CHEMICAL ENGINEERING BIBLIOGRAPHY (1967-1988)

MARTYN S. RAY

Curtin University of Technology
Chemical Engineering Dept.
P.O. Box U1987, Perth
Western Australia 6001

NOYES PUBLICATIONS
Park Ridge, New Jersey, U.S.A.
1990

Library of Congress Catalog Card Number: 89-77185
ISBN: 0-8155-1241-4
Printed in the United States

Published in the United States of America by
Noyes Publications
Mill Road, Park Ridge, New Jersey 07656

10 9 8 7 6 5 4 3 2 1

Library of Congress Cataloging-in-Publication Data

Ray, Martyn S., 1949-
Chemical engineering : bibliography (1967-1988) / Martyn S. Ray.
p. cm.
Includes bibliographical references
ISBN 0-8155-1241-4 :
1. Chemical engineering--Bibliography. I. Title.
Z7914.C4R35 1990
[TP155]
016.66--dc20 89-77185
CIP

Preface

This bibliography provides a valuable reference source for chemical engineers, and engineers and scientists from other disciplines. It is an essential reference source for practicing chemical engineers, students and academics. It contains nearly 20,000 references from the technical literature published during the period 1967-1988. The references are taken from 40 journals, these being the most relevant to the chemical engineering literature.

Other monthly and annual engineering indexes are available, however they are too expensive for personal purchase and too large for quick reference. The same comment applies to on-line computer databases. This bibliography provides chemical engineers with an "off-the-shelf" reference source for the most important literature published over the past 20 years, and at a price affordable to the individual. This book should occupy a place on the chemical engineer's bookshelf alongside a copy of *Perry's Chemical Engineers' Handbook.*

As with any publication, some readers will no doubt have suggestions regarding improvements, changes, additions, and errors. The author would welcome any correspondence which may improve the quality of any subsequent edition.

The table of contents is organized in such a way as to serve as a subject index, and provides easy access to the information contained in this bibliography. A more detailed subject cross-reference index is also included.

About the Author

Martyn S. Ray holds a B.Sc. and Ph.D. in Chemical Engineering from the University of Surrey, U.K. He has industrial experience as a chemical engineer with BOC International (UK) Ltd and has held positions as Lecturer in Chemical Engineering at Huddersfield Polytechic (UK), the University of the West Indies, and the Polytechnic of Wales. He is currently a Lecturer in Chemical Engineering at Curtin University of Technology, Western Australia. He is a Chartered Engineer, a Member of the Institution of Chemical Engineers (UK) and the American Institute of Chemical Engineers. He has published several papers and four textbooks, and is the holder of several patents.

Acknowledgements

The references in this bibliography were stored, and subsequently catalogued and sorted, using the *PC-FILE* database software program. This program was devised and distributed by Buttonware, Inc., U.S.A. I highly recommend this program for the preparation of bibliographies and other types of database applications.

My thanks to Brenda Tipping for her typing skills in transforming a mountain of index cards into a more manageable floppy disk format.

Finally, my thanks to my wife Cherry for her help, patience, understanding and encouragement during the preparation of this book.

This book is dedicated to my wife, Cherry, and our children, Frank, Harry, and Sally.

January 1990 *Martyn Ray*

Contents and Subject Index

About This Bibliography

This bibliography contains nearly 20,000 references taken from 40 journals published over the period 1967-1988. A list of the journals, the years covered, and the abbreviations used, is included at the end of this section. For the particular journals:
Chemical Engineering, The Chemical Engineer, Chemical Engineering Progress, Chemical Engineering Research and Development (formerly Transactions of the Institution of Chemical Engineers), and Hydrocarbon Processing;
all papers (excluding 'process' description-type papers) for the period 1967-1988 have been included. For other journals, I have exercised personal discretion and included those papers considered to be the most 'useful' to practicing engineers. The more obscure, esoteric, academic, and theoretical papers have been omitted mainly to keep this bibliography to one volume (and hence to a price acceptable to the individual). The researcher exploring a very specialist or theoretical topic not included here would probably refer to an on-line computer database or Engineering Index for a detailed literature search.

For the mainstream chemical engineering journals such as *Chemical Engineering Science* or the *AIChE Journal* approximately 80% of papers have been included, and a lower percentage for more specialist journals, e.g. *Energy & Fuels*. Papers detailing particular processes and flowsheets have generally been omitted, although some are included in Chapter 27: Miscellaneous Topics. Papers concerned with the design of particular process equipment such as an ammonia reactor, or reaction rate data, etc., are included in the relevant chapters. It is my intention to produce a companion volume detailing the literature of particular chemical processes and plant design in the future. It is also my intention to continue to compile chemical engineering bibliographies for future years, either as annual journal articles or in book form.

The reader may find it useful to refer to the book:
K Bourton, 'Chemical and Process Engineering Unit Operations: A Bibliographical Guide', MacDonald and Co. (Publishers) Ltd, London (1969).
Although this book is no longer in print, it may be available in engineering and technical libraries. This is a most useful book, and it provided the starting point (1967) for this current bibliography.

I have tried to avoid duplication of references in this bibliography and I have catalogued the entries in what I consider to be the most appropriate category, e.g. 'Control of a tubular reactor' in Chapter 18: Reactors, rather than Chapter 8: Process Control and Instrumentation. However, in some cases it has been necessary to include some duplicate references to make the task of information retrieval easier, e.g. 'Improved energy efficiency by heat exchanger design' in both Chapter 3: Energy Conservation and Chapter 17: Heat Exchangers. The extended subject index included at the end of this bibliography (containing nearly 1000 entries according to section reference rather than page number) provides a useful guide to the cross-referencing and cataloguing of entries.

Chapter 27 contains those references that could not be conveniently included in another chapter. Some of these references relate to particular processes and have been catalogued accordingly, e.g. Gasification (27.1) or Refining (27.6). Other papers which could not be placed in a particular category (or comprised too few references to merit a viable separate sub-section) have been included in Section 27.14: Miscellaneous Topics. These references are arranged primarily in chronological order and by author for each year, as with the other chapters. Chapter 27 provides a useful source of references for topics outside the mainstream chemical engineering subjects.

The references in this bibliography have been catalogued into a particular category (i.e. the chapter headings) and then into appropriate sub-topic headings. Each sub-section has then been organized chronologically (from 1967 to 1988), and the entries in each year arranged alphabetically by first author surname. Where only a few entries exist, individual year markers are omitted in favour of a year span, say 1967-1970, although entries are still included in sequential chronological and author order.

During the preparation of this bibliography every effort has been made to ensure the accuracy of the citations included. However, it is inevitable that some errors will occur with such a large number of references. Within the citation for a paper the *least likely* mistake is the journal title, due to the way in which the references were collected and processed. Any errors are probably concerning the volume (or part) number, or the page numbers, rather than the year of publication. Therefore, if a paper cannot be located from the reference given here, I suggest that you look initially to the author index for *Engineering Index* (or the journal issues directly) for the year given. I am confident that such errors have been kept to a minimum.

The main criticism of this bibliography will probably be that a list of references, even including the title, should also have included the abstract/summary for each paper. However, such an approach would have taken years to compile and the cost would have restricted purchase to libraries. It would also have required several volumes to compile such a bibliography. This bibliography is priced for individual purchase and is intended to provide an easy-to-use reference source in a single volume - for engineers to perform quick, initial literature searches. This book achieves these objectives.

List of Journals, Abbreviations Used, and Years Covered

Adsorption Science and Technology	Ads Sci Tech	84-87
American Institute of Chemical Engineers Journal	AIChEJ	67-88
Biochemical Engineering Journal (see Chemical Engineering Journal)	Biochem Eng J	71-88
British Chemical Engineering (see Processing)	Brit Chem Eng	66-70
Canadian Journal of Chemical Engineering	Can JCE	75-88
Catalysis Reviews	Catalysis Reviews	80-88
Chemical and Engineering News	C&E News	80-88
Chemical Engineer (UK), The*	Chem Engnr	67-88
Chemical Engineering (USA)*	Chem Eng	67-88
Chemical Engineering and Processing	Chem Eng & Proc	84-88
Chemical Engineering Communications	Chem Eng Commns	73-88
Chemical Engineering Education	Chem Eng Educ	76-88
Chemical Engineering in Australia	Chem Engng in Australia	88
Chemical Engineering Journal (including Biochemical Engineering Journal)	Chem Eng J	71-88
Chemical Engineering Progress*	Chem Eng Prog	67-88
Chemical Engineering Research and Development* (formerly Transactions of the Institution of Chemical Engineers)	CER&D	84-88
Chemical Engineering Science	Chem Eng Sci	70-88
Chemical Processing (see Processing)		
Chemistry and Industry	Chem & Ind	70-88
Chemtech	Chemtech	71-88
Computers and Chemical Engineering	Comput Chem Eng	77-88
Energy and Fuels	Energy & Fuels	87-88
Energy World	Energy World	73-88
Fuel	Fuel	75-88
Gas Separation and Purification	Gas Sepn & Purif	87-88
Hydrocarbon Processing*	Hyd Proc	67-88
Industrial and Engineering Chemistry Fundamentals	Ind Eng Chem Fund	67-86
Industrial and Engineering Chemistry Process Design and Development	Ind Eng Chem Proc Des Dev	67-86
Industrial and Engineering Chemistry Research	Ind Eng Chem Res	87-88
International Chemical Engineering	Int Chem Eng	67-88

International Journal of Heat and Mass Transfer	Int J Heat Mass Trans	80-88
Journal of Applied Chemistry	J Appl Chem	67-69
Journal of Applied Chemistry and Biotechnology	J Appl Chem Biotechnol	70-77
Journal of Chemical Technology and Biotechnology	J Chem Tech Biotechnol	78-88
Journal of the Institute of Energy	J Inst Energy	79-88
formerly:		
Journal of Institute of Fuel	J Inst Fuel	70-78
Periodica Polytechnica	Periodica Polytechnica	67-88
Plant/Operations Progress	Plant/Opns Prog	82-88
Powder Technology	Powder Tech	70-88
Process Economics International	Proc Econ Int	79-84
Process Engineering	Proc Engng	72-88
Processing	Processing	75-88
formerly:		
British Chemical Engineering	Brit Chem Eng	66-70
Brit Chem Eng and Process Technology	BCE & Proc Tech	71-72
Process Technology International	Proc Tech Int	72-73
Chemical Processing	Chem Procng	74
Separation and Purification Methods	Sepn & Purif Methods	74-88
Separation Science and Technology	Sepn Sci Technol	78-88
formerly:		
Separation Science	Sepn Sci	71-77
Solvent Extraction and Ion Exchange	Solv Extn & Ion Exch	83-88
Transactions of Institution of Chemical Engineers*	Trans IChemE	67-83

(see Chemical Engineering Research and Development)

* For these journals, **all** papers have been included for the years shown - excluding process description-type papers (see Chapter 27 for a selection of these papers).

Other Useful Journals

Advanced Materials and Processes
Ambient Energy
Ammonia Plant Safety (AIChE)
ASHRAE Journal
Biotechnology
Biotechnology Advances
Biotechnology Progress
British Ceramic Abstracts
British Ceramic Transactions and Journal
Bulk Solids Handling
Chemical Week
Corrosion
Corrosion Prevention and Control
Engineering (Design Council Journal, UK)
Engineering News Record (ENR) Construction Weekly
Environmental Progress (AIChE)
Filtration and Separation
Food Engineering International
Food Manufacturing News
Food Science and Technology Abstracts
Food Technology
Fuel and Energy Abstracts
Glass Technology
Hydrometallurgy
International Journal of Pressure Vessels and Piping
Journal of American Chemical Society
Journal of American Oil Chemist's Society (JAOCS)
Journal of Canadian Petroleum Technology
Journal of Chemical and Engineering Data
Journal of Energy Resources Technology (ASME)
Journal of Engineering for Industry (ASME)
Journal of Engineering Materials and Technology (ASME)
Journal of Food Engineering
Journal of Heat Treating
Journal of Petroleum Technology
Journal of Pressure Vessel Technology (ASME)
Journal of Solar Energy Engineering (ASME)
Journal of Water Pollution Control Federation
Light Metal Age
Manufacturing Chemist
Mechanical Energy (ASME)
Metallurgical Transactions
Nature
Nuclear Energy (Journal of British Nuclear Energy Society)
Oil and Gas Journal
Petroleum Times
Pipes and Pipelines International
Plastics Engineering
Plastics Technology
Process and Control Engineering
Progress in Material Science
Progress in Nuclear Energy
Refrigeration Air Conditioning and Heat Recovery
Science
Search (Science and Technology in Australia and New Zealand)
Solar Energy
Solar Energy Materials
Space Power
Technology Review
The Engineer
Water Engineering and Management

CHAPTER 1

DESIGN DATA: EQUATIONS OF STATE, PHASE EQUILIBRIA DATA, PHYSICAL AND THERMODYNAMIC PROPERTIES

1.1 Units and Information Sources

1966-1987

Chase, J.D., Searching for physical property data, Chem Eng, 12 Sept, 190-196 (1966).

Mullin, J.W., SI units in chemical engineering, Chem Engnr, Sept, CE176-178 (1967).

Lees, F.P., An SI unit conversion table for chemical engineering, Chem Engnr, Oct, CE341-344 (1968).

Parrish, A., Metrication in the process industries, Chem Engnr, March, CE34-37 (1968).

Dallaire, E.E., Adopting metrication, Chem Eng, 14 July, 111-114 (1969).

Martin, C.N.B., Using a comprehensive physical property estimation system, Chem Engnr, Sept, CE285-288 (1970).

Mullin, J.W., Preferred numbers in engineering, Chem Engnr, March, CE53-59 (1970).

Anon., SI unit conversion factors for chemical engineers, Brit Chem Eng, 16(9), 829-832 (1971).

Mullin, J.W., Recent developments in the changeover to SI units, Chem Engnr, Oct, 352-356 (1971).

Anon., Let's go metric (but not with SI), Chem Eng, 24 July, 141-144 (1972).

Biggert, E.C., State and city information sources, Chem Eng, 21 Aug, 94-103 (1972).

Biggert, E.C., Federal-government information sources, Chem Eng, 15 May, 103-114 (1972).

Canham, W.G., The international metric system, Chem Eng Prog, 68(7), 90-94 (1972).

Catlett, R.E.; Bush, W.D., and Hinkle, P.B., Metrication, Hyd Proc, 53(10), 181-196; 53(11), 232-238 (1974).

Glass, D.H., and Ponton, J.W., An interface to a physical property package, Chem Engnr, Dec, 760-766 (1975).

Holland, F.A., Metrication and chemical engineering, Chem Eng, 5 July, 77-80 (1976).

Loewer, H., Conversion of units: pascal, bar, newton, joule, watt, Int Chem Eng, 17(4), 719-740 (1977).

Winter, P., and Newell, R.G., Databases in design management, Chem Eng Prog, 73(6), 97-102 (1977).

Buck, E., Letter symbols for chemical engineering, Chem Eng Prog, 74(10), 73-80 (1978).

Dadyburjor, D.B., SI units for distribution coefficients, Chem Eng Prog, 74(4), 85-86 (1978).

Bondi, A., What thermophysical and other physical properties data are needed? Chem Eng Prog, 75(4), 70-74 (1979).

Bush, W.D., and Catlett, R.E., Whatever happened to metrication? Chem Eng Prog, 75(1), 61-66 (1979).

Hughson, R.V., SI metrication revisited, Chem Eng, 1 Jan, 91-92 (1979).

Jones, E.A., and James, R.B., Metrication in the process industries, Chem Eng Prog, 75(5), 76-78 (1979).

Youngquist, G.F., Update: SI on campus, Chem Eng Prog, 75(6), 87-88 (1979).

Baker, D.B., International information resources for chemical engineers, Chem Eng Prog, 77(4), 84-89 (1981).
Edmonds, B., Physical property data for chemical engineering design, Chem Engnr, Feb, 51 (1981).
Kharbanda, P.O., Nomograms: A tool for the design engineer, Chem Engnr, Feb, 52-53 (1981).
Norman, P., Physical property data banks for small computers, Chem Engnr, March, 111-113 (1981).
Stanley, W.G., Unique information resources for the chemical engineer, Chem Eng Prog, 77(6), 80-82 (1981).
Torres-Marchal, C., Getting chemical engineering information from non-English-speaking countries, Chem Eng, 13 July, 117-126 (1981).
Lowenstein, J.G., Conversion of pressure to pascals, Chem Eng, 8 March, 116 (1982).
Baltatu, M.E., On-line information, Chem Eng, 9 Jan, 69-72 (1984).
Kletz, T.A., Anglo-American glossary, Chem Engnr, Dec, 33-35 (1984).
Newman, S.A., Data prediction manual, Chem Eng Prog, 81(1), 16-21 (1985).
Short, W.A., International standards: How do they affect our industry? Chem Eng Prog, 81(11), 14-17 (1985).
Khan, M.S.; Pinkston, R.D., and Zaye, D.F., Chemical engineering information: What's available and how to get it, Chem Eng Prog, 82(1), 20-27 (1986).
Selover, T.B., Design Institute for Physical Property Data in its seventh year, Chem Eng Prog, 83(7), 18-21 (1987).

1.2 Thermodynamics Theory

1967-1979

Renoir, J.M., and Koppany, C.R., Equilibrium ratios by the Chao-Seader method, Hyd Proc, 46(11), 249-252 (1967).
Wheeler, J.D., and Smith, B.D., Application of molecular corresponding states theory to highly nonideal liquid mixtures, AIChEJ, 13(2), 303-311 (1967).
Kim, H.T., and Brodkey, R.S., Kinetic approach for polymer solution data, AIChEJ, 14(1), 61-68 (1968).
Hissong, D.W., and Kay, W.B., Calculation of critical locus curve of a binary hydrocarbon system, AIChEJ, 16(4), 580-587 (1970).
Edmister, W.C., Applied hydrocarbon thermodynamics, Hyd Proc, 47(9), 239-244; 47(11), 205-210, 47(12), 123-126 (1968); 48(5), 181-189 (1969); 48(6), 166-172; 48(8), 129-134 (1969); 50(2), 131-139 (1971).
Edmister, W.C., Applied hydrocarbon thermodynamics, Hyd Proc, 51(1), 109-112; 51(5), 145-152; 51(8), 97-101; 51(11), 165-170; 51(12), 93-101 (1972).
Gunn, R.D., Corresponding states for fluid mixtures, AIChEJ, 18(1), 183-193 (1972).
Winnick, J., Thermodynamics of simple polar liquids, Ind Eng Chem Fund, 11(2), 239-243 (1972).
Barner, H.E., and Adler, S.B., Calculation of solution nonideality from binary T-x data, Ind Eng Chem Proc Des Dev, 12(1), 71-75 (1973).

Edmister, W.C., Applied hydrocarbon thermodynamics, Hyd Proc, 52(3), 95-102; 52(4), 175-181; 52(5), 169-175; 52(6), 101-108; 52(7), 123-129; 52(8), 109-115 (1973).
Mollerup, J., and Fredenslund, A., Molecular thermodynamics of solutions in normal and critical regions, Chem Eng Sci, 28(6), 1285-1302 (1973).
Abrams, D.S., and Prausnitz, J.M., Statistical thermodynamics of liquid mixtures, AIChEJ, 21(1), 116-128 (1975).
Lee, B.I., and Kesler, M.G., Generalized thermodynamic correlation based on three-parameter corresponding states, AIChEJ, 21(3), 510-527 (1975).
Winnick, J., The state of affairs in thermo, Chemtech, March, 177-185; Dec, 756-762 (1975).
Prausnitz, J.M., Molecular thermodynamics for chemical process design, Chem Eng Educ, 10(2), 60-68 (1976).
Astarita, G., Historical and philosophical background of thermodynamics, Ind Eng Chem Fund, 16(1), 138-143 (1977).
Gutmann, V., Solvent concepts, Chemtech, April, 255-263 (1977).
Hanson, C.; Sohrabi, M., and Kaghazchi, T., Guestimate the behavior of binaries, Chemtech, Nov, 700-701 (1977).
Nitta, T.; Turek, E.A.; Greenkorn, R.A., and Chao, K.C., Group contribution molecular model of liquids and solutions, AIChEJ, 23(2), 144-160 (1977).
Tarakad, R.R., and Danner, R.P., Improved corresponding states method for polar fluids: Correlation of second virial coefficients, AIChEJ, 23(5), 685-695 (1977).
Anderson, T.F.; Abrams, D.S., and Grens, E.A., Evaluation of parameters for nonlinear thermodynamic models, AIChEJ, 24(1), 20-29 (1978).
Jenkins, J.D., and Gibson-Robinson, M., Thermodynamic consistency tests, Trans IChemE, 56, 43-49 (1978).
Lucas, K., Calculation of properties of gases and liquids by molecular theory: State of knowledge and application in chemical engineering, Int Chem Eng, 18(3), 408-417 (1978).
Martinez-Ortiz, J.A., and Manley, D.B., Direct solution of the isothermal Gibbs-Duhem equation for multicomponent systems, Ind Eng Chem Proc Des Dev, 17(3), 346-351 (1978).
McNeil, K.M., Experiments for teaching thermodynamics, Chem Eng Educ, 12(3), 130-135 (1978).
Nicolaides, G.L., and Eckert, C.A., Optimal representation of binary liquid mixture nonidealities, Ind Eng Chem Fund, 17(4), 331-340 (1978).
van Oss, C.J.; Neumann, A.W.; Omenyi, S.N., and Absolom, D.R., Role of repulsive van der Waals interactions in various separation methods, Sepn & Purif Methods, 7(2), 245-272 (1978).
Cropper, W.V., Standards and standardization, Chemtech, Sept, 550-559 (1979).
Gautam, R., and Seider, W.D., Computation of phase and chemical equilibrium, AIChEJ, 25(6), 991-1015 (1979).
Ginell, R., The meaning of entropy, Chemtech, July, 446-450 (1979).
Kesler, M.G.; Lee, B.I., and Sandler, S.I., Third parameter for use in generalised thermodynamic correlations, Ind Eng Chem Fund, 18(1), 49-54 (1979).

1980-1988

Mentzer, R.A.; Greenkorn, R.A., and Chao, K.C., Principle of corresponding states and prediction of gas-liquid separation factors and thermodynamic properties: A review, Sepn Sci Technol, 15(9), 1613-1678 (1980).

Snyder, L., Solvents and solubility, Chemtech, March, 188-193 (1980).

Ely, J.F., and Hanley, H.J.M., Prediction of transport properties, Ind Eng Chem Fund, 20(4), 323-332 (1981).

Spala, E.E., and Ricker, N.L., Thermodynamic model for solvating solutions with physical interactions, Ind Eng Chem Proc Des Dev, 21(3), 409-415 (1982).

Lebon, G., and Mathieu, P., Comparison of diverse theories of nonequilibrium thermodynamics, Int Chem Eng, 23(4), 651-663 (1983).

Buck, E., Applying phase equilibrium thermodynamics, Chemtech, Sept, 570-575 (1984).

Rangaiah, G.P., Estimation of frequency factor and activation energy in the Arrhenius equation, Chem Eng J, 29(3), 159-166 (1984).

Chimowitz, E.H., and Lee, C.S., Local thermodynamic models for high-pressure process calculations, Comput Chem Eng, 9(2), 195-200 (1985).

Debenedetti, P.G., and Reid, R.C., Diffusion and mass transfer in supercritical fluids, AIChEJ, 32(12), 2034-2046 (1986).

Kamlet, M.J.; Doherty, R.M.; Abboud, J.L.M.; Abraham, M.H., and Taft, R.W., Solubility: A new look, Chemtech, Sept, 566-576 (1986).

Michelsen, M.L., and Mollerup, J., Partial derivatives of thermodynamic properties, AIChEJ, 32(8), 1389-1392 (1986).

Olah, K., Entropy and temperature: Two simple thermodynamic models, Periodica Polytechnica, 30(1/2), 53-68 (1986).

Willman, B.T., and Teja, A.S., Continuous thermodynamics of phase equilibria using a multivariate distribution function and an equation of state, AIChEJ, 32(12), 2067-2078 (1986).

Farah, N., and Missen, R.W., The computer-derivation of thermodynamic equations, Can JCE, 64(1), 154-157 (1986); 65(1), 137-141 (1987).

Good, R.J., A contribution to the teaching of thermodynamics: A problem based on the Gibbs-Duhem equation, Chem Eng Educ, 21(2), 94-97 (1987).

Kalantar, A.H., Estimation of the parameters of Arrhenius-like equations, Chem Eng J, 34(3) 159-164 (1987).

Perry, M.B., and White, C.M., Correlations of molecular connectivity with critical volume and acentricity, AIChEJ, 33(1), 146-151 (1987).

Rangaiah, G.P., Adaptive methods for estimating Arrhenius parameters, Chem Eng J, 36(2), 123-128 (1987).

Waite, B.A., A simple molecular interpretation of entropy, Chem Eng Educ, 21(2), 98-100 (1987).

Haghtalab, A., and Vera, J.H., Nonrandom factor model for excess Gibbs energy of electrolyte solutions, AIChEJ, 34(5), 803-813 (1988).

Howell, W.J.; Lira, C.T., and Eckert, C.A., Linear chemical-physical theory model for liquid metal solution thermodynamics, AIChEJ, 34(9), 1477-1485 (1988).

Jonah, D.A., Interpolation between two excess Gibbs free energy-composition isotherms, Chem Eng Commns, 65, 95-108 (1988).

King, R.S.; Blanch, H.W., and Prausnitz, J.M., Molecular thermodynamics of aqueous two-phase systems, AIChEJ, 34(10), 1585-1594 (1988).
Kyle, B.G., The mystique of entropy, Chem Eng Educ, 22(2), 92-97 (1988).
McMahon, P.D.; Glandt, E.D., and Walker, J.S., Renormalization group theory in solution thermodynamics (review paper), Chem Eng Sci, 43(10), 2561-2586 (1988).
Teja, A.S., and Schaeffer, S.T., Research on thermodynamics and fluid properties, Chem Eng Educ, 22(4), 208-211, 222 (1988).

1.3 Equations of State

1968-1975

Edmister, W.C.; Vairogs, J., and Klekers, A.J., Generalized B-W-R equation of state, AIChEJ, 14(3), 479-482 (1968).
Auslander, G., Equation of state for real fluids, Brit Chem Eng, 14(4), 513-515 (1969).
Yorizane, M., and Masuoka, H., Gas-liquid equilibria from equations of state (high pressure), Int Chem Eng, 9(3), 532-540 (1969).
Zudkevitch, D., and Joffe, J., Correlation and prediction of vapor-liquid equilibria with the Redlich-Kwong equation of state, AIChEJ, 16(1), 112-119 (1970).
Lee, B.I., and Edmister, W.C., New three-parameter equation of state, Ind Eng Chem Fund, 10(1), 32-35 (1971).
Skamenca, D.G., and Tassios, D.P,, Improved Redlich-Kwong equation of state in supercritical region, Ind Eng Chem Proc Des Dev, 10(1), 59-64 (1971).
Wiehe, I.A.; Dorai, S., and Rader, C.G., Calculation of binary and multicomponent vapor-liquid equilibria with the two-parameter Flory-Huggins equation, Chem Eng Sci, 26(6), 901-914 (1971).
Yang, C.L., and Yendall, E.F., Equation of state for phase equilibrium, AIChEJ, 17(3), 596-607 (1971).
Bishnoi, P.R., and Robinson, D.B., Mixing rules improve use of BWR equation of state, Hyd Proc, 51(11), 152-156 (1972).
Cheng, S.I., and Chan, T.L., Estimation of Wilson parameters by nonlinear regression, Chem Eng J, 4(3), 282-286 (1972).
Horvath, A.L., Analysis of co-existing phases in Redlich-Kwong equation of state, Chem Eng Sci, 27(5), 1185-1190 (1972).
Soave, G., Equilibrium constants from modified Redlich-Kwong equation of state, Chem Eng Sci, 27(6), 1197-1204 (1972).
Vera, J.H., and Prausnitz, J.M., Generalized van der Waals theory for dense fluids, Chem Eng J, 3(1), 1-13 (1972).
Abbott, M.M., Cubic equations of state, AIChEJ, 19(3), 596-601 (1973).
Chaudron, J.; Asselineau, L., and Renon, H., Mixture properties and vapour-liquid equilibria by modified Redlich-Kwong equation of state, Chem Eng Sci, 28(11), 1991-2004 (1973).
Chaudron, J.; Asselineau, L., and Renon, H., New modification of Redlich-Kwong equation of state based on analysis of a large set of pure component data, Chem Eng Sci, 28(3), 839-846 (1973).

Heidemann, R.A., and Mandhane, J.M., Some properties of NRTL equation in correlating liquid-liquid equilibrium data, Chem Eng Sci, 28(5), 1213-1222 (1973).

Yamada, T., Improved generalized equation of state, AIChEJ, 19(2), 286-291 (1973).

Bishnoi, P.R.; Miranda, R.D., and Robinson, D.B., Equations of state for heavy compounds and hydrocarbon mixtures, Hyd Proc, 53(11), 197-201 (1974).

Lu, B.C.Y.; Yu, P., and Sugie, A.H., Prediction of vapour-liquid-liquid equilibria by modified regula-falsi method, Chem Eng Sci, 29(2), 321-326 (1974).

Mansoori, G.A., and Ali, I., Analytic equations of state of simple liquids and liquid mixtures, Chem Eng J, 7(3), 173-186 (1974).

Mattelin, A.C., and Vernhoeye, L.A.J., Correlation of binary miscibility data by NRTL equation, Chem Eng Sci, 30(2), 193-200 (1975).

McElroy, P.J., and Williamson, A.G., Simple equations of state for gas mixtures, Chem Eng Sci, 30(8), 819-824 (1975).

Nagata, I.; Ogura, M., and Nagashima, M., Extension of the Wilson equation to partially miscible systems, Ind Eng Chem Proc Des Dev, 14(4), 500-502 (1975).

Redlich, O., Three-parameter representation of the equation of state, Ind Eng Chem Fund, 14(3), 257-260 (1975).

1976-1979

De Santis, R.; Gironi, F., and Marrelli, L., Vapor-liquid equilibria from a hard-sphere equation of state, Ind Eng Chem Fund, 15(3), 183-189 (1976).

Epstein, L.F., Redlich-Kwong equation of state: Exact critical constant relations, Chem Eng Sci, 31(1), 87-88 (1976).

Fuller, G.G., Modified R-K-S equation of state for the liquid state, Ind Eng Chem Fund, 15(4), 254-257 (1976).

Gomez-Nieto, M., and Thodos, G., Benedict-Webb-Rubin parameters for ethanol, Can JCE, 54, 438-445 (1976).

Hederer, H.; Peter, S., and Wenzel, H., Calculation of thermodynamic properties from modified Redlich-Kwong equation, Chem Eng J, 11(3), 183-190 (1976).

Huron, M.J.; Vidal, J., and Asselineau, L., Predicting NRTL parameters for binary hydrocarbon mixtures for calculating liquid-vapor equilibriums of multicomponent systems, Chem Eng Sci, 31(6), 443-452 (1976).

Kato, M.; Cheng, W.K., and Lu, B.C.Y., Modified parameters for Redlich-Kwong equation of state, Can JCE, 54, 441-450 (1976).

Kato, M.; Chung, W.K., and Lu, B.C.Y., Binary interaction coefficients of Redlich-Kwong equation of state, Chem Eng Sci, 31(8), 733-736 (1976).

Peng, D.Y., and Robinson, D.B., New two-constant equation of state, Ind Eng Chem Fund, 15(1), 59-64 (1976).

Simonet, R., and Behar, E., Modified Redlich-Kwong equation of state for accurately representing pure components data, Chem Eng Sci, 31(1), 37-44 (1976).

Tassios, D., Limitations in correlating strongly nonideal binary systems with NRTL and LEMF equations, Ind Eng Chem Proc Des Dev, 15(4), 574-578 (1976).

Usdin, E., and McAuliffe, J.C., One parameter family of equations of state, Chem Eng Sci, 31(11), 1077-1084 (1976).

Chung, W.K., and Lu, B.C.Y., Representation of liquid properties of pure polar compounds by modified Redlich-Kwong equation of state, Can JCE, 55, 129-140 (1977).

Chung, W.K.; Haman, S.E.M., and Lu, B.C.Y., Modified RKS equation of state for the liquid state, Ind Eng Chem Fund, 16(4), 494-495 (1977).

Chung, W.K.; Kato, M., and Lu, B.C.Y., Generalized one-parameter Redlich-Kwong equation for vapour-liquid equilibrium calculations, Can JCE, 55, 701-710 (1977).

De Fre, R.M., and Verhoeye, L.A., Prediction of ternary liquid-liquid equilibrium by NRTL equation, J Appl Chem Biotechnol, 27, 667-679 (1977).

Horvath, C., and Lin, H.J., Simple three-parameter equation of state with critical compressibility factor correlation, Can JCE, 55, 450-460 (1977).

Silverman, N., and Tassios, D., Number of roots in the Wilson equation and its effect on vapor-liquid equilibrium calculations, Ind Eng Chem Proc Des Dev, 16(1), 13-20 (1977).

Anderson, T.F., and Prausnitz, J.M., Application of UNIQUAC equation to calculation of multicomponent phase equilibria, Ind Eng Chem Proc Des Dev, 17(4), 552-567 (1978).

Coward, I.; Gale, S.E., and Webb, D.R., Process engineering calculations with equations of state, Trans IChemE, 56, 19-27 (1978).

Flores Luna, J.L., and Barnes de Castro, F., Evaluation of various modifications of Redlich-Kwong equation, Int Chem Eng, 18(4), 611-627 (1978).

Gmehling, J.G.; Anderson, T.F., and Prausnitz, J.M., Solid-liquid equilibria using UNIFAC, Ind Eng Chem Fund, 17(4), 269-273 (1978).

Graboski, M.S., and Daubert, T.E., Modified Soave equation of state for phase equilibrium calculations, Ind Eng Chem Proc Des Dev, 17(4), 443-454 (1978).

Medani, M.S., and Hasan, M.A., Phase equilibria calculations with modified Redlich-Kwong equation of state, Can JCE, 56, 251-260 (1978).

Missen, R., Determination of van Laar parameters, Can JCE, 56, 126-135 (1978).

Nagata, I., and Ohta, T., Prediction of excess enthalpies of mixing of mixtures using the UNIFAC method, Chem Eng Sci, 33(2), 177-182 (1978).

Rasmussen, P., and Fredenslund, A., Review of prediction of separation factors using group contribution methods, Sepn & Purif Methods, 7(2), 147-182 (1978).

Robinson, D.B.; Peng, D.Y., and Ng, H.J., Capability of the Peng-Robinson programs, Hyd Proc, 57(4), 95-98 (1978).

Vidal, J., Mixing rules and excess properties in cubic equations of state, Chem Eng Sci, 33(6), 787-792 (1978).

Gmehling, J., and Onken, U., Calculation of activity coefficients from structural-group contributions, Int Chem Eng, 19(4), 566-571 (1979).

Martin, J.J., Review of cubic equations of state, Ind Eng Chem Fund, 18(2), 81-97 (1979).

Robinson, D.B.; Peng, D.Y., and Ng, H.J., Capabilities of the Peng-Robinson programs, Hyd Proc, 58(9), 269-273 (1979).

Skjold-Jorgensen, S., et al., Vapor-liquid equilibria by UNIFAC group contribution: Revision and extension, Ind Eng Chem Proc Des Dev, 18(4), 714-722 (1979).

Soave, G.S., Application of the R-K-S equation of state to solid-liquid equilibria calculations, Chem Eng Sci, 34(2), 225-230 (1979).

Tarakad, R.R.; Spencer, C.F., and Adler, S.B., Comparison of eight equations of state to predict gas-phase density and fugacity, Ind Eng Chem Proc Des Dev, 18(4), 726-739 (1979).

1980-1982

Harmens, A., and Knapp, H., Three-parameter cubic equation of state for normal substances, Ind Eng Chem Fund, 19(3), 291-294 (1980).

Ishikawa, T.; Chung, W.K., and Lu, B.C.Y., A cubic, perturbed, hard-sphere equation of state for thermodynamic properties and vapor-liquid equilibrium calculations, AIChEJ, 26(3), 372-378 (1980).

Jorgensen, S.S.; Rasmussen, P., and Fredenslund, A.A., Temperature dependence of the UNIQUAC/UNIFAC models, Chem Eng Sci, 35(12), 2389-2404 (1980).

Kikic, A.; Alessi, P.; Rasmussen, P., and Fredenslund, A.A., Combinatorial part of the UNIFAC and UNIQUAC models, Can JCE, 58, 253-258 (1980).

Krumins, A.E.; Rastogi, A.K.; Rusak, M.E., and Tassios, D., Prediction of binary vapor-liquid equilibrium from one-parameter equations, Can JCE, 58, 663-669 (1980).

Lin, C.T., and Daubert, T.E., Estimation of partial molar volume and fugacity coefficient of components in mixtures from the Soave and Peng-Robinson equations of state, Ind Eng Chem Proc Des Dev, 19(1), 51-59 (1980).

Lin, H.M., Modified Soave equation of state for phase equilibrium calculations, Ind Eng Chem Proc Des Dev, 19(3), 501-505 (1980).

Mauri, C., Unified procedure for solving multiphase-multicomponent vapor-liquid equilibrium calculation, Ind Eng Chem Proc Des Dev, 19(3), 482-489 (1980).

Mohanty, K.K.; Dombrowski, M., and Davis, H.T., Comparison of a pair of three parameter equations of state, Chem Eng Commns, 5(1), 85-92 (1980).

Montfort, J.P., and Hernandez, O., Calculation of critical constants of solutions with the NRTL solutions, Can JCE, 58, 271-273 (1980).

Raimondi, L., Modified Redlich-Kwong equation of state for vapour-liquid equilibrium calculations, Chem Eng Sci, 35(6), 1269-1277 (1980).

Schmidt, G., and Wenzel, H., Modified van der Waals-type equation of state, Chem Eng Sci, 35(7), 1503-1512 (1980).

Soave, G., Rigorous and simplified procedures for determining the pure-component parameters in the Redlich-Kwong-Soave equation of state, Chem Eng Sci, 35(8), 1725-1730 (1980).

Weber, J.H., Calculate equation-of-state variables, Chem Eng, 25 Feb, 93-100 (1980).

Ahrendts, J., and Baehr, H.D., Direct application of experimental values for any thermodynamic variables of state in establishing canonical equations of state, Int Chem Eng, 21(4), 557-572 (1981).

Ahrendts, J., and Baehr, H.D., Use of nonlinear regression processes in establishing thermodynamic equations of state, Int Chem Eng, 21(4), 572-580 (1981).

Herskowitz, M., and Gottlieb, M., UNIFAC group contribution method for silicone compounds, Ind Eng Chem Proc Des Dev, 20(2), 407-409 (1981).

Hiranuma, M., Significance and value of the third parameter in the modified Wilson equation, Ind Eng Chem Fund, 20(1), 25-28 (1981).

Joffe, J., Vapor-liquid equilibria and densities with the Martin equation of state, Ind Eng Chem Proc Des Dev, 20(1), 168-172 (1981).

Larrinaga, L., Graphically determine the Wilson parameters for vapor-liquid equilibria, Chem Eng, 6 April, 87-91 (1981).

Magnussen, T.; Rasmussen, P., and Fredenslund, A., UNIFAC parameter table for prediction of liquid-liquid equilibria, Ind Eng Chem Proc Des Dev, 20(2), 331-339 (1981).

Oellrich, L.; Ploker, U.; Prausnitz, J.M., and Knapp, H., Equation-of-state methods for computing phase equilibria and enthalpies, Int Chem Eng, 21(1), 1-17 (1981).

Poling, B.E.; Grens, E.A., and Prausnitz, J.M., Thermodynamic properties from a cubic equation of state, Ind Eng Chem Proc Des Dev, 20(1), 127-130 (1981).

Schmidt, G., and Wenzel, H., Estimation of critical data by equation of state, Can JCE, 59, 527-531 (1981).

Teja, A.S., and Patel, N.C., Applications of a generalized equation of state to correlation and prediction of phase equilibria, Chem Eng Commns, 13(1), 39-54 (1981).

Alessi, P.; Kikic, I.; Fredenslund, A., and Rasmussen, P., UNIFAC and infinite dilution activity coefficients, Can JCE, 60, 300-304 (1982).

Breivi, S.W., Simple correlations for UNIQUAC structure parameters, Ind Eng Chem Proc Des Dev, 21(3), 367-370 (1982).

Gmehling, J.; Rasmussen, P., and Fredenslund, A., Vapor-liquid equilibria by UNIFAC group contribution, Ind Eng Chem Proc Des Dev, 21(1), 118-127 (1982).

Gottlieb, M., and Herskowitz, M., External-degrees-of-freedom parameter in the UNIFAC-FV model, Ind Eng Chem Proc Des Dev, 21(3), 536-537 (1982).

Jorgensen, S.S.; Rasmussen, P., and Fredenslund, A., On the concentration dependence of the UNIQUAC/UNIFAC models, Chem Eng Sci, 37(1), 99-112 (1982).

Kumar, K.H., and Starling, K.E., The most general density-cubic equation of state: Application to nonpolar fluids, Ind Eng Chem Fund, 21(3), 255-262 (1982).

Patel, N.C., and Teja, A.S., A new cubic equation of state for fluids and fluid mixtures, Chem Eng Sci, 37(3), 463-474 (1982).

Sievers, U., and Schulz, S., Calculation of thermodynamic properties of fluid mixtures with a new equation of state, Chem Eng Commns, 17(1), 57-66 (1982).

Simonetty, J.; Yee, D., and Tassios, D., Prediction and correlation of liquid-liquid equilibria, Ind Eng Chem Proc Des Dev, 21(1), 174-180 (1982).

Stein, R.B., Modified Redlich-Kwong equation of state for phase equilibrium calculations, Ind Eng Chem Proc Des Dev, 21(4), 564-569 (1982).

Weber, J.H., Predict equation-of-state variables, Chem Eng, 11 Jan, 111-117 (1982).

1983-1985

Chien, C.H.; Greenkorn, R.A., and Chao, K.C., Chain-of-rotators equation of state, AIChEJ, 29(4), 560-571 (1983).

Georgeton, G.K., and Sommerfeld, J.T., Estimating vapor-liquid equilibria with UNIFAC, Chem Eng, 17 Oct, 61-65 (1983).

Lielmezs, J.; Howell, S.K., and Campbell, H.D., Modified Redlich-Kwong equation of state for saturated vapour-liquid equilibrium, Chem Eng Sci, 38(8), 1293-1302 (1983).

Macedo, E.A.; Weidlich, U.; Gmehling, J., and Rasmussen, P., Vapor-liquid equilibria by UNIFAC group contribution, Ind Eng Chem Proc Des Dev, 22(4), 676-678 (1983).

Marmur, A., Equations of state as conservation equations, Chem Eng Commns, 22(5), 299-302 (1983).

Mathias, P.M., A versatile phase equilibrium equation of state, Ind Eng Chem Proc Des Dev, 22(3), 385-391 (1983).

Ruzicka, V.; Fredenslund, A., and Rasmussen, P., Representation of petroleum fractions by group contribution, Ind Eng Chem Proc Des Dev, 22(1), 49-53 (1983).

Vetere, A., Vapor-liquid equilibria calculations using an equation of state, Chem Eng Sci, 38(8), 1281-1292 (1983).

Wilcox, R.F., Expanding the boundaries of UNIFAC, Chem Eng, 17 Oct, 66-68 (1983).

Larrinaga, L., Fitting activity coefficients to curves via Wilson model, Chem Eng, 16 April, 69-74 (1984).

Schotte, W., Temperature dependence of the UNIQUAC parameters, Chem Eng Sci, 39(1), 190-193 (1984).

Silverman, N., and Tassios, D., Prediction of multicomponent vapor-liquid equilibrium with Wilson equation, Ind Eng Chem Proc Des Dev, 23(3), 586-589 (1984).

Soave, G., Improvement of the Van der Waals equation of state, Chem Eng Sci, 39(2), 357-370 (1984).

Vera, J.H., and Vidal, J., Improved group method for hydrocarbon-hydrocarbon systems arising from a comparative study of ASOG and UNIFAC, Chem Eng Sci, 39(4), 651-662 (1984).

Vera, J.H.; Huron, M.J., and Vidal, J., Flexibility and limitations of cubic equations of state, Chem Eng Commns, 26(4), 311-318 (1984).

Vetere, A., Algorithm for fluid phase equilibria calculations using a multiconstant equation of state, Chem Eng Sci, 39(12), 1779-1784 (1984).

Dzialoszynski, L.; Richon, D., and Renon, H., Generalized semi-empirical non-cubic equation of state for representation of properties of pure substances, Chem Eng Commns, 39, 23-42 (1985).

Elliott, J.R., and Daubert, T.E., Revised procedures for phase equilibrium calculations with the Soave equation of state, Ind Eng Chem Proc Des Dev, 24(3), 743-748 (1985).

Marmur, A., Mathematical properties of equations of state, Chem Eng Sci, 40(10), 1881-1884 (1985).

Panaitescu, G., Parameter estimation in Wilson equation by method of maximum likelihood, Int Chem Eng, 25(4), 688-693 (1985).
Sampath, V.R., and Leipziger, S., Vapor-liquid-liquid equilibria computations, Ind Eng Chem Proc Des Dev, 24(3), 652-658 (1985).
Wong, J.O., and Prausnitz, J.M., Simple equation of state of the van der Waals form, Chem Eng Commns, 37(1), 41-54 (1985).

1986-1988
Arce, A.; Blanco, A., and Tojo, J., Prediction of vapor-liquid equilibria in binary systems using UNIFAC method: Comparison between OH and CCOH groups in alcohols, Int Chem Eng, 26(2), 278-290 (1986).
Bertucco, A.; Fermeglia, M., and Kikic, I., Modfied Carnahan-Starling-Van der Waals equation for supercritical fluid extraction, Chem Eng J, 32(1), 21-30 (1986).
Chao, Z., and Zhong, X., A four-parameter extension of modified Lee-Kesler equation of state, Chem Eng Commns, 43, 107-118 (1986).
Eckart, D.E.; Arnold, D.W.; Greenkorn, R.A., and Chao, K.C., A group contribution molecular model of liquids and solutions, AIChEJ, 32(2), 307-308 (1986).
Gupte, P.A., and Daubert, T.E., Prediction of low-pressure vapor-liquid equilibria of non-hydrocarbon-containing systems (ASOG or UNIFAC), Ind Eng Chem Proc Des Dev, 25(2), 481-487 (1986).
Gupte, P.A.; Rasmussen, P., and Fredenslund, A., A new group-contribution equation of state for vapor-liquid equilibria, Ind Eng Chem Fund, 25(4), 636-645 (1986).
Li, Y.K., and Nghiem, L.X., Phase equilibria of oil, gas and water/brine mixtures from a cubic equation of state and Henry's law, Can JCE, 64(3), 486-496 (1986).
Mathias, P.M., and Benson, M.S., Computational aspects of equations of state: Fact and fiction, AIChEJ, 32(12), 2087-2090 (1986).
Michelsen, M.L., Simplified flash calculations for cubic equations of state, Ind Eng Chem Proc Des Dev, 25(1), 184-188 (1986).
Orbey, H., and Vera, J.H., Rational construction of augmented hard core equation of state for pure compounds and study of its application to mixtures, Chem Eng Commns, 44, 95-106 (1986).
Ruiz, F., and Gomis, V., Correlation of quaternary liquid-liquid equilibrium data using UNIQUAC, Ind Eng Chem Proc Des Dev, 25(1), 216-221 (1986).
Sandarusi, J.A.; Kidnay, A.J., and Yesavage, V.F., Compilation of parameters for a polar fluid Soave-Redlich-Kwong equation of state, Ind Eng Chem Proc Des Dev, 25(4), 957-964 (1986).
Stryjek, R., and Vera, J.H., An improved Peng-Robinson equation of state with new mixing rules for strongly nonideal mixtures, Can JCE, 64(2), 334-340 (1986).
Stryjek, R., and Vera, J.H., A cubic equation of state for accurate vapor-liquid equilibria calculations, Can JCE, 64(5), 820-826 (1986).
Stryjek, R., and Vera, J.H., An improved Peng-Robinson equation of state for pure compounds and mixtures, Can JCE, 64(2), 323-333 (1986).

Wang, S.H., and Whiting, W.B., New algorithm for calculation of azeotropes from equations of state, Ind Eng Chem Proc Des Dev, 25(2), 547-552 (1986).
Agrawal, R.K., Activity coefficient from a generalized van der Waals equation, AIChEJ, 33(12), 2084-2086 (1987).
Czelej, M., Polynomial form of Wilson equation for binary and ternary systems, Int Chem Eng, 27(3), 535-539 (1987).
Fleming, P.D., and Brugman, R.J., Toward a molecular equation of state for real materials, AIChEJ, 33(5), 729-740 (1987).
Gupte, P.A., and Danner, R.P., Prediction of liquid-liquid equilibria with UNIFAC: A critical evaluation, Ind Eng Chem Res, 26(10), 2036-2042 (1987).
Joback, K.G., and Reid, R.C., Estimation of pure-component properties from group-contributions, Chem Eng Commns, 57, 233-244 (1987).
Larsen, B.L.; Rasmussen, P., and Fredenslund, A., A modified UNIFAC group-contribution model for prediction of phase equilibria and heats of mixing, Ind Eng Chem Res, 26(11), 2274-2286 (1987).
Mohamed, R.S.; Enick, R.M.; Bendale, P.G., and Holder, G.D., Empirical two-parameter mixing rules for a cubic equation of state, Chem Eng Commns, 59, 259-276 (1987).
Najjar, Y.S.H., and Mansour, A.R., Evaluation of Peng-Robinson equation of state in calculating thermophysical properties of combustion gases, Chem Eng Commns, 61, 327-346 (1987).
Ozokwelu, E.D., and Erbar, J.H., Improved Soave-Redlich-Kwong equation of state, Chem Eng Commns, 52, 9-20 (1987).
Sievers, U., and Schulz, S., Calculation of thermodynamic properties of fluid mixtures using equation of state with new mixing and combining rules, Int Chem Eng, 27(3), 379-405 (1987).
Tang, H., New cubic equation of state, Int Chem Eng, 27(1), 148-155 (1987).
Valderrama, J.O., et al., Applications of Patel-Teja equation of state to prediction of volumetric properties of mixtures, Chem Eng Commns, 54, 161-172 (1987).
Wang, S.H., and Whiting, W.B., A group contribution, continuous-thermodynamics framework for calculation of vapour-liquid equilibria, Can JCE, 65(3), 651-661 (1987).
Weidlich, U., and Gmehling, J., A modified UNIFAC model, Ind Eng Chem Res, 26(7), 1372-1382 (1987).
Bastos, J.C.; Soares, M.E., and Medina, A.G., Infinite dilution activity coefficients predicted by UNIFAC group contribution, Ind Eng Chem Res, 27(7), 1269-1277 (1988).
Bosse, M.A., and Reich, R., Correlation for the third virial coefficient, Chem Eng Commns, 66, 83-100 (1988).
Cairns, B.P., and Furzer, I.A., Sensitivity testing with the predictive thermodynamic models NRTL, UNIQUAC, ASOG and UNIFAC in multicomponent separations of methanol-acetone-chloroform, Chem Eng Sci, 43(3), 495-502 (1988).
Carmola, R., and Chimowitz, E.H., Cubic equations of state based upon infinite pressure limit, Chem Eng Commns, 73, 67-76 (1988).

Carrier, B.; Rogalski, M., and Peneloux, A., Correlation and prediction of physical properties of hydrocarbons with modified Peng-Robinson equation of state, Ind Eng Chem Res, 27(9), 1714-1720 (1988).

Gadalla, N.M., and Marsh, K.N., Prediction of thermophysical and thermodynamic properties using the BACK equation of state, Ind Eng Chem Res, 27(3), 536-540 (1988).

Georgeton, G.K., and Teja, A.S., A group contribution equation of state based on the simplified perturbed hard chain theory, Ind Eng Chem Res, 27(4), 657-664 (1988).

Han, S.J.; Lin, H.M., and Chao, K.C., Vapor-liquid equilibrium of molecular fluid mixtures by equation of state (review paper), Chem Eng Sci, 43(9), 2327-2368 (1988).

Hartounian, H., and Allen, D.T., Group contribution methods for coal-derived liquids: Hydrogen solubilities using UNIFAC approach, Fuel, 67(12), 1609-1614 (1988).

Jin, G., and Donohue, M.D., Equation of state for electrolyte solutions, Ind Chem Eng Res, 27(6), 1073-1084 (1988).

Jorgensen, S.S., Group contribution equation of state for phase equilibrium computations, Ind Eng Chem Res, 27(1), 110-118 (1988).

Lantagne, G.; Marcos, B., and Cayrol, B., Computation of complex equilibria by nonlinear optimization, Comput Chem Eng, 12(6), 589-600 (1988).

Lee, M.J., and Chao, K.C., Augmented BACK equation of state for polar fluids, AIChEJ, 34(5), 825-834 (1988).

Lee, M.J., and Chao, K.C., Augmented BACK equation of state, AIChEJ, 34(11), 1773-1780 (1988).

Macchietto, S.; Maduabueke, G.I., and Szczepanski, R., Efficient implementation of VLE procedures in equation-oriented simulators, AIChEJ, 34(6), 955-963 (1988).

Mainwaring, D.E.; Sadus, R.J., and Young, C.L., Deiters' equation of state and critical phenomena, Chem Eng Sci, 43(3), 459-466 (1988).

Mannan, M., and Starling, K.E., Equation-of-state vapour-liquid equilibrium prediction methodology for systems containing undefined fractions, Fuel, 67(6), 815-821 (1988).

Moshfeghian, M.; Shariat, A., and Maddox, R.N., Improved liquid densities with modified Peng-Robinson equation, Chem Eng Commns, 73, 205-216 (1988).

Orbey, H., Four parameter Pitzer-Curl type correlation of second virial coefficients, Chem Eng Commns, 65, 1-20 (1988).

Patel, M.R.; Joffrion, L.L., and Eubank, P.T., Simple procedure for estimating virial coefficients from Burnett PVT data, AIChEJ, 34(7), 1229-1232 (1988).

Topliss, R.J.; Dimitrelis, D., and Prausnitz, J.M., Computational aspects of non-cubic equation of state for phase-equilibrium calculations: Effect of density-dependent mixing rules, Comput Chem Eng, 12(5), 483-490 (1988).

1.4 Phase Equilibria Data

1967-1970

Chueh, P.L., and Prausnitz, J.M., Vapor-liquid equilibria at high pressures, AIChEJ, 13(6), 1099-1113 (1967).

Ellis, S.R.M.; McDermott, C., and Chiang, C.S., Predicting ternary isobaric vapour-liquid equilibria, Brit Chem Eng, 12(5), 727-729 (1967).

Guerreri, G., High-pressure vapor-liquid equilibria in ammonia-water-nitrogen-hydrogen system, AIChEJ, 13(5), 877-883 (1967).

Haughton, C.A., Vapour-liquid equilibria of benzene and acetic acid, Brit Chem Eng, 12(7), 1102-1103 (1967).

Ruhemann, M., and Harmens, A., Bibliography of vapour-liquid equilibrium of low boiling mixtures and notes on K-value correlations, Chem Engnr, Nov, CE254-258 (1967).

Sebastiani, E., and Lacquaniti, L., Acetic acid-water system thermodynamical correlation of vapor-liquid equilibrium data, Chem Eng Sci, 22(9), 1155-1162 (1967).

Cave, S.D., Nomogram for vapour-liquid equilibrium of ternary systems, Brit Chem Eng, 13(6), 849 (1968).

Fleck, R.N., and Prausnitz, J.M., Estimating binary vapor-liquid equilibria, Chem Eng, 20 May, 157-164 (1968).

Herington, E.F.G., Symmetrical-area tests for the consistency of vapour-liquid equilibrium data, J Appl Chem, 18, 285-291 (1968).

Hirata, M.; Hakuta, T., and Onoda, T., Vapor-liquid equilibria of propylene-propane at low temperatures, Int Chem Eng, 8(1), 175-179 (1968).

Leach, J.W.; Chappelear, P.S., and Leland, T.W., Use of molecular shape factors in vapor-liquid equilibrium calculations with the corresponding states principle, AIChEJ, 14(4), 568-576 (1968).

Sebastiani, E., Quick calculation of phase composition, Chem Eng, 11 March, 218-220 (1968).

Van Horn, L.D., and Kobayashi, R., Correlation of vapor-liquid equilibria of light hydrocarbons in paraffinic and aromatic solvents at low temperatures and elevated pressures, AIChEJ, 14(1), 92-100 (1968).

Boublikova, L., and Lu, B.C.Y., Isothermal vapour-liquid equilibria for ethanol/n-octane system, J Appl Chem, 19, 89-92 (1969).

Danciu, E., Vapor-liquid equilibrium: Equation for the boiling-point diagram of homogeneous binary mixtures, Int Chem Eng, 9(3), 426-434 (1969).

Ellis, S.R.M., Vapour-liquid equilibria: A review, Chem Engnr, July, CE289-302 (1969).

Manczinger, J.; Radnai, G., and Tettamanti, K., Vapour-liquid equilibrium of the system ethanol-dioxane, Periodica Polytechnica, 13, 189-206 (1969).

Nageshwar, G.D., and Mene, P.S., Vapour-liquid equilibrium data for n-butyl acetate/1,2 propylene glycol, J Appl Chem, 19, 195-196 (1969).

Popova, L.M.; Serafimov, L.A.,and Popov, V.V., Calculation of liquid-vapor phase equilibrium of non-ideal systems, Int Chem Eng, 9(4), 588-591 (1969).

Bowrey, R.G., and Marek, C.J., Calculation of liquid-vapour equilibrium data from vapour-phase measurements, Brit Chem Eng, 15(8), 1054-1055 (1970).

Furzer, I.A., and Ho, G.E., Vapour-liquid equilibrium for the system carbon tetrachloride-benzene, Brit Chem Eng, 15(1), 80-81 (1970).

Guerreri, G., Vapour-liquid equilibria in non-polar multicomponent systems, Brit Chem Eng, 15(8), 1049-1052 (1970).

Hudson, J.W., and Van Winkle, M., Multicomponent vapor-liquid equilibria in miscible systems from binary parameters, Ind Eng Chem Proc Des Dev, 9(3), 466-472 (1970).

Hutchison, H.P., and Fletcher, J.P., Orthogonal polynomials for fitting surfaces with application to ternery vapour-liquid equilibrium, Chem Engnr, Jan, CE29-31 (1970).

Jasinski, B., and Malanowski, S., Calculation of multicomponent vapour-liquid equilibrium from liquid boiling temperature data, Chem Eng Sci, 25(6), 913-920 (1970).

Lowell, P.S., and Van Winkle, M., Binary VLE correlation including heat of mixing data explicitly, Ind Eng Chem Proc Des Dev, 9(2), 289-292 (1970).

Tassios, D., Gas-liquid chromatography screens extractive distillation solvents, Hyd Proc, 49(7), 114-118 (1970).

Verhoeye, L.A.J., Determination of Wilson constants in correlation of vapour-liquid equilibrium data, Chem Eng Sci, 25(12), 1903-1910 (1970).

Wang, J.L.H.; Boublikova, L., and Lu, B.C.Y., Vapour-liquid equilibria of carbon tetrachloride-toluene system, J Appl Chem, 20, 172-174 (1970).

1971-1975

Bowrey, R.G., and Marek, C.J., Graphical solution for the conditions of vapor-liquid equilibrium, Brit Chem Eng, 16(1), 57-58 (1971).

Butcher, K.L.; Robinson, W.I., and Medani, M.S., Estimation of isothermal-isobaric vapour-liquid equilibrium data for methanol-ethanol mixtures, Brit Chem Eng, 16(10), 915-917 (1971).

Francesconi, R., et al., Liquid-vapour equilibrium in binary mixtures with associations of the components, Chem Eng Sci, 26(9), 1331-1356 (1971).

Mikolaj, P.G., and Dev, L., Prediction of vapor-liquid equilibria of petroleum fractions, AIChEJ, 17(2), 343-352 (1971).

Wang, J.L.H., and Lu, B.C.Y., Vapour-liquid equilibrium data for n-pentane/benzene system, J Appl Chem Biotechnol, 21, 297-299 (1971).

Wichterle, I.; Kobayashi, R., and Chappelear, P.S., Pinch points in phase composition (X-Y) diagrams, Hyd Proc, 50(11), 233-234 (1971).

Dakshinamurty, P.; Venkateswara, P.R., and Chiranjivi, C., Ternary vapour-liquid equilibria of acetone-chloroform-tetrachloroethylene, J Appl Chem Biotechnol, 22, 1217-1222 (1972).

Maripuri, V.C., and Ratcliff, G.A., Isothermal vapour-liquid equilibria in binary mixtures of ketones and alkanes, J Appl Chem Biotechnol, 22, 899-904 (1972).

Stitzell, J.A., Selecting the best vapor-pressure equation by computer, Chem Eng, 20 March, 136-138 (1972).

Dakshinamurty, P.; Subrahmanyam, V., and Rao, M.N., Ternary liquid equilibria: Acetic acid-water-toluene, J Appl Chem Biotechnol, 23, 323-328 (1973).

Dowling, G.R., and Todd, W.G., Comparing vapor-liquid equilibrium correlations, Chem Eng, 19 March, 115-120 (1973).

Guerreri, G., and Prausnitz, J.M., Vapour-liquid equilibria data (H2O-NH3-N2-H2-Ar-CH4), Proc Tech Int, April, 209-211 (1973).

Gugnoni, R.J.; Eldridge, J.W.; Okay, V.C., and Lee, T.J., Carbon dioxide-ethane system predictions, Hyd Proc, 52(9), 197-198 (1973).

Marina, J.M., and Tassios, D.P., Prediction of ternary liquid-liquid equilibrium from binary data, Ind Eng Chem Proc Des Dev, 12(3), 271-274 (1973).

Sinha, R., and Rao, R.J., Equilibrium distribution data for salicyclic acid-water-solvent systems, J Appl Chem Biotechnol, 23, 329-332 (1973).

Van Ness, H.C.; Byer, S.M., and Gibbs, R.E., Vapor-liquid equilibrium, AIChEJ, 19(2), 238-251 (1973).

Won, K.W., and Prausnitz, J.M., High-pressure vapor-liquid equilibria, Ind Eng Chem Fund, 12(4), 459-463 (1973).

Barton, P.; Holland, R.E., and McCormick, R.H., Correlation of vapor-liquid equilibria of C3-C5 hydrocarbons using solubility parameters, Ind Eng Chem Proc Des Dev, 13(4), 378-383 (1974).

Heidemann, R.A., Predict three-phase equilibria, Hyd Proc, 53(11), 167-170 (1974).

Nagata, I., and Ohta, T., Computation of vapor-liquid equilibrium data from binary and ternary vapor pressure and boiling point measurements, Ind Eng Chem Proc Des Dev, 13(3), 304-309 (1974).

Nitta, T.; Takeuchi, S., and Katayama, T., Effect of self-association on liquid-liquid equilibria, Chem Eng Sci, 29(11), 2213-2218 (1974).

van Zandijcke, F., and Verhoeye, L., Vapour-liquid equilibrium of ternary systems with limited miscibility at atmospheric pressure, J Appl Chem Biotechnol, 24, 709-730 (1974).

Abbott, M.M., and Van Ness, H.C., Vapor-liquid equilibrium, AIChEJ, 21(1), 62-76 (1975).

Arich, G.; Kikic, I., and Alessi, P., Liquid-liquid equilibrium for activity coefficient determination, Chem Eng Sci, 30(2), 187-192 (1975).

Fabries, J.F., and Renon, H., Method of evaluation and reduction of vapor-liquid equilibrium data of binary mixtures, AIChEJ, 21(4), 735-743 (1975).

Hall, K.R.; Eubank, P.T.; Myerson, A.S., and Nixon, W.E., New technique for collecting binary vapor-liquid equilibrium data without measuring composition: Method of intersecting isochores, AIChEJ, 21(6), 1111-1114 (1975).

Heidemann, R.A., and Mandhane, J.M., Ternary liquid-liquid equilibria: The van Laar equation, Chem Eng Sci, 30(4), 425-434 (1975).

Khanna, A., and Mukhopadhyay, M., Prediction of isobaric and isothermal vapour-liquid equilibria from limited experimental data, J Appl Chem Biotechnol, 25, 935-948 (1975).

Kohoutek, J., and Odstrcil, M., Determination of liquid-vapor equilibrium of complex hydrocarbon mixtures in industrial petrochemical plants, Int Chem Eng, 15(2), 269-274 (1975).

Nagata, I., and Gotoh, K., Palmer-Smith equation for correlation of VLE data, Ind Eng Chem Proc Des Dev, 14(1), 98-100 (1975).

Palmer, D.A., Predicting equilibrium relationships for maverick mixtures, Chem Eng, 9 June, 80-85 (1975).

Robinson, D.B., and Ng, H.J., Improved hydrate predictions, Hyd Proc, 54(12), 95-96 (1975).

Tamir, A., and Wisniak, J., Vapour-liquid equilibria in associating solutions, Chem Eng Sci, 30(3), 335-342 (1975).

1976-1980

George, B.; Brown, L.P.; Farmer, C.H.; Buthod, P., and Manning, F.S., Computation of multicomponent, multiphase equilibrium, Ind Eng Chem Proc Des Dev, 15(3), 372-377 (1976).

Hamam, S.E.M., and Lu, B.C.Y., Phase equilibria for system propane-ethane-carbon dioxide, Can JCE, 54, 333-340 (1976).

Joffe, J., Vapor-liquid equilibria by pseudocritical method, Ind Eng Chem Fund, 15(4), 298-303 (1976).

Lukacs, J.; Kemeny, S.,; Manczinger, J., and Tettamanti, K., Importance to the real behaviour of the vapour phase for the evaluation of vapour-liquid equilibrium data, Periodica Polytechnica, 20, 47-68 (1976).

Rod, V., Correlation of equilibrium data in ternary liquid-liquid systems, Chem Eng J, 11(2), 105-110 (1976).

Ronc, M., and Ratcliff, G.A., Measurement of vapor-liquid equilibria using semi-continuous total pressure static-equilibrium still, Can JCE, 54, 326-335 (1976).

Tatevosyan, A.V.; Khachatryan, S.S., and Kuleshova, Y.P., Modeling of vapor-liquid equilibrium process, Int Chem Eng, 16(2), 300-302 (1976).

Vega, R., and Vera, J.H., Phase equilibria of concentrated aqueous solutions containing volatile strong electrolytes, Can JCE, 54, 245-255 (1976).

Hamam, S.E.M.; Chung, W.K.; Elshayal, I.M., and Lu, B.C.Y., Generalized temperature-dependent parameters of the Redlich-Kwong equation of state for vapor-liquid equilibrium calculations, Ind Eng Chem Proc Des Dev, 16(1), 51-59 (1977).

Jenkins, J.D., and Gibson-Robinson, M., Vapour-liquid equilibrium in systems with association in both phases, Chem Eng Sci, 32(8), 931-938 (1977).

Kikic, I., and Alessi, P., Liquid-liquid equilibrium for activity coefficient determination, Can JCE, 55, 78-90 (1977).

Leach, M.J., An approach to multiphase vapor-liquid equilibria, Chem Eng, 23 May, 137-140 (1977).

Leesley, M.E., and Heyen, G., Dynamic approximation method of handling vapor-liquid equilibrium data in computer calculations for chemical processes, Comput Chem Eng, 1(2), 109-112 (1977).

Newsham, D.M.T., and Vahdat, N., Prediction of vapour-liquid-liquid equilibria from liquid-liquid equilibria, Chem Eng J, 13(1), 27-40 (1977).

Clark, F.G., and Koppany, C.R., Method tests V/L correlations, Hyd Proc, 57(11), 282-286 (1978).

Dingrani, J.G., and Thodos, G., Vapor-liquid equilibrium behavior of ethane/n-hexane system, Can JCE, 56, 616-625 (1978).

Edwards, T.J.; Newman, J., and Prausnitz, J.M., Thermodynamics of vapor-liquid equilibria for ammonia-water system, Ind Eng Chem Fund, 17(4), 264-269 (1978).

Francesconi, R., et al., Isobaric liquid-vapour equilibrium data correlation for ternary mixture exhibiting association: Acetic acid-acetone-trichloroethylene, Can JCE, 56, 364-370 (1978).

Gomez-Nieto, M., and Thodos, G., Vapor-liquid equilibrium behavior for propane-acetone system at elevated pressures, Chem Eng Sci, 33(12), 1589-1596 (1978).

Heller, H., Aqueous ammonia liquid-vapor composition nomograph, Chem Eng, 18 Dec, 131-132 (1978).

Kemeny, S., and Manczinger, J., Treatment of binary vapour-liquid equilibrium data, Chem Eng Sci, 33(1), 71-76 (1978).

Mapstone, G.E., Triangular or rectangular phase diagrams, Chem Engnr, Feb, 109 (1978).

Ozkardesh, H.; Tarakad, R.R., and Adler, S.B., Single-set treatment of all available binary system VLE data leads to simplified correlation procedures, Ind Eng Chem Fund, 17(3), 206-209 (1978).

Tasic, A., et al., Vapour-liquid equilibria of acetone-benzene-cyclohexane, and acetone-cyclohexane systems at 25 degC, Chem Eng Sci, 33(2), 189-198 (1978).

Van Ness, H.C.; Pedersen, F., and Rasmussen, P., Vapor-liquid equilibrium: Data reduction by maximum likelihood, AIChEJ, 24(6), 1055-1063 (1978).

Dincer, S.; Bonner, D.C., and Elefritz, R.A., Vapor-liquid equilibria in benzene-polybutadiene-cyclohexane system, Ind Eng Chem Fund, 18(1), 54-59 (1979).

Hanks, R.W.; O'Neill, T.K., and Christensen, J.J., Prediction of vapor-liquid equilibrium from heat of mixing data for binary hydrocarbon-alcohol mixtures, Ind Eng Chem Proc Des Dev, 18(3), 408-414 (1979).

Kusik, C.L.; Meissner, H.P., and Field, E.L., Estimation of phase diagrams and solubilities for aqueous multi-ion systems, AIChEJ, 25(5), 759-762 (1979).

Prausnitz, J.M., Practical applications of molecular thermodynamics for calculating phase equilibria, Int Chem Eng, 19(3), 401-410 (1979).

Sayegh, S.G.; Vera, J.H., and Ratcliff, G.A., Vapor-liquid equilibria for ternary system: n-heptane/n-propanol/1-chlorobutane and its constituent binaries at 298.15K, Can JCE, 57, 513-520 (1979).

Van Ness, H.C., and Abbott, M.M., Vapor-liquid equilibrium: Standard state fugacities for supercritical components, AIChEJ, 25(4), 645-653 (1979).

Abdel-Halim, T.A., Effect of steam on V-L of crude oil, Hyd Proc, 59(1), 115-119 (1980).

Anderson, T.F., and Prausnitz, J.M., Computational methods for high-pressure phase equilibria and other fluid-phase properties using a partition function (a review), Ind Eng Chem Proc Des Dev, 19(1), 1-14 (1980).

Barton, P.; Campbell, M.L.; McCormick, R.H., and Holland, R.E., Vapor-liquid equilibria for multicomponent C4 hydrocarbons, Ind Eng Chem Proc Des Dev, 19(2), 272-279 (1980).

Brandani, V.; Giacomo, G.D., and Foscolo, P.U., Isothermal vapor-liquid equilibria for the water-formaldehyde system: A predictive thermodynamic model, Ind Eng Chem Proc Des Dev, 19(1), 179-185 (1980).

Cavallotti, P.; Celeri, G.; Leonardis, B., and Gardini, L., Calculation of multicomponent multiphase equilibria, Chem Eng Sci, 35(11), 2297-2304 (1980).

Fredenslund, A.; Rasmussen, P., and Michelsen, M.L., Recent progress in computation of equilibrium ratios, Chem Eng Commns, 4(4), 485-500 (1980).

Howat, C.S., and Swift, G.W., New correlation of propene-propane vapor-liquid equilibrium, Ind Eng Chem Proc Des Dev, 19(2), 318-323 (1980).
Legret, D.; Richon, D., and Renon, H., Measuring vapor-liquid equilibria up to 50 bar, Ind Eng Chem Fund, 19(1), 122-126 (1980).
Reddy, K.V., and Husain, A., Vapor-liquid equilibrium relationship for ammonia in presence of other gases, Ind Eng Chem Proc Des Dev, 19(4), 580-586 (1980).
Sayegh, S.G., and Vera, J.H., Model-free methods for vapor-liquid equilibria calculations: Binary systems, Chem Eng Sci, 35(11), 2247-2256 (1980).

1981-1985
Castillo, J., and Grossmann, I.E., Computation of phase and chemical equilibria, Comput Chem Eng, 5(2), 99-108 (1981).
Cova, D.R., Vapor-liquid equilibria for stripper design, Ind Eng Chem Fund, 20(1), 99-100 (1981).
Goral, M.; Maczynski, A.; Schmidt, G., and Wenzel, H., Vapor-liquid equilibrium calculations in binary systems of hydrocarbons, Ind Eng Chem Fund, 20(3), 267-277 (1981).
Gothard, F.A., Simple model for predicting fluid-phase equilibria, Ind Eng Chem Fund, 20(3), 300-302 (1981).
Mentzer, R.A.; Greenkorn, R.A., and Chao, K.C., Principle of corresponding states and vapor-liquid equilibria of molecular fluids, and their mixtures with light gases, Ind Eng Chem Proc Des Dev, 20(2), 240-252 (1981).
Prausnitz, J.M., Calculation of phase equilibria for separation operations (review paper), Trans IChemE, 59, 3-16 (1981).
Saboungi, M.L., and Blander, M., Calculation of phase diagrams of ionic systems from fundamental solution theories, Chem Eng Commns, 11(6), 327-334 (1981).
Schotte, W., Collection of phase equilibrium data for separation technology, Ind Eng Chem Proc Des Dev, 19(3), 432-439 (1980); 20(3), 578 (1981).
Tamir, A., New correlations for fitting multicomponent vapor-liquid equilibria data and prediction of azeotropic behaviour, Chem Eng Sci, 36(9), 1453-1474 (1981).
Tochigi, K.; Lu, B.C.Y.; Ochi, K., and Kojima, K., Temperature dependence of ASOG parameters for VLE calculations, AIChEJ, 27(6), 1022-1024 (1981).
Arlt, W., and Onken, U., Liquid-liquid equilibria of organic compounds: Measurement, correlation and prediction, Chem Eng Commns, 15(1), 207-214 (1982).
Guy, J.L., Modeling the phase equilibria in dynamic systems, Chem Eng, 29 Nov, 75-79 (1982).
Malanowski, S., Rapid and accurate method for experimental determination of VLE, Chem Eng Commns, 18(1), 63-72 (1982).
Muir, R.F., and Howat, C.S., Predicting solid-liquid equilibrium data from vapor-liquid data, Chem Eng, 22 Feb, 89-92 (1982).
Mukhopadhyay, M., and Sahasranaman, K., Computation of multicomponent liquid-liquid equilibrium data for aromatics extraction systems, Ind Eng Chem Proc Des Dev, 21(4), 632-640 (1982).

Patino-Leal, H., and Reilly, P.M., Statistical estimation of parameters in vapor-liquid equilibrium, AIChEJ, 28(4), 580-587 (1982).
Pawlikowski, E.M.; Newman, J., and Prausnitz, J.M., Phase equilibria for aqueous solutions of ammonia and carbon dioxide, Ind Eng Chem Proc Des Dev, 21(4), 764-770 (1982).
Radosz, M.; Lin, H.M., and Chao, K.C., High-pressure vapor-liquid equilibria in asymmetric mixtures using new mixing rules, Ind Eng Chem Proc Des Dev, 21(4), 653-658 (1982).
Rice, V.L., Program performs vapor-liquid equilibrium calculations, Chem Eng, 28 June, 77-86 (1982).
Schotte, W., Vapor-liquid equilibrium calculations for polymer solutions, Ind Eng Chem Proc Des Dev, 21(2), 289-296 (1982).
Wozny, G., and Cremer, H., Phase equilibria of binary hydrocarbon mixtures, Int Chem Eng, 22(4), 611-619 (1982).
Adler, S.B., VLE predictions revisited, Hyd Proc, June, 93-97 (1983).
Aucejo, A.; Orchilles, A.; Berna, A., and Mulet, A., Vapor-liquid equilibrium calculations using the topological treatment of mixtures, Can JCE, 61(5), 745-752 (1983).
Hegazi, M.F., and Salem, A.B., Ternary data for acetic acid-water-mesityl oxide system, J Chem Tech Biotechnol, 33A(3), 145-149 (1983).
Hiranuma, M., Estimation of ternary liquid-liquid equilibria with binary Wilson parameters, Ind Eng Chem Fund, 22(4), 364-366 (1983).
Kertes, A.S., Methodology for processing numeric liquid-liquid distribution data, Solv Extn & Ion Exch, 1(1), 5-8 (1983).
Orlandini, M.; Fermeglia, M.; Kikic, I., and Alessi, P., Liquid-liquid equilibria for water-propanol and water-butanol-chlorocompound systems, Chem Eng J, 26(3), 245-250 (1983).
Ramasubramanian, J., and Srinivasan, D., Effect of dissolved sodium chloride salt on the liquid-liquid equilibria of the ternary system isopropyl ether-acetic acid-water, Chem Eng Commns, 19(4), 335-342 (1983).
Brule, M.R., and Corbett, R.W., Phase diagrams for critical-solvent processing (supercritical-gas extraction), Hyd Proc, June, 73-77 (1984).
Chimowitz, E.H.; Macchietto, S.; Anderson, T.F., and Stutzman, L.F., Local models for representing phase equilibria in multicomponent nonideal vapor-liquid and liquid-liquid systems, Ind Eng Chem Proc Des Dev, 22(2), 217-225 (1983); 23(3), 609-618 (1984).
Cooper, W.C., and Mak, Y.F., Modelling of equilibrium data for solvent extraction of copper and/or nickel from ammoniacal solutions, Solv Extn & Ion Exch, 2(7), 959-984 (1984).
Furzer, I.A., Liquid-liquid equilibria using UNIFAC for gasohol extraction systems, Ind Eng Chem Proc Des Dev, 23(2), 387-391 (1984).
Hunting, R.D., and Robinson, J.W., Predict composition of binary mixtures, Chem Eng, 9 July, 91-96 (1984).
Kuus, M., et al., Vapor-liquid equilibria prediction for systems containing unsaturated hydrocarbons: UNIFAC model for predicting boiling temperature, Chem Eng Commns, 26(1), 105-110 (1984).
Ohanomah, M.O., and Thompson, D.W., Computation of multicomponent phase equilibria, Comput Chem Eng, 8(3), 147-170 (1984).

Patel, A.N., Ternary phase-equilibrium studies of furfural-water-solvent systems, J Chem Tech Biotechnol, 34A(4), 161-164 (1984).
Poettmann, F.H., Butane hydrates equilibria, Hyd Proc, June, 111-112 (1984).
Rao, K.V., and Prasad, A.R., Measurement, correlation and prediction of binary vapor-liquid equilibria for alcohol-tetrachloroethene systems, Can JCE, 62(1), 142-148 (1984).
Sato, T., and Nakamura, T., Proposal for digitization of liquid-liquid distribution data, Solv Extn & Ion Exch, 2(3), 353-364 (1984).
Wong, D.S.H., and Sandler, S.I., Calculation of vapor-liquid-liquid equilibrium, Ind Eng Chem Fund, 23(3), 348-354 (1984).
Wong, D.S.H.; Sandler, S.I., and Teja, A., Vapor-liquid equilibrium calculations using a generalized corresponding states principle, Ind Eng Chem Fund, 23(1), 38-49 (1984).
Gothard, F.A., Predicting liquid-vapor equilibria by a model of the interaction equilibrium in restrained molecular systems, Ind Eng Chem Fund, 24(3), 330-339 (1985).
Herskowitz, A., Estimation of fluid properties and phase equilibria, Chem Eng Educ, 19(3), 148-149 (1985).
Hofman, T., Smoothing in the direct VLE data correlation methods, Chem Eng Sci, 40(4), 668-670 (1985).
Kabadi, V.N., and Danner, R.P., Modified SRK equation of state for water-hydrocarbon phase equilibria, Ind Eng Chem Proc Des Dev, 24(3), 537-541 (1985).
Langhorst, R.; Zeckland, S., and Knapp, H., A data bank for high pressure vapour-liquid equilibria, Chem Eng & Proc, 19(4), 205-210 (1985).
Lichtenbelt, J.H., and Schram, B.J., Vapor-liquid equilibrium of water-acetone-air at ambient temperatures and pressures (comparison of VLE methods), Ind Eng Chem Proc Des Dev, 24(2), 391-397 (1985).
Lin, H.M.; Kim, H.; Leet, W.A., and Chao, K.C., New vapor-liquid equilibrium apparatus for elevated temperatures and pressures, Ind Eng Chem Fund, 24(2), 260-262 (1985).
Moysan, J.M.; Paradowski, H., and Vidal, J., Correlation defines phase equilibria for hydrogen, methane, and nitrogen mixes, Hyd Proc, July, 73-76 (1985).
Paunovic, R.; Petkovska, M.; Krnic, Z., and Ciric, G., Simplified model for correlation of VLE data in the carbon dioxide-ammonia-water system, J Chem Tech Biotechnol, 35A(6), 311-319 (1985).
Roberts, B.E., and Tremaine, P.R., Vapour-liquid equilibrium calculations for dilute aqueous solutions of carbon dioxide, hydrogen sulfide, ammonia, and sodium hydroxide to 300 degC, Can JCE, 63(2), 294-300 (1985).
Walas, S.M., Estimate the eutectic conditions of binary mixtures, Chem Eng, 19 Aug, 86 (1985).

1986-1988

Charos, G.N.; Clancy, P., and Gubbins, K.E., Representation of highly non-ideal phase equilibria using computer graphics, Chem Eng Educ, 20(2), 88-91 (1986).
Desrosiers, R.E., Generation of a ternary phase diagram for vapour-liquid-liquid equilibrium, Chem Eng Educ, 20(2), 94-99,104 (1986).

Gautam, R., and Wareck, J.S., Computation of physical and chemical equilibria: Alternate specifications, Comput Chem Eng, 10(2), 143-151 (1986).
Kuczinski, M.; Hart, W., and Westerterp, K.R., Binary vapour-liquid equilibria of methanol with sulfolane, tetraethylene glycol dimethyl ether and 18-crown-6, Chem Eng & Proc, 20(1), 53-58 (1986).
Letcher, T.M.; Heyward, C.; Wootton, S., and Shuttleworth, B., Ternary phase diagrams for gasoline-water-alcohol mixtures, Fuel, 65(7), 891-894 (1986).
Liedo-Galindo, G., Multicomponent-mixture condensing curves, Chem Eng, 12 May, 89-95 (1986).
Maurer, G., Vapor-liquid equilibrium of formaldehyde- and water-containing multicomponent mixtures, AIChEJ, 32(6), 932-948 (1986).
Mock, W.; Evans, L.B., and Chen, C.C., Thermodynamic representation of phase equilibria of mixed-solvent electrolyte systems, AIChEJ, 32(10), 1655-1664 (1986).
Prasad, A.R., and Rao, K.V., Vapor-liquid equilibria of acetonitrile-1-propanol mixtures, Can JCE, 64(5), 813-819 (1986).
Ruff, W.A.; Glover, C.J., and Watson, A.T., Vapor-liquid equilibria from perturbation gas chromatography, AIChEJ, 32(12), 1948-1962 (1986).
Ruiz Bevia, F.; Zapata. R.R.; Gomis, A.F.M., and Rico, D.P., Model for phase equilibria correlation and prediction, Can JCE, 64(2), 311-322 (1986).
Weber, J.H., Predict parameters of correlations for liquid-phase composition, Chem Eng, 21 July, 57-62 (1986).
Wilcox, R.F., and White, S.L., Selecting the proper model to simulate vapor-liquid equilibrium, Chem Eng, 27 Oct, 141-144 (1986).
Wu, J.S., and Bishnoi, P.R., An algorithm for three-phase equilibrium calculations, Comput Chem Eng, 10(3), 269-276 (1986).
Atkinson, N., Getting V-L-E right, Proc Engng, Oct, 63 (1987).
Baes, C.F.; McDowell, W.J., and Bryan, S.A., Interpretation of equilibrium data from synergistic solvent-extraction systems, Solv Extn & Ion Exch, 5(1), 1-28 (1987).
Boduszynski, M.M., Composition of heavy petroleums, Energy & Fuels, 1(1), 2-12 (1987).
Du, P.C., and Mansoori, G.A., Phase equilibrium of multicomponent mixtures: Continuous-mixture Gibbs free energy minimization and phase rule, Chem Eng Commns, 54, 139-148 (1987).
Eubank, P.T., and Barrufet, M.A., General conditions of colinearity at the phase boundaries of fluid mixtures, AIChEJ, 33(11), 1882-1887 (1987).
Liedy, W.; Schlunder, E.U., and Turek, T., Vapor-liquid equilibrium of system benzene-hexamethyl disiloxane at normal pressure, Chem Eng & Proc, 22(3), 177-180 (1987).
Maddox, R.N., et al., Correlation of acid gas-ethanolamine equilibrium using ionic concentrations, Plant/Opns Prog, 6(2), 112-117 (1987).
Prange, M.M., and Riepe, W.H., Phase equilibria of multicomponent model mixture in supercritical carbon dioxide and trifluoromethane, Chem Eng & Proc, 22(4), 183-192 (1987).
Tan, T.C., Model for predicting the effect of dissolved salt on the vapour-liquid equilibrium of solvent mixtures, CER&D, 65(4), 355-366 (1987).
Abbasian, M.J., and Weil, S.A., Phase equilibria of continuous fossil fuel process oils, AIChEJ, 34(4), 574-582 (1988).

Abusleme, J.A., and Vera, J.H., Vapor-liquid equilibrium data for alcohol/n-amylamine binary systems at 333.15K: Reduction of experimental data by novel model-free method, Can JCE, 66(6), 964-969 (1988).

Adams, W.R.; Zollweg, J.A.; Streett, W.B., and Rizvi, S.S.H., New apparatus for measurement of supercritical fluid-liquid phase equilibria, AIChEJ, 34(8), 1387-1391 (1988).

Barbosa, D., and Doherty, M.F., Influence of equilibrium chemical reactions on vapor-liquid phase diagrams, Chem Eng Sci, 43(3), 529-540 (1988).

Blanco, A.; Correa, J.M.; Arce, A., and Correa, A., Liquid-liquid equilibria of the system water + acetic acid + methyl propyl ketone, Can JCE, 66(1), 137-141 (1988).

Choi, J.S., and Rhim, J.N., Ternary liquid-liquid equilibria of water, acetone and solvents, Int Chem Eng, 28(4), 698-707 (1988).

de Pablo, J.J., and Prausnitz, J.M., Thermodynamics of liquid-liquid equilibria including the critical region, AIChEJ, 34(10), 1595-1606 (1988).

Englezos, P., and Bishnoi, P.R., Prediction of gas hydrate formation conditions in aqueous electrolyte solutions, AIChEJ, 34(10), 1718-1721 (1988).

Hooper, H.H.; Michel, S., and Prausnitz, J.M., Correlation of liquid-liquid equilibria for some water-organic liquid systems, (20 - 250 degrees C), Ind Eng Chem Res, 27(11), 2182-2188 (1988).

Kvaalen, E., and Tondeur, D., Constraints on phase equilibrium equations, Chem Eng Sci, 43(4), 803-810 (1988).

Marco, J.M.; Galan, M.I., and Costa, J., Quaternary liquid-liquid equilibrium: Water-phosphoric acid-1 pentanol-3 pentanone at 25 degC, Solv Extn & Ion Exch, 6(1), 125-140 (1988).

Ortega, J., et al., Vapour-liquid equilibria of binary mixtures of alcohols, propyl ethanoate and ethyl propanoate, Can JCE, 65(6), 982-990 (1988).

Peters, C.J., et al., Global phase behavior of mixtures of short and long n-alkanes, AIChEJ, 34(5), 834-839 (1988).

Rajendran, M., et al., Effect of dissolved inorganic salts on vapor-liquid equilibria and enthalpy of mixing of methanol-ethyl acetate system, Chem Eng Commns, 74, 179-194 (1988).

Salem, A.B., and Sheirah, M.A., General correlations for liquid-liquid data systems, Sepn Sci Technol, 23(14), 2417-2444 (1988).

Sangster, J.; Talley, P.K.; Bale, C.W., and Pelton, A.D., Coupled optimization and evaluation of phase equilibria and thermodynamic properties of benzene-cyclohexane system, Can JCE, 66(6), 881-895 (1988).

Seckner, A.J.; McClellan, A.K., and McHugh, M.A., High-pressure solution behaviour of polystyrene-toluene-ethane system, AIChEJ, 34(1), 9-16 (1988).

Tan, T.C.; Teo, W.K., and Ti, H.C., Vapour-liquid equilibria of ethanol-water system saturated with glucose at subatmospheric pressures, CER&D, 66(1), 75-83 (1988).

Wang, S.H., and Whiting, W.B., Comparison of distribution functions for calculation of phase equilibria of continuous mixtures, Chem Eng Commns, 71, 127-144 (1988).

Wogatzki, H., and Gutsche, B., Binary and ternary vapor-liquid equilibrium calculations for methyl chloride, dimethyl ether and methanol, Chem Eng & Proc, 24(1), 57-62 (1988).

1.5 Physical and Thermodynamic Properties

1966-1967

Hadden, S.T., New correlation for surface tension of hydrocarbons, Hyd Proc, 45(10), 161-164 (1966).

Balakrishnan, S., and Krishnan, V., Nomogram for saturated vapour density of aliphatic monohydric alcohols, Brit Chem Eng, 12(10), 1617 (1967).

Chidambaram, S., Nomogram for thermal conductivities of organic liquids, Brit Chem Eng, 12(11), 1767 (1967).

Chidambaram, S., Nomogram for vapour heat capacity of n-alkyl benzenes, Brit Chem Eng, 12(3), 399 (1967).

Christie, A.O., and Crisp, D.J., Activity coefficients of the n-primary, secondary and tertiary aliphatic amines in aqueous solution, J Appl Chem, 17, 11-14 (1967).

Davis, D.S., Nomogram for viscosities of aqueous slurries, Brit Chem Eng, 12(4), 587 (1967).

Murti, P.S.; Sriram, M., and Narasimhamurty, G.S.R., Relative volatility, boiling points and heats of mixing charts for the ternary system: Benzene-cyclohexane-aniline, Brit Chem Eng, 12(12), 1882-1885 (1967).

Narsimhan, G., New correlation of latent heat of vaporisation, Brit Chem Eng, 12(6), 897-899 (1967).

Narsimhan, G., Temperature dependence of saturated liquid densities, Brit Chem Eng, 12(8), 1239-1240 (1967).

Nathan, D.I., Predictions of mixture enthalpies, Brit Chem Eng, 12(2), 223-226 (1967).

Pachaiyappan, V.; Ibrahim, S.H., and Kuloor, N.R., Simple correlation for determining viscosity of organic liquids, Chem Eng, 22 May, 193-196 (1967).

Pachaiyappan, V.; Ibrahim, S.H., and Kuloor, N.R., A new correlation for thermal conductivity, Chem Eng, 13 Feb, 140-144 (1967).

Pachaiyappan, V.; Ibrahim, S.H., and Kuloor, N.R., New correlation for surface tension, Chem Eng, 23 Oct, 172-174 (1967).

Pachaiyappan, V.; Ibrahim, S.H., and Kuloor, N.R., Simple correlation for liquid heat capacity, Chem Eng, 9 Oct, 241-243 (1967).

Reddy, C.C., and Murti, P.S., Enthalpy-concentration charts for alcohol-aromatic systems, Brit Chem Eng, 12(8), 1231-1235 (1967).

Renon, H., and Prausnitz, J.M., Thermodynamics of alcohol-hydrocarbon solutions, Chem Eng Sci, 22(3), 299-308 (1967).

Sastri, N.V.S., Heat capacity of liquids determined by simple equation, Chem Eng, 23 Oct, 200-202 (1967).

Tans, A.M.P., Nomogram for thermal conductivity and specific heat of aqueous solutions of ethylene glycol, Brit Chem Eng, 12(1), 93 (1967).

Van Vorst, W.D., Make your own diagram to estimate enthalpy changes of real gases, Chem Eng, 19 June, 229-231 (1967).

Viswanath, D.S., Thermal conductivities of liquids, AIChEJ, 13(5), 850-853 (1967).

Viswanath, D.S., and Kuloor, N.R., Ideal critical volume applied to Lennard-Jones potential energy parameters, Brit Chem Eng, 12(7), 1103 (1967).

Weintraub, M., and Corey, P.E., High-temperature viscosity of gases estimated quickly, Chem Eng, 23 Oct, 204 (1967).

1968

Anthony, R.G., and McKetta, J.J., Estimating water in hydrocarbons, Hyd Proc, 47(6), 131-134 (1968).

Austin, G.T., Quick charts for integrating heat-capacity equations, Chem Eng, 3 June, 128-130 (1968).

Balakrishnan, S., and Krishnan, V., Nomogram for densities of saturated vapours of 59 organic compounds, Brit Chem Eng, 13(12), 1741 (1968).

Butcher, K.L., and Medani, M.S., Thermodynamic properties of methanol-benzene mixtures at elevated temperatures, J Appl Chem, 18, 100-107 (1968).

Caplan, F., Nomographs to covert actual gas volume to standard volume, Chem Eng, 12 Feb, 168 (1968).

Caplan, F., Nomograph to determine humidity variables, Chem Eng, 1 July, 108 (1968).

Chidambaram, S., Nomogram for thermal conductivities of organic liquids, Brit Chem Eng, 13(3), 397 (1968).

Chidambaram, S., Nomogram for collision diameter, Brit Chem Eng, 13(10), 1459 (1968).

Chidambaram, S., Nomogram for vapour heat capacity of n-alkenes, Brit Chem Eng, 13(7), 1019 (1968).

Gallant, R.W., Physical properties of hydrocarbons, Hyd Proc, 45(10), 171-182 (1966); 47(6), 139-148 (1968); 47(8), 127-136; 47(9), 269-276; 47(11), 223-229; 47(12), 89-96 (1968).

Gonzalez, M.H., and Lee, A.L., Graphical viscosity correlation for hydrocarbons, AIChEJ, 14(2), 242-244 (1968).

Hinkamp, J.B., and Riggs, R.J., Estimate the Cetane number of fuels, Hyd Proc, 47(11), 233-237 (1968).

Procopio, J.M., and Su, G.J., Calculating latent heat of vaporization, Chem Eng, 3 June, 101-104 (1968).

Rao, A.V., and Ibrahim, S.H., Nomogram for specific heat and thermal conductivity of organic liquids, Brit Chem Eng, 13(3), 357 (1968).

Rao, A.V., and Ibrahim, S.H., Nomogram for diffusion coefficient of vapours in liquids at infinite dilution, Brit Chem Eng, 13(5), 699 (1968).

Rao, A.V.; Ibrahim, S.H., and Kuloor, N.R., Thermal conductivities of substituted organic liquids, Chem Eng, 16 Dec, 128-130 (1968).

Reddy, C.C., Enthalpy-concentration data for acetonitrile-water and acetonitrile-ethanol, Brit Chem Eng, 13(10), 1443-1445 (1968).

Roy, D., and Thodos, G., Thermal conductivity of gases, Ind Eng Chem Fund, 7(4), 529-534 (1968).

Sastri, S.R.S., Nomogram for heats of vaporisation of hydrocarbons, Brit Chem Eng, 13(8), 1159 (1968).

Sehgal, I.J.S.; Yesavage, V.F.; Mather, A.E., and Powers, J.E., Enthalpy predictions tested by data, Hyd Proc, 47(8), 137-143 (1968).

Shelton, R.J., A computer program for calculating flash equilibrium characteristics and heat contents of hydrocarbon systems, Chem Engnr, Nov, CE385-398 (1968).

Tans, A.M.P., Determine properties of n-mono-olefins, Hyd Proc, 47(11), 243-247 (1968).
Vaillant, A., Accurate determination of wet-bulb temperature, Chem Eng, 26 Aug, 134-136 (1968).
Viswanath, D.S., Ideal critical volume as a correlating parameter, Brit Chem Eng, 13(4), 532 (1968).
Zanker, A., Nomogram for temperature-viscosity relations of refinery products, Brit Chem Eng, 13(11), 1589 (1968).

1969

Adkhamov, A.A., and Shokirov, S., Statistical theory of viscous properties of simple liquids, Int Chem Eng, 9(3), 435-439 (1969).
Banerjee, S.C., Thermodynamic properties of chloroacetic acids, Brit Chem Eng, 14(5), 671 (1969).
Burgess, M.P., and Germann, R.P., Physical properties of hydrogen sulfide-water mixtures, AIChEJ, 15(2), 272-275 (1969).
Chidambaram, S., Nomogram for thermal conductivity of liquid refrigerants, Brit Chem Eng, 14(9), 1257 (1969).
Das, T.R.; Ibrahim, S.H., and Kuloor, N.R., Simple correlation predicts critical volume, Chem Eng, 1 Dec, 120-121 (1969).
Davis, D.S., Nomogram for mean molal specific heats of gaseous combustion products, Brit Chem Eng, 14(4), 535 (1969).
Davis, D.S., Nomogram for viscosities of gases, Brit Chem Eng, 14(7), 993 (1969).
Davis, D.S., Nomogram for vapour pressures of chlorobenzenes, Brit Chem Eng, 14(6), 849 (1969).
Gallant, R.W., Physical properties of hydrocarbons, Hyd Proc, 48(1), 153-160; 48(4), 151-162; 48(5), 143-151; 48(6), 141-148; 48(7), 135-141; 48(8), 117-126; 48(9), 199-205 (1969).
Gold, P.I., and Ogle, G.J., Estimating thermophysical properties of liquids, Chem Eng, 7 Oct, 152-154; 4 Nov, 185-190; 18 Nov, 170-174 (1968); 13 Jan, 119-122; 24 Feb, 109-112; 10 March, 122-129 (1969).
Gold, P.I., and Ogle, G.J., Estimating thermophysical properties of liquids, Chem Eng, 7 April, 130-132; 19 May, 192-194; 30 June, 129-131; 14 July, 121-123; 11 Aug, 97-100; 8 Sept, 141-146 (1969).
Hoffman, F.J., Relations between true boiling point and ASTM distillation curves, Chem Eng Sci, 24(1), 113-118 (1969).
Kojima, K.; Ochi, K., and Nakazawa, Y., Relationship between liquid activity coefficient and composition for ternary systems, Int Chem Eng, 9(2), 342-347 (1969).
Kuong, J.F., Nomogram for coefficient of thermal expansion for liquids, Brit Chem Eng, 14(12), 1729 (1969).
Lenoir, J.M., Predict K-values at low temperatures, Hyd Proc, 48(9), 167-172; 48(10), 121-124 (1969).
Major, C.J., Values of gas constant easily determined, Chem Eng, 8 Sept, 162-164 (1969).
Mapstone, G.E., Nomogram for rate of formation of biuret in urea solutions, Brit Chem Eng, 14(1), 81 (1969).

Mathur, B.C.; Ibrahim, S.H., and Kuloor, N.R., New, simple correlation predicts critical pressure, Chem Eng, 19 May, 212-216 (1969).
Mathur, B.C.; Ibrahim, S.H., and Kuloor, N.R., New, simple correlation predicts critical temperature, Chem Eng, 24 March, 182-184 (1969).
Naziev, Y.M., and Abasov, A.A., Thermal conductivities of gaseous unsaturated hydrocarbons, Int Chem Eng, 9(4), 631-633 (1969).
Osburn, J.O., and Markovic, P.L., Calculating Henry's law constant for gases in organic liquids, Chem Eng, 25 Aug, 105-108 (1969).
Ramaswamy, V., Nomogram for specific heats of liquids, Brit Chem Eng, 14(11), 1597 (1969).
Rowlinson, J.S., and Watson, I.D., Prediction of thermodynamic properties of fluids and fluid mixtures, Chem Eng Sci, 24(10), 1565-1580 (1969).
Sastri, S.R.S.; Rao, M.V.R.; Reddy, K.A., and Doraiswamy, L.K., Generalised method for estimating latent heat of vaporisation of organic compounds, Brit Chem Eng, 14(7), 959-963 (1969).
Singh, D.; Gupta, R.M., and Raju, K.S.N., H-x data for binary systems containing cumene, Brit Chem Eng, 14(4), 521 (1969).
Sisson, W., Nomograph for weight percent in ammonium nitrate or urea solutions, Hyd Proc, 48(7), 142 (1969).
Tans, A.M.P., Nomogram for specific volume of carbon dioxide, Brit Chem Eng, 14(5), 705 (1969).
Tans, A.M.P., Determine properties of n-acetylenes, Hyd Proc, 48(7), 155-159 (1969).
Tans, A.M.P., Determine thermodynamic properties of n-alkyl cyclopentanes, Hyd Proc, 48(10), 136-140 (1969).
Tans, A.M.P., Determine properties of n-aldehydes, Hyd Proc, 48(9), 156-160 (1969).
Tans, A.M.P., Determine properties of n-alcohols, Hyd Proc, 48(5), 168-172 (1969).
Williamson, A.G., Simplified correlation of vapour pressure of homologous series, Chem Engnr, Sept, CE158-159 (1969).
Zanker, A., Nomogram for viscosity index of lubricating oils, Brit Chem Eng, 14(2), 207; 14(3), 359 (1969).

1970

Bruin, S., Activity coefficient relations in miscible and partially miscible multicomponent systems, Ind Eng Chem Fund, 9(2), 305-314 (1970).
Carnahan, N.F., and Vadovic, C.J., Predict gas viscosity of alkanes, Hyd Proc, 49(5), 159-160 (1970).
Chidambaram, S., Nomogram for collision diameter, Brit Chem Eng, 15(7), 951 (1970).
Davis, D.S., Nomogram for thermal conductivity of nitric oxide, Brit Chem Eng, 15(6), 805 (1970).
Gallant, R.W., Physical properties of hydrocarbons, Hyd Proc, 49(1), 137-143; 49(2), 112-118; 49(3), 131-137; 49(4), 132-138 (1970).
Hilmi, A.K.; Ellis, S.R.M., and Barker, P.E., Methods of evaluating activity coefficient parameters, Brit Chem Eng, 15(10), 1321-1323; 15(11), 1453-1454 (1970).

Horvath, A.L., Correlating critical properties of halogenated hydrocarbons, Brit Chem Eng, 15(12), 1555 (1970).
Johnson, D.W., and Colver, C.P., Mixture properties by computer, Hyd Proc, 47(12), 79-83 (1968); 48(1), 127-133 (1969); 49(3), 113-122 (1970).
Joy, D.S., and Kyle, B.G., Evaluation of activity coefficients from ternary liquid-liquid equilibria, Ind Eng Chem Proc Des Dev, 9(2), 244-247 (1970).
King, F.G., and Naylor, J., Nomograph for fast estimation of heat of vaporization, Chem Eng, 13 July, 118; 7 Sept, 98 (1970).
Kreps, S.I., and Druin, M.L., Prediction of viscosity of liquid hydrocarbons, Ind Eng Chem Fund, 9(1), 79-83 (1970).
Mapstone, G.E., Nomogram for den loss in tripple superphosphate manufacture, Brit Chem Eng, 15(3), 387 (1970).
Oltmann, H.D., Practical formula for exact batch dilutions, Chem Eng, 20 April, 172-174 (1970).
Rao, S.S., and Satyanarayan, A., Latent heats by nomograph, Hyd Proc, 49(6), 102 (1970).
Roy, D., and Thodos, G., Thermal conductivity of gases, Ind Eng Chem Fund, 9(1), 71-79 (1970).
Ruel, M.J.M., A simple correlation for the thermal conductivity of n-alkanes, Chem Engnr, June, CE194-195 (1970).
Staples, B.G., and Procopio, J.M., Vapor-pressure data for common acids at high temperatures, Chem Eng, 16 Nov, 113-115 (1970).
Tans, A.M.P., Nomographs for thermodynamic properties of n-alkyl benzenes, Hyd Proc, 49(5), 156-158 (1970).
Thomas, H.L., and Smith, H., Correlation of vapour pressure, temperature, and latent heat of vaporisation, J Appl Chem, 20, 33-36 (1970).
Wisniak, J.; Eichholtz, C., and Fertilio, A., Solubility of fatty acids and their methyl esters in furfural and commercial grade sulfolane, Brit Chem Eng, 15(1), 76-77 (1970).

1971
Anon., Nomogram for density reduction of petroleum products, Brit Chem Eng, 16(11), 1059 (1971).
Basu, A., and Rafiuddin, S., Nomogram for liquid and vapour thermal conductivities of Freon refrigerants, Brit Chem Eng, 16(4), 409 (1971).
Basu, A., and Rafiuddin, S., Nomogram for viscosity of liquid Freon refrigerants, Brit Chem Eng, 16(6), 531 (1971).
Bogomolnyi, A.M., and Stankevich, T.S., Computer calculation of compressibility factor of gases and gas mixtures, Int Chem Eng, 11(4), 687-694 (1971).
Canfield, F.B., Estimate K-values by computer, Hyd Proc, 50(4), 137-138 (1971).
Chew, B., and Pye, J.W., Decomposition of R21 in the presence of steels, Brit Chem Eng, 16(12), 1136-1138 (1971).
Davis, D.S., Nomogram for friction factors of sulphuric acid solutions, Brit Chem Eng, 16(9), 843 (1971).
Eaton, E.O., Correlation and prediction of melting points, Chemtech, June, 362-366 (1971).

Grantcharov, I.N., Nomogram for estimating composition of nitrogen and phosphoric oxide based liquid fertilisers, Brit Chem Eng, 16(11), 1039 (1971).

Hall, K.R., and Yarborough, L., New simple correlation for predicting critical volume, Chem Eng, 1 Nov, 76-77 (1971).

Holland, C.D.; Hutton, A.E., and Pendon, G.P., Prediction of vaporization efficiencies for multicomponent mixtures by use of existing correlations for vapor and liquid film coefficients, Chem Eng Sci, 26(10), 1723-1736 (1971).

Jelinek, J., and Hlavacek, V., Compute boiling points faster, Hyd Proc, 50(8), 135-136 (1971).

Lee, B.I., and Edmister, W.C., Fugacity coefficients and isothermal enthalpy differences for pure hydrocarbon liquids, Ind Eng Chem Fund, 10(2), 229-236 (1971).

Loeb, M.B., Density of nitrogen, hydrogen and helium at high pressures, Chem Eng, 29 Nov, 78-81 (1971).

Loeb, M.B., Specific heat, gas constant values for nitrogen gas, Chem Eng, 4 Oct, 106 (1971).

Loeb, M.B., Properties of saturated liquid oxygen, Chem Eng, 12 July, 108-110 (1971).

Mapstone, G.E., Nomogram for entropies of gaseous compounds, Brit Chem Eng, 16(12), 1157 (1971).

Marathe, V.V., and Prausnitz, J.M., Correlation for helium solubilities in liquids, J Appl Chem Biotechnol, 21, 173 (1971).

Mathur, B.C., and Kuloor, N.R., Predict latent heat from molecular weight, Hyd Proc, 50(2), 106 (1971).

Mok, Y.I., and Yuan, S.C., Model for the heat capacity of solids, Brit Chem Eng, 16(2), 160-162 (1971).

Pachaiyappan, V., Thermal conductivity of organic liquids, Brit Chem Eng, 16(4), 382-384 (1971).

Pfeffer, J., Linear correlation approach to data estimation, Brit Chem Eng, 16(11), 1050-1051 (1971).

Ramaswamy, V., Enthalpy/concentration diagram for system acetone-carbon tetrachloride, Brit Chem Eng, 16(11), 1038 (1971).

Reddy, C.C., H-x data for alcohol-aliphatics, Brit Chem Eng, 16(11), 1036-1037 (1971).

Sappani, R., and Pitchumani, B., Nomogram for vapour pressure of alcohols, Brit Chem Eng, 16(10), 953 (1971).

Tamir, A., and Hasson, D., Evaporation and condensation coefficient of water, Chem Eng J, 2(3), 200-211 (1971).

Thinh, T.P.; Duran, J.L.; Ramalho, R.S., and Kaliaguine, S., Equations improve heat capacity predictions, Hyd Proc, 50(1), 98-104 (1971).

1972

Ambrose, D., and Lawrenson, I.J., Properties of saturated benzene, Proc Tech Int, 17(12), 967-969 (1972).

Caplan, F., Specific gravities of slurries or mixtures, Chem Eng, 21 Feb, 96-98 (1972).

Carli, A., Activity coefficients from boiling data by computer, Brit Chem Eng, 17(7), 649-651 (1972).
Chidambaram, S., Nomogram for ideal vapour heat capacity of amino-alkanes, Brit Chem Eng, 17(2), 156 (1972).
DiCave, S., and Carli, A., Heats of mixing of dimethylformamide and dimethylacetamide, Brit Chem Eng, 17(5), 439 (1972).
Dimoplon, W., Estimating specific heat of liquid mixtures, Chem Eng, 2 Oct, 64-66 (1972).
Frith, K.M., Your computer can help you estimate physical-property data, Chem Eng, 21 Feb, 72-75 (1972).
Hilado, C.J., and Clark, S.W., Autoignition temperatures of organic chemicals, Chem Eng, 4 Sept, 75-80 (1972).
Kahre, L.C., and Hankinson, R.W., Predict K-values for butadiene systems, Hyd Proc, 51(3), 94-96 (1972).
Larson, C.D., and Tassios, D.P., Prediction of ternary activity coefficients from binary data, Ind Eng Chem Proc Des Dev, 11(1), 35-38 (1972).
Manley, D.B., A better nomograph for calculating relative volatilities, Hyd Proc, 51(1), 113-114 (1972).
Medani, M.S., Viscosity of methanol-benzene mixtures, J Appl Chem Biotechnol, 22, 293-302 (1972).
Nagata, I., and Yamada, T., Correlation and prediction of heats of mixing of liquid mixtures, Ind Eng Chem Proc Des Dev, 11(4), 574-578 (1972).
Passut, C.A., and Danner, R.P., Correlation of ideal gas enthalpy, heat capacity, and entropy, Ind Eng Chem Proc Des Dev, 11(4), 543-546 (1972).
Paul, R.N., Nitrobenzene for fatty acid extraction, Proc Tech Int, 17(10), 800-801 (1972).
Paul, R.N., Ethylene dichloride is unselective for lower members of saturated fatty acid series, Brit Chem Eng, 17(3), 251-252 (1972).
Ramaswamy, V., Nomogram for vapour pressures of primary n-alkyl chlorides, Brit Chem Eng, 17(3), 250 (1972).
Seifert, W.F.; Jackson, L.L., and Sech, C.E., Organic fluids for high-temperature heat-transfer systems, Chem Eng, 30 Oct, 96-104 (1972).
Starling, K.E., et al., Thermodynamic data refined for LPG, Hyd Proc, 50(3), 101-104; 50(4), 139-145; 50(6), 116-122; 50(7), 115-121; 50(8), 170-176; 50(10), 90-96 (1971); 51(2), 86-92 (1972); 51(5), 129-132; 51(6), 107-115 (1972).
van Velzen, D.; Cardozo, R.L., and Langenkamp, H., Liquid viscosity-temperature-chemical constitution relation for organic compounds, Ind Eng Chem Fund, 11(1), 20-25 (1972).
Wichert, E., and Aziz, K., Calculate Z-values for sour gases, Hyd Proc, 51(5), 119-122 (1972).
Williamson, I.M., Computing properties of saturated steam, Chem Eng, 15 May, 128 (1972).
Zanker, A., Nomogram for net heat of combustion of aviation fuels, Brit Chem Eng, 17(2), 167 (1972).
Zanker, A., Nomogram for relative density and molecular weight of gases, Brit Chem Eng, 17(7), 657 (1972).
Zanker, A., Nomograph for calculation of viscosity of immiscible liquid mixtures, Chem Engnr, March, 121-122 (1972).

Zanker, A., Calculating the conductivity of granular beds, Chem Eng, 27 Nov, 112 (1972).

1973

Bhat, M.K., and Pitchumani, B., Nomogram for specific heats of n-alkyl phenols, Proc Tech Int, Nov, 429 (1973).

Chang, H.Y., Thermal conductivities of gases at atmospheric pressure, Chem Eng, 16 April, 122-123 (1973).

Coleman, J.V.; Greenkorn, R.A., and Chao, K.C., Thermodynamic properties of alkane liquids and group interactions, Ind Eng Chem Fund, 12(4), 452-458 (1973).

De Kee, D., and Laudie, H., Diffusivities of nonelectrolytes in water at atmospheric pressure, Chem Eng, 24 Dec, 74-75 (1973).

Gerrard, W., Significance of solubility of hydrocarbon gases in liquids in relation to intermolecular structure of liquids and the essential pattern of data for all gases, J Appl Chem Biotechnol, 23, 1-18 (1973).

Gubbins, K.E., Perturbation methods for calculating properties of liquid mixtures (review paper), AIChEJ, 19(4), 684-698 (1973).

Hayduk, W., and Laudie, H., Solubilities of gases in water and other associated solvents, AIChEJ, 19(6), 1233-1238 (1973).

Horvath, A.L., Physical properties of hydrogen chloride, Proc Tech Int, Jan, 67,69 (1973).

Horvath, A.L., Liquid thermal conductivity at boiling point, Proc Tech Int, March, 139 (1973).

Irving, J.B.; Jamieson, D.T., and Paget, D.S., The thermal conductivity of air at atmospheric pressure, Trans IChemE, 51, 10-13 (1973).

Lazaridis, A., Moisture content of air, Chem Eng, 3 Sept, 124 (1973).

Lee, B.I.; Erbar, J.H., and Edmister, W.C., Prediction of thermodynamic properties for low temperature hydrocarbon process calculations, AIChEJ, 19(2), 349-356 (1973).

Mahn, F.R., and Cosentino, J.P., Test for water in emulsions, Hyd Proc, 52(9), 177 (1973).

McWilliams, M.L., An equation to relate K-factors to pressure and temperature, Chem Eng, 29 Oct, 138-140 (1973).

Nisancioglu, K., and Newman, J., Diffusion in aqueous nitric acid solutions, AIChEJ, 19(4), 797-801 (1973).

Nothnagel, K.H.; Abrams, D.S., and Prausnitz, J.M., Generalized correlation for fugacity coefficients in mixtures at moderate pressures, Ind Eng Chem Proc Des Dev, 12(1), 25-35 (1973).

Parker, D.T., Measuring moisture in a vapor, Hyd Proc, 52(11), 215-216 (1973).

Rowlinson, J.S., et al., Prediction of thermodynamic properties of fluids and fluid mixtures, Chem Eng Sci, 28(2), 521-538 (1973).

Ruf, J.F.; Kurata, F., and McCall, T.F., Enthalpy prediction of mixtures using BWR equation of state with additional analytical functions, Ind Eng Chem Proc Des Dev, 12(1), 1-6 (1973).

Shipman, L.M., and Yen, L.C., A thermophysical data system, Chem Eng Prog, 69(7), 94-96 (1973).

Singh, D., and Upadhye, R.S., Enthalpy of sea water, Chem Eng, 19 March, 136 (1973).

Sissons, W., Calculating the weights of dry and wet flue gas, Chem Eng, 11 June, 134 (1973).
Spencer, C.F., and Daubert, T.E., Critical evaluation of methods for the prediction of critical properties of hydrocarbons, AIChEJ, 19(3), 482-486 (1973).
Spencer, C.F.; Daubert, T.E., and Danner, R.P., Critical review of correlations for critical properties of defined mixtures, AIChEJ, 19(3), 522-527 (1973).
Vosseller, W.P., Estimating specific heats of aqueous solutions, Chem Eng, 1 Oct, 100-102 (1973).
Wendt, J.O.L., and Frazier, G.C., Measurement of liquid-phase diffusivities, Ind Eng Chem Fund, 12(2), 239-243 (1973).
White, W.E.; Ferenczy, K.M., and Baudat, N.P., Shortcut to carbon dioxide solubility, Hyd Proc, 52(8), 107-108 (1973).
Winnick, J., Thermodynamics of mixtures containing polar liquids, Ind Eng Chem Fund, 12(2), 203-209 (1973).

1974

Abrams, D.S.; Massaldi, H.A., and Prausnitz, J.M., Vapor pressure of liquids as a function of temperature, Ind Eng Chem Fund, 13(3), 259-262 (1974).
Ambrose, D.; Broderick, B.E., and Townsend, R., Critical temperatures and pressures of thirty organic compounds, J Appl Chem Biotechnol, 24, 359-372 (1974).
Bonscher, F.S.; Shipman, L.M., and Yen, L.C., New charts for viscosity and conductivity (ethylene, ethane, propylene, propane), Hyd Proc, 53(1), 145-148; 53(2), 107-109; 53(3), 115-117; 53(4), 169-172 (1974).
Breslau, B.R.; Welsh, P.B., and Miller, I.F., Estimating the viscosities of aqueous electrolytic solutions, Chem Eng, 15 April, 112-113 (1974).
Brown, T.R., Computer calculation of conductivities and viscosities, Chem Eng, 18 March, 114 (1974).
Caplan, F., Properties of aqueous slurries, Chem Eng, 23 Dec, 84 (1974).
de Santis, R.; Breedveld, G.J.F., and Prausnitz, J.M., Thermodynamic properties of aqueous gas mixtures at advanced pressures, Ind Eng Chem Proc Des Dev, 13(4), 374-377 (1974).
Ghai, R.K.; Ertl, H., and Dullien, F.A.L., Liquid diffusion of nonelectrolytes (review paper), AIChEJ, 19(5), 881-900 (1973); 20(1), 1-20 (1974).
Grushka, E., and Maynard, V.R., Measuring diffusion coefficients by gas chromatography, Chemtech, Sept, 560-565 (1974).
Hildebrand, J.H., and Lamoreaux, R.H., Solubility of gases in liquids, Ind Eng Chem Fund, 13(2), 110-115 (1974).
Jacks, J.P.; Dluzniewski, J.H., and Adler, S.B., Predict enthalpy more accurately, Hyd Proc, 53(5), 133-136 (1974).
Johnston, J.C., Estimating flash points for organic aqueous solutions, Chem Eng, 25 Nov, 122 (1974).
Kee, D.D., and Laudie, H., Estimate diffusivity quickly, Hyd Proc, 53(9), 224-225 (1974).
Kumar, J., Quick visual comparison of fuel values, Chem Eng, 18 Feb, 156 (1974).
Miller, R.C., Estimating the densities of natural gas mixtures, Chem Eng, 28 Oct, 134-135 (1974).

Mo, K.C., and Gubbins, K.E., Molecular principle of corresponding states for viscosity and thermal conductivity of fluid mixtures, Chem Eng Commns, 1(6), 281-290 (1974).

Nagata, I., and Yamada, T., Correlation and prediction of excess thermodynamic functions of strongly nonideal liquid mixtures, Ind Eng Chem Proc Des Dev, 13(1), 47-53 (1974).

Pachaiyappan, V., et al., New correlation for liquid diffusivities, Chem Eng, 8 July, 108-110 (1974).

Pak, S.C., and Kay, W.B., Critical properties of binary hydrocarbon systems, Ind Eng Chem Fund, 11(2), 255-267 (1972); 13(3), 298 (1974).

Rowlinson, J.S., The use of the principle of corresponding states for prediction of thermodynamic properties, Chem Engnr, Nov, 718-721 (1974).

Saksena, M.P., and Harminder, A., Thermal conductivity of binary liquid mixtures, Ind Eng Chem Fund, 13(3), 245-247 (1974).

Swamy, K.M., and Narayana, K.L., Critical properties of halogenated hydrocarbons, Chemical Processing, June, 89 (1974).

Weber, T.W., How to calculate and use partial molar properties, Chem Eng, 11 Nov, 153-162 (1974).

Yaws, C.L., et al., Physical and thermodynamic properties of important industrial chemicals, Chem Eng, 10 June, 70-78; 8 July, 85-92; 19 Aug, 99-106; 30 Sept, 115-122; 28 Oct, 113-122; 25 Nov, 91-100; 23 Dec, 67-74 (1974).

Yeo, K.O., and Christian, S.D., Thermodynamic method for predicting solubilities of solutes in nonpolar solvents, Ind Eng Chem Fund, 13(3), 196-198 (1974).

1975

Bailey, R.G., and Chen, H.T., Predicting diffusion coefficients in binary gas systems, Chem Eng, 17 March, 86 (1975).

De Kee, D., and Laudie, H., Correlating diffusivity of n-alkanes in nonassociated solvents, Chem Eng, 14 April, 104-105 (1975).

Featherstone, W., Water-organic heterogeneous azeotropes: Composition and boiling point, Processing, Dec, 37 (1975).

Fish, L.W., and Lielmezs, J., General method for predicting latent heat of vaporization, Ind Eng Chem Fund, 14(3), 248-256 (1975).

Fredenslund, A.; Jones, R.L., and Prausnitz, J.M., Group-contribution estimation of activity coefficients in nonideal liquid mixtures, AIChEJ, 21(6), 1086-1100 (1975).

Ghosh, S.K., and Chopra, S.J., Activity coefficients from the Wilson equation, Ind Eng Chem Proc Des Dev, 14(3), 304-308 (1975).

Kyriakopoulos, G.B., Better fuel oil viscosity specs, Hyd Proc, 54(10), 40G-40MM (1975).

Lapina, R.P., Predict water thermodynamic properties, Hyd Proc, 54(2), 115-118 (1975).

Lenoir, J.M., Predict flash points accurately, Hyd Proc, 54(1), 95-99 (1975).

Luks, K.D.; Kohn, J.P.; Liu, P.H., and Kulkarni, A.A., Solubility of hydrocarbons in cryogenic NGL and LNG, Hyd Proc, 54(5), 181-184 (1975).

Mukhopadhyay, M., Prediction of binary azeotropes, Ind Eng Chem Proc Des Dev, 14(2), 195-196 (1975).
Narasimhan, K.S.; Swamy, K.M., and Narayana, K.L., New correlation for thermal conductivity, Chem Eng, 14 April, 83-84 (1975).
Ronc, M., and Ratcliff, G.A., Improved group solution model for prediction of excess free energies of liquid mixtures, Can JCE, 53, 329-340 (1975).
Sada, E.; Kito, S.; Oda, T., and Ito, Y., Diffusivities of gases in binary mixtures of alcohols and water, Chem Eng J, 10(2), 155-160 (1975).
Teja, A.S., and Kropholler, H.W., Critical states of mixtures in which azeotropic behaviour persists in the critical region, Chem Eng Sci, 30(4), 435-436 (1975).
Tyagi, K.P., Estimation of saturated liquid heat capacities, Ind Eng Chem Proc Des Dev, 14(4), 484-488 (1975).
Tyn, M.T., Estimating diffusion coefficients at any temperature, Chem Eng, 9 June, 106-107 (1975).
Tyn, M.T., and Calus, W.F., Self-diffusion coefficients from additive and constitutive parameters, Trans IChemE, 53, 44-49 (1975).
Yaws, C.L., et al., Physical and thermodynamic properties of important industrial chemicals, Chem Eng, 20 Jan, 99-106; 17 Feb, 87-94; 31 March, 101-109; 12 May, 89-97; 21 July, 113-122; 1 Sept, 107-115; 29 Sept, 73-81; 27 Oct, 119-127; 8 Dec, 119-128 (1975).

1976

Ambrose, D., Correlation of boiling points of alkanols, J Appl Chem Biotechnol, 26, 712-714 (1976).
Barnea, E., and Mizrahi, J., Effective viscosity of liquid-liquid dispersions, Ind Eng Chem Fund, 15(2), 120-125 (1976).
Chandak, B.S.; Nageshwar, G.D., and Mene, P.S., Excess enthalpy, volume and Gibbs free energy, and viscosity of methyl isobutyl ketone-methyl cellulose, Can JCE, 54, 647-655 (1976).
Cysewski, G.R., and Prausnitz, J.M., Estimation of gas solubilities in polar and nonpolar solvents, Ind Eng Chem Fund, 15(4), 304-309 (1976).
Duran, J.L.; Thinh, T.P.; Ramalho, R.S., and Kaliaguine, S., Predict heat capacity more accurately, Hyd Proc, 55(8), 153-156 (1976).
Eduljee, G.H., and Tiwari, K.K., Correlation of azeotropic data, Chem Eng Sci, 31(7), 535-540 (1976).
Ganapathy, V., Related properties of state for ammonia gas, Chem Eng, 2 Aug, 109 (1976).
Gothard, F.A., et al., Predicting parameters in Wilson equations for activity coefficients in binary hydrocarbon systems, Ind Eng Chem Proc Des Dev, 15(2), 333-337 (1976).
Green, L.E., Chromatography gives boiling point, Hyd Proc, 55(5), 205-207 (1976).
Grove, D.J., Moisture content of air at different pressures, Chem Eng, 30 Aug, 129 (1976).
Hafaz, M., and Hartland, S., Physical properties of three ternary systems, Chem Eng Sci, 31(3), 247-250 (1976).
Horvath, A.L., Liquid viscosities of halogenated hydrocarbons, Chem Eng, 29 March, 121-124 (1976).

Kesler, M.G., and Lee, B.I., Improve prediction of enthalpy of fractions, Hyd Proc, 55(3), 153-158 (1976).

Kikic, I., and Renon, H., Extension of chromatographic method of determination of thermodynamic properties, Sepn Sci, 11(1), 45-64 (1976).

King, M.B., The correlation, prediction and extrapolation of vapour pressure and related thermal data, Trans IChemE, 54, 54-60 (1976).

Lee, J.I.; Otto F.D., and Mather, A.E., Measurement and prediction of solubility of carbon dioxide and hydrogen sulfide mixtures in 2.5N monoethanolamine solution, Can JCE, 54, 214-220 (1976).

Lyman, T.J., and Danner, R.P., Correlation of liquid heat capacities with a four-parameter corresponding states method, AIChEJ, 22(4), 759-765 (1976).

Nakamura, R.; Breedveld, G.J.F., and Prausnitz, J.M., Thermodynamic properties of gas mixtures containing common polar and nonpolar components, Ind Eng Chem Proc Des Dev, 15(4), 557-564 (1976).

Palmer, D.J., Predict vapor pressure of 'undefined' fractions, Hyd Proc, 55(12), 121-122 (1976).

Plocker, U.J., and Knapp, H., Save time in computing density, Hyd Proc, 55(5), 199-201 (1976).

Purarelli, C., Estimating the atmospheric boiling points of liquids, Chem Eng, 30 Aug, 127-128 (1976).

Reid, R.C., and San Jose, J.L., Estimating liquid heat capacities, Chem Eng, 6 Dec, 161-164; 20 Dec, 67-71 (1976).

Schneider, A.J., Normality, molarity and molality, Chem Eng, 27 Oct, 140 (1975); 15 March, 110 (1976).

Singh, A., and Teja, A.S., Prediction of densities of n-alkanes, Chem Eng Sci, 31(5), 404-406 (1976).

Smith, G.; Winnick, J.; Abrams, D.S., and Prausnitz, J.M., Vapor pressures of high-boiling complex hydrocarbons, Can JCE, 54, 337-345 (1976).

Tamir, A., and Wisniak, J., Association effects in ternary vapour-liquid equilibria, Chem Eng Sci, 31(8), 625-630 (1976).

Tarakad, R.R., and Danner, R.P., Comparison of enthalpy prediction methods, AIChEJ, 22(2), 409-411 (1976).

Teng, T.T.; Sangster, J., and Lenzi, F., Additivity rules for prediction of density of aqueous solutions containing mixed-type solutes, Can JCE, 54, 600-610 (1976).

Thinh, T.P., and Trong, T.K., Accurate estimation of standard heats of formation, standard entropies of formation, standard free energies of formation, and absolute entropies of hydrocarbons from group contributions, Can JCE, 54, 344-355 (1976).

Thomas, L.H., Variation of liquid viscosity with temperature, Chem Eng J, 11(3), 201-206 (1976).

Thomas, L.H., Analytical expression for variation of vapour pressure of liquids with temperature up to critical conditions, Chem Eng J, 11(3), 191-200 (1976).

Tyn, M.T., Estimating self-diffusion coefficient at any temperature, Processing, Jan, 25 (1976).

Tyn, M.T., Estimation of diffusion coefficients of liquid mixtures at any temperature, Chem Eng J, 12(2), 149-150 (1976).

Wisniak, J., and Tamir, A., Correlation of boiling points of mixtures, Chem Eng Sci, 31(8), 631-636 (1976).
Yaws, C.L., et al., Physical and thermodynamic properties of important industrial chemicals, Chem Eng, 19 Jan, 107-115; 1 March, 107-115; 12 April, 129-137; 7 June, 119-127; 5 July, 81-89; 16 Aug, 79-87; 25 Oct, 127-135; 22 Nov, 153-162 (1976).
Zanker, A., Nonograph for heat of an equilibrium process, Chemtech, April, 234 (1976).

1977

Bissett, L., Equilibrium constants for shift reactions, Chem Eng, 24 Oct, 155-156 (1977).
Caplan, F., Adjusting pH with acid or caustic, Chem Eng, 4 July, 143-144 (1977).
Chappelear, P.S.; Chen, R.J.J., and Elliot, D.G., Select K-correlations carefully, Hyd Proc, 56(9), 215-217 (1977).
Corn, B.R.; Weber, J.H., and Tao, L.C., Chebychev polynomial correlation equations of composition and bubble/dew point temperature of binary mixtures containing ethane with propane, butane, and pentane, Ind Eng Chem Proc Des Dev, 16(1), 137-139 (1977).
Doherty, M.F., and Perkins, J.D., Properties of liquid-vapour composition surfaces at azeotropic points, Chem Eng Sci, 32(9), 1112-1114 (1977).
El-Sabaawi, M., and Pei, D.C.T., Moisture isotherms for hygroscopic porous solids, Ind Eng Chem Fund, 16(3), 321-326 (1977).
Ganapathy, V., Quick way to determine slurry densities, Chem Eng, 4 July, 144 (1977).
Ganapathy, V., et al., Physical properties of selected gas-streams, Chem Eng, 28 Feb, 195-199 (1977).
Goletz, E., and Tassios, D., Antoine-type equation for liquid viscosity dependency to temperature, Ind Eng Chem Proc Des Dev, 16(1), 75-79 (1977).
Gomez-Nieto, M., and Thodos, G., New vapor pressure equation and its application to normal alkanes, Ind Eng Chem Fund, 16(2), 254-259; 16(4), 495 (1977).
Gomez-Nieto, M., and Thodos, G., Generalized vapor pressure behavior of substances between their triple points and critical points, AIChEJ, 23(6), 904-913 (1977).
Katinas, T.G., and Danner, R.P., Easy method to predict latent heat of vaporization, Hyd Proc, 56(3), 157-160 (1977).
Kesler, M.G.; Lee, B.I.; Benzing, D.W., and Cruz, A., Method improves convergence pressure predictions, Hyd Proc, 56(6), 177-179 (1977).
Kesler, M.G.; Lee, B.I.; Fish, M.J., and Hadden, S.T., Correlation improves K-value predictions, Hyd Proc, 56(5), 257-262 (1977).
King, M.B.; Kassim, K., and Al-Najjar, H., Solubilities of carbon dioxide, hydrogen sulphide and propane in some normal alkane solvents, Chem Eng Sci, 32(10), 1241-1252 (1977).
Lawrenson, I.J., Recent publications on the physical and thermodynamic properties of organic compounds and their mixtures, Chem Engnr, Feb, 112 (1977).

Li, C.C., Rectilinear equation for binary azeotropes, AIChEJ, 23(2), 210-211 (1977).
Marreilli, L., and De Santis, R., Simple model to calculate solubility of liquid water in compressed air, Can JCE, 55, 183-190 (1977).
Medani, M.S., and Hasan, M.A., Viscosity of organic liquids at elevated temperatures and corresponding vapour pressures, Can JCE, 55, 203-210 (1977).
Peng, D.Y., and Robinson, D.B., Rigorous method for predicting critical properties of multicomponent systems from an equation of state, AIChEJ, 23(2), 137-144 (1977).
Purarelli, C., Heat-capacity ratios for real gases, Chem Eng, 14 March, 153 (1977).
Rao, A.K., Prediction of liquid activity coefficients, Chem Eng, 9 May, 143-147 (1977).
Raymont, M.E.D., Tests developed for evaluating sulfur, Hyd Proc, 56(6), 174-176 (1977).
Russell, R.A., Save computer time evaluating K-values, Hyd Proc, 56(2), 133-134 (1977).
Schneider, R.T., Calculate enthalpy with a pocket calculator, Chem Eng, 23 May, 145-152 (1977).
Seymour, K.M., et al., Empirical correlation among azeotropic data, Ind Eng Chem Fund, 16(2), 200-207; 16(4), 495 (1977).
Sridhar, T., and Potter, O.E., Predicting diffusion coefficients, AIChEJ, 23(4), 590-592 (1977).
Vera, J.H., Simple method for estimation of thermodynamic properties of concentrated and supersaturated aqueous solutions of NaCl and KCl, Can JCE, 55, 484-490 (1977).
Zanker, A., Nomographs for reduced temperatures and pressures, Proc Engng, May, 82-83 (1977).
Zanker, A., Nomograph for accurate prediction of vapour pressures, Proc Engng, Nov, 132-133 (1977).
Zanker, A., Nomograph for solubility of inorganic gases in petroleum liquids, Hyd Proc, 56(5), 255-256 (1977).

1978

Ackroyd, K., Refrigerants: Properties, selection and hazards, Chem Engnr, May, 366-370 (1978).
Asselineau, L.; Bogdanic, G., and Vidal, J., Calculation of thermodynamic properties and vapour-liquid equilibria of refrigerants, Chem Eng Sci, 33(9), 1269-1276 (1978).
Chen, S.A., and Lin, H.J., Estimation of critical solution temperature of an alcohol-aliphatic hydrocarbon mixture, Chem Eng J, 16(1), 69-72 (1978).
Douglas, J.M., Criteria for relative volatility, Hyd Proc, 57(2), 155-156 (1978).
Gaitonde, U.N.; Deshpande, D.D., and Sukhatme, S.P., Thermal conductivity of liquid mixtures, Ind Eng Chem Fund, 17(4), 321-325 (1978).
Gomez-Nieto, M., and Thodos, G., Generalized vapor pressure equation for nonpolar substances, Ind Eng Chem Fund, 17(1), 45-51 (1978).

Hsieh, M.W., and Cheh, H.Y., Binary diffusivities of nitrogen-methane and nitrogen-methyl chloride systems, Ind Eng Chem Fund, 17(3), 210-213 (1978).
Khoury, F.M., Calculate the right density for VLE predictions, Hyd Proc, 57(12), 155-157 (1978).
Lee, M.C., et al., Heat capacity determined for crude fractions, Hyd Proc, 57(1), 187-189 (1978).
Lin, C.T., and Daubert, T.E., Prediction of fugacity coefficients of nonpolar hydrocarbon systems from equations of state, Ind Eng Chem Proc Des Dev, 17(4), 544-549 (1978).
McGowan, J.C., Estimates of properties of liquids, J Appl Chem Biotechnol, 28, 599-607 (1978).
Mentzer, R.A.; Young, K.L.; Greenkorn, R.A., and Chao, K.C., Principle of corresponding states of liquid solutions: Excess enthalpy and the pseudocriticals, Chem Eng Sci, 33(2), 229-240 (1978).
Monfort, J.P., and Perez, J.L., Henry's constants of normal mixtures, Chem Eng J, 16(3), 205-210 (1978).
Mousa, A.H.N., Vapour pressure and critical constant of perfluorpropane, Can JCE, 56, 128-138 (1978).
O'Reilly, M.G., and Edmonds, B., Physical property data, Chem Engnr, Jan, 61-63 (1978).
Pesuit, D.R., Binary interaction constants for mixtures with a wide range in component properties, Ind Eng Chem Fund, 17(4), 235-242 (1978).
Rao, B.K.B., Glycols not best for smoke point, Hyd Proc, 57(12), 105-106 (1978).
Rodgers, M.A., Adapting humidity charts to brine, Chem Eng, 8 May, 212-214 (1978).
Santrach, D., and Lielmezs, J., Latent heat of vaporization prediction for binary mixtures, Ind Eng Chem Fund, 17(2), 93-96 (1978).
Sridhar, T., and Potter, O.E., Modification of Hildebrand's viscosity equation, Chem Eng J, 16(1), 57-60 (1978).
Tamir, A., and Wisniak, J., Activity coefficient calculations in multicomponent associating systems, Chem Eng Sci, 33(6), 651-656 (1978).
Tamir, A., and Wisniak, J., Correlation and prediction of boiling temperatures and azeotropic conditions in multicomponent systems, Chem Eng Sci, 33(6), 657-672 (1978).
Thomas, L.H.; Smith, H., and Davies, G.H., Viscosities of liquids close to freezing point, Chem Eng J, 16(3), 223-232 (1978).
Wild, J.D.; Sridhar, T., and Potter, O.E., Solubility of nitrogen and oxygen in cyclohexane, Chem Eng J, 15(3), 209-214 (1978).
Yorizane, M., and Miyano, Y., Generalized correlation for Henry's constants in nonpolar binary systems, AIChEJ, 24(2), 181-186 (1978).
Zanker, A., Nomograph for latent heat of vapourisation, Proc Engng, Sept, 72-73 (1978).
Zanker, A., Calculating viscosity of two-phase dispersions, Proc Engng, March, 100-101 (1978).
Zanker, A., Estimating the dew points of stack gases, Chem Eng, 10 April, 154-156 (1978).

Zia, T., and Thodos, G., Reduced vapor pressure equation for hydrocarbons, Chem Eng J, 16(1), 41-50 (1978).

1979

Battaerd, H.A.J., and Evans, D.G., Alternative representation of coal composition data, Fuel, 58(2), 105-108 (1979).

Brule, M.R.; Lee, L.L., and Starling, K.E., Predicting thermodynamic properties for fossil-fuel chemicals, Chem Eng, 19 Nov, 155-164 (1979).

Chao, J., Properties of alkylbenzenes, Hyd Proc, 58(11), 295-299 (1979).

Duhne, C.R., Viscosity-temperature correlations for liquids, Chem Eng, 16 July, 83-91 (1979).

Edmonds, B., Physical property data: Critical temperatures, Chem Engnr, June, 447 (1979).

Edmonds, B., Physical property data: Critical properties, Chem Engnr, Aug, 619 (1979).

Edmonds, B., Physical property data: Enthalpies of vaporisation, Chem Engnr, Feb, 109; March, 179, 181; May, 357 (1979).

Eduljee, G.H., and Tiwari, K.K., Prediction of ternary azeotropes, Chem Eng Sci, 34(7), 929-932 (1979).

Ganapathy, V., and Rajaram, S., Estimating mean densities for liquid-vapor mixtures, Chem Eng, 31 Dec, 71-73 (1979).

Green, K.A.; Tiffin, D.L.; Luks, K.D., and Kohn, J.P., Solubility of hydrocarbons in LNG/NGL, Hyd Proc, 58(5), 251-253 (1979).

Hankinson, R.W., and Thomson, G.H., New correlation for saturated densities of liquids and their mixtures, AIChEJ, 25(4), 653-663 (1979).

Hankinson, R.W., and Thomson, G.H., Calculate liquid densities accurately, Hyd Proc, 58(9), 277-283 (1979).

Heffington, W.M., Use the right heating value, Hyd Proc, 58(6), 141-142 (1979).

Heller, H., Aqua-ammonia density correction, Chem Eng, 4 June, 160 (1979).

Hikita, H.; Asai, S.; Ishikawa, H., Seko, M., and Kitajima, H., Diffusivities of carbon dioxide in aqueous mixed-electrolyte solutions, Chem Eng J, 17(1), 77-80 (1979).

Kabadi, V.N., and Danner, R.P., Nomograph solves for solubilities of hydrocarbons in water, Hyd Proc, 58(5), 245-246 (1979).

Kudchadker, S.A., Property data available for coal chemicals, Hyd Proc, 58(1), 169-171 (1979).

Mackay, M.E., and Paulaitis, M.E., Solid solubilities of heavy hydrocarbons in supercritical solvents, Ind Eng Chem Fund, 18(2), 149-153 (1979).

Macknick, A.B., and Prausnitz, J.M., Vapor pressures of heavy liquid hydrocarbons by a group-contribution method, Ind Eng Chem Fund, 18(4), 348-351 (1979).

Maher, P.J., and Smith, B.D., Infinite dilution activity coefficient values from total pressure VLE data, Ind Eng Chem Fund, 18(4), 354-357 (1979).

Meisen, A., and Bennet, H.A., Predict liquid sulfur vapor-pressure, Hyd Proc, 58(12), 131 (1979).

Ochi, K., and Lu, B.C.Y., Prediction of equilibrium liquid compositions from dew-point measurements for ternary systems, Chem Eng Sci, 34(2), 239-244 (1979).

Siman, J.E., and Vera, J.H., Heats of mixing of amine-alcohol systems: Analytical group solution model approach, Can JCE, 57, 355-365 (1979).
Sundaram, S., and Ibrahim, S.H., New correlation to predict latent heat of vaporization at normal boiling point, Can JCE, 57, 107-120 (1979).
Thompson, P.A., and Sullivan, D.A., Simple formula for saturated-vapor volume, Ind Eng Chem Fund, 18(1), 1-7 (1979).
Vetere, A., New correlations for predicting vaporization enthalpies of pure compounds, Chem Eng J, 17(2), 157-162 (1979).
Weber, J.H., Predicting viscosities of liquids and mixtures, Chem Eng, 30 July, 79-84 (1979).
Weber, J.H., Predict vapor-pressure vs. temperature for pure substances, Chem Eng, 5 Nov, 111-117 (1979).
Weber, J.H., Predicting volumes of pure liquids, Chem Eng, 26 March, 173-177 (1979).
Weber, J.H., Predict the viscosities of pure gases, Chem Eng, 18 June, 111-117 (1979).
Weber, J.H., Predicting volumes of saturated liquid mixtures, Chem Eng, 7 May, 95-99 (1979).
Zanker, A., Rapid calculation for binary gaseous diffusion, Proc Engng, Sept, 85-87 (1979).

1980

Amin, M.B., and Maddox, R.N., Estimate viscosity vs. temperature, Hyd Proc, 59(12), 131-135 (1980).
Bartlett, P.L., The solvent FC-113, Chemtech, June, 354-355 (1980).
Brelvi, S.W., Fugacities of supercritical hydrogen, helium, nitrogen, and methane in binary liquid mixtures, Ind Eng Chem Proc Des Dev, 19(1), 80-84 (1980).
Chao, J., Properties of elemental sulfur, Hyd Proc, 59(11), 217-223 (1980).
Chase, J.D., Enthalpy data for saturated lower fatty acids, Chem Eng, 24 March, 107-112 (1980).
Christensen, P.L., and Fredenslund, A.A., Corresponding states model for the thermal conductivity of gases and liquids, Chem Eng Sci, 35(4), 871-876 (1980).
Cohen, Y., and Sandler, S.I., Viscosity and thermal conductivity of simple dense gases, Ind Eng Chem Fund, 19(2), 186-188 (1980).
Cook, D., and Hamaker, J.D., Program calculates gas physical properties, Hyd Proc, 59(11), 213-215 (1980).
Cramer, S.D., Solubility of oxygen in brines from 0 to 300 degC, Ind Eng Chem Proc Des Dev, 19(2), 300-305 (1980).
Daubert, T.E., Property predictions: A review, Hyd Proc, 59(3), 107-112 (1980).
Dzialoszynski, L.; Fabries, J.F.; Renon, H., and Thiebault, D., Comparison of analytical representations of thermodynamic-properties of methane, ethane and ammonia, Ind Eng Chem Fund, 19(4), 329-337 (1980).
El-Twaty, A.I., and Prausnitz, J.M., Correlation of K-factors for mixtures of hydrogen and heavy hydrocarbons, Chem Eng Sci, 35(8), 1765-1768 (1980).
Guerrero, M.I., and Davis, H.T., Gradient theory for prediction of the surface tension of water, Ind Eng Chem Fund, 19(3), 309-311 (1980).

Heidemann, R.A., and Khalil, A.M., Calculation of critical points, AIChEJ, 26(5), 769-779 (1980).

Heinrich, J., et al., Calculation of saturated vapor pressure, Int Chem Eng, 20(1), 77-84 (1980).

Kato, M., Activity coefficients of pseudo-ternary mixtures, Ind Eng Chem Fund, 19(3), 253-259 (1980).

Kurtyka, Z.M., and Kurtyka, A., Azeotropic composition from the boiling points of components and of the azeotropes, Ind Eng Chem Fund, 19(2), 225-227 (1980).

Lazalde-Crabtree, H.; Breedveld, G.J.F., and Prausnitz, J.M., Solubility of methanol in compressed natural and synthetic gases, AIChEJ, 26(3), 462-470 (1980).

Lee, B.I., and Kesler, M.G., Improve vapor-pressure predictions, Hyd Proc, 59(7), 163-167 (1980).

Lin, C.T.; Brule, M.R.; Young, F.K.; Lee, L.L., Starling, K.E., and Chao, J., Data bank for synthetic fuels, Hyd Proc, 59(5), 229-233; 59(8), 117-123; 59(11), 225-232 (1980).

Mills, M.B.; Wills, M.J., and Bhirud, V.L., Calculation of density by the BWRS equation of state, AIChEJ, 26(6), 902-910 (1980).

Nelson, A., Material properties in SI units, Chem Eng Prog, 75(10), 95-97 (1979); 75(11), 70-71 (1979); 76(4), 86-88 (1980); 76(5), 83-85 (1980).

Neumann, K.K., and Ostertag, G., Computer systems to predict thermophysical properties of mixtures, Int Chem Eng, 20(1), 1-7 (1980).

O'Donnell, R.J., Predict thermal expansion of petroleum, Hyd Proc, 59(4), 229-231 (1980).

Ogiwara, K.; Arai, Y., and Saito, S., Thermal conductivities of liquid hydrocarbons and their binary mixtures, Ind Eng Chem Fund, 19(3), 295-300 (1980).

Prasad, R., and Gupta, A.K., Estimating vapor pressures of organic compounds, Chem Eng, 20 Oct, 152-153 (1980).

Rastogi, A., and Tassios, D., Estimation of thermodynamic properties of binary aqueous electrolytic solutions in range 25-100 degC, Ind Eng Chem Proc Des Dev, 19(3), 477-482 (1980).

Riazi, M.R., and Daubert, T.E., Prediction of the composition of petroleum fractions, Ind Eng Chem Proc Des Dev, 19(2), 289-294 (1980).

Riazi, M.R., and Daubert, T.E., Simplify property predictions, Hyd Proc, 59(3), 115-116 (1980).

Rowley, P.R., Calculator program solves gas equations, Chem Eng, 11 Feb, 97-100 (1980).

Santi, E., Calculating properties of chemical compounds, Chem Eng, 2 June, 75-77 (1980).

Teja, A.S., Corresponding states equation for saturated liquid densities, AIChEJ, 26(3), 337-345 (1980).

Weber, J.H., Predict thermodynamic properties of pure gases and binary mixtures, Chem Eng, 22 Sept, 155-158 (1980).

Weber, J.H., Predict properties of gas mixtures, Chem Eng, 19 May, 151-160 (1980).

Weber, J.H., Predict latent heat of vaporization, Chem Eng, 14 Jan, 105-110 (1980).

Zanker, A., Estimate surface tension changes for liquids, Hyd Proc, 59(4), 207-208 (1980).

1981

Allahwala, S.A., Compressibility of natural sour gas, Hyd Proc, 60(9), 271-272 (1981).

Bhasin, M.M., Applications of Auger and ESCA spectroscopies for solids analysis, Chem Eng Prog, 77(3), 60-67 (1981).

Bogart, M.J.P., A gas compression calculator program, Chem Eng Prog, 77(8), 66-70 (1981).

Chun, S.W.; Kay, W.B., and Teja, A.S., Critical states of propane-isomeric hexane mixtures, Ind Eng Chem Fund, 20(3), 278-280 (1981).

Eckart, C.A.; Newman, B.A.; Nicolaides, G.L., and Long, T.C., Measurement and application of limiting activity coefficients, AIChEJ, 27(1), 33-40 (1981).

Edwards, D.R., and Prausnitz, J.M., Estimation of vapor pressures of heavy liquid hydrocarbons containing nitrogen or sulfur by a group-contribution method, Ind Eng Chem Fund, 20(3), 280-283 (1981).

Hanna, J.G., Examination of the vapor pressure-temperature relationship, log p vs. log T/p, Ind Eng Chem Fund, 20(4), 376-381 (1981).

Jensen, T.; Fredenslund, A., and Rasmussen, P., Pure-component vapor pressures using UNIFAC group contribution, Ind Eng Chem Fund, 20(3), 239-246 (1981).

Kiang, Y.H., Predicting dewpoints of acid gases, Chem Eng, 9 Feb, 127 (1981).

Kurtyka, Z.M., and Kurtyka, A., Azeotropic composition from the activity coefficients of components, Ind Eng Chem Fund, 20(2), 177-180 (1981).

Lebrun, P., Program aids cryogenic solubility calculations, Chem Eng, 13 July, 127-131 (1981).

Lin, C.T., and Daubert, T.E, New correlation for the prediction of activity coefficients for binary nonpolar hydrocarbon mixtures, Ind Eng Chem Proc Des Dev, 20(4), 652-658 (1981).

Mullins, T.E., Compute API volume correction on calculator, Hyd Proc, 60(11), 295-296 (1981).

Newman, S.A., Correlations evaluated for coal-tar liquids, Hyd Proc, 60(12), 133-142 (1981).

Noor, A., Quick estimate of liquid densities, Chem Eng, 6 April, 111 (1981).

Pan, W.P., and Maddox, R.N., Determining properties of saturated liquids and vapors, Chem Eng, 2 Nov, 79-87 (1981).

Rizzi, A., and Huber, J.F.K., Comparative calculations of activity coefficients in binary liquid mixtures at infinite dilution using the 'solution of groups' method, Ind Eng Chem Proc Des Dev, 20(2), 204-210 (1981).

Roddy, J.W., Distribution of ethanol-water mixtures to organic liquids, Ind Eng Chem Proc Des Dev, 20(1), 104-108 (1981).

Roth, J.A., and Sullivan, D.E., Solubility of ozone in water, Ind Eng Chem Fund, 20(2), 137-140 (1981).

Rusin, M.H.; Chung, H.S., and Marshall, J.F., A 'transformation' method for calculating the research and motor octane numbers of gasoline blends, Ind Eng Chem Fund, 20(3), 195-204 (1981).

Schulze, G., and Prausnitz, J.M., Solubilities of gases in water at high temperatures, Ind Eng Chem Fund, 20(2), 175-177 (1981).
Sebastian, H.M.; Lin, H.M., and Chao, K.C., Correlation of the solubility of methane in hydrocarbon solvents, Ind Eng Chem Fund, 20(4), 346-349 (1981).
Sebastian, H.M.; Lin, H.M., and Chao, K.C., Correlation of solubility of carbon dioxide in hydrocarbon solvents, Ind Eng Chem Proc Des Dev, 20(3), 508-511 (1981).
Sebastian, H.M.; Lin, H.M., and Chao, K.C., Correlation of solubility of hydrogen in hydrocarbon solvents, AIChEJ, 27(1), 138-148 (1981).
Stournas, S., Viscosity index by computer, Hyd Proc, 60(8), 91-92 (1981).
Tamir, E.; Tamir, A., and King, M.B., Explaining characteristic azeotropic behaviour, Chem Eng Sci, 36(4), 759-764 (1981).
Tamir,A., Azeotropes prediction by 'sectionwise fitting' Chem Eng Sci, 36(1), 37-46 (1981).
Teja, A.S., and Rice, P., Generalized corresponding states method for prediction of thermal conductivity of liquids and liquid mixtures, Chem Eng Sci, 36(2), 417-422 (1981).
Teja, A.S., and Rice, P., A multifluid corresponding states principle for the thermodynamic properties of fluid mixtures, Chem Eng Sci, 36(1), 1-6 (1981).
Teja, A.S., and Rice, P., Generalized corresponding states method for viscosities of liquid mixtures, Ind Eng Chem Fund, 20(1), 77-81 (1981).
Twu, C.H., and Bulls, J.W., Viscosity-temperature relation for multicomponent mixtures, Hyd Proc, 60(4), 217-218 (1981).
Tyn, M.T., Temperature dependence of liquid phase diffusion coefficients, Trans IChemE, 59, 112-118 (1981).
Umesi, N.O., and Danner, R.P., Predicting diffusion coefficients in nonpolar solvents, Ind Eng Chem Proc Des Dev, 20(4), 662-665 (1981).
Weber, J.H., Predict thermal conductivities of liquid mixtures, Chem Eng, 1 June, 67-71 (1981).
Weber, J.H., Predict thermal conductivities of pure gases, Chem Eng, 12 Jan, 127-132 (1981).
Weber, J.H., Predict thermal conductivities of gas mixtures and liquids, Chem Eng, 9 March, 91-96 (1981).
Zabicky, J., Program predicts critical properties of organic compounds, Chem Eng, 23 Feb, 73-76 (1981).
Zanker, A., Nomograph for thermal conductivity of liquid mixtures, Hyd Proc, 60(3), 165-167 (1981).

1982

Baltatu, M.E., Prediction of the liquid viscosity for petroleum fractions, Ind Eng Chem Proc Des Dev, 21(1), 192-195 (1982).
Biarnes, R., Measurement of acid-dewpoint temperature, Plant/Opns Prog, 1(4), 230-236 (1982).
Dizechi, M., and Marschall, E., Correlation of viscosity data of liquid mixtures, Ind Eng Chem Proc Des Dev, 21(2), 282-289 (1982).
Dutcher, W.G., Shortcut program eases gas calculations (physical properties), Hyd Proc, Dec, 81-84 (1982).

Dutt, N.V.K., Estimation of vapor pressure from normal boiling point of hydrocarbons, Can JCE, 60, 707-710 (1982).

Fisher, C.H., Estimating liquid densities and critical properties of n-alkanes, Chem Eng, 11 Jan, 107-109 (1982).

Fisher, C.H., Equations correlate n-alkane physical properties with chain length, Chem Eng, 20 Sept, 111-113 (1982).

Gainer, J.L., and Brumgard, F.B., Using excess volume of mixing to correlate diffusivities in liquids, Chem Eng Commns, 15(5), 323-330 (1982).

Garvin, D.; Parker, V.B., and Wagman, D.D., Thermodynamic databanks, Chemtech, Nov, 691-697 (1982).

Gmehling, J., and Rasmussen, P., Flash points of flammable liquid mixtures using UNIFAC, Ind Eng Chem Fund, 21(2), 186-188; 21(3), 326 (1982).

Hankinson, R.W.; Coker, T.A., and Thomson, G.H., Accurate LNG densities, Hyd Proc, April, 207-208 (1982).

Hayduk, W., and Minhas, B.S., Correlations for prediction of molecular diffusivities in liquids, Can JCE, 60, 295-299 (1982).

Johnston, K.P.; Ziger, D.H., and Eckert, C.A., Solubilities of hydrocarbon solids in supercritical fluids: The augmented van der Waals treatment, Ind Eng Chem Fund, 21(3), 191-197 (1982).

Jou, F.Y.; Mather, A.E., and Otto, F.D., Solubility of water and carbon dioxide in aqueous methyldiethanolamine solutions, Ind Eng Chem Proc Des Dev, 21(4), 539-544 (1982).

Lapina, R.P., How to estimate compressibility factors and specific-heat ratios for hydrocarbon gases, Chem Eng, 8 Feb, 95-98 (1982).

Liley, P.E., Thermodynamic properties of methanol, Chem Eng, 29 Nov, 59-60 (1982).

Nash, J., Acentric factor and the critical volumes for normal fluids, Ind Eng Chem Fund, 21(3), 325-326 (1982).

Poltz, H., and Jugel, R., Thermal conductivity of some organic liquids between 30 and 190 degC, Int J Heat Mass Trans, 25(8), 1093-1102 (1982).

Pommersheim, J.M.; Seeley, J.T., and van Kirk, J., Use of the Stefan cell to obtain solvent diffusivities, Chem Eng J, 23(1), 105-110 (1982).

Srinivasan, P.; Abbas, S.P.; Ghosh, S.; Devotta, S., and Watson, F.A., A correlation between decomposition of fluorocarbons and viscosity changes of lubricants when exposed to elevated temperatures, Trans IChemE, 60, 380-382 (1982).

Teja, A.S., and Rice, P., Prediction of the thermal conductivity of binary aqueous mixtures, Chem Eng Sci, 37(5), 788-790 (1982).

Torquato, S., and Stell, G.R., An equation for the latent heat of vaporization, Ind Eng Chem Fund, 21(3), 202-205 (1982).

Weber, J.H., Predict gas-phase diffusion coefficients, Chem Eng, 3 May, 87-92 (1982).

Wormald, C.J., Thermo data for steam-hydrocarbons, Hyd Proc, May, 137-141 (1982).

Yu, W.C.; Lee, H.M., and Ligon, R.M., Predict high pressure properties, Hyd Proc, Jan, 171-178 (1982).

1983

Antunes, C., and Tassios, D., Modified UNIFAC model for prediction of Henry's constants, Ind Eng Chem Proc Des Dev, 22(3), 457-462 (1983).

Ashcroft, S.J.; Shearn, R.B., and Williams, G.J.J., A visual equilibrium cell for multiphase systems at pressures up to 690 bar, CER&D, 61, 51-55 (1983).

Baldauf, W., and Knapp, H., Measurements of diffusivities in liquids by the dispersion method, Chem Eng Sci, 38(7), 1031-1038 (1983).

Eckermann, R., Physical property databank for chemical engineering, Int Chem Eng, 23(1), 1-11 (1983).

Edmonds, B., Computer-based physical property storage and prediction (PPDS), Proc Engng, Oct, 51-53 (1983).

Fredrickson, A.G., Reference states and relative values of internal energy, enthalpy and entropy, Chem Eng Educ, 17(2), 64-69 (1983).

Ganapathy, V., Superheated-steam properties at a glance, Chem Eng, 7 Feb, 91 (1983).

Gasca-Ramirez, J., and Torres-Robles, R., Azeotrope of 2-methyl-1,3 butadiene/n-pentane system at different pressures, Chem Eng Commns, 22(3), 139-150 (1983).

Gooding, C.H., Estimating flash point and lower explosive limit, Chem Eng, 12 Dec, 88 (1983).

Krone, R.B., Viscosity-temperature relation for Newtonian liquids, Chem Eng Commns, 22(3), 161-180 (1983).

Luinstra, E.A., Blend viscosities by computer, Hyd Proc, April, 125-126; May, 104-106; June, 99-101 (1983).

Ma, P.; Jiang, B., and Zhang, J., Critical assessment and temperature correlations of viscosity data for gaseous substances at atmospheric pressure, Int Chem Eng, 23(1), 127-139 (1983).

McGarry, J., Correlation and prediction of the vapor pressures of pure liquids over large pressure ranges, Ind Eng Chem Proc Des Dev, 22(2), 313-322 (1983).

Newman, S.A., Coal char Cp correlations evaluated, Hyd Proc, March, 77-82 (1983).

Rao, K.V.K., Physical properties of common heat-transfer fluids, Chem Eng, 19 Sept, 89-90 (1983).

Rao, K.V.K., Viscosities of uncommon liquids, Chem Eng, 30 May, 90-91 (1983).

Rao, N.V.R., and Baird, M.H.I., Continuous measurement of surface and interfacial tension by stationary slug method, Can JCE, 61(4), 581-589 (1983).

Robinson, E.R., Calculate density of spiked crudes, Hyd Proc, May, 115-120 (1983).

Ruzicka, V., Estimation of vapor pressures by a group-contribution method, Ind Eng Chem Fund, 22(2), 266-267 (1983).

Sridhar, T., and Potter O.E., Diffusion coefficient of oxygen in liquids, Chem Eng Commns, 21(1), 47-54 (1983).

Tamir, A., Correlations for predicting azeotropic heat of vaporization of multicomponent mixtures, Ind Eng Chem Fund, 22(1), 83-86 (1983).

Teja, A.S.; Garg, K.B., and Smith, R.L., Method for calculating gas-liquid critical temperatures and pressures of multicomponent mixtures, Ind Eng Chem Proc Des Dev, 22(4), 672-676 (1983).

Utracki, L.A., Temperature and pressure dependence of liquid viscosity, Can JCE, 61(5), 753-758 (1983).

Weber, J.H., Predict liquid-phase diffusion coefficients, Chem Eng, 7 March, 161-163 (1983).

Yair, O.B., and Fredenslund, A., Pure-component vapor pressures by UNIFAC, Ind Eng Chem Proc Des Dev, 22(3), 433-436 (1983).

Yorizane, M., et al., Thermal conductivities of binary gas mixtures at high pressures, Ind Eng Chem Fund, 22(4), 458-463 (1983).

Zanker, A., P-V-T relationships for compressed air, Proc Engng, Sept, 73 (1983).

Zanker, A., Prediction of liquid density from critical properties, Proc Engng, Aug, 37 (1983).

Zanker, A., Predicting the flashpoint of binary blends, Proc Engng, Dec, 44-45 (1983).

1984

Alessi, P.; Alessandrini, A., and Orlandini, M., Activity coefficients at infinite dilution in solvents with two functional groups, Chem Eng Commns, 27(1), 59-68 (1984).

Allada, S.R., Solubility parameters of supercritical fluids, Ind Eng Chem Proc Des Dev, 23(2), 344-348 (1984).

Brule, M.R., and Starling, K.E., Thermophysical properties of complex systems, Ind Eng Chem Proc Des Dev, 23(4), 833-845 (1984).

Buck, E., and Frankl, E.M., Gaps in the pure component experimental physical property database, Chem Eng Prog, 80(3), 82-87 (1984).

Chase, J.D., The qualification of pure component physical property data, Chem Eng Prog, 80(4), 63-67 (1984).

Chung, T.H.; Lee, L.L., and Starling, K.E., Prediction of dilute gas viscosity and thermal conductivity, Ind Eng Chem Fund, 23(1), 8-13 (1984).

Coca, J., Diffusion coefficients in binary gas systems, Proc Engng, July, 57 (1984).

Feay, B.A.; Daubert, T.E.; Danner, R.P.; High, M.S., and Rhodes, C.L., Interactive computer for physical property data, Chem Eng Prog, 80(8), 55-57 (1984).

Fisher, C.H., and Huddle, B.P., Calculating properties of n-alkanes from boiling points, Chem Eng, 2 April, 102 (1984).

Ganapathy, V., Estimating the combustion temperature of fossil fuels, Proc Engng, Oct, 83 (1984).

Gandhidasan, P., Nomograph for moist-air properties, Chem Eng, 29 Oct, 118 (1984).

Hanna, J.G,, Entropy of vaporization from the log (T/p) vs. log p relationship, Ind Eng Chem Fund, 23(1), 101-104 (1984).

Klincewitz, K.M., and Reid, R.C., Estimation of critical properties with group contribution methods, AIChEJ, 30(1), 137-142 (1984).

Kudryavtseva, L., and Toome, M., Method for predicting ternary azeotropes, Chem Eng Commns, 26(4), 373-383 (1984).

Kung, J.K.; Nazario, F.N.; Joffe, J., and Tassios, D., Prediction of Henry's constants in mixed solvents from binary data, Ind Eng Chem Proc Des Dev, 23(1), 170-175 (1984).

Kuwairi, B., and Maddox, R.N., Generalized method for calculating latent heat of vaporization, Chem Eng Commns, 29(1), 337-352 (1984).

Maddox, R.N., and Erbar, J.H., Improve P-V-T predictions, Hyd Proc, Jan, 119-121 (1984).

McGovern, J.C., Estimation of solubility parameters and related properties of liquids, J Chem Tech Biotechnol, 34A(1), 38-42 (1984).

Pallady, P.H., and Henley, P.J., Relative humidity by direct calculation, Chem Eng, 29 Oct, 117 (1984).

Pedersen, K.S., Viscosity of crude oils, Chem Eng Sci, 39(6), 1011-1016 (1984).

Powell, L.J., and Murphy, T.J., Determining atomic weights, Chemtech, Dec, 726-730 (1984).

Rupp, W.; Hetzel, S.; Ojini, I., and Tassios, D., Prediction of enthalpies of mixing with group contribution models: Primary parameters, Ind Eng Chem Proc Des Dev, 23(2), 391-400 (1984).

Ruzicka, J.V.; Svab, L., and Novak, J.P., Estimation of densities of saturated liquid mixtures, Int Chem Eng, 24(1), 168-173 (1984).

Stoa, T.A., Formulas estimate data for dry saturated steam, Chem Eng, 10 Dec, 97 (1984).

Thomas, E.R., and Eckert, C.A., Prediction of limiting activity coefficients, Ind Eng Chem Proc Des Dev, 23(2), 194-209 (1984).

Wong, J.M., and Johnston, K.P., Thermodynamic models for nonrandom and strongly nonideal liquid mixtures, Ind Eng Chem Fund, 23(3), 320-326 (1984).

Zhang, G., and Hayduk, W., Solubility of isobutane and isobutylene in associated solvents, Can JCE, 62(5), 713-718 (1984).

Zhong, X., Improved generalized Watson equation for prediction of latent heat of vaporization, Chem Eng Commns, 29(1), 257-270 (1984).

Zhong, X., A reduced vapor pressure correlation and its application to refrigerants, Ind Eng Chem Proc Des Dev, 23(1), 7-11 (1984).

1985

Adams, J.T., and So, E.M.T., Automation of group-contribution techniques for estimation of thermophysical properties, Comput Chem Eng, 9(3), 269-284 (1985).

Adler, S.B., and Hall, K.R., Use correlation for oil properties, Hyd Proc, Nov, 71-75 (1985).

Adler, S.B., and Lin, T.C.T., K-constants: Water in hydrocarbons, Hyd Proc, April, 99-103 (1985).

Ali, M.F.; Hasan, M.U.; Bukhari, A.M., and Saleem, M., Arabian crude fractions analyzed, Hyd Proc, Feb, 83-86 (1985).

Ash, S.N.; Ray, P., and Dutta, B.K., Activity coefficients for binary and multicomponent liquid mixtures, AIChEJ, 31(5), 821-825 (1985).

Ashcroft, S.J., and Shearn, R.B., High pressure volumetric properties of ethene-carbon dioxide-propane at 294K and 311K, CER&D, 63, 283-290 (1985).

Baillagou, P.E., and Soong, D.S., Viscosity constitutive equation for PMMA-MMA solutions, Chem Eng Commns, 33(1), 125-134 (1985).

Banares-Alcantara, R.; Westerberg, A.W., and Rychener, M.D., Development of an expert system for physical property predictions, Comput Chem Eng, 9(2), 127-142 (1985).
Barduhn, A.J., Effect of pressure on azeotropes, Chem Eng Commns, 38(1), 9-16 (1985).
Carta, R.; Dernini, S., and Tola, G., Comparative analysis of derivation of infinite dilution activity coefficients from T-x and P-x data, Chem Eng Commns, 33(1), 231-236 (1985).
Diab, S., and Maddox, R.N., Calculating viscosity of mixtures by group contributions, Chem Eng Commns, 38(1), 57-66 (1985).
Dullien, F.A.L., and Asfour, A.F.A., Concentration dependence of mutual diffusion coefficients in regular binary solutions: A new predictive equation, Ind Eng Chem Fund, 24(1), 1-7 (1985).
Durand, A.A., Thermal conductivity of some insulating materials as a function of temperature, Chem Eng, 27 May, 153-154 (1985).
Gupte, P.A., and Daubert, T.E., New corresponding-states model for vapor pressure of non-hydrocarbon fluids, Ind Eng Chem Proc Des Dev, 24(3), 674-677 (1985).
Heidman, J.L.; Tsonopoulos, C.; Brady, C.J., and Wilson, G.M., High-temperature mutual solubilities of hydrocarbons and water, AIChEJ, 31(3), 376-384 (1985).
John, V.T.; Papadopoulos, K.D., and Holder, G.D., Generalized model for predicting equilibrium conditions for gas hydrates, AIChEJ, 31(2), 252-259 (1985).
Kerr, C.P., Calculate thermal conductivity for unassociated liquids, Chem Eng, 4 Feb, 90 (1985).
Mosey, F., Redox potentials in wastewater treatment, Chem Engnr, May, 21-24 (1985).
Przezdziecki, J.W., and Sridhar, T., Prediction of liquid viscosities, AIChEJ, 31(2), 333-335 (1985).
Ruiz-Bevia, F., et al., Liquid diffusion measurement by holographic interferometry, Can JCE, 63(5), 765-771 (1985).
Srinivasan, P.; Devotta, S., and Watson, F.A., Thermal stability of R11, R12B1, R113 and R114 and their compatibility with some lubricating oils, CER&D, 63, 230-234 (1985).
Stone, J.N., Determine the vapor pressure of HCl at low concentrations and temperatures, Chem Eng, 16 Sept, 102 (1985).
Teja, A., Correlation and prediction of diffusion coefficients using a generalized corresponding states principle, Ind Eng Chem Fund, 24(1), 39-44 (1985).
Vetere, A., Volumetric properties of pure compounds calculated by an equation of state, Chem Eng Sci, 40(3), 393-400 (1985).
Wagle, M.P., Predict saturation temperature as function of vapor pressure, Chem Eng, 10 June, 77-80 (1985).
Wagle, M.P., Estimate relative volatility quickly, Chem Eng, 29 April, 85 (1985).
Weber, J.H., Predict surface tensions of binary liquid mixtures, Chem Eng, 14 Oct, 87-90 (1985).
Weber, J.H., Predict surface tensions of pure liquids, Chem Eng, 4 Feb, 63-66 (1985).

Zhong, X., A reduced vapor pressure equation based on the theorem of corresponding states, Chem Eng Commns, 33(1), 107-124 (1985).

1986

Atkinson, G.; Kumar, A., and Atkinson, B.L., Modeling the PVT properties of concentrated electrolytes in water, AIChEJ, 32(9), 1561-1566 (1986).

Barr, R.S., and Newsham, D.M.T., Freezing temperatures of water, alkanoic acids and their mixtures, Chem Eng J, 33(2), 79-86 (1986).

Billingsley, D.S., and Lam, S., Critical point calculation with nonzero interaction parameters, AIChEJ, 32(8), 1393-1396 (1986).

Brucks, M.G., and Murad, S., Generalized corresponding states theory for surface tension of liquids and liquid mixtures, Chem Eng Commns, 40, 345-358 (1986).

Bures, M., Nonlinear equation describing molar heat capacities of gases as function of temperature, Int Chem Eng, 26(1), 160-165 (1986).

Cardozo, R.L., Prediction of the enthalpy of combustion of organic compounds, AIChEJ, 32(5), 844-848 (1986).

Dang, D., and Tassios, D.P., Prediction of enthalpies of mixing with a UNIFAC model, Ind Eng Chem Proc Des Dev, 25(1), 22-31 (1986).

de Pinto, N., and Graham, E.E., Evaluation of diffusivities in electrolyte solutions using Stefan-Maxwell equations, AIChEJ, 32(2), 291-296 (1986).

Debenedetti, P.G., and Kumar, S.K., Infinite dilution fugacity coefficients and the general behavior of dilute binary systems, AIChEJ, 32(8), 1253-1262 (1986).

Dong, W.G., and Lienhard, J.H., Corresponding states correlation of saturated and metastable properties, Can JCE, 64(1), 158-161 (1986).

England, C., Gas solubilities in physical solvents, Chem Eng, 28 April, 63-66 (1986).

Gonzalez-Pozo, V., Formulas estimate properties for dry, saturated steam, Chem Eng, 12 May, 123 (1986).

Guha, D.K., and De, P., Prediction of the volume correction factor in gas-liquid systems, Can JCE, 64(6), 1020-1022 (1986).

Herrera, R., The two-thirds power model for predicting diffusion coefficients in liquids, Chem Eng Commns, 40, 17-24 (1986).

Jalowka, J.W., and Daubert, T.E., Group contribution method to predict critical temperature and pressure of hydrocarbons, Ind Eng Chem Proc Des Dev, 25(1), 139-143 (1986).

Karandikar, B.M.; Morsi, B.I.; Shah, Y.T., and Carr, N.L., Effect of water on the solubility and mass transfer coefficients of carbon monoxide and hydrogen in a Fischer-Tropsch liquid, Chem Eng J, 33(3), 157-168 (1986).

Kolasinska, G., and Vera, J.H., Prediction of heats of vaporization of pure compounds, Chem Eng Commns, 43, 185-194 (1986).

Kumar, A., and Patwardhan, V.S., Prediction of vapour pressure of aqueous solutions of single and mixed electrolytes, Can JCE, 64(5), 831-838 (1986).

Lee, Z., Graphical representation of thermodynamic functions of state, Int Chem Eng, 26(1), 139-149 (1986).

Lielmezs, J., and Herrick, T.A., New surface tension correlation for liquids, Chem Eng J, 32(3), 165-170 (1986).

Luckas, M., and Lucas, K., Viscosity of liquids: An equation with parameters correlating with structural groups, AIChEJ, 32(1), 139-141 (1986).
Patrick, M., and Kaupisch, K.F., Properties of mixtures via parametric analogy, Chem Eng, 1 Sept, 104 (1986).
Patwardhan, V.S., and Kumar, A., A unified approach for prediction of thermodynamic properties of aqueous mixed-electrolyte solutions, AIChEJ, 32(9), 1419-1438 (1986).
Peng, D.Y., An empirical method for calculating vapor-liquid critical points of multicomponent mixtures, Can JCE, 64(5), 827-830 (1986).
Quinn, J.A.; Lin, C.H., and Anderson, J.L., Measuring diffusion coefficients by Taylor's method of hydrodynamic stability, AIChEJ, 32(12), 2028-2033 (1986).
Smith, R.L., and Teja, A.S., Critical point prediction using a multi-fluid generalized corresponding states principle, Chem Eng Commns, 43, 211-224 (1986).
Svoboda, J.; Repas, M., and Dykyj, J., Calculation of constants in Antione equation, Int Chem Eng, 26(2), 355-360 (1986).
Tamas, J., Determining physical properties for multicomponent hydrocarbon mixtures, Chem Eng, 29 Sept, 103-108 (1986).
Teja, A.S., and Thurner, P.A., Correlation and prediction of viscosities of mixtures over a wide range of pressure and temperature, Chem Eng Commns, 49, 69-80 (1986).
Twu, C.H., Generalized method for predicting viscosities of petroleum fractions, AIChEJ, 32(12), 2091-2094 (1986).
Vetere, A., An empirical correlation for the calculation of vapour pressures of pure components, Chem Eng J, 32(2), 77-86 (1986).
Wooley, R.J., Calculator program for finding values of physical properties, Chem Eng, 31 March, 109-118 (1986).

1987

Al-Najjar, H., and Al-Sammerrai, D., Thermogravimetric determination of the heat of vaporization of some highly polar solvents, J Chem Tech Biotechnol, 37(3), 145-152 (1987).
Arnett, E.M., et al., Master equations for calculating heterolysis energies in solution, Energy & Fuels, 1(1), 17-23 (1987).
Barduhn, A.J., Estimating relative volatilities of close-boiling species, Chem Eng Educ, 21(3), 144-145 (1987).
Bastawissi, A.E., Calculate kinematic viscosity of liquid water, Chem Eng, 14 Sept, 116 (1987).
Campanella, E.A.; Mathias, P.M., and O'Connell, J.P., Equilibrium properties of liquids containing supercritical substances, AIChEJ, 33(12), 2057-2066 (1987).
Castillo, C.A., An alternative method for the estimation of critical temperatures of mixtures, AIChEJ, 33(6), 1025-1027 (1987).
Chandar, S.C.R., and Singh, R.P., Estimation of vapor pressures of organic nitrogen and sulfur compounds by group-contribution method, Chem Eng Commns, 56, 107-116 (1987).
Cooney, W.R., and O'Connell, J.P., Correlation of partial molar volumes at infinite dilution of salts in water, Chem Eng Commns, 56, 341-350 (1987).

D'Souza, R., and Teja, A.S., Prediction of vapor pressures of carboxylic acids, Chem Eng Commns, 61, 13-22 (1987).
Eduljee, G.H., Correlating the fluidity of some n-alcohols using a modified free volume equation, Chem Eng J, 35(1) 1-8 (1987).
Fukai, J.; Watanabe, M.; Miura, T., and Ohtani, S., Simultaneous estimation of thermophysical properties by nonlinear least squares, Int Chem Eng, 27(3), 455-466 (1987).
Hart, D.R., Liquid activity coefficients, Chem Eng, 23 Nov, 131-138 (1987).
Iglesias-Silva, G.A., et al., Vapor pressure equation from extended asymptotic behavior, AIChEJ, 33(9), 1550-1556 (1987).
Liley, P.E., Improved Z charts, Chem Eng, 20 July, 123-126 (1987).
Ludmer, Z.; Shinnar, R., and Yakhot, V., Solubility in binary mixtures at the immiscibility critical point, AIChEJ, 33(11), 1776-1779 (1987).
Matthews, M.A., and Akgerman, A., Diffusion coefficients for binary alkane mixtures to 573K and 3.5 MPa, AIChEJ, 33(6), 881-885 (1987).
Meyer, E.C., Using vapor pressure information in a cubic equation of state, AIChEJ, 33(3), 503-505 (1987).
Moritz, P., Thermodynamic properties of pure fluid substances, Periodica Polytechnica, 31(3), 141-154 (1987).
Okkes, A.G., Get acid dew point of flue gas, Hyd Proc, 66(7), 53-55 (1987).
Raal, J.D., and Webley, P.A., Microflow calorimeter design for heats of mixing, AIChEJ, 33(4), 604-618 (1987).
Reid, R.C.; Prausnitz, J.M., and Poling, B.E., The Properties of Liquids and Gases, 4th Edn., McGraw-Hill, New York (1987).
Riazi, M.R., and Daubert, T.E., Predicting flash and pour points, Hyd Proc, 66(9), 81-83 (1987).
Said, A.S., and Al-Haddad, A.A., Curve fitting applied to relative volatility, bubble point, and dew point of ideal binary mixtures, Sepn Sci Technol, 22(4), 1199-1218 (1987).
Said, A.S.; Shaban, H.I., and Qader, S.A., Direct evaluation of bubble point of multicomponent ideal mixtures, Sepn Sci Technol, 22(5), 1439-1448 (1987).
Smith, R.L.; Teja, A.S., and Kay, W.B., Measurement of critical temperatures of thermally unstable n-alkanes, AIChEJ, 33(2), 232-238 (1987).
Takahashi, M.; Yokoyama, C., and Takahashi, S., Viscosities of gaseous R113, R114 and R115, Int Chem Eng, 27(1), 85-93 (1987).
Tsonopoulos, C., Critical constants of normal alkanes from methane to polyethylene, AIChEJ, 33(12), 2080-2083 (1987).
Vetere, A., Methods for the estimation of critical volumes, Chem Eng J, 34(3) 151-154 (1987).
Vetere, A., Methods for the estimation of critical volumes, Chem Eng J, 35(3), 215-217 (1987).
Vetere, A., Estimation of the critical temperatures and pressures of organic compounds by using the Rackett equation, Chem Eng J, 35(3), 211-214 (1987).

1988

Barker, I.K.; Bartle, K.D., and Clifford, A.A., Measurement of solubilities in fluids at supercritical temperatures and lower pressures using chromatographic retention, Chem Eng Commns, 68, 177-184 (1988).

Campbell, S.W., Initial estimate for pure-component vapor pressures in equation of state calculations, Ind Eng Chem Res, 27(7), 1333-1335 (1988).

Chung, T.H.; Ajlan, M.; Lee, L.L., and Starling, K.E., Generalized multiparameter correlation for nonpolar and polar fluid transport properties, Ind Eng Chem Res, 27(4), 671-679 (1988).

Cochran, H.D.; Pfund, D.M., and Lee, L.L., Theoretical models of thermodynamic properties of supercritical solutions, Sepn Sci Technol, 23(12), 2031-2048 (1988).

Colakyan, M., and Turton, R., Cramer's rule and partial thermodynamic properties. Ind Eng Chem Res, 27(4), 721-723 (1988).

Dickson, J.N., and Daubert, T.E., Consistency and extension of experimental vapor pressure and heat of vaporization data, Ind Eng Chem Res, 27(3), 523-527 (1988).

Diguilio, R., and Teja, A.S., Correlation and prediction of surface tensions of mixtures, Chem Eng J, 38(3), 205-208 (1988).

Gomes, J.F.P., Program calculates critical properties, Hyd Proc, 67(9), 110-112 (1988).

Harrison, B.K., and Seaton, W.H., Solution to missing group problem for estimation of ideal-gas heat capacities, Ind Eng Chem Res, 27(8), 1536-1540 (1988).

Horvath, A.L., Estimate properties of organic compounds, Chem Eng, 15 Aug, 155-158 (1988).

Jangkamolkulchai, A., and Luks, K.D., Isothermal compressibility of gas-saturated hydrocarbon liquids, Chem Eng Commns, 64, 197-206 (1988).

Jones, M.C., and Giarratano, P.J., Latent heats of supercritical fluid mixtures, AIChEJ, 34(12), 2059-2062 (1988).

Kopatsis, A.; Salinger, A., and Myers, A.L., Thermodynamics of solutions with solvent and solute in different pure states, AIChEJ, 34(8), 1275-1286 (1988).

Kramer, A., and Thodos, G., Adaptation of Flory-Huggins theory for modeling supercritical solubilities of solids, Ind Eng Chem Res, 27(8), 1506-1510 (1988).

Kumar, A., Prediction of activity coefficients in aqueous mixed electrolytes by McKay-Perring equation, Chem Eng Commns, 66, 201-206 (1988).

Le Lann, J.M.; Joulia, X., and Koehret, B., Computer program for prediction of thermodynamic properties and phase equilibria, Int Chem Eng, 28(1), 36-46 (1988).

Lee, H., and Thodos, G., Generalized viscosity behavior of fluids over complete gaseous and liquid states, Ind Eng Chem Res, 27(12), 2377-2384 (1988).

Lee, H., and Thodos,. G., Correlation for self-diffusivity, Ind Chem Eng Res, 27(6), 992-997 (1988).

Mallu, B.V., and Rao, Y.J., Estimating thermal conductivity, Hyd Proc, 67(1), 78 (1988).

Mazumdar, B.K., Correlation of physical and chemical properties of coal hydrogenation distillates, Energy & Fuels, 2(2), 230-234 (1988).

Narayan, J.; Wanchoo, R.K.; Raina, G.K., and Wani, G.A., Viscosity and surface tension of p-xylene-ethylacetate liquid mixtures, Can JCE, 66(6), 1021-1026 (1988).

Nass, K.K., Representation of solubility behavior of amino acids in water, AIChEJ, 34(8), 1257-1266 (1988).

Painter, P.C.; Park, Y., and Coleman, M.M., Thermodynamics of coal solutions, Energy & Fuels, 2(5), 693-702 (1988).

Roberts, B.E., and Mather, A.E., Solubility of carbon dioxide and hydrogen sulfide in mixed solvent, Chem Eng Commns, 72, 201-212 (1988).

Rodgers, P.A.; Creagh, A.L., and Prausnitz, J.M., Correlation of liquid heat capacities for fossil fuels using characterization data, Fuel, 67(1), 134-142 (1988).

Roy, R.; Liang, B., and Rosner, D.E., Simplified dewpoint predictions for N-trace salt-containing nonideal condensates in high temperature reactive vapor environments, Chem Eng Commns, 72, 35-46 (1988).

Rueff, R.M.; Sloan, E.D., and Yesavage, V.F., Heat capacity and heat of dissociation of methane hydrates, AIChEJ, 34(9), 1468-1476 (1988).

Sahimi, M., Determination of transport properties of disordered systems, Chem Eng Commns, 64, 177-196 (1988).

Shealy, G.S., and Sandler, S.I., Excess Gibbs free energy of aqueous nonelectrolyte solutions, AIChEJ, 34(7), 1065-1074 (1988).

Shigaki, Y., et al., Estimation of basic properties of fluorocarbon refrigerants (Freons), Int Chem Eng, 28(3), 447-455 (1988).

Shing, K.S., and Chung, S.T., Calculation of infinite-dilution partial molar properties by computer simulation, AIChEJ, 34(12), 1973-1980 (1988).

Shukla, K.P.; Chialvo, A.A., and Haile, J.M., Thermodynamic excess properties in binary fluid mixtures, Ind Eng Chem Res, 27(4), 664-671 (1988).

Siddiqi, S.A., and Teja, A.S., High-pressure densities of mixtures of coal chemicals, Chem Eng Commns, 72, 159-170 (1988).

Tayler, C., Physical property databases, Proc Engng, Feb, 57-60 (1988).

Teja, A.S., and Tardieu, G., Prediction of thermal conductivity of liquids and liquid mixtures including crude oil fractions, Can JCE, 66(6), 980-986 (1988).

Timar, L.; Siklos, J., and Edes, J., Computer generation of thermodynamic property charts for mixtures, Comput Chem Eng, 12(2/3), 123-126 (1988).

Toledo, P.G., and Reich, R., Comparison of enthalpy prediction methods for nonpolar and polar fluids and their mixtures, Ind Chem Eng Res, 27(6), 1004-1010 (1988).

Various, Surface chemistry of coals (symposium papers), Energy & Fuels, 2(2), 111-169 (1988).

Varsanyi, G., Revised reduced compressibility chart and fugacity diagram for fluids, Periodica Polytechnica, 32(4), 277-298 (1988).

Vetere, A., Empirical model for calculating second virial coefficients, Chem Eng Sci, 43(12), 3119-3128 (1988).

Wanchoo, R.K., et al., Viscosity and surface tension of toluene-ethylacetate liquid mixture, Chem Eng Commns, 69, 225-234 (1988).

Yaws, C.L., and Chiang, P.Y., Enthalpy of formation for 700 major organic compounds, Chem Eng, 26 Sept, 81-88 (1988).

Yaws, C.L.; Ni, H.M., and Chiang, P.Y., Heat capacity data for 700 compounds, Chem Eng, 9 May, 91-98 (1988).

CHAPTER 2

SITE CONSIDERATIONS AND PROJECT ASSESSMENT

2.1 Project and Process Assessment

1966-1970

Various, Materials management (feature report), Chem Eng, 5 Dec, 117-134 (1966).

Lee, E.S., and Gray, E.H., Optimizing complex chemical plants by mathematical modeling techniques, Chem Eng, 28 Aug, 131-138 (1967).

Spitz, P.H., Handling your process engineering requirements, Chem Eng, 23 Oct, 175-180 (1967).

Bridgwater, A.V., Long range process design and morphological analysis, Chem Engnr, April, CE75-81 (1968).

Cooke, P.J., A model for decision making capable of numerical analysis, Chem Engnr, April, CE82-85 (1968).

Elliott, D.M., Critical examination in process design, Chem Engnr, Nov, CE377-383 (1968).

Gilbertson, R., Automatic storage improves production control, Chem Eng, 4 Nov, 206-208 (1968).

Perry, R.H., and Singer, E., Practical guidelines for process optimization, Chem Eng, 26 Feb, 163-168 (1968).

Uchiyama, T., Best size for refinery and tankers, Hyd Proc, 47(12), 85-88 (1968).

Frazier, A.W., A seven-step approach to problem solving, Hyd Proc, 48(7), 189-190 (1969).

Klumpar, I.V., Process predesign by computer, Chem Eng, 22 Sept, 114-122 (1969).

Lewis, R.L., Managing an enginering project in Europe, Chem Eng, 6 Oct, 152-168 (1969).

Moore, J.F., Systems approach to refinery design and operations, Chem Eng Prog, 65(2), 71-75 (1969).

Pearson, A.W., and Gear, A.E., Project management in R & D, Brit Chem Eng, 14(11), 1537-1539 (1969).

Rothermel, T.W., Market simulation makes a science out of forecasting, Chem Eng, 24 March, 157-160 (1969).

Various, Technological forecasting (symposium papers), Chem Engnr, June, CE241-276 (1969).

Various, Market research and forecasting in the chemical industry (symposium papers), Chem Engnr, Nov, CE393-411 (1969).

Robbins, J., New approaches to process design, Chem Engnr, Oct, CE298-304 (1970).

Various, Process technology for license or sale (feature report), Chem Eng, 20 April, 114-144 (1970).

Weiss, A.S., New flowcharts provide process details at a glance, Chem Eng, 20 April, 170-172 (1970).

1971-1974

Fogel, I.M., Managing the small and medium-size project, Chem Eng, 13 Dec, 107-114 (1971).

Gregory, S.A., Creativity and innovation, Chem Engnr, June, 229-243 (1971).

Hartman, R.F., Project performance evaluation, Chem Eng Prog, 67(12), 42-46 (1971).
Kephart, H.L., Planning of refining/petrochemical complexes, Chem Eng Prog, 67(11), 68-74 (1971).
Klumpar, I.V., Determining economic and process variables of ventures, Chem Eng Prog, 67(4), 74-80 (1971).
Kovac, F.J., Technology forecasting: A case history of automobile tyres, Chemtech, Jan, 18-23 (1971).
Martino, J.P., Technological forecasting, Chem Eng, 27 Dec, 54-62 (1971).
McPherson, J.H., Decision-making techniques: A guide to selection, Hyd Proc, 50(8), 98-99 (1971).
Merckx, L.J., Compare large HPI plants by index of technical success, Hyd Proc, 50(8), 103-108 (1971).
Paradiso, A.J., The marketing of dyestuffs, Chemtech, May, 292-296 (1971).
Perry, W.H., Performance analysis in marketing planning, Brit Chem Eng, 16(8), 709-710 (1971).
Runyon, D.L., Move tanks without dismantling, Hyd Proc, 50(2), 112-114 (1971).
Shedden, I.W., Process design in Australia, Chem Engnr, Nov, 403-405 (1971).
Various, Effective process engineering design (symposium papers), Chem Engnr, Dec, 419-435 (1971).
Wilkes, A., and Pearson, A.W., Project management in research and development, Brit Chem Eng, 16(11), 1009-1011 (1971).
Birkhoff, J., One-step isometrics from models, Hyd Proc, 51(7), 95-97 (1972).
Blecker, H.G., and Nichols, T.M., Case study of process feasibility and economics of fish protein concentrate, Chemtech, Dec, 717-726 (1972).
Gregory, S.A., Transportation, Chem Engnr, Oct, 376-386 (1972).
Kerns, G.D., How to check process designs, Hyd Proc, 51(1), 100-102 (1972).
Mattson, R.H., Planning the expansion of titanium dioxide production, Chem Eng Prog, 68(5), 70-75 (1972).
Proctor, A., Planning for a medium-sized chemical plant, Proc Tech Int, 17(10), 781-783 (1972).
Various, Engineering design and construction using models, Chem Eng Prog, 68(6), 41-67 (1972).
Wilson, G.T., Rapid estimation of market size, Brit Chem Eng, 17(1), 25-26 (1972).
Hays, G.E., Profit-oriented research, Chem Eng Prog, 69(11), 70-71 (1973).
Jenett, E., Guidelines for successful project management, Chem Eng, 9 July, 70-82 (1973).
Klimpel, R.R., Operations research: Decision-making tool, Chem Eng, 16 April, 103-108; 30 April, 87-94 (1973).
Mapstone, G.E., Forecasting for sales and production, Chem Eng, 14 May, 126-132 (1973).
Coppen, J.L., Managing small design-construction projects, Chem Eng, 25 Nov, 85-90 (1974).
Various, Industrial chemical marketing (topic issue), Chem & Ind, 1 June, 426-436 (1974).
Wells, G., Selecting process routes, Proc Engng, March, 72-75 (1974).

1975-1979

Anon., Project independence: A critical assessment, Chem Eng, 6 Jan, 92-105 (1975).

Coates, W.H., Raw materials and the fertiliser industry, Chem Engnr, Oct, 589-592 (1975).

Folger, J.H., and Karr, A.E., Contract engineering, Chem Eng, 8 Dec, 130-132 (1975).

Lawrence, B., Preliminary project evaluation, Chemtech, Nov, 678-681 (1975).

Rudd, D.F., Modelling the development of the intermediate chemicals industry, Chem Eng J, 9(1), 1-20 (1975).

Rudkin, J., From bright idea to plant production, Chem Eng, 3 Feb, 69-71 (1975).

van Dalen, J.D., Problems in planning ethylene projects, Chem Eng Prog, 71(6), 91-98 (1975).

Wharton, F.D., and Craver, J.K., Technology assessment in product development, Chemtech, Sept, 547-551 (1975).

Coates, J.F., Technology assessment methods, Chemtech, June, 372-383 (1976).

Harris, J.S., New-product profile chart, Chemtech, Sept, 554-562 (1976).

Kladko, M., Designing for usable capacity, usable quality, Chem Eng, 2 Feb, 89-92 (1976).

Lee, D.A., Handy formulas for balancing materials, Chem Eng, 25 Oct, 156 (1976).

Pecoraro, C.A., Managing for effective design engineering, Chem Eng Prog, 72(1), 60-63 (1976).

Simmonds, W.H.C., Real criteria for chemical projects, Chemtech, Aug, 494-497 (1976).

Various, Project management (special report), Hyd Proc, 55(8), 73-99 (1976).

Weingast, M., Blending computations using a pocket calculator, Chem Eng, 25 Oct, 154-156 (1976).

Belanger, A., New isometric technique better than plant model, Hyd Proc, 56(10), 205-207 (1977).

Bergtraun, E.M., Contracting for new construction, Chem Eng, 20 June, 133-136 (1977).

King, R.A., How to achieve effective project control, Chem Eng, 4 July, 117-121 (1977).

Perrella, A.V., Pitfalls in evaluating R&D, Chem Eng, 1 Aug, 59-60 (1977).

Springmann, H., Plan large oxygen and nitrogen plants, Hyd Proc, 56(2), 97-101 (1977).

Woods, D.R., Teaching problem solving, Chem Eng Educ, 11(2), 86-94; 11(3), 140-144 (1977).

Cherry, D.H.; Grogan, J.C.; Holmes, W.A., and Perris, F.A., Availability analysis for chemical plants, Chem Eng Prog, 74(1), 55-60 (1978).

Ware, C.H., Improving R&D effectiveness, Chem Eng, 27 Feb, 99-108 (1978).

Aneja, A.P., and Aneja, V.P., Process options, feedstock selection and polyesters, Chemtech, April, 260-262 (1979).

Anon., Anatomy of a super-project, Proc Econ Int, 1(1), 20-28 (1979).

Anon., Experience curves for chemicals, Proc Econ Int, 1(1), 13-19 (1979).

Anon., Special considerations in design, logistics and construction of plants in Alaska, Proc Econ Int, 1(1), 29-34 (1979).

Anon., The contractor's view of consequential liability, Proc Econ Int, 1(1), 46-48 (1979).
Anon., An early appraisal of exploration projects, Proc Econ Int, 1(1), 42-45 (1979).
Bridge, G.L., The organisation of a process design department, Chem Engnr, May, 359, 361 (1979).
Kerridge, A.E., Check project progress with Bell and 'S' curves, Hyd Proc, 58(3), 189-202 (1979).
Picciotti, M., and Kaiser, V., Select process schemes for optimum petrochemicals, Hyd Proc, 58(6), 99-105 (1979).
Sommerfeld, J.T.; Sondhi, D.K.; Spurlock, J.M., and Ward, H.C., Usage of graph-theoretic methods in chemical venture analysis, Chem Eng J, 18(2), 117-124 (1979).
Various, Pilot plant operations (topic issue), Chem Eng Prog, 75(9), 45-63 (1979).
Various, Modular plant concepts (topic issue), Chem Eng Prog, 75(10), 49-91; 75(11), 35-45 (1979).
Woods, D.R., et al., What is problem solving? Chem Eng Educ, 13(3), 132-137 (1979).

1980-1984

Anon., Economics of ocean transportation, Proc Econ Int, 1(4), 11-16 (1980).
Bush, W.D., Project engineering, Hyd Proc, 59(11), 263-275 (1980).
Holding, J., What's ahead for process design offshore? Chem Engnr, Aug, 539-542 (1980).
Kauders, P.G., Process engineering design, Chem Engnr, Introduction, Jan, 51-52; Development of P&IDs, July, 475, 485; Hydraulic design, Aug, 530-531, Oct, 634, Dec, 753; Equipment design and pressure relief protection, April, 146-147 (1980).
Kurzawa, C.J., Megaprojects, Hyd Proc, 59(4), 283-294; 59(5), 291-293 (1980).
Law, J.A., Let's build a plant, Hyd Proc, 59(11), 249-253 (1980).
Polentz, L.M., The Monte Carlo method of predicting production time, Chem Eng, 11 Aug, 157-161 (1980).
Shanmugan, C., Estimating delivery times, Chem Eng, 3 Nov, 166-167 (1980).
Wojciechowski, B.W., Select a process by design, Hyd Proc, 59(3), 75-80 (1980).
Datz, M., Project management, Hyd Proc, 60(9), 161-177 (1981).
Glaser, L.B.; Kramer, J., and Siegelman, G.A., Modular and barge-mounted plants: An update, Chem Eng Prog, 77(9), 63-68 (1981).
Hayworth, H.C., A case of technology transfer, Chemtech, June, 342-346 (1981).
Malan, D.N., Commissioning, planning and execution, Chem Engnr, Oct, 442-444 (1981).
Various, Project management (special report), Hyd Proc, 60(12), 81-101 (1981).
Various, Pilot plants (topic issue), Chem Eng Prog, 77(8), 29-54 (1981).
Weaver, J.B., Project profitability and assessment, C&E News, 2 Nov, 37-46 (1981).
Zambon, D.M., Golden opportunities for modular construction, Chem Eng Prog, 77(8), 60-64 (1981).

Grave, M.J., The quality management of contract engineering research and development, Chem Engnr, April, 130-135 (1982).
Klein, R.L., Mega-projects: A growing management challenge, Hyd Proc, Jan, 225-228 (1982).
Martino, J.P., Methods for forecasting CPI technology change, Chem Eng, 11 Jan, 97-106 (1982).
Various, Project management (special report), Hyd Proc, Dec, 53-67 (1982).
Various, Modular technology offshore (topic issue), Chem Eng Prog, 78(11), 29-81 (1982).
Williams, J., A better way to revise P&IDs, Hyd Proc, May, 225-226 (1982).
Bolton, G.A., Project engineering modelling techniques, Chem Engnr, Dec, 26-28 (1983).
Glaser, L.B., and Kramer, J., Does modularization reduce plant investment? Chem Eng Prog, 79(10), 63-68 (1983).
Hulme, D., and La Trobe-Bateman, J., Take a closer look at modular construction, Hyd Proc, Jan, 34C-34R (1983).
Kliewer, V.D., Benefits of modular plant design, Chem Eng Prog, 79(10), 58-62 (1983).
Ostrovskii, G.M., et al., Steady-state simulation of chemical plants, Chem Eng Commns, 23(1), 181-190 (1983).
Various, Project management (special report), Hyd Proc, Dec, 63-80 (1983).
Waldheim, P.A.; Finneran, J.A., and Whittington, E.L., Productivity improvement in process design, Plant/Opns Prog, 2(4), 222-226 (1983).
Carlson, R.O., and Thorne, H.C., Avoiding the 'game' in project evaluation, Chem Eng Prog, 80(11), 11-13 (1984).
Kerridge, A.E., Integrate project controls with cost/schedule control systems criteria, Hyd Proc, Nov, 193-208 (1984).
Kerridge, A.E., How to develop a project schedule, Hyd Proc, Jan, 133-154 (1984).
Landis, R.L., and Hamilton, J.E., Technique evaluates engineering alternatives, Chem Eng, 1 Oct, 91-93 (1984).
Morrison, R.V., Analyzing batch process cycles, Chem Eng, 25 June, 175 (1984).
Reinhardt, J.P., Identification of key products and technologies, Proc Econ Int, 4(4), 12-15 (1984).
Stallworthy, E.A., and Kharbanda, O.P., New directions for project management, Proc Econ Int, 4(4), 25-29 (1984).

1985-1988

Bradford, M., Supply side process engineering, Hyd Proc, Jan, 111-120 (1985).
Copulsky, W., Buying and selling businesses, Chemtech, Oct, 591-593 (1985).
Veranth, J.M., Project scheduling techniques, Chem Eng, 15 April, 61-68 (1985).
Bradford, M.L., and Falconer, D., Process design follow-up: How to do it effectively, Hyd Proc, 65(11), 129-132 (1986).
Anon., Facts and figures for the chemical industry (1987), C&E News, 20 June, 34-82 (1988).
Anon., World chemical outlook, C&E News, 12 Dec, 26-44 (1988).

Aswani, A.G., Project planning and suppliers, Chem Eng, 10 Oct, 99-102 (1988).
Bem, J.Z., From concept to concrete, Chemtech, Jan, 42-46 (1988).
Contino, A.V., Checkout the product competition, Chem Eng, 20 June, 127-132 (1988).
Foveaux, M., and O'Connor, K.J., Assessing the US chemical industry, Chem Eng Prog, 84(7), 43-49; 56-66 (1988).
Gilles, E.D.; Holl, P., and Marquardt, W., Dynamic simulation of complex chemical processes, Int Chem Eng, 28(4), 579-593 (1988).
Harper, T., Pharmaceutical plant design, Processing, Nov, 25-34 (1988).
Landau, R., Harnessing innovation for growth, Chem Eng Prog, 84(7), 31-42 (1988).
Lesourd, J.B., and Hallegatte, R., Concept of process production function: Basic principles and application to calculation of efficiencies of irreversible thermodynamic cycles, Int Chem Eng, 28(4), 608-618 (1988).
Maiden, C.J., A project overview, Chemtech, Jan, 38-41 (1988).
Rawls, R.L., Facts and figures for chemical R & D, C&E News, 22 Aug, 29-64 (1988).
Sigurdsson, M., and Rudd, D.F., Trends in international production and trade in petrochemicals, Chem Eng Commns, 66, 125-169 (1988).
Stobaugh, R., and Gagne, J., Learning from petrochemical history: Strategies to the year 2000, Chem Eng Prog, 84(7), 25-30 (1988).
Storck, W.J., Top 100 chemicals 1987, C&E News, 2 May, 9-14 (1988).
Wyss, S.E., Guide to specification writing, Chem Eng, 9 May, 87-89 (1988).
Yamamura, K.; Nakajima, M., and Matsuyama, H., Detection of gross errors in process data using mass and energy balances, Int Chem Eng, 28(1), 91-99 (1988).

2.2 Plant Performance and Assessment

1967-1970

Corrigan, T.E., and Dean, M.J., Determining optimum plant size, Chem Eng, 14 Aug, 152-156 (1967).
Luecke, R.H.; McGuire, M.L., and Crosser, O.K., Dynamic flowsheeting, Chem Eng Prog, 63(2), 60-66 (1967).
Various, Chemical technology for sale or license (feature report), Chem Eng, 25 Sept, 135-175 (1967).
Campbell, K.S., Designing, building and operating a prototype plant, Chem Eng, 7 Oct, 163-166 (1968).
Goulcher, R., Application of queueing theory to batch manufacturing, Brit Chem Eng, 13(3), 377-380 (1968).
Katzen, R., When is the pilot plant necessary? Chem Eng, 25 March, 95-98 (1968).
Liddle, C.J., Preliminary evaluation of recycle processes, Brit Chem Eng, 13(6), 813-816 (1968).
Nishimura, H.; Hiraizumi, Y., and Yagi, S., Optimization of a process network by the linear model method, Int Chem Eng, 8(1), 186-194 (1968).

Wood, R.M., Optimisation of a processing unit, Brit Chem Eng, 13(5), 661-665 (1968).

Klima, B.B., and Youngblood, E.L., Constructing inexpensive plant models easily, Chem Eng, 24 Feb, 128-130 (1969).

Lofthouse, J.A., Large chemical plants and their problems, Chem Engnr, Sept, CE153-157 (1969).

Schwabe, G., and Purgand, J., Predicting optimum plant size, Brit Chem Eng, 14(2), 177-178 (1969).

Swager, W.L., Technological forecasting in R&D, Chem Eng Prog, 65(12), 39-46 (1969).

Estrup, C., and Nielsen, M.B., Chemical market forecasting, Brit Chem Eng, 15(7), 910-912 (1970).

Franzel, H.L., Guide to quality control of equipment design, Chem Eng, 1 June, 135-138 (1970).

Liddle, C.J., Modelling and analysing recycle streams, Brit Chem Eng, 15(1), 64-68; 15(3), 349-354 (1970).

1971-1979

Cinadr, B.F.; Curley, J.K., and Schooley, A.T., Miniplant design and use, Chem Eng, 25 Jan, 62-76 (1971).

Miller, R.E., Scale modeling of large and small plant projects, Chem Eng, 29 Nov, 69-73 (1971).

Various, Process flowsheets and descriptions (special edition; supplement), Brit Chem Eng, 16(12), 1-47 (1971).

Armstrong, R., Better ways to build process plants, Chem Eng, 17 April, 86-94 (1972).

Burklin, C.R., Safety standards, codes and practices for plant design, Chem Eng, 2 Oct, 56-63; 16 Oct, 113-120; 13 Nov, 143-155 (1972).

Gladstone, J., Project planning and control by network, Brit Chem Eng, 17(5), 401-406 (1972).

Johri, H.P.; Quon, D., and Reilly, P.M., Decision-making under uncertainty, Brit Chem Eng, 17(7), 597-600 (1972).

Potter, J.R., Generalized process evaluation, Chemtech, June, 343-349 (1972).

Powers, G.J., Heuristic synthesis in process development, Chem Eng Prog, 68(8), 88-95 (1972).

Robinson, E.R., Optimal charging rates for batch processes, Brit Chem Eng, 17(10), 807 (1972).

Strassburger, F., Polymer-plant engineering: Materials handling and compounding of plastics, Chem Eng, 3 April, 81-88 (1972).

Babcock, J.A., Engineering models, Chem Eng, 19 Feb, 112-118 (1973).

Erskine, J.B., Acoustic design for a petrochemical plant, Chem Engnr, June, 312-315 (1973).

Klumpar, I.V., Process feasibility by simplified simulation, Chemtech, Feb, 88-94 (1973).

Seth, N.D., and Subramanian, S.K., Graphical method to maximise batch efficiencies, Proc Tech Int, June, 260-264 (1973).

Szonyi, G., Factorial design methodology, Chemtech, Jan, 36-44 (1973).

Williams, R., Exclusion charts relate sales volume and unit price, Chemtech, Oct, 592-596 (1973).

Fruit, W.M.; Reklaitis, G.V., and Woods, J.M., Simulation of multiproduct batch chemical processes, Chem Eng J, 8(3), 199-212 (1974).
Rose, L.M.; Myhre, J., and Walter, O.H.D., Planning manufacturing capacity, Chemtech, Aug, 494-501 (1974).
Sherwin, M., Supply/demand and process technology, Chemtech, April, 225-229 (1974).
Freeman, R.A., and Gaddy, J.L., Quantitative overdesign of chemical processes, AIChEJ, 21(3), 436-440 (1975).
Weisman, J.; Pulido, H., and Khanna, A., Optimal process system design in accordance with demand and price forecasts, Ind Eng Chem Proc Des Dev, 14(1), 51-58 (1975).
Furzey, D.G., Economic feasibility studies for chemical plant, Chem Engnr, Jan, 33-35 (1976).
Anderson, M.C., Recovery calculation for a separation process, Chem Eng, 19 Dec, 106 (1977).
Farkas, D.F., Unit operations concepts optimize operations, Chemtech, July, 428-433 (1977).
Hlavacek, V., Analysis of steady-state and transient behavior of complex plants, Comput Chem Eng, 1(1), 75-100 (1977).
King, R., Unsuccessful processes: How to avoid and improve them (case studies), Proc Engng, March, 85-87 (1977).
Mahalec, V., and Motard, R.L., Procedures for the initial design of chemical processing systems, Comput Chem Eng, 1(1), 57-68 (1977).
Waggoner, R.C., and Loud, G.D., Algorithms for solution of material balance equations for nonconventional multistage operations, Comput Chem Eng, 1(1), 49-56 (1977).
Bisio, A., Investment recovery, Chemtech, Dec, 735-738 (1978).
Johns, W.R.; Marketos, G., and Rippin, D.W.T., The optimal design of chemical plant to meet time-varying demands in the presence of technical and commerical uncertainty, Trans IChemE, 56, 249-257 (1978).
Various, Symposium papers on integral design, J Appl Chem Biotechnol, 28, 40-68 (1978).
Allen, D.H., Project evaluation, Chemtech, July, 412-417 (1979).
Lamb, J.A., and Pomphrey, D.M., The design of plant under conditions of uncertainty, Chem Engnr, Feb, 102-105, 108 (1979).

1980-1988

Cran, J., Optimizing cycle-time calculations, Proc Engng, Dec, 73-75 (1980).
Shaw, R.J.; Sykes, J.A., and Ormsby, R.W., Plant-test manual, Chem Eng, 11 Aug, 126-132 (1980).
Various, Designing and operating batch process plants (feature report), Chem Eng, 25 Feb, 70-91 (1980).
Anon., Evaluation, design and construction of a new uranium plant in South Africa, Proc Econ Int, 2(4), 9-16 (1981).
Hoerner, D.R., and Crosser, O.K., Application of generalized geometric programming in preliminary plant design, Ind Eng Chem Proc Des Dev, 20(2), 210-219 (1981).

Paustian, J.E.; Puzio, J.F.; Stavropoulos, N., and Sze, M.C., A lesson in flowsheet design: Nicotinamide and acid, Chemtech, March, 174-178 (1981).

Pickwell, P., Nitric acid plant optimization, Chem & Ind, 21 Feb, 114-118 (1981).

Denn, M.M., and Lavie, R., Dynamics of plants with recycles, Chem Eng J, 24(1), 55-60 (1982).

Doherty, M.F., et al., Estimating bounds for process optimization problems: An analogy to rules-of-thumb and structural modifications, Ind Eng Chem Fund, 21(3), 289-298 (1982).

Knopf, F.C.; Okos, M.R., and Reklaitis, G.V., Optimal design of batch/semicontinuous processes, Ind Eng Chem Proc Des Dev, 21(1), 79-86 (1982).

Morari, M., et al., Design of resilient processing plants, Chem Eng Sci, 37(2), 245-270 (1982).

Suhami, I., and Mah, R.S.H., Optimal design of multipurpose batch plants, Ind Eng Chem Proc Des Dev, 21(1), 94-100 (1982).

Bonnell, L.W.; Heydorn, E.C., and Couch, J.R., Engineering information system for a demonstration plant facility, Plant/Opns Prog, 2(3), 160-165 (1983).

Henriksen, J.A., Systems engineering approach to pharmaceutical plant design, Plant/Opns Prog, 2(2), 79-84 (1983).

Kerridge, A.E., Predict project results with trending methods, Hyd Proc, July, 125-152 (1983).

Rippin, D.W.T., Simulation of single and multiproduct batch chemical plants for optimal design and operation (review paper), Comput Chem Eng, 7(3), 137-156 (1983).

Tayler, C., Improved process designs, Proc Engng, July, 31-32 (1983).

van Weenan, W.F., and Tielrooy, J., Optimizing hydrogen plant design, Chem Eng Prog, 79(2), 37-44 (1983).

Gonzalez, V.L., and Larder, K., Batch or continuous processing: A case history, Chemtech, Oct, 607-609 (1984).

Pacey, A.J., and Murray, G.R., Using decision trees, Chemtech, March, 156-161 (1984).

Takamatsu, T.; Hashimoto, I.; Hasebe, S., and O'Shima, M., Design of a flexible batch process with intermediate storage tanks, Ind Eng Chem Proc Des Dev, 23(1), 40-48 (1984).

Tayler, C., Determining the appropriate scale of production plants, Proc Engng, Sept, 29-35 (1984).

Anon., Ergonomics and engineering design, Proc Engng, July, 31,33 (1985).

Chaty, J.C., A general-purpose pilot-plant facility, Plant/Opns Prog, 4(2), 105-111 (1985).

Parakrama, R., Improving batch chemical processes, Chem Engnr, Sept, 24-25 (1985).

Shaw, J.H., Lab data check material balance, Hyd Proc, June, 83-84 (1985).

Tine, C.B.D., Evaluating errors in chemical plant mass balance, Can JCE, 63(2), 322-325 (1985).

Berkoff, C.E.; Kamholz, K.; Rivard, D.E.; Wellman, G., and Winicov, H., The process profile: Comparison and evaluation of chemical processes, Chemtech, Sept, 552-559 (1986).

Linnhoff, B., and Kotjabasakis, E., Downstream paths for operable process design (process optimization), Chem Eng Prog, 82(5), 23-28 (1986).
Nath, R.; Libby, D.J., and Duhon, H.J., Joint optimization of process units and utility system, Chem Eng Prog, 82(5), 31-38 (1986).
Tait, W.S., Optimizing the emulsion manufacturing process, Chem Eng Prog, 82(5), 29-30 (1986).
Fehr, M., Flowsheet is process language, Chem Eng Educ, 22(2), 88-90 (1988).
Fuge, C.; Eisele, P., and Whitehead, B.D., Optimization of operation of a gas terminal, Gas Sepn & Purif, 2(2), 103-106 (1988).
Lichtenstein, C.W., Bar-net schedule for project control, Chem Eng, 28 March, 53-56 (1988).
MacDonald, R.J., and Howat, C.S., Data reconciliation and parameter estimation in plant performance analysis, AIChEJ, 34(1), 1-8 (1988).
Malina, M.A., Optimizing plant capacity for inventory levels, Chem Eng, 23 May, 139-142 (1988).
Ostrovsky, G.M.; Ostrovsky, M.C., and Berezhinsky, T.A., Optimization of chemical plants with recycles, Comput Chem Eng, 12(4), 289-296 (1988).
Sinden, K., Modular approach to downstream processing, Chem Engnr, Aug, 32-33 (1988).
Smith, R., and Linnhoff, B., Design of separators in the context of overall processes, CER&D, 66(3), 195-228 (1988).

2.3 Equipment Design Data

1966-1969

Urban, W.J., and Holland, F.A., How to determine optimum plant size, Chem Eng, 28 March, 103-108 (1966).
Kehat, E., Thermodynamic analysis of chemical processes, Brit Chem Eng, 12(6), 877-878 (1967).
Mains, W.D., and Richenberg, R.E., Steam jet ejectors in pilot and production plants, Chem Eng Prog, 63(3), 84-88 (1967).
Pridgen, T.D., and Garcia, N., Structural design, Chem Eng, 8 May, 143-150; 22 May, 187-192 (1967).
Quigley, H.A., Nomograph simplifies selection of tank dimensions, Chem Eng, 10 April, 246 (1967).
Quigley, H.A., Raising operating level in horizontal tanks saves money, Chem Eng, 13 March, 200-202 (1967).
Rosen, A.M., and Krylov, V.S., Scaling-up of mass-transfer equipment and reactors: Use of hydraulic model experiments, Chem Eng Sci, 22(3), 407-416 (1967).
Shinnar, R., Sizing of storage tanks for off-grade material, Ind Eng Chem Proc Des Dev, 6(2), 263-264 (1967).
Watkins, R.N., Sizing separators and accumulators, Hyd Proc, 46(11), 253-256 (1967).
Bourne, J.R., Blending calculations for two-product specifications, Brit Chem Eng, 13(6), 834-836 (1968).
Hill, R.G., Drawing effective flowsheet symbols, Chem Eng, 1 Jan, 84-92 (1968).

Tang, S.S., Shortcut method for calculating tower deflections, Hyd Proc, 47(11), 230-232 (1968).
Various, Pilot plants (topic issue), Chem Eng Prog, 64(10), 33-58 (1968).
Witherspoon, D.L., Guidelines to bag and pallet sizes for bulk materials, Chem Eng, 17 June, 284-290 (1968).
Czerniak, E., Foundation design guide for stacks and towers, Hyd Proc, 48(6), 95-114 (1969).
Hughson, R.V., Process-equipment standards, Chem Eng, 17 Nov, 234-238 (1969).
Kern, D.Q., Scaleup and design, Chem Eng Prog, 65(7), 77-80 (1969).
McCabe, J.S., and Hickey, K.J., Heavy reactors now site assembled, Hyd Proc, 48(5), 133-136 (1969).
Moody, G.B., Mechanical design of a tall stack, Hyd Proc, 48(9), 173-178 (1969).
Timm, B., Application of high pressures in chemical technology, Brit Chem Eng, 14(5), 660-665 (1969).

1970-1974

Anon., Packaged-plant practice, Brit Chem Eng, 15(2), 181-186; 15(3), 369 (1970).
Boberg, I.E., Choosing tank designs, Chem Eng, 10 Aug, 134-136 (1970).
Crisi, J.S., Setting anchor bolts for tower foundations, Hyd Proc, 49(7), 119-120 (1970).
Various, Design of storage tanks, Brit Chem Eng, 15(10), 1332-1336 (1970).
Wills, J.S., Size vapor piping by computer, Hyd Proc, 49(5), 149-155 (1970).
Youness, A., New approach to tower deflection, Hyd Proc, 49(6), 121-126 (1970).
Badhwar, R.K., Shortcut design methods for piping, exchangers, towers, Chem Eng, 18 Oct, 112-122 (1971).
Brown, A.A., Easy way to compute tower foundation flexure, Hyd Proc, 50(1), 121-125 (1971).
Hashemi, H.T., and Wesson, H.R., LNG storage tanks, Hyd Proc, 50(8), 117-120 (1971).
Morgan, P.F., Move tanks on air cushions, Hyd Proc, 50(11), 229-230 (1971).
Various, USA and European design standards (special report), Hyd Proc, 50(6), 93-115 (1971).
White, C.H., Optimizing production rates for parallel equipment, Chem Eng, 14 June, 86-91 (1971).
Anon., Try a coordinated drafting system, Hyd Proc, 51(3), 131-138 (1972).
Hendry, J.E., and Hughes, R.R., Generating separation processes flowsheets, Chem Eng Prog, 68(6), 71-76 (1972).
Jope, J.A., Variable-unit scale-up methods applied to thermal systems, Chem Engnr, Aug, 291-294 (1972).
Jope, J.A., A variable-unit approach to scale-up, Chem Engnr, March, 112-113, 120 (1972).
Moody, G.B., Design of saddle supports for horizontal vessels, Hyd Proc, 51(11), 157-160 (1972).
Spence, J., and Findlay, G.E., Designing for pipe bends, Proc Engng, Dec, 62-67 (1972).

Brown, A.A., New foundation design method for pile supported tanks, Hyd Proc, 52(9), 175-176 (1973).
Durr, C.A., and Crawford, D.B., LNG terminal design, Hyd Proc, 52(11), 211-214 (1973).
Jope, J.A., Improved scale-up design using variable-unit analysis, Chem Engnr, July, 382-385 (1973).
Katell, S., Justifying pilot-plant operations, Chem Eng Prog, 69(4), 55-56 (1973).
Kowal, G., Quick calculation for holdups in horizontal tanks, Chem Eng, 11 June, 130-132 (1973).
Various, Storage in the HPI (special report), Hyd Proc, 52(8), 71-88 (1973).
Brown, A.A., New approach to tank foundation design, Hyd Proc, 53(10), 153-156 (1974).
Rosen, A.M., and Krylov, V.S., Theory of scaling up and hydrodynamic modelling of industrial mass-transfer equipment (review paper), Chem Eng J, 7(2), 85-98 (1974).

1975-1979

Arya, S.C.; Drewyer, R.P., and Pincus, G., Foundation design for vibrating machines, Hyd Proc, 54(11), 273-278 (1975).
Babcock, J.A., Design engineering of an arc acetylene plant, Chem Eng Prog, 71(3), 90-94 (1975).
Conn, A.L., Scaleup from lab to plant, Chemtech, March, 154-159 (1975).
Hahn, G.J., Designing experiments, Chemtech, Aug, 496-498; Sept, 561-562 (1975).
Hooper, W.B., Predicting flow patterns in plant equipment, Chem Eng, 4 Aug, 103-106 (1975).
Loeb, M.B., Locating bolt-holes for strength and convenience, Chem Eng, 17 March, 88 (1975).
Sparrow, R.E.; Forder, G.J., and Rippin, D.W.T., Choice of equipment sizes for multiproduct batch plants: Heuristics vs branch and bound, Ind Eng Chem Proc Des Dev, 14(3), 197-203 (1975).
Worley, N., Nuclear engineering power plant components, Chem Engnr, April, 242-244 (1975).
Zudkevitch, D., Imprecise data impacts plant design and operation, Hyd Proc, 54(3), 97-103 (1975).
Clark, J.P., Chemical plant design, Chemtech, Nov, 664-667 (1975); Jan, 23-26; April, 235-239 (1976).
Curran, D.F., Plotting product and raw-material storage, Chem Eng, 30 Aug, 128 (1976).
Sengupta, A.K., Adding acid to makeup water, Chem Eng, 2 Aug, 110 (1976).
Utley, C.O., Models of chemical plants, Chemtech, Aug, 488-493 (1976).
Various, Storage and transportation (feature report), Processing, Dec, 33-59 (1976).
Wachel, J.C., and Bates, C.L., Escape piping vibrations while designing, Hyd Proc, 55(10), 152-156 (1976).
de Rooij, A.H.; Dijkhuis, C., and Van Goolen, J.T.J., Scaling-up a process, Chemtech, May, 309-315 (1977).

Eddinger, R.T., Scaling-up using the cold model concept, Chemtech, Sept, 556-558 (1977).
Ganapathy, V., Estimating the holdup in dished heads, Chem Eng, 14 Feb, 108 (1977).
Ku, J.C.Z., and Bevan, D., Outdoor bulk storage for hydrophilic materials, Chem Eng, 29 April, 69-74 (1977).
Mah, R.S.H., Effects of thermophysical property estimation on process design, Comput Chem Eng, 1(3), 183-190 (1977).
Mahajan, K.K., Method for designing rectangular storage tanks, Chem Eng, 28 March, 107-112 (1977).
Markovitz, R.E., Process and project data pertinent to vessel design, Chem Eng, 10 Oct, 123-128 (1977).
Niida, K.; Yagi, H., and Umeda, T., An application of database management system (DBMS) to process design, Comput Chem Eng, 1(1), 33-40 (1977).
Sisson, W., Calculate fertilizer blends by nomograph, Chem Eng, 14 March, 156 (1977).
Getz, R., Stress analysis of steam-heated jacketed piping, Hyd Proc, 57(2), 139-141 (1978).
Kaferle, J.A., Sizing draw-off nozzles, Chem Eng, 5 June, 170 (1978).
Karcher, G.G., New design calculations for high temperature storage tanks, Hyd Proc, 57(10), 137-140 (1978).
El-Rifai, M., Estimate equipment weights, Hyd Proc, 58(9), 225-226 (1979).
Maitra, N., Graphs aid column base-plate design, Hyd Proc, 58(10), 155-158 (1979).
Mehra, Y.R., Liquid surge capacity in horizontal and vertical vessels, Chem Eng, 2 July, 87-88 (1979).
Santi, I.E., Finding volume in partially filled tanks, Chem Eng, 18 June, 144-147 (1979).
Streich, M., and Kistenmacher, H., Property inaccuracies influence low-temperature designs, Hyd Proc, 58(5), 237-241 (1979).
Ummarino, G., Computerize seal coolant flow calculations, Hyd Proc, 58(8), 137-141 (1979).
Various, Process drives (special report), Hyd Proc, 58(8), 73-91 (1979).
Various, Design and operation of batch processes (topic issue), Comput Chem Eng, 3(3), 9-13, 169-208 (1979).

1980-1984
Jayaraman, K., Estimating pipe expansion, Chem Eng, 11 Feb, 132 (1980).
Malina, M.A., Storage capacity: How big should it be? Chem Eng, 28 Jan, 121-124 (1980).
Sallenbach, H.G., Stepwise solution to vibrating equipment foundation design, Hyd Proc, 59(3), 93-100 (1980).
Sproesser, W.D., Models speed plant design, Chem Eng, 24 March, 113-116 (1980).
Williams, P., and Lawson, B., Structural-design update, Proc Engng, Dec, 33-39 (1980).
Burk, H.S., Conceptual design of refinery tankage, Chem Eng, 24 Aug, 107-110 (1981).

Gerunda, A., How to size liquid-vapor separators, Chem Eng, 4 May, 81-84 (1981).
Jordans, A., Simple weight calculations for steel towers and vessels, Hyd Proc, 60(8), 146 (1981).
Karcher, G.G., Simplified stress equations for elevated-temperature storage tanks, Hyd Proc, 60(7), 157-160 (1981).
Knap, J.E., Purchasing used equipment from other users, Chem Eng Prog, 77(3), 39-43 (1981).
Patel, C.R., Estimating the volumes of dished ends, Proc Engng, April, 85; June, 80 (1981).
Sproesser, W.D., Miniature models help to clarify design ideas, Chem Eng, 20 April, 145-146 (1981).
Zanker, A., Nomograph for tank truck loading by weight, Hyd Proc, 60(4), 197-198 (1981).
Zelnick, A.A., Avoiding oversized-flange problems, Chem Eng, 16 Nov, 253-257 (1981).
Buchmann, A.P.; Leesley, M.E., and Dale, A.G., The role, structure, and design of a database for disemination of process data during plant design, Chem Eng Commns, 16(1), 1-38 (1982).
Karamchandani, K.C.; Gupta, N.K., and Pattabiraman, J., Evaluation of percent critical damping of process towers, Hyd Proc, May, 205-208 (1982).
Purarelli, C., Optimum design of horizontal liquid-vapor separators, Chem Eng, 15 Nov, 127-129 (1982).
Schwartz, M., Structural design, Chem Eng, 17 May, 96-100; 9 Aug, 61-64; 1 Nov, 89-93; 27 Dec, 57-60 (1982).
Takamatsu, T.; Hashimoto, I., and Hasebe, S., Optimal design and operation of a batch process with intermediate storage tanks, Ind Eng Chem Proc Des Dev, 21(3), 431-440 (1982).
Various, Packaged plant (topic issue), Processing, Jan, 27-31 (1982).
Gallagher, T., Estimating horizontal-tank volumes, Chem Eng, 22 Aug, 100 (1983).
Gundzik, R.M., Standardized pilot plant procedures, Chem Eng Prog, 79(8), 29-34 (1983).
Kunesh, J.G., and Hollenack, W.R., Solving pilot-plant problems, Chem Eng Prog, 79(8), 51-54 (1983).
Livingston, A., Sizing process vessels, Chem Eng, 27 June, 70 (1983).
Piccinini, N., and Anatra, U., Intermediate tank to increase plant availability, Ind Eng Chem Fund, 22(2), 206-208 (1983).
Pinfold, G., Structural design of chimneys, Proc Engng, June, 40-43 (1983).
Power, V., Uniform method for drawings and documents, Chem Eng, 28 Nov, 73-75 (1983).
Ramshaw, C., and Arkley, K., Process intensification by miniature mass transfer (Higee system), Proc Engng, Jan, 29-31 (1983).
Blakey, P., and Orlando, G., Using inert gases for purging, blanketing and transfer, Chem Eng, 28 May, 97-102 (1984).
Constan, G.L., Pilot plants for medium-sized companies, Chem Eng Prog, 80(2), 56-57 (1984).
Feldman, J.M., Do electrical/control engineering early, Hyd Proc, Nov, 143-145 (1984).

Mullett, T.A., Manage modular projects, Hyd Proc, July, 93-94; Aug, 92-94; Sept, 185-187; Oct, 75-78 (1984).
Pfeffer, H.A.; Bhalla, S.K., and Dore, J.C., Pilot-plant design: How to avoid an expensive mistake, Plant/Opns Prog, 3(2), 98-101 (1984).
Schwartz, M., Structural design, Chem Eng, 7 March, 165-169; 2 May, 67-70; 11 July, 79-83; 5 Sept, 119-123; 31 Oct, 51-56 (1983); 23 Jan, 77-81 (1984).
Schwartz, M., and Koslov, J., Piping and instrumentation diagrams, Chem Eng, 9 July, 85-89 (1984).
Tan, M.A.; Kumar, R.P., and Kuilanoff, G., Modular design and construction, Chem Eng, 28 May, 89-96 (1984).
Whittaker, R., Onshore modular construction, Chem Eng, 28 May, 80-88 (1984).
Wu, F.H., Drum separator design: A new approach, Chem Eng, 2 April, 74-80 (1984).

1985-1988

Anon., Tank design, Processing, Aug, 41 (1985).
Bisio, A., and Kabel, R.L. (Eds), Scaleup of Chemical Processes, Wiley, New York (1985).
Fawbert, J., Open tendering for plant equipment, Proc Engng, March, 36-41 (1985).
Johnson, J.E., and Morgan, D.J., Graphical techniques for process engineering design calculations, Chem Eng, 8 July, 72-83 (1985).
Lowenstein, J.G., The pilot plant, Chem Eng, 9 Dec, 62-76 (1985).
von Brecht, R.C., Prefabricated control modules, Chem Eng, 13 May, 85-90 (1985).
Brown, T.R., Guidelines for preliminary selection of heat-exchanger type, Chem Eng, 3 Feb, 107-108 (1986).
Gerrard, M.; Puc, G., and Simpson, E., Optimize the design of wire-mesh separators, Chem Eng, 10 Nov, 91-93 (1986).
Heitmann, J.A., and Rhees, D.J., Scaling up, Chemtech, June, 344-350 (1986).
Lowenstein, J.G., Pilot-plant tests, Chem Eng, 12 May, 124 (1986).
Al-Abdulally, F.; Al-Shuwaib, S., and Gupta, B.L., Hazard analysis and safety considerations in refrigerated ammonia storage tanks, Plant/Opns Prog, 6(2), 84-88; 6(4), 06-08 (1987).
Martinez, O.A.; Madhavan, S., and Kellett, D.J., Damage to and replacement of ammonia storage-tank foundation, Plant/Opns Prog, 6(3), 129-141 (1987).
Marynowski, C.W., Scale-up: Three easy lessons, Chemtech, Sept, 560-563 (1987).
Walas, S.M., Rules of thumb: Selecting and designing equipment, Chem Eng, 16 March, 75-81 (1987).
Zlokarrik, M., Scale-up under conditions of partial similarity, Int Chem Eng, 27(1), 1-10 (1987).
Kabel, R.L., Instruction in scaleup, Chem Eng Educ, 22(3), 128-133 (1988).
Kaufman, N., Scale-up in the 1980s, Chemtech, May, 297-299 (1988).
Maha, A., Quick estimate of tank volume, Chem Eng., 19 Dec, 165 (1988).
McGuiness, W.N., Welding outlet fittings, Hyd Proc, 67(11), 101-103 (1988).

Muralidhara, H.S., The combined-fields approach to separations, Chemtech, April, 229-235 (1988).
Riggs, J.B., Simplified manifold design, Chem Eng, 10 Oct, 95-98 (1988).
Schwarzhoff, J.A., and Sommerfeld, J.T., Draining time of spheres, Chem Eng, 20 June, 158-160 (1988).
Various, Process synthesis and scale-up (topic issue), Chem & Ind, 15 Feb, 105-122 (1988).

2.4 Site Selection

1970-1982

Speir, W.B., Choosing and planning industrial sites, Chem Eng, 30 Nov, 69-75 (1970).
Mendel, O., How location affects U.S. plant-construction costs, Chem Eng, 11 Dec, 120-124 (1972).
Gushee. D.E., Plant siting and pollution control, Chemtech, Aug, 468-470 (1974).
Hyde, M.C., Locate new ethylene plants, Hyd Proc, 55(4), 53C-53F (1976).
Roskill, O.W., The location of chemical plants, Chem Engnr, Jan, 29-32 (1976).
Herzet, G.L., Construction site selection: A European viewpoint, Hyd Proc, 56(6), 116-120 (1977).
Maclean, W.D., Construction site selection: A US viewpoint, Hyd Proc, 56(6), 111-116 (1977).
Lambe, H.W., Plant cost and developing country location, Chemtech, Feb, 100-102 (1978).
Granger, J.E., Plantsite selection, Chem Eng, 15 June, 88-115 (1981).
Lovett, K.M.; Swiggett, G.E., and Cobb, C.B., When you select a plant site, Hyd Proc, May, 285-293 (1982).

2.5 Plant Layout

1967-1988

Mixon, G.M., Chemical plant lighting, Chem Eng, 5 June, 113-116 (1967).
Robertson, J.M., Plan small for expansion, Chem Eng Prog, 63(9), 87-90 (1967).
Gysemans, E.E., Piping layout, Chem Engnr, Oct, CE352-354 (1968).
Robertson, J.M., Design for expansion, Chem Eng, 22 April, 179-184; 6 May, 187-194 (1968).
House, F.F., An engineer's guide to process-plant layout, Chem Eng, 28 July, 120-128 (1969).
Constance, J.D., Plant-area design to prevent air-moisture condensation, Chem Eng, 5 Oct, 116-118 (1970).
Kaess, D., Guide to trouble-free plant layout, Chem Eng, 1 June, 122-134 (1970).
Sachs, G., Economic and technical factors in chemical plant layout, Chem Engnr, Oct, CE304-311 (1970).
Bush, M.J., and Wells, G.L., Unit plot plans for plant layout, Brit Chem Eng, 16(4), 325-327; 16(6), 514-517 (1971).

Bush, M., and Wells, G., Optimizing plant layout, Proc Engng, Sept, 135-137 (1972).
Thirkell, H., Optimization of storage farms and distribution terminals, Brit Chem Eng, 17(3), 235-241 (1972).
Friedrich, E.R., Floors for process areas, Chem Eng, 24 June, 157-162 (1974).
Spitzgo, C.R., Guidelines for overall chemical-plant layout, Chem Eng, 27 Sept, 103-107 (1976).
Kern, R., Plant layout, Chem Eng, 23 May, 130-136; 4 July, 123-129; 15 Aug, 153-160; 12 Sept, 169-177; 7 Nov, 93-99; 5 Dec, 131-140 (1977).
Teunissen, P.J.M., Development of a green field site, Hyd Proc, 56(11), 66EE-66J (1977).
Kern, R., Plant layout, Chem Eng, 30 Jan, 105-112; 27 Feb, 117-121; 10 April, 127-135; 8 May, 191-197; 17 July, 123-130; 14 Aug, 141-146 (1978).
Elton, R.L., Designing stormwater handling systems, Chem Eng, 2 June, 64-68 (1980).
Kaura, M.L., Plot plans must include safety, Hyd Proc, 59(7), 183-194 (1980).
Mecklenburgh, J.C., Process-plant layout, Proc Engng, Dec, 30-33 (1982).
Mecklenburgh, J.C. (Ed.), Process Plant Layout, Halstead Press, New York (1985).
Linnhoff, B., and Eastwood, A.R., Overall site optimisation by pinch technology, CER&D, 65(5), 408-414 (1987).
Nolan, P.F., and Bradley, W.J., Simple technique for optimization of layout and location for chemical plant safety, Plant/Opns Prog, 6(1), 57-61 (1987).
Hawk, C.W., Use of air-bubble enclosures, Chem Eng, 9 May, 83-85 (1988).
Madden, J., and Walters, K., Plant layout skills, Proc Engng, Sept, 59-61 (1988).

2.6 Pipework Design

1966-1975

Various, Piping (special report), Hyd Proc, 45(10), 113-139 (1966).
Elperin, I.T., et al., Decreasing the hydrodynamic resistance of pipelines, Int Chem Eng, 7(2), 276-279 (1967).
Various, Process piping systems (feature report), Chem Eng, 17 June, 190-256 (1968).
Daniel, P.T., and Hall, M., An integrated system of pipework estimating, detailing and control, Chem Engnr, May, CE169-178 (1969).
Gallant, R.W., Optimizing pipe sizes for liquids and vapors, Chem Eng, 24 Feb, 96-104 (1969).
Smith, B., Charts used for easier pipe sizing, Hyd Proc, 48(5), 173-180 (1969).
Various, Process pipelines: A design guide (Supplemental issues), Brit Chem Eng, 14(2), PPR3-PPR59; 14(9), PPR63-PPR123 (1969).
Birdwell, J.R., and Shull, W.W., Piping specifications by computer, Hyd Proc, 49(8), 103-106 (1970).
Moody, G.B., Structural tubing for pipe support, Hyd Proc, 49(4), 117-123 (1970).
Sisson, W., Sizing of manifold and branch lines made easy, Chem Eng, 18 May, 182 (1970).

Gay, B., and Middleton, P., Solution of pipe network problems, Chem Eng Sci, 26(1), 109-124 (1971).
Henry, J.P., and Louks, B.M., Economic study of pipeline gas production from coal, Chemtech, April, 238-247 (1971).
Loeb, M.B., Fast way of calculating piping weights, Chem Eng, 1 Nov, 96-98 (1971).
Various, Pipes and pipelines (special feature), Brit Chem Eng, 16(4), 305-336 (1971).
de Lesdernier, D.L., and Sommerfeld, J.T., Computer program sizes pipes, Hyd Proc, 51(3), 112-114 (1972).
Arcuri, K., Quiet-streamlining pipeline design, Chem Eng, 8 Jan, 134-138 (1973).
Benson, R.E., Analyzing piping flexibility, Chem Eng, 29 Oct, 102-109 (1973).
Various, Piping systems design (feature report), Proc Tech Int, Oct, 357-365 (1973).
Chimes, A.R., Confidence lines for sizing plant piping systems, Chem Eng, 5 Aug, 118-120 (1974).
Styer, R.F., and Wier, J.T., Revised piping code highlights, Hyd Proc, 53(3), 101-105 (1974).
Char, C.V., Check pipe support orientation, Hyd Proc, 54(9), 207-212 (1975).
Kern, R., Practical piping design, Chem Eng, 23 Dec, 58-66 (1974); 6 Jan, 115-120; 3 Feb, 72-78; 3 March, 161-168; 14 April, 85-93; 28 April, 119-126 (1975).
Kern, R., Practical piping design, Chem Eng, 26 May, 113-120; 23 June, 145-151; 4 Aug, 107-113; 15 Sept, 129-136; 13 Oct, 125-132; 10 Nov, 209-215 (1975).
Sisson, W., Thicknesses for pipe bends, Chem Eng, 22 Dec, 71 (1975).

1976-1987

Bickel, T.C.; Himmelblau, D.M., and Edgar, T.F., Optimal design of long gas pipelines, Chem Eng Prog, 72(9), 72-74 (1976).
Pothanikat, J.J., New approach to piping system weight analysis, Hyd Proc, 55(4), 195-197 (1976).
Stanley, G.M., and Mah, R.S.H., Estimation of flows and temperatures in process networks, AIChEJ, 23(5), 642-650 (1977).
Kent, G.R., Preliminary pipeline sizing, Chem Eng, 25 Sept, 119-120 (1978).
Various, Pipework (feature report), Processing, March, 53-74 (1978).
Various, Pipework (feature report), Processing, March, 53-74 (1978).
Char, C.V., Guide to pipe support design, Hyd Proc, 58(3), 133-139; 58(9), 241-248 (1979).
Duxbury, H.A., Relief line sizing for gases, Chem Engnr, Nov, 783-787; Dec, 851-852, 857 (1979).
Peng, L.C., Toward more consistent pipe stress analysis, Hyd Proc, 58(5), 207-211 (1979).
Simpson, L.L., Versatile calculator program eases piping design, Chem Eng, 29 Jan, 105-109 (1979).
Woods, D.R., and Dunn, R.W., Piping layout as a laboratory project, Chem Eng Educ, 13(2), 64-68 (1979).

Various, Pipework and piping systems (feature report), Processing, March, 57-88 (1980).
Kandell, P., Program sizes pipe and flare manifolds for compressible flow, Chem Eng, 29 June, 89-93 (1981).
Nowak, J.D., and Joye, D.D., Analytical equations for head vs. flow rate characteristics of branch/recycle piping systems, Ind Eng Chem Proc Des Dev, 20(3), 460-463 (1981).
Blackwell, W.W., Estimate equivalent line lengths of piping circuits, Chem Eng, 1 Nov, 69-72 (1982).
Hooper, W.B., Predict fittings for piping systems, Chem Eng, 17 May, 127-129 (1982).
Emery, S.J., Protecting against backflow in process lines, Chem Eng, 28 Nov, 81-84 (1983).
Hills, P.D., Designing piping for gravity flow, Chem Eng, 5 Sept, 111-114 (1983).
Silberring, L., Long distance pipelining of natural gas or its derivatives, Proc Econ Int, 4(1), 31-34 (1983).
Mikasinovic, M., and Marcucci, P.A., Sizing pipe for external pressure, Chem Eng, 30 April, 61-64 (1984).
Bell, N.J., Shortcut methods to determine optimum line sizing (SI units), Chem Eng, 14 Oct, 120-122 (1985).
Berryman, B., and Daniels, L., Pipeline technology, Chem Engnr, Nov, 17-19 (1985).
Brown, G.S., How to predict pressure drop before designing the piping, Chem Eng, 16 March, 85-86 (1987).
Calogero, M.V., and Brooks, A.W., Sizing pipe after steam traps, Chem Eng, 12 Oct, 138, 140 (1987).

2.7 Utilities

1967-1975

Partridge, E.P., and Paulson, E.G., Economic reuse of water via the closed cycle, Chem Eng, 9 Oct, 244-248 (1967).
Wardle, J.K.S., and Todd, G., Bulk storage of liquified gases, Chem Engnr, Nov, CE247-253, 260 (1967).
Kohli, J.P., Design best cooling water system, Hyd Proc, 47(12), 108-110 (1968).
Various, Utilities (special report), Hyd Proc, 47(6), 87-110 (1968).
Gunder, P.F., Specifying process furnaces, Hyd Proc, 48(10), 117-120 (1969).
McIlhenny, W.F., and Zeitoun, M.A., A chemical engineer's guide to seawater, Chem Eng, 3 Nov, 81-86; 17 Nov, 251-256 (1969).
Meador, L., and Shah, A., Steam lines designed for two-phase flow, Hyd Proc, 48(1), 143-146 (1969).
Scheel, L.F., Centrifugal or reciprocating compressors for refrigeration? Hyd Proc, 48(3), 123-129 (1969).
Sisson, W., Nomograph for flash steam from blowdown systems, Chem Eng, 19 May, 216 (1969).

Slack, J.B., Steam balance: A new exact method, Hyd Proc, 48(3), 154-156 (1969).
Weil, R.V., and Jackson, G., Water conservation in the petroleum industry, Chem Eng Prog, 65(11), 69-72 (1969).
Wohlk, W., Steam-jet refrigeration plants and their range of application, Brit Chem Eng, 14(3), 309-313 (1969).
Sisson, W., Boiler blowdown from nomograph, Chem Eng, 10 Aug, 138-140 (1970).
Troscinski, E.S., and Watson, R.G., Controlling deposits in cooling-water systems, Chem Eng, 9 March, 125-132 (1970).
Various, Process plant utilities (feature report), Chem Eng, 14 Dec, 130-146 (1970).
Various, Process cooling systems (special report), Hyd Proc, 49(10), 83-103 (1970).
Donohue, J.M., Making cooling water safe for steel and fish, too, Chem Eng, 4 Oct, 98-102 (1971).
Ellison, G.L., Steam-drum level stability factor, Hyd Proc, 50(5), 125-127 (1971).
Kearney, P.E., Deciding factors in the selection of closed and open circuit cooling systems, Chem Engnr, Sept, 333-337 (1971).
McCarthy, A.J., and Hopkins, M.E., Simplify refrigeration estimating, Hyd Proc, 50(7), 105-111 (1971).
Silverstein, R.M., and Curtis, S.D., Cooling water, Chem Eng, 9 Aug, 84-94 (1971).
Jones, F.A., Utilities consumption in natural gas plants, Chem Eng Prog, 68(10), 73-79 (1972).
Menicatti, S., and Cappiello, L., Find optimum furnace efficiency, Hyd Proc, 51(9), 226-230 (1972).
Thangappan, R., Nomogram for calculating furnace and boiler chimney height, Brit Chem Eng, 17(9), 721 (1972).
Walko, J.F., Controlling biological fouling in cooling systems, Chem Eng, 30 Oct, 128-132; 27 Nov, 104-108 (1972).
Askew, A., Selecting economic boiler-water pretreatment equipment, Chem Eng, 16 April, 114-120 (1973).
Gupton, P.S., and Krisher, A.S., Waste-heat boiler failures, Chem Eng Prog, 69(1), 47-50 (1973).
Kinard, G.E., and Gaumer, L.S., Mixed-refrigerant cascade cycle for LNG, Chem Eng Prog, 69(1), 56-61 (1973).
Anon., Review of coal-firing shell boilers, Energy World, April, 9-11 (1974).
Anon., Closed-circuit cooling systems, Proc Engng, Jan, 50-53 (1974).
Appleyard, C.J., and Shaw, M.G., Reuse and recycle of water in industry, Chem & Ind, 16 March, 240-246 (1974).
Downie, J.M., and Hoggarth, M.L., Review of industrial and commercial gas burner developments, J Inst Fuel, 47, 124-128 (1974).
Ferguson, B.C., Monitor boiler fuel density to control air/fuel ratio, Hyd Proc, 53(2), 89-93 (1974).
Thompson, G.A., Standby gas systems, Chemtech, Sept, 552-554 (1974).
Buffington, M.A., How to select package boilers, Chem Eng, 27 Oct, 98-106 (1975).

Chimes, A.R., Startup pressures for specifying steam traps, Chem Eng, 12 May, 114-116 (1975).
Field, A.A., Boilers, Energy World, Aug, 7-15; Oct, 12-27 (1975).
Grier, J.C., and Christensen, R.J., Biocides give flexibility in cooling water treatment, Hyd Proc, 54(11), 283-286 (1975).
Hinchley, P., Avoiding waste-heat boiler problems, Chem Eng, 1 Sept, 94-98 (1975).
How, M.E., Review of coal-fired water-tube boilers, Energy World, Feb, 7-10 (1975).

1976-1979

Anerousis, J.P., Softening water with sodium zeolite, Chem Eng, 7 June, 128-130 (1976).
Caplan, F., Is your water scaling or corrosive? Chem Eng, 1 Sept, 129 (1975); 2 Aug, 110 (1976).
James, E.W.; Maguire, W.F., and Harpel, W.L., Using wastewater as cooling-system makeup water, Chem Eng, 30 Aug, 95-100 (1976).
Kmiec, R.R., Distributing strong acid without using pumps or carboys, Chem Eng, 27 Sept, 135 (1976).
Kumar, J., and Fairfax, J.P., Rating alternatives to chromates in cooling-water treatment, Chem Eng, 26 April, 111-112 (1976).
Monroe, E.S., Install steam traps correctly, Chem Eng, 10 May, 121-126 (1976).
Robitaille, D.R., and Bilek, J.G., Molybdate cooling-water treatments, Chem Eng, 20 Dec, 77-80 (1976).
Roffman, H.K., and Roffman, A., Water that cools but does not pollute, Chem Eng, 21 June, 167-174 (1976).
Schweitzer, W.J., Hot water for process use: Storage vs. instantaneous heaters, Chem Eng, 8 Nov, 141-144 (1976).
Sisson, W., Estimating loads for unit heaters, Chem Eng, 5 July, 101 (1976).
Smith, R.D., and Scollon, R.B., Balancing boilers against plant loads, Chem Eng, 29 March, 125-128 (1976).
Stanbridge, D.W., Analyzing hydrogen generating plants, Chem Eng, 2 Aug, 110 (1976).
Various, Offsite facilities (special report), Hyd Proc, 55(12), 71-110 (1976).
Cooksey, D.L., Reducing refrigeration-system costs, Hyd Proc, 56(3), 131-132 (1977).
Ganapathy, V., Nomograph for basic data for steam generators, Chem Eng, 6 June, 197-198 (1977).
Kunz, R.G.; Yen, A.F., and Hess, T.C., Cooling-water calculations, Chem Eng, 1 Aug, 61-71 (1977).
Mikasinovic, M., and Dautovich, D.R., Designing steam transmission lines without steam traps, Chem Eng, 14 March, 137-139 (1977).
Sisson, W., Steam from flashing condensate, Chem Eng, 14 Feb, 105 (1977).
Various, Seawater cooling in chemical plant (topic issue), Chem & Ind, 2 July (supplement), pp.3-39 (1977).
Various, Water treatment (feature report), Processing, Aug, 41-53 (1977).
Various, Better furnace operations (special report), Hyd Proc, 56(2), 79-94 (1977).
Anon., Modular nitrogen-purge plants, Processing, July, 39-40 (1978).

Bergman, A., Inexpensive utilities metering and monitoring, Chem Eng, 28 Aug, 125-126 (1978).
Kaiser, V.; Salhi, O., and Pocini, C., Analyze mixed refrigerant cycles, Hyd Proc, 57(7), 163-167 (1978).
Kinsley, G.R., Controlling cross-connections keeps plant drinking-water safe, Chem Eng, 10 April, 121-126 (1978).
Krisher, A.S., Process water treatment, Chem Eng, 28 Aug, 78-98 (1978).
Krisher, A.S., Evaluation of low-toxicity cooling-water inhibitors, Chem Eng, 13 Feb, 115-116 (1978).
Larinoff, M.W.; Moles, W.E., and Reichhelm, R., Design and specification of air-cooled steam condensers, Chem Eng, 22 May, 86-94 (1978).
McWilliam, J.D., Protecting demineralizers from organic fouling, Chem Eng, 22 May, 80-84 (1978).
Puckorius, P.R., Controlling corrosive microorganisms in cooling-water systems, Chem Eng, 23 Oct, 171-174 (1978).
Various, Utility systems (feature report), Chem Eng, 18 Dec, 80-96 (1978).
Wilcox, J.C., Improving boiler efficiency, Chem Eng, 9 Oct, 127-130 (1978).
Howe, R.H.L., Boiler-water control for efficient steam production, Chem Eng, 26 Feb, 135-141 (1979).
Mehra, Y.R., Refrigerants, Chem Eng, 18 Dec, 97-104 (1978); 15 Jan, 131-139; 12 Feb, 95-101; 26 March, 165-172 (1979).
Miller, D.K., Sizing dual-suction risers in liquid overfeed refrigeration systems, Chem Eng, 24 Sept, 117-124 (1979).
Newman, V.G., Pumped-water storage, J Inst energy, 52, 178-184 (1979).
Omerod, W.G., and Read, A.W., Improved method of combustion control of coal-fired boilers using flue-gas carbon monoxide analysis, J Inst Energy, 52, 23-26 (1979).
Setchfield, P., Design of coil boilers, Processing, Oct, 40-41 (1979).
Stoa, T.A., Calculating boiler efficiency and economics, Chem Eng, 16 July, 77-81 (1979).
Subrahmanyan, N.V.; Pandian, G., and Ganapathy, V., Sizing waste-heat boilers, Hyd Proc, 58(9), 261-265 (1979).

1980-1984

Clark, J.K., and Helmick, N.E., How to optimize the design of steam systems, Chem Eng, 10 March, 116-128 (1980).
Crozier, R.A., Designing 'near optimum' cooling-water system, Chem Eng, 21 April, 118-127 (1980).
Gillis, E.A., Fuel cells for electric utilities, Chem Eng Prog, 76(10), 88-93 (1980).
Goyal, O.P., Guidelines for combustion design, Hyd Proc, 59(11), 205-211 (1980).
Highley, J., Design and control of fluidized-bed fired boilers, J Inst Energy, 53, 208-216 (1980).
Kemmer, F.N., Optimizing water-supply, treatment and recycle practices, Chem Eng, 6 Oct, 173-178 (1980).
Penninger, J.M.L., and Okazaki, J.K., Designing a high-pressure laboratory, Chem Eng Prog, 76(6), 65-71 (1980).
Various, Water treatment (feature report), Processing, July, 61-71 (1980).

Whittlesey, G., and Sommerfeld, J.T., Use liquified gases for refrigeration, Chem Eng, 22 Sept, 180 (1980).
Anon., Refrigeration systems, Processing, March, 15-16 (1981).
Barclay, F.W.; Nieman, R.E., and Hasinoff, M.P., Transient heat transfer and fluid mechanics of a recirculating pressurized water loop during blowdown and cold water injection, Can JCE, 59, 201-212 (1981).
Coe, W.W., How burners influence combustion, Hyd Proc, 60(5), 179-184 (1981).
Ganapathy, V., Evaluate the performance of waste-heat boilers, Chem Eng, 16 Nov, 291-292 (1981).
Ganapathy, V., Optimum design of waste heat boilers, Hyd Proc, 60(7), 167-168 (1981).
Gray, K.; Gunn, D.C.; Hopper, N.; Pollard, H., and williams, G., Economizers for modern boilers. J Inst Energy, 54, 151-157 (1981).
Howe, R.H.L., and Howe, R.C., Combining indexes for cooling-water evaluation, Chem Eng, 18 May, 157-158 (1981).
Monroe, E.S., Effects of carbon dioxide in steam systems, Chem Eng, 23 March, 209-212 (1981).
Sengupta, A.K., Side-stream filtration for cooling-water systems, Chem Eng, 23 Feb, 107-114 (1981).
van Duuren, F.A., Water utilisation in the process industries, Chem Engnr, Oct, 440-441 (1981).
Various, In-plant gas handling: Cryogenic gases, chlorine and ethylene (feature report), Chem Eng, 5 Oct, 114-138 (1981).
Various, Process plant utilities (special report), Hyd Proc, 60(8), 69-88 (1981).
Bogart, M.J.P., Ammonia absorption refrigeration, Plant/Opns Prog, 1(3), 147-152 (1982).
Holliday, A.D., Conserving and reusing water, Chem Eng, 19 April, 118-137 (1982).
Kempen, D., Manage water and energy in petrochemical plants, Hyd Proc, Aug, 109-112 (1982).
Livingston, A., Steam-line sizing, Chem Eng, 23 Aug, 119-120 (1982).
Marshall, A., Corrosion inhibitor for cooling systems, Proc Engng, Nov, 73-74 (1982).
Mehra, Y.R., Refrigeration systems for low-temperature processes, Chem Eng, 12 July, 94-103 (1982).
Price, N., Steam accumulators provide uniform loads on boilers, Chem Eng, 15 Nov, 131-135 (1982).
Richardson, D.S., Cooling-water system biofouling, Chem Eng, 13 Dec, 103-104 (1982).
Scott, W.J., Handling cooling-water systems during a low-pH excursion, Chem Eng, 22 Feb, 121-126 (1982).
Shah, J.M., Refrigerated ammonia storage tanks, Plant/Opns Prog, 1(2), 90-94 (1982).
Various, Process water, Chemtech, Jan, 22-23; Feb, 84-90; March, 145-149; June, 332-337; July, 413-419; Sept, 532-539; Oct, 602-603; Dec, 718-720 (1982).
Zanker, A., Estimating cooling-water scale, Proc Engng, Aug, 45 (1982).

Andrade, R.C.; Gates, J.A., and McCarthy, J.W., Controlling boiler carryover, Chem Eng, 26 Dec, 51-53 (1983).
Hawk, C.W., Packaged boilers, Chem Eng, 16 May, 89-92 (1983).
Horwitz, B.A., Estimate halogen flow from a cylinder, Chem Eng, 27 June, 68-70 (1983).
Patton, P.W., Vacuum systems in the chemical process industries, Chem Eng Prog, 79(12), 56-61 (1983).
Serna, M.C., Utilities guarantees: Contracts and performance, Hyd Proc, Nov, 34SS-34V (1983).
Sibley, H.W., Selecting refrigerants for process systems, Chem Eng, 16 May, 71-76 (1983).
Wetegrove, R.L., and Pocius, F.C., Controlling microorganisms in cooling-water systems, Chem Eng, 31 Oct, 75-78 (1983).
Baggio, J.L., Optimize refrigeration design, Hyd Proc, Jan, 97-99 (1984).
Barnet, P., Saving process water, Chemtech, Sept, 567-569 (1984).
Freedman, L., Chellant/phosphate treatment for boiler water, Chem Eng, 11 June, 105-108 (1984).
Hale, C.C., 1983 survey of refrigerated ammonia storage in USA and Canada, Plant/Opns Prog, 3(3), 147-159 (1984).
Hill, R., How to buy a water treatment plant, Chem Engnr, July, 17-21 (1984).
Rattan, I.S., and Pathak, V.K., Special problems in the design of closed-loop cooling systems, Chem Eng, 17 Sept, 127-128 (1984).

1985-1988

Beevers, A., De-aerating boiler feedwater, Proc Engng, Aug, 41 (1985).
Faust, C.A., Intermittent boiler operation, Chem Eng, 8 July, 115-118 (1985).
Ganapathy, V., Estimate boiler-blowoff steam flow during startup, Chem Eng, 22 July, 83 (1985).
Metzger, T.R., et al., On-site on-demand nitrogen generation, Plant/Opns Prog, 4(3), 168-172 (1985).
Peterson, J.F., and Mann, W.L., Steam-system design, Chem Eng, 14 Oct, 62-74 (1985).
Pring, E.J., The design of small to medium size cooling systems, Chem Engnr, March, 23-27 (1985).
Shelton, M.R., and Grossmann, I.E., A shortcut procedure for refrigeration systems, Comput Chem Eng, 9(6), 615-619 (1985).
Hurlbert, A.W., Airflow measurement for boilers, Chem Eng, 20 Jan, 87-88 (1986).
Nath, R.; Libby, D.J., and Duhon, H.J., Joint optimization of process units and utility system, Chem Eng Prog, 82(5), 31-38 (1986).
Spencer, J.L.; Connick, B.J., and Filippi, A.J., Optimal periodic control of boiler make-up and blowdown rates, Chem Eng Commns, 47, 329-344 (1986).
Chou, C.C., and Shih, Y.S., A thermodynamic approach to the design and synthesis of plant utility systems, Ind Eng Chem Res, 26(6), 1100-1109 (1987).
Graver, A., Control of steam systems, Processing, Dec, 15-16 (1987).
Hill, R., and Ward, K., High quality water, Chem Engnr, Dec, 17-18 (1987).
Hodgson, T., Attending to ancillary systems, Proc Engng, Sept, 55-57 (1987).

Royse, S., Sirofloc: A new approach to water treatment, Chem Engnr, June, 39 (1987).
Various, Utilities management (special report), Hyd Proc, 66(12), 27-37 (1987).
Adams, T.N., Effects of air jets on combustion in recovery boilers, Chem Eng Prog, 84(6), 45-50 (1988).
Bartok, W.; Lyon, R.K.; McIntyre, A.D.; Ruth, L.A., and Sommerlad, R.E., Combustors: Applications and design considerations, Chem Eng Prog, 84(3), 54-71 (1988).
Butt, A.R.; Bower, C.J.; Green, R.C.; Paterson, N.P., and Gale, J.J., Coal-fired appliances for process heating, J Inst Energy, 61(447), 59-72 (1988).
Garg, A., and Ghosh, H., Fired-heater design specifications, Chem Eng, 18 July, 77-80 (1988).
Graham, I., Control of boiler systems, Processing, Feb, 15-16 (1988).
Hooper, G., The operator's requirements for cooling water inhibitors, Chem & Ind, 7 Nov, 688-691 (1988).
Page, A., Specifying your next refrigeration plant, Chem Engnr, March, 32-34 (1988).
Proctor, A., Design and maintenance of refrigeration plant, Proc Engng, Nov, 57 (1988).
Sendelbach, M.G., Boiler-water treatment, Chem Eng, 15 Aug, 127-132 (1988).
Smith, J.D.; Spence, T.T.; Smith, P.J.; Blackham, A.U., and Smoot, L.D., Effects of coal quality on utility furnace performance, Fuel, 67(1),27-35 (1988).
Stevenson, J., Clean and dry compressed air, Processing, Jan, 21-22 (1988).

CHAPTER 3

ENERGY CONSERVATION

3.1 Energy Sources

1970-1974

Yellott, J.I., Solar energy in the 70s, Chem Eng, 29 June, 85-89 (1970).
Gambs, G.C., and Rauth, A.A., The energy crisis, Chem Eng, 31 May, 56-68 (1971).
Gaucher, L.P., Energy in perspective, Chemtech, March, 153-158 (1971).
Glaser, P.E., Concept for a satellite solar-power system, Chemtech, Oct, 606-614 (1971).
Murphy, J.J., New batteries, Chemtech, Aug, 487-494 (1971).
Thompson, T.J., Nuclear power, Chemtech, Aug, 495-501 (1971).
Robson, F.L., Production of electric power, Chemtech, April, 239-249 (1972).
Schoeppel, R.J., Prospects for hydrogen-fuelled vehicles, Chemtech, Aug, 476-480 (1972).
Wei, J., Energy: The ultimate raw material, Chemtech, March, 142-147 (1972).
Boer, K.W., The solar house, Chemtech, July, 394-400 (1973).
Klass, D.L., and Ghosh, S., Fuel gas from organic wastes, Chemtech, Nov, 689-698 (1973).
Mills, G.A., and Perry, H., Fossil fuel: Power and pollution, Chemtech, Jan, 53-63 (1973).
Szego, G.C., and Kemp, C.C., Energy forests and fuel plantations, Chemtech, May, 275-284 (1973).
Trammell, W.D., The energy crisis and the chemical industry, Chem Eng, 30 April, 68-82 (1973).
Winter, C., and Kohll, A., Energy imports: LNG vs. MeOH, Chem Eng, 12 Nov, 233-238 (1973).
Bacon, F.T., Fuel cells and the growing energy problem, J Inst Fuel, 47, 147-161 (1974).
France, D.H., US and UK substitute gas supplies, Energy World, Dec, 10-13 (1974).
Hendry, A.W., Fuel for transport, Energy World, April, 4-8 (1974).
Lewis, W.B., Nuclear fission energy, Chemtech, Sept, 531-536 (1974).
Marsham, T.N., Nuclear power in the UK, Energy World, Aug, 5-12 (1974).
Maude, C.W., and Jones, D.M., Practical approaches to power generation from low grade fuels, Chem Engnr, May, 298-303 (1974).
Parent, J.D., and Linden, H.R., Potential world crude oil supplies, Energy World, Jan, 3-9 (1974).
Richardson, R.W., Automotive engines for the 1980s, Chemtech, Nov, 660-669 (1974).
Saxton, J.C., et al., Federal findings on energy for industrial chemicals, Chem Eng, 2 Sept, 71-80 (1974).

1975-1977

Booth, H.R., et al., Nuclear power today, Chem Eng, 13 Oct, 102-118 (1975).
Brown, A.H., Bioconversion of solar energy, Chemtech, July, 434-437 (1975).
Chao, R.E., Thermochemical hydrogen production: Assessment of nonideal cycles, Ind Eng Chem Proc Des Dev, 14(3), 276-279 (1975).
Davison, R.R.; Harris,, W.B., and Martin, J.H., Storing sunlight underground, Chemtech, Dec, 736-741 (1975).

Field, A.A., Nuclear and geothermal energy for direct heat supply, Energy World, Dec, 5-8 (1975).
Field, A.A., District heating and total energy, Energy World, Feb, 11-15 (1975).
Gamburg, D.Y., and Semenov, V.P., Nuclear power and chemical technology, Int Chem Eng, 15(4), 616-621 (1975).
Hottel, H.C., Solar energy, Chem Eng Prog, 71(7), 53-65 (1975).
Hunt, S.E., Prospects for nuclear fusion reactors, Fuel, 54(1), 3-9 (1975).
Maple, J.H.C., Power from nuclear fusion, Energy World, April, 5-9 (1975).
Morgan, E.S., and Roughton, J.E., Estimation of calorific value of coal and heat input to a boiler, J Inst Fuel, 48, 10-15 (1975).
Rabl, A., and Nielsen, C.E., Solar ponds for space heating, Chemtech, Oct, 608-616 (1975).
Shinnar, R., Energy in perspective, Chemtech, April, 225-231 (1975).
Thring, M.W., UK energy policy, Proc Engng, Oct, 45-50 (1975).
Axtmann, R.C., and Peck, L.B., Geothermal chemical engineering (review paper), AIChEJ, 22(5), 817-828 (1976).
Capener, E.L., Economics of solar energy, Chemtech, March, 190-193 (1976).
Chubb, T.A., Chemical approach to solar energy, Chemtech, Oct, 654-657 (1976).
Glendenning, I., and Count, B.M., Wave power, Chem Engnr, Sept, 595-600, 604 (1976).
Heronemus, W.E., Wind power, Chemtech, Aug, 498-503 (1976).
Hill, J., Nuclear power: A review, J Inst Fuel, 49, 3-9 (1976).
Povich, M.J., Farming for fuels, Chemtech, July, 434-439 (1976).
Shaw, T.L., Tidal power, Chem Engnr, Sept, 592-594, 611 (1976).
Talbert, S.G.; Frieling, D.H.; Eibling, J.A.; and Nathan, R.A., Photochemical solar heating and cooling, Chemtech, Feb, 118-122 (1976).
Wallace, R.T., Energy: What are the best alternatives? Chem Eng, 1 March, 121-125 (1976).
Zener, C., and Rothfus, R.R., The solar sea power plant, Chemtech, Nov, 717-723 (1976).
Balzhiser, R.E., Energy options to the year 2000, Chem Eng, 3 Jan, 72-90 (1977).
Calvin, M., Energy of the future, Chemtech, June, 352-363 (1977).
Dickinson, W.C., Economics of process heat from solar energy, Chem Eng, 31 Jan, 101-104 (1977).
Doscher, T.M., Tertiary oil recovery and chemistry, Chemtech, April, 232-239 (1977).
Fernandes, J.H., Why not burn wood? Chem Eng, 23 May, 159-164 (1977).
Flowers, B., Nuclear power and public policy, Chemtech, Aug, 484-494 (1977).
Gicquel, R., Evaluating solar energy from a flat insolator, Int Chem Eng, 17(4), 575-583 (1977).
Goodger, E.M., Alcohol fuel prospects, J Inst Fuel, 50, 132-138 (1977).
Johnson, E.F., Chemical engineering in nuclear fusion power, AIChEJ, 23(5), 617-631 (1977).
Khandovletov, S., et al., Calculating capacity of solar thermoelectric generators, Int Chem Eng, 17(1), 105-109 (1977).
Lavallee, D., and Lavallee, C., Our solar home, Chemtech, April, 210-213 (1977).

Mills, G.A., Alternative fuels from coal, Chemtech, July, 418-423 (1977).
Pober, K., and Bauer, H., From garbage to oil, Chemtech, March, 164-169 (1977).
Singh, D., and Chauhan, R.S., Solar water-heater, Chem Eng, 14 Feb, 106 (1977).
Stott, J.B., New Zealand's energy supplies, J Inst Fuel, 50, 139-146 (1977).
Swearingen, J.S., Power from hot geothermal brines, Chem Eng Prog, 73(7), 83-86 (1977).
Tillman, D.A., Combustible renewable resources, Chemtech, Oct, 611-615 (1977).
Various, The nuclear fuel cycle, J Inst Fuel, 49, 87-91,130-134,210-217 (1976); 50, 33-40,153-160,173-178 (1977).
Wilson, J.S., A geothermal energy plant, Chem Eng Prog, 73(11), 95-98 (1977).

1978-1979

Armstead, H.C.H., Future of geothermal energy, J Inst Fuel, 51, 109-118 (1978).
Barnes, R.W., Btu's, goods, and the future, Chemtech, Jan, 30-36 (1978).
Bodle, W.W.; Punwani, D.V., and Mensinger, M.C., Uses of peat, Chemtech, Sept, 559-563 (1978).
Chiang, S.H.; Klinzing, G.E., and Cobb, J.T., Tomorrow's fuels will be less efficient, Hyd Proc, 57(11), 209-211 (1978).
Hearn, G.E., and Katory, M., Application of hydrodynamic analysis to wave power generators, J Inst Fuel, 51, 119-127 (1978).
Johnson, D.G., Thermo-hydraulic energy from the sea, J Inst Fuel, 51, 59-63 (1978).
Ledig, F.T., and Linzer, D.I.H., Fuel crop breeding, Chemtech, Jan, 18-27 (1978).
Morse, R.N., and Proctor, D., Soft energy for food processing, Chemtech, Aug, 478-483 (1978).
Opila, R.L., Energy for food processing, Chemtech, Feb, 104-107 (1978).
Root, D.H., Future energy sources, Hyd Proc, 57(4), 193-198 (1978).
Shinnar, R., and Shinnar, M., Cost of synthetic fuels, Chemtech, July, 418-423 (1978).
Starr, C., Energy and society, Chemtech, April, 248-255 (1978).
Various, Coal as a raw material for the chemical industry (topic issue), Chem & Ind, 5 Aug, 551-571 (1978).
Verma, A., From coal to gas, Chemtech, June, 372-381; Oct, 626-638 (1978).
Brennan, E., Biomass: Future feedstock, Processing, May, 38-39 (1979).
Cadwallader, E.A., and Westberg, J.E., Wind-powered processing, Chemtech, April, 254-259; May, 310-314 (1979).
Cameron, D., Fuel-cell energy generators, Chemtech, Oct, 633-637 (1979).
Franklin, N.L., Role and problems of breeder reactors, Energy World, Aug, 3-8 (1979).
Gicquel, R., Behavior of plane solar collectors under transient conditions, Int Chem Eng, 19(1), 51-66 (1979).
Hill, J., Future requirements for fast-reactor fuel reprocessing, Chem & Ind, 16 June, 406-409 (1979).
Hovel, H., Development of solar cells, Chemtech, March, 191-200 (1979).
Schachter, Y., Uses of oil shale, Chemtech, Sept, 568-570 (1979).

Various, Energy utilisation (feature report), Processing, Jan, 39-46 (1979).
Walters, P., Energy supplies, Chem & Ind, 7 July, 448-454 (1979).
White, N.A., International availability of energy minerals, Energy World, June, 6-16 (1979).

1980
Anon., Oil and alternative fossil energy resources: Trends and prospects, Proc Econ Int, 1(3), 18-22 (1980).
Dell, R.M., Hydrogen as a future alternative fuel, J Inst Energy, 53, 116-119 (1980).
Franklin, N.L., Nuclear energy in the 80s, J Inst Energy, 53, 120-123 (1980).
Gaensslen, H., Cost analysis of coal versus oil, Chemtech, Sept, 563-565 (1980).
Holmes, J.T., Sun power, Chemtech, Aug, 514-519 (1980).
Kadlec, E., A wind turbine, Chemtech, May, 324-326 (1980).
Lewis, C., European potential for bioenergy systems, Energy World, Jan, 2-8 (1980).
Lichtin, N.N., Storing solar energy abiotically, Chemtech, April, 252-260 (1980).
Longrigg, P., Resources in the southern ocean, Energy World, Dec, 2-5 (1980).
Lowes, T.M., and Lorimer, A.D., Waste as fuel, J Inst Energy, 53, 85-91 (1980).
Martin, J.F., Treat fuel gas and save energy, Hyd Proc, 59(3), 67-71 (1980).
Neal, W.E.J., Review of developments in solar-energy cooling, J Inst Energy, 53, 25-30,100 (1980).
Payne, P.R., Materials for solar energy systems, Chemtech, Sept, 550-557 (1980).
Usmani, I.H., An energy option, Chemtech, Feb, 94-98 (1980).
Various, Synthetic liquid fuels from coal and biomass (symposium papers), Can JCE, 58, 682-738 (1980).
Various, Developments in battery science (topic issue), Chem & Ind, 5 July, 519-531 (1980).
Various, Gasoline for the future (special report), Hyd Proc, 59(2), 57-75 (1980).

1981
Bronicki, Y.L., Electricity from solar ponds, Chemtech, Aug, 494-498 (1981).
Cazalet, P.G., UK energy resources: Problem or opportunity? Hyd Proc, 60(10), 68B-68P (1981).
Cook, E., Charting our energy future, Chemtech, July, 441-445 (1981).
Dawson, J.K., Prospects for UK renewable energy sources, Energy World, Nov, 3-9 (1981).
Duncan, W., Oil: An interlude in a century of coal, Chem & Ind, 2 May, 311-316 (1981).
Essenhigh, R.H., The uses of fuels, Chemtech, June, 351-359 (1981).
Fells, I., Nuclear energy: An assessment, Energy World, March, 2-6 (1981).
Goodger, E.M., Future trends in aviation fuels, Energy World, April, 2-11 (1981).
Goodger, E.M., Future trends in automotive fuels, Energy World, March, 8-18 (1981).
Goodger, E.M., Future trends in marine fuels, Energy World, May, 13-22 (1981).

Gregory, D.P.; Tsaros, C.L.; Arora, J.L., and Nevrekar, P., Economics of hydrogen as a fuel, Chemtech, July, 432-440 (1981).
Hawthorne, W., Uses of coal, Energy World, May, 2-12 (1981).
Manassen, J.; Hodes, G., and Cahen, D., Photoelectrochemical cells, Chemtech, Feb, 112-117 (1981).
May, E.K., Solar thermal energy for unit operations, Chem Eng Prog, 77(7), 60-64 (1981).
Morgan, R.P., and Shultz, E.B., Energy from seed oils, C&E News, 7 Sept, 69-77 (1981).
Myerly, R.C.; Nicholson, M.D.; Katzen, R., and Taylor, J.M., Fuel from forests, Chemtech, March, 186-192 (1981).
Rice, M.P., et al., Heat transfer analysis of receivers for solar concentrating collector, Chem Eng Commns, 8(4), 353-364 (1981).
Rose, A., Solar energy: A global view, Chemtech, Sept, 566-571; Nov, 694-697 (1981).
Schmitt, R.W., Coal-based electricity in the USA, J Inst Energy, 54, 63-75 (1981).
Scott, J.E., Solar water heating, Chemtech, June, 328-332 (1981).
Stotts, R.E.; Warrington, R.O., and Mussulman, R.L., Simulation and comparison of passive solar heating systems, Chem Eng Commns, 11(1), 81-98 (1981).
Vlitos, A., Natural products as feedstocks, Chem & Ind, 2 May, 303-310 (1981).
Walton, G.N., The temperature of nuclear power, Energy World, June, 7-9 (1981).
Yang, V.; Trindade, S.C., and Branco, J.R.C., Biomass for motor fuel, Chemtech, March, 168-172 (1981).

1982

Anderson, J., The nuclear power industry, C&E News, 20 Sept, 11-19 (1982).
Anon., Energy and economic growth: Past, present and future, Proc Econ Int, 3(1), 22-24 (1982).
Banal, M., and Bichon, A., Assessing use of tidal energy in France, J Inst Energy, 55, 86-91 (1982).
Dry, M.E., The Sasol route to fuels, Chemtech, Dec, 744-750 (1982).
Fells, I., Energy options to year 2030, Energy World, Feb, 15-21 (1982).
Geddes, K.A., and Deanin, R.D., Plastics in solar energy collectors, Chemtech, Dec, 736-740 (1982).
Hochman, J.M., Synthetic fuels, Chemtech, Aug, 500-505 (1982).
Kelley, J.H.; Escher, W.J.D., and van Deelen, W., Hydrogen uses and demands through the year 2025, Chem Eng Prog, 78(1), 58-61 (1982).
Koser, H.J.K.; Schmalstieg, G., and Siemers, W., Densification of water hyacinth as a fuel source, Fuel, 61(9), 791-798 (1982).
Locke, B., Feedstocks and fuels from coal: The need for process changes, Energy World, Feb, 5-14 (1982).
Multer, R.K., Solar ponds collect sun's heat, Chem Eng, 8 March, 87-89 (1982).
Noon, R., Power-grade butanol, Chemtech, Nov, 681-683 (1982).
Pollard, W.G., A theological view of nuclear energy, Chemtech, July, 420-423 (1982).
Pooley, D., Long-term energy options, Energy World, Dec, 12-15 (1982).

Russell, T.W.F., Photovoltaic unit operations, Chemtech, Sept, 540-545 (1982).
Sadoway, D.R., The materials-energy symbiosis, Chemtech, Oct, 625-627 (1982).
Shankland, R.V., Enhanced oil recovery, Chemtech, Nov, 684-688 (1982).
Tichener, A.L., Alternative fuels and methanol in New Zealand, Chem & Ind, 6 Nov, 841-846 (1982).
Various, Nuclear power (feature issue), Energy World, Aug, 2-15 (1982).
Various, Alternative energy sources, Energy World, May, 2-7 (1982).

1983
Cook, N.C.; Davis, G.C., and Wolfe, J.K., Cyclic process for making hydrogen, Chemtech, Dec, 755-757 (1983).
Davies, E., Hydrogen as an energy source, Proc Engng, 45-49 (1983).
Douglas, D.L., and Birk, J.R., Batteries for energy storage, Chemtech, Jan, 58-64; Feb, 120-126 (1983).
Golovoy, A., and Nichols, R.J., Natural-gas-powered vehicles, Chemtech, June, 359-363 (1983).
Gould, G., Converting to coal, Chemtech, May, 300-302 (1983).
Gray, J.A., Gas technology: Past and future, Energy World, Feb, 4-12 (1983).
Liang, C.C., Solid electrolytes and solid-state batteries, Chemtech, May, 303-305 (1983).
Mitchell, T.E.; Schroer, B.J.; Ziemke, M.C., and Peters, J.F., Biomass fuels: A national plan, Chemtech, April, 242-249 (1983).
Phillips, C.R., The uranium fuel cycle, Can JCE, 61(1), 3-19 (1983).
Sternlicht, B., Rebirth of the Stirling engine, Chemtech, Jan, 28-36 (1983).
Stobart, A.F., Wind energy: Collection, storage and application, Energy World, May, 4-6 (1983).
Various, Solar energy (feature report), Energy World, Dec supplement, pp.1-24 (1983).

1984
Anon., European demand, supply and fuel prices forecast to year 2000, Energy World, Feb, 9-14 (1984).
Comar, C.L., Plutonium, power and people, Chemtech, Nov, 660-663 (1984).
Dainton, A.D., Coal utilization in the UK: Problems and prospects, Energy World, Dec, 2-11 (1984).
Dainton, A.D., Prospects for coal liquifaction, Energy World, Feb, 6-8 (1984).
Davis, W.K., Problems and prospects for nuclear power, Chem Eng Prog, 80(6), 11-16 (1984).
Fells, I., Energy provision: Policies and technologies, Energy World, Oct, 2-10 (1984).
Klass, D.L., Renewable energy, Chemtech, Aug, 486-491; Oct, 610-615 (1984).
Linder, B., and Sjostrom, K., Operation of an internal combustion engine, Fuel, 63(11), 1485-1490 (1984).
Moore, J.S., and Shadis, W.J., Automotive fuels today and tomorrow, Chemtech, Sept, 554-561 (1984).
O'Sullivan, D.A., Pebble-bed nuclear reactor for power generation, C&E News, 5 March, 20-21 (1984).

Prentice, G., Fuel cells: Principles and prospects, Chemtech, Nov, 684-694 (1984).

Seefelder, M., Chemicals and energy, Chemtech, April, 240-242 (1984).

Spillman, R.W.; Spotnitz, R.M., and Lundquist, J.T., Making chemicals in fuel cells, Chemtech, March, 176-183 (1984).

Teggers, H., Conversion of brown coal to solid, gaseous and liquid products, Energy World, Jan, 6-19 (1984).

Ulman, M.; Blajeni, B.A., and Halmann, M., Fuel from carbon dioxide: An electrochemical study, Chemtech, April, 235-239 (1984).

Ushiba, K.K., Fuel cells, Chemtech, May, 300-307 (1984).

Various, Hydrogen: The next five years in the UK (topic issue), Chem & Ind, 16 Jan, 46-72 (1984).

1985

Abdul Hadi, M.I., Analysis and performance of a V-trough solar concentrator, Can JCE, 63(3), 399-405 (1985).

Beckman, D., and Elliott, D.C., Comparisons of the yields and properties of the oil products from direct thermochemical biomass-liquifaction processes, Can JCE, 63(1), 99-104 (1985).

Brodd, R.J., Advanced batteries, Chemtech, Oct, 612-621 (1985).

Langer, S.H., and Colucci-Rios, J.A., Chemicals and electric power, Chemtech, April, 226-233 (1985).

Owsley, D.C., and Bloomfield, J.J., Energy facts: A basis for decision, Chemtech, Feb, 94-98 (1985).

Perkins, R.P., and Manfred, R.K., Coal slurry fuel development program, Chem Eng Prog, 81(5), 69-76 (1985).

Various, The Sizewell B pressurized water reactor inquiry, Energy World, Dec, 2-13 (1985).

1986

Conn, A.L., The integrated gasification combined cycle power plant: Power from coal with minimum environmental problems in the USA, Energy World, Dec, 5-12,15 (1986).

Lede, J., et al., Preparation of hydrogen by thermolysis of water, Int Chem Eng, 26(4), 647-660 (1986).

Probert, W.R., Natural gas: Current trends and future prospects, Energy World, Jan, 8-18 (1986).

Various, Advances in battery technology (topic issue), Chem & Ind, 17 March, 192-209 (1986).

Zweibel, K., Photovoltaic cells, C&E News, 7 July, 34-48 (1986).

1987

Banks, R., and Isalski, W.H., Excess fuel gas? Recover hydrogen/LPG, Hyd Proc, 66(10), 47-51 (1987).

Diver, R.B., Transporting solar energy with chemistry, Chemtech, Oct, 606-611 (1987).

Hoffmann, P.; Hunter, D., and Ushio, S., Hydrogen research, Chem Eng, 26 Oct, 26-28 (1987).

Jones, C.A.; Leonard, J.J., and Sofranko, J.A., Fuels for the future: Remote gas conversion, Energy & Fuels, 1(1), 12-17 (1987).
Malik, V.A.; Lerner, S.L., and MacLean, D.L., Electricity, methane and liquid carbon dioxide production from landfill gas, Gas Sepn & Purif, 1(2), 77-83 (1987).

1988
Anon., World electricity prices, Energy World, June, 4-6 (1988).
Anon., European industrial energy consumption, Energy World, April, 2-4 (1988).
Bea, D.A, Jet fuels need more kerosine, Hyd Proc, 67(12), 51-52 (1988).
Butcher, C., What price energy? Chem Engnr, Oct, 20-22 (1988).
Elliott, D.C., et al., Production of liquid hydrocarbon fuels from peat, Energy & Fuels, 2(2), 234-235 (1988).
Ganapathy, V., Program computes turbine steam rates and properties, Hyd Proc, 67(11), 105-108 (1988).
Herapath, R.G., Innovation in process energy utilisation, Energy World, June, 10-11 (1988).
Lindstrom, O., Muscles, engines, and fuel cells, Chemtech, Nov, 686-693 (1988).
Lindstrom, O., Fuel-cell power plants, Chemtech, Sept, 553-559 (1988).
Lindstrom, O., Fuel cells, Chemtech, Aug, 490-497 (1988).
Moreau, D., The sweet smell of solar power, Chem Engnr, Oct, 24 (1988).
Spedding, P.J., Peat as a fuel (review paper), Fuel, 67(7), 883-900 (1988).
Swift-Hook, D.T., Progress in wind energy, Energy World, Feb, 13-15 (1988).
Taffe, P., Electroheat applications in UK, Processing, Dec, 17-21 (1988).
Takahashi, K., Progress in photovoltaics, Chemtech, Dec, 744-749 (1988).
Various, Advances in battery technology (topic issue), Chem & Ind, 1 Feb, 69-91 (1988).

3.2 Combustion

1973-1988
Reed, R.D., Save energy in furnaces, Hyd Proc, 52(7), 119-121 (1973).
Steenberg, L.R., Fuel recovery from flare systems, Chem Eng Prog, 70(7), 74-77 (1974).
Bond, A., Combustion control, Proc Engng, April, 74-77 (1975).
Gill, D.W., Review of pulverized-coal firing, Energy World, March, 4-6 (1975).
May, D.L., Cutting boiler fuel costs with combustion controls, Chem Eng, 22 Dec, 53-57 (1975).
Woodard, A.M., Control flue gas to improve furnace efficiency, Hyd Proc, 54(5), 165-166 (1975).
Reed, R.D., Recover energy from furnace stacks, Hyd Proc, 55(1), 127-128 (1976).
Wood, R., Burner design for fuel economy, Proc Engng, April, 73-75 (1977).
Salooja, K.C., Combustion with high efficiency and low pollutant emission, Energy World, Dec, 6-13 (1978).

Seebold, J.G., Conserving fuel in furnaces and flares, Hyd Proc, 60(11), 263-267 (1981).
Bonnet, C., Save energy in fired heaters, Hyd Proc, March, 131-137 (1982).
Challis, H., Conversion to coal-fired boilers, Proc Engng, Jan, 47-49 (1982).
Dattatreya, S., Include ambient temperature in combustion control for energy savings, Hyd Proc, Jan, 87-88 (1985).
Meunier, J.P., Find true furnace efficiency, Hyd Proc, Feb, 77-80 (1985).
Mahajani, V.V.; Kamat, A.B., and Mokashi, S.M., Recovering heat in fired heaters, Chem Eng, 18 Aug, 91-95 (1986).
Miller, D.B.; Soychak, T.J., and Gosar, D.M., Economics of recovering carbon dioxide from exhaust gases, Chem Eng Prog, 82(10), 38-46 (1986).
Miller, J.A., and Fisk, G.A., Combustion chemistry, C&E News, 31 Aug, 22-46 (1987).
Atkinson, N., Pressurised fluidized-bed combustion for power generation, Proc Engng, July, 28-31 (1988).
Bartok, W.; Lyon, R.K.; McIntyre, A.D.; Ruth, L.A., and Sommerlad, R.E., Combustors: Applications and design considerations, Chem Eng Prog, 84(3), 54-71 (1988).
Butt, A.R.; Bower, C.J.; Green, R.C.; Paterson, N.P., and Gale, J.J., Coal-fired appliances for process heating, J Inst Energy, 61(447), 59-72 (1988).
Fehr, M., An auditor's view of furnace efficiency, Hyd Proc, 67(11), 93-96 (1988).
Ferreira, M.A.; Carvalho, J.A., and Gill, W., Assessment of a combustion system by the concept of reversible work, Fuel, 67(4), 587-589 (1988).
Shook, J.R., Nonmetallic economizers improve combustion efficiency, Hyd Proc, 67(7), 41-43 (1988).

3.3 Cogeneration

1977-1984

Ganic, E., and Seider, W.D., Computer simulation of potassium-steam combined-cycle electrical power plants, Comput Chem Eng, 1(3), 161-170 (1977).
Marshall, W., Prospects for combined heat and power for district heating in the UK, Energy World, July, 3-9 (1977).
Bleay, J.A., and Fells, I., Linear programming optimization for design of combined heat and power schemes, J Inst Energy, 52, 125-139 (1979).
Bleay, J.A., and Fells, I., Analysis of some industrial combined heat and power schemes, Energy World, Oct, 12-17 (1979).
Holmes, J., and Lucas, N.J.D., The Marshall report on combined heat and power, Energy World, Oct 3-11 (1979).
Various, The future of cogeneration (feature report), Chem Eng, 26 Feb, 104-116 (1979).
Nishio, M.; Itoh, J.; Shiroko, K., and Umeda, T., Thermodynamic approach to steam-power system design, Ind Eng Chem Proc Des Dev, 19(2), 306-312 (1980).
Donnedu, M., Efficiency of an installation for combined production of electricity and heat, Int Chem Eng, 21(2), 311-323 (1981).

Garland, R.V., Gasification-based combined power plants, Chem Eng Prog, 77(1), 70-72 (1981).
Gartside, R.J., Cogenerate for energy efficiency, Hyd Proc, 60(12), 125-131 (1981).
Graybeal, P.E., and Manchester, A.H., Cogeneration, Chemtech, Jan, 48-51 (1982).
Marnet, C., Utilization of urban waste for cogeneration of electricity and district heat, J Inst Energy, 55, 144-152 (1982).
Kimber, A., Combined heat and power systems, Proc Engng, March, 47-49 (1983).
Roszkowski, T.R.; Grisso, J.R.; Klumpe, H.W., and Snyder, N.W., Gasification in combined/cogeneration cycles, Chem Eng Prog, 79(1), 9-15 (1983).
Townsend, D.W., and Linnhoff, B., Heat and power networks in process design, AIChEJ, 29(5), 742-771 (1983).
Ahner, D.J., Cogeneration economics for process plants, Chem Eng, 20 Aug, 177-182 (1984).
Iaquaniello, G.; Guerrini, S.; Pietrogrande, P., and Dreyer, H., Integrate gas turbine cogeneration with fired heaters, Hyd Proc, Aug, 57-60 (1984).
Nishio, M.; Koshijima, I.; Shiroko, K., and Umeda, T., A case study of heat and power supply systems optimization, Ind Eng Chem Proc Des Dev, 23(3), 450-456 (1984).
O'Shea, T.P., Electric power from coal-derived gas, Chem Eng Prog, 80(8), 71-76 (1984).
Ryan, F.J., and Cameron, D.S., Fuel cells for on-site cogeneration of heat and power, Energy World, Feb, 2-5 (1984).

1985-1988
Bancel, P.L., Cogeneration fuels: The high cost of lower heating values, Chem Eng, 11 Nov, 215-216 (1985).
Blasius, G.F., Municipal waste used for large scale cogeneration, Chem Eng Prog, 81(3), 64-69 (1985).
Gray, R.J., and Pesek, V., Petroleum-coke-fired cogeneration, Chem Eng Prog, 81(3), 70-77 (1985).
Kenson, R.E., Catalytic incineration in cogeneration systems, Chem Eng Prog, 81(11), 57-62 (1985).
Nishio, M.; Koshijima, I.; Shiroko, K., and Umeda, T., Synthesis of optimal heat and power supply systems for energy conservation, Ind Eng Chem Proc Des Dev, 24(1), 19-30 (1985).
Arnold, D.W.; Borzik, D.M., and Montgomery, L.K., Cogeneration in nitrogen fertilizer complex, Plant/Opns Prog, 5(2), 73-77 (1986).
Benz, A.D.; Degen, B.D., and McKibbin, J.R., Cogeneration in the petroleum refinery, Chem Eng Prog, 82(10), 21-27 (1986).
Peacock, N.F., Combined heat and power (small systems), Energy World, Dec, 17-19 (1986).
Sims, G.V., Better cogen economic analysis, Hyd Proc, 65(7), 61-64 (1986).
Strait, R., and Fischbach, M., Cut energy costs with cogeneration, Hyd Proc, 65(10), 55-57 (1986).
Zambo, R.A., Cogeneration economics, Chem Eng Prog, 82(10), 47-50 (1986).

Colmenares, T.R., and Seider, W.D., Heat and power integration of chemical processes, AIChEJ, 33(6), 898-915 (1987).
Garrett-Price, B.A., and Fassbender, L.L., Cogeneration: Right for your plant? Preliminary assessment of 'standard' configurations, Chem Eng, 27 April, 51-57 (1987).
Cooper, D., and Graves, A., Should you cogenerate electricity today? Hyd Proc, 67(7), 44-46 (1988).
Crossland, B., Combined heat and power and district heating, Energy World, Dec, 3-6, 25 (1988).
Ham, R.W.; Douglas, P.L., and Fulford, G., Effect of steam temperature on economics of cogeneration power plant, Can JCE, 66(6), 987-994 (1988).
Kunigita, E.; Nishitani, H., and Kutsawa, Y., Robustness of optimal operation of steam-and-power system, Int Chem Eng, 28(1), 75-84 (1988).
Royse, S., Assessing combined heat and power systems, Proc Engng, March, 61-63 (1988).

3.4 Energy Integration

1968-1980

Miller, R., Process energy systems, Chem Eng, 20 May, 130-148 (1968).
Canova, F., Matching turbomachinery to a process for direct energy recovery, Chem Eng, 2 June, 178-182 (1969).
Maikov, V.P.; Vilkov, G.G., and Galtsov, A.V., Thermo-economic optimum planning of multi-column rectification plants, Int Chem Eng, 12(2), 282-288; 12(3), 426-433 (1972).
Slack, J.B., Energy systems in large process plants, Chem Eng, 24 Jan, 107-111 (1972).
Rathore, R.N.S.; Van Wormer, K.A., and Powers, G.J., Synthesis strategies for multicomponent separation systems with energy integration, AIChEJ, 20(3), 491-502 (1974).
Rathore, R.N.S.; Vanwormer, K.A., and Powers, G.J., Synthesis of distillation systems with energy integration, AIChEJ, 20(5), 940-950 (1974).
Chimes, A.R., Graphical technique for energy balancing, Chem Eng, 29 Sept, 99-100 (1975).
Petterson, W.C., and Wells, T.A., Energy-saving schemes in distillation, Chem Eng, 26 Sept, 78-86 (1977).
Georgakis, C., and Worthey, D.J., Dynamical methods of heat integration design, AIChEJ, 24(6), 976-984 (1978).
Stacey, J.M., Improved heat integration, Processing, Oct, 37-38 (1979).
Umeda, T.; Niida, K., and Shiroko, K., Thermodynamic approach to heat integration in distillation systems, AIChEJ, 25(3), 423-429 (1979).
Various, Process integration and synthesis (topic issue), Comput Chem Eng, 3(5), 13-17, 241-306 (1979).
Manning, E., Design plant-wide heat recovery, Hyd Proc, 59(11), 245-247 (1980).
Morari, M., and Faith, D.C., Synthesis of distillation trains with heat integration, AIChEJ, 26(6), 916-928 (1980).

1981-1985

Kaiser, V., Energy optimization, Chem Eng, 23 Feb, 62-72 (1981).

Linnhoff, B., and Townsend, B.W., Designing total energy systems, Chem Eng Prog, 78(7), 72-80 (1982).

Naka, Y.; Terashita, M., and Takamatsu, T., Thermodynamic approach to multicomponent distillation system synthesis, AIChEJ, 28(5), 812-820 (1982).

Rathore, R.N.S., Process resequencing for energy conservation, Chem Eng Prog, 78(12), 75-82 (1982).

Townsend, D.W., and Linnhoff, B., Designing total energy systems by systematic methods, Chem Engnr, March, 91-97 (1982).

Chiang, T.P., and Luyben, W.L., Comparison of energy consumption in five heat-integrated distillation configurations, Ind Eng Chem Proc Des Dev, 22(2), 175-179 (1983).

Elaahi, A., and Luyben, W.L., Alternative distillation configurations for energy conservation in four-component separations, Ind Eng Chem Proc Des Dev, 22(1), 80-86 (1983).

Linnhoff, B., and Senior, P.R., Energy targets clarify scope for better heat integration, Proc Engng, March, 29-33 (1983).

Linnhoff, B.; Dunford, H., and Smith, R., Heat integration of distillation columns into overall processes, Chem Eng Sci, 38(8), 1175-1188 (1983).

Andrecovich, M.J., and Westerberg, A.W., Simple synthesis method based on utility bonding for heat-integrated distillation sequences, AIChEJ, 31(3), 363-375 (1985).

Andrecovich, M.J., and Westerberg, A.W., An MILP formulation for heat-integrated distillation sequence synthesis, AIChEJ, 31(9), 1461-1479 (1985).

Cheng, H.C., and Luyben, W.L., Heat-integrated distillation columns for ternary separations, Ind Eng Chem Proc Des Dev, 24(3), 707-713 (1985).

Elaahi, A., and Luyben, W.L., Control of an energy-conservative complex configuration of distillation columns for four-component separations, Ind Eng Chem Proc Des Dev, 24(2), 368-376 (1985).

Kaiser, V., and Gourlia, J.P., The ideal column concept: Applying exergy to distillation, Chem Eng, 19 Aug, 45-53 (1985).

Ramshaw, C., Process intensification, Chem Engnr, July, 30-33 (1985).

Steinmetz, F.J., and Chaney, M.O., Total plant process energy integration, Chem Eng Prog, 81(7), 27-32 (1985).

Tanaka, T.; Shimizu, K., and Matsubara, M., Exergy analysis of binary distillation columns with heat integration, Chem Eng Commns, 36(1), 223-232 (1985).

1986-1988

Bingzhen, C., and Westerberg, A.W., Structural flexibility for heat-integrated distillation columns, Chem Eng Sci, 41(2), 355-378 (1986).

Duran, M.A., and Grossmann, I.E., Simultaneous optimization and heat integration of chemical processes, AIChEJ, 32(1), 123-138 (1986).

Fidkowski, Z., and Krolikowski, L., Optimization of a thermally-coupled system of distillation columns, AIChEJ, 32(4), 537-546 (1986).

Itoh, J.; Shiroko, K., and Umeda, T., Extensive applications of the T-Q diagram to heat integrated system synthesis, Comput Chem Eng, 10(1), 59-66 (1986).
Kotas, T.J., Exergy method of thermal and chemical plant analysis (review paper), CER&D, 64(3), 212-229 (1986).
O'Reilly, A., Experiences in process integration, Chem Engnr, May, 56-59 (1986).
Vaselenak, J.A.; Grossmann, I.E., and Westerberg, A.W., Heat integration in batch processing, Ind Eng Chem Proc Des Dev, 25(2), 357-366 (1986).
Fidkowski, Z., and Krolikowski, L., Minimum energy requirements of thermally coupled distillation systems, AIChEJ, 33(4), 643-653 (1987).
Glenchur, T., and Govind, R., Study on a continuous heat-integrated distillation column, Sepn Sci Technol, 22(12), 2323-2338 (1987).
Handogo, R., and Luyben, W.L., Design and control of a heat-integrated reactor/column process, Ind Eng Chem Res, 26(3), 531-539 (1987).
Isla, M.A., and Cerda, J., Simultaneous synthesis of distillation trains and heat exchanger networks, Chem Eng Sci, 42(10), 2455-2464 (1987).
Isla, M.A., and Cerda, J., General algorithmic approach to optimal synthesis of energy-efficient distillation train designs, Chem Eng Commns, 54, 353-380 (1987).
Muraki, M., and Hayakawa, T., Evolutionary synthesis method of energy-integrated distillation separation process, Can JCE, 65(2), 250-255 (1987).
Viswanathan, M., and Evans, L.B., Studies in the heat integration of chemical process plants, AIChEJ, 33(11), 1780-1790 (1987).
Chiang, T.P., and Luyben, W.L., Comparison of the dynamic performance of three heat-integrated distillation configurations, Ind Eng Chem Res, 27(1), 99-105 (1988).
Floudas, C.A., and Paules, G.E., Mixed-integer nonlinear programming formulation for synthesis of heat-integrated distillation sequences, Comput Chem Eng, 12(6), 531-546 (1988).
Glavic, P.; Kravanja, Z., and Homsak, M., Modelling of reactors for process heat integration, Comput Chem Eng, 12(2/3), 189-194 (1988).
Glavic, P.; Kravanja, Z., and Homsak, M., Heat integration of reactors, Chem Eng Sci, 43(3), 593-608 (1988).
Isla, M.A., and Cerda, J., Heuristic method for synthesis of heat-integrated distillation systems, Chem Eng J, 38(3), 161-178 (1988).
Jelinek, J., and Ptacnik, R., Synthesis of heat-integrated rectification systems, Comput Chem Eng, 12(5), 427-432 (1988).
Kakhu, A.I., and Flower, J.R., Synthesising heat-integrated distillation sequences using mixed integer parameters, CER&D, 66(3). 241-254 (1988).
Lang, Y.D.; Biegler, L.T., and Grossmann, I.E., Simultaneous optimization and heat integration with process simulators, Comput Chem Eng, 12(4), 311-328 (1988).
Meszaros, I., and Fonyo, Z., Extensive state optimization for heat-integrated distillation columns, Comput Chem Eng, 12(2/3), 225-230 (1988).
Meszaros, I., and Fonyo, Z., Simple heuristic method to select heat-integrated distillation schemes, Chem Eng Sci, 43(11), 3109-3112 (1988).

Paterson, W.R., Heat integration in distillation columns, AIChEJ, 34(1), 147-149 (1988).

3.5 Pinch Technology

1980-1988

Townsend, D.W., Second law analysis in practice, Chem Engnr, Oct, 628-633 (1980).
Linnhoff, B., and Hindmarsh, E., Pinch design method for heat exchanger networks, Chem Eng Sci, 38(5), 745-764 (1983).
Linnhoff, B., and Vredeveld, D.R., Pinch technology has come of age, Chem Eng Prog, 80(7), 33-40 (1984).
Tjoe, T.N., and Linnhoff, B., Using pinch technology for process retrofit, Chem Eng, 28 April, 47-60 (1986).
Linnhoff, B., and Eastwood, A.R., Overall site optimisation by pinch technology, CER&D, 65(5), 408-414 (1987).
Smith, G., and Patel, A., Step-by-step through the pinch, Chem Engnr, Nov, 26-31 (1987).
Taffe, P., Pinch technology for batch processes, Processing, Dec, 20-21 (1987).
Deakin, A., and Kemp, I., Pinch technology for dryer efficiency, Processing, Oct, 49-56 (1988).
Linnhoff, B., and Polley, G., Stepping beyond the pinch, Chem Engnr, Feb, 25-32 (1988).
Linnhoff, B.; Polley, G.T., and Sahdev, V., General process improvements through pinch technology, Chem Eng Prog, 84(6), 51-58 (1988).
Obeng, E.D.A., and Ashton, G.J., Pinch technology-based procedures for design of batch processes, CER&D, 66(3). 255-259 (1988).

3.6 Heat Exchange Networks

1971-1979

Kobayashi, S.; Umeda, T., and Ichikawa, A., Synthesis of optimal heat-exchange systems, Chem Eng Sci, 26(9), 1367-1380; 26(11), 1841-1856 (1971).
Frith, J.F.; Bergen, B.M., and Shreehan, M.M., Optimize heat train design, Hyd Proc, 52(7), 89-91 (1973).
Ponton, J.W., and Donaldson, R.A.B., Fast method for synthesis of optimal heat exchanger networks, Chem Eng Sci, 29(12), 2375-2378 (1974).
Wright, J.D., and Bacon, D.W., Analysis of heat exchanger network using statistical time-series analysis methods, Ind Eng Chem Proc Des Dev, 14(4), 453-459 (1975).
Cena, V.; Mustacchi, C., and Natali, F., Synthesis of heat exchange networks: A non-iterative approach, Chem Eng Sci, 32(10), 1227-1232 (1977).
Kelahan, R.C., and Gaddy, J.L., Synthesis of heat exchange networks by mixed integer optimization, AIChEJ, 23(6), 816-822 (1977).
Wells, G., and Hodgkinson, M., Heat-content diagram for heat-exchanger networks, Proc Engng, Aug, 59-63 (1977).

Grossmann, I.E., and Sargent, R.W.H., Optimum design of heat-exchanger networks, Comput Chem Eng, 2(1), 1-8 (1978).
Linnhoff, B., and Flower, J.R., Synthesis of heat exchanger networks, AIChEJ, 24(4), 633-654 (1978).
Boland, D., and Linnhoff, B., The preliminary design of networks for heat exchange by systematic methods, Chem Engnr, April, 222-228 (1979).
Elshout, R.V., and Hohmann, E.C., The heat exchanger network simulator, Chem Eng Prog, 75(3), 72-77 (1979).

1980-1985

Flower, J.R., and Linnhoff, B., Thermodynamic-combinatorial approach to design of optimum heat exchanger networks, AIChEJ, 26(1), 1-9 (1980).
Stephanopoulos, G., and Westerberg, A.W., Modular design of heat exchanger networks, Chem Eng Commns, 4(1), 119-126 (1980).
Challand, T.B.; Colbert, R.W., and Venkatesh, C.K., Computerized heat exchanger networks, Chem Eng Prog, 77(7), 65-71 (1981).
Linnhoff, B., and Turner, J.A., Heat-recovery networks, Chem Eng, 2 Nov, 56-70 (1981).
Colbert, R.W., Industrial heat exchange networks, Chem Eng Prog, 78(7), 47-54 (1982).
Grimes, L.E.; Rychener, M.D., and Westerberg, A.W., Synthesis and evolution of networks of heat exchange that feature minimum number of units, Chem Eng Commns, 14(3), 339-360 (1982).
Grimes, L.E.; Rychener, M.D., and Westerberg, A.W., Synthesis and evolution of networks of heat exchange that feature minimum number of units, Chem Eng Commns, 14(3), 339-360 (1982).
Rev, E., and Fonyo, Z., Synthesis of heat exchanger networks, Chem Eng Commns, 18(1), 97-106 (1982).
Rev, E., and Fonyo, Z., Synthesis of heat exchanger networks, Chem Eng Commns, 18(1), 97-106 (1982).
Cerda, J., and Westerberg, A.W., Synthesizing heat exchanger networks having restricted stream/stream matches using transportation problem formulations, Chem Eng Sci, 38(10), 1723-1740 (1983).
Cerda, J.; Westerberg, A.W.; Mason, D., and Linnhoff, B., Minimum utility usage in heat exchanger network synthesis (a transportation problem), Chem Eng Sci, 38(3), 373-388 (1983).
Kleinschrodt, F., and Hammer, G.A., Exchanger networks for crude units, Chem Eng Prog, 79(7), 33-38 (1983).
Shiroko, K., and Umeda, T., A practical approach to optimal design of heat exchange systems, Proc Econ Int, 3(4), 44-49 (1983).
Boland, D., and Hindmarsh, E., Heat exchanger network improvements, Chem Eng Prog, 80(7), 47-54 (1984).
Su, J.L., and Motard, R.L., Evolutionary synthesis of heat-exchanger networks, Comput Chem Eng, 8(2), 67-80 (1984).
Wood, R.M.; Wilcox, R.J., and Grossmann, I.E., Minimum number of units for heat exchanger network synthesis, Chem Eng Commns, 39, 371-380 (1985).

1986-1988

Chato, J.C., and Damianides, C., Second-law-based optimization of heat exchanger networks using load curves, Int J Heat Mass Trans, 29(8), 1079-1086 (1986).

Floudas, C.A.; Ciric, A.R., and Grossmann, I.E., Automatic synthesis of optimum heat exchanger network configurations, AIChEJ, 32(2), 276-290 (1986).

Govind, R.; Mocsny, D.; Cosson, P., and Klei, J., Exchanger network synthesis on a microcumputer, Hyd Proc, 65(7), 53-57 (1986).

Jezwoski, J., and Hahne, E., Heat exchanger network synthesis by a depth-first method: A case study, Chem Eng Sci, 41(12), 2989-2998 (1986).

Jones, D.A.; Yilmaz, A.N., and Tilton, B.E., Synthesis techniques for retrofitting heat recovery systems, Chem Eng Prog, 82(7), 28-33 (1986).

Kotjabasakis, E., and Linnhoff, B., Sensitivity tables for design of flexible processes: How much contingency in heat exchanger networks is cost effective, CER&D, 64(3), 197-211 (1986).

Li, Y., and Motard, R.L., Optimal pinch approach temperature in heat-exchanger networks, Ind Eng Chem Fund, 25(4), 577-581 (1986).

Rev, E., and Fonyo, Z., Hidden and pseudo pinch phenomena and relaxation in the synthesis of heat-exchange networks, Comput Chem Eng, 10(6), 601-607 (1986).

Saboo, A.K.; Morari, M., and Colberg, R.D., An interactive software package for the synthesis and analysis of resilient heat-exchanger networks (RESHEX), Comput Chem Eng, 10(6), 577-599 (1986).

Atkinson, T.D., Second Law analysis of cryogenic processes using a three term model, Gas Sepn & Purif, 1(2), 84-89 (1987).

Dixon, A.G., Teaching heat exchanger network synthesis using interactive microcomputer graphics, Chem Eng Educ, 21(3), 118-121,156 (1987).

Floudas, C.A., and Grossmann, I.E., Automatic generation of multiperiod heat exchanger network configurations, Comput Chem Eng, 11(2), 123-142 (1987).

Floudas, C.A., and Grossmann, I.E., Synthesis of flexible heat exchanger networks with uncertain flowrates and temperatures, Comput Chem Eng, 11(4), 319-336 (1987).

Terrill, D.L., and Douglas, J.M., A T-H method for heat exchanger network synthesis, Ind Eng Chem Res, 26(1), 175-179 (1987).

Terrill, D.L., and Douglas, J.M., Heat-exchanger network analysis, Ind Eng Chem Res, 26(4), 685-696 (1987).

Terrill, D.L., and Douglas, J.M., Heat-exchanger network analysis, Ind Eng Chem Res, 26(4), 685-696 (1987).

Trivedi, K.K.; Roach, J.R.; and O'Neill, B.K., Shell targeting in heat exchanger networks, AIChEJ, 33(12), 2087-2090 (1987).

Engel, P., and Morari, M., Limitations of the primary loop-breaking method for synthesis of heat-exchanger networks, Comput Chem Eng, 12(4), 307-310 (1988).

Gundersen, T., and Naess, L., Synthesis of cost optimal heat exchanger networks: An industrial review of the state of the art, Comput Chem Eng, 12(6), 503-530 (1988).

Ptacnik, R., and Klemes, J., Application of mathematical optimization methods in heat-exchange network synthesis, Comput Chem Eng, 12(2/3), 231-236 (1988).
Qassim, R.Y., and Silveira, C.S., Heat-exchanger network synthesis: The goal programming approach, Comput Chem Eng, 12(11), 1163-1166 (1988).
Siegell, J.H., and Stachowicz, N.A., Heat-exchanger network response to process stream variations, Chem Eng Prog, 84(6), 37-44 (1988).
Trivedi, K.K.; O'Neill, B.K., and Roach, J.R., Synthesis of heat exchanger networks with designer imposed constraints, Chem Eng Commns, 69, 149-168 (1988).
Zhelev, T.K., and Boyadzhiev, K.B., Method for optimal synthesis of heat exchanger systems, Int Chem Eng, 28(3), 543-559 (1988).

3.7 Energy Conservation Methods

1967-1970

Caplow, S.D., and Bresler, S.A., Economics of gas turbine drives, Chem Eng, 27 March, 103-110 (1967).
Csathy, D., Evaluating boiler designs for process-heat recovery, Chem Eng, 5 June, 117-124 (1967).
Various, Low-cost power and the CPI (topic issue), Chem Eng Prog, 63(4), 35-67 (1967).
Jenett, E., Hydraulic power recovery systems, Chem Eng, 8 April, 159-164; 17 June, 257-262 (1968).
Various, Power for the CPI (topic issue), Chem Eng Prog, 64(3), 49-72 (1968).
Bouilloud, P., Compute steam balance by linear programming, Hyd Proc, 48(8), 127-128 (1969).
Dain, R.J., and Whitlock, D., Optimisation of total energy systems, Brit Chem Eng, 14(9), 1209-1212 (1969).
Dain, R.J., and Whitlock, D., Total-energy system design, Hyd Proc, 48(8), 175-178 (1969).
Isaacs, M., Selecting efficient, economical insulation, Chem Eng, 24 March, 143-150 (1969).
Rabb, A., Use entropy for quick evaluations of systems thermal efficiencies, Hyd Proc, 48(6), 133-135 (1969).
Taylor, F.M.H., Total energy for cost reduction, Brit Chem Eng, 14(7), 941-943 (1969).
Deeson, A.F.L., and Deeson, E., Thermal insulation in the chemical industry, Brit Chem Eng, 15(5), 621-626 (1970).
El Kabbani, A.S., Simplified method to estimate insulation thickness, Hyd Proc, 49(3), 145-148 (1970).
Kenyon, F., Insulation value of air spaces, Brit Chem Eng, 15(10), 1328-1329 (1970).

1971-1973

Bregman, J.I., Useful energy from unwanted heat, Chem Eng, 25 Jan, 83-87 (1971).

Conti, F., and Pistoia, G., Recent advances in high energy non-aqueous batteries, J Appl Chem Biotechnol, 21, 77-81,179 (1971).

Kodrashchenko, V.D., et al., Determining optimal fuel consumption in a process plant, Brit Chem Eng, 16(8), 705-706 (1971).

Polastri, F., Check pipe insulation thickness, Hyd Proc, 50(11), 231-232 (1971).

Butler, P., Gas turbines for energy savings, Proc Engng, Aug, 70-72 (1972).

Collins, C.E., Reducing fuel consumption in an expander plant, Chem Eng Prog, 68(11), 80-84 (1972).

Ediss, B.G., Steam-injection gas turbine cycle, Proc Tech Int, 17(11), 864-866 (1972).

Stettenbenz, L.M., Benefits of the power-recovery gas expander, Chem Eng, 10 Jan, 93-96 (1972).

Swearingen, J.S., Applications of turboexpanders, Chem Eng Prog, 68(7), 95-102 (1972).

Abadie, V.H., Turboexpanders recover energy, Hyd Proc, 52(7), 93-96 (1973).

Abgrall, R., and Kalaski, A., On-site power generation using gas turbines, Proc Tech Int, April, 193-199 (1973).

Brown, C.L., and Figenscher, D., Preheat process combustion air, Hyd Proc, 52(7), 115-116 (1973).

Kniel, L., Energy systems for LNG plants, Chem Eng Prog, 69(10), 77-84 (1973).

May, D.L., First steps in cutting steam costs, Chem Eng, 12 Nov, 228-232 (1973).

Mol, A., Which heat recovery system? Hyd Proc, 52(7), 109-112 (1973).

Soderlind, C., Tank insulation, Hyd Proc, 52(7), 122 (1973).

Steen-Johnsen, H., Turbine steam consumption, Hyd Proc, 52(7), 99-101 (1973).

Various, Energy management (special report), Hyd Proc, 52(7), 53-75 (1973).

Various, Heat recovery in process plants (feature report), Chem Eng, 28 May, 80-104 (1973).

1974

Austin, L.G., Fuel efficiency via the mass-energy balance, Chemtech, Oct, 631-638 (1974).

Barber, R.E., Rankine-cycle systems for waste heat recovery, Chem Eng, 25 Nov, 101-106 (1974).

Barker, R., Reactivity of calcium oxide towards carbon dioxide and its use for energy storage, J Appl Chem Biotechnol, 24, 221-228 (1974).

Barnes, F.J., and King, C.J., Synthesis of cascade refrigeration and liquefaction systems, Ind Eng Chem Proc Des Dev, 13(4), 421-433 (1974).

Cooke, B., Waste-heat recovery, Proc Engng, Dec, 81-83 (1974).

Cooper, A., Recover more heat with plate heat exchangers, Chem Engnr, May, 280-285 (1974).

David, T., Reducing steam loss, Proc Engng, Feb, 53-55 (1974).

Dickey, B.R.; Grimmett, E.S., and Kilian, D.C., Waste heat disposal via fluidized beds, Chem Eng Prog, 70(1), 60-64 (1974).

Fleming, J.B.; Lambrix, J.R., and Smith, M.R., Energy conservation in new-plant design, Chem Eng, 21 Jan, 112-122 (1974).

Hughes, R., and Deumaga, V., Insulation saves energy, Chem Eng, 27 May, 95-100 (1974).
MacDonald, J.O.S., Maximizing primary energy use, Chemical Processing, April, 99-101, 113 (1974).
Riekert, L., Efficiency of energy-utilization in chemical processes, Chem Eng Sci, 29(7), 1613-1620 (1974).
Robertson, J.C., Energy conservation in existing plants, Chem Eng, 21 Jan, 104-111 (1974).
Schumacher, C.E., and Girgis, B.Y., Conserving utilities' energy in new construction, Chem Eng, 18 Feb, 133-138 (1974).
Smithson, D.J., Energy savings in the UK iron and steel industry, Energy World, Nov, 2-5 (1974).
Tate, P., Fuel saving in industry, Energy World, July, 5-10 (1974).
Various, Energy management (special report), Hyd Proc, 53(7), 87-112 (1974).
Various, Insulation practices (topic issue), Chem Eng Prog, 70(8), 41-59 (1974).
Various, Energy saving ideas, Chem Eng, 30 Sept, 123-128 (1974).
Various, Energy-saving ideas, Chem Eng, 2 Sept, 50-61 (1974).

1975

Abramovitz, J.L., and Cordero, R., Selecting insulation thickness for hot pipes, Chem Eng, 21 July, 88-96 (1975).
Amir, S.J., Calculating heat transfer from a buried pipeline, Chem Eng, 4 Aug, 123-124; 27 Oct, 140 (1975).
Barlow, J.A., Energy recovery in a petrochemical plant, Chem Eng, 7 July, 93-97 (1975).
Briggs, M.A., Estimation of economic lagging thickness, Chem Engnr, Sept, 513-515 (1975).
Essenhigh, R.H., Fuel conservation, Chemtech, Feb, 112-116 (1975).
Field, A.A., Heat-recovery systems for buildings, Energy World, May, 3-5 (1975).
Field, A.A., Insulation and low-energy building design, Energy World, Nov, 2-9 (1975).
Field, A.A., Industrial space heating, Energy World, Jan, 9-14 (1975).
Franzke, A., Save energy with hydraulic power recovery turbines, Hyd Proc, 54(3), 107-110 (1975).
Furlong, L.E.; Bernstein, L.S., and Holt, E.L., Emission control and fuel economy, Chemtech, Jan, 34-38 (1975).
Holden, G.F., Strategic energy planning, Chem Eng, 14 April, 94-96 (1975).
Lewis, B.N., Fuel conservation cuts costs, Processing, Nov, 21-25 (1975).
Llovet, J.E.; Klooster, H.J., and Chapel, D.G., Refinery design for energy conservation, Chem Eng Prog, 71(6), 85-90 (1975).
Martin, R.B., Guide to better insulation, Chem Eng, 12 May, 98-100 (1975).
Monroe, E.S., Testing steam traps, Chem Eng, 1 Sept, 99-102 (1975).
Muller, R.G., Chemicals energy demand in year 2000? Hyd Proc, 54(12), 69-72 (1975).
Pettman, M.J., and Humphreys, G.C., Improve designs to save energy, Hyd Proc, 54(1), 77-81 (1975).

Rathore, R.N.S., and Powers, G.J., A forward-branching scheme for synthesis of energy-recovery systems, Ind Eng Chem Proc Des Dev, 14(2), 175-181 (1975).
Rex, M.J., Choosing equipment for process energy recovery, Chem Eng, 4 Aug, 98-102 (1975).
Shah, B.M., Saving energy with jet compressors, Chem Eng, 7 July, 106 (1975).
Slota, L., Improvement of evaporator steam economy by heat recovery, Chem Engnr, Feb, 84-90 (1975).
Valentine, A.C., and Wildman, S.V., Energy conservation from steam turbines, Chem Engnr, Sept, 516-519 (1975).
Various, Heat recovery and generation (feature report), Processing, May, 35-65 (1975).
Various, Energy conservation in the chemical industry (topic issue), Chem & Ind, 6 Sept, 717-728; 20 Sept, 771-785 (1975).
Various, Energy management (special report), Hyd Proc, 54(7), 73-100 (1975).
Various, Conservation of energy (topic issue), Chem Eng Prog, 71(10), 35-79 (1975).
Yates, W., Better steam trapping cuts energy waste, Hyd Proc, 54(11), 267-269 (1975).

1976

Abramovitz, J.L., Economic pipe insulation for cold systems, Chem Eng, 25 Oct, 105-112 (1976).
Briley, G.C., Conserve energy, refrigerate with waste heat, Hyd Proc, 55(5), 173-174 (1976).
Caplan, F., Converting boiler horsepower to steam, Chem Eng, 15 March, 107 (1976).
Curt, R.P., Economic insulation thickness, Hyd Proc, 55(3), 137-138 (1976).
Don, W.A., Efficiency of domestic oil-burning boilers at reduced loading, Energy World, Jan, 6-10 (1976).
Fleury, J., Transfer of energy by storage of compressed air, Int Chem Eng, 16(2), 308-315 (1976).
Freshwater, D.C., and Ziogou, E., Reducing energy requirements in unit operations, Chem Eng J, 11(3), 215-222 (1976).
Gunn, D.C., Waste-heat recovery in boilers, Energy World, Feb, 2-6 (1976).
Gushee, D.E., Energy accounting, Chemtech, Oct, 649-653 (1976).
Haring, W., Lower fuel bill, upgrade natural gas engines, Hyd Proc, 55(5), 211-214 (1976).
Hodgett, D.L., Efficient drying using heat pumps, Chem Engnr, July, 510-512 (1976).
Humphrey, R.C., Train for energy conservation, Hyd Proc, 55(10), 201-212 (1976).
Lucas, N.J.D., Space heating and energy supply, Energy World, March, 4-12 (1976).
Martin, D., Recuperator saves 60% of lost heat and washes exhaust gases, Proc Engng, June, 63 (1976).
Marx, A., Energy recovery from municipal refuse, Chem Engnr, Sept, 601-604 (1976).

Paros, S.V., Choosing cold insulation for piping and storage? Hyd Proc, 55(11), 257-259 (1976).
Reay, D., Energy conservation in industrial drying, Chem Engnr, July, 507-509 (1976).
Reynolds, J.A., Saving energy and costs in pumping systems, Chem Eng, 5 Jan, 135-138 (1976).
Ryder, C., Energy audits and case histories, Energy World, Nov, 2-6 (1976).
Summerell, H., Recovering energy from stacks, Chem Eng, 29 March, 147-148 (1976).
Teller, W.M.; Diskant, W., and Malfitani, L., Conserving fuel by heating with hot water instead of steam, Chem Eng, 21 June, 185-190 (1976).
Various, Energy management (special report), Hyd Proc, 55(7), 71-111 (1976).
Various, Conserving energy in plants and equipment (topic issue), Chem Eng Prog, 72(5), 49-82 (1976).
Vervalin, C.H., Information sources for energy trends, Hyd Proc, 55(11), 56G-56P; 55(12), 54JJ-54N (1976).
Wells, G.L.; Hodgkinson, M.G.; Al-Kadhi, H., and Wardle, I., Energy considerations during flowsheeting, Chem & Ind, 6 Nov, 943-947 (1976).
Williams, M.A., Organize for energy conservation, Hyd Proc, 55(4), 221-232 (1976).
Williams, M.A., Organizing an energy conservation program, Chem Eng, 11 Oct, 149-152 (1976).
Witt, J.A., and Aylott, G.W., Heat pump applications, Proc Engng, Nov, 81-83 (1976).
Witwer, J.G.; Ushiba, K.K., and Semrau, K.T., Energy conservation with LNG cold, Chem Eng Prog, 72(1), 50-55 (1976).
Wood, R., Heat-recovery equipment, Proc Engng, June, 58-61 (1976).

1977

Barr, D.J., Energy saving in drying, Processing, April, 33 (1977).
Barrows, G.L., Save energy with ceramic fiber insulation, Hyd Proc, 56(10), 187-189 (1977).
Don, W.A., Factors affecting the part-load efficiency of boilers, Energy World, Oct, 2-5 (1977).
Doolin, J.H., Select pumps to cut energy costs, Chem Eng, 17 Jan, 137-139 (1977).
Fuchs, W.; James, G.R., and Stokes, K.J., Economics of flue gas heat recovery, Chem Eng Prog, 73(11), 65-70 (1977).
Gaggioli, R.A., and Petit, P.J., Using the second law of thermodynamics, Chemtech, Aug, 496-506 (1977).
Ganapathy, V., Sizing piping insulation, Chem Eng, 21 Nov, 219 (1977).
Goyette, J., Estimating the costs of steam leaks, Chem Eng, 29 Aug, 95 (1977).
Harrison, M.R., and Pelanne, C.M., Cost-effective thermal insulation, Chem Eng, 19 Dec, 62-76 (1977).
Kelly, H., Stimulating energy technology, Chemtech, Jan, 32-35 (1977).
Nishio, M., and Johnson, A.I., Strategy for energy system expansion, Chem Eng Prog, 73(1), 73-80 (1977).
Perreault, E.A., and Prutzman, P.J., Strategies for reducing electric power consumption, Chem Eng, 23 May, 153-158 (1977).

Pigford, T.H., Fuel-cycle alternatives for nuclear power reactors, Ind Eng Chem Fund, 16(1), 75-81 (1977).
Pinto, A., and Rogerson, P.L., Impact of high fuel cost on plant design, Chem Eng Prog, 73(7), 95-100 (1977).
Reale, F.N., and Dillon, T.S., Investigation into the use of large-scale total-energy systems in mild and warm climates, Fuel, 56(3), 257-265 (1977).
Sayles, D., and Noakes, B., Energy savings from air:fuel ratio calculations, Proc Engng, Aug, 54-55 (1977).
Shepherd, D.G., Pick up energy from low heat sources, Hyd Proc, 56(12), 141-149 (1977).
Various, Energy and materials savings (feature report), Processing, April, 41-65 (1977).
Various, Energy management (special report), Hyd Proc, 56(7), 89-136 (1977).
Various, Techniques for saving energy in processes and equipment (feature report), Chem Eng, 4 July, 98-112 (1977).

1978

Anon., Energy recovery with gas turbines, Processing, Feb, 34-35 (1978).
Bannon, R.P., and Marple, S., Heat recovery in hydrocarbon distillation, Chem Eng Prog, 74(7), 41-45 (1978).
Beatty, R.E., and Krueger, R.G., How the steam trap saves plant energy, Chem Eng Prog, 74(9), 94-100 (1978).
Beckers, T., Design of steam traps, Processing, Jan, 30-31 (1978).
Bogart, M.J.P., Save energy in ammonia plants, Hyd Proc, 57(4), 145-151 (1978).
Brinsko, J.A., How to make a steam balance, Hyd Proc, 57(11), 227-235 (1978).
Caldwell, D.; Elfers, F., and Fankhanel, M., The vaporized fuel oil system, Chem Eng Prog, 74(11), 66-69 (1978).
Chiu, C.H., Energy efficiency in LNG plants, Hyd Proc, 57(9), 266-272 (1978).
Dziewulski, T.A., and Bews, J.H., Recover power from fluid-catalytic-cracking units, Hyd Proc, 57(12), 131-135 (1978).
Fisk, D.J., Economic value of conserving energy, J Inst Fuel, 51, 187-190 (1978).
Goldspink, K., Pipeline desuperheating, Processing, Feb, 37 (1978).
Goossens, G., Thermodynamic principles govern energy recovery, Hyd Proc, 57(8), 133-140 (1978).
Hall, D.O., Evaluation of solar energy conversion through biology as a practical energy source, Fuel, 57(6), 322-333 (1978).
Harker, J.H., Economic lagging thickness, Processing, July, 61 (1978).
Harker, J.H., Economic balancing of heat exchangers, Processing, June, 85 (1978).
Hickok, H.N., Save electrical energy, Hyd Proc, 57(8), 127-130 (1978).
Hlavecek, V., Synthesis in design of chemical processes (review paper), Comput Chem Eng, 2(1), 67-76 (1978).
Horr, K.S., Measure heat loss onstream, Hyd Proc, 57(12), 159 (1978).
Lock, J., Energy reduction in ammonia plant, Processing, July, 25-26 (1978).
Mix, T.J.; Dweck, J.S.; Weinburg, M., and Armstrong, R.C., Energy conservation in distillation, Chem Eng Prog, 74(4), 49-55 (1978).
Panesar, K.S., Select pumps to save energy, Hyd Proc, 57(10), 127-128 (1978).

Reay, D., Energy savings for drying, Proc Engng, July, 71-75 (1978).
Shelley, S.J., and Moore, A., Improving efficiency in fuel-fired industrial heating processes, J Inst Fuel, 51, 3-9 (1978).
Shinnar, R., Differential economic analysis of gasoline from coal, Chemtech, Nov, 686-693 (1978).
Shinskey, F.G., Saving energy with control systems, Proc Engng, May, 107-113 (1978).
Troyan, J.E., Energy conservation programs require accurate records, Chem Eng, 20 Nov, 189-195 (1978).
Various, Equipment/energy conservation (topic issue), Chem Eng Prog, 74(5), 43-89 (1978).
Various, Instruments save energy (special report), Hyd Proc, 57(2), 93-109 (1978).
Various, Energy management (special report), Hyd Proc, 57(7), 89-143 (1978).
Vernon, R., Economics of heat pumps, Proc Engng, May, 92-99 (1978).
Winter, E.F., Optimum industrial utilization of gaseous fuels, J Inst Fuel, 51, 46-58 (1978).

1979

Bojnowski, J.H., and Hanks, D.L., Low-energy separation processes, Chem Eng, 7 May, 67-71 (1979).
Brown, L.K., A strategy for saving electric power costs, Chem Eng, 12 March, 118 (1979).
Buividas, L.J., Cut energy costs in ammonia plants, Hyd Proc, 58(5), 257-259 (1979).
Christodoulou, A.P., Energy consumption in the chemical industry, Chemtech, Nov, 673-675 (1979).
Cole, W.S., and Suo, M., Waste heat recovery with fluidized beds, Chem Eng Prog, 75(12), 38-42 (1979).
Danekind, W.E., Steam management in a petroleum refinery, Chem Eng Prog, 75(2), 51-55 (1979).
DeJovine, J.M.; DeVries, D.L., and Keller, G.H., Fuel economy and real-world testing, Chemtech, June, 350-353 (1979).
Dev, L., Heat recovery in the forest products industry, Chem Eng Prog, 75(12), 25-29 (1979).
Fishel, F.D., and Howe, C.D., Cut energy costs with variable frequency motor drives, Hyd Proc, 58(9), 231-236 (1979).
Ganapathy, V., Best mix for two-steam sources, Hyd Proc, 58(11), 269-270 (1979).
Geyer, C.R., Conserving energy in a distillation train, Chem Eng Prog, 75(1), 41-45 (1979).
Givoni, B., Store energy in the ground, Chemtech, June, 384-390 (1979).
Harker, J.H., Economic balance in heat-recovery systems, Processing, June, 47 (1979).
Harker, J.H., and Hindmarsh, C.E., Energy storage in fluidized beds, J Inst Energy, 52, 45-48 (1979).
Hendrix, W.A., and Hoyos, G.H., Conserving boiler energy, Chem Eng, 31 Dec, 77-78 (1979).

Hendrix, W.A., and Moran, W.G., Save energy on compressed air, Chem Eng, 19 Nov, 177-178 (1979).
Kenney, W.F., Reducing the energy demand of separation processes, Chem Eng Prog, 75(3), 68-71 (1979).
Masters, J.; Webb, R.J., and Davies, R.M., Modelling techniques for design of recuperative burners, J Inst Energy, 52, 196-204 (1979).
Moran, W.G., and Hoyos, G.H, Boiler heat recovery, Chem Eng, 3 Dec, 111-112 (1979).
Robnett, J.D., Engineering approaches to energy conservation, Chem Eng Prog, 75(3), 59-67 (1979).
Sommerfeld, J.T., and White, R.H., Estimate energy consumption from heat of reaction, Chem Eng, 19 Nov, 140-147 (1979).
Stokes, K.J., Compression systems for ammonia plants, Chem Eng Prog, 75(7), 88-91 (1979).
Taniguchi, B., and Johnson, R.T., MTBE for octane improvement, Chemtech, Aug, 502-510 (1979).
Various, Thermal energy storage, J Inst Energy, 52, 185-196 (1979).
Various, Energy conservation (feature issue), Energy World, Oct supplement, (i)-(xx) (1979).
Various, Energy saving in the chemical industry (topic issue), Chem & Ind, 1 Sept, 566-591 (1979).
Various, Energy management (special report), Hyd Proc, 58(7), 105-162 (1979).
Venable, H.E., Floating balls insulate hot tanks, Chem Eng, 12 Feb, 136 (1979).
Whittingham, M.S., Storing energy by intercalation, Chemtech, Dec, 766-770 (1979).
Yamanouchi, N., and Nagasawa, H., Using LNG cold for air separation, Chem Eng Prog, 75(7), 78-82 (1979).
Zanker, A., Nomograph for economical pipe insulation, Proc Engng, March, 83-85 (1979).

1980

Anon., Heat pumps, Chemtech, July, 441-443 (1980).
Anon., Evaluation of a heat regenerator, Processing, April, 56 (1980).
Arscott, J.A.; Chew, P.E., and Lawn, C.J., Improvements in combustion efficiency through air/fuel matching in an oil-fired boiler, J Inst Energy, 53, 3-14 (1980).
Baker, R.E., Fuel consumption of 1980s cars, Chemtech, June, 375-381 (1980).
Baston, V.F., et al., Evaluating wastes as energy sources, Chemtech, July, 438-440 (1980).
Bate, D.J., Fuel-saving techniques in the petrochemicals industry, J Inst Energy, 53, 124-133 (1980).
Baudat, N.P., and Darrow, P.A., Power recovery in a closed cycle system, Chem Eng Prog, 76(2), 68-71 (1980).
Cheng, W.B., and Mah, R.S.H., Interactive synthesis of cascade refrigeration systems, Ind Eng Chem Proc Des Dev, 19(3), 410-420 (1980).
Cranfield, R.R., Fluidized-bed heat store for power generation, J Inst Energy, 53, 196-204 (1980).
Danton, S.A., Management techniques applied to energy conservation, Energy World, Aug, 17-20 (1980).

Davis, J.S., and Martin, J.R., Energy savings from cryogenics for syngas processing, Chem Eng Prog, 76(2), 72-79 (1980).
Finn, D.P., Select equipment drives to cut operating energy costs, Chem Eng, 24 March, 121-124 (1980).
Hearfield, F., Adipic acid reactor development: Benefits in energy and safety, Chem Engnr, Oct, 625-627, 633 (1980).
Hendrix, W.A., and Moran, W.G., Move electric motors from air-conditioned spaces, Chem Eng, 17 Nov, 299 (1980).
Hine, F., Energy saving in the chlor-alkali industry, Int Chem Eng, 20(4), 629-639 (1980).
Hoyos, G.H., and Muzzy, J.D., Use low-grade heat for refrigeration, Chem Eng, 5 May, 140-142 (1980).
Hunt, L., Evaluating energy savings by direct measurement, Proc Engng, Jan, 89-91 (1980).
Itoh, J.; Niida, K.; Shiroko, K., and Umeda., T., Analysis of available energy of a distillation system, Int Chem Eng, 20(3), 379-386 (1980).
Katzler, J., Tank insulation, Hyd Proc, 59(5), 197-198 (1980).
Keddy, E.S., and Ranken, W.A., Ceramic heat pipes for high temperature heat recovery, Chem Eng Commns, 4(2), 381-392 (1980).
Koenig, A.R., Choosing economic insulation thickness, Chem Eng, 8 Sept, 125-128 (1980).
Linnhoff, B., and Turner, J.A., Simple concepts in process synthesis give energy savings and elegant designs, Chem Engnr, Dec, 742-746 (1980).
MacLean, D.L.; Prince, C.E., and Chae, Y.C., Energy-saving modifications in ammonia plants, Chem Eng Prog, 76(3), 98-104 (1980).
Musgrave, G., et al., Utilization of industrial waste heat, J Inst Energy, 53, 137-141 (1980).
Neuzil, R.W., An energy-saving separation scheme, Chemtech, Aug, 498-503 (1980).
Paciotti, J.D., Unsteady-state heat losses from storage tanks, Chem Eng, 2 June, 104 (1980).
Pritchard, C., and Halfani, M., Heat economy in the Tanzanian oil refinery, Chem Engnr, Dec, 747-748 (1980).
Radway, J.E., Selecting and using fuel additives, Chem Eng, 14 July, 155-160 (1980).
Rodriguez, L., and Gaggioli, R.A., Second-law efficiency analysis of a coal gasification process, Can JCE, 58, 376-381 (1980).
Rush, F.E., Energy-saving alternatives to distillation, Chem Eng Prog, 76(7), 44-49 (1980).
Shah, J.V., and Westerberg, A.W., EROS: A program for quick evaluation of energy recovery systems, Comput Chem Eng, 4(1), 21-32 (1980).
Sommerfeld, J.T., Reuse hot wash water, Chem Eng, 30 June, 139-140 (1980).
Stockburger, D., and Bartmann, L., Exergetic evaluation of use of heat pumps (vapor compressors) in chemical plants, Int Chem Eng, 20(2), 197-203 (1980).
Sussman, M.V., Standard chemical availability for energy management, Chem Eng Prog, 76(1), 37-39 (1980).
Svec, O.J., and Palmer, J.H.L., Heat exchanger for in-ground heat storage, Chem Eng Commns, 5(5), 323-336 (1980).

Uppal, K.B., Tank insulation, Hyd Proc, 59(4), 165-168 (1980).
Various, Energy today (special report), Hyd Proc, 59(7), 57-98 (1980).
Various, Plant energy conservation (topic issue), Chem Eng Prog, 76(8), 35-71 (1980).
Verma, A., Use medium-Btu gas, Chemtech, June, 382-389 (1980).
Wade, D.W., and Moran, W.G., Steam system energy savers, Chem Eng, 11 Feb, 130-132 (1980).
Whitfield, M., Heat pumps for energy recovery, Proc Engng, Jan, 71-73 (1980).
Whittlesey, G., and Muzzy, J.D., Vapor recompression can reduce steam costs, Chem Eng, 28 July, 94-95 (1980).

1981

Anon., Energy outlook for the 80s, Hyd Proc, 60(2), 56D-56N (1981).
Anon., Energy savings in the cement industry, Proc Econ Int, 2(3), 30-32 (1981).
Bergman, A., Elapsed-time meters aid energy management, Chem Eng, 21 Sept, 155 (1981).
Bourguet, J.M., Problems with LNG cold, Hyd Proc, 60(1), 167-172 (1981).
Brunner, C.R., Program solves airstream energy balances, Chem Eng, 16 Nov, 265-269 (1981).
Catani, S.J., Control system cuts heating and cooling costs, Chem Eng, 24 Aug, 129 (1981).
Caton, J.A., and Heywood, J.B., Experimental and analytical study of heat transfer in an engine exhaust port, Int J Heat Mass Trans, 24(4), 581-596 (1981).
Chawla, O.P., and Khandwawala, A.I., Thermal performance of regenerators and waste-heat recovery, Int J Heat Mass Trans, 24(11), 1793-1800 (1981).
Elkotb, M.M., and El-Refaie, M.F., Multipass solar heater with heat-exchanging passes and exposed to non-uniform radiation, Chem Eng Commns, 11(1), 123-142 (1981).
Finlay, I.C., and Harrow, G.A., Future developments in car engines and fuels, Energy World, Aug, 2-12 (1981).
Gartside, R.J., Save energy with gas turbines, Hyd Proc, 60(1), 141-146 (1981).
Gay, R.L.; Barclay, K.M.; Grantham, L.F., and Yosim, S.J., Fuel from waste, Chemtech, Sept, 572-575 (1981).
Haas, J.R.; Holland, C.D.; Frederico, D.S., and Alejandro, G.M., Solution of systems of columns with energy exchange between recycle streams, Comput Chem Eng, 5(1), 41-50 (1981).
Harrison, M.R., Consider single layer insulation for high temperature piping, Hyd Proc, 60(9), 231-234 (1981).
Henze, R.H., and Humphrey, J.A.C., Enhanced heat conduction in phase-change thermal-energy storage devices, Int J Heat Mass Trans, 24(3), 459-474 (1981).
Huor, M.H., and Bugarel, R., Behavior of absorption heat pump operating under partial loading, Int Chem Eng, 21(4), 659-670 (1981).
Jacobs, W.P., Forecasting energy requirements, Chem Eng, 9 March, 97-99 (1981).
James, O.R., and Fan, S.K., Selecting an efficient fluid for recovering power from waste heat, Chem Eng, 28 Dec, 57-58 (1981).

Kawabata, J.I., et al., Performance of a pressurized two-stage fluidized gasification process for production of low-Btu gas from coal char, Chem Eng Commns, 11(6), 335-346 (1981).
Kragh, O.T., and Kraglund, A., Heat recovery in dryers, Chem Engnr, April, 149-153, 158 (1981).
Lauerhass, L.N., and Rudd, D.F., Thermodynamics of the chemical heat pump, Chem Eng Sci, 36(5), 803-808 (1981).
Laws, W., Waste heat as an energy source, Energy World, Nov, 10-19 (1981).
Miner, J.B., Using an energy audit to cut energy costs, Chem Eng, 2 Nov, 105-110 (1981).
Moore, A.G., and Nixon, K.A., Thermal efficiencies of electricity systems in EEC countries, J Inst Energy, 54, 103-112 (1981).
Mostafa, A., Thermodynamic availability analysis of fractional distillation with vapour compression, Can JCE, 59, 487-491 (1981).
Naphtali, L.M., and Shinnar, R., Effect of inflation on energy cost analyses, Chem Eng Prog, 77(2), 65-71 (1981).
Neal, J.E., and Clark, R.S., Saving heat energy in refractory-lined equipment, Chem Eng, 4 May, 56-70 (1981).
Salusinszky, A.L., Try adding work to recover heat, Hyd Proc, 60(3), 159-161 (1981).
Spiller, G.B., Energy use and conservation in a dairy, Energy World, Jan, 7-16 (1981).
Sundaram, S., and Eldridge, B.G., Solar production of process steam, Chem Eng Prog, 77(7), 50-54 (1981).
Tassou, S.A.; Green, R.K.; Wilson, D.R., and Searle, M., Energy conservation using capacity control of heat pumps, J Inst Energy, 54, 30-34 (1981).
Vallery, S.J., Are your steam traps wasting energy? Chem Eng, 9 Feb, 84-98 (1981).
Various, Heat recovery (topic issue), Processing, April, 27-33 (1981).
Various, Energy: Steam raising (topic issue), Processing, Oct, 33-49 (1981).
Various, Heat pumps, Energy World, Oct supplement, (i) - (xxxii) (1981).
Various, Energy conservation practices (topic issue), Chem Eng Prog, 77(10), 33-88 (1981).
Various, Energy management (special report), Hyd Proc, 60(7), 67-110 (1981).

1982

Antony, S.M., Save by improving power factor, Chem Eng, 28 June, 117 (1982).
Balzhiser, R.E., U.S. energy forecast, Chem Eng, 11 Jan, 74-96 (1982).
Bassett, J.W., Low-cost factory heating, Energy World, Nov, 2-7 (1982).
Bau, H.H., Temperature distribution in and around a buried heat-generating sphere, Int J Heat Mass Trans, 25(11), 1701-1708 (1982).
Bau, H.H., and Sadhal, S.S., Heat losses from a fluid flowing in a buried pipe, Int J Heat Mass Trans, 25(11), 1621-1630 (1982).
Bejan, A., and Schultz, W., Optimum flowrate history for cooldown and energy storage processes, Int J Heat Mass Trans, 25(8), 1087-1092 (1982).
Bommelburg, H.J., Use of ammonia in energy-related applications, Plant/Opns Prog, 1(3), 175-181 (1982).

Bumby, J.R., and Clarke, P.H., Development of internal combustion/battery electric hybrid vehicle in UK, Energy World, Dec, 2-12 (1982).
Clark, J.A., Analysis of technical and economic performance of a parabolic trough concentrator for solar industrial process heat application, Int J Heat Mass Trans, 25(9), 1427-1438 (1982).
Cross, P.H., The use of plate heat exchangers for energy economy, Chem Engnr, March, 87-90 (1982).
Davis, N., Waste-heat recovery, Proc Engng, Jan, 51 (1982).
de Nevers, N., Energy conservation and the second law, Chemtech, May, 306-317 (1982).
de Virgiliis, A., and Gerunda, A., Optimize energy usage in phthalic anhydride units, Hyd Proc, May, 173-175 (1982).
Deglise, X., and Lede, J., Upgrading biomass energy by thermal methods, Int Chem Eng, 22(4), 631-647 (1982).
Finkelstein, E., and Greenberg, A., Thermocompressor saves steam at phosphate plant, Chem Eng, 3 May, 113-114 (1982).
Gambera, S., and Lockett, W., Energy conservation in European refineries, Hyd Proc, Nov, 243-252 (1982).
Ganapathy, V., Evaluating waste heat recovery projects, Hyd Proc, Aug, 101-106 (1982).
Govindan, T.S., Energy management in the 1980s, Hyd Proc, May, 165-169 (1982).
Hansrani, S.P., Energy conservation in the steel industry, Energy World, Oct, 6-14 (1982).
Ishida, M., and Kawamura, K., Energy and exergy analysis of a chemical process system with distributed parameters based on the enthalpy-direction factor diagram, Ind Eng Chem Proc Des Dev, 21(4), 690-695 (1982).
Knoche, K.F., and Stehmeier, D., Exergetic criteria for development of absorption heat pumps, Chem Eng Commns, 17(1), 183-194 (1982).
Kotas, T.J., The exergy method, Proc Engng, Sept, 57-59 (1982).
Lach, J., and Pieczka, W., Heat transfer at the cladding-cooling fluid boundary, Int J Heat Mass Trans, 25(10), 1595-1604 (1982).
Laine, J., and Trimm, D.L., Conversion of heavy oils into more desirable feedstocks, J Chem Tech Biotechnol, 32, 813-833 (1982).
Le Goff, P., and Giuletti, M., Comparison of economic and energy optimizations for a heat exchanger, Int Chem Eng, 22(2), 252-269 (1982).
London, A.L., Economics and the second law: An engineering view and methodology, Int J Heat Mass Trans, 25(6), 743-752 (1982).
McAvoy, T.J., Integration of energy conservation principles into staged operations, Chem Eng Educ, 16(2), 88-93 (1982).
McKay, G., Cost effectiveness of energy conservation measures in buildings, Energy World, July, 2-6 (1982).
McKay, G.; Holland, C.R., and McConvey, I.F., Prediction of heat losses through furnace walls, Chem Engnr, March, 84-86 (1982).
Nelson, R.E., Vacuum pump aids ejectors, Hyd Proc, Dec, 95-96 (1982).
Neth, N.; Puhl, H., and Liebe, A., Improving ammonia plant production and energy usage, Chem Eng Prog, 78(7), 69-71 (1982).
Nishio, M.; Shiroko, K., and Umeda, T., Optimal use of steam and power in chemical plants, Ind Eng Chem Proc Des Dev, 21(4), 640-646 (1982).

Raine, J.H., Variations in energy saving calculations for industrial roof insulation, J Inst Energy, 55, 55-56 (1982).
Rathore, R.N.S., Reusing energy lowers fuel needs of distillation towers, Chem Eng, 14 June, 155-159 (1982).
Ringwald, R.M., Energy and the chemical industry, Chem & Ind, 1 May, 281-286 (1982).
Schmidt, P.S., Microcomputers in energy management, Chem Eng, 14 June, 169-170; 12 July, 123-126 (1982).
Sorotzkin, J., Train for energy conservation, Hyd Proc, Sept, 315-325 (1982).
Sunavala, P.D., Energy proficiency, J Inst Energy, 55, 153-159 (1982).
Various, Energy economy survey, Processing, Nov, 29-35 (1982).
Various, Energy supplement, Processing, Feb, 33-45 (1982).
Various, Energy management (special report), Hyd Proc, July, 85-156 (1982).
Vosseller, G.V., Estimate waste-gas heat savings, Chem Eng, 5 April, 138 (1982).

1983

Anon., Energy saving with hydraulic turbines, Processing, Jan, 21,23,27 (1983).
Anon., Insulation update, Processing, Dec, 21-24 (1983).
Anon., Energy conservation: A survey, Processing, Sept, 13-20 (1983).
Anon., Steamtraps, Processing, Jan, 37-39 (1983).
Baker, C., and Bahu, R., Reducing fuel consumption of dryers, Proc Engng, March, 42-45 (1983).
Boland, D., Energy management: Emphasis in the 80s, Chem Engnr, March, 24-28 (1983).
Challis, H., Equipment for steam savings, Proc Engng, April, 43-45 (1983).
Davies, D.D., Future markets for coal, Energy World, Aug, 4-12 (1983).
Davis, B.C., Flare efficiency studies, Plant/Opns Prog, 2(3), 191-199 (1983).
Dobis, O., Chemical cycle processes for improvement in conversion and rational use of energy, Periodica Polytechnica, 27, 205-222 (1983).
Ellingsen, W.R., Operating cost optimization using a dynamic process model, Chem Eng Prog, 79(1), 43-47 (1983).
Fish, J.D.; DeLaquil, P.; Faas, S.E., and Yang, C.L., Solar receiver steam systems, Chem Eng Prog, 79(1), 48-55 (1983).
Gerrard, A.M., Another look at forecasting energy requirements, Chem Eng, 30 May, 71-72 (1983).
Gluckman, R., Energy savings in refrigeration, Energy World, April, 8 (1983).
Hottel, H.C., The relative thermal value of tomorrow's fuels, Ind Eng Chem Fund, 22(3), 271-276 (1983).
Jackson, B., Energy conservation and investment problems, Energy World, Dec, 9-12 (1983).
Johnson, W.D., On entropy, efficiency and process design, Hyd Proc, Feb, 61-64 (1983).
Kenson, R.E., Emissions for energy conservation, Plant/Opns Prog, 2(3), 182-185 (1983).
Landauro-Paredes, J.M.; Watson, F.A., and Holland, F.A., Experimental study of the operating characteristics of a water-lithium bromide absorption cooler, CER&D, 61, 362-370 (1983).

Leatherman, H.R., Cost-effective mechanical vapor recompression, Chem Eng Prog, 79(1), 40-42 (1983).
Linnhoff, B., New concepts in thermodynamics for better chemical process design (review paper), CER&D, 61, 207-223 (1983).
Liptak, B.G., Optimizing controls for chillers and heat pumps, Chem Eng, 17 Oct, 40-51 (1983).
Marshall, D., Increasing boiler efficiency, Proc Engng, April, 39-41 (1983).
Monroe, E.S., Condensate-energy recovery systems, Chem Eng, 13 June, 117-120 (1983).
Murray, I., Waste-heat recovery equipment, Proc Engng, Jan, 41-43 (1983).
Reay, D., Heat-pump systems, Proc Engng, March, 35-37 (1983).
Rees, R., Energy pricing strategies, Energy World, July, 4-7 (1983).
Refsum, A., and Eghbali-Rashtkhari, A.R., Energy consumption and GDP, Energy World, Oct, 4-6 (1983).
Roberts, M., Key principles of energy management, Chem Engnr, Oct, 18-20 (1983).
Saez, A.E., and McCoy, B.J., Transient analysis of packed-bed thermal storage systems, Int J Heat Mass Trans, 26(1), 49-54 (1983).
Santoleri, J.J., Energy savings for process heaters, Plant/Opns Prog, 2(4), 233-238 (1983).
Schluter, L., and Schmidt, R., Energy saving in rectification, Int Chem Eng, 23(3), 427-439 (1983).
Schultz, W., and Bejan, A., Exergy conservation in parallel thermal insulation systems, Int J Heat Mass Trans, 26(3), 335-340 (1983).
Siddig-Mohammed, B.E.; Watson, F.A., and Holland, F.A., Study of the operating characteristics of a reversed absorption heat pump system (heat transformer), CER&D, 61, 283-289 (1983).
Standiford, F.C., and Weimer, L.D., Energy conservation in alcohol production, Chem Eng Prog, 79(1), 35-39 (1983).
Stephan, K., et al., An investigation of energy separation in a vortex tube, Int J Heat Mass Trans, 26(3), 341-348 (1983).
Swearingen, J.S., and Ferguson, J.E., Optimized power recovery from waste heat, Chem Eng Prog, 79(8), 66-70 (1983).
Tompeck, M.A., and Pfafflin, J.R., Statistical distribution of daytime temperatures, Energy World, Oct, 2-4 (1983).
Various, Energy-efficient process plant, Processing, March, 21-28 (1983).
Various, Modern gas-making plant and processes, Energy World, July supplement, pp.1-20 (1983).
Various, Energy management (special report), Hyd Proc, July, 57-108 (1983).
Zanker, A., Heat loss from insulated pipework, Proc Engng, Oct, 60 (1983).

1984

Aegerter, R., Energy conservation in process plants, Chem Eng, 3 Sept, 93-96 (1984).
Aird, R.J.; Bruffell, P.D., and Wooldridge, D.M., Energy consumption in a dye works, J Inst Energy, 57, 306-311 (1984).
Barr, J., Progress in energy auditing, monitoring and targeting, Energy World, Dec, 12-20 (1984).

Beasley, D.E., and Clark, J.A., Transient response of a packed bed for thermal energy storage, Int J Heat Mass Trans, 27(9), 1659-1670 (1984).
Cadogan, J.I.G., Energy and chemical technology: Problems and solutions, Chem & Ind, 5 Nov, 761-765 (1984).
Chadha, N., Use hydraulic turbines to recover energy, Chem Eng, 23 July, 57-61 (1984).
David, T., Steam trap design and operation, Proc Engng, Sept, 39 (1984).
Davis, N., Energy targeting and monitoring, Energy World, July, 5-7 (1984).
Debenedetti, P.G., Thermodynamic fundamentals of exergy, Chem Eng Educ, 18(3), 116-121 (1984).
Doane, R.C., Recovering low-level heat via expansion of natural gas, Chem Eng, 2 April, 89-91 (1984).
Essenhigh, R.H., Criteria of thermal efficiency, Chemtech, Jan, 58-63 (1984).
Fells, I., Future developments in energy supply and usage, Chem & Ind, 20 Aug, 566-567 (1984).
Fitzgerald, F., Process developments in the iron and steel industry, Energy World, Nov, 14-20 (1984).
Johnson, D.G., Selection criteria for steam turbine units in small power plants utilizing waste gas heat, J Inst Energy, 57, 332-339 (1984).
Langley, K.F., Chemical storage of energy, Chem & Ind, 20 Aug, 568-571 (1984).
Leprince, P., How conservation and substitution affect world energy demand, Hyd Proc, Aug, 98B-98M (1984).
Locke, B., Process rearrangement: Designing from new, Energy World, July, 2-5 (1984).
Molz, F.J., and Raymond, J.R., Store heat underground, Chemtech, Dec, 752-755 (1984).
More, A.I., Upgrading waste steam for process use, Proc Econ Int, 5(1), 42-43 (1984).
Nishio, M.; Koshijima, I.; Shiroko, K., and Umeda, T., A rational approach to a choice of energy conservation technologies in a total energy system, Ind Eng Chem Proc Des Dev, 23(3), 457-462 (1984).
Novak, R.G.; Troxler, W.L., and Dehnke, T.H., Recovering energy from hazardous waste incineration, Chem Eng, 19 March, 146-154 (1984).
Peate, J., Optimum boiler efficiency, Proc Engng, March, 43-46 (1984).
Pennington, R.J., Heat recovery from thermal oxidizers, Plant/Opns Prog, 3(2), 72-74 (1984).
Petroulas, T., and Reklaitis, G.V., Computer-aided synthesis and design of plant utility systems, AIChEJ, 30(1), 69-78 (1984).
Rotstein, E., Exergy change of reaction, Chem Eng Sci, 39(3), 413-418 (1984).
Ruggieri, R., Energy-saving projects, Proc Engng, March, 21-24 (1984).
Sharpe, C., Reducing steam wastage, Proc Engng, April, 47-51 (1984).
Shields, V.; Udengaard, N.B., and Berzins, V., Revamping hydrogen plants for improved yield and energy efficiency, Plant/Opns Prog, 3(2), 108-112 (1984).
Sieniutycz, S., Development of the relation between drying energy savings and thermodynamic irreversibility, Chem Eng Sci, 39(12), 1647-1660 (1984).
Storrar, A.M., A high-efficiency engine with potential for low emissions and wide fuel tolerance, J Inst Energy, 57, 266-272 (1984).

Supp, E., Convert methanol economically, Hyd Proc, July, 34C-34J (1984).
Thurlow, G.G., Case histories of some coal utilization R&D projects, Energy World, Nov, 2-13 (1984).
Various, Economic possibilities for fuel alcohol (topic issue), Chem & Ind, 18 June, 425-444 (1984).
Various, Developments in alternative-fuel propulsion, Energy World, Aug supplement, (i)-(xxviii), (1984).
Various, Developments in renewable energy resources, Energy World, Feb supplement, (i)-(xx) (1984).
Various, Better energy management (special report), Hyd Proc, July, 51-75 (1984).
Various, Energy management, conservation and retrofit (topic issue), Chem Eng Prog, 80(6), 33-62 (1984).
Wakefield, D., Recycling vent steam, Proc Engng, April, 53 (1984).

1985
Adler, D., and Ovshinsky, S.R., Amorphous photovoltaics for solar energy cells, Chemtech, Sept, 538-546 (1985).
Baker, R.E., and Shelef, M., Engine requirements for fuels and lubricants, Chemtech, Aug, 504-512 (1985).
Becker, F.E., and Zakak, A.I., Recover heat by mechanical vapor recompression, Hyd Proc, May, 77-80 (1985).
Becker, F.E., and Zakak, A.I., Recovering energy by mechanical vapor recompression, Chem Eng Prog, 81(7), 45-49 (1985).
Beevers, A., Energy-from-waste technology, Proc Engng, Oct, 51,53 (1985).
Braun, P.J.; James, S.A.; Catanese, G.A., and Shultis, W.C., Power recovery in a hydrocracker, Chem Eng Prog, 81(4), 45-47 (1985).
Brown, N.J., Local authorities' approach to energy management, Energy World, Feb, 14-17 (1985).
Campagne, W.V.L., Select gas turbines for energy efficiency, Hyd Proc, March, 77-80; April, 105-107 (1985).
Dahlstrom, D.W., Innovations in energy conservation for rotary calciners, Chem Eng Prog, 81(11), 43-47 (1985).
Daniels, L.C.; Bross, P.D., and Moyers, J.W., Try a real-time energy program, Hyd Proc, Aug, 73-76 (1985).
Eisa, M.A.R.; Devotta, S., and Holland, F.A., Study of economiser performance in a water-lithium bromide absorption cooler, Int J Heat Mass Trans, 28(12), 2323-2331 (1985).
Eisa, M.A.R.; Sane, M.G.; Devotta, S., and Holland, F.A., Experimental studies to determine the optimum flow ratio in a water-lithium bromide absorption cooler for high absorber temperatures, CER&D, 63, 267-270 (1985).
Foley, C., On-line monitoring of power station unit efficiencies, J Inst Energy, 58, 73-79 (1985).
Hampartsoumian, E.; Taylor, J.M., and Trangmar, D.T., Computer program for assessing energy efficiency of buildings, Energy World, Nov, 3-10 (1985).
Harker, J.H., and Kumar, V.G., Heat recovery by solid-gas transfer, J Inst Energy, 58, 86-93 (1985).
Harker, J.H., and Martyn, E.J., Energy storage in gravel beds, J Inst Energy, 58, 94-99 (1985).

Hauser, R.L.; McKeever, R.B., and Stull, D.P., Heat flux sensors for insulation economy and process control, Chem Eng Prog, 81(2), 24-27 (1985).
Hindmarsh, E.; Boland, D., and Townsend, D.W., Maximizing energy savings for heat engines in process plants, Chem Eng, 4 Feb, 38-47 (1985).
Holm, J., Energy recovery with turboexpander processes, Chem Eng Prog, 81(7), 63-68 (1985).
Laxton, J.R., Selecting high-temperature insulation, Proc Engng, July, 61,63 (1985).
Lehrer, I.H., Efficiency of energy conversion: Rational preliminary estimates, Ind Eng Chem Proc Des Dev, 24(3), 714-718 (1985).
Lindamood, D.M., Determine the most economical thickness for hot-pipe insulation, Chem Eng, 1 April, 96-98 (1985).
Moniotte, R.; Pouilliart, R., and Van Hecke, F., Make ammonium nitrate with net export steam, Hyd Proc, May, 109-113 (1985).
Monroe, E.S., Selecting steam traps, Chem Eng, 15 April, 73-75 (1985).
Monroe, R.C., Minimizing fan energy costs, Chem Eng, 27 May, 141-142; 24 June, 57-58 (1985).
Nishio, M.; Tanaka, H.; Shiroko, K., and Umeda, T., Optimal choice of energy conservation technologies for process systems, Chem Eng Sci, 40(8), 1539-1552 (1985).
O'Neill, P.S.; Wisz, M.W.; Ragi, E.G.; Page, E.H., and Antonelli, R., Vapor recompression systems with high efficiency components, Chem Eng Prog, 81(7), 57-62 (1985).
Peate, J., Improving boiler-plant thermal efficiency, Proc Engng, April, 49 (1985).
Randall, J.R., Power recovery for fluid catalytic cracking units, Chem Eng Prog, 81(2), 21-23 (1985).
Rao, P.R.; Shah, K.V., and Mahida, N.U., Energy conservation measures at a fertilizer company, J Inst Energy, 58, 80-85 (1985).
Reay, D.A., Heat recovery and heat pumps, Energy World, Feb, 11-14 (1985).
Skinner, A.C., Retrofitting increases energy recovery, Chem Eng Prog, 81(2), 28-29 (1985).
Stanley, G.T., and McAvoy, T.J., Dynamic energy conservation aspects of distillation control, Ind Eng Chem Fund, 24(4), 439-443 (1985).
Symsek, D.R., Use boiler feedwater for hot-water heating, Chem Eng, 19 Aug, 83 (1985).
Tye, R.P., Upgrading thermal insulation performance of industrial processes, Chem Eng Prog, 81(2), 30-34 (1985).
Various, Energy management (special report), Hyd Proc, July, 51-70 (1985).
Virr, M.J., and Williams, H.W., Heat recovery by shallow fluidized beds, Chem Eng Prog, 81(7), 50-56 (1985).
Webber, W.O., Retrofit insulation for profit, Hyd Proc, Nov, 79-80 (1985).

1986

Bhandari, V.A.; Paradis, R., and Saxena, A.C., Use performance indices for better control, Hyd Proc, 65(9), 59-61 (1986).
Bowrey, R.G.; Dang, V.B., and Sergeant, G.D., An energy model to minimize energy consumption in a low-temperature operation, steam ejector-cooling system, J Inst Energy, March, 45-48 (1986).

Bratley, J., How to tackle an $800M energy bill, Energy World, Dec, 13-15 (1986).
Doyle, J.B., Storing steam by changing water level in a deaerator, Chem Eng, 7 July, 77-78 (1986).
Elgal, G.M., Filtration with two-stage two-concentration solutions for energy conservation, Ind Eng Chem Proc Des Dev, 25(4), 872-878 (1986).
Fawkes, S., Energy management in Japan, Energy World, March, 2-4 (1986).
Foster-Pegg, R.W., Capital cost of gas-turbine heat-recovery boilers, Chem Eng, 21 July, 73-78 (1986).
Garg, D.R., and Yon, C.M., Adsorptive heat recovery drying system, Chem Eng Prog, 82(2), 54-60 (1986).
Guzdial, C.J., Energy-efficient location of blowdown, Chem Eng, 12 May, 123-124 (1986).
Holland, A., and Devotta, S., Prospects for heat pumps in process applications, Chem Engnr, May, 61-67 (1986).
Hopkins, P.D., Energy efficiency in industry: The electric answer, Energy World, Oct, 10-13 (1986).
Jones, C., and Webber, W.O., Why insulate if fuel costs drop? Hyd Proc, 65(8), 65-66 (1986).
Kelly, T.J., Energy savings from boiler blowdown, Plant/Opns Prog, 5(2), 110-115 (1986).
Kuppuraj, K., Estimate heat-recovery steam generator performance quickly, Hyd Proc, 65(7), 58-60; 65(11), 67-68 (1986).
Le Blanc, J.R., Retrofit ammonia plants to save energy and up capacity, Hyd Proc, 65(8), 39-44 (1986).
Linnhoff, B., Energy efficiency, Processing, March, 39,41 (1986).
Liss, V.M., Selecting thermal insulation, Chem Eng, 26 May, 103-105 (1986).
Magaeva, S., and Radnai, G., Evaluation of chemical exergy (utilizable energy) of multicomponent solutions of liquid nonelectrolytes by UNIFAC method, Int Chem Eng, 26(1), 78-82 (1986).
Martin, P.K., and Sweeting, J.T., Energy monitoring and targeting by microcomputer, Energy World, Nov, 12-15 (1986).
Mascone, C.F., and Short, H., Ammonia producers find new ways to cut energy, Chem Eng, 28 April, 14-17 (1986).
McAlister, D.R.; Corey, A.G., and Ziebold, S.A., Economically recovering sulfuric acid heat, Chem Eng Prog, 82(7), 34-38 (1986).
McKay, G., and Al-Duri, B., Energy recovery using flue gases from oil refinery furnaces, J Inst Energy, March, 57-60 (1986).
Nowakowski, G., Graphically determine payback/technical merit of energy-conservation projects, Chem Eng, 10 Nov, 136 (1986).
Pennink, H., Cost savings from evaporator vapor compression, Chem Eng, 6 Jan, 79-81 (1986).
Poje, J.B., and Smart, A.M., On-line energy optimization in a chemical complex, Chem Eng Prog, 82(5), 39-41 (1986).
Ramapriya, V.S., Estimate stack-gas heat loss, Chem Eng, 3 March, 129-131 (1986).
Russell, J.A.; Lyke, S.E.; Young, J.K., and Eberhardt, J.J., Redesign catalyst to save energy, Hyd Proc, 65(7), 65-68 (1986).

Saitoh, T., and Hirose, K., High-performance phase-change thermal energy storage using spherical capsules, Chem Eng Commns, 41, 39-58 (1986).
Serth, R.W., and Heenan, W.A., Gross error detection and data reconciliation in steam-metering systems, AIChEJ, 32(5), 733-742 (1986).
Strigle, R.F., and Fukuyo, K., Cut C4 recovery costs: Less reboiler duty for packed distillation columns, Hyd Proc, 65(6), 47-48 (1986).
van Ooyen, R., and Owen, D., Supplying plant energy at minimum cost, Proc Engng, June, 23,26,27,30 (1986).
Wagner, V.S., Retrofit ammonia plant steam turbines, Hyd Proc, 65(1), 67-69 (1986).
Webber, W.O., and Jones, C., Get better cost savings from enhanced insulation standards, Hyd Proc, 65(10), 59-61 (1986).

1987

Armer, A., Reducing steam costs in a flash, Chem Eng, 16 Feb, 185-188 (1987).
Bancheva, M.; Staeva, R., and Boyadzhiev, K., Estimation of thermodynamic efficiency of various chemical and physical processes, Int Chem Eng, 27(1), 121-125 (1987).
Cabano, L.J., Retrofit projects, Chem Eng Prog, 83(4), 27-31 (1987).
Cheng, C.S., and Shih, Y.S., Theoretical analysis of an absorption heat pump with continuous regeneration of working fluids by solvent extraction, CER&D, 65(5), 415-420 (1987).
Davidson, L.N., and Gullett, D.E., Gas turbine plant emissions, Chem Eng Prog, 83(3), 56-59 (1987).
Ganapathy, V., Evaluating gas-turbine heat-recovery boilers, Chem Eng, 7 Dec, 121-124 (1987).
Kemp, I.C., and Hart, D.R., Energy management: Obligation or opportunity? Chem Engnr, June, 26-29 (1987).
Klass, D.L., and Sen, C.T., Energy from waste, Chem Eng Prog, 83(7), 46-52 (1987).
Krane, R.J., A Second Law analysis of the optimum design and operation of thermal energy storage systems, Int J Heat Mass Trans, 30(1), 43-58 (1987).
Kuppuraj, K., Quickly estimate steam turbine power, Chem Eng, 7 Dec, 139-141 (1987).
Maclean, R., Energy saving techniques in the evaporation and drying of distillery effluent, Chem Engnr, July, 29-34 (1987).
Mercer, A., Fuel saving projects in the chemical industry, Proc Engng, Feb, 55-58 (1987).
Nelson, K.E., Forget about heat losses - stop wasting work, Chem Eng, 23 Nov, 143-146 (1987).
Nelson, K.E., and Cunningham, K., Checklist for reviewing energy-and-yield-saving projects, Chem Eng, 14 Sept, 115-116 (1987).
Smith, M., Understanding cooling towers: Route to energy savings, Plant/Opns Prog, 6(4), 181-184 (1987).
Strigle, R.F., and Nakano, T., Increasing efficiency in direct-contact heat transfer, Plant/Opns Prog, 6(4), 208-210 (1987).
Thompson, R.E., and King, C.J., Energy conservation in regenerated chemical absorption processes, Chem Eng & Proc, 21(3), 115-130 (1987).

Various, Energy control (special report), Hyd Proc, 66(7), 35-46 (1987).

1988

Aburas, R.; Lloyd, S., and Webster, M., Waste-heat recovery in Jordan petroleum refinery, Energy World, Nov, 6-9 (1988).

Armer, A., The right steam trap? Chem Eng, 15 Feb, 68-75 (1988).

Barton, J., Conserving steam, Chem Eng, 14 March, 159 (1988).

Cheah, C.W.; Hirt, D.E.; Liu, Y.A., and Squires, A.M., Vibrofluidized-bed heat exchanger for heat recovery from a hot gas, Powder Tech, 55(4), 257-276 (1988).

Collura, M.A., and Luyben, W.L., Energy-saving distillation designs in ethanol production, Ind Eng Chem Res, 27(9), 1686-1695 (1988).

de Silva, R., Flare-stack design, Proc Engng, Feb, 75-79 (1988).

Fitzgerald, F., Energy, high technology and economics in modern steelmaking, Energy World May, 3-12 (1988).

Fryer, T., Plant fuel selection and energy savings, Processing, Dec, 23-28 (1988).

Gauden, C.G., Air knives for energy efficiency in drying, Energy World, Dec, 7-10 (1988).

Halasz, G.; Toth, J., and Hangos, K.M., Energy-optimal operation conditions of a tunnel kiln, Comput Chem Eng, 12(2/3), 183-188 (1988).

Herbein, D.S., and Rohsenow, W.M., Comparison of entropy generation and conventional method of optimizing a gas turbine regenerator, Int J Heat Mass Trans, 31(2), 241-244 (1988).

Huang, H.M., and Govind, R., Studies on power plants using dissociating gases as working fluids, Chem Eng Commns, 72, 95-120 (1988).

Karaosmanoglu, F.; Aksoy, H.A., and Civelekoglu, H., Effects of isopropanol addition on gasoline-alcohol motor fuel blends, J Inst Energy, 61, 125-128 (1988).

Kesler, M.G., Retrofitting refinery and petrochemical plants to reduce heat losses, Chem Eng Prog, 84(6), 59-64 (1988).

Kim, W.S., and Song, H.O., Solidification heat transfer characteristics of heat storage system utilizing the PCM, Chem Eng Commns, 70, 157-170 (1988).

Koester, G.L., Improve efficiency of electric heating, Chem Eng, 15 Aug, 165-166 (1988).

Muller, C., and Flamant, G., Energy storage through magnesia sulfation in fluidized-bed reactor, AIChEJ, 34(3), 519-523 (1988).

Najjar, Y.S.H., and Othman, Fuel effect on vehicular gas-turbine engine performance, Fuel, 67(7), 994-996 (1988).

Royse, S., Steam or electric trace heating, Proc Engng, June, 57-59 (1988).

Smith, R., and Linnhoff, B., Design of separators in the context of overall processes, CER&D, 66(3), 195-228 (1988).

Thomas, A., and Hillis, D.L., Using low-grade thermal energy, Chemtech, Oct, 608-615 (1988).

Thompson, D., Biotechnology for energy conservation and a cleaner environment, Proc Engng, Dec, 39, 41 (1988).

Various, Energy conservation and integration (topic issue), Chem Eng Prog, 84(3), 35-71 (1988).

Various, Energy efficiency (supplement), Proc Engng, Nov, S1-S48 (1988).
Wolpert, V., Waste incineration for energy recovery, Processing, Oct, 37-38 (1988).

CHAPTER 4

ENVIRONMENTAL MANAGEMENT

4.1 Atmospheric Pollution

1966-1970

Weekley, G.H., and Sheehan, J.R., Jet compressors recover waste gases, Hyd Proc, 45(10), 165-170 (1966).

Specht, R.C., and Calaceto, R.R., Gaseous fluoride emissions from stationary sources, Chem Eng Prog, 63(5), 78-84 (1967).

Squires, A.M., Air pollution: Control of sulfur dioxide from power stacks, Chem Eng, 6 Nov, 260-268; 20 Nov, 133-140; 4 Dec, 188-196, 18 Dec, 101-109 (1967).

Teller, A.J., Control of gaseous fluoride emissions, Chem Eng Prog, 63(3), 75-79 (1967).

Willett, H.P., Cutting air pollution control costs, Chem Eng Prog, 63(3), 80-83 (1967).

Crocker, B.B., Minimizing air-pollution control costs, Chem Eng Prog, 64(4), 79-86 (1968).

Crocker, B.B., Opaque plumes caused by water vapor in effluent gases, Chem Eng, 15 July, 109-116 (1968).

Jaros, S., and Krizek, J., Catalytic reduction of waste nitrogen oxides, Int Chem Eng, 8(2), 261-267 (1968).

Kent, G.R., Determine radiation effect of flares, Hyd Proc, 47(6), 119-130 (1968).

Robins, D.L., and Mattia, M.M., Computer program helps design stacks for curbing air pollution, Chem Eng, 29 Jan, 119-122 (1968).

Vardi, J., and Biller, W.F., Thermal behavior of exhaust-gas catalytic convertor, Ind Eng Chem Proc Des Dev, 7(1), 83-90 (1968).

Various, Air pollution (topic issue), Chem Eng Prog, 64(9), 53-87 (1968).

Various, Air pollution techniques (topic issue), Chem Eng Prog, 64(1), 53-78 (1968).

Werner, K.D., Catalytic oxidation of industrial waste gases, Chem Eng, 4 Nov, 179-184 (1968).

Cortelyou, C.G., Commercial processes for sulfur dioxide removal, Chem Eng Prog, 65(9), 69-77 (1969).

Falkenberry, H.L., and Slack, A.V., Sulfur dioxide removal by limestone injection, Chem Eng Prog, 65(12), 61-66 (1969).

Oldenkamp, R.D., and Margolin, E.D., The molten carbonate process for sulfur oxide emissions, Chem Eng Prog, 65(11), 73-76 (1969).

Turk, A., Industrial odor control and its problems, Chem Eng, 3 Nov, 70-78 (1969).

Constance, J.D., Estimating exhaust-air requirements for processes, Chem Eng, 10 Aug, 116-118 (1970).

Ermenc, E.D., Controlling nitric oxide emission, Chem Eng, 1 June, 193-196 (1970).

Uno, T., et al., Scale-up of a sulfur dioxide control process, Chem Eng Prog, 66(1), 61-65 (1970).

Walther, J.E., and Amberg, H.R., Odor control in the Kraft pulp industry, Chem Eng Prog, 66(3), 73-80 (1970).

Yocom, J.E., and Duffee, R.A., Controlling industrial odors, Chem Eng, 15 June, 160-168 (1970).

1971-1975

Crynes, B.L., and Maddox, R.N., NOx control from combustion sources, Chemtech, Aug, 502-509 (1971).
Sem, G.J.; Borgos, J.A., and Olin, J.G., Monitoring particulate emissions, Chem Eng Prog, 67(10), 83-89 (1971).
Stairmand, C.J., The chemical engineers' contribution to air pollution control, Chem Engnr, Oct, 375-382 (1971).
Various, Nitrogen oxide pollution (topic issue), Chem Eng Prog, 67(2), 64-86 (1971).
Various, Sulfur dioxide emissions (topic issue), Chem Eng Prog, 67(5), 45-72 (1971).
Wilks, P.A., Infrared analysis of process emissions, Chem Eng, 13 Dec, 120-124 (1971).
Constance, J.D., Calculate effective stack height quickly, Chem Eng, 4 Sept, 81-83 (1972).
Haagen-Smit, A.J., The light side of smog, Chemtech, June, 330-335 (1972).
Iya, K.S., Reduce NOx in stack gases, Hyd Proc, 51(11), 163-164 (1972).
Morrow, N.L.; Brief, R.S., and Bertrand, R.R., Sampling and analyzing air pollution sources, Chem Eng, 24 Jan, 84-98 (1972).
Ross, S.R., Designing your plant for easier emission testing, Chem Eng, 26 June, 112-118 (1972).
Smith, R.S., Control stack-gas pollution, Hyd Proc, 51(9), 223-225 (1972).
Swithenbank, J., Ecological aspects of combustion devices (with reference to hydrocarbon flaring), AIChEJ, 18(3), 553-560 (1972).
Walters, R.M., Refinery air-pollution restrictions, Chem Eng Prog, 68(11), 85-88 (1972).
Bauerle, G.L., and Nobe, K., Two-stage catalytic exhaust converters: Transient operation, Ind Eng Chem Proc Des Dev, 12(4), 407-410 (1973).
Bauerle, G.L., and Nobe, K., Two-stage catalytic exhaust converter, Ind Eng Chem Proc Des Dev, 12(2), 137-141 (1973).
Butler, P., Instruments and processes for NOx removal, Proc Engng, April, 84-91 (1973).
D'Ambra, F.K., and Dobrowolski, Z.C., Pollution control for vacuum systems, Chem Eng, 25 June, 95-102 (1973).
del Grosso, R., Calculation of equilibrium composition of automotive exhaust gases, Ind Eng Chem Proc Des Dev, 12(3), 390-394 (1973).
Hales, J.M.; Wolf, M.A., and Dana, M.T., Linear model for predicting washout of pollutant gases from industrial plumes, AIChEJ, 19(2), 292-297 (1973).
Kyan, C.P., and Seinfeld, J.H., Real-time control of air pollution, AIChEJ, 19(3), 579-589 (1973).
McLain, L., Ozonation for effluent treatment, Proc Engng, Feb, 104-106 (1973).
Nichols, R.A., Hydrocarbon-vapor recovery, Chem Eng, 5 March, 85-92 (1973).
Querido, R., and Short, W.L., Removal of sulfur dioxide from stack gases by catalytic reduction to elemental sulfur with carbon monoxide, Ind Eng Chem Proc Des Dev, 12(1), 10-18 (1973).
Various, Emissions to the air (symposium papers), Chem Engnr, March, 121-150; April, 197-207 (1973).
Bierbower, R.G., and Van Sciver, J.H., Allied's sulfur dioxide reduction system, Chem Eng Prog, 70(8), 60-62 (1974).

Crocker, B.B., Monitoring plant air pollution, Chem Eng Prog, 70(1), 41-49 (1974).
Ferguson, N.B., and Finlayson, B.A., Transient modeling of a catalytic converter to reduce nitric oxide in automobile exhaust, AIChEJ, 20(3), 539-550 (1974).
Homolya, J.B., Continuous gas monitors connected to emission sources, Chemtech, July, 426-433 (1974).
Various, Factors affecting pesticide distribution throughout the atmosphere (topic issue), Chem & Ind, 2 March, 179-201 (1974).
Various, Odour and fume control (topic issue), Chem & Ind, 2 Nov, 846-861; 16 Nov, 895-908 (1974).
Various, Air pollution control techniques (topic issue), Chem Eng Prog, 70(5), 43-77 (1974).
Various, Sulfur dioxide and coal processing (topic issue), Chem Eng Prog, 70(6), 45-82 (1974).
Verhoff, F.H., and Banchero, J.T., Predicting dew points of flue gases, Chem Eng Prog, 70(8), 71-72 (1974).
Clay, D.T., and Lynn, S., Reduction and removal of sulfur dioxide and NOx from simulated flue gas using iron oxide as catalyst/adsorbent, AIChEJ, 21(3), 466-473 (1975).
Field, A.A., Chimneys for industrial boiler plant, Energy World, March, 7-9 (1975).
Kasthuri, R., and Reither, K., Reduction of gaseous emissions, Chem Engnr, Dec, 745-747 (1975).
Klooster, H.J.; Vogt, G.A., and Braun, G.F., Optimizing the design of relief and flare systems, Chem Eng Prog, 71(1), 39-44 (1975).
Lees, B., and Butcher, R.W., Reducing atmospheric pollution from oil-fired plant: Development of grit-arresting cyclones for multi-flue chimneys, J Inst Fuel, 48, 201-207 (1975).
Mahajan, K.K., Tall stack design simplified, Hyd Proc, 54(9), 217-220 (1975).
Mays, E.B., and Schwab, M.R., Elimination of NOx fumes, Chem Eng, 17 Feb, 112 (1975).
Various, Dust and fume control (feature report), Processing, Dec, 49-63 (1975).

1976-1980

Anderton, D., and Duggal, V.K., Diesel engine emissions and noise, J Inst Fuel, 49, 20-24 (1976).
Chapman, B., Stack monitoring, Proc Engng, Sept, 71-73 (1976).
Cooke, B., NOx and SOx emission control, Proc Engng, July, 62-63 (1976).
Hughes, R., Removal of contaminants from industrial gaseous effluents, Chem Engnr, Nov, 754-756 (1976).
Klooster, H.J.; Vogt, G.A., and Bernhart, D.G., Refinery odor control, Hyd Proc, 55(4), 121-123 (1976).
Sisson, W., Calculating the fly ash discharged to a stack, Chem Eng, 12 April, 154-156 (1976).
Tipton, D.F., Particle analyzer for stack emissions, Powder Tech, 14, 245-252 (1976).
Various, Measurements of particulate pollutants in gas streams (topic issue), Chem & Ind, 7 Aug, 621-640 (1976).

Various, Emission control (topic issue), Chem Eng Prog, 72(12), 33-57 (1976).
Yamaguchi, M.; Matsushita, K., and Takami, K., Remove NOx from nitric acid tail gas, Hyd Proc, 55(8), 101-106 (1976).
Harwood, C.F.; Siebert, P.C., and Oestreich, D.K., Optimizing baghouse performance to control asbestos emissions, Chem Eng Prog, 73(1), 54-56 (1977).
Klasens, H.A., Analyze stack gases via sampling or optically, in place, Chem Eng, 21 Nov, 201-205 (1977).
Tarbell, J.M., and Petty, C.A., Combustion modifications for control of NOx emissions, Chem Eng Sci, 32(10), 1177-1188 (1977).
Various, Control of gaseous emissions to the atmosphere (topic issue), Chem & Ind, 15 Jan, 50-64 (1977).
Various, Air pollution control (topic issue), Chem Eng Prog, 73(8), 31-73 (1977).
Various, Measuring pollutants in stack gases (feature report), Chem Eng, 31 Jan, 90-97 (1977).
Brough, A.; Parry, M.A., and Whittingham, C.P., Influence of aerial pollution on crop growth, Chem & Ind, 21 Jan, 51-53 (1978).
Pereira, C.J., and Varma, A., Uniqueness criteria of steady state in automotive catalysis, Chem Eng Sci, 33(12), 1645-1658 (1978).
Mogan, J.P.; Stewart, D.B., and Dainty, E.D., Oxidation of nitric oxide fraction of diluted diesel exhaust, Can JCE, 57, 378-388 (1979).
Wallace, M.J., Controlling fugitive emissions, Chem Eng, 27 Aug, 78-92 (1979).
England, G.C.; Heap, M.P., and Pershing, D.W., Control of NOx emissions, Hyd Proc, 59(1), 167-171 (1980).
Hegedus, L.L., and Gumbleton, J.J., Catalysts for automobile emissions, Chemtech, Oct, 630-642 (1980).
Kenson, R.E., and Hoffland, R.O., Control of toxic air emissions in chemical manufacture, Chem Eng Prog, 76(2), 80-83 (1980).
Macey, L.J., and Chandler, P., The low-temperature approach to airborne pollution control, Chem Engnr, April, 216-218, 224 (1980).
Marzo, L., and Fernandez, L., Destroy NOx catalytically, Hyd Proc, 59(2), 87-89 (1980).
Mitchell, D.V., System for treating furnace gases, Proc Engng, Sept, 99-103 (1980).
Murthy, K.S., Environmental assessment of fluidized-bed combustion, Chemtech, Jan, 58-63 (1980).
Scriven, R.A., Assessment of the global carbon dioxide problem and the greenhouse effect, J Inst Energy, 53, 15-16 (1980).
Various, Removal of gaseous pollutants (topic issue), Chem Eng Sci, 35(1), 145-194 (1980).
Wheeler, W.H., Chemical and engineering aspects of low NOx concentration, Chem Engnr, Nov, 693-699 (1980).

1981-1984

Duggan, M., Removing lead from petrol, Energy World, Feb, 10-11 (1981).
Ember, L.R., Acid pollutants, C&E News, 14 Sept, 20-31 (1981).
Ganapathy, V., Conversion of NOx units, Proc Engng, May, 81 (1981).

Gill, D.W., The potential of fluidised-bed combustion for emission control, Chem Engnr, June, 278-280 (1981).
Ireland, F.E., Emissions to air: UK standards for control, Chem Engnr, June, 281-284 (1981).
Sadakata, M., et al., Development of a low-NOx combustion system using a new type of continuous catalytic reactor and heat exchanger, Int Chem Eng, 21(2), 303-311 (1981).
Blackwood, T.R., An overview of fugitive particle emissions, Plant/Opns Prog, 1(4), 263-271 (1982).
Chameides, W.L., and Davis, D.D., Chemistry in the troposphere, C&E News, 4 Oct, 38-52 (1982).
Dayal, P., Recording ambient-air-quality monitoring data, Chem Eng, 25 Jan, 101-109 (1982).
Fumarola, G.; DeFaveri, D.M.; Palazzi, E., and Ferraiolo, G., Determine plume rise for elevated flares, Hyd Proc, Jan, 165-166 (1982).
Ganapathy, V., Figure particulate-emission rate quickly, Chem Eng, 26 July, 82 (1982).
Herfkens, A.H., Experience with tail gas treating, Hyd Proc, Nov, 199-203 (1982).
Horn, F.L., and Steinberg, M., Control of carbon dioxide emissions from a power plant (and use in enhanced oil recovery), Fuel, 61(5), 415-422 (1982).
Roland, L., Tail-gas scrubbing, Proc Engng, May, 55-57 (1982).
Benyon, A., Monitoring of smut emission from oil-fired boilers using microwave Doppler radar, J Inst Energy, 56, 45-51 (1983).
Burkinshaw, J.R., et al., Analysis of sulfur and nitrogen pollutants in three-phase coal combustion effluent samples, Ind Eng Chem Fund, 22(3), 292-298 (1983).
Capes, P., The cost of acid rain control, Proc Engng, May, 29-32 (1983).
Crane, A.J., Update on the carbon dioxide-climate problem, Energy World, Nov, 2-4 (1983).
Romberg, G., Gas-turbine combuster modelling for calculating pollutant emission, Int J Heat Mass Trans, 26(2), 197-210 (1983).
Anderson, A.P.; Foster, P.J., and Hedley, A.B., Measurement and characteristics of carbon monoxide emission from a small boiler at start-up, J Inst Energy, 57, 348-351 (1984).
Ember, L.R., Yellow rain, C&E News, 9 Jan, 8-34 (1984).
Fisher, B.E.A., Origin of acid rain, J Inst Energy, 57, 416-420 (1984).
Harbert, F., Detection of fugitive emissions using laser beams, Chem Engnr, Oct, 41-43 (1984).
Innes, W.B., It's raining nitrates and sulfates, Chemtech, July, 440-447 (1984).
Jones, A.L., Fugitive emissions of volatile hydrocarbons, Chem Engnr, Aug, 12-15 (1984).
Lees, B., Problems and policies of acid rain, Energy World, Aug, 2-5 (1984).
Redman, J., Acid rain, Chem Engnr, Oct, 10, 11 (1984).
Siddiqi, A.A., and Tenini, J.W., Coping with air pollution, Hyd Proc, Nov, 123-129 (1984).

1985-1988

Bergsma, F., Abatement of NOx from coal combustion (a review), Ind Eng Chem Proc Des Dev, 24(1), 1-7 (1985).

Buelt, J.L.; Fitzpatrick, V.F., and Timmerman, C.L., Electrical technique for in-place stabilization of contaminated soils, Chem Eng Prog, 81(3), 43-48 (1985).

Cooney, D.O., Modeling venturi scrubber performance for hydrogen sulfide removal from oil-shale retort gases, Chem Eng Commns, 35(1), 315-338 (1985).

Frank, N.W.; Kawamura, K., and Miller, G.A., Sulfur dioxide/NOx recovery and reuse, Chem Eng Prog, 81(6), 30-36 (1985).

Lachapelle, D.G., EPA SOx control program, Chem Eng Prog, 81(5), 56-62 (1985).

Lisauskas, R.A.; Itse, D.C.; Masser, C.C., and Abele, A.R., A prototype burner for NOx/SO2 control, Chem Eng Prog, 81(11), 51-56 (1985).

O'Sullivan, D.A., Acid rain in Europe, C&E News, 28 Jan, 12-18 (1985).

Beeley, G., On a clear day you can see for ever, Chem Engnr, March, 11 (1986).

Brauer, H., Biological purification of waste gases, Int Chem Eng, 26(3), 387-396 (1986).

Sheppard, S.V., Ionizing wet scrubber for air pollution control, Chem Eng Prog, 82(2), 40-43 (1986).

Various, The changing atmosphere (special report), C&E News, 24 Nov, 14-64 (1986).

Chynoweth, E., Laser-based techniques for emission control, Processing, Nov, 46-47 (1987).

Davidson, L.N., and Gullett, D.E., Gas turbine plant emissions, Chem Eng Prog, 83(3), 56-59 (1987).

Heisel, M.P., and Marold, F.J., CLINTOX Claus tailgas treatment, Gas Sepn & Purif, 1(2), 107-109 (1987).

Macgregor, S.A.; Syred, N., and Claypole, T.C., Minimisation of NOx emission by flame temperature control, Chem Eng Commns, 52, 163-172 (1987).

Zurer, P.S., The Antarctic ozone hole, C&E News, 2 Nov, 22-26 (1987).

Allen, J.K., Control of industrial emissions to air, Chem & Ind, 19 Dec, 777-781 (1988).

Barnes, F.J., et al., NOx emissions from radiant gas burners, J Inst Energy, Dec, 184-188 (1988).

Boardman, R.D., and Smoot, L.D., Prediction of nitric oxide in advanced combustion systems, AIChEJ, 34(9), 1573-1576 (1988).

Bohn, H., and Bohn, R., Soil beds remove air pollutants, Chem Eng, 25 April, 73-76 (1988).

Capper, A.J.; Davies, C., and Rees, G.J., Atmospheric pollution from coke ovens, J Inst Energy, Dec, 209-219 (1988).

Forrester, J.S., and Le Blanc, J.G., Vapor condensation to reduce emissions, Chem Eng, 23 May, 145-150 (1988).

Iborra, M.; Izquierdo, J.F.; Tejero, J., and Cunill, F., Getting the lead out of petrol with ethyl t-butyl ether, Chemtech, Feb, 120-122 (1988).

MacKerron, C.B., How to attain clean air, Chem Eng, 9 May, 35-39 (1988).

Oman, H., Controlling carbon dioxide buildup in the atmosphere, Chemtech, Feb, 116-119 (1988).

Royse, S., Alternatives to CFCs, Proc Engng, July, 33-34 (1988).
Royse. S., Biofilter removes volatile organic compounds from waste gases, Proc Engng, Sept, 65-69 (1988).
Rubin, E.S.; Salmento, J.S., and Frey, H.C., Cost-effective emission controls for coal-fired power plants, Chem Eng Commns, 74, 155-168 (1988).
Ruf, A., and Egli, S., Application of gas separation shown through purification of landfill gas, Gas Sepn & Purif, 2(2), 90-94 (1988).
Thompson, V.L., and Greenkorn, R.A., Non-Gaussian dispersion in model smokestack plumes, AIChEJ, 34(2), 223-228 (1988).
Thonchk, N.K., et al., Extraction of thiocyanate ions from coal gasification effluents by ion exchange, CER&D, 66(6), 503-517 (1988).
Zurer, P.S., Atmospheric ozone depletion, C&E News, 30 May, 16-25 (1988).
Zurer, P.S., Search for CFC alternatives, C&E News, 8 Feb, 17-20 (1988).

4.2 Wastewater Treatment

1967-1974

Inczedy, J., and Frankow, T., Contributions to the ion exchange treatment of waste waters containing cyanide, Periodica Polytechnica, 11, 53-60 (1967).
McIlhenny, W.F., Recovery of additional water from industrial wastewater, Chem Eng Prog, 63(6), 76-81 (1967).
Cecil, L.K., Water reuse and disposal, Chem Eng, 5 May, 92-104 (1969).
Gould, M., and Taylor, J., Temporary water clarification system, Chem Eng Prog, 65(12), 47-49 (1969).
Keith, F.W., and Little, T.H., Centrifuges in water and waste treatment, Chem Eng Prog, 65(11), 77-80 (1969).
Klein, L.A., Continuous sampling and flow measurement of wastewater, Chem Eng, 16 June, 116 (1969).
Trumbull, L.E., Method detects foreign waters in regular supply, Chem Eng, 8 Sept, 164 (1969).
Various, Water and waste treatment (topic issue), Chem Eng Prog, 65(6), 59-82 (1969).
Deeson, A.F.L., Water pollution and the chemical industry, Brit Chem Eng, 15(9), 1141-1146 (1970).
Fitzgerald, C.L.; Clemens, M.M., and Reilly, P.B., Coagulants for wastewater treatment, Chem Eng Prog, 66(1), 36-40 (1970).
Hiser, L.L., Selecting a wastewater treatment process, Chem Eng, 30 Nov, 76-80 (1970).
Various, Wastewater treatment and re-use (symposium papers), Chem Engnr, Jan, CE6-29 (1970).
Anon., Direct oxygenation of wastewater, Chem Eng, 29 Nov, 66-67 (1971).
Beychok, M.R., A review of wastewater treatment, Hyd Proc, 50(12), 109-112 (1971).
Mootz, E., Water-oil separator, Chem Eng, 6 Sept, 102-104 (1971).
Pitter, P., Possibility of biological decomposition of some hydroaromatic compounds, Int Chem Eng, 11(1), 19-25 (1971).
Carnes, W.A., Water-pollution control, Chem Eng, 11 Dec, 97-104 (1972).
Moores, C.W., Wastewater treatment, Chem Eng, 25 Dec, 63-66 (1972).

Nogaj, R.J., Selecting wastewater aeration equipment, Chem Eng, 17 April, 95-102 (1972).
Roberts, F.W., Pollution by warm water effluents, J Inst Fuel, 45, 558-561 (1972).
Sawyer, G.A., New trends in wastewater treatment and recycle, Chem Eng, 24 July, 120-128 (1972).
Butler, P., Chromate recovery from cooling water, Proc Engng, Jan, 74-75 (1973).
Levin, G.V.; Topol, G.J., and Tarnay, A.G., Biological removal of phosphates from wastewater, Chemtech, Dec, 739-744 (1973).
McGovern, J.G., Inplant wastewater control, Chem Eng, 14 May, 137-139 (1973).
Rabosky, J.G., and Koraido, D.L., Gaging and sampling industrial wastewaters, Chem Eng, 8 Jan, 111-120 (1973).
Various, Activated carbon for wastewater treatment (topic issue), Chem Eng Prog, 69(11), 45-69 (1973).
Various, Wastewater treatment (topic issue), Chem Eng Prog, 69(8), 71-84 (1973).
Bennett, J.E., Non-diaphragm electrolytic hypochlorite generators for water treatment, Chem Eng Prog, 70(12), 60-63 (1974).
Cadman, T.W., and Dellinger, R.W., Techniques for removing metals from process wastewater, Chem Eng, 15 April, 79-85 (1974).
Hawkes, H.A., Biological considerations of water quality, Chem & Ind, 21 Dec, 990-1000 (1974).
Krasnov, B.P.,; Paul, D.L., and Kirillova, T.V., Use of ozone for treatment of industrial wastewaters, Int Chem Eng, 14(4), 747-750 (1974).
Kuipers, E., Continuous sampler for plant waste waters, Chem Eng, 13 May, 128-130 (1974).
Pinto, A.P., Bypassing problems from the lime-clarification of wastewater, Chem Eng, 8 July, 111 (1974).
Various, Solids separation from industrial waters and effluents (topic issue), Chem & Ind, 19 Jan, 50-61 (1974).
Various, Treatment of aqueous effluents (topic issue), Chem Engnr, Feb, 91-107 (1974).

1975-1979

Milios, P., Water reuse at a coal gasification plant, Chem Eng Prog, 71(6), 99-104 (1975).
Miranda, J.G., Sump design for oil/water separators, Chem Eng, 24 Nov, 85 (1975).
Stevens, B.W., and Kerner, J.W., Recovering organic materials from wastewater, Chem Eng, 3 Feb, 84-87 (1975).
Thorsen, T., and Oen, R., How to measure industrial wastewater flow, Chem Eng, 17 Feb, 95-100 (1975).
Various, Developments in the treatment of metal-bearing effluents (topic issue), Chem & Ind, 2 Aug, 632-645 (1975).
Braunscheidel, D.E., and Gyger, R.G., UNOX system for wastewater treatment, Chem Eng Prog, 72(11), 71-72 (1976).

Bush, K.E., Refinery wastewater treatment and reuse, Chem Eng, 12 April, 113-118 (1976).
Chambers, D.B., and Cottrell, W.R.T., Flotation: Two new ways to treat effluent, Chem Eng, 2 Aug, 95-98 (1976).
Cummings-Saxton, J., Chemical-industry costs of water pollution abatement, Chem Eng, 8 Nov, 106-113 (1976).
Hebbel, G., Wastewater incineration, Int Chem Eng, 16(4), 603-614 (1976).
Redey, L.; Kovacs, I., and Zorkoczy, I., Electrochemical treatment of industrial waste solutions and sewage, Int Chem Eng, 16(1), 30-37 (1976).
Richter, H., Flotation process for wastewater treatment, Int Chem Eng, 16(4), 614-620 (1976).
Speece, R.E.; Siddiqi, R.H., and Aubert, R.P., Oxygen for wastewater treatment, Chem Eng, 15 March, 110 (1976).
Various, Waste treatment advances (topic issue), Chem Eng Prog, 72(10), 51-83 (1976).
Wiley, M.A., Water pollution control, Chemtech, Feb, 134-141 (1976).
Wilhelmi, A.R., and Ely, R.B., Two-step process for toxic wastewaters, Chem Eng, 16 Feb, 105-109 (1976).
Finelt, S., and Crump, J.R., Predict wastewater generation, Hyd Proc, 56(8), 159-166 (1977).
Kohn, P.M., Water treatment system cuts organics, Chem Eng, 15 Aug, 108-109 (1977).
Othmer, D.F., Oxygenation of aqueous wastes: The PROST system, Chem Eng, 20 June, 117-120 (1977).
Rizzo, J.L., and Shepherd, A.R., Treating industrial wastewater with activated carbon, Chem Eng, 3 Jan, 95-100 (1977).
Teale, J.M., In-plant recovery of spent solvents, Chem Eng, 31 Jan, 98-100 (1977).
Various, A decade of water-pollution control (feature report), Chem Eng, 15 Aug, 124-151 (1977).
Arden, T.V., and Forrest, R.D., Water treatment for industrial use, Chem Engnr, Dec, 919-922 (1978).
Bellew, E.F., Comparing chemical precipitation methods for water treatment, Chem Eng, 13 March, 85-91 (1978).
Bush, M.J., and Silveston, P.L., Computer simulation and optimal synthesis of wastewater treatment plants, Comput Chem Eng, 2(4), 143-160 (1978).
Creason, S.C., Selection and care of electrodes for effluent-stream pH analyzer, Chem Eng, 23 Oct, 161-163 (1978).
Howe, R.H.L., Aerobic-oxidation capability factor, Chem Eng, 31 July, 111-112 (1978).
Kovalcik, R.N., Single waste-treatment vessel both flocculates and clarifies, Chem Eng, 19 June, 117-120 (1978).
Lock, J., Cost-effective wastewater treatment, Processing, June, 37 (1978).
Nathan, M.F., Choosing a process for chloride removal from wastewater, Chem Eng, 30 Jan, 93-100 (1978).
Pallanich, P.J., Pure oxygen treatment of pesticide plant waste water, Chem Eng Prog, 74(4), 79-84 (1978).
Various, Separation of oil and water in oily wastes (topic issue), Chem & Ind, 4 Nov, 821-837 (1978).

Wang, S.T.; McMillan, A.F., and Chen, B.H., Effect of boundary conditions on pollutant dispersion in two-dimensional rivers, Can JCE, 56, 263-270 (1978).

White, M.J.D., Chemical engineering in wastewater treatment, Chem Engnr, Dec, 913-915, 918 (1978).

Wolfbauer, O.; Klettner, H., and Moser, F., Reaction engineering models of biological wastewater treatment and kinetics of activated sludge process, Chem Eng Sci, 33(7), 953-960 (1978).

Muratova, N.G., et al., Wastewater treatment by reverse osmosis, Int Chem Eng, 19(2), 350-352 (1979).

Various, Water: Environmental issues (topic issue), Chem & Ind, 17 March, 183-204 (1979).

1980-1984

Diesterweg, G., and Pascik, I., Treat waste water biologically, Hyd Proc, 59(11), 191-195 (1980).

Nishitani, H.; Ogaki, Y., and Kunugita, E., Multi-objective optimization of an aeration vessel for wastewater treatment, Chem Eng Commns, 5(1), 135-148 (1980).

Rowley, W.J., and Otto, F.D., Ozonation of cyanide with emphasis on gold mill wastewaters, Can JCE, 58, 646-653 (1980).

Russell, D.L., Monitoring and sampling liquid effluents, Chem Eng, 20 Oct, 108-120 (1980).

Various, Telemetry in water pollution control, Chem & Ind, 2 Aug, 610-623 (1980).

Ayling, G.W., and Castrantas, H.M., Waste treatment with hydrogen peroxide, Chem Eng, 30 Nov, 79-82 (1981).

Backhurst, J.R., and Matis, K.A., Electrolytic flotation in effluent treatment, J Chem Tech Biotechnol, 31, 431-434 (1981).

Bull, M.A.; Sterritt, R.M., and Lester, J.N., Treatment methods for dairy-industry wastewater, J Chem Tech Biotechnol, 31, 579-583 (1981).

Cooper, P.F., The use of biological fluidised beds for treatment of domestic and industrial wastewaters, Chem Engnr, Aug, 373-376 (1981).

Holiday, A.D., and Hardin, D.P., Activated carbon removes pesticides from wastewater, Chem Eng, 23 March, 88-89 (1981).

Howe, R.H.L.; Howe, R.C., and Howe, J.M., The complications of BOD tests, Chem Eng, 30 Nov, 99-100 (1981).

McKay, G., Design models for adsorption systems in wastewater treatment, J Chem Tech Biotechnol, 31, 717-731 (1981).

Aytimur, T.; Chen, B.H., and McMillan, A.F., Time dependence in a three-dimensional model of thermal pollution in rivers, Can JCE, 60, 699-703 (1982).

Bull, M.A.; Sterritt, R.M., and Lester, J.N., The effect of organic loading on the performance of anaerobic fluidised beds treating high strength wastewaters, Trans IChemE, 60, 373-376 (1982).

Kyuchoukov, G.; Hadjiev, D., and Boyadzhiev, L., Removal of heavy metal ions from industrial waste waters, Chem Eng Commns, 17(1), 219-226 (1982).

Olthof, M., and Oleszkiewicz, J., Anaerobic treatment of industrial wastewaters, Chem Eng, 15 Nov, 121-126 (1982).

Pedram, E.O.; Hines, A.L., and Cooney, D.O., Kinetics of adsorption of organics from an above-ground oil-shale retort water, Chem Eng Commns, 19(1), 167-176 (1982).

Various, Water treatment report, Processing, Feb, 13-19 (1982).

Wilson, D.J., et al., Removal of refractory organics from water by aeration, Sepn Sci Technol, 16(8), 907-936 (1981); 17(7), 897-924; 17(12), 1387-1396 (1982).

Busch, A.W., A practical approach to designing aeration systems, Chem Eng, 10 Jan, 76-84 (1983).

Langer, B.S., Wastewater reuse and recycle in petroleum refineries, Chem Eng Prog, 79(5), 67-76 (1983).

Bull, M.A.; Sterritt, R.M., and Lester, J.N., Developments in anaerobic treatment of high strength industrial wastewaters (review paper), CER&D, 62, 203-213 (1984).

Cotton, C., Use of water-treatment chemicals, Chemtech, June, 345-347 (1984).

Groves, F.R., Amine removal from waste water by ligand exchange, Chem Eng Commns, 31(1), 209-222 (1984).

Ibusuki, T., and Aneja, V.P., Mass transfer of ammonia into water at environmental concentrations, Chem Eng Sci, 39(7), 1143-1156 (1984).

1985-1988

Cloutier, J.N.; Leduy, A., and Ramalho, R.S., Peat adsorption of herbicide 2,4-D from wastewaters, Can JCE, 63(2), 250-257 (1985).

Duffy, G.J., and Kable, J.W., Comparative environmental impact of methods for disposal of coal washery tailings with particular reference to fluidized-bed combustion, J Inst Energy, 58, 31-48 (1985).

Eckenfelder, W.W.; Patoczka, J., and Watkin, A.T., Wastewater treatment, Chem Eng, 2 Sept, 60-74 (1985).

Johnson, G., Improved biological wastewater treatment, Proc Engng, Nov, 41-45 (1985).

Lie, L.X., Wastewater treating at Lanzhou refinery, Hyd Proc, June, 78-79 (1985).

Maeda, S., et al., Bioaccumulation of arsenic by freshwater algae and application to removal of inorganic arsenic from an aqueous phase, Sepn Sci Technol, 18(4), 375-386 (1983); 20(2), 153-162 (1985).

Mosey, F., Redox potentials in wastewater treatment, Chem Engnr, May, 21-24 (1985).

Naundorf, E.A., et al., Biological treatment of wastewater in compact reactor, Chem Eng & Proc, 19(5), 229-234 (1985).

Rockey, J.S., and Forster, C.F., Use of anaerobic expanded bed reactor for treatment of strong confectionary wastewater, CER&D, 63, 300-304 (1985).

Severin, B.F., and Suidan, M.T., Ultraviolet disinfection for municipal wastewater, Chem Eng Prog, 81(4), 37-44 (1985).

Tran, T.V., Advanced membrane filtration process treats industrial wastewater efficiently, Chem Eng Prog, 81(3), 29-33 (1985).

Forster, C.F.; Boyes, A.P.; Hay, B.A., and Butt, J.A., An aerobic fluidised bed reactor for wastewater treatment, CER&D, 64(6), 425-430 (1986).

Nonaka, M., Wastewater treatment system applying aeration-cavitation flotation mechanism, Sepn Sci Technol, 21(5), 457-474 (1986).

Wilson, D.J., et al., Removal of refractory organics by aeration, Sepn Sci Technol, 18(10), 941-968 (1983); 19(13), 1013-1024 (1984); 21(1), 57-78 (1986).

Anon., Progress in water treatment linked to environmental needs (technology focus), Chem Eng, 13 April, 33-42 (1987).

Armitage, S.J., and Nickrand, W.J., Statistical process control for water-treatment systems, Chem Eng, 25 May, 79-84 (1987).

Baker-Counsell, J., An integrated process water treatment system, Proc Engng, May, 55-57 (1987).

Basta, N., Better biological processes boost wastewater treatment, Chem Eng, 27 April, 14-15 (1987).

Kreevoy, M.M.; Kotchevar, A.T., and Aften, C.W., Decontamination of nitrate polluted water, Sepn Sci Technol, 22(2), 361-372 (1987).

Prasad, G., The application of solar energy in water reuse, J Chem Tech Biotechnol, 39(1), 29-36 (1987).

Proctor, A., Shock loadings in biological wastewater treatment, Proc Engng, Oct, 55-59 (1987).

Sales, D.; Valcarel, M.J.; Perez, L., and Ossa, E.M., Activated sludge treatment of wine-distillery wastewaters, J Chem Tech Biotechnol, 40(2), 85-100 (1987).

Solt, G., Removing nitrate from potable water, Chem Engnr, May, 33-36 (1987).

Various, Anaerobic water-treatment processes, Processing, June, 19-25 (1987).

Boaventura, R.A., and Rodrigues, A.E., Consecutive reactions in fluidized-bed biological reactors: Modeling and experimental study of wastewater denitrification, Chem Eng Sci, 43(10), 2715-2728 (1988).

Chen, S.J.; Li, C.T., and Shieh, W.K., Anaerobic fluidized-bed treatment of tannery wastewater, CER&D, 66(6), 518-523 (1988).

Huang, S.D., et al., Simultaneous removal of heavy metal ions from wastewater by foam separation techniques, Sepn Sci Technol, 23(4), 489-506 (1988).

Lankford, P.W.; Eckenfelder, W.W., and Torrens, K.D., Reducing wastewater toxicity, Chem Eng, 7 Nov, 72-82 (1988).

McKay, G., and McAleavey, G., Ozonation and carbon adsorption in three-phase fluidized bed for colour removal from peat water, CER&D, 66(6), 531-536 (1988).

Royse, S., Developments in liquid effluent processing, Proc Engng, Nov, 65-70 (1988).

Tipping, E., Colloids in the aqueous environment, Chem & Ind, 1 Aug, 485-490 (1988).

4.3 Activated Sludge Processes

1972-1979

Glinicki, Z.; Joncyk, E.; Sterninski, A., and Szymanska, Z., Biological treatment of wastes from pharmaceutical industry by activated sludge method, Int Chem Eng, 12(4), 644-649 (1972).

Horvath, I., Scale-up conditions in organic sludge sewage purification system, Int Chem Eng, 12(4), 606-616 (1972).

Pinto, A., How to measure a sludge blanket, Chem Eng, 30 Oct, 138 (1972).
McHarg, W.H., Designing the optimum system for biological-waste treatment, Chem Eng, 24 Dec, 46-49 (1973).
Butler, P., Developments in sewage and effluent treatment, Proc Engng, June, 85-87 (1975).
Thomson, S.J., Designing activated sludge units, Hyd Proc, 54(8), 99-102 (1975).
Harbold, H.S., How to control biological-waste-treatment processes, Chem Eng, 6 Dec, 157-160 (1976).
Ousby, J.C.; Walker, J., and Jones, R.T., Deep-shaft process treats domestic sewage, Proc Engng, Sept, 81-84 (1977).
Attir, U., and Denn, M.M., Dynamics and control of activated sludge wastewater process, AIChEJ, 24(4), 693-698 (1978).
Barber, N.R., and Dale, C.W., Increasing sludge-digester efficiency, Chem Eng, 17 July, 147-149 (1978).
Ramalho, R.S., Principles of activated sludge treatment, Hyd Proc, 57(10), 112-118; 57(11), 275-280; 57(12), 147-152 (1978).
Thibault, G.T., and Tracy, K.D., Controlling and monitoring activated-sludge units, Chem Eng, 11 Sept, 155-160 (1978).
Kucnerowicz, F., and Verstraete, W., Direct measurement of microbial ATP in activated sludge samples, J Chem Tech Biotechnol, 29, 707-712 (1979).
Versino, C.; Sarzanini, C.; Gigante, R., and Kodram, F., Removal of cupric ions by activated sludge: Kinetics, isotherms, and yields, Sepn Sci Technol, 14(10), 909-922 (1979).

1980-1988

Dixon, D.C., Effect of sludge funneling in gravity thickeners, AIChEJ, 26(3), 471-477 (1980).
Matsumoto, K.; Suganuma, A., and Kunii, D., Effect of cationic polymer on settling characteristics of activated sludge, Powder Tech, 25, 1-10 (1980).
Ramalho, R.S., Design of aerobic treatment units, Hyd Proc, 58(11), 285-292 (1979); 59(1), 159-164 (1980).
Scott, D.S.; Horlings, H., and Soupilas, A., Extraction of metals from sewage sludge, Can JCE, 58, 673-678 (1980).
Sherrard, J.H., Activated sludge wastewater treatment: Stoichiometric relationships, J Chem Tech Biotechnol, 30, 447-452 (1980).
Vilker, V.L.; Kamdar, R.S., and Frommhagen, L.H., Capacity of activated sludge solids for virus adsorption, Chem Eng Commns, 4(4), 569-576 (1980).
Lin, J.L., and Maa, J.R., Experimental study of design parameters of sludge blanket clarifier for water purification, Ind Eng Chem Proc Des Dev, 20(3), 456-459 (1981).
Carlson, C.H., Optimised aerobic and thermophilic treatment of municipal sewage sludges and night soils in a continuous operation, J Chem Tech Biotechnol, 32, 1010-1015 (1982).
Davis, N., Trends in sewage treatment, Proc Engng, May, 67-69 (1982).
Forster, C.F., Sludge surfaces and their relation to the rheology of sewage sludge suspensions, J Chem Tech Biotechnol, 32, 799-807 (1982).

Jones, G.L., and Paskins, A.R., Influence of high partial pressure of carbon dioxide and/or oxygen on nitrification, J Chem Tech Biotechnol, 32, 213-223 (1982).
Warden, J.H., Polymer treatment of waterworks coagulant sludges, Chem Engnr, Dec, 460-463 (1982).
Clarke, A.R., and Forster, C.F., Significance of ATP in settlement of activated sludge, J Chem Tech Biotechnol, 33B(2), 127-136 (1983).
Forster, C.F., Bound water im sewage sludges and its relationship to sludge surfaces and sludge viscosities, J Chem Tech Biotechnol, 33B(1), 76-84 (1983).
Paterson, R.B., and Denn, M.M., Computer-aided design and control of an activated sludge process, Biochem Eng J, 27(1), B13-B28 (1983).
Reddy, P.M.; Gaudy, A.F., and Manickam, T.S., Total oxidation process using an aerobic digester as source of recycle sludge, Chem Eng Commns, 23(1), 137-150 (1983).
Lawson, P.S.; Sterritt, R.M., and Lester, J.N., Adsorption and complexation mechanisms of heavy metal uptake in activated sludge, J Chem Tech Biotechnol, 34B(4), 253-262 (1984).
Takamatsu, T.; Shioya, S., and Kurome, H., Dynamics and control of an activated sludge process as a mixed culture system, AIChEJ, 30(3), 368-376 (1984).
van Esbroeck, H.; Schram, E., and Vereecken, J., An automatic ATP monitor for activated sludge characterisation, J Chem Tech Biotechnol, 34B(2), 76-86 (1984).
Giordano, P.M., Sewage sludge, Chemtech, Oct, 632-635 (1985).
Rappe, G., Deep-shaft waste treatment process, Chem Eng, 15 April, 44-45 (1985).
Hamer, G., and Zwiefelhofer, H.P., Aerobic thermophilic hygienization: A supplement to anaerobic mesophilic waste sludge digestion, CER&D, 64(6), 417-424 (1986).
Aulenbach, D.B., et al., Removal of heavy metals in activated sludge treatment systems, Chem Eng Commns, 60, 79-100 (1987).
Redman, J., Deep shaft treatment for sewage, Chem Engnr, Oct, 11-13 (1987).
Various, Sewage sludge disposal (topic issue), Chem & Ind, 2 May, 287-299 (1988).
Yoshida, H., and Yukawa, H., Theoretical analysis of electroosmotic dewatering of sludge, Int Chem Eng, 28(3), 477-486 (1988).
Yust, L.J., and Howell, J.A., Applications of a self-tuning regulator to control dissolved oxygen in an activated sludge plant, CER&D, 66(3), 260-264 (1988).

4.4 Incineration

1968-1979

Monroe, E.S., Burning waste waters, Chem Eng, 23 Sept, 215-220 (1968).
Various, Burning low-sulfur residuals (special report), Hyd Proc, 49(2), 89-100 (1970).

Bernadiner, M.N.; Rubinshtein, G.N., and Shurygin, A.P., Thermal methods of disposing of industrial wastewaters from chemical plants, Int Chem Eng, 13(2), 256-263 (1973).
Liebeskind, J.E., Pyrolysis for solid-waste management, Chemtech, Sept, 537-542 (1973).
McDonald, D., Waste incineration, Proc Engng, Feb, 76-80 (1973).
Searles, R.A., Catalyst for incineration odour removal, Proc Engng, Feb, 100-102 (1973).
Backhurst, J.R., and Harker, J.H., Drying and incineration of pig manure, Chem Engnr, July, 449-452 (1975).
Bond, G.C., and Sadeghi, N., Catalysed destruction of chlorinated hydrocarbons, J Appl Chem Biotechnol, 25, 241-248 (1975).
Wall, C.J.; Graves, J.T., and Roberts, E.J., How to burn salty sludges, Chem Eng, 14 April, 77-82 (1975).
Zimmer, J.C., and Guaitella, R., Incineration: Lowest cost HCl recovery, Hyd Proc, 55(8), 117-118 (1976).
Barr, W.H.; Strehlitz, F.W., and Dalton, S.M., Modifying large boilers to reduce nitric acid emissions, Chem Eng Prog, 73(7), 59-68 (1977).
Novak, R.G., et al., How sludge characteristics affect incinerator design, Chem Eng, 9 May, 131-136 (1977).
Pierce, R.R., Estimating acid dewpoints in stack gases, Chem Eng, 11 April, 125-128 (1977).
Fisher, M.J., Fluid-bed incineration for sludge disposal, Processing, May, 24-27 (1978).
Murcar, W., Effluent incinerators, Processing, Dec, 40-42 (1978).
Dunn, K.S., Problems and practicalities of incinerating chemical wastes, Chem Engnr, Nov, 779-782 (1979).
Fabian, H.W.; Reher, P., and Schoen, M., Waste incineration, Hyd Proc, 58(4), 183-192 (1979).

1980-1988

Ross, R.D., Burning hazardous waste, Chemtech, Nov, 708-712 (1980).
Anon., Effluent incineration for pharmaceuticals, Processing, June, 15-16 (1981).
Grace, C., Catalytic incineration, Proc Engng, May, 77-79 (1981).
Retallick, W.B., Design of transfer-limited catalytic incinerators, Chem Eng, 12 Jan, 123-125 (1981).
Seebold, J.G., Reduce heater NOx in the burner, Hyd Proc, Nov, 183-186 (1982).
Bell, C.T., and Warren, S., Experience with burner NOx reduction, Hyd Proc, Sept, 145-147 (1983).
Deneau, K.S., Pyrolytic destruction of hazardous waste, Plant/Opns Prog, 2(1), 34-38 (1983).
Monroe, E.S., Quicker, cheaper testing of incinerator performance, Chem Eng, 21 Feb, 69-71 (1983).
Underwood, J.G., Toxic smoke from burning plastics, Proc Econ Int, 4(2), 30-33 (1983).
Feeley, F.G., Burning wastes in steam boilers, Plant/Opns Prog, 3(1), 31-34 (1984).

Seebold, J.G., Practical flare design, Chem Eng, 10 Dec, 69-72 (1984).
McGowin, C.R., Municipal solid waste as a utility fuel, Chem Eng Prog, 81(3), 57-63 (1985).
Rickman, W.S.; Holder, N.D., and Young, D.T., Circulating bed incineration of hazardous wastes, Chem Eng Prog, 81(3), 34-38 (1985).
Tsai, T.C., Flare system design by microcomputer, Chem Eng, 19 Aug, 55-58 (1985).
Wong, S.L.S., and Tien, C., Biodegradation of organic compounds in fluidized bed reactors, Can JCE, 63(6), 954-962 (1985).
Zurer, P.S., Incineration of hazardous wastes at sea, C&E News, 9 Dec, 24-42 (1985).
Arai, N., et al., Partial combustion of surplus activated sludge in a countercurrent moving bed, Int Chem Eng, 26(1), 114-123 (1986).
Gupta, A.K., Combustion of chlorinated hydrocarbons, Chem Eng Commns, 41, 1-22 (1986).
Wallin, S., Incineration of waste, Chem Engnr, March, 18-21 (1986).
Baker-Counsell, J., Hazardous wastes: The future for incineration, Proc Engng, April, 25-29 (1987).
Brunner, C.R., Incineration: Today's hot option for waste disposal, Chem Eng, 12 Oct, 96-106 (1987).
Lightowlers, P., and Cape, J.N., PVC waste incineration and acid rain, Chem & Ind, 1 June, 390-393 (1987).
Vogg, H., Behavior of heavy metals in incineration of municipal wastes, Int Chem Eng, 27(2), 177-183 (1987).
Wiley, S.K., Incinerate your hazardous waste, Hyd Proc, 66(6), 51-54 (1987).
Anon., Burning the garbage, Chemtech, Nov, 656-657 (1988).
Arai, N.; Ninomiya, Y., and Hasatani, M., Emission of fuel-NOx from diffusional flame of fuel oil with preliminary gas-phase pyrolysis, Int Chem Eng, 28(1), 108-117 (1988).
de Silva, R., Safe way to efficient flaring, Processing, Nov, 37-38 (1988).
Hunter, D., Incinerating hazardous waste, Chem Eng, 14 March, 33-39 (1988).
Kolaczkowski, S., and Crittenden, B., Waste-gas combustion, Proc Engng, May, 109 (1988).
Lafleur, A.L., et al., Identification of aromatic alkynes and acyclic polyunsaturated hydrocarbons in output of jet-stirred combustor, Energy & Fuels, 2(5), 709-716 (1988).
Leite, O.C., Predict flare noise and spectrum, Hyd Proc, 67(12), 55-57 (1988).
Punjak, W.A., and Shadman, F., Aluminosilicate sorbents for control of alkali vapors during coal combustion and gasification, Energy & Fuels, 2(5), 702-709 (1988).
Tock, R.W., and Ethington, D., Transfer plasmas destroy PCB fluids, Chem Eng Commns, 71, 177-188 (1988).
Uchida, S.; Kamo, H., and Kubota, H., Source of HCl emission from municipal refuse incineration, Ind Eng Chem Res, 27(11), 2188-2191 (1988).

4.5 Radiation/Nuclear Waste

1970-1986

Lohse, G.E.; Modrow, R.D., and Wheeler, B.R., Nuclear-wastes disposal: Forming solids saves space, Chem Eng, 9 Feb, 94-96 (1970).

McBride, J.A., Regulatory controls on radioactive effluents, Chem Eng Prog, 66(4), 74-77 (1970).

Various, Radioactive wastes (topic issue), Chem Eng Prog, 66(2), 35-63 (1970).

Boback, M.W.; Davis, J.O.; Ross, K.N., and Stevenson, J.B., Disposal of low-level radioactive wastes from pilot plants, Chem Eng Prog, 67(4), 81-86 (1971).

McKay, H.A.C., By-products of nuclear power, Chem & Ind, 1 April, 275-280 (1972).

Bebbington, W.P., Environmental effect of a complex nuclear facility, Chem Eng Prog, 70(3), 85-86 (1974).

Lerch, R.E.; Cooley, C.R., and Atwood, J.M., Reducing radioactive wastes, Chemical Processing, Sept, 9-10 (1974).

Stubblefield, F.E., and Jackson, E.B., Improved control of radioactive wastes, Chem Eng Prog, 70(3), 87-88 (1974).

Slansky, C.M., Radioactive-waste management, Chemtech, March, 160-164 (1975).

Various, Handling nuclear wastes (topic issue), Chem Eng Prog, 72(3), 43-62 (1976).

Trevorrow, L.E., and Steindler, M.J., Technology assessment and nuclear waste management, Chemtech, Feb, 88-96 (1979).

Thomson, B.M., and Heggen, R.J., Uranium and water: Managing related resources, Chemtech, May, 294-299 (1983).

Zurer, P.S., Nuclear waste disposal in the US, C&E News, 18 July, 20-38 (1983).

Barlow, P., Handling waste from nuclear energy, Proc Engng, May, 25-29 (1984).

Bennett, W., Thermal oxide reprocessing at Sellafield, Chem Engnr, June, 35-38 (1985).

Taylor, R.F., Chemical engineering problems of radioactive waste fixation by vitrification (review article) Chem Eng Sci, 40(4), 541-570 (1985).

Ginniff, M.E., and Blair, I.M., The management and disposal of radioactive waste, Energy World, Jan, 5-7 (1986).

Short, H., Reprocessing of spent nuclear fuel, Chem Eng, 1 Sept, 19-24 (1986).

4.6 Flue Gas Desulfurization

1967-1979

Munro, A.J.E., and Masdin, E.,G., Atmospheric pollution: Method for desulphurising fuel gases, Brit Chem Eng, 12(3), 369-373 (1967).

Ludwig, S., Process for absorption of sulfur dioxide from flue gases, Chem Eng, 29 Jan, 70-72 (1968).

Newell, J.E., Making sulfur from sulfur dioxide in flue gas, Chem Eng Prog, 65(8), 62-66 (1969).

Stites, J.G.; Horlacher, W.R.; Bachofer, J.L., and Bartman, J.S., Removing sulfur dioxide from flue gas, Chem Eng Prog, 65(10), 74-79 (1969).
Davis, J.C., Sulfur dioxide absorbed from tail gas with sodium sulfite, Chem Eng, 29 Nov, 43-45 (1971).
Messman, H.C., Desulfurize coal? Chemtech, Feb, 114-116 (1971).
Various, Desulfurization processes (topic issue), Chem Eng Prog, 67(8), 57-91 (1971).
Welty, A.B., Flue-gas desulfurization technology, Hyd Proc, 50(10), 104-108 (1971).
Salooja, K.C., Some serious errors in sulphur trioxide determination in flue gases, Energy World, March, 10-11 (1974).
Jones, J.W., Disposal of flue-gas cleaning wastes, Chem Eng, 14 Feb, 79-85 (1977).
Reis, T., Taking sulfur out of coke, Chemtech, June, 366-373 (1977).
Cunningham, A.T.S., and Jackson, P.J., Reduction of atmospheric pollutants during burning of residual fuel oil in larger boilers, J Inst Fuel, 51, 20-30 (1978).
Engdahl, R.B., and Rosenberg, H.S., Status of flue gas desulfurization, Chemtech, Feb, 118-128 (1978).
McCarthy, J.E., Choosing a flue-gas desulfurization system, Chem Eng, 13 March, 79-84 (1978).
Ohtsuka, T., and Ishihara, Y., Emission control technology for sulphur oxides and nitrogen oxides from flue gases in Japan, J Inst Fuel, 51, 82-90 (1978).
Various, Flue gas desulfurization (topic issue), Chem Eng Prog, 74(2), 41-80 (1978).
Laseke, B.A., and Devitt, T.W., Status of flue gas desulfurization, Chem Eng Prog, 75(2), 37-50 (1979).

1980-1988

Various, Flue gas desulfurization (topic issue), Chem Eng Prog, 76(5), 45-77 (1980).
Bettelheim, J.; Kyte, W.S., and Littler, A., Fifty years' experience of flue gas desulphurisation, Chem Engnr, June, 275-277, 284 (1981).
Kyte, W.S., Some chemical and chemical engineering aspects of flue gas desulphurisation (review paper), Trans IChemE, 59, 219-228 (1981).
Hyne, J.B., Getting sulfur out of natural gas, Chemtech, Oct, 628-637 (1982).
Ting, A.P., and Fuchs, W., Sulfur-removal requirements, Chem Eng, 18 Oct, 131-132 (1982).
Chester, P.F., Coal and our atmosphere, Energy World, Feb, 2-12 (1986).
Simbeck, D.R., and Dickenson, R.L., Integrated gasification combined cycle for acid rain control, Chem Eng Prog, 82(10), 28-33 (1986).
Woodburn, E., Is flue gas desulphurisation the answer? Chem Engnr, Nov, 36-39 (1986).
Furimsky, E., et al., Role of iron in hydrogen sulphide removal from hot gas, Ads Sci Tech, 4(4), 230-240 (1987).
Gavalas, G.R., et al., Alkali-alumina sorbents for high temperature removal of sulfur dioxide, AIChEJ, 33(2), 258-266 (1987).
Halstead, D., Flue gas desulphurisation, Chem Engnr, Dec, 13-15 (1987).

Leidinger, B.; Natusch, K., and Scholl, G., Flue gas release from a cooling tower, J Inst Energy, June, 60-65 (1987).
Short, H., Acid rain spurs clean coal research in Europe, Chem Eng, 19 Jan, 12E-12I (1987).
Gollakota, S.V., and Chriswell, C.D., Study of an adsorption process using silicate for sulfur dioxide removal from combustion gases, Ind Eng Chem Res, 27(1), 139-143 (1988).
Lacey, J.A., Gasification for the clean use of coal, Energy World, March, 2-6, (1988).
Lees, B.,, Alternatives to desulphurisation of flue gases, Energy World, May, 3-4 (1988).
Redman, J., Flue gas desulphurisation, Chem Engnr, Oct, 29-36 (1988).
Redman, J., Flue gas desulphurisation on smaller plant, Chem Engnr, May, 33-44 (1988).
Royse, S., Flue gas desulphurisation, Proc Engng, Feb, 37-41 (1988).
Taffe, P., Programme of flue-gas desulphurisation, Processing, Oct, 21-24 (1988).

4.7 Accidents, Spills and Cleanups

1975-1988

Lindsey, A.W., Ultimate disposal of spilled hazardous materials, Chem Eng, 27 Oct, 107-114 (1975).
Mackay, D., and Mohtadi, M., Area affected by oil spills on land, Can JCE, 53, 140-151 (1975).
D'Alessandro, P.L., and Cobb, C.B., Oil spill control, Hyd Proc, 55(2), 121-124; 55(3), 145-148 (1976).
Wilson, D.C.; Smith, E.T., and Pearce, K.W., Uncontrolled hazardous waste sites in the UK, Chem & Ind, 3 Jan, 18-23 (1981).
Pizzi, F.P., Assess environmental risks, Hyd Proc, April, 253-267 (1982).
Mason, G.S., and Arnold, C., Contain liquid spills and improve safety with a flooded stormwater sewer, Chem Eng, 17 Sept, 105-109 (1984).
Various, Hazardous waste disposal and re-use of contaminated land (topic issue), Chem & Ind, 3 Sept, 602-631 (1984).
Lihou, D., Why did Bhopal ever happen? Chem Engnr, April, 18-19 (1985).
Various, Bhopal (topic issue), C&E News, 11 Feb, 14-65 (1985).
Deckert, A.H., Emergency responses to spills of sulfuric acid during transit, Chem Eng Prog, 82(7), 11-13 (1986).
Franklin, N., The accident at Chernobyl, Chem Engnr, Nov, 17-22 (1986).
Hohenemser, C.; Deicher, M.; Ernst, A.; Hofsass, H.; Linder, G., and Recknagel, E., Chernobyl, Chemtech, Oct, 596-605 (1986).
Anon., Clearing up cyanide, Chem Engnr, July, 15 (1987).
Keller, A., and Lamb, R., Quantify the risks from chemical spills into rivers, Proc Engng, July, 49-52 (1987).
Stallworthy, E.A., Chernobyl - was it simply mismanagement or more? Chem Engnr, Jan, 31 (1987).
Martins, K., Responding properly to hazardous-waste spills, Chem Eng, 18 Jan, 87-91 (1988).

Treworgy, E.D., Simulating hazardous-waste treatment, Chem Eng, 25 April, 95-97 (1988).

4.8 Legislation

1967-1980

Ross, L.W., Sizing up anti-pollution legislation, Chem Eng, 17 July, 191-196 (1967).
Popper, H., The cost of meeting environmental standards, Chem Eng, 23 Aug, 106-108 (1971).
Duprey, R.L., The status of sulfur oxides emission limitations, Chem Eng Prog, 68(2), 70-76 (1972).
Kyan, C.P., and Seinfeld, J.H., Meeting the provisions of the Clean Air Act, AIChEJ, 20(1), 118-127 (1974).
Harrison, E.B., Clean Water Act (1977), Hyd Proc, 57(2), 165-172 (1978).
Passow, N., US environmental regulations affecting the chemical process industries, Chem Eng, 20 Nov, 173-180 (1978).
Russell, D.L., and Tiede, J.J., Measurement uncertainties in the pollution-discharge permit system, Chem Eng, 9 Oct, 115-120 (1978).
Booth, R.L., Complying with National Pollution Discharge Elimination System (NPDES) requirements, Chem Eng Prog, 75(3), 82-85 (1979).
Hawkins, R.G.P., Recent legislation in the context of waste management and the role of the chemical engineer, Chem Engnr, Nov, 773-778, 782 (1979).
Wood, W.A., Premanufacturing notification (EPA regulations), Chemtech, July, 418-424 (1979).
Passow, N.R., Preparing the environmental impact statement, Chem Eng, 15 Dec, 69-74 (1980).
Various, Pollution control and the law (topic issue), Chem & Ind, 19 April, 328-347 (1980).

1983-1988

Ireland, F.E., Legislation and the environment, Energy World, June, 6-9 (1983).
Anon., Environmental law 1984, Chemtech, Feb, 121-127 (1984).
Regens, J.L., Acid rain and public policy, Chemtech, May, 310-316 (1984).
Various, Implementation of Part II of the Control of Pollution Act (COPA), UK, (topic issue), Chem & Ind, 16 July, 496-544; 20 Aug, 581-583 (1984).
Various, Odour nuisance, the law, impact and abatement (topic issue), Chem & Ind, 7 May, 320-347 (1984).
Davis, B.C., U.S. EPA's flare policy: Update and review, Chem Eng Prog, 81(4), 7-10 (1985).
Kahane, S.W.; Phinney, S.L., and Wright, A.P., Tightening regulations for hazardous waste management, Chem Eng Prog, 81(3), 11-15 (1985).
Fleckenstein, L.J., Federal government initiatives in environmental auditing, Chem Eng Prog, 82(10), 17-19 (1986).
Laing, I.G., Legislative developments in health, safety and environmental control, Chem & Ind, 7 April, 231-239 (1986).
Anderson, E.V., Fuel volatility regulations, C&E News, 6 July, 7-13 (1987).

Pollard-Cavalli, R.I., Environmental compliance without sacrificing customer requirements, Plant/Opns Prog, 6(4), 185-187 (1987).

Miles, J., Assessment of 12th Royal Commission report on environmental pollution. Chem & Ind, 1 Aug, 480-484 (1988).

Various, Government regulations in the process industries (topic issue), Chem Eng Prog, 84(12), 35-70 (1988).

4.9 Other Environmental Aspects

1967-1970

Ludwig, J.H., and Spaite, P.W., Control of sulfur oxide pollution, Chem Eng Prog, 63(6), 82-86 (1967).

Mencher, S.K., Minimizing waste in the petrochemical industry, Chem Eng Prog, 63(10), 80-88 (1967).

Chieffo, A.B., and McLean, R.H., Fast leak detection reduces hydrocarbon losses and pollution, Chem Eng, 15 July, 144-146 (1968).

Ross, W.L., Predict pollutants dispersion in air and water, Hyd Proc, 47(8), 144-150 (1968).

Various, Environmental engineering: A complete guide to pollution control (deskbook issue), Chem Eng, 14 Oct, 13-217 (1968).

Barker, W.G., and Schwarz, D., Engineering processes for waste control, Chem Eng Prog, 65(1), 58-61 (1969).

Dlouhy, P.E., and Dahlstrom, D.A., Food and fermentation waste disposal, Chem Eng Prog, 65(1), 52-57 (1969).

Richards, E.J., Noise generation in process plants, Chem Engnr, June, CE223-232 (1969).

Various, Problems of noise in chemical plant (symposium papers), Chem Engnr, April, CE104-129 (1969).

Driesen, M., Application of the thermal process technique to effluent problems, Brit Chem Eng, 15(9), 1156-1158 (1970).

Forster, E.J., and Stinson, L.E., Cars need more octane without lead, Hyd Proc, 49(12), 97-100 (1970).

Madonna, L.A., The pollution system: A chemical space model, Brit Chem Eng, 15(9), 1149-1152 (1970).

Novak, R.G., Eliminating or disposing of industrial solid wastes, Chem Eng, 5 Oct, 78-82 (1970).

Popper, H., and Hughson, R.V., Engineering ethics and environmental problems, Chem Eng, 2 Nov, 88-93 (1970).

Various, Pollution control (topic issue), Chem Eng Prog, 66(11), 31-78 (1970).

Various, A guide to industrial pollution control (Deskbook issue), Chem Eng, 27 April, 17-309 (1970).

1971-1973

Alonso, J.R.F., Handy form helps select pollution-control equipment, Chem Eng, 4 Oct, 104-106 (1971).

Dorsey, J.A., and Burckle, J.O., Particulate emissions and process monitors, Chem Eng Prog, 67(8), 92-96 (1971).

Eckenfelder, W.W., and Barnard, J.L., Treatment-cost relationship for industrial wastes, Chem Eng Prog, 67(9), 76-85 (1971).
Edwards, R., PCBs, occurrence and significance: A review, Chem & Ind, 20 Nov, 1340-1348 (1971).
Greenberg, J.H., Systems analysis of emissions: The iron foundry industry, Chemtech, Dec, 728-736 (1971).
Kolonics, Z., Disposal of wastes in the chemical industry, Int Chem Eng, 11(1), 1-6 (1971).
Meritt, A.D., and Golden, J.T., Low-cost continuous sampler for plant effluents, Chem Eng, 22 March, 122 (1971).
Rodriguez, F., Prospects for biodegradable plastics, Chemtech, July, 409-415 (1971).
Scher, J.A., Solid wastes characterization techniques, Chem Eng Prog, 67(3), 81-84 (1971).
Siegerman, H., Applications of electrochemistry to environmental problems, Chemtech, Nov, 672-679 (1971).
Various, Ocean pollution and marine waste disposal (feature report), Chem Eng, 8 Feb, 60-67 (1971).
Various, Environmental engineering: Pollution control law, technology and economics (deskbook issue), Chem Eng, 21 June, 9-266 (1971).
Witt, P.A., Disposal of solid wastes, Chem Eng, 4 Oct, 62-78 (1971).
Abert, J.G., and Zusman, M.J., Resource recovery from refuse (review paper), AIChEJ, 18(6), 1089-1106 (1972).
Caban, R., and Chapman, T.W., Review of pollution problem of mercury loss from chlorine plants, AIChEJ, 18(5), 892-913 (1972).
Downing, A.L., Chemical engineering and the hydrological cycle, Chem Engnr, April, 150-157 (1972).
Fair, J.R.; Crocker, B.B., and Null, H.R., Sampling and analyzing trace quantities, Chem Eng, 18 Sept, 146-154 (1972).
Fair, J.R.; Crocker, B.B., and Null, H.R., Trace-quality engineering, Chem Eng, 7 Aug, 60-74 (1972).
Herbert, W., Recycling municipal waste, Chem Eng, 10 Jan, 66-67 (1972).
Hoffmann, F., Selecting a pH control system for neutralizing waste acids, Chem Eng, 30 Oct, 105-110 (1972).
Metcalf, R.L.; Kapoor, I.P., and Hirwe, A.S., Development of biodegradable analogues of DDT, Chemtech, Feb, 105-109 (1972).
Pradt, L.A., Treating industrial waste by wet air oxidation, Chem Eng Prog, 68(12), 72-77 (1972).
Racine, W.J., Plant designed to protect the environment, Hyd Proc, 51(3), 115-119 (1972).
Snyder, H.D., Fungicides without mercury, Chemtech, Oct, 609-613 (1972).
Various, Environmental management (special report), Hyd Proc, 51(10), 79-176 (1972).
Various, Pollution abatement (topic issue), Chem Eng Prog, 68(8), 43-79 (1972).
Various, Environmental engineering (deskbook issue), Chem Eng, 8 May, 9-246 (1972).
Various, Effect of effluent restrictions upon industry (symposium papers), Chem Engnr, Nov, 415-428 (1972).

Allan, G.G., et al., Pesticides, pollution and polymers, Chemtech, March, 171-178 (1973).
Erskine, J.B., Acoustic design for a petrochemical plant, Chem Engnr, June, 312-315 (1973).
Fox, R.D., Pollution control at the source, Chem Eng, 6 Aug, 72-82 (1973).
Horvath, J.G., and MacDonald, J.O.S., Disposal of wastes from the chemical industry, Proc Tech Int, Jan, 61, 63 (1973).
Kumar, J., and Jedlicka, J.A., Selecting and installing synthetic pond-linings, Chem Eng, 5 Feb, 67-70 (1973).
Moulton, F.H., Environmental restrictions and chlorinated solvents, Chem Eng Prog, 69(10), 85-88 (1973).
Santoleri, J.J., Chlorinated hydrocarbon waste disposal and recovery systems, Chem Eng Prog, 69(1), 68-74 (1973).
Taylor, L.J., Polymer degradation, Chemtech, Sept, 552-559 (1973).
Various, Conserving our resources (topic issue), Chem & Ind, 16 June, 546-575 (1973).
Various, Sulfur developments and removal (topic issue), Chem Eng Prog, 69(12), 29-64 (1973).
Various, Disposal of solids wastes (symposium papers), Chem Engnr, Feb, 55-93 (1973).
Various, Environmental management (special report), Hyd Proc, 52(10), 71-101 (1973).
Various, Pollution control operations (topic issue), Chem Eng Prog, 69(6), 67-103 (1973).
Various, Industrial waste disposal (symposium papers), Chem Engnr, May, 251-270; June, 296-311 (1973).
Various, Environmental engineering (deskbook issue), Chem Eng, 18 June, 7-237 (1973).

1974-1975

Bennett, G.F., and Lash, L., Industrial waste disposal made profitable, Chem Eng Prog, 70(2), 75-85 (1974).
Darling, S.M.; Snyder, R.W., and Wotring, W.T., Influence of emission controls on fuel and lubricant additives, Chemtech, June, 356-365 (1974).
Furness, C.D., Biological treatment of effluents, Chem Engnr, Feb, 102-107 (1974).
Kuhn, A., Electroflotation for waste treatment, Chemical Processing, June, 9-12; July, 5-7 (1974).
Reiter, W.M., and Stocker, W.F., In-plant waste abatement, Chem Eng Prog, 70(1), 55-59 (1974).
Toms, E.H., and Minnis, H., Monitoring the flows through underground sewers, Chem Eng, 10 June, 108-110 (1974).
Various, Control of liquid effluents from chemical/petrochemical plants (topic issue), Chem & Ind, 5 Oct, 756-769 (1974).
Various, Environmental management (special report), Hyd Proc, 53(10), 95-132 (1974).
Various, Environmental engineering (deskbook issue), Chem Eng, 21 Oct, 7-203 (1974).

Walters, A.H., Microbial biodeterioration of materials, Chem & Ind, 4 May, 365-372 (1974).
Carter, L., and Moss, A.H., Disposal of untreated effluents to tidal waters, Chem Engnr, Feb, 96-99 (1975).
Coleman, A.K., New concepts in handling of industrial wastes, Chem & Ind, 5 July, 534-545 (1975).
Egan, H., Chemical analysis and environmental quality, Chem & Ind, 4 Oct, 814-820 (1975).
Graveland, A., and Heertjes, P.M., Removal of manganese from ground water by heterogeneous autocatalytic oxidation, Trans IChemE, 53, 154-164 (1975).
Hunt, R.G., and Franklin, W.E., Recycling of glass, Chemtech, Aug, 474-481 (1975).
Lacy, W.J., Point-source control of pollutants, Chemtech, Dec, 742-747 (1975).
Lederman, P.B.; Skovronek, H.S., and Des Rosiers, P.E., Pollution abatement in the pharmaceutical industry, Chem Eng Prog, 71(4), 93-97 (1975).
Saxton, J.C., and Narkus-Kramer, M., EPA findings on solid wastes from industrial chemicals, Chem Eng, 28 April, 107-112 (1975).
Shen, T.T., Online instruments expedite emissions tests, Chem Eng, 26 May, 109-112 (1975).
Various, Waste treatment (feature report), Processing, June, 53-72 (1975).
Various, Environmental protection survey (topic issue), Proc Engng, Feb, 85-143 (1975).
Various, Controlling pollution (feature report), Processing, Feb, 33-47 (1975).
Various, Environmental management (special report), Hyd Proc, 54(10), 73-111 (1975).
Various, Sulfur dioxide processing (topic issue), Chem Eng Prog, 71(5), 55-79 (1975).
Various, Vinyl chloride emission control (topic issue), Chem Eng Prog, 71(9), 41-62 (1975).
Various, Pollution control/materials of construction (topic issue), Chem Eng Prog, 71(3), 43-89 (1975).
Various, Noise control engineering (topic issue), Chem Eng Prog, 71(8), 31-59 (1975).
Various, Environmental engineering: Pollution-control equipment (deskbook issue), Chem Eng, 6 Oct, 9-210 (1975).
Wang, L.K., et al., Treatment of industrial effluents by activated carbon, J Appl Chem Biotechnol, 25, 475-502 (1975).

1976

Blakebrough, N., et al., Aerobic degradation of oil contamination in soils, J Appl Chem Biotechnol, 26, 550-558 (1976).
Brown, V.M., Advances in testing the toxicity of substances to fish, Chem & Ind, 21 Feb, 143-149 (1976).
Grutsch, J.F., and Mallatt, R.C., Optimize the effluent system, Hyd Proc, 55(3), 105-112; 55(4), 213-218; 55(5), 221-230; 55(6), 115-123; 55(7), 113-118; 55(8), 137-142 (1976).
Hoffman, E.R., and Fidler, R.K., Clean up urea plant effluent, Hyd Proc, 55(8), 111-112 (1976).

Horton, R., and Hawkes, D., Energy and fertilizer potential of natural organic wastes, Energy World, June, 3-7 (1976).

Horvath, A.L., Maximum permissible concentrations of halogenated hydrocarbons in reservoir waters, Chem & Ind, 3 Jan, 26-27 (1976).

Kiang, Y.H., Liquid waste disposal system, Chem Eng Prog, 72(1), 71-77 (1976).

Majer, D.J., Oil conservation and reclamation, Chem Engnr, Nov, 757-759, 764 (1976).

Middleton, A.H., Noise from chemical plant, Chem Engnr, Feb, 115-118 (1976).

Nishimura, H., Nitrogen cycles in a polluted sea area, Chem Engnr, Nov, 760-764 (1976).

Porter, J.J., and Brandon, C., Zero discharges from textile dyeing and finishing, Chemtech, June, 402-407 (1976).

Siddiqi, A.A.; Killion, L.D.; Tenini, J.W., and Adams, J.T., Tests quantify emissions from ship loadings, Hyd Proc, 55(4), 207-209 (1976).

Terry, R.C.; Berkowitz, J.B., and Porter, C.H., Waste clearinghouses and exchanges, Chem Eng Prog, 72(12), 58-62 (1976).

Unzelman, G.H., Emission vs conservation update, Hyd Proc, 55(6), 93-98 (1976).

Various, Environmental protection survey (topic issue), Proc Engng, Feb, 57-95 (1976).

Various, Environmental engineering (feature report), Processing, June, 37-59 (1976).

Various, Trade effluents, costs, charges and control, Chem & Ind, 2 Oct, 835-841 (1976).

Various, Recent advances in recovery of useful materials from industrial wastes (topic issue), Chem & Ind, 4 Sept, 709-729 (1976).

Various, Contract management and effluent treatment plant operation (topic issue), Chem & Ind, 6 Nov, 918-930 (1976).

Various, Sulfur compound cleanup (topic issue), Chem Eng Prog, 72(8), 80-103 (1976).

Various, Environmental management (special report), Hyd Proc, 55(10), 87-115 (1976).

Various, Environmental engineering: Toxic and flammable material protection (deskbook issue), Chem Eng, 18 Oct, 9-195 (1976).

Walters, J.K., and Wint, A., Process design and the environment, Chem Engnr, Nov, 751-753 (1976).

1977

Commoner, B., Carcinogens in the environment, Chemtech, Feb, 76-82 (1977).

Donovan, J.R.; Kennedy, E.D.; McAlister, D.R., and Smith, R.M., Analysis and control of sulfuric acid plant emissions, Chem Eng Prog, 73(6), 89-94 (1977).

Grover, P., A waste stream management system, Chem Eng Prog, 73(12), 71-73 (1977).

Mace, G.R., and Casaburi, D., Lime vs caustic for neutralizing power plant effluents, Chem Eng Prog, 73(8), 86-90 (1977).

Mandelik, B.G., and Turner, W., Selective oxidation in sulfuric and nitric acid plants, Chem Eng, 25 April, 123-130 (1977).

Miranda, J.G., Oil-water separation using parallel-plates, Chem Eng, 31 Jan, 105-107 (1977).
Mochizuki, S., Photochemical dechlorination of PCBs, Chem Eng Sci, 32(10), 1205-1210 (1977).
Seymour, R.B., and Sosa, J.M., Plastics from plastics, Chemtech, Aug, 507-511 (1977).
Street, E., The disposal of non-biodegradable hazardous wastes, Chem Engnr, April, 249-250 (1977).
Various, Physicochemical treatment of effluents (topic issue), Chem & Ind, 15 Oct, 808-836 (1977).
Various, Environmental protection survey (supplement issue), Proc Engng, Feb (1977).
Various, Waste recovery (topic issue), Chem Eng Prog, 73(5), 45-73 (1977).
Various, Environmental management (special report), Hyd Proc, 56(10), 101-135 (1977).
Various, Environmental engineering: Eliminating wastewater and air pollutants (deskbook issue), Chem Eng, 17 Oct, 10-207 (1977).

1978-1979
Abbas, K.B.; Knutsson, A.B., and Berglund, S.H., New thermoplastics from old, Chemtech, Aug, 502-508 (1978).
de Nevers, N., Strategies for pollution abatement, Chemtech, June, 344-352 (1978).
Elmore, C.L., New low-cost pumping technology, Chem Eng Prog, 74(1), 76-82 (1978).
Jones, J., Converting solid wastes and residues to fuel, Chem Eng, 2 Jan, 87-94 (1978).
Kittleman, T.A., and Akell, R.B., The cost of controlling organic emissions, Chem Eng Prog, 74(4), 87-91 (1978).
Various, Waste recovery (feature report), Processing, Feb, 49-70 (1978).
Various, Symposium papers on reclamation of solid wastes, J Appl Chem Biotechnol, 28, 229-340 (1978).
Various, Pollution control practices (topic issue), Chem Eng Prog, 74(12), 37-72 (1978).
Various, Environmental management (special report), Hyd Proc, 57(10), 91-126 (1978).
Bolme, D.W., and Horton, A., NOx removal in the nitric acid process, Chem Eng Prog, 75(3), 95-98 (1979).
Bowne, N.E., and Yocom, J.E., Meteorology and environmental problems, Chem Eng, 30 July, 56-68 (1979).
Flynn, B.L., Wet air oxidation of waste streams, Chem Eng Prog, 75(4), 66-69 (1979).
Goldman, M.I., Energy: What about the waste? Chem Eng Prog, 75(11), 65-69 (1979).
Paul, J.P., Reclaiming rubber, Chemtech, Feb, 104-108 (1979).
Taylor, L., Degradable plastics: Solution or illusion? Chemtech, Sept, 542-548 (1979).
Various, Waste treatment and recovery (feature report), Processing, Nov, 77-105 (1979).

Various, Environmental protection survey (supplement), Proc Engng, Feb (1979).
Various, Pollution control practices (topic issue), Chem Eng Prog, 75(8), 35-82 (1979).
Various, Environmental management (special report), Hyd Proc, 58(10), 89-121 (1979).
Various, Solid wastes: Treatment and disposal (feature report), Chem Eng, 29 Jan, 78-100; 26 Feb, 119-124; 9 April, 107-112; 21 May, 185-194; 2 July, 77-86; 13 Aug, 141-145 (1979).
Wilkinson, T.G., and Hamer, G., Microbial oxidation of mixtures of methanol, phenol, acetone, and isopropanol with reference to effluent purification, J Chem Tech Biotechnol, 29(1), 56-68 (1979).
Youn, K.C., Ammonia-Pt process promises NOx control, Hyd Proc, 58(2), 117-121 (1979).

1980-1981

Hawkes, D.L., and Horton, R., Potential of anaerobic digestion, Energy World, June, 10-13 (1980).
Smith, J.G., and Bubbar, G.L., Chemical destruction of PCBs by sodium naphthalenide, J Chem Tech Biotechnol, 30, 620-625 (1980).
Various, Reclamation of waste plastics (topic issue), J Chem Tech Biotechnol, 30, 151-186 (1980).
Various, Environmental problems (feature report), Processing, Dec, 27-41 (1980).
Various, Quality of tidal waters (topic issue), Chem & Ind, 19 April, 303-327 (1980).
Various, Environmental monitoring (topic issue), Chem & Ind, 2 Aug, 584-609 (1980).
Various, Environmental management (special report), Hyd Proc, 59(10), 59-86 (1980).
Various, Pollution control practices (topic issue), Chem Eng Prog, 76(10), 37-87 (1980).
Various, Environmental outlook for the 80s (feature report), Chem Eng, 30 June, 80-119; 28 July, 54-68 (1980).
Whitfield, M., Environmental protection survey (equipment review), Proc Engng, May, 49-101 (1980).
Anon., Economics of rubbish sorting to recover valuable materials, Proc Econ Int, 2(3), 26-29 (1981).
Briggs, R., Effluent control, Proc Engng, May, 45-50 (1981).
Lewis, R.E., Estuary mixing, Chem Engnr, Aug, 381-383 (1981).
Scharein, G., Recover products from chlorohydrocarbon residues, Hyd Proc, 60(9), 193-194 (1981).
Smith de Sucre, V., and Watkinson, A.P., Anodic oxidation of phenol for waste treatment, Can JCE, 59, 52-59 (1981).
Takahashi, T.; Kitamura, Y., and Nakada, K., Spreading of oil slicks on calm water surfaces: Effect of viscous drag on spreading rate, Int Chem Eng, 21(2), 244-251 (1981).
Various, Environment report, Processing, Dec, 29-33 (1981).
Various, Recovery and recycling (topic issue), Processing, March, 32-41 (1981).

Various, Toxic materials in industrial effluents (topic issue), Chem & Ind, 18 April, 267-290 (1981).
Various, Applications of biological treatment methods to industrial effluents (topic issue), Chem & Ind, 4 July, 446-469 (1981).
Various, Emission and effluent control (topic issue), Chem Eng Prog, 77(4), 37-75 (1981).
Various, Environmental management (special report), Hyd Proc, 60(10), 85-124 (1981).
Windle, M.R., and Chappell, T.E., A computer-based centralised control and supervisory system for an effluent treatment plant, Chem Engnr, Aug, 377-380, 383 (1981).

1982-1983

Anon., Cryogenic recycling of tyres: A new process, Proc Econ Int, 3(1), 64-67 (1982).
Baumer, A.R., Making environmental audits, Chem Eng, 1 Nov, 101-104 (1982).
Cantrell, C.J., Vapor recovery for refineries, Chem Eng Prog, 78(10), 56-60 (1982).
Edwards, P.R.; Campbell, I., and Milne, G.S., Impact of chloromethanes on the environment, Chem & Ind, 21 Aug, 574-578; 4 Sept, 619-622; 18 Sept, 714-718 (1982).
Hope, K., Safe disposal of industrial waste, Proc Engng, May, 43-45 (1982).
Hurford, N., Pollution and the chemical tanker, Chem Engnr, July, 287-290 (1982).
Payne, K., Chemistry and toxicology of polychlorodibenzo-p-dioxins, Chem & Ind, 1 May, 298-300 (1982).
Reed, S.B., and Wooley, B., Assessing the environmental impact of a new power station for Hong Kong, J Inst Energy, 55, 72-77 (1982).
Sachdev, A.K., and Narsimhan, G., Analysis of a filter bag reactor for gaseous pollution abatement, Chem Eng J, 23(2), 205-210 (1982).
Various, Controlled disposal of special wastes (topic issue), Chem & Ind, 17 April, 251-267 (1982).
Various, Pollution control practices (topic issue), Chem Eng Prog, 78(6), 35-82 (1982).
Various, Environmental management (special report), Hyd Proc, Oct, 75-89 (1982).
Wilkinson, T., An environmental programme for offshore oil operations, Chem & Ind, 20 Feb, 115-123 (1982).
Worthy, W., Hazardous-waste-treatment technology, C&E News, 8 March, 10-16 (1982).
Barbour, A.K., Environmental aspects of zinc, lead and cadmium production and use, Chem & Ind, 6 June, 409-415 (1983).
Block, R.M., and Kalinowski, T.W., Disposing of those old drums, Chem Eng, 21 March, 77-78; 18 April, 103-105 (1983).
Bown, M., pH control of industrial effluent, Proc Engng, Jan, 53-55; Feb, 49-51 (1983).
Burkhardt, C.W., Control pollution by air flotation, Hyd Proc, May, 59-61 (1983).

Edwards, R.E.; Speed, N.A., and Verwoert, D.E., Cleanup of chemically contaminated sites, Chem Eng, 21 Feb, 73-81 (1983).
Elkington, J., The case for pollution control, Proc Engng, May, 24-27 (1983).
Hobson, P.N., Kinetics of anaerobic digestion of farm wastes, J Chem Tech Biotechnol, 33B(1), 1-20 (1983).
Kitamura, Y., and Takahashi, T., Correlating the spread of oils slicks, Int Chem Eng, 23(4), 672-675 (1983).
Lawrence, A., Environmental-rule changes will affect facility planning, Chem Eng, 10 Jan, 85-89 (1983).
Nazly, N., et al., Detoxification of cyanide by immobilised fungi, J Chem Tech Biotechnol, 33B(2), 119-126 (1983).
Rich, L.G., Designing aerated lagoons to improve effluent quality, Chem Eng, 30 May, 67-70 (1983).
Syred, N.; Styles, A.C., and Sahatimehr, A., Emission control by cyclone combustor technology, J Inst Energy, 56, 125-130 (1983).
Various, Different waste disposal practices in Europe (topic issue), Chem & Ind, 6 June, 416-426 (1983).
Various, Lead in the working atmosphere and in the environment (topic issue), Chem & Ind, 4 April, 258-274 (1983).
Various, Chemical treatment methods for industrial effluents (topic issue), Chem & Ind, 18 July, 549-565 (1983).
Various, Reducing plant pollution exposure (topic issue), Chem Eng Prog, 79(3), 35-81 (1983).
Various, Environmental management (special report), Hyd Proc, Oct, 61-82 (1983).

1984-1985

Bucklin, R.W., and Mackey, J.D., Sulfur management in refineries, Chem Eng Prog, 80(6), 63-67 (1984).
Easterbrook, J., and Gagliardi, D.V., Sewers can pass on problems, Plant/Opns Prog, 3(1), 29-31 (1984).
Gittens, M.J., Energy and the environment, Energy World, July, 11-12 (1984).
Laughlin, B., Don't waste it, exchange it, Chemtech, Feb, 93-96 (1984).
Mackie, J.A., and Niesen, K., Hazardous-waste management: The alternatives, Chem Eng, 6 Aug, 50-64 (1984).
Various, Environmental management (special report), Hyd Proc, Oct, 39-51 (1984).
Various, Emission and pollution control (topic issue), Chem Eng Prog, 80(9), 33-76 (1984).
Various, A guide to groundwater contamination (feature report), Chem Eng, 26 Nov, 64-78 (1984).
Weston, A.F., Obtaining reliable priority-pollutant analyses, Chem Eng, 30 April, 54-60 (1984).
Zirschky, J., and Gilbert, R.O., Detecting hot spots at hazardous-waste sites, Chem Eng, 9 July, 97-100 (1984).
Aleksandrov, V.V., A Soviet view of nuclear winter, Chemtech, Nov, 658-665 (1985).
Baillod, C.R.; Lamparter, R.A., and Barna, B.A., Wet oxidation for industrial waste treatment, Chem Eng Prog, 81(3), 52-56 (1985).

Foller, P.C., and Goodwin, M.L., Electrochemical generation of high-concentration ozone for waste treatment, Chem Eng Prog, 81(3), 49-51 (1985).
Gilead, D., Plastics that self destruct, Chemtech, May, 299-301 (1985).
Holton, H.H., and Kutney, G.W., Wood pulp with less pollution, Chemtech, Sept, 568-571 (1985).
Keefer, G.B., and Sack, W.A., Iron recovery from neutralized acid mine drainage sludge, Chem Eng Commns, 35(1), 305-314 (1985).
Koerner, R.M., and Lord, A.E., Nondestructive testing methods of detecting buried wastes, Chem Eng Prog, 81(3), 39-42 (1985).
Mathews, J., Getting rid of nuclear debris, Chemtech, Dec, 728-730 (1985).
Reich, R.A., and Campbell, H.J., Organics, plastics, and synthetics industry discharges: Performance and regulations, Chem Eng Prog, 81(3), 16-21 (1985).
Russell, D.L., Environmental audits, Chem Eng, 24 June, 37-43 (1985).
Various, Recycling versus disposal (topic issue), Chem & Ind, 20 May, 327-336 (1985).
Various, Pollution control (topic issue), Chem Eng Prog, 81(10), 16-28, 33-56 (1985).
Various, Environmental control (special report), Hyd Proc, Oct, 43-53 (1985).
Wheatley, A., and Winstanley, I., Anaerobic digester systems for waste treatment, Chem Engnr, July, 42-46 (1985).

1986
Anderson, D., and Gaum, M., Conducting an environmental impact survey, Proc Engng, Feb, 53-55 (1986).
Anon., Hazardous-waste treatment undergoes dramatic improvement (equipment focus), Chem Eng, 10 Nov, 67-72 (1986).
Anon., Tasks of pollution control are difficult but manageable (equipment focus), Chem Eng, 12 May, 45-51 (1986).
Basta, N., Use electrodialytic membranes for waste recovery, Chem Eng, 3 March, 42-43 (1986).
Bell, M.M.G., and Allum, K.H., Upgrading treatment for effluent re-use, CER&D, 64(6), 409-416 (1986).
Chowdhury, J., CPI go below to remove groundwater pollutants, Chem Eng, 9 June, 14-19 (1986).
Colonna, G.R., Evaluating hazardous chemicals in the marine workplace, Chem Eng Prog, 82(4), 10-14 (1986).
Conner, J.R., Fixation and solidification of wastes, Chem Eng, 10 Nov, 79-85 (1986).
Cook, S.L., Groundwater monitoring at hazardous-waste facilities, Chem Eng, 13 Oct, 63-69 (1986).
Ditchfield, P., Effluent characterisation: The key to successful anaerobic treatment, Proc Engng, Nov, 39-43 (1986).
Edewor, J.O., A comparison of treatment methods for palm oil effluent wastes, J Chem Tech Biotechnol, 36(5), 212-218 (1986).
Goodfellow, H.D., and Berry, J., Clean-plant design, Chem Eng, 6 Jan, 55-61 (1986).

Herbert, P., et al., Occurrence of chlorinated solvents in the environment, Chem & Ind, 15 Dec, 861-869 (1986).
Ho, C.C.; Han, C.Y., and Khoo, K.H., Electrochemical treatment of effluents, J Chem Tech Biotechnol, 36(1), 7-14 (1986).
Kalbfus, W., Analyze the hydrocarbons in liquid refinery wastes, Hyd Proc, 65(1), 77-78 (1986).
Koltuniak, D.L., In-situ air stripping cleans contaminated soil, Chem Eng, 18 Aug, 30-31 (1986).
Moy, G.; Kenahan, T.; Prater, B., and Summerhill, D., Continuous measurement of pollutants, Chem Engnr, Oct, 28-30 (1986).
Pierce, V.E., and Bansal, B.B., Lead phase-out and octane enhancement, Chem Eng Prog, 82(3), 27-33 (1986).
Remirez, R., Looking into cheaper ways to deal with acid rain, Chem Eng, 7 July, 17-19 (1986).
Valais, M., and Bonnifay, P., Lead phase-out in Western Europe, Chem Eng Prog, 82(3), 34-38 (1986).
Various, Ecotoxicology: Effects of chemicals on the aquatic environment (topic issue), Chem & Ind, 3 Nov, 732-744 (1986).
Various, Management strategies for disposal of difficult industrial wastes (topic issue), Chem & Ind, 16 June, 410-422 (1986).
Various, Protecting the aquatic environment (topic issue), Chem & Ind, 2 June, 380-392 (1986).
Various, Pollution, politics and people (topic issue), Chem & Ind, 6 Jan, 9-24 (1986).
Vervalin, C.H., Environmental control, Hyd Proc, 65(10), 41-46 (1986).

1987

Anon., Solving pollution-control problems in the CPI, Chem Eng, 17 Aug, 45-49 (1987).
Anon., Hazardous waste management (technology focus), Chem Eng, 22 June, 95-101 (1987).
Anon., Waste management and pollution control 1987 (special advertising section), Chem Eng, 26 Oct, 53-87 (1987).
Basta, N., and MacKerron, C., Plastics recycling grows up, Chem Eng, 23 Nov, 22-27 (1987).
Bradford, M., Process plants as hazardous-waste facilities, Chem Eng, 2 March, 69-71 (1987).
Burger, R., Getting rid of asbestos: Removal from cooling towers, Chem Eng, 22 June, 167-170 (1987).
Chowdhury, J., New routes buoy efforts to trim heavy-metal wastes, Chem Eng, 12 Oct, 26-27 (1987).
Fromm, C.H.; Callahan, M.S.; Freeman, H.M., and Drabkin, M., Succeeding at waste minimization, Chem Eng, 14 Sept, 91-94 (1987).
Klass, D.L., and Sen, C.T., Energy from waste, Chem Eng Prog, 83(7), 46-52 (1987).
MacKerron, C.B., EPA tries techniques for hazardous-waste cleanup, Chem Eng, 26 Oct, 29-32 (1987).
Maclean, R., Energy saving techniques in the evaporation and drying of distillery effluent, Chem Engnr, July, 29-34 (1987).

Mailen, J.C., Secondary clean-up of Idaho Chemical Processing plant solvent, Sepn Sci Technol, 22(2), 335-346 (1987).
Martin, E.J., and Johnson, J.H. (Eds). Hazardous Waste Management Engineering, Van Nostrand Reinhold, New York (1987).
Matley, J., and Greene, R., Ethics of health, safety and environment, Chem Eng, 2 March, 40-50 (1987).
Matley, J.; Greene, R., and McCauley, C., Health, safety and environment (readers survey), Chem Eng, 28 Sept, 108-121 (1987).
Miller, J.A., and Fisk, G.A., Combustion chemistry, C&E News, 31 Aug, 22-46 (1987).
Ruckelshaus, W.D., Risk, science and democracy, Chemtech, Nov, 658-662; Dec, 738-741 (1987).
Russell, D.L., Understanding groundwater monitoring, Chem Eng, 26 Oct, 101-105 (1987).
Stinson, S.C., EPA to evaluate new technologies for cleaning up hazardous waste, C&E News, 25 May, 7-12 (1987).
Various, Chemicals and the environment (topic issue), Chem & Ind, 21 Sept, 641-653 (1987).
Various, Insurance and environmental protection (topic issue), Chem & Ind, 16 Nov, 775-792 (1987).
Various, Hazardous material control (topic issue), Chem Eng Prog, 83(11), 27-54 (1987).

1988

Anon., Pollution control case study, Chem Eng, 9 May, 107-108 (1988).
Attig., R.C.; Chapman, J.N.; Sheth, A.C., and Wu, S., Emission control by magnetohydrodynamics, Chemtech, Nov, 694-702 (1988).
Basta, N., Waste minimization, Chem Eng, 15 Aug, 34-37 (1988).
Coulson, J.A., Prevention and remediation of contaminated sites with linings, Chem & Ind, 7 Nov, 692-694 (1988).
Curtis, D.B., Checklist for waste treatment, Chem Eng, 23 May, 131-136 (1988).
de Ruiter, H., Pollution control and the European oil industry, Hyd Proc, 67(5), 108C-108F, 108N (1988).
Enander, R.T., and Nester, D.J., A structured approach to solid waste management, Chem Eng Prog, 84(4), 27-31 (1988).
Fromm, C.H., and Budaraju, S., Reducing equipment-cleaning wastes, Chem Eng, 18 July, 117-122 (1988).
Goldman, S.D., Software and environmental engineering, Chem Eng Prog, 84(12), 29-34 (1988).
Hileman, B., The Great Lakes cleanup effort, C&E News, 8 Feb, 22-39 (1988).
Hirschhorn, J.S., and Oldenburg, K.U., Prevent pollution upstream, Chemtech, May, 274-276 (1988).
Hollond, G.J., and Weber, W.F., Waste reduction at Du Pont, Plant/Opns Prog, 7(2), 111-113 (1988).
Kelsey, J., Specifying effluent treatment systems for optimum performance, Proc Engng, Aug, 55-56 (1988).
Langton, C.A., et al., Waste salt disposal at Savannah River plant, Chem Eng Commns, 66, 189-200 (1988).

Marco, G.J.; Hollingworth, R.M., and Durham, W.F., Silent Spring revisited, Chemtech, June, 350-353 (1988).
O'Sullivan, D.A., Hazardous waste exports, C&E News, 26 Sept, 24-27 (1988).
Palluzi, R.P., Asbestos removal, Chem Eng, 12 Sept, 95-99 (1988).
Prince, K., Waste disposal methods, Processing, Oct, 27-31 (1988).
Rice, S.C., Minimizing wastes from R & D activities, Chem Eng, 24 Oct, 85-88 (1988).
Skinner, M., Controlling vacuum without pollution, Processing, Dec, 35-37 (1988).
Thompson, D., Biotechnology for energy conservation and a cleaner environment, Proc Engng, Dec, 39, 41 (1988).
Turner, A., Electrical processes for waste treatment, Processing, July, 18-20 (1988).
Various, Safe handling and disposal of chlorinated solvent waste and residues, Chem & Ind, 15 Aug, 522-525 (1988).
Various, Environmental engineering (topic issue), CER&D, 66(6), 481-536 (1988).
Vervalin, C.H., Environmental control information resources, Hyd Proc, 67(10), 70-74 (1988).
Weiss, N.L., Hazardous-waste disposal in Arizona, Chemtech, Sept, 540-544 (1988).
Wilson, D.J.; Clarke, A.N., and Clarke, J.H., Soil clean-up by in-situ aeration: Mathematical modeling, Sepn Sci Technol, 23(10), 991-1038 (1988).

CHAPTER 5

ECONOMICS

5.1 Costing Data

1967-1970

Anon., Chemical engineering cost data, Brit Chem Eng, 11(5), 340-345; 11(6), 513,515; 11(7), 727,729; 11(8), 863,865; 11(11), 1401,1403 (1966); 12(5), 747 (1967); 12(6), 921; 12(7), 1119; 12(8), 1253; 12(9), 1419,1421; 12(10), 1619,1621; 12(11),1769 (1967).

Gallagher, J.T., Rapid estimation of plant costs, Chem Eng, 18 Dec, 89-96 (1967).

Haselbarth, J.E., Updated investment costs for 60 types of chemical plants, Chem Eng, 4 Dec, 214-215 (1967).

Hudig, J., Cost of multipurpose process units, Chem Eng Prog, 63(9), 79-81 (1967).

Smith, T.J., Reinforced-plastic pipe vs stainless: Comparing the actual costs, Chem Eng, 2 Jan, 110-113 (1967).

Deschner, W.W.; Gieck, J., and Potts, P., Economics of small ammonia plants, Hyd Proc, 47(9), 261-264 (1968).

Knox, W.G., Estimating the cost of process buildings via volumetric ratios, Chem Eng, 17 June, 292-294 (1968).

Sweeney, N., Ammonia costs, Hyd Proc, 47(9), 265-268 (1968).

Anon., Chemical engineering cost data, Brit Chem Eng, 12(12), 1915 (1967); 13(1), 117 (1968); 13(2), 403; 13(3), 551; 13(6), 851; 13(7), 1017 (1968); 14(1), 79 (1969); 14(2), 209; 14(3), 361,363; 14(4), 537; 14(5), 707; 14(6), 851; 14(7), 995 (1969).

Bolton, D.H., and Hanson, D., Economics of low pressures in methanol plants, Chem Eng, 22 Sept, 154-158 (1969).

Forbes, M.C., Cost accounting for pollution control, Hyd Proc, 48(10), 145-148 (1969).

Gallagher, J.T., Efficient estimating of worldwide plant costs, Chem Eng, 2 June, 196-202 (1969).

Guthrie, K.M., Data and techniques for preliminary capital cost estimating, Chem Eng, 24 March, 114-142 (1969).

Johnson, R.J., Cost of overseas plants, Chem Eng, 10 March, 146-152 (1969).

Mapstone, G.E., Find exponents for cost estimates, Hyd Proc, 48(5), 165-167 (1969).

Massey, D.J., and Black, J.H., Predicting chemical prices, Chem Eng, 20 Oct, 150-154 (1969).

Norden, R.B., CE cost indexes: Sharp rise since 1965, Chem Eng, 5 May, 134-138 (1969).

Chase, J.D., Plant cost vs. capacity: New way to use exponents, Chem Eng, 6 April, 113-118 (1970).

Freidman, W.F., Distribution costs, Chem Eng, 21 Sept, 169-174 (1970).

Guthrie, K.M., Capital and operating costs for 54 chemical processes, Chem Eng, 15 June, 140-156 (1970).

Hungerford, H.B., Controlling subcontractor costs, Chem Eng Prog, 66(1), 58-60 (1970).

1971-1975

Buehler, J.D., and Figge, G.J., Operating vs. capital costs: Evaluating tradeoff benefits, Chem Eng, 8 Feb, 96-102 (1971).
Jelen, F.C., and Cole, M.S., Optimum equipment life and replacement cost, Hyd Proc, 50(7), 97-100 (1971).
Lochmann, W.J., Corrosion allowances increase costs, Hyd Proc, 50(7), 101-104 (1971).
Wells, G.L., Determining interest rate on capital, Brit Chem Eng, 16(12), 1113 (1971).
Wilson, G.T., Capital investment for chemical plant, Brit Chem Eng, 16(10), 931-934 (1971).
Anon., Chemical engineering cost data, Brit Chem Eng, 14(8), 1115; 14(9), 1259; 14(10), 1459,1599; 14(12), 1731 (1969); 15(1), 111; 15(3), 389; 15(9), 1207; 15(10), 1347; 15(11), 1467 (1970); 16(2), 235; 16(4), 408; 16(6), 529 (1971); 17(2), 169 (1972).
Estrup, C., Six-tenths rule for capital cost estimation, Brit Chem Eng, 17(3), 213-214 (1972).
Mendel, O., How location affects U.S. plant-construction costs, Chem Eng, 11 Dec, 120-124 (1972).
Pitkin, D.W., Total-cost evaluation of mobile equipment, Chem Eng, 16 Oct, 128-130 (1972).
Thorsen, D.R., The seven year surge in the CE Cost Indexes, Chem Eng, 13 Nov, 168-170 (1972).
Yen, Y.C., Estimating plant costs in the developing countries, Chem Eng, 10 July, 89-92 (1972).
Blakeley, T.H., Nomograph for capital repayment, Chem Engnr, April, 196 (1973).
Cran, J., Location index for plant cost comparisons, Proc Engng, April, 109-111 (1973).
Cran, J., UK and USA plant construction cost indices, Proc Engng, March, 108-109 (1973).
May, V.T., Process development costs and experience, Chem Eng Prog, 69(2), 71-73 (1973).
Miller, C.A., Current concepts in capital cost forecasting, Chem Eng Prog, 69(5), 77-83 (1973).
Swaney, J.B., New approach to preliminary cost estimating, Hyd Proc, 52(4), 167-169 (1973).
Butler, P., More data required for capital cost estimation, Proc Engng, Dec, 84-89 (1974).
Hanna, J.H., Control the cost of new plants, Hyd Proc, 53(7), 183-188 (1974).
Various, Cost focus 1974 (feature report), Chem Eng, 24 June, 136-144 (1974).
Goyal, S.K., Economic plant expansion rate, Processing, Nov, 29 (1975).
Kearney, D.B., Ethylene plant economic size limitations, Chem Eng Prog, 71(11), 68-70 (1975).
Mapstone, G.E., Costing large and small plants, Hyd Proc, 54(1), 91-92 (1975).
Ricci, L.J., CE cost indexes accelerate 10-year climb, Chem Eng, 28 April, 117-118 (1975).
Whelan, T., Instrumentation costs in the HPI, Hyd Proc, 54(9), 185-188 (1975).

1976-1980

Various, Capital-cost estimating techniques, Processing, Jan, 17,19; Feb, 21-24; March, 19-22,23; April, 27-30; May, 31-32; June, 25,28; July, 21-22,34 (1976).

Wallace, D.M., Construction costs in Saudi Arabia, Hyd Proc, 55(11), 189-196 (1976).

Burford, C.L.; Liles. D.H., and Dryden, R.D., Effects of inflation on capital investments, Chemtech, Feb, 129-134 (1977).

Cordero, R., The cost of missing pipe insulation, Chem Eng, 14 Feb, 77-78 (1977).

Baltzell, H.J., Estimate refinery cost better, Hyd Proc, 57(1), 129-131 (1978).

Cran, J., Estimating factors for the food and drink industries, Proc Engng, Feb, 64-65 (1978).

Cran, J., Scaling-up chemical plant costs, Proc Engng, Sept, 154-161 (1978).

Hartzell, J., Chart for comparing energy costs, Chem Eng, 23 Oct, 180 (1978).

Kohn, P.M., CE cost indexes maintain 13-year ascent, Chem Eng, 8 May, 189-190 (1978).

Uri, N.D., Empirical findings of time-of-day price elasticities for electrical energy consumption, J Inst Fuel, 51, 199-201 (1978).

Anon., Process plant capital costs: Methanol, Proc Econ Int, 1(1), 12 (1979).

Anon., Process plant capital costs: Ethylene; VCM/PVC, Proc Econ Int, 1(2), 5 (1979).

Bridgwater, A.V., International construction cost location factors, Chem Eng, 5 Nov, 119-121 (1979).

Leonard, J.P., Synthetic gas and chemicals from coal: Economic appraisals, Chem Eng, 26 March, 183-186 (1979).

Miller, C.A., Converting construction costs from one country to another, Chem Eng, 2 July, 89-93 (1979).

Anon., Process plant capital costs: Ethylene glycol; ethylene oxide; LD polyethylene, Proc Econ Int, 1(3), 8-9 (1980).

Anon., Cost engineering in South East Asia, Proc Econ Int, 1(3), 11-17 (1980).

Anon., Process plant capital costs: Ammonia; HD polyethylene, Proc Econ Int, 1(4), 8-9 (1980).

Anon., Process plant capital costs: Oxygen; benzene, Proc Econ Int, 2(1), 3-4 (1980).

Beveridge, F.N., Economics and accounting of energy, Energy World, Aug, 12-16 (1980).

Gaensslen, H., Cost analysis of coal versus oil, Chemtech, Sept, 563-565 (1980).

Gonzalez, L., and Friedman, P., Use of simulation model of the torula yeast process for investment evaluation, Comput Chem Eng, 4(2), 123-132 (1980).

Kermode, R.I.; Nicholson, A.F., and Jones, J.E., Methanol from coal: Cost projections to 1990, Chem Eng, 25 Feb, 111-116 (1980).

Malina, M.A., Upgrading predictions of startup costs, Chem Eng, 11 Aug, 167-168 (1980).

Various, Controlling operating costs (feature report), Chem Eng, 24 March, 86-105 (1980).

1981-1984

Anon., Process plant capital costs: Acetic acid; nitric acid, Proc Econ Int, 2(4), 8-9 (1981).
Campagne, W.L., What's steam worth? Hyd Proc, 60(8), 117-122 (1981).
Salem, A.B., Estimate capital costs fast, Hyd Proc, 60(9), 199-201 (1981).
Anon., Process plant capital costs: Butadiene; glycerine; acetic acid; polyether polyols; polybutylene terephthalate; formaldehyde, Proc Econ Int, 3(1), 7-9 (1982).
Anon., Process plant capital costs: Hydrogen; ethanol, Proc Econ Int, 3(3), 5-6 (1982).
Kerridge, A.E., Evaluate project cost factors, Hyd Proc, July, 203-216 (1982).
Matley, J., Revised CE plant cost index, Chem Eng, 19 April, 153-156 (1982).
Parker, H.W., Comparing costs when selecting a synfuel process, Chem Eng, 14 June, 173-175 (1982).
Steinmeyer, D.E., Capital costs and energy conservation, Chemtech, March, 188-192 (1982).
Anon., Process plant capital costs: Propylene oxide; acrylonitrile, Proc Econ Int, 4(3), 7 (1983).
Anon., Process plant capital costs: Polymeric isocyanates (crude MDI); perchlorethylene, Proc Econ Int, 4(2), 5-6 (1983).
Anon., Process plant capital costs: Syngas; cyclohexane, Proc Econ Int, 4(1), 10-11 (1983).
Anon., Process plant capital costs: Sulphuric acid; methyl ethyl ketone, Proc Econ Int, 3(4), 6-7 (1983).
Cran, J., Cost indices for new plant costs, Proc Engng, April, 60-61 (1983).
Kay, S.R., Canadian process plant cost indicies: A comparison, Proc Econ Int, 4(1), 4-5 (1983).
Rolstadas, A., Cost study: Norwegian continental shelf, Proc Econ Int, 4(3), 15-23 (1983).
Anon., Process plant capital costs: Styrene; ethylbenzene, Proc Econ Int, 4(4), 7-8 (1984).
Anon., Process plant capital costs: Adipic acid; phthalic anhydride, Proc Econ Int, 5(1), 6-7 (1984).
Ceroke, C.J., Electric vs. hydraulic drives: An economic comparison, Chem Eng, 12 Nov, 133-134 (1984).
Chem Eng Magazine, Modern Cost Engineering: Methods and Data, Volume I(1979) and Volume II (1984), McGraw-Hill Publications, New York (1984).
Davis, B., Budgeting for maintenance, Proc Engng, Nov, 19-22 (1984).
Gerrard, M., and Ratanshi, K.K., Forecasting chemical price indices, Proc Engng, Oct, 65-66 (1984).
Lunde, K.E., Capacity exponents for field construction costs, Chem Eng, 5 March, 71-74 (1984).
Mudge, L.K.; Baker, E.G., and Brown, M.D., Economics of catalytic conversion of biomass to methanol, Proc Econ Int, 5(1), 30-34 (1984).
Vogel, G.A., and Martin, E.J., Estimating operating costs, Chem Eng, 5 Sept, 143-146; 17 Oct, 75-78; 28 Nov, 87-90 (1983); 9 Jan, 97-100; 6 Feb, 121-122 (1984).

Wagialla, K.M., Use of location factors in estimating total fixed investment of chemical plant, Proc Econ Int, 5(1), 37-41 (1984).
Ward, T.J., Predesign estimating of plant capital costs, Chem Eng, 17 Sept, 121-124 (1984).
Williams, L.F., and Gerrard, A.M., Computer-aided cost estimation: A survey, Proc Econ Int, 5(1), 28-30 (1984).

1985-1988
Douglas, J.M., and Woodcock, D.C., Cost diagrams and quick screening of process alternatives, Ind Eng Chem Proc Des Dev, 24(4), 970-976 (1985).
Fischer, A., Engineering manpower costs more for smaller projects, Hyd Proc, April, 133-134 (1985).
Klumpar, I.V., and Slavsky, S.T., Updated cost factors, Chem Eng, 22 July, 73-75; 19 Aug, 76-77; 16 Sept, 85-87 (1985).
Matley, J., CE cost indexes, Chem Eng, 29 April, 75-76 (1985).
Mendel, O., Estimating engineering cost, Chem Eng, 9 Dec, 117-118 (1985).
Ott, C.J., and Wu, T.K., Program speeds labor estimates, Hyd Proc, July, 84-85 (1985).
Rose, A., Parametric cost estimating, Proc Engng, Dec, 60-63 (1985).
Anon., Facts and figures for chemical R&D, C&E News, 28 July, 32-60 (1986).
Anon., Facts and figures for the chemical industry, C&E News, 9 June, 32-86 (1986).
Ledas, A.E., Determining manufacturing costs, Hyd Proc, 65(5), 55-56 (1986).
Webber, D., Top 50 chemicals production, C&E News, 21 April, 12-16 (1986).
Anon., Facts and figures for the chemical industry, C&E News, 8 June, 24-76 (1987).
Rawls, R.L., Facts and figures for chemical R&D, C&E News, 27 July, 32-62 (1987).
Anon., World electricity prices, Energy World, June, 4-6 (1988).
Matley, J., and Hick, A., 1987 cost indexes, Chem Eng, 11 April, 71-73 (1988).

5.2 Plant and Equipment Costing

1966-1970
Anon., Estimating costs of process dryers, Chem Eng, 31 Jan, 101 (1966).
Corrigan, T.E.; Lewis, W.E., and McKelvey, K.N., Costs of chemical reactors based on volume, Chem Eng, 22 May, 214-215 (1967).
Dean, M.J., and Corrigan, T.E., How plant size affects profits, Brit Chem Eng, 12(11), 1743-1744 (1967).
Derrick, G.C., Estimating the cost of jacketed, agitated and baffled reactors, Chem Eng, 9 Oct, 272 (1967).
Gallagher, J.T., Cost of direct-fired heaters, Chem Eng, 17 July, 232 (1967).
Kirk, M.M., Estimating costs of cranes, hoists and pulleys, Chem Eng, 27 Feb, 168 (1967).
Volkin, R.A., Economic piping of parallel equipment, Chem Eng, 27 March, 148-152 (1967).
Bosworth, D.A., Installed cost of outside piping, Chem Eng, 25 March, 132-133 (1968).

Cran, J.E., How to estimate piping costs, Chem Engnr, May, CE110-112 (1968).
Guthrie, K.M., Estimating the cost of high-pressure equipment, Chem Eng, 2 Dec, 144-148 (1968).
Hensley, E.F., and MacPhail, A.A., Minimizing investment in small scale plants, Chem Eng Prog, 64(9), 88-91 (1968).
Kronseder, J.G., Economics of phosphoric acid processes, Chem Eng Prog, 64(9), 97-102 (1968).
Cook, T.P., and Tennyson, R.N., Improved economics in synthesis gas plants, Chem Eng Prog, 65(11), 61-64 (1969).
Eckenfelder, W.W., and Ford, D.L., Economics of wastewater treatment, Chem Eng, 25 Aug, 109-118 (1969).
Guthrie, K.M., Cost of process modules from 'rapid calc' charts, Chem Eng, 13 Jan, 138-142 (1969).
Menicatti, S., Check tank-insulation economics, Hyd Proc, 48(4), 133-136 (1969).
Sokullu, E.S., Estimating piping costs from process flowsheets, Chem Eng, 10 Feb, 148-150 (1969).
Drew, J.W., and Ginder, A.F., How to estimate the cost of pilot-plant equipment, Chem Eng, 9 Feb, 100-110 (1970).
Liptak, B.G., Costs of process instruments, Chem Eng, 7 Sept, 60-76; 21 Sept, 175-179; 5 Oct, 83-86; 2 Nov, 94-100 (1970).

1971-1975

Alonso, J.R.F., Estimating the costs of gas-cleaning plants, Chem Eng, 13 Dec, 86-96 (1971).
Enyedy, G., Cost data for major equipment, Chem Eng Prog, 67(5), 73-81 (1971).
Epstein, L.D., Cost of standard-sized reactors and storage tanks, Chem Eng, 18 Oct, 160-161 (1971).
Hirschmann, W.B., Costs of new refineries, Chem Eng Prog, 67(8), 39-45 (1971).
Marshall, S.P., and Brandt, J.L., Installed cost of corrosion-resistant piping, Chem Eng, 23 Aug, 68-82; 4 Oct, 111-112 (1971).
Martinez, S., Equipment buying decisions, Chem Eng, 5 April, 146-150 (1971).
Carter, H.M., Planning in the chemical industry: The role of the equipment process manufacturer, Chem Engnr, Feb, 49-51 (1972).
Clark, F.D., and Terni, S.P., Cost of thick-wall pressure vessels, Chem Eng, 3 April, 112-116 (1972).
Epstein, L.D., Estimating costs of jacketed reactors, Hyd Proc, 51(12), 102 (1972).
Lerner, J.E., Simplified air-cooler cost estimating, Hyd Proc, 51(2), 93-100 (1972).
Robinson, J.D., and Loonkar, Y.R., Minimising capital investment for multi-product batch-plants, Proc Tech Int, 17(11), 861-863 (1972).
Sommerville, R.F., New method gives quick, accurate estimate of distillation costs, Chem Eng, 1 May, 71-76 (1972).
Finley, F., Total life-cycle costs of plant and equipment, Proc Tech Int, April, 177-180 (1973).

Grigsby, E.K.; Mills, E.W., and Collins, D.C., Cost of future refineries, Hyd Proc, 52(5), 133-135 (1973).
Marshall, S.P., and Brandt, J.L., Installed cost of corrosion-resistant piping, Chem Eng, 28 Oct, 94-106 (1974).
Allen, D.H., and Page, R.C., Revised technique for predesign cost estimating, Chem Eng, 3 March, 142-150 (1975).
Brownstein, A.M., Economics of ethylene glycol processes, Chem Eng Prog, 71(9), 72-76 (1975).
Featherstone, W., Estimating the cost of distillation columns, Processing, May, 21-25 (1975).
Maloney, K.F., Economic potential of steam turbines, Hyd Proc, 54(11), 261-264 (1975).
Nisenfeld, A.E., Cost comparisons of analog-control and computer-control systems, Chem Eng, 18 Aug, 104-106 (1975).
Woods, D.R., and Anderson, S.J., Evaluation of capital cost data: Drives, Can JCE, 53, 357-365 (1975).

1976-1979

Hoerner, G.M., Nomograph updates process equipment costs, Chem Eng, 24 May, 141-143 (1976).
Strohl, K.P., Comparing the costs of thermal and catalytic incinerators, Chem Eng, 7 June, 153-154 (1976).
Woods, D.R.; Anderson, S.J., and Norman, S.L., Evaluation of capital cost data: Heat exchangers, Can JCE, 54, 469-480 (1976).
Brown, T.R., Economic evaluation of future equipment needs, Chem Eng, 17 Jan, 125-127 (1977).
Davis, G.O., Economics of spare equipment, Chem Eng, 21 Nov, 187-191 (1977).
Edwards, M.F., and Stinchcombe, R.A., Cost comparison of gasketed plate heat exchangers and conventional shell and tube units, Chem Engnr, May, 338-341 (1977).
Holland, F.A., and Watson, F.A., Economic penalties of operating a process at reduced capacity, Chem Eng, 3 Jan, 91-94 (1977).
Johnnie, C.C., and Aggarwal, D.K., Calculating plant utility costs, Chem Eng Prog, 73(11), 84-88 (1977).
Kharbana, O.P., Capital cost of storage tanks and pressure vessels, Proc Engng, July, 61 (1977).
Miller, J.S., and Kapella, W.A., Installed cost of a distillation column, Chem Eng, 11 April, 129-133 (1977).
Phadke, P.S., and Kulkarni, P.D., Estimating the costs and weights of process vessels, Chem Eng, 11 April, 157-158 (1977).
Pikulik, A., and Diaz, H.E., Cost estimating for major process equipment, Chem Eng, 10 Oct, 106-122 (1977).
Harker, J.H., Economic pipe diameter, Processing, March, 74 (1978).
Harker, J.H., Economics of evaporator operation, Processing, Dec, 31-32 (1978).
Harker, J.H., Economic lagging thickness, Processing, July, 61 (1978).
Woods, D.R.; Anderson, S.J., and Norman, S.L., Evaluation of capital cost data: Gas-moving equipment, Can JCE, 56, 413-420 (1978).

Harker, J.H., Calculating vessel construction cost, Processing, Oct, 55, (1978); Feb, 35 (1979).
James, J.L., Economic evaluation of revamps, Hyd Proc, 58(11), 261-266 (1979).
Williams, L.F., Process plant and equipment cost estimation, Chem Engnr, Feb, 111; April, 249, 251; June, 448, 451; Dec, 855, 857 (1979).
Woods, D.R.; Norman, S.L., and Anderson, S.J., Economics of capital cost data: Liquid moving equipment and offsite utilities (supply), Can JCE, 57, 385-395; 533-543 (1979).

1980-1982
Chontos, L.W., Find economic pipe diameter via improved formula, Chem Eng, 16 June, 139-142 (1980).
Goodman, D.R., The economics of catalyst operation, Chem Engnr, Feb, 91-94 (1980).
Kharbanda, O.P., Nomographs for cost of coal slurry transport, Proc Engng, July, 41 (1980).
Mol, A., Reduce steam-cracker capital costs, Hyd Proc, 59(4), 233-239 (1980).
Williams, L.F., Process plant and equipment cost estimation, Chem Engnr, Jan, 52, 54; March, 170, 173; April, 230, 232, 236 (1980).
Anon., Costing procedures for heat exchangers and pressure vessels, Proc Econ Int, 2(3), 9-13 (1981).
Barrett, O.H., Installed cost of corrosion-resistant piping, Chem Eng, 2 Nov, 97-102 (1981).
Cran, J., and Viola, J.L., Charting routes to preliminary cost estimates (2 articles - feature report), Chem Eng, 9 April, 64-86 (1981).
Desai, M.B., Preliminary cost estimating of process plants, Chem Eng, 27 July, 65-70 (1981).
Epstein, L.D., Costs of standard vertical storage tanks and reactors, Chem Eng, 13 July, 141-142 (1981).
Flatz, W., Sizing equipment for minimum capital cost for multiproduct plants, Chem Eng, 13 July, 105-115 (1981).
Garrett, L.T., and McHenry, J.M., Analyzing costs of digital and analog control systems, Hyd Proc, 60(12), 103-108 (1981).
Konak, A.R., Predict cooling water treatment cost vs concentration cycles, Hyd Proc, 60(9), 247 (1981).
Mulet, A.; Corripio, A.B., and Evans, L.B., Estimate costs of distillation and absorption towers via correlations, Chem Eng, 28 Dec, 77-82 (1981).
Vatavuk, W.M., and Neveril, R.B., Estimating costs of air-pollution control systems, Chem Eng, 6 Oct, 165-168; 3 Nov, 157-162; 1 Dec, 111-115; 29 Dec, 71-73 (1980); 26 Jan, 127-132; 23 March, 223-228 (1981).
Anon., C4 recycling improves ethylene cracker economics, Proc Econ Int, 3(1), 39-42 (1982).
Anon., Sectional cost estimating of solids process plants, Proc Econ Int, 3(1), 15-21 (1982).
Corripio, A.B.; Chrien, K.S., and Evans, L.B., Estimate costs of centrifugal pumps and electric motors, Chem Eng, 22 Feb, 115-118 (1982).
Corripio, A.B.; Chrien, K.S., and Evans, L.B., Estimate costs of heat exchangers and storage tanks via corelations, Chem Eng, 25 Jan, 125-127 (1982).

Furze, J., Costing plant and equipment, Processing, Nov, 43-47 (1982).
Habit, J., Financing synfuels plants, Hyd Proc, March, 177-203 (1982).
Hall, R.S.; Matley, J., and McNaughton, K.J., Current costs of process equipment, Chem Eng, 5 April, 80-116 (1982).
Lonsdale, J.T., and Mundy, J.E., Estimating pipe heat-tracing costs, Chem Eng, 29 Nov, 89-93 (1982).
McChesney, M., and McChesney, P., Insulation without economics, Chem Eng, 3 May, 70-79 (1982).
Neal, G.W., Evaluating uncertainty in capital cost projections, Chem Eng, 6 Sept, 131-134 (1982).
Ruskan, R.P., Economic evaluation of synfuels plants, Hyd Proc, Jan, 197-220 (1982).
Vatavuk, W.M., and Neveril, R.B., Estimating costs of air-pollution control systems, Chem Eng, 18 May, 171-177; 15 June, 129-130; 7 Sept, 139-140; 30 Nov, 93-96 (1981); 22 March, 153-158; 12 July, 129-132; 4 Oct, 135-136 (1982).
Woods, D.R.; Anderson, S.J., and Norman-Sills, S.L., Evaluation of capital cost data: Onsite utilities (Industrial gases), Can JCE, 60, 173-201 (1982).
Zanker, A., Costing electrostatic precipitators, Proc Engng, Sept, 67 (1982).

1983-1984

Anon., Process equipment capital costs: Packaged boilers (oil-fired); heat exchangers, Proc Econ Int, 4(3), 8-9 (1983).
Anon., Costs of utilities for the process industries, Proc Econ Int, 3(1), 63-64 (1982); 4(2), 26-28 (1983).
Ganapathy, V., Life-cycle costing aids motor selection, Proc Engng, July, 51-52 (1983).
Livingstone, J.G., and Pinto, A., New ammonia process reduces costs, Chem Eng Prog, 79(5), 62-66 (1983).
Nendick, R.M., Economic comparison of carbon-in-pulp and Merill-Crowe processes for precious metal recovery, Proc Econ Int, 3(4), 38-43 (1983).
Purohit, G.P., Estimating costs of shell-and-tube heat exchangers, Chem Eng, 22 Aug, 56-67 (1983).
Seifert, W.F.; Beyrau, J.; Bogel, G., and Wuelpern, L.E., Economic evaluation of a waste-heat-recovery system, Chem Eng, 11 July, 105-110 (1983).
Shah, G.C., Optimizing the economics of cycle lengths for catalytic reactors, Chem Eng, 13 June, 123-126 (1983).
Wetherold, R.G.; Harris, G.E.; Steinmetz, J.I., and Kamas, J.W., Economics of controlling fugitive emissions, Chem Eng Prog, 79(11), 43-48 (1983).
Williams, L.F.; Gerrard, A.M.; Sebastian, D.J.G., and Wheeldon, D.H.V., Using the ECONOMIST package, Proc Econ Int, 4(3), 12-15 (1983).
Anon., Process equipment capital costs: Centrifuges; belt conveyors, Proc Econ Int, 5(1), 7-8 (1984).
Anon., Process equipment capital costs: Floating-head heat exchanger; kettle U-tube heat exchanger, Proc Econ Int, 4(4), 8-9 (1984).
Culler, D.W., Pipe-sizing economics, Chem Eng, 28 May, 113-116 (1984).
Kumana, J.D., Cost update on specialty heat exchangers, Chem Eng, 25 June, 169-172 (1984).

Kuri, C.J., and Corripio, A.B., Two computer programs for equipment cost estimation and economic evaluation of chemical processes, Chem Eng Educ, 18(1), 14-18 (1984).
Nobles, J.E., and Stover, J.C., Electric motor drives have lower operating costs for ammonia production, Chem Eng Prog, 80(1), 81-86 (1984).
Reeves, P., Maximising pay-back on process analyser investment, Proc Engng, Sept, 20-25 (1984).
Vatavuk, W.M., and Neveril, R.B., Estimating costs of air-pollution control systems, Chem Eng, 24 Jan, 131-132; 21 Feb, 89-90; 16 May, 95-98 (1983); 2 April, 97-99; 30 April, 95-98 (1984).

1985-1988
Harker, J.H., Economics of pipe insulation, Proc Engng, Jan, 60-61 (1985).
Purohit, G.P., Costs of double-pipe and multitube heat exchangers, Chem Eng, 4 March, 93-96; 1 April, 85-86 (1985).
Russell, H.W., Assessing costs for equipment selection, Chem Eng, 5 Aug, 103-106 (1985).
Burger, R., Retrofitting cuts cooling-tower costs, Chem Eng, 18 Aug, 117-119 (1986).
Foster-Pegg, R.W., Capital cost of gas-turbine heat-recovery boilers, Chem Eng, 21 July, 73-78 (1986).
Schuart, L., et al., Computer-aided development of quotations for fluidized-bed equipment, Int Chem Eng, 26(3), 419-423 (1986).
St John, B., Economics of atmospheric fluidized-bed boilers, Chem Eng, 8 Dec, 157-159 (1986).
Tolliver, T.L., Improving column control to reduce distillation operating cost, Chem Eng, 24 Nov, 99-101 (1986).
Tuthill, A.H., Installed cost of corrosion-resistant piping, Chem Eng, 3 March, 113-115; 31 March, 125-128; 28 April, 83-84; 26 May, 99-100; 23 June, 131-133 (1986).
Benning, M.A., Estimating costs of process development units, Chem Eng, 19 Jan, 139-141 (1987).
Anon., Estimating cost of instrumentation, Proc Engng, Oct, 16; Nov, 19; Dec, 15 (1988).
Hall, R.S.; Vatavuk, W.M., and Matley, J., Estimating process equipment costs, Chem Eng, 21 Nov, 66-75 (1988).
Raghunathan, S.N., Economics of energy saving systems, Chem Eng, 15 Feb, 109 (1988).

5.3 Economics: General Methods and Data

1967-1969
Allen, D.H., Two new tools for project evaluation, Chem Eng, 3 July, 75-78 (1967).
Anon., Three new economic indicators, Chem Eng, 10 April, 197-198 (1967).
Gallagher, J.T., Rapid estimating of engineering cost, Chem Eng, 19 June, 250-252 (1967).

Grumer, E.L., Selling price vs raw-material cost, Chem Eng, 24 April, 190-192 (1967).
Herron, D.P., Comparing investment evaluation methods, Chem Eng, 30 Jan, 125-132 (1967).
Lutz, J.H., Estimating project-completion costs, Chem Eng, 30 Jan, 164-168 (1967).
Nitchie, E.B., Accounting data and methods, Chem Eng, 19 Dec, 95-101 (1966); 2 Jan, 87-92; 16 Jan, 165-168 (1967).
Ransom, E.A., Guidelines for evaluating new projects, Chem Eng, 6 Nov, 286-292 (1967).
Sturgis, R.P., Control costs and define the scope for big savings, Chem Eng, 14 Aug, 188-190 (1967).
Teplitzky, G., Evaluating the business potential of R&D projects, Chem Eng, 5 June, 136-144 (1967).
Teplitzky, G., Using the profit-and-loss statement, Chem Eng, 19 June, 215-220 (1967).
Thorngren, J.T., Probability technique improves investment analysis, Chem Eng, 14 Aug, 143-151 (1967).
Various, Chemical technology for sale or license (feature report), Chem Eng, 25 Sept, 135-175 (1967).
Childs, J.F., Evaluating project proposals, Chem Eng, 26 Feb, 188-192 (1968).
DeCicco, R.W., Economic evaluation of research projects by computer, Chem Eng, 3 June, 84-90 (1968).
Gallagher, J.T., Analyzing 'cost plus' engineering bids, Chem Eng, 29 Jan, 140-146 (1968).
Gallagher, J.T., Analyzing field construction costs, Chem Eng, 20 May, 182-191 (1968).
Hegarty, W.P., Evaluating the incremental project: An illustrative example, Chem Eng, 9 Sept, 158-162 (1968).
Hegarty, W.P., Evaluating proposed ventures linked to existing facilities, Chem Eng, 12 Aug, 190-194 (1968).
Maristany, B.A., Repair or replace equipment? Chem Eng, 4 Nov, 210-212 (1968).
Moore, C.S., Supply and demand curves in profitability analysis, Chem Eng, 7 Oct, 198-205 (1968).
Murphy, G.H., Stockless purchasing, Chem Eng, 2 Dec, 140-142 (1968).
Park, W.R., and Jackson, D.E., Cash-flow analysis and discounted cash flow, Chem Eng, 1 Jan, 108-110 (1968).
Reul, R.I., Selecting an investment appraisal technique, Chem Eng, 22 April, 212-218 (1968).
Walton, P.R., Sources of error in operating-cost estimates, Chem Eng, 15 July, 150-152 (1968).
Allen, D.H., Economic evaluation and decision-making, Brit Chem Eng, 14(6), 790-793 (1969).
Champley, J.A., A 'business' approach to capital-budgeting decisions, Chem Eng, 22 Sept, 127-132 (1969).
Douglas, F.R., et al., Evaluating capital projects for profitability, Chem Eng, 17 Nov, 274-278 (1969).
Feldman, R.P., Economics of plant startups, Chem Eng, 3 Nov, 87-90 (1969).

Guthrie, K.M., Field-labor predictions for conceptual projects, Chem Eng, 7 April, 170-172 (1969).
Kapfer, W.H., Appraising rate of return methods, Chem Eng Prog, 65(11), 55-60 (1969).
Malloy, J.B., Instant economic evaluation, Chem Eng Prog, 65(11), 47-54 (1969).
Matthews, R.L., and Adams, J.F., Predicting manpower needs in engineering departments, Chem Eng, 30 June, 152-156 (1969).
Reilly, P.M., and Johri, H.P., Investment decision-making through opinion analysis, Chem Eng, 7 April, 122-129 (1969).
Various, Investment decisions (symposium papers), Chem Engnr, Dec, CE421-436 (1969).

1970-1972
Bobis, A.H., and Atkinson, A.C., Analyzing potential research projects, Chem Eng, 23 Feb, 95-100; 9 March, 133-141 (1970).
Congelliere, R.H., Correcting economic analyses, Chem Eng, 16 Nov, 109-112 (1970).
Dyremose, H.B., and Estrup, C., Turn-over ratios analysed, Brit Chem Eng, 15(1), 71-72 (1970).
Estrup, C., Realistic investment calculations, Brit Chem Eng, 15(3), 345-347 (1970).
Garbutt, D., Cost estimating, Chem & Ind, 11 July, 910-912 (1970).
Henney, A., Elements of stock control, Brit Chem Eng, 15(1), 59-62 (1970).
Jenckes, L.C., How to estimate operating costs and depreciation, Chem Eng, 14 Dec, 168-172 (1970).
Jones, L.R., Building-cost escalation, Chem Eng, 27 July, 170-174 (1970).
Klumpar, I.V., Process economics by computer, Chem Eng, 12 Jan, 107-116 (1970).
Klumpar, I.V., Project evaluation by computer, Chem Eng, 29 June, 76-84 (1970).
Kneale, M., A medium-sized chemical company's approach to project cost estimating and control, Chem Engnr, June, CE176-181 (1970).
Leung, T.K.Y., New nomograph for discounted cash flow, Chem Eng, 1 June, 208-209 (1970).
Loonkar, Y.R., and Robinson, J.D., Minimization of capital investment for batch processes, Ind Eng Chem Proc Des Dev, 9(4), 625-629 (1970).
Loring, R.J., Cost of preparing proposals, Chem Eng, 16 Nov, 126-127 (1970).
Prynn, P.J., Cost-benefit analysis, Chem & Ind, 3 Jan, 9-13 (1970).
Sommerfeld, J.T., and Lenk, C.T., Thermodynamics helps predict selling price, Chem Eng, 4 May, 136-138 (1970).
Sommerville, R.F., Estimating costs at low production rates, Chem Eng, 6 April, 148-150 (1970).
Syrett, W.P., and Kimber, F., Rapid estimation of DCF rates of return, Brit Chem Eng, 15(6), 765-767 (1970).
Estrup, C., Evaluation of the pay-back methods by DCF technique, Brit Chem Eng, 16(2), 171 (1971).
Estrup, C., Investment profitability and decentralisation, Brit Chem Eng, 16(4), 357-358 (1971).

Holzapfel, F.J., and Kuhn, W.A., Getting results from process improvements (case studies), Chem Eng, 18 Oct, 132-138 (1971).
Jenckes, L.C., Developing and evaluating a manufacturing-cost estimate, Chem Eng, 11 Jan, 168-170 (1971).
Keim, C.R., Meaningful production costs, Chem Eng, 15 Nov, 184-189 (1971).
Leibson, I., and Trischman, C.A., Economic analysis, Chem Eng, 31 May, 69-74; 14 June, 92-95; 28 June, 95-102; 9 Aug, 103-110; 6 Sept, 86-92; 4 Oct, 85-92; 1 Nov, 78-85; 13 Dec, 97-106 (1971).
Ohsol, E.O., Estimating marketing costs, Chem Eng, 3 May, 116-120 (1971).
Popper, H., The cost of meeting environmental standards, Chem Eng, 23 Aug, 106-108 (1971).
Ross, R.C., Uncertainty analysis helps in making business decisions, Chem Eng, 20 Sept, 149-155 (1971).
Various, Short-term planning of company operations (symposium papers), Chem Engnr, April, 145-168 (1971).
Betts, G.G., Investment appraisal, Chem Engnr, Feb, 78-81 (1972).
Caplan, F., Relating present worth, interest and time, Chem Eng, 20 March, 134 (1972).
Cappello, V.F., Simplifying scaleup cost estimation, Chem Eng, 7 Aug, 99 (1972).
Dobrow, P.V., Value engineering: A money-saving tool, Chem Eng, 21 Aug, 122-126 (1972).
Flanigan, O.; Wilson, W.W., and Sule, D.R., Process cost reduction through linear programming, Chem Eng, 7 Feb, 68-73 (1972).
Leibson, I., and Trischman, C.A., Economic analysis, Chem Eng, 24 Jan, 99-106; 21 Feb, 76-84; 20 March, 113-118; 17 April, 103-112 (1972).
Lyda, T.B., Determining working capital for a new project, Chem Eng, 18 Sept, 182-188 (1972).
Maristany, B.A., Consistent decision rules and methods for economic analysis, Brit Chem Eng, 17(2), 117-118 (1972).
Nathanson, D.M., Statistical techniques for forecasting petrochemical prices, Chem Eng Prog, 68(11), 89-96 (1972).
Simmonds, W.H., Demand forecasting for investment planning, Proc Engng, Aug, 61-62 (1972).
Stroup, R., Breakeven analysis, Chem Eng, 10 Jan, 122-124 (1972).
Walsh, M.J., Functional cost analysis for management, Brit Chem Eng, 17(4), 315-317 (1972).

1973-1975
Abbott, J.T.; Janssen, R.R., and Merz, C.M., Finance for engineers, Chem Eng, 25 June, 108-114 (1973).
Bridgwater, A.V., The build-up of costs, Chem Engnr, Nov, 538-544 (1973).
Eastman, N.S., Cost-improvement techniques spur widespread savings, Chem Eng, 10 Dec, 102-112 (1973).
Gregg, D.P., and Hill, G.W., Stock and production policy: Chances for gain and cost reduction, Chem Engnr, Nov, 544-551 (1973).
Hill, G.V., Costing in physical distribution management, Chem Engnr, Dec, 629-632 (1973).

Holland, F.A.; Watson, F.A., and Wilkinson, J.K., Engineering economics for chemical engineers, Chem Eng, 25 June, 103-107; 23 July, 118-121; 20 Aug, 139-144; 17 Sept, 123-126; 1 Oct, 80-86; 29 Oct, 115-119; 26 Nov, 83-89; 24 Dec, 61-66 (1973).
Jacobs, J.J., Cost-plus vs. fixed-fee construction contracts, Chem Eng, 22 Jan, 109-114 (1973).
Johnson, A.I., and Morgan, J.I., Modular approach to corporate business planning, Proc Tech Int, Jan, 25-32 (1973).
Liebeskind, D., Price forecasting via input/output techniques, Chemtech, Sept, 543-547 (1973).
Ohsol, E.O., Commercial evaluation of a new project, Chemtech, May, 285-289 (1973).
Richardson, J.A., and Templeton, L.R., The impact of plant unreliability upon cost, Chem Engnr, Dec, 625-629 (1973).
Holland, F.A.; Watson, F.A., and Wilkinson, J.K., Engineering economics for chemical engineers, Chem Eng, 7 Jan, 105-110; 4 Feb, 73-79; 4 March, 119-125; 1 April, 71-76; 15 April, 91-96; 13 May, 105-110; 10 June, 83-87; 8 July, 93-98; 5 Aug, 101-106; 2 Sept, 62-66; 16 Sept, 119-124; 28 Oct, 107-112 (1974).
Jelen, F.C., and Cole, M.S., Methods for economic analysis, Hyd Proc, 53(7), 133-139; 53(9), 227-233; 53(10), 161-163 (1974).
Lunger, J.W., Cost control through scope control, Chemtech, July, 413-417, (1974).
Malloy, J.B., Projecting chemical product prices, Chem Eng Prog, 70(9), 77-83 (1974).
Reuben, B., Accounting for inflation, Proc Engng, Sept, 120-125 (1974).
Abrams, H.J., Replace or renovate: Economic decisions, Proc Engng, Sept, 89-92 (1975).
Agarawal, J.C., and Klumpar, I.V., Profitability, sensitivity and risk analysis for project economics, Chem Eng, 29 Sept, 66-72 (1975).
Allen, D.H., Investment decisions: Evaluation techniques in perspective, Chem Engnr, Jan, 42-45 (1975).
Bechtel, L.R., How currency exchange affects investments, Hyd Proc, 54(5), 197-199 (1975).
Blake, A.B., Asset-disposal evaluation, Chemtech, May, 282-286 (1975).
Butler, P., Predicting project costs and inflation, Proc Engng, Oct, 73-77 (1975).
Mapstone, G.E., Calculate discount cash flow quickly, Hyd Proc, 54(12), 99-100 (1975).
Various, Changing economy: Its impact on equipment (feature report), Chem Eng, 17 March, 54-64 (1975).
Various, How to assess inflation of plant costs, Chem Eng, 7 July, 70-85 (1975).
Watson, F.A., Simplified calculation of discounted cash flow rate of return, Chem Engnr, July, 437-438 (1975).
Wells, G., Process optimization and economics, Proc Engng, July, 70-71 (1975).

1976-1979

Dolphin Development Company, Developments in capital cost estimating, Chem Engnr, Jan, 39-41 (1976).

Furzey, D.G., Economic feasibility studies for chemical plant, Chem Engnr, Jan, 33-35 (1976).
Herbert, V.D., and Bisio, A., The risk and the benefit, Chemtech, March, 174-179; July, 422-429; Nov, 691-693 (1976).
Jelen, F.C., Pitfalls in profitability analysis, Hyd Proc, 55(1), 111-115 (1976).
Kline, C., Maximizing profits in chemicals, Chemtech, Feb, 110-117 (1976).
Pollock, M.D., Earned value: Tool for measuring R&D? Chem Eng, 21 June, 175-176 (1976).
Steinmeyer, D., Energy price impacts designs, Hyd Proc, 55(11), 205-210 (1976).
Tacke, D.M., and Thorne, H.C., Predict prices with inflation, Hyd Proc, 55(4), 151-156 (1976).
Whiteley, F., The manager's influence on financial performance, Chem Engnr, Jan, 36-38 (1976).
Wild, N.H., Return on investment made easy, Chem Eng, 12 April, 153-154 (1976).
Carpenter, D.B., The use of custom processing, Chem Eng, 10 Oct, 129-132 (1977).
Childs, E.S., Markets for US thermoplastics, Chem Eng, 12 Sept, 163-168 (1977).
Cran, J., Estimating aids for the chemical industry, Proc Engng, Aug, 66-68 (1977).
Garcia-Borras, T., Research-project evaluation, Hyd Proc, 55(12), 137-146 (1976); 56(1), 171-186 (1977).
Goyette, J., Estimating the costs of steam leaks, Chem Eng, 29 Aug, 95 (1977).
Holland, F.A., and Watson, F.A., Putting inflation into profitability studies, Chem Eng, 14 Feb, 87-91 (1977).
Holland, F.A., and Watson, F.A., Project risk, inflation and profitability, Chem Eng, 14 March, 133-136 (1977).
Wild, N.H., Program for discounted-cash-flow return on investment, Chem Eng, 9 May, 137-142 (1977).
Various, Problems of cyclical investment (topic issue), Chem & Ind, 15 April, 246-253 (1978).
Various, Project cash flow (special report), Hyd Proc, 57(3), 77-98 (1978).
Wild, N.H., Logical calculation of DCF rate of return on investment projects, Chem Engnr, July, 575-579 (1978).
Anon., Cost inflation: The effect of time, Proc Econ Int, 1(2), 29-30 (1979).
Anon., Leasing in the process industries, Proc Econ Int, 1(1), 49-50 (1979).
Gaensslen, H., Short-cut to investment costs, Chemtech, May, 306-309 (1979).
Giles, E.M., Capital investment strategies in the chemical industry, Chem & Ind, 2 June, 373-377 (1979).
Griffith, J.W., and Keely, B.J., Life-cycle costing, Chemtech, April, 242-246 (1979).
Kasner, E., Break-even analysis evaluates investment alternatives, Chem Eng, 26 Feb, 117-118 (1979).
Kirkpatrick, D.M., Calculator program speeds up project financial analysis, Chem Eng, 27 Aug, 103-107 (1979).
Linsley, J., Return on investment: Discounted and undiscounted, Chem Eng, 21 May, 201-204 (1979).

Mascio, N.E., Predict costs reliably via regression analysis, Chem Eng, 12 Feb, 115-121 (1979).
Molleson, A.V.I., Investment control, Chem Engnr, Jan, 47-50 (1979).
Roth, J.E., Controlling construction costs, Chem Eng, 8 Oct, 88-100 (1979).
Various, Project evaluation (topic issue), Comput Chem Eng, 3(8), 21-27; 443-488 (1979).
Various, International project finance (special report), Hyd Proc, 58(12), 71-92 (1979).

1980-1982

Cevidalli, G., and Zaidman, B., Evaluate research projects rapidly, Chem Eng, 14 July, 145-152 (1980).
Cohen, R.L., and Zeftel, L., Materials requirements planning, Chem Eng Prog, 76(4), 59-63 (1980).
Horwitz, B.A., The mathematics of discounted cash flow analysis, Chem Eng, 19 May, 169-174 (1980).
Kerridge, A.E., Control project costs effectively, Hyd Proc, 59(2), 127-146 (1980).
Russell, T.W.F., and Bogaert, R.J., Applying microeconomics to process design, Ind Eng Chem Proc Des Dev, 19(2), 282-289 (1980).
Vatavuk, W.M., Levelized interest payments, Chem Eng, 2 June, 102-104 (1980).
Vervalin, C.H., Major factors in project financing, Hyd Proc, 59(1), 189-195 (1980).
Williams, J.E., Evaluating the cost of capital, Chemtech, April, 226-232 (1980).
Wynton, J.J.S., Control costs using earned value concept, Hyd Proc, 59(9), 289-298 (1980).
Gerrard, A.M., Some forecasting formulae, Chem Engnr, Feb, 70 (1981).
Horwitz, B.A., How does construction time affect return? Chem Eng, 21 Sept, 158 (1981).
Jelen, F.C., and Yaws, C.L., Unify interest compounding, Hyd Proc, 60(4), 223-226 (1981).
Kinsley, G.R., Hedging currency risk when buying foreign equipment, Chem Eng, 20 April, 171-174 (1981).
Naphtali, L.M., and Shinnar, R., Effect of inflation on energy cost analyses, Chem Eng Prog, 77(2), 65-71 (1981).
Pavone, A., and Patrick, G., Energy tax credit aids investment projects, Chem Eng, 23 Feb, 99-104 (1981).
Ryan, J.T., and Haugrud, B., Teaching market analysis, Chem Eng Educ, 15(1), 40-49 (1981).
Strickland, T.H., and Grady, E.C., Capital budgeting in project evaluation, Hyd Proc, 60(3), 179-204; 60(4), 247-269; 60(5), 233-250 (1981).
Various, Successful engineering procurement (topic issue), Chem Engnr, Nov, 501-512 (1981).
Anon., Realistic discounting of cash flows, Proc Econ Int, 2(4), 25-29 (1981); 3(1), 21-22 (1982).
Feldman, M.B., and Rangnow, D.G., Modern gasoline economics, Hyd Proc, Dec, 69-72 (1982).

Furze, J.E., Life cycle costing: An alternative approach to economic evaluation, Proc Econ Int, 3(3), 44-50 (1982).
Malloy, J.B., and Tacke, D.M., Forecasting demand for chemicals via econometric models, Chem Eng, 22 Feb, 101-105 (1982).
McIntyre, D.R., Evaluating the cost of corrosion-control methods, Chem Eng, 5 April, 127-132 (1982).
Monroe, K.B., Pricing decisions, Chemtech, Sept, 546-550 (1982).
Neal, G.W., Calculating net present value for varying cash flows, Chem Eng, 1 Nov, 107-109 (1982).
Neal, G.W., Discounted-cash-flow rates of return for varying cash flows, Chem Eng, 27 Dec, 73-75 (1982).
Pogue, G.A., and Lall, K., A primer on finance, Chemtech, Oct, 604-608 (1982).
Poland, G.F., Estimating cash flows for construction projects, Chem Eng, 9 Aug, 81-83 (1982).
Renshaw, T.A.; Sapakie, S.F., and Hanson, M.C., Concentration processes economics in the food industry, Chem Eng Prog, 78(5), 33-40 (1982).
Sommerfeld, J.T., and Roberts, R.S., Compare investments easier, Hyd Proc, March, 124 (1982).
Wessel, H.E., Improve profits in the 1980s, Hyd Proc, Sept, 327-340; Oct, 117-124 (1982).

1983-1985
Ellingsen, W.R., Operating cost optimization using a dynamic process model, Chem Eng Prog, 79(1), 43-47 (1983).
Hackney, J.W., How to explore the profitability of process projects, Chem Eng, 18 April, 99-101 (1983).
Lepeau, M., Evaluating the economics of heat recovery from burning refuse, Chem Eng, 21 March, 81-84 (1983).
Montfoort, A.G., and Meijer, F.A., Improved Lang factor approach to capital cost estimating, Proc Econ Int, 4(1), 20-22 (1983).
Salem, A.B., Predict profitability fast, Hyd Proc, April, 129-130 (1983).
Sancakter, S., Financial risk assessment at the plant design stage, Plant/Opns Prog, 2(3), 176-182 (1983).
Barna, B.A., Leverage, risk, and project economics, Chemtech, May, 295-297 (1984).
Fathi-Afshar, S., and Rippin, D.W.T., Game theory applied to chemical plant investment decisions in a competitive market, Chem Eng Commns, 26(1), 119-162 (1984).
Field, S., System predicts crude oil prices, Hyd Proc, Oct, 34G-34N (1984).
Hicks, C.L., and Bridgwater, A.V., Overview of modern investment appraisal, Proc Econ Int, 5(1), 12-21 (1984).
Kerridge, A.E., How to evaluate bids for major equipment, Hyd Proc, May, 141-154 (1984).
Konig, N., Financial engineering for turnkey industrial plant, Proc Econ Int, 4(4), 31-35 (1984).
Lunde, K.E., Joint-product costing, Chem Eng, 10 Dec, 89-92 (1984).
Martin, K.L., and Barna, B.A., Project economic analysis, Chem Eng, 23 July, 73-78 (1984).

McIlleron, W.G., and Bosman, J., Perspectives for developments and capital spending in the CPI, Proc Econ Int, 4(4), 16-21 (1984).
Pinches, G.E., Effective use of capital budgeting techniques, Chem Eng Prog, 80(11), 15-19 (1984).
Rubin, A.G., Choose process units via present-worth index, Chem Eng, 15 Oct, 129-132 (1984).
Tayler, C., Computer models for faster project cost estimates, Proc Engng, Dec, 20-21 (1984).
Frey, J.B., Pricing and product life cycle, Chemtech, Jan, 40-43 (1985).
Kerridge, A.E., Make your own econometric model, Hyd Proc, Nov, 217-223 (1985).
Lunde, K.E., Transfer pricing, Chem Eng, 4 Feb, 85-87 (1985).
Lunde, K.E., Joint-product costing, Chem Eng, 10 Dec, 89-92 (1984); 7 Jan, 95-100 (1985).
Powell, T.E., A review of recent developments in project evaluation, Chem Eng, 11 Nov, 187-194 (1985).
Taylor, J., Lease or buy, Chemtech, June, 334-336 (1985).
Woolgar, R., Improving productivity of capital plant, Proc Engng, April, 33-39 (1985).

1986-1988

Birkler, J.L.; Micklish, W.H., and Merrow, E.W., Comparing costs of oil refinery projects, Chem Eng, 29 Sept, 129-136 (1986).
Dunlap, J.L., Determining a proper parts inventory (ways to minimize costs), Chem Eng, 29 Sept, 93-97 (1986).
Hickman, W.E., and Moore, W.D., Managing the maintenance dollar, Chem Eng, 14 April, 68-77 (1986).
Intille, G.M., How refinery inventories threaten profitability, Hyd Proc, 65(7), 90-100 (1986).
Nowakowski, G., Graphically determine payback/technical merit of energy-conservation projects, Chem Eng, 10 Nov, 136 (1986).
Shinnar, R., and Dressler, O., Return on investments, Chemtech, Jan, 30-41 (1986).
Wilkinson, J., Life cycle costing, Proc Engng, Feb, 42-44 (1986).
Brestovansky, D.F., Rippin, D.W.T., and Russell, T.W.F., Microeconomic predictions for design strategy compared with case histories, Ind Eng Chem Res, 26(12), 2509-2515 (1987).
Finley, H.F., Reduce costs with knowledge-based maintenance, Hyd Proc, 66(1), 64-66 (1987).
Linsley, J., Quick graphical procedure for return on investment: Are underlying estimates good enough? Chem Eng, 22 June, 161-164 (1987).
Nelson, K.E., and Cunningham, K., Checklist for reviewing energy-and-yield-saving projects, Chem Eng, 14 Sept, 115-116 (1987).
Parkinson, G., New study provides clues to project-cost overruns, Chem Eng, 17 Aug, 33-37 (1987).
Proctor, A., Maximising plant performance, Proc Engng, Oct, 45-47 (1987).
Strauss, R., Use and misuse of manufacturing cost systems, Chem Eng, 16 March, 103-106; 27 April, 65-70; 25 May, 87-88 (1987).

Williams, K.A., and Holmes, J.M., Simulation of process economics for pioneer plants, Chem Eng Prog, 83(1), 31-36 (1987).
Anon., HPI spending outlook for 1989 (special report), Hyd Proc, 67(12), 70-91 (1988).
Kerridge, A.E., Measure project progress and performance effectively, Hyd Proc, 67(6), 69-74 (1988).
Kerridge, A.E., Project cost control, Hyd Proc, 67(5), 97-102 (1988).
Wirasinghe, E., Comparing project investments, Hyd Proc, 67(6), 66-67 (1988).

CHAPTER 6

SAFETY AND LOSS PREVENTION

6.1 Accidents

1967-1975

Gilmore, C.L., A statistical approach to preventing accidents, Chem Eng, 17 July, 224-230 (1967).
Masso, A.H., and Rudd, D.F., Disaster propagation, Ind Eng Chem Fund, 7(1), 131-141 (1968).
Emerick, R.H., A near-death day in the plant, Hyd Proc, 48(4), 163-164 (1969).
Flory, K.; Paoli, R., and Mesler, R., Molten metal-water explosions, Chem Eng Prog, 65(12), 50-54 (1969).
LeRoy, N.L., and Johnson, D.M., The story of a reservoir fire, Hyd Proc, 48(5), 129-132 (1969).
Webb, H.E., Accident investigation, Chem Eng, 24 Feb, 88-90 (1969).
Buehler, J.H., Report on explosion at Union Carbide's Texas City butadiene refining unit, Chem Eng, 7 Sept, 77-86 (1970).
Crocker, B.B., Preventing hazardous pollution during plant catastrophes, Chem Eng, 4 May, 97-102 (1970).
Griffith, S., and Keister, R.G., This butadiene unit exploded, Hyd Proc, 49(9), 323-327 (1970).
Constance, J.D., Pressure ventilate for explosion protection, Chem Eng, 20 Sept, 156-158 (1971).
Constance, J.D., Control of explosive or toxic air-gas mixtures, Chem Eng, 19 April, 121-124 (1971).
Katz, D.L., and Sliepcevich, C.M., LNG/water explosions: Cause and effect, Hyd Proc, 50(11), 240-244 (1971).
Munday, G., Detonations in vessels and pipelines, Chem Engnr, April, 135-144, 152 (1971).
Nakanishi, E., and Reid, R.C., LNG-water reactions and explosions, Chem Eng Prog, 67(12), 36-41 (1971).
Crouch, W.W., and Hillyer, J.C., What happens when LNG spills? Chemtech, April, 210-215 (1972).
O'Reilly, B.M., Dust explosions in factories, Chem & Ind, 1 Jan, 27-30 (1972).
de Oliveira, D.B., This paraffin chlorination unit exploded, Hyd Proc, 52(3), 112-126 (1973).
Drake, E.M.; Geist, J.M., and Smith, K.A., Prevent LNG 'rollover', Hyd Proc, 52(3), 87-90 (1973).
Lees, F.P., Some data on failure modes of instruments in chemical plant environment, Chem Engnr, Sept, 418-421 (1973).
Theimer, O.F., Cause and prevention of dust explosions in grain elevators and flour mills, Powder Tech, 8, 137-147 (1973).
Wesson, H.R.; Welker, J.R.; Brown, L.E., and Sliepcevich, C.M., Fight LNG fires, Hyd Proc, 52(10), 165-171; 52(11), 234-244 (1973).
Lawrence, W.E., and Johnson, E.E., Design for limiting explosion damage, Chem Eng, 7 Jan, 96-104 (1974).
Thompson, M.R., Methane explosions and their prevention, Energy World, Jan, 9-14 (1974).
Vervalin, C.H., API/NFPA report on HPI fire losses, Hyd Proc, 53(5), 182-184 (1974).
Ball, J.G., After the Flixborough report, Proc Engng, Dec, 39-46 (1975).

Butler, P., Flixborough explosion, Proc Engng, June, 83 (1975).
Drake, E.M., and Reid, R.C., Boiling of LNG spills, Hyd Proc, 54(5), 191-194 (1975).
Freeman, R., The Cherokee ammonia plant explosion, Chem Eng Prog, 71(11), 71-74 (1975).
King, R., Flixborough analysed, Proc Engng, Sept, 69-72 (1975).
Palmer, K.N., Explosions in dust collection plant, Chem Engnr, March, 136-137, 142 (1975).
Vervalin, C.H., Latest API/NFPA fire-loss report, Hyd Proc, 54(6), 149-154 (1975).
Warner, F., The Flixborough disaster, Chem Eng Prog, 71(9), 77-84 (1975).
Wirth, G.F., Preventing and dealing with in-plant hazardous spills, Chem Eng, 18 Aug, 82-96 (1975).

1976-1979

Ball, J.G., Flixborough explosion, Proc Engng, Jan, 53-60 (1976).
Kletz, T.A., Accidents caused by reverse flow, Hyd Proc, 55(3), 187-194 (1976).
Kletz, T.A., Preventing catastrophic accidents, Chem Eng, 12 April, 124-128 (1976).
Porteous, W.M., and Reid, R.C., Light hydrocarbon vapor explosions, Chem Eng Prog, 72(5), 83-89 (1976).
Various, Technical lessons of Flixborough (symposium papers), Chem Engnr, April, 266-281; May, 341-358 (1976).
Vervalin, C.H., Fire losses reported by API/NFPA, Hyd Proc, 55(5), 291-298 (1976).
Vervalin, C.H., Restrict vinyl chloride exposure, Hyd Proc, 55(2), 182-186 (1976).
Adno, H., et al., Pollution, safety, and disaster prevention, Int Chem Eng, 17(3), 443-468 (1977).
Bellingham, B., and Lees, F.P., The detection of malfunction using a process control computer; Part 1: A simple filtering technique for flow control loops; Part 2: A Kalman filtering technique for general control loops, Trans IChemE, 55, 1-16, 253-265 (1977).
Davenport, J.A., Prevent vapor cloud explosions, Hyd Proc, 56(3), 205-214 (1977).
Marshall, V.C., How lethal are explosions and toxic escapes? Chem Engnr, Aug, 573-577 (1977).
McKay, F.F.; Worrell, G.R.; Thornton, B.C., and Lewis, H.L., Ethylene pipeline ruptures, Hyd Proc, 56(11), 487-494 (1977).
Mecklenburgh, J.C., The investigation of major process disasters, Chem Engnr, Aug, 578-581, 586 (1977).
Monroy, A.D., and Majul, V.M.G., Stop fires in ethylene oxide plants, Hyd Proc, 56(9), 175-176 (1977).
Vervalin, C.H., Fire losses reported by NFPA, Hyd Proc, 56(2), 166-167 (1977).
Webb, H.E., What to do when disaster strikes, Chem Eng, 1 Aug, 46-58 (1977).
Schmit, K.H.; Lee, T.E.; Mitchen, J.H., and Spiteri, E.E., Clean-up of aluminum alkyl spills, Hyd Proc, 57(11), 341-353 (1978).

Vervalin, C.H., HPI loss-incident case histories, Hyd Proc, 57(2), 183-201 (1978).
Brown, L.E., and Romine, L.M., Liquified-gas fires: Which foam? Hyd Proc, 58(9), 321-332 (1979).
Cocks, R.E., Dust explosions: Prevention and control, Chem Eng, 5 Nov, 94-101 (1979).
Freshwater, D.C., Unconfined vapour cloud explosions: A review, Chem Engnr, Jan, 54-56 (1979).
Haddock, S.R., and Williams, R.J., Density of an ammonia cloud in the early stages of atmospheric dispersion, J Chem Tech Biotechnol, 29, 655-672 (1979).
Harvey, B., Flixborough: Five years later, Chem Engnr, Oct, 697-698 (1979).
Herzog, G.R., Tank-farm fires, Hyd Proc, 58(2), 165-168 (1979).
Kletz, T.A., Learn from HPI fires, Hyd Proc, 58(1), 243-250 (1979).
Kletz, T.A., A decade of safety lessons (case histories), Hyd Proc, 58(6), 195-204 (1979).

1980-1982
Ale, B.J.M., and Bruning, I.F., Unconfined vapour cloud explosions, Chem Engnr, Jan, 47-48 (1980).
Cobb, E.C., and Marshall, V.C., Three Mile Island, Harrisburg: The report of the President's Commission, Chem Engnr, Feb, 107-108 (1980).
Cronje, J.S.; Bishnoi, P.R., and Svrcek, W.Y., Application of characteristic method to shock tube data that simulate a gas pipeline rupture, Can JCE, 58, 289-294 (1980).
Gugan, K., Unconfined vapour cloud explosions, Chem Engnr, July, 491-494; Aug, 567-568 (1980).
Marshall, V.C., TNT equivalence: The Decatur anomaly, Chem Engnr, Feb, 108-109 (1980).
Marshall, V.C., Seveso: An analysis of the official report, Chem Engnr, July, 499-500, 516 (1980).
Stevens, J.J., What happened at Seveso? Chem & Ind, 19 July, 564-566 (1980).
Clark, S., Preventing dust explosions, Chem Eng, 5 Oct, 153-154 (1981).
Crain, S., Another tank farm fire, Hyd Proc, 60(1), 262-269 (1981).
McIntyre, D.R., Estimating temperature extremes of accidental fires, Chem Eng, 15 June, 123-126 (1981).
Baker, W.E., Explosion accident investigation, Plant/Opns Prog, 1(3), 144-147 (1982).
Delichatsios, M.A., Exposure of steel drums to external spill fire, Plant/Opns Prog, 1(1), 37-45 (1982).
Kletz, T.A., Three Mile Island: Lessons for the HPI, Hyd Proc, June, 187-192 (1982).
Lai, F.S.; Garrett, D.W., and Fan, L.T., Literature review of mechanisms of grain dust explosions as affected by particle size, Powder Tech, 32, 193-202 (1982).
Lees, F.P., The hazard warning structure of major hazards, Trans IChemE, 60, 211-221 (1982).
Marshall, V.C., Unconfined-vapor-cloud explosions, Chem Eng, 14 June, 149-154 (1982).

Norstrom, G.P., Fire/explosion losses in the chemical process industry, Chem Eng Prog, 78(8), 80-87 (1982).
Smith, D.W., Runaway reactions and thermal explosion, Chem Eng, 13 Dec, 79-84 (1982).
Solberg, D.M., Industrial gas explosion problems, Plant/Opns Prog, 1(4), 243-248 (1982).
Van Meerbeke, R.C., Accident at the Cove Point LNG facility, Chem Eng Prog, 78(1), 39-46 (1982).
Zilka, M.I., Assessing the cause of an accident, Chem Eng, 17 May, 121-124 (1982).

1983-1984

Cohn, D., Chain reaction power system failure in fertilizer complex, Plant/Opns Prog, 2(3), 168-173 (1983).
Elliot, W.H., The Chambers Works generic VOC bubble, Plant/Opns Prog, 2(1), 38-40 (1983).
Friedrichsen, F.G., An ammonia accident in Denmark, Plant/Opns Prog, 2(2), 122-123 (1983).
Heestand, J.; Shipman, C.W., and Meader, J.W., Predictive model for rollover in stratified LNG tanks, AIChEJ, 29(2), 199-207 (1983).
Lee, J.H.S., Explosion in vessels: Recent results, Plant/Opns Prog, 2(2), 84-89 (1983).
Lloyd, W.D., Methanol synthesis gas explosion, Plant/Opns Prog, 2(2), 120-122 (1983).
Prijatel, J., Accidental venting of liquid ammonia, Plant/Opns Prog, 2(2), 131-136 (1983).
Smith, M.E., et al., Atmospheric modeling for emergencies, Plant/Opns Prog, 2(1), 61-67 (1983).
Solberg, D.M., and Borgnes, O., Thermal response of process equipment to hydrocarbon fires, Plant/Opns Prog, 2(1), 50-58 (1983).
Sweat, M.E., An ammonia tank failure, Plant/Opns Prog, 2(2), 114-116 (1983).
Theofanous, T.G., Physiochemical origins of the Seveso accident, Chem Eng Sci, 38(10), 1615-1636 (1983).
Underwood, J.G., Persistant contamination from an industrial source: Seveso an example, Proc Econ Int, 4(1), 22-30 (1983).
Williams, G.P., and Hoehing, W.W., Causes of ammonia plant shutdowns, Chem Eng Prog, 79(3), 11-30 (1983).
Al-Ameeri, R.S.; Akashah, S.A.; Akbar, A.M., and Alawi, H.S.H., A $100-million vapour cloud fire, Hyd Proc, Nov, 181-188 (1984).
Anon., Case histories of plant design failures, Proc Engng, April, 57-59 (1984).
de Nevers, N., Pipe rupture by sudden fluid heating, Ind Eng Chem Proc Des Dev, 23(4), 669-674 (1984).
Elsworth, J.; Eyre, J., and Wayne, D., Liquified gas spillages in partially confined spaces, Chem Engnr, Feb, 26-31 (1984).
Grenier, M.L., Emergency response procedures for anhydrous ammonia vapor release, Plant/Opns Prog, 3(2), 66-71 (1984).
Harbert, F., Detection of fugitive emissions using laser beams, Chem Engnr, Oct, 41-43 (1984).

Hirano, T., Gas explosion processes in enclosures, Plant/Opns Prog, 3(4), 247-254 (1984).
Kletz, T.A., Prevention of major leaks: Better inspection after construction? Plant/Opns Prog, 3(1), 19-25 (1984).
Kletz, T.A., Accident investigation: How far should we go? Plant/Opns Prog, 3(1), 1-3 (1984).
Kletz, T.A., The Flixborough explosion: Ten years later, Plant/Opns Prog, 3(3), 133-136 (1984).
Lai, F.S.; Garrett, D.W., and Fan, L.T., Effect of particle size and composition on mechanisms of grain dust explosion, Powder Tech, 39, 263-278 (1984).
Markatos, N.C.; Rawnsley, S.M., and Spalding, D.B., Heat transfer during a small-break loss-of-coolant accident in a pressurized water reactor, Int J Heat Mass Trans, 27(8), 1379-1394 (1984).
Mullier, A.; Rustin, A., and Van Hecke, F., Fire in a compressor house, Plant/Opns Prog, 3(1), 46-50 (1984).
Rider, R.L., The Three Mile Island situation, Chemtech, Dec, 737-739 (1984).
Various, Flixborough - Ten years on, Chem Engnr, June, 25-33 (1984).

1985-1986

Anon., Chemical catastrophes continue, Processing, March, 43 (1985).
Anon., The chemical industry after Bhopal, Processing, Dec, 19-20 (1985).
Berenblut, B.J., and Whitehouse, S.M., Pemex: The forgotton disaster, Chem Engnr, Oct, 16-22 (1985).
Golec, R.A., The Milford Haven tank fire, Hyd Proc, Oct, 97-98 (1985).
Howard, W.B., Seveso: Cause and prevention, Plant/Opns Prog, 4(2), 103-105 (1985).
Kharbanda, O., Another Bhopal? Never again, Chem Engnr, July, 40 (1985).
King, R., Controlling major hazards, Proc Engng, May, 61,63 (1985).
Lepkowski, W., Bhopal: One year later, C&E News, 2 Dec, 18-32 (1985).
Lessinger, J.E., Medical management of anhydrous ammonia emergencies, Plant/Opns Prog, 4(1), 20-26 (1985).
McQuaid, J., Trials on dispersion of heavy gas clouds, Plant/Opns Prog, 4(1), 58-61 (1985).
Moodie, K., Use of water spray barriers to disperse spills of heavy gases, Plant/Opns Prog, 4(4), 234-242 (1985).
Naujokas, A.A., Spontaneous combustion of carbon beds, Plant/Opns Prog, 4(2), 120-126 (1985).
Pikaar, M.J., Unconfined vapour cloud dispersion and combustion: An overview of theory and experiments, CER&D, 63, 75-81 (1985).
Prugh, R.W., Mitigation of vapor cloud hazards, Plant/Opns Prog, 4(2), 95-103 (1985).
Various, The Sizewell B pressurized water reactor inquiry, Energy World, Dec, 2-13 (1985).
Dahn, C.J.; Ashum, M., and Williams, K., Contribution of low-level flammable concentrations to dust explosion output, Plant/Opns Prog, 5(1), 57-64 (1986).
Deckert, A.H., Emergency responses to spills of sulfuric acid during transit, Chem Eng Prog, 82(7), 11-13 (1986).

Goldwire, H.C., Large-scale ammonia spill tests, Chem Eng Prog, 82(4), 35-41 (1986).
Klem, T.J., Explosion in cold storage kills fire fighter, Plant/Opns Prog, 5(1), 27-30 (1986).
Lihou, D., 'Hazards IX' notebook, Chem Engnr, July, 41-43 (1986).
Lihou, D., Case studies of inadequate instrumentation, Chem Engnr, Nov, 41 (1986).
Martinsen, W.E.; Johnson, D.W., and Terrell, W.F., BLEVEs: Their causes, effects and prevention (BLEVE - boiling liquid, expanding vapor explosion), Hyd Proc, 65(11), 141-148 (1986).
McDaniel, J.T., Explosion of Benfield solution storage tank, Plant/Opns Prog, 5(1), 45-48 (1986).
Prugh, R.W., Mitigation of vapor cloud hazards, Plant/Opns Prog, 5(3), 169-174 (1986).
Roberts, R.H., and Handman, S.E., Minimize ammonia releases, Hyd Proc, 65(3), 58-64 (1986).
Sparrow, R.E., Firebox explosion in primary reformer furnace, Plant/Opns Prog, 5(2), 122-128 (1986).
Steinbrecher, L., Analysis of a tank-fire 'classic', Hyd Proc, 65(5), 85-91 (1986).

1987-1988

Brown, S.J., Effects of system dynamics on failure: Illustrative examples, Plant/Opns Prog, 6(1), 20-34 (1987).
Cartwright, P., Dust explosions, Processing, Feb, 31-32 (1987).
Costanza, P.A., et al., Estimation of catastrophic quantities of toxic chemicals by atmospheric dispersion modeling, Plant/Opns Prog, 6(4), 215-220 (1987).
Kletz, T.A., Another pipe failure, Chem Engnr, Jan, 32 (1987).
McRae, M.H., Anhydrous ammonia explosion in ice cream plant, Plant/Opns Prog, 6(1), 17-19 (1987).
Prugh, R.W., Guidelines for vapor release mitigation, Plant/Opns Prog, 6(3), 171-174 (1987).
Various, Control of industrial major accident hazards (topic issue), Chem & Ind, 2 Feb, 77-88 (1987).
Various, The Chernobyl accident and nuclear safety in the UK, Energy World, Feb, 7-9 (1987).
Anon., Analysis of Bhopal disaster, Chem Engnr, June, 6 (1988).
Anon., Piper Alpha explosion: Interim report, Chem Engnr, Nov, 9 (1988).
Atkinson, N., Why accidents happen, Proc Engng, June, 35-38 (1988).
de Nevers, N., Chemical engineers as expert witnesses in accident cases, Chem Eng Prog, 84(6), 22-27 (1988).
Garrison, W.G., Major fires and explosions analyzed for 30-year period, Hyd Proc, 67(9), 115-120 (1988).
Gibson, T.O., Loss prevention features perform in fired-heater loss incident, Plant/Opns Prog, 7(4), 258 (1988).
Holton, G.A., and Montague, D.A., Application of health effects data to chemical process accidents, Plant/Opns Prog, 7(3), 204-208 (1988).

Kletz, T.A., Fires and explosions of hydrocarbon oxidation plants, Plant/Opns Prog, 7(4), 226-230 (1988).
Koopman, R.P., Atmospheric dispersion of large-scale spills, Chem Eng Commns, 63, 61-86 (1988).
Prokop, J., The Ashland tank collapse, Hyd Proc, 67(5), 105-108 (1988).
Seltzer, R.J., Explosion of Nevada rocket oxidizer plant, C&E News, 8 Aug, 7-15 (1988).
Studer, D.W.; Cooper, B.A., and Doelp, L.C., Vaporization and dispersion modeling of contained refrigerated liquid spills, Plant/Opns Prog, 7(2), 127-135 (1988).
Swift, I., Developments in explosion protection, Plant/Opns Prog, 7(3), 159-168 (1988).
Various, Bhopal: The causes, Chem Engnr, Aug, 4, 6 (1988).

6.2 HAZOP and Assessment Methods

1969-1975

Browning, R.L., Estimating loss probabilities, Chem Eng, 15 Dec, 135-140 (1969).
Browning, R.L., Calculating loss exposures, Chem Eng, 17 Nov, 239-244 (1969).
Preddy, D.L., Guidelines for safety and loss prevention, Chem Eng, 21 April, 94-108 (1969).
Browning, R.L., Finding the critical path to loss, Chem Eng, 26 Jan, 119-124 (1970).
Allen, O.M., and Hanna, O.I., Arrow diagrams improve operational safety, Chem Eng, 20 Sept, 166-168 (1971).
Clancey, V.J., Assessment of explosion hazards, Chem & Ind, 19 Feb, 145-149 (1972).
Leone, L.C., Pinpointing losses in batch processes, Chem Eng, 19 March, 132-134 (1973).
Albrecht, A.R.,, Hazard classification and protection, Chemtech, Nov, 690-693 (1974).
Powers, G.J., and Tompkins, F.C., Fault-tree synthesis for chemical processes, AIChEJ, 20(2), 376-387 (1974).
Rivas, J.R., and Rudd, D.F., Synthesis of failure-safe operations, AIChEJ, 20(2), 320-325 (1974).
Various, Hazards and loss prevention (topic issue), Chem Engnr, Oct, 617-641, 652 (1974).
Browning, R.L., Analyze losses by diagram, Hyd Proc, 54(9), 253-260 (1975).
Henley, E.J., and Gandhi, S.L., Process reliability analysis, AIChEJ, 21(4), 677-686 (1975).
Wilson, R., Examples in risk-benefit analysis, Chemtech, Oct, 604-607 (1975).

1976-1980

Browning, R.L., Human factors in the fault tree, Chem Eng Prog, 72(6), 72-75 (1976).

Caceres, S., and Henley, E.J., Process failure analysis by block diagrams and fault trees, Ind Eng Chem Fund, 15(2), 128-134 (1976).
Gibson, S.B., Risk criteria in hazard analysis, Chem Eng Prog, 72(2), 59-62 (1976).
Hearfield, F., The philosophy of loss prevention, Chem Engnr, April, 257-259 (1976).
Lawley, H.G., Size up plant hazards this way, Hyd Proc, 55(4), 247-261 (1976).
Meadows, D.G., Hazard analysis, Processing, April, 13, 16 (1976).
Collacott, R.A., Loss prevention through fault diagnosis, Chem Engnr, Aug, 582-586 (1977).
Kolodner, H.J., Applications of fault-tree analysis, Hyd Proc, 56(9), 303-308 (1977).
Lambert, H.E., Fault trees for locating sensors in process systems, Chem Eng Prog, 73(8), 81-85 (1977).
Moores, C.W., Guidelines for pre-environmental-impact-assessment inspection, Hyd Proc, 56(3), 173-184 (1977).
Rowe, W.D., Risk, Chemtech, Aug, 477-483 (1977).
Shaeiwitz, J.A.; Lapp, S.A., and Powers, G.J., Fault-tree analysis of sequential systems, Ind Eng Chem Proc Des Dev, 16(4), 529-549 (1977).
Volkman, Y., Simple method for safety factor evaluation, AIChEJ, 23(2), 203-205 (1977).
Kumamoto, H., and Henley, E.J., Protective system hazard analysis, Ind Eng Chem Fund, 17(4), 274-276 (1978).
Hanna, R.C., Apply decision theory to hazard evaluation, Hyd Proc, 58(12), 133-150 (1979).
Krister, C.J., Safety evaluation, Chemtech, Nov, 668-672 (1979).
Menzies, R.M., and Strong, R., Some methods of loss prevention, Chem Engnr, March, 151-155, 160 (1979).
Slovic, P.; Fischhoff, B., and Lichtenstein, S., Rating the risks, Chemtech, Dec, 738-744 (1979).
Allen, D.J., and Rao, M.S.M., New algorithms for the synthesis and analysis of fault trees, Ind Eng Chem Fund, 19(1), 79-85 (1980).
Beyers, C.K., Failure analysis: Tool for HPI loss prevention, Hyd Proc, 59(8), 155-160 (1980).
Lees, F.P., Some aspects of hazard survey and assessment, Chem Engnr, Dec, 736-741 (1980).
Pilz, V., Fault tree analysis, Hyd Proc, 59(5), 275-283 (1980).
Prugh, R.W., Application of fault tree analysis, Chem Eng Prog, 76(7), 59-67 (1980).

1981-1984

Andow, P.K., Fault trees and failure analysis: Discrete state representation problems, Trans IChemE, 59, 125-128 (1981).
Kumamoto, H.; Inoue, K., and Henley, E.J., Computer-aided protective system hazard analysis, Comput Chem Eng, 5(2), 93-98 (1981).
Roach, J.R., and Lees, F.P., Some features of and activities in HAZOP studies, Chem Engnr, Oct, 456-462 (1981).
Wells, G.L., Safety reviews and plant design, Hyd Proc, 60(1), 241-261 (1981).

Calder, W., Intrinsic safety: Effects on loss prevention, Plant/Opns Prog, 1(1), 12-14 (1982).
Challis, H., Hazard assessment for process plant, Proc Engng, May, 37-39 (1982).
Doelp, L.C., and Brian, P.L.T., Reliability of pressure-protective systems: Markov analysis, Ind Eng Chem Fund, 21(2), 101-109 (1982).
Martin-Solis, G.A.; Andow, P.K., and Lees, F.P., Fault tree synthesis for design and real time applications, Trans IChemE, 60, 14-25 (1982).
O'Brien, G.J.; Gordon, M.D.; Hensler, C.J., and Marcali, K., Thermal stability hazards analysis, Chem Eng Prog, 78(1), 46-49 (1982).
Schreiber, A.M., Using event trees and fault trees, Chem Eng, 4 Oct, 115-120 (1982).
Bergmann, E.P., and Riegel, J.P., Deep Water Port: Fire and explosion hazards assessment, Plant/Opns Prog, 2(2), 89-99 (1983).
Freeman, R.A., Problems with risk analysis in the chemical industry, Plant/Opns Prog, 2(3), 185-191 (1983).
Hauptmanns, U., and Yllera, J., Fault-tree evaluation by Monte Carlo simulation, Chem Eng, 10 Jan, 91-97 (1983).
Holden, P., Quantitative risk assessment, Proc Engng, Dec, 32-35 (1983).
Kletz, T.A., Loss prevention: Some future problems, Chem Engnr, Feb, 33-35 (1983).
Watanabe, K., and Himmelblau, D.M., Fault diagnosis in nonlinear chemical processes, AIChEJ, 29(2), 243-261 (1983).
Willis, B.H., and Henebry, W.M., The EIS: A current perspective, Hyd Proc, Nov, 223-242 (1982); Jan, 135-152 (1983).
Allen, D.J., Digraphs and fault trees, Ind Eng Chem Fund, 23(2), 175-180 (1984).
Bendixen, L.M., and O'Neill, J.K., Chemical plant risk assessment using HAZOP and fault tree methods, Plant/Opns Prog, 3(3), 179-184 (1984).
Conrad, J., Total plant-safety audit, Chem Eng, 14 May, 83-86 (1984).
Doelp, L.C., et al., Quantitative fault tree analysis: Gate-by-gate method, Plant/Opns Prog, 3(4), 227-239 (1984).
Redmond, T., Hazard assessment for insurance purposes, Chem Engnr, Aug, 17-21 (1984).
Rodricks, J.V., and Tardiff, R.C., Risk assessment procedures, Chemtech, July, 394-397 (1984).

1985-1986

Gehring, P.J., Chemicals and risk, Chemtech, Sept, 522-525 (1985).
Gillett, J., Rapid ranking of process hazards, Proc Engng, Feb, 19-22 (1985).
Holden, P.L.; Lowe, D.R.T., and Opschoor, G., Risk analysis in the process industries: An ISGRA update, Plant/Opns Prog, 4(2), 63-68 (1985).
Kletz, T.A., Eliminating potential process hazards, Chem Eng, 1 April, 48-68 (1985).
Morgan, J., Cost-effective fault tree analysis for safer plants, Proc Engng, Jan, 28-31 (1985).
Ozog, H., Hazard identification analysis and control, Chem Eng, 18 Feb, 161-170 (1985).

Arendt, J.S.; Casada, M.L., and Rooney, J.J., Reliability and hazards analysis of cumene-hydroperoxide plant, Plant/Opns Prog, 5(2), 97-102 (1986).
Keey, R.B., Use of hazard-warning analysis, Chem Eng & Proc, 20(6), 289-296 (1986).
Lihou, D., Operability studies for busy people, Chem Engnr, May, 52-53 (1986).
Page, L.B., and Perry, J.E., A simple approach to fault-tree probabilities, Comput Chem Eng, 10(3), 249-257 (1986).
Vervalin, C.H., Hazard evaluation, Hyd Proc, 65(12), 35-40 (1986).

1987-1988
Al-Abdulally, F.; Al-Shuwaib, S., and Gupta, B.L., Hazard analysis and safety considerations in refrigerated ammonia storage tanks, Plant/Opns Prog, 6(2), 84-88; 6(4), 06-08 (1987).
Barry, T.F., Computer room risk assessment, Plant/Opns Prog, 6(2), 68-72 (1987).
Bendell, A., How to collect and use process plant reliability data, Proc Engng, June, 39-42 (1987).
Goldthwaite, W.H., et al., Guidelines for hazard evaluation procedures: Preparation and content, Plant/Opns Prog, 6(2), 63-67 (1987).
Kalinins, R.V., Emergency preparedness and response, Plant/Opns Prog, 6(1), 6-10 (1987).
Olsen, K.R., Hazard communication, Chem Eng, 13 April, 107-110 (1987).
Ozog, H., and Bendixen, L.M., Hazard identification and quantification, Chem Eng Prog, 83(4), 55-64 (1987).
Pitblado, R.M., and Lake, I.A., Guidelines for the application of hazard warning, CER&D, 65(4), 334-341 (1987).
Rich, S.H., and Venkatasubramanian, V., Model-based reasoning in diagnostic expert systems for chemical process plants, Comput Chem Eng, 11(2), 111-122 (1987).
Roberts, A.F., Health and Safety Executive (UK) research programme on loss prevention, CER&D, 65(4), 291-298 (1987).
Scheid, D.C., Health, safety and environmental protection: A new audit program, Plant/Opns Prog, 6(4), 211-214 (1987).
Various, Loss prevention review (a supplement), Chem Eng, Aug, 6-17 (1987).
Atkinson, N., Hazop evaluation in small companies, Proc Engng, March, 47-51 (1988).
Farquharson, G., Evaluating risks of pharmaceutical design, Processing, Nov, 19-22 (1988).
Finch, F.E., and Kramer, M.A., Narrowing diagnostic focus using functional decomposition, AIChEJ, 34(1), 25-36 (1988).
Huber, P.W., Private risk and public risk, Chemtech, Aug, 467-471 (1988).
Kletz, T.A., Indices of risk, Chem Engnr, Feb, 41 (1988).
Kletz, T.A., Computers and HAZOP, Chem Engnr, Oct, 52 (1988).
Lande, S.S., Risk assessment of products, Chemtech, Feb, 110-113 (1988).

6.3 Legislation

1973-1988

Ludwig, E.E., Designing process plants to meet OSHA requirements, Chem Eng, 3 Sept, 88-100 (1973).

Marshall, V.C., The Health and Safety at Work Act (UK), Chem Engnr, Nov, 661-663 (1975).

Smith, C.W., Toxic Substances Control Act, Hyd Proc, 56(1), 213-226 (1977).

Various, Occupational Safety and Health Administration (OSHA), Chem Eng, 11 April, 108-120 (1977).

Harrison, E.B., Clean Air Act 1977, Hyd Proc, 57(7), 255-262; 57(8), 173-188 (1978).

Vandergrift, E.F., Meeting OSHA regulations on toxic exposure, Chem Eng, 2 June, 69-73 (1980).

Benedetti, R.P., NFPA's impact on the chemical industry, Plant/Opns Prog, 1(1), 45-48 (1982).

Garside, R., Standards and approvals for intrinsic safety, Proc Engng, Oct, 34-35 (1982).

Ember, L.R., Legal action for toxics victims, C&E News, 28 March, 11-20 (1983).

Bierlein, L.W., Regulatory considerations in hazardous materials transportation, Plant/Opns Prog, 3(2), 112-113 (1984).

Sayle, A., Safety certification for North Sea topside processing modules, Proc Engng, Feb, 41-43 (1984).

Dewis, M., Regulations for control of industrial major accidents, Chem & Ind, 18 March, 187-189 (1985).

Lewis, P., Occupational exposure limits for toxic gas mixtures, Chem & Ind, 21 April, 268-271 (1986).

West, A.S., Safety evaluation of chemicals: The regulatory framework, Plant/Opns Prog, 5(1), 11-22 (1986).

Travis, C.C.; Richter, S.A.; Crouch, E.A.C.; Wilson, R., and Klema, E.D., Risk and regulation, Chemtech, Aug, 478-483 (1987).

Trowbridge, T.D., Permit space standards proposed by OSHA, Chem Eng Prog, 83(6), 68-73 (1987).

Various, Interpretation of toxicity data for worker protection (topic issue), Chem & Ind, 5 Oct, 680-693 (1987).

Blomquist, D.L., New LPG loss-control standards, Hyd Proc, 67(12), 58-61 (1988).

Burns, M.E.; Foster, R.H., and Shapiro, S.A., Lab safety and the law, Chemtech, May, 267-269 (1988).

Drake, E.M., and Croce, P.A., Guidelines for safe storage and handling of high toxic hazard materials, Plant/Opns Prog, 7(4), 259-264 (1988).

Various, Government regulations in the process industries (topic issue), Chem Eng Prog, 84(12), 35-70 (1988).

6.4 Equipment Safety

1967-1969

Buehler, C.A., Safety showers that work in winter, Chem Eng, 22 May, 204-212 (1967).

Burns, B.W., Design control centres to resist explosions, Hyd Proc, 46(11), 257-259 (1967).

Wardle, J.K.S., and Todd, G., Bulk storage of liquified gases, Chem Engnr, Nov, CE247-253, 260 (1967).

Galluzzo, J.F., Swing-joint connection avoids product cross-contamination, Chem Eng, 6 May, 216 (1968).

Rosenberg, R.H., Low-pressure alarm for gas cylinders in hazardous areas, Chem Eng, 29 July, 172 (1968).

Inkofer, W.A., Ammonia transport via pipeline, Chem Eng Prog, 65(3), 65-69 (1969).

Kauffmann, W.M., Safe operation of oxygen compressors, Chem Eng, 20 Oct, 140-148 (1969).

Mallinson, J.H., Grounding reinforced plastic process systems, Chem Eng, 25 Aug, 136-142 (1969).

Mallinson, J.H., Preventing fires in plastic ductwork, Chem Eng, 7 April, 162-168 (1969).

Tyler, D.A., How noisy is a refinery? Hyd Proc, 48(7), 173-174 (1969).

Watkins, R.A., Preventing ammonia-plant fires, Chem Eng Prog, 65(3), 69-72 (1969).

1970-1975

Bouilloud, P., Calculation of maximum flow rate through safety valves, Brit Chem Eng, 15(11), 1447-1449 (1970).

Chaffee, C.C., Transfer-line failure, Chem Eng Prog, 66(1), 54-57 (1970).

Nailen, R.L., Specifying motors for a quiet plant, Chem Eng, 18 May, 157-161 (1970).

Wittig, S.L.K., High-volume submillisecond pressure-relief emergency system, Ind Eng Chem Proc Des Dev, 9(4), 605-608 (1970).

Anyakora, S.N.; Engel, G.F.M., and Lees, F.P., Some data on the reliability of instruments in the chemical plant environment, Chem Engnr, Nov, 396-402 (1971).

Seebold, J.G., and Hersh, A.S., Control flare-steam noise, Hyd Proc, 50(2), 140 (1971).

Baudmann, H.D., Valve noise estimation, Brit Chem Eng, 17(1), 63-64 (1972).

Deloney, H.C., An evaluation of intrinsically safe instrumentation, Chem Eng, 29 May, 67-72 (1972).

Le Vine, R.Y., Electrical safety in process plants: Classes and limits of hazardous areas, Chem Eng, 1 May, 50-66 (1972).

Seebold, J G., Reducing control valve and furnace combustion noise, Hyd Proc, 51(3), 97-100 (1972).

Carne, C.M., and Bennett, J., Magnetic-logic systems for plant protection, Chem Eng, 29 Oct, 110-114 (1973).

De Heer, H.J., Analyzing failures in redundant safeguarding systems, Chem Eng, 9 July, 106 (1973).

Fields, J.B., Preventing mixing of bulk liquids with specific hose bibs, Chem Eng, 6 Aug, 108 (1973).

Kurz, G.R., Inert-gas protection, Chem Eng, 19 March, 112-114 (1973).

Rowe, G.D., Essentials of good industrial lighting, Chem Eng, 10 Dec, 113-122; 24 Dec, 50-60 (1973).

Elder, H.H., and Sommerfeld, J.T., Rapid estimation of tank leakage rates, Chemical Processing, April, 15-16 (1974).

Mattson, R.E., Improve electrical systems safety and reliability, Hyd Proc, 53(6), 146-150 (1974).

Moore, S.F., Gas-tight isolators: Safety and improved economy aspects, Energy World, Oct, 10-12 (1974).

Price, F.C., Are long-distance pipelines getting safer? Chem Eng, 14 Oct, 94-100 (1974).

Rivas, J.R.; Rudd, D.F., and Kelly, L.R., Computer-aided safety interlock systems, AIChEJ, 20(2), 311-319 (1974).

de Heer, H.J., Choosing an economical automatic protective system, Chem Eng, 17 March, 73-76 (1975).

Lawley, H.G., and Kletz, T.A., High-pressure-trip systems for vessel protection, Chem Eng, 12 May, 81-88 (1975).

Nailen, R.L., Applying electric motors in explosive atmospheres, Hyd Proc, 54(2), 101-105 (1975).

Rasmussen, E.J., Alarm and shutdown devices protect process equipment, Chem Eng, 12 May, 74-80 (1975).

Seebold, J.G., Reduce noise from pulsating combustion in elevated flares, Hyd Proc, 54(9), 225-227 (1975).

Zanker, A., Easy way to estimate the quantity of purge gas, Chem Eng, 22 Dec, 70 (1975).

1976-1978

Drake, E.M., Avoiding LNG rollover hazard, Hyd Proc, 55(1), 119-122 (1976).

Everett, W.S., Proper installation of exhaust-vent silencers, Chem Eng, 1 March, 116-120 (1976).

Ito, T., and Sawada, N., Ground flares aid safety, Hyd Proc, 55(6), 175-190 (1976).

Lees, F.P., Reliability of instruments, Chem & Ind, 6 March, 195-205 (1976).

McLarty, T.E., Selecting silencers to suppress plant noise, Chem Eng, 12 April, 104-112 (1976).

Babbidge, L.G.; Partridge, C.C., and D'Angelo, J.A., Fire-test valves, Hyd Proc, 56(12), 179-182 (1977).

Bloch, H.P., Improve safety and reliability of pumps and drivers, Hyd Proc, 56(1), 97-100; 56(2), 123-125; 56(3), 133-135; 56(4), 181-182; 56(5), 213-215 (1977).

Caplan, F., Space requirements for stairs, Chem Eng, 1 Aug, 84-85 (1977).

Cordes, R.J., Use compressors safely, Hyd Proc, 56(10), 227-247; 56(11), 469-484 (1977).

Grace, C., Safety in process heaters, Proc Engng, May, 85-88 (1977).

Howell, B.T., and Seymour, R.L., Protect reformer walls against burn out, Hyd Proc, 56(3), 147-148 (1977).

Kletz, T.A., Protect presure vessels from fire (BLEVEs), Hyd Proc, 56(8), 98-102 (1977).
Schmidt, T.R., Ground-level detector tames flare-stack flames, Chem Eng, 11 April, 121-124 (1977).
Wafelman, H.R., and Buhrmann, H., Maintain safer tank storage, Hyd Proc, 56(1), 229-236 (1977).
Weaver, F.L., Reliable overspeed protection for steam turbines, Hyd Proc, 56(4), 173-179 (1977).
Wood, R., Safe design of high-pressure bellows, Proc Engng, Nov, 42-43 (1977).
Zanker, A., Estimate tank vapor losses, Hyd Proc, 56(1), 117-120 (1977).
Bonilla, J.A., Estimate safe flare-headers quickly, Chem Eng, 10 April, 135-140 (1978).
Constance, J.D., How to pressure-ventilate large motors for corrosion, explosion and moisture protection, Chem Eng, 27 Feb, 113-116 (1978).
Hrycek, W., Optimum size for diked-in areas, Chem Eng, 13 March, 108 (1978).
Jackson, C., A practical vibration primer, Hyd Proc, 54(4), 161-163; 54(6), 109-111; 54(8), 109-111; 54(12), 251-258 (1975); 55(4), 171-179 (1976); 55(10), 141-149 (1976); 57(3), 119-124 (1978); 57(4), 209-215 (1978).
John, K.S., Use combustible-gas analyzers, Hyd Proc, 57(5), 282-292 (1978).
Nangia, S.C., Glycol-reboiler explosion protection, Chem Eng, 28 Aug, 128 (1978).
Vanchuk, J.T., Making safe compressed respiratory air, Chem Eng, 24 April, 109-112 (1978).

1979-1980

Baum, M.R., Blast waves generated by the rupture of gas pressurised ductile pipes, Trans IChemE, 57, 15-24 (1979).
Carriker, D., Steam-hose safety couplings, Chem Eng, 29 Jan, 121-125 (1979).
Eberhart, T.M., Designing fiberglass tanks for earthquake conditions, Chem Eng, 15 Jan, 147-150 (1979).
Hearfield, F., Thermal properties of reactors and some instabilities, Chem Engnr, March, 156-160 (1979).
Kinsley, G.R., Specifying sound levels for new equipment, Chem Eng, 18 June, 106-110 (1979).
Pesuit, D.R., Plant noise absorption by air, Chem Eng, 24 Sept, 125-129 (1979).
Rigard, J., and Vadot, L., Evaluate LNG storage hazard, Hyd Proc, 58(7), 267-268 (1979).
Rust, E.A., Explosion venting area for low-pressure equipment, Chem Eng, 5 Nov, 102-110 (1979).
Singh, J., Sizing vents for gas explosions, Chem Eng, 24 Sept, 103-109 (1979).
Verde, L., and Levy, G., Safety strategy for plant design, Hyd Proc, 58(3), 215-220 (1979).
Worrell, G.R., Flame-arrestors in ethylene pipes, Hyd Proc, 58(4), 255-258 (1979).
Zanker, A., Nomograph for diked-in area of tanks, Proc Engng, May, 67,69 (1979).
Andow, P.K., Real-time analysis of process plant alarms using a minicomputer, Comput Chem Eng, 4(3), 143-156 (1980).

Booth, A.D.; Karmarkar, M.; Knight, K., and Potter, R.C.L., Design of emergency venting system for phenolic resin reactors, Trans IChemE, 58, 75-90, 276 (1980).
Calmon, C., Explosion hazards of using nitric acid in ion-exchange equipment, Chem Eng, 17 Nov, 271-274 (1980).
Cherry, K.F., Prevent fluid mixups with a modified check valve, Chem Eng, 17 Nov, 275-276 (1980).
Clarkson, D.C., Reduction of vented pentane emissions during production of expandable polystyrene, Chem Engnr, Feb, 111, 114 (1980).
Hearfield, F., Adipic acid reactor development: Benefits in energy and safety, Chem Engnr, Oct, 625-627, 633 (1980).
Nomura, S.I., and Tanaka, T., Prediction of maximum rate of pressure rise due to dust explosion in closed spherical and nonspherical vessels, Ind Eng Chem Proc Des Dev, 19(3), 451-459 (1980).
Rajamani, S., Venting chemical reactions, Chem Eng, 15 Dec, 98 (1980).

1981

Arant, J.B., What is a 'fire-safe' valve? Hyd Proc, 60(5), 269-283 (1981).
Beckman, J.R., and Gilmer, J.R., Model for predicting emissions from fixed-roof storage tanks, Ind Eng Chem Proc Des Dev, 20(4), 646-651 (1981).
Gerardu, N.H., Safe blowoff conditions for storage vessels, Chem Eng Prog, 77(1), 49-52 (1981).
Hauptmanns, U., Fault-tree analysis of a proposed ethylene vaporization unit, Ind Eng Chem Fund, 19(3), 300-309 (1980); 20(3), 304-306 (1981).
Marshall, E.C.; Scanlon, K.E.; Shepherd, A., and Duncan, K.D., Panel diagnosis training for major-hazard continuous-process installations, Chem Engnr, Feb, 66-69 (1981).
McChesney, M., and McChesney, P., Preventing burns from insulated pipes, Chem Eng, 27 July, 59-64 (1981).
McCulloch, D.R., Preventing failure of bellows expansion-joints, Chem Eng, 16 Nov, 259-263 (1981).

1982

Aird, R.J., Reliability assessment of safety/relief valves, Trans IChemE, 60, 314-318 (1982).
Andow, P., Improving process safety with better monitoring and control systems, Chem Engnr, Aug, 325-328 (1982).
Anon., Firescreens, Processing, Jan, 35 (1982).
Badami, V.N., Safety valve protects vacuum lines, Chem Eng, 28 June, 116 (1982).
Bjorklund, R.A.; Kushida, R.O., and Flessner, M.F., Experimental evaluation of flashback flame arrestors, Plant/Opns Prog, 1(4), 254-263 (1982).
Brown, R.S., Damage detection, repair, and prevention in ammonia storage sphere, Plant/Opns Prog, 1(2), 97-101 (1982).
Caplan, K.J., Ventilation basics, Plant/Opns Prog, 1(3), 194-201 (1982).
Dransfield, P.B., and Greig, T.R., Decontaminating vessels containing hazardous materials, Chem Eng Prog, 78(1), 35-38 (1982).
Hare, J.E., Ammonia separator failure, Plant/Opns Prog, 1(3), 166-169 (1982).
Harrison, G.A., Emergency flair tip repair, Hyd Proc, July, 219-222 (1982).

Howard, W.B., Flame arresters and flashback preventers, Plant/Opns Prog, 1(4), 203-209 (1982).
Ilangovan, M.S., et al., Check smoke-detector performance in the plant, Chem Eng, 3 May, 114 (1982).
Kletz, T.A., Flame trap assembly for use with high melting-point materials, Plant/Opns Prog, 1(4), 252-254 (1982).
Lee, J.H.S., and Guirao, M., Pressure development in closed and vented vessels, Plant/Opns Prog, 1(2), 75-85 (1982).
Mikloucich, F.J., and Noronha, J.A., Heat transfer analysis of fire tests on water-filled drums, Plant/Opns Prog, 1(1), 65-69 (1982).
Noronha, J.; Merry, J.T., and Reid, W.C., Deflagration pressure containment for vessel safety design, Plant/Opns Prog, 1(1), 1-7 (1982).
Prescott, G.R., Cracking and nitriding in ammonia converters, Plant/Opns Prog, 1(2), 94-97 (1982).
Quraidis, S.I., Urea autoclave failure and repair, Plant/Opns Prog, 1(3), 159-166 (1982).
Rogerson, J.E., Entrance of dust into pressurized enclosures, Plant/Opns Prog, 1(1), 48-51 (1982).
Rosenhouse, G.; Lin, I.J., and Zimmels, Y., The use of sound-level maps in acoustic design, Chem Eng, 8 March, 72-78 (1982).
Short, W.A., Electrical equipment used in hazardous locations, Plant/Opns Prog, 1(1), 14-19 (1982).
Smith, A., Firescreens for process plant protection, Proc Engng, Aug, 32-33 (1982).
Walker, W., Venting panels for containing dust explosions, Proc Engng, Dec, 35-37 (1982).
Wassom, R.W., High-pressure heat exchanger ignition, Chem Eng Prog, 78(7), 55-58 (1982).

1983

Beychok, M.R., Calculate tank evaporation losses easier, Hyd Proc, March, 71-73 (1983).
Blanken, J.M., and Groefsema, T., Sudden pressure increase in vent header, Plant/Opns Prog, 2(3), 137-140 (1983).
Broschka, G.L., et al., Study of flame arrestors in piping systems, Plant/Opns Prog, 2(1), 5-13 (1983).
De Haven, E.S., Approximate hazard ratings and venting requirements from CSI-ARC data, Plant/Opns Prog, 2(1), 21-27 (1983).
Flores, A., Safety in design: An ethical viewpoint, Chem Eng Prog, 79(11), 11-14 (1983).
Harbert, F.C., On-line laser detection of gases, Plant/Opns Prog, 2(1), 58-61 (1983).
Heitner, I.; Trautmanis, T., and Morrissey, M., When gas-filled vessels are exposed to fire, Hyd Proc, Nov, 263-268 (1983).
Huston, W.C., Easy method for finding leaks in storage-tank floors, Hyd Proc, April, 133-134 (1983).
Johnson, O.W., An oil industry viewpoint on flame arrestors in pipe lines, Plant/Opns Prog, 2(2), 75-79 (1983).

Kirchoff, E.O., et al., Design and safe operation of a multifunctional monomer production facility, Plant/Opns Prog, 2(4), 238-241 (1983).
Lees, F.P., Process computer alarm and disturbance analysis: Review of the state of the art, Comput Chem Eng, 7(6), 669-694 (1983).
Novacek, D.A., Pump and motor failure in hot potassium carbonate system, Plant/Opns Prog, 2(4), 209-211 (1983).
O'Shea, S.M., Nitrogen blanketing of centrifuges reduces fire hazards, Chem Eng, 4 April, 73-76 (1983).
Ricca, P.M., IR techniques in fire-suppression systems, Plant/Opns Prog, 2(2), 99-101 (1983).
Robinson, G.F., and Holmes, M.L., Prediction of furnace pressure excursions, J Inst Energy, 56, 84-91 (1983).
Sanders, R.E., Plant modifications: Troubles and treatment, Chem Eng Prog, 79(2), 73-77 (1983).
Stevens, R., Emergency generators: A reliability study based on analysis of failure, Plant/Opns Prog, 2(4), 203-209 (1983).
Swift, I.; Fauske, H.K., and Grolmes, M.A., Emergency relief systems for runaway reactions, Plant/Opns Prog, 2(2), 116-120 (1983).
Tunkel, S.J., Barricade design criteria, Chem Eng Prog, 79(9), 50-55 (1983).
Various, Stress corrosion cracking in ammonia storage spheres, Plant/Opns Prog, 2(4), 247-260 (1983).
Wu, C.Y., Are your flare systems adequate? Chem Eng, 31 Oct, 41-44 (1983).

1984

Aldham, C.; Rhodes, N., and Tatchell, D.G., Three-dimensional calculations of explosion containment in fast reactors, Chem Eng Commns, 27(1), 79-100 (1984).
Allen, G., Do cone-roof tanks court disaster? Hyd Proc, Aug, 109-110 (1984).
Angelini, E., Protect plants with water spray, Hyd Proc, Dec, 89-95 (1984).
Bradford, M., and Durrett, D.G., Sizing distillation safety valves, Chem Eng, 9 July, 78-84 (1984).
Brahmbhatt, S.R., Are liquid thermal-relief valves needed? Chem Eng, 14 May, 69-71 (1984).
Burgoyne, J., Protecting exothermic reactors and pressure vessels, Proc Engng, Feb, 21-22 (1984).
Clarke, R.W., and Connaughton, G.E., Failure of 1500 psig steam line, Plant/Opns Prog, 3(3), 141-145 (1984).
DePaola, T.J., and Messina, C.A., Nitrogen blanketing, Plant/Opns Prog, 3(4), 203-210 (1984).
Fauske, H.K., Quick approach to reactor vent sizing, Plant/Opns Prog, 3(3), 145-147 (1984).
Fauske, H.K., Generalized vent sizing nomogram for runaway chemical reactions, Plant/Opns Prog, 3(4), 213-215 (1984).
Huff, J.E., Emergency venting requirements for gassy reactions from closed-system tests, Plant/Opns Prog, 3(1), 50-60 (1984).
Jones, M.R.O., and Bond, J., Electrostatic hazards associated with marine chemical tanker operations: Criteria of incendivity in tank cleaning operations, CER&D, 62, 327-333 (1984).

Kirby, D.C., and DeRoo, J.L., Water spray protection for a chemical processing unit, Plant/Opns Prog, 3(4), 254-258 (1984).
Kohan, D., The design of interlocks and alarms, Chem Eng, 20 Feb, 73-80 (1984).
Lawrence, G.M., Shell rupture of secondary reformer, Plant/Opns Prog, 3(1), 60-62 (1984).
Lees, F.P., Process computer alarm and disturbance analysis: Outline of methods for systematic synthesis of fault propagation structure, Comput Chem Eng, 8(2), 91-104 (1984).
Martel, J.T., Safety and reliability of microprocessor-based combustion-safeguard systems, Plant/Opns Prog, 3(2), 80-86 (1984).
Moore, P., Explosion suppression trials, Chem Engnr, Dec, 23-26 (1984).
Pankowski, R.A., Electrical circuits in hazardous locations, Chem Eng, 20 Feb, 93-94 (1984).
Piccinini, N., and Levy, G., Process safety analysis for better reactor cooling-system design in the ethylene oxide reactor, Can JCE, 62(4), 541-558 (1984).
Shafaghi, A.; Andow, P.K., and Lees, F.P., Fault tree synthesis based on control loop structure, CER&D, 62, 101-110 (1984).
Singh, J., Gas explosions in compartmented vessels: Pressure piling, CER&D, 62, 351-366 (1984).
Smith, A., Flame arrestors, Processing, April, 10 (1984).
Sonti, R.S., Practical design and operation of vapor-depressuring systems, Chem Eng, 23 Jan, 66-69 (1984).
Swift, I., Developments in emergency relief system design, Chem Engnr, Aug, 30-33 (1984).

1985

Andow, P.K., Fault diagnosis using intelligent knowledge based systems, CER&D, 63, 368-372 (1985).
Bartknecht, W., Effectiveness of explosion venting as protective measure for silos, Plant/Opns Prog, 4(1), 4-13 (1985).
Boix, J.A., Practical design features of emergency flare systems, Plant/Opns Prog, 4(4), 222-225 (1985).
Buck, M.E., and Belason, E.B., ASTM test for effects of large hydrocarbon pool fires on structural members, Plant/Opns Prog, 4(4), 225-230 (1985).
Cindric, D.T., Design a safe front-end vent for an ammonia plant, Plant/Opns Prog, 4(4), 242-246 (1985).
Coleman, C.R., Catastrophe theory and the incidence of furnace slagging, Energy World, Feb, 6 (1985).
Cory, J.M., and Riccioli, F.D., What are fire-safe valves, Chem Eng, 27 May, 147-148 (1985).
Crozier, R.A., Sizing relief valves for fire emergencies, Chem Eng, 28 Oct, 49-54 (1985).
Diliberto, M.C., and Gratzol, O.K., Fireproofing materials testing, Hyd Proc, April, 121-128 (1985).
Fauske, H.K., Emergency relief system design, Chem Eng Prog, 81(8), 53-56 (1985).

Gaudioso, C.P., Corrosion-resistant composite sleeve for the pipeline industry, Chem Eng Prog, 81(11), 40-42 (1985).
Gibson, N.; Harper, D.J., and Rogers, R.L., Evaluation of fire and explosion risk in drying powders, Plant/Opns Prog, 4(3), 181-189 (1985).
Grolmes, M.A., and Epstein, M., Vapor-liquid disengagement in atmospheric liquid-storage vessels subjected to external heat source, Plant/Opns Prog, 4(4), 200-207 (1985).
Grolmes, M.A.; Leung, J.C., and Fauske, H.K., Large-scale experiments of emergency relief systems, Chem Eng Prog, 81(8), 57-62 (1985).
Guth, D.C., and Clark, D.A., Inspection and repair of two ammonia spheres, Plant/Opns Prog, 4(1), 16-19 (1985).
Hart, R.L., Formulas for sizing explosion vents, Plant/Opns Prog, 4(1), 1-4 (1985).
Huff, J.E., Multiphase flashing flow in pressure-relief systems, Plant/Opns Prog, 4(4), 191-200 (1985).
Jones, M.R.O., and Bond, J., Electrostatic hazards associated with marine chemical tanker operations: Criteria of safety in tank cleaning operations, CER&D, 63, 383-390 (1985).
Kletz, T.A., Inherently safer plants, Plant/Opns Prog, 4(3), 164-168 (1985).
Kohlbrand, H.T., Reactive chemical screening for pilot-plant safety, Chem Eng Prog, 81(4), 52-56 (1985).
Lihou, D., and Kabir, Z., Sequential testing of safety systems, Chem Engnr, Dec, 35-36 (1985).
Lihou, D.A., Why screen modifications? Chem Engnr, Oct, 46-47 (1985).
Poole, G., Improved design of emergency relief systems, Proc Engng, May, 67,69 (1985).
Randhava, R., and Calderone, S., Hazard analysis of supercritical extraction, Chem Eng Prog, 81(6), 59-62 (1985).
Ringer, M.W., Detonation tests and response analysis of vessels and piping containing gas and aerated liquid, Plant/Opns Prog, 4(1), 26-47 (1985).
Shiozaki, J.; Matsuyama, H.; O'Shima, E., and Iri, M., Improved algorithm for diagnosis of system failures in the chemical process, Comput Chem Eng, 9(3), 285-294 (1985).
Swift, I., Emergency relief systems in context, Chem Engnr, July, 38-39 (1985).
Thompson, J.C., When a floating roof sinks, Hyd Proc, July, 119-120 (1985).
Vijayan, C.P.; Fontaine, J.C.,; Ghali, E.; Kaliaguine, S., and Galibois, A., Moisture evacuation during compressed natural gas discharge, CER&D, 63, 291-299 (1985).
Weiner, S., Use of programmable devices for safety, Plant/Opns Prog, 4(1), 47-49 (1985).

1986

Arendt, J.S., Determining heater retrofits through risk assessment, Plant/Opns Prog, 5(4), 228-231 (1986).
Bartknecht, W., Pressure venting of dust explosions in large vessels, Plant/Opns Prog, 5(4), 196-204 (1986).
Basta, N., Safety becomes a big issue at the pilot-plant level, Chem Eng, 12 May, 29-31 (1986).

Fauske, H.K., et al., Emergency relief-vent sizing for fire emergencies involving liquid-filled atmospheric storage vessels, Plant/Opns Prog, 5(4), 205-208 (1986).

Hansen, D.E., Boiler and machinery protection through control of piping reactions, Plant/Opns Prog, 5(3), 183-185 (1986).

Hub, L., and Jones, J.D., Early on-line detection of exothermic reactions, Plant/Opns Prog, 5(4), 221-224 (1986).

Jacobsen, C.T., High-temperature polymeric fire barriers, Plant/Opns Prog, 5(3), 148-154 (1986).

Kirby, G.N., and Siwek, R., Preventing failures of equipment subject to explosions, Chem Eng, 23 June, 125-128 (1986).

Klein, H.H., Analysis of DIERS venting tests: Validation of a tool for sizing emergency relief systems for runaway chemical reactions, Plant/Opns Prog, 5(1), 1-10 (1986).

Kletz, T.A., Plant modifications, Plant/Opns Prog, 5(3), 136-141 (1986).

Kobayashi, Y.; Kobayashi, T., and Coombs, K.M., Severe carry-over phenomena in 1200mtpd methanol plant, Plant/Opns Prog, 5(2), 103-107 (1986).

Lawrence, G.M., Multiple cracking and leakage of hot synthesis gas pipe, Plant/Opns Prog, 5(3), 175-178 (1986).

LeBlanc, R.W., Loss prevention through machinery vibration surveillance and analysis, Plant/Opns Prog, 5(3), 179-182 (1986).

Leung, J.C., Simplified vent sizing equations for emergency relief requirements in reactors and storage vessels, AIChEJ, 32(10), 1622-1634 (1986).

Lock, J., Standards for LNG storage tank design, Processing, June, 43-45 (1986).

McCoy, C.S.; Dillenback, M.D., and Truax, D.J., Major fire in steam-methane reformer furnace, Plant/Opns Prog, 5(3), 165-168 (1986).

Modarres, M., and Cadman, T., A method of alarm system analysis for process plants, Comput Chem Eng, 10(6), 557-565 (1986).

Moore, P., Towards large volume explosion suppression systems, Chem Engnr, Nov, 43-47 (1986).

Nieh, C.D., and Zengyan, H., Estimate exchanger vibration, Hyd Proc, 65(4), 61-65 (1986).

Prescott, G.R.; Blommaert, P., and Grisolia, L., Failure of high-pressure synthesis pipe, Plant/Opns Prog, 5(3), 155-159 (1986).

Rall, W., and Fromm, D., Syngas turbine damage by excessive steam superheating, Plant/Opns Prog, 5(2), 90-92 (1986).

Rao, S.; Wiltzen, R.C.A., and Jacobs, W., Investigation of damage and repair of 1000mtd horizontal ammonia converter, Plant/Opns Prog, 5(2), 116-121 (1986).

Schultz, N., Fire protection for cable trays in petrochemical facilities, Plant/Opns Prog, 5(1), 35-39 (1986).

Sotebier, D.L., and Rall, W., Leakage problems at large flanges in new BASF ammonia plant, Plant/Opns Prog, 5(2), 86-89 (1986).

Watanabe, K., and Himmelblau, D.M., Detection and location of a leak in a gas-transport pipeline by a new acoustic method, AIChEJ, 32(10), 1690-1701 (1986).

Weir, E.D.; Gravenstine, G.W., and Hoppe, T.F., Thermal runaways: Problems with agitation, Plant/Opns Prog, 5(3), 142-147 (1986).
Wiederuh, E., Compressing hydrogen-rich gases, Chem Eng, 9 June, 92 (1986).
Wrenn, C., Inerting for safety, Plant/Opns Prog, 5(4), 225-227 (1986).

1987
Brown, D.M., and Nolan, P.F., Interaction of shock waves on cylindrical structures, Plant/Opns Prog, 6(1), 52-56 (1987).
Castle, G.K., and Castle, G.G., Effect of fireproofing design on thermal performance of horizontal members with top flange exposed, Plant/Opns Prog, 6(4), 193-198 (1987).
Davis, G.D., Investigation and repair of an auxiliary boiler explosion, Plant/Opns Prog, 6(1), 42-45 (1987).
Fromm, D., and Rall, W., Fire at semi-lean pump by reverse motion, Plant/Opns Prog, 6(3), 162-164 (1987).
Green, S.J., Solving chemical and mechanical problems of pressurized water reactor steam generators, Chem Eng Prog, 83(7), 31-45 (1987).
Krisher, A.S., Plant integrity programs, Chem Eng Prog, 83(5), 20-24 (1987).
Leung, J.C., and Fauske, H.K., Runaway system characterization and vent sizing based on DIERS methodology, Plant/Opns Prog, 6(2), 77-83 (1987).
Madhavan, S., and Sathe, S.Y., Inspection of ammonia plants, Plant/Opns Prog, 6(1), 35-41 (1987).
Martinez, O.A.; Madhavan, S., and Kellett, D.J., Damage to and replacement of ammonia storage-tank foundation, Plant/Opns Prog, 6(3), 129-141 (1987).
Palluzi, R.P., Testing for leaks in pilot plants, Chem Eng, 9 Nov, 81-85 (1987).
Patel, K.; Sirl, D.; Thomas, W.J., and Ullah, U., Dynamic characteristics of a toxic vapour monitor, CER&D, 65(4), 326-333 (1987).
Russell, D.L., and Hart, S.W., Underground storage tanks: Potential for economic disaster, Chem Eng, 16 March, 61-69 (1987).
Stadler, H., Flame barrier valves, Plant/Opns Prog, 6(4), 175-180 (1987).
Straitz, J.F., Flare technology safety, Chem Eng Prog, 83(7), 53-62 (1987).
Swift, I., and Epstein, M., Performance of low-pressure explosion vents, Plant/Opns Prog, 6(2), 98-105 (1987).
Various, Hazard control in the oilseed solvent extraction industry, JAOCS, 64(1), 5-32 (1987).
Wilkie, F., Selecting a flame arrester, Proc Engng, March, 53-54 (1987).

1988
Barclay, D.A., Protecting process-safety interlocks, Chem Eng Prog, 84(2), 20-24 (1988).
Britton, L.G., Systems for electrostatic evaluation in industrial silos, Plant/Opns Prog, 7(1), 40-50 (1988).
Britton, L.G., and Smith, J.A., Static hazards of drum filling, Plant/Opns Prog, 7(1), 53-78 (1988).
Cartwright, P., Earthing electrostatic charge, Chem Engnr, Nov, 49 (1988).
Cobb, A.J., and Monier-Williams, S., Computerized ammonia plant trip system, Plant/Opns Prog, 7(4), 243-253 (1988).
Dent, B., Automatic switch-status monitoring for safety, Proc Engng, March, 59 (1988).

Dore, J.C., Process safety: An integral part of pilot plants, Plant/Opns Prog, 7(4), 223-225 (1988).
Fauske, H.K., Emergency relief-system design for reactive and nonreactive systems: Extension of DIERS methodology, Plant/Opns Prog, 7(3), 153-158 (1988).
Freeman, R.A., and Shaw, D.A., Sizing excess-flow valves, Plant/Opns Prog, 7(3), 176-182 (1988).
Friedel, L., Resistance to gas/vapor-liquid flow in safety valves, Int Chem Eng, 28(3), 406-424 (1988).
Fthenakis, V.M., and Moskowitz, P.D., Health and safety aspects of thin-film photovoltaic cell manufacturing technologies, Plant/Opns Prog, 7(4), 236-241 (1988).
Habermehl, R., Safe handling and disposal of spent catalysts, Chem Eng Prog, 84(2), 16-19 (1988).
Kletz, T.A., Plants should be friendly! Chem Engnr, March, 35 (1988).
Martin, T., Safety and pump shutdown, Chem Engnr, Dec, 96 (1988).
Mason, G.S., and Kumar, R., Algorithm for sizing flare piping, Chem Eng, 20 June, 99-102 (1988).
Rogers, J.M., Detecting toxic vapors early, Chem Eng, 11 April, 81-86 (1988).
Snyder, P.G., Brittle fracture of high-pressure heat exchanger, Plant/Opns Prog, 7(3), 148-152 (1988).
Swift, I., Designing explosion vents, Chem Eng, 11 April, 65-68 (1988).
Varey, P., Fire-proof valve testing, Chem Engnr, Nov, 33-38 (1988).
Venart, J.E., et al., Experiments on thermo-hydraulic response of pressure liquified gases in externally heated tanks with pressure relief, Plant/Opns Prog, 7(2), 139-144 (1988).
Williams, G.P.; Hoehing, W.W., and Byington, R.G., Causes of ammonia plant shutdowns, Plant/Opns Prog, 7(2), 99-110 (1988).
Yarbrough, R., Program for tower vibration, Hyd Proc, 67(5), 65-68 (1988).
Zirkle, W.D., Decommissioning nuclear reactors, Chemtech, July, 429 (1988).

6.5 Risk and Safety

1967-1969

Anon., Material factors: Safety features for particular chemicals, Chem Eng Prog, 62(8), 91-109 (1966); 63(1), 127-128 (1967); 63(2), 89-90; 63(3), 99-100; 63(4), 91-92; 63(5), 97-98; 63(6), 117-118; 63(7), 125-126 (1967).
Eichel, F.G., Static electricity, Chem Eng, 13 March, 153-167 (1967).
Johnson, R.T., Estimating concentration of system vapors in closed areas, Chem Eng, 13 March, 202-204 (1967).
Linton, F.L., and Brink, J.A., Safer preheating of air for ammonia plants, Chem Eng Prog, 63(2), 83-86 (1967).
Siegmund, J.M., Production, handling, and shipping of elemental fluorine, Chem Eng Prog, 63(6), 88-92 (1967).
Various, Loss prevention in the CPI (topic issue), Chem Eng Prog, 63(8), 43-78 (1967).
Various, Pilot-plant safety procedures (topic issue), Chem Eng Prog, 63(11), 49-71 (1967).

Baker, P.S., Radioisotopes in chemical processes, Chem Eng, 11 March, 179-186 (1968).
Benjaminsen, J.M., and van Wiechen, P.H., Calculate the mean time to electrical explosions! Hyd Proc, 47(8), 121-126 (1968).
Boiston, D.A., Safety study of acetone oxidation, Brit Chem Eng, 13(1), 85-88 (1968).
Chieffo, A.B., and McLean, R.H., Fast leak detection reduces hydrocarbon losses and pollution, Chem Eng, 15 July, 144-146 (1968).
Cook, G.A.; Dorr, V.A., and Shields, B.M., Region of noncombustion in nitrogen-oxygen and helium-oxygen diving atmospheres, Ind Eng Chem Proc Des Dev, 7(2), 308-311 (1968).
Leeah, C.J., Poor plant design courts disaster, Hyd Proc, 47(11), 248-251 (1968).
Methner, J.C., Hard sell for fire fighters, Chem Eng, 12 Feb, 112-116 (1968).
Miller, R.O., Explosions in condensed-phase nitric oxide, Ind Eng Chem Proc Des Dev, 7(4), 590-593 (1968).
Rasbash, D.J., Liquified gases as fire-fighting agents, Chem Engnr, April, CE89-93 (1968).
Stuhlbarg, D., Calculating the calculated risk, Chem Eng, 15 Jan, 152-154 (1968).
Tedeschi, R.J., et al., Safe liquifaction of acetylene, Ind Eng Chem Proc Des Dev, 7(2), 303-307 (1968).
Various, Estimating and reducing plant noise (special report), Hyd Proc, 47(12), 67-78 (1968).
Various, Loss prevention (topic issue), Chem Eng Prog, 64(6), 49-74 (1968).
Various, Handling plant emergencies (feature report), Chem Eng, 11 March, 164-178 (1968).
Anon., Pesticides: Present and future, Chem Eng, 7 April, 133-140 (1969).
Browning, R.L., Analyzing industrial risks, Chem Eng, 20 Oct, 109-114 (1969).
Freshwater, D.C., and Buffham, B.A., Reliability engineering for the process plant industries, Chem Engnr, Oct, CE367-369 (1969).
Lacey, B.G., Noise and its control in process plants, Chem Eng, 16 June, 74-84 (1969).
Lee, P.R., Safe storage and transportation of some potentially hazardous materials, J Appl Chem, 19, 345-351 (1969).
Lee, R.H., Electrical grounding: Safe or hazardous? Chem Eng, 28 July, 158-166 (1969).
Leeah, C.J., Better safety planning, Hyd Proc, 47(12), 136-138(1968); 48(1), 149-152 (1969).
McFarland, I., Preventing flange fires, Chem Eng Prog, 65(8), 59-61 (1969).
Richards, E.J., Noise generation in process plants, Chem Engnr, June, CE223-232 (1969).
Thompson, B.A., Chemicals in containers, Brit Chem Eng, 14(8), 1089-1090 (1969).
Various, Loss prevention (topic issue), Chem Eng Prog, 65(4), 29-52 (1969).
Various, Problems of noise in chemical plant (symposium papers), Chem Engnr, April, CE104-129 (1969).
Weismantel, G.E., A fresh look at intrinsic safety, Chem Eng, 30 June, 132-140 (1969).

1970-1972

Bigelow, C.R., A system for reporting transportation accidents, Chem Eng Prog, 66(4), 71-73 (1970).
Hilado, C.J., Predicting material flammability, Chem Eng, 14 Dec, 174-178 (1970).
Lawrence, W.W., and Cook, S.E., Initiation and propagation of explosive reactions in chlorine-ethane mixtures, Ind Eng Chem Proc Des Dev, 9(1), 47-49 (1970).
Polentz, L.M., Compressed water can be dangerous, Chem Eng, 23 March, 158-160 (1970).
Rose, H.E., Dust explosion hazards, Brit Chem Eng, 15(3), 371-375 (1970).
Rosival, L.; Vargova, M., and Uhnak, J., Toxic effects of heptachlor, Int Chem Eng, 10(4), 545-554 (1970).
Schuder, C.B., Coping with control-valve noise, Chem Eng, 19 Oct, 149-153 (1970).
Various, Plant loss prevention (topic issue), Chem Eng Prog, 66(9), 33-62 (1970).
Akita, K., Flame propagation in combustible fuel-air mixtures: Basic theory, Int Chem Eng, 11(4), 739-752 (1971).
Bowes, P.C., and Cameron, A., Self-heating and ignition of chemically activated carbon, J Appl Chem Biotechnol, 21, 244-250 (1971).
Fisher, J.N., and Phipps, A.J., Risk analysis for the HPI, Hyd Proc, 50(3), 63-70 (1971).
Frankton, H.M., Electrical problems in the chemical industry, Brit Chem Eng, 16(9), 785-792 (1971).
Hodnick, H.V., Thwarting maintenance deathtraps, Chem Eng, 5 April, 142-144 (1971).
Judd, S.H., Noise abatement in process plants, Chem Eng, 11 Jan, 139-145 (1971).
Malloy, J.B., Risk analysis of chemical plants, Chem Eng Prog, 67(10), 68-77 (1971).
Talmage, W.P., Safe vapor-phase oxidation, Chemtech, Feb, 117-119 (1971).
Various, Plant safety and loss prevention (topic issue), Chem Eng Prog, 67(6), 41-61 (1971).
Walls, W.L., Fire protection for LNG plants, Hyd Proc, 50(9), 205-208 (1971).
Constance, J.D., Calculating the masking effects of noise, Chem Eng, 17 April, 124-126 (1972).
Constance, J.D., Estimating fan noise from tip speeds, Chem Eng, 2 Oct, 92-94 (1972).
Croxford, B., Transporting cryogenic liquids, Brit Chem Eng, 17(3), 244-245 (1972).
Johnson, J.E., and Blair, E.H., Cost, time, and pesticide safety, Chemtech, Nov, 666-669 (1972).
Lauderback, W., Unique flare system retards smoke, Hyd Proc, 51(2), 127-128 (1972).
Palmer, K.N., and Butlin, R.N., Dust explosibility tests and their application, Powder Tech, 6, 149-157 (1972).
Searle, C.E., Chemical carcinogens, Chem & Ind, 5 Feb, 111-116 (1972).

Various, Fire protection and safety in the HPI (special report), Hyd Proc, 51(12), 49-66 (1972).
Various, Plant operations and loss prevention (topic issue), Chem Eng Prog, 68(5), 41-69 (1972).
Vervalin, C.H., Fire and safety information, Hyd Proc, 51(4), 163-165; 51(5), 162-164 (1972).
Vervalin, C.H., Information sources for health and safety, Hyd Proc, 51(2), 107 (1972).
Weisman, J., and Holzman, A.G., Optimal process system design under conditions of risk, Ind Eng Chem Proc Des Dev, 11(3), 386-397 (1972).
Woinsky, S.G., Predicting flammable-material classifications, Chem Eng, 27 Nov, 81-86 (1972).

1973-1974

Brand, R.R., and Burgess, F.L., Reducing hazards of handling reactive chemicals, Chem Eng, 12 Nov, 248-252 (1973).
de Heer, H.J., Calculating how much safety is enough, Chem Eng, 19 Feb, 121-128 (1973).
Deuschle, R., and Tiffany, F., Barrier intrinsic safety, Hyd Proc, 52(12), 111-116 (1973).
Imhof, H., Protecting process plants from power failures, Chem Eng, 2 April, 56-60 (1973).
Lawrence, F.E., Safety in the chemical industry, Chem Engnr, April, 211-214, 220 (1973).
Lou, S.C., Noise-control design for process plants, Chem Eng, 26 Nov, 77-82 (1973).
Mitsui, R., and Tanaka, T., Simple models of dust explosion, Ind Eng Chem Proc Des Dev, 12(3), 384-389 (1973).
Seebold, J.G., Does smooth piping reduce noise? Hyd Proc, 52(9), 189-191 (1973).
Stolin, A.M., and Merzhanov, A.G., Critical conditions for thermal detonation in presence of chemical and mechanical sources of heat, Int Chem Eng, 13(1), 91-96 (1973).
Various, Noise control (topic issue), Chem Eng Prog, 69(10), 51-66 (1973).
Various, Pilot operations and loss prevention (topic issue), Chem Eng Prog, 69(4), 37-54 (1973).
Various, Design for reliability (symposium papers), Chem Engnr, Oct, 457-484 (1973).
Vervalin, C.H., Information sources on fire and safety, Hyd Proc, 52(1), 128 (1973).
Wiley, S.K., Controlling vapor losses, Chem Eng, 17 Sept, 116-119 (1973).
Wood, W.S., Transporting, loading and unloading hazardous materials, Chem Eng, 25 June, 72-94 (1973).
Woodard, A.M., Design a plant firewater system, Hyd Proc, 52(10), 103-106 (1973).
Bently, D.E., Monitor machinery condition for safe operation, Hyd Proc, 53(11), 205-208 (1974).
Brown, L.E.; Wesson, H.R., and Welker, J.R., Predict LNG fire radiation, Hyd Proc, 53(5), 141-143 (1974).

Constance, J.D., Calculating sound levels from octave-band analysis, Chem Eng, 18 Feb, 152-154 (1974).
Davies, C.N., Deposition of inhaled particles in man, Chem & Ind, 1 June, 441-444 (1974).
Ecker, H.W.; James, B.A., and Toensing, R.H., Electrical safety: Designing purged enclosures, Chem Eng, 13 May, 93-97 (1974).
Snyder, I.G., Implementing a good safety program, Chem Eng, 27 May, 112-118 (1974).
Thumann, A., Interdisciplinary plant-noise control, Chem Eng, 19 Aug, 120-124 (1974).
Various, Loss prevention in plant operations (topic issue), Chem Eng Prog, 70(4), 45-84 (1974).
Various, Plant/equipment reliability (topic issue), Chem Eng Prog, 70(10), 53-76 (1974).
Welker, J.R.; Wesson, H.R., and Brown, L.E., Use foam to disperse LNG vapors? Hyd Proc, 53(2), 119-120 (1974).

1975-1976

Andow, P.K., and Lees, F.P., Process computer alarm analysis: Outline of a method based on list processing, Trans IChemE, 53, 195-208 (1975).
Bancroft, W.G.; Clark, K.R., and Corvini, G., Beware of oxygen in natural gas drying, Hyd Proc, 54(9), 203-205 (1975).
Bond, J., and Bryans, J.W., The two safeguard approach for minimising human failure injuries, Chem Engnr, April, 245-247 (1975).
Bruce, R.D., and Werchan, R.E., What does noise control cost? Hyd Proc, 54(5), 234-235 (1975).
Burgoyne, J.H., Risk management and chemical manufacture, Chem Engnr, March, 151-152 (1975).
Byrne, J.F., Good records control accidents, Hyd Proc, 54(11), 335-338 (1975).
Chakraborty, S.K.; Mukhopadhyay, B.N., and Chanda, B.C., Effect of inhibitors on flammability range of flames produced from LPG/air mixtures, Fuel, 54(1), 10-16 (1975).
Gadian, T., Carcinogens in industry: Dichlorobenzidine, Chem & Ind, 4 Oct, 821-831 (1975).
Kletz, T.A., Emergency isolation valves for chemical plants, Chem Eng Prog, 71(9), 63-71 (1975).
Marshall, V.C., Process-plant safety: A strategic approach, Chem Eng, 22 Dec, 58-60 (1975).
Various, Fire protection and safety (special report), Hyd Proc, 54(8), 65-86 (1975).
Vervalin, C.H., Trends in HPI fire/safety training, Hyd Proc, 54(2), 146-150; 54(3), 169-172 (1975).
Vervalin, C.H., HPI looks at toxicity, Hyd Proc, 54(4), 225-238 (1975).
Vervalin, C.H., Will your plant burn tomorrow? Hyd Proc, 54(12), 145-154 (1975).
Warren, J.H., and Corona, A.A., Method for testing fire protective coatings, Hyd Proc, 54(1), 121-137 (1975).
Barber, D.H., and Tibbetts, A.M., Developing safe workers in refineries, Chem Eng, 2 Feb, 117-119 (1976).

Bowen, J.H., Individual risk vs public risk criteria, Chem Eng Prog, 72(2), 63-67 (1976).

Broodo, A.; Gilmore, J., and Armstrong, A., Did arson cause that fire? Hyd Proc, 55(2), 163-179 (1976).

Coffee, R.D., Design your plant for survival, Hyd Proc, 55(5), 301-310 (1976).

Cooper, D.O., and Davidson, L.B., The parameter method for risk analysis, Chem Eng Prog, 72(11), 73-78 (1976).

Cox, N.D., Evaluating the profitability of standby components, Chem Eng Prog, 72(6), 76-79 (1976).

Dorsey, J.S., Eliminating static sparks, Chem Eng, 13 Sept, 203-205 (1976).

Driskell, L., Coping with high-pressure letdown, Chem Eng, 25 Oct, 113-118 (1976).

Fleming, J.B.; Lambrix, J.R., and Nixon, J.R., Safety in phenol-from-cumene process, Hyd Proc, 55(1), 185-196 (1976).

Gibson, S.B., Reliability engineering applied to the safety of new projects, Chem Engnr, Feb, 105-106 (1976).

Marshall, V.C., The strategic approach to safety, Chem Engnr, April, 260-262 (1976).

Middleton, A.H., Noise from chemical plant, Chem Engnr, Feb, 115-118 (1976).

Sandler, S., Apparatus to demonstrate explosive limits, Chem Eng Educ, 10(1), 40-43 (1976).

Taylor, H.D., What happens when loss prevention fails? Chem Engnr, April, 263-264 (1976).

Underwood, H.C.; Sourwine, R.E., and Johnson, C.D., Organize for plant emergencies, Chem Eng, 11 Oct, 118-130 (1976).

Various, Toxicity of polymers and polymer raw materials (topic issue), Chem & Ind, 5 June, 463-474 (1976).

Various, Plant loss prevention (topic issue), Chem Eng Prog, 72(11), 41-67 (1976).

Vervalin, C.H., Loss prevention in the HPI, Hyd Proc, 55(7), 205-216; 55(8), 182-184; 55(9), 321-337; 55(10), 215-222; 55(11), 305-306 (1976).

Walko, J., Process hazards, Proc Engng, Nov, 111-112 (1976).

Zanker, A., Nomograph for static electric charge from flowing hydrocarbons, Hyd Proc, 55(3), 133-135 (1976).

1977-1978

Archibald, R.G., Process sour gas safely, Hyd Proc, 56(3), 219-232 (1977).

Arney, H.E., Clean tanks and vessels safely, Hyd Proc, 56(5), 141-142 (1977).

Atallah, S., Security for chemical plants, Chem Eng, 10 Oct, 139-140 (1977).

Blair, D.A., Operate your refinery safely, Hyd Proc, 56(4), 241-256 (1977).

Crom, R.C.W., Safeguarding against electric shock hazards, Chem Eng, 28 March, 90-96 (1977).

Kletz, T.A., Evaluate risk in plant design, Hyd Proc, 56(5), 297-324 (1977).

Klunick, C.H., Aqueous film-forming foams, Hyd Proc, 56(9), 293-300 (1977).

Marinov, V.N., Self-ignition and mechanisms of interaction of coal with oxygen at low temperatures, Fuel, 56(2), 153-170 (1977).

Martinsen, W.E.; Muhlenkamp, S.P., and Olson, L.J., Disperse LNG vapors with water, Hyd Proc, 56(7), 261-266 (1977).

Morris, H., Inspect tank cars before shipment, Chem Eng, 7 Nov, 109-111 (1977).
Mukerji, A., Safe unloading and storage of vinyl chloride monomer, Chem Eng, 12 Sept, 155-160 (1977).
Rains, W.A., Testing for fire resistance, Chem Eng, 19 Dec, 97-100 (1977).
Robinson, G., Assess and control HPI plant noise, Hyd Proc, 56(6), 223-226 (1977).
Various, Fire protection design (special report), Hyd Proc, 56(8), 89-120 (1977).
Various, Plant loss prevention (topic issue), Chem Eng Prog, 73(9), 41-72, 80-81 (1977).
Ackroyd, K., Refrigerants: Properties, selection and hazards, Chem Engnr, May, 366-370 (1978).
Bergtraun, E.M., Safety during construction and alteration projects, Chem Eng, 30 Jan, 119-120; 27 Feb, 133-134 (1978).
Burgoyne, J.H., Testing of materials for dust explosion or fire, Chem & Ind, 4 Feb, 81-87 (1978).
Cocks, R.E., and Rogerson, J.E., Organizing a process safety program, Chem Eng, 23 Oct, 138-146 (1978).
Fothergill, C.D.H., A computerised reliability data store and its relevance to chemical plant operations, Chem Engnr, May, 381-382, 410 (1978).
Johnson, D.W., and Welker, J.R., Diked-in storage areas, Chem Eng, 31 July, 112-113 (1978).
Kletz, T.A., What you don't have can't leak, Chem & Ind, 6 May, 287-292 (1978).
Krienberg, M., Handling molten sulfur, Chem Eng, 4 Dec, 125-126 (1978).
Picciotti, M., Design for ethylene plant safety, Hyd Proc, 57(3), 191-202; 57(4), 261-282 (1978).
Pittom, A., Control of toxic hazards, Chem & Ind, 4 Feb, 77-80 (1978).
Sengupta, A.K., Unloading strong acid, Chem Eng, 8 May, 211 (1978).
Smith, I.W., Intrinsic reactivity of carbons to oxygen, Fuel, 57(7), 409-414 (1978).
Snow, K., Shipping hazardous material, Chem Eng, 6 Nov, 102-108 (1978).
Various, Safety (feature report), Processing, Nov, 69-97 (1978).
Various, Health and safety in the chemical industry (topic issue), Chem & Ind, 6 May, 293-314 (1978).
Various, Plant loss prevention (topic issue), Chem Eng Prog, 74(10), 45-72 (1978).
Various, Toxic substances, toxicology and the chemical engineer (feature report), Chem Eng, 24 April, 70-89 (1978).
Vervalin, C.H., Fire-loss report for the HPI, Hyd Proc, 57(1), 251-257 (1978).

1979

Anthony, E.J., and Powell, M.F., Peak flammability limits of hydrogen sulfide, carbon dioxide and air for upward propagation, Ind Eng Chem Fund, 18(3), 238-240 (1979).
Castle, G.K., A contractor views fireproofing, Hyd Proc, 58(8), 179-184 (1979).
Gay, P., Hazard warning signs for road tankers, Proc Engng, Jan, 34-37 (1979).
Gitzlaff, T.R., and Batton, F.E., Two inexpensive sampling systems reduce exposure, Chem Eng, 12 March, 117-118 (1979).

Goldfarb, A.S., Equation for API emissions chart, Chem Eng, 4 June, 160 (1979).
Hammock, A.A., Operate large plants safely, Hyd Proc, 58(4), 263-270 (1979).
Harrington, J.M., Safety and chemicals in the environment: Effects and epidemiology, Chem & Ind, 3 Nov, 722-728 (1979).
Kerr, D.J., Safety with the Du Pont company, Chem Engnr, March, 165-166 (1979).
Kletz, T.A., 'Layered' accident investigation, Hyd Proc, 58(11), 373-382 (1979).
Kletz, T.A., Consider vapor cloud dangers in plant design, Hyd Proc, 58(10), 205-212 (1979).
Kletz, T.A., Is there a simpler solution? Chem Engnr, March, 161-164 (1979).
Lamb, J.A., and Pomphrey, D.M., The design of plant under conditions of uncertainty, Chem Engnr, Feb, 102-105, 108 (1979).
Ling, K.C., Application of loss control to the fine chemicals industry, Chem Engnr, May, 342-344 (1979).
Lovelace, B.G., Safe sampling of liquid process streams, Chem Eng Prog, 75(11), 51-57 (1979).
Marshall, V.C., Hazard and risk: Occupational health and safety, Chem Engnr, March, 179 (1979).
Moore, P., Test methods for characterisation of dust explosibility, Chem & Ind, 7 July, 430-434 (1979).
Various, Safety and reliability of chemical processes, Comput Chem Eng, 3(9), 27-36; 489-512 (1979).
Vermeulen, T., Dynamics of runaway systems, Chem Eng Educ, 13(4), 156-158, 205-209 (1979).
Waldron, H.A., Solvents and occupational health, Chem Engnr, Jan, 39-40 (1979).

1980

Atallah, S., Assessing and managing industrial risk, Chem Eng, 8 Sept, 94-103 (1980).
Bull, D.C., Concentration limits to the initiation of unconfined detonation in fuel-air mixtures, Trans IChemE, 57, 220-227 (1979); 58, 281-284 (1980).
Gross, P., and Braun, D.C., Toxicology of fibers, Chemtech, July, 436-437 (1980).
Hardman, J.S.; Street, P.J., and Twamley, C.S., Spontaneous combustion in beds of activated carbon, Fuel, 59(3), 151-156; 213-214 (1980).
Kaura, M.L., Aid to firewater design, Hyd Proc, 59(12), 137-147 (1980).
Kletz, T.A., Seek intrinsically safe plants, Hyd Proc, 59(8), 137-151 (1980).
Le Cornu, M.J.P., Countering the fire hazards of plastic components, Chem Eng Prog, 76(1), 81-84 (1980).
Lewis, P., Controlling health hazards in the chemical industry, Chem & Ind, 1 Nov, 860-863 (1980).
Messing, A., Organizing a good safety/health program, Hyd Proc, 59(10), 193-198 (1980).
Rawls, R.L., Chemical transport: Coping with disasters, C&E News, 24 Nov, 20-30 (1980).
Schaeffer, J., Use flammable-vapor sensors? Hyd Proc, 59(1), 211-220 (1980).

Stein, T.N., Analyzing and controlling noise in process plants, Chem Eng, 10 March, 129-137 (1980).
Umeda, T.; Kuriyama, T.; O'Shima, E., and Matsuyama, H., Graphical approach to cause and effect analysis of chemical processing systems, Chem Eng Sci, 35(12), 2379-2388 (1980).
Various, Insulating materials and lung disease (topic issue), Chem & Ind, 1 March, 172-188 (1980).
Various, Loss prevention (topic issue), Chem Eng Prog, 76(11), 37-63 (1980).

1981
Chandler, P., Controlling noise in gas production plants, Proc Engng, Dec, 57-63 (1981).
Chandnani, M.K., Design HPI plants for safety, Hyd Proc, 60(11), 324-352 (1981).
Davenport, J.A., Examine your pre-emergency plan, Hyd Proc, 60(5), 287-294 (1981).
de Groot, J.J.; Groothuizen, T.M., and Verhoeff, J., Safety aspects of organic peroxides in bulk tanks, Ind Eng Chem Proc Des Dev, 20(1), 131-138 (1981).
Glassburn, L.E., Industrial noise: Properties, sources and solutions, Hyd Proc, 60(8), 127-130 (1981).
Hardy, C.J., and Collins, C., Testing of industrial chemicals for inhalation toxicity, Chem & Ind, 7 Feb, 89-92 (1981).
Kletz, T.A., Is plant safety too expensive? Hyd Proc, 60(12), 171-179 (1981).
Murphy, R.F., Fire fighting foam systems, Hyd Proc, 60(10), 213-226 (1981).
Skinner, G.A., Smoke: The hazard, the measurement, and the remedy, J Chem Tech Biotechnol, 31, 445-452 (1981).
Spranza, F.G., Improve your plant security system, Hyd Proc, 60(9), 331-338 (1981).
Various, Transport of chemicals by sea (topic issue), Chem & Ind, 7 Nov, 751-765 (1981).
Various, Fire protection (topic issue), Chem & Ind, 18 July, 481-491 (1981).
Various, Health and safety (topic issue), Chem & Ind, 5 Sept, 585-599 (1981).
Vervalin, C.H., Loss-prevention information sources, Hyd Proc, 60(3), 221-236 (1981).
Vervalin, C.H., Train for better fire protection, Hyd Proc, 60(4), 289-304; 60(7), 239-254 (1981).
Vervalin, C.H., Put loss experiences to work, Hyd Proc, 60(8), 175-181 (1981).

1982
Anon., Quantified risk analysis in the process industries, Chem Engnr, Oct, 385-386, 389 (1982).
Anon., Evolution of a hazardous materials safety policy in the Rijnmond area, Proc Econ Int, 3(1), 68-72 (1982).
Bartknecht, W., Containing air-dust explosions, Proc Engng, Sept, 46-47 (1982).
Brunner, R.L., Improve construction safety, Hyd Proc, May, 118-119 (1982).
Chiu, C.H., Apply depressuring analysis to cryogenic plant safety, Hyd Proc, Nov, 255-264 (1982).

Condiff, D.W., Propagation of thermal detonations through dispersions of hot liquid fuel in cooler volatile liquid coolants, Int J Heat Mass Trans, 25(1), 87-98 (1982).
Duch, M.W., et al., Thermal stability evaluation using differential scanning calorimetry and accelerating rate calorimetry, Plant/Opns Prog, 1(1), 19-27 (1982).
Fawcett, R.W., and Kletz, T.A., Use of computerised system to store and retrieve information on loss prevention, Plant/Opns Prog, 1(1), 7-12 (1982).
Gordon, M.D., et al., Mathematical modeling in thermal hazards evaluation, Plant/Opns Prog, 1(1), 27-33 (1982).
Grossmann, I.E.; Drabbant, R., and Jain, R.K., Incorporating toxicology in the synthesis of industrial chemical complexes, Chem Eng Commns, 17(1), 151-170 (1982).
Hale, C.C., Weather effects on ammonium storage, Plant/Opns Prog, 1(2), 107-114 (1982).
Hardman, R., Planning plant security, Proc Engng, Aug, 27-29 (1982).
Helmers, E.N., and Schaller, L.C., Calculated process risks and hazards management, Plant/Opns Prog, 1(3), 190-194 (1982).
Heperkan, H., and Greif, R., Heat transfer during the shock-induced ignition of an explosive gas, Int J Heat Mass Trans, 25(2), 267-276 (1982).
Howard, W.B., and Karabinis, A.H., Tests of explosion venting of buildings, Plant/Opns Prog, 1(1), 51-65 (1982).
Huberich, T., Safety and environmental protection in ammonia systems, Plant/Opns Prog, 1(2), 117-122 (1982).
Huff, J.E., Emergency venting requirements, Plant/Opns Prog, 1(4), 211-230 (1982).
Kletz, T.A., Minimize your product spillage, Hyd Proc, March, 207-215 (1982).
Kletz, T.A., Beware of the hazards in new technologies, Hyd Proc, May, 297-310 (1982).
Lee, R.S.; Aldis, D.F.; Garrett, D.W., and Lai, F.S., Improved diagnostics for determination of minimum explosive concentration, ignition energy and ignition temperature of dusts, Powder Tech, 31, 51-62 (1982).
McMillan, A.J., Why 'hazard zones' are inadequate, Proc Engng, Aug, 34-37 (1982).
McNaughton, K.J., Occupational skin disease, Chem Eng, 22 March, 147-150; 19 April, 149-150 (1982).
Monroe, C.B., Monitoring and education for a healthy workplace, Chemtech, Jan, 36-38 (1982).
Piccinini, N.; Anatra, U., and Malandrino, G., Safety analysis for allyl chloride plant, Plant/Opns Prog, 1(1), 69-74 (1982).
Pigford, T.H., Geological disposal of radioactive waste, Chem Eng Prog, 78(3), 18-26 (1982).
Ress, J.F., Safe production of bis(chloromethyl) ether, Plant/Opns Prog, 1(4), 248-254 (1982).
Rogerson, J.E., Fire tests of Class NFPA IIIA combustible liquids stored in drums, Plant/Opns Prog, 1(1), 33-37 (1982).
Various, Hazardous plant report, Processing, Dec, 19-25, 33 (1982).
Various, Biosafety (topic issue), Chem & Ind, 20 Nov, 876-903 (1982).

Various, Safety (topic issue), Chem & Ind, 16 Oct, 790-806 (1982).
Wiestling, C., Improve HPI loss investigation, Hyd Proc, Sept, 345-366 (1982).
Willette, D.J., Profile: Safety system in action, Hyd Proc, May, 120-121 (1982).

1983

Bailey, J.D., Is your plant secure from terrorists, Hyd Proc, May, 168-180 (1983).
Brown, C.C., Human toxicity from animal studies, Chemtech, June, 350-358 (1983).
Dagani, R., Toxicity testing, C&E News, 31 Oct, 7-13 (1983).
Duffy, J.L., Source control: Modification of a flame retardant chemical, Plant/Opns Prog, 2(4), 241-243 (1983).
Fritz, R.H., and Jack, G.G., Water in loss prevention (fire fighting agent), Hyd Proc, Aug, 77-90 (1983).
Hardman, J.S.; Lawn, C.J., and Street, P.J., Spontaneous ignition behaviour of activated carbon, Fuel, 62(6), 632-638 (1983).
Hess, J.L., and Kittleman, T.A., Predicting volatile emissions, Chem Eng Prog, 79(11), 40-42 (1983).
Jones, M.R.O., and Underwood, M.C., Calculation of the release rate of pressurized liquified gases, Chem Eng J, 26(3), 251-254 (1983).
Joschek, H.I., Risk assessment in the chemical industries, Plant/Opns Prog, 2(1), 1-5 (1983).
Kletz, T.A., A numerical comparison of short and long-term hazards, Chem Engnr, Nov, 9-13 (1983).
Kusnetz, H.L., and Phillips, C.F., Industrial hygiene evaluations for control design, Plant/Opns Prog, 2(3), 146-150 (1983).
Le Vine, R.Y., New concepts in classification of Class II hazardous locations, Plant/Opns Prog, 2(3), 140-144 (1983).
Lutzow, D., and Hemmer, G., Safe burning of explosive offgas, Plant/Opns Prog, 2(3), 150-153 (1983).
Macfie, R.A.B., Safety responsibilities of plant operators and specialist contractor, Plant/Opns Prog, 2(4), 226-233 (1983).
Martin, G., Formulating emergency plans, Proc Engng, Nov, 27-31 (1983).
Murray, R.W., and O'Neill, K.J., Automatic fire detection: Application and installation, Plant/Opns Prog, 2(1), 67-70 (1983).
Park, S., and Himmelblau, D.M., Fault detection and diagnosis via parameter estimation in lumped dynamic systems, Ind Eng Chem Proc Des Dev, 22(3), 482-487 (1983).
Riley, J.F., Selection and application of special extinguishing agents in industrial hazards, Plant/Opns Prog, 2(2), 101-108 (1983).
Russell, D., Safety in biochemical processing, Proc Engng, Aug, 18-21 (1983).
Schlichtharle, G., and Huberich, T., Tank-car loading station for liquid and aqueous ammonia, Plant/Opns Prog, 2(3), 165-168 (1983).
Various, Lead in the working atmosphere and in the environment (topic issue), Chem & Ind, 4 April, 258-274 (1983).
Various, Aspects of toxicity and toxicology (topic issue), Chem & Ind, 17 Jan, 52-62 (1983).
Various, Dioxin (feature report), C&E News, 6 June, 20-64 (1983).

Vervalin, C.H., Define loss prevention problems and remedies, Hyd Proc, Dec, 111-117 (1983).
Williams, L.F., Risk and insurance, Proc Econ Int, 4(3), 25-28 (1983).

1984

Ainsworth, J.B., and Ojeshina, A.O., Specify containment liners, Hyd Proc, Nov, 130-135 (1984).
Alspach, J., and Bianchi, R.J., Safe handling of phosgene in chemical processing, Plant/Opns Prog, 3(1), 40-42 (1984).
Arendt, J.S.; Campbell, D.J.; Casada, M.L., and Lorenzo, D.K., Risk analysis of a petroleum refinery, Chem Eng Prog, 80(8), 58-64 (1984).
Atwood, J.D., Energy pressures on ammonia, Plant/Opns Prog, 3(2), 95-98 (1984).
Austin, K., Pressure surges, Proc Engng, March, 35-38 (1984).
Beckman, J.R., Breathing losses from fixed-roof tanks by heat and mass transfer diffusion, Ind Eng Chem Proc Des Dev, 23(3), 472-479 (1984).
Bell, R., Safety of programmable electronic systems, Proc Engng, April, 21-24 (1984).
Castle, G.K., Improve control system fire protection, Hyd Proc, March, 113-125 (1984).
Colonna, G.R., Coast guard confined spaces safety procedures, Plant/Opns Prog, 3(3), 188-190 (1984).
Colt, W.J., Assessing plant damage, Hyd Proc, Oct, 85-91 (1984).
Cook, R.R., and Cartmill, J.B., Assessing the effects of dioxin, Chemtech, Sept, 534-537 (1984).
DiMaio, L.R., and Lange, R.F., Effect of water quality on fire fighting foams, Plant/Opns Prog, 3(1), 42-46 (1984).
Fauske, H.K., Scale-up for safety relief of runaway reactions, Plant/Opns Prog, 3(1), 7-12 (1984).
Fenlon, W.J., Comparison of ARC and other thermal stability test methods, Plant/Opns Prog, 3(4), 197-203 (1984).
Freeman, H.M., Role of incineration in remedial actions, Plant/Opns Prog, 3(2), 113-117 (1984).
Gray, B.F.; Griffiths, J.F., and Hasko, S.M., Spontaneous ignition hazards in stockpiles of cellulosic materials: Criteria for safe storage, J Chem Tech Biotechnol, 34A(8), 453-463 (1984).
Hagon, D.O., Use of frequency-consequence curves to examine the conclusions of published risk analyses and to define broad criteria for major hazard installations, CER&D, 62, 381-386 (1984).
Hamilton, T.B., Safety in combustion of unconventional fuels, Plant/Opns Prog, 3(2), 86-89 (1984).
Hamm, W., Safety and loss prevention in the food industry, Chem Engnr, Aug, 22-25 (1984).
Howard, W.B., Efficient time use to achieve safety of process, Plant/Opns Prog, 3(3), 129-133 (1984).
Janssens, H., Dust-explosion protection, Proc Engng, July, 47-49 (1984).
Jones, A.L., Fugitive emissions of volatile hydrocarbons, Chem Engnr, Aug, 12-15 (1984).
Kletz, T.A., Talking about safety, Chem Engnr, Jan, 33 (1984).

Lasseigne, A.H., The NTSB hazardous material spill-map program, Plant/Opns Prog, 3(2), 94-95 (1984).
Limb, D.I., and Healy, M.J., Loss analysis aids process selection, Hyd Proc, Aug, 87-89 (1984).
Mehne, P.H., Reducing risks at high pressures and temperatures, Plant/Opns Prog, 3(1), 37-40 (1984).
Nomura, S.I.; Torimoto, M., and Tanaka, T., Theoretical upper limit of dust explosion in relation to oxygen concentration, Ind Eng Chem Proc Des Dev, 23(3), 420-423 (1984).
Palazzi, E., et al., Flammability limits with short duration gas releases, Plant/Opns Prog, 3(3), 159-163 (1984).
Parker, R.V., Safety systems for continuous petrochemical process under vacuum, Plant/Opns Prog, 3(2), 119-121 (1984).
Pierson, K.L., How safe is safe enough? Plant/Opns Prog, 3(2), 71-72 (1984).
Plett, E.G., Flammability of gaseous mixtures of ethylene oxide, nitrogen and air, Plant/Opns Prog, 3(3), 190-194 (1984).
Roy, J.O., Explain those safety rules, Hyd Proc, Sept, 196-198 (1984).
Smith, D.W., Assessing the hazards of runaway reactions, Chem Eng, 14 May, 54-60 (1984).
Swift, I., Venting deflagrations: Theory and practice, Plant/Opns Prog, 3(2), 89-94 (1984).
Various, Process safety (topic issue), Chem Eng Prog, 80(3), 47-81 (1984).
Whitfield, D., Ergonomics problems in process operations, Chem Engnr, Dec, 36-37 (1984).

1985
Astleford, W.J.; Bass, R.L., and Colonna, G.R., Modeling of ship tank ventilation and occupational exposures to chemical vapors during tank entry, Plant/Opns Prog, 4(2), 90-95 (1985).
Bellamy, L., Human behaviour and plant safety, Proc Engng, July, 27-28 (1985).
Berhinig, R.M., Fire-resistance test for petrochemical facility structural elements, Plant/Opns Prog, 4(4), 230-234 (1985).
Buch, R.R., and Filsinger, D.H., Method for fire hazard assessment of fluid-soaked thermal insulation, Plant/Opns Prog, 4(3), 176-181 (1985).
Capizzani, R.E., An overview of flammable liquid drum storage and protection, Plant/Opns Prog, 4(3), 139-144 (1985).
Clarke, J.R.P., Monitoring gaseous emissions, Chem Engnr, July, 36-37 (1985).
Constance, J.D., Ventilating CPI buildings, Chem Eng, 9 Dec, 89-92 (1985).
Davies, P., and Hymes, I., Chlorine toxicity criteria for hazard assessment, Chem Engnr, June, 30-33 (1985).
Ember, L.R., Cleanup of acid pits, C&E News, 27 May, 11-21 (1985).
Fauske, H.K., Flashing flows: Some practical guidelines for emergency releases, Plant/Opns Prog, 4(3), 132-135 (1985).
Fauske, H.K., and Leung, J.C., New experimental technique for characterizing runaway chemical reactions, Chem Eng Prog, 81(8), 39-46 (1985).
Fisher, H.G., Emergency relief systems research program, Chem Eng Prog, 81(8), 33-36 (1985).

Franklin, N., The hazards of nuclear power: Myth or reality? Chem Engnr, Oct, 24-25 (1985).
Freeman, R.A., Use of risk assessment in the chemical industries, Plant/Opns Prog, 4(2), 85-90 (1985).
Gagliardi, D.V., Commonsense approach to loss prevention, Chem Eng Prog, 81(8), 26-28 (1985).
Gerritsen, H.G., and van't Land, C.M., Intrinsic continuous process safeguarding (a review), Ind Eng Chem Proc Des Dev, 24(4), 893-896 (1985).
Grolmes, M.A., and Leung, J.C., Code method for evaluating integrated relief phenomena, Chem Eng Prog, 81(8), 47-52 (1985).
Kirby, G.N., Explosion pressure shock resistance, Chem Eng Prog, 81(11), 48-50 (1985).
Kletz, T.A., Make plants inherently safe, Hyd Proc, Sept, 172-180 (1985).
Kletz, T.A., An atlas of safety thinking (copy in the colours!), Chem Engnr, May, 47-48 (1985).
Kletz, T.A., What is a protective system? Chem Engnr, July, 29 (1985).
Kletz, T.A., Knowledge in the wrong place, Chem Engnr, Sept, 51 (1985).
Kletz, T.A., Cheaper Safer Plants, I.Chem.E., U.K. (1985).
Kumar, R.K., and Tamm, H., Turbulent combustion of hydrogen in large volumes, Can JCE, 63(4), 662-667 (1985).
Lepkowski, W., Chemical safety in developing countries, C&E News, 8 April, 9-14 (1985).
Munson, R.E., Process hazards management in Du Pont, Plant/Opns Prog, 4(1), 13-16 (1985).
Nomura, S., and Callcott, T.G., Calculation of ignition sensitivity of dust clouds of varying size distributions, Powder Tech, 45, 145-154 (1985).
Rains, C.O., Control your contractor accident losses, Hyd Proc, July, 97-104 (1985).
Rains, W.A., Accelerated aging tests for evaluating fireproofing materials, Plant/Opns Prog, 4(4), 246-248 (1985).
Shiozaki, J., et al., Fault-diagnosis system for chemical plants, Int Chem Eng, 25(4), 651-668 (1985).
Soden, J.D., Basics of fire-protection design, Hyd Proc, May, 157-168; June, 101-110 (1985).
Stevens, R., Friedal-Crafts type reaction and explosion, Plant/Opns Prog, 4(2), 68-72 (1985).
Tyler, B.J., Using the Mond Index to measure inherent hazards, Plant/Opns Prog, 4(3), 172-176 (1985).
Various, Labelling chemicals for workplace and for transport (topic issue), Chem & Ind, 21 Jan, 40-55 (1985).
Various, Industrial accidents and risks to the community (topic issue), Chem & Ind, 1 July, 431-440; 15 July, 465-482 (1985).
Various, Control of substances hazardous to health (topic issue), Chem & Ind, 18 March, 180-185 (1985).
Various, Fire protection and safety (special report), Hyd Proc, Dec, 71-84 (1985).
Williams, J., Safety and reprocessing in nuclear power plants, Processing, Oct, 31-35 (1985).

Zurer, P.S., Hazards of asbestos, C&E News, 4 March, 28-41 (1985).

1986

Antal, L.; Hlavay, J., and Karpati, J., Analysis of respirable and sedimented dust samples, Periodica Polytechnica, 30(3/4), 209-212 (1986).

Blackburn, G.M., and Kellard, B., Chemical carcinogens, Chem & Ind, 15 Sept, 607-613; 20 Oct, 687-695; 17 Nov, 770-779 (1986).

Britton, L.G.; Taylor, D.A., and Wobser, D.C., Thermal stability of ethylene at elevated pressures, Plant/Opns Prog, 5(4), 238-251 (1986).

Brooks, K.S., and Glasser, D., A simplified model of spontaneous combustion in coal stockpiles, Fuel, 65(8), 1035-1041 (1986).

Buhrow, R.P., Know 'areas of vulnerability'. Hyd Proc, 65(8), 99-102 (1986).

Chowdhury, J., Troubleshooting comes online in the CPI, Chem Eng, 13 Oct, 14-19 (1986).

Gallagher, G.A., and McCone, S.W., Lessons in hazardous material transportation based on case histories, Plant/Opns Prog, 5(3), 186-191 (1986).

Ghosh, R., Spontaneous combustion of certain Indian coals: Some physico-chemical considerations, Fuel, 65(8), 1042-1046 (1986).

Guldemond, C.P., Behaviour of denser than air ammonia in presence of obstacles: Wind tunnel experiments, Plant/Opns Prog, 5(2), 93-96 (1986).

Harron, J.G., Loss prevention challenge: How one refinery meets it, Hyd Proc, 65(9), 167-175 (1986).

Jebens, W.L., Reduce plant operating risks, Hyd Proc, 65(12), 65-66 (1986).

Kletz, T.A., Inherent safety and the nuclear industry, Chem Engnr, July, 35 (1986).

Kletz, T.A., Will cold petrol explode in the open air? Chem Engnr, June, 63 (1986).

Kletz, T.A., Tests should be like real life, Chem Engnr, Oct, 39 (1986).

Kletz, T.A., Package deals, Chem Engnr, Jan, 37 (1986).

Kletz, T.A., Transportation of hazardous substances: The UK scene, Plant/Opns Prog, 5(3), 160-164 (1986).

Lihou, D., The illusion of double standards, Chem Engnr, Feb, 45 (1986).

Matthiessen, R.C., Estimating chemical exposure levels in the workplace, Chem Eng Prog, 82(4), 30-34 (1986).

McAlister, J., Surveillance - the forgotten maintenance tool? Chem Eng, 13 Oct, 87-89 (1986).

McMillan, A., Flammable hazards: Area classification, Proc Engng, July, 35-39 (1986).

Narasimhan, N.D., Predict flare noise, Hyd Proc, 65(4), 133-136 (1986).

Norstrom, G.P., Property insurance considerations in loss prevention expenditures, Plant/Opns Prog, 5(4), 209-220 (1986).

Ross, C., Good vibes from anti-sound, Chem Engnr, Feb, 22-24 (1986).

Stevenson, F.D.; Maher, S.T.; Sharp, D.R., and Sloane, B.D., Process safety management of a fuel gas conditioning facility using fault tree analysis, Can JCE, 64(5), 848-853 (1986).

Various, In-house self regulation of health and safety (topic issue), Chem & Ind, 15 Sept, 598-606 (1986).

Young, E.K., Controlling and eliminating perceived risks, Chem Eng Prog, 82(7), 14-16 (1986).

1987

Baker-Counsell, J., How long will your plant run safely, Proc Engng, May, 61-63 (1987).

Bradford, M., Process plants as hazardous-waste facilities, Chem Eng, 2 March, 69-71 (1987).

Chowdhury, J., Chemical-plant safety: An international drawing card, Chem Eng, 16 March, 14-17 (1987).

Dave, N.D., Loss-control representatives: What they do, Chem Eng, 11 May, 49-51 (1987).

Davenport, J.A., Gas plant and fuel handling facilities: An insurer's view, Plant/Opns Prog, 6(4), 199-202 (1987).

Guymer, P.; Kaiser, G.D.; McKelvey, T.C., and Hannaman, G.W., Probabilistic risk assessment in the CPI, Chem Eng Prog, 83(1), 37-45 (1987).

Hoyle, E.R., and Stricoff, R.S., Functional requirements and design criteria for design of high hazard containment laboratories, Plant/Opns Prog, 6(3), 146-150 (1987).

Ichniowski, T., Making it safer to move hazardous materials, Chem Eng, 20 July, 14-19 (1987).

Kauffman, G.J., Combustion safety inspection, Chem Eng Prog, 83(7), 23-26 (1987).

Keller, C.L., Future requirements for unrestricted entry into confined spaces, Plant/Opns Prog, 6(3), 142-145 (1987).

Kemp, H.S., Process safety - a commitment? Chem Eng Prog, 83(5), 46-47 (1987).

Latino, C.J., Solving human-caused failure problems, Chem Eng Prog, 83(5), 42-45 (1987).

Martin, E.J., and Johnson, J.H. (Eds). Hazardous Waste Management Engineering, Van Nostrand Reinhold, New York (1987).

Matley, J., and Greene, R., Ethics of health, safety and environment, Chem Eng, 2 March, 40-50 (1987).

Matley, J.; Greene, R., and McCauley, C., Health, safety and environment (readers survey), Chem Eng, 28 Sept, 108-121 (1987).

McNaughton, D.J.; Worley, G.G., and Bodner, P.M., Evaluating emergency response models for the chemical industry, Chem Eng Prog, 83(1), 46-51 (1987).

Sims, E.R., Safety store hazardous and flammable materials, Chem Eng, 28 Sept, 129-132 (1987).

Smith, J.B., System approach for plant reliability, Chem Eng Prog, 83(4), 47-54 (1987).

Spiegelman, A., Gas plant problems and the insurance industry, Plant/Opns Prog, 6(4), 190-192 (1987).

Taylor, R.M., and Lewis, W.I., Using robots to reduce personnel radiation exposure at Savannah River plant, Plant/Opns Prog, 6(3), 165-170 (1987).

Various, Science, technology and industrial security (topic issue), Chem & Ind, 15 June, 412-415 (1987).

Various, Handling of chemical carcinogens (topic issue), Chem & Ind, 15 June, 404-411 (1987).
Whalley, S., What can cause human error? Chem Engnr, Feb, 37 (1987).
Wierzba, I.; Karim, G.A.; Cheng, H., and Hanna, M., The flammability of rich mixtures of hydrogen and ethylene in air, J Inst Energy, March, 3-7 (1987).

1988
Allman, W.F., Staying alive in the 20th century, Chemtech, Dec, 720-724 (1988).
Atkinson, N., Assessing human reliability, Proc Engng, Oct, 35-37 (1988).
Bartknecht, W., Ignition capabilities of hot surfaces and mechanically generated sparks in flammable gas and dust/air mixtures, Plant/Opns Prog, 7(2), 114-121 (1988).
Berkey, B.D.; Pratt, T.H., and Williams, G.M., Review of literature related to human spark scenarios, Plant/Opns Prog, 7(1), 32-36 (1988).
Brooks, K.; Balakotaiah, V., and Luss, D., Effect of natural convection on spontaneous combustion of coal stockpiles, AIChEJ, 34(3), 353-365 (1988).
Brooks, K.; Svanas, N., and Glasser, D., Evaluating the risk of spontaneous combustion in coal stockpiles, Fuel, 67(5), 651-656 (1988).
Bunn, A.R., and Lees, F.P., Expert design of plant handling hazardous materials: Design expertise and CAD methods with illustrative examples, CER&D, 66(5), 419-444 (1988).
Ciolek, W.H., Laboratory shielding for projectiles, Plant/Opns Prog, 7(2), 79-86 (1988).
Commission Report, Static electricity: Rules for plant safety, Plant/Opns Prog, 7(1), 1-22 (1988).
Crowl, D.A., and Louvar, J.F., Safety and loss prevention undergraduate syllabus, Chem Eng Educ, 22(2), 74-79 (1988).
DiMaio, L.R., and Norman, E.C., Performance of aqueous Hazmat foams on selected hazardous materials, Plant/Opns Prog, 7(3), 195-198 (1988).
Ducatman, A.; Crawl, J.R., and Conwill, D.E., Cancer clusters: Correlation, causation and common sense, Chemtech, April, 204-210 (1988).
Fleischman, M., Incorporating health and safety into the curriculum, Chem Eng Educ, 22(1), 30-34 (1988).
Gittus, J., and Gunning, A., Safety and nuclear power in perspective, Chem Engnr, May, 12-18 (1988).
Hawksley, J.L., Process safety management: A UK approach, Plant/Opns Prog, 7(4), 265-269 (1988).
Kletz, T.A., Ammonia can explode, Chem Engnr, Feb, 41 (1988).
Kletz, T.A., Safety topics, Chem Engnr, May, 47 (1988).
Kletz, T.A., Some hazards arn't obvious, Chem Engnr, Aug, 34 (1988).
Le, N.B.; Santay, A.J., and Zabrenski, J.S., Laboratory safety design criteria, Plant/Opns Prog, 7(2), 87-94 (1988).
Liang, H., and Tanaka, T., Simulation of spontaneous heating for evaluation of ignition temperature and induction time of combustible dust, Int Chem Eng, 28(4), 652-661 (1988).
Mall, R.D., Safety in multistream ammonia plants utilizing common utility services, Plant/Opns Prog, 7(2), 136-138 (1988).

Mancini, R.A., Use and misuse of bonding for control of static ignition hazards, Plant/Opns Prog, 7(1), 23-31 (1988).
Mascone, C.F.; Gordon, H.S., and Vagi, D.L., CPI safety survey results, Chem Eng, 10 Oct, 74-86 (1988).
Morris, G.D.L., Fire protection report, Chem Eng, 10 Oct, 30-37 (1988).
Nazario, F.N., Preventing or surviving explosions, Chem Eng, 15 Aug, 102-109 (1988).
Owens, J.E., Spark ignition hazards caused by charge induction, Plant/Opns Prog, 7(1), 37-39 (1988).
Rosenthal, L.A., Static electricity and plastic drums, Plant/Opns Prog, 7(1), 51-52 (1988).
Schwab, R.F., Consequences of solvent ignition during drum filling operation, Plant/Opns Prog, 7(4), 242 (1988).
Singh, J., Assessing the hazards of runaway reactions, Processing, July, 43-44 (1988).
Spence, J.P., and Noronha, J.A., Reliable detection on runaway reaction precursors in liquid-phase reactions, Plant/Opns Prog, 7(4), 231-235 (1988).
Stallworthy, E.A., Coping with catastrophe, Chem Engnr, July, 59 (1988).
Torrent, J.G.; Armada, I.S., and Pedreira, R.A., Correlation between composition and explosibility index for coal dust, Fuel, 67(12), 1629-1632 (1988).
Various, Safety issues in biotechnology (topic issue), J Chem Tech Biotechnol, 43(4), 245-378 (1988).
Various, Hazard assessment for ecotoxicological testing of chemicals (topic issue), Chem & Ind, 4 Jan, 8-24 (1988).
Various, Chemical plant safety (feature issue), Chem Eng Prog, 84(9), 25-70 (1988).
Various, Static electricity and safety (topic issue), Plant/Opns Prog, 7(1), 1-78 (1988).
Wampler, F.M., Formation of diacrylic acid during acrylic acid storage, Plant/Opns Prog, 7(3), 183-189 (1988).
Wilson, D.K., Failure mode management: Loss prevention philosophy for programmable logic controllers, Plant/Opns Prog, 7(4), 254-257 (1988).
Winkless, N., Defining toxic materials, Chemtech, Nov, 658-660 (1988).
Yapijakis, C., Chemical spills, Chemtech, Oct, 604-607 (1988).

CHAPTER 7

PLANT OPERATIONS

7.1 Startup and Shutdown

1967-1979

Hettig, S.B., Emergency shutdown systems, Chem Eng, 27 Feb, 141-143 (1967).
Various, Plant startup problems (topic issue), Chem Eng Prog, 63(12), 33-63 (1967).
Finneran, J.A.; Sweeney, N.J., and Hutchinson, T.G., Startup performance of large ammonia plants, Chem Eng Prog, 64(8), 72-77 (1968).
Feldman, R.P., Economics of plant startups, Chem Eng, 3 Nov, 87-90 (1969).
Goyal, S.K., How much manpower for turnarounds? Hyd Proc, 48(5), 155-156 (1969).
Grieve, P., Plant startups, Chem Eng, 8 Sept, 148-150 (1969).
Matley, J., Keys to successful plant startups, Chem Eng, 8 Sept, 110-130 (1969).
Clark, M.E.; DeForest, E.M., and Steckley, L.R., Problems with plant startups, Chem Eng Prog, 67(12), 25-28 (1971).
Parsons, R.H., Guidelines for plant startups, Chem Eng Prog, 67(12), 29-31 (1971).
Swain, R.T., and Hopper, B.J., Contractor problems during startup, Chem Eng Prog, 67(12), 32-35 (1971).
Various, Startups made easier (special report), Hyd Proc, 50(12), 87-95 (1971).
Goldman, R., Plant startups, Chem Eng, 25 Dec, 72-74 (1972).
Ryan, G.T., Managing the project startup, Chem Eng Prog, 68(12), 65-71 (1972).
Fisher, J.T., Designing emergency shutdown systems, Chem Eng, 10 Dec, 138-144 (1973).
Galluzzo, J.F., Maintenance aspects of a pharmaceutical-plant startup, Chem Eng, 2 April, 82-86 (1973).
Godard, K.E., Gas plant startup problems, Hyd Proc, 52(9), 151-155 (1973).
Various, Plant startup problems (topic issue), Chem Eng Prog, 69(8), 85-104 (1973).
Butler, P., Commissioning chemical plants, Proc Engng, June, 88-89 (1974).
Stainthorp, F.P., and West, B., Computer controlled plant start-up, Chem Engnr, Sept, 526-530 (1974).
Arndt, J.H., and Kiddoo, D.B., Special purpose gearing: Avoiding startup problems, Hyd Proc, 54(3), 113-118 (1975).
Clifton, R.H., Reducing plant downtime, Proc Engng, Feb, 78-81 (1975).
Henderson, P.E., Optimum boiler start-up control, Energy World, Feb, 3-6 (1975).
Martin, D., Causes of plant stoppages, Proc Engng, Oct, 86-93 (1975).
Butzert, H.E., Ammonia/methanol plant start-up, Chem Eng Prog, 72(1), 56-59 (1976).
Dolle, J., and Gilbourne, D., Start-up of the Skikda LNG plant, Chem Eng Prog, 72(1), 39-43 (1976).
Gans, M., The A to Z of plant startup, Chem Eng, 15 March, 72-82 (1976).
Bress, D.F., and Packbier, M.W., The startup of two major urea plants, Chem Eng Prog, 73(5), 80-84 (1977).
Goyal, S.K., Optimum time between shutdowns, Processing, May, 41; Oct, 33 (1977).

Mandelik, B.G., and Turner, W., How to operate nitric acid plants, Hyd Proc, 56(7), 175-177 (1977).
Williams, G.P., Causes of ammonia plant shutdowns, Chem Eng Prog, 74(9), 88-93 (1978).
Vargas, K.J., Managing plant turnarounds, Chem Eng, 4 June, 116-130 (1979).

1980-1988
Kao, Y.K., A simple start-up policy for a catalytic reactor with catalyst deactivation, Chem Eng J, 20(3), 237-246 (1980).
Britton, C.F., Preventing unscheduled shutdowns, Chem Eng, 1 June, 83-85 (1981).
Bell, H.V., Shutdown and start-up of a cryogenic tank, Chem Eng Prog, 78(2), 74-77 (1982).
Fulks, B.D., Planning and organizing for less troublesome plant startups, Chem Eng, 6 Sept, 96-106 (1982).
Meier, F.A.. Is your control system ready to start-up? Chem Eng, 22 Feb, 76-87 (1982).
Rhodes, T., Shutdown systems for better managed process operations, Hyd Proc, May, 151-154 (1982).
Anderson, A.P., and Hedley, A.B., Start-up characteristics of a small oil-fired boiler, J Inst Energy, 56, 111-118,225 (1983).
Anderson, G.D., Initial controller settings to use at plant startup, Chem Eng, 11 July, 113-116 (1983).
Gans, M.; Kiorpes, S.A., and Fitzgerald, F.A., Plant startup - step by step by step, Chem Eng, 3 Oct, 74-100 (1983).
Max, D.A., and Jones, S.T., Flareless ethylene plant: Start-up and shut-down procedures, Hyd Proc, Dec, 89-90 (1983).
Badder, E.E., and Brooks, B.W., Start-up procedures for continuous-flow emulsion-polymerization reactors, Chem Eng Sci, 39(10), 1499-1510 (1984).
Facer, J.R., and Rich, R.W., Shutdown winterization of 1500 ton per day ammonia plant, Plant/Opns Prog, 3(1), 12-14 (1984).
Fisch, E., Winterizing process plants, Chem Eng, 20 Aug, 128-143 (1984).
Prijatel, J., Failure analysis of ammonia plant shutdown instrumentation and control, Plant/Opns Prog, 3(1), 25-29 (1984).
Braye, J., Planning for commissioning and start-up, Proc Engng, March, 45-48 (1985).
Rattan, I.S., and Pathak, V.K., Startup of centrifugal pumps in flashing or cryogenic liquid service, Chem Eng, 1 April, 95-96 (1985).
Ruziska, P.A., et al., Exxon low-energy ammonia process start-up experience, Plant/Opns Prog, 4(2), 79-85 (1985).
Toering, W., Installation and start-up experiences of synthesis gas dryers in existing ammonia plant, Plant/Opns Prog, 4(3), 127-132 (1985).
Twigg, R.J., Preventing corrosion during mothballing, Chem Eng, 16 Sept, 91-94 (1985).
Kletz, T.A., When is a shutdown not a shutdown? Chem Engnr, April, 55 (1986).
Onderdonk, J.K., Understanding shutdown systems, Chem Eng, 7 July, 45-50 (1986).

Tropp, R.I., More efficient turnarounds, Hyd Proc, 65(1), 55-57 (1986).
Wetherill, F.E., and Wallsgrove, C.S., Operator training for start-up, Plant/Opns Prog, 5(2), 78-80 (1986).
Miller, R.G., and King, R.A., Mothballing your plant, Chem Engnr, March, 41-43 (1987).
Velasco, J.R.G.; Ortiz, M.A.G.; Pelayo, J.M.C., and Marcos, J.A.G., Improvements in batch distillation startup, Ind Eng Chem Res, 26(4), 745-750 (1987).
Merrow, E.W., Estimating startup times for solids-processing plants, Chem Eng, 24 Oct, 89-92 (1988).

7.2 Commissioning, Revamps and Retrofits

1967-1980

House, F.F., Winterizing chemical plants, Chem Eng, 11 Sept, 173-180 (1967).
Marcon, J.M., and Neidhart, J.F., Old plants revived with new control centre, Chem Eng, 14 July, 115-120 (1969).
Crowley, M.S., Guide to refractory lining repair, Hyd Proc, 49(4), 149 (1970).
Thorngren, J.T., Predict exchanger tube damage, Hyd Proc, 49(4), 129-131 (1970).
Voelker, C.H., and Zeis, L.A., Repair welding HK-40 furnace tubes, Hyd Proc, 51(4), 121-124 (1972).
Brussel-Smith, I.E., Uses of photography in plant construction, Chem Eng, 28 May, 124-128 (1973).
Fair, E.W., Replacing old equipment, Chem Eng, 15 Oct, 124-126 (1973).
Thorne, H.C., Post-installation appraisals, Hyd Proc, 52(4), 203-211 (1973).
Taube, A.T., Decoke furnace tubes faster, Hyd Proc, 53(4), 151-156 (1974).
Fertilio, A., and Princip, B., Inspection of reformer tubes for repair, Hyd Proc, 54(9), 174-180 (1975).
Sullivan, J.A., Defining work scope for turnarounds, Chem Eng, 8 Dec, 141-144 (1975).
Grimes, T.L.; Schadewald, F.H.; Pauli, W.A.; Fonner, D.E., and Griffin, J.C., Automated cleaning in pharmaceuticals processing, Chem Eng Prog, 72(2), 68-73 (1976).
Mol, A., Why revamp older steam crackers? Hyd Proc, 57(7), 179-184 (1978).
Houghton, J.D., Correct way to overhaul turbomachinery, Hyd Proc, 58(6), 129-136 (1979).
Minet, R.G., and Tsai, F., Design and retrofit of thermal cracking coils, Chem Eng Prog, 76(7), 31-35 (1980).

1981-1985

Malan, D.N., Commissioning, planning and execution, Chem Engnr, Oct, 442-444 (1981).
Nielsen, A., et al., Revamp of ammonia plants, Plant/Opns Prog, 1(3), 186-190 (1982).
Wang, S.I.; Smith, D.D.; Patel, N.M., and DiMartino, S.P., Performance tests for steam-methane reformers, Hyd Proc, Aug, 89-92 (1982).

Berzins, V., and Udengaard, N.R., Revamp your hydrogen plant, Hyd Proc, May, 65-67 (1983).
Slear, D.G.; Long, R.L., and Jones, J.D., Repair of TMI-1 OTSG tube failures, Plant/Opns Prog, 2(3), 173-176 (1983).
Sumner, C., and Fernandez-Baujin, J.M., Retrofitting olefin cracking plants, Chem Eng Prog, 79(12), 39-44 (1983).
Tsao, U., Troubleshooting the new plant, Chemtech, Dec, 750-754 (1983).
Breuer, C.T., Retrofitting coal-fired boilers, Chem Eng, 17 Sept, 97-103 (1984).
Combs, J.R.; Copeland, B.G.; Abbott, K.W., and Bachus, J.B., Manage plant modifications effectively during turnaround, Hyd Proc, Sept, 213-226 (1984).
Edgar, M.D.; Johnson, A.D.; Pistorius, J.T., and Varadi, T., Troubleshooting made easy, Hyd Proc, May, 65-70 (1984).
Shields, V.; Udengaard, N.B., and Berzins, V., Revamping hydrogen plants for improved yield and energy efficiency, Plant/Opns Prog, 3(2), 108-112 (1984).
Wang, S.I., and Patel, N.M., Revamping ammonia plants can improve profitability, Plant/Opns Prog, 3(2), 101-108 (1984).
Zakrzewski, A., Successful commissioning of computer-controlled plant, Proc Engng, Oct, 34-38 (1984).
Baker-Counsell, J., Process-plant revamping, Proc Engng, April, 21,23 (1985).
Hyde, J.M., New developments in cleaning practices, Chem Eng Prog, 81(1), 39-41 (1985).
Ramshaw, C., Process intensification, Chem Engnr, July, 30-33 (1985).
Schillmoller, C.M., Use these materials to retrofit ethylene furnaces, Hyd Proc, Sept, 101-104 (1985).
Skinner, A.C., Retrofitting increases energy recovery, Chem Eng Prog, 81(2), 28-29 (1985).
Speedie, I., Engineering for successful revamps, Chem Engnr, April, 49-54 (1985).
Tsai, F.W.; Che, S.C., and Minet, R.G., Why retrofit furnaces? Hyd Proc, Aug, 41-47 (1985).
Webber, W.O., Retrofit insulation for profit, Hyd Proc, Nov, 79-80 (1985).

1986-1988

Arendt, J.S., Determining heater retrofits through risk assessment, Plant/Opns Prog, 5(4), 228-231 (1986).
Burger, R., Retrofitting cuts cooling-tower costs, Chem Eng, 18 Aug, 117-119 (1986).
Danos, R.J., Retrofitting ammonium phosphate granulation plants, Chem Eng Prog, 82(5), 50-54 (1986).
Le Blanc, J.R., Retrofit ammonia plants to save energy and up capacity, Hyd Proc, 65(8), 39-44 (1986).
Marschner, F., and Moertel, H.G., Revamps increase efficiency, Hyd Proc, 65(1), 63-66 (1986).
Schillmoller, C.M., Consider these alloys for ammonia plant retrofit, Hyd Proc, 65(9), 63-65 (1986).
Smet, E.J., Sequential completion of plant construction to allow early production, Plant/Opns Prog, 5(2), 108-109 (1986).

Tjoe, T.N., and Linnhoff, B., Using pinch technology for process retrofit, Chem Eng, 28 April, 47-60 (1986).
Wagner, V.S., Retrofit ammonia plant steam turbines, Hyd Proc, 65(1), 67-69 (1986).
Westerterp, K.R., and Kuczynski, M., Retrofit methanol plants with this converter system, Hyd Proc, 65(11), 80-83 (1986).
Atkinson, R., Planning and implementing plant revamps, Chem Eng, 2 March, 51-56 (1987).
Austin, J.F., and Sultan, M.S., Effective documentation can improve revamps, Hyd Proc, 66(11), 111-121 (1987).
Cabano, L.J., Retrofit projects, Chem Eng Prog, 83(4), 27-31 (1987).
Fisher, W.R.; Doherty, M.F., and Douglas, J.M., Screening of process retrofit alternatives, Ind Eng Chem Res, 26(11), 2195-2204 (1987).
Gibson, G.J., Efficient test runs: Procedures when planning and implementing process-plant test runs, Chem Eng, 11 May, 75-78 (1987).
Jaske, C.E., Benefits of remaining life assessment, Chem Eng Prog, 83(4), 37-46 (1987).
Lockett, W., and Plumstead, J.A., Improving refinery efficiencies, Chem Eng Prog, 83(4), 33-36 (1987).
Taylor, W.K., and Pinto, A., Commissioning of an ammonia plant, Plant/Opns Prog, 6(2), 106-111 (1987).
Vaselenak, J.A.; Grossman, I.E., and Westerberg, A.W., Optimal retrofit design of multiproduct batch plants, Ind Eng Chem Res, 26(4), 718-726 (1987).
White, R.L., On-line troubleshooting of chemical plants, Chem Eng Prog, 83(5), 33-38 (1987).
Elshout, R.V., and Kilstrom, M.S., Tips for better retrofits, Chem Eng, 15 Aug, 95-99 (1988).
Glazer, J.L.; Schott, M.E., and Stapf, L.A., Improved hydrocracking by upgrading recycle, Hyd Proc, 67(10), 61-62 (1988).
Granelli, F., Ways to revamp urea units, Hyd Proc, 67(6), 59-63 (1988).
Mancuso, J.R., Retrofitting gear couplings with diaphragm couplings, Hyd Proc, 67(10), 47-51 (1988).
Pistikopoulos, E.N., and Grossmann, I.E., Optimal retrofit design for improving process flexibility in linear systems, Comput Chem Eng, 12(7), 719-732 (1988).
Savage, P., Demothballing plants, Chem Eng, 23 May, 26-29 (1988).
Various, Maintenance and retrofitting (special report), Hyd Proc, 67(1), 35-61 (1988).

7.3 Maintenance

1967-1972

Sawers, D.S., Piping configuration cuts down maintenance time, Chem Eng, 18 Dec, 130 (1967).
Bresler, S.A., and Hertz, M.J., Equipment warranties, Chem Eng, 25 March, 86-94; 8 April, 137-142 (1968).
Jordan, J.H., Evaluating the advantages of contract maintenance, Chem Eng, 25 March, 124-130 (1968).

Sarappo, J.W., Contract maintenance, Chem Eng, 17 Nov, 264-272 (1969).
Various, Maintenance (special report), Hyd Proc, 48(1), 89-121 (1969).
Various, Plant engineering and maintenance (topic issue), Chem Eng Prog, 65(1), 33-51 (1969).
Hopkins, C.D., Preventive maintenance of gas plant equipment, Chem Eng Prog, 66(6), 60-61 (1970).
Maten, S., Program machine maintenance by measuring vibration velocity, Hyd Proc, 49(9), 291-296 (1970).
Matley, J., Trends in maintenance, Chem Eng, 23 March, 112-126 (1970).
Trotter, J.A., Techniques of predictive maintenance, Chem Eng, 24 Aug, 66-70 (1970).
Various, Maintenance (special report), Hyd Proc, 49(1), 99-136 (1970).
Buhrow, R.P., Scheduling pipe inspections, Chem Eng, 26 July, 120-122 (1971).
Jumper, C., New approach to inventory control, Hyd Proc, 50(8), 137-138 (1971).
Mack, W.C., Predict remaining service life of tubing, Chem Eng, 9 Aug, 124-126 (1971).
Simon, E.L., and Gonzalez, H.S., Maintenance control for medium-size plants, Chem Eng, 23 Aug, 102-104 (1971).
Tator, K.B., Engineered painting pays off, Chem Eng, 27 Dec, 84-87 (1971).
Various, Maintenance (special report), Hyd Proc, 50(1), 73-97 (1971).
Various, Contract maintenance and plant operations (topic issue), Chem Eng Prog, 67(4), 43-49 (1971).
Wrasman, T.J., Plastic-piping maintenance, Chem Eng, 8 Feb, 88-94 (1971).
Brown, P.J., Compressor and engine maintenance, Chem Eng Prog, 68(6), 77-80 (1972).
Donahoe, P.G., Preventive maintenance for gas turbines, Chem Eng Prog, 68(7), 87-89 (1972).
Drost, J.G., Preventive maintenance through oil analysis, Chem Eng Prog, 68(8), 84-87 (1972).
King, C.F., and Rudd, D.F., Design and maintenance of economically failure-tolerant processes (review paper), AIChEJ, 18(2), 257-269 (1972).
Pilborough, L., Planned maintenance versus unscheduled shutdown, Proc Engng, Aug, 54-59 (1972).
Prescott, J.H., Thermography for identifying imminent plant problems, Chem Eng, 18 Sept, 178-180 (1972).
Various, Maintenance (special report), Hyd Proc, 51(1), 73-92 (1972).
Various, Guide to trouble-free plant operation: Filtration, instrumentation, computer-process interface, Chem Eng, 26 June, 88-111 (1972).

1973-1979

Kauber, G.A., Formal operating instructions and operations auditing, Chem Eng, 17 Sept, 146-148 (1973).
McCullough, D.G., Comparing maintenance performance, Chem Eng, 29 Oct, 120-124 (1973).
Powley, C., Terotechnology: Reducing total cost of maintenance, Proc Engng, June, 140-142 (1973).
Various, Maintenance (special report), Hyd Proc, 52(1), 65-91 (1973).

Various, Plant maintenance and engineering (deskbook issue), Chem Eng, 26 Feb, 13-155 (1973).
de la Mare, R.F., Terotechnology, Chemical Processing, April, 103-106; May, 93-96 (1974).
Fair, E.W., What about maintenance costs? Chem Eng, 4 March, 132-134 (1974).
Various, Maintenance: Diagnozing equipment problems (special report), Hyd Proc, 53(1), 85-114 (1974).
Perkins, R.L., Maintenance costs and equipment reliability, Chem Eng, 31 March, 126-128 (1975).
Roseman, W.W., Defining maintenance work load, Hyd Proc, 54(4), 187-188 (1975).
van Eijk, F.P., Instrument trip system maintenance and improvement program, Chem Eng Prog, 71(1), 48-53 (1975).
Various, Maintenance (special report), Hyd Proc, 54(1), 53-76 (1975).
Goyal, S.K., Planning of optimum maintenance frequency, Processing, July, 32-34 (1976).
Gumm, W.G., and Turner, J.E., Nondestructive testing methods, Chem Eng, 16 Aug, 64-78 (1976).
Pippitt, R.R., Maintenance of kilns, calciners and dryers, Chem Eng Prog, 72(2), 41-45 (1976).
Santini, F., Maintenance by objectives, Hyd Proc, 55(1), 159-162 (1976).
Various, Maintenance of rotating equipment (topic issue), Chem Eng Prog, 72(2), 35-58 (1976).
Various, Maintenance (special report), Hyd Proc, 55(1), 73-93 (1976).
Bergtraun, E.M., In-plant vs. contract maintenance, Chem Eng, 28 March, 131-133 (1977).
Collacott, R.A., Condition monitoring prevents plant failure, Proc Engng, Jan, 39-41 (1977).
Martin, D., Maintenance investment pays back operating costs with interest, Proc Engng, Jan, 50-52 (1977).
Various, Maintenance (special report), Hyd Proc, 56(1), 77-112 (1977).
Weiss, W.H., Management by exception in operations and maintenance, Chem Eng, 5 Dec, 151-154 (1977).
Clark, R.L., How to prepare inspection reports, Hyd Proc, 57(2), 159-161; 57(3), 159-168 (1978).
Various, Maintenance (feature report), Processing, May, 53-76 (1978).
Various, Maintenance (special report), Hyd Proc, 57(1), 85-120 (1978).
Weiss, W.H., Coordinating maintenance and production operations, Chem Eng, 6 Nov, 129-130 (1978).
Various, Maintenance (special report), Hyd Proc, 58(1), 89-125 (1979).

1980-1983

Aird, R.J., The application of reliability engineering to process plant maintenance, Chem Engnr, May, 301-305, 311 (1980).
Anon., Condition monitoring in the process industries, Chem Engnr, May, 315, 318 (1980).
Bader, M.E., Quality assurance and quality control, Chem Eng, 11 Feb, 86-92; 7 April, 89-93; 16 June, 123-129; 25 Aug, 95-97 (1980).

Edwards, J.D., Management of the chemical plant maintenance function, Chem Engnr, May, 316-318 (1980).
Idelson, I.V., and Hemelryk, S.O., What consultants can contribute to the management of maintenance, Chem Engnr, May, 306-307, 311 (1980).
Partington, E.V., Applying terotechnology to chemical plant maintenance, Chem Engnr, May, 312-314 (1980).
Shaw, R.J.; Sykes, J.A., and Ormsby, R.W., Plant-test manual, Chem Eng, 11 Aug, 126-132 (1980).
Sims, I.G., Condition monitoring of chemical plants, Chem Engnr, May, 308-311 (1980).
Varagas, K.J., Maintenance records, Chem Eng, 21 April, 161-168 (1980).
Vargas, K.J., Setting up a spare-parts program, Chem Eng, 28 Jan, 133-136 (1980).
Various, Maintenance (special report), Hyd Proc, 59(1), 71-110 (1980).
Baldin, A.E., Condition-based maintenance, Chem Eng, 10 Aug, 89-95 (1981).
Hellhake, F.J., Setting up a preventive maintenance program, Chem Eng, 13 July, 145-150 (1981).
Nolting, H.F., Computer control of maintenance operations, Chem Eng Prog, 77(1), 53-55 (1981).
Peters, L.A., Checklist for verifying shop drawings, Chem Eng, 20 April, 177-178 (1981).
Various, Maintenance and condition monitoring (topic issue), Processing, Feb, 21-29, 39 (1981).
Various, Maintenance (special report), Hyd Proc, 60(1), 97-136 (1981).
Cramp, J.H.W., Laser monitoring of process plant, Chem & Ind, 20 Feb, 124-128 (1982).
de Matteis, U., Planning and budgeting for maintenance, Chem Eng, 15 Nov, 105-112 (1982).
Evans, R.K., Condition monitoring for plant maintenance, Proc Engng, June, 59-63 (1982).
Various, Maintenance (special report), Hyd Proc, Jan, 77-130 (1982).
Bailey, P., Developing an equipment-inspection program, Chem Eng, 27 June, 30-36 (1983).
Charlton, J.S., and Polarski, M., Radioisotope techniques for plant and process problems, Chem Eng, 24 Jan, 125-128; 21 Feb, 93-98 (1983).
Mann, L., and Bostock, H.H., Short-range maintenance planning/scheduling using network analysis, Hyd Proc, March, 97-101 (1983).
Peters, L.A., Specifying a maintenance information system, Chem Eng, 8 Aug, 91-92 (1983).
Various, Maintenance: Special report, Hyd Proc, Jan, 63-97 (1983).

1984-1986

McAlister, J., and Elder, R.T., Cut maintenance expenses with deficiency tagging, Chem Eng, 1 Oct, 109-110 (1984).
Proctor, J., Maintenance management by computer, Proc Engng, June, 29-33 (1984).
Various, Maintenance and retrofitting (special report), Hyd Proc, Jan, 61-86 (1984).

Wells, A., Micro-aided maintenance for raising efficiency and cutting costs, Chem Engnr, July, 34-36 (1984).
Backert, W., and Rippin, D.W.T., Determination of maintenance strategies for plants subject to breakdown, Comput Chem Eng, 9(2), 113-126 (1985).
Various, Maintenance and retrofitting (special report), Hyd Proc, Jan, 51-77 (1985).
Wilkie, F., Maintenance planning, Proc Engng, Nov, 59-60 (1985).
Baguley, P., Condition-based maintenance, Chem Engnr, July, 38-40 (1986).
Chowdhury, J., Troubleshooting comes online in the CPI, Chem Eng, 13 Oct, 14-19 (1986).
Hickman, W.E., and Moore, W.D., Managing the maintenance dollar, Chem Eng, 14 April, 68-77 (1986).
McAlister, J., Surveillance - the forgotten maintenance tool? Chem Eng, 13 Oct, 87-89 (1986).
Pierce, F.R., Maintenance: Do more with less, Hyd Proc, 65(6), 101-107 (1986).
Woodruff, D.M., and Phillips, F.M., The Pareto chart: Tool for problem solving, Chem Eng, 14 April, 111-114 (1986).

1987-1988

Anon., Tackling plant-maintenance problems (equipment focus and new products), Chem Eng, 19 Jan, 69-73 (1987).
Bagadia, K., Microcomputer-Aided Maintenance Management, Marcel Dekker, New York (1987).
Bond, A., and Cooke, D., Managing instrumentation and plant maintenance on a PC, Proc Engng, May, 67-69 (1987).
Finley, H.F., Reduce costs with knowledge-based maintenance, Hyd Proc, 66(1), 64-66 (1987).
Green, S.J., Solving chemical and mechanical problems of pressurized water reactor steam generators, Chem Eng Prog, 83(7), 31-45 (1987).
Grosshandler, S., Predictive maintenance of pressure vessels, Proc Engng, Aug, 27-31 (1987).
Krisher, A.S., Plant integrity programs, Chem Eng Prog, 83(5), 20-24 (1987).
Polk, M., Better couplings reduce pump maintenance, Hyd Proc, 66(1), 62-63 (1987).
Tayler, C., Planning and scheduling for optimising production, Proc Engng, Sept, 35-37 (1987).
Cowick, R.M., Develop a maintenance resource management program, Hyd Proc, 67(4), 100A-100L (1988).
Egol, L., Computerized maintenance management, Chem Eng, 26 Sept, 107-109 (1988).
Feit, E., Preventive maintenance using ultrasonic scanning, Chem Eng., 19 Dec, 158-160 (1988).
Feuless, S.C., and Madhaven, S., Seven steps to shorter turnarounds, Chem Eng, 14 March, 126-131 (1988).
Fromm, C.H., and Budaraju, S., Reducing equipment-cleaning wastes, Chem Eng, 18 July, 117-122 (1988).
Gibson, G.L., Plant inspection for routine maintenance, Chem Eng, 26 Sept, 95-98 (1988).

Madhavan, S., and Kirsten, R.A., Anatomy of a plant audit, Hyd Proc, 67(12), 92-95 (1988).
Madhaven, S., and Kirsten, R.A., Anatomy of a plant audit, Hyd Proc, 67(11), 123-130 (1988).

7.4 Operation and Equipment Problems

1967-1970

Baker, R.C., Determine water vapor in compressed air, Chem Eng, 19 June, 246-248 (1967).
Fox, C.T., Blast it clean with water, Chem Eng, 14 Aug, 180-186 (1967).
Kellstrom, A.A., Guard against air-pressure losses, Chem Eng, 19 June, 242-244 (1967).
Kirchner, R.W., Equipment testing, Chem Eng, 28 Aug, 107-119 (1967).
Miller, R.L., Solving practical problems in big ammonia plants, Chem Eng, 5 June, 125-127 (1967).
Roehrs, R.J., Leak testing of process vessels, Chem Eng, 28 Aug, 120-130 (1967).
Thielsch, H., How to pressure-test process equipment for safety, Chem Eng, 25 Sept, 198-200 (1967).
Buscarello, R.T., Practical solutions for vibration problems, Chem Eng, 12 Aug, 157-166 (1968).
Elshout, R., Graphs determine time to drain vessels, Chem Eng, 23 Sept, 246-250 (1968).
Fleming, R.C., Reduce piping maintenance with T-strainer, Chem Eng, 18 Nov, 196 (1968).
Miller, N.H., Modern lubrication practices, Chem Eng, 26 Feb, 155-162; 11 March, 193-198; 25 March, 105-111 (1968).
Munro, H.P.; Martin, F.W., and Roberts, M.C., How to use simulation techniques to determine optimum manning levels for continuous process plants, Chem Engnr, Oct, CE355-358 (1968).
Ostrofsky, B., Nondestructive tests for on- and off-stream inspections, Chem Eng, 20 May, 174-180 (1968).
Van Amerongen, L., A guide to rigging and lifting equipment, Chem Eng, 22 April, 202-210 (1968).
Coopey, W., Spring-loaded packings for pump leaks, Hyd Proc, 48(9), 215-216 (1969).
Hattiangadi, U.S., Relocate equipment to cure cavitation problems, Chem Eng, 1 Dec, 118-120 (1969).
Evans, D.J., Non-destructive testing in the field, Chem Eng Prog, 66(9), 66-69 (1970).
Ostrofsky, B., and Heckler, N.B., Ultrasonics check reformer headers, Hyd Proc, 49(7), 99-102 (1970).
Various, Plant/equipment reliability (topic issue), Chem Eng Prog, 66(12), 29-58 (1970).
Yaki, S.J., and Carpenter, R., Cost of pump packing: A case history, Hyd Proc, 49(7), 136 (1970).

1971-1972

Baudmann, H.D., Control-valve noise: Causes and cure, Chem Eng, 17 May, 120-126 (1971).

Engle, J.P., Cleaning boiler tubes chemically, Chem Eng, 18 Oct, 154-158 (1971).

Hodnick, H.V., Thwarting maintenance deathtraps, Chem Eng, 5 April, 142-144 (1971).

Isaacs, M., and Setterlund, R.B., Pipe joining methods, Chem Eng, 3 May, 74-90 (1971).

Jackson, C., Alignment of barrel-type centrifugal compressors, Hyd Proc, 50(9), 189-194 (1971).

Mueller, E.R., Synthetics: Rx for lubrication problems, Chem Eng, 28 June, 91-94 (1971).

Tustin, W., Measurement and analysis of machinery vibration, Chem Eng Prog, 67(6), 62-69 (1971).

Various, Plant/equipment reliability (topic issue), Chem Eng Prog, 67(1), 33-76 (1971).

Anyakora, S.N., and Lees, F.P., Detection of instrument malfunction by the process operator, Chem Engnr, Aug, 304-309 (1972).

Bradley, W.A., Care and feeding of gears, Chem Eng, 1 May, 84-92 (1972).

Brown, R.N., Can specifications improve compressor reliability? Hyd Proc, 51(7), 89-91 (1972).

Cates, J.H., Electric motors: Principles and applications in the chemical process plant, Chem Eng, 11 Dec, 82-87 (1972).

Imgram, A.G., and McCandless, J.B., Check furnace tube condition with in-place metallography, Hyd Proc, 51(4), 125-126 (1972).

Lee, R., Fast cure for failures, Chem Eng, 24 Jan, 118-122 (1972).

Palm, G., How to lift the disc of a stuck gate valve, Chem Eng, 30 Oct, 136 (1972).

Pritchett, D.H., Electric motors: A guide to standards, Chem Eng, 11 Dec, 88-92 (1972).

Ruckstuhl, R.E., Specifying compressor lube and seal systems, Hyd Proc, 51(5), 127-128 (1972).

Sohre, J.S., Causes and cures for silica deposits in steam turbines, Hyd Proc, 51(12), 87-89 (1972).

Various, Polyethylene plant operations (topic issue), Chem Eng Prog, 68(11), 51-79 (1972).

Various, Plant/equipment reliability (topic issue), Chem Eng Prog, 68(3), 47-81 (1972).

Walko, J.F., Controlling biological fouling in cooling systems, Chem Eng, 30 Oct, 128-132; 27 Nov, 104-108 (1972).

1973-1974

Beisel, R.O., Spare parts: Lifeblood of overseas plants, Chem Eng, 19 Feb, 119-120 (1973).

Bushar, T.A., Bearings - why they fail, Chem Eng, 5 Feb, 92-98 (1973).

Campbell, M.E., Solid lubricants, Chem Eng, 1 Oct, 56-66 (1973).

Essinger, J.N., Turbomachinery alignment, Hyd Proc, 52(9), 185-188 (1973).

Jackson, C., Prevent pump cavitation, Hyd Proc, 52(5), 157-160 (1973).

Kletz, T.A., Uses, availability and pitfalls of data on reliability, Proc Tech Int, March, 111-113 (1973).
Margetts, R.J., Flow problems in feed-water coils, Chem Eng Prog, 69(1), 51-55 (1973).
Miller, J.W., Super-lube systems eliminate shaft-seal leakage, Chem Eng, 9 July, 88-90 (1973).
Neale, D.F., Better mechanical testing can improve compressor reliability, Hyd Proc, 52(9), 165-171 (1973).
Richardson, J.A., and Templeton, L.R., The impact of plant unreliability upon cost, Chem Engnr, Dec, 625-629 (1973).
Shah, B.M., Repairing reactor manhole covers without unloading catalyst, Chem Eng, 1 Oct, 98 (1973).
Turner, B., How reliable is your high-pressure seal-oil system? Hyd Proc, 52(11), 205-209 (1973).
Berger, D.M., Preparing for painting, Chem Eng, 28 Oct, 130-132 (1974).
Berger, D.M., Choosing and applying paint, Chem Eng, 25 Nov, 112-116 (1974).
Budris, A.R., Try filled-TFE bearings for problem services, Chem Eng, 8 July, 102-107 (1974).
Hancock, W.P., Controlling pump vibration, Hyd Proc, 53(3), 107-113 (1974).
Levi, E.J., New developments in the basics of cooling-water treatment, Chem Eng, 10 June, 88-92 (1974).
Lewis, P., Increase compressor reliability by better component design, Hyd Proc, 53(5), 117-120 (1974).
Murray, M.G., Guidelines for machinery alignment, Hyd Proc, 53(10), 139-145 (1974).
von Nimitz, W.; Wachel, J.C., and Szenasi, F.R., Case histories of specialized turbomachinery problems, Hyd Proc, 53(4), 141-146 (1974).

1975-1976
Berger, D.M., Inspecting coatings for film thickness, Chem Eng, 17 Feb, 106-110 (1975).
Berger, D.M., Detecting film flaws in coatings, Chem Eng, 17 March, 79-83 (1975).
Britt, C.H., Hot tapping and stoppling in cryogenic service, Chem Eng Prog, 71(11), 59-62 (1975).
Brown, C.W., Electric pipe tracing, Chem Eng, 23 June, 172-178 (1975).
Diehl, G.M., Controlling compressor noise, Hyd Proc, 54(7), 157-159 (1975).
Mapes, W.H., How to keep motors running, Chem Eng, 21 July, 107-112 (1975).
Pattison, D.A., Practical lubrication for process plants, Chem Eng, 28 April, 98-106 (1975).
Pekrul, P.J., Vibration monitoring increases equipment availability, Chem Eng, 18 Aug, 109-114 (1975).
Phelan, J.V., and Gelosa, L.R., How to control boiler iron deposits, Chem Eng, 3 March, 174-178 (1975).
Raynesford, J.D., Dynamic vibration absorbers, Hyd Proc, 54(4), 167-171 (1975).

Schanzenbach, G.P., Sensors for machinery monitoring, Hyd Proc, 54(2), 85-88 (1975).
Story, G., and MacFarland, I., Sealing defective heat exchanger tubes, Chem Eng Prog, 71(7), 94-96 (1975).
Various, Air cooler corrosion, Hyd Proc, 54(7), 169-174 (1975).
Whittaker, D.M., Improving valve life in reciprocating compressors, Hyd Proc, 54(5), 143-144 (1975).
Wright, J., Lubricated or non-lubricated shaft couplings? Hyd Proc, 54(4), 191-193 (1975).
Barnes, E.F., New materials solve reciprocating compressor problems, Hyd Proc, 55(7), 147-149 (1976).
Bell, A.W., and Breen, B.P., Converting gas boilers to oil and coal, Chem Eng, 26 April, 93-101 (1976).
Cox, N.D., Evaluating the profitability of standby components, Chem Eng Prog, 72(6), 76-79 (1976).
Cronenwett, R.H., Low-cost automation of steam tracing, Chem Eng, 10 May, 141 (1976).
Massey, M.J.; McMichael, F.C., and Dunlap, R.W., Strategies for keeping coal-gas lines clean, Chem Eng, 30 Aug, 109-113 (1976).
Maukonen, D.W., Spot heating with portable heating systems, Chem Eng, 24 May, 163-166 (1976).
Monroe, E.S., Select the right steam trap, Chem Eng, 5 Jan, 129-134 (1976).
Murray, M.G., Select O-rings carefully, Hyd Proc, 55(4), 191-192 (1976).
Neale, D.F., Shop testing for improved compressor reliability, Hyd Proc, 55(6), 135-137 (1976).
Sisson, W., Cooling and air circulation for electric motors, Chem Eng, 5 July, 104 (1976).
Sisson, W., Calculate the paint needed for tanks, Chem Eng, 16 Feb, 122 (1976).

1977-1978

Ganapathy, V., Determine the natural vibration frequency of heat exchanger tubes, Chem Eng, 26 Sept, 122-124 (1977).
Henley, E.J., and Hoshino, H., Effect of storage tanks on plant availability, Ind Eng Chem Fund, 16(4), 439-443 (1977).
Murray, M.G., Electric-motor failures, Hyd Proc, 56(2), 127-128 (1977).
Pollack, M.J., Plant monitoring using conductivity measurement, Chem Eng, 12 Sept, 161-162 (1977).
Sharp, E.C., Operation and maintenance records for heating equipment, Chem Eng, 25 April, 141-142; 23 May, 171-172 (1977).
Turner, B., Learn from equipment failure, Hyd Proc, 56(11), 317-320 (1977).
Cloudt, W.O., Tensioning bolts in the field, Chem Eng, 27 March, 147-152 (1978).
Epstein, J.P., Buying and selling used equipment, Chem Eng, 20 Nov, 185-188 (1978).
Farr, W., Preventing pumps from running dry, Chem Eng, 18 Dec, 133 (1978).
Morrison, W.G., Improving the reliability of electrical distribution systems, Chem Eng, 25 Sept, 102-112 (1978).
Nailen, R.L., Protect motors by heat sensing, Hyd Proc, 57(4), 175-179 (1978).
Nicholas, M.P., Reconditioning old valves, Chem Eng, 22 May, 117-118 (1978).

Pengelly, R., Select expansion joints properly, Hyd Proc, 57(3), 141-144 (1978).
Roebuck, A.H., Safe chemical cleaning with organics, Chem Eng, 31 July, 107-110 (1978).
Salot, W.J., Mystery leaks in a waste-heat boiler, Chem Eng, 11 Sept, 177-182 (1978).
Various, Phosphoric acid plant problems, Chem Eng Prog, 74(11), 37-65 (1978).
Various, Trouble-free equipment for process plants (feature report), Chem Eng, 5 June, 114-142 (1978).
Various, Seals and packings for rotating shafts (feature report), Chem Eng, 9 Oct, 96-110 (1978).
Whitney, J.B., Detecting faults in glass-lined reactors, Chem Eng Prog, 74(10), 81-83 (1978).

1979-1980
Chase, G., Flexible connectors for corrosive fluids, Chem Eng, 10 Sept, 149-155 (1979).
Jackson, C., and Leader, M.E., Turbomachines: Avoiding operating problems, Hyd Proc, 58(11), 281-284 (1979).
McNaughton, K.J., Relining deteriorated pipelines with polyolefin pipe, Chem Eng, 19 Nov, 173-174 (1979).
Messer, J., Selecting and maintaining reciprocating-compressor piston rods, Chem Eng, 21 May, 209-217 (1979).
Smith, A.P., Unexplained wear in large centrifugal pumps, Chem Eng, 13 Aug, 153-155 (1979).
Smith, A.P., Avoid pipe stresses in pumps, Chem Eng, 16 July, 121-123 (1979).
Snow, M., Maintaining pressure vessels, Chem Eng, 1 Jan, 109-112 (1979).
Anon., Lifting and shifting heavy loads, Proc Econ Int, 2(1), 37-40 (1980).
Constance, J.D., Solve gas purging problems graphically, Chem Eng, 29 Dec, 65-68 (1980).
Constance, J.D., Removing moisture from buildings by exhaust ventilation, Chem Eng, 19 May, 177-178 (1980).
Finley, R.W., Incipient failure detection in rotating equipment, Chem Eng, 14 July, 104-112 (1980).
Kletz, T.A., Plant instruments: Which ones don't work and why, Chem Eng Prog, 76(7), 68-71 (1980).
Kraus, M.N., How to protect mechanical equipment and motors, Chem Eng, 15 Dec, 59-68 (1980).
Lee, J.W., Pretreatment of cooling water systems, Chem Eng Prog, 76(7), 56-58 (1980).
Mueller, J.F., Electrical-distribution-system survey, Chem Eng, 8 Sept, 135-138 (1980).

1981-1983
Churchman, C.G., Washbox design cuts splashing and saves money, Chem Eng, 21 Sept, 156 (1981).
Granek, K., and Heckenkamp, F.W., Spiral-wound gaskets - panacea or problem? Chem Eng, 23 March, 231-238 (1981).
Jackson, C., Guidelines for improving rotating equipment reliability, Hyd Proc, 60(9), 223-228 (1981).

Kenson, R.E., Controlling odors, Chem Eng, 26 Jan, 94-100 (1981).
Martin, H.W., AIChE equipment testing procedures, Chem Eng Prog, 77(3), 44-48 (1981).
Mazlack, S.W., Avoid problems with steam turbine carbon ring seals, Hyd Proc, 60(8), 143-145 (1981).
Sangerhausen, C.R., Reduce seal failures, Hyd Proc, 60(3), 119-122 (1981).
Scully, W.A., Safety-relief-valve malfunctions, Chem Eng, 10 Aug, 111-114 (1981).
Shanmugam, C., Instrument-installation manhours, Chem Eng, 26 Jan, 124-125 (1981).
Warner, D., Beware of safety-meter inaccuracies, Chem Eng, 4 May, 103 (1981).
Anon., Estimating the reliability and availability of new process plant, Proc Econ Int, 3(1), 42-49 (1982).
Barsness, D., Adhesives and sealants make maintenance easier, Chem Eng, 8 Feb, 129-131 (1982).
Bass, C.D., Safe governing of turbine speed, Plant/Opns Prog, 1(2), 101-107 (1982).
Bilic, M., and Milakovic, T., Manual liquid-filling system, Chem Eng, 5 April, 136 (1982).
Carmer, J.P., and Woods, J.A., Cleaning debris out of process lines, Chem Eng, 31 May, 150 (1982).
Cichowski, J.S., Turndown for nitric acid production, Chem Eng, 15 Nov, 149-150 (1982).
Clark, R.L., Improve equipment-source inspections, Hyd Proc, April, 213-242 (1982).
Crozier, R.A., Increase flow to cut fouling, Chem Eng, 8 March, 115 (1982).
Dye, R.J., Testing of ammonia converter, Plant/Opns Prog, 1(3), 169-175 (1982).
Raghavachari, S., Machine-maintenance pointers, Chem Eng, 23 Aug, 120 (1982).
Ridenhour, L.W., Process and equipment monitoring by computer, Plant/Opns Prog, 1(4), 236-243 (1982).
Robitaille, D.R., Molybdate inhibitors for problem cooling waters, Chem Eng, 4 Oct, 139-142 (1982).
Roodman, R.G., Chemical plant operation, Chem Eng, 17 May, 131-133 (1982).
Simmons, P.E., The optimum provision of installed spares, Hyd Proc, April, 189-192 (1982).
Adams, W.V., Troubleshooting mechanical seals, Chem Eng, 7 Feb, 48-57 (1983).
Bayler, H., Guide to conveyor belt cleaners, Proc Engng, Feb, 45 (1983).
Bertagnolio, M., Modernizing a lube plant, Hyd Proc, March, 103-106 (1983).
Capes, P., Seals, packings and gaskets, Proc Engng, Aug, 34-35 (1983).
Ivanus, G., Operational flexibility of chemical plants, Proc Econ Int, 4(3), 28-33 (1983).
LoPinto, L., Designing and writing operating manuals, Chem Eng, 11 July, 77-78 (1983).
Martin, H.W., Operation at less than design rate, Plant/Opns Prog, 2(3), 144-146 (1983).

Mundis, J.A., Problems in pressurized water reactor steam generators, Chem Eng Prog, 79(7), 39-46 (1983).
Raghavachari, S., and Ranganathan, A., Longer life of motor windings in dusty atmospheres, Chem Eng, 30 May, 92 (1983).
Shorthouse, B.O., Boiler circulation during chemical circulation, Chem Eng, 22 Aug, 75-79 (1983).
Treleaven, W.D., Acoustic emission testing for chemical plants, Chem Eng, 7 Feb, 87-89 (1983).
Tung, P.C., and Mikasinovic, M., Eliminating cavitation from pressure-reducing orifices, Chem Eng, 12 Dec, 69-71 (1983).

1984-1985

Hoose, J., Use liquid nitrogen to cool reactors, Hyd Proc, Oct, 71-72 (1984).
Kauders, P., Designing for plant upset conditions, Chem Engnr, Jan, 9-11 (1984).
Koehler, F.H., Draining elliptical vessel heads, Chem Eng, 14 May, 90-92 (1984).
Krishnaswamy, R., and Parker, N.H., Corrective maintenance and performance optimization, Chem Eng, 16 April, 93-98 (1984).
Lewis, M.W.J., Failure diagnosis, Chem Eng, 9 July, 117-120; 6 Aug, 87-90 (1984).
Stus, T.F., Writing operating instructions, Chem Eng, 26 Nov, 105-106 (1984).
Various, Better service from rotating equipment (special report), Hyd Proc, Aug, 51-68 (1984).
Armstrong, D.J., Investigation of ammonia plant performance using radioactive tracers, Plant/Opns Prog, 4(2), 72-79 (1985).
Batterham, R.J., External corrosion: How long before pipework inspection? Plant/Opns Prog, 4(3), 154-161 (1985).
Cohen, E.M., Fault-tolerant processes, Chem Eng, 16 Sept, 73-78 (1985).
Irhayem, A.Y.M., A simple, automatic method to reduce scale buildup in sewer-pump systems, Chem Eng, 16 Sept, 100 (1985).
King, J.D., and Egan, K.M., Modify your batch material-transfer system, Chem Eng, 27 May, 153 (1985).
Kipin, P., Cleaning pipelines: A pigging primer, Chem Eng, 4 Feb, 53-58 (1985).
Kremers, J., Fighting water hammer, Processing, Jan, 21-22 (1985).
Leslie, V.J., and Ferguson, D., Radioisotope techniques for solving ammonia plant problems, Plant/Opns Prog, 4(3), 144-149 (1985).
Valdes, E.C., and Svoboda, K.J., Estimating relief loads for thermally blocked-in liquids, Chem Eng, 2 Sept, 77-82 (1985).
Vivian, B., Identifying and evaluating valve problems, Proc Engng, Oct, 33-36 (1985).

1986-1987

Antalffy, L., Convert hot wall fluid catalytic cracking lines to cold wall design, Hyd Proc, 65(1), 58-62 (1986).
Artus, C., Fluid transfer: Consider thermoplastic hose, Chem Eng, 8 Dec, 123-127 (1986).

Chynoweth, E., Troubleshooting with gamma rays on the plant, Proc Engng, Aug, 44-45 (1986).
Coats, L., Monitoring site productivity, Proc Engng, Sept, 27-30 (1986).
Dooner, R., Looking for trouble, Proc Engng, March, 51-53 (1986).
Dunlap, J.L., Determining a proper parts inventory (ways to minimize costs), Chem Eng, 29 Sept, 93-97 (1986).
Egli, U.M., and Rippin, D.W.T., Short-term scheduling for multiproduct batch chemical plants, Comput Chem Eng, 10(4), 303-325 (1986).
Goldman, S., A two-channel spectrum analyzer used to diagnose rotating machinery problems, Hyd Proc, 65(1), 51-54 (1986).
Nassauer, J., and Kessler, H.G., Effect of electrostatic phenomena on cleaning of surfaces, Chem Eng & Proc, 20(1), 27-32 (1986).
Palazzolo, A.B., Combine computer simulation and field testing to solve problems, Hyd Proc, 65(5), 59-62 (1986).
Riance, X.P., More learning - the hard way, Chem Eng, 13 May, 143; 10 June, 123 (1985); 17 Feb, 161-162 (1986).
Adams, W.V., Better high temperature sealing, Hyd Proc, 66(1), 53-59 (1987).
Conroy, J.A., A computer-based inspection system, Chem Eng Prog, 83(5), 39-41 (1987).
Contino, A.V., Improve plant performance via statistical process control, Chem Eng, 20 July, 95-102 (1987).
Flitney, R.K., Reliability of seals in centrifugal process pumps, Proc Engng, July, 39-42 (1987).
Fusillo, R.H., and Powers, G.J., A synthesis method for chemical plant operating procedures, Comput Chem Eng, 11(4), 369-382 (1987).
Ganapathy, V., Avoid heat transfer equipment vibration, Hyd Proc, 66(6), 61-63 (1987).
Goggin, D.S., Vibration monitoring: A programmed approach, Plant/Opns Prog, 6(1), 1-5 (1987).
Ku, H.; Rajagopalan, D., and Karimi, I., Scheduling in batch processes, Chem Eng Prog, 83(8), 35-45 (1987).
Reeves, G.G., Avoid self-priming centrifugal pump problems, Hyd Proc, 66(1), 60-61 (1987).

1988

Fort, J., and Jehl, J., Magnetic bearings and dry seals improve compressor operation, Hyd Proc, 67(10), 53-56 (1988).
Fusillo, R.H., and Powers, G.J., Computer-aided planning of purge operations, AIChEJ, 34(4), 558-566 (1988).
Junique, J.C., Flush or blow lines adequately, Hyd Proc, 67(7), 55-56 (1988).
Kubic, W.L., and Stein, F.P., Theory of design reliability using probability and fuzzy sets, AIChEJ, 34(4), 583-601 (1988).
Miller, J.E., Selecting chemical treatment programs, Hyd Proc, 67(9), 97-100 (1988).
Pathak, V.K., and Rattan, I.S., Turndown limit sets heater control, Chem Eng, 18 July, 103-105 (1988).
Ranade, S.M.; Robert, W.E., and Zapata, A., Evaluate electric drive replacements, Hyd Proc, 67(10), 41-45 (1988).
Reeves, C., Vibration monitoring, Proc Engng, June, 45-50 (1988).

Schlager, S.T., Process troubleshooting with expert systems, Chemtech, Dec, 750-758 (1988).
Van Blarcom, P.P., Valves for vessel draining and sampling, Chem Eng, 21 Nov, 76-80 (1988).
Various, Quality control (topic issue), Chem Eng Prog, 84(4), 25-64 (1988).

CHAPTER 8

PROCESS CONTROL AND INSTRUMENTATION

8.1 Control Systems and Strategies

1967

Casciano, R.M., and Staffin, H.K., Model-reference adaptive control system, AIChEJ, 13(3), 485-491 (1967).

Davies, W.D.T., A self-adjusting controller, Brit Chem Eng, 12(8), 1227-1230 (1967).

Ewing, R.W.; Glahn, G.L.; Larkins, R.P., and Zartman, W.N., Generalized process control programming system, Chem Eng Prog, 63(1), 104-110 (1967).

Foster, R.D., and Stevens, W.F., Method for noninteracting control of a class of linear multivariable systems, AIChEJ, 13(2), 334-340 (1967).

Franks, R.G.E., and Schiesser, W.E., Evolution of digital simulation programs, Chem Eng Prog, 63(4), 68-78 (1967).

Hays, J.R.; Clements, W.C., and Harris, T.R., Frequency domain evaluation of mathematical models for dynamic systems, AIChEJ, 13(2), 374-378 (1967).

Lapidus, L., Control of nonlinear systems via second-order approximations (review), Chem Eng Prog, 63(12), 64-71 (1967).

Latour, P.R.; Koppel, L.B., and Coughanowr, D.R., Time-optimum control of chemical processes for set-point changes, Ind Eng Chem Proc Des Dev, 6(4), 452-460 (1967).

Liu, S.L., Noninteracting process control, Ind Eng Chem Proc Des Dev, 6(4), 460-468 (1967).

Lombardo, J.M., The case for digital backup in direct digital control systems, Chem Eng, 3 July, 79-84 (1967).

McAvoy, T.J., and Johnson, E.F., Quality of control problem for dead-time plants, Ind Eng Chem Proc Des Dev, 6(4), 440-446 (1967).

Mosler, H.A.; Koppel, L.A., and Coughanowr, D.R., Application of conventional loop tuning to sampled-data systems, Ind Eng Chem Proc Des Dev, 6(1), 101-105 (1967).

Mosler, H.A.; Koppel, L.B., and Coughanowr, D.R., Process control by digital compensation, AIChEJ, 13(4), 768-778 (1967).

Rijnsdorp, J.E., Chemical process systems and automatic control (review), Chem Eng Prog, 63(7), 97-116 (1967).

Rybak, T.B.M., and Colliss, B.A., The status of computer control, Brit Chem Eng, 12(4), 549-553 (1967).

Whitman, K., Digital computer control, Chem Eng, 5 Dec, 135-138(1966); 2 Jan, 81-86 (1967).

Wilde, D.J., and Passy, U., Partial control of linear inventory systems, AIChEJ, 13(2), 236-240 (1967).

1968

Cadman, T.W., and Smith, T.G., Learn about analog computers, Hyd Proc, 47(2), 140-144; 47(3), 129-132; 47(4), 141-144; 47(5), 171-176; 47(6), 149-154; 47(7), 121-126; 47(8), 151-156; 47(10), 125-132; 47(11), 213-218 (1968).

Distefano, G.P., and Richards, W., Hybrid computers, Chem Eng, 6 May, 195-202 (1968).

Greenfield, G.G., and Ward, T.J., Feedforward and dynamic uncoupling control of linear multivariable systems, AIChEJ, 14(5), 783-789 (1968).
Itahara, S., Direct digital control for batch processes, Chem Eng, 18 Nov, 159-164 (1968).
Latour, P.R.; Koppel, L.B., and Coughanowr, D.R., Feedback time-optimum process controllers, Ind Eng Chem Proc Des Dev, 7(3), 345-353 (1968).
Luecke, R.H., and McGuire, M.L., Analysis of optimal composite feedback-feedforward control, AIChEJ, 14(1), 181-189 (1968).
Maurer, C.J., and Garlid, K.L., Stability of naturally bounded nonlinear systems, AIChEJ, 14(1), 3-8 (1968).
Noon, D.W., Nomograph solves rate-limit problems in control devices, Chem Eng, 12 Feb, 164 (1968).
Rafal, M.D., and Stevens, W.F., Discrete dynamic optimization applied to on-line optimal control, AIChEJ, 14(1), 85-91 (1968).
Seinfeld, J.H., and Lapidus, L., Computational aspects of optimal control of distributed-parameter systems, Chem Eng Sci, 23(12), 1461-1484 (1968).
Sienfeld, J.H., and Lapidus, L., Singular solutions in optimal control of lumped and distributed parameter systems, Chem Eng Sci, 23(12), 1485-1500 (1968).
Steymann, E.H., Justifying process computer control, Chem Eng, 12 Feb, 124-129 (1968).
Various, Computer process control (topic issue), Chem Eng Prog, 64(4), 33-78 (1968).

1969

Ast, P.A.; Cugini, J.C., and Davis, R.S., New trends in handling computer control projects, Chem Eng, 28 July, 129-132 (1969).
Baker, W.J., and Weber, J.C., Direct digital control of batch processes, Chem Eng, 15 Dec, 121-128 (1969).
Bell, C.A., and Ray, D.J., Design of a digital controller, Brit Chem Eng, 14(6), 807-808 (1969).
Davies, W.D.T., Identification of a multidimensional process, Brit Chem Eng, 14(2), 194-198 (1969).
Gibilaro, L.G., and Lees, F.P., Reduction of complex transfer function models to simple models using the method of moments, Chem Eng Sci, 24(1), 85-94 (1969).
Koppel, L.B., and Aiken, P.M., A general process controller, Ind Eng Chem Proc Des Dev, 8(2), 174-184 (1969).
Lim, H.C., Classical approach to bang-bang control of linear processes, Ind Eng Chem Proc Des Dev, 8(3), 334-342 (1969).
Lopez, A.M.; Smith, C.L., and Murrill, P.W., An advanced tuning method, Brit Chem Eng, 14(11), 1553-1555 (1969).
McCoy, R.E., Comparing flow-ratio control systems, Chem Eng, 16 June, 92-96 (1969).
Miller, J.A.; Murrill, P.W., and Smith, C.L., Applying feedforward control, Hyd Proc, 48(7), 165-172 (1969).
Rao, A.M., and Corrigan, T.E., Bode diagrams for some chemical reactors, Brit Chem Eng, 14(10), 1406-1407 (1969).

Ray, W.H., Optimal control of processes modeled by transfer functions containing pure time delays, Chem Eng Sci, 24(2), 209-216 (1969).
Sealey, C.J., Introduction to linear control theory, Brit Chem Eng, 14(5), 651-655; 14(8), 1063-1069; 14(11), 1558-1560; 14(12), 1698-1702 (1969).
Shih, Y.P., Optimal control of distributed-parameter systems with integral equation constraints, Chem Eng Sci, 24(4), 671-681 (1969).
Storey, C., Liapunov methods in chemical engineering, Brit Chem Eng, 13(11), 1585-1588 (1968); 14(7), 965-967 (1969).
Various, Computer control (topic issue), Chem Eng Prog, 65(8), 33-52 (1969).
Various, Computer process control (topic issue), Chem Eng Prog, 65(5), 45-67 (1969).
Various, Process control (feature report), Chem Eng, 2 June, 94-164 (1969).
West, H.H., and McGuire, M.L., Optimal feedforward-feedback control of dead-time systems, Ind Eng Chem Fund, 8(2), 253-257 (1969).
Yung-Cheng, C., New approach to response surface generation by analogue computer, Brit Chem Eng, 14(10), 1422-1424 (1969).

1970

Almasy, G.A.; Gertler, J.J., and Pallai, I.M., Adaptive model for dynamic optimal control of a chemical plant, Brit Chem Eng, 15(2), 213-215 (1970).
Amrehn, H., and Winkler, O., Backup for process-control computers, Chem Eng, 9 March, 116-124 (1970).
Coggan, G.C., and Noton, A.R.M., Discrete-time sequential state and parameter estimation in chemical engineering, Trans IChemE, 48, T255-264 (1970).
Dube, J.H.; Smith, C.L., and Murrill, P.W., Controller settings improved, Hyd Proc, 49(12), 82-84 (1970).
Fitzpatrick, T.J., and Law, V.J., Noninteracting control for multivariable sampled data systems: Transform method design of decoupling controllers, Chem Eng Sci, 25(5), 867-874 (1970).
Martin, R.L., and Webber, W.O., Combination control: Better response, Hyd Proc, 49(3), 149-152 (1970).
Mellichamp, D.A., Model predictive time-optimal control of second-order processes, Ind Eng Chem Proc Des Dev, 9(4), 494-502 (1970).
Ray, W.H., and Soliman, M.A., Optimal control of processes containing pure time delays, Chem Eng Sci, 25(12), 1911-1926 (1970).
Ritter, A.B., and Douglas, J.M., Frequency response of nonlinear systems, Ind Eng Chem Fund, 9(1), 21-28 (1970).
Ruxton-Davies, W.D.T., Introduction to analog computation, Brit Chem Eng, 15(1), 83-87; 15(2), 222-224 (1970).
Soule, L.M., Automatic control, Chem Eng, 22 Sept, 133-138; 20 Oct, 115-119; 1 Dec, 101-104 (1969); 12 Jan, 103-106; 26 Jan, 130-134; 23 Feb, 113-116; 9 March, 142-144 (1970).
Wells, J.F., Automatic control in a dye-works, Brit Chem Eng, 15(5), 655-657 (1970).

1971

Ackerman, C.D.; Huling, G.P., and Metzger, K.J., Effects and benefits of pilot-plant automation, Chem Eng Prog, 67(4), 50-53 (1971).

Ahlgren, T.D., and Stevens, W.F., Adaptive control of a chemical process system, AIChEJ, 17(2), 428-435 (1971).
Brambilla, A., et al., Study of dynamic behaviour of chemical plants by digital simulation, Chem Eng Sci, 26(7), 1101-1112 (1971).
Elsey, J.I., and Bruley, D.F., JASON procedure for design of multivariable control systems, Ind Eng Chem Proc Des Dev, 10(4), 431-441 (1971).
Holmes, W.S., An introduction to electrical control circuits, Chem Eng, 15 Nov, 176-182 (1971).
Ryan, P.J., and Crandall, E.D., Multiparameter adaptive process control via constrained objective functions, AIChEJ, 17(2), 326-335 (1971).
Schlossmacher, E.J., and Lapidus, L., Suboptimal control of nonlinear systems using Liapunov-like functions, AIChEJ, 17(6), 1330-1341 (1971).
Steve, E.H., Logic diagram boosts process-control efficiency, Chem Eng, 8 March, 96-104 (1971).
Various, Computer control (topic issue), Chem Eng Prog, 67(10), 41-67 (1971).
Weber, A.P.J., and Lapidus, L., Suboptimal control of nonlinear systems, AIChEJ, 17(3), 641-658 (1971).
Weinrich, S.D., and Lapidus, L., Optimally sensitive and adaptive control systems, AIChEJ, 17(6), 1471-1480 (1971).

1972
Arant, J.B., Applying ratio control to chemical processing, Chem Eng, 18 Sept, 155-158 (1972).
Edgar, T.F., and Lapidus, L., Computation of optimal singular bang-bang control, AIChEJ, 18(4), 774-785 (1972).
Friedmann, P.G., and Moore, J.A., For process control, select the key variable, Chem Eng, 12 June, 85-90 (1972).
Hileman, J.R., Justifying a minicomputer for process control, Chem Eng, 29 May, 61-66 (1972).
Howarth, B.R.; Grens, E.A., and Foss, A.S., Root-locus interpretation of modal control, Ind Eng Chem Fund, 11(3), 403-406 (1972).
Jolls, K.R., and Riedinger, R.L., Applied electronics, Chem Eng, 12 May, 95-98; 15 June, 101-106; 24 July, 137-140; 21 Aug, 104-108; 18 Sept, 165-170; 2 Oct, 67-71; 30 Oct, 117-123 (1972).
Kegerreis, J.E., and Weigand, W.A., Improved feedback control and system design for second-order dead-time processes, Ind Eng Chem Proc Des Dev, 11(2), 246-253 (1972).
Luyben, W.L., Damping-coefficient design charts for sampled-data control of processes with deadtime, AIChEJ, 18(5), 1048-1052 (1972).
McAvoy, T.J., Time optimal and Ziegler-Nichols control, Ind Eng Chem Proc Des Dev, 11(1), 71-78 (1972).
Newell, R.B.; Fisher, D.G., and Seborg, D.E., Computer control using optimal multivariable feedforward-feedback algorithms, AIChEJ, 18(5), 976-984 (1972).
O'Connor, G.E., and Denn, M.M., Three mode control as a optimum control, Chem Eng Sci, 27(1), 121-128 (1972).
Ray, W.H., and Barney, J.R., Application of differential game theory to process control problems, Chem Eng J, 3(3), 237-244 (1972).

Soliman, M.A., and Ray, W.H., Optimal control of processes containing pure time delays, Chem Eng Sci, 27(12), 2183-2188 (1972).

Steadman, J.F., and Koppel, L.B., Bang-bang control is faster, Hyd Proc, 51(7), 101-105 (1972).

Various, Process controls and computers (topic issue), Chem Eng Prog, 68(4), 45-56 (1972).

Weigand, W.A., and Kegerreis, J.E., Comparison of controller-setting techniques as applied to second-order dead-time processes, Ind Eng Chem Proc Des Dev, 11(1), 86-90 (1972).

1973

Buelens, P.F., and Hellinckx, L.J., Time-optimal control of linear multivariable systems, Chem Eng J, 5(2), 137-144 (1973).

Edgar, T.F.; Vermeychuk, J.G., and Lapidus, L., The linear-quadratic control problem: Review of theory and practice, Chem Eng Commns, 1(2), 57-76 (1973).

Foss, A.S., Critique of chemical process control theory, AIChEJ, 19(2), 209-214 (1973).

Hutchinson, J.F., and McAvoy, T.J., On-line control of a nonlinear multivariable process, Ind Eng Chem Proc Des Dev, 12(3), 226-231 (1973).

Jolls, K.R., and Riedinger, R.L., Applied electronics, Chem Eng, 27 Nov, 93-98; 25 Dec, 67-70 (1972); 8 Jan, 121-128; 19 Feb, 129-134; 2 April, 61-68; 14 May, 140-148 (1973).

King, D., Logic diagrams for plant operation, Chem Eng, 20 Aug, 154-156 (1973).

Luyben, W.L., Parallel cascade control, Ind Eng Chem Fund, 12(4), 463-467 (1973).

Lynch, E.P., Process logic diagrams, Chem Eng, 28 May, 105-109 (1973).

Narsimhan, G., Control of steady-state through concentration signal, Chem Eng Sci, 28(11), 2102-2105 (1973).

Nieman, R.E., and Fisher, D.G., Experimental evaluation of optimal multivariable servo control in conjunction with conventional regulatory control, Chem Eng Commns, 1(2), 77-88 (1973).

Niemann, R.E., and Fisher, D.G., Experimental evaluation of time-optimal, open-loop control, Trans IChemE, 51, 132-140 (1973).

Various, Process control (topic issue), Chem Eng Prog, 69(9), 45-61 (1973).

Vermeychuk, J.G., and Lapidus, L., Suboptimal feedback control of distributed systems, AIChEJ, 19(1), 123-137 (1973).

1974

Alevisakis, G., and Seborg, D.E., Control of multivariable systems containing time delays using multivariable Smith predictor, Chem Eng Sci, 29(2), 373-381 (1974).

Ball, J.; Brez, C., and Cassiday, J., Batch-control improvement through computers, Chem Eng, 5 Aug, 93-95 (1974).

Chintapalli, P.S., and Douglas, J.M., Controllability considerations for multivariable processes, Chem Eng Sci, 29(2), 403-410 (1974).

Harrison, R.E.; Felder, R.M., and Rousseau, R.W., Accuracy of parameter estimation by frequency response analysis, Ind Eng Chem Proc Des Dev, 13(4), 389-391 (1974).
Knudsen, J.K.H., and Kummel, M., Frequency domain solution of optimal control problem for linear distributed systems, Chem Eng Sci, 29(2), 521-532 (1974).
Luus, R., Optimal control by direct search on feedback gain matrix, Chem Eng Sci, 29(4), 1013-1018 (1974).
Luus, R., Practical approach to time-optimal control of nonlinear systems, Ind Eng Chem Proc Des Dev, 13(4), 405-408 (1974).
Lynch, E.P., Using Boolean algebra and logic diagrams, Chem Eng, 19 Aug, 107-113; 16 Sept, 111-118; 14 Oct, 101-104 (1974).
Mayfield, R., Backup for batch systems, Chem Eng, 10 June, 79-82 (1974).
Mutharasan, R., and Coughanowr, D.R., Feedback direct digital control algorithms for a class of distributed-parameter systems, Ind Eng Chem Proc Des Dev, 13(2), 168-176 (1974).
Oliver, W.K.; Seborg, D.E., and Fisher, D.G., Model reference adaptive control based on Liapunov's direct method, Chem Eng Commns, 1(3), 125-140 (1974).
Scotting, D.J.; Cowsley, C.W.; Mitchell, F.R.G., and Kenney, C.N., Computer controlled experimentation, Trans IChemE, 52, 349-353 (1974).
Stainthorp, F.P., and Benson, R.S., Computer aided design of process control systems, Chem Engnr, Sept, 531-535 (1974).

1975

Bullin, J.A., and Dukler, A.E., Hybrid computer modelling of stochastic nonlinear dynamic system, Chem Eng Sci, 30(5), 631-636 (1975).
Cheng, Y.C., and Ward, T.J., Noninteracting control, Ind Eng Chem Proc Des Dev, 14(2), 193-195 (1975).
Chintapalli, P.S., and Douglas, J.M., Use of economic performance measures to synthesize optimal control systems, Ind Eng Chem Fund, 14(1), 1-10 (1975).
Cusset, B.F., and Mellichamp, D.A., On-line identification of process dynamics: Multi-frequency response method, Ind Eng Chem Proc Des Dev, 14(4), 359-368 (1975).
MacGregor, J.F.; Wright, J.D., and Hong, H.M., Optimal tuning of digital PID controllers using dynamic-stochastic models, Ind Eng Chem Proc Des Dev, 14(4), 398-402 (1975).
Turrie, B.D.; Clinton, J.H., and Weigand, W.A., Investigation of a method for time optimal control of linear systems with multiple inputs, Chem Eng Commns, 2(1), 5-18 (1975).
Waller, K.V.T., and Gustafsson, S.E., Performance indices for multivariable PI-control with incomplete feedback, Chem Eng Sci, 30(10), 1265-1272 (1975).
Waller, K.V.T., and Nygardas, C.G., Inverse response in process control, Ind Eng Chem Fund, 14(3), 221-223 (1975).
Yang, C.H., and Ward, T.J., Secondary-variable control, Ind Eng Chem Fund, 14(3), 171-174 (1975).

1976

Bizarro, L.A., Networking computers for process control, Chem Eng, 6 Dec, 151-156 (1976).

Bohl, A.H., and McAvoy, T.J., Linear feedback vs. time optimal control, Ind Eng Chem Proc Des Dev, 15(1), 24-33 (1976).

Brez, C.; Draves, R., and Gore, F., Operator-oriented design in computer-controlled batch processes, Chem Eng, 20 Dec, 58-62 (1976).

Brodmann, M.T., and Smith, C.L., Computer control of batch processes, Chem Eng, 13 Sept, 191-198 (1976).

Dunn, I.J.; Prenosil, J.E., and Ingham, J., Digital simulation experiments, Chem Eng Educ, 10(1), 23-27 (1976).

Graham, G.E., Analyzing process control loops, Chem Eng, 2 Aug, 72-78 (1976).

Hammett, J.L., and Lindsay, L.A., Advanced computer control of ethylene plants pays off, Chem Eng, 8 Nov, 115-120 (1976).

Hatfield, J.M., Process control by computer: A review, Chem Engnr, March, 171-174, 177 (1976).

Jeffreson, C.P., Controllability of process systems: Application of Harriott's index of controllability, Ind Eng Chem Fund, 15(3), 171-179 (1976).

Kestenbaum, A.; Shinnar, R., and Thau, F.E., Design concepts for process control, Ind Eng Chem Proc Des Dev, 15(1), 2-13 (1976).

Khandheria, J., and Luyben, W.L., Experimental evaluation of digital algorithms for anti-reset windup, Ind Eng Chem Proc Des Dev, 15(2), 278-285 (1976).

Lee, W., and Weekman, V.W., Advanced control practice in the chemical process industry: A view from industry (review paper), AIChEJ, 22(1), 27-38 (1976).

Meyer, C.; Seborg, D.E., and Wood, R.K., Comparison of the Smith predictor and conventional feedback control, Chem Eng Sci, 31(9), 775-778 (1976).

Mutharasan, R., and Coughanowr, D.R., Sampled-data control of a distributed-parameter process, Ind Eng Chem Proc Des Dev, 15(3), 378-381 (1976).

Mutharasan, R., and Luus, R., Linear direct digital control algorithms for a class of distributed processes, Ind Eng Chem Proc Des Dev, 15(1), 137-141 (1976).

Nishida, N.; Liu, Y.A.; Lapidus, L., and Hiratsuka, S., Effective computational algorithm for suboptimal singular and/or bang-bang control, AIChEJ, 22(3), 505-523 (1976).

Oh, S.H., and Luus, R., Optimal feedback control of time-delay systems, AIChEJ, 22(1), 140-147 (1976).

Palmenberg, R.E., and Ward, T.J., Derivative decoupling control, Ind Eng Chem Proc Des Dev, 15(1), 41-47 (1976).

Prasad, C.C., and Krishnaswamy, P.R., Analysis of a predictor-regulator system, Chem Eng Sci, 31(1), 1-8 (1976).

Stephanopoulos, G., and Schuelke, L.M., Process design in a dynamic environment, AIChEJ, 22(5), 855-867 (1976).

Various, Process control practices (topic issue), Chem Eng Prog, 72(4), 72-93 (1976).

Watson, J.M., Total distributed control, Chem Engnr, March, 167-170 (1976).

1977

Bellingham, B., and Lees, F.P., The detection of malfunction using a process control computer; Part 1: A simple filtering technique for flow control loops; Part 2: A Kalman filtering technique for general control loops, Trans IChemE, 55, 1-16, 253-265 (1977).

Berenblut, B.J., and Whitehouse, H.B., A method for monitoring process plant based on a decision table analysis, Chem Engnr, March, 175-181 (1977).

Bourne, E.J., and Herring, W.M., Interfacing pilot plants with process computers, Chem Eng Prog, 73(12), 56-61 (1977).

Wengrow, H.R., A systems control simulator, Chem Eng Educ, 11(1), 32-33 (1977).

Williams, D.L., Microprocessors enhance computer control of plants, Chem Eng, 18 July, 95-99 (1977).

1978

Cox, W.E., Economics of computer control, Hyd Proc, 57(4), 169-171 (1978).

Gupta, S.R., and Coughanowr, D.R., On-line gain identification of flow processes with application to adaptive pH control, AIChEJ, 24(4), 654-664 (1978).

Joseph, B., and Brosilow, C.B., Inferential control of processes, AIChEJ, 24(3), 485-509 (1978).

Krishnaswamy, P.R.; Chandraprasad, C., and Mohandas, K.P., A multiloop compensator for dead time, Chem Eng J, 16(3), 177-184 (1978).

Maggioli, V.J., Applying programmable controllers, Hyd Proc, 57(12), 137-142 (1978).

Martin, R.L., Simple solutions to control-problems, Chem Eng, 22 May, 103-111 (1978).

Mellefont, D.J., and Sargent, R.W.H., Selection of measurements for optimal feedback control, Ind Eng Chem Proc Des Dev, 17(4), 549-552 (1978).

Schork, F.J., and Deshpande, P.B., Double-cascade controller tested, Hyd Proc, 57(6), 113-117 (1978).

Sundaresan, K.R., and Krishnaswamy, P.R., Estimation of time-delay time-constant parameters in time, frequency, and Laplace domains, Can JCE, 56, 257-267 (1978).

Van Horn, L.D., Implementing computer control, Hyd Proc, 57(9), 243-245 (1978).

Various, Process control in the CPI (topic issue), Chem Eng Prog, 74(6), 49-88 (1978).

1979

Becker, J.V., and Hill, R., Fundamentals of interlock systems, Chem Eng, 8 Oct, 101-108 (1979).

Bhalodia, M., and Weber, T.W., Feedback control of two-input, two-output interacting process, Ind Eng Chem Proc Des Dev, 18(4), 599-607 (1979).

Brown, T., and Cadick, J.L., Electrical energy, Chem Eng, 1 Jan, 72-76; 29 Jan, 111-115; 12 March, 85-91; 7 May, 89-94; 18 June, 119-122; 16 July, 107-112 (1979).

Gray, J.O.; El-Dhuwaib, Z., and Hassapis, G., Process control with sampled feedback, Trans IChemE, 57, 73-83 (1979).

Kumamoto, H., and Henley, E.J., Safety and reliability synthesis of systems with control loops, AIChEJ, 25(1), 108-113 (1979).

Mahood, R.F., and Martin, J.E., Improve automatic control system reliability, Hyd Proc, 58(5), 215-221 (1979).

Nisenfeld, E., Beware of mistuning control loops, Chem Eng, 24 Sept, 139-140 (1979).

Ogunnaike, B.A., and Ray, W.H., Multivariable controller design for linear systems having multiple time delays, AIChEJ, 25(6), 1043-1057 (1979).

Palmor, Z.J., and Shinnar, R., Design of sampled data controllers, Ind Eng Chem Proc Des Dev, 18(1), 8-30, (1979).

Perkins, P.R., The oilfield: A distributed data system, Chem Engnr, May, 334-337 (1979).

Various, Process control by computer (topic issue), Comput Chem Eng, 3(11), 573-592 (1979).

Various, Computer optimization (special report), Hyd Proc, 58(6), 71-92 (1979).

Various, Process control (topic issue), Chem Eng Prog, 75(5), 37-67 (1979).

Weber, T.W., and Bhalodia, M., Optimum behavior of third-order process under feedback control, Ind Eng Chem Proc Des Dev, 18(2), 217-223 (1979).

1980

Anon., Process control update, Proc Engng, Nov, 45-49 (1980).

Bittman, G.V., and Ward, T.J., Digital control of time-delay processes, AIChEJ, 26(2), 295-297 (1980).

Blevins, T.L., and Langley, K., Computer models aid process-control system design, Hyd Proc, 59(11), 197-201 (1980).

Bond, A., and Hunt, L., Glossary of microcomputing terms, Proc Engng, April, 66-69 (1980).

Bristol, E.H., After direct digital control: Idiomatic control, Chem Eng Prog, 76(11), 84-89 (1980).

Brown, T., and Cadick, J.L., Electrical energy, Chem Eng, 10 Sept, 137-140; 22 Oct, 127-130; 3 Dec, 101-103 (1979); 28 Jan, 117-120; 10 March, 149-152; 21 April, 145-149 (1980).

Casa, P.D.; Barbieri, R., and Albertini, E., Better batch processes control, Hyd Proc, 59(3), 101-105 (1980).

Deshpande, P.B., Process identification of open-loop unstable systems, AIChEJ, 26(2), 305-308 (1980).

Eyermann, V.W., Cascade controller returns condensate, Chem Eng, 7 April, 115 (1980).

Fuhrman, J.E.; Mutharasan, R., and Coughanowr, D.R., Computer control of a distributed parameter system, Ind Eng Chem Proc Des Dev, 19(4), 537-546 (1980).

Harris, S.L., and Mellichamp, D.A., On-line identification of process dynamics: Use of multifrequency binary sequences, Ind Eng Chem Proc Des Dev, 19(1), 166-174 (1980).

Hsiao-Ping, H., and Yung-Cheng, C., Parameter identification and self-tuning adaptive control by a sensitivity approach, Chem Eng Commns, 6(6), 313-332 (1980).

Lee, P.L., and Newell, R.B., Use of Krasovskii's stability technique for control strategy evaluation and system design, Can JCE, 58, 389-392 (1980).

McCool, J., Implementing adaptive control, Proc Engng, June, 93,95 (1980).
Mesniaeff, P.G., Timers are the key elements in process control systems, Chem Eng, 1 Dec, 87-93 (1980).
Morari, M., Stability of model reference adaptive control systems, Ind Eng Chem Proc Des Dev, 19(2), 279-281 (1980).
Morari, M.; Arkun, Y., and Stephanopoulos, G., Studies in the synthesis of control structures for chemical processes, AIChEJ, 26(2), 220-260; 26(6), 975-991 (1980).
Murty, B.S.N.; Gangiah, K., and Husain, A., Performance of various methods in computing optimal control policies, Chem Eng J, 19(3), 201-208 (1980).
Sandefur, M.J., Implementing digital control, Hyd Proc, 59(7), 115-129 (1980).
Schagrin, E.F., Evolution of a distributed control system, Chem Eng Prog, 76(6), 72-75 (1980).
Shah, G.C., Understanding minicomputer control systems, Hyd Proc, 59(4), 153-157 (1980).
Various, Computers and process control (topic issue), Chem Eng Prog, 76(4), 31-56 (1980).

1981

Anderson, M., and Whitcomb, P., How to document process-monitoring-and-control systems, Chem Eng, 7 Sept, 117-120 (1981).
Arkun, Y., and Stephanopoulos, G., Studies in the synthesis of control structures for chemical processes, AIChEJ, 27(5), 779-793 (1981).
Brantley, R.O.; Leffew, K.W., and Deshpande, P.B., Do you need cascade control? Hyd Proc, 60(3), 139-142 (1981).
Frankland, P.B., Wide turndown flow control, Hyd Proc, 60(4), 187-189 (1981).
Groves, F.R., Application of comparison theorems to determination of response of nonlinear systems, Chem Eng Commns, 8(4), 341-352 (1981).
Guy, J.L., Process control loops, Chem Eng, 24 Aug, 111-117 (1981).
Guy, J.L., Solving the mathematical models for dynamic systems, Chem Eng, 16 Nov, 271-277 (1981).
Guy, J.L., Modeling process systems on analog/hybrid computers, Chem Eng, 28 Dec, 63-68 (1981).
Guy, J.L., Fundamentals of chemical process dynamics, Chem Eng, 29 June, 74-80 (1981).
Harris, S.L., and Mellichamp, D.A., Frequency domain adaptive controller, Ind Eng Chem Proc Des Dev, 20(2), 188-196 (1981).
Love, L., LOOP: A process control system design package, Chem Engnr, Feb, 56-57, 65 (1981).
Martin, G.D., Long-range predictive control, AIChEJ, 27(5), 748-753 (1981).
McAvoy, T.J., Connection between relative gain and control loop stability and design, AIChEJ, 27(4), 613-619 (1981).
Palmor, Z.J., and Shinnar, R., Design of advanced process controllers, AIChEJ, 27(5), 793-805 (1981).
Petkovski, D.B., Design of decentralized P&I controllers for multivariable systems, Comput Chem Eng, 5(1), 51-56 (1981).
Ray, W.H., Process control: Practice, research, and education, Chemtech, May, 300-304 (1981).

Taggart, G., Problems with feedforward control for Claus units, Hyd Proc, 60(3), 155-157 (1981).
Various, Process control survey (topic issue), Processing, Aug, 21-35 (1981).
Various, Distributed computer control (special report), Hyd Proc, 60(6), 91-114 (1981).
Various, Putting computers to work (topic issue), Chem Eng Prog, 77(11), 33-68 (1981).

1982
Abd-El-Bary, M.F., and Chari, S., Digital computer application in process control, Chem Eng Educ, 16(3), 118-121 (1982).
Barth, P.W., Silicon sensors meet integrated circuits, Chemtech, Nov, 666-673 (1982).
Brantley, R.O.; Schaefer, R.A., and Deshpande, P.B., On-line process identification, Ind Eng Chem Proc Des Dev, 21(2), 297-301 (1982).
Dartt, S.R., Distributed digital control, Chem Eng, 6 Sept, 107-114 (1982).
DiBiano, R.; Hales, G., and Autenreith, A., Advantages of third generation computer control, Hyd Proc, June, 117-121 (1982).
Gagnepain, J.P., and Seborg, D.E., Analysis of process interactions with applications to multiloop control system design, Ind Eng Chem Proc Des Dev, 21(1), 5-11 (1982).
Garcia, C.E., and Morari, M., Internal model control, Ind Eng Chem Proc Des Dev, 21(2), 308-323 (1982).
Govind, R., and Powers, G.J., Control system synthesis strategies, AIChEJ, 28(1), 60-73 (1982).
Graboski, M.S.; Kabel, R.L.; Danner, R.P., and Al-Ameeri, R.S., Process input analysis, Chem Eng Commns, 17(1), 137-150 (1982).
Harris, T.L.; MacGregor, J.F., and Wright, J.D., Overview of discrete stochastic controllers: Generalized PID algorithms with dead-time compensation, Can JCE, 60, 425-432 (1982).
Huang, C.T., and Clements, W.C., Parameter estimation for the second-order plus dead-time model, Ind Eng Chem Proc Des Dev, 21(4), 601-603 (1982).
Huang, H.P., and Chao, Y.C., Optimal tuning of practical digital PID controller, Chem Eng Commns, 18(1), 51-62 (1982).
Kennedy, J.P., Using process computers for refining operations, Hyd Proc, May, 160-164 (1982).
Kletz, T.A., Human problems with computer control, Plant/Opns Prog, 1(4), 209-211 (1982).
Lin, C.S., Tsai, T.H., and Lane, J.W., Coordinated control and its implementation, Plant/Opns Prog, 1(2), 85-90 (1982).
Meier, F.A.. Is your control system ready to start-up? Chem Eng, 22 Feb, 76-87 (1982).
Morari, M., and Fung, A.K.W., Nonlinear inferential control, Comput Chem Eng, 6(4), 271-282 (1982).
Ogunnaike, B.A., and Ray, W.H., Computer-aided multivariable control system design for processes with time delays, Comput Chem Eng, 6(4), 311-326 (1982).
Rinard, I.H., Control-system design, Chem Eng, 29 Nov, 46-58 (1982).

Various, Distributed process-control systems, Processing, Aug, 21-31 (1982).
Various, Programmable controllers (topic issue), Processing, Aug, 33-41 (1982).
Various, Putting computers to work (topic issue), Chem Eng Prog, 78(9), 39-90 (1982).
Weber, R.; StAubin, R.J., and Halberg, M.R., Graphics-based process interface, Chem Eng Prog, 78(1), 50-53 (1982).
Yuwana, M., and Seborg, D.E., New method for on-line controller tuning, AIChEJ, 28(3), 434-440 (1982).

1983
Carter, E.R., Operator interrupts for programmable controllers, Chem Eng, 26 Dec, 39-42 (1983).
Giannici B., and Galluzzo, M., Feedforward control schemes for chemical processes: An algorithmic approach, Chem Eng Commns, 21(4), 279-292 (1983).
Gordon, L.M., Feedback control modes, Chem Eng, 8 Aug, 79-85 (1983).
Gordon, L.M., Basic concepts, terminology, and techniques for process control, Chem Eng, 30 May, 58-66 (1983).
Harding, M., Data transfer in distributed systems, Proc Engng, Sept, 41-44 (1983).
Herring, W.M., Bench-top data acquisition and control system, Chem Eng Prog, 79(8), 46-50 (1983).
Kinney, T.B., Tuning process controllers, Chem Eng, 19 Sept, 67-72 (1983).
Koppel, L.B., Input multiplicities in process control, Chem Eng Educ, 17(2), 58-63, 89-92 (1983).
Kusnetz, H.L., and Phillips, C.F., Industrial hygiene evaluations for control design, Plant/Opns Prog, 2(3), 146-150 (1983).
Leigh, R., Improving process-control signals, Proc Engng, Aug, 24-25 (1983).
Leigh, R., Distributed process-control systems, Proc Engng, July, 22-27 (1983).
Leister, M.E., and Sanders, R.R., Convert to microprocessor controls without shutdown, Hyd Proc, April, 101-104 (1983).
Marchetti, J.L.; Mellichamp, D.A., and Seborg, D.E., Predictive control based on discrete convolution models, Ind Eng Chem Proc Des Dev, 22(3), 488-495 (1983).
Martinovic, A., Architectures of distributed digital control systems, Chem Eng Prog, 79(2), 67-72 (1983).
McAvoy, T.J., Dynamic interaction analysis of complex control systems, Ind Eng Chem Proc Des Dev, 22(1), 42-49 (1983).
Mehta, G.A., The benefits of batch process control, Chem Eng Prog, 79(10), 47-52 (1983).
Myron, T.J., Feedback methods for process control systems, Chem Eng, 14 Nov, 233-238 (1983).
Ollila, A., Operating experience with distributed digital control, Hyd Proc, Oct, 87-91 (1983).
Palmer, D., Microprocessors in distributed control systems, Chem Engnr, May, 91-93 (1983).
Ray, W.H., Multivariable process control: A survey, Comput Chem Eng, 7(4), 367-394 (1983).

Seitz, D.R., Profile of a successful computer control project, Chem Eng Prog, 79(10), 44-46 (1983).
Severns, G., and Hedrick, J., Planning control methods for batch processes, Chem Eng, 18 April, 69-78 (1983).
Shinskey, F.G., Uncontrollable processes and what to do about them, Hyd Proc, Nov, 179-182 (1983).
Srinivasan, R., and Mellichamp, D.A., Stability and response properties of the analytical predictor, Ind Eng Chem Proc Des Dev, 22(4), 571-576 (1983).
Tompkins, D., Function block programming, Proc Engng, April, 27-30 (1983).
VanHorn, L.D., and Crosby, J.E., Impact of advanced applications technology, Hyd Proc, Aug, 34A-34G (1983).
Various, Process systems engineering (topic issue), Comput Chem Eng, 7(4), 201-566 (1983).
Various, Digital process control (special report), Hyd Proc, June, 59-82 (1983).
Various, Process control: Optimization and simulation (topic issue), Chem Eng Prog, 79(6), 39-89 (1983).
Wang, R.K., et al., Advanced distributed computer system for pilot plants, Chem Eng Prog, 79(8), 55-58 (1983).
Williams, T.J., Robots in the CPI, Chem Eng, 26 Dec, 29-34 (1983).
Wu, W.T., and Liu, W.C., Suboptimal control of nonlinear dynamic systems via moving model, Comput Chem Eng, 7(1), 35-42 (1983).

1984

Arkun, Y., and Ramakrishnan, S., Structural sensitivity analysis in the synthesis of process control systems, Chem Eng Sci, 39(7), 1167-1180 (1984).
Arkun, Y.; Manousiouthakis, B., and Palazoglu, A., Robustness analysis of process control systems, Ind Eng Chem Proc Des Dev, 23(1), 93-101 (1984).
Badavas, P.C., Feedforward methods for process control systems, Chem Eng, 15 Oct, 103-108 (1984).
Badavas, P.C., Direct-synthesis and adaptive controls, Chem Eng, 6 Feb, 99-103 (1984).
Banks, W.W., and Cerven, F., Predictor displays: Application of human engineering in process control systems, Plant/Opns Prog, 3(4), 215-222 (1984).
Berkowitz, P.N., Process control projects need these precautions, Hyd Proc, Aug, 76-78 (1984).
Chang, H.C., and Chen, L.H., Bifurcation characteristics of nonlinear systems under conventional PID control, Chem Eng Sci, 39(7), 1127-1142 (1984).
Dartt, S.R., Improve project management of a distributed digital control system, Hyd Proc, Aug, 73-75,78 (1984).
Fay, L., Feedforward process control in the Bayer process, Can JCE, 62(5), 661-683 (1984).
Friedly, J.C., Use of the Bristol array in designing noninteracting control loops (a limitation and extension), Ind Eng Chem Proc Des Dev, 23(3), 469-472 (1984).
Haggin, J., Process control: An update, C&E News, 2 April, 7-16; 21 May, 7-13; 4 June, 7-13 (1984).

Joseph, B., and Elliott, D.L., Microcomputer based laboratory for teaching computer process control, Chem Eng Educ, 18(3), 136-139 (1984).

Juska, D.W., Simplify controller tuning, Hyd Proc, Dec, 67-68 (1984).

Jutan, A., and Rodriguez, E., Calculator program for new controller-tuning method, Chem Eng, 3 Sept, 69-73 (1984).

Jutan, A., and Rodriguez, E.S., New method for on-line controller tuning, Can JCE, 62(6), 802-808 (1984).

Kurth, T.C., How to use feedback loops to meet process conditions, Chem Eng, 30 April, 77-83 (1984).

McAlister, J., Control disturbances? Check grounding, Chem Eng, 24 Dec, 77-78 (1984).

McGee, H.A., Laser applications in chemical engineering, Chem Eng Prog, 80(9), 11-14 (1984).

Moore, J.F.; Truesdale, P.B., and Sipowicz, W.W., Trends in refinery operations monitoring and control, Hyd Proc, March, 38B-38N (1984).

Olson, D.H., and Scharringhausen, D., Effectively using distributed control concepts, Hyd Proc, Oct, 59-61 (1984).

Qinwei, W., Process control in the People's Republic of China, Hyd Proc, March, 160R-160X (1984).

Rogers, J.M., and Edgar, T.F., Specifying control and isolation dampers, Hyd Proc, Jan, 125-129 (1984).

Rys, R.A., Advanced control methods, Chem Eng, 20 Aug, 151-158 (1984).

Shaw, R., The microprocessor in process control, Chem Engnr, Feb, 34-39 (1984).

Smith, C.L., and Hirche, J., Integrated process information and control, Chem Eng, 12 Nov, 113-120 (1984).

Song, H.K.; Fisher, D.G., and Shah, S.L., Experimental evaluation of a robust self-tuning PID controller, Can JCE, 62(6), 755-763 (1984).

Srinivas, S.P.; Deshpande, P.B., and Krishnaswamy, P.R., Extension of Nichols chart for identification of open-loop unstable systems, AIChEJ, 30(4), 684-686 (1984).

Various, Computerized control and operation of chemical plants (topic issue), Comput Chem Eng, 8(5), 253-314 (1984).

Various, Practical process control (special report), Hyd Proc, June, 51-69 (1984).

Vervalin, C.H., Training by simulation, Hyd Proc, Dec, 41-50 (1984).

Zakrzewski, A., Installing advanced control and optimisation strategies, Proc Engng, Nov, 63-66 (1984).

1985

Abell, M., Building programmable-controller systems, Proc Engng, Nov, 81-84 (1985).

Arkun, Y., et al., Computer-aided analysis and design of robust multivariable control systems for chemical processes, Comput Chem Eng, 9(1), 27-60 (1985).

Bader, F.P., Process controller/computer connection, Hyd Proc, Sept, 109-112 (1985).

Balakotaiah, V., and Luss, D., Input-multiplicity in lumped-parameter systems, Chem Eng Commns, 39, 309-322 (1985).

Balakotaiah, V.; Luss, D., and Keyfitz, B.L., Steady-state multiplicity analysis of lumped-parameter systems described by a set of algebraic equations, Chem Eng Commns, 36(1), 121-148 (1985).

Barsamian, J.A., Distributed computer systems for the plant, Chem Eng, 18 Feb, 179-182 (1985).

Baxter, R.A., Systems approach to process control design, Proc Engng, July, 45-50 (1985).

Brignole, E.A.; Gani, R., and Romagnoli, J.A., Simple algorithm for sensitivity and operability analysis of separation processes, Ind Eng Chem Proc Des Dev, 24(1), 42-48 (1985).

Buzzi-Ferraris, G., and Tronconi, E., Operational optimization and sensitivity analysis of multistage separators, Ind Eng Chem Proc Des Dev, 24(1), 112-118 (1985).

Cohen, E.M., and Fehervari, W., Sequential control, Chem Eng, 29 April, 61-66 (1985).

Cordova, G.J.; Hertanu, H.I., and Doyle, G.T., Microprocessor-based distributed control systems, Chem Eng, 21 Jan, 86-95 (1985).

Drott, D., and Freshwater, D., Teaching process control using micro-computers, Chem Engnr, Sept, 46-49 (1985).

Fisher, W.R.; Doherty, M.F., and Douglas, J.M., Steady-state control as a prelude to dynamic control, CER&D, 63, 353-357 (1985).

Grosdidier, P.; Morari, M., and Holt, B.R., Closed-loop properties from steady-state gain information, Ind Eng Chem Fund, 24(2), 221-235 (1985).

Gunkler, A.A., Putting computers in control, Chem Eng, 11 Nov, 177-181 (1985).

Harris, S.L., and Mellichamp, D.A., Controller tuning using optimization to meet multiple closed-loop criteria, AIChEJ, 31(3), 484-487 (1985).

Herman, D.J.; Sullivan, G.R., and Thomas, S., Integration of process design, simulation, and control systems, CER&D, 63, 373-377 (1985).

Huang, H.P.; Chao, Y.C., and Liu, P.H., Predictive adaptive control system for unmeasured disturbances, Ind Eng Chem Proc Des Dev, 24(3), 666-673 (1985).

Johnston, R.D.; Barton, G.W., and Brisk, M.L., Single-input-single-output (SISO) control system synthesis, Comput Chem Eng, 9(6), 547-566 (1985).

Larrabee, G.B., Microelectronics, Chem Eng, 10 June, 51-59 (1985).

McGee, N.F., Microprocessor controllers, Chem Eng, 4 Feb, 67-74 (1985).

Mijares, G., et al., Analysis and evaluation of the relative gains for nonlinear systems, Comput Chem Eng, 9(1), 61-70 (1985).

Moore, J.A., Retrofitting control systems, Chem Eng, 11 Nov, 172-176 (1985).

O'Kelly, D., and Mullins, M., Programming control algorithms, Chem Eng, 18 March, 183-186 (1985).

Ogunnaike, B.A., Optimal controller for discrete time delay systems requiring no prediction, Chem Eng Commns, 37(1), 249-264 (1985).

Palazoglu, A., and Arkun, Y., Robust tuning of process control systems using singular values and their sensitivities, Chem Eng Commns, 37(1), 315-332 (1985).

Palmor, Z.J., and Powers, D.V., Improved dead-time compensator controllers, AIChEJ, 31(2), 215-221 (1985).

Perkins, J.D., and Wong, M.P.F., Assessing controllability of chemical plants, CER&D, 63, 358-362 (1985).
Ralston, P.A.S.; Watson, K.R., Patwardhan, A.A., and Deshpande, P.B., Computer algorithm for optimized control, Ind Eng Chem Proc Des Dev, 24(4), 1132-1136 (1985).
Sargent, R.W.H., Trends in development of process control systems, CER&D, 63, 349-352 (1985).
Schweber, W.L., Microcomputer process control, Chem Eng, 5 Aug, 109-111 (1985).
Shaw, J.A., Human factors aspects of advanced process control, Plant/Opns Prog, 4(2), 111-116 (1985).
Tan, L.Y., and Weber, T,W., Controllability and observability of time-varying linear system using block-pulse functions, Chem Eng Commns, 36(1), 149-160 (1985).
Tan, L.Y., and Weber, T.W., Controller tuning of a third-order process under proportional-integral control, Ind Eng Chem Proc Des Dev, 24(4), 1155-1160 (1985).
Tayler, C., Pattern recognition in self-organising control, Proc Engng, Feb, 34-39 (1985).
Tayler, C., Distributed-intelligence control systems, Proc Engng, Aug, 36-39 (1985).
Various, Process control (topic issue), Chem Eng Prog, 81(12), 11-14, 25-51 (1985).
Various, Better process control (special report), Hyd Proc, June, 53-71 (1985).
Watson, K.R.; Wong, J.P., and Deshpande, P.B., Simulation of simple controlled processes with dead time, Chem Eng Educ, 19(1), 44-45 (1985).

1986

Anon., Programmable controllers (equipment focus), Chem Eng, 3 Feb, 47-52 (1986).
Ara Barsamian, J., Justifying plantwide computer control, Chem Eng, 12 May, 105-108 (1986).
Arulalan, G.R., and Deshpande, P.B., New algorithm for multivariable control, Hyd Proc, 65(6), 51-54 (1986).
Berger, R.W., and Hart, T.H., Statistical Process Control: A Guide for Implementation, Marcel Dekker, New York (1986).
Berk, A.A., Microprocessors in Process and Product Control, McGraw-Hill, New York (1986).
Bhandari, V.A.; Paradis, R., and Saxena, A.C., Use performance indices for better control, Hyd Proc, 65(9), 59-61 (1986).
Bozenhardt, H., and Dybeck, M., Estimating savings from upgrading process control, Chem Eng, 3 Feb, 99-102 (1986).
Chowdhury, J., Troubleshooting comes online in the CPI, Chem Eng, 13 Oct, 14-19 (1986).
Chowdhury, J., New programmable controllers widen process-control choices, Chem Eng, 4 Aug, 22-27 (1986).
Cooper, D.J.; Ramirez, W.F., and Clough, D.E., Comparison of linear distributed-parameter filters to lumped approximations, AIChEJ, 32(2), 186-194 (1986).

Davidson, R.S., Stop redesigning pilot plant systems: Use distributed, hierarchical control systems, Chem Eng Prog, 82(12), 18-21 (1986).
Fay, C.R., Project management for a pilot plant control scheme, Chem Eng Prog, 82(12), 44-49 (1986).
Garcia, C.E., and Morshedi, A.M., Quadratic programming solution of dynamic matrix control, Chem Eng Commns, 46, 73-88 (1986).
Hide, D., Computer control: A customised system, Proc Engng, April, 42-49 (1986).
Husain, A., Chemical Process Simulation, Halstead Press, New York (1986).
Jensen, N.; Fisher, D.G., and Shah, S.L., Interaction analysis in multivariable control systems, AIChEJ, 32(6), 959-970 (1986).
Jerome, N.F., and Ray, W.H., High-performance multivariable control strategies for systems having time delays, AIChEJ, 32(6), 914-931 (1986).
Jovic, F., Process Control Systems: Principles of Design and Operation, Gulf Publishing Co., Texas (1986).
Jovic, F., Process Control Systems, Gulf Publishing Co., Texas (1986).
Kane, L., Advanced process control handbook, Hyd Proc, 65(2), 35-100 (1986).
Kimber, A., Programmable controllers update, Processing, Sept, 27-37 (1986).
Leach, D.B., Specifying a batch-process control system, Chem Eng, 8 Dec, 115-122 (1986).
Lucas, M.P., Distributed Control Systems: Evaluation and Design, Van Nostrand Reinhold, New York (1986).
Luyben, W.L., Simple method for tuning SISO controllers in multivariable systems, Ind Eng Chem Proc Des Dev, 25(3), 654-660 (1986).
Margetts, A., Design and test your PLC control logic this way, Proc Engng, March, 41-45 (1986).
Mijares, G.; Cole, J.D.; Naugle, N.W.; Preisig, H.A., and Holland, C.D., New criterion for the pairing of control and manipulated variables, AIChEJ, 32(9), 1439-1449 (1986).
Moore, J.A., Digital Control Devices: Equipment and Applications, ISA Press, North Carolina (1986).
Ogunnaike, B.A., and Adewale, K.E.P., Dynamic matrix control for process systems with time varying parameters, Chem Eng Commns, 47, 295-314 (1986).
Palm, W.J., Control Systems Engineering, Wiley, New York (1986).
Postlethwaite, I., New perspectives in control system design, Proc Engng, March, 31-33 (1986).
Rangaiah, G.P., and Krishnaswamy, P.R., Dynamics of probabilistic time delay models, Chem Eng J, 33(3), 175-182 (1986).
Saunders, I., Programmable logic controllers in the food and drink industries, Chem Engnr, April, 50-51 (1986).
Seborg, D.E.; Edgar, T.F., and Shah, S.L., Adaptive control strategies for process control: A survey, AIChEJ, 32(6), 881-913 (1986).
Stephanopoulos, G. and Townsend, D.W., Synthesis in process development, CER&D, 64(3), 160-174 (1986).
Strickler, G.R., Control of pilot plants and laboratories using microcomputers, Chem Eng Prog, 82(12), 50-56 (1986).
Tayler, C., Computer control systems, Proc Engng, Dec, 29-32 (1986).
Tayler, C., Computers in control, Proc Engng, April, 51-55 (1986).

Tayler, C., Process control review (1965-1986): From panel boards to computer screens, Proc Engng, Oct, 59-63 (1986).
Thomas, H.W., Adaptive control, J Inst Energy, Dec, 213-215 (1986).
Tsai, T.H.; Lane, J.W., and Lin, C.S., Modern Control Techniques for the Process Industries, Marcel Dekker, New York (1986).
Various, Teaching undergraduate process control (special feature), Chem Eng Educ, 20(2), 70-83, 106-108 (1986).
Yu, C.C., and Luyben, W.L., Conditional stability in cascade control, Ind Eng Chem Fund, 25(1), 171-174 (1986).
Yu, C.C., and Luyben, W.L., Design of multiloop SISO controllers in multivariable processes, Ind Eng Chem Proc Des Dev, 25(2), 498-504 (1986).
Zgurovskii, M.Z., Algorithm for optimal control of distributed technological process as described in multidimensional Euclidian space, Int Chem Eng, 26(1), 165-171 (1986).

1987

Agarwal, M., and Seeborg, D.E., Multivariable nonlinear self-tuning controller, AIChEJ, 33(8), 1379-1386 (1987).
Agemennoni, O.E.; Desages, A.C., and Romagnoli, J.A., Adaptive control scheme for SISO processes with delays, Ind Eng Chem Res, 26(4), 774-782 (1987).
Anon., Process automation 1987 (special advertising section), Chem Eng, 28 Sept, 45-107 (1987).
Bashan, Y., and Handelsman, M., Microcomputer-based control of a batch process, Chem Eng, 19 Jan, 123-125 (1987).
Basta, N., Process-control techniques make rapid advances, Chem Eng, 26 Oct, 22-25 (1987).
Beevers, A., Distributed control systems, Processing, May, 33-37 (1987).
Brown, S., Graphics mapping in process control, Processing, Feb, 26-28; March, 22-23 (1987).
Chien, I.L.; Seborg, D.E., and Mellichamp, D.A., Self-tuning control with decoupling, AIChEJ, 33(7), 1079-1088 (1987).
Chowdhury, J., Process data go digital, Chem Eng, 16 Feb, 22-24 (1987).
Contino, A.V., Improve plant performance via statistical process control, Chem Eng, 20 July, 95-102 (1987).
Copulsky, W., and Weinreb, H.G., Process computer control, Chem Eng, 9 Nov, 93-95 (1987).
Espino, R.L., Problem solving by computer simulation, Chem Eng Prog, 83(8), 20-24 (1987).
Fan, L.T.; Gharpuray, M.M., and Huang, Y.W., Sequencing a separation via fuzzy heuristics, Chem Eng, 2 March, 57-59 (1987).
Foss, A.S., Multiloop computer control program (UC ONLINE), Chem Eng Educ, 21(3), 122-125, 154-156 (1987).
Gallier, P.W., and Kisala, T.P., Process optimization by simulation, Chem Eng Prog, 83(8), 60-66 (1987).
Georgiou, A.; Caston, L., and Georgakis, C., Dynamic properties of the extensive variable control structures, Chem Eng Commns, 60, 119-144 (1987).

Gerry, J.P., and Hansen, P.D., Choosing the right controller: Choice between P&I or model-predictive controllers, Chem Eng, 25 May, 65-68 (1987).
Gore, F.E., Plantwide computerization: Where are we now? Hyd Proc, 66(10), 26A-26D (1987).
Grosdidier, P., and Morari, M., A computer-aided methodology for the design of decentralised controllers, Comput Chem Eng, 11(4), 423-433 (1987).
Harris, T.J., and MacGregor, J.F., Design of multivariable linear-quadratic controllers using transfer functions, AIChEJ, 33(9), 1481-1495 (1987).
Johnston, R.D., and Barton, G.W., Design and performance assessment of control systems using singular-value analysis, Ind Eng Chem Res, 26(4), 830-840 (1987).
Kane, L., Advanced process control handbook II, Hyd Proc, 66(3), 55-109 (1987).
Koenig, D.M., Tuning rules for non-interacting multidimensional PI control algorithm, Chem Eng Commns, 50, 177-184 (1987).
Kramer, M.A., and Palowitch, B.L., Rule-based approach to fault diagnosis using signed directed graph, AIChEJ, 33(7), 1067-1078 (1987).
Kravaris, C., and Chung, C.B., Nonlinear state feedback synthesis by global input-output linearization, AIChEJ, 33(4), 592-603 (1987).
Krishnaswamy, P.R.; Chan, B.E.M., and Rangaiah, G.P., Closed-loop tuning of process control systems, Chem Eng Sci, 42(9), 2173-2182 (1987).
Lipowicz, M., More process-control software, Chem Eng, 11 May, 63-65 (1987).
Love, J., Confidence in control, Chem Engnr, Dec, 36-38 (1987).
Love, J., Batch process control, Chem Engnr, June, 34-35 (1987).
Love, J., Strategies for batch control, Chem Engnr, Sept, 29-31 (1987).
Lyberatos, G., and Tsiligiannis, C.A., Analysis of feedback-induced Hopf bifurcations via the Carleman approximation, Chem Eng Commns, 57, 1-14 (1987).
Papadoulis, A.V.; Tsiligiannis, C.A., and Svoronos, S.A., A cautious self-tuning controller for chemical processes, AIChEJ, 33(3), 401-409 (1987).
Patwardhan, A.A.; Karim, M.N., and Shah, R., Controller tuning by least-squares method, AIChEJ, 33(10), 1735-1737 (1987).
Procyk, L.M., and Berndt, P.W., Batch distributed control: Tips on installation, Chem Eng, 17 Aug, 139-141 (1987).
Ray, W.H., et al., CONSYD: Integrated software for computer-aided control system design and analysis, Comput Chem Eng, 11(2), 187-203 (1987).
Roat, S.D., and Melsheimer, S.S., Microcomputer-aided control systems design, Chem Eng Educ, 21(1), 34-39 (1987).
Roffel, B., and Chin, P.A., Controlling processes with inverse response and dead time, Hyd Proc, 66(12), 40-42 (1987).
Rugh, W.J., Design of nonlinear PID controllers, AIChEJ, 33(10), 1738-1742 (1987).
Sanathanan, C.K., and Quinn, S.B., Design of set point regulators for processes involving time delay, AIChEJ, 33(11), 1873-1881 (1987).
Saucier, M.F.; Chang, H.C., and Seborg, D.E., Bifurcation analysis of multivariable feedback control systems, Chem Eng Commns, 57, 215-232 (1987).

Svoronos, S.A., Linear model-dependent control, AIChEJ, 33(3), 394-400 (1987).
Sztraka, L.; Boros, B., and Molnar-Paal, E., A computer program for simulating some signal-to-noise enhancement methods used in high performance instruments, Periodica Polytechnica, 31(1/2), 67-82 (1987).
Tan, L.Y., and Weber, T.W., New solution to inverse problem of optimal regulator control, AIChEJ, 33(1), 36-42 (1987).
Various, Process control (special report), Hyd Proc, 66(6), 37-49 (1987).
Various, Operability and control (special topic issue), CER&D, 65(6), 449-489 (1987).
Wilby, R., Programmable controllers in distributed control, Chem Engnr, April, 13-15 (1987).
Wong, P.M.; Taylor, P.A., and Wright, J.D., An experimental evaluation of saturation algorithms for advanced digital controllers, Ind Eng Chem Res, 26(6), 1117-1127 (1987).

1988

Agamennoni, O.E.; Desages, A.C., and Romagnoli, J.A., Robust controller design methodology for multivariable chemical processes, Chem Eng Sci, 43(11), 2937-2950 (1988).
Aluko, M.E., Linear feedback equivalence and nonlinear control of a class of two-dimensional systems, Chem Eng Commns, 68, 31-42 (1988).
Arkun, Y., Relative sensitivity: A dynamic closed-loop interaction measure and design tool, AIChEJ, 34(4), 672-675 (1988).
Arkun, Y.; Charos, G.N., and Reeves, D.E., Model predictive control experiments, Chem Eng Educ, 22(4), 178-183, 187 (1988).
Ayral, T.E.; Conley, R.C.; England, J., and Antis, K., Advanced control documentation for operators, Hyd Proc, 67(9), 103-104 (1988).
Bisio, A., The changing face of process control, Chem Engnr, Oct, 16 (1988).
Bozenhardt, H., Integrated approach to microprocessor-based control, Chem Eng Prog, 84(12), 23-28 (1988).
Callaghan, P.J., and Lee, P.L., Experimental investigation of predictive controller design by principal component analysis, CER&D, 66(4), 345-356 (1988).
Calvet, J.P., and Arkun, Y., Feedforward and feedback linearization of nonlinear systems and implementation using internal model control (IMC), Ind Eng Chem Res, 27(10), 1822-1832 (1988).
Clatworthy, S., Programmable electronic systems guidelines, Chem Engnr, Aug, 16-20 (1988).
Collins, R.S., and Clements, W.C., Evaluation of Gautam-Mutharasan control algorithm, Chem Eng Commns, 71, 145-156 (1988).
Deming, S.N., Statistical process control, Chemtech, Sept, 560-566 (1988).
Deshpande, P.B., Multivariable control methods, Chem Eng Educ, 22(4), 188-191 (1988).
Dybeck, M., and Bozenhardt, H., Justifying automation projects, Chem Eng, 20 June, 113-116 (1988).
Fisher, W.R.; Doherty, M.F., and Douglas, J.M., The interface between design and control, Ind Eng Chem Res, 27(4), 597-615 (1988).

Florence, D., Specifying programmable controllers, Chem Engnr, July, 46-51 (1988).
Galloway, P.J., and Holt, B.R., Multivariable time delay approximations for analysis and control, Comput Chem Eng, 12(7), 637-650 (1988).
Golden, M.P.; Chesna, A., and Ydstie, B.E., Adaptive nonlinear model control, Chem Eng Commns, 63, 17-38 (1988).
Grimm, W.M., and Lee, P.L., Robust closed-loop log modulus design of SISO Smith predictors for a special class of processes, Chem Eng Commns, 64, 217-232 (1988).
Guilandoust, M.T.; Morris, A.J., and Tham, M.T., Adaptive estimation algorithm for inferential control, Ind Eng Chem Res, 27(9), 1658-1664 (1988).
Gupta, Y.P., Suboptimal control though model reduction, Can JCE, 66(1), 142-146 (1988).
Harriott, P., Optimum controller settings for processes with dead-time: Effects of type and location of disturbance, Ind Eng Chem Res, 27(11), 2060-2063 (1988).
Heck, G.F., Simulating feedback control, Chem Eng, 9 May, 99-101 (1988).
Karimi, I., and Ku, H., Modified heuristic for initial sequence in flowshop scheduling, Ind Eng Chem Res, 27(9), 1654-1657 (1988).
Kravaris, C., Input-output linearization: A nonlinear analog of placing poles at process zeros, AIChEJ, 34(11), 1803-1812 (1988).
Kravaris, C., and Palanki, S., Robust nonlinear state feedback under structured uncertainty, AIChEJ, 34(7), 1119-1127 (1988).
Kutsuwa, Y.; Nishitani, H., and Kunugita, E., Design and analysis of flexible process system with a fixed tolerance using mixed-integer linear-programming model, Int Chem Eng, 28(2), 314-330 (1988).
Lee, P.L., and Sullivan, G.R., Generic model control (GMC), Comput Chem Eng, 12(6), 573-580 (1988).
Lewin, D.R., and Scali, C., Feedforward control in presence of uncertainty, Ind Eng Chem Res, 27(12), 2323-2330 (1988).
Li, W.C., and Biegler, L.T., Process control strategies for constrained nonlinear systems, Ind Eng Chem Res, 27(8), 1421-1433 (1988).
Li, W.C., and Biegler, L.T., Process control strategies for constrained nonlinear systems, Ind Eng Chem Res, 27(8), 1421-1433 (1988).
Love, J., Trends and issues in batch control, Chem Engnr, April, 24-26 (1988).
Luyben, W.L., The concept of 'eigenstructure' in process control, Ind Eng Chem Res, 27(1), 206-208 (1988).
Mamzic, C.L., and Tucker, T.W., Statistical process control in the process industries, Hyd Proc, 67(11), 132B-132P (1988).
Mamzic, C.L., and Tucker, T.W., Statistical process control in the process industries, Hyd Proc, 67(11), 132B-132P; 67(12), 34C-34H (1988).
Maurath, P.R.; Laub, A.J.; Seborg, D.E., and Mellichamp, D.A., Predictive controller design by principal components analysis, Ind Eng Chem Res, 27(7), 1204-1212 (1988).
Maurath, P.R.; Mellichamp, D.A., and Seborg, D.E., Predictive controller design for single input-single output systems, Ind Chem Eng Res, 27(6), 956-963 (1988).

May, D.L., Back-up systems for process control computers, Chem Eng, 25 April, 64-70 (1988).
Miller, D.L., Interactive process-control panel (direct manipulation interface for process simulations), Comput Chem Eng, 12(12), 1257-1262 (1988).
Monica, T.J.; Yu, C.C., and Luyben, W.L., Improved multiloop single input-single output controllers for multivariable processes, Ind Chem Eng Res, 27(6), 969-973 (1988).
Nguyen, T.C.; Barton, G.W.; Perkins, J.D., and Johnston, R.D., Condition number scaling policy for stability robustness analysis, AIChEJ, 34(7), 1200-1206 (1988).
Parrish, J.R., and Brosilow, C.B., Nonlinear inferential control, AIChEJ, 34(4), 633-644 (1988).
Phillips, S.F.; Seborg, D.E., and Legal, K.J., Adaptive control strategies for achieving desired temperature profiles during process start-up, Ind Eng Chem Res, 27(8), 1434-1449 (1988).
Proctor, A., Self-tuning controllers, Proc Engng, Feb, 63-66 (1988).
Ray, A., Distributed data communication networks for real-time process control, Chem Eng Commns, 65, 139-154 (1988).
Reeves, D.E., and Schork, F.J., Simulation problems in digital process control, Chem Eng Educ, 22(3), 154-157 (1988).
Roberts, A., Conception and design of process supervision and control, Manufacturing Chemist, June, 37-40 (1988).
Shen, G.C., and Lee, W.K., Multivariable adaptive inferential control, Ind Eng Chem Res, 27(10), 1863-1873 (1988).
Shen, G.C., and Lee, W.K., Generalized analytical predictor, AIChEJ, 34(4), 676-678 (1988).
Takamatsu, T., et al., Model predictive control in terms of structure and degrees of freedom, Int Chem Eng, 28(4), 661-669 (1988).
Vaidya, C.M., and Deshpande, P.B., Single-loop simplified-model predictive control, Hyd Proc, 67(6), 53-57 (1988).
Various, Advanced process control handbook III, Hyd Proc, 67(3), 53-106 (1988).
Various, Computer control (topic issue), Chem Eng Prog, 84(10), 19-67 (1988).
Yeo, Y.K., and Williams, D.C., Adaptive model predictive control for SISO bilinear systems with stable inverses, Chem Eng Commns, 65, 79-94 (1988).
Yethiraj, A.; Kuszta, B., and Smith, C.B., Identification and feedback control of nonlinear systems, Chem Eng Commns, 71, 1-22 (1988).
Yu, C.C., Design of parallel cascade control for disturbance-rejection, AIChEJ, 34(11), 1833-1838 (1988).

8.2 Expert Systems/Artificial Intelligence

1983-1986

Leigh, R., and Wetton, M., Fuzzy logic, Proc Engng, Nov, 36-37 (1983).
Kumamoto, H.; Ikenchi, K.; Inoue, K., and Henley, E.J., Application of expert system techniques to fault diagnosis, Chem Eng J, 29(1), 1-10 (1984).
Peate, J., Applications of expert systems, Proc Engng, Jan, 21-23 (1984).

Andow, P.K., Fault diagnosis using intelligent knowledge based systems, CER&D, 63, 368-372 (1985).
Berkovitch, I., and Baker-Counsell, J., Applications of expert systems, Proc Engng, Dec, 25-28 (1985).
Dohnal, M., Fuzzy flowsheeting, Chem Eng J, 30(2), 71-80 (1985).
Henry, M., Expert systems and all that, Chem Engnr, Dec, 32-34 (1985).
Voller, V., and Knight, B., Expert systems, Chem Eng, 10 June, 93-96 (1985).
Higham, E., Expert systems in self-tuning controllers, Chem Engnr, March, 41-43 (1986).
Kane, L.A., Artificial intelligence and manufacturing automation protocol in the processing industries, Hyd Proc, 65(6), 55-58 (1986).
King, R.A., Expert systems for materials selection and corrosion, Chem Engnr, Dec, 42-43 (1986).
Niida, K.; Itoh, J.; Umeda, T.; Kobayashi, S., and Ishikawa, A., Some expert system experiments in process engineering, CER&D, 64(5), 372-380 (1986).
Venkatasubramanian, V., Artificial intelligence in process engineering, Chem Eng Educ, 20(4), 188-192 (1986).

1987-1988

Atkinson, N., Applications of expert systems, Proc Engng, Oct, 33-34 (1987).
Barnwell, J., and Ertl, B., Expert systems and the chemical engineer, Chem Engnr, Sept, 41-43 (1987).
Kerridge, A.E., Operators can use expert systems, Hyd Proc, 66(9), 97-105 (1987).
Kramer, M.A., Malfunction diagnosis using quantitative models with non-Boolean reasoning in expert systems, AIChEJ, 33(1), 130-140 (1987).
Moore, A., Use of expert systems, Proc Engng, Jan, 25-29 (1987).
Tayler, C., What benefit expert systems for supervising control? Proc Engng, April, 45-46 (1987).
Various, Expert systems (topic issue), Chem Eng Prog, 83(9), 21-75 (1987).
Zadeh, L.A., Fuzzy logic, Chemtech, June, 340-344; July, 406-410 (1987).
Andow, P., Developments in expert systems, Proc Engng, Oct, 63-66 (1988).
Basta, N., Expert systems for the CPI, Chem Eng, 14 March, 26-29 (1988).
Bunn, A.R., and Lees, F.P., Expert design of plant handling hazardous materials: Design expertise and CAD methods with illustrative examples, CER&D, 66(5), 419-444 (1988).
Erdmann, H.H., et al., Use of expert systems in chemical engineering, Chem Eng & Proc, 23(2), 125-134 (1988).
Evitt, S.D., and Mukaddam, W.A., Expert systems in the HPI, Hyd Proc, 67(1), 92B-92J (1988).
Lewin, D.R., and Morari, M., ROBEX: An expert system for robust control synthesis, Comput Chem Eng, 12(12), 1187-1198 (1988).
Norman, P.W., and Naveed, S., Knowledge acquisition analysis and structuring for construction of real-time supervisory expert systems, CER&D, 66(5), 470-480 (1988).
Schlager, S.T., Process troubleshooting with expert systems, Chemtech, Dec, 750-758 (1988).
Tayler, C., Expert systems in process plants, Proc Engng, April, 63-66 (1988).

Various, Artificial intelligence in chemical engineering: Research and development (topic issue), Comput Chem Eng, 12 (9/10), 853-1074 (1988).

8.3 Alarm Systems

1984-1988

Renton, A., Specification and realization of industrial control alarm systems, Chem Engnr, April, 33-35 (1984).
Andow, P., Alarm system and alarm analysis, Plant/Opns Prog, 4(2), 116-120 (1985).
Andow, P., Alarm systems, Chemtech, Feb, 124-128 (1986).
Margetts, A., Programmable logic controllers for alarm and shutdown systems, Chem Engnr, July, 36-37 (1986).
Martini, R.A.; Sullivan, C., and Cinar, A., An undergraduate experiment in alarm system design, Chem Eng Educ, 22(1), 22-25 (1988).

8.4 Equipment Control

1967-1970

Dor, M.N., On-line computer control of fixed-bed chemical reactors, Brit Chem Eng, 12(6), 888-891 (1967).
Foster, R.D., and Stevens, W.F., Application of noninteracting control to a continuous flow stirred-tank reactor, AIChEJ, 13(2), 340-345 (1967).
Corrigan, T.E., and Shaefer, R., Frequency response for N stirred tanks with recycle, Brit Chem Eng, 13(5), 680-683 (1968).
Luyben, W.L., Nonlinear feedforward control of chemical reactors, AIChEJ, 14(1), 37-45 (1968).
Ogunye, A.F., and Ray, W.H., Non-simple control policies for reactors with catalyst decay, Trans IChemE, 46, T225-231 (1968).
Lenoir, J.M., Measurement of surface temperature of furnace tubes, Hyd Proc, 48(10), 97-101 (1969).
Prescott, J.H., Computer-controlled catalytic cracking pays off, Chem Eng, 5 May, 128-132 (1969).
Buckley, P.S., Protective controls for a chemical reactor, Chem Eng, 20 April, 145-150 (1970).
Cadman, T.W., and Hsu, C.K., Dynamics and control of multistage liquid extraction, Trans IChemE, 48, T209-226 (1970).
Landis, D.M., Process control of centrifuge operations, Chem Eng Prog, 66(1), 51-53 (1970).

1971-1979

Kleinpeter, J.A., and Weaver, R.E.C., Multivariate control of phase separation process, AIChEJ, 17(3), 513-519 (1971).
Kortlandt, D., and Zwart, R.L., Find optimum timing for sampling product quality, Chem Eng, 1 Nov, 66-72 (1971).
Azpitarte, J.L., and Trevino, J.A., Controlling process plant deaerators, Chem Eng, 25 Dec, 85 (1972).

Huang, I.D., and Sonn, M., Computer control of batch processes, Brit Chem Eng, 17(6), 507-512 (1972).
Joffe, B.L., and Sargent, R.W.H., The design of an on-line control scheme for a tubular catalytic reactor, Trans IChemE, 50, 270-282 (1972).
Jowett, A., Mathematical models of mineral processing operations in process control, Chem Engnr, Dec, 459-465 (1972).
Liptak, B.G., Control of heat exchangers, Brit Chem Eng, 17(7), 637-645 (1972).
Marroquin, G., and Luyben, W.L., Experimental evaluation of nonliner cascade controllers for batch reactors, Ind Eng Chem Fund, 11(4), 552-556 (1972).
Dashevskii, L.N., Determination of effectiveness of automatic control systems for industrial processes, Int Chem Eng, 13(4), 695-698 (1973).
Matsubara, M.; Nishimura, Y., and Takahashi, N., Periodic operation and control of CSTR, Chem Eng Sci, 28(7), 1369-1386 (1973).
Various, Instrumentation/control of pilot plants (topic issue), Chem Eng Prog, 70(11), 53-88 (1974).
Davis, J.J., and Kermode, R.I., Optimum control of simplified polymerization process from its stochastic process model, Ind Eng Chem Proc Des Dev, 14(4), 459-466 (1975).
Scull, W.L., Selecting temperature controls for heaters, Chem Eng, 26 May, 128-132 (1975).
Bruns, D.D., and Bailey, J.E., Nonlinear feedback control for operating a nonisothermal CSTR near an unstable steady state, Chem Eng Sci, 32(3), 257-264 (1977).
Hsu, E.H., and Bacher, S., Interacting feedback control system for gas-liquid reactor, Ind Eng Chem Fund, 16(2), 259-263 (1977).
Various, Process control systems and equipment (feature report), Processing, March, 53-67 (1977).
Brisk, M.L., and Barton, G.W., On-line optimal control of catalytic reactors: A simplified approach, Trans IChemE, 56, 113-119 (1978).
Economopoulos, A.P., Acetic acid dehydration: Process and control, Chem Engnr, March, 199-202 (1978).
Luyben, W.L., and Melcic, M., Consider reactor control lags, Hyd Proc, 57(3), 115-117 (1978).
Nisenfeld, A.E., and Cho, C.H., Considering parallel compressor control, Hyd Proc, 57(2), 147-150 (1978).
Cheung, T.F., and Luyben, W.L., Liquid-level control in single tanks and cascades of tanks with proportional-only and proportional-integral feedback controllers, Ind Eng Chem Fund, 18(1), 15-21 (1979).
Cheung, T.F., and Luyben, W.L., PD control improves reactor stability, Hyd Proc, 58(9), 215-218 (1979).
Danziger, R., Distillation columns with vapor recompression, Chem Eng Prog, 75(9), 58-64 (1979).
Litchfield, R.J.; Campbell, K.S., and Locke, A., The application of several Kalman filters to the control of a real chemical reactor, Trans IChemE, 57, 113-120 (1979).

1980-1983

Foss, A.S.; Edwards, J.M., and Kouvaritakis, B., Multivariable control system for two-bed reactors by the characteristic locus method, Ind Eng Chem Fund, 19(1), 109-117 (1980).

Giger, G.K.; Coughanowr, D.R., and Richarz, W., Implementation of a feedback direct-digital control algorithm for a heat exchanger, Ind Eng Chem Proc Des Dev, 19(4), 546-550 (1980).

Horak, J., and Jiracek, F., Control of stirred tank reactors in open-loop unstable states, Chem Eng Sci, 35(1), 483-491 (1980).

Van Horn, L.D., Evaluating crude unit computer control, Hyd Proc, 59(4), 145-148 (1980).

Bertram, C.G., Sizing and specifying level-controlled condensate pots, Hyd Proc, 60(8), 151-154 (1981).

Bromley, J.A., and Ward, T.J., Fluidized catalytic cracker control: A structural analysis approach, Ind Eng Chem Proc Des Dev, 20(1), 74-81 (1981).

Lee, P.L.; Newell, R.B., and Agnew, J.B., A comparative study of control techniques applied to an unstable steady-state of a chemical reactor, Trans IChemE, 59, 105-111 (1981).

Mukerji, A., Specifying process chromatography systems, Hyd Proc, 60(5), 187-192 (1981).

Dattatreya, S., Adaptive gain improves coal-fired boiler control, Hyd Proc, Sept, 261-263 (1982).

Dohnal, M., Fuzzy models of unit operations, Chem Eng Commns, 19(1), 129-140 (1982).

Paulissen, G.T., Computer control for industrial utilities, Chem Eng Prog, 78(5), 56-58 (1982).

Adams, R.M.S.; Nourse, J.C., and Calmer, J.C., Process control for dechlorination, Chem Eng, 4 April, 113-114 (1983).

Asgari, M.; Yang, C.H.; Minarik, Z., and Zelenka, J., Computerization of a modern ethylene plant, Chem Eng Prog, 79(11), 27-33 (1983).

Brice, J.C., and Krikorian, K.V., Improve fluid catalytic cracking profitability with better control, Hyd Proc, May, 83-87 (1983).

Hinze, P., and Jenkins, P., Control of a power plant, Proc Engng, April, 31-37 (1983).

Honka, I.; Kuukkanen, K.K.; Malmivuori, P., and Aksela, R., Computer control of sulfite recovery boiler in the pulp and paper industry, Chem Eng Prog, 79(9), 31-35 (1983).

Hulbert, D.G., and Woodburn, E.T., Multivariable control of a wet-grinding circuit, AIChEJ, 29(2), 186-191 (1983).

Mukesh, D., and Cooper, A.R., Partial simulation and control of a continuous stirred-tank reactor with a digital computer, Ind Eng Chem Fund, 22(1), 145-149 (1983).

Swanson, K., An advanced combustion control system, Plant/Opns Prog, 2(1), 30-34 (1983).

1984-1986

Cebuhar, W.A., and Costanza, V., Nonlinear control of CSTRs, Chem Eng Sci, 39(12), 1715-1722 (1984).

Guffey, C.G., and Heenan, W.A., Process control of turboexpander plants, Hyd Proc, May, 71-74 (1984).
Martin, G.D.; Van Horn, L.D., and Cassaday, K.M., Experience with hydrotreater computer control, Hyd Proc, March, 66-70 (1984).
Watanabe, N.; Kurimoto, H., and Matsubara, M., Periodic control of continuous stirred tank reactors, Chem Eng Sci, 39(1), 31-36 (1984).
Baker-Counsell, J., Automating batch control, Proc Engng, April, 45,47 (1985).
Cluett, W.R.; Shah, S.L., and Fisher, D.G., Adaptive control of a batch reactor, Chem Eng Commns, 38(1), 67-78 (1985).
Eigenberger, G., Dynamics and stability of chemical engineering processes. Int Chem Eng, 25(4), 595-611 (1985).
Gomes, V.G., Controlling fired heaters, Chem Eng, 7 Jan, 63-68 (1985).
Graham, B.P.; Lee, P.L., and Newell, R.B., Simulation of multivariable control of a cement kiln, CER&D, 63, 363-367 (1985).
Belanger, R.R.; Rochon, L.; Dumont, G.A., and Gendron, S., Self-tuning control of chip level in a Kamyr digester, AIChEJ, 32(1), 65-74 (1986).
Doyle, J.B., Storing steam by changing water level in a deaerator, Chem Eng, 7 July, 77-78 (1986).
Guthrie, J.E.; Braunstein, B.A.; Harrington, M.T., and Merrick, R.D., Monitoring corrosion with a microcomputer, Chem Eng, 1 Sept, 81-84 (1986).
Liptak, B.G., Controlling and optimizing chemical reactors, Chem Eng, 26 May, 69-81 (1986).
Mandler, J.A.; Morari, M., and Seinfeld, J.H., Robust multivariable control system design for a fixed-bed reactor, Ind Eng Chem Fund, 25(4), 645-656 (1986).
Montgomery, D.P., Onstream calibration of pressurized gas-flow controller, Chem Eng, 9 June, 91-92 (1986).
Ryskamp, C.J.; McGee, N.F., and Badavas, P.C., Better alkylation control, Hyd Proc, 65(11), 113-118 (1986).

1987-1988

Armitage, S.J., and Nickrand, W.J., Statistical process control for water-treatment systems, Chem Eng, 25 May, 79-84 (1987).
Graver, A., Control of steam systems, Processing, Dec, 15-16 (1987).
Kozub, D.J.; MacGregor, J.F., and Wright, J.D., Application of LQ and IMC controllers to a packed-bed reactor, AIChEJ, 33(9), 1496-1506 (1987).
Najim, K.; Lann, M.U.L., and Casamatta, G., Learning control of a pulsed liquid-liquid extraction column, Chem Eng Sci, 42(7), 1619-1628 (1987).
Veland, L.H.; Hoyland, J.; Aronson, C.R., and White, D.C., Unique features improved crude unit advanced control, Hyd Proc, 66(9), 73-78 (1987).
Anon., Equipment update, Chem Eng, 26 Sept, 55-68 (1988).
Armer, A., Efficient condensate removal, Hyd Proc, 67(1), 81-83 (1988).
Bernard, J.W., MAP in the process industries, Hyd Proc, 67(6), 51-52 (1988).
Birch, J.R., Controlling small-scale liquid additions, Chem Eng, 18 July, 128-130 (1988).
Callaghan, P.J.; Lee, P.L., and Newell, R.B., Pilot-scale heat-recovery system for computer process control teaching and research, Chem Eng Educ, 22(2), 68-71 (1988).

Fish, N., Control of fermenters, Processing, May, 51-52 (1988).
Fryer, T., Advances in control of batch processes, Processing, Jan, 14-18 (1988).
Onderwater, D.; MacGregor, J.F., and Wright, J.D., Use of nonlinear transformations and a self-tuning regulator to develop an algorithm for catalytic-reactor temperature control, Can JCE, 66(3), 478-484 (1988).
Pan, D.F.; Schnitzlein, K., and Hofmann, H., Design of the control scheme of a concentration-controlled recycle reactor, Ind Eng Chem Res, 27(1), 86-93 (1988).
Phillips, S.F.; Seborg, D.E., and Legal, K.J., Adaptive control strategies for achieving desired temperature profiles during process start-up, Ind Eng Chem Res, 27(8), 1434-1449 (1988).
Proctor, A., Systems approach to improving plant efficiency, Proc Engng, Sept, 43-46 (1988).
Wensley, J.H., and Harclerode, C.S., Chemical reactor control by fault-tolerant computer, Manufacturing Chemist, June, 65-68 (1988).
Yust, L.J., and Howell, J.A., Applications of a self-tuning regulator to control dissolved oxygen in an activated sludge plant, CER&D, 66(3), 260-264 (1988).

8.5 Control Valves

1967-1980

Boger, H.W., Methods for sizing control valves, Chem Eng, 6 Nov, 247-250 (1967).
Canon, J., Guidelines for selecting process-control valves, Chem Eng, 21 April, 109-114; 5 May, 105-109 (1969).
Driskell, L.R., New approach to control valve sizing, Hyd Proc, 48(7), 131-134 (1969).
Simon, H., and Whelan, T.H., Select optimum control valve by computer, Hyd Proc, 49(7), 103-106 (1970).
Walton, P., Control valve actuators, Brit Chem Eng, 15(8), 1015-1020 (1970).
Arant, J.B., Special control valves reduce noise and vibration, Chem Eng, 6 March, 92-98 (1972).
Baumann, H.D., How to estimate pressure drop across liquid-control valves, Chem Eng, 29 April, 137-142 (1974).
Chalfin, S., Specifying control valves, Chem Eng, 14 Oct, 105-114 (1974).
Constance, J.D., Sizing steam control valves, Chem Eng, 2 Sept, 94-96 (1974).
Various, Valves (feature report), Chemical Processing, Sept, 77-106 (1974).
Constance, J.D., Piping around control valves, Chem Eng, 5 Nov, 129-131 (1979).
Durand, A.A., Sizing hot-vapors bypass valve, Chem Eng, 25 Aug, 111-112 (1980).
Hancock, V., Control-valve design, Proc Engng, Feb, 85-87 (1980).
Searle, D., Actuators, Proc Engng, April, 59-61 (1980).
Various, Valves (special report), Hyd Proc, 59(8), 61-77 (1980).

1981-1988

Baumann, H.D., Control valve vs. variable-speed pump, Chem Eng, 29 June, 81-84 (1981).

Monsen, J.F., Program sizes control valves for liquids, Chem Eng, 18 May, 159-163 (1981).

Various, Valves survey (topic issue), Processing, Dec, 9-21 (1981).

Mukaddam, W.M., Pneumatic memory smoothes valve opening, Chem Eng, 28 June, 115 (1982).

Pechey, R., Matching valves and actuators, Proc Engng, Feb, 31-33 (1982).

Fordham, R., Specifying control valves, Proc Engng, April, 47-49 (1983).

Monsen, J.F., Program sizes control valves for gas and vapor, Chem Eng, 31 Oct, 45-49 (1983).

Overend, M., Valves: Plant and equipment survey, Processing, Feb, 17-29 (1983).

Various, Control valves (feature report), Chem Eng, 5 Sept, 94-110 (1983).

Adams, M., Don't overspecify control valves, Chem Eng, 29 Oct, 121-124 (1984).

Adams, M., and Boyd, D., Control valves: Time for review, Hyd Proc, May, 87-91 (1984).

Anon., Control valves survey, Processing, Jan, 25-27 (1984).

Choquette, A.E., Evaluating fire-tested valves: A critical look, Hyd Proc, July, 85-87 (1984).

Fagerlund, A., Predicting aerodynamic noise from control valves, Chem Eng, 14 May, 65-67 (1984).

Symalla, M., How to effectively select fire-safe valves and actuators, Hyd Proc, July, 83-84 (1984).

Anon., Valves: A survey, Processing, Feb, 10-16 (1985).

Beevers, A., Guide to actuators, Proc Engng, Dec, 59 (1985).

Merrick, R.C., A guide to selecting manual valves, Chem Eng, 1 Sept, 52-64 (1986).

Montana, W.M., Butterfly valves for control, Chem Eng, 3 March, 123-126 (1986).

Connell, J.R., Realistic control-valve pressure drops, Chem Eng, 28 Sept, 123-127 (1987).

Driskell, L., Select the right control valve for difficult service, Chem Eng, 17 Aug, 123-127 (1987).

Wood, A., Developments in valve technology, Processing, Jan, 19-27 (1987).

Anon., Expert advice on control valve selection, Chem Engnr, Jan, 27-30 (1988).

Anon., Valves and actuators update (advertising feature), Chem Eng, 9 May, 59-66 (1988).

Evans, R., Control valves for steam sterilisation, Proc Engng, Dec, 37 (1988).

8.6 Instrumentation

1967-1970

Gries, W.H., Thermocouples calibrated through simple set-up, Chem Eng, 18 Dec, 130-132 (1967).

McConnell, J.A., and Smuck, W.W., Gamma backscatter technique for level and density detection, Chem Eng Prog, 63(8), 79-82 (1967).
Regenczuk, T.J., Selecting temperature monitoring and control systems, Chem Eng, 4 Dec, 202-208 (1967).
Angel, D., Chemical engineering aspects of integrated circuits, Chem Eng Prog, 64(5), 90-95 (1968).
Bartz, A.M., and Ruhl, H.D., Process control via infrared analyzers, Chem Eng Prog, 64(8), 45-49 (1968).
Brown, J.E., Onstream process analyzers, Chem Eng, 6 May, 164-176 (1968).
Buckley, C.F., et al., In-line chromatograph analyzer solves process control problems, Chem Eng Prog, 64(8), 50-52 (1968).
Considine, D.M., Process instrumentation, Chem Eng, 29 Jan, 84-113; 12 Feb, 137-144 (1968).
Krell, G.H., Sensing pressure through diaphragms, Chem Eng, 1 July, 87-92 (1968).
Maier, H.F., Infrared radiometry for temperature control, Chem Eng, 7 Oct, 188-196 (1968).
Various, Contents gauging and level indication (Supplement issue), Brit Chem Eng, 13(9), 1-14 (1968).
Weiland, R.H., Simple level-controller works over wide range of flowrates, Chem Eng, 12 Feb, 162 (1968).
Rowe, S., and Cook, H.L., Nuclear gages for density and level control, Chem Eng, 27 Jan, 159-166 (1969).
Rowton, E.E., Digital meters for process instruments, Chem Eng, 9 Feb, 111-113 (1970).
Topham, W.H., Simple plant chromatograph for process control in the petroleum industry, Brit Chem Eng, 15(2), 209-212 (1970).

1971-1975

Marcus, G.M., Precise control with ordinary temperature indicator, Chem Eng, 19 April, 134-136 (1971).
Anyakora, S.N., and Lees, F.P., Detection of instrument malfunction by the process operator, Chem Engnr, Aug, 304-309 (1972).
Elsworth, R., The value and use of dissolved oxygen measurement in deep culture, Chem Engnr, Feb, 63-71 (1972).
Masek, J.A., Location of thermowells, Hyd Proc, 51(4), 147-151 (1972).
Standen, G., and Fortier, R., Fluidic controls, Chemtech, Oct, 620-626 (1972).
Various, Process analyzers (special report), Hyd Proc, 51(2), 69-84 (1972).
Various, Analytical monitoring of processes (topic issue), Chem Eng Prog, 68(10), 39-55 (1972).
Various, Instrumentation and process control (deskbook issue), Chem Eng, 11 Sept, 7-140 (1972).
Lawford, V.N., How to select liquid-level instruments, Chem Eng, 15 Oct, 109-118 (1973).
Lees, F.P., Some data on failure modes of instruments in chemical plant environment, Chem Engnr, Sept, 418-421 (1973).
Picker, S., Measuring levels of freezing liquids, Chem Eng, 19 Feb, 146 (1973).
Winn, B.M., Batch temperature control near freezing point, Chem Eng, 16 April, 124 (1973).

Boyer, F.E., Current views on fluidics technology, Chem Eng, 9 Dec, 117-124 (1974).
Dolan, T.J., Traversing thermocouple system, Chem Eng, 25 Nov, 118 (1974).
Harshe, B.L., Simplified approach to estimating instrument accuracy, Chem Eng, 18 March, 93-96 (1974).
Various, Level measurement and control (feature report), Chemical Processing, July, 57-65 (1974).
Foster, R.A., Guidelines for selecting online process analyzers, Chem Eng, 17 March, 65-72 (1975).
Liptak, B.G., Higher profits via advanced instrumentation, Chem Eng, 23 June, 152-162 (1975).
Pierre, A.F.S., Plant instrument zeroing by process computer, Chem Eng, 10 Nov, 223-224 (1975).
Ryan, J.B., Pressure control, Chem Eng, 3 Feb, 63-68 (1975).
Various, Instrumentation and measurement techniques (topic issue), Chem Engnr, May, 297-315 (1975).
Warren, C.W., Interpreting instrument flow sheets, Hyd Proc, 54(7), 163-165; 54(9), 191-193 (1975).

1976-1980

Flanagan, T.P., Choosing the right instruments, Chem Engnr, March, 178-179 (1976).
Hoyle, D.L., Designing for pH control, Chem Eng, 8 Nov, 121-126 (1976).
Various, Process-plant control and instrumentation (feature report), Processing, April, 45-60 (1976).
Various, Perspective on instrumentation and process control (feature report), Chem Eng, 21 June, 128-166 (1976).
Wallace, L.M., Sighting in on level instruments, Chem Eng, 16 Feb, 95-104 (1976).
Balasubramanian, G.R., and Sivasankaran, K., How to design a metering system, Chem Eng, 29 Aug, 96 (1977).
Kardos, P.W., Response of temperature-measuring elements, Chem Eng, 29 Aug, 79-83 (1977).
Lieberman, N.P., Plant instrumentation for smooth operation, Chem Eng, 12 Sept, 140-154 (1977).
McDonough, R., Selecting sight flow indicators, Chem Eng, 4 July, 113-116 (1977).
Riedmuller, G.F., Calibrate transmitters faster, Hyd Proc, 56(7), 179-180 (1977).
Hammett, J.L., Remote multiplexing, Hyd Proc, 57(9), 253-254 (1978).
Jones, C., and Reed, J.E., Near-infrared analyzers refine process control, Chem Eng, 9 Oct, 111-114 (1978).
McWhorter, E.W., Correct analyzer steam-tracing, Hyd Proc, 57(12), 125-128 (1978).
Various, On-line analysis (feature report), Processing, Sept, 69-91 (1978).
Anderson, G., Zero suppression and elevation primer, Chem Eng, 4 June, 155-158 (1979).
Astbury, G.R., Sampling lines under vacuum or pressure, Chem Engnr, June, 451 (1979).

Garrett, L.T., Multiplexing systems, Hyd Proc, 58(3), 145-149 (1979).
Raghavachari, S., Quick tips on instrument installation, Chem Eng, 22 Oct, 159-160 (1979).
Schenck, A.E., Automatic pressure/vacuum water seal, Chem Eng, 30 July, 98 (1979).
Stanton, B.D., and Sterling, M.A., Better way to control composition, Hyd Proc, 58(11), 275-280 (1979).
Subramanian, P., and Arunachalam, V.R., An easy-to-make viscometer, Chem Eng, 12 Feb, 138 (1979).
Various, Instrumentation and process control (deskbook issue), Chem Eng, 15 Oct, 11-160 (1979).
Walls, W.R., High integrity flow metering, Chem Engnr, May, 331-333 (1979).
Bobeck, R.F., Electrochemical sensors for oxygen analysis, Chem Eng, 14 July, 113-117 (1980).
Cheung, T.F., and Luyben, W.L., Nonlinear and nonconventional liquid-level controllers, Ind Eng Chem Fund, 19(1), 93-98 (1980).
Kletz, T.A., Plant instruments: Which ones don't work and why, Chem Eng Prog, 76(7), 68-71 (1980).
Lazenby, B., Level monitoring and control, Chem Eng, 14 Jan, 88-96 (1980).
Martin, R.E., Hydratect: Detection of the presence or absence of water in steam plant, Chem Engnr, Nov, 687-690 (1980).
Medlock, R.S., State of the art and trends in industrial instrumentation, Chem Engnr, Nov, 679-682 (1980).
Picker, S., A substitute for limit switches, Chem Eng, 7 April, 116 (1980).
Various, Process analyzer systems (special report), Hyd Proc, 59(6), 45-68 (1980).
Verdin, A., Process analysers, Chem Engnr, Nov, 683-686, 690 (1980).

1981-1982

Bartran, D.S., Testing the response of online analyzers, Chem Eng, 9 March, 115-116 (1981).
Cobb, R.F., Portable gage to measure average specific gravity, Chem Eng, 9 Feb, 128 (1981).
Gumtz, G.D., and Gray, D.M., A closer look at pH estimates, Chem Eng, 29 June, 111-112 (1981).
Howe, R.H.L., and Howe, R.C., Avoid errors in averaging pH values, Chem Eng, 6 April, 109 (1981).
Masek, J.A., Designing pressure-gage connections, Chem Eng, 4 May, 71-79 (1981).
Mowery, R.A., Online process liquid chromatography, Chem Eng, 18 May, 145-152 (1981).
Tarbutton, J., Selecting easy-to-maintain analytical instrumentation, Chem Eng, 28 Dec, 85-87 (1981).
Vannah, W.E., and Calder, W., Effects of service conditions on electronic instrumentation, Chem Eng, 7 Sept, 143-146 (1981).
Asher, R.C., Ultrasonic techniques for non-invasive instrumentation on chemical and process plant, Chem Engnr, July, 317-319 (1982).
Faries, G., and Gross, W., Instrument design for economic installation, Hyd Proc, Feb, 101-104 (1982).

Masek, J.A., Installing pressure switches, Chem Eng, 15 Nov, 113-120 (1982).
Raghavachari, S., Instrument-maintenance hints, Chem Eng, 18 Oct, 134 (1982).
Settle, F.A.; Pleva, M.A., and Jackson, L.L., Instrument selection, Chemtech, July, 444-448 (1982).
Various, Pressure measurement survey, Processing, July, 31-39 (1982).
Various, Temperature control survey, Processing, Jan, 11-21 (1982).
Various, On-line instrumentation survey, Processing, May, 28-43 (1982).
Various, Applying new process control instrumentation (special report), Hyd Proc, Aug, 67-86 (1982).
Vivona, M.A., Use telemetry for monitoring and control, Chem Eng, 8 March, 91-95 (1982).

1983-1984
Adb-El-Bary, M.F., Direct digital-control liquid-level experiment, Chem Eng Educ, 17(1), 28-31,47 (1983).
Clarke, J.R.P., Analytical instruments: A survey, Processing, June, 25-33,51 (1983).
Dooley, G.D., and Szybist, D.J., Accessing the analog world, Chem Eng, 22 Aug, 81-83 (1983).
Kennedy, R.H., Selecting temperature sensors, Chem Eng, 8 Aug, 54-71 (1983).
Kimmel, E., Temperature-sensing materials, Chem Eng, 5 Sept, 125-138 (1983).
Langdon, R.M., Vibrating transducers for chemical plant, Chem Engnr, Nov, 33-35 (1983).
Love, J., pH control using a DDC system, Chem Engnr, July, 24-27 (1983).
Masek, J.A., Improved instrument valves for pressure switches and pressure gages, Chem Eng, 18 April, 89-93 (1983).
Nair, V., Pressure measurement and control, Processing, Oct, 22-27 (1983).
Pitt, G.D., Fibre-optic devices for process control, Proc Engng, Oct, 30-33 (1983).
Smith, R., Electronic control: A survey, Processing, July, 9-23 (1983).
Thomas, P., Temperature control in batch reactors, Processing, April, 19-25 (1983).
Anon., Level controllers, Processing, Oct, 25-27 (1984).
Anon., Instrumentation and control systems (special advertising section), Chem Eng, 25 June, 57-134 (1984).
Bajek, W.A.; Kuchar, P.J., and Remec, A.A.B., How boiling point monitors can increase profits, Hyd Proc, March, 77-81 (1984).
Charlton, S., The application of nucleonic instruments in measurement and control, Chem Engnr, Aug, 49-51 (1984).
Dealy, J.M., Viscometers for online measurement and control, Chem Eng, 1 Oct, 62-70 (1984).
Ditcham, S., A practical guide to liquid density transducers, Chem Engnr, July, 29-31 (1984).
Gordon, L.M., Scaling converts process signals to instrument ones, Chem Eng, 25 June, 141-146 (1984).
Hoeppner, C.H., Online measurement of liquid density, Chem Eng, 1 Oct, 71-78 (1984).
Love, J., The management of instrumentation and control projects, Chem Engnr, May, 17-21 (1984).

Peate, J., Liquid-level measurement and instrumentation, Proc Engng, Nov, 70-71 (1984).
Petersen, H.J.S., and Haastrup, P., Fault-tree evaluation shows importance of testing instruments and controls, Chem Eng, 26 Nov, 85-87 (1984).
Priestman, G.H., and Tippetts, J.R., Development and potential of power fluidics for process flow control (review paper), CER&D, 62, 67-80 (1984).
Puzniak, T.J., On-line analysis for process optimization and control, Chem Eng Prog, 80(8), 29-32 (1984).
Various, Temperature control survey, Processing, Oct, 11-16 (1984).
Various, Process analyser survey, Processing, June, 5-9 (1984).
Warmoth, D., Electrolytic conductivity: Its use in control, Chem Engnr, Dec, 18-20 (1984).

1985
Anon., Pressure measurement survey, Processing, March, 11-16 (1985).
Bartran, D.S., and Nelson, K., Noncontact temperature measurement in kilns, Chem Eng, 24 June, 65-66 (1985).
Borer, J., Instrumentation and Control for the Process Industries, Elsevier, New York (1985).
Bown, M., Theory and application of electrolytic conductivity measurement, Proc Engng, May, 73-76 (1985).
Briscoe, B., and Mahgerefteh, H., Continuous gas density measurement at high temperature and pressure, Chem Engnr, March, 28-29 (1985).
Crocker, R., Ultrasound for diagnosis and measurement, Chem Engnr, Dec, 27-30 (1985).
Demorest, W.J., Pressure measurement, Chem Eng, 30 Sept, 56-68 (1985).
Hampartsoumian, E., and Williams, A., Principles and applications of fibre optic sensors for process instrumentation and control, J Inst Energy, 58, 159-168 (1985).
Higham, E., Optical fibres as sensors, Chem Engnr, June, 45-47 (1985).
Lawson, F.J., Process-signal transmitters, Chem Eng, 1 April, 73-76 (1985).
Leigh, R., Temperature-measuring equipment, Proc Engng, Dec, 48-51 (1985).
May, D.L., Plant instrumentation, Chem Eng, 4 March, 67-71 (1985).
Narayanaswamy, R., Analytical instruments update, Processing, Aug, 27-38 (1985).
Rogers, N., On-line sodium ion analysis, Chem Engnr, March, 32-35 (1985).
Seebold, J.G., Tube skin thermocouples, Chem Eng Prog, 81(12), 57-59 (1985).
Smith, C.L., Smart instrumentation (special advertising report), Chem Eng, 27 May, 53-110 (1985).
Stein, T.N., Select thermowell dimensions to avoid vibration resonance from gas flow, Chem Eng, 22 July, 84-86 (1985).
Turner, B., High-accuracy measurement of process temperature, Proc Engng, Feb, 63-67 (1985).

1986
Bartran, D.S., and Nelson, K., Noncontact temperature measurement, Chem Eng, 6 Jan, 90 (1986).
Challoner, A., Sensing the pressure, Proc Engng, Aug, 36-39 (1986).

Cheremisinoff, N.P., Instrumentation for Complex Fluid Flows, Technomic Publishing Co., Pennsylvania (1986).
Clevett, K.J., Process Analyzer Technology, Wiley, New York (1986).
DeVries, E.A., Experience with ultrasonic flowmeters, Hyd Proc, 65(6), 65-66 (1986).
Earle, W.; Hall, F.; Williams, R.A., and Villalobos, R., Analyzer communications with control systems, Chem Eng Prog, 82(12), 40-43 (1986).
Fetty, C., Intrinsically safe temperature measurement, Chem Eng, 3 March, 95-97 (1986).
Gordon, L.M., Simple optimization for dual composition control, Hyd Proc, 65(6), 59-62 (1986).
Hulley, B.J., Measuring pH on-line, Proc Engng, Sept, 67-68 (1986).
Lihou, D., Case studies of inadequate instrumentation, Chem Engnr, Nov, 41 (1986).
Liptak, B.G. (Ed.), Instrument Engineers' Handbook: Process Control Revised Edition, Chilton Book Co., Pennsylvania (1986).
Liptak, B.G., On-line instrumentation and process control (special advertising section), Chem Eng, 31 March, 49-96 (1986).
McDonald, K.A.; McAvoy, T.J., and Tits, A., Optimal averaging level control, AIChEJ, 32(1), 75-86 (1986).
Morris, A., A fresh approach to plant instrumentation, Proc Engng, Jan, 32-35 (1986).
Peacock, G.R., Commissioning and verifying radiation thermometers, Chem Eng, 23 June, 103-108 (1986).
Smith, L., Self-tuning and adaptive tuners for temperature control, Processing, Dec, 11,13 (1986).
Tayler, C., Process control electronics, Proc Engng, June, 36-43 (1986).
Turner, R., Process temperature controllers, Chem Engnr, June, 83-85 (1986).
Various, Analytical instruments update, Processing, Sept, 47-55 (1986).
Yeh, J.T.Y., Online composition analysers, Chem Eng, 20 Jan, 55-68 (1986).
Zanetti, R., IR thermometers zero in on new CPI targets, Chem Eng, 21 July, 22-25 (1986).

1987

Anon., Temperature measurement update, Processing, Oct, 25-35 (1987).
Beevers, A., Pressure transducers, Processing, March, 37-42 (1987).
Fowler, T.J., Acoustic emission testing of vessels and piping, Chem Eng Prog, 83(5), 25-32 (1987).
Haskins, D.E.; Wick, D.J., and Oburn, H.W., Successful handling of instrumentation-modernization projects, Chem Eng, 19 Jan, 84-94 (1987).
Liedy, W., Development of equation for calibration curve of infrared gas analyzer, Chem Eng & Proc, 22(2), 117-120 (1987).
Lipowicz, M., Process-control software, Chem Eng, 8 Dec, 129-131 (1987).
Liptak, B.G., Pressure regulators, Chem Eng, 13 April, 69-76 (1987).
Mascone, C.F., Flowmetering gets boost from novel designs, Chem Eng, 16 Feb, 37-39 (1987).
Reeves, P., Direct in-process analysis using optical fibre technology, Proc Engng, Nov, 67-70 (1987).

Tayler, C., A multipurpose ultrasonic level meter, Proc Engng, March, 47-51 (1987).
Varey, P., Purifying brine: On-line analysis, Chem Engnr, Sept, 22-23 (1987).
Watmough, N., and Flower, J., Ultrasonics for level measurement, Processing, Dec, 23-24 (1987).

1988

Anon., Instrumentation and process control 1988 (special advertising section), Chem Eng, 14 March, 77-122 (1988).
Balcombe, A., New standards for pressure measurement, Processing, Feb, 22-29 (1988).
Christian, G.D., and Ruzicka, J., Analysis technique for on-line control, Chem Eng, 28 March, 57-60 (1988).
Early, P.L., Hydrostatic tank gauging control and instrumentation, Hyd Proc, 67(11), 89-91 (1988).
Fryer, T., Developments in mass flowmeters, Processing, March, 17-24 (1988).
Gay, P., Continuous gas analysis optimises acid removal, Processing, Jan, 24-25 (1988).
Hazlewood, E., Removing electrical noise from instrumentation, Chem Eng, 21 Nov, 105-108 (1988).
Kern, A.G., Simplify batch temperature control, Chem Eng, 28 March, 61-63 (1988).
Milnes, J., Measurement of pH, Processing, May, 33-34 (1988).
Pathak, V.K., and Rattan, I.S., Turndown limit sets heater control, Chem Eng, 18 July, 103-105 (1988).
Proctor, A., Ultrasonic systems for level measurement, Proc Engng, Nov, 73, 76 (1988).
Royse, S., On-line viscometry for process control, Proc Engng, Dec, 53 (1988).
Various, Temperature measurement update, Processing, Nov, 41-52 (1988).

CHAPTER 9

MATHEMATICAL METHODS

9.1 Modeling and Simulation

1967-1976

Koenig, D.M., Invariant imbedding: New design method in unit operations, Chem Eng, 11 Sept, 181-184 (1967).

Tierney, J.W., and Bruno, J.A., Equilibrium stage calculations, AIChEJ, 13(3), 556-563 (1967).

Murrill, P.W.; Pike, R.W., and Smith, C.L., Dynamic mathematical models, Chem Eng, 9 Sept, 117-120; 7 Oct, 177-182; 18 Nov, 165-169; 16 Dec, 103-106 (1968).

Murrill, P.W.; Pike, R.W., and Smith, C.L., Dynamic mathematical models, Chem Eng, 27 Jan, 167-172; 24 Feb, 105-108; 10 March, 111-116; 7 April, 151-154; 19 May, 195-200; 16 June, 97-100; 28 July, 139-142; 25 Aug, 125-128 (1969).

Vajta, L., and Valoczy, I., Application of network analysis on modelling of a chemical industry complex, Periodica Polytechnica, 13, 1-8 (1969).

Tomich, J.F., New simulation method for equilibrium-stage processes, AIChEJ, 16(2), 229-232 (1970).

Chung, B.T.F.; Fan, L.T., and Hwang, C.L., Surface renewal and penetration models in the transient state, AIChEJ, 17(1), 154-160 (1971).

Naphtali, L.M., and Sandholm, D.P., Multicomponent separation calculations by linearization, AIChEJ, 17(1), 148-153 (1971).

Sinibaldi, F.J.; Koehler, T.L., and Bobis, A.H., Transformed data simplify and confirm math models, Chem Eng, 17 May, 139-146 (1971).

Orbach, O.; Crowe, C.M., and Johnson, A.I., Multicomponent separation calculations by modified Newton's method, Chem Eng J, 3(2), 176-186 (1972).

Umeda, T., and Nishio, M., Comparison between sequential and simultaneous approaches in process simulation, Ind Eng Chem Proc Des Dev, 11(2), 153-160 (1972).

Hutchison, H.P., Plant simulation by linear methods, Trans IChemE, 52, 287-290 (1974).

Aris, R., Mathematical modelling, Chem Eng Educ, 10(3), 114-124 (1976).

1977-1983

Browne, D.W.; Ishii, Y., and Otto, F.D., Solving multicolumn equilibrium stage operations by total linearization, Can JCE, 55, 307-315 (1977).

Gunn, D.J., Analysis of physical separations by staged processes at steady state, Chem Eng Sci, 32(1), 19-22 (1977).

Huber, W.F., Matrix method for distillation, Hyd Proc, 56(8), 121-125 (1977).

Murphy, T.D., Design and analysis of industrial experiments, Chem Eng, 6 June, 168-182 (1977).

Mann, U.; Rubinovitch, M., and Crosby, E.J., Characterization and analysis of continuous recycle systems, AIChEJ, 25(5), 873-882 (1979).

Parlagh, G.; Szekely, G., and Racz, G., Mathematical simulation of continuous gas chromatography, Periodica Polytechnica, 20, 205-222 (1976); 23, 75-86 (1979).

Tatrai, F.; Barkai, L., and Vamos, E., Analysis of chemical technological networks, Periodica Polytechnica, 24, 189-200 (1980).

Thambynayagam, R.K.M.; Winter, P., and Branch, S.J., Film penetration model for mass and heat transfer with high fluxes, Trans IChemE, 58, 277-280 (1980).

Wasek, K., and Socha, J., Steam stripper calculation by simplified Newton-Raphson, technique, Chem Eng Sci, 35(3), 623-626 (1980).

Do, D.D., and Bailey, J.E., Useful representations of series solutions for transport problems, Chem Eng Sci, 36(11), 1811-1818 (1981).

Guy, J.L., Solving the mathematical models for dynamic systems, Chem Eng, 16 Nov, 271-277 (1981).

Li, M.C., and Frost, R.J., Flexible solution method for generalized equilibrium stage columns, Can JCE, 59, 388-394 (1981).

Guy, J.L., Modelling process systems via digital computers, Chem Eng, 8 March, 97-103 (1982).

Hittle, D.C., and Bishop, R., Improved root-finding procedure for use in calculating transient heat flow through multilayered slabs, Int J Heat Mass Trans, 26(11), 1685-1694 (1983).

1984-1988

Cussler, E.L., Teaching mass transfer, Chem Eng Educ, 18(3), 124-127, 149-152 (1984).

Dunne, A., and Lacey, L., Model discrimination: Posterior probabilities, Can JCE, 62(4), 535-540 (1984).

Kutateladze, S.S., The mixing length hypothesis in the turbulence theory, Int J Heat Mass Trans, 27(11), 1947-1952 (1984).

Stewart, W.E., Simulation and estimation by orthogonal collocation, Chem Eng Educ, 18(4), 204-212 (1984).

Ravichandran, V., A modelling problem, Chem Eng Educ, 19(3), 140-142 (1985).

Shacham, M., Comparing software for the solution of systems of nonlinear algebraic equations arising in chemical engineering, Comput Chem Eng, 9(2), 103-112 (1985).

Duran, M.A., and Grossmann, I.E., A mixed-integer nonlinear programming algorithm for process systems synthesis, AIChEJ, 32(4), 592-606 (1986).

Bonvin, D., and Mellichamp, D.A., Scaling procedure for structural and interaction analysis of dynamic models, AIChEJ, 33(2), 250-257 (1987).

Johnston, P.R., and Do, D.D., New method for solving a large class of heat and mass transfer problems, Chem Eng Commns, 49, 247-272 (1987).

Krishna, R., Unified theory of separation processes based on irreversible thermodynamics, Chem Eng Commns, 59, 33-64 (1987).

Klein, G.F.,, Modeling three-variable equations painlessly, Chem Eng, 15 Feb, 110 (1988).

Riggs, J.B., A systematic approach to modeling, Chem Eng Educ, 22(1), 26-29 (1988).

Whitaker, S., Levels of simplification in engineering analysis, Chem Eng Educ, 22(2), 104-108 (1988).

9.2 Statistics

1967-1974

Balasubramanian, S.; Shantarum, R., and Doraiswamy, L.K., Statistical approach to Indanthrone production, Brit Chem Eng, 12(3), 377-380 (1967).

Avriel, M., and Wilde, D.J., Engineering design under uncertainty, Ind Eng Chem Proc Des Dev, 8(1), 124-131 (1969).

Lee, S.M., and White, H.R., Analyzing time-based data, Chem Eng, 20 Oct, 104-108 (1969).

Pavelic, V., and Saxena, U., Basics of statistical experimental design, Chem Eng, 6 Oct, 175-180 (1969).

Isaacson, W.B., Statistical analyses for multivariable systems, Chem Eng, 29 June, 69-75 (1970).

Mapstone, G.E., Tie-line interpolation and checking, Brit Chem Eng, 15(6), 778-779 (1970).

Steiger, F.H., Practical applications of the Weibull distribution function, Chemtech, April, 225-231 (1971).

Umeda, T.; Nishio, M., and Komatsu, S., Method for plant data analysis and parameters estimation, Ind Eng Chem Proc Des Dev, 10(2), 236-243 (1971).

Gibilaro, L.G., and Drinkenburg, A.A.H., Numerical evaluation of mean and variance from the Laplace transform, Chem Eng Sci, 27(2), 445-447 (1972).

Gupta, A.S., A quick visual summary of experimental results, Chem Eng, 27 Nov, 110-112 (1972).

Michelsen, M.L., Least-squares method for residence-time distribution analysis, Chem Eng J, 4(2), 171-179 (1972).

Nogita, S., Statistical test and adjustment of process data, Ind Eng Chem Proc Des Dev, 11(2), 197-200 (1972).

Hahn, G.J., Coefficient of correlation, Chemtech, Oct, 609-612 (1973).

Jackisch, P.F., Shortcuts to small-sample statistics problems, Chem Eng, 11 June, 107-110 (1973).

Churchill, S.W., and Usagi, R., Standardized procedure for production of correlations in the form of common empirical equations, Ind Eng Chem Fund, 13(1), 39-44 (1974).

Felder, R.M.; Harrison, R.E., and Rousseau, R.W., Parameter estimation by curve-fitting techniques and method of moments, Chem Eng Commns, 1(4), 187-190 (1974).

Hahn, G.J., Random sampling and statistical significance, Chemtech, Jan, 16-17, 55 (1974).

Kleinfeld, J.M., Easier, more accurate interpolation of graphs, Chem Eng, 15 April, 116 (1974).

Kowalski, B.R., Pattern recognition techniques for predicting performance, Chemtech, May, 300-304 (1974).

Mitton, P.B., Statistical examination of testing methods, Chemtech, April, 248-253 (1974).

Rowe, P.N., Correlating data, Chemtech, Jan, 9-14 (1974).

Sherwood, T.K., Treatment and mistreatment of data, Chemtech, Dec, 736-740 (1974).

1975-1979

Hahn, G.J., Random samplings, Chemtech, Jan, 61-62; March, 186-187 (1975).
Kim, Y.G., Use of statistical methods in chemical engineering, Int Chem Eng, 15(2), 319-325 (1975).
Hah, G.J., The normal distribution, Chemtech, Aug, 530-532 (1976).
Hahn, G.J., Evaluating probabilities, Chemtech, Nov, 712-713 (1976).
Hahn, G.J., Sample sizes in statistics, Chemtech, Feb, 142-143 (1976).
Rowe, P.N., Correlating data, Chemtech, April, 266-270 (1976).
Hahn, G.J., Randomization, Chemtech, Oct, 630-632 (1977).
Hahn, G.J., Beware of nonindependent observations, Chemtech, Feb, 117-118 (1977).
Hahn, G.J., Estimating sources of variability, Chemtech, Sept, 580-582 (1977).
Lemcoff, N.O., Pseudo nonlinear regression, Chem Eng J, 13(1), 71-74 (1977).
Tettamanti, K.; Stomfai, R.; Kemeny, S., and Manczinger, J., Application of weighted regression, Periodica Polytechnica, 21, 333-343 (1977).
Weiner, P.H., Solve problems by factor analysis, Chemtech, May, 321-328 (1977).
Campbell, J.R., and Alonso, J.R.F., Getting curves from data points, Hyd Proc, 57(1), 123-126 (1978).
Hahn, G.J., Hazards of extrapolation, Chemtech, Nov, 699-701 (1978).
Hahn, G.J., Comparisons and statistics, Chemtech, May, 317-318 (1978).
Hahn, G.J., Randomization, Chemtech, March, 164-168 (1978).
Zanker, A., Use of nomographs, Proc Engng, June, 108-111 (1978).
Hahn, G.J., Smoothing data, Chemtech, Aug, 492-493 (1979).
Hahn, G.J., Hazards of extrapolation, Chemtech, Jan, 46-49 (1979).
Hahn, G.J., Sample size determines precision, Chemtech, May, 294-295 (1979).
Hendrix, C.D., Experimental design, Chemtech, March, 167-174 (1979).
Volk, W., Calculator programs to correlate data, Chem Eng, 23 April, 128-132; 4 June, 133-138; 10 Sept, 131-135; 19 Nov, 149-153; 17 Dec, 93-98 (1979).

1980-1984

Brosens, J.R., Correlating the hyperbolic function, Chem Eng, 7 April, 97-99 (1980).
Chapin, J.F., Calculator program for normal and log-normal distributions, Chem Eng, 15 Dec, 75-78 (1980).
Hahn, G.J., Retrospective studies versus planned experimentation, Chemtech, June, 372-373 (1980).
Hahn, G.J., Planning experiments: An annotated bibliography, Chemtech, Jan, 36-39 (1980).
Hurlburt, H.Z., Curve fitting by computer, Hyd Proc, 59(8), 107-110 (1980).
Ingels, R.M., Analyzing test data by computer regression analysis, Chem Eng, 11 Aug, 145-156 (1980).
Kollar-Hunek, K.; Hunek, J., and Sawinsky, J., Use of spline functions to analyze countercurrent extraction processes, Periodica Polytechnica, 24, 259-268 (1980).

Machej, K., Use of transformation in method of least squares, Int Chem Eng, 20(1), 110-117 (1980).
Payne, C.S., Program correlates data, Chem Eng, 28 July, 75-77 (1980).
Copulsky, W.; Flicek, J., and Wang, D., Regression analysis on a hand calculator, Chem Eng, 29 June, 86-87 (1981).
Djukic, D.S., and Atanackovic, T.M., Least squares method: Kantorovich approach, Int J Heat Mass Trans, 24(3), 443-448 (1981).
Prahl, W.H., Fitting linear equations to sets of experimental data, Chem Eng, 10 Aug, 85-88 (1981).
Thomas, W.B., Calculate statistics from a histogram, Chem Eng, 27 July, 95-96 (1981).
Vatuvuk, W.M., Interpolating on logarithmic coordinates, Chem Eng, 9 March, 118 (1981).
Clare, B.W., Nonlinear regression on pocket calculator, Chem Eng, 23 Aug, 83-89 (1982).
Clare, B.W., Polynomial regression on a pocket calculator, Chem Eng, 4 Oct, 121-124 (1982).
Fisher, C.H., and Huddle, B.P., Reciprocal equations: Four basic components and curve fitting, Chem Eng, 8 Feb, 99-101 (1982).
Hahn, G.J., Statistical comparison of on-line process alternatives, Chemtech, Dec, 741-743 (1982).
Hahn, G.J., Demonstrating performance with 'high statistical confidence', Chemtech, May, 286-289 (1982).
Kratochvil, B., and Taylor, J.K., Sampling for chemical analysis, Chemtech, Sept, 564-570 (1982).
Weisz, P.B., Correlations, Chemtech, March, 180-181 (1982).
Clare, B.W., Curve fitting via orthogonal polynomials, Chem Eng, 18 April, 85-88 (1983).
Hendrix, C.D., Evaluating numerical data, Chemtech, Oct, 598-605 (1983).
Pogany, G.A., You can still go wrong using statistics, Chemtech, Feb, 90-93 (1983).
Prahl, W.H., Regression lines through a given point, Chem Eng, 2 May, 65-66 (1983).
Taitel, Y., and Tamir, A., Avoiding unwarranted inflection points in fitting of data, AIChEJ, 29(1), 153-157 (1983).
Bennett, W.R., Program performs polynomial regression, Chem Eng, 30 April, 65-68 (1984).
Kuns, R.C. Multiple regression on programmable calculator, Chem Eng, 1 Oct, 85-89 (1984).
Totman, R.S., Rapid curve-fitting technique, Chem Eng, 24 Dec, 63-65 (1984).
Viswanathan, K., Curve-fitting monotonic functions, AIChEJ, 30(4), 657-660 (1984).

1985-1988

Graham, B.P., and Jutam, A., Time series analysis, Chem Eng Educ, 19(4), 186-189 (1985).
Jonas, L.A.; Sansone, E.B., and Conlon, J.C., Regression equations forced through a coordinate point, Chem Eng, 27 May, 129-131 (1985).

Miller, R.E., Statistics for chemical engineers, Chem Eng, 23 July, 40-44; 17 Sept, 111-115; 26 Nov, 89-93 (1984); 21 Jan, 107-110; 18 March, 173-178; 10 June, 85-90 (1985).
Miller, R.E., Statistics for chemical engineers (tutorial series), Chem Eng, 23 July, 40-44; 17 Sept, 111-115; 26 Nov, 89-93 (1984); 21 Jan, 107-110; 18 March, 173-178; 10 June, 85-90; 30 Sept, 71-75 (1985); 3 Feb, 77-80; 14 April, 85-88; 23 June, 113-117; 1 Sept, 73-76 (1986).
Miller, R.E., Statistics for chemical engineers, Chem Eng, 30 Sept, 71-75 (1985); 3 Feb, 77-80; 14 April, 85-88; 23 June, 113-117; 1 Sept, 73-76 (1986).
Narasimhan, S.; Mah, R.S.H.; Tamhane, A.C.; Woodward, J.W., and Hale, J.C., A composite statistical test for detecting changes of steady states, AIChEJ, 32(9), 1409-1418 (1986).
Orofino, T.A., Be careful with correlation coefficients, Chem Eng, 3 March, 129 (1986).
Cheremisinoff, N.P., Statistical regression routines on spreadsheets, Chem Eng, 17 Aug, 153-157 (1987).
Said, A.S., and Al-Ameeri, R.S., Curve fitting in science and technology, Sepn Sci Technol, 22(1), 65-84 (1987).
Said, A.S., and Al-Haddad, A.A., Curve fitting applied to relative volatility, bubble point, and dew point of ideal binary mixtures, Sepn Sci Technol, 22(4), 1199-1218 (1987).
Atkinson, N., Statistics for process analysis, Proc Engng, April, 41, 43 (1988).
Rippin, D.W.T., Statistical methods for experimental planning in chemical engineering, Comput Chem Eng, 12(2/3), 109-116 (1988).
Taylor, J.K., Taking relevant samples, Chemtech, May, 294-296 (1988).

9.3 Dimensional Analysis

1978-1984

Krug, R.R., Fitting data to dimensionless groups, Ind Eng Chem Fund, 17(4), 306-308 (1978).
Hirose, T., Dimensional analysis by dimensional constants, Int Chem Eng, 20(1), 28-36 (1980).
Quraishi, M.S., and Fahidy, T.Z., Simplified procedure for dimensional analysis employing SI units, Can JCE, 59, 563-567 (1981).
Quraishi, M.S., and Fahidy, T.Z., A generalized method of dimensional analysis, Can JCE, 61(1), 116-120 (1983).
Andrews, G.F., Dimensional analysis, Chem Eng Educ, 18(3), 112-115 (1984).

9.4 Optimization

1966-1970

Nemec, D., and Salusinszky, A.L., Linear programming, Hyd Proc, 45(10), 193-196 (1966).
Kermode, R.I., Geometric programming: A simple and efficient optimization technique, Chem Eng, 18 Dec, 97-100 (1967).

Lee, E.S., Quasilinearization in optimization: A numerical study, AIChEJ, 13(6), 1043-1051 (1967).
Aronofsky, J.S., Linear programming models, Chem Eng Prog, 64(4), 87-92 (1968).
Lee, E.S., Iterative techniques in optimization, AIChEJ, 14(6), 908-916 (1968).
Murphy, J.S., Resolving problems with OR, Chem Eng, 29 Jan, 114-118 (1968).
Norman, J.R., and Bowrey, R.C., Review of linear programming techniques, Brit Chem Eng, 13(11), 1575-1579 (1968).
Vrba, J., and Rod, V., Optimization of stepwise separating processes, Int Chem Eng, 8(4), 717-724 (1968).
Jenett, E., Experience with and evaluation of critical path methods, Chem Eng, 10 Feb, 96-106 (1969).
Mudar, P.J., Heuristic programming, Chem Eng Prog, 65(12), 20-24 (1969).
Norris, W.E., Linear programming: Advantages and shortcomings, Chem Eng, 1 Dec, 95-100 (1969).
Barneson, R.A.; Brannock, N.F.; Moore, J.G., and Morris, C., Picking optimization methods, Chem Eng, 27 July, 132-142 (1970).
Christensen, J.H., Structuring of process optimization, AIChEJ, 16(2), 177-184 (1970).
Heck, R.M., and Smith, T.G., Linear programming for process engineers, Brit Chem Eng, 15(9), 1171-1173 (1970).
Wheeler, J.M., and Aris, R., Studies in optimization, Chem Eng Sci, 25(3), 445-462 (1970).

1971-1978

Allen, D.H., Linear programming models for plant operations planning, Brit Chem Eng, 16(8), 685-691 (1971).
Umeda, T., and Ichikawa, A., Modified complex method for optimization, Ind Eng Chem Proc Des Dev, 10(2), 229-236 (1971).
Woolsey, R.E.D.; Kochenburger, G.A., and Linck, K.R., Geometric programming, Hyd Proc, 50(8), 133-134 (1971).
Adelman, A., and Stevens, W.F., Process optimization by the 'complex' method, AIChEJ, 18(1), 20-24 (1972).
Alonso, J.R.F., Dynamic programming formulation of the generalised multistage problem, Proc Tech Int, 17(12), 935-939 (1972).
Caillaud, J.B., and Padmanabhan, L., Optimization technique for a class of discrete allocation problems, Chem Eng J, 3(1), 14-21 (1972).
Lamonte, R.R., and Lederman, P.B., Uses and limitations of geometric programming, Brit Chem Eng, 17(1), 34-37 (1972).
Glass, R.W., and Bruley, D.F., REFLEX method for empirical optimization, Ind Eng Chem Proc Des Dev, 12(1), 6-10 (1973).
Keefer, D.L., SIMPAT: Self-bounding direct-search method for optimization, Ind Eng Chem Proc Des Dev, 12(1), 92-99 (1973).
Klimpel, R.R., Operations research: Decision-making tool, Chem Eng, 16 April, 103-108; 30 April, 87-94 (1973).
Luus, R., and Jaakola, T.H.I., Optimization of nonlinear functions subject to equality constraints, Ind Eng Chem Proc Des Dev, 12(3), 380-383 (1973).
Luus, R., and Jaakola, T.H.I., Optimization by direct search and systematic reduction of the size of search region, AIChEJ, 19(4), 760-766 (1973).

Allen, D.H., How to use mixed-integer programming, Chem Eng, 29 March, 114-120 (1976).
Morshedi, A.M., and Luecke, R.H., On-line optimization of stochastic processes, Ind Eng Chem Proc Des Dev, 16(4), 473-478 (1977).
Grossmann, I.E., and Sargent, R.W.H., Optimum design of chemical plants with uncertainty parameters, AIChEJ, 24(6), 1021-1028 (1978).
Schweitzer, O.R., Reverse synthesis simplifies problem solving, Chem Eng, 19 June, 111-116 (1978).

1980-1988

Berna, T.J., Locke, M.H., and Westerberg, A.W., A new approach to optimization of chemical processes, AIChEJ, 26(1), 37-43 (1980).
Hendrix, C., Practical optimization methods, Chemtech, Aug, 488-497 (1980).
Kilikas, A.C., and Hutchison, H.P., Process optimisation using linear models, Comput Chem Eng, 4(1), 39-48 (1980).
Kondelik, P., et al., Optimization of chemical engineering equipment, Int Chem Eng, 22(4), 699-729 (1982).
Schweitzer, O.R., Systematic trial-and-error problem solving, Chem Eng, 17 May, 109-110 (1982).
Locke, M.H.; Westerberg, A.W., and Edahl, R.H., Improved successive quadratic programming optimization algorithm for engineering design problems, AIChEJ, 29(5), 871-874 (1983).
Lu, C.Y., and Weisman, J., Close approximations of global optima of process design problems, Ind Eng Chem Proc Des Dev, 22(3), 391-396 (1983).
Papoulias, S.A., and Grossmann, I.E., A structural optimization approach in process synthesis: Utility systems, heat-recovery networks, and total processing systems, Comput Chem Eng, 7(6), 695-734 (1983).
Various, Large scale optimization (topic issue), Comput Chem Eng, 7(5), 567-668 (1983).
Biegler, L.T.; Grossmann, I.E., and Westerberg, A.W., Approximation techniques for process optimization, Comput Chem Eng, 9(2), 201-206 (1985).
Crosser, O.K., Adjoint variables and their role in optimal problems, Chem Eng Educ, 19(1), 68-71 (1985).
Vickery, D.J., and Taylor, R., Path-following approaches to the solution of multicomponent multistage separation process problems, AIChEJ, 32(4), 547-556 (1986).
Cuthrell, J.E., and Biegler, L.T., Optimization of differential-algebraic process systems, AIChEJ, 33(8), 1257-1270 (1987).
Jang, S.S.; Joseph, B., and Mukai, H., On-line optimization of contrained multivariable chemical processes, AIChEJ, 33(1), 26-35 (1987).
Rhinehart, R.R., and Beasley, J.D., Dynamic programming for chemical engineering applications, Chem Eng, 7 Dec, 113-119 (1987).
Ali, L., Structured programming, Chem Eng, 10 Oct, 111-116 (1988).
Bohl, A.H., Optimizing plastics formulations, Chemtech, May, 284-289 (1988).
Kocis, G.R., and Grossmann, I.E., Global optimization of nonconvex mixed-integer nonlinear programming (MINLP) problems in process synthesis, Ind Eng Chem Res, 27(8), 1407-1421 (1988).

9.5 Experimental Error

1967-1988

de Chazal, M., How to discard invalid measurements, Chem Eng, 13 Feb, 182-184 (1967).

Strauss, R., The sensitivity chart for estimating uncertainties, Chem Eng, 25 March, 112-116 (1968).

Rigby, L.V., The nature of human error, Chemtech, Dec, 712-718 (1971).

Stanecki, J.W., How many check points verify a formula? Chem Eng, 30 Oct, 138 (1972).

Karlsson, H.T., and Bjerle, I., Simple approximation of the error function, Comput Chem Eng, 4(1), 67-69 (1980).

Park, S.W., and Himmelblau, D.M., Error in the propagation of error formula, AIChEJ, 26(1), 168-170 (1980).

Rod, V., and Hancil, V., Iterative estimation of model parameters when measurements of all variables are subject to error, Comput Chem Eng, 4(1), 33-38 (1980).

Romagnoli, J.A., and Stephanopoulos, G., Rectification of process measurement data in the presence of gross errors, Chem Eng Sci, 36(11), 1849-1865 (1981).

Mah, R.S.H., and Tamhane, A.C., Detection of gross errors in process data, AIChEJ, 28(5), 828-830 (1982).

Almasy, G.A., and Mah, R.S.H., Estimation of measurement error variances from process data, Ind Eng Chem Proc Des Dev, 23(4), 779-784 (1984).

Hahn, G.J., Calculating overall error, Chemtech, Nov, 696-697 (1984).

Zanker, A., Detection of outliers by Nalimov's test, Chem Eng, 6 Aug, 74-75 (1984).

Green, J.C.A., An error analysis for complex separation systems, Comput Chem Eng, 9(2), 143-152 (1985).

Asbjornsen, O.A., Error in the propagation of error formula, AIChEJ, 32(2), 332-334 (1986).

Heenan, W.A., and Serth, R.W., Detecting errors in process data, Chem Eng, 10 Nov, 99-103 (1986).

Heenan, W.A., and Serth, R.W., Detecting errors in process data, Chem Eng, 10 Nov, 99-103 (1986).

Vatavuk, W.M., How significant are your figures? Chem Eng, 18 Aug, 97-98 (1986).

Narasimhan, S., and Mah, R.S.H., Generalized likelihood ratio method for gross error identification, AIChEJ, 33(9), 1514-1521 (1987).

Crowe, C.M., Recursive identification of gross errors in linear data reconciliation, AIChEJ, 34(4), 541-550 (1988).

Narasimhan, S., and Mah, R.S.H., Generalized likelihood ratios for gross error identification in dynamic processes, AIChEJ, 34(8), 1321-1331 (1988).

Pai, C.C.D., and Fisher, G.D., Application of Broyden's method to reconciliation of nonlinearly constrained data, AIChEJ, 34(5), 873-876 (1988).

9.6 Engineering Mathematics

1966-1969

Koenig, D.M., Numerical differentiation of experimental data, Chem Eng, 12 Sept, 186-189 (1966).

Bopp, G.R., The role of electrodynamics in transport phenomena, Chem Eng Prog, 63(10), 74-79 (1967).

Richards, J.W., Nomograms without equations, Brit Chem Eng, 12(10), 1587-1590 (1967).

Villadsen, J.V., and Stewart, W.E., Solution of boundary-value problems by orthogonal collocation, Chem Eng Sci, 22(11), 1483-1502 (1967).

Buckheimer, M., and Sterling, R.C., Mathematics for engineers: A new approach, Brit Chem Eng, 13(12), 1727-1730 (1968).

Davison, E.J., Numerical solution of large systems of linear differential equations, AIChEJ, 14(1), 46-50 (1968).

Distefano, G.P., Stability of numerical integration techniques, AIChEJ, 14(6), 946-955 (1968).

Hedstrom, B., Computation techniques in chemical engineering, Brit Chem Eng, 13(4), 528-531 (1968).

Wong, A., Practical method of constructing nomographs, Chem Eng, 21 Oct, 176 (1968).

Finlayson, B.A., Applications of the method of weighted residuals and variational methods, Brit Chem Eng, 14(1), 53-57; 14(2), 179-182 (1969).

Johnson, E.E., Curve fitting easily done with hyperbolic equation, Chem Eng, 11 Aug, 122-126 (1969).

Seinfeld, J.H., Nonlinear estimation and identification of parameters in partial differential equations, Chem Eng Sci, 24(1), 65-84 (1969).

1970-1974

Brailovskaya, I.Y.; Kuskova, T.V., and Chudov, L.A., Difference methods of solving the Navier-Stokes equations, Int Chem Eng, 10(2), 228-237 (1970).

Hanson, D.T., Linear algebra, Chem Eng, 6 April, 119-122; 18 May, 153-156; 15 June, 169-174; 27 July, 154-156; 24 Aug, 83-86; 21 Sept, 180-184 (1970).

Seinfeld, J.H.; Lapidus, L., and Hwang, M., Review of numerical integration techniques for stiff ordinary differential equations, Ind Eng Chem Fund, 9(2), 266-275 (1970).

Caillaud, J.B., and Padmanabhan, L., Improved semi-implicit Runge-Kutta method for stiff systems, Chem Eng J, 2(4), 227-232 (1971).

Hanson, D.T., Linear algebra, Chem Eng, 19 Oct, 154-158; 30 Nov, 81-86 (1970); 11 Jan, 133-138; 8 March, 117-122; 5 April, 123-126 (1971).

Kubicek, M., and Hlavacek, V., Method with third-order convergence for solution of nonlinear algebraic equations, Chem Eng J, 2(2), 100-103 (1971).

Caplan, F., Graphical solution for the general quadratic equation, Chem Eng, 4 Sept, 98 (1972).

Corbo, V.J., and Lapidus, L., An invariant imbedding difference equation algorithm, Chem Eng J, 3(1), 35-51 (1972).

Davies, I.J., and Wood, R.M., Some properties of the matrices associated with the transition behaviour of multicomponent systems, Trans IChemE, 50, 176-178 (1972).
Halsted, D.J., and Brown, D.E., Zakian's technique for inverting Laplace-transforms, Chem Eng J, 3(3), 312-313 (1972).
Hohmann, E.C., and Lockhart, F.J., The hyperbola: Curve fitting and iteration with a three-constant equation, Chemtech, Oct, 614-619 (1972).
Hwang, M., and Seinfeld, J.H., New algorithm for estimation of parameters in ordinary differential equations, AIChEJ, 18(1), 90-93 (1972).
Lockhart, F.J., Calculating log mean averages from arithmetic averages, Chem Eng, 12 June, 120 (1972).
Michelsen, M.L., and Villadsen, J., Convenient computational procedure for collocation constants, Chem Eng J, 4(1), 64-68 (1972).
Burgess, W.P., and Lapidus, L., Composite numerical solutions of partial differential equations, Chem Eng Commns, 1(1), 33-56 (1973).
Flynn, B.P., Substituting straight lines for a curve in computer calculation, Chem Eng, 19 Feb, 150 (1973).
Friedland, A.J., Estimating the peak value of a curve, Chem Eng, 19 Feb, 148 (1973).
Leung, V.P., and Padmanabhan, L., Improved estimation algorithms using smoothing and relinearization, Chem Eng J, 5(3), 197-208 (1973).
Shacham, M., and Kehat, E., Converging interval methods for iterative solution of a nonlinear equation, Chem Eng Sci, 28(12), 2187-2195 (1973).
Aiken, R.C., and Lapidus, L., Numerical integration method for typical stiff systems, AIChEJ, 20(2), 368-375 (1974).
Bosch, B.V.D., and Hellinckx, L., Estimation of parameters in differential equations, AIChEJ, 20(2), 250-255 (1974).
Dang, N.D.P., and Gibilaro, L.G., Numerical inversion of Laplace transforms by simple curve-fitting technique, Chem Eng J, 8(2), 157-163 (1974).
King, R.P., Applications of stochastic differential equations to chemical engineering problems: An introductory review, Chem Eng Commns, 1(5), 221-238 (1974).
Kubicek, M., and Visnak, K., Nonlinear explicit algorithm for efficient integration of stiff systems of ordinary differential equations, Chem Eng Commns, 1(6), 291-296 (1974).
Lynch, E.P., Using Boolean algebra and logic diagrams, Chem Eng, 19 Aug, 107-113; 16 Sept, 111-118; 14 Oct, 101-104 (1974).
Van den Bosch, B., and Hellinckx, L.J., Solution of nonlinear initial value problems by orthogonal collocation method, Chem Eng J, 7(1), 73-78 (1974).

1975-1979

Carey, G.F., and Finlayson, B.A., Orthogonal collocation on finite elements, Chem Eng Sci, 30(5), 587-596 (1975).
Levy, A.V., and Montalvo, A., Comparison of multiplier and quasilinearization methods, Ind Eng Chem Proc Des Dev, 14(4), 385-391 (1975).
Rajakumar, A., and Krishnaswamy, P.R., Time to frequency domain conversion of step response data, Ind Eng Chem Proc Des Dev, 14(3), 250-256 (1975).

Arva, P., and Szeiffert, F., Solution of boundary-value problems in unit operation calculations, Int Chem Eng, 16(1), 37-43 (1976).

Book, N.L., and Ramirez, W.F., Selection of design variables in systems of algebraic equations, AIChEJ, 22(1), 55-66 (1976).

Cuddihy, E.F., Determine area under a curve without integration, Chem Eng, 16 Feb, 120 (1976).

de Kerf, J., Interpolation methods of Sprague and Karup, Int Chem Eng, 16(1), 47-54 (1976).

Towler, B.F., and Yang, R.Y.K., Numerical solution of nonlinear parabolic PDEs by asymmetric finite-difference formulae, Chem Eng J, 12(2), 81-88 (1976).

Zanker, A., Developing empirical equations, Chem Eng, 19 Jan, 101-106 (1976).

Gabbrielli, E., and Spadoni, G., Iterative solution of large systems of linear equations: Some new empirical criteria and tests, Comput Chem Eng, 1(2), 113-124 (1977).

Michelsen, M.L., Application of semi-implicit Runge-Kutta methods for integration of ordinary and partial differential equations, Chem Eng J, 14(2), 107-112 (1977).

Rivas, O.R., Analytical solution of cyclic mass-transfer operations, Ind Eng Chem Proc Des Dev, 16(3), 400-405 (1977).

Tan, H., Calculation of J functions by pocket calculator, Chem Eng, 24 Oct, 158 (1977).

Birnbaum, I., and Lapidus, L., Studies in approximation methods (5 parts), Chem Eng Sci, 33(4), 415-470 (1978).

Carr, W.A., Sorting out yield problems, Chem Eng, 6 Nov, 119-120 (1978).

Chan, Y.N.I.; Birnbaum, I., and Lapidus, L., Solution of stiff differential equations and use of imbedding techniques (a review), Ind Eng Chem Fund, 17(3), 133-148 (1978).

Douglas, J.M., Derive short-cut correlations, Hyd Proc, 57(12), 145 (1978).

Goulcher, R., and Casares-Long, J.J., Solution of steady-state chemical engineering optimisation problems using a random-search algorithm, Comput Chem Eng, 2(1), 33-36 (1978).

Heller, H., Choosing the right formula for calculator curve fitting, Chem Eng, 13 Feb, 119-120 (1978).

Horwitz, B.A., Finding the slope of a curve on log and semi-log paper, Chem Eng, 23 Oct, 178 (1978).

Shacham, M., and Mah, R.S.H., Newton-type linearization method for solution of nonlinear equations, Comput Chem Eng, 2(1), 64-66 (1978).

White, R.E., Newman's numerical technique for solving boundary value problems, Ind Eng Chem Fund, 17(4), 367-369 (1978).

Deliquet, A., Computer program derives equation, Hyd Proc, 58(3), 140-142 (1979).

Lavery, J.E., The perils of differentiating engineering data numerically, Chem Eng, 15 Jan, 121-123 (1979).

Michelsen, M.L., Fast solution technique for a class of linear PDEs, Chem Eng J, 18(1), 59-66 (1979).

Prahl, W.H., Solving graphical problems on a calculator, Chem Eng, 22 Oct, 118-126 (1979).

Silverberg, P.M., Finding a quadratic equation, Chem Eng, 4 June, 158 (1979).
Weimer, A.W., and Clough, D.E., Critical evaluation of semi-implicit Runge-Kutta methods for stiff systems, AIChEJ, 25(4), 730-732 (1979).
Weiss, G.H., Transport equations with quadratic nonlinearities, Sepn Sci Technol, 14(3), 243-246 (1979).

1980-1982

Cash, J.R., A semi-implicit Runge-Kutta formula for the integration of stiff systems of ordinary differential equations, Chem Eng J, 20(3), 219-224 (1980).
Mah, R.S.H., and Lin, T.D., Comparison of modified Newton's methods, Comput Chem Eng, 4(1), 75-78 (1980).
Van Zee, J.; Edmund, M.A., and White, R.E., Application of Newman's technique to coupled, nonlinear partial differential equations, Ind Eng Chem Fund, 19(4), 438-440 (1980).
Westerberg, A.W., Solving sets of nonlinear algebraic equations, Chem Eng Educ, 14(2), 72-77 (1980).
Cerny, J., Comparison of methods for speeding up convergence in the simulation of chemical engineering processes at steady states, Int Chem Eng, 21(4), 694-703 (1981).
Ferraris, G.B., Newton-Raphson method for multistage separation problems, AIChEJ, 27(1), 163-166 (1981).
Finlayson, B.A., Finite element methods, Chem Eng Educ, 15(1), 20-25 (1981).
Kerr, C.P., Aid to iterative calculations, Hyd Proc, 60(12), 145-149 (1981).
Prokopakis, G.J., and Seider, W.D., Adaptive semi-implicit Runge-Kutta method for solution of stiff ordinary differential equations, Ind Eng Chem Fund, 20(3), 255-266 (1981).
Radeke, K.H., Critical remarks on using moments method, Ind Eng Chem Fund, 20(3), 302-303 (1981).
Soliman, M.A., Linearization of nonlinear dynamic systems, Comput Chem Eng, 5(2), 111-114 (1981).
Baden, N., and Villadsen, J., Collocation-based methods for parameter estimation in differential equations, Chem Eng J, 23(1), 1-14 (1982).
Burka, M.K., Solution of stiff ordinary differential equations by decomposition and orthogonal collocation, AIChEJ, 28(1), 11-20 (1982).
Donati, G.; Marini, L., and Marziano, G.L., A comprehensive approach to chemical engineering computational problems, Chem Eng Sci, 37(8), 1265-1282 (1982).
Heydweiller, J.C., and Patel, H.S., Improved biased difference scheme for solution of hyperbolic equations by method of lines, Comput Chem Eng, 6(2), 101-110 (1982).
Hlavacek, V., and Van Rompay, P., Calculation of parametric dependence and finite-difference methods, AIChEJ, 28(6), 1033-1036 (1982).
Ponton, J.W., Numerical evaluation of analytical derivatives, Comput Chem Eng, 6(4), 331-334 (1982).
Taylor, R., Solution of the linearized equations of multicomponent mass transfer, Ind Eng Chem Fund, 21(4), 407-413 (1982).

1983-1984

Bhatia, Q.S., and Hlavacek, V., Integration of difficult initial and boundary value problems by arc-length strategy, Chem Eng Commns, 22(5), 287-298 (1983).

Fairweather, G.; Davis, M.E., and Yamanis, J., Non-iterative solution of nonlinear parabolic and mixed-type problems arising in reactor design, Chem Eng Commns, 23(1), 89-100 (1983).

Hwang, C., and Shih, Y.P., Solutions of stiff differential equations via generalized block pulse functions, Chem Eng J, 27(2), 81-86 (1983).

Jerri, A.J., Application of transform-iterative method to nonlinear concentration boundary value problems, Chem Eng Commns, 23(1), 101-114 (1983).

Kalogerakis, N., and Luus, R., Improvement of Gauss-Newton method for parameter estimation using an information index, Ind Eng Chem Fund, 22(4), 436-445 (1983).

Krasnov, V.I.; Volin, Y.M., and Ostrovskii, G.M., Application of graph theory methods in solution of sparse systems of equations, Int Chem Eng, 23(4), 752-766 (1983).

Mehra, R.K.; Heidemann, R.A., and Aziz, K., An accelerated successive substitution algorithm, Can JCE, 61(4), 590-596 (1983).

Shibata, T., and Kugo, M., Generalization and application of Laplace transformation formulas for diffusion, Int J Heat Mass Trans, 26(7), 1017-1028 (1983).

Snellenberger, R.W., and Petty, C.A., Estimates of average mass transfer rates using an approximate hydrodynamic Green's function, Chem Eng Commns, 20(5), 311-334 (1983).

Billingsley, D.S., Accelerated convergence for Newton-type iterations applied to multicomponent distillation problems, AIChEJ, 30(4), 686-688 (1984).

Chen, H.S., and Stadtherr, M.A., Solving large sparse nonlinear equation systems, Comput Chem Eng, 8(1), 1-8 (1984).

Crowe, C.M., Relationship between quasi-Newton and dominant eigenvalue methods for numerical solution of nonlinear equations, Comput Chem Eng, 8(1), 35-42 (1984).

Feng, A.; Holland, C.D., and Gallun, S.E., Development and comparison of generalized semi-implicit Runge-Kutta method with Gear's method for systems of coupled differential and algebraic equations, Comput Chem Eng, 8(1), 51-60 (1984).

Gjumbir, M., and Olujic, Z., Effective ways to solve single nonlinear equations, Chem Eng, 23 July, 51-56 (1984).

Haber, S., and Brenner, H., Symbolic operator solutions of Laplace's and Stokes' equations, Chem Eng Commns, 27(5), 283-312 (1984).

Johnson, N.L.; Graham, A.L., and Gort, G.E., Comparison of methods for solving nonlinear finite-element equations in heat transfer, Chem Eng Commns, 26(1), 269-284 (1984).

Kim, D.H., and Chang, K.S., Numerical solution of unsteady-state problems, Chem Eng J, 29(1), 11-18 (1984).

Ricker, N.L., Comparison of methods for nonlinear parameter estimation, Ind Eng Chem Proc Des Dev, 23(2), 283-286 (1984).

Shacham, M., Decomposition of systems of nonlinear algebraic equations, AIChEJ, 30(1), 92-99 (1984).

Tan, K.S., and Spinner, I.H., Numerical methods of solution of continuous countercurrent processes in the nonsteady state, AIChEJ, 30(5), 770-786 (1984).

1985-1986

Kuru, S., and Westerberg, A.W., Newton-Raphson based strategy for exploiting latency in dynamic simulation, Comput Chem Eng, 9(2), 175-182 (1985).

Leis, J.R., and Kramer, M.A., Sensitivity analysis of systems of differential and algebraic equations, Comput Chem Eng, 9(1), 93-96 (1985).

Levinson, W., Automatic solution of single variable equations, Chem Eng, 25 Nov, 43-46 (1985).

Seader, J.D., Computers and equilibrium stage operations, Chem Eng Educ, 19(2), 88-103 (1985).

Soliman, M.A., On the convergence acceleration of iterative processes, Comput Chem Eng, 9(1), 97-98 (1985).

Soliman, M.A., A new update for the solution of nonlinear algebraic equations, Comput Chem Eng, 9(4), 407-410 (1985).

Biegler, L.T.; Damiano, J.J., and Blau, G.E., Nonlinear parameter estimation: A case study comparison, AIChEJ, 32(1), 29-45 (1986).

Davis, H.T., Completeness theorem for a product of self-adjoint matrices, Chem Eng Commns, 41, 267-270 (1986).

Do, D.D., and Bailey, J.E., Solution method for a class of nonlinear boundary value problems, Chem Eng Commns, 43, 1-24 (1986).

Kumar, S.K., and Reid, R.C., Derivation of the relationships between partial derivatives of Legendre transforms, AIChEJ, 32(7), 1224-1226 (1986).

Laura, P.A.A., and Cortinez, V.H., Optimization of eigenvalues when using the Galerkin method, AIChEJ, 32(6), 1025-1026 (1986).

Maddox, R.N., and Peterson, E.R., Estimating equation coefficients, Chem Eng, 17 Feb, 127-129 (1986).

McGregor, D.R., et al., Simple solution models applied to virial coefficients, AIChEJ, 32(7), 1221-1223 (1986).

Singh, V., Equation-solvers for chemical engineers, Chem Eng, 17 Feb, 149-152 (1986).

Srinivasan, V., and Aiken, R.C., Stagewise parameter estimation for stiff differential equations, AIChEJ, 32(2), 195-199 (1986).

1987-1988

Hughson, R.V., Equation-solving programs, Chem Eng, 12 Oct, 123-126 (1987).

McCoy, B.J., Approximate polynomial expansion method for inverting Laplace transforms of impulse responses, Chem Eng Commns, 52, 93-104 (1987).

Audry-Sanchez, J., Numerical solution of differential algebraic equations, Can JCE, 66(6), 1031-1035 (1988).

Byrne, G.D., and Ponzi, P.R., Differential-algebraic systems: Applications and solutions, Comput Chem Eng, 12(5), 377-382 (1988).

Cameron, I.T., and Gani, R., Adaptive Runge-Kutta algorithms for dynamic simulation, Comput Chem Eng, 12(7), 705-718 (1988).

Cook, T.E., Newton-Raphson technique, Chem Eng, 15 Aug, 169-172 (1988).

Cremades. L.; Pibouleau, L., and Koehret, B., Fitting of discontinuous functions for chemical engineering applications, Chem Eng J, 39(1), 17-26 (1988).

Ehrhardt, K.; Klusacek, K., and Schneider, P., Finite-difference scheme for solving dynamic multicomponent diffusion problems, Comput Chem Eng, 12(11), 1151-1156 (1988).

Hahn, G.J., and Morgan, C.B., Design experiments by computer, Chemtech, Nov, 664-669 (1988).

Hanna, O.T., New explicit and implicit 'improved Euler' methods for integration of ODEs, Comput Chem Eng, 12(11), 1083-1086 (1988).

Hillion, P., Solutions of the diffusion equation, Int J Heat Mass Trans, 31(6), 1173-1176 (1988).

Kuno, M., and Seader, J.D., Computing all real solutions to systems of nonlinear equations with a global fixed-point homotopy, Ind Eng Chem Res, 27(7), 1320-1329 (1988).

Poloschi, J.R., and Perkins, J.D., Implementation of quasi-Newton methods for solving sets of nonlinear equations, Comput Chem Eng, 12(8), 767-776 (1988).

Shacham, M., and Cutlip, M.B., Applications of a microcomputer computation package, Chem Eng Educ, 22(1), 18-21, 34-35 (1988).

Sieniutycz, S., Variational expressions underlying equations for heat, mass, and momentum transport in highly unsteady-state processes, Int Chem Eng, 28(2), 353-362 (1988).

Simandi, B.; Balint, A., and Sawinsky, J., Application of method of attenuated moments to evaluation of curves for residence-time distribution, Int Chem Eng, 28(2), 362-369 (1988).

Soyama, R.; Gotoh, M.; Matsumoto, S., and Suzuki, M., Real-time identification of system parameters by integral transform, Int Chem Eng, 28(1), 117-125 (1988).

Sun, E.T., and Stadtherr, M.A., Nonlinear equation solving in chemical engineering, Comput Chem Eng, 12(11), 1129-1140 (1988).

Sun, E.T., and Stadtherr, M.A., Sparse finite-difference schemes applied to chemical process engineering problems, Comput Chem Eng, 12(8), 849-851 (1988).

Tao, T.M., and Watson, A.T., Adaptive algorithm for fitting with splines, AIChEJ, 34(10), 1722-1725 (1988).

Vrentas, J.S., and Vrentas, C.M., Green's function method for the solution of diffusion-reaction problems, AIChEJ, 34(2), 347-348 (1988).

Wait, R., and Landauro, J., Parallel algorithms for multicomponent separation calculations, AIChEJ, 34(6), 964-968 (1988).

CHAPTER 10

COMPUTER-AIDED DESIGN

10.1 Computing

1969-1980

Meador, L., Computer time-sharing, Chem Eng, 13 Jan, 109-114 (1969).
Phelan, P.F., Writing and debugging computer programs, Chem Eng, 11 June, 98-106 (1973).
Rapier, P.M., The engineer's guide to electronic calculators, Chem Eng, 20 Aug, 114-120 (1973).
Evans, L.B., and Seider, W.D., The requirements of an advanced computing system, Chem Eng Prog, 72(6), 80-83 (1976).
Hall, M., and Feinstein, A.M., Database management systems, Chem Eng Prog, 73(10), 79-82 (1977).
Ridlon, S., How to verify computer programs, Chem Eng, 20 June, 121-123 (1977).
Russell, R.A., Improve your efficiency in writing computer programs, Chem Eng, 25 April, 111-116 (1977).
Waligura, C.L., and Motard, R.L., Data management on engineering and construction projects, Chem Eng Prog, 73(12), 62-70 (1977).
Winter, P., and Newell, R.G., Databases in design management, Chem Eng Prog, 73(6), 97-102 (1977).
Smith, C.L., FORTRAN 77: New programming capability, Chem Eng, 25 Sept, 113-118 (1978).
Various, Databases and library programs (topic issue), Comput Chem Eng, 3(7), 419-442 (1979).
Leesley, M.E., and Buchmann, A.P., Databases for computer-aided process plant design, Comput Chem Eng, 4(2), 79-84 (1980).

1981-1985

Various, Application of computer graphics in chemical engineering (topic issue), Comput Chem Eng, 5(4), 187-306 (1981).
Abrahamson, E.A., Time and labor saving technical computer services, Chem Eng Prog, 78(1), 54-57 (1982).
Cherry, D.H.; Grogan, J.C.; Knapp, G.L., and Perris, F.A., Use of databases in engineering design, Chem Eng Prog, 78(5), 59-67 (1982).
Brody, G.L., and Fisher, T.B., Workstations enhance productivity, Hyd Proc, May, 55-58 (1983).
Shammas, N.C., BASIC, FORTRAN and Pascal, Chem Eng, 25 July, 81-84 (1983).
Baltatu, M.E., On-line information, Chem Eng, 9 Jan, 69-72 (1984).
Finlayson, B.A., The impact of computers on undergraduate education, Chem Eng Prog, 80(2), 70-74 (1984).
Lipowicz, M.A., Integrated computing, Chem Eng, 9 Jan, 58-65 (1984).
Shammas, N.C., FORTRAN, Ada and Modula-2, Chem Eng, 1 Oct, 95-98 (1984).
Various, Putting computers to work (topic issue), Chem Eng Prog, 80(8), 29-70 (1984).
Craft, J., Designing a database for the process engineer, Proc Engng, May, 47-51 (1985).

Gupta, A.K., and Dimarzo, M., Computer applications to combustion research, Chem Eng Commns, 39, 175-192 (1985).
Herman, D.J.; Sullivan, G.R., and Thomas, S., Integration of process design, simulation, and control systems, CER&D, 63, 373-377 (1985).
Ollero, P., Tearing algorithm for recycle process networks, CER&D, 63, 264-266 (1985).
Various, Engineering with personal computers (special report), Hyd Proc, Feb, 51-63 (1985).

1986-1988

Englund, R., Computers for decision making, Chem Eng Prog, 82(9), 21-23 (1986).
Forster, A.V., PC/Mainframe interactions enhance productivity, Chem Eng Prog, 82(9), 37-42 (1986).
Grulke, E.A., Using spreadsheets for teaching design, Chem Eng Educ, 20(3), 128-131,153 (1986).
Miquel, J., and Castells, F., Curve fitting made easy, Hyd Proc, 65(11), 121-124 (1986).
Various, Man-machine interface (topic issue), Chem & Ind, 5 May, 304-324 (1986).
Vegeais, J.A.; Coon, A.B., and Stradther, M.A., Advanced computer architectures: An overview, Chem Eng Prog, 82(12), 23-31 (1986).
Basta, N., New microelectronics add computing power, Chem Eng, 12 Oct, 22-25 (1987).
Benayoune, M., and Preece, P.E., Review of information management in computer-aided engineering, Comput Chem Eng, 11(1), 1-6 (1987).
Hughson, R.V., Computer languages, Chem Eng, 14 Sept, 105-107 (1987).
Various, Pitfalls in computer application (topic issue), Chem & Ind, 2 Aug, 506-520 (1987).
Edgar, T.F.; Mah, R.S.H.; Reklaitis, G.V., and Himmelblau, D.M., Computer aids in chemical education, Chemtech, May, 277-283 (1988).
Egol, L., Systems manage chemical data, Chem Eng, 24 Oct, 111-114 (1988).
Evans, H., Guide to information technology, Proc Engng, March, 37-42 (1988).
Seider, W.D., Computing and chemical engineering, Chem Eng Educ, 22(3), 134-138 (1988).

10.2 Flowsheeting

1968-1980

Andrew, S.M., Computer flowsheeting using NETWORK 67: An example, Trans IChemE, 46, T123-132 (1968).
Forder, G.J., and Hutchison, H.P., Analysis of chemical plant flowsheets, Chem Eng Sci, 24(4), 771-786 (1969).
Flower, J.R., and Whitehead, B.D., Computer-aided design: A survey of flowsheeting, Chem Engnr, April, 208-210, 223; May, 271-277 (1973).
Gaddy, J.L., and Culberson, O.L., Prediction of variable process performance by stochastic flowsheet simulation, AIChEJ, 19(6), 1239-1243 (1973).

Peters, N., and Barker, P.E., PEETPACK: A new non-proprietary flowsheeting program, Chem Engnr, Dec, 763-768 (1974).

Gaines, L.D., and Gaddy, J.L., Process optimization by flowsheet simulation, Ind Eng Chem Proc Des Dev, 15(1), 206-211 (1976).

Ballman, S.H., and Gaddy, J.L., Optimization of methanol process by flowsheet simulation, Ind Eng Chem Proc Des Dev, 16(3), 337-341 (1977).

Metcalfe, S.R., and Perkins, J.D., Information flow in modular flowsheeting systems, Trans IChemE, 56, 210-213 (1978).

Westerberg, A.W., and Berna, T.J., Decomposition of very large-scale Newton-Raphson based flowsheeting problems, Comput Chem Eng, 2(1), 61-64 (1978).

Sood, M.K.; Reklaitis, G.V., and Woods, J.M., Solution of material balances for flowsheets modelled with elementary modules, AIChEJ, 25(2), 209-229 (1979).

Various, Flowsheeting (topic issue), Comput Chem Eng, 3(6), 17-21, 307-418 (1979).

Doering, F.J., and Gaddy, J.L., Optimization of sulfuric acid process with a flowsheet simulator, Comput Chem Eng, 4(2), 113-122 (1980).

Hanyak, M.E., Textual expansion of chemical process flowsheets into algebraic equation sets, Comput Chem Eng, 4(4), 223-240 (1980).

1981-1988

Winter, P., Process flowsheeting, Proc Engng, June, 33-38 (1981).

Shacham, M.; Macchietto, S.; Stutzman, L.F., and Babcock, P., Equation oriented approach to process flowsheeting (review paper), Comput Chem Eng, 6(2), 79-96 (1982).

Stadtherr, M.A., and Hilton, C.M., Efficient solution of large-scale Newton-Raphson-based flowsheeting problems in limited core, Comput Chem Eng, 6(2), 115-120 (1982).

Ponton, J.W., Dynamic process simulation using flowsheet structure, Comput Chem Eng, 7(1), 13-18 (1983).

Rippin, D.W.T., Sample problems for flowsheeting programs, Comput Chem Eng, 7(2), 65-72 (1983).

Stadtherr, M.A., and Wood, E.S., Sparse matrix methods for equation-based chemical process flowsheeting: Reordering and numerical phases, Comput Chem Eng, 8(1), 9-34 (1984).

Trevino-Lozano, R.A.; Kisala, T.P., and Boston, J.F., Simplified absorber model for nonlinear simultaneous modular flowsheet calculations, Comput Chem Eng, 8(2), 105-116 (1984).

Winter, P., Process flowsheeting systems, Proc Engng, March, 29-31 (1984).

Beazley, W.G., Transferring CAD flowsheets, Chem Eng, 28 Oct, 61-63; 9 Dec, 113-114 (1985).

Chen, H.S., and Stadtherr, M.A., A simultaneous-modular approach to process flowsheeting and optimization, AIChEJ, 31(11), 1843-1881 (1985).

Kjaer, J., Solution of process flowsheets by formal gauss elimination, Comput Chem Eng, 9(2), 153-166 (1985).

Cuthrell, J.E., and Biegler, L.T., Simultaneous solution and optimization of process flowsheets with differential equation models, CER&D, 64(5), 341-346 (1986).

Hutchison, H.P.; Jackson, D.J., and Morton, W., The development of an equation-oriented flowsheet simulation and optimization package (QUASILIN), Comput Chem Eng, 10(1), 19-47 (1986).

Milani, S.M., and Findley, M.E., Linear process calculations as convergence accelerators in flowsheet-sequenced programs, AIChEJ, 32(4), 624-631 (1986).

Sawicki, S.J.; Young, R.D., and Sund, S.E., A new program (GRAF) to create chemical process flow diagrams, Comput Chem Eng, 10(3), 297-301 (1986).

Chimowitz, E.H., and Bielinis, R.Z., Analysis of parallelism in modular flowsheet calculations, AIChEJ, 33(6), 976-986 (1987).

Ganesh, N., and Biegler, L.T., Reduced Hessian strategy for sensitivity analysis of optimal flowsheets, AIChEJ, 33(2), 282-296 (1987).

Bogle, I.D.L., and Perkins, J.D., Sparse Newton-like methods in equation oriented flowsheeting, Comput Chem Eng, 12(8), 791-806 (1988).

Kirkwood, R.L.; Locke, M.H., and Douglas, J.M., A prototype expert system for synthesizing chemical process flowsheets, Comput Chem Eng, 12(4), 329-344 (1988).

Sillett, C., Flowsheeting for food product safety, Chem Engnr, July, 21-25 (1988).

Zitney, S.E., and Stadtherr, M.A., Computational experiments in equation-based chemical process flowsheeting, Comput Chem Eng, 12(12), 1171-1186 (1988).

10.3 Models

1969-1975

Christensen, J.H., and Rudd, D.F., Structuring design computations, AIChEJ, 15(1), 94-100 (1969).

Gilmour, R.H., Simulation of an alkylation plant, Brit Chem Eng, 14(3), 315-316 (1969).

Gerdes, F.O.; Hoftyzer, P.J.; Kemkes, J.F.; Van Loon, M., and Schweigman, C., Mathematical models for chemical process plant, Chem Engnr, Sept, CE267-278 (1970).

Johnson, A.I., et al., Modular approach to simulation and design, Brit Chem Eng, 16(10), 923-929 (1971).

Siirola, J.J., and Rudd, D.F., Computer-aided synthesis of chemical process designs, Ind Eng Chem Fund, 10(3), 353-362 (1971).

Friedman, P., and Pinder, K.L., Optimization of a simulation model of a chemical plant, Ind Eng Chem Proc Des Dev, 11(4), 512-520 (1972).

Johnson, A.I., Computer-aided process analysis and design: A modular approach, Brit Chem Eng, 17(1), 28-33; 17(2), 119-122; 17(3), 217-223 (1972).

King, C.J.; Gantz, D.W., and Barnes, F.J., Systematic evolutionary process synthesis, Ind Eng Chem Proc Des Dev, 11(2), 271-283 (1972).

Ramirez, W.F.,and Vestal, C.R., Algorithms for structuring design calculations, Chem Eng Sci, 27(12), 2243-2254 (1972).

Branch, J., Problems in simulation of chemical processes, Chem & Ind, 4 Oct, 832-835 (1975).

Hutchison, H.P., and Kilikas, A.C., Linear simulation with parameter variation, Chem Engnr, Dec, 753-756, 759 (1975).

Motard, R.L.; Shacham, M., and Rosen, E.M., Steady-state chemical process simulation (review paper), AIChEJ, 21(3), 417-436 (1975).

Umeda, T., and Ichikawa, A., Rational approach to process synthesis, Chem Eng Sci, 30(7), 699-708 (1975).

Wells, G., Simulation strategy for process plant projects, Proc Engng, Jan, 60-64 (1975).

1976-1980

Jayaraman, K., and Lapidus, L., Practical realizations in process modelling, AIChEJ, 22(2), 298-315 (1976).

Nishida, N.; Liu, Y.A., and Ichikawa, A., Studies in chemical process design and synthesis, AIChEJ, 22(3), 539-549 (1976).

Stephanopoulos, G., and Westerberg, A.W., Studies in process synthesis, Chem Eng Sci, 30(8), 963-972 (1975); 31(3), 195-204 (1976).

Evans, L.B.; Joseph, B., and Seider, W.D., System structures for process simulation, AIChEJ, 23(5), 658-666 (1977).

Heydweiller, J.C.; Sincovec, R.F., and Fan, L.T., Dynamic simulation of chemical processes described by distributed and lumped parameter models, Comput Chem Eng, 1(2), 125-132 (1977).

Lamb, J.A., and Pomphrey, D.M., Simulation aids plant design, Processing, Feb, 27-29 (1977).

Nishida, N.; Liu, Y.A., and Lapidus, L., Studies in chemical process design and synthesis, AIChEJ, 23(1), 77-93 (1977).

Fowler, J.R., and Harvey, D.J., Dynamic simulation of a PVC process, Chem Eng Prog, 74(1), 61-66 (1978).

Overtuft, B.W.; Reklaitis, G.V., and Woods, J.M., Simulation of batch/semicontinuous operations using the GASP IV language, Ind Eng Chem Proc Des Dev, 17(2), 161-175 (1978).

Grossmann, I.E., and Sargent, R.W.H., Optimum design of multipurpose chemical plants, Ind Eng Chem Proc Des Dev, 18(2), 343-348 (1979).

Seider, W.D., et al., Routing of calculations in process simulation, Ind Eng Chem Proc Des Dev, 18(2), 292-297 (1979).

Various, Mathematical methods and optimization (topic issue), Comput Chem Eng, 3(12), 593-625 (1979).

Various, Simulation of unit operations (topic issue), Comput Chem Eng, 3(10), 513-572 (1979).

Various, Advances in model building (topic issue), Comput Chem Eng, 3(2), 61-161 (1979).

Grossmann, I.E., and Santibanez, J., Applications of mixed-integer linear programming in process synthesis, Comput Chem Eng, 4(4), 205-214 (1980).

Patterson, G.K., and Rozsa, R.B., DYNSYL: A general-purpose dynamic simulator for chemical processes, Comput Chem Eng, 4(1), 1-20 (1980).

1981-1985

Ford, J.R., Polymer production simulation: A generalized plant program, Chem Eng Prog, 77(9), 74-79 (1981).

Nishida, N.; Stephanopoulos, G., and Westerberg, A.W., Review of process synthesis, AIChEJ, 27(3), 321-351 (1981).

Gundersen, T., Numerical aspects of implementation of cubic equations of state in flash calculation routines, Comput Chem Eng, 6(3), 245-256 (1982).

Murtagh, B.A., Simultaneous solution and optimization of large-scale engineering systems, Comput Chem Eng, 6(1), 1-6 (1982).

van Deemter, J.J., Basics of process modeling (review paper), Chem Eng Sci, 37(5), 657-664 (1982).

Van Deemter, J.J., Basics of process modeling (review paper), Chem Eng Sci, 37(5), 657-663 (1982).

Felder, R.M., Simulation: A tool for optimizing batch-process production, Chem Eng, 18 April, 79-84 (1983).

Ostrovskii, G.M., et al., Steady-state simulation of chemical plants, Chem Eng Commns, 23(1), 181-190 (1983).

Tayler, C., Plant simulators, Proc Engng, July, 38-41 (1984).

Alcock, P., Dynamic simulation, Proc Engng, Sept, 40-43 (1985).

Douglas, J.M., Hierarchical decision procedure for process synthesis, AIChEJ, 31(3), 353-362 (1985).

Various, Computer-aided process design (topic issue), Comput Chem Eng, 9(5), 413-546 (1985).

1986-1988

Ferrall, J.F.; Pappano, A.W., amd Jennings, C.N., Process analysis on a spreadsheet, Chem Eng, 3 March, 101-104 (1986).

Glaser, D.C., The PC simulator, Chem Eng Prog, 82(9), 45-48 (1986).

Ostrovsky, G.M.; Mikhailova, Y.M., and Berzhinsky, T.A., Large-scale system optimization, Comput Chem Eng, 10(2), 123-128 (1986).

Palazzolo, A.B., Combine computer simulation and field testing to solve problems, Hyd Proc, 65(5), 59-62 (1986).

Ponton, J.W., and Vasek, V., A two-level approach to chemical plant and process simulation, Comput Chem Eng, 10(3), 277-286 (1986).

Stephenson, G.R., and Shewchuk, C.F., Reconciliation of process data with process simulation, AIChEJ, 32(2), 247-254 (1986).

Tomiak, A., Simulate countercurrent washing on belt filters, Chem Eng, 4 Aug, 61-64 (1986).

Various, Modelling chemical engineering systems (special issue), Chem Eng Sci, 41(6), 1371-1698 (1986).

Fairclough, M.P., Modelling conceptual designs for nuclear reprocessing, Proc Engng, June, 33-35 (1987).

Floudas, C.A., Separation synthesis of multicomponent feed streams into multicomponent product streams, AIChEJ, 33(4), 540-550 (1987).

Hunter, D., Process modeling: When bigger is better, Chem Eng, 16 March, 12E-12H (1987).

Wilson, J.A., Make batch process improvements using models, Proc Engng, June, 45-49 (1987).

Cheng, S.H., and Liu, Y.A., Studies in chemical process design and synthesis, Ind Eng Chem Res, 27(12), 2304-2322 (1988).

Floquet, P.; Pibouleau, L., and Domenech, S., Mathematical programming tools for chemical engineering process design synthesis, Chem Eng & Proc, 23(2), 99-114 (1988).

Gilles, E.D.; Holl, P., and Marquardt, W., Dynamic simulation of complex chemical processes, Int Chem Eng, 28(4), 579-593 (1988).

Harrison, B.K., Not enough data for simulation? Chem Eng, 18 July, 133-134 (1988).

Holmes, J.W., Fluid-flow analysis with spreadsheets, Chem Eng., 19 Dec, 166-168 (1988).

Oyeleye, O.O., and Kramer, M.A., Qualitative simulation of chemical process systems: Steady-state analysis, AIChEJ, 34(9), 1441-1454 (1988).

Pinto, J.C., and Biscaia, E.C., Order-reduction strategies for models of staged separation systems, Comput Chem Eng, 12(8), 821-832 (1988).

Schultheisz, D.J., and Sommerfeld, J.T., Discrete-event simulation in chemical engineering, Chem Eng Educ, 22(2), 98-102 (1988).

Seider, W.D., Computing and process design, Chem Eng Educ, 22(4), 212-217 (1988).

Various, Papers from Use of Computers in Chemical Engineering (1987): Simulation and modelling of processes, Comput Chem Eng, 12(5), 357-370, 383-426, 449-460, 469-474 (1988).

10.4 Software

1967-1983

Hodge, B., and Mantey, J.P., Computers for engineering applications, Chem Eng, 25 Sept, 180-184; 23 Oct, 167-171; 20 Nov, 141-146; 18 Dec, 110-113 (1967).

Hodge, B., and Mantey, J.P., Computers for engineering applications, Chem Eng, 1 Jan, 79-83; 26 Feb, 169-172; 11 March, 187-192; 22 April, 185-188; 20 May, 165-167; 17 June, 271-276; 29 July, 151-156 (1968).

Hughson, R.V., and Steymann, E.H., Computer programs for chemical engineers, Chem Eng, 20 Aug, 121-132; 17 Sept, 127-140 (1973).

Kehat, E., and Shacham, M., Chemical process simulation programs, Proc Tech Int, Jan, 35-37; March, 115-118; April, 181-184 (1973).

Smith, J.M., Engineering analysis on the pocket calculator, Chem Eng, 26 April, 80-92 (1976).

Benenati, R.F., Solving engineering problems on programmable pocket calculators, Chem Eng, 28 Feb, 201-206; 14 March, 129-132 (1977).

Peterson, J.N.; Chen, C.C., and Evans, L.B., Computer programs for chemical engineers (1978), Chem Eng, 5 June, 145-154; 3 July, 69-82; 31 July, 79-86; 28 Aug, 107-115 (1978).

Various, Microcomputing for chemical engineers: Hardware and software (feature report), Chem Eng, 31 May, 104-115 (1982).

Garrett, J.R., Published calculator programs for chemical engineers, Chem Eng, 7 March, 149-160 (1983).

Selk, S., Spreadsheet software solves engineering problems, Chem Eng, 27 June, 51-53 (1983).

1984-1988

Dunhill, S., Integration of design packages, Proc Engng, June, 45-46 (1984).
Lipowicz, M.A., Desktop software, Chem Eng, 9 Jan, 66-68 (1984).
Schmidt, W.P., and Upadhye, R.S., Material balances on a speadsheet, Chem Eng, 24 Dec, 67-70 (1984).
Wynne, D.R., Microcomputer data-acquisition systems, Chem Eng, 26 Nov, 95-96 (1984).
Adams, A., Microcomputer software for process engineers, Chem Engnr, Jan, 13-15 (1985).
Goldfarb, S.M., and Hirschel, R., Spreadsheets for chemical engineers, Chem Eng, 15 April, 91-94 (1985).
Ferrall, J.F.; Pappano, A.W., and Jennings, C.N., Process analysis on a spreadsheet, Chem Eng, 3 March, 101-104 (1986).
Lipowicz, M., Process engineering software, Chem Eng, 14 April, 95-98 (1986).
Lipowicz, M., Selecting personal-computer software, Chem Eng, 9 June, 75-77 (1986).
Lipowicz, M.A., Environmental, distillation, and control-valve-sizing software, Chem Eng, 20 Jan, 73-76 (1986).
Preece, P., The making of PFG and PIG (graphical software packages), Chem Engnr, June, 87-88 (1986).
Sawyer, P., Integrated software for process engineering calculations, Chem Engnr, Nov, 32-34 (1986).
Singh, V., Equation solvers for chemical engineers, Chem Eng, 17 Feb, 149-152 (1986).
Wilkie, F., Customised software for CAD, Proc Engng, June, 59-63 (1986).
Winter, P., and Rosen, S.M., Process-engineering databases, Chem Eng, 7 July, 65-67 (1986).
Cheremisinoff, N.P., Statistical regression routines on spreadsheets, Chem Eng, 17 Aug, 153-157 (1987).
Cifuentes, L., Suggestions for evaluating software, Chem Eng, 7 Dec, 135-137 (1987).
Sowa, C.J., Engineering calculations on a spreadsheet, Chem Eng, 2 March, 61-63 (1987).
Goldman, S.D., Software and environmental engineering, Chem Eng Prog, 84(12), 29-34 (1988).
Hughson, R.V., Chemical engineering software survey, Chem Eng, 12 Sept, 113-116 (1988).
Hughson, R.V., Computer packages for chemical engineering design, Chem Eng, 7 Nov, 109-112 (1988).
Lang, C., Software documentation, Chem Eng, 14 March, 149-152 (1988).

10.5 Computer-Aided Design

1969-1980

Andrew, S.M., Digital computers in the design of complete chemical processes, Brit Chem Eng, 14(8), 1057-1062 (1969).

Carter, A.G., Computers and chemical plant engineering, Brit Chem Eng, 15(11), 1427-1432 (1970).

Westerberg, A.W., and Edie, F.C., Computer-aided design, Chem Eng J, 2(1), 9-25; 2(2), 114-124 (1971).

Leesley, M.E., Process plant design by computer, Proc Tech Int, Nov, 403-405 (1973).

Pho, T.K., and Lapidus, L., Topics in computer-aided design, AIChEJ, 19(6), 1170-1189 (1973).

Clark, J.P., and Sommerfeld, J.T., Use of FLOWTRAN simulation, Chem Eng Educ, 10(2), 90-92 (1976).

Coscia, M., and Lord, R.C., Consider advantages of exchanger drawings by computer, Hyd Proc, 56(1), 157-159 (1977).

Rosen, E.M., and Pauls, A.C., Computer-aided chemical process design: The FLOWTRAN system, Comput Chem Eng, 1(1), 11-22 (1977).

Cheng, W.B., and Mah, R.S.H., Interactive synthesis of pipeline networks using PIGRAPH, Comput Chem Eng, 2(4), 133-142 (1978).

Economopoulos, A.P., General design of stage operations solution packages, Chem Engnr, Nov, 841-844 (1978).

Gay, P., CAD for process plant engineering, Proc Engng, May, 42-43 (1979).

Russell, R.A., Chemical process simulation: The GMB system, Comput Chem Eng, 4(3), 167-190 (1980).

Various, Piping-system design by CAD, Proc Engng, Feb, 77-79,89-91 (1980).

1981-1985

Lauher, V.A., Computerized drafting: Problems and benefits, Chem Eng Prog, 77(3), 49-54 (1981).

Thambynayagam, R.K.M.; Wood, R.K., and Winter, P., Dynamic process simulator: An engineer's tool for dynamic process analysis, Chem Engnr, Feb, 58-65 (1981).

Hope, K., Computer plant design, Proc Engng, July, 35-37 (1982).

Thomasson, J.M., and Lebosse, M., Experience with CAD, Hyd Proc, June, 125-131; July, 161-165 (1982).

Various, Computer-aided design (topic issue), Chem Eng Prog, 78(12), 33-60 (1982).

Carell, R., and Kuroczko, M., Train effectively for CAD/D, Hyd Proc, April, 145-151 (1983).

Chimowitz, E.H.; Bielinis, R.Z., and Jobsky, R.W., Speed up microcomputer design calculations, Chem Eng, 26 Dec, 43-45 (1983).

Robinson, E., Microcomputing systems in chemical engineering, Chem Engnr, March, 43-45 (1983).

Sucksmith, I., Microcomputer systems for process design, Chem Eng, 10 Jan, 115-118 (1983).

Williams, L.F.; Gerrard, A.M.; Sebastian, D.J.G., and Wheeldon, D.H.V., Using the ECONOMIST package, Proc Econ Int, 4(3), 12-15 (1983).

Branch, J., Review of CAD packages, Processing, Aug, 33-34 (1984).
Goldman, G.S., Design and drafting on a microcomputer, Chem Eng, 3 Sept, 81-84 (1984).
O'Reilly, M.G., The role of desk top microcomputers in process simulation, Chem Engnr, April, 16-17 (1984).
Peters, R.A., Computer-aided engineering, Chem Eng, 11 June, 91-94; 9 July, 107-109 (1984).
Seider, W.D., Use of FLOWTRAN simulation, Chem Eng Educ, 18(1), 26-29, 41-43 (1984).
Bakker, B., CAD of process plant, Processing, Oct, 26,29 (1985).
Branch, J., Integrated process design, Processing, March, 37-39 (1985).
Leigh, R., CAD packages for control systems, Proc Engng, March, 55-56 (1985).
Peters, R.A., Integrated computer-aided engineering, Chem Eng, 13 May, 95-98 (1985).
Tayler, C., CAD piping database, Proc Engng, Nov, 33-36 (1985).
Various, Computer-aided engineering (special topic, 9 papers), Chem Eng Prog, 81(9), 14-17, 21-61 (1985).
Wright, V.E., Microcomputer CAD systems, Chem Eng, 8 July, 99-102 (1985).

1986-1988

Basta, N., Computer-aided design for chemical engineers, Chem Eng, 1 Sept, 14-17 (1986).
Knight, F., 3-D CAD: A user's opinion, Proc Engng, March, 36-39 (1986).
Sawyer, P., Getting a grip on the nature of CAD, Chem Engnr, Oct, 22-23 (1986).
Winter, P., Computer aided process design at a watershed (review paper), CER&D, 64(5), 329-331 (1986).
Atkinson, N., 3D modelling in CAD, Proc Engng, Nov, 53-55 (1987).
Briggs, R., CADD for project engineering, Proc Engng, July, 44-46 (1987).
Bullock, R.H., Computer-aided design on personal computers, Chem Eng, 22 June, 143-146 (1987).
Rosen, E.M., and Adams, R.N., A review of spreadsheet usage in chemical engineering calculations, Comput Chem Eng, 11(6), 723-736 (1987).
Stephanopoulos, G., et al., DESIGN-KIT: An object oriented environment for process engineering, Comput Chem Eng, 11(6), 629-638 (1987).
Walton, A., Is PC CADD the answer? Chem Engnr, July, 36-37 (1987).
Wood, A., CAD for process design, Processing, June, 32-34 (1987).
Atkinson, N., Contracting CAD, Proc Engng, April, 53-57 (1988).
Egol, L., Software for desktop design, Chem Eng, 21 Nov, 111-114 (1988).
McLeod, N., Developments in CAD, Processing, June, 18-22 (1988).
Morrison, J., Personal computing in the process industries, Chem Engnr, March, 36-37 (1988).
Pantelides, C.C., Speedup: Recent advances in process simulation, Comput Chem Eng, 12(7), 745-756 (1988).
Pericleous, K.A., Computer modelling for the analysis of fluid flow, heat transfer and combustion in industry, Energy World, Aug, 9-11 (1988).
Rose, M., Process engineering on the PC, Proc Engng, Nov, 61-63 (1988).
Ross, M., CADCAM packages, Chem Engnr, March, 13-16 (1988).

Savage, P., Computer-integrated manufacturing, Chem Eng, 28 March, 20-21 (1988).
Sawyer, P., A personal workstation for integrated process engineering, Chem Engnr, Feb, 22-24 (1988).
Tayler, C., Computerised 3D plant models, Proc Engng, Dec, 42-43 (1988).
Various, Process design and CAD (topic issue), CER&D, 66(5), 396-444 (1988).
Wright, A., and Bramfitt, V., Development and application of a CAD package for batch processing industry, Chem & Ind, 15 Feb, 114-118 (1988).

10.6 Equipment and Plant Design

1967-1970

Sargent, R.W.H., Integrated design and optimization of processes, Chem Eng Prog, 63(9), 71-78 (1967).
Goldman, M.R., and Robinson, E.R., Computer simulation of batch distillation processes, Brit Chem Eng, 13(12), 1713-1716 (1968).
Johnson, A.I.; Aizawa, M., and Petryschuk, W.F., Simulation of a synthetic rubber plant, Brit Chem Eng, 13(10), 1432-1438 (1968).
Lederman, P.B., Process design with computers, Chem Eng, 23 Sept, 221-226; 21 Oct, 151-154; 2 Dec, 127-132; 16 Dec, 107-112 (1968).
Youle, P.V., Computing for petrochemicals, Brit Chem Eng, 13(2), 225-228 (1968).
Andrew, S.M., Computer modelling and optimisation in the design of a complete chemical process, Trans IChemE, 47, T79-84 (1969).
Armstrong, M., and Schofield, A.E., The design of air separation distillation columns using a computer, Chem Engnr, May, CE184-189 (1969).
Cocks, A.M., Plate heat exchanger design by computer, Chem Engnr, May, CE193-198 (1969).
Daniel, P.T., and Hall, M., An integrated system of pipework estimating, detailing and control, Chem Engnr, May, CE169-178 (1969).
Emerson, W.H., Application of digital computer to design of surface condensers, Chem Engnr, May, CE178-184 (1969).
Klumpar, I.V., Process predesign by computer, Chem Eng, 22 Sept, 114-122 (1969).
Lepper, A.M., and Houtby, D.K., Algorithmic models for the specification of heat-transfer equipment involving condensation, Chem Engnr, May, CE189-193 (1969).
Ludwig, W.R., and Peterson, R.P., Computer design helps small chemical company, Chem Eng, 10 March, 98-105 (1969).
Katell, S., and Jones, P.R., Optimum heat exchanger design, Brit Chem Eng, 15(4), 491-494 (1970).
Lee, W., and Tayyabkhan, M.T., Increase profits by computer optimization, Hyd Proc, 49(9), 286-290 (1970).
Mijolaj, P.G., Computer aid for crude oil distillation unit designs, Brit Chem Eng, 15(5), 638-643 (1970).

1971-1979

Various, Philosophy of computer-aided process and plant design (symposium papers), Chem Engnr, Aug, 293-311 (1971).

Yamada, I., Computer-aided distillation calculations, Int Chem Eng, 13(1), 106-122 (1973).

Briddell, E.T., Process design by computer, Chem Eng, 4 Feb, 60-63; 4 March, 113-118; 1 April, 77-84; 15 April, 5 (1974).

Hutchison, H.P., Plant simulation by linear methods, Trans IChemE, 52, 287-290 (1974).

Mah, R.S.H., Pipeline network calculations using sparse computation techniques, Chem Eng Sci, 29(7), 1629-1638 (1974).

Sparrow, R.E.; Rippin, D.W.T., and Forder, G.J., Multi-batch: A computer package for the design of multi-product batch plants, Chem Engnr, Sept, 520-525 (1974).

Stainthorp, F.P., and Benson, R.S., Computer aided design of process control systems, Chem Engnr, Sept, 531-535 (1974).

Beychok, M.R., Program calculators for design study, Hyd Proc, 55(9), 261-263 (1976).

Madden, J., and Winter, P., Pipework design by CAD, Processing, Oct, 13-16 (1976).

Anon., Computer design of heat exchangers, Processing, Sept, 64 (1977).

Ewell, R.B., and Gadmer, G., Design catalytic crackers by computer, Hyd Proc, 57(4), 125-134 (1978).

Johns, W.R.; Marketos, G., and Rippin, D.W.T., The optimal design of chemical plant to meet time-varying demands in the presence of technical and commerical uncertainty, Trans IChemE, 56, 249-257 (1978).

Latour, P.R., Online computer optimization, Hyd Proc, 58(7), 219-223 (1979).

1981-1985

Al-Zakri, A.S., and Bell, K.J., Estimating performance when uncertainties exist, Chem Eng Prog, 77(7), 39-49 (1981).

Biegler, L.T., and Hughes, R.R., Approximation programming of chemical processes with Q/LAP, Chem Eng Prog, 77(4), 76-83 (1981).

Domenech, S., and Enjalbert, M., Program for simulating batch rectification as a unit operation, Comput Chem Eng, 5(3), 181-185 (1981).

Singh, C.P.P., and Saraf, D.N., Process simulation of ammonia plant, Ind Eng Chem Proc Des Dev, 20(3), 425-433 (1981).

Thurston, C.W., Computer-aided design of distillation column controls, Hyd Proc, 60(7), 128-130; 60(8), 135-140 (1981).

Various, Computer modeling of separation processes (symposium papers), Sepn Sci Technol, 16(10), 1299-1428 (1981).

Yang, R.Y.K., and Colton, C.K., Computer-aided analysis and design of an ATP regeneration reactor, Chem Eng Commns, 16(1), 227-244 (1982).

Morris, R.C., Simulating batch processes, Chem Eng, 16 May, 77-81 (1983).

Ollero, P., and Amselem, C., Decomposition algorithm for chemical process simulation, CER&D, 61, 303-307 (1983).

Singh, C.P.P., and Carr, N.L., Process simulation of an SRC II plant, Ind Eng Chem Proc Des Dev, 22(1), 104-118 (1983).

Dokurno, M.G., and Douglas, P.L., Simulation of an ethylene oligomerization reactor system using ASPEN, Can JCE, 62(6), 818-824 (1984).
Jacobs, O.L.R.,; Badran, W.A., and Proudfoot, C.G., Computer-aided design of systems for regulating pH, Chem Engnr, March, 19-21 (1984).
Madhavan, S., Ammonia process simulator, Plant/Opns Prog, 3(1), 14-18 (1984).
Williams, L.F., and Gerrard, A.M., Computer-aided cost estimation: A survey, Proc Econ Int, 5(1), 28-30 (1984).
Gostoli, C., and Spadoni, C., Linearization of the head-capacity curve in the analysis of pipe networks including pumps, Comput Chem Eng, 9(1), 89-92 (1985).
Kenefick, J.F., and Chirillo, R.D., Photogrammetry and computer-aided piping design, Chem Eng, 18 Feb, 173-176 (1985).
Salisbury, A., Integrated pipeline-design systems, Proc Engng, Feb, 27-32 (1985).

1986-1988

Govind, R.; Mocsny, D.; Cosson, P., and Klei, J., Exchanger network synthesis on a microcumputer, Hyd Proc, 65(7), 53-57 (1986).
Guthrie, J.E.; Braunstein, B.A.; Harrington, M.T., and Merrick, R.D., Monitoring corrosion with a microcomputer, Chem Eng, 1 Sept, 81-84 (1986).
Jones, D.A.; Yilmaz, A.N., and Tilton, B.E., Synthesis techniques for retrofitting heat recovery systems, Chem Eng Prog, 82(7), 28-33 (1986).
Neil, J., and Stone, S.J., Using CAD systems to prepare McCabe-Thiele diagrams, Chem Eng, 10 Nov, 135-136 (1986).
Palen, J.W., Designing heat exchangers by computer, Chem Eng Prog, 82(7), 23-27 (1986).
Pase, G.K., Computer programs for heat exchanger design, Chem Eng Prog, 82(9), 53-56 (1986).
Wachel, L.J., Exchanger simulator: Guide to less fouling, Hyd Proc, 65(11), 107-110 (1986).
White, J.R., Use spreadsheets for better refinery operation, Hyd Proc, 65(10), 49-52 (1986).
Leonard, R.A., Electronic worksheets for calculation of stagewise solvent extraction processes, Sepn Sci Technol, 22(2), 535-556 (1987).
Lin, W.J.; Seader, J.D., and Wayburn, T.L., Computing multiple solutions to systems of interlinked separation columns, AIChEJ, 33(6), 886-897 (1987).
Anon., Piping software, Chem Eng, 20 June, 145-146 (1988).
Bunn, A.R., and Lees, F.P., Expert design of plant handling hazardous materials: Design expertise and CAD methods with illustrative examples, CER&D, 66(5), 419-444 (1988).
Kerlin, M., Spreadsheet tray hydraulic design, Hyd Proc, 67(3), 49-51 (1988).
Leone, H.; Scenna, N., and Vecchietti, A., An iso-propanol distillery revamping using IMBAD simulator, Comput Chem Eng, 12(8), 807-820 (1988).
Morris, C.G.; Sim, W.D.; Vysniauskas, T., and Svrcek, W.Y., Crude-tower simulation on a personal computer, Chem Eng Prog, 84(11), 63-68 (1988).

Pibouleau, L.; Floquet, P., and Domenech, S., Optimal synthesis of reactor separator systems by nonlinear programming method, AIChEJ, 34(1), 163-166 (1988).

Ranzi, E.; Rovaglio, M.; Faravelli, T., and Biadri, G., Role of energy balances in dynamic simulation of multicomponent distillation columns, Comput Chem Eng, 12(8), 783-786 (1988).

Tayler, C., Database for process instrument selection, Proc Engng, Oct, 57-59 (1988).

CHAPTER 11

MATERIALS

11.1 Traditional Materials

1966-1970

Various, Materials of construction (22nd biennial report), Chem Eng, 7 Nov, 187-238 (1966).

Burst, J.F., and Spieckerman, J.A., A guide to selecting modern refractories, Chem Eng, 31 July, 85-104 (1967).

Various, Nonferrous metals, Chem Eng, 21 Nov, 109-114; 5 Dec, 139-146; 19 Dec, 105-112 (1966); 16 Jan, 159-164; 30 Jan, 139-146 (1967).

Briggs, A., The commonly used austenitic steels, Brit Chem Eng, 13(3), 369-371 (1968).

Samans, C.H., Making the most of contemporary steels, Chem Eng, 12 Feb, 150-160 (1968).

Various, Materials (topic issue), Chem Eng Prog, 64(2), 35-63 (1968).

Various, Materials of contruction (23rd biennial report), Chem Eng, 4 Nov, 126-178 (1968).

Husen, C., and Samans, C.H., Avoiding the problems of stainless steels, Chem Eng, 27 Jan, 178-186 (1969).

Loginow, A.W., and Brickner, K.G., Designing with a new stainless steel, Chem Eng, 8 Sept, 152-161 (1969).

Price, F.C., Iron and steel today, Chem Eng, 11 Aug, 76-88 (1969).

Tyson, S.E., Shortcut to stainless-steel specification, Chem Eng, 6 Oct, 188-192 (1969).

Kies, F.K.; Franson, I.A., and Coad, B., New uses for ferritic stainless steel, Chem Eng, 23 March, 150-156 (1970).

Knoth, R.J.; Lasko, G.E., and Matejka, W.A., New Ni-free stainless bids to oust austenitic, Chem Eng, 18 May, 170-176 (1970).

Landels, H.H., and Stout, E., Glassed steel equipment: A guide to current technology, Brit Chem Eng, 15(10), 1289-1295 (1970).

Schwartz, C.D.; Franson, I.A., and Hodges, R.J., Inventing a new ferritic stainless steel, Chem Eng, 20 April, 164-169 (1970).

Various, Engineering materials (deskbook issue), Chem Eng, 12 Oct, 15-221 (1970).

1971-1975

Anon., Making sense from alloy compositions, Chem Eng, 1 Nov, 92-95 (1971).

Hogan, R.E., Solder glasses, Chemtech, Jan, 41-43 (1971).

Molineux, K.J., Glass for chemical plant equipment, Brit Chem Eng, 16(9), 796-798; 16(12), 1128-1130 (1971).

Stookey, S.D., Glass chemistry, Chemtech, Aug, 458-465 (1971).

Buckley, D.H., and Johnson, R.L., Solid lubricants, Chemtech, May, 302-310 (1972).

Gaugh, R.R., and Perry, D.C., A new stainless steel for the CPI, Chem Eng, 2 Oct, 84-90 (1972).

Molineux, K.J., Glass as a material of construction in chemical plant, Brit Chem Eng, 17(2), 150-151 (1972).

Skalny, J., and Daugherty, K.E., Portland cement, Chemtech, Jan, 38-45 (1972).

Various, Metallurgy for process and mechanical design engineers (special report), Hyd Proc, 51(8), 65-80 (1972).

Various, Engineering materials (deskbook issue), Chem Eng, 4 Dec, 19-125 (1972).
Baum, B.M., Flame-retardant fabrics, Chemtech, March, 167-170; May, 311-316; July, 416-421 (1973).
Kopecki, E.S., Stainless steel for saline-water service, Chem Eng, 22 Jan, 124-128 (1973).
Mack, W.C., Selecting steel tubing, Chem Eng, 26 Nov, 94-96 (1973).
Various, Copper technology (feature report), Chem Eng, 16 April, 94C-94HHH (1973).
Wagner, H.B., Polymer modification of Portland cement systems, Chemtech, Feb, 105-108 (1973).
Copeland, J.F., and Pense, A.W., Hardenability: Key to vessel plate strength, Hyd Proc, 53(4), 157-160 (1974).
Hentschel, R.A.A., Spunbonded sheet products, Chemtech, Jan, 32-41 (1974).
Nordin, S., Stainless special steels for the chemical industry, Chem Engnr, Nov, 724-727 (1974).
Hobson, C.G., and Christenson, A., Using castable refractories with success, Chem Eng, 29 Sept, 97-98 (1975).

1976-1980

Fryatt, J., Low thermal-mass furnace linings, Energy World, Feb, 7-13 (1976).
Margat, E.E., and Morrison, R.E., Evolution of man-made fibres, Chemtech, Nov, 702-709 (1976).
Martin, D., Wood for process plant, Proc Engng, Sept, 82-83 (1976).
McCandless, J.B., and Imgram, A.G., Evaluate materials using new nondestructive test method, Hyd Proc, 55(10), 159-161 (1976).
Ross, A.P., Rediscovering lead, Chem Eng, 24 Nov, 79-82 (1975); 2 Aug, 107-108; 22 Nov, 175-178 (1976).
Various, Chemicals in processing (feature report), Processing, July, 37-47 (1976).
Baker, D.S., Flame-retarding wood and timber products, Chem & Ind, 15 Jan, 74-79 (1977).
Arnold, J.L., Heat treatment protects steel alloys from hot gases, Chem Eng, 8 May, 205-208 (1978).
Bernett, F.E., and Wagner, H.B., Inorganic adhesives, Chemtech, Nov, 684-685 (1978).
Hanson, B.H., The use of titanium in the chemical industry, Chem Engnr, April, 276-280 (1978).
Lyman, W.S., and Cohen, A., Engineering with copper alloys, Chem Eng, 13 March, 99-102; 10 April, 147-150 (1978).
Sellers, T., From forest to plywood, Chemtech, Oct, 622-626 (1979).
Giragosian, N.H., Commercial development of Loctite, Chemtech, Oct, 604-609 (1980).
Gupta, N., Refractory materials, Proc Engng, Oct, 62-63 (1980).
Various, Guide to process-plant materials, Proc Engng, Oct, 69-77 (1980).

1981-1984

Herzegh, F., The tubeless tire, Chemtech, April, 224-228 (1981).
Knill, R.B., Trends in tire technology, Chemtech, Nov, 688-692 (1981).
Marra, G.G., and Youngquist, J.A., Wood composites, Chemtech, July, 418-421 (1981).
Schumacher, W.J., A stainless steel alternative to cobalt wear alloys, Chem Eng, 21 Sept, 149-152 (1981).
Various, Cements (topic issue), Chem & Ind, 19 Sept, 620-645 (1981).
Capes, P., Applications of titanium, Proc Engng, Nov, 52-53 (1982).
Gray, R.L., and Parham, R.A., Structure of wood, Chemtech, April, 232-241 (1982).
Redmond, J.D., and Miska, K.H., The basics of stainless steels (30th biennial report on materials of construction), Chem Eng, 18 Oct, 78-118 (1982).
Various, Adhesive bonding survey, Processing, June, 37-43 (1982).
Various, Blended cements (topic issue), Chem & Ind, 6 Nov, 829-840 (1982).
Various, Carbon and graphite (topic issue), Chem & Ind, 18 Sept, 675-713 (1982).
Wright, J.B., Cast austenitic alloys for valves, Chem Eng Prog, 78(12), 61-63 (1982).
Bensted, J., Oilwell cements, Chem & Ind, 17 Oct, 776-781 (1983).
Bucsko, R.T., Glass as a material of construction, Chem Eng Prog, 79(2), 82-85 (1983).
Capes, P., Advances in cement technology, Proc Engng, Nov, 41-43 (1983).
Davies, R.E., Blanket lining for refractories, Proc Engng, Jan, 49 (1983).
Ganapathy, V., Estimate maximum allowable pressures for steel piping, Chem Eng, 25 July, 99 (1983).
Rosenfelder, W.J., Industrial development of amorphous metal, Chem & Ind, 15 Aug, 639-641 (1983).
Smith, T., and Hoult, D., Glass linings for steels, Proc Engng, Nov, 51-53 (1983).
Various, Recent developments in materials used in the water supply industry (topic issue), Chem & Ind, 5 Sept, 659-678 (1983).
Various, Fifty years of road and building materials (topic issue), Chem & Ind, 5 Dec, 878-895 (1983).
Fletcher, J.R., New stainless steels for the process industries, Chem Engnr, May, 39-41 (1984).
Schillmoller, C.M., and Althoff, H.J., How to avoid failures of stainless steels, Chem Eng, 28 May, 119-122 (1984).
Schillmoller, C.M., and van den Bruck, U.W., Furnace alloys update, Hyd Proc, Dec, 55-59 (1984).
Seelinger, S.M., Structural steel: Galvanizing vs. painting? Chem Eng, 16 April, 101-104 (1984).
Various, 31st biennial report on materials of construction, Chem Eng, 29 Oct, 69-103 (1984).

1985-1988

Allcock, H.R., Inorganic macromolecules, C&E News, 18 March, 22-36 (1985).
Cheng, F.L., Materials for strong and light cars, Chemtech, Sept, 550-557 (1985).

Chruma, J.L., and Chapman, R.D., Nylon properties and applications, Chem Eng Prog, 81(1), 49-54 (1985).
Moretti, C., Glass of the past, Chemtech, June, 340-344 (1985).
Morimoto, T., Porous aluminium, Chemtech, Feb, 112-117 (1985).
Rave, T.W., Synthetic pulp, Chemtech, Jan, 54-62 (1985).
Anon., Materials of construction: Current literature, Chem Eng, 24 Nov, 65-84 (1986).
Broxterman, W.E., Adhesives, Chemtech, Jan, 44-47 (1986).
Redmond, J.D., Selecting second-generation duplex stainless steels, Chem Eng, 27 Oct, 153-155; 24 Nov, 103-105 (1986).
Reichle, W.T., Anionic clay minerals, Chemtech, Jan, 58-63 (1986).
Seagle, S.R., and Thomas, D.E., Status of titanium technology, Chem Eng Prog, 82(6), 63-68 (1986).
Mark, H.F., Textile science and engineering, Chem Eng Prog, 83(12), 44-54 (1987).
Endo, M., Uses and preparation of carbon fibres, Chemtech, Sept, 568-576 (1988).
McKay, G., Lignite: A versatile resource, Chem Engnr, Feb, 34-36 (1988).
Moslemi, A.A., Inorganically bonded wood composites, Chemtech, Aug, 504-510 (1988).
Zweben, C., The limitless world of composites, Chemtech, Dec, 733-737 (1988).

11.2 New Materials

1968-1980

Benson, K.E., Silicon crystals for microelectronic applications, Chem Eng Prog, 64(4), 93-101 (1968).
Roth, D.W.H.; Twilley, I.C., and Harvie, L.K., Synthetic fiber technology, Chem Eng, 16 Dec, 86-94 (1968).
Various, Textile engineering (topic issue), Chem Eng Prog, 65(10), 41-54 (1969).
Anon., Ferralium: A new corrosion-resisting alloy, Brit Chem Eng, 16(11), 1032-1033 (1971).
Kammermeyer, K., Biomaterials: Developments and applications, Chemtech, Dec, 719-727 (1971).
Kane, P.F., Semiconductors: A challenge in trace analysis, Chemtech, Sept, 532-539 (1971).
Castellano, J.A., and Brown, G.H., Thermotropic liquid crystals, Chemtech, Jan, 47-52; April, 229-235 (1973).
Fitzer, E., and Heym, M., Carbon fibres: The outlook, Chem & Ind, 21 Aug, 663-676 (1976).
Gregory, E., Practical superconductors, Chemtech, Aug, 516-522 (1976).
Sleight, A.W., Newer superconductors, Chemtech, July, 468-470 (1976).
Ball, J.R., Carbon fibres, Chem Engnr, May, 333-336 (1977).
Bruck, S.D., Smooth flexible biomaterials, Chemtech, April, 240-246 (1977).
Hess, D.W., Process technology of silicon integrated-circuits, Chemtech, July, 432-445 (1979).

Knittel, D.R., Zirconium, Chem Eng, 2 June, 95-98 (1980).

1982-1988

Seagle, S.R., and Bannon, B.P., Titanium: Its properties and uses, Chem Eng, 8 March, 111-113 (1982).

Brown, G.H., nd Crooker, P.P., Liquid crystals, C&E News, 31 Jan, 24-37 (1983).

Ubbelode, A.R., Well-ordered graphites for technical applications of materials with very high anisotropy, Chem & Ind, 15 Aug, 636-638 (1983).

Dotts, R.L.; Curry, D.M., and Tillian, D.J., Space shuttle materials, Chemtech, Oct, 616-626 (1984).

Fensom, D.H., and Clark, B., Tantalum: Its uses in chemical industry, Chem Engnr, Aug, 46-48 (1984).

Wallis, B., Current trends in seals materials, Chem Engnr, Nov, 10-13 (1984).

Bechgaard, K., and Jerome, D., Superconducting organic solids, Chemtech, Nov, 682-685 (1985).

Kurnik, R.T., Chemical vapor deposition in microelectronics, Chem Eng Prog, 81(5), 30-35 (1985).

Larrabee, G.B., Chemical technology of microelectronics, Chemtech, March, 168-174 (1985).

Lashway, R.W., Sintered alpha-silicon carbide, Chem Eng, 9 Dec, 121-122 (1985).

Pogge, H.B., Material aspects of semiconductors, Chemtech, Aug, 497-503 (1985).

Shriver, D.F., and Farrington, G.C., Solid ionic conductors, C&E News, 20 May, 42-57 (1985).

Asphahani, A.I., Overview of advanced materials technology, Chem Eng Prog, 82(6), 33-40 (1986).

King, R., New materials for old problems, Chem Engnr, May, 17-21 (1986).

Martin, C.; Rives, V., and Malet, P., Texture properties of titanium dioxide, Powder Tech, 46(1), 1-12 (1986).

Dagani, R., Superconductivity: A revolution in electricity is taking shape, C&E News, 11 May, 7-16 (1987).

Margolis, J.M., Advanced composites, Chem Eng Prog, 83(12), 30-43 (1987).

Parkinson, G., New ways to make crystals for semiconductor uses, Chem Eng, 25 May, 14-17 (1987).

Parkinson, G., CPI keeping close watch on superconductors craze (developments in high-temperature superconductors), Chem Eng, 17 Aug, 25,27,29 (1987).

Reisch, M.S., High-performance fibers, C&E News, 2 Feb, 9-14 (1987).

Dagani, R., New superconductors for higher temperatures, C&E News, 16 May, 24-29 (1988).

Fasth, R., and Eckert, C.H., Advanced materials markets, Chemtech, July, 408-412 (1988).

Forney, R.C., Advanced materials and technological innovation, Chemtech, March, 178-183 (1988).

Matsuda, H.S., Reinforcing fibers for advanced composites, Chemtech, May, 310-313 (1988).

McCartney, B., The semiconductor industry, Chem Engnr, July, 71 (1988).

Werschky, D.E., and Williams, C.E., Uses of
g Prog, 64(10), 74-78 (1968).
ceramics and coatings (symposium papers),
(1974).
Ceramic pipes for furnace heat recovery, Chem
79).
iber furnace linings, Hyd Proc, 60(4), 169-172

ics? Chem Eng, 20 Sept, 123-126 (1982).
Chemtech, April, 230-239 (1983).
, C&E News, 9 July, 26-40 (1984).
Ceramics heat up, Chemtech, April, 221-225

.A., Engineering ceramics, Chemtech, March,

ic issue), Chem & Ind, 6 Oct, 630-667 (1986).
, Ceramic films and coatings, Chem Eng Prog,

., Fine ceramics from glass, Chemtech, Nov,

as advanced structural materials, C&E News, 1

ne filters, Chem Engnr, Feb, 15-16 (1988).
nemical service (33rd biennial report), Chem

mechanical seals, Proc Engng, May, 77-78

Pattison, D.A., Picking the right elastomer to fit
ec, 118-128 (1967).
for the automotive industry, Chem Eng Prog,

Schoneman, D.P., Elastomers for high-impact
g, 63(7), 118-124 (1967).
om rigid polyurethane foams, Chem Eng, 25

ing, Chem Eng Prog, 63(6), 94-100 (1967).
ropylenes for automotive applications, Chem
7).
es and resins, Chem Eng, 18 Nov, 182-192

Vaill, E.W., Applications of thermoset molding, Chem Eng Prog, 64(12), 50-55 (1968).
Bott, T.R., and Barker, A.J., Behaviour of model composites in contact with different environments, Trans IChemE, 47, T188-193 (1969).
Spencer, F.J., Engineered plastics today, Hyd Proc, 48(7), 123-130 (1969).
Slama, W.R., and McMahon, R.E., Testing plastics for fire resistance, Chem Eng, 2 Nov, 120-125 (1970).
Platzer, N., Performance characteristics of thermoplastics, Chemtech, March, 165-175 (1971).
Roberts, R., Designing for fluoroplastic linings, Chem Eng, 22 Feb, 138-144 (1971).
Rodriguez, F., Prospects for biodegradable plastics, Chemtech, July, 409-415 (1971).
Shadduck, A.K., Designing for reinforced plastics, Chem Eng, 9 Aug, 116-122 (1971).
De Falco, J.J., High-speed fans of reinforced polyester, Chem Eng, 4 Sept, 88-92 (1972).
Griffith, J.R.; O'Rear, J.G., and Reines, S.A., Fluorinated epoxy resins, Chemtech, May, 311-316 (1972).
Peters, E.M., and Gervasi, J.A., Nylon 4 - between synthetic and natural fibers, Chemtech, Jan, 16-25 (1972).
Anon., Using large-diameter polyethylene pipe, Chem Eng, 29 Oct, 132-136 (1973).
Hyland, J., Teflon tank-linings, Chem Eng, 11 June, 124-128 (1973).
Kirk, D.N., Selectivity in reactions of epoxides, Chem & Ind, 3 Feb, 109-117 (1973).
Luce, W.A., TFE or FEP for corrosion-resistant linings, Proc Engng, April, 80-81 (1973).
Sheard, E.A., Commercial status of liquid elastomers, Chemtech, May, 298-303 (1973).
Brookman, R.S., Analysis of PVC resins, Chemtech, Dec, 741-743 (1974).
Conte, A.A., Painting with polymer powders, Chemtech, Feb, 99-103 (1974).
Masson, C.R., Polymer theory for silicate melts, Chemtech, Jan, 56-62 (1974).
Morton, E.R., Fiber-glass-reinforced plastics for corrosion resistance, Chem Eng, 24 Dec, 70-72 (1973); 21 Jan, 140-144 (1974).
Platzer, N., Copolymers, polyblends, and composites, Chemtech, Feb, 90-95 (1974).
Robertson, A.B., and Miller, W.A., Fluorinated polymers for coating CPI equipment, Chem Eng, 30 Sept, 138-146 (1974).
Samuels, R.J., Structured polymer properties, Chemtech, March, 169-177 (1974).
Seymour, R., Role of fillers in plastics composites, Chemtech, July, 422-425 (1974).

1975-1979

Allcock, H.R., Poly (organophosphazenes), Chemtech, Sept, 552-560 (1975).
Bruns, L.E., Plastics in nuclear processing plants, Chem Eng Prog, 71(1), 59-62 (1975).

Dorsey, J.S., Use of reinforced plastics for process equipment, Chem Eng, 15 Sept, 104-114 (1975).
Leslie, V.J.; Rose, J.; Rudkin, G.O., and Feltzin, J., Polyethersulfones, Chemtech, July, 426-432 (1975).
Rastogi, A.K., Fiber-reinforced plastics, Chemtech, June, 349-355 (1975).
Various, Polyethylene mini-feature (topic issue), Chem Eng Prog, 71(2), 73-80 (1975).
Harper, C.A., What you should know about plastics processing, Chem Eng, 10 May, 100-114 (1976).
O'Connor, M., Plastics in chemical plant fabrication, Chem Engnr, Feb, 95-96 (1976).
Patton, J.T., Applications of polyether-based polyurethanes, Chemtech, Dec, 780-784 (1976).
Pitman, J.S., Rubber linings, Chem Engnr, Feb, 97-98 (1976).
Puckett, D.B., Fiberglass-reinforced plastic tanks, Chem Eng, 27 Sept, 129-132 (1976).
Shannon, J., Polybutylene: A new thermoplastic for industrial piping, Chem Eng, 30 Aug, 121-124 (1976).
Shaw, J.F.G., Paint and protective coatings, Chem Engnr, Feb, 99-102 (1976).
Steinberg, M., Substituting polyketones and polysulfones for polyethylene, Chem Eng Prog, 72(9), 75-79 (1976).
Various, Plastics for process plant (feature report), Processing, Jan, 29-45 (1976).
Furman, H.N., The synergism of Teflon-lined fiberglass-reinforced plastic, Chem Eng Prog, 73(11), 92-94 (1977).
Platzer, N., Elastomers for toughening styrene polymers, Chemtech, Oct, 634-641 (1977).
Surtess, L.S., and Rooney, P., Specifying fiberglass-reinforced plastic piping, Chem Eng, 21 Nov, 215-216 (1977).
Todd, D.B., and Baumann, D.K., Compounding glass into plastics, Chem Eng Prog, 73(1), 65-68 (1977).
Downing, P.A., Glass fibre reinforced furane resins, Chem Engnr, April, 272-274 (1978).
Shen, M., and Kawai, H., Properties and structure of polymeric alloys (review paper), AIChEJ, 24(1), 1-20 (1978).
Various, Plasticisers for polyvinyl chloride and co-polymers (topic issue), Chem & Ind, 19 Aug, 610-621 (1978).
Various, Use of plastic pipelines in the process industries (topic issue), Chem & Ind, 3 June, 361-377 (1978).
Eberhart, T.M., Designing fiberglass tanks for earthquake conditions, Chem Eng, 15 Jan, 147-150 (1979).
Gegner, P.J., Using reinforced plastics, Chemtech, Nov, 676-681 (1979).
Hatch, L.F., and Matar, S., Introduction to polymer chemistry, Hyd Proc, 58(3), 165-172 (1979).
Hatch, L.F., and Matar, S., Thermoplastics, Hyd Proc, 58(9), 175-187 (1979).
Kraus, M., and Patchornik, A., Polymeric reagents, Chemtech, Feb, 118-128 (1979).
Morgan, P.W., Aromatic polyamides, Chemtech, May, 316-326 (1979).

Stevens, T.C., and Littlewood, M.J., Plastics pipes for water transport, Chem & Ind, 17 March, 205-210 (1979).
Various, Plastics in processing (feature report), Processing, April, 61-76 (1979).

1980-1983

Bauer, R.S., Versatile epoxies, Chemtech, Nov, 692-700 (1980).
Bedson, J.H., Elastomeric expansion joints, Chem Eng, 29 Dec, 34-40 (1980).
Fowler, T.J., and Scarpellini, R.S., Non-destructive testing of fibre-reinforced plastic equipment, Chem Eng, 20 Oct, 145-148; 17 Nov, 293-296 (1980).
Hatch, L.F., and Matar, S., Thermosetting resins, Hyd Proc, 59(1), 141-151 (1980).
Hatch, L.F., and Matar, S., Synthetic fibers, Hyd Proc, 59(4), 211-219 (1980).
Hatch, L.F., and Matar, S., Synthetic rubber, Hyd Proc, 59(5), 207-213 (1980).
Ohm, R.F., Polynorbornene: The porous polymer, Chemtech, March, 183-187 (1980).
Rolston, J.A., Fiberglass composite materials and fabrication processes, Chem Eng, 14 Jan, 96-110 (1980).
Owen, M.J., Behavior of silicones, Chemtech, May, 288-292 (1981).
Platzer, N., Commodity and engineering plastics, Chemtech, Feb, 90-94 (1981).
Senich, G.A., Migration to and from plastics, Chemtech, June, 360-365 (1981).
Various, Advances in polymer technology (topic issue), Chem & Ind, 21 Nov, 788-804 (1981).
Castro, G.O., Selecting and installing plastic-lined pipe, Chem Eng, 22 March, 112-118 (1982).
Eise, K., and Mielcarek, D.F., Compounding of additives and fillers, Chem Eng Prog, 78(1), 62-64 (1982).
Griffith, J.R., Epoxy resins containing fluorine, Chemtech, May, 290-293 (1982).
Johnsen, R.E., Specifying plastic-lined pipe, Chem Eng, 22 March, 119-125 (1982).
Kaeding, W.W.; Young, L.B., and Prapas, A.G., Para-methylstyrene, Chemtech, Sept, 556-562 (1982).
Kossoff, R.M., Engineering plastics industry, Chemtech, Sept, 552-555 (1982).
Margus, E.A., Engineered plastics for pumps, Chem Eng Prog, 78(12), 69-74 (1982).
Arkles, B., Uses of silicones, Chemtech, Sept, 542-555 (1983).
Brode, G.L.; Jones, T.R., and Chow, S.W., Phenol-formaldehyde resins, Chemtech, Nov, 676-681 (1983).
Legge, N.R., Thermoplastic elastomers. Chemtech, Oct, 630-639 (1983).
Robinson, T.L., Thermoplastic valves, Chem Eng Prog, 79(1), 62-67 (1983).
Sherwood, M., Polythene and its origins, Chem & Ind, 21 March, 237-242 (1983).
Vandenberg, E.J., Development of epichlorohydrin elastomers, Chemtech, Aug, 474-477 (1983).

1984-1987

Baines, D., Glass reinforced plastics in the process industries, Chem Engnr, July, 24-27 (1984).
Dibbo, A., Plastic process plant, Chem Engnr, Nov, 38-41 (1984).

Kardos, J.L., Bonding polymer composites, Chemtech, July, 430-434 (1984).
Mark, H.F., New elastomers, Chemtech, April, 220-228 (1984).
Bergenn, W.R., and Rigby, R.B., PES and PEEK: Tough engineering thermoplastics, Chem Eng Prog, 81(1), 36-38 (1985).
Dennis, R., Latex in the construction industry, Chem & Ind, 5 Aug, 505-511 (1985).
Dix, J.S., PPS: The versatile engineering plastic, Chem Eng Prog, 81(1), 42-44 (1985).
Galvin, T.J.; Chaudhari, M.A., and King, J.J., High-performance matrix resin systems, Chem Eng Prog, 81(1), 45-48 (1985).
Browning, C.E., Processing science of graphite/epoxy composites, Chem Eng Prog, 82(6), 41-44 (1986).
Ender, D.H., Elastomeric seals, Chemtech, Jan, 52-56 (1986).
Moseley, J.D., and Nowak, R.M., Engineering thermoplastics, Chem Eng Prog, 82(6), 49-54 (1986).
Niesse, J.E., Innovations in organic linings, Chem Eng Prog, 82(6), 55-62 (1986).
Webber, D., Engineering plastics, C&E News, 18 Aug, 21-46 (1986).
Greek, B.F., Plastics in 1987, C&E News, 24 Aug, 27-65 (1987).
Scott, G., Polymer durability: An essential design parameter, Chem & Ind, 21 Dec, 841-845 (1987).
Stupp, S.I., Thermotropic liquid crystal polymers, Chem Eng Prog, 83(12), 17-22 (1987).
Worthy, W., Polymer composites, C&E News, 16 March, 7-13 (1987).

1988

Baney, R.H., Designing preceramic polymers, Chemtech, Dec, 738-742 (1988).
Bradbury, J.H., and Pereva, M.C.S., Advances in epoxidation of unsaturated polymers, Ind Eng Chem Res, 27(12), 2196-2202 (1988).
Greek, B.F., Use of plastics additives, C&E News, 13 June, 35-57 (1988).
Kauffman, G.B., Development of Nylon, Chemtech, Dec, 725-731 (1988).
Keller, T.M., High-performance electrically conductive polymers, Chemtech, Oct, 635-639 (1988).
Lantos, P.R., Performance plastics: Why not thermosets? Chemtech, Feb, 99-102 (1988).
Lepenye, G., et al., Improved use-value of polyester-based fabrics, Periodica Polytechnica, 32(1), 167-168 (1988).
Massingill, J.L., High-solids, epoxy-based coatings, Chemtech, April, 236-241 (1988).
Mertzel, E.A.; Perchak, D.R.; Ritchey, W.M., and Keonig, J.L., Modeling of polymer network systems, Ind Eng Chem Res, 27(4), 580-586 (1988).
Nauman, E.B., et al., Compositional quenching: Process for forming polymer-in-polymer microdispersions and cocontinuous networks, Chem Eng Commns, 66, 29-56 (1988).
Proctor, A., Plastics for corrosion-resistant linings, Proc Engng, Aug, 37-38 (1988).
Rader, C.P., Thermoplastic elastomers, Chemtech, Jan, 54-59 (1988).
Reynolds, J.R., Electrically conductive polymers, Chemtech, July, 440-447 (1988).

Sacks, W., Multilayer plastic packaging containers, Chemtech, Aug, 480-483 (1988).
Smith, C.P., High performance polymers, Chemtech, May, 290-291 (1988).
Sperling, L.H., Interpenetrating polymer networks, Chemtech, Feb, 104-109 (1988).
Various, Underground applications of plastics, Chem & Ind, 4 July, 414-432 (1988).
Various, Polymer engineering (topic issue), Chem Eng Prog, 84(11), 35-62 (1988).
Wright, M.A.; Hamblin, N.R., and Rader, C.P., Thermoplastic elastomers for cars, Chemtech, June, 354-357 (1988).

11.5 Materials Applications

1967-1969

Black, S., Reduce equipment costs by correct design-in aluminium, Brit Chem Eng, 12(2), 233-236 (1967).
Boschma, L.G., Lubricants for future automotive engines, Chem Eng Prog, 63(5), 99-102 (1967).
Campbell, R.W., and Browning, J.E., Materials of construction for cryogenic processes, Chem Eng, 23 Oct, 188-197 (1967).
Darden, J.F., Selecting materials by service experience, Chem Eng, 13 Feb, 174-178 (1967).
Dell, G.J., Construction materials for Phos-acid manufacture, Chem Eng, 10 April, 234-242 (1967).
Newman, D.J., and Miller, R., Making nitric acid in all-stainless plants, Chem Eng, 31 July, 138-141 (1967).
Pike, J.J., Choosing materials for Phos-acid concentration, Chem Eng, 13 March, 192-198 (1967).
Rinckhoff, J.B., Making acid in iron and steel equipment, Chem Eng, 28 Aug, 160-162 (1967).
Salot, W.J., Safe and economical aluminum storage tanks, Chem Eng, 8 May, 172-178 (1967).
Anon., New materials data for high-pressure design, Chem Eng, 3 June, 122-124 (1968).
Avery, R.E., and Valentine, H.L., Materials for high-temperature piping systems, Chem Eng Prog, 64(1), 89-92 (1968).
Dukes, R.R., and Schwarting, C.H., Choosing materials for making chlorine and caustic, Chem Eng, 11 March, 206-214; 8 April, 172-176 (1968).
Irving, G.M., Construction materials for breweries, Chem Eng, 1 July, 100-104 (1968).
Lederman, P.B., and Kallas, D.H., Materials: Key to exploiting the oceans, Chem Eng, 3 June, 105-113 (1968).
Miller, R., Materials for formaldehyde, Chem Eng, 15 Jan, 182-188 (1968).
Miller, R., Matching materials to temperatures, Chem Eng, 6 May, 210-214 (1968).
Brown, R.W., and Sandmeyer, K.H., Applications of fused-cast refractories, Chem Eng, 16 June, 106-114 (1969).

Hines, J.G., The selection of materials for chemical plant, Trans IChemE, 47, T173-176 (1969).
Rappleyea, L., Stress-grading lumber, Chem Eng, 21 April, 126-130 (1969).
Skavdahl, R.E., and Zebroski, E.L., Finding materials for fast reactors of the future, Chem Eng, 11 Aug, 114-120 (1969).
Steensland, O., and Pulkkinen, R., Application of stainless steel in phosphoric acid manufacture, Brit Chem Eng, 14(4), 516-519 (1969).

1970-1974

Desensy, M.G.J., Materials for seawater cooling, Chem Eng, 15 June, 182-188 (1970).
Hilado, C.J., Predicting material flammability, Chem Eng, 14 Dec, 174-178 (1970).
Loeb, M.B., Thermal-contraction graphs for cryogenic temperatures, Chem Eng, 15 June, 194 (1970).
Various, Materials of construction (symposium papers), Chem Engnr, Dec, CE419-436 (1970).
Various, Materials, design and fabrication of chemical plant (symposium papers), Chem Engnr, Oct, CE312-350 (1970).
Adams, L., Supporting cryogenic equipment with wood, Chem Eng, 17 May, 156-158 (1971).
Fuchs, E., Essential welding knowledge for the project engineer, Brit Chem Eng, 16(10), 905-911; 16(11), 1033 (1971).
Hines, J.G., Selection of materials for chemical plants: Economic and communication aspects, Chem Engnr, March, 101-104, 106 (1971).
Montrone, E.D., and Long, W.P., Choosing materials of construction for carbon dioxide absorption systems, Chem Eng, 25 Jan, 94-99 (1971).
Hoffman, C.H., Wood-tank engineering, Chem Eng, 17 April, 120-122 (1972).
Hoffman, C.H., Consider wood for process-plant uses, Chem Eng, 20 March, 126-132 (1972).
Sheppard, W.L., Membranes behind brick, Chem Eng, 15 May, 122-126; 12 June, 110-116 (1972).
Bhat, V.K., and Carpenter, W.W., Thermomechanical properties of glass-lined equipment, Chem Eng, 3 Sept, 118-122 (1973).
Hart, G.L., A review of resins, reinforcements and composite systems for use in chemical plant applications, Chem Engnr, April, 215-220 (1973).
Pierce, R.R., and Bressi, V., Vessel linings for the process industry, Chem Eng Prog, 69(6), 104-109 (1973).
Tesmen, A.B., Materials of construction for process plants, Chem Eng, 19 Feb, 140-144; 19 March, 126-130; 14 May, 158-162 (1973).
Assini, J., Choosing welding fittings and flanges, Chem Eng, 2 Sept, 90-91 (1974).
Cameron, J.A., and Danowski, F.M., Selecting materials for centrifugal compressors, Hyd Proc, 53(6), 115-125 (1974).
Hoodbhoy, A.I., Urethane uses in automobiles, Chemtech, April, 238-240 (1974).
McDowell, D.W., Specifications for acidproof brick, Chem Eng, 10 June, 100-104 (1974).

Schley, J.R., Impervious graphite for process equipment, Chem Eng, 18 Feb, 144-150; 18 March, 102-110 (1974).
Various, Handling mineral acids (26th biennial materials of construction report), Chem Eng, 11 Nov, 118-146 (1974).
Verink, E.D., Aluminum alloys for saline waters, Chem Eng, 15 April, 104-110 (1974).

1975-1979
Kuhlkamp, A., Developments in industrial coatings, Chem & Ind, 16 Aug, 693-699 (1975).
MacNab, A.J., Design and materials requirements for coal gasification, Chem Eng Prog, 71(11), 51-58 (1975).
McDowell, D.W., Materials for handling phosphoric acid and phosphate fertilizers, Chem Eng, 4 Aug, 119-121; 1 Sept, 121-124 (1975).
Waterman, N.A., Selecting materials for process plant, Proc Engng, Jan, 68-69 (1975).
Artus, C.H., All about chemical hose, Chem Eng, 26 April, 121-124 (1976).
Berger, D.M., Preparing concrete surfaces for painting, Chem Eng, 25 Oct, 141-148 (1976).
Elgee, H., Using wood tanks, Chem Eng, 5 July, 95-98 (1976).
Mack, W.C., Selecting steel tubing for high-temperature service, Chem Eng, 7 June, 145-150 (1976).
Maukonen, D.W., and Wagner, J., In-place annealing of high-temperature furnace tubes, Chem Eng, 19 July, 173-174 (1976).
Various, Materials for high-temperature service (special report), Hyd Proc, 55(6), 75-90 (1976).
McDowell, D.W., Choosing materials for sulfuric-acid services, Chem Eng, 4 July, 137-140 (1977).
Morgan, J.D., and Kirby, R.C., Chemical engineers in the metals field, Chem Eng, 20 June, 111-116 (1977).
O'Hara, J.B.; Lochmann, W.J., and Jentz, N.E., Materials for coal liquefaction plants, Chem Eng, 11 April, 147-154 (1977).
Hackman, L.E., and Chambers, H., New refractories solve old problems, Hyd Proc, 57(11), 259-265 (1978).
Kobrin, G., and Kopecki, E.S., Choosing alloys for ammonia services, Chem Eng, 18 Dec, 115-128 (1978).
McDowell, D.W., Choosing materials for ethylbenzene services, Chem Eng, 16 Jan, 159-160 (1978).
Michels, H.T., and Hoxie, E.C., Alloys for sulfur dioxide scrubbers, Chem Eng, 5 June, 161-166 (1978).
Crowley, M.S., Design better vessel linings, Hyd Proc, 58(12), 127-130 (1979).
Dawson, J.D., Designing high-temperature piping, Chem Eng, 9 April, 127-132 (1979).
Lerman, M.J., Extended life for glass-lined equipment, Chem Eng, 27 Aug, 113-114; 24 Sept, 135-136 (1979).
Timmer, W., Materials for contact lenses, Chemtech, March, 175-179 (1979).

1980-1983

Berger, D., and Border, F., How to specify coatings, Chem Eng, 14 Jan, 123-124 (1980).
Deck, D.C., High-temperature effects on metallic materials, Chem Eng, 30 June, 131-136 (1980).
Kirby, G.N., How to select materials (29th biennial report on materials of construction), Chem Eng, 3 Nov, 86-149 (1980).
Ostrofsky, B., Materials identification in the plant, Chem Eng, 15 Dec, 91-92 (1980).
Brown, R.S., Selecting stainless steel for pumps, valves and fittings, Chem Eng, 9 March, 109-112 (1981).
Dhingra, A.K., Fibers in metal castings, Chemtech, Oct, 600-608 (1981).
Evans, L., Materials selection tips for process plants, Chem Eng, 4 May, 99-100 (1981).
Hill, R.B., Materials of construction in the electricity supply industry, Chem Engnr, May, 226-228 (1981).
Horowitz, N.C., Selecting materials for chlorine-gas neutralization, Chem Eng, 6 April, 105-108 (1981).
Marshall, W.W., Construction materials for chemical process industries, Chem Engnr, May, 221-225, 228 (1981).
McAusland, D.D., and Webb, J.A., Development of materials and fabrication requirements for oilfield production valves in sour (hydrogen sulfide/chloride) service, Chem Engnr, May, 229-232 (1981).
Ostrofsky, B., Materials identification in the plant, Chem Eng, 15 Dec, 92 (1980); 12 Jan, 141-142; 9 Feb, 119-124 (1981).
Cole, S.A., Materials for tubular filters, Chem Eng Prog, 78(10), 70-74 (1982).
Marsch, H.D., Nitriding of steel, Plant/Opns Prog, 1(3), 152-159 (1982).
Nagl, G.J., Effects of insulation on refractory structures, Chem Eng, 18 Oct, 127-128 (1982).
Rolston, J.A., Fiberglass composites in filter applications, Chem Eng Prog, 78(10), 75-79 (1982).
Schiefer, H.M., and Pape, P.G., Use of silicones in process plants, Chem Eng, 8 Feb, 123-128 (1982).
Sheppard, W.L., Avoiding problems with acid-resistant brick, Chem Eng, 3 May, 107-110 (1982).
Hoult, D., Glass-lined steel equipment can solve process problems, Chem Engnr, Nov, 32-33 (1983).
Kirchner, R.W., Materials of construction for flue-gas-desulphurization systems, Chem Eng, 19 Sept, 81-86 (1983).
Malcolmson, R.W., Elastomers for cars, Chemtech, May, 286-292 (1983).
McRae, R.C., Weld-ells or cold bent pipe? Hyd Proc, Sept, 139-142 (1983).
Redmond, J.D., and Miska, K.H., High-performance stainless steels for high-chloride service, Chem Eng, 25 July, 93-96; 22 Aug, 91-94 (1983).
Shuker, F.S., When to use refractory metals and alloys in the plant, Chem Eng, 2 May, 81-84 (1983).

1984-1986

Hergenrother, P.M., High-temperature adhesives, Chemtech, Aug, 496-502 (1984).

Pierce, R.R., and Semler, C.E., Ceramic and refractory linings for acid condensation, Chem Eng, 12 Dec, 81 (1983); 23 Jan, 101-104 (1984).
Sennik, L., Selecting elastomers for plate heat exchanger gaskets, Chem Engnr, Aug, 41-45 (1984).
Webb, W.P., and Gupta, S.C., Metals for hydrogen service, Chem Eng, 1 Oct, 113-116 (1984).
Baker-Counsell, J., Improved surface coatings, Proc Engng, Nov, 65-68 (1985).
Ireland, N., Pipework in sulphuric acid service, Chem Engnr, Nov, 21-22 (1985).
Kirby, G.N., Selecting alloys for chloride service, Chem Eng, 4 Feb, 81-83; 4 March, 99-102 (1985).
Klein, R.L., and Rancombe, A.J., Performance of water pipeline materials, Chem & Ind, 3 June, 353-358 (1985).
Schillmoller, C.M., Use these materials to retrofit ethylene furnaces, Hyd Proc, Sept, 101-104 (1985).
Silence, W.L.; Kolts, J., and Wu, J.B.C., Using tests and service histories to select metals and alloys, Chem Eng, 29 April, 79-82 (1985).
Caprio, J.A., Third generation fluid catalytic cracking unit refractories, Hyd Proc, 65(3), 51-52 (1986).
Hirschfeld, T., Microengineering applications, Chemtech, Feb, 118-123 (1986).
Hoffman, A.S., Materials for biotech, Chemtech, July, 426-432 (1986).
Jones, J.E., and Olson, D.L., Selecting arc-welding processes, Chem Eng, 3 March, 117-120; 31 March, 131-134 (1986).
Niebur, D.R., Using borosilicate glass in the process plant, Chem Eng, 21 July, 81-84 (1986).
Schillmoller, C.M., Consider these alloys for ammonia plant retrofit, Hyd Proc, 65(9), 63-65 (1986).
Schillmoller, C.M., Solving high-temperature problems, Chem Eng, 6 Jan, 83-87 (1986).
Tuthill, A.H., Fabrication and post-fabrication cleanup of stainless steels, Chem Eng, 29 Sept, 141-146 (1986).

1987-1988

Dekumbis, R., Surface treatment of materials by lasers, Chem Eng Prog, 83(12), 23-29 (1987).
Lerman, M.J., Extending the life of glass-lined equipment, Chem Eng, 27 April, 40-49 (1987).
Lowrie, R., Materials for oxygen service: Selection criteria and careful design, Chem Eng, 27 April, 75-80 (1987).
MacNab, A.J., Alloys for ethylene-cracking furnace tubes, Hyd Proc, 66(12), 43-45 (1987).
Nowell, D., Explosive metal cladding process, Proc Engng, Aug, 47-48 (1987).
Turner, M., Limitations of materials, Chem Engnr, Nov, 17-18 (1987).
Anon., Material reduces scrubber corrosion, Chem Eng, 18 July, 126 (1988).
Beardmore, P., Automobile materials of the future, Chemtech, Oct, 599-603 (1988).
Carmen, R., Glass-lined equipment: Replace or reglass? Chem Eng, 24 Oct, 99-102 (1988).

Crowe, C.R., and Cooper D., Brick/membrane linings for acid service, Chem Eng, 18 July, 83-86 (1988).
Ember, L.R., Preserving and restoring historical objects, C&E News, 14 Nov, 10-19 (1988).
Jordan, G., Material selection for peristaltic pumps, Processing, Dec, 39-40 (1988).
Parkinson, G., Rapid solidification of metals, Chem Eng, 11 April, 18-21 (1988).
Ulrich, D.R., Sol-gel processing for homogeneous materials, Chemtech, April, 242-249 (1988).

11.6 Corrosion and In-service Conditions

1967-1969

Leonard, R.B., Preventing corrosion in Phos-acid concentration, Chem Eng, 5 June, 158-162 (1967).
Reilly, A.F., Uses of urethane slab foam, Chem Eng Prog, 63(5), 104-108 (1967).
Rinckhoff, J.B., Controlling corrosion in wet-gas sulfuric acid plants, Chem Eng, 20 Nov, 158-162 (1967).
Sisler, C.W., Zinc paints yield optimum corrosion resistance, Chem Eng, 16 Jan, 182-189 (1967).
Burkhalter, L.C.; Shelton, M.F., and Tomlinson, E.H., Cathodic cure for corrosion, Chem Eng, 21 Oct, 164-170 (1968).
Iverson, W.P., Microbiological corrosion, Chem Eng, 23 Sept, 242-244 (1968).
Lichtenberg, E.B., and Katona, E., Corrosion resistance of aluminium and its alloys in the chemical industry, Int Chem Eng, 8(2), 313-318 (1968).
Mapstone, G.E., Nomogram for pressure of hydrogen in steel, Brit Chem Eng, 13(4), 547 (1968).
Pierce, R.R., Protecting concrete floors from chemicals, Chem Eng, 16 Dec, 118-124 (1968).
Sorell, G., Controlling corrosion by process design, Chem Eng, 29 July, 162-170 (1968).
Wagner, J., Corrosion problems with sea-water cooling, Chem Eng Prog, 64(10), 59-66 (1968).
Anon., Setting a value on creep strength, Chem Eng, 19 May, 208-210 (1969).
Barker, A.J., and Bott, T.R., Corrosion studies of glass fibre surfaces using a scanning electron microscope, Trans IChemE, 47, T212-221 (1969).
Cornet, I., and Kappesser, R., Cathodic protection of a rotating cylinder, Trans IChemE, 47, T194-197 (1969).
Heckler, N.B., Failure analysis and selection of materials, Chem Eng, 3 Nov, 100-106 (1969).
Landrum, R.J., Designing for corrosion resistance, Chem Eng, 24 Feb, 118-124; 24 March, 172-180 (1969).
Leonard, R.B., New nickel-based corrosion-resistant alloys, Chem Eng Prog, 65(7), 84-86 (1969).
Obrecht, M.F., Scale and corrosion control in aqueous systems from the environmental engineer's viewpoint, Trans IChemE, 47, T183-187 (1969).

Postlethwaite, J., and Sharp, D.M., Interfacial concentrations in aqueous corrosion processes, Trans IChemE, 47, T198-203 (1969).
Rama Char, T.L., and Padma, D.K., Corrosion inhibitors in industry, Trans IChemE, 47, T177-182 (1969).
Roberts, K.J., and Shemilt, L.W., Strain effects in the corrosion of copper in a flowing electrolyte, Trans IChemE, 47, T204-211 (1969).
Staehle, R.W., Effects of fabrication and processing on stress corrosion cracking of Fe-Ni-Cr alloys, Trans IChemE, 47, T227-240 (1969).
Thomas, B., Designing brick linings to resist hot chemicals, Chem Eng, 1 Dec, 110-116 (1969).
Various, Corrosion in the chemical industry (symposium papers), Chem Engnr, March, CE60-95 (1969).

1970-1971

Brooke, J.W., Inhibitors: New demands for corrosion control, Hyd Proc, 49(3), 138-144; 49(8), 107-110; 49(9), 299-302; 49(10), 117-122 (1970).
Dunlop, A.K., Using corrosion inhibitors, Chem Eng, 5 Oct, 108-114 (1970).
Estefan, S.L., Design guide to metallurgy and corrosion in hydrogen processes, Hyd Proc, 49(12), 85-92 (1970).
McCoy, J.D., and Hamel, F.B., New corrosion rate data for hydrodesulfurizing units, Hyd Proc, 49(6), 116-120 (1970).
Most, C.R., Comparing coatings for wear and corrosion-resistance, Chem Eng, 26 Jan, 140-145 (1970).
Schillings, R.C., Protection for chlorine load cells, Chem Eng Prog, 66(2), 68-69 (1970).
Siebert, O.W., Failure analysis: A materials engineering tool, Chem Eng Prog, 66(9), 63-65 (1970).
Various, Corrosion control practices (topic issue), Chem Eng Prog, 66(10), 33-65 (1970).
Various, Protective coatings (topic issue), Chem Eng Prog, 66(8), 31-53 (1970).
Weaver, P.E., Specifying coatings, Hyd Proc, 49(2), 127-128 (1970).
Anon., Process-corrosion testing, Chem Eng, 22 March, 116-120 (1971).
Battilana, R.E., Fretting corrosion under mechanical seals, Chem Eng, 8 March, 130-132 (1971).
Freyling, E.N., Adding automatic control to corrosion protection, Chem Eng, 19 April, 130-133 (1971).
Henthorne, M., Fundamentals of corrosion, Chem Eng, 17 May, 127-132; 14 June, 102-106; 26 July, 99-104; 23 Aug, 89-94; 20 Sept, 159-164; 18 Oct, 139-146; 15 Nov, 163-166; 27 Dec, 73-79 (1971).
Krystow, P.E., Materials and corrosion problems in urea plants, Chem Eng Prog, 67(4), 59-64 (1971).
McDonald, D.P., Corrosion-resistant materials and their applications, Brit Chem Eng, 16(9), 801-804 (1971).
Pierce, R.R., Protecting metal from corrosive atmospheres, Chem Eng, 11 Jan, 160-166 (1971).
Ross, T.K., Fundamentals of corrosion protection in chemical plant, Chem Engnr, March, 95-101 (1971).
Shaw, M.C., Fundamentals of wear, Chemtech, July, 432-439 (1971).

Wismer, M., and Bosso, J.F., Make the part the cathode: Key to resistant coatings, Chem Eng, 14 June, 114-118 (1971).

1972-1975

Atkinson, H.E., Fluoroplastic linings for corrosive service, Chem Eng, 25 Dec, 76-80 (1972).

Bravenec, E.V., Why steels fracture, Chem Eng, 10 July, 100-104 (1972).

Catlett, R.E., Specifications and the corrosion engineer, Chem Eng, 7 Aug, 90-94 (1972).

Cooper, C.M., Specifying corrosion allowances, Hyd Proc, 51(5), 123-126 (1972).

Henthorne, M., Fundamentals of corrosion, Chem Eng, 10 Jan, 103-108; 7 Feb, 82-87; 6 March, 113-118; 3 April, 97-102 (1972).

McGill, W.A., and Weinbaum, M.J., Aluminum reducing pipe and tubing corrosion? Hyd Proc, 51(6), 127-128 (1972).

Butwell, K.F.; Hawkes, E.N., and Mago, B.F., Corrosion control in carbon dioxide removal systems, Chem Eng Prog, 69(2), 57-61 (1973).

Ewald, G.W., Increasing the life of synthetic pond-linings, Chem Eng, 1 Oct, 67-70 (1973).

Maukonen, D., and Vest, G., Heat-treating welds, Chem Eng, 9 July, 100-102 (1973).

McDowell, D.W., Corrosion: Back to basics, Chem Eng, 1 Oct, 92-96 (1973).

Mills, J.F., and Oakes, B.D., Bromine chloride: Less corrosive than bromine, Chem Eng, 6 Aug, 102-106 (1973).

Anon., Predicting and preventing brittle failures, Chem Eng, 5 Aug, 114-116 (1974).

Bunsell, A.R., Fiber failure and fatigue, Chemtech, May, 292-299 (1974).

Creamer, E.L., Stress corrosion cracking of reformer tubes, Chem Eng Prog, 70(8), 69-70 (1974).

Evans, F.L., Corrosion in refineries, Hyd Proc, 53(4), 109-112 (1974).

Lambertin, W.J., and Vaughan, F.H., Equipment failure by catastrophic brittle failure, Hyd Proc, 53(9), 217-221 (1974).

McDowell, D.W., Corrosion in urea-synthesis reactors, Chem Eng, 13 May, 118-124 (1974).

Phelan, J.V., and Mandel, S.B., Cutting condensate corrosion with catalyzed hydrazine, Chem Eng, 30 Sept, 148-150 (1974).

Szymanski, W.A., and Kloda, D.W., Polyester and furfural alcohol resins for corrosion control, Chem Eng Prog, 70(1), 51-54 (1974).

Berger, D.M., How to test linings for corrosion-resistant tanks, Chem Eng, 14 April, 100-101 (1975).

Cangi, J.W., How 'carbon pickup' can cause casting failures, Chem Eng, 12 May, 106-110 (1975).

Margus, E., Polyvinylidene fluoride for corrosion-resistant pumps, Chem Eng, 27 Oct, 133-134 (1975).

Zeis, L.A., and Paul, G.T., Minimizing stress cracking, Hyd Proc, 54(9), 229 (1975).

1976-1977

Berger, D.M., Liquid-applied linings for steel tanks, Chem Eng, 22 Dec, 65-67 (1975); 19 Jan, 123-126 (1976).
Cracknell, A., The effects of hydrogen on steel, Chem Engnr, Feb, 92-94 (1976).
Daly, J.J., Controlled shotpeening prevents stress-corrosion cracking, Chem Eng, 16 Feb, 113-116 (1976).
Lee, R.P., Tracing the causes of metal failures, Chem Eng, 13 Sept, 213-220 (1976).
Various, Plant corrosion problems, Hyd Proc, 55(8), 145-151 (1976).
Various, Corrosion-resistant stainless steels and high-nickel alloys (27th biennial materials of construction report), Chem Eng, 22 Nov, 118-152 (1976).
Berger, D.M., Applicator's guide to zinc-rich primers, Chem Eng, 14 March, 147-150 (1977).
Berger, D.M., How corrosion theory relates to protective coatings, Chem Eng, 1 Aug, 77-80; 29 Aug, 89-94 (1977).
Brautigam, F.C., Welding practices that minimize corrosion, Chem Eng, 17 Jan, 145-147; 14 Feb, 97-102 (1977).
Bravery, A.F., Biodeterioration of solid and constructional timbers, Chem & Ind, 20 Aug, 675-678 (1977).
Hawk, C.W., Do you understand galvanic corrosion? Chem Eng, 6 June, 193-195 (1977).
Lee, R.P., Systematized failure analysis, Chem Eng, 16 Aug, 105-106 (1976); 3 Jan, 107-108; 31 Jan, 129-134 (1977).
Miksic, B.A., Volatile corrosion-inhibitors find a new home, Chem Eng, 26 Sept, 115-118 (1977).
Morris, P.E., and Kain, R.M., An electrochemical approach to alloy protection, Chem Eng Prog, 73(6), 103-104 (1977).
Schumacher, W.J., Wear and galling of materials, Chem Eng, 9 May, 155-160 (1977).
Sengupta, A.K., Scaling-corrosion test made easy, Chem Eng, 1 Aug, 83 (1977).
Various, Plant corrosion (feature report), Processing, July, 41-59 (1977).

1978-1979

Amos, R.S., and Townsend, D.W., Stainless steels for corrosion prevention in process plant (review paper), Chem Engnr, April, 270-271 (1978).
Berger, D.M., Six stages in the corrosion of coated steel, Chem Eng, 28 Aug, 121-122 (1978).
Clark, C.C., Damage to alloy steels by high-temperature hydrogen, Chem Eng, 3 July, 87-90 (1978).
Dawson, J.L., Methods of corrosion control, Chem Engnr, April, 266-269 (1978).
Evans, L.S., Corrosion in the chemical and process industries, Chem Engnr, April, 263-265 (1978).
Feist, W.C., Protecting wooden structures, Chemtech, March, 160-162 (1978).
Hodge, F.G., High performance alloys make wet scrubbers work, Chem Eng Prog, 74(10), 84-88 (1978).
Miller, R.M., Control initial aqueous condensate corrosion, Hyd Proc, 57(6), 135-137 (1978).

Moreland, P.J., and Hines, J.G., Corrosion monitoring: Select the right system, Hyd Proc, 57(11), 251-255 (1978).
Pritchard, G.; Swampillai, G., and Taneja, N., Degradation of polyester-glass laminates by hot water: Some experimental data, Trans IChemE, 56, 96-100 (1978).
Sheppard, W.L., Using chemical-resistant masonry in air-pollution-control equipment, Chem Eng, 20 Nov, 203-211 (1978).
Various, Coatings and surface treatment (feature report), Processing, July, 43-61 (1978).
Various, Predictability and integrity of materials preservation (topic issue), Chem & Ind, 2 Sept, 643-651 (1978).
Various, Process corrosion control (topic issue), Chem Eng Prog, 74(3), 37-74 (1978).
Various, Use of lined pipe and equipment for corrosive applications (28th biennial materials of construction report), Chem Eng, 20 Nov, 117-171 (1978).
Armour, A.W., and Robitaille, D.R., Corrosion inhibition by sodium molybdate, J Chem Tech Biotechnol, 29, 619-628 (1979).
Burrill, K.A., Corrosion product transport in water-cooled nuclear reactors, Can JCE, 55, 54-64 (1977); 56, 79-89 (1978); 57, 211-220 (1979).
Cross, T.A., How to evaluate urethane coatings, Chem Eng, 22 Oct, 153-156 (1979).
Figg, J., Corrosion of steel in concrete, Chem & Ind, 20 Jan, 39-43 (1979).
Flanders, R.B., Tantalum for corrosion resistance, Chem Eng, 17 Dec, 109-110 (1979).
Moore, R.E., Selecting materials to meet environmental conditions, Chem Eng, 2 July, 101-103; 30 July, 91-94 (1979).
Newman, J., Fighting corrosion with titanium castings, Chem Eng, 4 June, 149-154 (1979).
Romano, F.J., Corrosion and related problems in glass-lined vessels, Chem Eng Prog, 75(3), 92-94 (1979).
Rothwell, G.P., Measuring corrosion as it happens, Chem Eng, 7 May, 107-112 (1979).
Various, Corrosion and surface treatment (feature report), Processing, May, 73-87 (1979).
Various, Plant corrosion (topic issue), Proc Engng, Nov, 52-111 (1979).

1980-1981

Britton, C., Corrosion update, Proc Engng, Oct, 45-49 (1980).
Crombie, D.J.; Moody, G.J., and Thomas, J.D.R., Corrosion of iron by sulphate-reducing bacteria Chem & Ind, 21 June, 500-504 (1980).
Dickie, R.A., and Smith, A.G., How paint arrests rust, Chemtech, Jan, 31-35 (1980).
Gallagher, R., Beat corrosion with rubber hose, Chem Eng, 8 Sept, 105-118 (1980).
Kirby, G.N., Corrosion performance of carbon steel, Chem Eng, 12 March, 72-84 (1979); 17 Nov, 5 (1980).
Kraus, N.J., Fiberglass-reinforced plastic underground horizontal tanks for corrosive chemicals, Chem Eng, 11 Feb, 125-128 (1980).

Mackay, W.B., Cathodic protection material and design for offshore applications, Chem Engnr, Aug, 546-548 (1980).
McIntyre, D.R., How to prevent stress-corrosion cracking in stainless steels, Chem Eng, 7 April, 107-112; 5 May, 131-136 (1980).
Schillmoller, C.M., Alloys to resist chlorine, hydrogen chloride and hydrochloric acid, Chem Eng, 10 March, 161-164 (1980).
Various, Corrosion control (feature report), Processing, Nov, 43-65 (1980).
Various, Materials for equipment protection and linings (feature report), Processing, May, 67,69,72,97,101-105 (1980).
Verhoff, F.H., and Choi, M.K., Effects of sulphuric acid condensation on stack-gas equipment, J Inst Energy, 53, 92-98 (1980).
Berger, D.M., Specifying zinc-rich primers, Chem Eng, 14 Dec, 101-104 (1981).
Britton, C., Controlling corrosion, Proc Engng, March, 35-38 (1981).
Danilov, B., Examples of corrosion control, Hyd Proc, 60(3), 115-118 (1981); 60(4), 192-194 (1981).
Ludema, K.C., How to select material properties for wear resistance, Chem Eng, 27 July, 89-92 (1981).
Magee, T.R.A., and McKay, G., Corrosion: Monitoring and prevention, Chemtech, Feb, 104-107 (1981).
Rehrig, P., Selecting centrifugal compressor materials for harsh environments, Hyd Proc, 60(10), 137-139 (1981).
Strutt, J.E.; Robinson, M.J., and Turner, W.H., Recent developments in electrochemical corrosion monitoring techniques, Chem Engnr, Dec, 567-572 (1981).
Treseder, R.S., Guarding against hydrogen embrittlement, Chem Eng, 29 June, 105-108 (1981).
Various, Corrosion report, Processing, Oct, 25-31 (1981).
Wallace, A.E., and Webb, W.P., Cut vessel costs with realistic corrosion allowances, Chem Eng, 24 Aug, 123-126 (1981).
Weismantel, G.E., Paints and coatings for CPI plants and equipment, Chem Eng, 20 April, 130-143 (1981).
Wiles, D.M., and Carlsson, D.J., Photodegradation, Chemtech, March, 158-161 (1981).

1982-1983

Anon., Corrosion report, Processing, Dec, 27, 31 (1982).
Berger, D.M., Electrochemical and galvanic corrosion of coated steel surfaces, Chem Eng, 28 June, 103-109 (1982).
Gossett, J.L., Stop stress-corrosion cracking, Chem Eng, 15 Nov, 143-146 (1982).
Imoto, Y.; Terada, S., and Maki, K., Predicting creep damage, Plant/Opns Prog, 1(2), 127-134 (1982).
Kawai, T., et al., Creep-rupture properties of HK40 spun cast tubes, Plant/Opns Prog, 1(3), 181-186 (1982).
Kirby, G.N., What you need to know about maintenance paints, Chem Eng, 26 July, 75-78; 23 Aug, 113-116 (1982).
Konoki, K., et al., Creep rupture of steam reforming tube due to thermal stress, Plant/Opns Prog, 1(2), 122-127 (1982).

Mallinson, J.H., Abrasion of fiber-reinforced plastics in corrosive environments, Chem Eng, 31 May, 143-146 (1982).
Takemoto, M., Stress corrosion cracking of austenitic stainless steel: Effect of environmental conditions on crack propagation, Int Chem Eng, 22(2), 338-346 (1982).
Various, Corrosion and materials report, Processing, Nov, 13-23 (1982).
Berger, D.M., How to inspect paints and coatings, Chem Eng, 7 March, 177-180; 4 April, 105-110 (1983).
Chen, Y.S., Scaling and corroding tendencies of water, Chem Eng, 14 Nov, 253-254 (1983).
Crook, P., and Asphahani, A., Alloys to protect against corrosion and wear, Chem Eng, 10 Jan, 127-132 (1983).
El-Dahshan, M.E., The role of carbides in the accelerated attack of cobalt-based alloys, CER&D, 61, 13-20 (1983).
Figg, J., Chloride and sulphate attack on concrete, Chem & Ind, 17 Oct, 770-775 (1983).
France, P.E., High-temperature silicone-based coatings, Chem Eng, 27 June, 61-63 (1983).
Mayenkar, K.V., Quick way to determine scaling or corrosive tendencies of water, Chem Eng, 30 May, 85-86 (1983).
Miller, D.R.; Begeman, S.R., and Lintner, E.L., Current and future applications of on-line surveillance and monitoring systems in the petroleum industry, Chem & Ind, 17 Oct, 782-785 (1983).
Spencer, G.R., Program for cooling-water corrosion and scaling, Chem Eng, 19 Sept, 61-65 (1983).
Sugitani, J., et al., Ultrasonic determination of creep damage in catalyst tubes, Plant/Opns Prog, 2(1), 40-47 (1983).

1984-1985

Dorsey, J.S., Controlling external underground corrosion, Chem Eng, 5 March, 77-84 (1984).
Elliott, P., Controlling process-plant corrosion, Proc Engng, Nov, 43-46 (1984).
Hullcoop, R., Paints and coatings: A survey, Processing, April, 13-16 (1984).
Sheppard, W.L., Failure analysis of chemically resistant monolithic surfacings, Chem Eng, 9 July, 123-125 (1984).
Various, Corrosion-resistant materials, Processing, Nov, 26-29; Dec, 24 (1984).
Anon., Corrosion update, Processing, Feb, 21-24 (1985).
Baker-Counsell, J., Radioactivation for corrosion monitoring, Proc Engng, Feb, 47,49 (1985).
Beevers, A., Spray-on coating protection, Proc Engng, April, 40-41 (1985).
Brandsema, W., Constructing with corrosion-resistant lead, Chem Eng, 14 Oct, 111-114 (1985).
Dahlberg, E.P., and Zipp, R.D., Fracture and the scanning electron microscope, Chemtech, Feb, 118-122 (1985).
Ganainy, O.E., Failure of dissimilar metals weld in reformer tubes, Plant/Opns Prog, 4(3), 149-154 (1985).
Hale, K., Detecting cracks with optical-fibre sensors, Proc Engng, Aug, 27 (1985).

Lee, T.S., Preventing galvanic corrosion in marine environments, Chem Eng, 1 April, 89-92 (1985).
Mattsson, E., Corrosion: An electrochemical problem, Chemtech, April, 234-243 (1985).
Pelosi, P.F., and Cappabianca, C.J., Corrosion control in steam and condensate lines, Chem Eng, 24 June, 61-63 (1985).
Pickard, S.S., Sulfur concrete for acid resistance, Chem Eng, 22 July, 77-80 (1985).
Rolston, J.A., Dual-polymer laminates for corrosion-resistant equipment, Chem Eng, 7 Jan, 89-92 (1985).

1986-1987

Anon., On-line monitors reduce process corrosion, Processing, Feb, 34-35 (1986).
Baker-Counsell, J., Database/CAD link-up organises corrosion data gathering, Proc Engng, June, 33-35 (1986).
Bowen, J.M., and Campbell, J.D., Metallurgical examination of existing piping in ammonia plant, Plant/Opns Prog, 5(2), 81-85 (1986).
DeClerck, D.H., and Patarcity, A.J., Guidelines for selecting corrosion-resistant materials (32nd biennial report on materials of construction), Chem Eng, 24 Nov, 47-63 (1986).
Dillon, C.P., Corrosion Control in the Chemical Process Industries, McGraw-Hill, New York (1986).
Guthrie, J.E.; Braunstein, B.A.; Harrington, M.T., and Merrick, R.D., Monitoring corrosion with a microcomputer, Chem Eng, 1 Sept, 81-84 (1986).
King, R.A., Expert systems for materials selection and corrosion, Chem Engnr, Dec, 42-43 (1986).
Martin, J.R., and Noon, D.W., Protecting electronic systems in corrosive environments, Chem Eng, 9 June, 85-88 (1986).
Patton, D.M., and Clemente, M.J., Hard-facing protects against wear, Chem Eng, 3 Feb, 93-94 (1986).
Schillmoller, C.M., Amine stress cracking: Causes and cures, Hyd Proc, 65(6), 37-39 (1986).
Taffe, P., Magnetic 'pig' detects corrosion problems, Processing, Sept, 19-20 (1986).
Turner, M., and King, R., Should you worry about corrosion monitoring? Chem Engnr, Feb, 17-21 (1986).
Tuthill, A.H., Installed cost of corrosion-resistant piping, Chem Eng, 3 March, 113-115; 31 March, 125-128; 28 April, 83-84; 26 May, 99-100; 23 June, 131-133 (1986).
Beckwith, B.M., Evaluating cladding processes, Chem Eng, 16 Feb, 191-193 (1987).
Chowdhury, J., Coatings get a new look, Chem Eng, 19 Jan, 14-19 (1987).
King, R.A., and Miller, R.G., Sulphide: The unwelcome part of the sulphur cycle, Chem Engnr, July, 38-39 (1987).
Lock, J., Corrosion update, Processing, Feb, 17-20 (1987).

Lunde, L., and Nyborg, R., Effect of oxygen and water on stress corrosion cracking of mild steel in liquid and vaporous ammonia, Plant/Opns Prog, 6(1), 11-16 (1987).

Mitchell Liss, V., Preventing corrosion under insulation, Chem Eng, 16 March, 97-100 (1987).

Steele, D.F., Corrosion control in nuclear fuel reprocessing plants, CER&D, 65(5), 375-380 (1987).

Various, Corrosion and Abrasion Resistant Materials '87 (special advertising section), Chem Eng, 17 Aug, 53-100 (1987).

Wilkie, F., Linings for corrosion resistance in aggressive media, Proc Engng, Feb, 49-53 (1987).

1988

Anon., Assessing the coating condition of buried pipelines, Proc Engng, Dec, 47 (1988).

Atkinson, A., Reducing waterside corrosion in boiler plant, Proc Engng, April, 59, 61 (1988).

Chowdhury, J., Cathodic protection applications, Chem Eng, 12 Sept, 25-29 (1988).

Jones, T.N., New interpretation of alkali-silica reaction and expansion mechanisms in concrete, Chem & Ind, 18 Jan, 40-44 (1988).

Muller, H.M., and Branch, C.A., Comparison of indices for scaling and corrosion tendency of water, Can JCE, 66(6), 1005-1007 (1988).

Schofield, M., Corrosion case studies, Chem Engnr, March, 39-40 (1988).

Schofield, M., and King, R., Corrosion control, Chem Engnr, Dec, 83-84 (1988).

Turner, M., Corrosion testing, Chem Engnr, July, 52-54 (1988).

Various, Corrosion (feature report, plus 18 page supplement), Processing, April, 30-38 (1988).

CHAPTER 12

BIOTECHNOLOGY

12.1 Medical Applications

1967-1979

Guccione, E., Biomedical engineering, Chem Eng, 30 Jan, 107-124 (1967).

Guccione, E., How the artificial kidney cleans blood, Chem Eng, 30 Jan, 94-96 (1967).

Flower, J.R., Chemical engineering in medicine: The artificial kidney and lung machines, Chem Engnr, May, CE120-139 (1968).

Grassmann, P., Chemical engineering and medicine, Chem Engnr, June, CE233-240 (1969).

Hills, B.A., Chemical engineering principles in medicine and biology, Brit Chem Eng, 16(8), 700-703 (1971).

Rogers, A., Engineering and medicine: The interdisciplinary relationship, Chem Engnr, July, 258-260 (1972).

Huckaba, C.E.; Hansen, L.W.; Downey, J.A., and Darling, R.C., Calculation of temperature distribution in the human body, AIChEJ, 19(3), 527-532 (1973).

Davis, E.J.; Cooney, D.O., and Chang, R., Mass transfer between capillary blood and tissues, Chem Eng J, 7(3), 213-226 (1974).

Various, Cardiac pacemakers (topic issue), Chem & Ind, 20 July, 562-567 (1974).

White, P.A.F., A chemical engineering contribution to medical technology? Chem Engnr, July, 465-466, 474 (1974).

Huckaba, C.E.; Tam, H.S.; Darling, R.C., and Downey, J.A., Prediction of dynamic temperature distributions in the human body, AIChEJ, 21(5), 1006-1012 (1975).

Michaels, A.S.; Chandrasekaran, S.K., and Shaw, J.E., Drug permeation through human skin: Theory and in-vitrio experimental measurement, AIChEJ, 21(5), 985-996 (1975).

Lovinger, A.J., and Gryte, C.C., Simulation of the cardiopulmonary circulation, Chem Eng Educ, 10(1), 28-32,39 (1976).

Pasternack, A., Engineering aspects of artificial kidney development and operation, Int Chem Eng, 16(1), 1-10 (1976).

Sitharamayya, S., and Mathur, V.K., The artificial kidney: A review, Chem Engnr, Feb, 112-114 (1976).

Crick, F., Predictions in biology, Chemtech, May, 298-305 (1979).

1981-1988

Artandi, C., Fibers in medicine, Chemtech, Aug, 476-481 (1981).

Henley, E.J., Engineering + medicine = fluidotherapy, Chemtech, April, 215-220 (1982).

Sanders, H.J., Drugs to combat heart disease, C&E News, 12 July, 26-38 (1982).

Krassner, M.B., Brain chemistry, C&E News, 29 Aug, 22-33 (1983).

Zurer, P.S., The chemistry of vision, C&E News, 28 Nov, 24-35 (1983).

Weathersby, P.K., Engineering analysis of diver decompression sickness, Chem Eng Commns, 30(3), 183-190 (1984).

Kambic, H.E.; Murabayashi, S., and Nose, Y., Biomaterials in artificial organs, C&E News, 14 April, 30-48 (1986).

Various, Biomedical engineering applications (topic issue), Chem Eng Commns, 47, 1-183 (1986).
Anon., Biotechnology and biomedicine, Chem Eng Prog, 84(8), 28-29 (1988).
Bryant, R.J., Manufacture of medicinal alkaloids from the opium poppy: Review of a traditional biotechnology, Chem & Ind, 7 March, 146-153 (1988).
King, W.E.; Schultz, D.S., and Gatenby, R.A., Analysis of systemic tumor oxygenation using multi-region models, Chem Eng Commns, 64, 137-154 (1988).
Sakai, K.; Ohashi, H., and Naitoh, A., Effects of blood contact on the properties of tubular dialysis membranes, Biochem Eng J, 38(1), B1-B6 (1988).
Zurer, P.S., Understanding cocaine dependency, C&E News, 21 Nov, 7-13 (1988).

12.2 Biotechnology Applications and Principles

1967-1970

Blakebrough, N., Mass transfer in aerobic microbial systems, Brit Chem Eng, 12(1), 78-80 (1967).
Atkinson, B., and Daoud, I.S., The analogy between micro-biological reactions and heterogeneous catalysis, Trans IChemE, 46, T19-24 (1968).
Elsworth, R., et al., Production of E-Coli as a source of nucleic acids, J Appl Chem, 18, 157-166 (1968).
Iverson, W.P., Microbiological corrosion, Chem Eng, 23 Sept, 242-244 (1968).
Lilly, M.D., and Sharp, A.K., The kinetics of enzymes attached to water-insoluble polymers, Chem Engnr, Jan, CE12-18 (1968).
Solomons, G.L., Some aspects of research on industrial enzymes, Chem Engnr, Jan, CE9-12 (1968).
Anon., Pesticides: Present and future, Chem Eng, 7 April, 133-140 (1969).
Wang, D.I.C., and Humphrey, A.E., Biochemical engineering, Chem Eng, 15 Dec, 108-120 (1969).
Wingard, L.B., Bioengineering problems, Chem Eng Prog, 65(1), 69-79 (1969).
Atkinson, B., and Daoud, I.S., Diffusion effects within microbial films, Trans IChemE, 48, T245-254 (1970).
Callihan, C.D., Applications of microbes, Chem Eng, 21 Sept, 160-164 (1970).
Kuliev, A.M., et al., Microbiological oxidation of various fractions of diesel fuels, Int Chem Eng, 10(2), 241-244 (1970).
Meshkov, A.N., et al., Biosynthesis of proteolytic enzymes in some antibiotic-producing actinomycetes, Int Chem Eng, 10(2), 216-219 (1970).
Pozsarickaja, L.M., and Gradova, N.B., Thermodynamic problems of cultures of microorganisms, Int Chem Eng, 10(3), 362-365 (1970).

1971-1973

Aiba, S., and Endo, I., Statistical analysis of growth of microorganisms, AIChEJ, 17(3), 608-612 (1971).
Bergmann, E.D., Insect control, Chemtech, Dec, 740-744 (1971).
Calam, C.T.; Ellis, S.H., and McCann, M.J., Mathematical models of fermentations and simulation of the Griseofulvin fermentation, J Appl Chem Biotechnol, 21, 181-189 (1971).

Greer, F.; Ignoffo. C.M., and Anderson, R.F., The first viral pesticide: A case history, Chemtech, June, 342-347 (1971).
Perlman, D., Antibiotics, Chemtech, Sept, 540-547 (1971).
Carbonell, R.G., and Kostin, M.D., Enzyme kinetics and engineering (review paper), AIChEJ, 18(1), 1-12 (1972).
Elsworth, R., The value and use of dissolved oxygen measurement in deep culture, Chem Engnr, Feb, 63-71 (1972).
Emery, A.N.; Hough, J.S.; Novais, J.M., and Lyons, T.P., Some applications of solid-phase enzymes in biological engineering, Chem Engnr, Feb, 71-76 (1972).
Various, Papers from symposium on continuous culture of microorganisms, J Appl Chem Biotechnol, 22, 55-148, 217-292, 345-440, 509-558 (1972).
Fogarty, W.M., and Griffin, P.J., Device for production of microbial extracellular enzymes in concentrated form, J Appl Chem Biotechnol, 23, 401-406 (1973).
Meienhofer, J., Why peptides are synthesized and how, Chemtech, April, 242-254 (1973).
Sargeant, K., Pilot-scale preparative microbiology: Cell culture, extraction, and purification, J Appl Chem Biotechnol, 23, 151-158 (1973).
Wilkinson, C.F., Insecticide synergism, Chemtech, Aug, 492-497 (1973).
Woods, G.A., Bacteria: Friends or foe? Chem Eng, 5 March, 81-84 (1973).

1974-1975

Atkinson, B.; Davies, I.J., and How, S.Y., The overall rate of substrate uptake (reaction) by microbial films, Trans IChemE, 52, 248-268 (1974).
Coulter, P.R., and Potter, O.E., Rate at which starch becomes susceptible to hydrolysis by enzymes, Ind Eng Chem Proc Des Dev, 13(4), 324-327 (1974).
Furness, C.D., Biological treatment of effluents, Chem Engnr, Feb, 102-107 (1974).
Pirt, S.J., Theory of fed batch culture with reference to penicillin fermentation, J Appl Chem Biotechnol, 24, 415-424 (1974).
Royer, G.P., Supports for immobilized enzymes and affinity chromatography, Chemtech, Nov, 694-700 (1974).
Tanner, R.D., and De Angelis, L.H., Sigmoidal and growth rate kinetic hysteresis in biochemical systems, Chem Eng J, 8(2), 113-124 (1974).
Vieth, W.R., and Venkatasubramanian, K., Enzyme engineering, Chemtech, Nov, 677-687 (1973); Jan, 47-55 (1974); May, 309-320; July, 434-444 (1974).
Baker, R.W., and Lonsdale, H.K., Controlled drug delivery: New use for membranes, Chemtech, Nov, 668-674 (1975).
Buyske, D.A., Drugs from nature, Chemtech, June, 361-369 (1975).
Cardarelli, N., Controlled-release pest control agents, Chemtech, Aug, 482-485 (1975).
Chang, T.M.S., Artificial cells, Chemtech, Feb, 80-85 (1975).
Dunn, I.J., and Einsele, A., Oxygen transfer coefficients by the dynamic method, J Appl Chem Biotechnol, 25, 707-720 (1975).
Emery, A.N., Biochemical engineering, Chem Engnr, Nov, 682-683, 697 (1975).

Fernandes, P.M.; Constantinides, A.; Vieth, W.R., and Venkatasubramanian, K., Enzyme engineering, Chemtech, July, 438-445 (1975).
Frear, D.S., Herbicide metabolism and its significance, Chemtech, Oct, 629-632 (1975).
Greenhalgh, S.H.; McManamey, W.J., and Potter, K.E., Comparison of oxygen mass transfer into sodium sulphite solution with a biological system, J Appl Chem Biotechnol, 25, 143-160 (1975).
MacLaren, D.D., Single-cell protein: An overview, Chemtech, Oct, 594-597 (1975).
Pirt, S.J., and Mancini, B., Inhibition of penicillin production by carbon dioxide, J Appl Chem Biotechnol, 25, 781-784 (1975).
Shoda, M.; Nagai, S., and Aiba, S., Simulation of growth of methane-utilising bacteria in batch culture, J Appl Chem Biotechnol, 25, 305-318 (1975).
Wain, R.L., Selective herbicidal activity, Chemtech, June, 356-360 (1975).
Weiner, M., Drug testing and use, Chemtech, April, 205-209 (1975).
Zinkel, D.F., Chemicals from trees, Chemtech, April, 235-241 (1975).

1976-1977

Fritz, J.C., Bioavailability of mineral nutrients, Chemtech, Oct, 643-648 (1976).
Gunstone, F.D., Natural and unnatural unsaturated fatty acids, Chem & Ind, 20 March, 243-251 (1976).
Hedin, P., Using a sex attractant for insect control, Chemtech, July, 444-451 (1976).
Khan, M.; Gassman, M., and Haque, R., Biodegradation of pesticides, Chemtech, Jan, 62-69 (1976).
Kuhr, R., Insecticide metabolites in and on plants, Chemtech, May, 316-321 (1976).
McCann, J., Mutagenesis, carcinogenesis, and the Salmonella test, Chemtech, Nov, 682-687 (1976).
Watts, H., Single-cell protein, Chem & Ind, 3 July, 537-540 (1976).
Alder, E.F.; Wright, W.L., and Klingman, G.C., Man against weeds, Chemtech, June, 374-380 (1977).
Denson, D., Modern anesthetics, Chemtech, July, 446-453 (1977).
Hansch, C., and Fukunaga, J., Designing biologically active materials, Chemtech, Feb, 120-128 (1977).
Horsfall, J.G., Fungicides: Past, present and future, Chemtech, May, 302-305 (1977).
Ladisch, M.R.; Emery, A., and Rodwell, V.W., Economic implications of purification of glucose isomerase prior to immobilization, Ind Eng Chem Proc Des Dev, 16(3), 309-313 (1977).
Paca, J., and Gregr, V., Determination of oxygen transfer coefficients (KLa) with correction for actual cultivation conditions, J Appl Chem Biotechnol, 27, 155-164 (1977).
Petrow, V., Chemistry of contraceptives, Chemtech, Sept, 563-569 (1977).
Pitcher, W.H., Enzyme immobilisation processes, Proc Engng, July, 77-80 (1977).
Rushton, A., and Koo, H.E., Filtration characteristics of yeast, J Appl Chem Biotechnol, 27, 99-109 (1977).

Shabi, F.A., and Ilett, K.J., The role of biological processes in the treatment of effluents from the chemical industry, Chem Engnr, April, 247-248, 250 (1977).
Weisburger, J.H., Cancer prevention, Chemtech, Dec, 734-740 (1977).

1978-1979

Archer, M.D., Photosynthesis 'in vitro', J Inst Fuel, 51, 100-108 (1978).
Bailey, J.E., et al., Measurement of structured microbial population dynamics by flow microfluorometry, AIChEJ, 24(4), 570-577 (1978).
Black, W., Marketing biomass, Chemtech, Oct, 606-608 (1978).
Endo, I.; Ohtaguchi, K.; Nagamune, T., and Inoue, I., Functional representation of yeast batch-culture systems, Int Chem Eng, 18(4), 634-642 (1978).
Hoffmann, C.E., Virus chemotherapy, Chemtech, Dec, 726-732 (1978).
Hoffmann, O.L., Herbicide antidotes, Chemtech, Aug, 488-492 (1978).
Keller, R., and Dunn, I.J., Computer simulation of biomass production rate of cyclic fed batch continuous culture, J Appl Chem Biotechnol, 28, 784-790 (1978).
Keller, R., and Dunn, I.J., Fed-batch microbial culture: Models, errors and applications, J Appl Chem Biotechnol, 28, 508-514 (1978).
Loo, A.C.; Tanner, R.D., and Crooke, P.S., Simplifying enzyme and fermentation kinetic models, Chem Eng J, 16(2), 137-150 (1978).
Papoutsakis, E.; Lim, H.C., and Tsao, G.T., Single-cell protein production on C1 compounds, AIChEJ, 24(3), 406-417 (1978).
Samour, C.M., Polymeric drugs, Chemtech, Aug, 494-501 (1978).
Various, Pesticides packaging (topic issue), Chem & Ind, 18 Feb, 107-123 (1978).
Bylinsky, G., Cloning, Chemtech, Dec, 722-727 (1979).
Cantell, K.J., Clinical use of interferon, Chemtech, Sept, 537-541 (1979).
Cape, R.E., Microbial genetics and the pharmaceutical industry, Chemtech, Oct, 638-644 (1979).
Greenstein, J.S., Studies on a peerless contraceptive, Chemtech, April, 217-221 (1979).
Kovaly, K.A., Nitrogen fixation, Chem Engnr, June, 417-418 (1979).
Lin, S.H., Effects of mass-transfer resistances on encapsulated enzymatic reaction systems, Chem Eng J, 17(1), 55-62 (1979).
Neil, G.L., Antitumor compounds from fermentation, Chemtech, July, 428-431 (1979).
Ohtaguchi, K.; Endo, I., and Inoue, I., Functional representation of yeast and bacterial mixed culture, Int Chem Eng, 19(4), 591-600 (1979).
Oliver, J.E., Nitrosamines from pesticides, Chemtech, June, 366-371 (1979).
Roelofs, W., Electroantennograms, Chemtech, April, 222-227 (1979).
Whetstone, R., Vinyl phosphate insecticides, Chemtech, June, 360-364 (1979).
Wolstenholme, G., Genetic manipulation, Chemtech, June, 354-358 (1979).

1980-1981

Crooke, P.S.; Wei, C.J., and Tanner, R.D., Effect of specific growth rate and yield expressions on extension of oscillatory behavior of continuous fermentation model, Chem Eng Commns, 6(6), 333-348 (1980).
Giurgea, C., A drug for the mind, Chemtech, June, 360-365 (1980).

Langer, R., Polymeric delivery systems for controlled drug release (review paper), Chem Eng Commns, 6(1), 1-48 (1980).
Parliment, T.H., Chemistry of aroma, Chemtech, May, 284-289 (1980).
Royer, G.P., Immobilized enzymes catalysis reviews (1978), Catalysis Reviews, 22(1), 29-74 (1980).
Skalsky, M., and Farrell, P.C., Uptake characteristics of selected biochemicals on coated and uncoated activated charcoal, Trans IChemE, 58, 91-97 (1980).
Stephanopoulos, G., Dynamics of mixed cultures of microorganisms: Some topological considerations, AIChEJ, 26(5), 802-816 (1980).
Various, Fertiliser recommendations: Formulation and use (topic issue), Chem & Ind, 6 Sept, 677-696 (1980).
Various, Assessment of in-vitro methods of detecting mutagens and carcinogens (topic issue), Chem & Ind, 1 Nov, 844-859 (1980).
Various, Chemical aspects of pro-drug design, Chem & Ind, 7 June, 433-461 (1980).
Westheimer, F.H., Models for enzyme systems, Chemtech, Dec, 748-754 (1980).
Blanch, H.W., Microbial growth kinetics (review paper), Chem Eng Commns, 8(4), 181-212 (1981).
Kalogerakis, N., and Boyle, T.J., Implementation and demonstration of a quasi-steady-state controller for yeast fermentation, Can JCE, 59, 377-381 (1981).
Magee, P.S., Discovering bioactive materials, Chemtech, June, 378-384 (1981).
Moellering, R.C., and Murray, B.E., Antibiotic resistance, Chemtech, May, 280-282 (1981).
Papoutsakis, E., and Lim, H.C., Bioefficiency of single-cell protein production on C1 compounds, Ind Eng Chem Fund, 20(4), 307-314 (1981).
Ramachandran, P.A.; Kulkarni, B.D., and Sadana, A., Analysis of multiple steady states of complex biochemical reactions, J Chem Tech Biotechnol, 31, 546-552 (1981).
Sadana, A., Simple probabilistic approach to enzyme deactivations, Ind Eng Chem Fund, 20(4), 336-340 (1981).
Sanders, H.J., Herbicides, C&E News, 3 Aug, 20-35 (1981).
Slater, N.K.H.; Powell, M.S., and Johnson, P., Relevance of bacterial mobility to fermenter contaminations: An experimental study for Bacillus cereus, Trans IChemE, 59, 170-176, 286-287 (1981).
Toda, K., Induction and repression of enzymes in microbial culture (review paper), J Chem Tech Biotechnol, 31, 775-790 (1981).
Various, Enzyme mechanisms and design of enzyme inhibitors (topic issue), Chem & Ind, 7 March, 131-156 (1981).
Various, Joint action of mixtures of drugs or pesticides (topic issue), Chem & Ind, 1 Aug, 518-537 (1981).
Various, Biotechnology (topic issue), Chem & Ind, 4 April, 204-247 (1981).

1982

Adler-Nissen, J., Limited enzymic degradation of proteins, J Chem Tech Biotechnol, 32, 138-156 (1982).

Ashley, M.H.J., Continuous sterilisation of media, Chem Engnr, Feb, 54-58 (1982).
Birch, J.R., and Cartwright, T., Environmental factors influencing the growth of animal cells in culture, J Chem Tech Biotechnol, 32, 313-317 (1982).
Challenger, J.G., Contractors can help to bridge the biotechnology gap, Chem Engnr, April, 127-129 (1982).
Cushman, D.W., and Ondetti, M.A., Inhibitors of angiotensin-converting enzyme, Chemtech, Oct, 620-624 (1982).
Davis, N., In-house enzymes from the mash, Proc Engng, March, 39 (1982).
Dobbs, A.J.; Peleg, M.; Mudgett, R.E., and Rufner, R., Some physical characteristics of active dry yeast, Powder Tech, 32, 63-69 (1982).
Fox, J.L., Biotechnology investments, C&E News, 29 March, 10-15 (1982).
Geankoplis, C.J., and Hu, M.C., Interaction effects on diffusion in protein solutions, Ind Eng Chem Fund, 21(2), 135-141 (1982).
Harris, B.L., Chemical warfare, Chemtech, Jan, 28-35 (1982).
Heden, C.G., Global aspects of biotechnology, J Chem Tech Biotechnol, 32, 18-24 (1982).
Hoare, M., Protein precipitation and precipitate ageing, Trans IChemE, 60, 79-87, 157-163 (1982).
Humphrey, A.E., Biotechnology: The way ahead, J Chem Tech Biotechnol, 32, 25-33 (1982).
King, P.P., Biotechnology: An industrial view, J Chem Tech Biotechnol, 32, 2-8 (1982).
Kroner, K.H.; Schutte, H.; Stach, W., and Kula, M.R., Scale-up of formate dehydrogenase by partition, J Chem Tech Biotechnol, 32, 130-137 (1982).
Leegwater, M.P.M.; Neijssel, O.M., and Tempest, D.W., Aspects of microbial physiology in relation to process control, J Chem Tech Biotechnol, 32, 92-99 (1982).
Lilly, M.D., Two-liquid-phase biocatalytic reactions, J Chem Tech Biotechnol, 32, 162-169 (1982).
Luong, J.H.T., and Volesky, B., Indirect determination of biomass concentration in fermentation processes, Can JCE, 60, 163-168 (1982).
Melchior, J.L., et al., Biomethanation, J Chem Tech Biotechnol, 32, 189-197 (1982).
Nisbet, L.J., Search for bioactive microbial metabolites, J Chem Tech Biotechnol, 32, 251-270 (1982).
Pirt, S.J., Microbial photosynthesis in the harnessing of solar energy, J Chem Tech Biotechnol, 32, 198-202 (1982).
Pym, D., Biotechnology in outer space, Chem Engnr, April, 143-144 (1982).
Rebeiz, C.A., Chlorophyll: Anatomy of a discovery, Chemtech, Jan, 52-63 (1982).
Rehm, H.J., Biotechnological research in Europe, J Chem Tech Biotechnol, 32, 9-13 (1982).
Reub, M.; Bajpai, R.K., and Berke, W., Effective oxygen consumption rates in fermentation broths with filamentous organisms, J Chem Tech Biotechnol, 32, 81-91 (1982).
Reuss, M.; Debus, D., and Zoll, G., Rheological properties of fermentation fluids, Chem Engnr, June, 233-236 (1982).

Smith, I.H., and Pace, G.W., Recovery of microbial polysaccharides, J Chem Tech Biotechnol, 32, 119-129 (1982).
Spier, R.E., Animal cell technology: An overview, J Chem Tech Biotechnol, 32, 304-312 (1982).
Thomas, D., and Gellf, G., Enzyme technology and molecular biology, J Chem Tech Biotechnol, 32, 14-17 (1982).
Thomas, T., Chemistry and biology of oils, Chem & Ind, 17 July, 484-488 (1982).
van den Berg, L., and Kennedy, K.J., Comparison of intermittent and continuous loading of stationary fixed-film reactors for methane production from wastes, J Chem Tech Biotechnol, 32, 427-432 (1982).
Various, Growth potential of biotechnology, Processing, May, 21, 25 (1982).
Various, Opportunities in biotechnology (topic issue), Chem & Ind, 7 Aug, 508-537 (1982).
Various, Biosafety (topic issue), Chem & Ind, 20 Nov, 876-903 (1982).
Volesky, B.; Yerushalmi, L., and Luong, J.H.T., Metabolic-heat relation for aerobic yeast respiration and fermentation, J Chem Tech Biotechnol, 32, 650-659 (1982).
Voser, W., Isolation of hydrophilic fermentation products by adsorption chromatography, J Chem Tech Biotechnol, 32, 109-118 (1982).

1983

Abeles, R.H., Suicide enzyme inactivators, C&E News, 19 Sept, 48-56 (1983).
Barry, B.W., Drug delivery systems, Chemtech, Jan, 38-44 (1983).
Bott, T.R., and Miller, P.C., Mechanisms of biofilm formation on aluminium tubes, J Chem Tech Biotechnol, 33B(3), 177-184 (1983).
Carlson, P.S., Plant genetic engineering, Chemtech, Dec, 744-746 (1983).
Henriksen, J., Water-piping systems for pharmaceuticals, Proc Engng, Dec, 28-31 (1983).
Putnam, A.R., Allelopathic chemicals: Nature's herbicides, C&E News, 4 April, 34-45 (1983).
Russell, D., Safety in biochemical processing, Proc Engng, Aug, 18-21 (1983).
Schmidt, S., Guidelines for painting food plants, Proc Engng, June, 55-59 (1983).
Various, Symposium papers on continuous culture, J Chem Tech Biotechnol, 33B(3), 195-202 (1983).
Various, The role of pesticides in industry (topic issue), Chem & Ind, 20 June, 465-474 (1983).
Various, Political and economic aspects of fermentation raw materials markets (topic issue), Chem & Ind, 7 Feb, 95-104 (1983).
Various, Crop protection in China (topic issue), Chem & Ind, 20 June, 453-464 (1983).
Webb, C.; Black, G.M., and Atkinson, B., Liquid fluidisation of highly porous particles, CER&D, 61, 125-134 (1983).
Wilkie, K.C.B., Hemicellulose, Chemtech, May, 306-319 (1983).
Zeelen, F.J., Drug development: A systematic approach, Chemtech, July, 419-425 (1983).

1984

Akelah, A., Biological applications of functionalised polymers (review paper), J Chem Tech Biotechnol, 34A(6), 263-286 (1984).

Allegretto, B., Using polymers for agricultural frost protection, Chemtech, March, 152-155 (1984).

Andrews, G.F.; Fonta, J.P.; Marrotta, E., and Stroeve, P., Effects of cells on oxygen transfer coefficients, Biochem Eng J, 29(3), B39-B55 (1984).

Atkinson, B.; Cunningham, J.D., and Pinches, A., Biomass hold-ups and overall rates of substrate (glucose) uptake of support particles containing a mixed microbial culture, CER&D, 62, 155-164 (1984).

Barnes, G., Agricultural fungicides, Chem & Ind, 19 Nov, 799-802 (1984).

Bodor, N., The soft drug approach, Chemtech, Jan, 28-38 (1984).

Chase, H.A., Affinity separations utilising immobilised monoclonal antibodies: A new tool for the biochemical engineer (review paper), Chem Eng Sci, 39(7), 1099-1125 (1984).

Chase, H.A., Affinity separations utilising monoclonal antibodies: A new tool for the biochemical engineer (review paper), Chem Eng Sci, 39(7), 1099-1126 (1984).

Davey, K.R., and Wood, D.G., Laboratory evaluation of Lin's model of continuous sterilisation of micro-organisms, CER&D, 62, 117-122 (1984).

Domach, M.M., Specifying immobilized enzyme performance parameters, Biochem Eng J, 29(1), B1-B8 (1984).

Douglas, K.T., Anticancer drugs, Chem & Ind, 1 Oct, 693-698; 15 Oct, 738-742; 5 Nov, 766-771 (1984).

Dunnill, Biotechnology and British industry, Chem & Ind, 2 July, 470-475 (1984).

Garrett, G.W., and Salladay, D., Fluid fertilizers, Chemtech, April, 250-253 (1984).

Higgins, B., Pharmaceutical manufacture by continuous processing, Proc Engng, Feb, 35-39 (1984).

Krieger, J.H., Plant biotechnology, C&E News, 29 Oct, 16-19 (1984).

Lee, Y.K., and Pirt, S.J., Effect of pH on carbon dioxide absorption rate in an algal culture, J Chem Tech Biotechnol, 34B(1), 28-32 (1984).

Michaels, A.S., The impact of genetic engineering, Chem Eng Prog, 80(4), 9-15 (1984).

Michaels, A.S., Adapting modern biology to industrial practice, Chem Eng Prog, 80(6), 19-25 (1984).

Murray, J.R., The biotechnology dilema, Chemtech, Aug, 482-485 (1984).

Oyama, K., and Kihara, K., New developments for enzyme technology, Chemtech, Feb, 100-105 (1984).

Randerson, D., Hybridoma technology and the process engineer, Chem Engnr, Dec, 12-15 (1984).

Rawls, R.L., Oncogenes and the origin of cellular cancer, C&E News, 10 Dec, 11-15 (1984).

Rosevear, A., Immobilised biocatalysts: A critical review, J Chem Tech Biotechnol, 34B(3), 127-150 (1984).

Schenke, E., and Giorgio, R.J., Pharmaceuticals by recombinant DNA technology, Proc Engng, April, 27-30 (1984).

van Dijk, H.J.M.; Walstra, P., and Schenk, J., Theoretical and experimental study of one-dimensional syneresis of a protein gel, Biochem Eng J, 28(3), B43-B50 (1984).
Various, Plant biotechnology (topic issue), Chem & Ind, 3 Dec, 817-849 (1984).
Various, Chemical inputs and agricultural productivity (topic issue), Chem & Ind, 17 Sept, 645-665 (1984).
Various, Mycotoxins (topic issue), Chem & Ind, 6 Aug, 530-552 (1984).
Various, Chemical engineering in biotechnology (topic issue), Chem Eng Prog, 80(12), 7-21, 37-63 (1984).
Various, Genetic engineering (feature report), C&E News, 13 Aug, 10-63 (1984).
Webber, D., Biotechnology: An assessment, C&E News, 16 April, 11-19 (1984).
Zurer, P.S., Drugs in sport, C&E News, 30 April, 69-78 (1984).

1985
Bailey, J.E., and Ollis, D.F., Biochemical engineering fundamentals, Chem Eng Educ, 19(4), 168-171 (1985).
Birkner, J.H., Biotechnology transfer: National security implications, Chemtech, Dec, 734-737 (1985).
Bjurstrom, E., Biotechnology, Chem Eng, 18 Feb, 126-158 (1985).
Bowden, C., Recovery of micro-organisms from fermented broth, Chem Engnr, June, 50-54 (1985).
Bowden, C.P., Novel processes for cell recovery technology developments, J Chem Tech Biotechnol, 35B(4), 253-265 (1985).
Donaldson, E.C., and Grula, E.A., Bacteria in oil, Chemtech, Oct, 602-604 (1985).
Dreikorn, B.A., and O'Doherty, G.O.P., Case study of discovery and development of a rodenticide, Chemtech, July, 424-430 (1985).
Fauquex, P.F.; Hustedt, H., and Kula, M.R., Phase equilibrium in agitated vessels during extractive enzyme recovery, J Chem Tech Biotechnol, 35B(1), 51-59 (1985).
Fukuto, T.R., Development of pesticides, Chemtech, June, 362-367 (1985).
Hanson, D.J., Effects of animal drugs on human health, C&E News, 7 Oct, 7-11 (1985).
Helmes, C.T.; Sigman, C.C., and Papa, P.A., Predicting carcinogenicity from chemical structure, Chemtech, Jan, 48-53 (1985).
Karel, S.F.; Libicki, S.B., and Robertson, C.R., The immobilization of whole cells (review article), Chem Eng Sci, 40(8), 1321-1354 (1985).
Karel, S.F.; Libicki, S.B., and Robertson, C.R., The immobilization of whole cells: Engineering principles (review paper), Chem Eng Sci, 40(8), 1321-1354 (1985).
Kollerup, F., and Daugulis, A.J., Screening and identification of extractive fermentation solvents using a database, Can JCE, 63(6), 919-927 (1985).
Leonhardt, B.A., Phermones for pest control, Chemtech, June, 368-374 (1985).
Marshall, A., Biotechnology trends, Processing, Feb, 27-28 (1985).
Oosterhuis, N.M.G.; Sweere, A.P.J., and Kossen, N.W.F., Determination of the liquid side oxygen transfer coefficient in a biological medium, CER&D, 63, 203-205 (1985).

Parikh, I., and Cuatrecasas, P., Affinity chromatography, C&E News, 26 Aug, 17-32 (1985).
Phillips, J.A., Uses of biomass, Chemtech, June, 376-384 (1985).
Rawls, R.L., Biological response modifiers for treating cancer, C&E News, 17 June, 10-16 (1985).
Rawls, R.L., Understanding metastasis is key to controlling cancer, C&E News, 25 Feb, 10-17 (1985).
Sanders, H.J., Controlled release of drugs, C&E News, 1 April, 30-48 (1985).
Sillett, C., User specifications for hygenic process plant, Chem Engnr, April, 28-30 (1985).
Smith, M.D., and Ho, C.S., Dissolved carbon dioxide in penicillin fermentations, Chem Eng Commns, 37(1), 21-28 (1985).
Spark, L., Biotechnology industrial trends, Proc Engng, July, 35-39 (1985).
Turunen, I., et al., Fuzzy modelling in biotechnology: Sucrose inversion, Biochem Eng J, 30(3), B51-B60 (1985).
Various, Pesticides in the Third World (topic issue), Chem & Ind, 16 Sept, 609-625; 7 Oct, 654-659 (1985).
Webber, D., Commercialization of biotechnology, C&E News, 18 Nov, 25-60 (1985).
Wilson, T.; Klausner, A., and Payson, C., Networking in biotechnology, Chemtech, Nov, 666-670 (1985).
Yoshioka, H., Development of the insecticide fenvalerate, Chemtech, Aug, 482-486 (1985).

1986

Baier, R.E., and Meyer, A.E., Biosurface chemistry, Chemtech, March, 178-185 (1986).
Baum, R.M., AIDS research, C&E News, 1 Dec, 7-12 (1986).
Baum, R.M., Enzyme chemistry, C&E News, 14 July, 7-14 (1986).
Berkovitch, I., Sterile plant, Proc Engng, July, 41-42 (1986).
Boyles, D.T., Biomass for energy: A review, J Chem Tech Biotechnol, 36(11), 495-511 (1986).
Bungay, H.R., Biochemical engineering with extensive use of personal computers, Chem Eng Educ, 20(3), 122-123,155 (1986).
Dagani, R., Drugs and vaccines to combat AIDS, C&E News, 8 Dec, 7-14 (1986).
Foster, P.R.; Dickson, A.J.; Stenhouse, A., and Walker, E.P., A process control system for the fractional precipitation of human plasma proteins, J Chem Tech Biotechnol, 36(10), 461-466 (1986).
Fox, J.L., Applied science and biotechnology in Cuba, C&E News, 12 May, 31-39 (1986).
Hayes, M.E.; Nestaas, E., and Hrebenar, K.R., Microbial surfactants, Chemtech, April, 239-243 (1986).
Hossain, M.M.; Do, D.D., and Bailey, J.E., Immobilization of enzymes in porous solids under restricted diffusion conditions, AIChEJ, 32(7), 1088-1098 (1986).
Jacobsson, S.; Jamison, A., and Rothman, H., The Biotechnological Challenge, Cambridge University Press, U.K. (1986).

Klibanov, A.M., Enzymes that work in organic solvents, Chemtech, June, 354-359 (1986).
Layman, P.L., Industrial enzymes, C&E News, 15 Sept, 11-14 (1986).
Monbouquette, H.G., and Ollis, D.F., Even live catalysts die, Chemtech, Sept, 542-551 (1986).
Moo-Young, M., Biochemical engineering and industrial biotechnology, Chem Eng Educ, 20(4), 194-197 (1986).
Rawls, R.L., Light-activated pesticides, C&E News, 22 Sept, 21-24 (1986).
Robins, R.K., Synthetic antiviral agents, C&E News, 27 Jan, 28-40 (1986).
Rothstein, M., Biochemical studies of aging, C&E News, 11 Aug, 26-39 (1986).
Sharma, S.K., Recovery of genetically engineered proteins from Escherichia coli, Sepn Sci Technol, 21(8), 701-726 (1986).
Stinson, S.C., Antibacterial drugs, C&E News, 29 Sept, 33-67 (1986).
Tsezos, M.; Baird, M.H.I., and Schemilt, L.W., The kinetics of radium biosorption, Biochem Eng J, 33(2), B35-B42 (1986).
Various, Biotechnology symposium papers, Can JCE, 64(4), 529-638 (1986).
Various, Biochemical engineering and its applications (topic issue), Chem Eng Commns, 45, 1-309 (1986).
Yue, P.L., and Lowther, K., Enzymatic oxidation of C1 compounds in a biochemical fuel cell, Biochem Eng J, 33(3), B69-B77 (1986).

1987

Datar, R., and Rosen, C.G., Centrifugal separation in the recovery of intracellular protein from E. coli, Biochem Eng J, 34(3) B49-B56 (1987).
Davey, K.R., Determining thermal survivor data of micro-organisms: The Lin model studies, CER&D, 65(3), 234-237 (1987).
Dunnill, P., Biochemical engineering and biotechnology, CER&D, 65(3), 211-217 (1987).
Honda, H.; Mano, T.; Taya, M.; Shimizu, K.; Matsubara, M., and Kobayashi, T., A general framework for the assessment of extractive fermentations, Chem Eng Sci, 42(3), 493-498 (1987).
Iijima, S., Use of novel turbidimeter to monitor microbial growth and control glucose concentration, J Chem Tech Biotechnol, 40(3), 203-213 (1987).
Kenney, A.C., and Chase, H.A., Automated production-scale affinity purification of monoclonal antibodies, J Chem Tech Biotechnol, 39(3), 173-182 (1987).
Kingdon, C., Developments in biosensors, Processing, Nov, 35,38 (1987).
Lorenz, T.; Schmidt, W., and Schugerl, K., Sampling devices in fermentation technology: A review, Biochem Eng J, 35(2), B15-B22 (1987).
Matsuhisa, S.; Takesawa, S., and Sakai, K., Binary-solute adsorption of dosed drugs on serum albumin, Biochem Eng J, 34(2) B21-B28 (1987).
Mavituna, F., and Park, J.M., Size distribution of plant cell aggregates in batch culture, Biochem Eng J, 35(1) B9-B14 (1987).
McConvey, I.F., Sorption of vitamin B12 from aqueous solution, CER&D, 65(3), 231-233 (1987).
Park, D.H.; Malaney, G.W., and Tanner, R.D., Kinetics of protein secretion and fractionation in baker's yeast fermentations, J Chem Tech Biotechnol, 39(2), 85-92 (1987).

Skalak, R., and Chien, S., Handbook of Bioengineering, McGraw-Hill, New York (1987).
Storck, W.J., Home and garden pesticides, C&E News, 6 April, 11-17 (1987).
Tsezos, M.; Baird, M.H.I., and Shemilt, L.W., The elution of radium adsorbed by microbial biomass, Biochem Eng J, 34(3) B57-B64 (1987).
Various, Biotechnology developments, Processing, May, 38-50 (1987).
Various, Pest control in Australia (topic issue), Chem & Ind, 20 April, 258-293 (1987).
Various, Biomanufacturing (topic issue), Chem Eng Prog, 83(10), 21-70 (1987).
Wu, W.T.; Jang, W.D., and Wu, S.C., On-line control for cultivation of Saccharomyces cerevisiae, Biochem Eng J, 36(1), B1-B6 (1987).

1988

Allen, G., Biotechnology for the non-petrochemical industry, Chem Engnr, April, 13-22 (1988).
Barton, J.K., Recognizing DNA structures, C&E News, 26 Sept, 30-42 (1988).
Blank, M., Biological switches, Chemtech, July, 434-438 (1988).
Builder, S.E., and Hancock, W.S., Analytical and process chromatography in pharmaceutical protein production, Chem Eng Prog, 84(8), 42-46 (1988).
Dochain, D., and Pauss, A., On-line estimation of microbial specific growth-rates: An illustrative case study, Can JCE, 66(4), 626-631 (1988).
George, N., and Davies, J.T., Adsorption of microorganisms on activated charcoal cloth with biotechnology applications, J Chem Tech Biotechnol, 43(2), 117-130 (1988).
Georgiou, G., Optimizing the production of recombinant proteins in microorganisms (review paper), AIChEJ, 34(8), 1233-1248 (1988).
Heitz, J.R., Photoactivated pesticides, Chemtech, Aug, 484-488 (1988).
Herrett, R.A., Low-dosage pesticides: Less is better, Chemtech, April, 220-225 (1988).
Huang, H.P., et al., On-line optimal feed of substrate during fed-batch culture of cell mass, Chem Eng Commns, 68, 221-236 (1988).
Murray, K., Aquacultural engineering, Chem Engnr, Jan, 42 (1988).
Ng, T.K.L.; Gonzalez, J.F., and Hu, W.S., Teaching biochemical engineering, Chem Eng Educ, 22(4), 202-207 (1988).
Parkinson, G., The next wave of bioprocessing arrives, Chem Eng, 19 Dec, 37-44 (1988).
Randolph, T.W.; Blanch, H.W., and Prausnitz, J.M., Enzyme-catalyzed oxidation of cholesterol in supercritical carbon dioxide, AIChEJ, 34(8), 1354-1360 (1988).
Rebeiz, C.A., Biocides from first principles, Chemtech, Feb, 90-93 (1988).
Royse, S., Antibody engineering for the 1990s, Proc Engng, May, 83-88 (1988).
Ruggeri, B.; Specchia, V.; Sassi, G., and Gianetto, A., Numerical estimation of biokinetic parameters, Biochem Eng J, 39(2), B17-B24 (1988).
Sato, N., et al., Cyclohexanedione herbicides, Chemtech, July, 430-433 (1988).
Sawyer, D.T., Oxygen in biological processes, Chemtech, June, 369-375, (1988).
Schaeffer, J.R.; Burdick, B.A., and Abrams, C.T., Thin-film biocatalysts, Chemtech, Sept, 546-550 (1988).

Smithson, L.H., Biotechnology: Now and soon, Chemtech, March, 168-173 (1988).
Sonnet, P.E., Enzymes for chiral synthesis, Chemtech, Feb, 94-98 (1988).
Specchia, V.; Ruggeri, B., and Gianetto, A., Mechanisms of activated carbon bioremoval, Chem Eng Commns, 68, 99-118 (1988).
Trezl, L., et al., Protein and amino acid chemical studies, Periodica Polytechnica, 32(1), 151-159 (1988).
Van der Hoek, J.P.; Griffioen, A., and Klapwijk, A.B., Biological regeneration of nitrate-loaded anion-exchange resins by denitrifying bacteria, J Chem Tech Biotechnol, 43(3), 213-222 (1988).
Vardar-Sukan, F., Efficiency of natural oils as antifoaming agents in bioprocesses, J Chem Tech Biotechnol, 43(1), 39-48 (1988).
Various, Use of fine chemicals in flavours and fragrances (topic issue), Chem & Ind, 19 Sept, 575-596 (1988).
Various, Safety issues in biotechnology (topic issue), J Chem Tech Biotechnol, 43(4), 245-378 (1988).
Various, Advances in biotechnology (topic issue), Chem Eng Prog, 84(8), 16-74 (1988).
Various, Use of garden pesticides (topic issue), Chem & Ind, 21 Nov, 711-726 (1988).
Various, Extremophiles, (topic issue), J Chem Tech Biotechnol, 42(4), 289-322 (1988).
Various, Biotechnology (topic issue), Chem Eng Prog, 84(8), 16-74 (1988).
Ward, M., Biotech faces market challenge, Processing, May, 44-48 (1988).
Wyke, A., The state of biotechnology, Chem Eng Prog, 84(8), 16-27 (1988).

12.3 Biotechnology Processes

1968-1975

Various, Pharmaceutical engineering (topic issue), Chem Eng Prog, 64(2), 64-79 (1968).
Wang, D.I.C., Proteins from petroleum, Chem Eng, 26 Aug, 99-108 (1968).
Fuchs, R.; Ryu, D.D.Y., and Humphrey, A.E., Effect of surface aeration on scale-up procedures for fermentation processes, Ind Eng Chem Proc Des Dev, 10(2), 190-196 (1971).
Johnson, M.J., Fermentation: A review, Chemtech, June, 338-341 (1971).
Nyiri, L., Preparation of enzymes by fermentation, Int Chem Eng, 11(3), 447-458 (1971).
Royston, M.G., Continuous fermentation, Chem & Ind, 6 Feb, 170-171 (1971).
Sinclair, C.G.; Topiwala, H.H., and Brown, D.E., An experimental investigation of a growth model for Aerobacter aerogenes in continuous culture, Chem Engnr, May, 198-201 (1971).
Various, Production of large molecules by biochemical processes (symposium papers), Chem Engnr, July, 259-264 (1971).
Calam, C.T., and Russell, D.W., Microbial aspects of fermentation process development, J Appl Chem Biotechnol, 23, 225-238 (1973).
Loucaides, R., and McManamey, W.J., Mass transfer into simulated fermentation media, Chem Eng Sci, 28(12), 2165-2179 (1973).

Ryu, D.D.Y., and Humphrey, A.E., Examples of computer-aided fermentation systems, J Appl Chem Biotechnol, 23, 283-296 (1973).
Humphrey, A.E., Current developments in fermentation, Chem Eng, 9 Dec, 98-112 (1974).
Perlman, D., Prospects for fermentation industries (1974-1983), Chemtech, April, 210-216 (1974).
Halligan, J.E.; Herzog, K.L., and Parker, H.W., Synthesis gas from bovine wastes, Ind Eng Chem Proc Des Dev, 14(1), 64-69 (1975).
Tsao, G.T., and Lee, D.D., Oxygen transfer in fermentation, AIChEJ, 21(5), 979-985 (1975).
Various, Palm oil (topic issue), Chem & Ind, 1 Nov, 892-913 (1975).

1976-1979

Bamford, C., From polyethylene to protein, Chem & Ind, 18 Sept, 755-762 (1976).
Holloway, J.W., and Burrows, S., The technology of yeast production, Chem Engnr, June, 435-439 (1976).
Imanaka, T., and Aiba, S., Estimation of rate of heat evolution in fermentation, J Appl Chem Biotechnol, 26, 559-567 (1976).
Laine, B.M.; Snell, R.C., and Peet, W.A., Production of single cell protein from n-paraffins, Chem Engnr, June, 440-443, 446 (1976).
Lewis, C.W., Energy requirements for single cell protein production, J Appl Chem Biotechnol, 26, 568-575 (1976).
Hoogerheide, J.C., Traditional fermentation, Chemtech, Feb, 94-97 (1977).
Humphrey, A.E., Fermentation technology, Chem Eng Prog, 73(5), 85-91 (1977).
Perlman, D., Fermentation industries, Chemtech, July, 434-443 (1977).
Trilli, A., Prediction of costs in continuous fermentations, J Appl Chem Biotechnol, 27, 251-259 (1977).
Russell, R.M., and Tanner, R.D., Multiple steady states in continuous fermentation processes with bimodal growth kinetics, Ind Eng Chem Proc Des Dev, 17(2), 157-161 (1978).
Tong, G.E., Fermentation routes to C3 and C4 chemicals, Chem Eng Prog, 74(4), 70-74 (1978).
Weijenberg, D.C.; Mulder, J.J.; Drinkenburg, A.A.H., and Stemerding, S., Recovery of protein from potato juice wastewater by foam separation, Ind Eng Chem Proc Des Dev, 17(2), 209-213 (1978).
Zabriskie, D.W., and Humphrey, A.E., Real-time estimation of aerobic batch fermentation biomass concentration by component balancing, AIChEJ, 24(1), 138-146 (1978).
Chapman, V.J., Organic chemicals from the sea, Chemtech, Aug, 484-490 (1979).
Moo Young, M.; Moreira, A.R., and Daugulis, A.J., Economics of fermentation processes for SCP production from agricultural wastes, Can JCE, 57, 741-750 (1979).
Nichols, D., and Blouin, G.M., Ammonia fertilizer from coal, Chemtech, Aug, 512-518 (1979).
Shioya, S., and Dunn, I.J., Analysis of fed-batch fermentation processes by graphical methods, J Chem Tech Biotechnol, 29, 180-192 (1979).

Spear, M., Chemicals from renewable raw materials, Chem Engnr, June, 419-421 (1979).

1980-1982

Chain, E., Discovery of penicillin, Chemtech, Aug, 474-481 (1980).

Hobson, P.N., and McDonald, I., Methane production from acids in piggery-waste digesters, J Chem Tech Biotechnol, 30, 405-408 (1980).

Luong, J.H.T., and Volesky, B., Determination of the heat of some aerobic fermentations, Can JCE, 58, 497-504 (1980).

Various, Gashol/biomass developments (topic issue), Chem Eng Prog, 76(9), 39-64 (1980).

Ericsson, M.; Ebbinghaus, L., and Lindblom, M., Single-cell protein from methanol: Economic aspects of the Norprotein process, J Chem Tech Biotechnol, 31, 33-43 (1981).

Karthigesan, J., and Brown, B.S., Conversion of waste plastics to single-cell protein by pyrolysis followed by fermentation, J Chem Tech Biotechnol, 31, 55-65 (1981).

Takamatsu, T.; Shioya, S., and Furuya, T., Mathematical model of gluconic acid fermentation, J Chem Tech Biotechnol, 31, 697-704 (1981).

Various, Brewing industry report (topic issue), Processing, Oct, 13-21 (1981).

Various, Gasohol developments (topic issue), Chem Eng Prog, 77(6), 35-70 (1981).

Atkinson, B., and Sainter, P., Development of downstream processing in biotechnology, J Chem Tech Biotechnol, 32, 100-108 (1982).

Atkinson, B., and Sainter, P., Downstream biological process engineering, Chem Engnr, Nov, 410-419 (1982).

Brown, D.E., Industrial-scale operation of microbial processes, J Chem Tech Biotechnol, 32, 34-46 (1982).

Chen, T.S., Industrial fermentation in China, J Chem Tech Biotechnol, 32, 669-673 (1982).

Cho, G.H.; Choi, C.Y.; Choi, Y.D., and Han, M.H., Ethanol production by immobilised yeast and its carbon dioxide gas effects in a packed bed reactor, J Chem Tech Biotechnol, 32, 959-967 (1982).

Hawtin, P., Downstream processing in biochemical technology, Chem Engnr, Jan, 11-13 (1982).

Jones, B.E., The manufacture of hard gelatin capsules, Chem Engnr, May, 174-177 (1982).

Kovaly, K.A., Biomass to chemicals, Chemtech, Aug, 486-489 (1982).

Maiorella, B.L., Fermentation alcohol: Better to convert to fuel, Hyd Proc, Aug, 95-97 (1982).

Martin, S.R., The production of fuel ethanol from carbohydrates, Chem Engnr, Feb, 50-53, 58 (1982).

Nyeste, L., and Sevella, B., Mathematical modelling of fermentation systems, Int Chem Eng, 22(4), 729-736 (1982).

Schreiner, H., Laboratory experiments on methane digestion, Chem Engnr, June, 221-223, 229 (1982).

Various, Pharmaceutical industry survey, Processing, Sept, 13-19 (1982).

1983-1984

Cabib, G.; Silva, H.J.; Giulietti, A., and Ertola, R., Use of cane sugar stillage for single-cell protein production, J Chem Tech Biotechnol, 33B(1), 21-28 (1983).

Chen, N.Y., Making gasoline by fermentation, Chemtech, Aug, 488-492 (1983).

Clements, L.D.; Beck, S.R., and Heintz, C., Chemicals from biomass feedstocks, Chem Eng Prog, 79(11), 59-62 (1983).

Cundy, V.A.; Maples, D., and Tauzin, C., Combustion of bagasse: Use of an agricultural-derived waste, Fuel, 62(7), 775-780 (1983).

Karapinar, M., and Worgan, J.T., Bioprotein production from waste products of olive oil extraction, J Chem Tech Biotechnol, 33B(3), 185-188 (1983).

Lee, K.J., and Rogers, P.L., Fermentation kinetics of ethanol production, Biochem Eng J, 27(2), B31-B38 (1983).

Powell, G.E.; Hilton, M.G.; Archer, D.B., and Kirsop, B.H., Kinetics of methanogenic fermentation of acetate, J Chem Tech Biotechnol, 33B(4), 209-215 (1983).

Radhika, L.G.; Seshardi, S.K., and Mohandas, P.N., Study of biogas generation from coconut pith, J Chem Tech Biotechnol, 33B(3), 189-194 (1983).

Wennersten, R., Extraction of citric acid from fermentation broth using a solution of a tertiary amine, J Chem Tech Biotechnol, 33B(2), 85-94 (1983).

Anon., Distillery industry survey, Processing, Nov, 9-11 (1984).

Dawson, P.S.S., Continuous fermentation (review paper), Can JCE, 62(3), 293-300 (1984).

Felming, H.P., Developments in cucumber fermentation, J Chem Tech Biotechnol, 34B(4), 241-252 (1984).

Hallberg, D.E., Fermentation ethanol, Chemtech, May, 308-309 (1984).

Hartolegen, F.J.; Coburn, J.M., and Roberts, R.L., Microbial desulfurization of petroleum, Chem Eng Prog, 80(5), 63-67 (1984).

Jaworski, E.G., Biotechnology and the chemical industry, Processing, March, 17-21 (1984).

Nystrom, J.M.; Greenwald, C.G.; Harrison, F.G., and Gibson, E.D., Making ethanol from cellulosics, Chem Eng Prog, 80(5), 68-74 (1984).

Scheper, T.; Halwachs, W., and Schugerl, K., Production of L-amino acid by liquid membrane technique, Biochem Eng J, 29(2), B31-B38 (1984).

Shah, R.B.; Clausen, E.C., and Gaddy, J.L., Production of chemical feedstocks from biomass, Chem Eng Prog, 80(1), 76-80 (1984).

1985

Admassu, W.; Korus, R.A., and Heimsch, R.C., Ethanol fermentation with a flocculating yeast, Biochem Eng J, 31(1), B1-B8 (1985).

Arnold, F.H.; Blanch, H.W., and Wilke, C.R., Analysis of affinity separations, Biochem Eng J, 30(2), B9-B36 (1985).

Bjurstrom, E., Biotechnology: Fermentation and downstream processing, Chem Eng, 18 Feb, 126-158 (1985).

Chao, J.F.; Hollein, H.C., and Huang, C.R., Semicontinuous protein separation via variable-field-strength electrophoresis, Ind Eng Chem Fund, 24(4), 489-497 (1985).

Esplin, G.J.; Fung, D.P.C., and Hsu, C.C., Development of sampling and analytical procedures for biomass gasifiers, Can JCE, 63(6), 946-953 (1985).

Godia, F.; Casas, C., and Sola, C., Alcoholic fermentation by immobilised yeast cells, J Chem Tech Biotechnol, 35B(2), 139-144 (1985).

Hasegawa, S.; Shimizu, K.; Kobayashi, T., and Matsubara, M., Efficiency of repeated batch penicillin fermentation using two fermenters: Computer simulation and optimisation, J Chem Tech Biotechnol, 35B(1), 33-40 (1985).

Ho, C.C., and Tan, Y.K., Anaerobic treatment of palm-oil mill effluent by tank digesters, J Chem Tech Biotechnol, 35B(2), 155-164 (1985).

Kar, R., and Viswanathan, L., Ethanolic fermentation by thermotolerant yeasts, J Chem Tech Biotechnol, 35B(4), 235-238 (1985).

McGregor, W.C. (Ed.), Membrane Separations in Biotechnology, Marcel Dekker, New York (1985).

Nagamune, T.; Endo, I., and Inoue, I., Optimization method for successive batch-fed fermentation, Int Chem Eng, 25(4), 668-680 (1985).

Reschke, M., and Schugerl, K., Reactive extraction of penicillin, Biochem Eng J, 28(1), B1-B20 (1984); 29(2), B25-B30 (1984); 31(3), B19-B26 (1985).

Sadler, A.M.; Winkler, M.A., and Wiseman, A., Recovery of microsomal cytochrome P-450 from yeast using low-speed centrifugation, Biochem Eng J, 30(3), B43-B50 (1985).

Sekhar, C., Guidelines for large-scale r-DNA fermentations, Chem Eng, 29 April, 57-59 (1985).

Tanner, R.D.; Dunn, I.J.; Bourne, J.R., and Klu, M.K., Effect of imperfect mixing on an idealized fermentation model, Chem Eng Sci, 40(7), 1213-1220 (1985).

1986

Baker-Counsell, J., Improving control during protein production, Proc Engng, March, 61-63 (1986).

Basta, N., Cell culturing emerges as a leading biotech route, Chem Eng, 17 Feb, 23-28 (1986).

Girard, P.; Scharer, J.M., and Moo-Young, M., Two-stage anaerobic digestion for the treatment of cellulosic wastes, Biochem Eng J, 33(1), B1-B10 (1986).

Hoare, M., and Dunnill, P., Processing with proteins, Chem Engnr, Sept, 23-25 (1986).

Hoare, M., and Dunnill, P., Protein processing - new prospects, Chem Engnr, Dec, 39-41 (1986).

Jurasek, L., and Paice, M., Pulp, paper and biotechnology, Chemtech, June, 360-365 (1986).

Mata-Alvarez, J., and Viturtia, A.M., Laboratory simulation of municipal solid waste fermentation with leachate recycle, J Chem Tech Biotechnol, 36(12), 547-556 (1986).

Olson, G.J., and Brinckman, F.E., Bioprocessing of coal, Fuel, 65(12), 1638-1646 (1986).

Phillips, A.; Ball, G.; Fantes, K.; Finter, N., and Johnston, M., Growing cells on a 2000 gallon scale, Chemtech, July, 433-435 (1986).

1987

Baum, R.M., Biotech industry moving pharmaceutical products to market, C&E News, 20 July, 11-32 (1987).

Byerley, J.J.; Scharer, J.M., and Charles, A.M., Uranium (VI) biosorption from process solutions, Biochem Eng J, 36(3), B49-B59 (1987).

Mascone, C.F., Separations are key to biotech scaleup, Chem Eng, 19 Jan, 21-25 (1987).

Nicolaidis, A.A., Microbial mineral processing: The opportunities for genetic manipulation, J Chem Tech Biotechnol, 38(3), 167-186 (1987).

Parkinson, G., New techniques may squeeze more chemicals from algae, Chem Eng, 11 May, 19-23 (1987).

Royse, S., Scaling up for mammalian cell culture, Chem Engnr, Nov, 12-13 (1987).

Schmidt, W., and Schugerl, K., Continuous ethanol production by Zymomonas mobilis on a synthetic medium, Biochem Eng J, 36(3), B39-B48 (1987).

Shimizu, K., and Matsubara, M., A solvent screening criterion for multicomponent extractive fermentation, Chem Eng Sci, 42(3), 499-504 (1987).

Stead, C.V., The use of reactive dyes in protein separation processes, J Chem Tech Biotechnol, 38(1), 55-71 (1987).

Wilson, B.W., et al., Microbial conversion of low-rank coal: Characterization of biodegraded product, Energy & Fuels, 1(1), 80-84 (1987).

1988

Akiyama, A., et al., Enzymes in organic synthesis, Chemtech, Oct, 627-634 (1988).

Bar, R., Effect of interphase mixing on a water-organic solvent two-liquid phase microbial system: Ethanol fermentation, J Chem Tech Biotechnol, 43(1), 49-62 (1988).

Bjurstrom, E.E., and Smelser, B.J., Commercializing biotechnology processes, Chem Eng, 18 Jan, 81-84 (1988).

Cross, J., Production of apyrogenic water, Chem Engnr, July, 69 (1988).

Eligwe, C.A., Microbial desulphurization of coal, Fuel, 67(4), 451-458 (1988).

Fish, N., and Thornhill, N., Monitoring and control of fermentation, Chem Engnr, Sept, 31-33 (1988).

Godia, F.; Casas, C., and Sola, C., Batch alcoholic fermentation modelling by simultaneous integration of growth and fermentation equations, J Chem Tech Biotechnol, 41(2), 155-165 (1988).

Hileman, B., Fluoridation of water, C&E News, 1 Aug, 26-42 (1988).

Hubbard, D.W.; Harris, L.R., and Wierenga, M.K., Scaleup for polysaccharide fermentation, Chem Eng Prog, 84(8), 55-61 (1988).

Jones, J.L.; Fong, W.S.; Hall, P., and Cometta, S., Bioprocesses for volume chemicals, Chemtech, May, 304-309 (1988).

Likidis, Z., and Schugerl, K., Simulation of the continuous re-extraction of penicillin G from the solution of its ion-pair complex in three different types of bench-scale column, Chem Eng Sci, 43(6), 1243-1247 (1988).

Likidis, Z., and Schugerl, K., Reextraction of Penicillin G and V from organic phase in bench-scale Karr column after reactive extraction by LA-2, Chem Eng & Proc, 23(1), 61-64 (1988).

Record, P., Pharmaceutical tabletting, Chem Engnr, May, 48-50 (1988).
Rubio, F.C., et al., Enzymatic hydrolysis of carboxymethyl cellulose, Int Chem Eng, 28(4), 618-627 (1988).
Satoh, H.; Yoshizawa, J., and Kametani, S., Bacteria help desulfurize gas, Hyd Proc, 67(5), 76D-76F (1988).
Schiweck, H.; Rapp, K., and Vogel, M., Utilisation of sucrose as an industrial bulk chemical: State of the art and future implications, Chem & Ind, 4 April, 228-234 (1988).
Sikes, C.S., and Wheeler, A.P., Regulators of biomineralization Chemtech, Oct, 620-626 (1988).
Thompson, D., Biotechnology for energy conservation and a cleaner environment, Proc Engng, Dec, 39, 41 (1988).
Various, Crop protection in Brazil (topic issue) Chem & Ind, 21 March, 175-199 (1988).
Wright, J.D., Ethanol from biomass by enzymatic hydrolysis, Chem Eng Prog, 84(8), 62-74 (1988).

12.4 Equipment Design

1967-1972

Atkinson, B.; Swilley, E.L.; Busch, A.W., and Williams, D.A., Kinetics, mass transfer, and organism growth in a biological film reactor, Trans IChemE, 45, T257-264 (1967).
Atkinson, B.; Daoud, I.S., and Williams, D.A., A theory for the biological film reactor, Trans IChemE, 46, T245-250 (1968).
Irving, G.M., Construction materials for breweries, Chem Eng, 1 July, 100-104 (1968).
Lengyel, Z.L., Problems of mixing and aeration in submerged culture processes, Int Chem Eng, 10(2), 252-257 (1970).
Atkinson, B., and Williams, D.A., The performance characteristics of a trickling filter with hold-up of microbial mass controlled by periodic washing, Trans IChemE, 49, 215-224 (1971).
Babayants, A.V., et al., Parameters for thermophilic methane fermentor with continuous liquid feed, Int Chem Eng, 11(4), 585-588 (1971).
Greenshields, R.N., and Smith, E.L., Tower fermentation systems and their applications, Chem Engnr, May, 182-190 (1971).
Atkinson, B., and Davies, I.J., The completely mixed microbial film fermenter, Trans IChemE, 50, 208-216 (1972).
Blakebrough, N., Mixing effects in biological systems, Chem Engnr, Feb, 58-63 (1972).
Erickson, L.E.; Lee, S.S., and Fan, L.T., Modelling and analysis of tower fermentation processes, J Appl Chem Biotechnol, 22, 199-216 (1972).

1973-1975

Choi, P.S.K., and Fan, L.T., Transient behaviour of encapsulated enzyme reactor systems, J Appl Chem Biotechnol, 23, 531-548 (1973).
Hockenhull, D.J.D., Fermentation pilot plants, Chem & Ind, 19 May, 461-464 (1973).

Katoh, S., and Yoshida, F., Rate of blood oxygenation in a flat plate membrane oxygenator, Chem Eng J, 6(1), 51-58 (1973).

Pirt, S.J., Quantitative theory of action of microbes attached to a packed column, J Appl Chem Biotechnol, 23, 389-400 (1973).

Saunders, P.T., and Bazin, M.J., Attachment of microorganisms in a packed column, J Appl Chem Biotechnol, 23, 847-854 (1973).

Vieth, W.R., et al., Mass transfer and biochemical reaction in enzyme membrane reactor systems, Chem Eng Sci, 28(4), 1013-1020 (1973).

Ramachandran, P.A., General model for packed bed encapsulated enzyme reactor, J Appl Chem Biotechnol, 24, 265-276 (1974).

Smith, E.L., and Greenshields, R.N., Tower fermentation systems and their application to aerobic processes, Chem Engnr, Jan, 28-34 (1974).

Starkie, G.L., Design of fermentation equipment, Chem & Ind, 16 Feb, 142-146 (1974).

Topiwala, H.H., and Hamer, G., Mass transfer and dispersion properties in a fermenter with a gas-inducing impeller, Trans IChemE, 52, 113-120 (1974).

Cooper, P.G., and Silver, R.S., Basin fermentor for single cell protein, Chem Eng Prog, 71(9), 85-88 (1975).

Klei, H.E.; Sundstrom, D.W., and Molvar, A.E., Control of well-mixed biological reactors subject to variations in feed concentration and flowrate, J Appl Chem Biotechnol, 25, 535-548 (1975).

Mashelkar, R.A., and Ramachandran, P.A., New model for hollow-fibre enzyme reactor, J Appl Chem Biotechnol, 25, 867-880 (1975).

Nishikawa, A.H., Affinity purification of enzymes, Chemtech, Sept, 564-571 (1975).

Shaffer, A.G., and Hamrin, C.E., Enzyme separation by parametric pumping, AIChEJ, 21(4), 782-786 (1975).

1976-1977

Atkinson, B., and Ali, M.E.A.R., Wetted area, slime thickness and liquid phase mass transfer in packed bed biological film reactors (trickling filters), Trans IChemE, 54, 239-250 (1976).

Hseih, F.; Davidson, B., and Vieth, W.R., Modelling of combined mass transfer kinetic effects in an enzyme membrane reactor system, J Appl Chem Biotechnol, 26, 631-644 (1976).

Paca, J.; Ettler, P., and Gregr, V., Hydrodynamic behaviour and oxygen transfer rate in a pilot plant fermenter, J Appl Chem Biotechnol, 26, 309-317 (1976).

Spruytenburg, R., Experience with a computer-coupled bioreactor, Chem Engnr, June, 447-449 (1976).

Brown, D.E., Measurement of fermenter power input, Chem & Ind, 20 Aug, 684-688 (1977).

Chen, H.T.; Hsieh, T.K.; Lee, H.C., and Hill, F.B., Separation of proteins via semicontinuous pH parametric pumping, AIChEJ, 23(5), 695-701 (1977).

Knowlton, H.E., Biological rotating treatment units, Hyd Proc, 56(9), 227-230 (1977).

Lin, S.H., Analysis of immobilized enzymatic reaction in a packed-bed reactor with enzyme denaturation, Chem Eng J, 14(2), 129-136 (1977).

Pasquali, G., and Magelli, F., Analysis of tower fermentation systems, Chem Eng J, 14(2), 147-152 (1977).
Pollard, R., and Shearer, C.J., The application of chemical engineering concepts to the design of fermenters, Chem Engnr, Feb, 106-110 (1977).
Schneider, K., and Frischknecht, K., Determination of oxygen and carbon dioxide KLa values in fermenters by the dynamic method measuring step responses in the gas phase, J Appl Chem Biotechnol, 27, 631-642 (1977).

1978-1979

Brauer, H.; Schmidt, H., and Thiele, H., Fluid-dynamic investigations of cascaded mixer-fermenter, Int Chem Eng, 18(4), 549-558 (1978).
Jackson, M.L., and Shen, C.C., Aeration and mixing in deep tank fermentation systems, AIChEJ, 24(1), 63-71 (1978).
Lin, S.H., Performance characteristics of a CSTR containing immobilised enzyme particles, J Appl Chem Biotechnol, 28, 677-686 (1978).
McNaughton, G.S.; Robins, J.C., and Zwaaneveld, C.H., Continuous countercurrent ion-exchange recovery of catalase from beef liver extract, Sepn & Purif Methods, 7(1), 31-54 (1978).
Shioya, S.; Dang, N.D.P., and Dunn, I.J., Bubble column fermenter modeling, Chem Eng Sci, 33(8), 1025-1030 (1978).
Various, Trends in fermenter design (topic issue), Chem & Ind, 21 Oct, 782-789 (1978).
Chen, H.T.; Wong, Y.W., and Wu, S., Continuous fractionation of protein mixtures by pH parametric pumping, AIChEJ, 25(2), 320-327 (1979).
Jarai, M., Factors affecting scale-up of aerated fermentation processes, Int Chem Eng, 19(4), 701-708 (1979).
Laine, J., and Kuoppamaki, R., Development of the design of large-scale fermenters, Ind Eng Chem Proc Des Dev, 18(3), 501-506 (1979).
Pirt, S.J.; Panikov, N., and Lee, Y.K., The miniloop: A small-scale air-lift microbial culture vessel and photobiological reactor, J Chem Tech Biotechnol, 29, 437-441 (1979).
Stephanopoulos G.; Fredrickson, A.G., and Aris, R., Growth of competing microbial populations in CSTR with periodically varying inputs, AIChEJ, 25(5), 863-872 (1979).

1980-1981

Chen, H.T., et al., Separation of proteins via pH parametric pumping, Sepn Sci Technol, 15(6), 1377-1392 (1980).
Chen, H.T.; Yang, W.T.; Pancharoen, U., and Parisi, R., Separation of proteins via multicolumn pH parametric pumping, AIChEJ, 26(5), 839-849 (1980).
Paca, J., Elimination of ethanol inhibition of yeast growth by a multistream ethanol feed in a multistage tower fermenter, J Chem Tech Biotechnol, 30, 764-771 (1980).
Prasad, R.; Gupta, A.K., and Bajpai, R.K., Adsorption of Streptomycin on ion exchange resins, J Chem Tech Biotechnol, 30, 324-331 (1980).
Schwartzberg, H.G., Continuous counter-current extraction in the food industry, Chem Eng Prog, 76(4), 67-85 (1980).
Various, Biochemical reactor engineering (topic issue), Chem Eng Sci, 35(1), 99-144 (1980).

Vyas, S.N.; Patwardhan, S.R., and Padhye, V.M., Ion exchangers for recovery of penicillin from its waste, Sepn Sci Technol, 15(2), 111-122 (1980).

Buchholtz, H.; Luttman, R,; Zakrzewski, W., and Schugerl, K., Performance of a tower bioreactor with external loop, J Chem Tech Biotechnol, 31, 435-444 (1981).

Chen, H.T., et al., Protein separations using semicontinuous pH parametric pumping, Sepn Sci Technol, 16(1), 43-62 (1981).

Chen, H.T.; Ahmed, Z.M., and Rollen, V., Enzyme purification by parametric pumping, Ind Eng Chem Fund, 20(2), 171-174 (1981).

Sakoda, A.; Sadakata, M.; Koya, T.; Furusawa, T., and Kunii, D., Gasification of biomass in a fluidized bed, Chem Eng J, 22(3), 221-228 (1981).

Shieh, W.K.; Mulcahy, L.T., and LaMotta, E.J., Fluidised bed biofilm reactor effectiveness factor expressions, Trans IChemE, 59, 129-133 (1981).

Wick, E., Bubble pumping in cocurrent gas-liquid flow and its effect upon conversion in continuous fermentation, Can JCE, 59, 297-302 (1981).

1982

Adler, I.; Deckwer, W.D., and Schugerl, K., Performance of tower loop bioreactors, Chem Eng Sci, 37(2), 271-276; 37(3), 417-424 (1982).

Agrawal, P.; Lee, C.; Lim, H.C., and Ramkrishna, D., Theoretical investigations of dynamic behaviour of isothermal continuous stirred-tank biological reactors, Chem Eng Sci, 37(3), 453-462 (1982).

Anderson, C.; LeGrys, G.A., and Solomons, G.L., Concepts in the design of large-scale fermenters for viscous culture broths, Chem Engnr, Feb, 43-49 (1982).

Andrew, S.P.S., Gas-liquid mass-transfer in microbiological reactors (review paper), Trans IChemE, 60, 3-13 (1982).

Bajpai, R.K., and Reuss, M., Coupling of mixing and microbial kinetics for evaluating performance of bioreactors, Can JCE, 60, 384-392 (1982).

Brauer, H., Development and improvement of bioreactors, Chem Engnr, June, 224-229 (1982).

Bull, M.A.; Sterritt, R.M., and Lester, J.N., The effect of organic loading on the performance of anaerobic fluidised beds treating high strength wastewaters, Trans IChemE, 60, 373-376 (1982).

Dussap, G., and Gros, J.B., Energy consumption and interfacial mass-transfer area in an air-lift fermenter, Chem Eng J, 25(2), 151-162 (1982).

Graham, E.E., and Fook, C.F., Rate of protein absorption and desorption on cellulosic ion exchangers, AIChEJ, 28(2), 245-250 (1982).

Hollein, H.C.; Ma, H.C.; Huang, C.R., and Chen, H.T., Protein separations by parametric pumping, Ind Eng Chem Fund, 21(3), 205-214 (1982).

Kargi, F., and Park, J.K., Optimal biofilm thickness for fluidised-bed biofilm reactors, J Chem Tech Biotechnol, 32, 744-748 (1982).

Luttmann, R.; Thoma, M.; Buchholz, H., and Schugerl, K., Nonsteady-state simulation of extended cultures in tower loop reactors, Chem Eng Sci, 37(12), 1771-1783 (1982).

Rizzuti, L.; Augugliaro, V.; Dardanoni, L., and Torregrossa, M.V., Kinetic parameter determination in monoculture and monosubstrate biological reactors, Can JCE, 60, 608-612 (1982).

Roels, J.A., Mathematical models and design of biochemical reactors, J Chem Tech Biotechnol, 32, 59-72 (1982).
Schugerl, K., New bioreactors for aerobic processes, Int Chem Eng, 22(4), 591-611 (1982).
Schugerl, K., Basic principles for layout of tower bioreactors, J Chem Tech Biotechnol, 32, 73-80 (1982).
Sittig, W., The present state of fermentation reactors, J Chem Tech Biotechnol, 32, 47-58 (1982).
Sittig, W., and Faust, U., Biochemical applications of the airlift loop reactor, Chem Engnr, June, 230-232 (1982).

1983

Blakebrough, N.; McManamey, W.J., and Tart, K.R., Heat transfer to fermentation systems in an air-lift fermenter, Trans IChemE, 56, 127-135 (1978); CER&D, 61, 264-266, 383-387 (1983).
Brown, D.E., Scaling-up microbial processes, Chemtech, March, 164-169 (1983).
Bull, M.A.; Sterritt, R.M., and Lester, J.N., Influence of COD, hydraulic, temperature and pH shocks on stability of an unheated fluidised-bed bioreactor, J Chem Tech Biotechnol, 33B(4), 221-230 (1983).
Fields, P.R.; Fryer, P.J.; Slater, N.K.H., and Woods, G.P., Adsorptive bubble fractionation in a bubble column fermenter, Biochem Eng J, 27(1), B3-B12 (1983).
Fox, D.; Dunn, N.W.; Gray, P.P., and Marsden, W.L., Saccharification of bagasse using a countercurrent plug-flow reactor, J Chem Tech Biotechnol, 33B(2), 114-118 (1983).
Luttmann, R.; Thoma, M.; Buchholtz, H., and Schugerl, K., Model development, parameter identification, and simulation of SCP production processes in air-lift tower bioreactors with external loop, Comput Chem Eng, 7(1), 43-64 (1983).
Meiners, M., et al., Analysis of process dynamics in bioreactors, J Chem Tech Biotechnol, 33B(3), 164-176 (1983).
Oosterhuis, N.M.G., and Kossen, N.W.F., Oxygen transfer in a production scale bioreactor, CER&D, 61, 308-312 (1983).
Pirt, S.J., et al., Design and performance of a tubular bioreactor for photosynthetic production of biomass from carbon dioxide, J Chem Tech Biotechnol, 33B(1), 35-58 (1983).
Popovic, M.; Papalexiou, A., and Reuss, M., Gas residence time distribution in stirred-tank bioreactors, Chem Eng Sci, 38(12), 2015-2026 (1983).
Rodrigues, A.; Grasmick, A., and Elmaleh, S., Modelling of biofilm reactors, Biochem Eng J, 27(2), B39-B48 (1983).
Scott, C.D., Fermentation in a fluidized-bed reactor, Chemtech, June, 364-365 (1983).
Sittig, W., Fermentation reactors, Chemtech, Oct, 606-613 (1983).
Walach, M.R., et al., Computer control of an algal bioreactor with simulated diurnal illumination, J Chem Tech Biotechnol, 33B(1), 59-75 (1983).

1984

Andrews, G.F., and Elkcechen, S., Solid adsorbents in batch fermentations, Chem Eng Commns, 29(1), 139-152 (1984).

Bulock, J.D.; Comberbach, D.M., and Ghommidh, C., Continuous ethanol production using a highly flocculent yeast in a gas-lift tower fermenter, Biochem Eng J, 29(1), B9-B24 (1984).

Do, D.D., Enzyme deactivation studies in a continuous stirred basket reactor, Biochem Eng J, 28(3), B51-B60 (1984).

Glasgow, L.A., et al., Wall pressure fluctuations and bubble size distributions at several positions in an airlift fermentor, Chem Eng Commns, 29(1), 311-336 (1984).

Heijnen, J.J., and Van't Riet, K., Mass transfer, mixing and heat transfer phenomena in low-viscosity bubble column reactors (review paper), Biochem Eng J, 28(2), B21-B42 (1984).

Jurecic, R.; Berovic, M.; Steiner, W., and Koloini, T., Mass transfer in aerated fermentation broths in a stirred tank reactor, Can JCE, 62(3), 334-339 (1984).

McManamey, W.J.; Wase, D.A.J.; Raymahasay, S, and Thayanithy. K., Influence of gas inlet design on gas hold-up values for water and various solutions in a loop-type air-lift fermenter, J Chem Tech Biotechnol, 34B(3), 151-164 (1984).

O'Sullivan, T.J., Epstein, A.C.; Korchin, S.R., and Beaton, N.C., Applications of ultrafiltration in biotechnology, Chem Eng Prog, 80(1), 68-75 (1984).

Ohkawa, A.; Sakai, N.; Imai, H., and Endoh, K., Mechanical foam control in a stirred draft-tube reactor, J Chem Tech Biotechnol, 34B(2), 87-96 (1984).

Park, Y.H.; Han, M.H., and Rhee, H.K., Effect of external mass transfer in a packed-bed reactor system for a reversible enzyme reaction, J Chem Tech Biotechnol, 34B(1), 57-69 (1984).

Schugerl, K., Influence of tower-reactor design and operation on culturing single-cell organisms, Int Chem Eng, 24(4), 603-618 (1984).

Shieh, W.K., and Chen, C.Y., Biomass hold-up correlations for a fluidised bed biofilm reactor, CER&D, 62, 133-136 (1984).

1985

Baker-Counsell, J., Fermenters for biotechnology, Proc Engng, July, 43 (1985).

Beg, S.A., and Hassan, M.M., Biofilm model for packed bed reactors, Biochem Eng J, 30(1), B1-B8 (1985).

Chen, S.J.; Li, C.T., and Shieh, W.K., Performance evaluation of the anaerobic fluidised-bed system, J Chem Tech Biotechnol, 35B(2), 101-109; (3), 183-190; (4), 229-234 (1985).

Muth, W.L., Scaleup biotechnology safely, Chemtech, June, 356-361 (1985).

Pickett, A., and Ladwa, N., Updating fermentation process control, Processing, Oct, 23-24 (1985).

Samson, R., and Leduy, A., Multistage continuous cultivation of blue-green alga in flat tank photobioreactors with recycle, Can JCE, 63(1), 105-112 (1985).

Shuler, M.L., Use of chemically structured models for bioreactors (review paper), Chem Eng Commns, 36(1), 161-190 (1985).

Tayler, C., and Opie, R., Understanding and improving fermenter design, Proc Engng, Nov, 55 (1985).

Thomas, C.R., and Yates, J.G., Expansion index for biological fluidised beds, CER&D, 63, 67-70 (1985).
Vaidya, R.N., and Pangarkar, V.G., Hydrodynamics and mass transfer in rotating biological contactor, Chem Eng Commns, 39, 337-354 (1985).
Verschoor, H., Developments in bioreactors, Chem Engnr, June, 39 (1985).
Wheatley, A., and Winstanley, I., Anaerobic digester systems for waste treatment, Chem Engnr, July, 42-46 (1985).

1986

Bauer, W., A comparison between a conventional submerged culture fermenter and a new concept gas/solid fluid bed bioreactor for glutathione production, Can JCE, 64(4), 561-566 (1986).
Belanger, R.R.; Rochon, L.; Dumont, G.A., and Gendron, S., Self-tuning control of chip level in a Kamyr digester, AIChEJ, 32(1), 65-74 (1986).
Cabral, J.M.S.; Novais, J.M.; Cardoso, J.P., and Kennedy, J.F., Design of immobilised glucoamylase reactors using a simple kinetic model for the hydrolysis of starch, J Chem Tech Biotechnol, 36(6), 247-254 (1986).
Cheryan, M., and Mehaia, M.A., Membrane bioreactors, Chemtech, Nov, 676-681 (1986).
Dekker, M.; Van Riet, K.; Weijers, S.R.; Baltussen, W.A.; Laane, C., and Bijsterbosch, B.H., Enzyme recovery by liquid-liquid extraction using reversed micelles, Biochem Eng J, 33(2), B27-B34 (1986).
Hoffman, A.S., Materials for biotech, Chemtech, July, 426-432 (1986).
Kawase, Y., and Moo-Young, M., Mixing and mass transfer in concentric-tube airlift fermenters: Newtonian and non-Newtonian media, J Chem Tech Biotechnol, 36(11), 527-538 (1986).
Loh, V.Y.; Richards, S.R., and Richmond, P., Particle suspension in a circulating bed fermenter, Biochem Eng J, 32(2), B39-B41 (1986).
Narendranathan, T.J., Designing fermentation equipment, Chem Engnr, May, 23-31 (1986).
Patel, S.A.; Glasgow, L.A.; Erickson, L.E., and Lee, C.H., Characterization of downflow section of airlift column using bubble size distribution measurements, Chem Eng Commns, 44, 1-20 (1986).
Pucci, A.; Mikitenko, P., and Asselineau, L., Three-phase distillation: Simulation and application to the separation of fermentation products, Chem Eng Sci, 41(3), 485-494 (1986).
Reschke, M., and Schugerl, K., Simulation of the continuous reactive extraction of penicillin G in a Karr column, Biochem Eng J, 32(1), B1-B6 (1986).
Schoutens, G.H.; Guit, R.P.; Zieleman, G.J.; Luyben, K.C.A.M., and Kossen, N.W.F., A comparative study of a fluidised bed reactor and a gas-lift loop reactor for the IBE process, J Chem Tech Biotechnol, 36(7), 335-343; 36(9), 415-426; 36(12), 565-576 (1986).
Schugerl, K., Reaction engineering fundamentals relating to design and operation of bioreactors, Int Chem Eng, 26(2), 204-231 (1986).
Sterbackova, M., and Sterbacek, Z., Non-ideal flow characteristics, power input and oxygen transfer in a sectioned bubble column fermenter provided with twin dispersing disc agitators, Biochem Eng J, 33(1), B11-B18 (1986).

Varlaan, P.; Tramper, J.; Van Riet, K., and Luyben, K.C.M., Hydrodynamic model for airlift-loop bioreactor with external loop, Biochem Eng J, 33(2), B43-B53 (1986).

Walter, J.F., and Blanch, H.W., Bubble break-up in gas-liquid bioreactors (turbulent flows), Biochem Eng J, 32(1), B7-B17 (1986).

1987

Beck, C.; Stiefel, H., and Stinnett, T., Cell-culture bioreactors, Chem Eng, 16 Feb, 121-129 (1987).

Beevers, A., Adsorption for biotech separations, Processing, Sept, 41,43 (1987).

Bovonsombut, S.; Wilhelm, A.M., and Riba, J.P., Influence of gas distributor design on the oxygen transfer characteristics of an airlift fermenter, J Chem Tech Biotechnol, 40(3), 167-176 (1987).

Brink, L.E.S., and Tramper, J., Facilitated mass transfer in a packed-bed immobilized-cell reactor by using an organic solvent as substrate reservoir, J Chem Tech Biotechnol, 37(1), 21-44 (1987).

Cameron, W., Mechanical seals for bioreactors, Chem Engnr, Nov, 41-42 (1987).

Elmaleh, S.; Papaconstantinou, S.; Rios, G.M., and Grasmick, A., Organic carbon conversion in a large-particle spouted bed, Biochem Eng J, 34(2) B29-B34 (1987).

Hassan, M.M., and Beg, S.A., Theoretical analysis of a packed-bed biological reactor for various reaction kinetics, Biochem Eng J, 36(2), B15-B28 (1987).

Hoffmann, H.; Scheper, T.; Schugerl, K., and Schmidt, W., Use of membranes to improve bioreactor performance, Biochem Eng J, 34(1) B13-B19 (1987).

Hossain, M.M., and Do, D.D., Effects of nonuniform immobilized enzyme distribution in porous solid supports on the performance of a continuous reactor, Biochem Eng J, 34(2) B35-B47 (1987).

Hu, T.T., and Wu, J.Y., Characteristics of biological fluidized bed in magnetic field, CER&D, 65(3), 238-242 (1987).

Kafarov, V.V.; Vinarov, A.Y., and Gordeev, L.S., Modeling of bioreactors, Int Chem Eng, 27(4), 615-642 (1987).

Kavanagh, P.R., and Brown, D.E., Cross-flow separation of yeast cell suspensions using a sintered stainless steel filter tube, J Chem Tech Biotechnol, 38(3), 187-200 (1987).

La Nauze, R.D., A review of the fluidised bed combustion of biomass, J Inst Energy, June, 66-76 (1987).

Likidis, Z., and Schugerl, K., Continuous reactive extraction of penicillin G and its re-extraction in three different column types, Chem Eng Sci, 43(1), 27-32 (1987).

Lorenz, T.; Diekman, J.; Freueh, K.; Hiddessen, R., and Moeller, J., On-line measurement and control of penicillin V production in a tower loop reactor, J Chem Tech Biotechnol, 38(1), 41-54 (1987).

Mehrotra, I.; Alibhai, K.R.K., and Forster, C.F., Removal of heavy metals in anaerobic upflow sludge blanket reactors, J Chem Tech Biotechnol, 37(3), 195-202 (1987).

Meldrum, A., Hollow-fibre membrane bioreactors, Chem Engnr, Oct, 28-31 (1987).
Menawat, A.; Mutharasan, R., and Coughanowr, D.R., Singular optimal control strategy for a fed-batch bioreactor: Numerical approach, AIChEJ, 33(5), 776-783 (1987).
Meyer, H.D.; Kuhlman, W.; Lubbert, A., and Schugerl, K., Development of a microcomputer-based system for single-stirred bioreactors and their cascade, J Chem Tech Biotechnol, 40(1), 19-32 (1987).
Miller, R., and Melick, M., Modeling bioreactors, Chem Eng, 16 Feb, 112-120 (1987).
Qureshi, N.; Pai, J.S., and Tamhane, D.V., Reactors for ethanol production using immobilised yeast cells, J Chem Tech Biotechnol, 39(2), 75-84 (1987).
Rosen, C.G., Biotechnology: Time to scale up and commercialize, Chemtech, Oct, 612-618 (1987).
Salmon, P.M., and Robertson, C.R., A theoretical analysis of a hollow-fibre reactor with two substrates, Biochem Eng J, 35(1) B1-B8 (1987).
Shackleton, R., The application of ceramic membranes to the biological industries, J Chem Tech Biotechnol, 37(1), 67-69 (1987).
Shinizu, K., and Matsubara, M., Product formation patterns and performance improvement for multistage continuous stirred tank fermenters, Chem Eng Commns, 52, 61-74 (1987).
Tang, W.T., and Fan, L.S., Steady state phenol degradation in draft-tube gas-liquid-solid fluidized-bed bioreactor, AIChEJ, 33(2), 239-249 (1987).

1988
Anon., Liquid-liquid extraction for biotechnology, Chem Engnr, Feb, 33 (1988).
Chisti, Y., and Moo-Young, M., Gas holdup behaviour in fermentation broths and other non-Newtonian fluids in pneumatically agitated reactors, Biochem Eng J, 39(3), B31-B36 (1988).
Chisti, Y., and Moo-Young, M., Prediction of liquid circulation velocity in airlift reactors with biological media, J Chem Tech Biotechnol, 42(3), 211-220 (1988).
Dahuron, L., and Cussler, E.L., Protein extractions with hollow fibres, AIChEJ, 34(1), 130-136 (1988).
Fish, N., Control of fermenters, Processing, May, 51-52 (1988).
Fryer, T., Valves for biotech, Processing, May, 41-42 (1988).
Fryer, T., One-step beer filtration process, Processing, July, 29-37 (1988).
Hamdane, M.; Wilhelm, A.M., and Riba, J.P., Modelling of a fluidized bed immobilized enzyme reactor, Biochem Eng J, 39(2), B25-B30 (1988).
Harris, T.A.J.; Reuben, B.G.; Cox, D.J.; Vaid, A.K., and Carvel, J., Cross-flow filtration of an unstable b-lactam antibiotic fermentation broth, J Chem Tech Biotechnol, 42(1), 19-30 (1988).
Hess, W.F., and Kalwa, M., High-pressure filtration for downstream processing of biological dispersions, Chem Eng & Proc, 23(3), 179-188 (1988).
Kafarov, V.V.; Vinarov, A.Y., and Gordeev, L.S., Modeling of bioreactors, Int Chem Eng, 28(1), 14-36 (1988).
Krebser, U.; Meyer, H.P., and Fiechter, A., Comparison of performance of continuously stirred-tank bioreactors and a TORUS bioreactor with respect

to highly viscous culture broths, J Chem Tech Biotechnol, 43(2), 107-116 (1988).
Merchuk, J.C., and Siegel, M.H., Air-lift reactors in chemical and biological technology, J Chem Tech Biotechnol, 41(2), 105-120 (1988).
Moresi, M., and Patete, M., Prediction of Kla in conventional stirred fermenters, J Chem Tech Biotechnol, 42(3), 197-210 (1988).
Nelson, T.B., and Skaates, J.M., Attrition in a liquid fluidized-bed bioreactor, Ind Eng Chem Res, 27(8), 1502-1505 (1988).
Papathansiou, T.D.; Kalogerakis, N., and Behie, L.A., Dynamic modelling of mass transfer phenomena with chemical reaction in immobilized-enzyme bioreactors, Chem Eng Sci, 43(7), 1489-1498 (1988).
Park, S., and Ramirez, W.F., Optimal production of secreted protein in fed-batch reactors, AIChEJ, 34(9), 1550-1558 (1988).
Rechnitz, G.A., Biosensors, C&E News, 5 Sept, 24-36 (1988).
Short, H., Biotech fermenters, Chem Eng, 28 March, 22-26 (1988).
Steinmeyer, D.; Cho, T.; Efthymiou, G., and Shuler, M.L., A multimembrane bioreactor with integral product recovery, Chemtech, Nov, 680-685 (1988).
Various, Bioseparations (topic issue), Sepn Sci Technol, 23(8), 759-943 (1988).

12.5 Food and Brewing

1968-1974

Clarke, R.J., Chemical engineering in the food industry, Chem Engnr, Nov, CE374-376 (1968).
Iyengar, M.S., and Baruah, J.N., Protein from petroleum hydrocarbons, Brit Chem Eng, 13(5), 684-686 (1968).
Shore, D.T., and Royston, M.G., Chemical engineering of the continuous brewing process, Chem Engnr, May, CE99-109 (1968).
Various, World protein resources (topic issue), Chem Eng Prog, 65(9), 20-36 (1969).
Various, Food from petroleum (special report), Hyd Proc, 48(3), 95-112 (1969).
Holdsworth, S.D., Heat transfer in the freezing of fruit and vegetables, Chem Engnr, May, CE127-134 (1970).
Warman, K.G., and Reichel, A.J., Development of a freeze-drying process for food, Chem Engnr, May, CE134-139 (1970).
Yamamoto, T., Food from natural gas, Int Chem Eng, 10(3), 478-484 (1970).
Hetherington, P.J.; Follows, M.; Dunnill, P., and Lilly, M.D., Release of protein from bakers yeast by disruption in an industrial homogeniser, Trans IChemE, 49, 142-148 (1971).
Haughey, D.P., Chemical engineering in the meat industry, Chem Engnr, Dec, 472-475 (1972).
Lasztity, R., Recent results in cereal protein research, Periodica Polytechnica, 16, 331-346 (1972).
Scott-Blair, G.W., Rheology of foodstuffs, Periodica Polytechnica, 16, 81-84 (1972).
Whitman, W.E., Foods and the future, Chem Engnr, Oct, 372-376 (1972).
Bathory, J., and Vamos, E., Microbiological processing of petroleum for production of food, Int Chem Eng, 13(3), 490-496 (1973).

Hall, R.L., Faith, fad, fear, and food, Chemtech, July, 412-415 (1973).
Hawthorn, J., Present-day problems in food science, Periodica Polytechnica, 17, 39-46 (1973).
Hepner, L., Chemical engineering methods applied to food and drink processing, Proc Engng, June, 137-138 (1973).
Rosenfield, D., Spun-fiber vegetable protein products, Chemtech, June, 352-355 (1974).
Various, Sulphur dioxide in food processing (topic issue), Chem & Ind, 21 Sept, 716-722 (1974).

1975-1976

Fidgett, M., and Smith, E.L., Model of sugar utilisation during batch beer fermentation, J Appl Chem Biotechnol, 25, 355-366 (1975).
Franzen, K.L., and Kinsella, J.E., Physicochemical aspects of food flavouring, Chem & Ind, 21 June, 505-509 (1975).
Goldenberg, N., and Matheson, H.R., 'Off-flavours' in foods (1948-1974), Chem & Ind, 5 July, 551-557 (1975).
Hedrick, T.I., Aeromicrobiological contamination control in dairy plants, Chem & Ind, 18 Oct, 868-872 (1975).
Redfern, R., A new generation food processor, Chem Engnr, Sept, 521-524 (1975).
Solbett, J.M., and Hepner, L., Recovery of by-products from the treatment of whey, Chem Engnr, July, 447-448, 452 (1975).
Aguirre, F., et al., Protein from waste, Chemtech, Oct, 636-642 (1976).
Anton, J.J., Soy: Protein of the decade, Chemtech, Feb, 90-93 (1976).
Clarke, R.J., Food engineering and coffee, Chem & Ind, 17 April, 362-365 (1976).
Harrison, D.E.F., Protein from methane, Chemtech, Sept, 570-574 (1976).
Jebson, R.S., Chemical engineering in the dairy industry, Chem Engnr, June, 450-451, 457 (1976).
Lin, S.H., Continuous high temperature, short-time sterilization of liquid foods with steam-injection heating, Chem Eng Sci, 31(1), 77-82 (1976).
Scott, R., Mechanical cheesemaking, Chem Engnr, June, 444-446 (1976).
Swift, J.R., Analysis of liquids in the food industry, Chem Engnr, March, 175-177 (1976).
Various, Novel feeds and foods (topic issue), Chem & Ind, 17 July, 581-598 (1976).
Various, Recent advances on the chemistry and technology of oils and fats (topic issue), Chem & Ind, 18 Sept, 763-779 (1976).
Various, Roles of water in food (topic issue), Chem & Ind, 18 Dec, 1039-1060 (1976).

1977-1978

Bailey, C., and James, S.J., The prediction of process design data for meat refrigeration, Chem Engnr, Nov, 788-791 (1977).
Burton, H., The direct-heating process for the UHT sterilisation of milk, Chem Engnr, Nov, 792-795 (1977).
Clausen, E.C., Sitton, O.C., and Gaddy, J.L., Converting crops into methane, Chem Eng Prog, 73(1), 71-72 (1977).

Elson, C.R., Increased design efficiency through improved product characterisation: Heat transfer with food processing, Chem Engnr, Nov, 783-787 (1977).
Emery, A.N.; Barker, A.J., and Hargrave, A.L., Processing of microbial protein for food use, Chem Engnr, July, 506-509 (1977).
Jowitt, R., Heat transfer in some food-processing applications of fluidisation, Chem Engnr, Nov, 779-782 (1977).
Patterson, J.T., and Stewart, D.B., Microbiological quality of exported food, Chem & Ind, 7 May, 349-353 (1977).
Reymond, D., Flavor chemistry: Coffee, cocoa, and tea, Chemtech, Nov, 664-670 (1977).
Various, Advances in production and application of processed food commodities (topic issue), Chem & Ind, 4 June, 425-453; 18 June, 489-500 (1977).
Various, Scale-up in the food industry (topic issue), Chem & Ind, 5 Feb, 98-110 (1977).
Davey, K.R.; Lin, S.H., and Wood, D.G., Effect of pH on continuous high-temperature/short-time sterilization of liquid foods, AIChEJ, 24(3), 537-540 (1978).
Kolodny, S., Economics of sweeteners, Chemtech, May, 292-296 (1978).
Litchfield, J.H., Microbial cells for food, Chemtech, April, 218-223 (1978).
Various, Oils and fats (topic issue), Chem & Ind, 16 Sept, 698-722 (1978).
Wingard, R.E.; Crosby, G.A., and DuBois, G.E., Non-absorbable sweeteners, Chemtech, Oct, 616-621 (1978).

1979-1980

Booth, E., History of the seaweed industry, Chem & Ind, 2 July, 528-534 (1977); 4 Nov, 838-840 (1978); 20 Jan, 52-55 (1979); 2 June, 378-383 (1979).
Hesseltine, C.W., and Wang, H.L., Fermented foods, Chem & Ind, 16 June, 393-399 (1979).
Holzberg, I., Engineered foods, Chemtech, Feb, 110-113 (1979).
Various, Fermentation and brewing (topic issue), Chem & Ind, 15 Dec, 887-898 (1979).
Various, Crop protection (topic issue), Chem & Ind, 17 Nov, 769-795 (1979).
Various, Food packaging health safety (topic issue), Chem & Ind, 19 May, 328-345 (1979).
Various, Carcinogenic risks from food (topic issue), Chem & Ind, 3 Feb, 73-86 (1979).
Baxter, A.G., et al., Reverse osmosis concentration of flavor components in apple-juice and grape-juice waters, Chem Eng Commns, 4(4), 471-484 (1980).
Dalgleish, J. McN., and Lamb, J., Food engineering education for the food and beverage process industries, Chem Engnr, March, 152-154 (1980).
Kidger, P.A., Flash pasteurisation of beer, Chem Engnr, March, 149-151 (1980).
Mussinan, C., Analytical chemistry and flavor creation, Chemtech, Oct, 618-622 (1980).
Sanders, H.J., Tooth decay, C&E News, 25 Feb, 30-42 (1980).

1981-1982

Beausejour, D.; Leduy, A., and Ramalho, R.S., Batch cultivation in cheese whey, Can JCE, 59, 522-526 (1981).

Chopra, C.L., et al., Production of citric acid by submerged fermentation: Effect of medium sterilisation at pilot-plant level, J Chem Tech Biotechnol, 31, 122-126 (1981).

Flinck, K.E., Optimisation of food processes, Chem & Ind, 16 May, 351-354 (1981).

Grabenbauer, G.C., and Glatz, C.E., Protein precipitation: Analysis of particle size distribution and kinetics, Chem Eng Commns, 12(1), 203-220 (1981).

Holman, R.T., Essential fatty acids in nutrition and disease, Chem & Ind, 17 Oct, 704-709 (1981).

Melrose, D.R., Use of anabolic agents in meat production, Chem & Ind, 7 Nov, 766-770 (1981).

North, W.J., Kelp, Chemtech, May, 294-298 (1981).

O'Brien, L., and Gelardi, R.C., Alternative sweeteners, Chemtech, May, 274-278 (1981).

Sanders, H.J., Diabetes, C&E News, 2 March, 30-45 (1981).

Sharon, N., and Lis, H., Glycoproteins, C&E News, 30 March, 21-44 (1981).

Baird, D.G., and Labropoulos, A.E., Food dough rheology (review paper), Chem Eng Commns, 15(1), 1-26 (1982).

Brin, M., Nutrition and vitamin C, Chemtech, July, 428-433 (1982).

Dutton, H.J., Hydrogenated fats: Processing, analysis and biological implications, Chem & Ind, 2 Jan, 9-17 (1982).

Edelman, J., Food, chemistry and cuisine, Chem & Ind, 17 July, 481-483 (1982).

Golberg, L., New directions in food toxicology, Chem & Ind, 5 June, 354-357 (1982).

Jarvis, B., and Holmes, A.W., Biotechnology in the food industry, J Chem Tech Biotechnol, 32, 224-232 (1982).

Jarvis, B., and Paulus, K., Food preservation, J Chem Tech Biotechnol, 32, 233-250 (1982).

Renshaw, T.A.; Sapakie, S.F., and Hanson, M.C., Concentration processes economics in the food industry, Chem Eng Prog, 78(5), 33-40 (1982).

Various, Meat processing survey, Processing, March, 49-54 (1982).

Various, Food technology (topic issue), Chem & Ind, 20 March, 177-196 (1982).

Various, Agriculture and food (topic issue), Chem & Ind, 4 Sept, 647-659 (1982).

Various, Commercial aspects of oils and fats (topic issue), Chem & Ind, 3 July, 428-459 (1982).

Various, Food safety and toxins (topic issue), Chem & Ind, 18 Dec, 972-987 (1982).

Wheatley, A.D.; Mitra, R.I., and Hawkes, H.A., Protein recovery from dairy industry wastes with aerobic biofiltration, J Chem Tech Biotechnol, 32, 203-212 (1982).

1983-1984

Bakal, A.I., Functionality of combined sweeteners in several food applications, Chem & Ind, 19 Sept, 700-708 (1983).

Bataille, M.P., and Bataille, P.F., Extraction of proteins from shrimp processing waste, J Chem Tech Biotechnol, 33B(4), 203-208 (1983).

Bishop, M., Food in, energy out, Chemtech, Aug, 494-496 (1983).

Danehy, J.P., Taste and flavour and the Maillard reaction, Chemtech, July, 412-418 (1983).

Porter, A.B., Effectiveness of multiple sweeteners and other ingredients in food formulation, Chem & Ind, 19 Sept, 696-699 (1983).

Various, Agriculture and the food industry (topic issue), Chem & Ind, 1 Aug, 581-614 (1983).

Various, Trace elements in food and drink products (topic issue), Chem & Ind, 4 July, 496-508 (1983).

Various, Quantitative structure activity relationships (QSAR) in taste and olfaction (topic issue), Chem & Ind, 3 Jan, 10-42 (1983).

Weatherley, L.R., Protein via fish farming, Chemtech, Oct, 614-620 (1983).

Williams, A.A., Defining sensory quality in food and beverages, Chem & Ind, 3 Oct, 740-745 (1983).

Jones, D.P., Food, taste and culture, Chem & Ind, 21 May, 361-366 (1984).

Mellor, J.W., and Adams, R.H., Food for third world countries, C&E News, 23 April, 32-39 (1984).

Padley, F.B., New developments in oils and fats, Chem & Ind, 19 Nov, 788-792 (1984).

Various, Symposium papers on new techniques for processing and purification of proteins, J Chem Tech Biotechnol, 34B(3), 176-228 (1984).

Various, Food intolerance (topic issue), Chem & Ind, 6 Feb, 87-107 (1984).

Wood, D.A., Microbial processes in mushroom cultivation: A large-scale solid substrate fermentation, J Chem Tech Biotechnol, 34B(4), 232-240 (1984).

1985-1986

Appel, C.E., Taste and flavour, Chemtech, July, 420-423 (1985).

Bown, G., Microcomputer-controlled batch sterilisation in the food industry, Chem & Ind, 3 June, 359-365 (1985).

O'Leary, J.W., Saltwater crops, Chemtech, Sept, 562-566 (1985).

Redman, J., Grappling with the grape, Chem Engnr, Dec, 18-19 (1985).

Senouci, A.; Smith, A., and Richmond, P., Extrusion cooking: Food for thought, Chem Engnr, Sept, 30-33 (1985).

Various, Food processing: A survey, Processing, July, 15-19 (1985).

Various, Use of pesticides to control post-harvest losses in food crops (topic issue), Chem & Ind, 4 Feb, 70-90 (1985).

Ammar, K.A.; El-Kady, S.A.; El-Nemer, K., and Lasztity, R., Production of orange juice powder as affected by method of drying, packaging and storage, Periodica Polytechnica, 30(1/2), 3-10 (1986).

Beevers, A., Food irradiation, Proc Engng, June, 55 (1986).

Kimber, A., Process control in the food industry, Processing, Jan, 32 (1986).

Lamb, J., Getting the lump out of food handling, Chem Engnr, April, 18-20 (1986).

Lasztity, R.; Ammar, K.A., and El-Kady, S.A., Biochemical changes of green peas during processing and storage, Periodica Polytechnica, 30(1/2), 11-20 (1986).
O'Toole, C.; Richmond, P., and Reynolds, J., Extracting foodstuffs using supercritical carbon dioxide, Chem Engnr, June, 73-79 (1986).
Various, Crop protection in Japan (topic issue), Chem & Ind, 20 Jan, 46-67; 3 March, 164-167 (1986).
Wren, J.J., Future of food processing, Chem & Ind, 7 April, 227-230 (1986).
Zurer, P.S., Food irradiation, C&E News, 5 May, 46-56 (1986).

1987-1988

Allen, G., The impact of biotechnology on the food industry, Chem Engnr, April, 28-29 (1987).
Baum, R.M., Agricultural biotechnology advances toward commercialization, C&E News, 10 Aug, 9-14 (1987).
Baum, R.M., Agricultural biotechnology, C&E News, 10 Aug, 9-14 (1987).
Boey, S.C.; Garcia del Cerro, M.C., and Pyle, D.L., Extraction of citric acid by liquid membrane extraction, CER&D, 65(3), 218-223 (1987).
Bramwell, A., The role of flavourings in the food supply, Chem & Ind, 1 June, 380-385 (1987).
David, P., Savoury flavourings, Chem & Ind, 1 June, 386-389 (1987).
Le, M.S., Recovery of beer from tank bottoms with membranes, J Chem Tech Biotechnol, 37(1), 59-66 (1987).
Ough, C.S., Chemicals used in making wine, C&E News, 5 Jan, 19-28 (1987).
Richardson, P.; Gaze, J., and Holdsworth, D., Aseptic processing of particulate foods, Proc Engng, April, 65-71 (1987).
Skudder, P., and Biss, C., Aseptic processing of food products using ohmic heating, Chem Engnr, Feb, 26-28 (1987).
Various, Food processing update, Processing, Oct, 37-48 (1987).
Ackman, R.G., Technology of fish oils, Chem & Ind, 7 March, 139-145 (1988).
Burrows, C., Food from crude oil vapor, Chemtech, July, 422-423 (1988).
Chowdhury, J., Freeze concentration for food processing, Chem Eng, 25 April, 24-31 (1988).
Redman, J., Low-alcohol brewing, Chem Engnr, Dec, 92-93 (1988).
Sillett, C., Flowsheeting for food product safety, Chem Engnr, July, 21-25 (1988).
Various, Food processing (topic issue), Chem Eng Prog, 84(5), 19-69 (1988).
Zaritzky, N.E., and Bevilacqua, A.E., Oxygen diffusion in meat tissues, Int J Heat Mass Trans, 31(5), 923-930 (1988).

CHAPTER 13

PRESSURE VESSELS

13.1 Pressure Vessel Design

1967-1975

Wardle, J.K.S., and Todd, G., Bulk storage of liquified gases, Chem Engnr, Nov, CE247-253, 260 (1967).

Anon., New materials data for high-pressure design, Chem Eng, 3 June, 122-124 (1968).

Strelzoff, S., and Pan, L.C., Designing pressure vessels, Chem Eng, 4 Nov, 191-198 (1968).

Various, High-pressure technology (feature report), Chem Eng, 23 Sept, 194-214; 21 Oct, 143-150 (1968).

Adams, L., Comparing pressure-vessel steels, Chem Eng, 15 Dec, 150-151 (1969).

Arthur, I.P., Prestressed concrete pressure vessels, Chem Eng Prog, 65(5), 84-88 (1969).

Fowler, D.W., New analysis method for pressure vessel column supports, Hyd Proc, 48(5), 157-162 (1969).

Tate, R.W., Estimating liquid discharge from pressurized containers, Chem Eng, 2 Nov, 126-128 (1970).

Various, Pressure vessels (special feature), Brit Chem Eng, 15(7), 871-890; 15(8), 1057; 15(9), 1169-1170 (1970).

Cheers, R.F., and Furman, T.T., Algorithms for the optimal design of pressure vessels, Brit Chem Eng, 16(11), 1027-1030; 16(12), 1125-1128 (1971).

Markovitz, R.E., Choosing the most economical vessel head, Chem Eng, 12 July, 102-106 (1971).

Sood, P., Nomographs for nozzle reinforcement design (ASME pressure vessel code), Hyd Proc, 50(8), 139-141 (1971).

Strauss, W., Experimentation at very high pressures, Brit Chem Eng, 16(2), 169-171 (1971).

Strauss, W., and Pollard, L.J., Pressure vessels for very high pressures, Brit Chem Eng, 16(1), 49-51 (1971).

Various, Pressure vessels (special feature), Brit Chem Eng, 16(6), 473-503 (1971).

Clark, F.D., and Terni, S.P., Cost of thick-wall pressure vessels, Chem Eng, 3 April, 112-116 (1972).

McLeish, R.D., Pressure vessel cost variation depends on shape and size, Proc Engng, Aug, 65-68 (1972).

Voelker, C.H., Practical repair welding of pressure vessels, Hyd Proc, 52(2), 95-96 (1973).

Dimoplon, W., Determine the geometry of pressure vessel heads, Hyd Proc, 53(8), 71-74 (1974).

Ganapathy, V., Estimating weights of pressure vessels, Chem Eng, 23 Dec, 86 (1974).

Loeb, M.B., Optimum vessel design for gas storage, Chem Eng, 24 June, 170-174 (1974).

Pilorz, B.H., Quick estimates of vessel-head volumes and areas, Chem Eng, 10 June, 110 (1974).

Kaferle, J.A., Calculating pressures for dimple jackets, Chem Eng, 24 Nov, 86 (1975).

1976-1987

Logan, P., Pressure vessel thickness from new ASME code addenda, Hyd Proc, 55(5), 217-218 (1976).

Logan, P.J., Simplified approach to pressure vessel head design, Hyd Proc, 55(11), 265-266 (1976).

Singh, K.P., Design of skirt-mounted supports for pressure vessels, Hyd Proc, 55(4), 199-203 (1976).

Wood, R., Material problems in pressure vessel fabrication, Proc Engng, Dec, 39-40 (1976).

Mahajan, K.K., Size vessel stiffeners quickly, Hyd Proc, 56(4), 207-208 (1977).

Various, High pressure operations (topic issue), Chem Eng Prog, 73(12), 33-55 (1977).

Kent, G.R., Selecting gaskets for flanged joints, Chem Eng, 27 March, 125-128 (1978).

Baum, M.R., Blast waves generated by the rupture of gas pressurised ductile pipes, Trans IChemE, 57, 15-24 (1979).

Heinze, A.J., Pressure vessel design for process engineers, Hyd Proc, 58(5), 181-191 (1979).

Stippick, J., Design weld-neck flanges fast, Hyd Proc, 58(5), 201-204 (1979).

Whited, J.D., and Wells, D.D., Moving pressure vessels from shop to site, Hyd Proc, 58(5), 195-198 (1979).

Ganapathy, V., and Elango, R., Designing vessels and tubes for external pressure, Chem Eng, 19 May, 143-146 (1980).

Hagel, W.C., and Miska, K.H., How to select alloy steels for pressure vessels, Chem Eng, 28 July, 89-91; 25 Aug, 105-108 (1980).

Jawadekar, S.P., Corrosion in pressure vessel calculations, Chem Eng, 15 Dec, 96-98 (1980).

Spear, M., Quality assurance of pressure vessels, Proc Engng, Aug, 59,61 (1980).

Fitt, J.S., Pressure piling, Chem Engnr, May, 237-239 (1981).

Litchfield, A.B., LNG terminal for storage and regasification, Chem Eng Prog, 77(1), 83-88 (1981).

Mulet, A.; Corripio, A.B., and Evans, L.B., Estimate costs of pressure vessels via correlations, Chem Eng, 5 Oct, 145-150 (1981).

Tankha, A., Selecting formed heads for cylindrical vessels, Chem Eng, 1 June, 89 (1981).

Knight, G.L.B., and Smith, R.J., Optimize material requirements for prismatic vessels, Hyd Proc, Feb, 95-96 (1982).

Sivasankaran, S., and Gupta, J.P., Computer program for flange design, Hyd Proc, June, 122-124 (1982).

Ricord, N.J., Maintain maximum liquid level in pressurized vessels, Chem Eng, 14 Nov, 254 (1983).

Sivasankaran, S., and Gupta, J.P., Computer program for pressure vessel head design, Hyd Proc, Aug, 67-69 (1983).

Sivasankaran, S., and Gupta, J.P., Computer program for nozzle opening reinforcement design, Hyd Proc, May, 99-103 (1983).

Conte, P., and Felici, M., Optimum pipe thickness for vacuum, Hyd Proc, Aug, 96-98 (1984).

Baker-Counsell, J., Pressure-vessel quality assurance, Proc Engng, Feb, 43,45 (1985).
Taylor, C., The ins and outs of bellows design, Chem Engnr, Sept, 40-45 (1985).
Bednar, H.H., Pressure Vessel Design Handbook, 2nd End., Van Nostrand Reinhold, New York (1986).
Chao, Y.J., and Sutton, M.A., Program determines vessel weights, Hyd Proc, 65(9), 71-73 (1986).
Megyesy, E.F., Pressure Vessel Handbook, 7th Ed., PVH Publishers, Oklahoma (1986).
Grosshandler, S., Safe and cost-effective design of pressure vessels, Proc Engng, Oct, 51-53 (1987).
Kirkpatrick, H.L., Program designs and prices vessels, Hyd Proc, 66(3), 41-44 (1987).
Mascone, C.F., New gauging system (for process vessels), Chem Eng, 14 Sept, 25-29 (1987).
Russell, D.L., and Hart, S.W., Underground storage tanks: Potential for economic disaster, Chem Eng, 16 March, 61-69 (1987).

13.2 Pressure Vessel Codes

1968-1986

Witkin, D.E., The new ASME pressure vessel code, Chem Eng, 26 Aug, 124-130 (1968).
Anon., Explaining Division 2 of the ASME vessel code, Hyd Proc, 48(1), 147-148 (1969).
Anon., New code for fiber-glass-reinforced plastic pressure vessels, Chem Eng, 14 July, 128-130 (1969).
Hopkins, C.H.R., The Pressure Vessel Authority, Brit Chem Eng, 14(4), 508-509 (1969).
Dall-Ora, F., European pressure vessel codes, Hyd Proc, 50(6), 93-96 (1971).
McGrath, R.V., and Palmer, J.R., USA pressure vessel and tank codes and standards, Hyd Proc, 50(6), 96-101 (1971).
Stevens, B., International pressure vessel standards, Proc Engng, Jan, 67-71 (1973).
Various, Guide to world pressure vessel codes (special report), Hyd Proc, 54(12), 53-68 (1975).
Caplan, F., ASME requirements for vessels, Chem Eng, 10 May, 139-141 (1976).
Morris, M., BS5500 for tubesheet design, Proc Engng, April, 78-80 (1976).
Morris, M., Pressure vessel design using BS5500, Proc Engng, Feb, 43 (1977).
Various, Guide to worldwide pressure vessel codes, Hyd Proc, 57(12), 89-101 (1978).
Capes, P., The need for statutory standards for pressure vessels, Proc Engng, Nov, 65-67 (1982).
Smolen, A.M., and Mase, J.R., ASME pressure-vessel code: Which division to choose? Chem Eng, 11 Jan, 133-136 (1982).
Boyer, C.B., High pressure codes and standards, Plant/Opns Prog, 3(1), 3-7 (1984).

Yokell, S., Understanding the pressure vessel codes, Chem Eng, 12 May, 75-85 (1986).

13.3 Pressure Vessel Inspection and Testing

1970-1988

McFarland, I., Safety in pressure vessel design, Chem Eng Prog, 66(6), 56-58 (1970).
Kirby, N., and Bentley, P.G., Pressure vessel inspection by acoustic emission, Brit Chem Eng, 16(10), 912-913 (1971).
Ferge, D.T., Dynamic pressure testing, Chem Eng Prog, 68(9), 85-87 (1972).
Witt, P.A., Pressure-vessel monitoring by acoustic emissions, Chem Eng Prog, 68(1), 56-57 (1972).
Parry, D.L., Qualify pressure vessel integrity with acoustic emission analysis, Hyd Proc, 55(12), 132-134 (1976).
Snow, M., Maintaining pressure vessels, Chem Eng, 1 Jan, 109-112 (1979).
Polentz, L.M., Calculate the energy in a pressurized vessel, Chem Eng, 15 Dec, 95-96 (1980).
Mischiatti, M., and Ripamonti, B., Burst test yields vessel data, Hyd Proc, Nov, 83-86 (1985).
Fowler, T.J., Acoustic emission testing of vessels and piping, Chem Eng Prog, 83(5), 25-32 (1987).
Grosshandler, S., Predictive maintenance of pressure vessels, Proc Engng, Aug, 27-31 (1987).
Gwynn, J.E., Positive pressure-testing system, Plant/Opns Prog, 7(3), 169-172 (1988).

13.4 Pressure Relief Systems

1967-1979

Grote, S.H., Calculating pressure-release times, Chem Eng, 17 July, 203-206 (1967).
Saunders, M.J., Evacuation time of gaseous systems calculated quickly, Chem Eng, 20 Nov, 166-168 (1967).
Rearick, J.S., Designing pressure-relief systems, Hyd Proc, 48(9), 161-166 (1969).
Isaacs, M., Pressure-relief systems, Chem Eng, 22 Feb, 113-124 (1971).
Anderson, F.E., Pressure-relieving devices, Chem Eng, 24 May, 128-134 (1976).
Cheng, W.B., and Mah, R.S.H., Optimal design of pressure relieving piping networks by discrete merging, AIChEJ, 22(3), 471-476 (1976).
Kern, R., Pressure-relief valves for process plants, Chem Eng, 28 Feb, 187-194 (1977).
Willis, R.P., Sizing of vessel nozzles for safety-valve service, Chem Eng, 6 June, 200 (1977).
Frankland, P.B., Correct location of relief valves, Hyd Proc, 57(4), 189-191 (1978).

Duxbury, H.A., Relief line sizing for gases, Chem Engnr, Nov, 783-787; Dec, 851-852, 857 (1979).

1980-1988

Crozier, R.A., Pressure relief to prevent heat-exchanger failure, Chem Eng, 15 Dec, 79-83 (1980).

Duxbury, H.A., The sizing of relief systems for polymerisation reactions, Chem Engnr, Jan, 31-37 (1980).

Mukerji, A., How to size relief valves, Chem Eng, 2 June, 79-86 (1980).

Aird, R.J., Reliability assessment of safety/relief valves, Trans IChemE, 60, 314-318 (1982).

Copigneaux, P., Calculating theoretical flowrates of relief valves, Hyd Proc, Nov, 209-213 (1982).

Friedal, L., and Lohr, G., Design of pressure-relief devices for gas-liquid reaction systems, Int Chem Eng, 22(4), 619-631 (1982).

Uchiyama, T., Effective pressure relief of offsite piping, Hyd Proc, May, 213-214 (1982).

Van Boskirk, B.A., Sensitivity of relief valves to inlet and outlet line lengths, Chem Eng, 23 Aug, 77-82 (1982).

Fauske, H.K.; Grolmes, M.A., and Henry, R.E., Emergency relief systems: Sizing and scale-up, Plant/Opns Prog, 2(1), 27-30 (1983).

Papa, D.M., How back pressure affects safety relief valves, Hyd Proc, May, 79-81 (1983).

Andrew, B.E., Relieving overpressure in a bulging drum, Chem Eng, 14 May, 90 (1984).

Moore, A., Pressure-relieving systems, Chem Engnr, Oct, 13-16 (1984).

Sonti, R.S., Practical design and operation of vapor-depressuring systems, Chem Eng, 23 Jan, 66-69 (1984).

Constantinescu, S., Sizing gas pressure-relief nozzles, Chem Eng, 29 April, 85-86 (1985).

Emerson, G.B., Selecting pressure relief valves, Chem Eng, 18 March, 195-200 (1985).

Jones, B.G., and Duckett, R.C., Thermographic survey of integrity of process plant pressure-relief system, Plant/Opns Prog, 4(3), 161-164 (1985).

Sallett, D.W., and Somers, G.W., Flow capacity and response of safety relief valves to saturated water flow, Plant/Opns Prog, 4(4), 207-217 (1985).

Chen, Z.; Govind, R., and Weisman, J., Vapor-liquid flow through spring-loaded relief valves, Chem Eng Commns, 49, 23-34 (1986).

Simpson, L.L., and Woinsky, S.G., Computer nomographs for pressure-relief devices, Plant/Opns Prog, 5(1), 49-51 (1986).

Stikvoort, W.J., Piping reactions on pressure-vessel nozzles, Chem Eng, 7 July, 51-53 (1986).

Fullarton, D.; Evripidis, J., and Schlunder, E.U., Influence of product vapour condensation on venting of storage tanks, Chem Eng & Proc, 22(3), 137-144 (1987).

Van Boskirk, B.A., Sizing and selecting conservation vents, Chem Eng, 23 Nov, 173-180 (1987).

Woolfolk, W.H., and Sanders, R.E., Dynamic testing and maintenance of safety relief valves, Chem Eng, 26 Oct, 119-124 (1987).

Emerson, G.B., Pressure relief valve types and selection, Hyd Proc, 67(5), 71-72 (1988).
Huff, J.E., Pressure-relief system design, Chem Eng Prog, 84(9), 44-51 (1988).
Mayinger, F., Design and layout of safety and pressure relief valves: Two-phase flow phenomena with depressurization, Chem Eng & Proc, 23(1), 1-12 (1988).

13.5 Rupture Discs

1970-1988

Alba, C., Size rupture discs by nomograph, Hyd Proc, 49(9), 297-298 (1970).
Brodie, G.W., Dome bursting discs, Proc Engng, Dec, 92-93 (1973).
Li, K.W., A simple leak-detecting device for rupture discs, Chem Eng, 23 Dec, 84-86 (1974).
Various, Simple leak-detecting device for rupture disks, Chem Eng, 17 March, 88 (1975).
Ganapathy, V., Rupture disks for gases and liquids, Chem Eng, 25 Oct, 152-154 (1976).
Zook, R.J., Rupture disks for low-burst pressures, Chem Eng, 1 March, 131-136 (1976).
Hoffman, B., Guide to bursting discs, Proc Engng, Nov, 95 (1980).
Whitfield, M., Selecting bursting discs, Proc Engng, Aug, 51,53 (1980).
Anon., Bursting discs, Processing, Feb, 31,33 (1983).
Harris, L.R., Select the right rupture disk, Hyd Proc, May, 75-78 (1983).
Mathews, T., Bursting discs for over-pressure protection, Chem Engnr, Aug, 21-23 (1983).
Phadke, P.S., Calculator program for rupture disc design, Hyd Proc, Dec, 34B-34L (1983).
Beveridge, H., Improving efficiency of rupture-disk systems, Proc Engng, Jan, 35,37 (1985).
Walker, W., The new British Standard BS 2915: 1984 for bursting discs, Chem Engnr, Feb, 18-21 (1985).
Nazario, F.N., Rupture discs: A primer, Chem Eng, 20 June, 86-96 (1988).

CHAPTER 14

MIXING

14.1 Theory

1966-1970

Campbell, H., and Bauer, W.C., Cause and cure of demixing in solid-solid mixers, Chem Eng, 12 Sept, 179-185 (1966).

Bourne, J.R., Statistical analysis of batchwise blending processes, Trans IChemE, 45, T280-284 (1967).

Gren, U., Solids mixing: Review of present theory, Brit Chem Eng, 12(11), 1733-1737 (1967).

Hagedorn, D., and Salamone, J.J., Batch heat-transfer coefficients for pseudoplastic fluids in agitated vessels, Ind Eng Chem Proc Des Dev 6(4), 469-475 (1967).

Hoogendoorn, C.J., and den Hartog, A.P., Model studies on mixers in viscous flow region, Chem Eng Sci, 22(12), 1689-1700 (1967).

Johnson, R.T., Batch mixing of viscous liquids, Ind Eng Chem Proc Des Dev, 6(3), 340-345 (1967).

Uhl, V.W., and Root, W.L., Heat transfer to granular solids in agitated units, Chem Eng Prog, 63(7), 81-92 (1967).

Ciborowski, J., and Wolny, A., Influence of moisture content of loose materials on dynamics of their mixing, Int Chem Eng, 8(2), 199-204 (1968).

Lehrer, L.H., Gas agitation of liquids, Ind Eng Chem Proc Des Dev, 7(2), 226-239 (1968).

Nepomnyashchii, E.A., Kinetics of separation and mixing of dispersed materials, Int Chem Eng, 8(1), 86-91 (1968).

Narayanan, S.; Bhatia, V.K.; Guha, D.K., and Rao, M.N., Suspension of solids by mechanical agitation, Chem Eng Sci, 24(2), 223-230 (1969).

Pollard, J., and Kantyka, T.A., Heat transfer to agitated non-Newtonian fluids, Trans IChemE, 47, T21-27 (1969).

Todd, D.B., and Irving, H.F., Axial mixing in a self-wiping reactor, Chem Eng Prog, 65(9), 84-89 (1969).

Blasinski, H.; Kochanski, B., and Rzyski, E., The Froude number in the mixing process, Int Chem Eng, 10(2), 176-180 (1970).

Coyle, C.K.; Hirschland, H.E.; Michel, B., and Oldshue, J.Y., Mixing in viscous liquids, AIChEJ, 16(6), 903-906 (1970).

Hobler, T., and Palugniok, H., Minimum Reynolds number required in a mixer to obtain a suspension of immiscible liquids, Int Chem Eng, 10(1), 15-21 (1970).

Inoue, I., and Yamaguchi, K., Particle motion in a two-dimensional V-type mixer, Int Chem Eng, 10(3), 490-497 (1970).

Lodh, B.B.; Murthy, G.S.R.N., and Murti, P.S., Turbulence promotion for improved heat transfer to gas-solids mixtures, Brit Chem Eng, 15(1), 73-75 (1970).

Schofield, C., Assessing mixtures by autocorrelation, Trans IChemE, 48, T28-34 (1970).

Smith, J.M., Secondary flow phenomena in mixing viscous and visco-elastic liquids, Chem Engnr, March, CE45-49 (1970).

1971-1972

Bridgwater, J., and Ingram, N.D., Rate of spontaneous inter-particle percolation, Trans IChemE, 49, 163-169 (1971).

Hubbard, D.W., and Patel, H., Hydrodynamic measurements for imperfect mixing processes: Newtonian fluids, AIChEJ, 17(6), 1387-1393 (1971).

Mancott, A., Fast blending calculations, Chem Eng, 19 April, 136 (1971).

Rogers, A.R., and Clements, J.A., Examination of segregation of granular materials in a tumbling mixer, Powder Tech, 5, 167-178 (1971).

Abouzeid, A.Z.M., and Fuerstenau, D.W., Effect of humidity on mixing of particulate solids, Ind Eng Chem Proc Des Dev, 11(2), 296-301 (1972).

Blasinski, H., and Rzyski, E., Mixing of non-Newtonian liquids, Int Chem Eng, 12(1), 24-36 (1972).

Chen, S.J.; Fan, L.T., and Watson, C.A., Stochastic approach to solid particles mixing in a motionless mixer, AIChEJ, 18(5), 984-989 (1972).

Ford, D.E.; Mashelkar, R.A., and Ulbrecht, J., Mixing times in Newtonian and non-Newtonian fluids, Proc Tech Int, 17(10), 803-807 (1972).

Kelkar, J.V.; Mashelkar, R.A., and Ulbrecht, J., On the rotational viscoelastic flows around simple bodies and agitators, Trans IChemE, 50, 343-352 (1972).

Nagata, S.; Nishikawa, M.; Katsube, T., and Takaishi, K., Mixing of highly viscous non-Newtonian liquids, Int Chem Eng, 12(1), 175-182 (1972).

Smith, J.M., Alternative flow regimes in agitated vessels, Chem Engnr, May, 182-185 (1972).

Various, Intensive mixing, Brit Chem Eng, 17(4), 330-335 (1972).

1973-1975

Orr, N.A., and Shotton, E., The mixing of cohesive powders, Chem Engnr, Jan, 12-19 (1973).

Powley, C., Homogenisation, Proc Engng, April, 101-103 (1973).

Rao, D.P., and Edwards, L.L., Mixing effects in stirred tank reactors: Comparison of models, Chem Eng Sci, 28(5), 1179-1192 (1973).

Shearer, C.J., Mixing of highly viscous liquids, Chem Eng Sci, 28(4), 1091-1098 (1973).

Stevens, B., Static mixing, Proc Engng, April, 76-78 (1973).

Various, Handling viscous materials (feature report), Chem Eng, 19 March, 98-111 (1973).

Williams, J.C., and Khan, M.I., The mixing and segregation of particulate solids of different particle size, Chem Engnr, Jan, 19-25 (1973).

Bridgwater, J., and Scott, A.M., Statistical models of packing: Application to gas absorption and solids mixing, Trans IChemE, 52, 317-324 (1974).

Landau, J.I., and Petersen, E.E., Theoretical development and experimental verifiction of a novel well-mixed vessel, AIChEJ, 20(1), 166-171 (1974).

Rees, L.H., Evaluating homogenizers for emulsions and dispersions, Chem Eng, 13 May, 86-92 (1974).

Simpson, L.L., Turbulence and industrial mixing, Chem Eng Prog, 70(10), 77-79 (1974).

Chavan, V.V.; Ford, D.E., and Arumugan, M., Influence of fluid rheology on circulation, mixing and blending, Can JCE, 53, 628-640 (1975).

Fan, L.T.; Gelves-Arocha, H.H.; Walawender, W.P., and Lai, F.S., Mechanistic kinetic model of rate of mixing segregating solid particles, Powder Tech, 12, 139-156 (1975).

Gunkel, A.A., and Weber, M.E., Flow phenomena in stirred tanks, AIChEJ, 21(5), 931-949 (1975).

Hersey, J.A., Ordered mixing: A new concept in powder mixing practice, Powder Tech, 11, 41-44 (1975).

Mashelkar, R.A.; Kale, D.D., and Ulbrecht, J., Rotational flows of non-Newtonian fluids, Trans IChemE, 53, 143-153 (1975).

1976-1977

Akao, Y.; Kunisawa, H.; Fan, L.T.; Lai, F.S., and Wang, R.H., Degree of mixedness and contact number: Study of mixture of particulate solids and structure of solid mixtures, Powder Tech, 15, 267-277 (1976).

Bridgwater, J., Fundamental powder mixing mechanisms, Powder Tech, 15, 215-236 (1976).

Cooke, M.H.; Stephens, D.J., and Bridgwater, J., Powder mixing: A literature survey, Powder Tech, 15, 1-20 (1976).

Ford, D.E., and Ulbrecht, J., Influence of rheological properties of polymer solutions on mixing and circulation times, Ind Eng Chem Proc Des Dev, 15(2), 321-326 (1976).

Khang, S.J., and Levenspiel, O., Mixing-rate number for agitator-stirred tanks, Chem Eng, 11 Oct, 141-143 (1976).

Kristensen, H.G., Characterization of non-random mixtures: A survey, Powder Tech, 13, 103-113 (1976).

Various, Liquid agitation, Chem Eng, 8 Dec, 110-114 (1975); 5 Jan, 139-145; 2 Feb, 93-100; 26 April, 102-110; 24 May, 144-150 (1976).

Various, Liquid agitation, Chem Eng, 19 July, 141-148; 2 Aug, 89-94; 30 Aug, 101-108; 27 Sept, 109-112; 25 Oct, 119-126; 8 Nov, 127-133; 6 Dec, 165-170 (1976).

King, G.T., A gravimetric continuous mixing system, Processing, March, 31 (1977).

Martin, D., Solutions to mixing problems, Proc Engng, May, 53-55 (1977).

Quraishi, A.Q.; Mashelkar, R.A., and Ulbrecht, J.J., Influence of drag reducing additives on mixing and dispersing in agitated vessels, AIChEJ, 23(4), 487-492 (1977).

Wang, R.H., and Fan, L.T., Stochastic modeling of segregation in a motionless mixer, Chem Eng Sci, 32(7), 695-702 (1977).

1978-1979

Baldi, G.; Conti, R., and Alaria, E., Complete suspension of particles in mechanically agitated vessels, Chem Eng Sci, 33(1), 21-26 (1978).

Laidler, P., and Ulbrecht, J.J., Numerical analysis of flow in close-clearance mixers, Chem Eng Sci, 33(12), 1615-1622 (1978).

Ries, H.B., Mixing quality: Problems, test methods, and results, Int Chem Eng, 18(3), 426-443 (1978).

Various, Mixing (topic issue), Proc Engng, May, 61-71 (1978).

Fan. L.T., and Shin, S.H., Stochastic diffusion model of non-ideal mixing in a horizontal drum mixer, Chem Eng Sci, 34(6), 811-820 (1979).

Nauman, E.B., Enhancement of heat transfer and thermal homogeneity with motionless mixers, AIChEJ, 25(2), 246-258 (1979).
Nienow, A.W.; Chapman, C.M., and Middleton, J.C., Gas recirculation rate through impeller cavities and surface aeration in sparged agitated vessels, Chem Eng J, 17(2), 111-118 (1979).
Okamoto, Y.; Nishikawa, M.; Matsuda, K., and Hashimoto, K., Measurement of turbulent diffusion coefficients in mixing vessels, Int Chem Eng, 19(4), 639-646 (1979).
Reed, R.D., and Narayan, B.C., Mixing fluids under turbulent flow conditions, Chem Eng, 4 June, 131-132 (1979).
Rieger, F.; Ditl, P., and Novak, V., Vortex depth in mixed unbaffled vessels, Chem Eng Sci, 34(3), 397-404 (1979).
Sawant, S.B., and Joshi, J.B., Critical impeller speed for onset of gas induction in gas-inducing types of agitated contactors, Chem Eng J, 18(1), 87-92 (1979).
Snee, R.D., Experimenting with mixtures, Chemtech, Nov, 702-710 (1979).
Tasucher, W.A., and Streiff, F.A., Static mixing of gases, Chem Eng Prog, 75(4), 61-65 (1979).
Various, Mixing (topic issue), Proc Engng, June, 85-95 (1979).

1980-1981
El-Shawarby, S.I., and Eissa, S.H., Determination of interfacial area in stirred tanks by chemical absorption and by power measurement, Ind Eng Chem Proc Des Dev, 19(3), 469-477 (1980).
Farritor, R.E., and Hughmark, G.A., Interfacial area and mass transfer with gas-liquid systems in turbine-agitated vessels, Chem Eng Commns, 4(1), 143-148 (1980).
Grosz-Roll, F., Assessing homogeneity in motionless mixers, Int Chem Eng, 20(4), 542-550 (1980).
Heim, A., Modeling momentum and heat transfer in mixers with close-clearance agitators, Int Chem Eng, 20(2), 271-289 (1980).
McManamey, W.J., A circulation model for batch mixing in agitated, baffled vessels, Trans IChemE, 58, 271-276 (1980).
Middleton, J.C.; Edwards, M.F., and Stewart, I., Recommended standard terminology and nomenclature for mixing, Chem Engnr, Aug, 557-562 (1980).
Schofield, C., and Stewart, I.W., Pigment deagglomeration in a model mixer, Chem Engnr, July, 486-489 (1980).
Yuan, H.H.S., and Tatterson, G.B., Estimation of fluid forces in an agitated vessel, Chem Eng Commns, 4(4), 531-538 (1980).
Yuu, S., and Oda, T., Measurement of turbulence parameters in a non-baffled stirred tank with high rotation speeds, Chem Eng J, 20(1), 35-42 (1980).
Blasinski, H., and Kuncewicz, C., Heat transfer during mixing of pseudoplastic fluids with ribbon agitators, Int Chem Eng, 21(4), 679-684 (1981).
Conti, R.; Sicardi, S., and Specchia, V., Effect of the stirrer clearance in agitated vessels on the particle suspension, Chem Eng J, 22(3), 247-250 (1981).
Hiraoka, S., and Fan, L.T., Two-dimensional model analysis of turbulent flow in an agitated vessel with paddle impeller, Chem Eng Commns, 10(1), 149-164 (1981).

Mann, R.; Mavros, P.P., and Middleton, J.C., A structured stochastic flow model for interpreting flow-follower data from a stirred vessel, Trans IChemE, 59, 271-278 (1981).
Oldshue, J.Y., Understanding mixing, Chemtech, Sept, 554-561 (1981).
Ottino, J.M.; Ranz, W.E., and Macosko, C.W., Description of mechanical mixing of fluids, AIChEJ, 27(4), 565-577 (1981).
Sverak, S., and Hruby, M., Gas entrainment from liquid surface of vessels with mechanical agitation, Int Chem Eng, 21(3), 519-527 (1981).
Tatterson, G.B., et al., Liquid-dispersion mechanisms in agitated tanks, Chem Eng Commns, 10(4), 205-222 (1981).
Tojo, K.; Miyanami, K., and Mitsui, H., Vibratory agitation in solids-liquid mixing, Chem Eng Sci, 36(2), 279-284 (1981).
Tucker, C.L., Sample variance measurement of mixing, Chem Eng Sci, 36(11), 1829-1840 (1981).
Wichterle, K., and Wein, O., Threshold of mixing of non-Newtonian liquids, Int Chem Eng, 21(1), 116-121 (1981).

1982-1984

Fajner, D.; Magelli, F., and Pasquali, G., Modelling of non-standard mixers stirred with multiple impellers, Chem Eng Commns, 17(1), 285-296 (1982).
Kang, I.S., and Chang, H.N., Effect of turbulence promoters on mass transfer: Numerical analysis and flow visualization, Int J Heat Mass Trans, 25(8), 1167-1182 (1982).
Kuriyama, M.; Inomata, H.; Arai, K., and Saito, S., Numerical solution for the flow of highly viscous fluid in agitated vessel with anchor impeller, AIChEJ, 28(3), 385-391 (1982).
Lane, A.G.C., and Rice, P., An investigation of liquid jet mixing employing an inclined side entry jet, Trans IChemE, 60, 171-176 (1982).
Tojo, K., and Miyanami, K., Solids suspension in mixing tanks, Ind Eng Chem Fund, 21(3), 214-220 (1982).
Bourne, J.R., Mixing on the molecular scale (micromixing), Chem Eng Sci, 38(1), 5-8 (1983).
Ducla, J.M.; Desplanches, H., and Chevalier, J.L., Effective viscosity of non-Newtonian fluids in a mechanically stirred tank, Chem Eng Commns, 21(1), 29-36 (1983).
Lintz, H.G., and Weber, W., Experimental study of mixing in CSTR using an autocatalytic reaction, Int Chem Eng, 23(4), 618-634 (1983).
Mordarski, J., and De Kee, D., Pneumatic cylinders inject gas into liquid, Chem Eng, 7 March, 187 (1983).
Shintre, S.N., and Ulbrecht, J.J., Model of mixing in a motionless mixer, Chem Eng Commns, 24(1), 115-138 (1983).
Glatt, I., and Kafri, O., Analysis of the turbulent mixing of liquids by moire deflectometry, Chem Eng Sci, 39(11), 1637-1639 (1984).
Harnby, N., Trends in powder mixing, Chem Engnr, July, 22-23 (1984).
Ou, J.J., and Ranz, W.E., Mixing and chemical reaction: Thermal effects, Chem Eng Sci, 39(12), 1735-1740 (1984).
Takase, H.; Unno, H., and Akehata, T., Oxygen transfer in surface-aeration tank with square cross section, Int Chem Eng, 24(1), 128-135 (1984).

1985-1986

Ahmad, S.W.; Latto, B., and Baird, M.H.I., Mixing of stratified liquids, CER&D, 63, 157-167 (1985).

Arimond, J., and Erwin, L., Simulation of a motionless mixer, Chem Eng Commns, 37(1), 105-126 (1985).

King, R., Fluid/structure interactions in mixing processes, Proc Engng, Feb, 50-51 (1985).

Mamleev, R.A., Modeling of turbulent mixing in two-phase dispersed liquid/liquid systems, Int Chem Eng, 25(3), 566-570 (1985).

Ulbrecht, J.J., and Patterson, G.K. (Eds), Mixing of Liquids by Mechanical Means, Gordon and Breach, New York (1985).

Wheeler, D., Problems of powder unmixing, Processing, Nov, 16,27 (1985).

Baker-Counsell, J., Lasers for a better understanding of mixing, Proc Engng, April, 39 (1986).

Brodberger, J.F.; Valentin, G., and Storck, A., Use of conducimetric microprobe to study mixing within agitated reactors, Int Chem Eng, 26(1), 69-78 (1986).

Cloete, F.L.D., and Coetzee, M.C., Calculating minimum powder required for complete suspension of solids in a mixer, Powder Tech, 46(2), 239-244 (1986).

Kuboi, R., and Nienow, A.W., Intervortex mixing rates in high-viscosity liquids agitated by high-speed dual impellers, Chem Eng Sci, 41(1), 123-134 (1986).

Shamlou, P.A., and Edwards, M.F., Heat transfer to viscous Newtonian and non-Newtonian fluids for helical ribbon mixers, Chem Eng Sci, 41(8), 1957-1968 (1986).

Tanaka, M.; Noda, S., and Oshima, E., Effect of location of a submerged impeller on enfoldment of air bubbles from free surface of a stirred vessel, Int Chem Eng, 26(2), 314-327 (1986).

Uhl, V.W., and Gray, J.B., (Eds), Mixing Theory and Practice, Vol. 3, Academic Press, Florida (1986).

1987-1988

Barresi, A., and Baldi, G., Solid dispersion in an agitated vessel, Chem Eng Sci, 42(12), 2949-2956 (1987).

Beevers, A., Liquid mixing developments, Processing, Feb, 34-39 (1987).

Cooker, B., and Nedderman, R.M., Theory of mechanics of helical ribbon powder agitator, Powder Tech, 50(1), 1-14 (1987).

Dackson, K., and Nauman, E.B., Fully developed flow in twisted tapes: Model for motionless mixer, Chem Eng Commns, 54, 381-395 (1987).

Various, Dry mixing update, Processing, Sept, 53,55 (1987).

Hancil, V., and Rod, V., Break-up of a drop in stirred tank, Chem Eng & Proc, 23(3), 189-193 (1988).

Nienow, A.W., and Elson, T.P., Aspects of mixing in rheologically complex fluids, CER&D, 66(1), 5-15 (1988).

Smith, J.M., and Schoenmakers, A.W., Blending of liquids of differing viscosity, CER&D, 66(1), 16-21 (1988).

Tadmor, Z., Number of passage distribution functions with application to dispersive mixing, AIChEJ, 34(12), 1943-1948 (1988).

14.2 Design

1967-1970

Peters, D.C., and Smith, J.M., Fluid flow in the region of anchor agitator blades, Trans IChemE, 45, T360-366 (1967).

Strek, F., and Masiuk, S., Heat transfer in liquid mixers, Int Chem Eng, 7(4), 693-702 (1967).

Blasinski, H., and Tyczkowski, A., Hydrodynamic evaluation of certain geometric parameters of mixers, Int Chem Eng, 8(1), 43-53 (1968).

Hall, K.R., and Godfrey, J.C., Mixing rates of highly viscous Newtonian and non-Newtonian fluids in a laboratory sigma-blade mixer, Trans IChemE, 46, T205-212 (1968).

Harrell, J.E., and Perona, J.J., Mixing of fluids in tanks of large length-to-diameter ratio by recirculation, Ind Eng Chem Proc Des Dev, 7(4), 464-468 (1968).

Bourne, J.R., and Butler, H., An analysis of the flow produced by helical ribbon impellers, Trans IChemE, 47, T11-17 (1969).

Connolly, J.R., and Winter, R.L., Approaches to mixing operation scale-up, Chem Eng Prog, 65(8), 70-78 (1969).

Dykman, M., and Michel, B.J., Comparing mechanical aerator designs, Chem Eng, 10 March, 117-121 (1969).

Leggett, E., Multistage mixing solved graphically, Hyd Proc, 48(5), 141-142 (1969).

Berresford, H.I.; Gibilaro, L.G.; Spikins, D.J., and Kropholler, H.W., Continuous blending of low viscosity fluids (scale-up criteria), Trans IChemE, 48, T21-27 (1970).

Bowers, R.H., Some aspects of agitator design, Chem Engnr, March, CE50-52 (1970).

Penney, W.R., Guide to trouble-free mixers, Chem Eng, 1 June, 171-180 (1970).

1971-1975

Anderson, K.S., Plant requirements for the production of pigment masterbatches and other concentrates in batch and continuous mixers, Chem Engnr, July, 269-278 (1971).

Miller, D.N., Scale-up of agitated vessels, Ind Eng Chem Proc Des Dev, 10(3), 365-375 (1971).

Blakebrough, N., Mixing effects in biological systems, Chem Engnr, Feb, 58-63 (1972).

Casto, L.V., Practical tips on designing turbine-mixer systems, Chem Eng, 10 Jan, 97-102 (1972).

Cheng, D.C.H., and Schofield, C., A scale-up procedure for a new mixer for paste-like materials, Trans IChemE, 50, 6-11 (1972).

Edwards, M.F., and Wilkinson, W.L., Heat transfer in agitated vessels, Chem Engnr, Aug, 310-319; Sept, 328-335 (1972).

Chavan, V.V., and Ulbrecht, J., Internal circulation in vessels agitated by screw impellers, Chem Eng J, 6(3), 213-224 (1973).

Ho, F.C., and Kwong, A., A guide to designing special agitators, Chem Eng, 23 July, 94-104 (1973).

Miller, D.N., Scale-up of agitated vessels for gas-liquid mass transfer, AIChEJ, 20(3), 445-453 (1974).
Nienow, A.W., Constant turnover time as scale-up criterion for agitated tanks, Chem Eng Sci, 29(4), 1043-1045 (1974).
Pravdin, V.G., et al., Hydraulic resistance of paddle-type vortexers, Int Chem Eng, 14(3), 501-503 (1974).
Ulbrecht, J., Mixing of viscoelastic fluids by mechanical agitation, Chem Engnr, June, 347-353, 367 (1974).
Wang, R.H., and Fan, L.T., Methods for scaling-up tumbling mixers, Chem Eng, 27 May, 88-94 (1974).
Fan, L.T., and Wang, R.H., Mixing indices, Powder Tech, 11, 27-32 (1975).
Koen, C., Practical mixing as a rate process, Chem Engnr, Feb, 91-95 (1975).
Vant Riet, K., and Smith, J.M., Trailing vortex system produced by Rushton turbine agitators, Chem Eng Sci, 30(9), 1093-1106 (1975).
Wien, O.; Wichterle, K., and Klohna, J., Arrangement of an agitator for homogenizing suspensions, Int Chem Eng, 15(1), 1-7 (1975).

1976-1980

Blasinski, H., and Rzyski, E., Mixing of non-Newtonian fluids with turbine propeller and paddle agitators, Int Chem Eng, 16(4), 751-755 (1976).
Brennan, D.J., Vortex geometry in unbaffled vessels with impeller agitation, Trans IChemE, 54, 209-217 (1976).
Brennan, D.J., and Lehrer, I.H., Impeller mixing in vessels: A general mixing time equation, Trans IChemE, 54, 139-152 (1976).
Khang, S.J., and Levenspiel, O., New scale-up and design method for stirrer-agitated batch-mixing vessels, Chem Eng Sci, 31(7), 569-578 (1976).
Anon., Reducing mixing time, Processing, June, 25 (1977).
Irving, S.; King, R., and Bull, D., Mixing studies in a large-scale sludge digester, Chem Engnr, Nov, 831-832, 837 (1978).
Nienow, A.W., and Miles, D., Effect of impeller/tank configurations on fluid-particle mass transfer, Chem Eng J, 15(1), 13-24 (1978).
Pace, G.W., Mixing of highly viscous fermentation broths, Chem Engnr, Nov, 833-837 (1978).
Rieger, F., and Novak, V., Axial thrust acting on agitators in non-Newtonian fluids, Chem Eng J, 16(1), 27-34 (1978).
Smith, J.M., Gas dispersion in viscous liquids with a static mixer, Chem Engnr, Nov, 827-830 (1978).
Lopes, M.M., and Calderbank, P.H., Scale-up of aerated mixing vessels for specified oxygen dissolution rates, Chem Eng Sci, 34(11), 1333-1338 (1979).
Joshi, J.B., Modifications in design of gas-inducing impeller, Chem Eng Commns, 5(1), 109-114 (1980).
Nigam, K.D.P., and Vasudeva, K., Residence time distribution in static mixer, Can JCE, 58, 543-545 (1980).
Sato, K., Characteristics of rectangular mixing vessels, Chem Eng Commns, 7(1), 45-56 (1980).
Tojo, K.; Mitsui, H., and Miyanami, K., Mixing performance of vibrating disk tank, Chem Eng Commns, 6(4), 305-311 (1980).

1981-1985

Chandrasekharan, K., and Calderbank, P.H., Scale-up of aerated mixing vessels, Chem Eng Sci, 36(5), 819-824 (1981).

Bathija, P.R., Jet mixing design and applications, Chem Eng, 13 Dec, 89-94 (1982).

Boss, J., and Czastkiewicz, W., Principles of scale-up for laminar mixing processes of Newtonian fluids in static mixers, Int Chem Eng, 22(2), 362-368 (1982).

Nauman, E.B., Reactions and residence time distributions in motionless mixers, Can JCE, 60, 136-140 (1982).

Tojo, K., and Miyanami, K., Mixing performance of high shear impellers, Chem Eng Commns, 16(1), 159-174 (1982).

Al Taweel, A.M., and Walker, L.D., Liquid dispersion in static in-line mixers, Can JCE, 61(4), 527-533 (1983).

Garrison, C.M., How to design and scale mixing pilot-plants, Chem Eng, 7 Feb, 63-70 (1983).

Oldshue, J.Y., Fluid mixing technology and practice, Chem Eng, 13 June, 82-108 (1983).

Chudacek, M.W., Does your mixing tank bottom have the right shape? Chem Eng, 1 Oct, 79-83 (1984).

Greaves, M., and Kobbacy, K.A.H., Measurement of bubble size distribution in turbulent gas-liquid dispersions, CER&D, 62, 3-12 (1984).

Guerin, P.; Carreau, P.J.; Patterson, W.I., and Paris, J., Characterization of helical impellers by circulation times, Can JCE, 62(3), 301-309 (1984).

Benayad, S.; David, R., and Cognet, G., Measurement of coupled velocity and concentration fluctuations in discharge flow of Rushton turbine in stirred tank, Chem Eng & Proc, 19(3), 157-166 (1985).

Bowen, R.L., Agitation intensity (scaleup), Chem Eng, 18 March, 159-168 (1985).

Nigam, K.D.P., and Nauman, E.B., Residence time distributions of power law fluids in motionless mixers, Can JCE, 63(3), 519-521 (1985).

Rzyski, E., Mixing time (to homogenization) in the transition region of mixing, Chem Eng J, 31(2), 75-82 (1985).

Wichterle, K.; Zak, L., and Mitschka, P., Shear stresses on the walls of agitated vessels, Chem Eng Commns, 32(1), 289-306 (1985).

1986-1988

Carpenter, K.J., Fluid processing in agitated vessels (review paper), CER&D, 64(1), 3-10 (1986).

Nienow, A.W.; Konno, M., and Bujalski, W., Studies on three-phase mixing: A review and recent results, CER&D, 64(1), 35-42 (1986).

Pustelnik, P., Investigation of residence time distribution in Kenics static mixer, Chem Eng & Proc, 20(3), 147-154 (1986).

Xanthopoulos, C., and Stamatoudis, M., Turbulent-range impeller power numbers in closed cylindrical and square vessels, Chem Eng Commns, 46, 123-128 (1986).

Bourne, J.R., and Dell'Ava, P., Micro- and macro-mixing in stirred tank reactors of different sizes, CER&D, 65(2), 180-186 (1987).

Parkinson, G., Made a mess of mixing? Good design and proper testing can eliminate some mixing problems, Chem Eng, 22 June, 32-35 (1987).
Raidoo, A.D.; Rao, K.S.; Sawant, S.B., and Joshi, J.B., Improvements in gas-inducing impeller design, Chem Eng Commns, 54, 241-264 (1987).
Ziolkowski, D., and Morawski, J., Flow characteristic of liquid streams inside tubular apparatus equipped with new type of static mixing elements, Chem Eng & Proc, 21(3), 131-140 (1987).
Berkman, P.D., and Calabrese, R.V., Dispersion of viscous liquids by turbulent flow in static mixer, AIChEJ, 34(4), 602-609 (1988).
Bertrand, J., and Couderc, J.P., Numerical and experimental study of flow induced by an anchor in viscous, Newtonian and pseudo-plastic fluids, Int Chem Eng, 28(2), 257-271 (1988).
Deckert, A., and O'Brien, E.E., Analysis of statistically steady multiple-reaction turbulent mixers, Chem Eng Commns, 74, 85-94 (1988).
Komori, S., and Murakami, Y., Turbulent mixing in baffled stirred tanks with vertical-blade impellers, AIChEJ, 34(6), 932-937 (1988).
Kroezen, A.B.J., et al., Foam generation in rotor-stator mixer, Chem Eng & Proc, 24(3), 145-156 (1988).
Lakin, M.B., How to mix a reactor, Chemtech, May, 300-303 (1988).
Mann, R., Recent progress in modelling of mixers, Proc Engng, Aug, 40-41 (1988).
Nauman, E.B.; Etchells, A.W., and Tatterson, G.B., Mixing: The state-of-the-art, Chem Eng Prog, 84(5), 58-69 (1988).
Platzer, B., and Noll, G., Modelling of local distributions of velocity components and turbulence parameters in agitated vessels, Chem Eng & Proc, 23(1), 13-32 (1988).
Rao, K.S., and Joshi, J.B., Liquid-phase mixing in mechanically agitated vessels, Chem Eng Commns, 74, 1-26 (1988).
Rao, K.S.M.; Rewatkar, V.B., and Joshi, J.B., Critical impeller speed for solid suspension in mechanically agitated contactors, AIChEJ, 34(8), 1332-1340 (1988).
Robinson, J.E.; Debelak, L.A., and Tanner, R.D., Two-zone unequal volume model to describe mixing in chemical and biochemical processes, Chem Eng Commns, 63, 143-156 (1988).
Wichterle, K.; Mitschka, P.; Hajek, J., and Zak, L., Shear stresses on the walls of vessels with axial impellers, CER&D, 66(1), 102-106 (1988).

14.3 Equipment

1969-1975

Oldshue, J.Y., Specifying mixers, Hyd Proc, 48(10), 73-80 (1969).
Anon., Mixers and mixing operations, Brit Chem Eng, 15(1), 45-51 (1970).
Miles, J.E.P., and Schofield, C., Performance of several industrial mixers using non-segregating free flowing powders, Trans IChemE, 48, T85-89 (1970).
Barrett, D., Rotating mixer reactor for kinetic studies, Trans IChemE, 49, 80-82 (1971).
Penny, W.R., Trends in mixing equipment, Chem Eng, 22 March, 86-98 (1971).

Burghardt, A., and Lipowska, L., Mixing phenomena in a CFSTR, Chem Eng Sci, 27(10), 1783-1796 (1972).
Connolly, J.R., How to make mixers meet process needs, Chem Eng, 23 July, 128-132 (1973).
Burghardt, A., and Lipowska, L., Mixing phenomena in continuous tank reactor, Int Chem Eng, 13(2), 227-235 (1973); 14(1), 1-8, 94-102 (1974).
Cook, P., and Hersey, J.A., Evaluation of a Nauta mixer for preparing a multicomponent powder mixture, Powder Tech, 9, 257-261 (1974).
Efremov, Y.V.; Poddubnyi, V.V., and Krasnova, I.E., Self-aligning mixer, Int Chem Eng, 14(2), 342-343 (1974).
Morris, W.D., and Misson, P., Experimental study of mass transfer and flow resistance in the Kenics static mixer, Ind Eng Chem Proc Des Dev, 13(3), 270-275 (1974).
Udaltsov, V.V., Trends in development of modern mixing equipment, Int Chem Eng, 14(2), 242-245 (1974).
Wichterle, K., and Riha, P., Optimum parameters of a screw agitator for laminar processes, Chem Eng J, 7(2), 105-110 (1974).
Butler, P., Mixing technology: Current problems, Proc Engng, Jan, 38-39 (1975).
Gates, L.E.; Henley, T.L., and Fenic, J.G., Selecting the optimum turbine agitator, Chem Eng, 8 Dec, 110-114 (1975).
Harwood, C.F.; Walanski, K.; Luebcke, E., and Swanstrom, C., Performance of continuous mixers for dry powders, Powder Tech, 11, 289-296 (1975).
Redfern, R., A new generation food processor, Chem Engnr, Sept, 521-524 (1975).
Sheridan, L.A., The practical side of mixing in extruders, Chem Eng Prog, 71(2), 83-84 (1975).
Todd, D.B., Mixing in starved twin-screw extruders, Chem Eng Prog, 71(2), 81-82 (1975).
Wesselingh, J.A., Mixing of liquids in cylindrical storage tanks with side-entering propellers, Chem Eng Sci, 30(8), 973-982 (1975).

1976-1979

Carreau, P.J.; Patterson, I., and Yap, C.Y., Mixing of viscoelastic fluids with helical-ribbon agitators, Can JCE, 54, 135-145 (1976).
Mersmann, A.; Einenkel, W.D., and Kappel, M., Design and scale-up of mixing equipment, Int Chem Eng, 16(4), 590-603 (1976).
Morris, W.D., and Benyon, J., Turbulent mass transfer in the Kenics static mixer, Ind Eng Chem Proc Des Dev, 15(2), 338-342 (1976).
Thyn, J., and Duffek, K., Powder mixing in a horizontal batch mixer, Powder Tech, 15, 193-197 (1976).
Morris, W.D., and Proctor, R., Effect of twist ratio on forced convection in the Kenics static mixer, Ind Eng Chem Proc Des Dev, 16(3), 406-412 (1977).
Walko, J., Viscous-product mixers, Proc Engng, May, 61-63 (1977).
Walko, J., Sabre-blade mixer, Proc Engng, May, 59 (1977).
Blasinski, H., and Rzyski, E., Mixing of non-Newtonian fluids with anchor, screw and ribbon mixers, Int Chem Eng, 18(4), 708-712 (1978).
Oldshue, J.Y., and Mady, O.B., Flocculation performance of mixing impellers, Chem Eng Prog, 74(8), 103-108 (1978).

Oldshue, J.Y., and Mady, O.B., Flocculator impellers: A comparison, Chem Eng Prog, 75(5), 72-75 (1979).
Patterson, W.I.; Carreau, P.J., and Yap, C.Y., Mixing with helical ribbon agitators, AIChEJ, 25(3), 508-521 (1979).

1980-1985

Fasano, J.B., and Eberhart, T.M., Designing fiberglass-reinforced-plastic vessels for agitator service, Chem Eng, 5 May, 115-125 (1980).
Murakami, Y.; Hirose, T., and Ohshima, M., Mixing with an up and down impeller, Chem Eng Prog, 76(5), 78-82 (1980).
Pollard, G., Designing shaft mixers, Proc Engng, June, 81-83 (1980).
Sasakura, T.; Kato, Y.; Yamamuro, S., and Ohi, N., Mixing process in a stirred vessel, Int Chem Eng, 20(2), 251-258 (1980).
Searle, D., Fluidized mixing, Proc Engng, March, 79 (1980).
Swanborough, A., Plastics compounding extruders, Chem Engnr, July, 482-485 (1980).
Williams, G.D., Static-mixer design, Proc Engng, June, 85,87 (1980).
Various, Mixers: Plant and equipment survey, Processing, Jan, 18-32; Sept, 29-41 (1981).
Weetman, R.J., and Salzman, R.N., Impact of side flow on mixing impeller, Chem Eng Prog, 77(6), 71-75 (1981).
Pahl, M., and Muschelknautz, E., Static mixers and their applications, Int Chem Eng, 22(2), 197-206 (1982).
Capes, P., Mixing design and research, Proc Engng, Oct, 57-59 (1983).
Chavan, V.V., Close-clearance helical impellers, AIChEJ, 29(2), 177-186 (1983).
Dickey, D.S., Computer program chooses agitator, Chem Eng, 9 Jan, 73-81 (1984).
Wheeler, D.A., Mixers: An update, Processing, Jan, 15-23; May, 23-26; Aug, 9,12 (1984).
Williams, G., How to buy a static mixer, Chem Engnr, Oct, 30-33 (1984).
Chella, R., and Ottino, J.M., Fluid mechanics of mixing in a single-screw extruder, Ind Eng Chem Fund, 24(2), 170-180 (1985).
Chudacek, M.W., Solids suspension behaviour in profiled bottom and flat bottom mixing tanks, Chem Eng Sci, 40(3), 385-392 (1985).
Heywood, N., Selecting a motionless mixer, Processing, March, 19-25 (1985).
Jones, R.L., Mixing equipment for powders and pastes, Chem Engnr, Nov, 41-43 (1985).
Placek, J., and Tavlarides, L.L., Turbulent flow in stirred tanks, AIChEJ, 31(7), 1113-1120 (1985).

1986-1988

Anon., Mixing and size reduction (special advertising section), Chem Eng, 23 June, 53-92 (1986).
Bourne, J.R., and Garcia-Rosas, J., Rotor-stator mixers for rapid micromixing, CER&D, 64(1), 11-17 (1986).
Cybulski, A., and Werner, K., Criteria for applications and selection of static mixers. Int Chem Eng, 26(1), 171-181 (1986).

Mutsakis, M., and Streiff, F.A., Advances in static mixing technology, Chem Eng Prog, 82(7), 42-48 (1986).
Various, Mixers update, Processing, Sept, 41-43 (1986).
Anon., Mixing and Size Reduction '87 (special advertising section), Chem Eng, 22 June, 36-55 (1987).
McDonagh, M., Mixers for powder/liquid dispersion, Chem Engnr, March, 29-32 (1987).
Pandit, A.; Niranjan, K., and Davidson, J., A multipurpose impeller, Chem Engnr, Sept, 21 (1987).
Tatterson, G.B., and Morrison, G.L., Effect of tank to impeller diameter ratio on flooding transition for disc turbines, AIChEJ, 33(10), 1751-1753 (1987).
Anon., Mixing: Equipment focus, Chem Eng, 12 Sept, 57-70 (1988).
Chen, K.Y.; Hajduk, J.C., and Johnson, J.W., Laser-Doppler anemometry in baffled mixing tank, Chem Eng Commns, 72, 141-158 (1988).

14.4 Efficiency and Power Consumption

1967-1974

Hruby, M., Relationship between mechanical energy dissipation and heat transfer in agitated vessels, Int Chem Eng, 7(1), 86-91 (1967).
Bourne, J.R., and Butler, H., Power consumption of helical ribbon impellers in viscous liquids, Trans IChemE, 47, T263-270 (1969).
Hall, K.R., and Godfrey, J.C., Power consumption by helical ribbon impellers, Trans IChemE, 48, T201-208 (1970).
Ando, K.; Hara, H., and Endoh, K., Flow behavior and power consumption in horizontal stirred vessels, Int Chem Eng, 11(4), 735-739 (1971).
McPhee, A.D., and Brown, N.L., Power consumption in solid-liquid slurries, Ind Eng Chem Proc Des Dev, 10(4), 456-459 (1971).
Nienow, A.W., and Miles, D., Impeller power numbers in closed vessels, Ind Eng Chem Proc Des Dev, 10(1), 41-43 (1971).
Seichter, P., Efficiency of the screw mixers with a draught tube, Trans IChemE, 49, 117-123 (1971).
Williams, J.C., and Rahman, M.A., Prediction of performance of continuous mixers for particulate solids using residence time distributions, Powder Tech, 5, 87-92, 307-316 (1971).
Chavan, V.V., and Ulbrecht, J., Power correlation for helical ribbon impellers in inelastic non-Newtonian fluids, Chem Eng J, 3(3), 308-311 (1972).
Chavan, V.V.; Jhaveri, A.S., and Ulbrecht, J., Power consumption for mixing of inelastic non-Newtonian fluids by helical screw agitators, Trans IChemE, 50, 147-155 (1972).
Moo-Young, M.; Tichar, K., and Dullien, F.A.L., Blending efficiencies of some impellers in batch mixing, AIChEJ, 18(1), 178-182 (1972).
Rieger, F., and Novak, V., Scale-up method for power consumption of agitators in creeping flow regime, Chem Eng Sci, 27(1), 39-44 (1972).
Chavan, V.V., and Ulbrecht, J., Power correlation for off-centred helical screw impellers in highly viscous Newtonian and non-Newtonian liquids, Trans IChemE, 51, 349-354 (1973).

Kelkar, J.V.; Mashelkar, R.A., and Ulbrecht, J., Scale-up method for power consumption of agitators in creeping flow regime, Chem Eng Sci, 28(2), 664-666 (1973).
Rieger, F., and Novak, V., Power consumption of agitators in highly viscous non-Newtonian liquids, Trans IChemE, 51, 105-111 (1973).
Bruin, W.; van Riet, K., and Smith, J.M., Power consumption with aerated Rushton turbines, Trans IChemE, 52, 88-104 (1974).
Chavan, V.V., and Ulbrecht, J., Power correlations for close-clearance helical impellers in non-Newtonian liquids, Ind Eng Chem Proc Des Dev, 12(4), 472-476 (1973); 13(3), 309 (1974).
Rieger, F., and Novak, V., Power consumption scale-up in agitating non-Newtonian fluids, Chem Eng Sci, 29(11), 2229-2234 (1974).
Rieger, F., and Novak, V., Power consumption for agitating viscoelastic liquids in the viscous regime, Trans IChemE, 52, 285-286 (1974).

1975-1980

Novak, V., and Rieger, F., Homogenization efficiency of helical ribbon and anchor agitators, Chem Eng J, 9(1), 63-70 (1975).
Connolly, J.R., Energy conservation in fluid mixing, Chem Eng Prog, 72(5), 52-55 (1976).
Sawinsky, J.; Havas, G., and Deak, A., Power requirement of anchor and helical ribbon impellers for agitation of Newtonian and pseudo-plastic liquids, Chem Eng Sci, 31(6), 507-510 (1976).
Van Riet, K.; Boom, J.M., and Smith, J.M., Power consumption, impeller coalescence and reciriculation in aerated vessels, Trans IChemE, 54, 124-131 (1976).
Hassan, I.T.M., and Robinson, C.W., Stirred-tank mechanical power requirements and gas holdup in aerated aqueous phases, AIChEJ, 23(1), 49-56 (1977).
Loiseau, B.; Midoux, N., and Charpentier, J.C., Hydrodynamics and power input data for mechanically agitated gas-liquid contactors, AIChEJ, 23(6), 931-935 (1977).
Novak, V., and Rieger, F., Influence of vessel to screw diameter ratio on efficiency of screw agitators, Trans IChemE, 55, 202-206 (1977).
Chowdhury, R., and Tiwari, K.K., Power consumption studies of helical-ribbon screw mixers, Ind Eng Chem Proc Des Dev, 18(2), 227-231 (1979).
Ito, R.; Hirata, Y.; Sakata, K., and Nakahara, I., Power consumption of the impellar in agitated vessel in continuous flow system, Int Chem Eng, 19(4), 605-611 (1979).
Nishikawa, M., et al., Agitation power and mixing time in off-centred mixing, Int Chem Eng, 19(1), 153-159 (1979).
Sawinsky, J.; Deak, A., and Havas, G., Power consumption of screw agitators in Newtonian liquids of high viscosity, Chem Eng Sci, 34(9), 1160-1163 (1979).
Zundelevich, Y., Power consumption and gas capacity of self-inducing turbo aerators, AIChEJ, 25(5), 763-773 (1979).
Blasinski, H., and Rzyski, E., Power requirements of helical ribbon mixers, Chem Eng J, 19(2), 157-160 (1980).

Blasinski, H.; Nowicki, J., and Rzyski, E., Mixing power and mixing times of propeller agitators introduced laterally, Int Chem Eng, 20(1), 92-98 (1980).
Einsele, A., and Finn, R.K., Influence of gas flow rates and gas holdup on blending efficiency in stirred tanks, Ind Eng Chem Proc Des Dev, 19(4), 600-603 (1980).
Hughmark, G.A., Power requirements and interfacial area in gas-liquid turbine agitated systems, Ind Eng Chem Proc Des Dev, 19(4), 638-641 (1980).
Medek, J., Power characteristics of agitators with flat inclined blades, Int Chem Eng, 20(4), 664-673 (1980).
Ottino, J.M., and Macosko, C.W., Efficiency parameter for batch mixing of viscous fluids, Chem Eng Sci, 35(6), 1454-1458 (1980).

1981-1984

Garrison, C.M., How to cut agitation costs, Chem Eng, 30 Nov, 73-78 (1981).
Okamoto, Y.; Nishikawa, M., and Hashimoto, K., Energy dissipation rate distribution in mixing vessels and effects on liquid-liquid dispersion and solid-liquid mass transfer, Int Chem Eng, 21(1), 88-95 (1981).
Gray, D.J.; Treybal, R.E., and Barnett, S.M., Mixing of single and two-phase systems: Power consumption of impellers, AIChEJ, 28(2), 195-199 (1982).
Lehtola, S., and Kuoppamaki, R., Measurement of mixing efficiency of continuous mixers, Chem Eng Sci, 37(2), 185-192 (1982).
Sir, J., and Lecjaks, Z., Pressure drop and homogenization efficiency of a motionless mixer, Chem Eng Commns, 16(1), 325-334 (1982).
Greaves, M.; Kobbacy, K.A.H., and Millington, G.C., Gassed power dynamics of disc turbine impeller, Chem Eng Sci, 38(11), 1909-1916 (1983).
Kuboi, R.; Nienow, A.W., and Allsford, K., Multipurpose stirred tank facility for flow visualisation and dual impeller power measurement, Chem Eng Commns, 22(1), 29-40 (1983).
Nienow, A.W., et al., Effect of rheological complexities on power consumption in an aerated agitated vessel, Chem Eng Commns, 19(4), 273-294 (1983).
Ottino, J.M., Mechanical mixing efficiency parameter for static mixers, AIChEJ, 29(1), 159-161 (1983).
Pasquali, G.; Fajner, D., and Magelli, F., Effect of suspension viscosity on power consumption in the agitation of solid-liquid systems, Chem Eng Commns, 22(5), 371-376 (1983).
Ismail, A.F.; Nagase, Y., and Imon, J., Power characteristics and cavity formation in aerated agitations, AIChEJ, 30(3), 487-489 (1984).
Oliver, D.R.; Nienow, A.W.; Mitson, R.J., and Terry, K., Power consumption in mixing of Boger fluids, CER&D, 62, 123-127 (1984).

1985-1988

Bertrand, J., and Couderc, J.P., Evaluation of power consumption in agitation of viscous Newtonian or pseudoplastic liquids by two-bladed, anchor or gate agitators, CER&D, 63, 259-263 (1985).
Collias, D.J., and Prudhomme, R.K., Effect of fluid elasticity on power consumption and mixing times in stirred tanks, Chem Eng Sci, 40(8), 1495-1506 (1985).
Deak, A.; Havas, G., and Sawinsky, J., Power requirements for anchor, ribbon and helical-screw agitators, Int Chem Eng, 25(3), 558-566 (1985).

Shamlou, P.A., and Edwards, M.F., Power consumption of helical ribbon mixers in viscous Newtonian and non-Newtonian fluids, Chem Eng Sci, 40(9), 1773-1782 (1985).
Rieger, F.; Novak, V., and Havelkova, D., Homogenization efficiency of helical ribbon agitators, Chem Eng J, 33(3), 143-150 (1986).
Sestak, J.; Zitny, R., and Houska, M., Anchor-agitated systems: Power input correlation for pseudoplastic and thixotropic fluids in equilibrium, AIChEJ, 32(1), 155-158 (1986).
Sinevic, V.; Kuboi, R., and Nienow, A.W., Power numbers, Taylor numbers and Taylor vortices in viscous Newtonian and non-Newtonian fluids, Chem Eng Sci, 41(11), 2915-2924 (1986).
Valderrama, J.O., et al., Experimental study on mixing: Power consumption and degree of suspension, Chem Eng Commns, 44, 331-346 (1986).
Cooker, B., and Nedderman, R.M., Circulation and power consumption in helical ribbon powder agitators, Powder Tech, 52(2), 117-130 (1987).
Masiuk, S., Power consumption, mixing time and attrition action for solid mixing in ribbon mixer, Powder Tech, 51(3), 217-230 (1987).
King, R.L.; Hiller, R.A., and Tatterson, G.B., Power consumption in a mixer, AIChEJ, 34(3), 506-509 (1988).
Rao, K.S.R., and Joshi, J.B., Liquid-phase mixing and power consumption in mechanically agitated solid-liquid contactors, Chem Eng J, 39(2), 111-124 (1988).
Rieger, F.; Novak, V., and Havelkova, D., Influence of geometrical shape on power requirement of ribbon impellers, Int Chem Eng, 28(2), 376-383 (1988).

14.5 Applications and Systems

1969-1975

Novak, V., and Rieger, F., Homogenization with helical screw agitators, Trans IChemE, 47, T335-340 (1969).
Poroshin, V.V., and Streltsov, V.V., Distinctive features of the mixing of loess-like loams in paddle mixers, Int Chem Eng, 9(1), 27-29 (1969).
Skelland, A.H.P., and Dimmick, G.R., Heat transfer between coils and non-Newtonian fluids with propeller agitation, Ind Eng Chem Proc Des Dev, 8(2), 267-274 (1969).
Butters, J.R., Recent developments in the mixing of dry solids, Brit Chem Eng, 15(1), 41-43 (1970).
Bridgwater, J., Mixing of cohesionless powders, Powder Tech, 5, 257-260 (1971).
Robertson, W.S., Use of mechanical aerators in the activated sludge process, Chem Engnr, May, 176-181 (1971).
Sethuraman, K.J., and Davies, G.S., Studies on solids mixing in a double-cone blender, Powder Tech, 5, 115-118 (1971).
Chen, S.J.; Fan, L.T., and Watson, C.A., Mixing of solid particles in a motionless mixer: Axial-dispersed plug-flow model, Ind Eng Chem Proc Des Dev, 12(1), 42-47 (1973).

Various, Mixing and blending (feature report), Chemical Processing, June, 65-85 (1974).
Chen, S.J., Static mixing of polymers, Chem Eng Prog, 71(8), 80-83 (1975).
Hicks, R.W., and Gates, L.E., Fluid agitation in polymer reactors, Chem Eng Prog, 71(8), 74-79 (1975).
Schott, N.R.; Weinstein, B., and LaBombard, D., Motionless mixers in plastic processing, Chem Eng Prog, 71(1), 54-58 (1975).

1976-1980

Hersey, J.A., Powder mixing: Theory and practice in pharmacy, Powder Tech, 15, 149-153 (1976).
Various, Mixing and blending (feature report), Processing, Oct, 49-74 (1976).
Wang, R.H., and Fan, L.T., Axial mixing of grains in a motionless Sulzer (Koch) mixer, Ind Eng Chem Proc Des Dev, 15(3), 381-388 (1976).
Williams, J.C., Continuous mixing of solids: A review, Powder Tech, 15, 237-243 (1976).
Oldshue, J.Y., and Connelly, F.L., Gas-liquid contacting with impeller mixers, Chem Eng Prog, 73(3), 85-89 (1977).
Wadsworth, J.K., Mixing with ultrasonics, Processing, Sept, 21-23 (1977).
Havas, G.; Sawinsky, J.; Deak, A., and Fekete, A., Investigation of homogenization efficiency of agitators and impellers for mixing of high-viscosity Newtonian liquids, Periodica Polytechnica, 21, 315-344 (1978).
Yano, T.; Sato, M., and Terashita, K., Recent work in Japan on the mixing of solids, Powder Tech, 20, 9-14 (1978).
Various, Mixing and blending (feature report), Processing, March, 61-76 (1979).
Yung, C.N.; Wong, C.W., and Chang, C.L., Gas holdup and aerated power consumption in mechanically stirred tanks, Can JCE, 57, 672-682 (1979).
Jarvis, J.E., Dispersion mixing in the paint industry, Chem Engnr, July, 477-481 (1980).
Keirstead, K.F.; DeKee, D., and Carreau, P.J., Effect of mixers on formation and stability of aerated ammonium nitrate gels, Can JCE, 58, 549-552 (1980).
Oldshue, J.Y., Mixing in hydrogenation processes, Chem Eng Prog, 76(6), 60-64 (1980).

1981-1984

Chapman, C.M.; Nienow, A.W., and Middleton, J.C., Particle suspension in gas sparged Rushton-turbine agitated vessel, Trans IChemE, 59, 134-137 (1981).
Stamatoudis, M., and Tavlarides, L.L., Effect of impeller rotational speed on the drop size distributions of viscous liquid-liquid dispersions in agitated vessels, Chem Eng J, 21(1), 77-78 (1981).
Joshi, J.B.; Pandit, A.B., and Sharma, M.M., Mechanically agitated gas-liquid reactors (review paper), Chem Eng Sci, 37(6), 813-844 (1982).
Lane, A.G.C., and Rice, P., Comparative assessment of performance of three designs for liquid-jet mixing, Ind Eng Chem Proc Des Dev, 21(4), 650-653 (1982).
Oldshue, J.Y.; Mechler, D.O., and Grinnell, D.W., Fluid mixing variables in suspension and emulsion polymerization, Chem Eng Prog, 78(5), 68-74 (1982).

Bowsher, M.E., and Hooley, D.F., Optimizing reactor agitation in heat-transfer-limited situations, Chem Eng, 27 June, 45-50 (1983).
Chapman, C.M.; Nienow, A.W.; Cooke, M., and Middleton, J.C., Particle-gas-liquid mixing in stirred vessels, CER&D, 61, 71-95, 167-185 (1983).
Harvey, P.S., and Greaves, M., Turbulent flow in an agitated vessel, Trans IChemE, 60, 195-210 (1982); 61, 136-137 (1983).
Nagase, Y., and Yasui, H., Fluid motion and mixing in a gas-liquid contactor with turbine agitators, Chem Eng J, 27(1), 37-48 (1983).
Rounsley, R.R., Oil dispersion with a turbine mixer, AIChEJ, 29(4), 597-603 (1983).
Akiyama, T., and Tada, I., Mixing characteristics of the positive and negative pressure air mixers, Ind Eng Chem Proc Des Dev, 23(4), 737-741 (1984).
Prudhomme, R.K., and Shaqfeh, E., Effect of elasticity on mixing torque requirements for Rushton turbine impellers, AIChEJ, 30(3), 485-486 (1984).
Shiue, S.J., and Wong, C.W., Studies on homogenization efficiency of various agitators in liquid blending, Can JCE, 62(5), 602-609 (1984).

1985-1988

Chudacek, M.W., Impeller power numbers and impeller flow numbers in profiled bottom tanks, Ind Eng Chem Proc Des Dev, 24(4), 858-867 (1985).
Dobby, G.S., and Finch, J.A., Mixing characteristics of industrial flotation columns, Chem Eng Sci, 40(7), 1061-1069 (1985).
Harnby, N.; Edwards, M.F., and Nienow, A.W., Mixing in the Process Industries, Butterworth Publishers, Massachusetts (1985).
Raghavachari, S., Extend the life of slurry agitators, Chem Eng, 1 April, 98 (1985).
Bowen, R.L., Shear-sensitive mixing systems, Chem Eng, 9 June, 55-63 (1986).
Gosman, A.D., and Simitovic, R., Experimental study of confined jet mixing, Chem Eng Sci, 41(7), 1853-1872 (1986).
Kimber, A., Mixing update, Processing, March, 27-33 (1986).
Lu, W.M., and Chen, H., Flooding and critical impeller speed for gas dispersion in aerated turbine-agitated vessels, Chem Eng J, 33(2), 57-62 (1986).
Wong, C.W., and Shiuan, J.H., Effect of additives on mass transfer in an aerated mixing vessel, Chem Eng Commns, 43, 133-146 (1986).
Al-Ameeri, R.S., Influence of drag reducing additives on power consumption in agitated vessels, Chem Eng Commns, 59, 1-14 (1987).
Benbow, J.J.; Oxley, E.W., and Bridgwater, J., The extrusion of pastes: The influence of paste formulation on extrusion parameters, Chem Eng Sci, 42(9), 2151-2162 (1987).
Hold, P., Mixing for polymer processing, Chem Eng Prog, 83(6), 51-55 (1987).
Mielcarek, D.F., Twin-screw compounding for mixing of plastics, Chem Eng Prog, 83(6), 59-67 (1987).
Saito, F., and Kamiwano, M., A technique for prediction of the mixing time of high-viscosity liquid-mixing systems with negligible diffusivity of solute, Chem Eng J, 36(2), 93-100 (1987).
Butcher, C., Lasers for mixing research, Chem Engnr, June, 39-42 (1988).

Lal, P.; Kumar, S.; Upadhyay, S.N., and Upadhya, Y.D., Solid-liquid mass transfer in agitated Newtonian and non-Newtonian fluids, Ind Eng Chem Res, 27(7), 1246-1259 (1988).

Nocentini, M.; Magelli, F., and Pasquali, G., Flooding-loading transition for gas-liquid vessels stirred with Rushton turbines, CER&D, 66(4), 378-381 (1988).

Smith, J.M., and Verbeek, D.G.F., Impeller cavity developments in nearly boiling liquids, CER&D, 66(1), 39-46 (1988).

CHAPTER 15

FLUID FLOW

15.1 Theory and Applications

1966-1968

Collier, J.G., and Hewitt, G.F., Measurement of liquid entrainment, Brit Chem Eng, 11(11), 1375-1379 (1966).
Collier, J.G., and Hewitt, G.F., Experimental techniques in two-phase flow, Brit Chem Eng, 11(12), 1526-1531 (1966).
Chidambaram, S., Nomogram for minimum entrainment velocity of a liquid jet, Brit Chem Eng, 12(6), 919 (1967).
Collier, J.G., and Hewitt, G.F., Film thickness measurement in two-phase flow, Brit Chem Eng, 12(5), 709-715 (1967).
Gomezplata, A., and Nichols, C.R., Estimate holdup in vertical two-phase flow, Chem Eng, 13 Feb, 182 (1967).
Hamielec, A.E.; Hoffman, T.W., and Ross, L.L., Numerical solution of the Navier-Stokes equation for flow past spheres, AIChEJ, 13(2), 212-224 (1967).
Paige, P.M., Estimating the pressure drop of flashing mixtures, Chem Eng, 14 Aug, 159-164 (1967).
Bjorklund, I.S., and Dygert, J.C., Small scale tests for attrition resistance of solids in slurry systems, AIChEJ, 14(4), 553-557 (1968).
Brown, F.C., and Kranich, W.L., Model for prediction of velocity and void fraction profiles in two-phase flow, AIChEJ, 14(5), 750-758 (1968).
Buffham, B.A., Laminar flow in open circular channels and symmetrical lenticular tubes, Trans IChemE, 46, T152-157 (1968).
Capps, D.O., and Rehm, T.R., Empirical expression for turbulent-flow velocity distribution, Ind Eng Chem Proc Des Dev, 7(2), 311-313 (1968).
Clump, C.W., and Kwasnoski, D., Turbulent flow in concentric annuli, AIChEJ, 14(1), 164-168 (1968).
Gloyer, W., Correlation for two-phase flow, Chem Eng, 1 Jan, 93-95 (1968).
Hodossy, L., Review of momentum, heat and mass transfer in two-phase flow, Int Chem Eng, 8(3), 427-438 (1968).
Huber, O., and Penzkofer, A., Rheological behaviour of coating clays and aqueous clay dispersions, Int Chem Eng, 8(1), 92-99 (1968).
Jenson, V.G.; Horton, T.R., and Wearing, J.R., Drag coefficients and transfer factors for spheres in laminar flow, Trans IChemE, 46, T177-184 (1968).
Oliver, D.R., and Young Hoon, A., Two-phase non-Newtonian flow, Trans IChemE, 46, T106-122 (1968).
Wohl, M.H., Designing for non-Newtonian fluids, Chem Eng, 15 Jan, 148-151; 12 Feb, 130-136; 25 March, 99-104; 8 April, 143-146; 6 May, 183-186; 3 June, 95-100; 1 July, 81-86; 15 July, 127-132; 26 Aug, 113-118 (1968).

1969-1970

Chandler, J.L., Rapid solution of the Dittus-Boelter equation for a fluid flowing in a circular duct, Brit Chem Eng, 14(7), 983 (1969).
Gay, E.C.; Nelson, P.A., and Armstrong, W.P., Flow properties of suspensions with high solids concentration, AIChEJ, 15(6), 815-822 (1969).
Gutierrez, A., and Lynn, S., Minimum critical velocity for one-phase flow of liquids, Ind Eng Chem Proc Des Dev, 8(4), 486-491 (1969).

Halwagi, M.M.E., and Eissa, S., Determine the void fraction in gas-liquid flow, Chem Eng, 15 Dec, 158-160 (1969).
Kambe, H., Rheology of concentrated suspensions, Int Chem Eng, 9(1), 164-171 (1969).
Loeb, M.B., New graphs for solving compressible flow problems, Chem Eng, 19 May, 179-184 (1969).
Rushton, E., and Leslie, D.C., Evaluation of saturated water flow through nozzles, Brit Chem Eng, 14(3), 319-323 (1969).
Savchenko, I.V., Comparison of rheological characteristics of non-Newtonian systems, Int Chem Eng, 9(2), 207-210 (1969).
Singh, K.; Pierre, C.C.S.; Crago, W.A., and Moeck, E.O., Liquid film flowrates in two-phase flow of steam and water at 1000 psia, AIChEJ, 15(1), 51-56 (1969).
Various, Liquids handling (deskbook issue), Chem Eng, 14 April, 11-282 (1969).
Azizov, A.; Zysina, L.M.; Kuznetsova, V.M., and Soskova, I.N., Effect of temperature factor on transition from laminar to turbulent flow conditions in a boundary layer, Int Chem Eng, 10(1), 23-27 (1970).
Boothroyd, R.G., and Goldberg, A.S., Measurements in flowing gas-solids suspensions, Brit Chem Eng, 14(12), 1705-1708 (1969); 15(3), 357-362 (1970).
Jones, M.L., Method solves fluid-flow problems simply and accurately, Chem Eng, 23 March, 160-162 (1970).
Zanker, A., Nomogram for liquid film thickness in turbulent flow regime (falling films), Brit Chem Eng, 15(12), 1585 (1970).
Zanker, A., Nomograph for calculating flow in rectangular weirs, Chem Eng, 10 Aug, 142 (1970).

1971-1972

Boyadzhiev, K., Hydrodynamics of certain two-phase flows, Int Chem Eng, 11(3), 464-488 (1971).
Constance, J.D., Estimating fluid friction in ducts of nonstandard shapes, Chem Eng, 22 Feb, 146-149 (1971).
DeGance, A.E., and Atherton, R.W., Aspects of two-phase flow, Chem Eng, 23 March, 135-139; 20 April, 151-158; 4 May, 113-120; 13 July, 95-103; 10 Aug, 119-126; 5 Oct, 87-94; 2 Nov, 101-108 (1970); 22 Feb, 125-132 (1971).
Jameson, G.J., The contribution of disturbance waves to the overall pressure drop in annular two-phase flow, Trans IChemE, 49, 42-48 (1971).
Kasturi, G.; Stepanek, J., and Holland, F.A., Review of two-phase flow literature, Brit Chem Eng, 16(4), 333-336; 16(6), 511-514 (1971).
Various, Non-Newtonian flow (topic issue), Chem Engnr, Jan, 16-34 (1971).
Baum, M.R., An experimental study of adiabatic choked gas-liquid bubble flow in a convergent-divergent nozzle, Trans IChemE, 50, 293-299 (1972).
Dombrowski, N., and Wolfsohn, D.L., The atomisation of water by swirl spray pressure nozzles, Trans IChemE, 50, 259-269 (1972).
Fouda, A.E., and Rhodes, E., Two-phase annular flow stream division, Trans IChemE, 50, 353-363 (1972).

Kasturi, G., and Stepanek, J.B., Two-phase flow, Chem Eng Sci, 27(10), 1871-1892 (1972).

Lihou, D.A.; Lowe, W.D., and Hattangady, K.S., Studies of carbon dioxide drops in a stream of high-pressure carbon dioxide gas, Trans IChemE, 50, 217-223 (1972).

Lynch, E.P., Timing gravity flow from vertical tanks, Chem Eng, 15 May, 130 (1972).

1973-1975

Boyadjiev, C., and Krylov, V.S., Analysis of stability of laminar liquid film flow, Chem Eng J, 6(3), 225-232 (1973).

Grace, J.R., Shapes and velocities of bubbles rising in infinite liquids, Trans IChemE, 51, 116-120 (1973).

Hallett, V.A., Asymmetric flow of a falling viscous film, Proc Tech Int, June, 279-283 (1973).

Harris, J.B., and Pittman, J.F.T., Applicability of the similarity solution for converging channel flow to a practical case, Trans IChemE, 51, 369-373 (1973).

Little, J., Calculating flows and pressure drops by square roots, Chem Eng, 19 Feb, 148 (1973).

Mewis, J., Variable-spectrum approach in steady-state shear flow, Chem Eng J, 6(3), 205-212 (1973).

Sande, E., and Cordemans, W., Large scale turbulence characteristics of a submerged water jet, Trans IChemE, 51, 247-250 (1973).

Denham, M.K., and Patrick, M.A., Laminar flow over a downstream-facing step in a two-dimensional flow channel, Trans IChemE, 52, 361-367 (1974).

Fouda, A.E., and Rhodes, E., Two-phase annular flow stream division in a simple tee, Trans IChemE, 52, 354-360 (1974).

Noyer, J.J., Calculating pressure drops by calculator or computer, Chem Eng, 18 Feb, 154-156; 10 June, 110 (1974).

Austin, P.P., Simplified fluid flow calculations, Hyd Proc, 54(9), 197-201 (1975).

Baum, M.R., and Cook, M.E., The modelling of cavitation inception, Trans IChemE, 53, 131-135 (1975).

Lin, O.C.C., Rheology and surface coatings, Chemtech, Jan, 51-60 (1975).

Payne, G.J., and Prince, R.G.H., Transition from jetting to bubbling at a submerged orifice, Trans IChemE, 53, 209-223 (1975).

Ruskin, R.P., Calculating line sizes for flashing steam-condensate, Chem Eng, 18 Aug, 101-103; 24 Nov, 88 (1975).

Zangger, R., and Bell, D., Nomogram for flow characteristics of Bingham bodies, Proc Engng, May, 86-87 (1975).

1976-1977

Markatos, N.C.G., Stochastic analysis of turbulent air-water wavy interfaces, Trans IChemE, 54, 184-195 (1976).

Taitel, Y., and Dukler, A.E., Model for predicting flow regime transitions in horizontal and near horizontal gas-liquid flow, AIChEJ, 22(1), 47-55 (1976).

Taylor, M.A., Program for compressible flow, Proc Engng, Oct, 83-84 (1976).

Thwaites, G.R.; Kulov, N.N., and Nedderman, Liquid film properties in two-phase annular flow, Chem Eng Sci, 31(6), 481-486 (1976).
Virk, P.S., Drag reduction fundamentals (review paper), AIChEJ, 21(4), 625-656 (1975); 22(2), 398 (1976).
Yagawa, G.; Ishida, Y., and Ando, Y., Finite-element method applied to Navier-Stokes equation, Int Chem Eng, 16(2), 253-258 (1976).
Adler, P.M., Formation of an air-water two-phase flow, AIChEJ, 23(2), 185-191 (1977).
Churchill, S.W., Comprehensive correlating equations for heat, mass and momentum transfer in fully developed flow in smooth tubes, Ind Eng Chem Fund, 16(1), 109-116 (1977).
Dukler, A.E., The role of waves in two-phase flow, Chem Eng Educ, 11(3), 108-117 (1977).
Gay, B., and Preece, P.E., Matrix methods for the solution of fluid network problems, Trans IChemE, 53, 12-15 (1975); 55, 38-45, 285 (1977).
Hanik, P.P., and Sterba, V.J., Computer simulation of fluid flow networks, Chem Eng Prog, 73(4), 100-104 (1977).
Otten, L., and Fayed, A.S., Slug velocity and slug frequency measurements in concurrent air/non-Newtonian slug flow, Trans IChemE, 55, 64-67 (1977).
Various, In-plant slurry handling (feature report), Chem Eng, 25 April, 94-110 (1977).

1978-1979

Markatos, N.C.G.; Sala, R., and Spalding, D.B., Flow in an annulus of non-uniform gap, Trans IChemE, 56, 28-35 (1978).
Various, Fluids handling (feature report), Processing, Oct, 65-101 (1978).
Various, Liquids handling: Solving problems in piping, pumps, valves, tanks and transportation (deskbook issue), Chem Eng, 3 April, 9-215 (1978).
Ackermann, N.L., and Shen, H.T., Rheological characteristics of solid-liquid mixtures, AIChEJ, 25(2), 327-332 (1979).
Carreau, P.J., and De Kee, D., Review of some useful rheological equations, Can JCE, 57, 3-15 (1979).
Castillo, C., and Williams, M.C., Rheology of very concentrated coal suspensions, Chem Eng Commns, 3(6), 529-546 (1979).
Eisenberg, F.G., and Weinberger, C.B., Annular two-phase flow of gases and non-Newtonian liquids, AIChEJ, 25(2), 240-246 (1979).
Markatos, N.C.G., and Moult, A., The computation of steady and unsteady turbulent, chemically reacting flows in axi-symmetrical domains, Trans IChemE, 57, 156-162 (1979).
Moore, S.W., Bingham-plastics pressure drop, Chem Eng, 24 Sept, 142 (1979).
Moult, A.; Spalding, D.B., and Markatos, N.C.G., The solution of flow problems in highly irregular domains by the finite-difference method, Trans IChemE, 57, 200-204 (1979).
Rodriguez, F., Analogy between fluid flow and electric circuitry, Chem Eng Educ, 13(2), 96-98 (1979).
Srinivasan, S.; Bobba, K.M., and Stenger, L.A., Viscous flow in rectangular open channels, Ind Eng Chem Fund, 18(2), 130-133 (1979).
Toda, M.; Yonehara, J., and Maeda, S., Horizontal solid-liquid two-phase flow at low flowrates, Int Chem Eng, 19(4), 646-650 (1979).

Toda, M.; Yonehara, J.; Kimura, T., and Maeda, S., Transition velocities in horizontal solid-liquid two-phase flow, Int Chem Eng, 19(1), 145-153 (1979).

Various, Fluids handling (feature report), Processing, Oct, 93-111 (1979).

1980-1981

Cheng, D.C.H., Viscosity-concentration equations and flow curves for suspensions, Chem & Ind, 17 May, 403-407 (1980).

Edwards, M.F., and Smith, R., The integration of the energy equation for fully developed turbulent pipe-flow, Trans IChemE, 58, 260-264 (1980).

Embaby, M.H., and Verba, A., Mean flow properties in the developing region of a circular pipe for turbulent flow at maximum drag reduction, Periodica Polytechnica, 24, 83-93 (1980).

Farooqi, S.I., and Richardson, J.F., Rheological behaviour of kaolin suspensions in water and water-glycerol mixtures, Trans IChemE, 58, 116-124 (1980).

Hodge, S.A.; Sanders, J.P., and Klein, D.E., Slope and intercept of dimensionless velocity profile for artificially roughened surfaces, Int J Heat Mass Trans, 23(2), 135-140 (1980).

Markatos, N.C.G., Turbulent air flow over water waves, Trans IChemE, 58, 251-259 (1980).

Stein, M.A.; Kessler, D.P., and Greenkorn, R.A., Empirical model of velocity profiles for turbulent flow in smooth pipes, AIChEJ, 26(2), 308-310 (1980).

Various, Fluid handling (feature report), Processing, Oct, 48-83 (1980).

Wall, T.F.; Nguyen, H.; Subramanian, V.; Mai-Viet, T., and Howley, P., Direct measurements of the entrainment by single and double concentric jets in the regions of transition and flow establishment, Trans IChemE, 58, 237-241 (1980).

Al-Hayes, R.A.M., and Winterton, R.H.S., Bubble growth and bubble diameter on detachment in flowing liquids, Int J Heat Mass Trans, 24(2), 213-230 (1981).

Athey, R.D., System flow properties, Chemtech, May, 308-314 (1981).

Darby, R., and Melson, J., How to predict the friction factor for flow of Bingham plastics, Chem Eng, 28 Dec, 59-61 (1981).

Foster, T.C., Time required to empty a vessel, Chem Eng, 4 May, 105 (1981).

Kadambi, V., Void fraction and pressure drop in two-phase stratified flow, Can JCE, 59, 584-589 (1981).

Kawase, Y., and Ulbrecht, J.J., Formation of drops and bubbles in flowing liquids, Ind Eng Chem Proc Des Dev, 20(4), 636-640 (1981).

Kraus, M.N., Use V-trough as distributor, Chem Eng, 9 Feb, 127-128 (1981).

Lehrer, I.H., A new model for free turbulent jets of miscible fluids of different density and a jet mixing time criterion, Trans IChemE, 59, 247-252 (1981).

Norouzian, M., How to build a constant-flow liquid feeder, Chem Eng, 14 Dec, 107 (1981).

Srivastava, R.P.S., and Narasimhamurthy, G.S.R., Void fraction and flow pattern during two-phase flow of pseudoplastic fluids, Chem Eng J, 21(2), 165-176 (1981).

Thome, J.R., and Davey, G., Bubble growth rates in liquid nitrogen, argon and their mixtures, Int J Heat Mass Trans, 24(1), 89-98 (1981).

Verhas, J., Viscosity of colloids, Periodica Polytechnica, 25, 53-61 (1981).
Wang, C.C.C., and Charles, M.E., Cocurrent stratified flow of immiscible liquids, Can JCE, 59, 668-676 (1981).

1982-1983

Anderson, P.C.; Veal, C.J., and Withers, V.R., Rheology of coal-oil dispersions, Powder Tech, 32, 45-54 (1982).
Andersson, B., Fluid to particle mass transport in slurries, Chem Eng Sci, 37(1), 93-98 (1982).
Anwar, M.M.; Bright, A.; Das, T.K., and Wilkinson, W.L., Laminar liquid jets in immiscible liquid systems, Trans IChemE, 60, 306-313 (1982).
Baird, D.G., and Labropoulos, A.E., Food dough rheology (review paper), Chem Eng Commns, 15(1), 1-26 (1982).
Conder, J.R.; Gunn, D.J., and Shaikh, M.A., Heat and mass transfer in two-phase flow: Mathematical model for laminar film flow and its experimental validation, Int J Heat Mass Trans, 25(8), 1113-1126 (1982).
Hewitt, G.F., Applications of two-phase flow, Chem Eng Prog, 78(7), 38-46 (1982).
Landis, R.L., Solve flashing-fluid critical-flow problems, Chem Eng, 8 March, 79-82 (1982).
Lane, A.G.C., and Rice, P., The flow characteristics of a submerged bounded jet in a closed system, Trans IChemE, 60, 245-248 (1982).
Lasztity, R., Correlation between the chemical structure and rheological properties of gluten, Periodica Polytechnica, 26, 3-25 (1982).
Lin, S.H., Pressure drop for slurry transport, Chem Eng, 17 May, 115-117 (1982).
Sawinsky, J., and Simandi, B., The residence time distribution for laminar flow of a non-Newtonian fluid in a straight circular tube, Trans IChemE, 60, 188-190 (1982).
Shenoy, A.V., and Saini, D.R., New velocity profile model for turbulent pipe-flow of power-law fluids, Can JCE, 60, 694-697 (1982).
Wall, T.F.; Subramanian, V., and Howley, P., An experimental study of the geometry, mixing and entrainment of particle-laden jets up to ten diameters from the nozzle, Trans IChemE, 60, 231-239 (1982).
Awasthi, R.C., and Vasudeva, K., Mean residence times in flow systems, Chem Eng Sci, 38(2), 313-320 (1983).
Crozier, R.A., Solving incompressible fluid flow problems, Chem Eng, 28 Nov, 57-60 (1983).
Fairhurst, P., Modelling of multiphase flow, Proc Engng, July, 34-36 (1983).
King, R., Tackling fluids handling problems for the process industries, Chem Engnr, Aug, 40-46 (1983).
Kremers, J., Avoid water hammer, Hyd Proc, March, 67-68 (1983).
Liu, T.Y.; Soong, D.S., and Kee, D.D., Model for structured fluids, Chem Eng Commns, 22(5), 273-286 (1983).
Murty, K.N., Assessing temperature effects on incompressible fluid flowrate, Chem Eng, 25 July, 101 (1983).
Roco, M.C., and Shook, C.A., Modeling of slurry flow: Effect of particle size, Can JCE, 61(4), 494-503 (1983).

1984-1985

Clark, N.N., and Flemmer, R.L.C., Vertical downward two-phase flow, Chem Eng Sci, 39(1), 170-173 (1984).

Gonzalez-Velasco, J.R., and Elorriaga, J.B., A nonideal flow experiment, Chem Eng Educ, 18(2), 74-77, 93 (1984).

Khatib, Z., and Richardson, J.F., Vertical co-current flow of air and shear thinning suspensions of kaolin, CER&D, 62, 139-154 (1984).

Serghides, T.K., Estimate friction factor accurately, Chem Eng, 5 March, 63-64 (1984).

Soliman, R., Two-phase pressure drop computed, Hyd Proc, April, 155-157 (1984).

Wene, D.G., Measuring the flow of water-saturated gases, Chem Eng, 2 April, 101-102 (1984).

Alessandrini, A.; Lapasin, R., and Papo, A., A class of rheological models for concentrated suspensions: Application to clay/kaolin aqueous systems, Chem Eng Commns, 37(1), 29-40 (1985).

Clark, N.N., and Flemmer, R.L., Gas-liquid contacting in vertical two-phase flow, Ind Eng Chem Proc Des Dev, 24(2), 231-236 (1985).

Constantinescu, S., Designing the optimum trough (open-channel flow), Chem Eng, 9 Dec, 131 (1985).

Lin, S.H., Preventing cavitation with multistage orifices, Hyd Proc, July, 93-95 (1985).

McNulty, G., Studying multi-phase flow, Proc Engng, Oct, 45,47 (1985).

Nasr-El-Din, H.; Shook, C.A., and Esmail, M.N., Wall sampling in slurry systems, Can JCE, 63(5), 746-753 (1985).

Quemada, D.; Flaud, P., and Jezequel, P.H., Rheological properties and flow of concentrated disperse media, Chem Eng Commns, 32(1), 61-100 (1985).

Sutterby, J.L., Explaining viscosity, Chemtech, July, 416-419 (1985).

Tadros, T.F., Rheology of concentrated suspensions, Chem & Ind, 1 April, 210-218 (1985).

Tandon, T.N.; Varma, H.K., and Gupta, C.P., Void-fraction model for annular two-phase flow, Int J Heat Mass Trans, 28(1), 191-198 (1985).

Wilson, K.C., and Thomas, A.D., New-analysis of the turbulent flow of non-Newtonian fluids, Can JCE, 63(4), 539-546 (1985).

Zigrang, D.J., and Sylvester, N.D., Turbulent-flow friction-factor equations, Chem Eng, 29 April, 86-88 (1985).

1986-1987

Mascone, C.F., Flow injection analysis moves into CPI plants, Chem Eng, 7 July, 14-15 (1986).

Oren, M.J., and MacKay, G.D.M., Rheological and calorific properties of peat-in-oil slurries, Fuel, 65(5), 644-646 (1986).

Palmer, M., Scale modelling of flow problems, Chem Engnr, Jan, 28-30 (1986).

Stein, W.A., Approximate calculation of characteristic shear gradient for aerated non-Newtonian liquids, Chem Eng & Proc, 20(3), 137-146 (1986).

Tsutsumi, A., and Yoshida, K., Rheological behaviour of coal-solvent slurries, Fuel, 65(7), 906-909 (1986).

Yamashiro, C.E.; Espiell, L.G.S., and Farina, I.H., Program determines two-phase flow, Hyd Proc, 65(12), 46-47 (1986).

Andritsos, N., and Hanratty, T.J., Influence of interfacial waves in stratified gas-liquid flows, AIChEJ, 33(3), 444-454 (1987).
Hanzevack, E.L.; Bowers, C.B., and Ju, C.H., Two-phase flow analysis by laser image processing, AIChEJ, 33(12), 2003-2007 (1987).
Knarr, P.S., Equations improve flow measurement, Hyd Proc, 66(1), 58-59 (1987).
Sami, S.M., and Lakis, A.A., Characteristics of turbulent two-phase flow, Chem Eng Commns, 54, 173-210 (1987).
Wace, P.F.; Morrell, M.S., and Woodrow, J., Bubble formation in transverse horizontal liquid flow, Chem Eng Commns, 62(1), 93-106 (1987).

1988

Bruno, K., and McCready, M.J., Origin of roll waves in horizontal gas-liquid flows, AIChEJ, 34(9), 1431-1440 (1988).
Butcher, C., Multiphase pumping, Chem Engnr, Dec, 87-89 (1988).
Carpenter, K.J., Recent developments and new challenges in fluid processing, CER&D, 66(1), 2-4 (1988).
Cieslicki, K., Flows in channels with periodic abrupt expansions, Chem Eng Commns, 64, 67-82 (1988).
Clarke, D., Introduction to water hammer, Chem Engnr, July, 44-45 (1988).
Clarke, D., Water hammer, Chem Engnr, Sept, 34-35 (1988).
El-Genk, M.S., et al., Free convection experiments of atmospheric air in vertical open annuli, Chem Eng Commns, 63, 225-244 (1988).
Ghim, Y.S., and Chang, H.N., Free-surface boundary conditions for analysis of steady film flow, Int Chem Eng, 28(4), 684-690 (1988).
Ghoniem, S.A.A., Effect of deformation sequence on rheological behavior and mechanical degradation of polymer solutions, Chem Eng Commns, 63, 129-142 (1988).
Hanzevack, E.L., et al., Inexpensive experimental system for study of two-phase flow by laser image processing, Chem Eng Commns, 65, 161-168 (1988).
Hasan, A.R., Void fraction in bubbly, slug, and churn flow in vertical two-phase up-flow, Chem Eng Commns, 66, 101-112 (1988).
Heywood, N., Handling of slurries, Proc Engng, Oct, 71-74 (1988).
Irvine, T.F., Generalized Blasius equation for power law fluids, Chem Eng Commns, 65, 39-48 (1988).
Kimoto, H., Method of skin friction determination of circular cylinder in water flow field, Chem Eng Commns, 74, 63-72 (1988).
Kulshreshtha, A.K., and Caruthers, J.M., Unsteady motion of a sphere in Newtonian fluid, Chem Eng Commns, 72, 1-24 (1988).
Leung, J.C., and Epstein, M., Generalized critical flow model for nonideal gases, AIChEJ, 34(9), 1568-1572 (1988).
Leung, J.C., and Grolmes, M.A., Generalized correlation for flashing choked flow of initially subcooled liquid, AIChEJ, 34(4), 688-691 (1988).
Reza, J., and Martin, H., Analytical and numerical studies of separated laminar two-phase flow in elliptical ducts of arbitrary axis ratio, Chem Eng & Proc, 24(3), 121-132 (1988).
Wang, M.L.; Chang, R.Y., and Chung, I.Y., Analysis of non-Newtonian fluid flowing around two horizontal cylinders in tandem arrangement by finite element methods, Chem Eng Commns, 70, 19-38 (1988).

Zabaras, G.J., and Dukler, A.E., Countercurrent gas-liquid annular flow, including the flooding state, AIChEJ, 34(3), 389-396 (1988).

15.2 Pipe Flow

1967-1969

Caplan, F., Find pressure drop of air in steel pipe from nomograph, Chem Eng, 31 July, 148 (1967).
Chisholm, D., Pressure gradients during the flow of incompressible two-phase mixtures through pipes, venturis and orifice plates, Brit Chem Eng, 12(9), 1368-1371 (1967).
Harris, J., Turbulent flow of non-Newtonian fluids through round tubes, Chem Engnr, Nov, CE243-246 (1967).
Michiyoshi, I.; Shirataki, K., and Takitani, K., The steam volume fraction in two-phase flow, Int Chem Eng, 7(1), 159-167 (1967).
Roberts, R.N., Pipelines for process slurries, Chem Eng, 31 July, 125-130 (1967).
Solveev, A.V.; Preobrazhenskii, E.I., and Semenov, P.A., Hydraulic resistance in two-phase flow, Int Chem Eng, 7(1), 59-63 (1967).
Giraldo, J., Generalized method for vapor-phase pressure drop in pipes, Hyd Proc, 47(11), 219-222 (1968).
Hamilton, W., and MacDonald, B.B., Flow of fluids in pipes, Brit Chem Eng, 13(8), 1143-1144 (1968).
Robertson, J.M.; Martin, J.D., and Burkhart, T.H., Turbulent flow in rough pipes, Ind Eng Chem Fund, 7(2), 253-265 (1968).
Serwinski, M., et al., Pressure drop of pulp suspensions in pipelines, Int Chem Eng, 8(3), 453-462 (1968).
Cramer, S.D., and Marchello, J.M., Design procedure for laminar, isothermal, non-Newtonian flow in pipes and annuli, Ind Eng Chem Proc Des Dev, 8(3), 293-298 (1969).
Kern, R., Sizing process piping for two-phase flow, Hyd Proc, 48(10), 105-116 (1969).
Narasimhamurty, G.S.R., and Prasad, S.S., Effect of turbulence-promoters in two-phase gas-liquid flow in horizontal pipes, Chem Eng Sci, 24(2), 331-342 (1969).
Templeton, L., How to convert flow data to various pipe schedules, Chem Eng, 16 June, 118-120 (1969).

1970-1971

Alves, G.E., Cocurrent liquid-gas pipeline contactors, Chem Eng Prog, 66(7), 60-67 (1970).
Srinivasan, P.S.; Nandapurkar, S.S., and Holland, F.A., Friction factors for coils, Trans IChemE, 48, T156-161 (1970).
White, D.A., Correlation of pressure drop data in pipe flow of dilute polymer solutions, Chem Eng Sci, 25(7), 1127-1132 (1970).
Anaya, A., and Garritz, A., Design sonic flowchart for compressible fluids in pipes, Chem Eng, 1 Nov, 98-100 (1971).

Aude, T.C.; Cowper, N.T.; Thompson, T.L., and Wasp, E.J., Slurry piping systems: Trends, design methods, guidelines, Chem Eng, 28 June, 74-90 (1971).

Bonnecaze, R.H.; Erskine, W., and Greskovich, E.J., Holdup and pressure drop for two-phase slug flow in inclined pipelines, AIChEJ, 17(5), 1109-1113 (1971).

Castro, W.E., and Neuwirth, J.G., Polymer additions reduce fluid flow friction, Chemtech, Nov, 697-701 (1971).

Edwards, M.F., and Wilkinson, W.L., Review of potential applications of pulsating flow in pipes, Trans IChemE, 49, 85-94 (1971).

Greskovich, E.J., and Shrier, A.L., Pressure drop and holdup in horizontal slug flow, AIChEJ, 17(5), 1214-1219 (1971).

Hasson, D., Friction factor of coils, Chem Engnr, March, 105-106 (1971).

Ibragimov, M.K.; Subbotin, V.I., and Taranov, G.S., Velocity and temperature fluctuations and correlations for turbulent air flow in pipes, Int Chem Eng, 11(4), 659-665 (1971).

Shaheen, E.I., Rheological study of viscosities and pipeline flow of concentrated slurries, Powder Tech, 5, 245-256 (1971).

Turian, R.M.; Yuan, T.F., and Mauri, G., Pressure drop correlation for pipeline flow of solid-liquid suspensions, AIChEJ, 17(4), 809-817 (1971).

Wasp, E.J.; Thompson, T.L., and Snoek, P.E., Design of slurry pipelines, Chemtech, Sept, 552-562 (1971).

1972-1974

Chimes, A.R., Fast way to choose pipe diameters, Chem Eng, 27 Nov, 114 (1972).

Cowper, N.T.; Thompson, T.L.; Aude, T.C., and Wasp, E.J., Processing steps: Keys to successful slurry-pipeline systems, Chem Eng, 7 Feb, 58-67 (1972).

Edwards, M.F.; Nellist, D.A., and Wilkinson, W.L., Pulsating flow of non-Newtonian fluids in pipes, Chem Eng Sci, 27(3), 545-554 (1972).

Kern, R., How discharge piping affects pump performance, Hyd Proc, 51(3), 89-93 (1972).

Soo, S.L., and Tung, S.K., Deposition and entrainment in pipe flow of a suspension, Powder Tech, 6, 283-294 (1972).

Bending, M.J., and Hutchison, H.P., Calculation of steady-state incompressible flow in large networks of pipes, Chem Eng Sci, 28(10), 1857-1864 (1973).

Bertela, M.; Santarelli, F., and Foraboschi, F.P., Some experimental results on non-isothermal laminar flow in vertical tubes, Trans IChemE, 51, 168-171 (1973).

Ivanov, A.P., Turbulent friction for fluid flow in pipes and channels, Int Chem Eng, 13(2), 327-332 (1973).

Srivastava, R.P.S., and Naraisimhamurty, G.S.R., Hydrodynamics of non-Newtonian two-phase flow in pipes, Chem Eng Sci, 28(2), 553-558 (1973).

Burnykh, V.S., Excess pressure losses in a gas pipeline on reduction of hydraulic efficiency coefficient, Int Chem Eng, 14(4), 723-725 (1974).

Guzhov, A.I.; Medvedev, V.F., and Savelev, V.A., Movement of gas-water-oil mixtures through pipelines, Int Chem Eng, 14(4), 713-715 (1974).

Tsaturyan, S.I., and Markelov, S.S., Unsteady-state gas flow in pipelines, Int Chem Eng, 14(4), 719-723 (1974).

Zanker, A., Nomograph for calculation of transport velocities in solids/water suspensions, Chem Engnr, Nov, 727-728 (1974).

1975-1977

Cheng, D.C.H., Pipeline design for non-Newtonian fluids, Chem Engnr, Sept, 525-528, 532; Oct, 587-588, 595 (1975).

Evgenev, A.E.; Ivannikov, V.G., and Rozenberg, G.D., Determination of equivalent roughness of piping, Int Chem Eng, 15(3), 429-431 (1975).

Greskovich, E.J., and Cooper, W.T., Correlation and prediction of gas-liquid holdups in inclined upflows, AIChEJ, 21(6), 1189-1192 (1975).

Joseph, B.; Smith, E.P., and Adler, R.J., Numerical treatment of laminar flow in helically coiled tubes of square cross section, AIChEJ, 21(5), 965-979 (1975).

Larson, A.M., Streamlining head-loss calculations, Chem Eng, 27 Oct, 115-118 (1975).

Oliver, D.R., and Asghar, S.M., The laminar flow of Newtonian and viscoelastic liquids in helical coils, Trans IChemE, 53, 181-186 (1975).

Caplan, F., Flow through rectangular channels, Chem Eng, 22 Nov, 181-182 (1976).

Horwitz, B., Quick method of manifold sizing, Chem Eng, 10 May, 142 (1976).

Kopalinsky, E.M., and Bryant, R.A.A., Friction coefficients for bubbly two-phase flow in horizontal pipes, AIChEJ, 22(1), 82-86 (1976).

Churchill, S.W., Friction-factor equation spans all fluid-flow regimes, Chem Eng, 7 Nov, 91-92 (1977).

Levenspiel, O., Discharge of gases from a reservoir through a pipe, AIChEJ, 23(4), 402-403 (1977).

Mashelkar, R.A., and Devarajan, G.V., Secondary flows of non-Newtonian fluids in coiled tubes, Trans IChemE, 54, 100-114 (1976); 55, 29-37 (1977).

McFetridge, R.H., and Klinzing, G.E., Experimentally determined effect of artificially roughened surfaces on hydraulic loss coefficients, Ind Eng Chem Proc Des Dev, 16(2), 176-180 (1977).

Nauman, E.B., Residence time distribution for laminar flow in helically coiled tubes, Chem Eng Sci, 32(3), 287-294 (1977).

Shilimkan, R.V., and Stepanek, J.B., Interfacial area in cocurrent gas-liquid upward flow in tubes of various size, Chem Eng Sci, 32(2), 149-154 (1977).

Turian, R.M., and Yuan, T.F., Flow of slurries in pipelines, AIChEJ, 23(3), 232-243 (1977).

1978-1979

Branan, C.R., Estimating pressure drop for turbulent flow in steel pipes, Chem Eng, 28 Aug, 126 (1978).

Ganapathy, V., Estimating air and flue-gas velocities, Chem Eng, 5 June, 169 (1978).

Kubie, J., and Oates, H.S., Aspects of two-phase frictional pressure drop in tubes, Trans IChemE, 56, 205-209 (1978).

Mujawar, B.A., and Rao, M.R., Flow of non-Newtonian fluids through helical coils, Ind Eng Chem Proc Des Dev, 17(1), 22-27 (1978).
Nguyen, H.X., Simplify calculation of economic pipe size, Hyd Proc, 57(2), 143-144 (1978).
Taitel, Y.; Lee, N., and Dukler, A.E., Transient gas-liquid flow in horizontal pipes, AIChEJ, 24(5), 920-934 (1978).
Blass, E., Gas/film flow in tubes, Int Chem Eng, 19(2), 183-196 (1979).
Heywood, N.I., and Richardson, J.F., Slug flow of air-water mixtures in a horizontal pipe, Chem Eng Sci, 34(1), 17-30 (1979).
Jayaraman, K., Estimating pressure drop in gas ducts, Chem Eng, 9 April, 136-138 (1979).
Mishra, P., and Gupta, S.N., Momentum transfer in curved pipes, Ind Eng Chem Proc Des Dev, 18(1), 130-142, (1979).
Pham, Q.T., Explicit equations for the solution of turbulent pipe-flow problems, Trans IChemE, 57, 281-283 (1979).
Shah, G.C., Pipeline pressure drop: A Cv approach, Chem Eng, 30 July, 99 (1979).
Verma, C.P., Solve pipe flow problems directly, Hyd Proc, 58(8), 122-124 (1979).
Zisselmar, R., and Molerus, O., Investigation of solid-liquid pipe flow for turbulence modification, Chem Eng J, 18(3), 233-240 (1979).

1980-1981

Alshamani, K.M.M., A study of turbulent flow in ducts, Chem Eng J, 20(1), 7-20 (1980).
Balasubramanian, G.R., et al., Pressure drop for gas-liquid flow, Chem Eng, 2 June, 101-102 (1980).
Chen, N.H., An explicit equation for friction factor in pipes, Ind Eng Chem Fund, 18(2), 296-303 (1979); 19(2), 228-230 (1980).
Farooqi, S.I.; Heywood, N.I., and Richardson, J.F., Drag reduction by air injection for suspension flow in a horizontal pipeline, Trans IChemE, 58, 16-27 (1980).
Friedel, L., Pressure drop during gas/vapor-liquid flow in pipes, Int Chem Eng, 20(3), 352-368 (1980).
Irasga, M.A., Calculate the friction factor, Chem Eng, 21 April, 129-130 (1980).
Meyer, J.M., Versatile program for pressure-drop calculations, Chem Eng, 10 March, 139-142 (1980).
Moore, R.G.; Bishnoi, P.R., and Donnelly, J.K., Rigorous design of high pressure natural gas pipelines using BWR equation of state, Can JCE, 58, 103-112 (1980).
Oroskar, A.R., and Turian, R.M., Critical velocity in pipeline flow of slurries, AIChEJ, 26(4), 550-558 (1980).
Thorley, A.R.D., Surge suppression and control in slurry pipelines, Chem Engnr, April, 222-224 (1980).
Various, Piping design (special report), Hyd Proc, 59(12), 87-96 (1980).
Blackwell, W.W., Calculating two-phase pressure drop, Chem Eng, 7 Sept, 121-125 (1981).
Frankel, I., Figuring fluegas Reynolds number, Chem Eng, 24 Aug, 132 (1981).

Hooper, W.B., The two-K method predicts head losses in pipe fittings, Chem Eng, 24 Aug, 96-100 (1981).

Lang, F.D., and Miller, B.L., Use of friction-factor correlation in pipe-network problems, Chem Eng, 29 June, 95-97 (1981).

Mujawar, B.A., and Rao, M.R., Gas/non-Newtonian liquid two-phase flow in helical coils, Ind Eng Chem Proc Des Dev, 20(2), 391-397 (1981).

Olujic, Z., Compute friction factors fast for flow in pipes, Chem Eng, 14 Dec, 91-93 (1981).

Rajaram, S., Quick estimate of steam-system pressure drop, Chem Eng, 9 Feb, 130 (1981).

Van de Sande, E.; Sloof, R., Hiemstra, W., Turbulent velocity profiles in a straight circular pipe at high temperatures, Trans IChemE, 59, 283-285 (1981).

1982-1983

Blackwell, W.W., Sizing condensate (2 phase-flow) return lines, Chem Eng, 12 July, 105-108 (1982).

Brunner, C.R., Program predicts pressure drop for steam flow, Chem Eng, 22 Feb, 97-99 (1982).

Durand, A.A., Polymer additives cut pipe friction, Chem Eng, 20 Sept, 130 (1982).

El Telbany, M.M.M., and Reynolds, A.J., Empirical description of turbulent channel flows, Int J Heat Mass Trans, 25(1), 77-86 (1982).

Farooqi, S.I., and Richardson, J.F., Horizontal flow of air and liquid in a smooth pipe, Trans IChemE, 60, 292-305, 323-333 (1982).

Kurmarao, P.S.V., Steam flow through critical nozzles, Chem Eng, 8 Feb, 132 (1982).

Rao, K.V.K., Head losses in fittings, Chem Eng, 8 Feb, 132-133 (1982).

Shook, C.A., Flow of stratified slurries through horizontal venturi meters, Can JCE, 60, 342-345 (1982).

Wilson, K.C., and Brown, N.P., Analysis of fluid friction in dense-phase pipeline flow, Can JCE, 60, 83-86 (1982).

Azzopardi, B.J., and Gibbons, D.B., Annular two-phase flow in large diameter tube, Chem Engnr, Dec, 19,21,30,31 (1983).

Burfoot, D., and Rice, P., Heat transfer and pressure drop characteristics of short lengths of swirl flow inducers interspaced along a circular duct, CER&D, 61, 253-258 (1983).

Chhabra, R.P.; Farooqi, S.I.,; Richardson, J.F., and Wardle, A.P., Co-current flow of air and shear thinning suspensions in pipes of large diameter, CER&D, 61, 56-61 (1983).

Delhaye, J.M., Two-phase pipe flow, Int Chem Eng, 23(3), 385-411 (1983).

Liu, T.J., Fully developed flow of power-law fluids in ducts, Ind Eng Chem Fund, 22(2), 183-186 (1983).

Pigford, R.L.; Ashraf, M., and Miron, Y.D., Flow distribution in piping manifolds, Ind Eng Chem Fund, 22(4), 463-471 (1983).

Rajaram, S., Quick estimate of air flow in ducts, Chem Eng, 2 May, 88-90 (1983).

Sargent, J.B., Check liquid-line size by mental arithmetic, Chem Eng, 17 Oct, 82'(1983).

Verba, A.; Embaby, M.H., and Angyal, I., Experimental measurements on non-Newtonian and drag reduction flows in pipes, Periodica Polytechnica, 27, 57-70 (1983).

1984-1985

Bradford, M.L., Tables simplify pressure-drop calculations, Chem Eng, 25 June, 137-140 (1984).

Burfoot, D., and Rice, P., Heat transfer and pressure drop characteristics of alternate rotation swirl-flow inducers in a circular duct, CER&D, 62, 128-132 (1984).

Chhabra, R.P., and Richardson, J.F., Prediction of flow pattern for the cocurrent flow of gas and non-Newtonian liquid in horizontal pipes, Can JCE, 62(4), 449-454 (1984).

Chhabra, R.P.; Farooqi, S.I., and Richardson, J.F., Isothermal two-phase flow of air and aqueous polymer solutions in a smooth horizontal pipe, CER&D, 62, 22-32, 398-400 (1984).

Fastenakels, M., and Campana, H., Find optimum pipe size, Hyd Proc, Sept, 163-165 (1984).

Hallett, W.L.H., and Gunther, R., Flow and mixing in swirling flow in a sudden expansion, Can JCE, 62(1), 149-155 (1984).

Maten, S., Field criteria for pipe vibration, Hyd Proc, July, 107-108 (1984).

Patel, V.K., Friction loss in piping, Hyd Proc, Jan, 100-102 (1984).

Roco, M.C., and Shook, C.A., Computational method for coal slurry pipelines with heterogeneous size distribution, Powder Tech, 39, 159-176 (1984).

Shipley, D.G., Two-phase flow in large diameter pipes, Chem Eng Sci, 39(1), 163-165 (1984).

Troniewski, L., and Ulbrich, R., Two-phase gas-liquid flow in rectangular channels, Chem Eng Sci, 39(4), 751-766 (1984).

Troniewski, L., and Ulbrich, R., Analysis of flow regime maps of two-phase gas-liquid flow in pipes, Chem Eng Sci, 39(7), 1213-1224 (1984).

Chen, J.J.J., Predict gas/liquid-flow pressure drop, Chem Eng, 22 July, 83-84 (1985).

Edwards, M.F.; Jadallah, M.S.M., and Smith, R., Head losses in pipe fittings at low Reynolds numbers, CER&D, 63, 43-50 (1985).

Grundmann, R., Friction diagram of helically coiled tube, Chem Eng & Proc, 19(2), 113-116 (1985).

Gyori, I., Calculator program for compressible flow in pipes, Chem Eng, 28 Oct, 55-60 (1985).

Ippolito, M., and Sabatino, C., Velocity profiles of non-Newtonian suspensions in smooth pipes, Chem Eng Commns, 39, 127-146 (1985).

Olujic, Z., Predicting two-phase-flow friction loss in horizontal pipes, Chem Eng, 24 June, 45-50 (1985).

Watts, P., Vent system design on or offshore, Chem Engnr, Jan, 34-35 (1985).

1986-1987

Gnielinski, V., Correlations for pressure drop in helically coiled tubes, Int Chem Eng, 26(1), 36-45 (1986).

Kapoor, B.S.; Garde, R.J., and Raju, K.G.R., Discharge characteristics of orifice meters in sediment-laden flows, Can JCE, 64(1), 36-41 (1986).

Muller, H., and Heck, K., Simple friction pressure drop correlation for two-phase flow in pipes, Chem Eng & Proc, 20(6), 297-308 (1986).
Nokay, R., Quick determination of resistance coefficients for reducers and enlargers, Chem Eng, 4 Aug, 91-92 (1986).
Rao, K.V.K., Determine laminar-flow head losses for fittings, Chem Eng, 3 Feb, 108-109 (1986).
Brown, G.S., How to predict pressure drop before designing the piping, Chem Eng, 16 March, 85-86 (1987).
Cindric, D.T.; Gandhi, S.L. and Williams, R.A., Designing piping systems for two-phase flow, Chem Eng Prog, 83(3), 51-55 (1987).
Ebner, L., et al., Characterization of hydrodynamic regimes in horizontal two-phase flow, Chem Eng & Proc, 22(1), 39-52 (1987).
Ellul, I.R., and Issa, R.I., Prediction of the flow of interspersed gas and liquid phases through pipe bends, CER&D, 65(1), 84-96 (1987).
Kale, D.D., Drag reduction in two-phase gas-liquid flows, AIChEJ, 33(2), 351-352 (1987).
Katsaounis, A., Flow pattern and pressure drop in tees, Chem Eng Commns, 54, 119-138 (1987).
Kowalski, J.E., Wall and interfacial shear stress in stratified flow in horizontal pipe, AIChEJ, 33(2), 274-281 (1987).
Mavridis, H.; Hyrmak, A.N., and Vlachopoulos, T., Finite-element simulation of stratified multiphase flows, AIChEJ, 33(3), 410-422 (1987).
Turian, R.M.; Hsu, F.L., and Ma, T.W., Estimation of critical velocity in pipeline flow of slurries, Powder Tech, 51(1), 35-48 (1987).

1988

Hooper, W.B., Calculate head loss caused by change in pipe size, Chem Eng, 7 Nov, 89-92 (1988).
MacKay, M.E.; Yeow, Y.L., and Boger, D.V., Pressure drop in pipe contractions: Experimental measurement vs. finite element simulation, CER&D, 66(1), 22-25 (1988).
Ogawa, K.; Kuroda, C., and Yoshikawa, S., Shape of turbulent lump in circular pipe flow determined by spatial-dependence matrix, Chem Eng Commns, 66, 113-124 (1988).
Salman, A.D., New approach to calculation of variation of mixing length over the pipe diameter, Periodica Polytechnica, 32(4), 269-276 (1988).
Sultan, A.A., Sizing pipe for non-Newtonian flow, Chem Eng, 19 Dec, 140-146 (1988).
Various, Piping and valves (feature report), Hyd Proc, 67(8), 37-47 (1988).
Vigneaux, P.; Chenais, P., and Hulin, J.P., Liquid-liquid flows in inclined pipe, AIChEJ, 34(5), 781-789 (1988).
Vrentas, J.S., and Vrentas, C.M., Dispersion in laminar tube flow at low Peclet numbers or short times, AIChEJ, 34(9), 1423-1430 (1988).
Waliullah, S., Design of vortex breakers, Chem Eng, 9 May, 108-109 (1988).

15.3 Pumps and Compressors

1966-1969

Holland, F.A., and Chapman, F.S., Positive-displacement pumps, Chem Eng, 14 Feb, 129-152 (1966).

Magliozzi, T.L., Control system prevents surging in centrifugal-flow compressors, Chem Eng, 8 May, 139-142 (1967).

Montgomery, R., Centrifugal pump troubleshooting, Chem Eng, 24 April, 182-184 (1967).

Newman, E.F., How to specify steam-jet ejectors, Chem Eng, 10 April, 203-208 (1967).

Paige, P.M., Shortcuts to optimum-size compressor piping, Chem Eng, 13 March, 168-172 (1967).

Quigley, H.A., Pump horsepower calculations made easy, Chem Eng, 8 May, 184 (1967).

Chodnowsky, N.M., Centrifugal compressors for high-pressure service, Chem Eng, 2 Dec, 110-116 (1968).

Hernandez, L.A., Controlled-volume pumps, Chem Eng, 21 Oct, 124-136 (1968).

Morrow, D.R., Compressor performance curves, Chem Eng Prog, 64(12), 56-61 (1968).

Scheel, L.F., New ideas on centrifugal compressors, Hyd Proc, 47(9), 253-260; 47(10), 161-165; 47(11), 238-242 (1968).

Vetter, G., and Bohm, O., The design characteristics and uses of glandless metering pumps, Brit Chem Eng, 13(9), 1280-1286 (1968).

Whillier, A., Pump efficiency determination in chemical plant from temperature measurements, Ind Eng Chem Proc Des Dev, 7(2), 194-196 (1968).

Alack, C.S., Pumping abrasive slurries, Chem Eng, 15 Dec, 146-148 (1969).

Baird, M.H.I., Graph for calculating pump output with bypass, Chem Eng, 8 Sept, 162 (1969).

Carter, G., The application of helical gear pumps to the handling of non-Newtonian fluids, Chem Engnr, Jan, CE12-16 (1969).

Evghenide, C., Effect of viscosity on characteristics of centrifugal pumps, Int Chem Eng, 9(1), 69-74 (1969).

Gibbs, B.H., Development of centrifugal pumps for industry, Brit Chem Eng, 14(1), 61-66 (1969).

Scheel, L.F., New ideas on fan selection, Hyd Proc, 48(6), 125-130 (1969).

1970-1971

Bresler, S.A., Guide to trouble-free compressors, Chem Eng, 1 June, 161-170 (1970).

Franzke, A., Benefits of energy-recovery turbines, Chem Eng, 23 Feb, 109-112 (1970).

Hattiangadi, U.S., Diagram to speed up pump specification, Chem Eng, 13 July, 114-116 (1970).

Hattiangadi, U.S., Specifying centrifugal and reciprocating pumps, Chem Eng, 23 Feb, 101-108 (1970).

Lady, E.R., Compressor efficiency, Chem Eng, 10 Aug, 113-115 (1970).

Nailen, R.L., Guide to trouble-free electric-motor drives, Chem Eng, 1 June, 181-186 (1970).
Payne, D.C., Reliable viscous output with positive-displacement pumps, Chem Eng, 23 Feb, 128-132 (1970).
Scheel, L.F., A guide to gas expanders, Hyd Proc, 49(2), 105-111 (1970).
Various, Design of pumping systems (topic issue), Chem Eng Prog, 66(5), 43-72 (1970).
Walkden, A.J., and Eveleigh, B.D., Experimental and theoretical study of reciprocating-jet pump using pneumo-hydraulic drive, Trans IChemE, 48, T121-128 (1970).
Cooke, E.F., Selecting adjustable-speed drives, Chem Eng, 6 Sept, 70-82 (1971).
Farrow, J.F., User guide to steam turbines, Hyd Proc, 50(3), 71-75 (1971).
Green, C.F.A., Liquid-ring vacuum pumps, Brit Chem Eng, 16(1), 37 (1971).
Kusay, R.G.P., Vacuum equipment for chemical processes, Brit Chem Eng, 16(1), 29-35 (1971).
Moens, J.P.C., Adapting the process to suit the centrifugal compressor, Hyd Proc, 50(12), 96-100 (1971).
Nuttall, L.C., Recent developments in centrifugal pumps, Chem Engnr, March, 121-129 (1971).
Paul, K.S., Selecting a mechanical vacuum pump, Brit Chem Eng, 16(2), 202-204 (1971).
Stafford, J.D., High-pressure centrifugal compressor loop, Chem Eng Prog, 67(4), 54-58 (1971).
Various, Specification of centrifugal compressors (special report), Hyd Proc, 50(10), 69-88 (1971).
Various, Pump and valve selector (deskbook issue), Chem Eng, 11 Oct, 29-208 (1971).

1972-1973

Ibragimov, G.Z., and Khisamutdinov, N.I., Conditions determining stalling of a submerged centrifugal pump, Int Chem Eng, 12(2), 312-314 (1972).
Kern, R., How to size pump suction piping, Hyd Proc, 51(4), 119-120 (1972).
MacDonald, J.O.S., Economising compressor installations, Brit Chem Eng, 17(5), 431-437; 17(7), 647 (1972).
Mehta, D.D., Design of compressors and converters, Hyd Proc, 51(6), 129-135 (1972).
Rabb, A., Selection of gas and steam turbines, Proc Tech Int, 17(12), 965-967 (1972).
Sharipov, A.G., et al., Method of determining intake pressure of a submerged centrifugal pump, Int Chem Eng, 12(2), 306-309 (1972).
Various, Specifying process pumps (special report), Hyd Proc, 51(6), 85-106 (1972).
Various, Pumps and unit operations review, Chem Engnr, Sept, 338-359 (1972).
Vetter, G., and Fritsch, H., Cavitation experiments on reciprocating positive-displacement pumps, Brit Chem Eng, 17(2), 133-137 (1972).
White, M.H., Surge control for centrifugal compressors, Chem Eng, 25 Dec, 54-62 (1972).

Abraham, R.W., Reliability of rotating equipment, Chem Eng, 15 Oct, 96-108 (1973).

Baker, C.D., Predicting pump flows from system characteristics, Chem Eng, 26 Nov, 102 (1973).

Barker, M.L., and Simper, J.I., A unified nomenclature for jet pumps, Chem Engnr, April, 195, 207 (1973).

Boyd, O.W., A single graph for gas-compression horsepower, Chem Eng, 6 Aug, 110 (1973).

Cavaliere, G.F., and Gyepes, R.A., Cost evaluation of intercooler systems for air compressors, Hyd Proc, 52(10), 107-110 (1973).

Doolin, J.H., Updating standards for chemical pumps, Chem Eng, 11 June, 117-119 (1973).

Gatehouse, L.W., Pumping corrosive and erosive fluids: A manufacturer's viewpoint, Chem Engnr, July, 370-376 (1973).

Kern, R., Nomographs to quickly size pump piping and components, Hyd Proc, 52(3), 81-86 (1973).

Pollak, R., Selecting fans and blowers, Chem Eng, 22 Jan, 86-100 (1973).

1974-1975

Birk, J.R., and Peacock, J.H., Pump requirements, Chem Eng, 18 Feb, 116-124 (1974).

Brown, R.N., Testing compressor performance, Hyd Proc, 53(4), 133-140 (1974).

Li, K.W., Pumping viscous liquids to recycled operating units, Chem Eng, 30 Sept, 152 (1974).

Miller, J.E., Pumping abrasive slurries: The Miller number, Chem Eng, 22 July, 103-106 (1974).

Neerken, R.F., Pump selection, Chem Eng, 18 Feb, 104-115 (1974).

Van Blarcom, P.P., Bypass systems for centrifugal pumps, Chem Eng, 4 Feb, 94-98 (1974).

Various, Pumps and pumping (feature report), Chemical Processing, May, 59-76 (1974).

Yedidiah, S., Improving pump performance, Hyd Proc, 53(4), 165-167 (1974).

Caplan, F., Estimating minimum required flows through pumps, Chem Eng, 17 March, 84 (1975).

Field, A.A., Pumps in heating and cooling systems, Energy World, June, 13-14 (1975).

Haigh, J.C., Air operated pumps for handling special chemicals, Chem Engnr, Jan, 32-34 (1975).

Lapina, R.P., Can you rerate your centrifugal compressor? Chem Eng, 20 Jan, 95-98 (1975).

Moore, J.C., Electric motor drivers for centrifugal compressors, Hyd Proc, 54(5), 133-139 (1975).

Neerken, R.F., Guide to compressor selection, Chem Eng, 20 Jan, 78-94 (1975).

Nisenfeld, A.E.; Miyasaki, R.; Liem, T., and Eskes, J.M., Multistage compressor control, Hyd Proc, 54(4), 153-156 (1975).

Sisson, W., Easy way to get compression temperatures, Chem Eng, 9 June, 104 (1975).

Sparks, B.E., Gas turbine pumps for difficult fuels, Chem Engnr, Jan, 40-41 (1975).

Trent, A.W., Metering with gear pumps, Chem Eng, 20 Jan, 107-111 (1975).

Various, Pumps and pumping (feature report), Processing, March, 37-59 (1975).

Various, Centrifugal compressors and blowers (special report), Hyd Proc, 54(6), 57-78 (1975).

Winstanley, J.A., Canned pumps reduce costs, Chem Engnr, Jan, 38-39 (1975).

1976-1977

Anderson, L.M., New steam-extraction control system for steam turbines, Hyd Proc, 55(2), 111-113 (1976).

Bloch, H.P., Use keyless couplings for large compressor shafts, Hyd Proc, 55(4), 181-186 (1976).

Cohen, H., and Wolf, D., Pumping gases at low rates of flow, Chem Eng, 16 Feb, 120-122 (1976).

De Santis, G.J., How to select a centrifugal pump, Chem Eng, 22 Nov, 163-168 (1976).

Huff, G.A., Selecting a vacuum system, Chem Eng, 15 March, 83-86 (1976).

James, R., Pump maintenance, Chem Eng Prog, 72(2), 35-40 (1976).

McKelvey, J.M.; Maire, U., and Haupt, F., Performance of gear pumps and screw pumps in polymer-processing applications, Chem Eng, 27 Sept, 94-102 (1976).

Patton, P.W., and Joyce, C.F., Selecting the lowest-cost vacuum system, Chem Eng, 2 Feb, 84-88 (1976).

Sayyed, S., How compressor components affect performance, Hyd Proc, 55(9), 273-277 (1976).

Sayyed, S., Aerodynamics for compressor performance, Hyd Proc, 55(12), 125-128 (1976).

Various, Pumps and valves (feature report), Processing, Nov. 37-73 (1976).

Buse, F., Effects of dimensional variations on centrifugal pumps, Chem Eng, 26 Sept, 93-100 (1977).

Doolin, J.H., Select pumps to cut energy costs, Chem Eng, 17 Jan, 137-139 (1977).

Karrasik, I.J., Tomorrow's centrifugal pump, Hyd Proc, 56(9), 247-251 (1977).

Margus, E., Plastic centrifugal pumps for corrosive service, Chem Eng, 28 Feb, 213-215 (1977).

Various, Pumps and valves (feature report), Processing, Nov, 53-68 (1977).

Various, Prime movers (feature report), Processing, Jan, 33-41 (1977).

Various, Turbomachinery (special report), Hyd Proc, 56(12), 77-95 (1977).

Walko, J., Nomograph for pumping costs, Proc Engng, June, 40-42 (1977).

Yedidiah, S., Diagnosing troubles of centrifugal pumps, Chem Eng, 24 Oct, 124-128; 21 Nov, 193-199; 5 Dec, 141-143 (1977).

1978-1979

Dimoplon, W., A guide to compressors, Hyd Proc, 57(5), 221-227 (1978).

Hundy, G.F., The development of screw compressors for refrigeration duties, Chem Engnr, May, 375-377 (1978).

Jackson, C., Install compressors for high availability, Hyd Proc, 57(11), 243-249 (1978).

Jeelani, S.A.K.; Rajkumar, A., and Rao, K.V.K., Designing air-jet ejectors, Chem Eng, 25 Sept, 135-136 (1978).
Penney, W.R., Inert gas in liquids reduces pump performance, Chem Eng, 3 July, 63-68 (1978).
Richter, S.H., Size relief systems for two-phase flow, Hyd Proc, 57(7), 145-152 (1978).
Sayyed, S., Compressor shop tests, Hyd Proc, 57(4), 201-206 (1978).
Shah, G.C., Inert gas injection reduces pump noise, Chem Eng, 3 July, 93 (1978).
Simmons, P.E., Turbomachinery: New acceptance criteria proposed, Hyd Proc, 57(1), 169-171 (1978).
Various, Electric motor drives (topic issue), Proc Engng, Feb, 34-43 (1978).
Various, Centrifugal pumps (special report), Hyd Proc, 57(6), 93-108 (1978).
Willoughby, W.W., Steam rate: The key to turbine selection, Chem Eng, 11 Sept, 146-154 (1978).
Winters, R.G., Lubricating air compressors, Chem Eng, 14 Aug, 157-160 (1978).
Grohmann, M., Extend pump performance with inducers, Hyd Proc, 58(12), 121-124 (1979).
Henshaw, T.L., Improve power pump suction systems, Hyd Proc, 58(10), 161-162 (1979).
Jeelani, S.A.K., et al., Designing steam-jet ejectors, Chem Eng, 9 April, 135-136 (1979).
Lightle, J., and Hohman, J., Keep pumps operating efficiently, Hyd Proc, 58(9), 227-229 (1979).
Odrowaz-Pieniazek, S., Solids handling pumps: A guide to selection, Chem Engnr, Feb, 94-101 (1979).
Poynton, J.P., Basics of reciprocating metering pumps, Chem Eng, 21 May, 156-165 (1979).
Reed, R.D., Shortcut method for sizing air blowers, Hyd Proc, 58(6), 147-148 (1979).
Smith, A.P., Unexplained wear in large centrifugal pumps, Chem Eng, 13 Aug, 153-155 (1979).
Staroselsky, N., and Ladin, L., Improved surge control for centrifugal compressors, Chem Eng, 21 May, 175-184 (1979).
Van Ormer, H., Include power costs in air compressor selection, Hyd Proc, 58(9), 219-221 (1979).
Webster, G.R., The canned pump in the petrochemical environment, Chem Engnr, Feb, 91-93 (1979).
Wong, G.S.; Davis, D.E., and Gilman, H.H., Development of a high-pressure centrifugal slurry pump, Chem Eng Prog, 75(12), 58-65 (1979).

1980-1981

Blackwell, W.W., Rapid calculation of centrifugal-pump hydraulics, Chem Eng, 14 Jan, 111-115 (1980).
Budris, A.R., Preventing cavitation in rotary gear pumps, Chem Eng, 5 May, 109-112 (1980).
Heller, H., Running gear pumps in reverse, Chem Eng, 28 July, 95 (1980).

Krienberg, M., Pump selection for fuel oil tank farms, Chem Eng, 22 Sept, 179 (1980).
Monroe, R.C., Consider variable-pitch fans, Hyd Proc, 59(12), 122-128 (1980).
Natarajan, V.K., Venting ejectors for centrifugal compressors, Chem Eng, 30 June, 142 (1980).
Neerken, R.F., How to select and apply positive-displacement rotary pumps, Chem Eng, 7 April, 76-87 (1980).
Rajarm, S., Estimate volume of flow from a fan, Hyd Proc, 59(10), 134-135 (1980).
Van Blarcom, P.P., Recirculation systems for cooling centrifugal pumps, Chem Eng, 24 March, 131-134 (1980).
Van Ormer, H.P., Better service from rotary screw air compressor packages, Hyd Proc, 59(5), 181-187 (1980).
Various, Steam and gas turbines (feature report), Chem Eng, 25 Aug, 62-89 (1980).
Vetter, G., and Hering, L., Leakfree pumps for the chemical process industries, Chem Eng, 22 Sept, 149-154 (1980).
Buse, F., Using centrifugal pumps as hydraulic turbines, Chem Eng, 26 Jan, 113-117 (1981).
Eads, D.K., Establishing a centrifugal fan performance curve, Chem Eng, 23 March, 201-208 (1981).
Ekstrum, J.D., Sizing pulsation dampeners for reciprocating pumps, Chem Eng, 12 Jan, 111-118 (1981).
Henshaw, T.L., and Bristol, J.M., Positive displacement pumps (2 articles - feature report), Chem Eng, 21 Sept, 104-140 (1981).
Johnson, J.D., Variable-speed drives can cut pumping costs, Chem Eng, 10 Aug, 107-108 (1981).
King, R., Pumps update, Proc Engng, Feb, 29-34 (1981).
Kraus, M.N., Protect motors and rotating equipment, Chem Eng, 6 April, 110-111 (1981).
Mikasinovic, M., and Tung, P.C., Sizing centrifugal pumps for safety service, Chem Eng, 23 Feb, 83-85 (1981).
Murphy, J., Pumping slurries and suspensions, Proc Engng, Feb, 55-57 (1981).
Peters, K.L., Applying multiple inlet compressors, Hyd Proc, 60(5), 171-176 (1981).
Poynton, J.P., Metering pumps: Types and applications, Hyd Proc, 60(11), 279-284 (1981).
Ryans, J.L., and Croll, S., Selecting vacuum systems, Chem Eng, 14 Dec, 72-90 (1981).
Summerell, H.M., Consider axial-flow fans when choosing a gas mover, Chem Eng, 1 June, 59-62 (1981).
Tinney, W.S., Venting deepwell pumps, Chem Eng, 30 Nov, 83-85 (1981).
Various, Compressors (special report), Processing, Aug, 15-19 (1981).

1982-1983

Baker, D.F., Surge control for multistage centrifugal compressors, Chem Eng, 31 May, 117-122 (1982).
Burke, P.Y., Compressor intercoolers and aftercoolers: Predicting off-performance, Chem Eng, 20 Sept, 107-109 (1982).

Doll, T.R., Making the proper choice of adjustable-speed drives, Chem Eng, 9 Aug, 46-60 (1982).
Gaston, J.R., Antisurge control schemes for turbocompressors, Chem Eng, 19 April, 139-147 (1982).
Hallam, J.L., Centrifugal pumps: Which suction specific speeds are acceptable? Hyd Proc, April, 195-197 (1982).
Kannappan, S., Determining centrifugal compressor piping loads, Hyd Proc, Feb, 91-93 (1982).
Karassik, I.J., Centrifugal pumps and system hydraulics, Chem Eng, 4 Oct, 84-106 (1982).
Kraus, M.N., Reducing centrifugal fan-wheel capacity, Chem Eng, 3 May, 116 (1982).
Lapina, R.P., How to use the performance curves to evaluate behavior of centrifugal compressors, Chem Eng, 25 Jan, 86-93 (1982).
McLean, M.G., How to select and apply flexible-impeller pumps, Chem Eng, 20 Sept, 101-106 (1982).
Pri-Or, A., and Winter, Y., Calibrate your metering pump, Chem Eng, 26 July, 81 (1982).
Simmons, P.E., A new approach to pump flange loading, Hyd Proc, March, 150-154 (1982).
Smith, C.A., Selecting an air compressor, Proc Engng, Oct, 68-69 (1982).
Steuber, A., Factors for optimum vacuum system selection, Hyd Proc, Sept, 267-269 (1982).
Tsai, M.J., Accounting for dissolved gases in pump design, Chem Eng, 26 July, 65-69 (1982).
Various, Pumps survey (topic issue), Processing, Nov, 11-49 (1981); Jan, 37-39 (1982).
Anon., Compressors: A survey, Processing, March, 41-47 (1983).
Antony, S.M., and Mani, S., Suction saturator raises compressor capacity, Chem Eng, 22 Aug, 98-100 (1983).
Capes, P., Rotodynamic pump systems, Proc Engng, Feb, 56-57 (1983).
Davis, H., Evaluating multistage centrifugal compressors, Chem Eng, 26 Dec, 35-38 (1983).
Evans, J.E., Locating pump-discharge pressure gages for valid readings, Chem Eng, 25 July, 100 (1983).
Goyal, S., Choosing between induced- and forced-draft fans, Chem Eng, 7 Feb, 92-93 (1983).
Jewett, R.L., Metering pump operation, Plant/Opns Prog, 2(4), 211-216 (1983).
Lapina, R.P., Calculator method for steam-turbine efficiency, Chem Eng, 27 June, 37-42 (1983).
Maceyka, T.D., Detect and evaluate surge in multistage compressors, Hyd Proc, Nov, 203-204 (1983).
Martino, E., Materials for reciprocating process gas compressors, Chem Eng Prog, 79(4), 77-83 (1983).
McKelvey, J.M., Gear pumps, Processing, Jan, 25-35 (1983).
McKelvey, J.M., and Rice, W.T., Retrofitting plasticating extruders with gear pumps, Chem Eng, 24 Jan, 89-94 (1983).
Shilston, P., Variable-speed pumps, Proc Engng, June, 49-51 (1983).

Simnett, R.W., and Anderson, E., Air-motor drives for small pumps, Chem Eng, 12 Dec, 73-75 (1983).
Taggart, G.W., Vary speed of Claus blowers, Hyd Proc, Dec, 83-86 (1983).
Talwar, M., Analyzing centrifugal-pump circuits, Chem Eng, 22 Aug, 69-73 (1983).
Thompson, J.E., and Trickler, C.J., Fans and fan systems, Chem Eng, 21 March, 48-63 (1983).
Yates, M., Pumps report, Processing, Dec, 11-13 (1983).

1984

Anon., Report on pumps and valves, Processing, July, 25-30 (1984).
Anon., Pumps and piping systems (special advertising section), Chem Eng, 19 March, 71-143 (1984).
Dobrowolski, Z.C., Mechanical vacuum pumps in the process industry, Chem Eng Prog, 80(7), 75-83 (1984).
Fisher, B., Operating centrifugal compressors in parallel, Proc Engng, June, 59-62 (1984).
Gadsden, C., Squeezing the best from peristaltic pumps, Chem Engnr, June, 42-43 (1984).
George, J.R., Transient-damping improves flow control of centrifugal pumps, Chem Eng, 6 Aug, 69-71 (1984).
Harvest, J., Recent developments in gear pumps, Chem Engnr, May, 28, 29 (1984).
Hornsby, I., How to buy a pump, Chem Engnr, Jan, 30-31 (1984).
Lightfoot, E.N.; Thorne, P.S., and Stoll, L.L., Dimensionless presentation of performance data for fans and blowers, AIChEJ, 30(2), 341-345 (1984).
Margus, E.A., Choosing plastic pumps, Chem Eng, 12 Nov, 137-140 (1984).
Paugh, J.J., Head vs. capacity characteristics of centrifugal pumps, Chem Eng, 15 Oct, 91-93 (1984).
Various, Better service from rotating equipment (special report), Hyd Proc, Aug, 51-68 (1984).
Witton, J., Gas turbines, Proc Engng, Aug, 18-21 (1984).
Yedidiah, S., Multistage centrifugal pumps, Chem Eng, 26 Nov, 81-83 (1984).

1985

Anon., Pumps survey, Processing, Jan, 14-19 (1985).
Anon., Pumps update, Processing, Dec, 21-29 (1985).
Beevers, A., Applications of peristaltic pumps, Proc Engng, Dec, 45,47 (1985).
Bloch, H.P., and Johnson, D.A., Downtime prompts upgrading of centrifugal pumps, Chem Eng, 25 Nov, 35-41 (1985).
Cody, D.J.; Vandell, C.A., and Spratt, D., Selecting positive-displacement pumps, Chem Eng, 22 July, 38-52 (1985).
Dillon, M.L.; St Clair, K.A., and Kline, P.H., Predicting flowrates from positive-displacement rotary pumps, Chem Eng, 22 July, 57-60 (1985).
Etheridge, R., Control of centrifugal pumps, Chem Engnr, Feb, 46-48 (1985).
Hughes, S., Improving reliability of rotodynamic pumps, Proc Engng, Aug, 21-25 (1985).
Johnson, S.R., Determine possible compressor icing conditions, Chem Eng, 4 Feb, 89-90 (1985).

Lobanoff, V.S., and Ross, R.R., Centrifugal Pumps: Design and Application, Gulf Publishing Co., Texas (1985).
Magnani, I.; Nutini, G., and Tosi, G., Aspects of gas compression equipment - onshore and offshore, Chem Engnr, Jan, 28-31 (1985).
Marshall, P., Positive displacement pumps: A brief survey, Chem Engnr, Oct, 52-55 (1985).
Nevill, K., Glandless centrifugal pumps for critical process duties, Proc Engng, July, 56-57 (1985).
Polonyi, M., Operating performance of reciprocating or positive-displacement compressors, Chem Eng, 9 Dec, 132 (1985).
Rana, S.A.Z., Understand multistage compressor antisurge control, Hyd Proc, April, 69-74 (1985).
Rattan, I.S., and Pathak, V.K., Startup of centrifugal pumps in flashing or cryogenic liquid service, Chem Eng, 1 April, 95-96 (1985).
Sayyed, S., Selecting high-performance centrifugal compressors, Hyd Proc, Oct, 57-60 (1985).
Tadmor, Z.; Mehta, P.S.; Valsamis, L.N., and Yang, J.C., Corotating disc pumps for viscous liquids, Ind Eng Chem Proc Des Dev, 24(2), 311-320 (1985).
Various, Pumps and compressors developments, Processing, July, 37-41 (1985).
Wilkie, F., Gas compressor update, Proc Engng, Dec, 55,57 (1985).

1986

Bacchetti, J.A., Pumps and piping (special advertising section), Chem Eng, 17 Feb, 49-113 (1986).
Brown, R.N., Compressors: Selection and Sizing, Gulf Publishing Co., Texas (1986).
Clark, N.N., and Dabolt, R.J., General design equation for air lift pumps operating in slug flow, AIChEJ, 32(1), 56-64 (1986).
Frings, A., and Kasthuri, R., Fans for the process industries, Chem Engnr, Jan, 20-22 (1986).
Ganapathy, V., Check pump performance from motor data, Chem Eng, 13 Oct, 91-92 (1986).
Karassik, I.J.; Krutzsch, W.C.; Fraser, W.H., and Messina, J.P. (Eds), Pump Handbook, 2nd Edn., McGraw-Hill, New York (1986).
Lowe, R.E., Specifying, evaluating and procuring dynamic compressors, Hyd Proc, 65(8), 46-50 (1986).
Nasr, A.M., When to select a sealless pump, Chem Eng, 26 May, 85-89 (1986).
Odrowaz-Pieniazek, S., and Steele, K., Advances in slurry pumps, Chem Engnr, Feb, 34-37; March, 30-33 (1986).
Rattan, I.S., and Pathak, V.K., Startup of centrifugal pumps, Chem Eng, 6 Jan, 89 (1986).
Reynolds, J.A., Standard pumps are not obsolete, Chem Eng, 12 May, 119-120 (1986).
Royse, S., Switched reluctance drives for pump control, Chem Engnr, May, 49-51 (1986).
Stadler, E.L., Understand centrifugal compressor stage curves, Hyd Proc, 65(8), 51-53 (1986).
Stanecki, J.W., Program performs calculations for polytropic compression, Chem Eng, 6 Jan, 63-68 (1986).

Taffe, P., Compressors update, Processing, July, 31-35 (1986).
Various, Pumps update, Processing, Dec, 17-23 (1986).
Yedidiah, S., Unusual problems with centrifugal pumps, Chem Eng, 8 Dec, 143-145 (1986).
Zafar, R.S.A., Design for compressors with direct quench recycle, Chem Engnr, March, 35-37 (1986).

1987

Brauer, R., Diaphragm metering pumps, Chem Eng Prog, 83(4), 18-24 (1987).
Brown, R.N., Compressors: Selection and Sizing, Gulf Publishing Co., Texas (1987).
Chynoweth, E., Pumps update, Processing, Dec, 29-35 (1987).
Chynoweth, E., Mixing and matching modular pumps: The CAD approach, Proc Engng, Sept, 77-79 (1987).
Constantinescu, S., Alternative to Gaede's formula for vacuum pumpdown time, Chem Eng, 11 May, 81-82 (1987).
Edwards, R.M., Improve your acceptance criteria for high energy pumps, Proc Engng, April, 33-36 (1987).
Feldman, E.J., Specifying electric motors, Chem Eng, 11 May, 37-48 (1987).
Fischer, S.M., Designing centrifugal pump systems, Chem Eng, 16 Feb, 147-150 (1987).
Grandjean, B.P.A.; Ajersch, F.; Carreau, P.J., and Patterson, I., Study of an air-lift system, Can JCE, 65(3), 430-432 (1987).
Hughes, S., A new market for centrifugal pumps, Chem Engnr, May, 40-42 (1987).
Jackson, C., Put the screws to your compressor applications, Chem Eng, 28 Sept, 137-139 (1987).
Liptak, B.G., Integrating compressors into one system, Chem Eng, 19 Jan, 97-103 (1987).
Mathews, J., Low-pressure-steam ejectors: Operation and maintenance, Chem Eng, 22 June, 155-158 (1987).
Neerken, R.F., Progress in pumps, Chem Eng, 14 Sept, 76-88 (1987).
Polk, M., Better couplings reduce pump maintenance, Hyd Proc, 66(1), 62-63 (1987).
Reeves, G.G., Avoid self-priming centrifugal pump problems, Hyd Proc, 66(1), 60-61 (1987).
Roco, M.C., and Addie, G.R., Erosion wear in slurry pumps and pipes, Powder Tech, 50(1), 35-46 (1987).
Taylor, I., Pump bypasses are now more important: Installation of automatic bypasses on large centrifugal pumps, Chem Eng, 11 May, 53-57 (1987).
Various, Rotating equipment (special report), Hyd Proc, 66(10), 33-42 (1987).

1988

Ablitt, I., Developments in gas compressors, Processing, June, 37-40 (1988).
Anon., Plant and equipment survey for pumps, Processing, Dec, 42-49 (1988).
Bloch, H.P., Low-maintenance compressors, Chem Eng, 18 July, 97-101 (1988).
Martin, T., Pump control, Chem Engnr, Sept, 37 (1988).

Rao,, S.P.R., and Singh, R.P., Performance characteristics of single-stage steam jet ejectors using two simple models, Chem Eng Commns, 66, 207-220 (1988).
Rendell, J., Disc pumps for abrasive and viscous fluids, Proc Engng, April, 49, 51 (1988).
Stark, B., and Taylor, G., Centrifugal pump vibration problems at part-load, Proc Engng, Oct, 79-87 (1988).
Tuthill, A.H., Selecting materials for saline water pumps, Chem Eng, 12 Sept, 88-92 (1988).
Yedidiah, S., Performance of centrifugal pumps, Chem Eng, 20 June, 139-142 (1988).

15.4 Seals and Bearings

1967-1988

Coopey, W., Spring-loaded packing, Chem Eng, 6 Nov, 278-284 (1967).
Lindsey, M.H., Mechanical seals: Carbon's key role, Chem Eng, 27 Feb, 160-166 (1967).
Samoiloff, A.A., Mechanical seals: Longer runs, less maintenance, Chem Eng, 29 Jan, 130-138 (1968).
Spear, M., Shaft sealing: Art or science? Processing, Feb, 7-10 (1975).
Terry, E.S., Guide to mechanical sealing, Processing, Sept, 8,10 (1975).
Phillips, J., The solution to difficult shaft sealing problems, Chem Engnr, Feb, 87-90 (1979).
Hawk, C.W., Why mechanical seals fail, Chem Eng, 11 Aug, 171-174 (1980).
Jones, P.T., Resolving seal and bearing failures in process pumps, Chem Eng, 16 June, 145-148 (1980).
Kerklo, P., Oil seals for centrifugal compressors, Hyd Proc, Oct, 112-114 (1982).
Wallis, B., Current trends in seals materials, Chem Engnr, Nov, 10-13 (1984).
Buse, F.W., Power consumption of double mechanical seals, Chem Eng, 9 Dec, 125-128 (1985).
Riance, X.P., Expansion and nonexpansion bearings, Chem Eng, 13 May, 109-110; 10 June, 105-106 (1985).
Budrow, J.S., Seals for abrasive slurries, Chem Eng, 1 Sept, 67-71 (1986).
Flitney, R., Mechanical seals for the process industries, Chem Engnr, Sept, 37-41 (1986).
Parkinson, G., Call for higher quality is heeded by seal makers, Chem Eng, 12 May, 22-27 (1986).
Abrams, P., and Olson, R., Specifying mechanical seals, Hyd Proc, 67(11), 98-99 (1988).
Anon., Mechanical seal repairs, Chem Eng, 18 July, 125 (1988).
Ferland, R.H., Packing options for rotating equipment, Hyd Proc, 67(4), 39-41 (1988).
Newby, T., Sealing options for pumps, Chem Engnr, April, 33-38 (1988).
Royse, S., Secondary sealing, Proc Engng, June, 67-68 (1988).
Taylor, I.; Cameron, W., and Wong, W., User's guide to mechanical seals, Chem Eng, 12 Sept, 81-87 (1988).

Wallace, N., Secondary sealing systems for safer operation, Processing, Oct, 33-34 (1988).

15.5 Valves

1973-1988

Bean, D.W., How to select high-performance valves, Chem Eng, 5 Feb, 62-66 (1973).
Various, Valves (special report), Hyd Proc, 53(6), 87-99 (1974).
Various, Valves, pipes and fittings (feature report), Processing, Oct, 37-63 (1975).
Wier, J.T., Selecting valves, Chem Eng, 24 Nov, 62-71 (1975).
Bertrem, B.E., Butterfly valves for flow of process fluids, Chem Eng, 20 Dec, 63-66 (1976).
Pikulik, A., Selecting and specifying valves for new plants, Chem Eng, 13 Sept, 168-190 (1976).
Farley, H.E., Choosing and applying line-blind valves, Chem Eng, 8 Oct, 121-124 (1979).
Whitfield, M., Developments in control valves, Proc Engng, April, 68-71 (1979).
Cook, D.T., Selecting hand-operated valves for process plants, Chem Eng, 14 June, 126-140 (1982).
Kalsi, M.S., and Guerrero, D.C., Finite element analysis aids noncircular valve specification, Hyd Proc, May, 195-201 (1982).
Various, Control valves (feature report), Chem Eng, 5 Sept, 94-110 (1983).
Anon., Valves and actuators (special advertising section), Chem Eng, 20 Aug, 69-123 (1984).
Thorley, D., The dynamic response of check valves, Chem Engnr, April, 12-15 (1984).
Anon., Valves and actuators (special advertising section), Chem Eng, 18 Feb, 69-124 (1985).
Baumann, H.D., Preventing cavitation in butterfly valves, Chem Eng, 18 March, 149-153 (1985).
Irhayem, A., Prevent plug valves from sticking and jamming, Chem Eng, 29 April, 88 (1985).
Merrifield, R., The development of ball valves, Chem Engnr, Oct, 36-41 (1985).
Pittman, E.D., Specifying rotary valves, Chem Eng, 8 July, 89-95 (1985).
Dibbo, A., Valves update, Processing, Jan, 21-30 (1986).
Ellis, J., and Mualla, W., Selecting check valves, Proc Engng, Nov, 47-53 (1986).
Merrick, R.C., A guide to selecting manual valves, Chem Eng, 1 Sept, 52-64 (1986).
Anon., New valves and actuators offer greater versatility (equipment focus), Chem Eng, 16 Feb, 76A-76D (1987).
Anon., Valves and actuators for the CPI (special advertising section), Chem Eng, 7 Dec, 53-92D (1987).
Morley, P., and Heasman, W., Selecting a cryogenic valve, Chem Engnr, Feb, 13-16 (1987).

Anderson, V.R., Rotary actuators for quarter-turn valves, Chem Eng, 15 Aug, 143-151 (1988).
Husu, M., Variable pressure recovery: A rotary valve advantage, Hyd Proc, 67(9), 107-108 (1988).
Various, Valves and actuators (feature report), Processing, April, 17-28 (1988).
Vivian, B., Selecting process valves for in-service reliability, Proc Engng, April, 33-39 (1988).

15.6 Flowmeters and Control

1967-1972

Kumar, R., and Kuloor, N.R., Small to large flowrates measured by new meter, Chem Eng, 25 Sept, 200-202 (1967).
Liptak, B.G., Process instrumentation for slurries and viscous materials, Chem Eng, 30 Jan, 133-138 (1967).
Liptak, B.G., Instruments to measure and control slurries and viscous materials, Chem Eng, 13 Feb, 151-158 (1967).
Platt, G., How to control small flows in high-pressure streams, Chem Eng, 5 June, 168 (1967).
Evans, G.V., Triangular-plate vortex generators in flow measurement, Brit Chem Eng, 13(3), 375-376 (1968).
Spolidoro, E.F., Comparing positive-displacement meters, Chem Eng, 3 June, 91-94 (1968).
Dieterich, P.D., New approach for flow metering, Hyd Proc, 48(4), 115-116 (1969).
Kumar, R., Flow-pressure drop linear relation with orifice meter, Chem Eng, 14 July, 134-136 (1969).
Mandersloot, W.G.B.; Hicks, R.E., and Langejan, J.J.D., Flow rates from multipoint measurements with velocity probes (Pitot tubes), Chem Engnr, Oct, CE372-378 (1969).
Pack, G.E., Correcting orifice meters, Hyd Proc, 48(7), 160-162 (1969).
Plaskowski, A., and Beck, M., Indirect measurement of liquid flowrates, Int Chem Eng, 9(3), 418-422 (1969).
Brade, W.R., Rotameter gas flow converted to standard conditions, Chem Eng, 7 Sept, 98 (1970).
Jacoby, R.H., and Tracht, J.H., Fluid sampling technique, Hyd Proc, 49(2), 101-102 (1970).
May, D.L., Accurate flow measurements with turbine meters, Chem Eng, 8 March, 105-108 (1971).
Nedoborov, Y.P., and Semenov, P.A., Mathematical correlation of two-phase flow data in rectangular venturi tubes, Int Chem Eng, 11(4), 638-641 (1971).
Adams, W.C., Flow control in pilot plants, Chem Eng Prog, 68(1), 41-42 (1972).
Felton, G.L., Low-flow measurement with the integral orifice, Chem Eng Prog, 68(1), 43-47 (1972).
Harris, J., and Magnall, A.N., The use of orifice plates and venturi meters with non-Newtonian fluids, Trans IChemE, 50, 61-68, 292, 394-395, (1972).

Moore, D.C., Easy way to measure slurry flowrates, Chem Eng, 2 Oct, 96 (1972).
Smith, J.R., Measuring flows through vents, Chem Eng, 25 Dec, 82-84 (1972).

1973-1979

Chavan, V.V., A flow meter for thick liquids, Chem Eng, 1 Oct, 102 (1973).
McNulty, F.G., Designing orifices for flow regulation, Chem Eng, 12 Nov, 239-241 (1973).
Simo, F.E., Which flow control: Valve or pump? Hyd Proc, 52(7), 103-106 (1973).
Zacharias, E.M., and Franz, D.W., Sound velocimeters monitor process systems, Chem Eng, 22 Jan, 101-108 (1973).
McNulty, F., Designing critical flow orifices, Chem Eng, 5 Aug, 122-124 (1974).
Yard, J., A practical guide to low-flow measurement, Chem Eng, 15 April, 74-78 (1974).
Coe, G.H., The accurate pumping of liquids, Chem Engnr, Jan, 35-37, 39 (1975).
Corcoran, W.S., and Honeywell, J., Practical methods for measuring flows, Chem Eng, 7 July, 86-92 (1975).
Klapper, W.T., Estimating pipe sizes for orifice runs, Chem Eng, 4 Aug, 126 (1975).
Loeb, M.B., Determining chlorine-gas flowrates from cylinders, Chem Eng, 6 Jan, 126-130 (1975).
Various, Flow measurement and control (feature report), Processing, July, 33-43 (1975).
Monroe, E.S., How to size and rate steam traps, Chem Eng, 12 April, 119-123 (1976).
Nguyen, H.X., Estimating venturi or orifice diameter, Chem Eng, 3 July, 92 (1978).
Various, Flow and level control (feature report), Processing, April, 49-71 (1978).
Edwards, J.E., Flow measurement practice, Chem Engnr, May, 325-330 (1979).
Nayak, K.R., Sizing orifice restriction, Chem Eng, 17 Dec, 114 (1979).
Nguyen, H.X., Maximize orifice meter performance, Hyd Proc, 58(4), 217-222 (1979).
Nguyen, H.X., Sizing critical-flow orifices, Chem Eng, 2 July, 105-106 (1979).
Polentz, L.M., Compressible flow through orifices, Chem Eng, 27 Aug, 118-120 (1979).
Walls, W.R., High integrity flow metering, Chem Engnr, May, 331-333 (1979).

1980-1984

Harrison, P., Flow measurement: A review, Chem Eng, 14 Jan, 97-104 (1980).
Heller, H., Use weir to measure fluid flow, Chem Eng, 17 Nov, 300-302 (1980).
Mink, W.H., Program calculates orifice sizes for gas, Chem Eng, 25 Aug, 91-94 (1980).
Vaux, W.G., Calculating flow through gas rotameters, Chem Eng, 1 Dec, 119-120 (1980).
Ganapathy, V., Orifice-meter sizing, Chem Eng, 24 Aug, 130 (1981).

Hayward, A.T.J., Selecting flowmeters, Proc Engng, April, 45-50 (1981).
Nangia, S.C., Adjust diaphragm meters correctly, Chem Eng, 4 May, 104-105 (1981).
Raghavachari, S., Check turbine-flowmeter calibration fast, Chem Eng, 16 Nov, 292-294 (1981).
Various, Flowmeters: Plant and equipment survey (topic issue), Processing, May, 23-37 (1981).
Hayward, A., Flowmeters: Plant and equipment survey, Processing, Dec, 35-47 (1982).
Radhakrishnan, R.; Kumar, S.N.K., and Rajaram, S., Pressure tube anemometers for air flow measurement, Hyd Proc, April, 201-203 (1982).
Stephens, F.A., Program predicts pressure drop for gas flow across an orifice meter, Chem Eng, 29 Nov, 69-74 (1982).
Tung, P.C., and Mikasinovic, M., Sizing orifices for flow of gases and vapors, Chem Eng, 8 March, 83-85 (1982).
King, C., Power fluidics: Fluid control without moving parts, Proc Engng, Oct, 37-39 (1983).
Levin, H., and Escorza, M.M., Gas flow through rotameters, Ind Eng Chem Fund, 22(2), 163-166 (1983).
Noor, A., Sizing orifice and venturi meters, Chem Eng, 22 Aug, 97 (1983).
Anon., Gas and liquid metering, Processing, Feb, 9-13 (1984).
Baker, R., and Deacon, J., The behaviour of turbine, vortex and electromagnetic flowmeters, Chem Engnr, March, 13-15 (1984).
Brown, K., Pitot tube calculations with a TI-59, Hyd Proc, June, 91-92 (1984).
Ditcham, S., A practical guide to liquid density transducers, Chem Engnr, July, 29-31 (1984).
Hills, J.H., Investigation into the suitability of a transverse pitot tube for two phase flow measurements, CER&D, 61, 371-376 (1983); 62, 269-271 (1984).
Hobbs, J.M., Calibrating flowmeters, Processing, Dec, 9-10 (1984).

1985-1988

Heywood, N., Selecting a viscometer, Chem Engnr, June, 16-23 (1985).
Rao, K.V.K., Orifice-meter design for steam flow, Chem Eng, 19 Aug, 83-86 (1985).
Reeves, G.G., Use pressure gage to measure flow, Chem Eng, 14 Oct, 119 (1985).
Zigrang, D.J., and Sylvester, N.D., Equations for orifice-meter sizing, Chem Eng, 4 Feb, 91 (1985).
De Vries, E.A., Experience with ultrasonic flowmeters, Hyd Proc, 65(6), 65-66 (1986).
Hall, K.R.; Eubank, P.T.; Holste, J.C., and Marsh, K.N., Performance equations for compressible flow through orifices, AIChEJ, 32(3), 517-519 (1986).
Hayward, A., Flowmeters update, Processing, May, 61-71 (1986).
Moore, A., Selecting a flowmeter, Chem Engnr, April, 39-45 (1986).
Zigrang, D.J., and Sylvester, N.D., Equations for orifice-meter sizing, Chem Eng, 6 Jan, 94 (1986).
Anon., Fluid flow: Products and literature (special advertising section), Chem Eng, 16 Feb, 83-109 (1987).

Chynoweth, E., Direct mass flowmeters, Proc Engng, Aug, 35-37 (1987).
Fadel, T.M., The safe way to install restriction orifices, Chem Eng, 13 April, 114 (1987).
Ginesi, D., and Grebe, G., Flowmeters: A performance review, Chem Eng, 22 June, 102-118 (1987).
Hobbs, M., A guide to the selection of orifice plates, Proc Engng, Feb, 33-35 (1987).
Humphreys, J.M., Improve the performance of orifice plates, Proc Engng, Feb, 37-41 (1987).
Mascone, C.F., Flowmetering gets boost from novel designs, Chem Eng, 16 Feb, 37-39 (1987).
Various, Flowmeters update, Processing, June, 39-45 (1987).
Baker, R., Measuring multiphase flow, Chem Engnr, Oct, 39-45 (1988).
Morrison, G.L.; Sheth, K.K., and Tatterson, G.B., Elbow flowmeter calibrations for slurries, Chem Eng Commns, 63, 39-48 (1988).
Wilbeck, K., Vortex-shedding flowmeters, Hyd Proc, 67(8), 53-55 (1988).

CHAPTER 16

FLUID AND PARTICULATE SYSTEMS

16.1 Theory

1967-1975

Browning, J.E., Agglomeration techniques, Chem Eng, 4 Dec, 147-170 (1967).

Bridgwater, J.; Sharpe, N.W., and Stocker, D.C., Particle mixing by percolation, Trans IChemE, 47, T114-119 (1969).

Marrucci, G., A theory of coalescence, Chem Eng Sci, 24(6), 975-986 (1969).

Han, C.D., and Wilenitz, I., Mathematical modeling of steady-state behavior in industrial granulation, Ind Eng Chem Fund, 9(3), 401-411 (1970).

Capes, C.E., Correlation of agglomerate strength with size, Powder Tech, 5, 119-125 (1971).

Holland-Batt, A.B., Behaviour of particles accelerating in fluids, Trans IChemE, 50, 12-20 (1972).

Plachco, F.P., and Krasuk, J.H., Cocurrent solid-liquid extraction with initial concentration profile and constant diffusivity, Chem Eng Sci, 27(2), 221-226 (1972).

Various, Papers from 2nd annual meeting of the Fine Particle Society (1971), Powder Tech, 6, 3-37 (1972).

Barnea, E., and Mizrahi, J., Generalized approach to fluid dynamics of particulate systems, Chem Eng J, 5(2), 171-190 (1973).

Saileswaran, N., and Panchanathan, V., Compaction of grains: General parameter evaluation, Powder Tech, 8, 19-26 (1973).

Various, Papers from 4th annual meeting of the Fine Particle Society (1973), Powder Tech, 10, 101-128 (1974).

Lai, F.S., and Fan, L.T., Application of discrete mixing model to study of mixing of multicomponent solid particles, Ind Eng Chem Proc Des Dev, 14(4), 403-411 (1975).

Van Brakel, J., Pore space models for transport phenomena in porous media: Review and evaluation with special emphasis on capillary liquid transport, Powder Tech, 11, 205-236 (1975).

1976-1981

Capes, C.E., Basic research in particle technology and some novel applications, Can JCE, 54, 3-12 (1976).

Various, Papers from 5th annual meeting of the Fine Particle Society (1974), Powder Tech, 13, 1-31 (1976).

Ballesteros, F., Three-phase separation in one stage process, Proc Engng, Sept, 78-79 (1977).

Messman, H.C., Agglomeration, Chemtech, July, 424-427 (1977).

Parfitt, G.D., Dispersion of powders in liquids: An introduction, Powder Tech, 17, 157-162 (1977).

Svarovsky, L., Errors in measurement of efficiency of particle-fluid separators, Powder Tech, 17, 139-143 (1977).

Barnea, E., and Mednick, R.L., Generalized approach to fluid dynamics of particulate systems, Chem Eng J, 15(3), 215-228 (1978).

Tomi, D.T., and Bagster, D.F., The behaviour of aggregates in stirred vessels, Trans IChemE, 56, 1-18 (1978).

Cross, M., Transverse motion of solids moving through rotary kilns, Powder Tech, 22, 187-190 (1979).

Spaninks, J.A.M., and Bruin, S., Mathematical simulation of performance of solid-liquid extractors, Chem Eng Sci, 34(2), 199-216 (1979).

Abouzeid, A.Z.M.; Fuerstenau, D.W., and Sastry, K.V.S., Transport behavior of particulate solids in rotary drums, Powder Tech, 27, 241-250 (1980).

French, R.M., and Wilson, D.J., Fluid mechanics-foam flotation interactions, Sepn Sci Technol, 15(5), 1213-1228 (1980).

Lee, Y.; Beddow, J.K.; Vetter, A.F., and Lenth, R., Morphological analysis of fine particle mixtures, Powder Tech, 25, 137-145 (1980).

Herndl, G., and Mersmann, A.B., Fluid dynamics and mass transfer in stirred suspensions, Chem Eng Commns, 13(1), 23-38 (1981).

Schubert, H., Principles of agglomeration, Int Chem Eng, 21(3), 363-378 (1981).

Various, Effects of powder characteristics on final compact properties (special issue), Powder Tech, 30, 1-94 (1981).

1982-1984

Cooke, M.H., and Bridgwater, J., Simulation of a particle disperser, Powder Tech, 33, 239-247 (1982).

Downton, G.E.; Flores-Luna, J.L., and King, C.J., Mechanism of stickiness in hygroscopic, amorphous powders, Ind Eng Chem Fund, 21(4), 447-451 (1982).

Graichen, K., Modelling of blending processes, Int Chem Eng, 22(1), 68-74 (1982).

Luerkens, D.W.; Beddow, J.K., and Vetter, A.F., Morphological Fourier descriptors of particle profile representation, Powder Tech, 31, 209-220 (1982).

Murthy, D.V.S., and Ananth, M.S., A one-parameter model for granulation, Chem Eng J, 23(2), 177-184 (1982).

Ouchiyama, N., and Tanaka, T., Kinetic analysis and simulation of batch granulation, Ind Eng Chem Proc Des Dev, 21(1), 29-35 (1982).

Ouchiyama, N., and Tanaka, T., Physical requisite to appropriate granule growth rate, Ind Eng Chem Proc Des Dev, 21(1), 35-37 (1982).

Krycer, I.; Pope, D.G., and Hersey, J.A., Evaluation of tablet binding agents, Powder Tech, 34, 39-56 (1983).

Paramanathan, B.K., and Bridgwater, J., Attrition of solids, Chem Eng Sci, 38(2), 197-224 (1983).

Smith, P.G., and Nienow, A.W., Particle growth mechanisms in fluidised bed granulation, Chem Eng Sci, 38(8), 1223-1240 (1983).

Hiestand, H.E.N., and Smith, D.P., Indices of tableting performance, Powder Tech, 38, 145-159 (1984).

Jones, G.L., Simulating the effects of changing particle characteristics in solids processing, Comput Chem Eng, 8(6), 329-338 (1984).

Rietema, K., Powders, what are they? Powder Tech, 37, 5-24 (1984).

van Brakel, J. (Ed), Coal fineparticle technology (special issue), Powder Tech, 40, 1-362 (1984).

1985-1988

Stiess, M., Survey of criteria for separations of two component mixtures of solids. Int Chem Eng, 25(2), 234-240 (1985).

Datye, A.K.; Smith, D.M., and Williams, F.L., Characterization of powders and porous materials, Chem Eng Educ, 20(4), 198-201 (1986).
Akiyama, T., and Naito, T., Vibrated beds of powders: A new mathematical formulation, Chem Eng Sci, 42(6), 1305-1312 (1987).
Khan, A.R., and Richardson, J.F., Resistance to motion of solid sphere in a fluid, Chem Eng Commns, 62(1), 135-150 (1987).
Woodcock, L.V., Powder processing research, Proc Engng, Oct, 41-43 (1987).
Salman, A.D., and Verba, A., New approximate equations to estimate the drag coefficient of different particles of regular shape, Periodica Polytechnica, 32(4), 261-268 (1988).
Wang, P.T., Constitutive model for consolidation of granular materials, Powder Tech, 54(2), 107-118 (1988).
Wasowski, T., and Blass, E., Wake phenomena behind solid and liquid particles, Int Chem Eng, 28(4), 593-608 (1988).
Yu, A.B., and Standish, N., Analytical-parametric theory of random packing of particles, Powder Tech, 55(3), 171-186 (1988).

16.2 Equipment/Separation Techniques

1966-1972

Various, Solid-liquid separation (feature report), Chem Eng, 20 June, 139-202 (1966).
Bullock, H.L., Advice on processing dry particles, Chem Eng, 24 April, 186-188 (1967).
Lieberman, A., Fine-particle technology, Chem Eng, 27 March, 97-102; 10 April, 209-218; 24 April, 163-166 (1967).
Grieves, R.B., Studies on foam separation processes, Brit Chem Eng, 13(1), 77-82 (1968).
Anon., Mechanical separation of solids from liquids, Brit Chem Eng, 14(10), 1362-1368 (1969).
Pugh, G., and Morris, G., Equipment selection for handling abrasive and friable materials, Chem Engnr, May, CE160-162 (1969).
Stoev, S.M., and Watson, D., Pelletising on a flat surface using three-dimensional vibrations, Brit Chem Eng, 14(3), 325-327 (1969).
Various, Solids handling (deskbook issue), Chem Eng, 13 Oct, 7-148 (1969).
Chen, N.H., Optimum theoretical stages in countercurrent leaching, Chem Eng, 24 Aug, 71-74 (1970).
Roberts, E.J., et al., Solids concentration, Chem Eng, 29 June, 52-68 (1970).
Waldie, B., Preparation of powders at high temperatures, Trans IChemE, 48, T90-93 (1970).
Bridgwater, J., and Ingram, N.D., Rate of spontaneous inter-particle percolation, Trans IChemE, 49, 163-169 (1971).
Various, Solids separations (deskbook issue), Chem Eng, 15 Feb, 11-139 (1971).
Fish, G.F., The solids-handling jet pump, Brit Chem Eng, 17(5), 423-427 (1972).
Pierson, H.G.W., Selection of liquid/solid separation systems, Brit Chem Eng, 17(6), 524-527 (1972).

1973-1976

Campbell, A.P., and Bridgwater, J., The mixing of dry solids by percolation, Trans IChemE, 51, 72-74 (1973).

Linkson, P.B.; Glastonbury, J.R., and Duffy, G.J., The mechanism of granule growth in wet pelletising, Trans IChemE, 51, 251-259 (1973).

Powell, T.E., Granulation in the fertilizer industry, Proc Tech Int, June, 271-278 (1973).

Masliyah, J., and Bridgwater, J., Particle percolation: A numerical study, Trans IChemE, 52, 31-42 (1974).

Various, Selection of solid-liquid separation equipment (feature report), Chem Eng, 29 April, 116-136 (1974).

Roberts, A.G., and Shah, K.D., The large scale application of prilling, Chem Engnr, Dec, 748-750 (1975).

Various, Mineral separation (topic issue), Chem & Ind, 18 Jan, 54-70 (1975).

Agarwal, J.C., and Klumpar, I.V., Multistage-leaching simulation, Chem Eng, 24 May, 135-140 (1976).

Krambrock, W., Mixing and homogenizing of granular bulk materials in a pneumatic mixer unit, Powder Tech, 15, 199-206 (1976).

Leaver, R.H., Evaluating industrial pelleting, Chem Eng, 5 Jan, 155-156 (1976).

Ruskan, R.P., Prilling vs. granulation for nitrogen fertilizer production, Chem Eng, 7 June, 114-118 (1976).

Sisson, W., Determine conveyor belt length, Chem Eng, 7 June, 154 (1976).

Various, Powder technology (feature report), Processing, Feb, 29-47 (1976).

Wes, G.W.J.; Drinkenburg, A.A.H., and Stemerding, S., Solids mixing, residence time distribution, and heat transfer in a horizontal rotary drum reactor, Powder Tech, 13, 177-192 (1976).

1977-1979

Parnaby, J., The design and control of homogenisation systems for minerals processing, Trans IChemE, 55, 104-113 (1977).

Roe, L.A., Mineral processing methods: Review and forecast, Chem Eng, 20 June, 102-110 (1977).

Tiller, F.M., and Crump, J.R., Solid-liquid separation: An overview, Chem Eng Prog, 73(10), 65-75 (1977).

Wilson, M.F., and Roberts, A.G., Chemical engineering techniques in the study of granulation processes, Chem Engnr, Dec, 860-862 (1977).

Bridgwater, J.; Cooke, M.H., and Scott, A.M., Inter-particle percolation: Equipment development and mean percolation velocities, Trans IChemE, 56, 157-167 (1978).

Gottfried, B.S., and Abara, J., Computer simulation of coal preparation plants, Comput Chem Eng, 2(2), 99-108 (1978).

Lieberman, A., Optical instruments monitor liquid-borne solids, Chem Eng, 18 Dec, 105-109 (1978).

Miura, M., and Williams, T.M., Advances in liquid-solid separation, Chem Eng Prog, 74(4), 66-69 (1978).

Nguyen, H.X., Calculating actual stages in countercurrent leaching, Chem Eng, 6 Nov, 121-122 (1978).

Purchas, D.B., Solid/liquid separation equipment: A preliminary experimental selection programme, Chem Engnr, Jan, 47-49 (1978).

Various, Materials handling (deskbook issue), Chem Eng, 30 Oct, 9-148 (1978).
Robinson, T., and Waldie, B., Dependency of growth on granule size in a spouted bed granulator, Trans IChemE, 57, 121-127 (1979).
Svarovsky, L., Advances in solid-liquid separation, Chem Eng, 2 July, 62-76; 16 July, 93-105; 30 July, 69-78 (1979).

1980-1985

Neville, J.M., and Seider, W.D., Coal pretreatment: Extensions of FLOWTRAN to model solids-handling equipment, Comput Chem Eng, 4(1), 49-62 (1980).
Nienow, A.W., and Naimer, N.S., Continuous mixing of two particulate species of different density in a gas fluidised bed, Trans IChemE, 58, 181-186 (1980).
Various, Fluid-solid separations (symposium papers), Sepn Sci Technol, 15(3), 165-370 (1980).
Bozzay, J.; Dombai, Z.; Rusznak, I., and Torok, L., New process for granular pesticide manufacture, Periodica Polytechnica, 25, 103-110 (1981).
Lyne, C.W., and Johnston, H.G., Selection of pelletisers, Powder Tech, 29, 211-216 (1981).
Heywood, N.I.; Carleton, A.J., and Bransby, P.L., Problems in handling and processing wet solids, Chem Engnr, Dec, 465-471 (1982).
Ridgway, K., Pharmaceutical tablet making, Chem Engnr, May, 169-173 (1982).
Various, Application of solid separation systems to industrial effluents (topic issue), Chem & Ind, 6 Feb, 80-98 (1982).
Wells, I.S., Wet separation of paramagnetic minerals, Chem Engnr, Nov, 424-427 (1982).
Bozzay, J.; Dombai, Z.; Rusznak, I., and Torok, L., Development of a new granulation technique, Periodica Polytechnica, 27, 5-12; 187-194 (1983).
Mittal, A.K., Uprating existing belt conveyors, Chem Eng, 4 April, 114-116 (1983).
Anon., Solid materials handling (special advertising section), Chem Eng, 15 Oct, 61-76 (1984).
Eberts, D.H., Program calculates losses for countercurrent decantation, Chem Eng, 12 Nov, 121-124 (1984).
Kaye, L.A., and Fiocco, R.J., Fine-particle separations technology, Sepn Sci Technol, 19(11), 783-800 (1984).
Colijn, H., Mechanical Conveyors for Bulk Solids, Elsevier, New York (1985).
Grist, J., Fine-chemicals process plants, Proc Engng, March, 19-23 (1985).
Holm, P.; Schaefer, T., and Kristensen, H.G., Granulation in high-speed mixers, Powder Tech, 43, 213-233 (1985).
Purchas, D., Selecting solid/liquid separation equipment, Proc Engng, Sept, 53-57 (1985).
Reece, E.V., Bulk solids handling, Chem Eng, 29 April, 38-52 (1985).

1986-1987

Akiyama, T., et al., Densification of powders by air, vibratory and mechanical compactions, Powder Tech, 46(2), 173-180 (1986).

Anon., Material-handling equipment (equipment focus), Chem Eng, 9 June, 47-51 (1986).
Elban, W.L., and Chiaito, M.A., Quasi-static compaction study of coarse HMX explosive, Powder Tech, 46(2), 181-194 (1986).
Nystrom, C., et al., Direct compression of tablets, Powder Tech, 46(1), 67-76; 47(3), 201-210 (1986).
Rademacher, F.J.C., and Haaker, G., Possible deviations in determination of bulk solid characteristics, caused by loading mechanism of the Jenike shear cell, Powder Tech, 46(1), 33-44 (1986).
Rossiter, A.P., and Douglas, J.M., Design and optimisation of solids processes (3 parts), CER&D, 64(3), 175-196 (1986).
Stovall, T.; de Larrard, F., and Buil, M., Linear packing density model of grain mixtures, Powder Tech, 48(1), 1-12 (1986).
Wakeman, R.J., Developments in solid-liquid separation, CER&D, 64(2), 80-82 (1986).
Herron, D.J., Choosing purge vessels for mass transfer and solids handling, Chem Eng, 7 Dec, 107-110 (1987).
Hobson, L., and Gupta, D., Transporting ferromagnetic powders using the linear induction motor, Proc Engng, April, 39-42 (1987).
Jain, R.C.; Acharjee, D.K., and Gupta, P.S., Some studies in the leaching of rock phosphate in a jet contactor, Chem Eng J, 35(2), 105-114 (1987).
Jaraiz, E., and Estevez, A.M., Design concepts for collar-type magnetic valve for small-size solids: Theory and experiments, Powder Tech, 53(1), 1-10 (1987).
Knaff, G., and Schlunder, E.U., Mass transfer for dissolving solids in supercritical carbon dioxide, Chem Eng & Proc, 21(3), 151-162; 21(4), 193-198 (1987).
Lee, W.K., Rheological nature of solid pressures of granular media, Powder Tech, 51(3), 261-266 (1987).
Various, Advances in particulate technology (special topic issue), Chem Eng Sci, 42(4), 591-922 (1987).
Various, Fine particle research (topic issue), Powder Tech, 51(1), 1-134 (1987).

1988

Alonso, M.; Satoh, M., and Miyanami, K., Powder coating in a rotary mixer with rocking motion, Powder Tech, 56(2), 135-141 (1988).
Bronkala, W.J., Selecting magnetic separators, Chem Eng, 10 Oct, 89-92 (1988).
Bronkala, W.J., Magnetic separations, Chem Eng, 14 March, 133-138 (1988).
Fang, C.S., et al., Microwave demulsification, Chem Eng Commns, 73, 227-240 (1988).
Huang, C.C., and Kono, H.O., Mathematical coalescence model in batch fluidized-bed granulator, Powder Tech, 55(1), 35-50 (1988).
Huang, C.C., and Kono, H.O., Granulation of partially prewetted alumina powder: New concept in coalescence mechanism, Powder Tech, 55(1), 19-34 (1988).
Johanson, J.R., and Johanson, K.D., Improve solids processing, Chem Eng, 15 Feb, 77-81 (1988).

Kim, N.K.; Srivastava, R., and Lyon, J., Simulation of industrial rotary calciner with Trona ore decomposition, Ind Eng Chem Res, 27(7), 1194-1198 (1988).
Knight, P.C., and Johnson, S.H., Measurement of powder cohesive strength with penetration test, Powder Tech, 54(4), 279-284 (1988).
Litster, J.D., and Waters, A.G., Influence of material properties of iron ore sinter feed on granulation effectiveness, Powder Tech, 55(2), 141-152 (1988).
Liu, Y.A., et al., Studies in magnetochemical engineering, Powder Tech, 56(4), 259-292 (1988).
Nikolakakis, I., and Pilpel, N., Effects of particle shape and size on tensile strengths of powders, Powder Tech, 56(2), 95-104 (1988).
Siegell, J.H., Magnetically frozen beds, Powder Tech, 55(2), 127-132 (1988).
Tampy, G.K., et al., Wettability measurements of coal using a modified Washburn technique, Energy & Fuels, 2(6), 782-787 (1988).

16.3 Weighing and Powder Metering

1968-1988

Pryadkin, N.M.; Shklyarenko, Y.V., and Sobolov, G.P., Continuous bulk doser for metering powdery materials with high accuracy, Int Chem Eng, 8(4), 724-726 (1968).
Chatters, K.E., Using load cell weighing systems, Brit Chem Eng, 14(1), 49-50 (1969).
Kravitz, S., Estimating weight of natural solids piles, Chem Eng, 8 Sept, 166 (1969).
Kumar, R., Measuring device for mass flowrate of solids, Chem Eng, 6 Oct, 196-198 (1969).
Parkins, E.G., and Alcorn, J.B., Process control and automatic weighing, Proc Tech Int, 17(11), 877-882 (1972).
Burke, A.J., Weighing bulk materials in the process industries, Chem Eng, 5 March, 66-80 (1973).
Turner, G.A., The conditions for linearity and reproducibility of a belt weigher, Trans IChemE, 51, 1-3 (1973).
Various, Methods for conveying and weighing solids, Chem Eng, 28 Feb, 176-186; 28 March, 97-106 (1977).
Various, Weighing (feature report), Processing, Sept, 93-106 (1979).
Various, Process weighing (topic issue), Proc Engng, Oct, 63-79 (1979).
Anon., Process weighing (feature report), Processing, Sept, 29-41 (1980).
Belsham, P., Agitated weighing, Processing, Jan, 37 (1981).
Al-Din, N., and Gunn, D.J., Metering of solids by a rotary valve feeder, Powder Tech, 36, 25-31 (1983).
Zecchin, P., Process weighing for the chemical industry, Chem Engnr, Nov, 28-29 (1983).
Kravitz, M., Choosing weigh-belt and loss-in-weight feeders, Chem Eng, 6 Aug, 65-68 (1984).
Pettit, J.R., Feeding solids accurately, Chem Eng, 20 Aug, 159-162 (1984).
Tayler, C., Automation of process weighing systems, Proc Engng, March, 51 (1984).

Peate, J., Batchweighing technology, Proc Engng, Jan, 55,57 (1985).
Turland, J., Safety and control in process weighing, Chem Engnr, May, 54-56 (1985).
Ford, D., and Smith, D., Advances in loadcell weighing, Chem Engnr, Nov, 49-51 (1986).
McKenzie, G., Measuring mass directly, Chem Engnr, Oct, 33-36 (1987).
Proctor, A., Process weighing equipment and systems, Proc Engng, Sept, 69-72 (1987).
Zecchin, P., Process weighing in pharmaceutical production, Chem Engnr, July, 42-43 (1987).
Redman, J., Weighing up weighing, Chem Engnr, Nov, 51-56 (1988).

16.4 Particle Size Analysis

1967-1972

Richards, J.C., A method to assess efficiency of size separation of industrial air-swept classifiers, Trans IChemE, 45, T385-391 (1967).
De Chazal, L.E.M., and Hung, Y.C., Effect of sample size on analysis of powder mixtures, AIChEJ, 14(1), 169-173 (1968).
Lapple, C.E., Particle-size analysis and analyzers, Chem Eng, 20 May, 149-156 (1968).
Rose, H.E., and English, J.E., Effect of particle size characteristics on differential thermal analyses of powdered materials, Brit Chem Eng, 13(8), 1135-1140 (1968).
Allen, T., and Khan, A.A., Critical evaluation of powder sampling procedures, Chem Engnr, May, CE108-112 (1970).
Garrett, K.H., and James, R.H., Methods for controlling the properties of powders, Chem Engnr, April, CE74-78 (1970).
Heywood, H., Communications and definitions in particle technology research, Chem Engnr, April, CE70-73, 78 (1970).
Olivier, J.P.; Hickin, G.K., and Orr, C., Rapid automatic particle size analysis in the subsieve range, Powder Tech, 4, 257-263 (1970).
Pietsch, W., Improving powders by agglomeration, Chem Eng Prog, 66(1), 31-35 (1970).
Rosen, H.N., and Hulburt, H.M., Size analysis of irregular-shaped particles in sieving, Ind Eng Chem Fund, 9(4), 658-661 (1970).
Gebbett, J.G., Sieving, straining and grading, Brit Chem Eng, 16(7), 613-614 (1971).
Harris, C.C., A multi-purpose Alyavdin-Rosin-Rammler-Weibull chart, Powder Tech, 5, 39-42 (1971).
Caplan, F., Calculating openings in wire mesh screens, Chem Eng, 15 May, 132 (1972).
English, J.E., Detection of small differences in the specific surface area of powders, Brit Chem Eng, 17(9), 715-716 (1972).
Matthews, C.W., Size reduction of solids by screening, Chem Eng, 10 July, 76-83 (1972).
Zanker, A., Nomogram for calculation of open area of square mesh cloth, Proc Tech Int, 17(12), 969 (1972).

1973-1976

Lines, R.W., Sampling for particle size analysis with the Coulter counter, Powder Tech, 7, 129-136 (1973).

Records, F.A., Sieving practice and the gyrating screen, Proc Tech Int, Jan, 47-53 (1973).

Rose, H.E., and English, J.E., The influence of blinding material on the results of test sieving, Trans IChemE, 51, 14-21 (1973).

Saksena, R.K., and Mitra, C.R., Batch classification of mixtures, Periodica Polytechnica, 17, 277-294 (1973).

Patel, D.J., Determining the efficiency of classifying equipment, Chem Eng, 18 March, 112 (1974).

Various, Mercury intrusion porosimetry (topic issue), Powder Tech, 9, 157-211 (1974).

Bucsky, G., Particle-size measurement techniques, Proc Engng, July, 52-54 (1975).

Gupta, V.S.; Fuerstenau, D.W., and Mika, T.S., Investigation of sieving in presence of attrition, Powder Tech, 11, 257-271 (1975).

Narasimhan, K.S., and Sastri, S.R.S., Estimating the size distribution of crushed products, Chem Eng, 9 June, 77-79 (1975).

Pulvermacher, B., and Ruckenstein, E., Time evolution of the size spectrum in granulation, Chem Eng J, 9(1), 21-30 (1975).

Karuhn, R.; Davies, R.; Kaye, B.H., and Clinch, M.J., Studies on the Coulter counter, Powder Tech, 11, 157-171 (1975); 12, 157-166 (1975); 13, 193-201 (1976).

Konowalchuk, H.; Naylor, A.G., and Kaye, B.H., Rapid method for assessing the quality of sieves, Powder Tech, 13, 97-101 (1976).

Williams, J.C., Segregation of particulate materials: A review, Powder Tech, 15, 245-251 (1976).

Zanker, A., Estimating equivalent diameters of solids, Chem Eng, 5 July, 101-104 (1976).

1977-1979

Gibson, K.R., Particle classification efficiency calculations by geometry, Powder Tech, 18, 165-170 (1977).

King, E.H., How to determine plant screening requirements, Chem Eng Prog, 73(5), 74-79 (1977).

Small, H., Measure particle size by hydrodynamic chromatography, Chemtech, March, 196-200 (1977).

Bernotat, S., and Gregor, W., Aspects of separation in a cross-flow air-classifier, Chem Eng Sci, 33(6), 751-758 (1978).

Johanson, J.R., What to do about particle segregation, Chem Eng, 8 May, 183-188 (1978).

Vose, J.R., Separating grain components by air classification, Sepn & Purif Methods, 7(1), 1-30 (1978).

Zanker, A., Size distribution of ground products, Chem Eng, 31 July, 113 (1978).

Angelidou, C.; Psimopoulos, M., and Jameson, G.J., Size distribution functions of dispersions, Chem Eng Sci, 34(5), 671-676 (1979).

Bright, D., and Chabay, I., Measuring aerosol particles, Chemtech, Nov, 694-699 (1979).
Harrison, L., Guide to process screening, Proc Engng, March, 103,105 (1979).
Stone, L.H., Upgrading circular vibratory screen separators, Chem Eng, 15 Jan, 125-130 (1979).
Tsubaki, J., and Jimbo, G., Proposed new characterization of particle shape and its application, Powder Tech, 22, 161-178 (1979).
Various, Analytical sieving technology (topic issue), Powder Tech, 24, 113-166 (1979).
von Wolff, W.T.E., and Gaigher, J.L., Particle-size analysis of fine grit from a pulverised-coal-fired power station, J Inst Energy, 52, 11-14 (1979).

1980-1984

Kaye, B.H., Standard fine-particles, Chemtech, Jan, 40-43 (1980).
Beddow, J.K., Dry separation techniques, Chem Eng, 10 Aug, 70-84 (1981).
Beeckmans, J.M., Inclined-plane particle classifier, Powder Tech, 28, 129-134 (1981).
Falivene, P.J., Graph paper for sieve analyses, Chem Eng, 23 Feb, 87-89 (1981).
Pham, M.L., Comparison of the three calibration techniques for Coulter counters, Powder Tech, 28, 217-220 (1981).
Various, Mercury porosimetry (special issue), Powder Tech, 29, 1-208 (1981).
Smithwick, R.W., Generalized analysis for mercury porosimetry, Powder Tech, 33, 201-209 (1982).
Stone, L.H., Protecting screening machines from products they process, Chem Eng Prog, 78(12), 64-68 (1982).
Viswanathan, K., and Mani, B.P., A new particle size distribution, Ind Eng Chem Proc Des Dev, 21(4), 776-778 (1982).
Beeckmans, J.M., and Hill, J., Probability screening, Powder Tech, 35, 263-269 (1983).
Diehl, D.S., Fiber savings by pressure-fed static screens, Chem Eng Prog, 79(9), 23-26 (1983).
Harris, C.C., Mineral sample preparation in sub-sieve size ranges, Powder Tech, 34, 131-134 (1983).
Meloy, T.P., and Makino, K., Characterizing residence times of powder samples on sieves, Powder Tech, 36, 253-258 (1983).
Various, Particle-size analysis and pore-size measurement in the construction industry (topic issue), Chem & Ind, 15 Aug, 626-635 (1983).
Zenz, F.A., Particulate solids, Chem Eng, 28 Nov, 61-67 (1983).
Erdesz, K., Separation of two-component particulate material mixture by vibro-impacting separator, Sepn Sci Technol, 19(4), 241-260 (1984).
Harnby, N., Trends in powder mixing, Chem Engnr, July, 22-23 (1984).
Johnson, D.L., et al., Particle analysis to bulk analysis, Chemtech, Nov, 678-683 (1984).
Muldoon, J., Screens and screening, Chem Eng, 6 Feb, 90-94 (1984).
Viswanathan, K.; Aravamudhan, S., and Mani, B.P., Solids separation based on shape, Powder Tech, 39, 83-98 (1984).

1985-1986

Beeckmans, J.M.; Germain, E.H.R., and McIntyre, A., Performance characteristics of a probability screening machine, Powder Tech, 43, 249-256 (1985).

Durst, F., and Macagno, M., Experimental particle size distributions and their representation by log-hyperbolic functions, Powder Tech, 45, 223-244 (1985).

Hou, T.H., Evaluation of separator performance by number-size distribution data, Powder Tech, 41, 99-104 (1985).

Jimbo, G.; Yamazaki, M.; Tsubaki, J., and Suh, T.S., Mechanism of classification in Sturtevant-type air classifier, Chem Eng Commns, 34(1), 37-48 (1985).

Novak, J.W., and Thompson, J.R., Extending the use of particle sizing instrumentation to calculate particle shape factors, Powder Tech, 45, 159-169 (1985).

Rich, G.A., Do you need a fourth particulate-sampling test? Chem Eng, 5 Aug, 117-119 (1985).

Standish, N., Kinetics of batch sieving, Powder Tech, 41, 57-67 (1985).

Zimmels, Y., Theory of density separations of particulate systems, Powder Tech, 43, 127-139 (1985).

Beeckmans, J.M., and Thielen, S., Performance characteristics of roller-gap particle classifier, Powder Tech, 48(2), 181-186 (1986).

Chen, Y.M., and Doo, S.W., Experimental investigation of particle size analysis by modified Andreasen pipet, Powder Tech, 48(1), 23-30 (1986).

Clark, N.N., Three techniques for implementing digital fractal analysis of particle shape, Powder Tech, 46(1), 45-52 (1986).

Gerrard, M.; Puc, G., and Simpson, E., Optimize the design of wire-mesh separators, Chem Eng, 10 Nov, 91-93 (1986).

Haff, P.K., and Werner, B.T., Computer simulation of mechanical sorting of grains, Powder Tech, 48(3), 239-246 (1986).

Hostomsky, J., et al., Size analysis of non-spherical particles, Powder Tech, 49(1), 45-52 (1986).

Iinoya, K., Performance evaluation of particle size classifiers, Ind Eng Chem Fund, 25(4), 701-704 (1986).

Klumpar, I.V.; Currier, F.N., and Ring, T.A., Air classifiers, Chem Eng, 3 March, 77-92 (1986).

Rosato, A.; Prinz, F.; Standburg, K.J., and Swendsen, R., Monte Carlo simulation of particulate matter segregation, Powder Tech, 49(1), 59-70 (1986).

Standish, N.; Bharadwaj, A.K., and Hariri-Akbari, G., Effect of operating variables on efficiency of vibrating screen, Powder Tech, 48(2), 161-172 (1986).

1987-1988

Benbow, J.J.; Ouchiyama, N., and Bridgwater, J., Prediction of extrudate pore structure from particle size, Chem Eng Commns, 62(1), 203-220 (1987).

Furuuchi, M., et al., Optimal performance of a shape classifier for binary mixtures of granular materials, Powder Tech, 50(2), 137-146 (1987).

Luerkens, D.W.; Beddow, J.K., and Vetter, A.F., Structure and morphology: The science of form applied to particle characterization (review paper), Powder Tech, 50(2), 93-102 (1987).

Yamada, Y.; Yasuguchi, M., and Iinoya, K., Effect of particle dispersion and circulation systems on classification performance, Powder Tech, 50(3), 275-280 (1987).

Anon., Particle-size analyzers, Chem Eng, 7 Nov, 95-98 (1988).

Cameron, P.W., and Frey, R., Shortcut for screening area, Chem Eng, 11 April, 90-92 (1988).

Chin, A.D.; Butler, P.B., and Luerkens, D.W., Influence of particle shape on the size measured by light-blockage technique, Powder Tech, 54(2), 99-106 (1988).

Clark, N.N., and Meloy, T.P., Particle-sample shape description with an automated cascadograph particle analyzer, Powder Tech, 54(4), 271-278 (1988).

D'Amore, M., et al., Apparent particle density of a fine powder, Powder Tech, 56(2), 129-134 (1988).

Furuuchi, M., and Gotoh, K., Continuous shape separation of binary mixture of granular particles, Powder Tech, 54(1), 31-38 (1988).

Glaves, C.L.; Davis, P.J., and Smith, D.M., Surface area determination via NMR: Fluid and frequency effects, Powder Tech, 54(4), 261-270 (1988).

Islam, M.N., and Matzen, R., Size distribution analysis of ground wheat by a hammer mill, Powder Tech, 54(4), 235-242 (1988).

Popplewell, L.M., et al., Description of normal, log-normal, and Rosin-Rammler particle populations by modified beta distribution function, Powder Tech, 54(2), 119-126 (1988).

Popplewell, L.M.; Campanella, O.H., and Peleg, M., Comparison of modified beta and normal distribution functions for description of populations with finite size range, Powder Tech, 54(2), 157-160 (1988).

Whiteman, M., and Ridgway, K., Comparison between two methods of shape-sorting particles, Powder Tech, 56(2), 83-94 (1988).

Yamamoto, H.; Utsumi, R., and Kushida, A., Aperture size in a screen of plain Dutch weave, Int Chem Eng, 28(3), 455-461 (1988).

16.5 Comminution

1967-1970

Maroudas, N.G., Electrohydraulic crushing, Brit Chem Eng, 12(4), 558-562 (1967).

Barlow, C.G., The granulation of powders, Chem Engnr, July, CE196-201 (1968).

Clarke, B., and Kitchener, J.A., Influence of pulp viscosity on fine grinding in a ball mill, Brit Chem Eng, 13(7), 991-995 (1968).

Halasyamani, P.; Venkatachalam, S., and Mallikarjunan, R., Influence of pH on kinetics of comminution of quartz, Ind Eng Chem Proc Des Dev, 7(1), 79-83 (1968).

Suzuki, A., and Tanaka, T., Crushing efficiency in relation to some operational variables and material constraints, Ind Eng Chem Proc Des Dev, 7(2), 161-166 (1968).

Garside, J., and Wildsmith, J.A., Prediction of performance of swing hammer crushers used for fertilisers, Trans IChemE, 47, T285-291 (1969).

Kapur, P.C., and Fuerstenau, D.W., Coalescence model for granulation, Ind Eng Chem Proc Des Dev, 8(1), 56-62 (1969).

Yigit, E.; Johnston, H.A., and Maroudas, N.G., Selective breakage in electrohydraulic comminution, Trans IChemE, 47, T332-334 (1969).

Hiorns, F.J., Advances in comminution, Brit Chem Eng, 15(12), 1565-1572 (1970).

Kapur, P.C., and Agrawal, P.K., Approximate solutions to the discretized batch grinding equation, Chem Eng Sci, 25(6), 1111-1113 (1970).

Owe Berg, T.G., and Avis, L.E., Exploratory experiments on kinetics of comminution, Powder Tech, 4, 27-31 (1970).

Tamura, K., and Tanaka, T., Rate of ball milling and vibration milling on basis of comminution law, Ind Eng Chem Proc Des Dev, 9(2), 165-173 (1970).

Vederaman, R.; Raghavendra, N.M., and Venkateswarlu, D., Studies in vibration milling, Powder Tech, 4, 313-321 (1970).

1971-1972

Austin, L.G., Introduction to mathematical description of grinding as a rate process (review paper), Powder Tech, 5, 1-17 (1971).

Austin, L.G., and Bhatia, V.K., Experimental methods for grinding studies in laboratory mills, Powder Tech, 5, 261-266 (1971).

Austin, L.G., and Luckie, P.T., Methods for determination of breakage distribution parameters, Powder Tech, 5, 215-222 (1971).

Austin, L.G., and Luckie, P.T., Estimation of non-normalized breakage distribution parameters from batch grinding tests, Powder Tech, 5, 267-271 (1971).

Furuya, M.; Nakajima, Y., and Tanaka, T., Theoretical analysis of closed-circuit grinding system based on comminution kinetics, Ind Eng Chem Proc Des Dev, 10(4), 449-456 (1971).

Kapur, P.C., Energy-size reduction relationships in comminution of solids, Chem Eng Sci, 26(1), 11-16 (1971).

Matsui, K.; Kurihara, T., and Sekiguchi, T., Fundamental relations in ball mill crushing, Int Chem Eng, 11(1), 162-167 (1971).

Sastri, S.R.S., and Narasimhan, K.S., Nomogram for equilibrium size distribution of grinding media in tumbling mills, Brit Chem Eng, 16(2), 233 (1971).

Snow, R.H., Annual review of size reduction, Powder Tech, 5, 351-364 (1971).

Johnson, R.T., Quick check for attrition of fine particles, Chem Eng, 15 May, 130 (1972).

Pebworth, J.T., Selecting grinding mills for heat-sensitive materials, Chem Eng, 7 Aug, 81-85 (1972).

Prescott, T.W.L., and Webb, F.C., Size distribution produced in a hammer mill, Trans IChemE, 50, 21-25 (1972).

Ratcliffe, A., Size reduction of solids by crushing and grinding, Chem Eng, 10 July, 62-75 (1972).

Somasundaran, P., and Lin, I.J., Effect of the nature of environment on comminution processes, Ind Eng Chem Proc Des Dev, 11(3), 321-331 (1972).

Tanaka, T., Scale-up theory of jet mills on basis of comminution kinetics, Ind Eng Chem Proc Des Dev, 11(2), 238-241 (1972).

1973-1974

Austin, L.G., Comments on Kick, Rittinger, and Bond laws of grinding, Powder Tech, 7, 315-317 (1973).

Austin, L.G., Understanding ball mill sizing, Ind Eng Chem Proc Des Dev, 12(2), 121-129 (1973).

Austin, L.G.; Shoji, K., and Everett, M.D., Explanation of abnormal breakage of large particle sizes in laboratory mills, Powder Tech, 7, 3-7 (1973).

Furuya, M.; Nakajima, Y., and Tanaka, T., Design of closed-circuit grinding system with tube mill and nonideal classifier, Ind Eng Chem Proc Des Dev, 12(1), 18-23 (1973).

McLain, L., Cyrogrinding, Proc Engng, June, 62-67 (1973).

Nakajima, Y., and Tanaka, T., Solution of batch grinding equation, Ind Eng Chem Proc Des Dev, 12(1), 23-25 (1973).

Reed, R.M., and Reynolds, J.C., The spherodizer granulation process, Chem Eng Prog, 69(2), 62-66 (1973).

Rumpf, H., Physical aspects of comminution and new formulation of a law of comminution, Powder Tech, 7, 145-159 (1973).

Sastri, S.R.S., and Narasimhan, K.S., Nomogram for make-up ball size for grinding, Proc Tech Int, Oct, 383 (1973).

Snow, R.H., Annual review of size reduction (1972), Powder Tech, 7, 69-83 (1973).

Berg, S., Comminution and particle size analysis, Powder Tech, 10, 1-8 (1974).

Gupta, V.K., and Kapur, P.C., Simple mill matrix for grinding mills, Chem Eng Sci, 29(2), 634-637 (1974).

Ouchiyama, N., and Tanaka, T., Mathematical model in the kinetics of granulation, Ind Eng Chem Proc Des Dev, 13(4), 383-389 (1974).

Shoji, K., and Austin, L.G., Model for batch rod milling, Powder Tech, 10, 29-35 (1974).

Snow, R.H., and Luckie, P.T., Annual review of size reduction (1973), Powder Tech, 10, 129-142 (1974).

Whiten, W.J., Matrix theory of comminution machines, Chem Eng Sci, 29(2), 589-600 (1974).

1975-1976

Obeng, D.M., and Trezek, G.J., Simulation of comminution of heterogeneous mixtures of brittle and nonbrittle materials in a swing hammermill, Ind Eng Chem Proc Des Dev, 14(2), 113-117 (1975).

Ouchiyama, N., and Tanaka, T., Probability of coalescence in granulation kinetics, Ind Eng Chem Proc Des Dev, 14(3), 286-289 (1975).

Sadler, L.Y.; Stanley, D.A., and Brooks, D.R., Attrition mill operating characteristics, Powder Tech, 12, 19-28 (1975).

Sastri, S.R.S., and Narasimhan, K.S., Predicting grinding-mill energy use, Chem Eng, 1 Sept, 103-106 (1975).

Zanker, A., Estimating power for crushing and grinding, Chem Eng, 14 April, 105 (1975).
Austin, L.G.; Shoji, K., and Luckie, P.T., Effect of ball size on mill performance, Powder Tech, 14, 71-79 (1976).
Bond, A., Developments in comminution, Proc Engng, March, 55-57 (1976).
Jindal, V.K., and Austin, L.G., Kinetics of hammer milling of maize, Powder Tech, 14, 35-39 (1976).
Menyhart, M., and Miskiewicz, L., Comminution and structural changes in a jet mill, Powder Tech, 15, 261-266 (1976).
Mika, T.S., Solution of distributed parameter model of continuous grinding mill at steady state, Chem Eng Sci, 31(4), 257-262 (1976).
Pokorny, J., and Zaloudik, P., Dry grinding in agitated ball mills, Powder Tech, 15, 181-186 (1976).
Snow, R.H., and Luckie, P.T., Annual review of size reduction (1974), Powder Tech, 13, 33-48 (1976).
Wary, J., and Davis, R.B., Cryopulverizing, Chemtech, March, 200-203 (1976).
Zanker, A., Nomograph for ball diameters in grinding mills, Proc Engng, Dec, 42-43 (1976).
Zanker, A., Sizing balls and rods for use in grinding mills, Chem Eng, 20 Dec, 86 (1976).
Zanker, A., Calculating the optimum and critical speeds for ball mills, Chem Eng, 2 Aug, 111 (1976).

1977-1979

Cutting, G.W., Grindability assessments using laboratory rod mill tests, Chem Engnr, Oct, 702-704 (1977).
Le Houillier, R.; Van Neste, A., and Marchand, J.C., Influence of charge on the parameters of the batch grinding equation and its implications in simulation, Powder Tech, 16, 7-15 (1977).
Lowrison, G.C., The future for comminution, Chem Engnr, Oct, 699-701 (1977).
Opoczky, L., Fine grinding and agglomeration of silicates, Powder Tech, 17, 1-7 (1977).
Smith, E.A., Comminution problems, Processing, Nov, 35-37 (1977).
Solomon, J.A., and Mains, G.J., A mild, protective and efficient procedure for grinding coal: Cryocrushing, Fuel, 56(3), 302-304 (1977).
Stevens, D.J., Size reduction problems and processes in flour-milling, Chem Engnr, Oct, 705-708 (1977).
Klimpel, R.R., and Manfroy, W., Chemical grinding aids for increasing throughput in wet grinding of ores, Ind Eng Chem Proc Des Dev, 17(4), 518-523 (1978).
Ryason, P.R., and England, C., New method of feeding coal: Continuous extrusion of fully plastic coal, Fuel, 57(4), 241-244 (1978).
Zanker, A., Shortcut technique gives size distribution of comminuted solids, Chem Eng, 24 April, 101-103 (1978).
Austin, L.G.; Jindal, V.K., and Gotsis, C., Model for continuous grinding in a laboratory hammer mill, Powder Tech, 22, 199-204 (1979).
Cutting, G.W., The characterisation of mineral release during comminution processes, Chem Engnr, Dec, 845-849 (1979).

Hersey, J.A., and Krycer, I., Fine grinding and the production of coarse particulates, Chem Engnr, Dec, 837-840 (1979).

Prem, H., and Prior, M., Impact mills for fine size reduction, Chem Engnr, Dec, 841-844 (1979).

Snow, R.H., and Luckie, P.T., Annual review of size reduction (1975), Powder Tech, 23, 31-46 (1979).

1980-1981

Biddulph, M.W., Coolers for cryogenic grinding, Chem Eng, 11 Feb, 93-96 (1980).

Hunt, L., Equipment for grinding, Proc Engng, March, 68-71 (1980).

Krogh, S.R., Crushing characteristics, Powder Tech, 27, 171-181 (1980).

Krycer, I., and Hersey, J.A., Comparative study of commminution in rotary and vibratory ball mills, Powder Tech, 27, 137-141 (1980).

Ouchiyama, N., and Tanaka, T., Stochastic model for compaction of pellets in granulation, Ind Eng Chem Proc Des Dev, 19(4), 555-560 (1980).

Shoji, K.; Lohrasb, S., and Austin, L.G., Variation of breakage parameters with ball and powder loading in dry ball milling, Powder Tech, 25, 109-114 (1980).

Waldie, B., and Robinson, T., Granulation in spouted beds: Attrition and other mechanisms, Powder Tech, 27, 163-169 (1980).

Zanker, A., Nomograph for grinding energy, Proc Engng, March, 63,65 (1980).

Austin, L.G., and Bagga, P., Analysis of fine dry grinding in ball mills, Powder Tech, 28, 83-90 (1981).

Austin, L.G.; Bagga, P., and Celik, M., Breakage properties of some materials in a laboratrory ball mill, Powder Tech, 28, 235-243 (1981).

Austin, L.G.; Van Orden, D.; McWilliams, B., and Perez, J.W., Breakage parameters of some materials in smooth roll crushers, Powder Tech, 28, 245-251 (1981).

Bruynseels, J.P., Granulate in fluid bed, Hyd Proc, 60(9), 203-208 (1981).

Crisi, J.S., Selecting balls and liners for grinding mills, Chem Eng, 7 Sept, 127-129 (1981).

Holt, C.B., Shape of particles produced by comminution: A review, Powder Tech, 28, 59-63 (1981).

Kapur, P.C.; Sastry, K.V.S., and Fuerstenau, D.W., Mathematical models of open-circuit balling or granulating devices, Ind Eng Chem Proc Des Dev, 20(3), 519-524 (1981).

Krycer, I., and Hersey, J.A., Grinding and granulation in a vibratory ball mill, Powder Tech, 28, 91-95 (1981).

Shinozaki, M., and Senna, M., Effects of number and size of milling balls on the mechanochemical activation of fine crystalline solids, Ind Eng Chem Fund, 20(1), 59-62 (1981).

Swaroop, S.H.R.; Abouzeid, A.Z.M., and Fuerstenau, D.W., Flow of particulate solids through tumbling mills, Powder Tech, 28, 253-260 (1981).

1982-1983

Austin, L.G., et al., Analysis of ball-and-race milling, Powder Tech, 29, 263-275 (1981); 33, 113-134 (1982).

Austin, L.G.; Shoji, K., and Ball, D., Rate equations for non-linear breakage in mills due to materials effects, Powder Tech, 31, 127-133 (1982).
Cross, M., Method for extracting product size distributions from empirical comminution models, Powder Tech, 31, 233-237 (1982).
Dowding, C.H., and Lytwynyshyn, G., Point load-determination relationships and design of jaw crusher plates, Powder Tech, 31, 277-286 (1982).
Gupta, V.K.; Hodouin, D., and Everell, M.D., Analysis of wet grinding operation using linearized population balance model for pilot scale grate-discharge ball mill, Powder Tech, 32, 233-244 (1982).
Klimpel, R., Laboratory studies of grinding and rheology of coal-water slurries, Powder Tech, 32, 267-277 (1982).
Klimpel, R.R., and Austin, L.G., Chemical additives for wet grinding of minerals, Powder Tech, 31, 239-253 (1982).
Mohanty, B., and Narasimhan, K.S., Fluid energy grinding, Powder Tech, 33, 135-141 (1982).
Shoji, K.; Austin, L.G.; Smaila, F.; Brame, K., and Luckie, P.T., Further studies of ball and powder filling effects in ball milling, Powder Tech, 31, 121-126 (1982).
Wheeler, D.A., Size reduction: Plant and equipment survey, Processing, Nov, 55-59 (1982).
Austin, L.G., and Brame, K., Comparison of Bond method for sizing wet tumbling ball mills with a size-mass balance simulation model, Powder Tech, 34, 261-274 (1983).
Austin, L.G.; Rogovin, Z.; Rogers, R.S.C., and Trimarchi, T., Axial mixing model applied to ball mills, Powder Tech, 36, 119-126 (1983).
Chan, S.Y.; Pilpel, N., and Cheng, D.C.H., Tensile strengths of single powders and binary mixtures, Powder Tech, 34, 173-189 (1983).
Hulbert, D.G., and Woodburn, E.T., Multivariable control of a wet-grinding circuit, AIChEJ, 29(2), 186-191 (1983).
Peate, J., Developments in grinding, Proc Engng, May, 41-43 (1983).
Rogers, R.S.C., The double-roll crusher model applied to a two-component feed material, Powder Tech, 35, 131-134 (1983).
Rogers, R.S.C., Generalized transfer parameter treatment of crushing and grinding circuit simulation, Powder Tech, 36, 137-143 (1983).
Rogers, R.S.C., and Shoji, K., A double-roll crusher model applied to a full scale crusher, Powder Tech, 35, 123-129 (1983).
Voller, V.R., Energy-size reduction relationships in comminution, Powder Tech, 36, 281-286 (1983).

1984

Ansems, A., and Bloebaum, R.K., Chemical comminution, Powder Tech, 40, 265-268 (1984).
Austin, L.G.; Luckie, P.J.; Shoji, K.; Rogers, R.S.C., and Brame, K., Simulation model of air-swept ball mill grinding coal, Powder Tech, 38, 255-266 (1984).
Cottaar, W., and Rietema, K., Effect of interstitial gas on milling, Powder Tech, 38, 183-194 (1984).
El-Shall, H., and Somasundaran, P., Mechanisms of grinding modification by chemical additives (organic reagents), Powder Tech, 38, 267-273 (1984).

El-Shall, H., and Somasundaran, P., Physico-chemical aspects of grinding: Review of use of additives, Powder Tech, 38, 275-293 (1984).
Khoe, G.K.; Ruda, M.M., and Epstein, N., Batch comminution of coal in a spouted bed, Powder Tech, 39, 249-262 (1984).
Klimpel, R.R., and Austin, L.G., Back-calculation of specific rates of breakage from continuous mill data, Powder Tech, 38, 77-91 (1984).
Lytle, J.M., and Prisbrey, K.A., Material-dependent non-linear modeling of fine coal grinding, Powder Tech, 38, 93-97 (1984).
Moir, D.N., Size reduction, Chem Eng, 16 April, 54-68 (1984).
Opoczky, L., and Farnady, F., Fine grinding and states of equilibrium, Powder Tech, 39, 107-115 (1984).
Petela, R., Exergitic efficiency of comminution of solid substances, Fuel, 63(3), 414-418 (1984).
Revnivtsev, V.I., et al., Selective liberation of minerals in inertial cone crushers, Powder Tech, 38, 195-203 (1984).
Stehr, N., Residence time distributions in stirred ball mill and effect on comminution, Chem Eng & Proc, 18(2), 73-84 (1984).

1985

Austin, L.G., and Klimpel, R.R., Ball wear and ball size distributions in tumbling ball mills, Powder Tech, 41, 279-286 (1985).
Austin, L.G., and Rogers, R.S.C., Powder technology in industrial size reduction, Powder Tech, 42, 91-109 (1985).
Cottaar, W.; Rietema, K., and Stemerding, S., Effect of interstitial gas on milling, Powder Tech, 43, 189-198 (1985).
Gotsis, C., and Austin, L.G., Batch grinding kinetics in presence of dead space (hammer mill), Powder Tech, 41, 91-98 (1985).
Gotsis, C., and Trass, O., Performance of Szego mill in dry grinding coal and wheat, Powder Tech, 41, 287-294 (1985).
Gotsis, C.; Austin, L.G.; Luckie, P.T., and Shoji, K., Modeling of grinding circuit with swing-hammer mill and twin-cone classifier, Powder Tech, 42, 209-216 (1985).
Gupta, V.K.; Zouit, H., and Hodouin, D., Effect of ball and mill diameters on grinding rate parameters in dry grinding operation, Powder Tech, 42, 199-208 (1985).
Heim, A.; Leszczyniecki, R., and Amanowicz, K., Determination of parameters for wet-grinding model in Perl mills, Powder Tech, 41, 173-179 (1985).
Koka, V.R.; Papachristodoulou, G., and Trass, O., Settling stability of coal slurries prepared by wet grinding in the Szego mill, Can JCE, 63(4), 585-590 (1985).
Kuga, Y., et al., Measurement and statistical analysis of shape of particles comminuted by a screen mill, Powder Tech, 44, 281-290 (1985).
Papachristodoulou, G., and Trass, O., Grinding of coal-oil slurries with a Szego mill, Can JCE, 63(1), 43-50 (1985).
Peterson, T.W.; Scotto, M.V., and Sarofim, A.F., Comparison of comminution data with analytical solutions of the fragmentation equation, Powder Tech, 45, 87-93 (1985).
Robinson, G.F., Model of transient operation of a coal pulverizer, J Inst Energy, 58, 51-63 (1985).

Standish, N., and Meta, I.A., Kinetic aspects of continuous screening, Powder Tech, 41, 165-171 (1985).

Various, Milling and grinding: A survey, Processing, June, 43-49 (1985).

Various, Milling and grinding developments, Processing, Nov, 39-45 (1985).

Viswanathan, K., Theoretical expressions for the distribution function of comminution kinetics, Ind Eng Chem Fund, 24(3), 339-343 (1985).

Weismantal, G., and Sresty, G., Mixing and size reduction (special advertising section), Chem Eng, 24 June, 71-109 (1985).

Zanker, A., Design parameters for ball mills, Proc Engng, March, 63 (1985).

1986

Anon., Mixing and size reduction (special advertising section), Chem Eng, 23 June, 53-92 (1986).

Austin, L.G., et al., Improved simulation model for semi-autogenous grinding, Powder Tech, 47(3), 265-284 (1986).

Austin, L.G.; Barahona, C.A., and Menacho, J.M., Fast and slow chipping fracture and abrasion in autogenous grinding, Powder Tech, 46(1), 81-88 (1986).

Belardi, G.; Bonifazi, G., and Massacci, P., Particle breakage and mineral liberation, CER&D, 64(5), 381-391 (1986).

Bignell, J.D., and Newton, S., Comminution: A review, CER&D, 64(2), 91-93 (1986).

Cottaar, W., and Rietema, K., Effect of interstitial gas on milling, Powder Tech, 46(1), 89-98 (1986).

Dutkiewicz, R.K., et al., Energy-size reduction relationship in impact crushing of coal, Powder Tech, 49(1), 83-86 (1986).

Kanda, Y.; Sano, S., and Yashima, S., Grinding limit based on fracture mechanics, Powder Tech, 48(3), 263-268 (1986).

Mankosa, M.J.; Adel, G.T., and Yoon, R.H., Effect of media size in stirred ball mill grinding of coal, Powder Tech, 49(1), 75-82 (1986).

Menacho, J., and Concha, F., Mathematical model of ball wear in grinding mills, Powder Tech, 47(1), 87-96 (1986).

Menacho, J.M., Some solutions for kinetics of combined fracture and abrasion breakage, Powder Tech, 49(1), 87-96 (1986).

Ogawa, K., Effectiveness of information entropy for evaluation of grinding efficiency, Chem Eng Commns, 46, 1-10 (1986).

Opoczky, L.O.; Verdes, S., and Mrakovics, K., Grinding technology for producing high-strength cement of high slag content, Powder Tech, 48(1), 91-98 (1986).

Yuregir, K.R.; Ghadiri, M., and Clift, R., Impact attrition of granular solids, Powder Tech, 49(1), 53-58 (1986).

1987

Anon., Mixing and Size Reduction '87 (special advertising section), Chem Eng, 22 June, 36-55 (1987).

Austin, L.G., Approximate calculation of specific fracture energies for grinding, Powder Tech, 53(2), 145-150 (1987).

Austin, L.G., and Tangsathitkulchai, C., Comparison of methods for sizing ball mills using open-circuit wet grinding of phosphate ore as a test example, Ind Eng Chem Res, 26(5), 997-1003 (1987).
Bemrose, C.R., and Bridgwater, J., Review of attrition and attrition test methods, Powder Tech, 49(2), 97-126 (1987).
Hang, T., et al., Explosive comminution of bituminous coal using steam, Energy & Fuels, 1(6), 529-535 (1987).
Koka, V.R., and Trass, O., Determination of breakage parameters and modelling of coal breakage in Szego mill, Powder Tech, 51(2), 201-214 (1987).
Koka, V.R.; Hohmann, R., and Trass, O., Flow of dry particulates in the Szego mill, Powder Tech, 51(2), 189-200 (1987).
Mamaghani, A.H.; Beddow, J.K., and Vetter, A.F., Chemical comminution of coal, AIChEJ, 33(2), 319-321 (1987).
Menacho, J.M., and Concha, F.J., Mathematical model of ball wear in grinding mills, Powder Tech, 52(3), 267-278 (1987).
Scieszka, S.F., Grinding behaviour of coal in different size-reduction systems, Powder Tech, 49(2), 191-192 (1987).
Yoda, M.; Tamura, K.; Hashimoto, A., and Sato, Y., Method of crushing using triaxial stress apparatus, Powder Tech, 52(2), 171-178 (1987).

1988

Austin, L.G.; Stubican, J.; Rogers, R.S.C., and Brame, K.A., Rapid computational procedure for unsteady-state ball mill circuit simulation, Powder Tech, 56(1), 1-12 (1988).
Celik, M.S., Acceleration of breakage rates of anthracite during grinding in a ball mill, Powder Tech, 54(4), 227-234 (1988).
Celik, M.S., Comparison of dry and wet fine grinding of coals in a ball mill, Powder Tech, 55(1), 1-10 (1988).
Devaswithin, A.; Pitchumani, B., and de Silva, S.R., Modified back-calculation method to predict particle size distributions for batch grinding in a ball mill, Ind Eng Chem Res, 27(4), 723-726 (1988).
Howat, D.D., and Vermeulen, L.A., Fineness of grind and consumption and wear rates of metallic grinding media in tumbling mills, Powder Tech, 55(4), 231-240 (1988).
Kanda, Y., et al, Fundamental study of dry and wet grinding with breaking strength, Powder Tech, 56(1), 57-62 (1988).
Kanda, Y.; Abe, Y., and Sasaki, H., Examination of ultra-fine grinding by preferential grinding, Powder Tech, 56(3), 143-148 (1988).
Kapur, P.C.; Berloiz, L.M., and Fuerstenau, D.W., Effect of fluid-medium density on ball mill torque, energy distribution and grinding kinetics, Powder Tech, 54(3), 217-224 (1988).
Meloy, T.P., Geometry for characterizing fractured particle shape, Powder Tech, 55(4), 285-292 (1988).
Pauw, O.G., Minimization of overbreakage during repetitive impact breakage of single ore particles, Powder Tech, 56(4), 251-258 (1988).
Pauw, O.G., Optimization of individual events in grinding mills during which breakages occur, Powder Tech, 55(4), 247-256 (1988).
Pauw, O.G., and Mare, M.S., Determination of optimum impact-breakage routes for an ore, Powder Tech, 54(1), 3-14 (1988).

Rogovin, Z., and Hogg, R., Internal classification in tumbling grinding mills, Powder Tech, 56(3), 179-190 (1988).
Tangsathitkulchai, C., and Austin, L.G., Rheology of concentrated slurries of particles of natural size distribution produced by grinding, Powder Tech, 56(4), 293-300 (1988).
Various, Developments in grinding (feature report), Processing, June, 25-34 (1988).
Wen, S.B.; Chen, C.K., and Liu, H.S., Size reduction of magnetic sand to nanometre powder in laboratory vibration mill, Powder Tech, 55(1), 11-18 (1988).

16.6 Particulate Removal

1967-1970

Botterill, J.S.M., and Aynsley, E., Collection of airbourne dusts, Brit Chem Eng, 12(10), 1593-1597; 12(12), 1899-1903 (1967).
Elenkov, D., and Boyadzhiev, K., Hydrodynamics and mass transfer in nozzleless venturi absorber: Absorption of sulfur dioxide in water and aqueous solutions of surfactants, Int Chem Eng, 7(2), 191-194 (1967).
Morash, N.; Krouse, M., and Vosseller, W.P., Removing solid and mist particles from exhaust gases, Chem Eng Prog, 63(3), 70-74 (1967).
Svanda, J., Cyclone calculations and some design parameters affecting cyclone efficiency, Int Chem Eng, 7(2), 238-246 (1967).
Bolmosov, V.I., and Bichkov, A.D., Hydrocyclone operating efficiencies, Int Chem Eng, 8(4), 629-631 (1968).
Srimathi, C.R.; Singh, B., and Bhat, G.N., Transient-state theory of centrifuges, Brit Chem Eng, 13(12), 1725-1727 (1968).
Brink, J.A., and Porthouse, J.D., Efficient dust control via new sampling technique, Chem Eng, 10 March, 106-110 (1969).
Hanson, D.N., and Wilke, C.R., Electrostatic precipitator analysis, Ind Eng Chem Proc Des Dev, 8(3), 357-364 (1969).
Sargent, G.D., Dust collection equipment, Chem Eng, 27 Jan, 130-150 (1969).
Anon., Dust separation trends, Brit Chem Eng, 15(3), 315-326; 15(4), 536 (1970).
Calvert, S., Venturi and other atomizing scrubbers efficiency and pressure drop, AIChEJ, 16(3), 392-396 (1970).
Mukhopadhyay, S.N., and Chowdhury, K.C.R., Collection efficiency of a cyclone separator, Brit Chem Eng, 15(4), 529 (1970).
Walling, J.C., Ins and outs of gas filter bags, Chem Eng, 19 Oct, 162-167 (1970).

1971-1974

Hanf, E.B., Entrainment separator design, Chem Eng Prog, 67(11), 54-59 (1971).
Lancaster, B.W., and Strauss, W., Study of steam injection into wet scrubbers, Ind Eng Chem Fund, 10(3), 362-369 (1971).
Thompson, B.W., and Strauss, W., Application of vortex theory to design of cyclone collectors, Chem Eng Sci, 26(1), 125-132 (1971).

Wagner, J., and Murphy, R.S., Miniature liquid cyclones, Ind Eng Chem Proc Des Dev, 10(3), 346-352 (1971).
Zanker, A., Dust-collector performance calculated with nomograph, Chem Eng, 22 March, 124 (1971).
Vandenhoeck, P., Cooling hot gases before baghouse filtration, Chem Eng, 1 May, 67-70 (1972).
Zanker, A., Nomogram for collection efficiency in electrical dust precipitators, Brit Chem Eng, 17(4), 343 (1972).
Anon., Nomogram for estimating electrostatic precipitator performance, Proc Tech Int, Aug, 339 (1973).
Bloor, M.I.G., and Ingham, D.B., Efficiency of the industrial cyclone, Trans IChemE, 51, 173-176 (1973).
Bloor, M.I.G., and Ingham, D.B., Theoretical investigation of the flow in a conical hydrocyclone, Trans IChemE, 51, 36-41 (1973).
Boll, R.H., Particle collection and pressure drop in venturi scrubbers, Ind Eng Chem Fund, 12(1), 40-50 (1973).
Dullien, F.A.L., and Munro, T.S., Fractional mass efficiency measurements on a wet dust scrubber, Powder Tech, 8, 57-68 (1973).
Peters, J.M., Predicting efficiency of fine-particle collectors, Chem Eng, 16 April, 99-102 (1973).
Gupta, J.P., and Grover, P.D., Optimum design of hydrocyclones, Chemical Processing, June, 38-39 (1974).
Kutepov, A.M., Performance indices for separation processes in hydrocyclones, Int Chem Eng, 14(4), 697-700 (1974).
Owen, L.T., Selecting in-plant dust-control systems, Chem Eng, 14 Oct, 120-126 (1974).
Stamp, J.L., Valve operation system for a cyclone, Chem Eng, 28 Oct, 136 (1974).

1975-1976
Bloor, M.I.G., and Ingham, D.B., The leakage effect in the industrial cyclone, Trans IChemE, 53, 7-11 (1975).
Bloor, M.I.G., and Ingham, D.B., Turbulent spin in a cyclone, Trans IChemE, 53, 1-6 (1975).
Gerrard, A.M., and Liddle, C.J., The optimal selection of multiple hydrocyclone systems, Chem Engnr, May, 295-296 (1975).
Hollands, K.G.T., and Goel, K.C., General method for predicting pressure loss in venturi scrubbers, Ind Eng Chem Fund, 14(1), 16-22 (1975).
Rao, K.N., and Rao, T.C., Estimating hydrocyclone efficiency, Chem Eng, 26 May, 121-122 (1975).
Rothwell, E., Fabric dust filtration, Chem Engnr, March, 138-142 (1975).
Schneider, G.G.; Horzella, T.I.; Cooper, J., and Striegl, P.J., Selecting and specifying electrostatic precipitators, Chem Eng, 26 May, 94-108; 18 Aug, 97-100 (1975).
Svarovsky, L., Gas cyclone selection procedure, Chem Engnr, March, 133-135, 152 (1975).
Swift, P., Wet dedusting: Design and application, Chem Engnr, March, 146-150 (1975).

Swift, P., Dust control related to the bulk delivery of particulate materials, Chem Engnr, March, 143-145, 150 (1975).
Taheri, M., and Sheih, C.M., Mathematical modeling of atomizing scrubbers, AIChEJ, 21(1), 153-157 (1975).
Yoshida, T.; Kousaka, Y.; Inake, S., and Nakai, S., Pressure drop and collection efficiency of an irrigated bag filter, Ind Eng Chem Proc Des Dev, 14(2), 101-105 (1975).
Zenz, F.A., Size cyclone diplegs better, Hyd Proc, 54(5), 125-128 (1975).
Bloor, M.I.G., and Ingham, D.B., Boundary layer flows on the side walls of conical cyclones, Trans IChemE, 54, 276-280 (1976).
Ciliberti, D.F., and Lancaster, B.W., Fine dust collection in a rotary flow cyclone, Chem Eng Sci, 31(6), 499-504 (1976).
Ciliberti, D.F., and Lancaster, B.W., Performance of rotary flow cyclones, AIChEJ, 22(2), 394-398 (1976).
Dorman, R.G., and Maggs, F.A.P., Filtration of fine particles and vapours from gases, Chem Engnr, Oct, 671-674 (1976).
Douglas, P.L.; Dullien, F.A.L., and Spink, D.R., Operating parameters of low-energy wet scrubber for fine particulates, Can JCE, 54, 173-180 (1976).
Gerrard, A.M., and Liddle, C.J., Optimal choice of multiple cyclones, Powder Tech, 13, 251-254 (1976).
Gerrard, A.M., and Liddle, C.J., Designing hydrocyclone systems, Proc Engng, June, 105-107 (1976).
Marietta, M.G., and Swan, G.W., Particle diffusion in electrostatic precipitators, Chem Eng Sci, 31(9), 795-802 (1976).

1977-1978

Beeckmans, J.M., and Kim, C.J., Analysis of efficiency of reverse-flow cyclones, Can JCE, 55, 640-650 (1977).
Billings, C.E., Fabric filter installations for flue-gas fly ash control, Powder Tech, 18, 79-110 (1977).
Bump, R.L., Electrostatic precipitators in industry, Chem Eng, 17 Jan, 129-136 (1977).
Calvert, S., How to choose a particulate scrubber, Chem Eng, 29 Aug, 54-68; 24 Oct, 133-148 (1977).
Doerschlag, C., and Miczek, G., How to choose a cyclone dust collector, Chem Eng, 14 Feb, 64-72 (1977).
Goel, K.C., and Hollands, K.G.T., General method for predicting particulate collection efficiency of venturi scrubbers, Ind Eng Chem Fund, 16(2), 186-193 (1977).
Harwood, C.F.; Siebert, P.C., and Oestreich, D.K., Optimizing baghouse performance to control asbestos emissions, Chem Eng Prog, 73(1), 54-56 (1977).
Koch, W.H., and Licht, W., New design approach boosts cyclone efficiency, Chem Eng, 7 Nov, 80-88 (1977).
Loeffler, F., Fine-particle dust collection, Int Chem Eng, 17(2), 208-217 (1977).
Semrau, K.T., Practical process design of particulate scrubbers, Chem Eng, 26 Sept, 87-91 (1977).
Sheng, H.P., Separation of liquids in a conventional hydrocyclone, Sepn & Purif Methods, 6(1), 89-128 (1977).

Vincent, J.H., Particle dynamics in a grid-type electrostatic precipitator, Chem Eng Sci, 32(9), 1077-1082 (1977).

Zanker, A., Hydrocyclones: Dimensions and performance, Chem Eng, 9 May, 122-125 (1977).

Abrahamson, J.; Martin, C.G., and Wong, K.K., The physical mechanisms of dust collection in a cyclone, Trans IChemE, 56, 168-177 (1978).

Calvert, S., Field evaluation of fine particle scrubbers, Chem Engnr, June, 485-490 (1978).

Calvert, S., Guidelines for selecting mist eliminators, Chem Eng, 27 Feb, 109-112 (1978).

Caplan, F., Sizing electrostatic precipitators, Chem Eng, 10 April, 153 (1978).

Frenkel, D.I., Improving electrostatic precipitator performance, Chem Eng, 19 June, 105-110 (1978).

Gerrard, A.M., and Liddle, C.J., Numerical optimisation of multiple hydrocyclone systems, Chem Engnr, Feb, 107-109 (1978).

Horzella, T.I., Selecting, installing and maintaining cyclone dust collectors, Chem Eng, 30 Jan, 84-92 (1978).

Kelly, W.J., Maintaining venturi-tray scrubbers, Chem Eng, 4 Dec, 133-137 (1978).

Oglesby, S., and Gooch, J.P., Electrostatic precipitation theory: A review, Chem Engnr, June, 473-476, 479 (1978).

Puri, R., Automatic control of reagent feed boosts wet-scrubber efficiency, Chem Eng, 23 Oct, 157-159 (1978).

Rothwell, E., The collection of fine particulates in fabric filters, Chem Engnr, Feb, 115-119 (1978).

Sager, F., Cut polymer costs with cyclones, Hyd Proc, 57(2), 119-121 (1978).

Shah, Y.M., and Price, R.T., Calculator program solves cyclone efficiency, Chem Eng, 28 Aug, 99-102 (1978).

Vincent, J.H., and Humphries, W., Collection of airbourne dusts by bluff bodies, Chem Eng Sci, 33(8), 1147-1156 (1978).

Wrotnowski, A.C., Final filtration with felt bag strainers, Chem Eng Prog, 74(10), 89-93 (1978).

Yuu, S.; Jotaki, T.; Tomita, Y., and Yoshida, K., Reduction of pressure drop due to dust loading in a conventional cyclone, Chem Eng Sci, 33(12), 1573-1580 (1978).

1979-1981

Bundy, R.P., and Plunkett, T.P., Outside bag collectors for industrial boilers, Chem Eng Prog, 75(12), 43-45 (1979).

Fan, K.C., and Gentry, J.W., Effect of packing density on collection efficiencies of charged fibers, Ind Eng Chem Fund, 18(4), 306-311 (1979).

Harker, J.H., and Taha, T.S.A., Use of conditioning agents in electrostatic precipitation, J Inst energy, 52, 150-152 (1979).

Kraus, M.N., Baghouses: Separating and collecting industrial dusts, Chem Eng, 9 April, 94-106; 23 April, 133-142 (1979).

Lund, I.E.; Mills, B., and Moyes, A.J., Practical problems with dust generators and dust collectors, J Inst Energy, 52, 32-44 (1979).

Mozley, R., Hydrocyclones for particle recovery, Processing, Dec, 21-23 (1979).

Laverack, S.D., The effect of particle concentration on the boundary layer flow in a hydrocyclone, Trans IChemE, 58, 33-42 (1980).
Parida, A., and Chand, P., Turbulent swirl with gas-solid flow in a cyclone, Chem Eng Sci, 35(4), 949-954 (1980).
Purchas, D.B., Scaling-up multicomponent shaker bag filters, Chem Engnr, Aug, 551-554 (1980).
Severson, S.D., et al., The economics of fabric filters and precipitators, Chem Eng Prog, 76(1), 68-73 (1980).
Bergmann, L., Baghouse filter fabrics, Chem Eng, 19 Oct, 177-178 (1981).
Dietz, P.W., Collection efficiency of cyclone separators, AIChEJ, 27(6), 888-892 (1981).
Furlong, D.A., and Shevlin, T.S., Fabric filtration at high temperatures, Chem Eng Prog, 77(1), 89-91 (1981).
Kanagawa, A., Errors associated with measurement of dust collection efficiencies with a light-scattering photometer, Int Chem Eng, 21(2), 236-244 (1981).
Maxwell, M.A., Consider baghouses for product recovery, Chem Eng, 28 Dec, 54-56 (1981).
Merrill, F.H., Program calculates hydrocyclone efficiency, Chem Eng, 2 Nov, 71-78 (1981).
Placek, T.D., and Peters, L.K., Analysis of particulate removal in venturi scrubbers: Effect of operating variables on performance, AIChEJ, 27(6), 984-993 (1981).

1982-1983

Boysan, F.; Ayers, W.H., and Swithenbank, J., A fundamental mathematical modelling approach to cyclone design, Trans IChemE, 60, 222-230 (1982).
Dietz, P.W., Electrostatically enhanced cyclone separators, Powder Tech, 31, 221-226 (1982).
Ernst, M., et al., Evaluation of a cyclone dust collector for high-temperature high-pressure particulate control, Ind Eng Chem Proc Des Dev, 21(1), 158-161 (1982).
Klimpel, R.R., Influence of chemical dispersant on sizing performance of a 24 inch hydrocyclone, Powder Tech, 31, 255-262 (1982).
Lammers, G.C., Venturi scrubber scaleup, Chem Eng, 26 July, 80-81 (1982).
Moore, S.J., Hydrocyclone balances, Chem Eng, 31 May, 149-150 (1982).
Rennhack, R., Dust extraction from hot gases, Int Chem Eng, 22(3), 405-415 (1982).
Stanton, C., Dust control techniques, Proc Engng, Sept, 50-53 (1982).
Various, Dust control: Plant and equipment survey, Processing, Oct, 37-45 (1982).
Bryant, H.S.; Silverman, R.W., and Zenz, F.A., How dust in gas affects cyclone pressure drop, Hyd Proc, June, 87-90 (1983).
Casal, J., and Martinez-Benet, J.M., A better way to calculate cyclone pressure drop, Chem Eng, 24 Jan, 99-100 (1983).
Chironna, R.J., and Voepel, L.P., Double duty scrubber for TDI fumes, Plant/Opns Prog, 2(4), 243-247 (1983).
Colman, D.A., and Thew, M.T., Correlation of separation results from light dispersion hydrocyclones, CER&D, 61, 233-240 (1983).

Gaunt, G.N., Effect of high particle concentration on the boundary layer flow in a hydrocyclone, CER&D, 61, 271-281 (1983).
Holmes, T.L.; Meyer, C.F., and DeGarmo, J.L., Reverse-jet scrubber for control of fine particulates, Chem Eng Prog, 79(2), 60-66 (1983).
Lindsey, W.E., Obtain accurate samples to determine scrubber efficiency, Chem Eng, 17 Oct, 81 (1983).
Reid, K.J., and Voller, V.R., Reconciling hydrocyclone particle-size data, Chem Eng, 27 June, 43-44 (1983).
van Duijn, G., and Rietema, K., Performance of a large-cone-angle hydrocyclone, Chem Eng Sci, 38(10), 1651-1674 (1983).
Williamson, R.D., et al., Use of hydrocyclones for small particle separation, Sepn Sci Technol, 18(12), 1395-1416 (1983).

1984

Bayvel, L.P., The effect of polydispersity of drops on the efficiency of a venturi scrubber, Trans IChemE, 60, 31-34 (1982); CER&D, 62, 60-61 (1984).
Biffin, M.; Syred, N., and Sage, P., Enhanced collection efficiency for cyclone dust separators, CER&D, 62, 261-265 (1984).
Busnaina, A.A., and Lilley, D.G., Numerical simulation of swirling flow in a cyclone chamber, Chem Eng Commns, 26(1), 73-88 (1984).
Constantinescu, S., Sizing gas cyclones, Chem Eng, 20 Feb, 97-98 (1984).
Forsyth, R.A., Cyclone separation in natural gas transmission systems, Chem Engnr, June, 37-41 (1984).
Hofer, H.H., and Wolter, A., Dust resistance and dust separation behaviour of electrostatic precipitations, Powder Tech, 37, 95-104 (1984).
Holmes, T.L., and Chen, G.K., Design and selection of spray/mist elimination equipment, Chem Eng, 15 Oct, 82-89 (1984).
King, R.P., and Juckes, A.H., Cleaning of fine coals by dense-medium hydrocyclone, Powder Tech, 40, 147-160 (1984).
Kuramarao, P.S.V., Critical-velocity plots for moisture separators, Chem Eng, 25 June, 178 (1984).
Lamb, G.E.R., Trapping dust electrically, Chemtech, Sept, 562-566 (1984).
Lee, R.W., Selection of equipment for removing particulates from gases, Chem & Ind, 2 July, 476-480 (1984).
Martinez-Benet, J.M., and Casal, J., Optimization of parallel cyclones, Powder Tech, 38, 217-221 (1984).
Ollero, P., Program calculates venturi-scrubber efficiency, Chem Eng, 28 May, 103-105 (1984).
Reijnen, K., and van Brakel, J., Gas cleaning at high temperatures and high pressures: A review, Powder Tech, 40, 81-112 (1984).
Swift, P., A user's guide to dust control, Chem Engnr, May, 22-26 (1984).
Tawari, T.D., and Zenz, F.A., Evaluating cyclone efficiencies from stream compositions, Chem Eng, 30 April, 69-73 (1984).
Trasi, P.R., and Licht, W., Effect of recycle on cyclone performance, Ind Eng Chem Proc Des Dev, 23(3), 479-482 (1984).
Various, Removal of particulate matter from air (topic issue), Chem & Ind, 20 Feb, 121-143 (1984).
Witbeck, W.O., and Woods, D.R., Pressure drop and separation efficiency in a flooded hydrocyclone, Can JCE, 62(1), 91-98 (1984).

Zanker, A., Determining air inlet-velocity for cyclones, Chem Eng, 19 March, 159-160 (1984).

1985-1986

Ahmed, A.A.; Ibraheim, G.A., and Doheim, M.A., Influence of apex diameter on pattern of solid/liquid ratio distribution within a hydrocyclone, J Chem Tech Biotechnol, 35A(8), 395-402 (1985).

Holzer, K., Wet separation of fine dusts and aerosols, Int Chem Eng, 25(2), 223-234 (1985).

Moir, D.N., Selection and use of hydrocyclones, Chem Engnr, Jan, 20-27 (1985).

Peate, J., Dust and fume control systems, Proc Engng, July, 59 (1985).

van Santen, A., and Allen, R., Selection of wet dedusting plant, Chem Engnr, Dec, 45-48 (1985).

Williams, R., and Nosker, R.W., Dust, Chemtech, July, 434-439 (1985).

Dabir, B., and Petty, C.A., Measurements of mean velocity profiles in a hydrocyclone using laser doppler anemometry, Chem Eng Commns, 48, 377-388 (1986).

Valdez, M.G.; Garcia, I., and Beato, B., Sizing gas cyclones for efficiency, Chem Eng, 14 April, 119-120 (1986).

1987-1988

Beeckmans, J.M., and Morin, B., Effect of particulate solids on pressure drop across a cyclone, Powder Tech, 52(3), 227-232 (1987).

Bloor, M.I.G., and Ingham, D.B., The reduced efficiency of the industrial cyclone, Chem Eng Commns, 57, 31-40 (1987).

Coury, J.R.; Thambimuthu, K.V., and Clift, R., Capture and rebound of dust in granular-bed gas filters, Powder Tech, 50(3), 253-266 (1987).

Doheim, M.A.; Ibraheim, G.A., and Ahmed, A.A., Modelling of hydrocyclones at high feed-solids concentrations, Chem Eng J, 34(2) 81-88 (1987).

Jaasund, S.A., Electrostatic precipitators: Better wet than dry, Chem Eng, 23 Nov, 159-163 (1987).

Li, P.M.; Lin, S., and Vatistas, G.H., Predicting collection efficiency of separation cyclones: A momentum analysis, Can JCE, 65(5), 730-735 (1987).

Sumner, R.J.; Briens, C.L., and Bergougnou, M.A., Study of a novel uniflow cyclone design, Can JCE, 65(3), 470-475 (1987).

Davidson, M.R., Similarity solutions for flow in hydrocyclones, Chem Eng Sci, 43(7), 1499-1506 (1988).

Harker, J.H., and Pimparkar, P.M., Effect of additives on electrostatic precipitation of fly ash, J Inst Energy, 61, 134-142 (1988).

Kanaoka, C.; Emi, H.; Hiragai, S., and Myojo, T., Morphology of particle agglomerates on a fiber when inertia and interception are predominant mechanisms of collection, Int Chem Eng, 28(3), 512-520 (1988).

Kwasniak, J., Application of Multvir method to separation of droplets and solid particles from gases, Chem Eng & Proc, 24(4), 211-216 (1988).

Li, Z.; Zisheng, Z., and Kuotsung, Y., Study of structure parameters of cyclones, CER&D, 66(2), 114-120 (1988).

Montz, K.W.; Beddow, J.K., and Butler, P.B., Adhesion and removal of particulate contaminants in high-decibel acoustic field, Powder Tech, 55(2), 133-140 (1988).
Mothes, H., and Loffler, F., Prediction of particle removal in cyclone separators, Int Chem Eng, 28(2), 231-241 (1988).
Petroll, J., and Fodisch, H., Modelling of dust particle collection in plate-type electrostatic precipitators, Chem Eng & Proc, 24(2), 105-118 (1988).
Ramarao, B.V., and Tien, C., Stochastic simulation of aerosol deposition in model filters, AIChEJ, 34(2), 253-262 (1988).
Spink, D.R., Handling mists and dusts, Chemtech, June, 364-368 (1988).

16.7 Sedimentation

1967-1972

Ratcliff, G.A.; Blackadder, D.A., and Sutherland, D.N., Compressibility of sediments, Chem Eng Sci, 22(2), 201-208 (1967).
Dollimore, D., and McBride, G.B., Alternative methods of calculating particle size from hindered settling measurements, J Appl Chem, 18, 136-140 (1968).
George, D.R.; Riley, J.M., and Ross, J.R., Potassium recovery by chemical precipitation and ion exchange, Chem Eng Prog, 64(5), 96-99 (1968).
Manchanda, K.D., and Woods, D.R., Significant design variables in continuous gravity decantation, Ind Eng Chem Proc Des Dev, 7(2), 182-187 (1968).
Scott, K.J., Experimental study of continuous thickening of a flocculated silica slurry, Ind Eng Chem Fund, 7(4), 582-595 (1968).
Barskii, M.D.; Shteinberg, A.M., and Dolganov, E.A., Effect of material concentration in a stream on efficiency of gravitational classification, Int Chem Eng, 9(1), 43-45 (1969).
Gondo, S., and Kusunoki, K., New look at gravity settlers, Hyd Proc, 48(9), 209-210 (1969).
Davies, G.A.; Jeffreys, G.V., and Ali, F., Design and scale-up of gravity settlers, Chem Engnr, Nov, CE378-385 (1970).
Kwatra, B.; Ahuja, L.D., and Ramakrishna, V., Cluster formation of kaolin using the hindered settling technique, J Appl Chem, 20, 123-126 (1970).
Scott, K.J., Continuous thickening of flocculated suspensions, Ind Eng Chem Fund, 9(3), 422-427 (1970).
Turner, G.A., and Fayed, M.E., Sedimentation of fine powders in air, Powder Tech, 4, 241-249 (1970).
Davies, R., and Kaye, B.H., Experimental investigation into settling behaviour of suspensions, Powder Tech, 5, 61-68 (1971).
Fitch, B., Batch tests predict thickener performance, Chem Eng, 23 Aug, 83-88 (1971).
Barfod, N., Concentration dependence of sedimentation rate of particles in dilute suspensions, Powder Tech, 6, 39-43 (1972).
Bikerman, J.J., Foam fractionation and drainage, Sepn Sci, 7(6), 647-652 (1972).
Charewicz, W., and Walkowiak, W., Selective flotation of inorganic ions, Sepn Sci, 7(6), 631-646 (1972).

Fitch, B., Calculating terminal settling rates in sedimentation, Chem Eng, 7 Aug, 96-98 (1972).
Goldberg, M., and Rubin, E., Foam fractionation in a stripping column, Sepn Sci, 7(1), 51-74 (1972).
Holland-Batt, A.B., Two-dimensional motion of particles accelerating in fluids, Trans IChemE, 50, 156-167 (1972).
Zanker, A., Nomographs to assist in the more rapid estimation of the sedimentation velocity, Chem Engnr, Oct, 387-391, 396 (1972).
Zanker, A., Nomograph for determination of viscosity of liquid-solid suspensions, Chem Engnr, Feb, 76-77 (1972).

1973-1975

Dolimore, D.; Goddard, J.A., and Nicklin, T., Settling of powder slurries with reference to sulphur slurries obtained from the Stretford process, Trans IChemE, 51, 309-314 (1973).
Dollimore, D., and McBride, G.B., Application of hindered settling methods to suspensions of oxysalts, Powder Tech, 8, 207-212 (1973).
Lockett, M.J., and Al-Habbooby, H.M., Differential settling by size of two particle species in a liquid, Trans IChemE, 51, 281-292 (1973).
Wacholder, E., Sedimentation in a dilute emulsion, Chem Eng Sci, 28(7), 1447-1454 (1973).
Himsworth, J.R., Hindered settling classifiers: The by-pass particle size controller, Powder Tech, 10, 181-187 (1974).
Hung, N.X., Computing the terminal velocity of spherical particles, Chem Eng, 25 Nov, 120 (1974).
Lockett, M.J., and Al-Habbooby, H.M., Relative particle velocities in two-species settling, Powder Tech, 10, 67-71 (1974).
Mehrotra, S.P., and Kapur, P.C., Optimal-suboptimal synthesis and design of flotation circuits, Sepn Sci, 9(3), 167-184 (1974).
Rice, R.G.;, Oliver, A.D.; Newman, J.P., and Wiles, R.J., Reduced dispersion using baffles in column flotation, Powder Tech, 10, 201-210 (1974).
Tarrer, A.R.; Lim, H.C.; Koppel, L.B., and Grady, C.P.L., Model for continuous thickening, Ind Eng Chem Proc Des Dev, 13(4), 341-346 (1974).
Ahmad, S.I., Laws of foam formation and foam fractionation, Sepn Sci, 10(6), 649-700 (1975).
Eklund, L.G., and Jernqvist, A., Experimental study of dynamics of vertical continuous thickener, Chem Eng Sci, 30(5), 597-606 (1975).
Goodarz-Nia, I., and Sutherland, D.N., Floc simulation: Effects of particle size and shape, Chem Eng Sci, 30(4), 407-412 (1975).
Harris, C.C.; Somasundaran, P., and Jensen, R.R., Sedimentation of compressible materials: Analysis of batch sedimentation curve, Powder Tech, 11, 75-84 (1975).
Kleeman, J.R., Equations for terminal settling-velocities of spheres, Chem Eng, 14 April, 102 (1975).
Petty, C.A., Continuous sedimentation of a suspension with nonconvex flux law, Chem Eng Sci, 30(12), 1451-1458 (1975).
Sansone, E.B., and Civic, T.M., Liquid sedimentation analysis, Powder Tech, 12, 11-18 (1975).

Zahavi, E., and Rubin, E., Settling of solid suspensions under and between inclined surfaces, Ind Eng Chem Proc Des Dev, 14(1), 34-41 (1975).

1976-1977

Collins, G.L., and Jameson, G.J., Flotation of fine particles: Influence of particle size and charge, Chem Eng Sci, 31(11), 985-992 (1976).

Davies, L.; Dollimore, D., and Sharp, J.H., Sedimentation of suspensions: Implications of theories of hindered settling, Powder Tech, 13, 123-132 (1976).

Dixon, D.C.; Souter, P., and Buchanan, J.E., Study of inertial effects in sedimentation, Chem Eng Sci, 31(9), 737-740 (1976).

Eklund, L.G., Working conditions in continuous pilot thickener at optimal load, Chem Eng Sci, 31(10), 881-892 (1976).

McKay, R.B., Hindered settling of organic pigment dispersions in hydrocarbon liquids, J Appl Chem Biotechnol, 26, 55-66 (1976).

Smiles, D.E., Sedimentation: Integral behaviour, Chem Eng Sci, 31(4), 273-276 (1976).

Smiles, D.E., Sedimentation and filtration equilibria, Sepn Sci, 11(1), 1-16 (1976).

Turner, J.P.S., and Glasser, D., Continuous thickening in a pilot plant, Ind Eng Chem Fund, 15(1), 23-30 (1976).

Abernathy, M.W., Design horizontal gravity settlers, Hyd Proc, 56(9), 199-202 (1977).

Allen, T., and Baudet, M.G., The limits of gravitational sedimentation, Powder Tech, 18, 131-138 (1977).

Barnea, E., New plot enhances value of batch-thickening tests, Chem Eng, 29 Aug, 75-78 (1977).

Blake, J.R., and Colombera, P.M., Sedimentation: Comparison between theory and experiment, Chem Eng Sci, 32(2), 221-228 (1977).

Collins, G.L., and Jameson, G.J., Double-layer effects in flotation of fine particles, Chem Eng Sci, 32(3), 239-246 (1977).

Joosten, G.E.H.; Schilder, J.G.M., and Broere, A.M., The suspension of floating solids in stirred vessels, Trans IChemE, 55, 220-222 (1977).

Kos, P., Fundamentals of gravity thickening, Chem Eng Prog, 73(11), 99-105 (1977).

Richmond, P., Fundamental concepts in flotation, Chem & Ind, 1 Oct, 792-796 (1977).

Tory, E.M., and Pickard, D.K., Three-parameter Markov model for sedimentation, Can JCE, 55, 655-665 (1977).

Wouda, T.W.M.; Rietema, K., and Ottengraf, S.P.P., Continuous sedimentation theory: Effects of density gradients and velocity profiles on sedimentation efficiency, Chem Eng Sci, 32(4), 351-358 (1977).

1978-1979

Bishop, P.L., Removal of powdered activated carbon from water by foam separation, Sepn Sci Technol, 13(1), 47-58 (1978).

Clarke, A.N., and Wilson, D.J., Separation by flotation, Sepn & Purif Methods, 7(1), 55-98 (1978).

Davies, L., and Dollimore, D., Sedimentation of suspensions, Powder Tech, 16, 45-49 (1977); 16, 59-61 (1977); 17, 147-152 (1977); 18, 285-287 (1977); 19, 1-6 (1978).

Dixon, D.C., Momentum-balance aspects of free-settling theory, Sepn Sci, 12(2), 171-204 (1977); 13(9), 753-766 (1978).

Mace, G.R., and Laks, R., Developments in gravity sedimentation, Chem Eng Prog, 74(7), 77-83 (1978).

Maljian, M.V., and Howell, J.A., Dynamic response of a continuous thickener to overloading and underloading, Trans IChemE, 56, 56-61 (1978).

Musil, L., and Vlk, J., Suspending solid particles in an agitated conical-bottom tank, Chem Eng Sci, 33(8), 1123-1132 (1978).

Sadowski, Z.; Mager, J., and Laskowski, J., Hindered settling of coagulating suspensions, Powder Tech, 21, 73-79 (1978).

Slagle, D.J.; Shah, Y.T.; Klinzing, G.E., and Walters, J.G., Settling of coal in coal-oil slurries, Ind Eng Chem Proc Des Dev, 17(4), 500-504 (1978).

Wilson, D.J., Kinetic and equilibrium aspects of floc coagulation, Sepn Sci Technol, 13(1), 25-39 (1978).

Adler, P.M., Study of disaggregation effects in sedimentation, AIChEJ, 25(3), 487-493 (1979).

Biddulph, M.W., Mixing in water elutriators, Can JCE, 57, 268-280 (1979).

Blake, J.R.; Colombera, P.M., and Knight, J.H., One-dimensional model of sedimentation using Darcy's law, Sepn Sci Technol, 14(4), 291-304 (1979).

Charewicz, W.A., and Strzelbicki, J., Foam separation and precipitate floatation, J Chem Tech Biotechnol, 29, 149-153 (1979).

Cordoba-Molina, J.F.; Hudgins, R.R., and Silveston, P.L., Gravity clarifier as a stratified flow phenomenon, Can JCE, 57, 249-259 (1979).

Dublanc, E.A., Fast calculation of change in terminal settling velocity with temperature, Chem Eng, 12 Feb, 135-136 (1979).

Fitch, B., Sedimentation of flocculent suspensions (review paper), AIChEJ, 25(6), 913-930 (1979).

Grace, J.R., and Tuot, J., A theory for cluster formation in vertically conveyed suspensions of intermediate density, Trans IChemE, 57, 49-54 (1979).

Graves, E.C.; Schnelle, K.B., and Wilson, D.J., Experimental verification of a mathematical model for quiescent settling of a flocculating slurry, Sepn Sci Technol, 14(10), 923-934 (1979).

Khatib, Z., and Howell, J.A., Batch and continuous sedimentation behaviour of flocculated china clay slurries, Trans IChemE, 57, 170-175 (1979).

Machej, K., and Niemiec, W., Dimensioning of rectangular settling tanks with provision for inlet disturbances, Int Chem Eng, 19(1), 93-96 (1979).

Masliyah, J.H., Hindered settling in multispecies particle system, Chem Eng Sci, 34(9), 1166-1168 (1979).

Mirza, S., and Richardson, J.F., Sedimentation of suspensions of particles of two or more sizes, Chem Eng Sci, 34(4), 447-454 (1979).

Shirato, M.; Aragaki, T.; Manabe, A., and Takeuchi, N., Electroforced sedimentation of thick clay suspensions in consolidation region, AIChEJ, 25(5), 855-863 (1979).

Silveston, P.L., et al., Residence time distributions in design of clarifiers, Can JCE, 57, 83-91 (1979).

Weiland, R.H., and McPherson, R.R., Accelerated settling by addition of buoyant particles, Ind Eng Chem Fund, 18(1), 45-49 (1979).

1980-1981

Cheng, D.C.H., Sedimentation of suspensions and storage stability, Chem & Ind, 17 May, 408-414 (1980).

Clarke, J.H.; Clarke, A.N., and Wilson, D..J., Theory of clarifier operation, Sepn Sci Technol, 13(9), 767-790; 13(10), 881-916 (1978); 14(1), 1-12 (1979); 15(7), 1429-1444 (1980).

Kiefer, J.E., and Wilson, D., Electrical aspects of adsorbing colloid flotation, Sepn Sci Technol, 15(1), 57-74 (1980).

Puracelli, C., Direct sizing of gravity settlers, Chem Eng, 22 Sept, 182 (1980).

Reed, C.C., and Anderson, J.L., Hindered settling of suspensions at low Reynolds number, AIChEJ, 26(5), 816-827 (1980).

Various, Concentrated suspensions (topic issue), Chem & Ind, 15 March, 211-227 (1980).

Zanker, A., Calculating countercurrent decantation efficiency, Proc Engng, Sept, 95,97 (1980).

Zanker, A., Nomographs determine settling velocities for solid-liquid systems, Chem Eng, 19 May, 147-150 (1980).

Austin, L.G., and Klimpel, R.R., Improved method for analyzing classifier data, Powder Tech, 29, 277-281 (1981).

Dixon, D.C., Thickener dynamic analysis, accounting for compression effects, Chem Eng Sci, 36(3), 499-508 (1981).

Graves, E.C.; Schnelle, K.B., and Wilson, D.J., Initial particle growth effect on settling of a flocculating slurry, Sepn Sci Technol, 16(3), 263-274 (1981).

Kawase, Y., and Ulbrecht, J.J., Sedimentation of particles in non-Newtonian fluids, Chem Eng Commns, 13(1), 55-64 (1981).

Lai, R.W.M., Get more information from flotation-rate data, Chem Eng, 19 Oct, 181-182 (1981).

Masliyah, J.H.; Kwong, T.K., and Seyer, F.A., Theoretical and experimental studies of a gravity separation vessel, Ind Eng Chem Proc Des Dev, 20(1), 154-160 (1981).

Tiller, F.M., Revision of Kynch sedimentation theory, AIChEJ, 27(5), 823-829 (1981).

Warren, L.J., Shear flocculation, Chemtech, March, 180-185 (1981).

1982-1984

Attia, Y.A., Fine-particle separation by selective flocculation (review paper), Sepn Sci Technol, 17(3), 485-494 (1982).

Hickey, T.J., Developments in flotation, Chem Engnr, Nov, 420-423 (1982).

Matsumoto, K.; Kutowy, O., and Capes, C.E., Theoretical aspects of electrophoretic sedimentation of particles, Powder Tech, 31, 197-207 (1982).

Al Taweel, A.M.; Farag, H.A.; Fadaly, O., and Mackay, G.D.M., Method for determining the settling behaviour of dense suspensions, Can JCE, 61(4), 534-540 (1983).

Biddulph, M.W., Separating efficiency of a water elutriator, AIChEJ, 29(6), 956-961 (1983).

Carpenter, C.R., Calculate settling velocities for unrestricted particles or hindered settling, Chem Eng, 14 Nov, 227-231 (1983).
Fitch, B., Kynch theory of sedimentation and compression zones, AIChEJ, 29(6), 940-947 (1983).
Selim, M.S.; Kothari, A.C., and Turian, R.M., Sedimentation of multisized particles in concentrated suspensions, AIChEJ, 29(6), 1029-1038 (1983).
Smith, T.N., Limitations to the capacity of continuous thickeners, CER&D, 61, 45-50 (1983).
Strang, R.M.; Schnelle, K.B., and Wilson, D.J., Theory of clarifier operation, Sepn Sci Technol, 18(3), 253-292 (1983).
Zanker, A., Nomograph for continuous gravity decanters, Proc Engng, June, 37 (1983).
Biddulph, M.W., Water elutriators in materials recycling, Can JCE, 62(3), 357-362 (1984).
Kipke, K., Suspension by side entering agitators, Chem Eng & Proc, 18(4), 233-238 (1984).
Klimpel, R.R., Use of chemical reagents in flotation, Chem Eng, 3 Sept, 75-79 (1984).
Nonaka, M., and Uchio, T., Microhydrodynamic model of sedimentation process, Sepn Sci Technol, 19(4), 337-356 (1984).
Papavergos, P.G., and Hedley, A.B., Particle deposition behaviour from turbulent flows (review paper), CER&D, 62, 275-295 (1984).
Stewart, R.F., and Sutton, D., Structure of flocculated suspensions, Chem & Ind, 21 May, 373-378 (1984).
Zanker, A., Liquid requirements in agglomeration processes, Proc Engng, Nov, 79 (1984).

1985-1986

Bos, A.S., and Zuiderweg, F.J., Kinetics of continuous agglomeration in suspension, Powder Tech, 44, 43-51 (1985).
El-Genk, M.S.; Kim, S.H., and Erickson, D., Sedimentation of binary mixtures of particles of unequal densities and of different sizes, Chem Eng Commns, 36(1), 99-120 (1985).
Ferrell, D.P., and Huang, C.P., Removal of fine coal particles from water by flotation, Chem Eng Commns, 35(1), 351-372 (1985).
Harris, C.C., and Khandrika, S.M., Flotation machine design, Powder Tech, 43, 243-248; 273-278 (1985).
Hirtzel, C.S., and Rajagopalan, R., Stability of colloidal dispersions (review paper), Chem Eng Commns, 33(5), 301-324 (1985).
Kneule, F., Scale-up in suspension of solids in agitated vessels, Int Chem Eng, 25(2), 214-223 (1985).
Merta, H., and Ziolo, J., Calculation of thickener area and depth based on the data of a batch-settling test, Chem Eng Sci, 40(7), 1301-1304 (1985).
Pasquali, G.; Fajner, D., and Magelli, F., Behaviour of multistage mechanically stirred columns with neutrally buoyant solid-liquid suspensions, CER&D, 63, 51-58 (1985).
Patwardhan, V.S., and Tien, C., Sedimentation and liquid fluidization of solid particles of different sizes and densities, Chem Eng Sci, 40(7), 1051-1060 (1985).

Vargha-Butler, E.I., Surface tension effects in sedimentation of coal particles in various liquid mixtures, Chem Eng Commns, 33(5), 255-276 (1985).
Venkataraman, S., and Weiland, R.H., Buoyant-particle-promoted settling of industrial suspensions, Ind Eng Chem Proc Des Dev, 24(4), 966-970 (1985).
Androutsopoulos, G.P., Froth flotation performance evaluation: A schematic illustration and a numerical separation index, Fuel, 65(7), 968-974 (1986).
Bhatty, J.I., Clusters formation during sedimentation of dilute suspensions, Sepn Sci Technol, 21(9), 953-968 (1986).
Chin, A.D., et al., Shape-modified size correction for terminal settling velocity in the intermediate region, Powder Tech, 48(1), 59-66 (1986).
Davies, J.T., Particle suspension and mass transfer rates in agitated vessels, Chem Eng & Proc, 20(4), 175-182 (1986).
Hsu, J.P., Floc breakage analysis: A simulation approach, Chem Eng Commns, 44, 21-32 (1986).
Iordache, O., and Corbu, S., Stochastic approach to sedimentation, Chem Eng Sci, 41(10), 2589-2594 (1986).
Kawashima, Y., et al., Spherical agglomeration of calcium carbonate dispersed in aqueous medium containing sodium oleate, Powder Tech, 46(1), 61-66 (1986).
Lev, O.; Rubin, E., and Sheintuch, M., Steady state analysis of a continuous clarifier-thickener system, AIChEJ, 32(9), 1516-1525 (1986).
Merta, H., and Ziolo, J., Method of thickener-area calculation based on the data of batch-settling tests, Chem Eng Sci, 41(7), 1918-1921 (1986).
Yamazaki, H.; Tojo, T., and Miyanami, K., Concentration profiles of solids suspended in a stirred tank, Powder Tech, 48(3), 205-216 (1986).

1987

Choi, C.; Dyrkacz, G.R., and Stock, L.M., Density separation of alkylated coal macerals, Energy & Fuels, 1(3), 280-287 (1987).
Concha, F., and Bustos, M.C., Modification of the Kynch theory of sedimentation, AIChEJ, 33(2), 312-315 (1987).
Dabak, T., and Yucel, O., Modeling of concentration and particle size distribution effects on rheology of highly concentrated suspensions, Powder Tech, 52(3), 193-206 (1987).
Flynn, S.A., and Woodburn, E.T., A froth ultra-fine model for selective separation of coal from mineral in dispersed-air flotation cell, Powder Tech, 49(2), 127-142 (1987).
Galvin, K.P., and Waters, A.G., Effect of sedimentation feed flux on solids flux curve, Powder Tech, 53(2), 113-120 (1987).
Kesavan, S.K., Behaviour of coal particles suspended in coal liquids, Powder Tech, 50(1), 15-24 (1987).
Law, H.S.; Masliyah, J.H.; MacTaggart, R.S., and Nandakumar, K., Gravity separation of bidisperse suspensions (light and heavy particle species), Chem Eng Sci, 42(7), 1527-1538 (1987).
Lin, I.J., Hydrocycloning thickening: Dewatering and densification of fine particulates, Sepn Sci Technol, 22(4), 1327-1348 (1987).
Pickard, D.K.; Tuckman, B.A., and Tory, E.M., Three-parameter Markov model for sedimentation, Powder Tech, 49(3), 227-240 (1987).

Seifert, J.A., Selecting thickeners and clarifiers, Chem Eng, 12 Oct, 111-118 (1987).

Shih, Y.T.; Gidaspow, D., and Wasan, D.T., Hydrodynamics of sedimentation of multisized particles, Powder Tech, 50(3), 201-216 (1987).

Turton, R., and Clark, N.N., Explicit relationship to predict spherical particle terminal velocity, Powder Tech, 53(2), 127-130 (1987).

Wesselingh, J.A., Velocity of particles, drops and bubbles, Chem Eng & Proc, 21(1), 9-14 (1987).

1988

Aoki, M., et al., New separator for sediments and floating solid matter, Int Chem Eng, 28(2), 337-344 (1988).

Bustos, M.C., and Concha, F., Simulation of batch sedimentation with compression, AIChEJ, 34(5), 859-861 (1988).

Davis, R.H., and Birdsell, K.H., Hindered settling of semidilute monodisperse and polydisperse suspensions, AIChEJ, 34(1), 123-129 (1988).

Font, R., Compression zone effect in batch sedimentation, AIChEJ, 34(2), 229-238 (1988).

Fuerstenau, D.W., et al., Assessing wettability and degree of oxidation of coal by film flotation, Energy & Fuels, 2(3), 237-241 (1988).

Hirosue, H.; Yamada, N.; Abe, E., and Tateyama, H., Coagulation of red mud suspensions with an inorganic coagulent, Powder Tech, 54(1), 27-30 (1988).

Jean, R.H., and Fan, L.S., Particle terminal velocity in a gas-liquid medium with liquid as the continuous phase, Can JCE, 65(6), 881-886 (1988).

Kim, S., and Lawrence, C.J., Suspension mechanics for particle contamination control (review article), Chem Eng Sci, 43(5), 991-1016 (1988).

Koziol, K., and Glowacki, P., Determination of free settling parameters of spherical particles in power law fluids, Chem Eng & Proc, 24(4), 183-188 (1988).

Landman, K.A.; White, L.R., and Buscall, R., Continuous-flow gravity thickener: Steady-state behavior, AIChEJ, 34(2), 239-252 (1988).

Li, A.C., et al., Flows of neutrally buoyant suspensions in several viscometers, Chem Eng Commns, 73, 95-120 (1988).

McKay, G.; Murphy, W.R., and Hillis, M., Settling characteristics of cylindrical and disc-shaped particles, CER&D, 66(1), 107-112 (1988).

Moir, D.N., A guide to sedimentation centrifuges, Chem Eng, 28 March, 42-51 (1988).

Nasr-El-Din, H.; Masliyah, J.H.; Nandakumar, K., and Law, D.H.S., Continuous gravity separation of a bidisperse suspension in a vertical column, Chem Eng Sci, 43(12), 3225-3234 (1988).

Sell, N.J.; Doshi, M.R., and Hawes, J.M., Sedimentation behavior of various pulp fibres, Chem Eng Commns, 73, 217-226 (1988).

Shiragami, N., et al., Enhancement of settling in a tank by inclined plates, Int Chem Eng, 28(4), 669-677 (1988).

Stamatakis, K., and Tien, C., Dynamics of batch sedimentation of polydispersed suspensions, Powder Tech, 56(2), 105-118 (1988).

Tiller, F.M., and Chen, W., Limiting operating conditions for continuous thickeners, Chem Eng Sci, 43(7), 1695-1704 (1988).

Tory, E.M., and Kamel, M.T., Divergence problem in calculating particle velocities in dilute dispersions of identical spheres, Powder Tech, 55(3), 187-192 (1988).

Tory, E.M., and Kamel, M.T., Divergence problem in calculating particle velocities in dilute dispersions of identical spheres, Powder Tech, 55(1), 51-60 (1988).

Tsai, S.C., Staged flotation of fine coal and effects of mineral size and distribution, Ind Eng Chem Res, 27(9), 1669-1674 (1988).

Webb, J., Method for observation of fine particles in aqueous suspension by cryo-SEM, Powder Tech, 55(1), 71-73 (1988).

Zimmels, Y., Simulation of nonsteady sedimentation of polydisperse particle mixtures, Powder Tech, 56(4), 227-250 (1988).

16.8 Filtration and Centrifuges

1967-1969

Fierstine, B.A., Continuous bulk filtration centrifugals, Chem Eng Prog, 63(10), 115-118 (1967).

Han, C.D., and Bixler, H.J., Washing of the liquid retained by granular solids, AIChEJ, 13(6), 1058-1066 (1967).

Heertjes, P.M., and Lerk, C.F., The functioning of deep-bed filters, Trans IChemE, 45, T129-145 (1967).

Kearney, R.D., Applications of bulk centrifuges, Chem Eng Prog, 63(12), 72-78 (1967).

Kehat, E.; Lin, A., and Kaplan, A., Clogging of filter media, Ind Eng Chem Proc Des Dev, 6(1), 48-55 (1967).

Michaels, A.S.; Baker, W.E.; Bixler, H.J., and Vieth, W.R., Permeability and washing characteristics of flocculated kaolinite filter cakes, Ind Eng Chem Fund, 6(1), 25-40 (1967).

Smith, G.R.S., Improve your filter-aid filtration, Chem Eng, 30 Jan, 154-161 (1967).

Weissman, B.J., and Elsken, J.C., Countercurrent washing in a continuous screening centrifuge, Chem Eng Prog, 63(9), 91-94 (1967).

Armour, J.C., and Cannon, J.N., Fluid flow through woven screens, AIChEJ, 14(3), 415-420 (1968).

Crook, M.D., and Jones, W.D., Assessing the filterability characteristics of industrial sludges using an expanded form of the Carman equation, Brit Chem Eng, 13(1), 94-98 (1968).

Dlouhy, P.E., and Dahlstrom, D.A., Continuous filtration in pharmaceutical production, Chem Eng Prog, 64(4), 116-121 (1968).

Gould, D.Z., Filter eliminates screening of ferrite slurry, Chem Eng, 3 June, 126-128 (1968).

Kouloheris, A.P., and Meek, R.L., Centrifugal washing of solids, Chem Eng, 9 Sept, 121-126 (1968).

Shephard, D., and Grice, M.A.K., Strips on filter cloth prevent filter-cake cracking, Chem Eng, 23 Sept, 246 (1968).

Singer, S., and Hacker, C.H., Testing of ultrafine filters, Chem Eng Prog, 64(6), 75-77 (1968).

Ambler, C.M., Selecting the optimum centrifuge, Chem Eng, 20 Oct, 96-103 (1969).
Dahlstrom, D.A., and Davis, S.S., Plastics in continuous filtration equipment, Chem Eng Prog, 65(10), 80-85 (1969).
Dollinger, L.L., Specifying filters, Hyd Proc, 48(10), 88-92 (1969).
Galluzzo, J.F., Portable filter design for flexible operation, Chem Eng, 24 March, 182 (1969).
Hooley, F.V., An approach to centrifuge evaluation, Brit Chem Eng, 14(11), 1540 (1969).
Kaji, A.T., and Sharma, M.M., Mass-transfer characteristics of the bowl and cone contactor, Brit Chem Eng, 14(10), 1416-1419 (1969).
Li, N.N., Separation of wax crystals from oil by centrifugation, Ind Eng Chem Proc Des Dev, 8(1), 89-92 (1969).
Murkes, J., Effect of suspension characteristics in centrifugal separation, Brit Chem Eng, 14(12), 1692-1697 (1969).
Shirato, M.; Sambuichi, M.; Kato, H., and Aragaki, T., Internal flow mechanism in filter cakes, AIChEJ, 15(3), 405-409 (1969).
Smick, K.F., Filter-cloth porosity tester easily assembled, Chem Eng, 19 May, 212 (1969).
Thrush, R.E., and Honeychurch, R.W., Specifying centrifuges, Hyd Proc, 48(10), 81-87 (1969).
Zubkov, V.A., and Golovko, Y.D., Flow of liquid in the rotor outlet of a tubular centrifuge, Int Chem Eng, 9(3), 403-406 (1969).

1970-1971

Cherry, G.B.; Moss, A.A.H., and Scott, E., Comparison of fixed and variable volume chamber mechanised filters for the isolation of dyestuffs, Chem Engnr, April, CE95-100 (1970).
Ives, K.J., Advances in deep-bed filtration, Trans IChemE, 48, T94-100 (1970).
Kozicki, W.; Rao, A.R.K., and Tiu, C., Correction for transient flow in the initial stage of constant-rate filtration, Ind Eng Chem Fund, 9(2), 261-265 (1970).
Kuchinskii, M.K., Calculation of mean throughput of vacuum drum filters allowing for filter cloth plugging, Int Chem Eng, 10(3), 427-431 (1970).
Kuo, M.T., and Barrett, E.C., Continuous filter-cake washing performance, AIChEJ, 16(4), 633-638 (1970).
Perry, M.G., and Dobson, B., Model study of rotary vacuum filtration, Chem Engnr, April, CE83-87, 94 (1970).
Purchas, D.B., A non-guide to filter selection for liquids, Chem Engnr, April, CE79-82 (1970).
Rushton, A., Effect of filter cloth structure on flow resistance, bleeding, blinding and plant performance, Chem Engnr, April, CE88-94 (1970).
Schnittger, J.R., Integrated theory of separation for bulk centrifuges, Ind Eng Chem Proc Des Dev, 9(3), 407-413 (1970).
Smiles, D.E., Theory of constant pressure filtration, Chem Eng Sci, 25(6), 985-996 (1970).
Sokolov, V.I., and Shamsutdinov, U.G., Selection of dynamic parameters of filters with vibrational removal of the deposit, Int Chem Eng, 10(1), 100-103 (1970).

Hauslein, R.H., Ultra-fine filtration of bulk fluids, Chem Eng Prog, 67(5), 82-88 (1971).
Raichel, D.R., and Schaefer, R.B., Performance parameters of centrifuges, Sepn Sci, 6(4), 599-610 (1971).
Rousselet, C., Pressure filtration or centrifugal filtration, Brit Chem Eng, 16(11), 1031 (1971).
Rushton, A., and Griffiths, P., Fluid flow in monofilament filter media, Trans IChemE, 49, 49-59 (1971).
Sutherland, K.S., Advances in centrifugal equipment, Chem Engnr, March, 114-120 (1971).
Various, Centrifugal separations (topic issue), Chem Eng Prog, 67(9), 45-68 (1971).

1972-1974

Dombe, A.I., and Shkoropad, D.E., Separation of suspensions in filtering screw centrifuges, Int Chem Eng, 12(1), 46-50 (1972).
Freshwater, D.C., and Stenhouse, J.I.T., Retention of large particles in fibrous filters, AIChEJ, 18(4), 786-791 (1972).
Glass, J.S., Filter efficiency studies using image analyzing techniques, Chem Eng Prog, 68(1), 58-61 (1972).
Maloney, G.F., Selecting and using pressure leaf filters for saturated solutions, Chem Eng, 15 May, 88-94 (1972).
Maracek, J.; Krpata, M., and Moudry, F., Effect of particle size distribution on removal of liquid phase from centrifuged filter cake, Int Chem Eng, 12(3), 384-389 (1972).
Purchas, D.B., Cake filtration: A standard test method for any filter, Chem Eng, 21 Aug, 86-93 (1972).
Reinhardt, H., Centrifuging and centrifuges, Chem & Ind, 6 May, 363-366 (1972).
Zenz, F.A., and Krockta, H., Granular beds for filtration and adsorption, Brit Chem Eng, 17(3), 224-227 (1972).
McLain, L., Advances in filter media, Proc Engng, Jan, 80-82 (1973).
Mirokhin, A.M.; Dorokhov, I.N., and Kafarov, V.V., Design of filtration equipment for countercurrent multi-stage washing of a cake, Int Chem Eng, 13(3), 517-522 (1973).
Pierson, H.G.W., Low-pressure filtration avoids crystal collapse, Proc Engng, April, 82-83 (1973).
Tomiak, A., Theoretical recoveries in filter cake reslurrying and washing, AIChEJ, 19(1), 76-84 (1973).
Various, Centrifuges (topic issue), Chem Eng Prog, 69(9), 62-74 (1973).
Various, Filtration (symposium papers), Chem Engnr, Dec, 579-602 (1973).
Various, Filtration report (IChemE Working Party Report), Chem Engnr, Nov, Supplement, 40 pages, (1973).
Day, R.W., Techniques for selecting centrifuges, Chem Eng, 13 May, 98-104 (1974).
Kemper, E.A., Cover design gives access to basket centrifuges, Chem Eng, 18 March, 112 (1974).
Machej, J., Filtration study: Practical application of test data, Int Chem Eng, 14(1), 27-33 (1974).

Records, F.A., The continuous scroll discharge decanting centrifuge, Chem Engnr, Jan, 41-47 (1974).
Shpanov, N.V., Design of filtration processes with cake formation during constant-rate regime, Int Chem Eng, 14(4), 715-719 (1974).
Various, Filtration/separation techniques (topic issue), Chem Eng Prog, 70(12), 33-59 (1974).
Vesilind, P.A., Estimating centrifuge capacities, Chem Eng, 1 April, 54-57 (1974).
Wakeman, R.J., and Rushton, A., Structural model for filter cake washing, Chem Eng Sci, 29(9), 1857-1866 (1974).

1975-1976

Rymarz, T.M., Specifying pulse-jet filters, Chem Eng, 31 March, 97-100 (1975).
Various, Filtration (feature report), Processing, Aug, 29-44 (1975).
Various, Filtration/separation (topic issue), Chem Eng Prog, 71(12), 37-80 (1975).
Wnek, W.J.; Gidaspow, D., and Wasan, D.T., The role of colloid chemistry in modelling deep-bed liquid filtration, Chem Eng Sci, 30(9), 1035-1048 (1975).
Cleasby, J.L., Filtration with granular beds, Chem Engnr, Oct, 663-667, 682 (1976).
Henry, J.D.; Lui, A.P., and Kuo, C.H., Dual functional solid-liquid separation process based on filtration and settling, AIChEJ, 22(3), 433-441 (1976).
Kai, T., Analysis of concentration distribution in centrifuges, Int Chem Eng, 16(2), 240-249 (1976).
Spiewok, L., Scrapper-centrifuge control system, Proc Engng, Oct, 69-71 (1976).
Suttle, H.K., Development of industrial filtration, Chem Engnr, Oct, 675-682 (1976).
Various, Filtration: Advances and guidelines (feature report), Chem Eng, 16 Feb, 80-94 (1976).
Wakeman, R.J., Filtration research in UK universities (dated, but good bibliography!), Chem Engnr, Oct, 683-685, 703 (1976).
Wakeman, R.J.; Rushton, A., and Brewis, L.N., Residual saturation of dewatered filter cakes, Chem Engnr, Oct, 668-670 (1976).
White, D.A., Prediction of leaching and filtration performance of drum filters, Chem Eng Sci, 31(6), 419-426 (1976).
Wronski, S.K.; Bin, A.K., and Laskowski, L.K., Anomalous behaviour during initial stage of constant-pressure filtration, Chem Eng J, 12(2), 143-148 (1976).
Zastrow, J., Theoretical calculation of performance of disc centrifuges, Int Chem Eng, 16(3), 515-518 (1976).

1977-1978

Basso, A.J., Getting the most out of filteraids, Chem Eng, 12 Sept, 185-190 (1977).
Farr, K., Emptying techniques for filtration, Proc Engng, Oct, 52-53 (1977).

Kelly, J.J., and O'Donnell, P., Residence time model for rotary drums, Trans IChemE, 55, 243-252 (1977).
Lamb, G., and Constanza, P., Improving the performance of fabric filters, Chem Eng Prog, 73(1), 51-53 (1977).
Molerus, O., and Brunner, K., Design of oscillating-screen centrifuges, Int Chem Eng, 17(2), 230-237 (1977).
Purchas, D., Safety margins for pressure filtration, Proc Engng, Sept, 72-75 (1977).
Smith, T.N., Recovery fractions in centrifuges, Chem Eng J, 13(1), 22-26 (1977).
Stenhouse, J.I.T., and Freshwater, D.C., Particle adhesion in fibrous air filters, Trans IChemE, 54, 95-99 (1976); 55, 285 (1977).
Tiller, F.M., and Crump, J.R., Increasing filtration rates in continuous filters, Chem Eng, 6 June, 183-187 (1977).
Various, Filtration and separation (feature report), Processing, May, 53-75 (1977).
Various, Filtration and separation (topic issue), Chem Eng Prog, 73(4), 57-91 (1977).
Ademondi, B., Better filtration of waste-oil particles, Hyd Proc, 57(12), 103 (1978).
Bender, W., Filtering, washing and deliquoring of finely dispersed solids, Int Chem Eng, 18(1), 48-59 (1978).
Bonem, J.M., Plant testing for improved filter performance, Chem Eng, 13 Feb, 107-109 (1978).
Chen, N.H., Liquid-solid filtration: Generalized design and optimization equations, Chem Eng, 31 July, 97-101 (1978).
Fokina, Z.A., et al., Determination of process parameters for treatment of cake with steam at a final production filter, Int Chem Eng, 18(1), 136-139 (1978).
Heertjes, P.M., and Zuideveld, P.L., Clarification of liquids using filter aids, Powder Tech, 19, 17-64 (1978).
Rushton, A., Pressure variation effects in rotary drum filtration with incompressible cakes, Powder Tech, 20, 39-46 (1978).
Rushton, A., Design throughputs in rotary-disc vacuum filtration with incompressible cakes, Powder Tech, 21, 161-169 (1978).
Wakeman, R.J., A numerical integration of the differential equations describing the formation of and flow in compressible filter cakes, Trans IChemE, 56, 258-265 (1978).
Zanker, A., Finding solids buildup in a filter, Chem Eng, 8 May, 214 (1978).
Zeitsch, K., Effect of the feed rate on the active acceleration of overflow centrifuges, Trans IChemE, 56, 281-284 (1978).

1979-1980

Cooke, M.H., and Bridgwater, J., Interparticle percolation: A statistical mechanical interpretation, Ind Eng Chem Fund, 18(1), 25-27 (1979).
Crosby, H.L., Suppressing centrifuge chatter, Chem Eng Prog, 75(10), 92-94 (1979).
Harker, J.H., Improving batch filtration, Processing, Feb, 34-35 (1979).

Hauslein, R.H., and Simkins, R.H., Filter cartridge tests, Chem Eng Prog, 75(1), 46-51 (1979).

Kaplan, S.J.; Morland, C.D., and Hsu, S.C., Predict non-Newtonian fluid pressure drop across random-fibre filters, Chem Eng, 27 Aug, 93-98 (1979).

Macdonald, I.F.; El-Sayed, M.S.; Mow, K., and Dullien, F.A.L., Flow through porous media: The Ergun equation revisited (a review) Ind Eng Chem Fund, 18(3), 199-208 (1979).

Tien, C., and Payatakes, A.C., Advances in deep-bed filtration (review paper), AIChEJ, 25(5), 737-759 (1979).

Tien, C.; Turian, R.M., and Pendse, H., Simulation of dynamic behavior of deep-bed filters, AIChEJ, 25(3), 385-395 (1979).

Tomiak, A., Predict performance of belt-filter washing, Chem Eng, 23 April, 143-146 (1979).

Tomiak, A., Solve complex filtration-washing problems, Chem Eng, 8 Oct, 109-113 (1979).

Various, Filtration and dust control (feature report), Processing, Aug, 37-52 (1979).

Clarke, J.W., and Rantell, T.D., Filtration in coal liquifaction: Influence of digestion conditions in the filtration of non-hydrogenated coal digests, Fuel, 59(3), 208-212 (1980).

Clarke, J.W., and Rantell, T.D., Filtration in coal liquifaction: Influence of filtration conditions in non-hydrogenated systems, Fuel, 59(1), 35-41 (1980).

Rothwell, E., Recent advances in filtration and separation practice, Chem Engnr, Jan, 24-26 (1980).

Rushton, A., Filtration and separation update, Proc Engng, Sept, 49-55 (1980).

Various, Filtration and separation developments, Proc Engng, Sept, 73-87 (1980).

Ward, A.S., Filtration research in Japan: The work of Mopei Shirato, Chem Engnr, Jan, 27-30, 37 (1980).

Willis, M.S., and Tosun, I., A rigorous cake filtration theory, Chem Eng Sci, 35(12), 2427-2438 (1980).

Yoshimura, Y.; Ueda, K.; Mori, F., and Yoshioka, N., Initial particle collection mechanism in clean, deep-bed filtration, Int Chem Eng, 20(4), 600-609 (1980).

1981-1982

Greenkorn, R.A., Steady flow through porous media (a review), AIChEJ, 27(4), 529-545 (1981).

Gupta, S.K., Scale-up procedures for disc-stack centrifuges, Chem Eng J, 22(1), 43-50 (1981).

Maracek, J., Filter cloth plugging and its effect on the filter performance, Ind Eng Chem Proc Des Dev, 20(4), 693-698 (1981).

Sai, C.R.M., Get more flow through a filter press, Chem Eng, 14 Dec, 108 (1981).

Various, Filtration (special report), Processing, July, 24-29 (1981).

Wakeman, R.J., Thickening and filtration: A review and evaluation of recent research, Trans IChemE, 59, 147-160 (1981).

Wakeman, R.J., The formation and properties of apparently incompressible filter cakes under vacuum on downward facing surfaces, Trans IChemE, 59, 260-270 (1981).

Basso, A.J., Vacuum filtration using filteraids, Chem Eng, 19 April, 159-162 (1982).

Blosse, P., Advantages of membrane filters, Proc Engng, Sept, 34-37 (1982).

Cheape, D.W., Leaf tests can establish optimum rotary-vacuum-filter operation, Chem Eng, 14 June, 141-148 (1982).

Clarke, J.W.; Rantell, T.D., and Parsons, D.A., Filtration in coal liquifaction: Preparation of filter aid from filter cake, Fuel, 61(4), 364-368 (1982).

Freeman, M.P., Vacuum electrofiltration, Chem Eng Prog, 78(8), 74-79 (1982).

Nepomnyashchy, E.A., Kinetics of fine powder centrifugation, Powder Tech, 33, 25-29 (1982).

Purchas, D.B., The mysterious origins of the rotary vacuum filter, Chem Engnr, Aug, 335-336 (1982).

Shirato, M.; Murase, T., and Atsumi, K., Optimization of operation of a filter press with membrane-compression mechanism, Int Chem Eng, 22(4), 689-699 (1982).

Smiles, D.E.; Raats, P.A.C., and Knight, J.H., Constant pressure filtration: Effect of a filter membrane, Chem Eng Sci, 37(5), 707-714 (1982).

Tettamanti, B., Modelling and scaling of filtration in industry, Int Chem Eng, 22(3), 561-572 (1982).

Tomiak, A., Washing loss calculations for dynamic filters, Ind Eng Chem Proc Des Dev, 21(3), 500-505 (1982).

Various, Batch pressure filtration (feature report), Chem Eng, 26 July, 46-63 (1982).

1983-1984

Campbell, R.W., Liquid-solid separation: Cake filters in batch filtration, Plant/Opns Prog, 2(4), 216-222 (1983).

Hermia, J., Constant pressure blocking filtration laws: Application to power-law non-Newtonian fluids, Trans IChemE, 60, 183-187 (1982); 61, 68 (1983).

Jaisinghani, R.A., Interfacial phenomena in the filtration/separation of petroleum products, Sepn Sci Technol, 18(12), 1295-1322 (1983).

Lecey, R.W., and Pietila, K.A., Improving belt-filter-press performance, Chem Eng, 28 Nov, 69-72 (1983).

Leu, W., and Tiller, F.M., Experimental study of mechanism of constant-pressure cake filtration: Clogging the filter media, Sepn Sci Technol, 18(12), 1351-1370 (1983).

MacKiewicz, J., Development of flocculation effects in filter theory, Chem Eng Commns, 23(4), 305-314 (1983).

Massiah, T.F., Estimate product in perforate basket centrifuge, Chem Eng, 2 May, 90 (1983).

Shirato, M.; Murase, T., and Mori, H., Centrifugal dehydration of a packed particulate bed, Int Chem Eng, 23(2), 298-307 (1983).

Tiller, F.M., and Horng, L.L., Hydraulic deliquoring of compressible filter cakes, AIChEJ, 29(2), 297-305 (1983).

Ward, A., Liquid filtration: A survey, Processing, March, 31-35 (1983).

Willis, M.S.; Collins, R.M., and Bridges, W.G., Complete analysis of non-parabolic filtration behaviour, CER&D, 61, 96-109 (1983).
Anon., Liquid-solids separation and filtration (special advertising section), Chem Eng, 6 Feb, 45-86B (1984).
Cook, L.N., Laboratory approach optimizes filter-aid addition, Chem Eng, 23 July, 45-50 (1984).
Dunhill, S., Developments in filtration, Proc Engng, Sept, 53-55 (1984).
Hsu, E.H., and Fan, L.T., Experimental study of deep-bed filtration: Stochastic treatment, AIChEJ, 30(2), 267-273 (1984).
Kiesewetter, W., and Alt, C., Clarifying/filtering centrifuge: A novel separating unit, Int Chem Eng, 24(4), 629-639 (1984).
Lindley, J., Centrifuges: Guidelines on selection, Chem Engnr, Dec, 28-31 (1984).
Millington, P., Liquid filtration survey, Processing, Dec, 12-16 (1984).

1985-1986

Anon., Filtration and separation (special advertising report), Chem Eng, 5 Aug, 39-94 (1985).
Avery, L., Filtration plant scaleup, Chemtech, Oct, 622-631 (1985).
Carleton, A.J., Choosing a compression filter, Chem Engnr, April, 20-23 (1985).
Fan, L.T.; Hwang, S.H.; Chou, S.T., and Nassar, R., Birth-death modeling of deep-bed filtration: Sectional analysis, Chem Eng Commns, 35(1), 101-122 (1985).
Lan, L.T.; Hwang, S.H.; Nassar, R., and Chou, S.T., Experimental study of deep-bed filtration: Stochastic analysis, Powder Tech, 44, 1-11 (1985).
Lindley, J., Centrifuges, Chem Engnr, Part 1: Guidelines on selection, Dec, 28-31(1984); Part 2: Safe operation, Feb, 41-44 (1985).
Macdonald, D., Evaluation of separation problems for disc bowl centrifuges, Chem Engnr, March, 15-17 (1985).
Purchas, D., Quantifying the filterability of slurries, Proc Engng, July, 53,55 (1985).
Tosun, I., Mathematical formulation of cake filtration for deformable solid particles, Chem Eng Sci, 40(4), 673-674 (1985).
Waterman, M.J., Microcomputer-controlled facility for phase analysis from AKUFVE mixer-centrifuge unit, J Chem Tech Biotechnol, 35A(2), 83-88 (1985).
West, J., Disc-bowl centrifuges, Chem Eng, 7 Jan, 69-73 (1985).
Willis, M.S.; Tosun, I., and Collins, R.M., Filtration mechanics, CER&D, 63, 175-183 (1985).
Anon., Filtration and separation (special advertising section), Chem Eng, 27 Oct, 57-108 (1986).
Berger, M.H., Rule-of-thumb for binary isotope separations in a gas centrifuge, Sepn Sci Technol, 21(4), 383-392 (1986).
Carleton, A., Centrifuges for slurry separations, Processing, April, 29,31,35-37 (1986).
Conlisk, A.T., Effect of aspect ratio and feed flow rate on separative power in a gas centrifuge, Chem Eng Sci, 41(10), 2639-2650 (1986).

Coyne, K.; O'Brien, R.; Conner, W.C., and Rucinski, K., Filter morphology and performance: Porosimetry and microscopy of oil filter media compared with filtration, Chem Eng J, 32(1), 53-62 (1986).

Elgal, G.M., Filtration with two-stage two-concentration solutions for energy conservation, Ind Eng Chem Proc Des Dev, 25(4), 872-878 (1986).

Horng, J.S., and Maa, J.R., The effect of a surfactant on mixer-settler operation, J Chem Tech Biotechnol, 36(1), 15-26 (1986).

Morris, N., High-efficiency filters in process filtration, Chem & Ind, 3 Nov, 745-748 (1986).

Nassar, R.; Chou, S.T., and Fan, L.T., Modelling and simulation of deep-bed filtration: A stochastic compartmental model, Chem Eng Sci, 41(8), 2017-2028 (1986).

Tomiak, A., Simulate countercurrent washing on belt filters, Chem Eng, 4 Aug, 61-64 (1986).

Tosun, I., Formulation of cake filtration, Chem Eng Sci, 41(10), 2563-2568 (1986).

Various, Filtration update, Processing, Feb, 17-29 (1986).

Wakeman, R.J., Transport equations for filter cake washing, CER&D, 64(4), 308-319 (1986).

1987

Allen, T., Photocentrifuges, Powder Tech, 50(3), 193-200 (1987).

Anon., Filtration and separation (special advertising section), Chem Eng, 12 Oct, 45-92 (1987).

Batigun, A., and Tosun, I., Washing theory for unsaturated filter cakes using capillary model, Chem Eng Commns, 61, 89-106 (1987).

Fairclough, A.R.N.; Chan, K.L., and Davies, G.A., An analysis of fibre filters for filtration of dilute suspensions, CER&D, 65(5), 396-407 (1987).

Havsteen, B., Kinetics of formation of a flow-inhibiting boundary layer in liquid-solid filtration, Chem Eng J, 35(2), 123-136 (1987).

Lawton, D.J., and Constantino, A., What makes cartridge filters perform effectively? Chem Eng Prog, 83(11), 20-26 (1987).

Mackie, R.I.; Horner, R.M.W., and Jarvis, R.J., Dynamic modeling of deep-bed filtration, AIChEJ, 33(11), 1761-1775 (1987).

Nakanishi, K., et al., Specific resistance of cakes of microorganisms, Chem Eng Commns, 62(1), 187-202 (1987).

Purchas, D., Filtration problems, Proc Engng, Sept, 47-50 (1987).

Saata, A.M., and Halilsoy, M., A new solution of the deep bed filter equations, Chem Eng J, 34(3) 147-150 (1987).

Sambuichi, M., et al., Theory of batchwise centrifugal filtration, AIChEJ, 33(1), 109-120 (1987).

Sharma, M.M., and Yortsos, Y.C., Network model for deep bed filtration processes, AIChEJ, 33(10), 1644-1653 (1987).

Smiles, D.E., and Kirby, J.M., Aspects of one-dimensional filtration, Sepn Sci Technol, 22(5), 1405-1424 (1987).

Tiller, F.M., and Yeh, C.S., Role of porosity in filtration, AIChEJ, 33(8), 1241-1256 (1987).

Tiller, F.M.; Yeh, C.S., and Leu, W.F., Compressibility of particulate structures in relation to thickening, filtration, and expression: A review, Sepn Sci Technol, 22(2), 1037-1064 (1987).
Tosun, I., and Sahinoglu, S., On the constancy of average porosity in filtration, Chem Eng J, 34(2) 99-106 (1987).
Tosun, I., and Willis, M.S., The effect of filter cake geometry on average porosity, Chem Eng J, 35(1) 65-66 (1987).

1988
Anon., Filtration (special advertising section) Chem Eng, 23 May, 79-116 (1988).
Anon., Solid-liquid separation: Equipment focus, Chem Eng, 10 Oct, 59-71 (1988).
Carleton, A., Selecting improved equipment for solid-liquid separation, Proc Engng, Sept, 35-41 (1988).
Coelho, M.A.N., and Guedes de Carvalho, J.R.F., Transverse dispersion in granular beds, CER&D, 66(2), 165-189 (1988).
Euston, J.A., Low-turbulence flocculation as an aid to vacuum filtration, Powder Tech, 55(1), 61-70 (1988).
Hess, W.F., and Kalwa, M., High-pressure filtration for downstream processing of biological dispersions, Chem Eng & Proc, 23(3), 179-188 (1988).
Johnston, P.R., Liquid filtration, Chem Eng Prog, 84(11), 18-26 (1988).
Mackay, D., and Salusbury, T., Choosing between centrifugation and crossflow microfiltration, Chem Engnr, April, 45-50 (1988).
Nield, P., Approaches to cake filtration, Chem Engnr, Nov, 45-48 (1988).
Ramirez, W.F.; Smith, M.F., and Feerer, J.L., Velocity and temperature effects on dispersion in porous media: Applications to unconsolidated lignite, Chem Eng Commns, 64, 13-26 (1988).
Rege, S.D., and Fogler, H.S., Network model for deep-bed filtration of solid particles and emulsion drops, AIChEJ, 34(11), 1761-1772 (1988).
Shucosky, A.C., Select the right cartridge filter, Chem Eng, 18 Jan, 72-77 (1988).
Templin, B.R., and Leith, D., Effect of operating conditions on pressure drop in pulse-jet cleaned fabric filter, Plant/Opns Prog, 7(4), 215-222 (1988).
Tosun, I.; Willis, M.S., and Batigun, A., Parameter estimation in cake washing, Chem Eng Commns, 73, 151-162 (1988).

16.9 Particle Storage

1967-1970
Perry, M.G., and Handley, M.F., The dynamic arch in free flowing granular material discharging from a model hopper, Trans IChemE, 45, T367-371 (1967).
Walker, D.M., and Blanchard, M.H., Pressures in experimental coal hoppers, Chem Eng Sci, 22(12), 1713-1746 (1967).
Shinohara, K., Mechanism of gravity flow of particles from a hopper, Ind Eng Chem Proc Des Dev, 7(3), 378-383 (1968).

Bosley, J.; Schofield, C., and Shook, C.A., An experimental study of granule discharge from model hoppers, Trans IChemE, 47, T147-153 (1969).

Holland, J.; Miles, J.E.P.; Schofield, C., and Shook, C.A., Fluid drag effects in the discharge of granules from hoppers, Trans IChemE, 47, T154-159 (1969).

McDougall, I.R., Ambient fluid influence on solids discharge from hoppers, Brit Chem Eng, 14(8), 1079-1082 (1969).

McDougall, I.R., and Knowles, G.H., Flow of particles through orifices, Trans IChemE, 47, T73-78 (1969).

Colijin, H., and Carroll, P.J., Feeders for bins and hoppers, Brit Chem Eng, 15(8), 1029-1042 (1970).

Kotchanova, I.I., Experimental and theoretical investigations on the discharge of granular materials from bins, Powder Tech, 4, 32-37 (1970).

Lvin, J.B., Analytical evaluation of pressures of granular materials on silo walls, Powder Tech, 4, 280-285 (1970).

Perry, M.G., and Jangda, H.A.S., Pressures in flowing and static sand in model bunkers, Powder Tech, 4, 89-96 (1970).

Shinohara, K.; Shoji, K., and Tanaka, T., Mechanism of segregation and blending of particles flowing out of mass-flow hoppers, Ind Eng Chem Proc Des Dev, 9(2), 174-180 (1970).

Shook, C.A.; Carleton, A.J., and Flain, R.J., Effect of fluid drag on the flow-rate of granular solids from hoppers, Trans IChemE, 48, T173-175 (1970).

Toyama, S., Flow of granular materials in moving beds, Powder Tech, 4, 214-220 (1970).

1971-1973

Johanson, J.R., and Jenike, A.W., Effect of gaseous phase on pressures in a cylindrical silo, Powder Tech, 5, 133-145 (1971).

Carleton, A.J., Effect of fluid-drag forces on discharge of free-flowing solids from hoppers, Powder Tech, 6, 91-96 (1972).

Pikon, J.; Sasiadek, B., and Drozdz, M., Arching in storage bins for loose materials, Proc Tech Int, 17(11), 888-891 (1972).

Resnick, W., Particle flow through orifices, Trans IChemE, 50, 289-291 (1972).

Shinohara, K.; Shoji, K., and Tanaka, T., Mechanism of size segregation of particles in filling a hopper, Ind Eng Chem Proc Des Dev, 11(3), 369-376 (1972).

Shook, C.A.; Garrett, G.C., and Zabel, T., A study of the discharge of solid particles under water, Trans IChemE, 50, 125-131 (1972).

Sutton, H.M., Design for bulk storage bottleneck problems, Proc Engng, Aug, 84-86 (1972).

Yuasa, Y., and Kuno, H., Effects of an efflux tube on the rate of flow of glass beads from a hopper, Powder Tech, 6, 97-102 (1972).

Davidson, J.F., and Nedderman, R.M., The hour-glass theory of hopper flow, Trans IChemE, 51, 29-35 (1973).

Kemper, E.A., Calculating volumes of rectangular bins, Chem Eng, 9 July, 104-106 (1973).

Leung, L.S., and Wilson, L.A., Downflow of solids in standpipes, Powder Tech, 7, 343-349 (1973).

Matsen, J.M., Flow of fluidized solids and bubbles in standpipes and risers, Powder Tech, 7, 93-96 (1973).
McDougall, I.R., and Pullen, R.J.F., Effect on solids mass flowrate of an expansion chamber between a hopper outlet and a vertical standpipe, Powder Tech, 8, 231-242 (1973).
Parnaby, J.; Battye, P.G., and Waite, G.S., Optimal design of homogeneous systems incorporating layered stockpiles and fluidised silos for the control of raw materials quality, Trans IChemE, 51, 323-330 (1973).
Rao, V.L., and Venkateswarlu, D., Determination of velocities and flow patterns of particles in mass flow hoppers, Powder Tech, 7, 263-265 (1973).
Ring, R.J.; Buchanan, R.H., and Doig, I.D., Discharge of granular material from hoppers submerged in water, Powder Tech, 8, 117-125 (1973).
Stainforth, P.T., and Ashley, R.C., Analytical hopper design method for cohesive powders, Powder Tech, 7, 215-243 (1973).
Sutton, H.M., and Richmond, R.A., Improving the storage conditions of fine powders by aeration, Trans IChemE, 51, 97-104 (1973).
Wahl, E.A., Bin activators: Key to practical storage and flow of solids, Chem Eng Prog, 69(1), 62-67 (1973).
Walters, J.K., Theoretical analysis of stresses in axially-symmetric hoppers and bunkers, Chem Eng Sci, 28(3), 779-790 (1973).

1974-1976

Bransby, P.L., and Blair-Fish, P.M., Wall stresses in mass-flow bunkers, Chem Eng Sci, 29(5), 1061-1074 (1974).
Eckhoff, R.K., and Leversen, P.G., Design of mass flow hoppers, Powder Tech, 10, 51-58 (1974).
Hancock, A.W., and Nedderman, R.M., Prediction of stresses on vertical bunker walls, Trans IChemE, 52, 170-179 (1974).
Kemper, E.A., Convenient scale-tank calibration, Chem Eng, 8 July, 110 (1974).
Lapidot, H., Weight of contents in cylindrical storage tanks, Chem Eng, 21 Jan, 123-126 (1974).
Rao, V.L., and Venkateswarlu, D., Static and dynamic wall pressures in experimental mass flow hoppers, Powder Tech, 10, 143-152 (1974).
Shinohara, K., and Tanaka, T., Approximate consideration on mechanism of gravity flow particles through aperture of storage vessel on basis of block-flow model, Chem Eng Sci, 29(9), 1977-1990 (1974).
Bransby, P.L., and Blair-Fish, P.M., Initial deformations during mass flow from a bunker, Powder Tech, 11, 273-288 (1975).
Caplan, F., Finding valley angles for hoppers and bins, Chem Eng, 27 Oct, 138 (1975).
de Jong, J.A.H., Aerated solids flow through a vertical standpipe below a pneumatically discharged bunker, Powder Tech, 12, 197-208 (1975).
Enstad, G., Theory of arching in mass flow hoppers, Chem Eng Sci, 30(10), 1273-1284 (1975).
Kurz, H.P., and Rumpf, H., Flow processes in aerated silos, Powder Tech, 11, 147-156 (1975).
Shinohara, K., and Tanaka, T., Effect of tapping on flow of particles from storage vessels, Ind Eng Chem Proc Des Dev, 14(1), 1-11 (1975).

Darton, R.C., Structure and dispersion of jets of solid particles falling from a hopper, Powder Tech, 13, 241-250 (1976).
Kurz, H.P., Stability of material bridges in aerated silos, Powder Tech, 13, 57-72 (1976).
Perry, M.G.; Rothwell, E., and Woodfin, W.T., Model studies of mass-flow bunkers, Powder Tech, 12, 51-56 (1975); 14, 81-92 (1976).
Rao, V.L., and Venkateswarlu, D., Internal pressures in flowing granular materials from mass-flow hoppers, Powder Tech, 11, 133-146 (1975); 13, 151-155 (1976).
Ratkai, G., Particle flow and mixing in vertically vibrated beds, Powder Tech, 15, 187-192 (1976).

1977-1979

Bagster, D.F., Effect of compressibility on the assessment of the contents of a stockpile, Powder Tech, 16, 193-196 (1977).
Craven, G.C., Storage and discharge of bulk solids in the chemical industry, Chem Engnr, March, 171-174 (1977).
Crewdson, B.J.; Ormond, A.L., and Nedderman, R.M., Air-impeded discharge of fine particles from a hopper, Powder Tech, 16, 197-207 (1977).
Enstad, G., Stresses and dome formation in axially symmetric mass-flow hoppers, Chem Eng Sci, 32(3), 337-339 (1977).
Leung, L.S., Design of fluidized gas-solids flow in standpipes, Powder Tech, 16, 1-6 (1977); 17, 291-293 (1977).
Levinson, M.; Shmutter, B ., and Resnick, W., Displacement and velocity fields in hoppers, Powder Tech, 16, 29-43 (1977).
Midgley, G., Bulk storage of urea fertilisers, Chem Engnr, Dec, 865-866 (1977).
Molerus, O., and Schoneborn, P.R., Bunker design based on experiments in a bunker-centrifuge, Powder Tech, 16, 265-272 (1977).
Williams, J.C., Rate of discharge of coarse granular materials from conical mass-flow hoppers, Chem Eng Sci, 32(3), 247-256 (1977).
Winters, R.J., Improving material flow from bins, Chem Eng Prog, 73(9), 82-86 (1977).
Canon, R.M., and Medlin, G.W., Calculating solids inventory, Chem Eng, 3 July, 91 (1978).
Gilbert, N., and Handman, S.E., Continuous recirculation system for in-bin blending of granular solids, Chem Eng, 23 Oct, 147-152 (1978).
Horne, R.M., and Nedderman, R.M., Stress distribution in hoppers, Powder Tech, 19, 243-254 (1978).
Kurylchek, A.L., Bin discharging problems and solutions, Chem Eng Prog, 74(7), 84-87 (1978).
Morrison, H.L., One-dimensional analysis of granular flow in bunkers, Chem Eng Sci, 33(2), 241-253 (1978).
Spink, C.D., and Nedderman, R.M., Gravity discharge rate of fine particles from hoppers, Powder Tech, 21, 245-261 (1978).
Dumbaugh, G.D., Induced vertical flow from storage bins, Chem Eng, 23 April, 159-167 (1979).
Dumbaugh, G.D., Flow from bulk-solids storage bins, Chem Eng, 26 March, 189-193 (1979).

Mickiewicz, A., and Wlodarski, A., Composition of bulk solid discharged from a bin with continuous recirculation, Powder Tech, 22, 97-100 (1979).
Sundaram, V., and Cowin, S.C., Reassessment of static bin pressure experiments, Powder Tech, 22, 23-32 (1979).
Yuu, S., and Jotaki, T., Experimental analysis of static pressure in bins, Chem Eng Sci, 34(7), 913-918 (1979).

1980-1982
Abouzeid, A.Z.M., and Fuerstenau, D.W., Scale-up of particulate hold-up in rotary drums, Powder Tech, 25, 65-70 (1980).
Abouzeid, A.Z.M., and Fuerstenau, D.W., Study of hold-up in rotary drums with discharge end constrictions, Powder Tech, 25, 21-29 (1980).
Altiner, H.K., Fluid pressure distribution in an aerated hopper, AIChEJ, 26(2), 297-299 (1980).
Gardner, J.E., and Freeburn, F.D., Update of lockhopper valve development, Chem Eng Prog, 76(2), 84-92 (1980).
Murfitt, P.G., and Bransby, P.L., Deaeration of powders in hoppers, Powder Tech, 27, 149-162 (1980).
Smith, J.C., and Hattiangadi, U.S., Profiling solids flow from bins, Chem Eng Commns, 6(1), 105-116 (1980).
Altenkirch, R.A., and Eichhorn, R., Effect of fluid drag on low Reynolds number discharge of solids from a circular orifice, AIChEJ, 27(4), 593-598 (1981).
Koehler, F.H., Estimate the solids inventory in a silo, Chem Eng, 27 July, 98 (1981).
Lemlich, R., and Doshi, V.K., Operating bins and hoppers, Chemtech, Jan, 48-49 (1981).
Pitcher, G., Problems with discharge of powders from bins, Proc Engng, March, 47-49 (1981).
Akiyama, T., et al., Mixing characteristics of particulate material in a negative-pressure air mixer, Ind Eng Chem Proc Des Dev, 21(4), 664-670 (1982).
Allen, T., and Macsporran, W.C., Permeametry: Correction for gas expansion as it flows through a bed of powder, Powder Tech, 33, 195-200 (1982).
Kaza, K.R., and Jackson, R., Rate of discharge of coarse granular material from a wedge-shaped mass flow hopper, Powder Tech, 33, 223-237 (1982).
Murfitt, P.G., and Bransby, P.L., Pressures in hoppers filled with fine powders, Powder Tech, 31, 153-174 (1982).
Nedderman, R.M., Theoretical prediction of stress distributions in hoppers (review paper), Trans IChemE, 60, 259-275 (1982).
Tsunakawa, H., Use of partition plates and circular cones to reduce stresses on particulate solids in hoppers, Int Chem Eng, 22(2), 280-287 (1982).
Tuzun, U., and Nedderman, R.M., Investigation of flow boundary during steady-state discharge from a funnel-flow bunker, Powder Tech, 31, 27-43 (1982).
Zanker, A., Estimating bulk material volumes, Proc Engng, Oct, 61-63 (1982).

1983-1985
Emanuel, J.H.; Best, J.L.; Mahmoud, M.H., and Hasanain, G.S., Parametric study of silo-material interaction, Powder Tech, 36, 223-243 (1983).

Fan, L.T.; Toda, M., and Satija, S., Hold-up of fine particles in the packed dense bed of the multisolid pneumatic transport bed, Powder Tech, 36, 107-114 (1983).

Kraus, M.N., Maintain flow from bin activators, Chem Eng, 2 May, 87 (1983).

Martin, P.D., and Davidson, J.F., Flow of powder through an orifice from a fluidised bed, CER&D, 61, 162-166 (1983).

Nedderman, R.M.; Tuzun, U., and Thorpe, R.B., Effect of interstitial air pressure gradients on discharge from bins, Powder Tech, 35, 69-81 (1983).

Fleming, A., Practical aspects of silo design and installation, Chem Engnr, March, 34-36 (1984).

Jenike, A.W., Analysis of solids densification during pressurization of lock hoppers, Powder Tech, 37, 131-144 (1984).

Kaza, K.R., and Jackson, R., Boundary conditions for a granular material flowing out of a hopper or bin, Chem Eng Sci, 39(5), 915-917 (1984).

Michalowski, R.L., Flow of granular material through a plane hopper, Powder Tech, 39, 29-40 (1984).

Peterson, E.R., Constructing conical bottoms of storage bins for dry powder or pellets, Chem Eng, 25 June, 176 (1984).

Shinohara, K., and Miyata, S.I., Mechanism of density segregation of particles in filling vessels, Ind Eng Chem Proc Des Dev, 23(3), 423-428 (1984).

Stacey, R., An overview of powder valve design, Chem Engnr, June, 45-47 (1984).

Weare, F., and Wright, H., Bulk storage design and maintenance, Proc Engng, Jan, 33-35 (1984).

Zanker, A., Calculating voids percentage, real density and bulk density of solids, Proc Engng, March, 73 (1984).

Chen, Y.M.; Rangachari, S., and Jackson, R., Theoretical and experimental investigation of fluid and particle flow in a vertical standpipe, Ind Eng Chem Fund, 23(3), 354-370 (1985).

Cowin, S.C., Compensating corrections for static bin pressures, Powder Tech, 43, 169-173 (1985).

Drescher, A., and Vgenopoulou, I., Theoretical analysis of channeling in bins and hoppers, Powder Tech, 42, 181-191 (1985).

Ducker, J.R.; Ducker, M.E., and Nedderman, R.M., Discharge of granular materials from unventilated hoppers, Powder Tech, 42, 3-14 (1985).

Geldart, D., and Williams, J.C., Fine-powder flooding from hoppers, Powder Tech, 43, 181-183 (1985).

Kalson, P.A., and Resnick, W., Angles of repose and drainage for granular materials in a wedge-shaped hopper, Powder Tech, 43, 113-116 (1985).

Kirby, J.M., Deaeration of powders in hoppers, Powder Tech, 44, 69-75 (1985).

Moore, B.A., and Arnold, P.C., Alternative presentation of design parameters for mass flow hoppers, Powder Tech, 42, 79-89 (1985).

Prieto, R.F., Shortcut methods for estimating solids inventory, Chem Eng, 11 Nov, 229-230 (1985).

Standish, N., Studies of size segregation in filling and emptying a hopper, Powder Tech, 45, 43-56 (1985).

Sundaram, V., and Cowin, S.C., Experimental evaluation of effect of material consolidation on static bin pressures, Powder Tech, 42, 241-247 (1985).

Tardos, G.I., et al., Unsteady flow of compressible gas through consolidated porous media: Application to deaeration of hoppers, Powder Tech, 41, 135-146 (1985).

Wilms, H., and Schwedes, J., Analysis of the active stress field in hoppers, Powder Tech, 42, 15-25 (1985).

1986-1988

Cooke, S., Intermediate bulk containers for bulk solids, Chem Engnr, March, 22-24 (1986).

Gudehus, G.; Kolymbas, D., and Tejchman, J., Behaviour of granular materials in cylindrical silos, Powder Tech, 48(1), 81-90 (1986).

Molerus, O., and Egerer, B., Predictive characterization of particulate solids in design of storage and transport systems, Chem Eng & Proc, 20(2), 61-72 (1986).

Parbery, R.D., and Roberts, A.W., Equivalent friction for accelerated gravity flow of granular materials in chutes, Powder Tech, 48(1), 75-80 (1986).

Pitman, E.B., Stress and velocity fields in two- and three-dimensional hoppers, Powder Tech, 47(3), 219-232 (1986).

Schofield, C., Recent research on the storage of particulate materials in hoppers, CER&D, 64(2), 89-90 (1986).

Graham, D.P.; Tait, A.R., and Wadmore, R.S., Measurement and prediction of flow patterns of granular solids in cylindrical vessels, Powder Tech, 50(1), 65-76 (1987).

Leung, L.S.; Chong, Y.O., and Lottes, J., Operation of V-valves for gas-solid flow, Powder Tech, 49(3), 271-276 (1987).

Williams, J.C.; Al-Salman, D., and Birks, A.H., Measurement of static stresses on wall of cylindrical container for particulate solids, Powder Tech, 50(3), 163-176 (1987).

Malhotra, K., et al., Fundamental particle mixing studies in agitated bed of granular materials in cylindrical vessel, Powder Tech, 55(2), 107-114 (1988).

Moriyama, R., and Jimbo, G., Pulsating pressure on wall in hopper section of a bin, Int Chem Eng, 28(1), 84-91 (1988).

Nedderman, R.M., Measurement of velocity profile in granular material discharging from a conical hopper, Chem Eng Sci, 43(7), 1507-1516 (1988).

Standish, N.; Liu, Y.N., and McLean, A.G., Quantification of the degree of mixing in bins, Powder Tech, 54(3), 197-208 (1988).

16.10 Particle Conveying

1966-1969

Kraus, M.N., Starting up pneumatic conveyors, Chem Eng, 31 Jan, 94-100 (1966).

Condolios, E.; Chapus, E.E., and Constans, J.A., New trends in solids pipelines, Chem Eng, 8 May, 131-138 (1967).

Decker, E.B., and Vaarst, W.V., Gravity flow of solids to weighfeeders, Chem Eng Prog, 63(2), 92-95 (1967).

Gasparyan, A.M., and Mirzakhanyan, R.M., Pneumatic transport of coarse-grained materials in compact bed, Int Chem Eng, 7(2), 330-333 (1967).

Haag, A., Velocity losses in pneumatic conveyor pipe bends, Brit Chem Eng, 12(1), 65-66 (1967).

Jones, J.H.; Braun, W.G.; Daubert, T.E., and Allendorf, H.D., Estimation of pressure drop for vertical pneumatic transport of solids, AIChEJ, 13(3), 608-611 (1967).

Knafelc, F.M., Troublefree diverters for pneumatic conveying systems, Chem Eng, 25 Sept, 176-179 (1967).

Weisselberg, E., Troubleshooting chemical solids handling, Chem Eng Prog, 63(11), 72-76 (1967).

Hickerson, W.L., Vibratory conveyor and feeder systems, Brit Chem Eng, 13(6), 817-819 (1968).

Hickerson, W.L., Vibratory conveyor and feeder systems, Brit Chem Eng, 13(7), 995-999 (1968).

McCarthy, H.E., and Olson, J.H., Turbulent flow of gas-solids suspensions, Ind Eng Chem Fund, 7(3), 471-483 (1968).

Anon., Conveyor practice, Brit Chem Eng, 14(8), 1093-1094,1142 (1969).

Bates, L., Handling of bulk solids by helical screw equipment, Brit Chem Eng, 14(8), 1072-1076 (1969).

Bennett, B.A.; Cloete, F.L.D.; Miller, A.I., and Streat, M., Accurate hydraulic metering of solids in dense phase flow, Chem Engnr, Nov, CE412-414 (1969).

Corrigan, T.E.; Dean, M.J., and Denton, D., Feeding pneumatic conveyors, Brit Chem Eng, 14(3), 297-300 (1969).

Leung, L.S.; Wiles, R.J., and Nicklin, D.J., Transition from fluidised to packed bed flow in vertical hydraulic conveying, Trans IChemE, 47, T271-278 (1969).

Maddocks, K.L., Conveying by vibration, Brit Chem Eng, 14(8), 1084-1085 (1969).

Murty, J.S., and Dey, D.N., Nomogram for air consumption and pipe diameter in pneumatic conveying, Brit Chem Eng, 14(8), 1113 (1969).

Reddy, K.V.S., and Pei, D.C.T., Particle dynamics in solids-gas flow in a vertical pipe, Ind Eng Chem Fund, 8(3), 490-497 (1969).

Toda, M.; Konno, H.; Saito, S., and Maeda, S., Hydraulic conveying of solids through horizontal and vertical pipes, Int Chem Eng, 9(3), 553-560 (1969).

Toda, M.; Konno, H.; Saito, S., and Maeta, S., Hydraulic conveying of solids through horizontal and vertical pipes, Brit Chem Eng, 14(8), 1077 (1969).

Yoshida, T., and Kousaka, Y., Flow of granular solids through a vibrating orifice, Int Chem Eng, 9(1), 177-180 (1969).

1970-1972

Bena, J.; Lodes, A., and Sefcik, J., Aerodynamic properties of granular materials, Int Chem Eng, 10(1), 89-100 (1970).

Boothroyd, R.G., and Haque, H., Experimental investigation of heat transfer in the entrance region of a heated duct conveying fine particles, Trans IChemE, 48, T109-120 (1970).

Carr, R.L., Particle behaviour, storage and flow, Brit Chem Eng, 15(12), 1541-1549 (1970).

Heertjes, P.M.; Verloop, J., and Willems, R., Measurement of local mass flowrates and particle velocities in fluid-solids flow, Powder Tech, 4, 38-40 (1970).
Kolpakov, V.M., and Donat, E.V., Investigation of pressure drop in acceleration zone of a vertical pipeline for conveying solid particles, Int Chem Eng, 10(3), 394-398 (1970).
Letan, R., Effect of restraints on vertical transport of solids, Trans IChemE, 48, T178-179 (1970).
Papazoglou, C.S., and Pyle, D.L., Air-assisted flow from a bed of particles, Powder Tech, 4, 9-18 (1970).
Stainforth, P.T.; Ashley, R.C., and Morley, J.N.B., Computer analysis of powder flow characteristics, Powder Tech, 4, 250-256 (1970).
Various, Solids materials handling (topic issue), Chem Eng Prog, 66(6), 31-55 (1970).
Carnell, D.W., Solids feeder for charging through small openings, Chem Eng, 1 Nov, 100 (1971).
DeCamps, F.; Dumont, G., and Goossens, W., Vertical pneumatic conveyer with a fluidized bed as mixing zone, Powder Tech, 5, 299-306 (1971).
Johanson, J.R., Modeling flow of bulk solids, Powder Tech, 5, 93-99 (1971).
Leung, L.S.; Wiles, R.J., and Nicklin, D.J., Correlation for predicting choking flowrates in vertical pneumatic conveying, Ind Eng Chem Proc Des Dev, 10(2), 183-189 (1971).
Botterill, J.S.M.; Van der Kolk, M.; Elliott, D.E., and McGuigan, S., Flow of fluidised solids, Powder Tech, 6, 343-351 (1972).
Flain, R.J., Pneumatic conveying: Matching system to materials, Proc Engng, Nov, 88-90 (1972).
Foster, E.P., and Bonk, D.L., Controlling small flows of solids and slurries, Chem Eng, 30 Oct, 134 (1972).
Jayasinghe, S.S., Cohesion and flow of particulate solids, Proc Tech Int, 17(11), 872-874 (1972).
Lanchester, F.G., Pneumatic conveying of fine powders, Proc Engng, Nov, 114-116 (1972).
Mason, J.S., and Smith, B.V., Erosion of bends by pneumatically conveyed suspensions of abrasive particles, Powder Tech, 6, 323-335 (1972).
Paulson, C.A.J., and Philipp, D.H., Feeding solids into gas streams, Chem Eng, 4 Sept, 94-96 (1972).
Sowden, R.E., The art of solids handling technology, Proc Tech Int, 17(11), 882-887 (1972).

1973-1974

Botterill, J.S.M., and Bessant, D.J., Flow properties of fluidized solids, Powder Tech, 8, 213-222 (1973).
Bransby, P.L.; Blair-Fish, P.M., and James, R.G., Investigation of flow of granular materials, Powder Tech, 8, 197-206 (1973).
Khan, J.I., and Pei, D.C., Pressure drop in vertical solid-gas suspension flow, Ind Eng Chem Proc Des Dev, 12(4), 428-431 (1973).
Kostyuk, G.F., and Dzyadzio, A.M., Pressure losses during vertical high-density pneumatic transport, Int Chem Eng, 13(4), 611-613 (1973).

Leung, L.S., and Towler, B.F., Design of vertical pneumatic conveyor with a fluidized bed as mixing zone, Powder Tech, 8, 27-32 (1973).
Manieh, A.A., and Spink, D.R., Purging cures corrosion and condensation in solids feed, Chem Eng, 16 April, 124 (1973).
Wang, P.Y., and Heldman, D.R., Pneumatic transport of dry milk powder, Ind Eng Chem Proc Des Dev, 12(4), 424-427 (1973).
Yang, W.C., Estimating the solid particle velocity in vertical pneumatic conveying lines, Ind Eng Chem Fund, 12(3), 349-352 (1973).
Augenstein, D.A., and Hogg, R., Friction factors for powder flow, Powder Tech, 10, 43-49 (1974).
Cowin, S.C., A theory for the flow of granular materials, Powder Tech, 9, 61-69 (1974).
Ferretti, E.J., Feeding coal to pressurized systems, Chem Eng, 9 Dec, 113-116 (1974).
Hogg, R.; Shoji, K., and Austin, L.G., Axial transport of dry powders in horizontal rotating cylinders, Powder Tech, 9, 99-106 (1974).
Mullins, W.W., Experimental evidence for the stochastic theory of particle flow under gravity, Powder Tech, 9, 29-37 (1974).
Rademacher, F.J.C., Characteristics of vertical screw conveyors for granular material, Powder Tech, 9, 71-89 (1974).
Van Zuilichem, D.J.; Van Egmond, N.D., and De Swart, J.G., Density behaviour of flowing granular material, Powder Tech, 10, 161-169 (1974).
Various, Solids handling (special report), Hyd Proc, 53(3), 75-93 (1974).
Yang, W.C., Correlations for solid friction factors in vertical and horizontal pneumatic conveyings, AIChEJ, 20(3), 605-607 (1974).

1975-1976

Galimov, Z.F.; Latypov, M.G., and Levinter, M.E., Operation of vertical pressurized standpipes and overflow tubes under conditions of backpressure of gas phase on a stream of granular material, Int Chem Eng, 15(2), 266-269 (1975).
Gerchow, F.J., Specifying components of pneumatic-conveying systems, Chem Eng, 31 March, 88-96 (1975).
Gerchow, F.J., How to select a pneumatic-conveying system, Chem Eng, 17 Feb, 72-86 (1975).
Johanson, J.R., Flow of bulk powders, Chemtech, Sept, 572-576 (1975).
Kopko, R.J.; Barton, P., and McCormick, R.H., Hydrodynamics of vertical liquid-solids transport, Ind Eng Chem Proc Des Dev, 14(3), 264-269 (1975).
Radin, I.; Zakin, J.L., and Patterson, G.K., Drag reduction in solid-fluid systems, AIChEJ, 21(2), 358-371 (1975).
Shinohara, K., and Tanaka, T., Effect of air pressure on solids flow from storage vessels, Chem Eng Sci, 30(4), 369-378 (1975).
Various, Solids flow techniques (topic issue), Chem Eng Prog, 71(2), 53-72 (1975).
Yang, W.C., Choking in vertical pneumatic conveying, AIChEJ, 21(5), 1013-1015 (1975).
Zanker, A., Nomograph for flow of solids through openings, Proc Engng, July, 66-67 (1975).

Birchenough, A., and Mason, J.S., Local particle velocity measurements with a laser anemometer in upward flowing gas-solid suspension, Powder Tech, 14, 139-152 (1976).
Botterill, J.S.M., and Bessant, D.J., Flow properties of fluidized solids, Powder Tech, 14, 131-137 (1976).
Gutman, R.G., Vibrated beds of powders, Trans IChemE, 54, 174-183, 251-257 (1976).
Leung, L.S., and Wiles, R.J., Quantitative design procedure for vertical pneumatic conveying systems, Ind Eng Chem Proc Des Dev, 15(4), 552-557 (1976).
Marcus, R.D.; Dickson, A.J., and Rallis, C.J., Drag reduction and pressure pulsations in pneumatic conveying, Powder Tech, 15, 107-116 (1976).
Sato, I., Predicting powder flow properties, Processing, Oct, 38-39 (1976).
Shook, C.A., Developments in hydrotransport, Can JCE, 54, 13-20 (1976).
Various, Pneumatic conveying of materials (topic issue), Chem Eng Prog, 72(3), 63-80 (1976).

1977-1978

Bates, L., The handling of bulk material, Chem Engnr, March, 165-170 (1977).
Bransby, P.L., Current work in materials handling, Chem Engnr, March, 161-164 (1977).
Coughlin, R.W., Improved gravity feeding of granular solids, Powder Tech, 17, 65-71 (1977).
Mann, U., and Crosby, E.J., Flow measurement of coarse particles in pneumatic conveyers, Ind Eng Chem Proc Des Dev, 16(1), 9-13 (1977).
Mills, D., and Mason, J.S., Particle concentration effects in bend erosion for pneumatic conveying, Powder Tech, 17, 37-53 (1977).
Payen, M., Pneumatic materials handling in the aluminium industry, Chem Engnr, April, 255-257 (1977).
Swift, D.L., Transport and effects of fine particles, Powder Tech, 18, 49-52 (1977).
Bandrowski, J., and Kaczmarzyk, G., Gas-to-particle heat transfer in vertical pneumatic conveying of granular materials, Chem Eng Sci, 33(10), 1303-1310 (1978).
Caloine, R., and Clayton, C.G.A., Drying and handling of polyester granules, Chem Engnr, March, 185-188 (1978).
Cawley, J., Some aspects of powder handling, Chem Engnr, Jan, 59-60 (1978).
Glikin, P.G., Transport of solids through flighted rotating drums, Trans IChemE, 56, 120-126 (1978).
Jones, P.J., and Leung, L.S., Comparison of correlations for saltation velocity in horizontal pneumatic conveying, Ind Eng Chem Proc Des Dev, 17(4), 571-575 (1978).
Knowlton, T.M., and Hirsan, I., L-values characterized for solids flow, Hyd Proc, 57(3), 149-156 (1978).
Leung, L.S., and Jones, P.J., Flow of gas-solid mixtures in standpipes: A review, Powder Tech, 20, 145-160 (1978).
Leung, L.S.; Jones, P.J., and Knowlton, T.M., Analysis of moving-bed flow of solids down standpipes and slide valves, Powder Tech, 19, 7-15 (1978).

Ottjes, J.A., Digital simulation of pneumatic particle transport, Chem Eng Sci, 33(6), 783-786 (1978).
Rademacher, F.J.C., Are inclined screw blades for vertical grain augers advantageous? Powder Tech, 21, 135-145 (1978).
Scott, A.M., Pneumatic transport of granules, Chem Engnr, March, 189-190 (1978).
Smith, T.N., Limiting volume fractions in vertical pneumatic transport, Chem Eng Sci, 33(6), 745-750 (1978).
Spedding, P.L., and Nguyen, V.T., Holdup in conveying systems for fluid-solid two-phase flow, Chem Eng J, 15(2), 131-146 (1978).
Various, Solids handling (feature report), Processing, Jan, 41-55 (1978).
Yang, W.C., Correlation for solid friction factor in vertical pneumatic conveying lines, AIChEJ, 24(3), 548-552 (1978).

1979-1980

Arastoopour, H., and Gidaspow, D., Vertical pneumatic conveying using four hydrodynamic models, Ind Eng Chem Fund, 18(2), 123-130 (1979).
Botterill, J.S.M., and Abdul-Halim, B.H., Open-channel flow of fluidized solids, Powder Tech, 23, 67-78 (1979).
Carstensen, J.T., and Laughlin, S.M., Dynamic flowrates of granular particles, Powder Tech, 23, 79-84 (1979).
Ganapathy, V., Solids discharge rates from vertical pipes, Chem Eng, 22 Oct, 161 (1979).
Johanson, J.R., Two-phase-flow effects in solids processing and handling, Chem Eng, 1 Jan, 77-86 (1979).
Judd, M.R., and Dixon, P.D., The effect of aeration on the flowability of powders, Trans IChemE, 57, 67-69 (1979).
Klinzing, G.E., Damping solid-flow fluctuations, Chem Eng Sci, 34(7), 971-974 (1979).
Klinzing, G.E., Vertical pneumatic transport of solids in minimum pressure drop region, Ind Eng Chem Proc Des Dev, 18(3), 404-408 (1979).
Nedderman, R.M., and Tuzun, U., Kinematic model for flow of granular materials, Powder Tech, 22, 243-253 (1979).
Rademacher, F.J.C., Non-spill discharge characteristics of bucket elevators, Powder Tech, 22, 215-241 (1979).
Televantos, Y., et al., Flow of slurries of coarse particles at high solids concentrations, Can JCE, 57, 255-265 (1979).
Vanasse, R.; Coupal, B., and Boulos, M.I., Hydraulic transport of peat moss suspensions, Can JCE, 57, 238-248 (1979).
Various, Powder handling (feature report), Processing, Feb, 53-81 (1979).
Yang, W.C.; Vaux, W.G.; Keairns, D.L., and Vojnovich, T., High-temperature pneumatic transport line test facility, Ind Eng Chem Proc Des Dev, 18(4), 695-703 (1979).
Zenz, F.A., and Zenz, F.E., Gravity flow of gases, liquids and bulk solids, Ind Eng Chem Fund, 18(4), 345-348 (1979).
Chen, T.Y.; Walawender, W.P., and Fan, L.T., Moving-bed solids flow in an inclined pipe leading into a fluidized bed, AIChEJ, 26(1), 24-36 (1980).
Ginestra, J.C.; Rangachari, S., and Jackson, R., A one-dimensional theory of flow in a vertical standpipe, Powder Tech, 27, 69-84 (1980).

Kaczmarzyk, G., and Bandrowski, J., Gas-solid heat-transfer coefficient in vertical pneumatic transport, Int Chem Eng, 20(1), 98-110 (1980).
Klinzing, G.E., Comparison of pressure losses in bends between recent data and models for gas-solid flow, Can JCE, 58, 670-673 (1980).
Klinzing, G.E., A simple light-sensitive meter for measurement of solid/gas loadings and flow steadiness, Ind Eng Chem Proc Des Dev, 19(1), 31-33 (1980).
Leung, L.S., Vertical pneumatic conveying: A flow regime diagram and review of choking versus non-choking systems, Powder Tech, 25, 185-190 (1980).
McKay, G., and McLain, H.D., The transportation of cuboid particles in horizontal pipelines, Trans IChemE, 58, 175-180 (1980).
Molerus, O., Description of pneumatic conveying, Int Chem Eng, 20(1), 7-19 (1980).
Staub, F.W., Steady-state and transient gas-solids flow characteristics in vertical transport lines, Powder Tech, 26, 147-159; 27, 124 (1980).
Tomita, Y.; Jotaki, T., and Fukushima, K., Feed rate characteristics of a blow-tank solids conveyor in transport of granular materials, Powder Tech, 26, 29-33 (1980).
Tomita, Y.; Yutani, S., and Jotaki, T., Pressure drop in vertical pneumatic transport lines of powdery material at high solids loading, Powder Tech, 25, 101-107 (1980).

1981

Bachovchin, D.M.; Mulik, P.R.; Newby, R.A., and Keairns, D.L., Pulsed transport of bulk solids between adjacent fluidized beds, Ind Eng Chem Proc Des Dev, 20(1), 19-26 (1981).
Bandrowski, J., and Kaczmarzyk, G., Operation and design of vertical pneumatic conveying, Powder Tech, 28, 25-33 (1981).
Ganapathy, V., Estimation of screw-conveyor capacity, Proc Engng, July, 47 (1981).
Hwang, L.Y.; Wood, D.J., and Kao, D.T., Capsule hoist system for vertical transport of coal and other mineral solids, Can JCE, 59, 317-324 (1981).
Jackson, R., and Judd, M.R., Effect of aeration on the flowability of powders, Trans IChemE, 59, 119-121 (1981).
Jafari, J.F.; Clarke, B., and Dyson, J., Characteristics of dense phase pneumatic transport of grains in horizontal pipes, Powder Tech, 28, 195-199 (1981).
Klinzing, G.E., and Mathur, M.P., Dense and extrusion flow regime in gas-solid transport, Can JCE, 59, 590-594 (1981).
Molerus, O., Prediction of pressure drop with steady-state pneumatic conveying of solids in horizontal pipes, Chem Eng Sci, 36(12), 1977-1984 (1981).
Molerus, O., and Wellmann, P., A new concept for the calculation of pressure drop with hydraulic transport of solids in horizontal pipes, Chem Eng Sci, 36(10), 1623-1632 (1981).
Parzonka, W.; Kenchington, J.M., and Charles, M.E., Hydrotransport of solids in horizontal pipes: Effects of solids concentration and particle size on the deposit velocity, Can JCE, 59, 291-296 (1981).
Wahl, R., Effect of upstream solids discharge on downstream processing, Chem Eng Prog, 77(6), 76-79 (1981).

1982

Arastoopour, H.; Lin, S.C., and Weil, S.A., Analysis of vertical pneumatic conveying of solids using multiphase flow models, AIChEJ, 28(3), 467-473 (1982).

Carmichael, G.R., Estimation of the drag coefficient of regularly shaped particles in slow flows from morphological descriptors, Ind Eng Chem Proc Des Dev, 21(3), 401-403 (1982).

Chan, S.M.; Remple, D.; Shook, C.A., and Esmail, M.N., One-dimensional model of plug-flow pneumatic conveying, Can JCE, 60, 581-588 (1982).

Davis, N., Pneumatic conveying of coal, Proc Engng, Feb, 37-39 (1982).

Grikitis, K., Mechanised materials handling, Proc Engng, Feb, 41-43 (1982).

Matsen, J.M., Mechanisms of choking and entrainment in pneumatic transport systems, Powder Tech, 32, 21-33 (1982).

Molerus, O., Flow behaviour of cohesive materials (review paper), Chem Eng Commns, 15(5), 257-290 (1982).

Nedderman, R.M.; Tuzun, U.; Savage, S.B., and Houlsby, G.T., Flow of granular materials (review paper), Chem Eng Sci, 37(11), 1597-1610; 37(12), 1691-1710 (1982).

Nedderman, R.M.; Tuzun, U.; Savage, S.B., and Houlsby, G.T., The flow of granular materials: Discharge rates from hoppers (review paper), Chem Eng Sci, 37(11), 1597-1609 (1982).

Rietema, K., Dispersed two-phase systems (review paper), Chem Eng Sci, 37(8), 1125-1150 (1982).

Rogers, R.S.C., Closed-form analytical solutions for models of closed-circuit roll crushers, Powder Tech, 32, 125-127 (1982).

Scott, D.S., and Piskorz, J., Low-rate entrainment feeder for fine solids, Ind Eng Chem Fund, 21(3), 319-322 (1982).

Shih, Y.T.; Arastoopour, H., and Well, S.A., Hydrodynamic analysis of horizontal solids transport, Ind Eng Chem Fund, 21(1), 37-43 (1982).

Singh, B., Analysis of pressure drop in vertical pneumatic conveying, Powder Tech, 32, 179-191 (1982).

Smeltzer, E.E.; Waever, M.L., and Klinzing, G.E., Individual electrostatic particle interaction in pneumatic transport, Powder Tech, 33, 31-42 (1982).

Smeltzer, E.E.; Weaver, M.L., and Klinzing, G.E., Pressure drop losses due to electrostatic generation in pneumatic transport, Ind Eng Chem Proc Des Dev, 21(3), 390-394 (1982).

Straker, E., Materials handling, Processing, Oct, 21-25 (1982).

Tomita, Y.; Jotaki, T.; Makimoto S., and Fukushima, K., Similarity of granular flow in a blow tank solids conveyor, Powder Tech, 32, 1-8 (1982).

Tuzun, U.; Houlsby, G.T.; Nedderman, R.M., and Savage, S.B., The flow of granular materials: Velocity distributions in slow flow (review paper), Chem Eng Sci, 37(12), 1691-1709 (1982).

Various, Conveyors survey, Processing, March, 23-45 (1982).

Yang, W.; Jaraiz, E.; Levenspiel, O., and Fitzgerald, T.J., A magnetic control valve for flowing solids, Ind Eng Chem Proc Des Dev, 21(4), 717-721 (1982).

1983

Ahmadi, G., and Shahinpoor, M., Turbulent modeling of rapid flow of granular materials, Powder Tech, 35, 241-248 (1983).

Baker, P., Pipeline transport of solids: Economics and design, Proc Engng, Feb, 29-34 (1983).

Bates, L., Bulk handling survey, Processing, July, 43-47 (1983).

Brennen, C.E.; Sieck, K., and Paslaski, J., Hydraulic jumps in granular material flow, Powder Tech, 35, 31-37 (1983).

Briscoe, B.J.; Radwan, H., and Streat, M., Model experiments on sliding friction for application in hydraulic conveying of solids, Can JCE, 61(6), 769-775 (1983).

Capes, P., Vibratory conveying, Proc Engng, Feb, 41-43 (1983).

Chhabra, R.P., and Richardson, J.F., Hydraulic transport of coarse gravel particles in a smooth horizontal pipe, CER&D, 61, 313-317 (1983).

Dry, R.J.; Judd, M.R., and Shingles, T., Two-phase theory and fine powders, Powder Tech, 34, 213-223 (1983).

Irons, G.A., and Chang, J.S., Dispersed powder flow through vertical pipes, Powder Tech, 34, 233-242 (1983).

Savage, S.B.; Nedderman, R.M.; Tuzun, U., and Houlsby, G.T., The flow of granular materials, Chem Eng Sci, 38(2), 189-196 (1983).

Savage, S.B.; Nedderman, R.M.; Tuzun, U., and Houlsby, G.T., The flow of granular materials: Rapid shear flows (review paper) Chem Eng Sci, 38(2), 189-195 (1983).

Takami, A., Mechanism of discharge flow of powder from a vertical tube, Powder Tech, 34, 1-8 (1983).

Yang, W.C., Criteria for choking in vertical pneumatic conveying lines, Powder Tech, 35, 143-150 (1983).

1984

Al-Din, N., and Gunn, D.J., Flow of non-cohesive solids through orifices, Chem Eng Sci, 39(1), 121-128 (1984).

Anon., Solids handling: A survey, Processing, July, 7-16; Aug, 44 (1984).

Bienstock, D., Development of coal-liquid mixtures, Chem Engnr, Feb, 18-23 (1984).

Fan, L.S.; Satija, S.; Kim, B.C., and Nack, H., Limestone/dolomite sulfonation in a vertical pneumatic transport reactor, Ind Eng Chem Proc Des Dev, 23(3), 538-545 (1984).

Kano, T.; Takeuchi, F.; Sugiyama, H., and Yamazaki, E., Optimum conditions for plug-type pneumatic conveying of granular materials, Int Chem Eng, 24(4), 702-710 (1984).

Konrad, K., and Davidson, J.F., Gas-liquid analogy in horizontal dense-phase pneumatic conveying, Powder Tech, 39, 191-198 (1984).

Larouere, P.J.; Joseph, S., and Klinzing, G.E., Stability concepts in relation to electrostatics and pneumatic transport, Powder Tech, 38, 1-6 (1984).

Mathur, M.P., and Klinzing, G.E., Flow measurement in pneumatic transport of pulverized coal, Powder Tech, 40, 309-322 (1984).

Taubmann, H., Technology of bulk materials, Int Chem Eng, 24(3), 432-441 (1984).

Teo, C.S., and Chong, Y.O., Pressure drop in vertical pneumatic conveying, Powder Tech, 38, 175-179 (1984).
Viswanathan, K., and Mani, B.P., Hold-up studies in the hydraulic conveying of solids in horizontal pipelines, AIChEJ, 30(4), 682-684 (1984).
Yoshioka, S.; Hirato, M.; Satomi, Y., and Ozaki, H., Control of flow of solid particles from a throat by addition of gas above the throat, Int Chem Eng, 24(1), 97-104 (1984).

1985
Ahmadi, G., Turbulence model for rapid flows of granular materials, Powder Tech, 44, 261-279 (1985).
Bello, R.A.; Robinson, C.W., and Moo-Young, M., Prediction of the volumetric mass transfer coefficient in pneumatic contactors, Chem Eng Sci, 40(1), 53-58 (1985).
Bohnet, M., Advances in design of pneumatic conveyors, Int Chem Eng, 25(3), 387-406 (1985).
Briens, C.L., and Bergougnou, M.A., Simple calculation technique to solve the Leung/Wiles equations for the choking load of multisize particles in pneumatic transport lines, Can JCE, 63(6), 995-996 (1985).
Chhabra, R.P., and Richardson, J.F., Hydraulic transport of coarse particles in viscous Newtonian and non-Newtonian media in a horizontal pipe, CER&D, 63, 390-397 (1985).
Maunder, A., Pneumatic conveying of hazardous materials, Chem Engnr, March, 11-13 (1985).
Peirce, T., The challenge of coal-liquid mixtures, Chem Engnr, Sept, 17-20 (1985).
Satija, S., and Fan, L.S., Terminal velocity of dense particles in the multisolid pneumatic transport bed, Chem Eng Sci, 40(3), 259-268 (1985).
Satija, S.; Young, J.B., and Fan, L.S., Pressure fluctuations and choking criterion for vertical pneumatic conveying of fine particles, Powder Tech, 43, 257-271 (1985).
Tuzun, U., and Nedderman, R.M., Gravity flow of granular materials around obstacles, Chem Eng Sci, 40(3), 325-352 (1985).
Various, Solids handling update, Processing, July, 25-35 (1985).
Wirth, K.E., and Molerus, O., Influence of pipe geometry on critical velocity of horizontal pneumatic conveying of coarse particles, Powder Tech, 42, 27-34 (1985).
Wolny, A., and Kabata, M., Mixing of solid particles in vertical pneumatic transport, Chem Eng Sci, 40(11), 2113-2118 (1985).

1986
Adewumi, M.A., and Arastoopour, H., Two-dimensional steady-state hydrodynamic analysis of gas-solids flow in vertical pneumatic conveying systems, Powder Tech, 48(1), 67-74 (1986).
Anon., Solids handling update, Processing, Jan, 17-19 (1986).
Antonin, L.; Otto, M., and Jozef, M., Generalized equation of pressure losses in dispersed powder flow through vertical pipes, Chem Eng Commns, 41, 151-162 (1986).

Briens, C.L., and Bergougnou, M.A., New model to calculate the choking variety of monosize and multisize solids in vertical pneumatic transport lines, Can JCE, 64(2), 196-204 (1986).

Chellappan, S., and Ramaiyan, G., Experimental study of design parameters of gas-solid injector feeder, Powder Tech, 48(2), 141-144 (1986).

Chong, Y.O., and Leung, L.S., Comparison of choking velocity correlations in vertical pneumatic conveying, Powder Tech, 47(1), 43-50 (1986).

Enick, R.M., and Klinzing, G.E., Correlation for acceleration length in vertical gas-solid transport, Chem Eng Commns, 49, 127-132 (1986).

Golda, J., Hydraulic transport of coal in pipes with drag-reducing additives, Chem Eng Commns, 43, 53-68 (1986).

Hilgraf, P., Investigation of pneumatic dense-phase conveying, Chem Eng & Proc, 20(1), 33-42 (1986).

Kim, H.J., et al., Minimum velocity for transport of a sand-water slurry through a pipeline, Int Chem Eng, 26(4), 731-738 (1986).

Kimber, A., Conveyors update, Processing, June, 28-33 (1986).

Konrad, K., Dense-phase pneumatic conveying through long pipelines: Effect of significantly compressible air flow on pressure drop, Powder Tech, 48(3), 193-204 (1986).

Konrad, K., Dense-phase pneumatic conveying (review paper), Powder Tech, 49(1), 1-36 (1986).

Kraus, M.N., Pneumatic conveying systems, Chem Eng, 13 Oct, 50-61 (1986).

Leung, L.S. (Ed.), Standpipe flow and pneumatic conveying (various papers), Powder Tech, 47(2), 103-200 (1986).

Shu, M.T.; Hamshar, J.A., and Weinberger, C.B., Cocurrent flow of gas-particulate mixtures through helical tubular coils, AIChEJ, 32(4), 529-536 (1986).

Tashiro, H., and Tomita, Y., Influence of diameter ratio on sudden expansion of circular pipe in gas-solid two-phase flow, Powder Tech, 48(3), 227-232 (1986).

Venkataraman, K.S., and Fuerstenau, D.W., Effect of lifter shape and configuration on material transport Powder Tech, 46(1), 23-32 (1986).

Weinberger, C.B., and Shu, M.T., Helical gas-solids flow, Powder Tech, 48(1), 13-22 (1986).

Yin, M.J.; Beddow, J.K., and Vetter, A.F., Effects of particle shape on two-phase flow in pipes, Powder Tech, 46(1), 53-60 (1986).

1987

Ayazi Shamlou, P., Hydraulic transport of particulate solids, Chem Eng Commns, 62(1), 233-250 (1987).

Borzone, L.A., and Klinzing, G.E., Dense-phase transport: Vertical plug flow, Powder Tech, 53(3), 273-284 (1987).

Chynoweth, E., A new pneumatic conveying system, Proc Engng, April, 63 (1987).

Davies, C.E., and Spedding, N.B., Mass flow determination from pressure drop across fluidized solids in fluidized-bed feeder, Powder Tech, 53(2), 131-136 (1987).

Gullett, B.K., and Gillis, G.R., Low flowrate laboratory feeders for agglomerative particles, Powder Tech, 52(3), 257-260 (1987).

Kim, M.H., et al., Flow of suspension through a vertical tube, Chem Eng Commns, 50, 31-50 (1987).
Klinzing, G.E.; Rohatgi, N.D.; Zaltash, A., and Myler, C.A., Generalized phase diagram approach to pneumatic transport (review paper), Powder Tech, 51(2), 135-150 (1987).
Kmiec, A., and Leschonski, K., Acceleration of the solid phase during pneumatic conveying in vertical pipes, Chem Eng J, 36(1), 59-70 (1987).
Majumdar, A., et al., Experimental study on solid particle dynamics in shear flow, Powder Tech, 49(3), 217-226 (1987).
Ouyang, C.J.P., and Tatterson, G.B., Effect of distributors on two-phase and three-phase flows in vertical columns, Chem Eng Commns, 49, 197-216 (1987).
Pittman, E.D., Properly maintain pneumatic conveying systems (tips on maintenance and troubleshooting), Chem Eng, 22 June, 123-127 (1987).
Thomas, B., et al., Dynamics of vibrated beds of granular solids, Powder Tech, 52(1), 77-92 (1987).
Tsubaki, J., and Tien, C., Solid velocity in crossflow moving beds, Powder Tech, 53(2), 105-112 (1987).
Wypich, P.W., and Arnold, P.C., Improving scale-up procedures for pneumatic conveying design, Powder Tech, 50(3), 281-294 (1987).
Yang, W.C., et al., Pneumatic transport in 10cm horizontal loop, Powder Tech, 49(3), 207-216 (1987).
Yianneskis, M., Velocity, particle sizing and concentration measurement techniques for multiphase flow, Powder Tech, 49(3), 261-270 (1987).

1988

Aziz, Z.B., and Klinzing, G.E., Plug-flow transport of cohesive coal: Horizontal and inclined flow, Powder Tech, 55(2), 97-106 (1988).
Byrne, M., Update on solids conveyors, Proc Engng, Dec, 49, 51 (1988).
Erdesz, K., and Nemeth, J., Methods of calculation of vibrational transport rate of granular materials, Powder Tech, 55(3), 161-170 (1988).
Erdesz,, K., and Szalay, A., Experimental study on vibrational transport of bulk solids, Powder Tech, 55(2), 87-96 (1988).
Hidaka, J.; Miwa, S., and Makino, K., Mechanism of sound generation in shear flow of granular materials, Int Chem Eng, 28(1), 99-108 (1988).
Kikkawa, H., et al., Effect of adsorption characteristics of dispersant on flow and storage properties of coal-water mixtures, Powder Tech, 55(4), 277-284 (1988).
Kitano, K.; Wisecarver, K.D.; Satija, S., and Fan, L.S., Holdup of fine particles in the fluidized dense bed of the multisolid pneumatic transport bed, Ind Eng Chem Res, 27(7), 1259-1264 (1988).
Ma, D., and Ahmadi, G., Kinetic model for rapid granular flows of nearly elastic particles including interstitial fluid effects, Powder Tech, 56(3), 191-208 (1988).
McDonagh, M., Screw conveying of wet materials, Chem Engnr, Feb, 19-20 (1988).
O'Dea, D.P.; Chong, Y.O., and Leung, L.S., Experimental study of transitional packed-bed flow in standpipe, Powder Tech, 55(3), 223-225 (1988).

Spedding, P.L., and Chen, J.J.J., Application of the Lockhart-Martinelli theory to fluid-solid transport, Chem Eng J, 37(2), 123-130 (1988).

CHAPTER 17

HEAT EXCHANGERS

17.1 Heat Transfer Theory and Data

1967-1969

Carr, A.D., and Balzhiser, R.E., Temperature profiles and eddy diffusivities in liquid metals, Brit Chem Eng, 12(1), 53-57 (1967).

Davis, D.S., Nomogram for heat-transfer coefficients from single horizontal cylinders to any fluid, Brit Chem Eng, 12(2), 248 (1967).

Gambill, W.R., An evaluation of recent correlations for high-flux heat transfer, Chem Eng, 28 Aug, 147-154 (1967).

Holt, A.D., Heating and cooling of solids, Chem Eng, 23 Oct, 145-166 (1967).

Starczewski, J., Economy gained by use of low-fin tubes, Brit Chem Eng, 12(2), 239-241 (1967).

Valakhonova, V.I., Heat and mass transfer processes during liquid evaporation from free surface into rarefied gas, Int Chem Eng, 7(3), 471-475 (1967).

Balakrishna, K., Nomogram for heat transfer to non-Newtonian liquids in agitated vessels, Brit Chem Eng, 13(9), 1304 (1968).

Caplan, F., Nomograph for radiant heat-transfer coefficient, Chem Eng, 18 Nov, 198 (1968).

De Voe, D.H., Nomograph for heat-tracing coil, Chem Eng, 8 April, 182 (1968).

Dent, J.C., An electrical network method for combined free convective and radiative transfer from annular finned surfaces, Brit Chem Eng, 13(1), 90-93 (1968).

Fells, I., and Harker, J.H., Temperature and heat transfer measurements within the discharge zone of a propane-air flame, Trans IChemE, 46, T236-242 (1968).

Hooper, W.B., Calculate corrected mean temperature difference, Chem Eng, 21 Oct, 174 (1968).

Kasper, S., Selecting heat-transfer media, Chem Eng, 2 Dec, 117-120 (1968).

Mathews, R.T., Economic applications of air cooling to process industries, Brit Chem Eng, 13(10), 1425-1432 (1968).

Murti, P.S., and Rao, K.B., Heat transfer in mixers for non-Newtonian liquids, Brit Chem Eng, 13(10), 1441-1442 (1968).

Porter, J.W.; Goren, S.L., and Wilke, C.R., Direct-contact heat transfer between immiscible liquids in turbulent pipe flow, AIChEJ, 14(1), 151-158 (1968).

Sherwin, K., Laminar convection in uniformly heated vertical concentric annuli, Brit Chem Eng, 13(11), 1580-1585 (1968).

Yurkanin, R.M., and Classon, E.O., Impedance heating, Chem Eng, 12 Aug, 182-188 (1968).

Bott, T.R., and Nair, B., The behaviour of some pseudo-plastic solutions in scraped-film heat transfer equipment, Chem Engnr, Oct, CE361-362 (1969).

Everett, M., Forced convection heat transfer inside tubes (turbulent flow), Chem Engnr, Sept, CE159 (1969).

Gay, B., and Cameron, P.T., Computer methods for the heat conduction equation, Brit Chem Eng, 14(9), 1222-1223; 14(10), 1409-1412 (1969).

Kneale, M., Design of vessels with half coils, Trans IChemE, 47, T279-284 (1969).

Oliver, D.R., Non-Newtonian heat transfer (non-circular tubes), Trans IChemE, 47, T18-20 (1969).

Osburn, J.O., and Sollami, B.J., Simplified calculation of thermal radiation shields, Chem Eng, 22 Sept, 139-144 (1969).
Rachkov, V.I., and Morozov, V.H., Designing curved tube plates, Brit Chem Eng, 14(5), 670 (1969).
Sherwin, K., Combined natural and forced laminar flows, Brit Chem Eng, 14(9), 1215-1217 (1969).

1970-1972
Agnew, J.B., and Potter, O.E., Heat transfer properties of packed tubes of small diameter, Trans IChemE, 48, T15-20 (1970).
Boothroyd, R.G., and Haque, H., Experimental investigation of heat transfer in the entrance region of a heated duct conveying fine particles, Trans IChemE, 48, T109-120 (1970).
Bradshaw, A.V.; Johnson, A.; McLachlan, N.H., and Chiu, Y.T., Heat transfer between air and nitrogen and packed beds of non-reacting solids, Trans IChemE, 48, T77-84 (1970).
Brdlik, P.M., Heat and mass transfer in a binary boundary layer with natural convection, Int Chem Eng, 10(1), 49-56; 119-124 (1970).
Chernousko, F.L., Solution of nonlinear heat conduction problems in media with phase changes, Int Chem Eng, 10(1), 42-49 (1970).
Dey, D.N., Nomogram for volumetric heat transfer coefficient of granular materials in beds, Brit Chem Eng, 15(9), 1205 (1970).
Lykov, A.V.; Berkovskii, B.M., and Fertman, V.E., Experimental investigation of convection during heating from above, Int Chem Eng, 10(1), 30-34 (1970).
Oosthuizen, P.H., Laminar combined convection from an isothermal circular cylinder to air, Trans IChemE, 48, T227-231 (1970).
Sergeeva, L.A., Comparison of heat-transfer rates under unsteady-state and steady-state conditions, Int Chem Eng, 10(3), 457-461 (1970).
Starczewski, J., Short-cut to tubeside heat-transfer coefficient, Hyd Proc, 49(2), 129-130 (1970).
Barrow, H., and Sitharamarao, T.L., Effect of variation in the volumetric expansion coefficient on free convection heat transfer, Brit Chem Eng, 16(8), 704-705 (1971).
Porter, J.E., Heat transfer at low Reynolds number (highly viscous liquids in laminar flow), Trans IChemE, 49, 1-29 (1971).
Sherwin, K., Forced-convection heat transfer, Brit Chem Eng, 16(7), 593-596 (1971).
Basu, A., and Narasimham, K.S., Convective heat transfer between air and a cloud of falling particles, Brit Chem Eng, 17(1), 67-69 (1972).
Bertram, C.G.; Desai, V.J., and Interess, E., Designing steam tracing, Chem Eng, 3 April, 74-80 (1972).
Eastop, T.D., and Smith, C., Heat transfer from a sphere to an air stream at sub-critical Reynolds number, Trans IChemE, 50, 26-31 (1972).
Herbert, L.S., and Sterns, U.J., Heat transfer in vertical tubes: Interaction of forced and free convection, Chem Eng J, 4(1), 46-52 (1972).
Madsen, N., Comparing parallel flow with counterflow operation, Chem Eng, 10 July, 106-108 (1972).

Seifert, W.F.; Jackson, L.L., and Sech, C.E., Organic fluids for high-temperature heat-transfer systems, Chem Eng, 30 Oct, 96-104 (1972).

Various, Heat transfer (topic issue), Chem Eng Prog, 68(7), 53-86 (1972).

Various, Heat transfer (topic issue), Chem Eng Prog, 68(2), 49-69 (1972).

1973-1975

Davies, J.T., and Shawki, A.M., The enhancement of turbulent heat transfer, Chem Engnr, Nov, 528-532 (1973).

Edney, H.G.S.; Edwards, M.F., and Marshall, V.C., Heat transfer to a cooling coil in an agitated vessel, Trans IChemE, 51, 4-9 (1973).

Edwards, M.F.; Nellist, D.A., and Wilkinson, W.L., Heat transfer to viscous fluids in pulsating flow in pipes, Chem Engnr, Nov, 532-537 (1973).

Forrest, G., and Wilkinson, W.L., Laminar heat transfer to power law fluids in tubes with constant wall temperature, Trans IChemE, 51, 331-338 (1973).

Lynch, E.P., Calculating the linear feet in a flat-spiral coil, Chem Eng, 3 Sept, 128 (1973).

Mishra, P., and Tripathi, G., Heat and momentum transfer to purely viscous non-Newtonian fluids flowing through tubes, Trans IChemE, 51, 141-150 (1973).

Otazo, J., Designing heating coils for tanks by nomograph, Chem Eng, 24 Dec, 73-74 (1973).

Starczewski, J., Short cut to shell-side heat-transfer coefficient, Hyd Proc, 52(4), 155-157 (1973).

Various, Heat transfer (topic issue), Chem Eng Prog, 69(7), 57-85 (1973).

Young, J., and McCutcheon, A.R.S., The performance of Ranque-Hilsch vortex tubes, Chem Engnr, Nov, 522-528 (1973).

Forrest, G., and Wilkinson, W.L., Laminar heat transfer to power law fluids in tubes with constant wall heat flux, Trans IChemE, 52, 10-16 (1974).

Frank, O., Estimating overall heat transfer coefficients, Chem Eng, 13 May, 126-128 (1974).

Manka, H., and Bandrowski, J., Equivalent heat-transfer coefficient for radiating gases, Int Chem Eng, 14(2), 343-347 (1974).

Phadke, P.S., Empirical formula for tube-side heat transfer coefficients, Chem Eng, 23 Dec, 88 (1974).

Various, New directions in heat transfer (feature report), Chem Eng, 19 Aug, 82-98 (1974).

Dury, T., Heat transfer literature review 1974-75, Proc Engng, Nov, 62-67 (1975).

Frikken, D.R.; Rosenberg, K.S., and Steinmeyer, D.E., Understanding vapor-phase heat-transfer media, Chem Eng, 9 June, 86-90 (1975).

Smyth, R., Some experimental results on heat transfer in two-dimensional diffusers, Trans IChemE, 53, 50-54 (1975).

1976-1978

Basiulis, A., What good is the heat pipe? Chemtech, March, 208-211 (1976).

Caplan, F., Calculating radiant heat transfer, Chem Eng, 27 Sept, 136-137 (1976).

Duncan, P., Heat transfer literature review 1975-76, Proc Engng, Nov, 63-69 (1976).

Edney, H.G.S., and Edwards, M.F., Heat transfer to non-Newtonian and aerated fluids in stirred tanks, Trans IChemE, 54, 160-166 (1976).
Ganapathy, V., Quick estimation of gas heat-transfer coefficients, Chem Eng, 13 Sept, 199-202 (1976).
Lemaire, R., Special aspects of biological heat transfer, Int Chem Eng, 16(4), 571-576 (1976).
Lin, S., Ice-pile formation by free convection of moist air in a freezer, Trans IChemE, 54, 287-288 (1976).
Oliver, D.R., and Asghar, S.M., Heat transfer to Newtonian and viscoelastic liquids during laminar flow in helical coils, Trans IChemE, 54, 218-224 (1976).
Szpiro, O., Bibliography of laminar-flow heat transfer in horizontal ducts, Proc Engng, Nov, 100-104 (1976).
Various, Heat transfer (topic issue), Chem Eng Prog, 72(7), 53-95 (1976).
Brisbane, T., Heat transfer literature review 1976-1977, Proc Engng, Nov, 63-71 (1977).
Dodd, R., Free convection effects for heat transfer to fluids in laminar flow inside vertical tubes with constant wall temperature, Trans IChemE, 55, 223-224 (1977).
Ganapathy, V., Estimating heat-transfer coefficients, Hyd Proc, 56(10), 139-141; 56(11), 303-306; 56(12), 105-108 (1977).
Ganapathy, V., Relating heat emission to surface temperature, Chem Eng, 19 Dec, 106 (1977).
Martignon, D.R., Heat more efficiently using electric immersion heaters, Chem Eng, 23 May, 141-144 (1977).
Withee, E., Vaporization and condensation: Equilibrium-flash calculations on a calculator, Chem Eng, 26 Sept, 121-122 (1977).
Brown, A., Relative importance of viscous dissipation and pressure stress effects on laminar free convection, Trans IChemE, 56, 77-80 (1978).
Economides, M.J., and Maloney, J.O., Two experiments for estimating free convection and radiation heat-transfer coefficients, Chem Eng Educ, 12(3), 122-126 (1978).
Oliver, D.R., and Rao, S.S., Heat transfer to viscous Newtonian liquids in laminar flow in straight horizontal circular tubes, Trans IChemE, 56, 62-66 (1978).

1979-1980

Kohli, I.P., Steam tracing of pipelines, Chem Eng, 26 March, 156-163 (1979).
Oliver, D.R., and Rao, S.S., Heat transfer to Newtonian liquids in laminar flow in straight horizontal elliptical tubes, Trans IChemE, 57, 104-112 (1979).
Polley, G.T., Correlations for forced convection heat transfer, Chem Engnr, April, 233-234 (1979).
Siddall, R.G., and Selcuk, N., Evaluation of a new six-flux model for radiative transfer in rectangular enclosures, Trans IChemE, 57, 163-169 (1979).
Stephan, K., Heat transfer with phase change in chemical engineering processes, Int Chem Eng, 19(4), 557-566 (1979).
Tagg, D.J.; Patrick, M.A., and Wragg, A.A., Heat and mass transfer downstream of abrupt nozzle expansions in turbulent flow, Trans IChemE, 57, 176-181 (1979).

Bandrowski, J., and Ziolo, J., Heat and mass transfer bibliography: Polish works (1977-1978), Int J Heat Mass Trans, 23(3), 387-392 (1980).

Brunner, C.R., Program calculates heat transfer through composite walls, Chem Eng, 16 June, 119-122 (1980).

Campo, A., Comparison of three algorithms for nonlinear heat conduction problems, Comput Chem Eng, 4(2), 139-142 (1980).

Chisholm, D., Heat transfer literature review 1978-79, Proc Engng, Jan, 43-53 (1980).

Horwitz, B., Can log-mean temperature difference equal zero? Chem Eng, 5 May, 142 (1980).

Mori, Y., Heat transfer bibliography: Japanese works, Int J Heat Mass Trans, 23(1), 123-126 (1980).

Mori, Y., Heat transfer bibliography: Japanese works, Int J Heat Mass Trans, 23(11), 1589-1594 (1980).

Murdock, D.L., Electrical heat-tracing is simple and inexpensive, Chem Eng, 20 Oct, 151 (1980).

Pierce, W.L., Heat transfer Colburn-factor equation, Chem Eng, 17 Dec, 113 (1979); 10 March, 170 (1980).

Shah, M.M., General correlation for critical heat flux in annuli, Int J Heat Mass Trans, 23(2), 225-234 (1980).

Shibayama, S., and Morooka, S., Study on a heat pipe, Int J Heat Mass Trans, 23(7), 1003-1014 (1980).

Sim, W.J., and Daubert, T.E, Prediction of vapor-liquid equilibria of undefined mixtures, Ind Eng Chem Proc Des Dev, 19(3), 386-393 (1980).

Soloukhin, R.I., and Martynenko, O.G., Heat and mass transfer bibliography: Soviet works, Int J Heat Mass Trans, 23(2), 235-246; 23(5), 713-722; 23(7), 1033-1042; 23(8), 1147-1156 (1980).

Sparrow, E.M., and Laing, L.R., Heat transfer bibliography, Int J Heat Mass Trans, 23(3), 393-404; 23(10), 1385-1398 (1980).

1981

Aly, A.M.M., Flow regime boundaries for an interior subchannel of a horizontal 37-element bundle, Can JCE, 59, 158-163 (1981).

Aziz, A., and Na, T.Y., Periodic heat transfer in fins with variable thermal parameters, Int J Heat Mass Trans, 24(8), 1397-1404 (1981).

Bandrowski, J., and Ziolo, J., Heat and mass transfer bibliography: Polish works (1979-1980), Int J Heat Mass Trans, 24(10), 1733-1738 (1981).

Eckert, E.R.G., et al., Heat transfer: A review of 1979 literature, Int J Heat Mass Trans, 24(1), 1-34 (1981).

Eckert, E.R.G., et al., Heat transfer: Review of 1980 literature, Int J Heat Mass Trans, 24(12), 1863-1902 (1981).

Ganapathy, V., Estimate nonluminous radiation heat transfer coefficients, Hyd Proc, 60(4), 235-237 (1981).

Grissom, W.M., and Wierum, F.A., Liquid spray cooling of a heated surface, Int J Heat Mass Trans, 24(2), 261-272 (1981).

Horwitz, B.A., How to cool your beer more quickly, Chem Eng, 10 Aug, 97-98 (1981).

Liapis, A.I., and McAvoy, T.J., Transient solutions for a class of hyperbolic countercurrent distributed heat and mass transfer systems, Trans IChemE, 59, 89-94 (1981).

McNaught, J., Heat-transfer literature review 1979-80, Proc Engng, Jan, 37-41 (1981).

Mori, Y., Heat transfer bibliography: Japanese works, Int J Heat Mass Trans, 24(11), 1753-1758 (1981).

Rabas, T.J.; Eckels, P.W., and Sabatino, R.A., Effect of fin density on heat transfer and pressure drop performance of low-finned tube banks, Chem Eng Commns, 10(1), 127-149 (1981).

Sampson, P., and Gibson, R.D., Mathematical model of nozzle blockage by freezing, Int J Heat Mass Trans, 24(2), 231-242 (1981).

Singh, J., Selecting heat-transfer fluids for high-temperature service, Chem Eng, 1 June, 53-58 (1981).

Soloukhin, R.I., and Martynenko, O.G., Heat and mass transfer bibliography: Soviet works, Int J Heat Mass Trans, 24(1), 171-180; 24(6), 959-968; 24(7), 1277-1286 (1981).

Sparrow, E.M., and Hsu, C.F., Analysis of two-dimensional freezing on the outside of a coolant-carrying tube, Int J Heat Mass Trans, 24(8), 1345-1358 (1981).

1982

Blackwell, W.W., Estimate heat-tracing requirements for pipelines, Chem Eng, 6 Sept, 115-118 (1982).

Burfoot, D., and Rice, P., Turbulent forced convection heat transfer enhancement using pall rings in a circular duct, Ind Eng Chem Proc Des Dev, 21(4), 646-650 (1982).

Chu, H.S.; Chen, C.K., and Weng, C.I., Applications of Fourier series technique to transient heat transfer problem, Chem Eng Commns, 16(1), 215-226 (1982).

Eastop, T.D., and Turner, J.R., Air flow around three cylinders at various pitch-to-diameter ratios for both a longitudinal and a transverse arrangement, Trans IChemE, 60, 359-363 (1982).

Eckert, E.R.G., et al., Heat transfer literature review (1981), Int J Heat Mass Trans, 25(12), 1783-1812 (1982).

Fletcher, D.F.; Maskell, S.J., and Patrick, M.A., Theoretical investigation of the Chilton-Colburn analogy, Trans IChemE, 60, 122-125 (1982).

Guy, J.L., Modeling heat-transfer systems, Chem Eng, 3 May, 93-98 (1982).

Havas, G.; Deak, A., and Sawinsky, J., Heat-transfer coefficients in an agitated vessel with vertical tube baffles, Chem Eng J, 23(2), 161-166 (1982).

Hieber, C.A., Laminar mixed convection in an isothermal horizontal tube: Correlation of heat transfer data, Int J Heat Mass Trans, 25(11), 1737-1746 (1982).

Kurmarao, P.S.V., How efficient is the circular tube in heat transfer? Chem Eng, 23 Aug, 122 (1982).

Marzi, M., Solving problems of varying heat-transfer areas in batch processes, Chem Eng, 17 May, 101-107 (1982).

McMurray, R., Flare radiation estimated, Hyd Proc, Nov, 175-181 (1982).

Mori, Y., Heat transfer bibliography: Japanese works, Int J Heat Mass Trans, 25(12), 1813-1818 (1982).
Murray, I., Heat exchanger literature review 1980-81, Proc Engng, Jan, 27-31 (1982).
Pick, A.E., Consider direct steam injection for heating liquids, Chem Eng, 28 June, 87-89 (1982).
Rabadi, N.J.; Chow, J.C.F., and Simon, H.A., Heat transfer in curved tubes with pulsating flow, Int J Heat Mass Trans, 25(2), 195-204 (1982).
Schroder, F.S., Calculating heat loss or gain by an insulated pipe, Chem Eng, 25 Jan, 111-114 (1982).
Sharma, A., and Minkowycz, W.J., KNOWTRAN: Artificial intelligence system for solving heat transfer problems, Int J Heat Mass Trans, 25(9), 1279-1290 (1982).
Soloukhin, R.I., and Martynenko, O.G., Heat and mass transfer bibliography: Soviet works, Int J Heat Mass Trans, 25(2), 151-160; 25(4), 439-448; 25(7), 899-908; 25(8), 1075-1086 (1982).

1983

Bondy, F., and Lippa, S., Heat transfer in agitated vessels, Chem Eng, 4 April, 62-71 (1983).
Danko, G., The possibility of determining and using a new local heat-transfer coefficient, Int J Heat Mass Trans, 26(11), 1679-1684 (1983).
Desplanches, H.; Bruxelmane, M.; Chevalier, J.L., and Ducla, J., Characteristic variable, prediction and scale-up for heat transfer to coils in agitated vessels, CER&D, 61, 3-12 (1983).
Eckert, E.R.G., et al., Heat transfer literature review (1982), Int J Heat Mass Trans, 26(12), 1733-1770 (1983).
Escoe, A.K., Heat transfer in vessels and piping, Hyd Proc, Jan, 107-112 (1983).
Hasan, R., and Rhodes, E., Two-phase flow heat transfer in a horizontal steam/water system, Chem Eng Commns, 22(3), 205-220 (1983).
Linnhoff, B., New concepts in thermodynamics for better chemical process design (review paper), CER&D, 61, 207-223 (1983).
Murray, I., Heat transfer publications review for 1982, Proc Engng, Jan, 36-39 (1983).
Rao, K.V.K., Physical properties of common heat-transfer fluids, Chem Eng, 19 Sept, 89-90 (1983).
Soloukhin, R.I., and Martynenko, O.G., Heat and mass transfer bibliography: Soviet works (1981-82), Int J Heat Mass Trans, 26(1), 1-10; 26(3), 323-334; 26(6), 795-804; 26(12), 1771-1782 (1983).
Zimmels, Y., Application of elementary analytic functions for solution of heat transfer from arrays of pipes, Chem Eng Commns, 21(1), 1-22 (1983).

1984

Bandrowski, J., and Ziolo, J., Heat and mass transfer bibliography: Polish works (1981-82), Int J Heat Mass Trans, 27(2), 157-162 (1984).
Dang, V.D., Forced convection heat transfer of power law fluid at low Peclet number flow, CER&D, 62, 367-372 (1984).

Del Villar, R.; Carreau, P.J., and Patterson, W.I., Heat transfer and drag reduction in pipes of various geometries, Chem Eng Commns, 25(1), 321-332 (1984).

Eckert, E.R.G., et al., Heat-transfer literature review (1983), Int J Heat Mass Trans, 27(12), 2179-2215 (1984).

Ganapathy, V., Estimate convective heat losses from surfaces, Chem Eng, 20 Feb, 100 (1984).

Ganapathy. V., Nomograph for draft pressure loss across tube bundles, Proc Engng, Feb, 61 (1984).

Lower, C.G., Heat-transfer coefficient depends on tubeside flowrate, Chem Eng, 9 July, 101-103 (1984).

Mori, Y., and Echigo, R., Heat transfer bibliography: Japanese works (1982-1983), Int J Heat Mass Trans, 27(4), 479-486 (1984).

Paradowski, H.; Kaiser, V., and Gourguechon, F., Use mixed cycle for gas cooling, Hyd Proc, March, 73-75 (1984).

Said, M.N.A., and Trupp, A.C., Predictions of turbulent flow and heat transfer in internally finned tubes, Chem Eng Commns, 31(1), 65-100 (1984).

Schlunder, E.U., Heat transfer to packed and stirred beds from the surface of immersed bodies, Chem Eng & Proc, 18(1), 31-54 (1984).

Venkatesh, C.K., Estimate errors in exchanger heat transfer coefficients, Hyd Proc, July, 56-58 (1984).

1985

Bandrowski, J., and Ziolo, J., Heat and mass transfer bibliography: Polish works (1983-84), Int J Heat Mass Trans, 28(12), 2229-2234 (1985).

De Faveri, D.M.; Fumarola, G.; Zonato, C., and Ferraiolo, G., Estimate flare radiation intensity, Hyd Proc, May, 89-91 (1985).

Eckert, E.R.G., et al., Heat transfer literature review (1984), Int J Heat Mass Trans, 28(12), 2181-2228 (1985).

Foord, A., and Mason, G., Recycle-heating experiment, Chem Eng Educ, 19(3), 136-139, 143 (1985).

Ganapathy, V., Find surface heat loss and flue gas density quickly, Hyd Proc, April, 82-83 (1985).

Ganapathy, V., Estimate surface area of finned tubes, Chem Eng, 27 May, 156 (1985).

Kalinin, E.K., and Dreitser, G.A., Unsteady convective heat transfer for turbulent flows of gases and liquids in tubes, Int J Heat Mass Trans, 28(2), 361-370 (1985).

Kumar, P.; Devotta, S., and Holland, F.A., Experimental heat and mass transfer studies on the solar generator of an open cycle absorption tank, CER&D, 63, 139-148 (1985).

Kumar, P.; Sane, M.G.; Devotta, S., and Holland, F.A., Experimental studies with an absorption system for simultaneous cooling and heating, CER&D, 63, 133-136 (1985).

Mori, Y., and Echigo, R., Heat transfer bibliography: Japanese works 1983/84, Int J Heat Mass Trans, 28(1), 1-6; 28(10), 1795-1804 (1985).

Mroz, W., and Nowakowska, H., Influence of heat flux on minimum liquid flowrate required for complete wetting of horizontal tube surface, Chem Eng & Proc, 19(6), 329-336 (1985).

Soloukhin, R.I., and Martynenko, O.G., Heat and mass transfer bibliography: Soviet works (1983-84), Int J Heat Mass Trans, 28(12), 2235-2246 (1985).

Vilemas J.V., and Simonis, V.M., Heat transfer and friction of rough ducts carrying gas flow with variable physical properties, Int J Heat Mass Trans, 28(1), 59-68 (1985).

Warrington, R.O., and Powe, R.E., Heat transfer by natural convection between bodies and their enclosures, Int J Heat Mass Trans, 28(2), 319-330 (1985).

1986

Anon., Heat transfer: A technology update (special advertising section), Chem Eng, 18 Aug, 45-79 (1986).

Bondy, F.; Mesagno, J., and Schwartz, M., An easy way to design steam tracing for pipes, Chem Eng, 4 Aug, 65-70 (1986).

Cross, W.T., and Ramshaw, C., Process intensification: Laminar-flow heat transfer, CER&D, 64(4), 293-301 (1986).

Eckert, E.R.G., et al., Heat transfer: Review of 1985 literature, Int J Heat Mass Trans, 29(12), 1767-1842 (1986).

Kroger, D.G., Performance characteristics of industrial finned tubes presented in dimensional form, Int J Heat Mass Trans, 29(8), 1119-1126 (1986).

Martinez, O.M., et al., Estimation of pseudohomogeneous one-dimensional heat transfer coefficient in fixed bed, Chem Eng & Proc, 20(5), 245-254 (1986).

Mascone, C.F., CPI strive to improve heat transfer in tubes, Chem Eng, 3 Feb, 22-25 (1986).

Vines, H.L., Upgrading natural gas (cryogenic processes), Chem Eng Prog, 82(11), 46-50 (1986).

1987

Bandrowski, J., and Ziolo, J., Heat and mass transfer bibliography: Polish works (1985-1986), Int J Heat Mass Trans, 30(12), 2533-2538 (1987).

Crittenden, B.D.; Hout, S.A., and Alderman, N.J., Model experiments of chemical reaction fouling, CER&D, 65(2), 165-170 (1987).

Deshpande, S.D., and Bishop, A.A., Heat transfer to non-Newtonian liquids flowing through horizontal tubes, Chem Eng Commns, 52, 339-354 (1987).

Eckert, E.R.G., et al., Heat transfer: A review of 1986 literature, Int J Heat Mass Trans, 30(12), 2449-2524 (1987).

Famularo, J., A computer-controlled heat exchange experiment, Chem Eng Educ, 21(2), 84-88 (1987).

Fletcher, P., Heat transfer coefficients for stirred batch reactor design, Chem Engnr, April, 33-37 (1987).

Mori, Y., and Echigo, R., Heat transfer bibliography: Japanese works (1985-1986), Int J Heat Mass Trans, 30(3), 417-426 (1987).

Murty, K.N., Optimize thermal performance of spiral coils, Chem Eng, 13 April, 113-114 (1987).

Soloukhin, R.I., and Martynenko, O.G., Heat and mass transfer bibliography: Soviet works (1985-1986), Int J Heat Mass Trans, 30(12), 2525-2532 (1987).

Soloukhin, R.I., and Martynenko, O.G., Heat and mass transfer bibliography: Soviet works (1984-1985), Int J Heat Mass Trans, 30(1), 3-13 (1987).

Soloukhin, R.I., and Martynenko, O.G., Heat and mass transfer bibliography: Soviet works (1985-1986), Int J Heat Mass Trans, 30(5), 819-826 (1987).
Various, Heat transfer (special report), Hyd Proc, 66(8), 37-50 (1987).
Ziolkowski, D., and Legawiec, B., Thermokinetic parameters of mathematical modes of heat transfer in packed bed columns, Chem Eng & Proc, 21(2), 65-76 (1987).

1988

Aksan, D.; Borak, F., and Onsan, Z.I., Heat-transfer coefficients in coiled stirred-tank systems, Can JCE, 65(6), 1013-1017 (1988).
Annamalai, K., et al., Application of nonintegral method to combustion and natural convection problems, Chem Eng Commns, 65, 231-242 (1988).
Dixon, A.G., Wall and particle-shape effects on heat transfer in packed beds, Chem Eng Commns, 71, 217-238 (1988).
Eckert, E.R.G., et al., Heat transfer: Review of 1987 literature, Int J Heat Mass Trans, 31(12), 2401-2488 (1988).
Futagami, K., and Aoyama, Y., Laminar heat transfer in a helically coiled tube, Int J Heat Mass Trans, 31(2), 387-396 (1988).
Klimenko, V.V., A generalized correlation for two-phase forced-flow heat transfer, Int J Heat Mass Trans, 31(3), 541-552 (1988).
Martynenko, O.G., Heat and mass transfer bibliography: Soviet works (1986), Int J Heat Mass Trans, 31(12), 2489-2504 (1988).
Miyatake, O., and Iwashita, H., The Graetz-Nusselt problem for fluid flowing axially between heated cylinders, Int Chem Eng, 28(3), 461-469 (1988).
Nakayama, W.; Kuwahara, H., and Hirasawa, S., Heat transfer from tube banks to air/water mist flow, Int J Heat Mass Trans, 31(2), 449-460 (1988).
Neale, A.J.; Babus-Haq, R.F., and Probert, S.D., Steady-state heat transfers across an obstructed air-filled rectangular cavity, CER&D, 66(5), 458-462 (1988).
Oliver, D.J., and Aldington, R.W.J., Heat transfer enhancement in round tubes using wire matrix turbulators: Newtonian and non-Newtonian liquids, CER&D, 66(6), 555-565 (1988).
Parikh, R.S., and Mahalingham, R., A collocational approach to laminar-flow heat transfer in non-Newtonian fluids, Chem Eng J, 38(1), 1-8 (1988).
Pawlowski, M., and Suszek, E., Heat transfer by perpendicular impact of an air stream on a flat surface, Int Chem Eng, 28(2), 369-376 (1988).
Raina, G.K., and Grover, P.D., Direct contact heat transfer with change of phase: Experimental technique, AIChEJ, 34(8), 1376-1380 (1988).
Sarma, P.K.; Subrahmanyam, T., and Rao, V.D., Natural convection from a radiating fin in air: A conjugate problem, Chem Eng Commns, 71, 23-38 (1988).
Shelton, S.V.; Wepfer, W.J., and Miles, D.J., External fluid heating of porous bed, Chem Eng Commns, 71, 39-52 (1988).
Skiepko, T., Effect of matrix longitudinal heat conduction on temperature fields in rotary heat exchanger, Int J Heat Mass Trans, 31(11), 2227-2238 (1988).
Tanasawa, I., and Echigo, R., Heat transfer bibliography: Japanese works (1986-87), Int J Heat Mass Trans, 31(7), 1335-1344 (1988).
Wang, B.X.; Guo, Z.Y., and Ren, Z.P., Review of Chinese literature (1983-1986) on heat transfer, Int J Heat Mass Trans, 30(11), 2215-2224 (1988).

Wang, C.Y., Buoyancy effects on stagnation flow on heated vertical plate, Chem Eng Commns, 68, 237-244 (1988).
White, I., Single-medium systems for heating/cooling cycles, Proc Engng, Nov, 41-42 (1988).
Yang, T.Y., et al., Heat transfer in partially miscible ternary systems, Chem Eng Commns, 64, 1-12 (1988).
Yao, H., and Kovenklioglu, S., Heat transfer enhancement with chemical reaction: Model comparisons, Chem Eng Commns, 71, 205-216 (1988).
Zhukauskas, A.A., et al., Local characteristics of liquid flow in staggered bundles of roughened tubes, Int Chem Eng, 28(3), 535-543 (1988).
Zhukauskas, A.A., et al., Characteristics of flow and local heat transfer in staggered bundles of finned tubes, Int Chem Eng, 28(3), 527-535 (1988).

17.2 Heat Exchanger Design

1966-1969

Bergman, D.J., High-temperature exchanger problems, Hyd Proc, 45(10), 158-160 (1966).
Gilmour, C.H., Troubleshooting heat exchanger design, Chem Eng, 19 June, 221-228 (1967).
Kroehle, T.P., Rules for reducing exchanger costs, Chem Eng, 14 Aug, 157-158 (1967).
Trommelen, A.M., Heat transfer in a scraped-surface heat exchanger, Trans IChemE, 45, T176-181 (1967).
Various, Design of heat exchangers (topic issue), Chem Eng Prog, 63(7), 47-80 (1967).
Anon., The new TEMA standards, Hyd Proc, 47(9), 277-281 (1968).
Blanco, J.A.; Gill, W.N., and Nunge, R.J., Computational procedures for recent analyses of counterflow heat exchangers, AIChEJ, 14(3), 505-507 (1968).
Bott, T.R., and Azoory, S., Scraped-surface heat transfer with KnitMesh scrapers, Brit Chem Eng, 13(3), 372-374 (1968).
Bott, T.R.; Azoory, S., and Porter, K.E., Scraped-surface heat exchangers, Trans IChemE, 46, T33-42 (1968).
Kern, D.Q., Misuse of low-finned tubing, Chem Eng Prog, 64(12), 49 (1968).
Parker, R.O., and Mok, Y.I., Shell-side pressure loss in baffled heat exchangers, Brit Chem Eng, 13(3), 366-368 (1968).
Rubin, F.L., Practical heat-exchanger design, Chem Eng Prog, 64(12), 44-48 (1968).
Rubin, F.L., Specifying heat exchangers, Chem Eng, 8 April, 130-136 (1968).
Samoshka, P.S., et al., Heat transfer and pressure drop for closely spaced tube banks in water flows, Int Chem Eng, 8(3), 388-392 (1968).
Srinivasan, P.S.; Nandapurkar, S.S., and Holland, F.A., Pressure drop and heat transfer in coils, Chem Engnr, May, CE113-119 (1968).
Balasubramanyam, P.R., Nomograph for double-pipe heat exchanger design, Brit Chem Eng, 14(2), 175 (1969).
Dent, J.C., An automatic system for determining heat exchanger heat-transfer coefficients, Brit Chem Eng, 14(10), 1397-1399 (1969).

Fisher, J., and Parker, R.O., New ideas on heat exchanger design, Hyd Proc, 48(7), 147-154 (1969).
Horn, G., and Atherton, A., Statistical techniques for evaluating experimental performance results on a batch of finned heat exchanger elements, Trans IChemE, 47, T43-51 (1969).
Jenssen, S.K., Heat exchanger optimization, Chem Eng Prog, 65(7), 59-66 (1969).
Kehat, E., and Letan, R., Design of a spray-column heat exchanger, Brit Chem Eng, 14(6), 803-805 (1969).
Lohrisch, F.W., Shortcut heat exchanger tube-side rating, Hyd Proc, 48(4), 125-132 (1969).
Messa, C.J.; Foust, A.S., and Poehlein, G.W., Shell-side heat-transfer coefficients in helical-coil heat exchangers, Ind Eng Chem Proc Des Dev, 8(3), 343-347 (1969).
Mottram, J.A., Mean temperature difference found quickly and accurately, Chem Eng, 16 June, 116-118 (1969).
Tien, C., and Srinivasan, S., Approximate solution for countercurrent heat exchangers, AIChEJ, 15(1), 39-46 (1969).

1970-1972

Anon., Advances in heat-exchanger design, Brit Chem Eng, 15(4), 473-477 (1970).
Dodd, R., Constant or linear overall heat transfer coefficient in tubular exchangers? Chem Engnr, June, CE190-193 (1970).
Gay, B., and Williams, T.A., Heat transfer on the shell-side of a cylindrical shell-and-tube heat exchanger fitted with segmental baffles, Trans IChemE, 46, T95-100 (1968); 48, T3-6 (1970).
Jamin, B., Exchanger stages solved graphically, Hyd Proc, 49(7), 137-144 (1970).
Katell, S., and Jones, P.R., Optimum heat exchanger design, Brit Chem Eng, 15(4), 491-494 (1970).
Lord, R.C.; Minton, P.E., and Slusser, R.P., Design of heat exchangers, Chem Eng, 26 Jan, 96-118 (1970).
Minton, P.E., Designing spiral-tube heat exchangers, Chem Eng, 18 May, 145-152 (1970).
Stuhlbarg, D., Calculating optimum exchanger size for forced circulation, Hyd Proc, 49(1), 149-152 (1970).
Dimoplon, W., Compressible flow in heat exchangers, Hyd Proc, 50(9), 195-200 (1971).
Markovitz, R.E., Picking the best vessel jacket, Chem Eng, 15 Nov, 156-162 (1971).
Messa, C.J.; Poehlein, G.W., and Foust, A.S., Heat exchanger modeling by conservative scalar pulse testing, Ind Eng Chem Proc Des Dev, 10(4), 466-472 (1971).
Starczewski, J., Short-cut method to predict exchanger shell-side pressure drop, Hyd Proc, 50(6), 147-150 (1971).
Starczewski, J., Short-cut method to exchanger tube-side pressure drop, Hyd Proc, 50(5), 122-124 (1971).

Tarrer, A.R.; Lim, H.C., and Koppel, L.B., Finding the economically optimum heat exchanger, Chem Eng, 4 Oct, 79-84 (1971).
Trommelen, A.M., and Beek, W.J., Flow phenomena in scraped-surface heat exchanger (Votator-type), Chem Eng Sci, 26(11), 1933-1942; 26(12), 1977-2003 (1971).
Cooper, A., and Cocks, A.M., Computer design of heat exchangers, Proc Engng, Dec, 76-77 (1972).
Peters, D.L., and Nicole, F.J.L., Efficient programming for cost-optimised heat exchanger design, Chem Engnr, March, 98-111 (1972).
Petrosky, J.T., Direct calculation of exchanger exit temperatures, Chem Eng, 17 April, 128 (1972).
Thompson, J.W., How not to buy heat exhangers! Hyd Proc, 51(12), 83-85 (1972).
Yim, Y.J.; Wellman, P., and Katell, S., Importance of temperature approach in heat exchangers, Chemtech, March, 167-172 (1972).

1973-1975

Bandrowski, J., and Rybski, W., Shell-side pressure calculation procedure in baffled heat exchangers, Int Chem Eng, 13(4), 676-681 (1973).
Berryman, J.E., and Himmelblau, D.M., Influence of stochastic inputs and parameters on heat exchanger design, Ind Eng Chem Proc Des Dev, 12(2), 165-171 (1973).
Dobryakov, B.A., et al., Heat exchange equipment with cross-flow of heat transfer agents, Int Chem Eng, 13(1), 81-84 (1973).
Kalinin, E.K.; Dreitser, G.A., and Kozlov, A.K., Heat transfer intensification during lengthwise flow along tube bundles with various relative pitches, Int Chem Eng, 13(1), 1-5 (1973).
Malek, R.G., Improved exchanger design, Hyd Proc, 52(5), 128-130 (1973).
Somayajulu, K.R., Quick calculation for cross-flow area, Chem Eng, 14 May, 168 (1973).
Walker, R.A., and Bott, T.R., Effect of roughness on heat transfer in exchanger tubes, Chem Engnr, March, 151-156 (1973).
Yokell, S., Double-tubesheet heat-exchanger design stops shell-tube leakage, Chem Eng, 14 May, 133-136 (1973).
Bevevino, J.W., Tube to tubesheet welding techniques, Chem Eng Prog, 70(7), 71-73 (1974).
Mathur, J., Performance of steam heat-exchangers, Chem Eng, 3 Sept, 101-106 (1973); 18 March, 86 (1974).
Roetzel, W., Iteration-free calculation of heat transfer coefficient in air-cooled crossflow heat exchangers, Chem Eng J, 7(1), 79-82 (1974).
Singh, K.P., Location of impingement plates in tubular heat exchangers, Hyd Proc, 53(10), 147-149 (1974).
Various, Heat exchanger design (feature report), Chemical Processing, April, 65-88 (1974).
Char, C.V., Heat exchanger foundation design, Hyd Proc, 54(3), 121-126 (1975).
Cowie, R., Relative costs of shell and tube heat exchangers, Proc Engng, Nov, 85-87 (1975).

Dagsoz, A.K., Determining the heat transfer on tube-bundle support plates, Chem & Ind, 2 Aug, 656-658 (1975).
Ganapathy, V., Quick calculation for exchanger tubesheet thickness, Chem Eng, 12 May, 114 (1975).
Hills, D.E.G., Graphite heat exchangers, Chem Eng, 23 Dec, 80-83 (1974); 20 Jan, 116-119 (1975).
Kazmerovich, V., et al., Application of the method of modelling for the construction of parametric series of heat exchangers, Int Chem Eng, 15(4), 700-704 (1975).
Knight, W.P., Plant operating data improve heat exchanger design, Hyd Proc, 54(5), 151-154 (1975).
Kobalskii, B.S.; Silantev, A.V., and Marchenko, P.S., Design of spiral heat exchangers, Int Chem Eng, 15(1), 90-95 (1975).
Murray, I., Developments in heat transfer equipment, Proc Engng, Nov, 68-71 (1975).

1976-1977

Foxall, D.H., and Gilbert, P.T., Selecting tubes for heat exchangers, Chem Eng, 15 March, 99-104; 12 April, 147-150; 10 May, 133-136 (1976).
Matsuyama, H., and Oshima, E., Dynamic model of multipass heat exchangers, Int Chem Eng, 16(1), 154-162 (1976).
Morris, M., BS5500 for tubesheet design, Proc Engng, April, 78-80 (1976).
Rodriguez, F., Approximate LMTD, Hyd Proc, 55(2), 125 (1976).
Tucker, W.H., Temperature approach in counterflow heat exchanger experiment, Chem Eng Educ, 10(1), 36-39 (1976).
Various, Heat exchanger design (feature report), Processing, March, 41-59 (1976).
Various, Designing shell-and-tube heat exchangers (feature report), Chem Eng, 5 July, 62-76 (1976).
Anon., Spiral exchangers, Processing, June, 19, 23 (1977).
Anon., Computer design of heat exchangers, Processing, Sept, 64 (1977).
Ganapathy, V., Charts simplify spiral finned-tube calculations, Chem Eng, 25 April, 117-122 (1977).
Gardner, K., and Taborek, J., Mean temperature difference: A reappraisal, AIChEJ, 23(6), 777-786 (1977).
Gasior, S., et al., Approximate mean temperature difference for heat-exchanger calculations, Can JCE, 55, 741-749 (1977).
Golan, L.P., and Borushko, G., Multi-phase flow in the annulus of a double-pipe exchanger, Chem Eng Prog, 73(2), 79-83 (1977).
Kulkarni, P.D., and Phadke, P.S., Tubeside heat-transfer, Chem Eng, 14 Feb, 108 (1977).
Leenaerts, R., Calculating tube bundles with longitudinally finned tubes, Int Chem Eng, 17(1), 140-164 (1977).
Malek, R.G., New approach to exchanger tubesheet design, Hyd Proc, 56(1), 163-169 (1977).
Murray, I., Developments in heat exchangers, Proc Engng, Nov, 57-59 (1977).
Roetzel, W., Iteration-free calculation of heat-transfer coefficients in heat exchangers, Chem Eng J, 13(3), 233-238 (1977).

Russell, J.J., and Carnavos, T.C., Air cooling of internally finned tubes, Chem Eng Prog, 73(2), 84-88 (1977).
Simonson, J.R., Transient and steady-state analysis of crossflow heat exchangers by programs in Fortran, Trans IChemE, 55, 53-58 (1977).
Walko, J., Rod-baffle heat exchanger, Proc Engng, Nov, 60-61 (1977).
Weierman, C., Pressure drop data for heavy-duty finned tubes, Chem Eng Prog, 73(2), 69-72 (1977).

1978-1979

Dimoplon, W., Calculate the length of helical heating coils, Chem Eng, 23 Oct, 177 (1978).
Tan, K.S., and Spinner, I.H., Dynamics of shell-and-tube heat exchanger with finite tube-wall heat capacity and finite shell-side resistance, Ind Eng Chem Fund, 17(4), 353-358 (1978).
Various, Heat exchanger design (feature report), Processing, June, 53-85 (1978).
Various, Heat-exchanger design developments (topic issue), Proc Engng, Nov, 35-77 (1978).
Various, Design of air-cooled exchangers (feature report), Chem Eng, 27 March, 106-124 (1978).
Davidson, R.M., and Miska, K.H., Stainless-steel heat exchangers, Chem Eng, 12 Feb, 129-133; 12 March, 111-114 (1979).
Deliquet, A., Simple method predicts heat exchanger outlet temperature, Hyd Proc, 58(4), 213 (1979).
Ernst, W.R., and Gerrard, A.M., Optimisation of single-pipe heat exchangers, Proc Engng, March, 95,97 (1979).
Gnielinski, V., Equations for heat transfer in single tube rows and banks of tubes in transverse flow, Int Chem Eng, 19(3), 380-391 (1979).
Gnielinski, V., and Gaddis, E.S., Calculation of mean heat-transfer coefficients on shell side of shell and tube heat exchangers with segmental baffles, Int Chem Eng, 19(3), 391-401 (1979).
Gutierrez, H.J., and Cooper, A.R., Heat exchanger process dynamics review, Chem Eng J, 17(1), 13-18 (1979).
Karamercan, O.E., and Gainer, J.L., Effect of pulsations on heat transfer in exchangers, Ind Eng Chem Fund, 18(1), 11-15 (1979).
Patrickson, P., Pipe heat exchangers, Processing, Nov, 37-39 (1979).
Petchonka, J.J., Specifying exchanger surface area, Chem Eng, 30 July, 97-98 (1979).
Rubin, F.L., and Gainsboro, N.R., Latest TEMA standards for shell-and-tube exchangers, Chem Eng, 24 Sept, 111-116 (1979).
Zhukauskas, A.A.; Ulinskas, R.V., and Shvegzhda, A.A., Heat transfer efficiency of heat exchangers, Int Chem Eng, 19(4), 711-714 (1979).

1980-1981

Aiba, S.; Ota, T., and Tsuchida, H., Heat transfer of tubes closely spaced in an in-line bank, Int J Heat Mass Trans, 23(3), 311-320 (1980).
Baker, W.J., Selecting and specifying air-cooled heat exchangers, Hyd Proc, 59(5), 173-177 (1980).
Chaksh, S.A., Increased heat-transfer rate on shell side of shell-and-tube heat exchangers using cross baffles, Int Chem Eng, 20(3), 498-503 (1980).

Crosser, O.K., and Park, K.Y., Analysis of batch scraped-surface heat exchange, Int J Heat Mass Trans, 23(12), 1683-1686 (1980).
Dodd, R., Mean temperature difference and temperature efficiency for shell and tube heat exchangers connected in series with two tube passes per shell pass, Trans IChemE, 58, 9-15, 276 (1980).
Franklin, W.C., and Cocks, R.E. Determine the resistance that controls heat exchanger cost, Chem Eng, 28 July, 93-94 (1980).
Gutterman, G., Specifying heat exchangers, Hyd Proc, 59(4), 161-163 (1980).
Heggs, P.J., and Stones, P.R., Improved design methods for finned tube heat exchangers, Trans IChemE, 58, 147-154 (1980).
Karanth, N.G., Predict heat exchanger outlet temperatures, Hyd Proc, 59(9), 262-263 (1980).
London, A.L., and Seban, R.A., Generalization of the methods of heat exchanger analysis, Int J Heat Mass Trans, 23(1), 5-16 (1980).
Rubin, F.L., TEMA exchanger classifications, Hyd Proc, 59(6), 92 (1980).
Various, Heat exchanger design (feature report), Proc Engng, Jan, 50-63 (1980).
Various, Heat exchangers (feature report), Chem Eng, 6 Oct, 120-151 (1980).
Watkinson, P., Process heat transfer: Some practical problems, Can JCE, 58, 553-558 (1980).
Blackwell, W.W., and Haydu, L., Calculating the corrected LMTD in shell-and-tube heat exchangers, Chem Eng, 24 Aug, 101-106 (1981).
Gupta, J.P., and Sivasankaran, S., Computer program for tubesheet design, Hyd Proc, 60(11), 273-276 (1981).
Reay, D.A., Heat-pipe heat exchangers, Chem Engnr, April, 154-158 (1981).
Urbicain, M.J., and Paloschi, J., Simulation of air-cooled heat exchangers, Comput Chem Eng, 5(2), 75-82 (1981).
Wales, R.E., Mean temperature difference in heat exchangers, Chem Eng, 23 Feb, 77-81 (1981).
Webb, R.L., Performance evaluation criteria for use of enhanced heat transfer surfaces in heat exchanger design, Int J Heat Mass Trans, 24(4), 715-726 (1981).

1982

Ahuja, A.S., Thermal design of a heat exchanger with laminar flow of particle suspensions, Int J Heat Mass Trans, 25(5), 725-728 (1982).
Brown, K., Ease thermal stresses in shell-and-tube heat exchangers, Chem Eng, 8 Feb, 131 (1982).
Corripio, A.B.; Chrien, K.S., and Evans, L.B., Estimate costs of heat exchangers and storage tanks via corelations, Chem Eng, 25 Jan, 125-127 (1982).
Dodd, R., Temperature efficiency of heat exchangers with one shell pass and even number of tube passes, Trans IChemE, 60, 364-368 (1982).
Fehr, M., Exchanger temperatures: Estimates for new services, Hyd Proc, Nov, 215-216 (1982).
Ievlev, V.M.; Dzyubenko, B.V.; Dreitser, G.A., and Vilemas, Y.V., In-line and cross-flow helical tube heat exchangers, Int J Heat Mass Trans, 25(3), 317-324 (1982).
Patil, R.K.; Shende, B.W., and Ghosh, P.K., Designing a helical-coil heat exchanger, Chem Eng, 13 Dec, 85-88 (1982).

Pigorini, A.; De Pascale, T., and Milanesi, F., Program simplifies log mean temperature difference calculations, Hyd Proc, Nov, 205-207 (1982).

Sodha, M.S., Performance of countercurrent heat exchanger with periodic inlet temperatures, Int J Heat Mass Trans, 25(10), 1609-1611 (1982).

Various, Heat exchangers: Plant and equipment survey, Processing, Feb, 23-29 (1982).

Webb, R.L., Performance evaluation criteria for air-cooled finned-tube heat-exchanger surface geometries, Int J Heat Mass Trans, 25(11), 1770-1771 (1982).

Yokell, S., Heat-exchanger tube-to-tubesheet connections, Chem Eng, 8 Feb, 78-94 (1982).

Ziolkowska, I., and Dolata, M., Application of turbulence promoters for optimization of a gas heat exchanger, Chem Eng Commns, 18(1), 121-136 (1982).

1983

Cizmar, L.E., Mechanical design of exchangers, Chem Eng Prog, 79(7), 47-50 (1983).

Kasza, K.E.; Bobis, J.P., and Lawrence, W.P., Overview of thermal transient induced buoyancy phenomena in pipe and heat exchanger flows, Chem Eng Commns, 19(4), 295-316 (1983).

Kolenda, Z.; Szmyd, J.; Slupek, S., and Baez, L.M., Numerical modelling of heat transfer processes with supplementary data, Can JCE, 61(5), 627-634 (1983).

Kroger, D.G., Design optimization of an air-oil heat exchanger, Chem Eng Sci, 38(2), 329-334 (1983).

Krupiczka, R., et al., Influence of shell-side structures on heat transfer in shell and tube heat exchangers, Chem Eng Commns, 19(4), 325-334 (1983).

Kurmarao, P.S.V., Determine LMTD for a 1-1 split-flow exchanger, Chem Eng, 7 March, 188-190 (1983).

Murty, K.N., Analysis of 1-2 split flow heat exchanger, Int J Heat Mass Trans, 26(10), 1571-1574 (1983).

Parry, W., Heat exchangers survey, Processing, May, 31-35 (1983).

Pase, G.K., and Yokell, S., Interpass temperature effects in multipass straight tube exchangers, Chem Eng Prog, 79(7), 51-53 (1983).

Purohit, G.P., Estimating costs of shell-and-tube heat exchangers, Chem Eng, 22 Aug, 56-67 (1983).

Purohit, G.P., Thermal and hydraulic design of hairpin and finned-bundle exchangers, Chem Eng, 16 May, 62-70 (1983).

Roach, G.H., and Wood, R.M., Flexible heat exchanger designs, Chem Engnr, Oct, 25-27 (1983).

Shaikh, N.M., Estimate air-cooler size, Chem Eng, 12 Dec, 65-68 (1983).

Sivasankaran, S., and Gupta, J.P., Computer program for flanged and flued expansion joints, Hyd Proc, Dec, 93-96 (1983).

Sparrow, E.M., and Comb, J.W., Effect of interwall spacing and fluid flow inlet conditions on a corrugated-wall heat exchanger, Int J Heat Mass Trans, 26(7), 993-1006 (1983).

Various, Perspective on shell-and-tube heat exchangers, Chem Eng, 25 July, 46-84 (1983).

Walczyk, H., Enhancement of heat transfer in finned-tube heat exchangers by water injection into air streams, Chem Eng Commns, 19(4), 317-324 (1983).

Zaleski, T., and Thullie, J., Mathematical model of a multichannel cross-flow heat exchanger (temperature profiles), Int Chem Eng, 23(3), 561-568 (1983).

Zijl, W., and De Bruijn, H., Continuum equations for the prediction of shell-side flow and temperature patterns in heat exchangers, Int J Heat Mass Trans, 26(3), 411-424 (1983).

Zozulya, N.V.; Khavin, A.A., and Leonova, V.I., Influence of sparseness of finned tube bundles on their thermal and aerodynamic characteristics, Int Chem Eng, 23(3), 558-561 (1983).

1984

Cai, Z.H.; Li, M.L.; Wu, Y.W., and Ren, H.S., Modified selected-point matching technique for testing compact heat exchanger surfaces, Int J Heat Mass Trans, 27(7), 971-978 (1984).

Horwitz, B.A., Evaluating heat-exchanger bypass flow, Chem Eng, 10 Dec, 98-100 (1984).

Johnston, D., Modelling techniques for shell-and-tube heat exchangers, Proc Engng, May, 33-36 (1984).

Murty, K.N., Calculate LMTD for 1-1 divided-flow heat exchangers, Chem Eng, 20 Feb, 98-100 (1984).

Norman, C., Advances in heat-exchanger design, Proc Engng, Jan, 25-29 (1984).

Phadke, P.S., Determining tube counts for shell-and-tube exchangers, Chem Eng, 3 Sept, 65-68 (1984).

Tanaka, O., Analysis of simultaneous heat and water vapor exchange through a flat-plate crossflow total heat exchanger, Int J Heat Mass Trans, 27(12), 2259-2266 (1984).

Various, Heat exchangers: A survey, Processing, Nov, 12-21 (1984).

Zaleski, T., General mathematical model of parallel-flow, multichannel heat exchangers and analysis of its properties, Chem Eng Sci, 39(7), 1251-1260 (1984).

1985

Berryman, R., and Russell, C., Troubleshooting air-cooled heat exchangers, Proc Engng, April, 25-29 (1985).

Chowdhury, K., et al., Analytical studies on temperature distribution in spiral plate heat exchangers: Straightforward design formulae for efficiency and mean temperature difference, Chem Eng & Proc, 19(4), 183-190 (1985).

Gaddis, E.S., and Gnielinski, V., Pressure drop in crossflow across tube bundles, Int Chem Eng, 25(1), 1-16 (1985).

Goyal, O.P., Guidelines on exchangers, Hyd Proc, Aug, 55-60 (1985).

Huang, B.J., and Tsuei, J.T., Method of analysis for heat-pipe heat exchangers, Int J Heat Mass Trans, 28(3), 553-562 (1985).

Murty, K.N., Quickly calculate temperature cross in 1-2 parallel-counterflow heat exchangers, Chem Eng, 16 Sept, 99-100 (1985).

Purohit, G.P., Costs of double-pipe and multitube heat exchangers, Chem Eng, 4 March, 93-96; 1 April, 85-86 (1985).
Saatdjian, E., and Large, J.F., Heat transfer simulation in a raining packed bed exchanger, Chem Eng Sci, 40(5), 693-698 (1985).
Smith, R., Up-date on shellside two-phase heat transfer, Chem Engnr, May, 16-19 (1985).
Sparrow, E.M., and Kang, S.S., Longitudinally finned cross-flow tube banks and their heat transfer and pressure drop characteristics, Int J Heat Mass Trans, 28(2), 339-350 (1985).
Sparrow, E.M., and Samie, F., Heat transfer and pressure drop results for one- and two-row arrays of finned tubes, Int J Heat Mass Trans, 28(12), 2247-2260, 2379-2381 (1985).
Suzuki, K.; Hirai, E.; Miyake, T., and Sato, T., Numerical and experimental studies on a two-dimensional model of an offset-strip-fin type compact heat exchanger used at low Reynolds number, Int J Heat Mass Trans, 28(4), 823-836 (1985).
Tadrist, L., et al., Experimental and numerical study of direct-contact heat exchangers, Int J Heat Mass Trans, 28(6), 1215-1228 (1985).
Tammami, B., Weighted mean-temperature-difference for exchangers with phase changes, Chem Eng, 11 Nov, 230-232 (1985).

1986

Berryman, R., and Russell, C., Assessing air-side performance of air-cooled heat exchangers, Proc Engng, April, 59-63 (1986).
Brown, T.R., Guidelines for preliminary selection of heat-exchanger type, Chem Eng, 3 Feb, 107-108 (1986).
Johnston, A., Miniaturized heat exchangers for chemical processing, Chem Engnr, Dec, 36-38 (1986).
Lipowicz, M., Heat-exchanger software, Chem Eng, 4 Aug, 73-76 (1986).
Maingonnat, J.F., and Corrieu, G., Thermal performance of scraped-surface heat exchanger, Int Chem Eng, 26(1), 45-69 (1986).
Margittai, T.B., New heat exchanger geometry, Chem Eng Prog, 82(7), 39-41 (1986).
Merker, G.P., and Hanke, H., Heat transfer and pressure drop on the shell-side of tube-banks having oval-shaped tubes, Int J Heat Mass Trans, 29(12), 1903-1910 (1986).
Murty, K.N., How effective are finned tubes in heat exchangers? Chem Eng, 14 April, 120-122 (1986).
Palen, J.W., Designing heat exchangers by computer, Chem Eng Prog, 82(7), 23-27 (1986).
Pase, G.K., Computer programs for heat exchanger design, Chem Eng Prog, 82(9), 53-56 (1986).
Quick, K., Direct calculation of exchanger exit temperatures in cocurrent flow, Chem Eng, 13 Oct, 92 (1986).
Rohsenow, W.M.; Hartnett, J.P., and Ganic, E.N., Handbook of Heat Transfer Fundamentals, 2nd Edn.; Handbook of Heat Transfer Applications, 2nd Edn., McGraw-Hill, New York (1986).

Sparrow, E.M., and Reifschneider, L.G., Effect of interbaffle spacing on heat transfer and pressure drop in a shell-and-tube heat exchanger, Int J Heat Mass Trans, 29(11), 1617-1628 (1986).

Tammami, B., Weighted mean-temperature-difference for exchangers with phase changes, Chem Eng, 6 Jan, 92 (1986).

Turton, R., and Frederick, N., Analyze cross-flow heat exchangers, Chem Eng, 7 July, 55-58 (1986).

Turton, R.; Ferguson, D., and Levenspiel, O., Charts for performance and design of heat exchangers, Chem Eng, 18 Aug, 81-88 (1986).

Wachel, L.J., Exchanger simulator: Guide to less fouling, Hyd Proc, 65(11), 107-110 (1986).

1987

Crane, R., and Arrazola, R., Design and optimization of heat exchangers for batch heating by NTU-effectiveness method, Chem Eng Commns, 50, 103-112 (1987).

Diaz, M., and Aguayo, A.T., How flow dispersion affects exchanger performance, Hyd Proc, 66(4), 57-60 (1987).

Ganapathy, V., Simplified approach to designing heat-transfer equipment, Chem Eng, 13 April, 81-87 (1987).

Idem, S.A.; Jung, C.; Gonzalez, G.J., and Goldschmidt, V.W., Performance of air-to-water copper finned-tube heat exchangers at low air side Reynolds numbers, including effects of baffles, Int J Heat Mass Trans, 30(8), 1733-1742 (1987).

Joshi, H.M., and Webb, R.L., Heat transfer and friction in the offset strip-fin heat exchanger, Int J Heat Mass Trans, 30(1), 69-84 (1987).

Large, J.F.; Guignon, P., and Molodtsof, Y., Hydrodynamics of raining-particle heat exchangers and their applications, Int Chem Eng, 27(4), 607-615 (1987).

McDonough, M.J., Hairpin exchangers: Double-pipe and multitube, Chem Eng, 20 July, 87-90 (1987).

Proctor, A., Heat exchanger integrated design packages, Proc Engng, Sept, 61-62 (1987).

Ratnasamy, F., Exchanger design using disc and donut baffles, Hyd Proc, 66(4), 63-65 (1987).

Wonchala, E.P., and Wynnyckyj, J.R., The phenomenon of thermal channelling in countercurrent gas-solid heat exchangers, Can JCE, 65(5), 736-743 (1987).

Zaleski, T., Mathematical modelling of cross-flow heat exchangers, Chem Eng Sci, 42(7), 1517-1526 (1987).

Zaleski, T., and Lachowski, A., Unsteady temperature profiles in parallel-flow spiral heat exchangers, Int Chem Eng, 27(3), 556-566 (1987).

1988

Abichandani, H., and Sarma, S.C., Heat transfer and power requirements in horizontal thin-film scraped-surface heat exchangers, Chem Eng Sci, 43(4), 871-882 (1988).

Baclic, B.S.; Romie, F.E., and Herman, C.V., Galerkin method for two-pass crossflow heat exchanger problem, Chem Eng Commns, 70, 177-198 (1988).

Fryer, P.J.; Hobin, P.J., and Mawer, S.P., Optimal design of heat exchanger undergoing reaction fouling, Can JCE, 66(4), 558-562 (1988).

Grazzini, G., and Gori, F., Entropy parameters for heat exchanger design, Int J Heat Mass Trans, 31(12), 2547-2554 (1988).

Guzman, P.E.R., Speed up heat-exchanger design, Chem Eng, 14 March, 143-146 (1988).

Kuye, A., and Ogboja, O., Mathematical correlations of some shell and tube heat exchanger design parameters, Chem Eng Commns, 74, 39-46 (1988).

Prasad, R., Improved LMTD approximation, Chem Eng, 10 Oct, 110 (1988).

Prasad, R.C., Generalized solution and effectiveness for concentric-tube heat exchangers, Int J Heat Mass Trans, 31(12), 2571-2578 (1988).

Redman, J., Compact future for heat exchangers, Chem Engnr, Sept, 12-16 (1988).

Schrage, D.S.; Hus, J.T., and Jensen, M.K., Two-phase pressure drop in vertical crossflow across a horizontal tube bundle, AIChEJ, 34(1), 107-115 (1988).

17.3 Heat Exchanger Operation

1968-1973

Fleming, R.C., Reduce tube plugging in heat exchangers, Chem Eng, 16 Dec, 128 (1968).

Fuller, N.C., and Hoover, B.O., Leak-free exchangers using electric stud heaters, Hyd Proc, 47(8), 118-120 (1968).

Various, Heat exchanger problems (topic issue), Chem Eng Prog, 64(3), 73-89 (1968).

Wild, N.H., Noncondensable gas eliminates hammering in heat exchanger, Chem Eng, 21 April, 132-134 (1969).

Imaeda, M.; Honda, S., and Sugiyama, S., Dynamics of heat exchangers subject to simultaneous changes in flowrate and inlet fluid temperature, Int Chem Eng, 10(2), 303-309 (1970).

Impagliazzo, A.M., and Murphy, J.J., Re-evaluation of external water jackets, Chem Eng Prog, 66(3), 66-72 (1970).

Lord, R.C.; Minton, P.E., and Slusser, R.P., Guide to trouble-free heat exchangers, Chem Eng, 1 June, 153-160 (1970).

Steensland, O., and Magnusson, L., Analysis of damage to heat exchangers, Brit Chem Eng, 15(4), 485-490 (1970).

Bott, T.R., and Walker, R.A., Fouling in heat transfer equipment, Chem Engnr, Nov, 391-395 (1971).

Manlove, J.C., Early detection of heat exchanger leaks, Chem Eng, 17 May, 162 (1971).

Wigham, I., Designing optimum cooling systems, Chem Eng, 9 Aug, 95-102 (1971).

Hendrickson, R.B., and Lashmet, P.K., Effect of random plugging on performance of multipassaged cryogenic heat exchanger, Ind Eng Chem Proc Des Dev, 11(1), 53-59 (1972).

Liptak, B.G., Control of heat exchangers, Brit Chem Eng, 17(7), 637-645 (1972).
Moore, J.A., Development of a low-maintenance heat-recovery exchanger, Chem Eng Prog, 69(1), 43-46 (1973).
Walker, R.A., and Bott, T.R., Prediction of fouling in heat exchanger tubes from existing data, Trans IChemE, 51, 165-167 (1973).

1974-1977

Franklin, G.M., and Munn, W.B., Problems with heat exchangers in low temperature environments, Chem Eng Prog, 70(7), 63-67 (1974).
Konak, A.R., Prediction of fouling curves in heat transfer equipment, Trans IChemE, 51, 377-378 (1973); 52, 386 (1974).
Rozenman, T., and Pundyk, J., Reducing solidification in air-cooled heat exchangers, Chem Eng Prog, 70(10), 80-85 (1974).
Shipes, K.V.; Brown, J.W., and Benkly, G.J., Air coolers in cold climates, Hyd Proc, 53(5), 147-150 (1974).
Bott, T.R., Fouling in heat exchangers, Proc Engng, Nov, 76-81 (1975).
Brown, P.M.M., and France, D.W., Protecting air-cooled heat exchangers against overpressure, Hyd Proc, 54(8), 103-106 (1975).
Donohue, J.M., and Nathan, C.C., Unusual heat exchanger problems with cooling water treatment, Chem Eng Prog, 71(7), 88-93 (1975).
Fischer, P.; Suitor, J.W., and Ritter, R.B., Fouling measurement techniques, Chem Eng Prog, 71(7), 66-72 (1975).
Kern, W.I., Continuous tube cleaning improves performance of condensers and heat exchangers, Chem Eng, 13 Oct, 139-144 (1975).
Haluska, J.L., Effective heat exchanger fouling control, Hyd Proc, 55(7), 153-162 (1976).
Schwartz, G.W., Preventing vibration in shell-and-tube heat exchangers, Chem Eng, 19 July, 134-140 (1976).
Troup, D.H., and Richardson, J.A., Scale nucleation on a heat transfer surface and its prevention, Chem Eng Commns, 2(2), 167-180 (1976).
Bott, T.R., and Gudmundsson, J.S., Deposition of paraffin wax from kerosene in cooled heat-exchanger tubes, Can JCE, 55, 381-390 (1977).
Suitor, J.W.; Marner, W.J., and Ritter, R.B., Fouling of heat exchangers in cooling water service (review paper), Can JCE, 55, 374-385 (1977).
Wilkins, C., Spiral heat exchangers avoid fouling, Proc Engng, March, 89-90 (1977).

1978-1980

Barrington, E.A., Cure exchanger acoustic vibration, Hyd Proc, 57(7), 193-198 (1978).
Bestcherevnykh, A., Relating heat-exchanger fouling factors to coefficients of conductivity, Chem Eng, 13 Feb, 122 (1978).
Spencer, R.A., Predicting heat-exchanger performance by successive summation, Chem Eng, 4 Dec, 121-124 (1978).
Zanker, A., Predict fouling by nomograph, Hyd Proc, 57(3), 145-148 (1978).
Boland, D., and Linnhoff, B., The preliminary design of networks for heat exchange by systematic methods, Chem Engnr, April, 222-228 (1979).

Hill, A.B., and Bevers, D.V., Detecting equipment hot-spots, Chemtech, April, 247-253 (1979).

Silvestrini, R., Heat exchanger fouling and corrosion, Chem Eng Prog, 75(12), 29-35 (1979).

Crozier, R.A., Pressure relief to prevent heat-exchanger failure, Chem Eng, 15 Dec, 79-83 (1980).

Giger, G.K.; Coughanowr, D.R., and Richarz, W., Implementation of a feedback direct-digital control algorithm for a heat exchanger, Ind Eng Chem Proc Des Dev, 19(4), 546-550 (1980).

Herndon, R.C.; Hubble, P.E., and Gainer, J.L., Two pulsators for increasing heat transfer, Ind Eng Chem Proc Des Dev, 19(3), 405-410 (1980).

Rubin, F.L., Winterizing air-cooled heat exchangers, Hyd Proc, 59(10), 147-149 (1980).

Vukadinovic, M., What to do when heat exchangers plug, Chem Eng Prog, 76(7), 38-40 (1980).

1981-1983

Bott, T.R., Heat-exchanger fouling, Proc Engng, Jan, 27-30 (1981).

Chambers, A.K.; Wynnyckyj, J.R., and Rhodes, E., Development of a monitoring system for ash deposits on boiler-tube surfaces, Can JCE, 59, 230-235 (1981).

El-Shobokshy, M.S., Method for reducing deposition of small particles from turbulent fluid by creating a thermal gradient at the surface, Can JCE, 59, 155-157 (1981).

Pasteris, R.M., Pretreating mild-steel water-cooled heat exchangers, Chem Eng, 16 Nov, 285-288 (1981).

Rothernberg, D.H., and Nicholson, R.L., Interacting controls for air coolers, Chem Eng Prog, 77(1), 80-82 (1981).

Chiappetta, L.M., and Szetela, E.J., Heat exchanger computational procedure for temperature-dependent fouling, Chem Eng Commns, 16(1), 189-204 (1982).

Hwu, M.C., and Foster, R.D., Detection of fouling in a tubular reactor, Chem Eng Prog, 78(7), 62-68 (1982).

Miller, P.C., and Bott, T.R., Effects of biocide and nutrient availability on microbial contamination of heat exchanger surfaces, J Chem Tech Biotechnol, 32, 538-546 (1982).

Vargas, K.J., Troubleshooting compression refrigeration systems, Chem Eng, 22 March, 137-143 (1982).

Brookman, J., Using petroleum heat-transfer oils, Proc Engng, Sept, 67-69 (1983).

Croke, R., and Russell, C., Control of air-cooled heat exchangers, Proc Engng, Oct, 44-45 (1983).

Lukas, M.P., and Kaya, A., Adaptive control of heat exchanger using function blocks, Chem Eng Commns, 24(4), 259-274 (1983).

Shipes, K., Correcting air-cooled exchanger problems, Chem Eng Prog, 79(7), 56-57 (1983).

Stegelman, A.F., and Renfftlen, R., On-line mechanical cleaning of heat exchangers, Hyd Proc, Jan, 95-97 (1983).

1984-1986

Antony, S.M., and Joshi, G.H., Use a test heat exchanger to monitor scaling and corrosion, Chem Eng, 2 April, 103 (1984).

Chang, S.C., Simulating heat-exchanger performance, Chem Eng, 2 April, 81-88 (1984).

Chen, C.C., Predicting the performance of a heat-exchanger train, Chem Eng, 19 March, 155-158 (1984).

Knox, A.C., Venting requirements for deaerating heaters, Chem Eng, 23 Jan, 95-98 (1984).

Murty, K.N., and Ganapathy, V., Evaluate heat exchanger fouling, Chem Eng, 6 Aug, 93-96 (1984).

Fryer, P.J., and Slater, N.K.H., Direct simulation procedure for chemical reaction fouling in heat exchangers, Chem Eng J, 31(2), 97-108 (1985).

Grantom, R.L., Tube wall temperature monitoring technique, Chem Eng Prog, 81(7), 41-44 (1985).

Knudsen, J.G., Fouling of heat exchangers, Chem Eng Prog, 80(2), 63-69 (1985).

Orbons, H.G., and Huurdemann, T.L., Stress corrosion cracking in syngas heat exchangers, Plant/Opns Prog, 4(1), 49-58 (1985).

Starczewski, J., Better refrigerant exchanger design, Hyd Proc, April, 93-97 (1985).

Calandranis, J., and Stephanopoulos, G., Structural operability analysis of heat exchanger networks, CER&D, 64(5), 347-364 (1986).

Floudas, C.A., and Grossmann, I.E., Synthesis of flexible heat exchanger networks for multiperiod operation. Comput Chem Eng, 10(2), 153-168 (1986).

Fryer, P.J., and Slater, N.K.H., Simulation of heat exchanger control with tube-side chemical reaction fouling, Chem Eng Sci, 41(9), 2363-2372 (1986).

Jones, D.A.; Yilmaz, A.N., and Tilton, B.E., Synthesis techniques for retrofitting heat recovery systems, Chem Eng Prog, 82(7), 28-33 (1986).

Klaren, D.C., A fluid end to fouling? Proc Engng, Jan, 45-47 (1986).

Kotjabasakis, E., and Linnhoff, B., Sensitivity tables for design of flexible processes: How much contingency in heat exchanger networks is cost effective, CER&D, 64(3), 197-211 (1986).

Miyasugi, T.; Yoshioka, S.; Kosaka, S., and Suzuki, A., Effect of carbon deposition on heat exchanger-type steam reformer with low steam/carbon ratio, Int Chem Eng, 26(1), 130-139 (1986).

Nieh, C.D., and Zengyan, H., Estimate exchanger vibration, Hyd Proc, 65(4), 61-65 (1986).

1987-1988

Crittenden, B.D.; Kolaczkowski, S.T., and Hout, S.A., Modelling hydrocarbon fouling, CER&D, 65(2), 171-179 (1987).

Fryer, P., Modelling heat exchanger fouling, Chem Engnr, Oct, 20-22 (1987).

Fryer, P.J.; Paterson, W.R., and Slater, N.K.H., Robustness of fouling heat exchanger networks, CER&D, 65(3), 267-271 (1987).

Stuhlbarg, D., and Szurgot, A.M., Eliminating height and wind effects on air-cooled heat-exchangers, Chem Eng, 23 Nov, 151-154 (1987).

Weaver, D.S., Avoid vibration problems on heat exchanger tubes, Proc Engng, Nov, 33-38 (1987).
Wu, W.T.; Chu, Y.T., and Tsao, J.H., Bounded disturbances adaptive control of a heat exchanger, Chem Eng Commns, 59, 173-184 (1987).
Yokell, S., Extending the life of tubular heat exchangers, Chem Eng, 20 July, 74-86 (1987).
Calandranis, J., and Stephanopoulus, G., Structural approach to design of control systems in heat exchanger networks, Comput Chem Eng, 12(7), 651-670 (1988).
Forsyth, J., Liquid fouling problems in heat exchangers, Processing, Feb, 19-20 (1988).
Latif, N.A.; Al-Madfai, S.H.F., and Ghanim, A.N., Removal of scale deposited on heat-transfer surfaces using chemical methods, Ind Eng Chem Res, 27(8), 1548-1551 (1988).
Mukherjee, R., How to debottleneck exchangers, Hyd Proc, 67(7), 47-49 (1988).
Muller, H.M., and Branch, C.A., Influence of thermal boundary conditions on calcium carbonate fouling in double pipe heat exchangers, Chem Eng & Proc, 24(2), 65-74 (1988).
Snyder, P.G., Brittle fracture of high-pressure heat exchanger, Plant/Opns Prog, 7(3), 148-152 (1988).

17.4 Heat Exchanger Applications

1967-1979

Bodman, S.W., and Cortez, D.H., Heat transfer to agitated two-phase liquids in jacketed vessels, Ind Eng Chem Proc Des Dev, 6(1), 127-133 (1967).
Hood, R.R., Designing heat exchangers in teflon, Chem Eng, 22 May, 181-186 (1967).
Letan, R., and Kehat, E., Mechanism of heat transfer in a spray-column heat exchanger, AIChEJ, 14(3), 398-405 (1968).
Garside, J.; Francis, J.C., and Powell, T.E., The 'waterfall' cooler: A new type of product cooler for fertiliser plants, Brit Chem Eng, 14(2), 191-193 (1969).
Preece, R.J., and Hitchcock, J.A., Comparative performance characteristics of some extended surfaces for air-cooler applications, Chem Engnr, June, 238-244 (1972).
Doyle, P.T., and Benkly, G.J., Use fanless air cooler, Hyd Proc, 52(7), 81-86 (1973).
Rao, K.B., and Murti, P.S., Heat transfer in mechanically agitated gas-liquid systems, Ind Eng Chem Proc Des Dev, 12(2), 190-197 (1973).
Brown, J.W., and Benkly, G.J., Heat exchangers in cold service, Chem Eng Prog, 70(7), 59-62 (1974).
Newell, R.G., Air-cooled heat exchangers in low temperature environments, Chem Eng Prog, 70(10), 86-91 (1974).
Shipes, K.V., Air-cooled exchangers in cold climates, Chem Eng Prog, 70(7), 53-58 (1974).
Kakabaev, A., et al., Test results from a large-scale solar air-conditioning pilot plant, Int Chem Eng, 16(1), 60-65 (1976).

Harker, J.H., Optimizing water coolers, Processing, Nov, 54 (1978).
Umeda, T.; Itoh, J., and Shiroko, K., Heat exchange system synthesis, Chem Eng Prog, 74(7), 70-76 (1978).
Berntsson, T., Heat transfer in a new type of heat exchanger for divided solids, Powder Tech, 22, 101-111 (1979).
Raju, K.S.N., and Rattan, V.K., Heat-pipe construction, Chem Eng, 17 Dec, 99-101 (1979).
Turner, J.R., and Eastop, T.D., A hot wire anemometry method for the flow patterns in an array of heat exchanger tubes, Trans IChemE, 57, 139-142 (1979).

1980-1988
Degnan, T.F., and Wei, J., The co-current reactor-heat exchanger, AIChEJ, 26(1), 60-67 (1980).
Malone, R.J., Sizing external heat exchangers for batch reactors, Chem Eng, 1 Dec, 95-101 (1980).
Herkenhoff, R.G., A new way to rate an existing heat exchanger, Chem Eng, 23 March, 213-215 (1981).
Martin, H., Fluid-bed heat exchangers: A new model for particle convective energy transfer, Chem Eng Commns, 13(1), 1-22 (1981).
Timmerhaus, K.D., Fundamental concepts and applications of cryogenic heat transfer, Chem Eng Educ, 15(2), 68-72, 98-104 (1981).
Bolliger, D.H., Assessing heat transfer in process-vessel jackets, Chem Eng, 20 Sept, 95-100 (1982).
Ganapathy, V., Evaluating waste heat recovery projects, Hyd Proc, Aug, 101-106 (1982).
Kumana, J.D., and Kothari, S.P., Predict storage-tank heat transfer precisely, Chem Eng, 22 March, 127-132 (1982).
Mehra, Y.R., Refrigeration systems for low-temperature processes, Chem Eng, 12 July, 94-103 (1982).
Murray, I., Nonmetallic materials for heat transfer, Proc Engng, Jan, 34-37 (1982).
Noble, R.D., Experiment on solar hot-water heating by natural convection, Chem Eng Educ, 17(1), 20-23 (1983).
Sjogren, S., and Grueiro, W., Applying heat exchangers in hydrocarbon processing, Hyd Proc, Sept, 133-136 (1983).
Miyasugi, T.; Kosaka, S.; Kawai, T., and Suzuki, A., A heat-exchanger type steam reformer for ammonia production, Chem Eng Prog, 80(7), 41-45 (1984).
Zinemanas, D.; Hasson, D., and Kehat, E., Simulation of heat exchangers with change of phase, Comput Chem Eng, 8(6), 367-376 (1984).
Wagner,, R.L., and Sjogren, S., Optimizing heat exchanger design for crude oil stabilization, Chem Eng Prog, 81(2), 46-51 (1985).
Anon., Fluid bed heat exchanger to minimise fouling, Chem Engnr, Feb, 28 (1986).
Govind, R.; Mocsny, D.; Cosson, P., and Klei, J., Exchanger network synthesis on a microcomputer, Hyd Proc, 65(7), 53-57 (1986).
Chynoweth, E., Thermal fluids for process heating, Proc Engng, Oct, 37-39 (1987).

Havas, G.; Deak, A., and Sawinsky, J., Heat transfer to helical coils in agitated vessels, Chem Eng J, 35(1) 61-64 (1987).
Gruver, M.E., and Pike, R., Getting jacketed reactors to perform better, Chem Eng., 19 Dec, 149-152 (1988).
Zhelev, T.K., and Boyadzhiev, K.B., Method for optimal synthesis of heat exchanger systems, Int Chem Eng, 28(3), 543-559 (1988).

17.5 Plate Heat Exchangers

1969-1979

Buonopane, R.A., and Troupe, R.A., Effects of internal rib and channel geometry in plate heat exchangers, AIChEJ, 15(4), 585-596 (1969).
Cocks, A.M., Plate heat exchanger design by computer, Chem Engnr, May, CE193-198 (1969).
Minton, P.E., Designing spiral-plate heat exchangers, Chem Eng, 4 May, 103-112 (1970).
Usher, J.D., Evaluating plate heat-exchangers, Chem Eng, 23 Feb, 90-94 (1970).
Marriott, J., Where and how to use plate heat exchangers, Chem Eng, 5 April, 127-134 (1971).
Clark, D.F., Plate heat exchanger design and recent development, Chem Engnr, May, 275-279, 285 (1974).
Cooper, A., Recover more heat with plate heat exchangers, Chem Engnr, May, 280-285 (1974).
Edwards, M.F.; Vaie, A.A.C., and Parrott, D.L., Heat transfer and pressure drop characteristics of a plate heat exchanger using Newtonian and non-Newtonian liquids, Chem Engnr, May, 286-288, 293 (1974).
Harvey, D.C., and Glass, G.E., Uses of inflated-plate heat exchangers, Chem Eng, 11 Nov, 170-174 (1974).
Kullendorff, A., Transfer characteristics of plate heat exchangers, Proc Engng, June, 56-59 (1974).
Wilkinson, W.L., Flow distribution in plate heat exchangers, Chem Engnr, May, 289-293 (1974).
Cowan, C.T., Choosing materials of construction for plate heat exchangers, Chem Eng, 9 June, 100-103; 7 July, 102-104 (1975).
Edwards, M.F., and Stinchcombe, R.A., Cost comparison of gasketed plate heat exchangers and conventional shell and tube units, Chem Engnr, May, 338-341 (1977).
Marriott, J., Performance of an Alfaflex plate heat exchanger, Chem Eng Prog, 73(2), 73-78 (1977).
Price, A.F., and Fattah, A.F.M.A., Hydrodynamic characteristics of a plate heat exchanger channel, Trans IChemE, 56, 217-228 (1978).
Cross, P.H., Prevent fouling in plate heat exchangers, Chem Eng, 1 Jan, 87-90 (1979).

1980-1988

Raju, K.S.N., and Chand, J., Consider the plate heat exchanger, Chem Eng, 11 Aug, 133-144 (1980).

Bond, M.P., Plate heat exchangers for effective heat transfer, Chem Engnr, April, 162-167 (1981).
Cross, P.H., The use of plate heat exchangers for energy economy, Chem Engnr, March, 87-90 (1982).
Mills, S.H., and Munson, W.H., Graphite tubes vs plate exchangers in quench acid service, Chem Eng Prog, 78(7), 33-37 (1982).
Tishchenko, Z.V., and Bondarenko, V.N., Comparison of efficiency of smooth-finned plate heat exchangers, Int Chem Eng, 23(3), 550-558 (1983).
Bassiouny, M.K., and Martin, H., Flow distribution and pressure drop in plate heat exchangers, Chem Eng Sci, 39(4), 693-704 (1984).
Sennik, L., Selecting elastomers for plate heat exchanger gaskets, Chem Engnr, Aug, 41-45 (1984).
Jarzebski, A.B., and Wardas-Koziel, E., Dimensioning of plate heat exchangers to give minimum annual operating costs, CER&D, 63, 211-218 (1985).
Gregory, E., Plate and fin heat exchangers, Chem Engnr, Sept, 33, 35, 37-39 (1987).
Lines, J.R., Asymmetric plate heat exchangers, Chem Eng Prog, 83(7), 27-30 (1987).
Lowe, R.E., Plate and fin heat exchangers for cryogenic service, Chem Eng, 17 Aug, 131-135 (1987).
Atkinson, N., Plate heat exchangers for viscous fluids, Proc Engng, Jan, 41-45 (1988).
Khan, A.R.; Baker, N.S., and Wardle, A.P., Dynamic characteristics of a countercurrent plate heat exchanger, Int J Heat Mass Trans, 31(6), 1269-1278 (1988).
Pignotti, A., and Tamborenea, P.I., Thermal effectiveness of multipass plate exchangers, Int J Heat Mass Trans, 31(10), 1983-1992 (1988).

17.6 Evaporators

1967-1969

Baum, V.A.; Bairamov, R., and Toiliev, K., Industrial calculation of unsteady-state thermal regime in glass-enclosed solar evaporators, Int Chem Eng, 7(4), 569-572 (1967).
Cheng, C.Y., and Cheng, S.W., Constant total-pressure evaporation with heat reuse by a built-in engine, AIChEJ, 13(3), 528-534 (1967).
Findley, M.E., Membrane evaporators, Ind Eng Chem Proc Des Dev, 6(2), 226-230 (1967).
Gouw, T.H., and Jentoft, R.E., Efficiency measurements on an all-glass wiped-film evaporator, Ind Eng Chem Proc Des Dev, 6(1), 62-67 (1967).
Kharisov, M.A., and Kogan, V.B., A distribution device for thin-film evaporation equipment, Int Chem Eng, 7(2), 189-191 (1967).
Mizushima, T.; Ito, R., and Miyashita, H., Experimental study of evaporative cooler, Int Chem Eng, 7(4), 727-732 (1967).
Oden, E.C., Charts speed computations for multiple-effect evaporator problems, Chem Eng, 24 April, 159-162 (1967).
Partridge, G.C., Redesign of a malt extract evaporation plant, Brit Chem Eng, 12(3), 374-376 (1967).

Skocylas, A., Thin-film evaporator construction and performance, Brit Chem Eng, 12(8), 1235-1239 (1967).

Unterberg, W., and Edwards, D.K., Effect of dissolved solid on wiped-film evaporation, Ind Eng Chem Proc Des Dev, 6(3), 268-276 (1967).

Dickinson, D.R., and Marshall, W.R., Rates of evaporation of sprays, AIChEJ, 14(4), 541-552 (1968).

Guerreri, G., Recovery of liquid substances from dilute solutions, Brit Chem Eng, 13(4), 524-527 (1968).

Hornby, J., and Taylor, R.F., Entrainment removal from climbing-film evaporators, Brit Chem Eng, 13(3), 361-365 (1968).

Itahara, S., and Stiel, L.I., Optimal design of multiple-effect evaporators with vapor bleed streams, Ind Eng Chem Proc Des Dev, 7(1), 6-11 (1968).

Lu, C.H., and Fabuss, B.M., Calcium sulfate scaling in sea water evaporation, Ind Eng Chem Proc Des Dev, 7(2), 206-212 (1968).

Moore, J.G., and Pinkel, E.B., Using single-pass evaporators, Chem Eng Prog, 64(7), 39-44 (1968).

Newman, H.H., Testing evaporator performance, Chem Eng Prog, 64(7), 33-38 (1968).

Starmer, R., and Lowes, F., Nuclear desalting, Chem Eng, 9 Sept, 127-134 (1968).

Clemons, D.B., Axial mixing of a low-viscosity liquid in a wiped-film evaporator, Ind Eng Chem Fund, 8(2), 279-281 (1969).

Findley, M.E.; Tanna, V.V.; Rao, Y.B., and Yeh, C.L., Mass and heat transfer relations in evaporation through porous membranes, AIChEJ, 15(4), 483-489 (1969).

Gelperin, N.I., and Shur, V.A., Calculation of heating surface of multistage evaporators, Int Chem Eng, 9(3), 406-410 (1969).

Kovalev, E.M., and Ulianova, L.A., A natural circulation evaporator with a channel heating chamber, Brit Chem Eng, 14(10), 1413 (1969).

Reiser, C.O., System for controlling water evaporation from reservoirs, Ind Eng Chem Proc Des Dev, 8(1), 63-69 (1969).

Robbins, J., and Davies, P.J., Design of non-scaling evaporators, Brit Chem Eng, 14(9), 1204-1207 (1969).

1970-1971

Barba, D., and Giona, A., Pressure loss in a falling-film evaporator, Brit Chem Eng, 15(11), 1436-1437 (1970).

Bruin, S., Analysis of heat transfer in a centrifugal film evaporator, Chem Eng Sci, 25(9), 1475-1486 (1970).

Delyannis, A.A., and Delyannis, E.A., Solar desalting, Chem Eng, 19 Oct, 136-140 (1970).

Kelleher, J., Critical entrainment rate in the evaporator of a phospheric acid plant, Brit Chem Eng, 15(10), 1324-1326 (1970).

Mandil, M.A., and Ghafour, E.E., Optimization of multistage flash evaporation plants, Chem Eng Sci, 25(4), 611-622 (1970).

Osborn, O.; Schrieber, C.F., and Smith, H.G., Metals performance in evaporator-desalination plants, Chem Eng Prog, 66(7), 74-79 (1970).

Skoczylas, A., Heat-transfer coefficients for a hinged-blade wiped-film evaporator, Brit Chem Eng, 15(2), 221-222 (1970).

Thomas, D.G., and Young, G., Thin-film evaporation enhancement by finned surfaces, Ind Eng Chem Proc Des Dev, 9(2), 317-323 (1970).
Wetherhorn, D., Guide to trouble-free evaporators, Chem Eng, 1 June, 187-192 (1970).
Yoshida, T., and Hyodo, T., Evaporation of water in air, humid air, and superheated steam, Ind Eng Chem Proc Des Dev, 9(2), 207-214 (1970).
Burdett, J.W., and Holland, C.D., Dynamics of a multiple-effect evaporator system, AIChEJ, 17(5), 1080-1089 (1971).
Hammond, R.P., New type of flash evaporator for seawater distillation, Chemtech, Dec, 754-757 (1971).
Porteous, A., The theory, practice and economics of solar distillation, Chem Engnr, Nov, 406-411 (1971).
Porteous, A., and Muncaster, R., Model for equilibration rates in flashing flow through open channels, nozzles and short tubes, Brit Chem Eng, 16(1), 59-64 (1971).
Various, Papers from symposium on evaporation of heat-sensitive foodstuff liquids, J Appl Chem Biotechnol, 21, 349-377 (1971).
Williams, J.S., and Hodgson, A.S., Multistage flash desalination utilizing diesel-generator waste heat, Ind Eng Chem Proc Des Dev, 10(4), 460-466 (1971).

1972-1975

Godau, H.J., Improved heat transfer performance in film evaporators, Proc Tech Int, 17(12), 945-946 (1972).
Gull, H.C., Applied physics in evaporation plants, Brit Chem Eng, 17(1), 39-42; 17(2), 123-132 (1972).
Kovalev, E.M., and Kostenko, Z.F., An evaporating apparatus with a reversed circulation path, Int Chem Eng, 12(2), 252-254 (1972).
Newell, R.B., and Fisher, D.G., Model development, reduction, and experimental evaluation for an evaporator, Ind Eng Chem Proc Des Dev, 11(2), 213-221 (1972).
Pancharatnam, S., Transient behavior of a solar pond and prediction of evaporation rates, Ind Eng Chem Proc Des Dev, 11(2), 287-292 (1972).
Fisher, D.G., and Jacobson, B.A., Computer control of a pilot plant evaporator, Chem Engnr, Nov, 552-558, 562 (1973).
Hajdu, H., and Tettamanti, K., Heat transfer in vertical tube evaporators, Periodica Polytechnica, 16, 347-366 (1972); 17, 321-334 (1973).
Pugh, O., Desalination (a review), Proc Tech Int, Jan, 57-59 (1973).
Rodionov, A.I., and Degtyarev, V.V., Mass transfer in gas phase during water evaporation from aqueous solutions with various viscosities in grid-plate columns, Int Chem Eng, 13(4), 661-664 (1973).
Waltrich, P.F., Sizing vacuum equipment for evaporative coolers, Chem Eng, 14 May, 164-166 (1973).
Fletcher, L.S.; Sernas, V., and Galowin, L.S., Evaporation from thin water films on horizontal tubes, Ind Eng Chem Proc Des Dev, 13(3), 265-269 (1974).
Gray, R.M., Reducing evaporation costs, Chemical Processing, Sept, 35-41 (1974).
Barduhn, A.J., The status of freeze-desalination, Chem Eng Prog, 71(11), 80-87 (1975).

Cole, J.W., Mechanical vapour recompression for evaporators, Chem Engnr, Feb, 76-78, 81 (1975).
Dickson, A.N., The optimisation of multi-stage flash distillation plant, Chem Engnr, Feb, 79-81 (1975).
Fletcher, L.S.; Sernas, V., and Parken, W.H., Evaporation heat-transfer coefficients for thin sea-water films on horizontal tubes, Ind Eng Chem Proc Des Dev, 14(4), 411-416 (1975).
Fokin, V.S., Design of multiple-effect evaporation equipment for concentration of saturated solutions, Int Chem Eng, 15(1), 122-125 (1975).
Godau, H.J., Flow processes in thin-film evaporators, Int Chem Eng, 15(3), 445-450 (1975).
Guttridge, D., and Anderson, B., Thin-film evaporators, Proc Engng, Jan, 49-51; Feb, 69-71 (1975).
Houghton, J., and Bailes, R.K., Economics of multiple effect evaporators, Chem Engnr, Feb, 82-83, 90 (1975).
Sheth, N.J.; Peck, R.E., and Wasan, D.T., Desalination of sea water using solar radiation under retarded evaporation conditions, Ind Eng Chem Proc Des Dev, 14(4), 351-358 (1975).
Slota, L., Improvement of evaporator steam economy by heat recovery, Chem Engnr, Feb, 84-90 (1975).
Smirnov, N., et al., Evaporation of water-alcohol solutions of ascorbic acid in a falling-film evaporator, Int Chem Eng, 15(4), 704-707 (1975).

1976-1978

Bairamov, R.B., et al., Analysis of thermal schemes for evaporative desalination plants, Int Chem Eng, 16(3), 387-392 (1976).
Dockendorff, J.D., and Cheng, P.J., Energy-conscious evaporators, Chem Eng Prog, 72(5), 56-61 (1976).
Farin, W.G., Low-cost evaporation method saves energy by reusing heat, Chem Eng, 1 March, 100-106 (1976).
Rozycki, J., Energy conservation via recompression evaporation, Chem Eng Prog, 72(5), 69-72 (1976).
Van der Mast, V.C., and Bromley, L.A., Interfacial phenomena in falling-film evaporation of natural seawater, AIChEJ, 22(3), 533-538 (1976).
Various, Putting evaporators to work (topic issue), Chem Eng Prog, 72(4), 41-71 (1976).
White, I., Effect of surfactants on evaporation of water close to 100 degC, Ind Eng Chem Fund, 15(1), 53-59 (1976).
Anon., Thin-film evaporators, Processing, Sept, 57, 58, 62 (1977).
Burrows, M.J., and Beveridge, G., Glass lining for thin-film evaporators, Proc Engng, Nov, 84-86 (1977).
Chintapalli, P.; Seborg, D.E., and Fisher, D.G., Model reference identification of state space models for double-effect evaporator, Can JCE, 55, 213-225 (1977).
Cole, J., Developments in evaporation, Proc Engng, Nov, 102-103 (1977).
Fellows, S.K., and Rothbaum, H.P., Industrial brine production using power station waste heat to assist solar evaporation of sea water, J Appl Chem Biotechnol, 27, 685-695 (1977).

Frank, J.T., and Lutcha, J., Residence time of materials treated in a rotary-film evaporator, Int Chem Eng, 17(3), 511-520 (1977).
Murthy, V.N., and Sarma, P.K., Falling film evaporators: Design equation for heat transfer rates, Can JCE, 55, 732-740 (1977).
Saaski, E.W., and Franklin, J.L., Performance of an evaporative heat transfer wick, Chem Eng Prog, 73(7), 74-77 (1977).
Stewart, G., and Beveridge, G.S.G., Steady-state cascade simulation in multiple-effect evaporation, Comput Chem Eng, 1(1), 3-10 (1977).
Bennett, R.C., Recompression evaporation, Chem Eng Prog, 74(7), 67-71 (1978).
Bucher, F., Improved design of thin-film evaporators, Processing, June, 73-75 (1978).
Casten, J.W., Mechanical recompression evaporators, Chem Eng Prog, 74(7), 61-67 (1978).
Dev, L., and Kelso, P.W., Steam stripping of Kraft Mill evaporator condensate, Chem Eng Prog, 74(1), 72-75 (1978).
Harker, J.H., Economics of evaporator operation, Processing, Dec, 31-32 (1978).
Kalishevich, Y.I., et al., Temperature depression in evaporation equipment with film-type vaporizers, Int Chem Eng, 18(1), 134-136 (1978).
Kleinman, G., Double effect evaporation of crude phosphoric acid, Chem Eng Prog, 74(11), 37-40 (1978).
Seraq El-Din, S.G.; Darwish, M.A., and El-Dessouky, H.T., Interactions in a single-stage flash unit, Ind Eng Chem Proc Des Dev, 17(4), 381-388 (1978).
Swartz, A., Guide to troubleshooting multiple-effect evaporators, Chem Eng, 8 May, 175-182 (1978).
Westbrook, N.J., Improved evaporator design, Processing, Sept, 39-41 (1978).

1979-1980

Burrows, M.J., and Beveridge, G.S.G., The centrifugally agitated wiped film evaporator, Chem Engnr, April, 229-232 (1979).
Epstein, N., Optimum evaporator cycles with scale formation, Can JCE, 57, 659-669 (1979).
Freese, H.L., and Glover, W.B., Mechanically agitated thin-film evaporators, Chem Eng Prog, 75(1), 52-58 (1979).
Radovic, L.R., et al., Computer design and analysis of operation of multiple-effect evaporator system in the sugar industry, Ind Eng Chem Proc Des Dev, 18(2), 318-323 (1979).
Sato, M., et al., Dynamics of desalting multistage flash evaporator for a partial load, Int Chem Eng, 19(3), 463-470; 19(4), 631-639 (1979).
Beesley, A.H., and Rhinesmith, R.D., Energy conservation by vapor compression evaporation, Chem Eng Prog, 76(8), 37-41 (1980).
Bennett, R.C., and Fakatselis, T.E., The elbow separator evaporator, Chem Eng Prog, 76(11), 64-69 (1980).
Cole, J., Saving energy with MVR falling-film evaporators, Proc Engng, Jan, 82-83 (1980).
Frank, J.T., and Lutcha, J., Material film thickness in film-type rotary evaporators, Int Chem Eng, 20(1), 65-77 (1980).

Klaren, D.G., and Halberg, N., Development of a multi-stage flash/fluidized bed evaporator, Chem Eng Prog, 76(7), 41-43 (1980).
Newell, R.B., Comparative study of model and goal coordination in multilevel optimization of double-effect evaporator, Can JCE, 58, 275-278 (1980).
Peak, W.E., Desalting seawater by flash evaporation, Chem Eng Prog, 76(7), 50-53 (1980).
Stern, G.; Bayles, B.J., and Chukumerije, O.H., Choosing materials for desalting by flash distillation, Chem Eng, 22 Sept, 171-176 (1980).
Tang, C.L., Two-phase flow in a climbing-film evaporator, Can JCE, 58, 425-430 (1980).
Weimer, L.D.; Dolf, H.R., and Austin, D.A., A systems engineering approach to vapor recompression evaporators, Chem Eng Prog, 76(11), 70-77 (1980).
Zimmer, A., Developments in energy-efficient evaporation, Chem Eng Prog, 76(8), 50-56 (1980).

1981-1983

Alkidas, A.C., Influence of size-distribution parameters on the evaporation of polydisperse dilute sprays, Int J Heat Mass Trans, 24(12), 1913-1924 (1981).
Hoffman, D., Low-temperature evaporation plants, Chem Eng Prog, 77(10), 59-62 (1981).
Hughes, C.H., and Emmermann, D.K., Vertical tube vapour-compression evaporation for sea water, Chem Eng Prog, 77(7), 72-73 (1981).
Sephton, H.H., Vertical tube foam evaporation, Chem Eng Prog, 77(10), 83-86 (1981).
Yundt, B., and Rhinesmith, R., Horizontal spray-film evaporation, Chem Eng Prog, 77(9), 69-73 (1981).
Bukacek, R.F., and Tahir, T.B., The refrigerant sets the process, Hyd Proc, Oct, 109-110 (1982).
Droz, N.A.R., Urea evaporator entrainment separator, Chem Eng Prog, 78(3), 62-65 (1982).
Hadley, G.R., Theoretical treatment of evaporation front drying, Int J Heat Mass Trans, 25(10), 1511-1522 (1982).
Izumi, K., et al., Alkaline scale formation in multistage flash-evaporation-type desalination plant, Int Chem Eng, 22(1), 82-91 (1982).
Izumi, K.; Yamada, A.; Sawa, T., and Takahashi, S., Iron-sludge formation in multistage flash-evaporator-type desalination plant. Int Chem Eng, 22(2), 301-309 (1982).
Katsaros, K.B., and Garrett, W.D., Effects of organic surface films on evaporation and thermal structure of water in free and forced convection, Int J Heat Mass Trans, 25(11), 1661-1670 (1982).
Nakamura, K., and Watanabe, T., Flow in agitated thin-film horizontal evaporator, Chem Eng Commns, 18(1), 173-190 (1982).
Taylor, D.C., The history of evaporation (1750 to 1900), Chem Engnr, May, 187-190 (1982).
Various, Evaporation and evaporators: Plant and equipment survey, Processing, April, 23-31 (1982).
Angell, C.W.; Baird, J.L., and Vivian, J.E., Thin-film evaporators for processing radioactive wastes, Chem Eng Prog, 79(5), 52-55 (1983).

Anthony, D., Improved evaporators for radioactive wastes, Chem Eng Prog, 79(7), 58-63 (1983).
Arlidge, D.B., Wiped film evaporators as pilot plants, Chem Eng Prog, 79(8), 35-40 (1983).
Barde, D.K., and Patel, J.C., Optimum efficiency in pulp and paper concentrators, Chem Eng Prog, 79(9), 27-30 (1983).
Chow, L.C., and Chung, J.N., Evaporation of water into a laminar stream of air and superheated steam, Int J Heat Mass Trans, 26(3), 373-380 (1983).
Esplugas, S., and Mata, J., Calculator design of multistage evaporators, Chem Eng, 7 Feb, 59-61 (1983).
Gropp. U.; Schnabel, G., and Schlunder, E.U., Effect of liquid-side mass transfer resistance on selectivity during partial evaporation of binary mixtures in a falling film, Int Chem Eng, 23(1), 11-18 (1983).
Kasyanenko, M.K., et al., Heat transfer in concentrators with external boiling zone and natural circulation, Int Chem Eng, 23(4), 766-769 (1983).
Leatherman, H.R., Cost-effective mechanical vapor recompression, Chem Eng Prog, 79(1), 40-42 (1983).
Logsdon, J.D., Evaporator applications/trends in the pulp and paper industry, Chem Eng Prog, 79(9), 36-40 (1983).
Streit, D.E., Free falling evaporation system, Chem Eng Prog, 79(9), 41-45 (1983).
Wagner, W.M., and Finnegan, D.R., Select a seawater-desalting process, Chem Eng, 7 Feb, 71-75 (1983).

1984-1986

Cole, J., A guide to selection of evaporation plant, Chem Engnr, June, 20-23 (1984).
Fakatselis, T.E., Direct contact air cooled evaporation. Plant/Opns Prog, 3(3), 136-141 (1984).
Yundt, B., Troubleshooting vapor compression evaporators, Chem Eng, 24 Dec, 46-55 (1984).
Awerbuch, L.; Van der Mast, V., and Weekes, M., The geothermal flash evaporation process, Chem Eng Prog, 81(2), 40-45 (1985).
Han, J.C., and Fletcher, L.S., Falling-film evaporation and boiling in circumferential and axial grooves on horizontal tubes, Ind Eng Chem Proc Des Dev, 24(3), 570-575 (1985).
Harker, J., Optimizing evaporator operation, Proc Engng, Aug, 45-47 (1985).
Loughlin, K.F.; Coates, L.H., and Halhouli, K., High mass-flux evaporation, Chem Eng Sci, 40(7), 1263-1272 (1985).
Manganaro, J.L., and Schwartz, J.C., Simulation of an evaporative solar salt pond, Ind Eng Chem Proc Des Dev, 24(4), 1245-1251 (1985).
Clegg, G.T., and Papadakis, G., Rates of evaporation accompanying the depressurization of a pool of saturated Freon-11, Chem Eng Sci, 41(12), 3037-3044 (1986).
Gault, T., Evaporation enhancement through use of sprays, Plant/Opns Prog, 5(1), 23-26 (1986).
Helal, A.M.; Medani, M.S.; Soliman, M.A., and Flower, J.R., A tridiagonal matrix model for multistage flash desalination plants, Comput Chem Eng, 10(4), 327-342 (1986).

Lerner, B.J., High-tech mist elimination in multistage evaporators, Plant/Opns Prog, 5(1), 52-56 (1986).
MacDonald, E., and Kemp, I., Process integration gives new insights on evaporators, Proc Engng, Nov, 25-27 (1986).
Mehra, D.K., Selecting evaporators, Chem Eng, 3 Feb, 56-72 (1986).
Minton, P.E., Handbook of Evaporation Technology, Noyes Publications, New Jersey (1986).
Murray, A.J., Practical and economic benefits of falling-film evaporation, Plant/Opns Prog, 5(1), 31-34 (1986).
Pennink, H., Cost savings from evaporator vapor compression, Chem Eng, 6 Jan, 79-81 (1986).

1987-1988

Inuzuka, M., et al., Evaporation of binary mixtures under vacuum, Int Chem Eng, 27(1), 100-107 (1987).
Lambert, R.N.; Joye, D.D., and Koko, F.W., Design calculations for multiple-effect evaporators, Ind Eng Chem Res, 26(1), 100-107 (1987).
Maclean, R., Energy saving techniques in the evaporation and drying of distillery effluent, Chem Engnr, July, 29-34 (1987).
Ramakrishna, P., Estimate optimum number of effects for multi-effect evaporation, Chem Eng, 11 May, 82 (1987).
Tsotsas, E., and Schlunder, E.U., Heat transfer during evaporation and condensation of binary mixtures, Chem Eng & Proc, 21(4), 209-216 (1987).
Various, Thin film evaporators, Chem Engnr, Sept, 17-22 (1987).
Banvolgyi, G.; Valko, P.; Vajda, S., and Fulop, N., Estimation of influential parameters in a steady-state evaporator model, Comput Chem Eng, 12(2/3), 117-122 (1988).
Calu, M.P., and Lameloise, M.L., Interpretation of residence-time distribution measurements in flow of variable density: Applications in modeling of climbing-film evaporator for sugar industry, Int Chem Eng, 28(3), 424-435 (1988).
Hillenbrand, J.B., and Westerberg, A.W., Synthesis of multiple-effect evaporator systems using minimum utility insights, Comput Chem Eng, 12(7), 611-636 (1988).
Joye, D.D., and Koko, F.W., Simpler method of multiple-effect evaporator calculations, Chem Eng Educ, 22(1), 52-56 (1988).
Kocamustafaogullari, G., and Chen, I.Y., Falling-film heat transfer analysis on a bank of horizontal tube evaporator, AIChEJ, 34(9), 1539-1549 (1988).
Nguyen, V.T.; Furzeland, R.M., and Ijpelaar, M.J.M., Rapid evaporation at the superheat limit, Int J Heat Mass Trans, 31(8), 1687-1700 (1988).
Sandall, O.C.; Hanna, O.T., and Ruiz, G., Heating and evaporation of turbulent falling liquid films, AIChEJ, 34(3), 502-505 (1988).
Taffe, P., Refinery use of MVR evaporators, Processing, Dec, 31-32 (1988).
Wilkins, E.S., and Ramachandran, R.S., Gelled solar ponds, Chemtech, July, 414-421 (1988).

17.7 Cooler/Condensers

1967-1969

Avriel, M., and Wilde, D.J., Optimal condenser design by geometric programming, Ind Eng Chem Proc Des Dev, 6(2), 256-263 (1967).

Dobratz, C.J., and Oldershaw, C.F., Desuperheating vapors in condensers unnecessary, Chem Eng, 31 July, 146 (1967).

Henderson, C.L., and Marchello, J.M., Role of surface tension and tube diameter in film condensation on horizontal tubes, AIChEJ, 13(3), 613-614 (1967).

Leonard, W.K., and Estrin, J., Effect of vapor velocity on condensation on a vertical surface, AIChEJ, 13(2), 401-402 (1967).

O'Bara, J.T.; Killian, E.S., and Roblee, L.H.S., Dropwise condensation of steam at atmospheric and above atmospheric pressures, Chem Eng Sci, 22(10), 1305-1314 (1967).

Keller, J.B., Condensing subcooling service from old condenser, Chem Eng, 1 July, 108 (1968).

Mizushina, T.; Ito, R., and Miyashita, H., Characteristics and thermal design methods of evaporative coolers, Int Chem Eng, 8(3), 532-538 (1968).

Schrodt, J.T., and Gerhard, E.R., Simultaneous condensation of methanol and water from noncondensing gas on a vertical tube bank, Ind Eng Chem Fund, 7(2), 281-285 (1968).

Stern, F., and Votta, F., Condensation from superheated gas-vapor mixtures, AIChEJ, 14(6), 928-933 (1968).

Thomas, D.G., Enhancement of film condensation rate on vertical tubes by longitudinal fins, AIChEJ, 14(4), 644-649 (1968).

Williams, A.G.; Nandapurkar, S.S., and Holland, F.A., A review of methods for enhancing heat transfer rates in surface condensers, Chem Engnr, Nov, CE367-373 (1968).

Dehne, M.F., Air-cooled overhead condensers, Chem Eng Prog, 65(7), 51-58 (1969).

Dolloff, J.B.; Metzger, N.H., and Roblee, L.H.S., Dropwise condensation of steam at elevated pressures, Chem Eng Sci, 24(3), 571-584 (1969).

Emerson, W.H., Application of digital computer to design of surface condensers, Chem Engnr, May, CE178-184 (1969).

Grant, I.D.R., Condenser performance: Effect of different arrangements for venting non-condensing gases, Brit Chem Eng, 14(12), 1709-1711 (1969).

Krupiczka, R., Cooler-condenser design, Brit Chem Eng, 14(11), 1550-1551 (1969).

Lepper, A.M., and Houtby, D.K., Algorithmic models for specification of heat-transfer equipment involving condensation, Chem Engnr, May, CE189-193 (1969).

Schwartz, A., and Goldschmidt, S., A new type of water-cooled condenser, Brit Chem Eng, 14(4), 497-498 (1969).

Trub, I.A., Design of a tray-type barometric condenser, Brit Chem Eng, 14(8), 1069 (1969).

1970-1974

Gloyer, W., Thermal design of mixed-vapor condensers, Hyd Proc, 49(6), 103-108; 49(7), 107-110 (1970).

Lord, R.C.; Minton, P.E., and Slusser, R.P., Design parameters for condensers and reboilers, Chem Eng, 23 March, 127-134 (1970).

Various, Air-cooled condensers - some comments, Chem Eng Prog, 66(4), 78-84 (1970).

Kals, W., Wet-surface aircoolers, Chem Eng, 26 July, 90-94 (1971).

Kosky, P.G., and Staub, F.W., Local condensing heat transfer coefficients in the annular flow regime, AIChEJ, 17(5), 1037-1043 (1971).

Withers, J.G., and Young, E.H., Steam condensing on vertical rows of horizontal corrugated and plain tubes, Ind Eng Chem Proc Des Dev, 10(1), 19-30 (1971).

Arrowsmith, R.M., Design of flash cooling systems, Proc Engng, Aug, 50-52 (1972).

Barnea, E., and Mizrahi, J., Heat transfer coefficient in the condensation of a hydrocarbon-steam mixture, Trans IChemE, 50, 286-288 (1972).

Fair, J.R., Designing direct-contact coolers/condensers, Chem Eng, 12 June, 91-100 (1972).

Schrodt, J.T., Analytical solutions for predicting vapor-gas behavior in cooler-condensers, Ind Eng Chem Proc Des Dev, 11(1), 20-26 (1972).

Archambault, J.P.; Jauffret, J.P., and Luyben, W.L., Experimental study of condensate subcooling control in a vertical condenser, AIChEJ, 19(5), 923-928 (1973).

Goncharenko, G.K., and Moshiashvili, M.D., Condensation of mixtures of vapors of immiscible liquids, Int Chem Eng, 13(2), 225-227 (1973).

Schrodt, J.T., Simultaneous heat and mass transfer from multicomponent condensing vapor-gas systems, AIChEJ, 19(4), 753-759 (1973).

Starczewski, J., Graphical method for partial-condenser design, Proc Engng, March, 72-77 (1973).

Wilkins, D.G., and Bromley, L.A., Dropwise condensation phenomena, AIChEJ, 19(4), 839-845 (1973).

Wilkins, D.G.; Bromley, L.A., and Read, S.M., Dropwise and filmwise condensation of water vapour on a pure gold tube, AIChEJ, 19(1), 119-123 (1973).

Billet, R., Partial counter-current direct condensation, Chemical Processing, July, 11-17 (1974).

Sharp, A.K., Transient dropwise condensation of water vapour from air on to cylindrical food cans, Trans IChemE, 52, 17-30 (1974).

Various, Why condensers don't operate as they are supposed to, Chem Eng Prog, 70(7), 78-82 (1974).

Volejnik, M., Industrial-scale condensers, Int Chem Eng, 14(2), 247-256 (1974).

Wood, D.G., and Subrahmaniyam, S., Pressure drop in annular condensation, Trans IChemE, 52, 379-380 (1974).

1975-1977

Kennedy, R., Optimization of air cooler design, Proc Engng, Nov, 73-75 (1975).

Lihou, D.A., and Hattangady, K.S., Studies of drops formed by condensation from a stream of high-pressure gas, Trans IChemE, 53, 242-246 (1975).

Lovett, D.A., Condensation control by porous insulating materials, Trans IChemE, 53, 112-116 (1975).
Salov, V.S., and Danilov, O.L., Condensation of binary mixtures of vapors of immiscible liquids on nonisothermal surfaces, Int Chem Eng, 15(1), 39-43 (1975).
Standart, G.L., and Akhtar, N., Condensation of a vapour with variable coolant temperature, Chem Eng J, 10(3), 165-188 (1975).
Swamy, M.S.K., Reliable spray-type desuperheaters, Chem Eng, 17 March, 86 (1975).
Velichko, G.N.; Stefanovskii, V.M., and Socherbakov, A.Z., Heat transfer during condensation of binary vapor mixtures, Int Chem Eng, 15(4), 625-627 (1975).
Chandran, R., and Watson, F.A., Condensation on static and rotating pinned tubes, Trans IChemE, 54, 65-72 (1976).
Detz, C.M., and Vermesh, R.J., Nucleation effects in the dropwise condensation of steam on electroplated gold surfaces, AIChEJ, 22(1), 87-93 (1976).
Izumi, R.; Ishimaru, T., and Aoyagi, W., Heat transfer and pressure drop for refrigerant R-12 condensing in horizontal tube, Int Chem Eng, 16(3), 500-506 (1976).
Russell, C., Designing air coolers for less noise, Proc Engng, Nov, 77-79 (1976).
Wu, W.H., and Maa, J.R., Heat transfer in dropwise condensation, Chem Eng J, 12(3), 225-232 (1976).
Krishna, R., and Pancha, C.B., Condensation of a binary vapour mixture in the presence of an inert gas, Chem Eng Sci, 32(7), 741-746 (1977).

1978-1980
Butterworth, D., Shell-and-tube condenser design, Proc Engng, Nov, 62-69 (1978).
Knight, W.P., Improve sulfur condensers, Hyd Proc, 57(5), 239-241 (1978).
Larinoff, M.W.; Moles, W.E., and Reichhelm, R., Design and specification of air-cooled steam condensers, Chem Eng, 22 May, 86-94 (1978).
Maa, J.R., Drop size distribution and heat flux of dropwise condensation, Chem Eng J, 16(3), 171-176 (1978).
Mizushina, T.; Ito, R.; Yamashita, S., and Kamimura, H., Film condensation of a pure superheated vapor inside a vertical tube, Int Chem Eng, 18(4), 672-680 (1978).
Nguyen, H.X., Optimize water outlet temperature from cooler/condensers, Hyd Proc, 57(5), 245-246 (1978).
Razavi, M.D., and Damle, A.S., Heat transfer coefficients for turbulent filmwise condensation, Trans IChemE, 56, 81-85 (1978).
Abdul-Hadi, M.I., Investigation into dropwise condensation of different steam-air mixtures on substrates of various materials, Can JCE, 57, 451-469 (1979).
Jesch, J.; Lutcha, J., and Michal, V., Use of air coolers for viscous media, Int Chem Eng, 19(4), 680-689 (1979).
Standiford, F.C., Effect of non-condensibles on condenser design and heat transfer, Chem Eng Prog, 75(7), 59-62 (1979).
Tamir, A., 'Mixed' pattern condensation of multicomponent mixtures, Chem Eng J, 17(2), 141-156 (1979).

Tamir, A., and Merchuk, J.C., Verification of a theoretical model for multicomponent condensation, Chem Eng J, 17(2), 125-140 (1979).
Chan, S.H.; Cho, D.H., and Condiff, D.W., Heat transfer from a high-temperature condensable mixture, Int J Heat Mass Trans, 23(1), 63-72 (1980).
Christman, J., Air vs water-cooled steam condensers, Hyd Proc, 59(9), 257-259 (1980).
Dobran, F., and Thorsen, R.S., Forced-flow laminar filmwise condensation of a pure saturated vapour in a vertical tube, Int J Heat Mass Trans, 23(2), 161-178 (1980).
Foxall, D.H., and Chappell, H.R. Superheated vapor condensation in heat exchanger design, Chem Eng, 29 Dec, 41-50 (1980).
Kotake, S., and Oswatitsch, K., Parameters of binary-mixture film condensation, Int J Heat Mass Trans, 23(11), 1405-1416 (1980).
Kurmarao, P.S.V., Heat transfer coefficients of steam condensers, Chem Eng, 25 Aug, 112-114 (1980).

1981-1982

Abdelmessih, A.H.; Rotenberg, Y., and Neumann, A.W., Experimental study of effects of surface characteristics and material thermal properties on dropwise condensation, Can JCE, 59, 138-148 (1981).
Baker, W.J., How to specify and select surface condensers, Hyd Proc, 60(9), 239-245 (1981).
Bandrowski, J., and Kubaczka, A., Condensation of multicomponent vapours in presence of inert gases, Int J Heat Mass Trans, 24(1), 147-154 (1981).
Haydu, L.J., Calculator program for a steam condenser, Chem Eng, 9 Feb, 99-102 (1981).
Kistler, R.S., and Kassem, A.E., Stepwise rating of condensers, Chem Eng Prog, 77(7), 55-59 (1981).
Kurmarao, P.S.V., Avoid oversizing desuperheater-condensers, Chem Eng, 19 Oct, 182-184 (1981).
Prasad, V., and Jaluria, Y., Film condensation on a horizontal isothermal surface, Chem Eng Commns, 11(4), 231-240 (1981).
Rabas, T.J., Tubeside exit condition with complete condensation in multitube crossflow heat exchangers, Chem Eng Commns, 10(1), 13-24 (1981).
Razavi, M.D., and Clutterbuck, E.K., Pressure drop and heat transfer prediction for total condensation inside horizontal tube, Ind Eng Chem Proc Des Dev, 19(4), 715-717 (1980); 20(2), 410 (1981).
Rose, J.W., Dropwise condensation theory, Int J Heat Mass Trans, 24(2), 191-194 (1981).
Schroppel, J., Impact of wall temperature variation on the film condensation of binary gas-vapour mixtures, Int J Heat Mass Trans, 24(1), 165-170 (1981).
Starczewski, J.., Simplify design of partial condensers, Hyd Proc, 60(3), 131-136 (1981).
Kutateladze, S.S., Semi-empirical theory of film condensation of pure vapours, Int J Heat Mass Trans, 25(5), 653-660 (1982).
LoPinto, L., Fog formation in low-temperature condensers, Chem Eng, 17 May, 111-113 (1982).

Maron, D.M., and Sideman, S., Condensation inside near horizontal tubes in cocurrent and countercurrent flow, Int J Heat Mass Trans, 25(9), 1439-1444 (1982).

Murty, K.N., Vertical vs. horizontal condensers, Chem Eng, 31 May, 152 (1982).

Prasad, V., and Jaluria, Y., Transient film condensation on a finite horizontal plate, Chem Eng Commns, 13(4), 327-343 (1982).

Rubin, F.L., Multizone condensers: Tabulate design data to prevent errors, Hyd Proc, June, 133-135 (1982).

Seban, R.A., and Hodgson, J.A., Laminar film condensation in a tube with upward vapor flow, Int J Heat Mass Trans, 25(9), 1291-1300 (1982).

Soliman, H.M., Annular-to-wavy flow pattern transition during condensation inside horizontal tubes, Can JCE, 60, 475-481 (1982).

1983-1985

Bell, K.J., Coping with an improperly vented condenser, Chem Eng Prog, 79(7), 54-55 (1983).

Ganapathy, V., Nomograph for gas cooler design, Proc Engng, April, 63-65 (1983).

Jorda, J.L., Estimate condensate flow quickly, Chem Eng, 7 March, 190 (1983).

Owen, R.G., and Lee, W.C., Some recent developments in condensation theory (review paper), CER&D, 61, 335-361 (1983).

Rao, D.S., and Murthy, M.S., Heat transfer during dropwise condensation, Chem Eng J, 26(1), 1-12 (1983).

Krebs, R.G., and Schlunder, E.U., Condensation with non-condensing gases inside vertical tubes with turbulent gas and film flow, Chem Eng & Proc, 18(6), 341-356 (1984).

Leach, R.P., and Rajguru, A., Design for 'free' chilling, Hyd Proc, Aug, 80-81 (1984).

Paschke, L.F., Condensing heat exchangers save heat, Chem Eng Prog, 80(7), 70-74 (1984).

Pathak, V.K., and Rattan, I.S., How much condensate will flash? Chem Eng, 14 May, 89-90 (1984).

Rose, J.W., Effect of pressure gradient in forced-convection film condensation on a horizontal tube, Int J Heat Mass Trans, 27(1), 39-48 (1984).

Bowie, G.E., Specifying and operating desuperheaters, Chem Eng, 27 May, 119-124 (1985).

Burghardt, A., and Dubis, A., Computational design method for condensation of binary vapours of immiscible liquids, Chem Eng & Proc, 19(5), 243-256 (1985).

Kotake, S., Effects of a small amount of noncondensable gas on film condensation of multicomponent mixtures, Int J Heat Mass Trans, 28(2), 407-414 (1985).

Krupiczka, R., Effect of surface tension on laminar film condensation on horizontal cylinder, Chem Eng & Proc, 19(4), 199-204 (1985).

Mangnall, K., and Webb, D., Vacuum condensation, Chem Engnr, Dec, 37-40 (1985).

McNaught, J., Developments in condenser design, Proc Engng, Jan, 38-43 (1985).

Rahman, M.M.; Fathi, A.M., and Soliman, H.M., Flow pattern boundaries during condensation: New experimental data, Can JCE, 63(4), 547-552 (1985).
Zondlo, J.W., and Rothfus, R.R., Prediction of coolant effects in vertical condensers, Ind Eng Chem Proc Des Dev, 24(3), 621-625 (1985).

1986-1987

Asano, K., and Matsuda, S., Total and partial condensation of binary vapor mixtures of methanol and water on vertical flat plates, Int Chem Eng, 26(4), 671-680 (1986).
Fullarton, D., and Schlunder, E.U., Approximate calculation of heat exchanger area for condensation of gas-vapor mixtures, Int Chem Eng, 26(3), 408-419 (1986).
Honda, H.; Nozu, S.; Uchima, B., and Fujii, T., Effect of vapour velocity on film condensation of R-113 on horizontal tubes in crossflow, Int J Heat Mass Trans, 29(3), 429-438 (1986).
Marto, P.J.; Looney, D.J.; Rose, J.W., and Wanniarachi, A.S., Evaluation of organic coatings for promotion of steam dropwise condensation, Int J Heat Mass Trans, 29(8), 1109-1118 (1986).
McAlister, J., Graphical method for when to retube steam condensers, Chem Eng, 1 Sept, 103 (1986).
Fullarton, D.; Schlunder, E.U., and Yuksel, L., Condensation of mixtures of isopropanol and steam, Int Chem Eng, 27(4), 597-607 (1987).
Krupiczka, R., and Kozak, A., Mathematical model of condensation process on horizontal tube with rectangular fins, Chem Eng & Proc, 22(1), 53-58 (1987).
Morsy, M.G.; Wassef, F.M.; Morcos, V.H., and El Biblawy, H.A.M., Overall heat-transfer coefficient for a multi-tube rotating condenser, Chem Eng Commns, 57, 41-50 (1987).
Murty, K.N., How vacuum affects condenser heat-transfer, Chem Eng, 7 Dec, 141 (1987).
Rashtchian, D., and Webb, D.R., Condensation of steam from mixtures with air in shell and tube exchanger at atmospheric and reduced pressures, CER&D, 65(2), 157-164 (1987).
Tsotsas, E., and Schlunder, E.U., Heat transfer during evaporation and condensation of binary mixtures, Chem Eng & Proc, 21(4), 209-216 (1987).
Webb, D.R., and Panagoulias, D., An improved approach to condenser design using film models, Int J Heat Mass Trans, 30(2), 373-378 (1987).

1988

Burghardt, A., and Berezowski, M., Computational design method for multicomponent condensation, Chem Eng & Proc, 24(4), 189-202 (1988).
Crick, G., Choice of process chillers, Processing, March, 26-28 (1988).
Irhayem, A.Y.N., Purging prevents condenser corrosion, Chem Eng, 15 Aug, 178 (1988).
Ogino, F.; Kanzaki, S., and Mizushina, T., Condensation of binary vapors of immiscible liquids, Int J Heat Mass Trans, 31(2), 245-250 (1988).

Oh, C.H., and Wadkins, R.P., Heat transfer in low-pressure lithium condensation, Chem Eng Commns, 73, 141-150 (1988).

Rashwan, F.A., and Soliman, H.M., Onset of slugging in horizontal condensers, Can JCE, 65(6), 887-898 (1988).

Rifert, V.G., Heat transfer and flow modes of phases in laminar-film vapour condensation inside a horizontal tube, Int J Heat Mass Trans, 31(3), 517-524 (1988).

Unsal, M., Forced vapour-flow condensation, Int J Heat Mass Trans, 31(8), 1613-1626 (1988).

Wang, C.Y., and Tun, C.J., Effects of non-condensable gas on laminar film condensation in vertical tube, Int J Heat Mass Trans, 31(11), 2339-2346 (1988).

17.8 Boiling, Boilers and Vaporizers

1967-1968

Bennett, A.W.; Hewitt, G.F.; Kearsey, H.A.; Keeys, R.K.F., and Pulling, D.J., Studies of burnout in boiling heat transfer, Trans IChemE, 45, T319-333 (1967).

Cole, R., Bubble frequencies and departure volumes at subatmospheric pressures in nucleate pool boiling, AIChEJ, 13(4), 779-783 (1967).

Csathy, D., Evaluating boiler designs for process-heat recovery, Chem Eng, 5 June, 117-124 (1967).

Frost, W., and Dzakowic, G.S., Graphical estimation of nucleate-boiling heat transfer, Ind Eng Chem Proc Des Dev, 6(3), 346-347 (1967).

Kesselring, R.C.; Rosche, P.H., and Bankoff, S.G., Transition and film boiling from horizontal strips, AIChEJ, 13(4), 669-675 (1967).

Lee, D.H., Boiling burnout: A review of recent work, Brit Chem Eng, 12(9), 1363-1368 (1967).

Van Stralen, S.J.D., Bubble growth rates in boiling binary mixtures, Brit Chem Eng, 12(3), 390-394 (1967).

Beaver, P.R., and Hughmark, G.A., Heat-transfer coefficients and circulation rates for thermosyphon reboilers, AIChEJ, 14(5), 746-749 (1968).

Fortuna, G., and Sideman, S., Direct contact heat transfer between immiscible liquid layers with simultaneous boiling and stirring, Chem Eng Sci, 23(9), 1105-1120 (1968).

Grigorev, L.N.; Khairullin, I.K., and Usmanov, A.G., Experimental study of critical heat flux in boiling of binary mixture, Int Chem Eng, 8(1), 39-42 (1968).

Grigorev, L.N.; Sarkisyan, L.A., and Usmanov, A.G., Experimental study of heat transfer in boiling of three-component mixtures, Int Chem Eng, 8(1), 76-78 (1968).

Hodgson, A.S., Forced convection, sub-cooled boiling heat transfer with water in an electrically heated tube at 100-550 psi, Trans IChemE, 46, T25-31 (1968).

Kern, R., Thermosyphon reboiler piping simplified, Hyd Proc, 47(12), 118-122 (1968).

Kosky, P.G., and Lyon, D.N., Pool boiling heat transfer to cryogenic liquids, AIChEJ, 14(3), 372-387 (1968).
Zemaitis, J.F., and Kermode, R.I., Experimental analysis of forced-convection film boiling from a flat horizontal plate, Ind Eng Chem Proc Des Dev, 7(3), 354-359 (1968).

1969-1972

Eckhart, R.A., Dynamic simulation of an LPG vaporizer, Ind Eng Chem Proc Des Dev, 8(4), 491-495 (1969).
Gelperin, N.I., and Korobkov, E.I., Experimental investigation of heat transfer during boiling of liquids. Int Chem Eng, 9(4), 627-630 (1969).
Hospeti, N.B., and Mesler, R.B., Vaporization at the base of bubbles of different shape during nucleate boiling of water, AIChEJ, 15(2), 214-219 (1969).
Hughmark, G.A., Designing thermosyphon reboilers, Chem Eng Prog, 65(7), 67-70 (1969).
Kermode, R.I., and Zemaitis, J.F., Experimental analysis of pool boiling from a flat horizontal plate, Brit Chem Eng, 14(5), 675-677 (1969).
Van Gasselt, M.L.G., Boiling and burn-out heat transfer and pressure drop for biphenyl, Brit Chem Eng, 14(5), 679-680 (1969).
Glushchenko, L.F., Generalization of experimental data on critical thermal loadings during boiling with underheating, Int Chem Eng, 10(2), 219-223 (1970).
Michiyoshi, I.; Nakajima, T.; Inoue, M., and Ito, Y., Forced-convection boiling heat transfer to diphenyl in annuli, Int Chem Eng, 10(1), 150-156 (1970).
van Stralen, S.J.D., Boiling paradox in binary liquid mixtures, Chem Eng Sci, 25(1), 149-172 (1970).
Yaglov, V.V.; Gorodov, A.K., and Labuntsov, D.A., Heat transfer during boiling of liquids at reduced pressures under conditions of free flow, Int Chem Eng, 10(4), 607-612 (1970).
Matsui, G., Flow instability in boiling channel systems, Int Chem Eng, 11(3), 554-560 (1971).
Wright, R.D.; Clements, L.D., and Colver, C.P., Nucleate and film pool boiling of ethane-ethylene mixtures, AIChEJ, 17(3), 626-630 (1971).
Ded, J.S., and Lienhard, J.H., Peak pool-boiling heat flux from a sphere, AIChEJ, 18(2), 337-342 (1972).
Kozitskii, V.I., Heat-transfer coefficients during boiling of n-butane on surfaces of various roughnesses, Int Chem Eng, 12(4), 685-687 (1972).
Kutateladze, S.S., and Malenkov, I.G., Vapor compressibility effect on stability criterion of nucleate boiling, Int Chem Eng, 12(1), 81-85 (1972).
Moles, F.D., and Shaw, J.F.G., Boiling heat transfer to sub-cooled liquids under conditions of forced convection, Trans IChemE, 50, 76-84 (1972).
Rice, P., and Calus, W.F., Pool boiling of single component and binary liquid mixtures, Chem Eng Sci, 27(9), 1677-1698 (1972).
Thorngren, J.T., Reboiler computer evaluation, Ind Eng Chem Proc Des Dev, 11(1), 39-43 (1972).
Turner, D.J., Feasibility of a new method of maintaining clean conditions in once-through boilers, J Appl Chem Biotechnol, 22, 983-992 (1972).

1973-1974

Calus, W.F.; di Montegnacco, A., and Denning, R.K., Heat transfer in natural circulation single-tube reboiler, Chem Eng J, 6(3), 233-264 (1973).

Frank, O., and Prickett, R.D., Designing vertical thermosyphon reboilers, Chem Eng, 3 Sept, 107-110 (1973).

Gogonin, I.I., and Svorkova, I.N., Heat transfer during boiling of Freon-21 on finned surfaces, Int Chem Eng, 13(4), 698-701 (1973).

Graham, J.P.; Fulton, J.W.; Kuk, M.S., and Holland, C.D., Predictive methods for determination of vaporization efficiencies, Chem Eng Sci, 28(2), 473-488 (1973).

Haigh, C.P., and Ponter, A.B., Pool-boiling correlation of Rice and Calus, Chem Eng Sci, 28(12), 2265-2267 (1973).

Malek, R.G., Predict nucleate boiling transfer rates, Hyd Proc, 52(2), 89-92 (1973).

Orrell, W.H., Physical considerations in designing vertical thermosyphon reboilers, Chem Eng, 17 Sept, 120-122 (1973).

Roizen, L.I., and Rubin, G.R., Heat transfer during boiling of liquids on finned surface, Int Chem Eng, 13(1), 37-41 (1973).

Sobolev, O.B., and Platonov, V.M., Selection of optimum dimensions of ribbed heat-transfer surfaces in boiling, Int Chem Eng, 13(1), 68-71 (1973).

Ahmadzadeh, J., and Harker, J.H., Evaporation from liquid droplets in free fall, Trans IChemE, 52, 108-111 (1974).

Kirichenko, Y.A., Separation of vapor bubbles during nucleate boiling, Int Chem Eng, 14(2), 265-270 (1974).

Loeb, M.B., From liquid nitrogen to purge gas (vaporization and superheating), Chem Eng, 9 Dec, 134-138 (1974).

Marchenko, A.N.; Solyanik, O.N., and Nerubatskaya, V.D., Thin-film rotary vaporizer with graduated heat-transfer surface, Int Chem Eng, 14(2), 232-234 (1974).

Sarma, N.V.L.S.; Reddy, P.J., and Murti, P.S., Computer design method for vertical thermosyphon reboilers, Ind Eng Chem Proc Des Dev, 12(3), 278-290 (1973); 13(3), 303 (1974).

Seshadri, C.V., and Shanbhag, P.K., Condensation and boiling of binary azeotrope on porous boiling surface, Chem Eng Commns, 1(4), 175-180 (1974).

Smith, J.V., Improving the performance of vertical thermosyphon reboilers, Chem Eng Prog, 70(7), 68-70 (1974).

Vishnev, I.P., General relationships for heat-transfer during boiling of cryogenic liquids, Int Chem Eng, 14(1), 8-15 (1974).

Yue, P.L., and Weber, M.E., Minimum film boiling flux of binary mixtures, Trans IChemE, 52, 217-222 (1974).

1975-1977

Blander, M., and Katz, J.L., Bubble nucleation in liquids, AIChEJ, 21(5), 833-848 (1975).

Golovinskii, G.P. Heat transfer from finned and corrugated tubes to boiling liquids, Int Chem Eng, 15(2), 258-261 (1975).

Zeugin, L.; Donovan, J., and Mesler, R., Microlayer evaporation for three binary mixtures during nucleate boiling, Chem Eng Sci, 30(7), 679-684 (1975).

Collins, G.K., Horizontal thermosyphon reboiler design, Chem Eng, 19 July, 149-152 (1976).
Condiff, D.W., and Epstein, M., Transient volumetric pool boiling, Chem Eng Sci, 31(12), 1139-1162 (1976).
Dwyer, O.E., Growth rates of hemispherical bubbles in nucleate boiling of liquid metals, Chem Eng Sci, 31(3), 187-194 (1976).
Mesler, R., High heat fluxes during nucleate boiling, AIChEJ, 22(2), 246-252 (1976).
Mirzaev, I.T.; Kutepov, A.M., and Rustamov, K.R., Critical heat flux during boiling of aqueous solutions at low pressures, Int Chem Eng, 16(2), 368-370 (1976).
Mohan Rao, P.K., and Andrews, D.G., Effect of heater diameter on critical heat flux from horizontal cylinders in pool boiling, Can JCE, 54, 403-410 (1976).
Tailby, S.R.; Clutterbuck, E.K., and Bakshi, A.S., Heat transfer and pressure drop for vaporisation of water in a vertical tube with particular reference to entrance effects, Trans IChemE, 54, 84-88 (1976).
Barr, W.H.; Strehlitz, F.W., and Dalton, S.M., Modifying large boilers to reduce nitric acid emissions, Chem Eng Prog, 73(7), 59-68 (1977).
Frea, W.J.; Knapp, R., and Taggart, T.D., Flow boiling and pool boiling critical-heat-flux in water and ethylene glycol mixtures, Can JCE, 55, 37-50 (1977).
Helzner, A.E., Operating performance of steam-heated reboilers, Chem Eng, 14 Feb, 73-76 (1977).
Hinchley, P., Waste heat boilers: Problems and solutions, Chem Eng Prog, 73(3), 90-96 (1977).
Mesler, R., and Mailen, G., Nucleate boiling in thin liquid films, AIChEJ, 23(6), 954-957 (1977).
Mesler, R.B., Forced-convection boiling heat transfer, AIChEJ, 23(4), 448-453 (1977).

1978-1980

Ganapathy, V., Short-cut calculation for steam heaters and boilers, Chem Eng, 13 March, 105-106 (1978).
Gordon, E.; Hashemi, M.H.; Dodge, R.D., and La Rosa, J., A versatile steam balance program, Chem Eng Prog, 74(7), 51-56 (1978).
Holden, B.S., and Katz, J.L., Homogeneous nucleation of bubbles in superheated binary liquid mixtures, AIChEJ, 24(2), 260-267 (1978).
Lowry, J.A., Evaluate reboiler fouling, Chem Eng, 13 Feb, 103-106 (1978).
Sandru, E., and Chiriac, F., Ammonia vaporization heat transfer during flow through horizontal pipe systems at low vapor concentrations, Int Chem Eng, 18(4), 692-700 (1978).
Various, Coal-fired boilers (feature report), Chem Eng, 16 Jan, 110-128 (1978).
Hinchley, P., Specifying and operating reliable waste-heat boilers, Chem Eng, 13 Aug, 120-134 (1979).
Johnson, D.L., and Yukawa, Y., Vertical thermosyphon reboilers in vacuum service, Chem Eng Prog, 75(7), 47-52 (1979).
Leong, L.S., and Cornwell, K., Heat transfer coefficients in a reboiler tube bundle, Chem Engnr, April, 219-221 (1979).

Shah, G.C., Troubleshooting reboiler systems, Chem Eng Prog, 75(7), 53-58 (1979).
Volejnik, M., Investigation of industrial reboilers, Int Chem Eng, 19(4), 689-696 (1979).
Bennett, D.L., and Chen, J.C., Forced convective boiling in vertical tubes for saturated pure components and binary mixtures, AIChEJ, 26(3), 454-461 (1980).
Guglielmini, G., and Nannei, E., Note on Kutateladze's equation for partial nucleate boiling, Int J Heat Mass Trans, 23(5), 729-731 (1980).
Kao, Y.K.; Rahrooh, G., and Weisman, J., Transition boiling heat transfer in vertical round tube, Chem Eng Commns, 4(2), 219-236 (1980).
Messina, A.D., and Park, E.L., A new tool for nucleate boiling research, Chem Eng Commns, 4(1), 69-76 (1980).
Prakash, S., and Sirignano, W.A., Theory of convective droplet vaporization with unsteady heat transfer in circulating liquid phase, Int J Heat Mass Trans, 23(3), 253-268 (1980).
Stephan, K., and Abdelsalam, M., Heat-transfer correlations for natural convection boiling, Int J Heat Mass Trans, 23(1), 73-88 (1980).
Taghavi-Tafreshi, K., and Dhir, V.K., Taylor instability in boiling, melting, and condensation or evaporation, Int J Heat Mass Trans, 23(11), 1433-1446 (1980).

1981

Achard, J.L.; Drew, D.A., and Lahey, R.T., Effect of gravity and friction on stability of boiling flow in a channel, Chem Eng Commns, 11(1), 59-80 (1981).
Avedisian, C.T., and Glassman, I., Superheating and boiling of water in hydrocarbons at high pressures, Int J Heat Mass Trans, 24(4), 695-706 (1981).
Chen, C.C.; Loh, J.V., and Westwater, J.W., Prediction of boiling heat transfer duty in a compact plate-fin heat exchanger using the improved local assumption, Int J Heat Mass Trans, 24(12), 1907-1912 (1981).
Cooper, M.G., and Stone, C.R., Boiling of binary mixtures: Study of individual bubbles, Int J Heat Mass Trans, 24(12), 1937-1950 (1981).
Csathy, D., Design of waste-heat boilers, Chem Engnr, April, 159-161 (1981).
Ganapathy, V., Quick estimate of boiling-heat-transfer coefficient, Chem Eng, 4 May, 103-104 (1981).
Gray, K.; Gunn, D.C.; Hopper, N.; Pollard, H., and williams, G., Economizers for modern boilers. J Inst Energy, 54, 151-157 (1981).
Katto, Y., Relation between critical heat flux and outlet flow pattern of forced convection boiling in uniformly heated vertical tubes, Int J Heat Mass Trans, 24(4), 541-544 (1981).
Kenning, D.B.R., and Del Valle, V.H., Fully developed nucleate boiling: Overlap of areas of influence and interference between bubble sites, Int J Heat Mass Trans, 24(6), 1025-1032 (1981).
Klimenko, V.V., Film boiling on a horizontal plate: New correlation, Int J Heat Mass Trans, 24(1), 69-80 (1981).

Mertol, A.; Greif, R., and Zvirin, Y., Transient, steady-state and stability behavior of a thermosyphon with throughflow, Int J Heat Mass Trans, 24(4), 621-634 (1981).

Messina, A.D., and Park, E.L., Effects of precise arrays of pits on nucleate boiling, Int J Heat Mass Trans, 24(1), 141-146 (1981).

Ramamohan, T.R.; Rao, A.S., and Srinivas, N.S., Forced-convection heat transfer to boiling binary mixtures, Can JCE, 59, 400-403 (1981).

Sedler, B., and Mikielewicz, J., Simplified model of the boiling crisis, Int J Heat Mass Trans, 24(3), 431-438 (1981).

Shekriladze, I.G., Developed boiling heat transfer, Int J Heat Mass Trans, 24(5), 795-802 (1981).

Shoukri, M.; Yanchis, R.J., and Rhodes, E., Effect of heat flux on pressure drop in low-pressure flow boiling in a horizontal tube, Can JCE, 59, 149-154 (1981).

Stephan, K., and Auracher, H., Correlations for nucleate-boiling heat transfer in forced convection, Int J Heat Mass Trans, 24(1), 99-108 (1981).

Ueda, T., and Isayama, Y., Critical heat flux and exit film flowrate in a flow boiling system, Int J Heat Mass Trans, 24(7), 1267-1276 (1981).

Ueda, T.; Inoue, M., and Nagatome, S., Critical heat flux and droplet entrainment rate in boiling of falling liquid films, Int J Heat Mass Trans, 24(7), 1257-1266 (1981).

Unal, H.C.; van Gasselt, M.L.G., and van't Verlaat, P.M., Dryout and two-phase flow pressure drop in sodium heated-helically coiled steam-generator tubes at elevated pressures, Int J Heat Mass Trans, 24(2), 285-298 (1981).

1982

Abbott, W.F., Thermosyphon reboiler piping designed by computer, Hyd Proc, March, 127-129 (1982).

Aoki, S.; Inoue, A.; Aritomi, M., and Sakamoto, Y., Experimental study on the boiling phenomena within a narrow gap, Int J Heat Mass Trans, 25(7), 985-990 (1982).

Bau, H.H., and Torrance, K.E., Boiling in low-permeability porous materials, Int J Heat Mass Trans, 25(1), 45-56 (1982).

Bjorge, R.W.; Hall, G.R., and Rohsenow, W.M., Correlation of forced-convection boiling heat-transfer data, Int J Heat Mass Trans, 25(6), 753-758 (1982).

Cornwell, K., and Schuller, R.B., Study of boiling outside a tube bundle using high-speed photography, Int J Heat Mass Trans, 25(5), 683-690 (1982).

Dhar, P.L., and Jain, V.K., Evaluation of correlations for prediction of heat transfer coefficient in nucleate flow boiling, Int J Heat Mass Trans, 25(8), 1250-1252 (1982).

France, D.M.; Chiang, T.; Carlson, R.D., and Priemer, R., Experimental evidence supporting two-mechanism critical boiling heat flux, Int J Heat Mass Trans, 25(5), 691-698 (1982).

Hasan, R., and Rhodes, E., Boiling two-phase flow heat transfer in a horizontal bend, Chem Eng Commns, 18(1), 191-210 (1982).

Kao, T.T.; Cho, S.M., and Pai, D.H., Thermal modeling of steam-generator tubing under CHF-induced temperature oscillations, Int J Heat Mass Trans, 25(6), 781-790 (1982).

Katto, Y., Analytical investigation on CHF of flow boiling in uniformly heated vertical tubes with special reference to governing dimensionless groups, Int J Heat Mass Trans, 25(9), 1353-1362 (1982).

Lazarek, G.M., and Black, S.H., Evaporative heat transfer, pressure drop and critical heat flux in a small vertical tube with R-113, Int J Heat Mass Trans, 25(7), 945-960 (1982).

Morford, P.S., and Messina, A.D., Convective currents in nucleate pool boiling, Chem Eng Commns, 18(1), 149-162 (1982).

Palen, J.W.; Shih, C.C., and Taborek, J., Mist flow in thermosyphon reboilers, Chem Eng Prog, 78(7), 59-61 (1982).

Schuller, R.B., and Cornwell, K., Theoretical model for recirculating flow in a boiler, Chem Eng Commns, 13(4), 271-288 (1982).

Witte, L.C., and Lienhard, J.H., Existence of two 'transition' boiling curves, Int J Heat Mass Trans, 25(6), 771-780 (1982).

1983-1984

Fair, J.R., and Klip, A., Thermal design of horizontal reboilers, Chem Eng Prog, 79(3), 86-96 (1983).

Freedman, L., Boiler treatment for high purity feedwater, Hyd Proc, Jan, 101-104 (1983).

Hahne, E., and Muller, J., Boiling on a finned tube and a finned tube bundle, Int J Heat Mass Trans, 26(6), 849-860 (1983).

Haramura, Y., and Katto, Y., New hydrodynamic model of critical heat flux, applicable widely to both pool and forced convection boiling on submerged bodies in saturated liquids, Int J Heat Mass Trans, 26(3), 389-400 (1983).

Miyasaka, Y.; Inada, S., and Izumi, T., Boiling-characteristic curves in subcooled pool-boiling of water, Int Chem Eng, 23(1), 48-56 (1983).

Schlunder, E.U., Heat transfer in nucleate boiling of mixtures, Int Chem Eng, 23(4), 589-600 (1983).

Thome, J.R., Prediction of binary-mixture boiling heat-transfer coefficients using only phase equilibrium data, Int J Heat Mass Trans, 26(7), 965-974 (1983).

Bier, K.; Schmadl, J., and Gorenflo, D., Effect of heat flux density and boiling pressure on heat transfer in pool boiling of binary mixtures, Int Chem Eng, 24(2), 227-232 (1984).

Ganapathy, V., Size or check waste heat boilers quickly, Hyd Proc, Sept, 169-170 (1984).

Garcia-Borras, T., Chemicals that clean boilers, Chemtech, June, 381-384 (1984).

Popiel, C.O., Heat-transfer coefficient for nuclear pool boiling of chlorine, Chem Eng Commns, 31(1), 185-192 (1984).

Srinivasan, J., and Rao, N.S., Numerical study of heat transfer in laminar film boiling by the finite-difference method, Int J Heat Mass Trans, 27(1), 77-84 (1984).

Wang, B.X., and Shi, D.H., Film boiling in laminar boundary-layer flow along a horizontal plate surface, Int J Heat Mass Trans, 27(7), 1025-1030 (1984).

Yang, Y.M., and Maa, J.R., Dynamic surface effect on boiling of mixtures, Chem Eng Commns, 25(1), 47-62 (1984).

1985-1986

Afgan, N.H.; Jovic, L.A.; Kovalev, S.A., and Lenykov, V.A., Boiling heat transfer from surfaces with porous layers, Int J Heat Mass Trans, 28(2), 415-422 (1985).

Chawla, T.C., and Bingle, J.D., Downward heat transfer in heat-generating boiling pools. Int J Heat Mass Trans, 28(1), 81-90 (1985).

Chowdhury, S.K.R., and Winterton, R.H.S., Surface effects in pool boiling, Int J Heat Mass Trans, 28(10), 1881-1890 (1985).

Del Valle, V.H., and Kenning, D.B.R., Subcooled flow boiling at high heat flux, Int J Heat Mass Trans, 28(10), 1907-1920 (1985).

Derewnicki, K.P., Experimental studies of heat transfer and vapour formation in fast transient boiling, Int J Heat Mass Trans, 28(11), 2085-2092 (1985).

Dhuga, D.S., and Winterton, R.H.S., Measurement of surface contact in transition boiling, Int J Heat Mass Trans, 28(10), 1869-1880 (1985).

Fodemski, T.R., Influence of liquid viscosity and system pressure on stagnation point vapor thickness during forced-convection film boiling, Int J Heat Mass Trans, 28(1), 69-80 (1985).

Hui, T.O., and Thorne, J.R., Study of binary mixture boiling: Boiling site density and subcooled heat transfer, Int J Heat Mass Trans, 28(5), 919-928 (1985).

Sgheiza, J.E., and Myers, J.E., Behavior of nucleation sites in pool boiling, AIChEJ, 31(10), 1605-1613 (1985).

Singh, R.L.; Saini, J.S., and Varma, H.K., Effect of cross-flow on boiling heat transfer of R-12, Int J Heat Mass Trans, 28(2), 512-514 (1985).

Bennett, D.L.; Hertzler, B.L., and Kalb, C.E., Down-flow shell-side forced convective boiling, AIChEJ, 32(12), 1963-1970 (1986).

Foster-Pegg, R.W., Capital cost of gas-turbine heat-recovery boilers, Chem Eng, 21 July, 73-78 (1986).

Gropp, U., and Schlunder, E.U., Influence of liquid-side mass transfer on heat transfer and selectivity during surface and nucleate boiling of liquid mixtures in a falling film, Chem Eng & Proc, 20(2), 103-114 (1986).

Gungor, K.E., and Winterton, R.H.S., General correlation for flow boiling in tubes and annuli, Int J Heat Mass Trans, 29(3), 351-358 (1986).

Hurlbert, A.W., Airflow measurement for boilers, Chem Eng, 20 Jan, 87-88 (1986).

Hwang, T.H., and Yao, S.C., Forced convective boiling in horizontal tube bundles, Int J Heat Mass Trans, 29(5), 785-796 (1986).

Mayinger, F., Boiling: The stabilizer and destabilizer of safe operation, Int Chem Eng, 26(3), 373-387 (1986).

Muller, H., and Steiner, D., Heat transfer at flow boiling of argon within a horizontal tube, Chem Eng & Proc, 20(4), 201-212 (1986).

Nakayama, A., and Koyama, H., Analysis of combined free and forced convection film boiling, AIChEJ, 32(1), 142-148 (1986).

Peterson, E.R., and Maddox, R.N., Vaporizer system smooths flow from gas cylinders, Chem Eng, 29 Sept, 115-118 (1986).

Sheehan, E., Heat-transfer oil boilers, Plant/Opns Prog, 5(4), 193-195 (1986).

Sinai, Y.L., Pool height for transient pool boil-up, Chem Eng Sci, 41(10), 2507-2516 (1986).

St John, B., Economics of atmospheric fluidized-bed boilers, Chem Eng, 8 Dec, 157-159 (1986).

1987-1988

Andrews, P.R., and Cornwell, K.J., Cross-sectional and longitudinal heat transfer variations in a reboiler tube bundle section, CER&D, 65(2), 127-130 (1987).

Gungor, K.E., and Winterton, R.H.S., Simplified general correlation for saturated flow boiling and comparisons of correlations with data, CER&D, 65(2), 148-156 (1987).

Liptak, B.G., Improving boiler efficiency, Chem Eng, 25 May, 49-60 (1987).

Lu, S.M., and Chang, R.H., Pool boiling from a surface with porous layer, AIChEJ, 33(11), 1813-1828 (1987).

Mudawwar, I.A.; Incropera, T.A., and Incropera, F.P., Boiling heat transfer and critical heat flux in liquid films falling on vertically mounted heat sources, Int J Heat Mass Trans, 30(10), 2083-2096 (1987).

Nishio, S., Prediction technique for minimum heat flux point condition of saturated pool boiling, Int J Heat Mass Trans, 30(10), 2045-2058 (1987).

Ross, H.; Radermacher, R.; Marzo, M., and Didion, D., Horizontal flow boiling of pure and mixed refrigerants, Int J Heat Mass Trans, 30(5), 979-992 (1987).

Xin, M.D., and Chao, Y.D., Analysis and experiment of boiling heat transfer on T-shaped finned surfaces, Chem Eng Commns, 50, 185-200 (1987).

Yilmaz, S.B., Horizontal shellside thermosyphon reboilers, Chem Eng Prog, 83(11), 64-70 (1987).

Casarosa, C., and Dobran, F., Experimental investigation and analytical modeling of a closed two-phase thermosyphon with imposed convection boundary conditions, Int J Heat Mass Trans, 31(9), 1815-1834 (1988).

Maracy, M., and Winterton, R.H.S., Hysteresis and contact angle effects in transition pool boiling of water, Int J Heat Mass Trans, 31(7), 1443-1450 (1988).

Muller, H., et al., Effect of dissolved gases on subcooled flow boiling heat transfer, Chem Eng & Proc, 23(2), 115-124 (1988).

Papastathopoulou, H.S., and Luyben, W.L., Design and control of condensate-throttling reboilers, Ind Eng Chem Res, 27(12), 2293-2303 (1988).

Tzan, Y.L., and Yang, Y.M., Pool boiling of binary mixtures, Chem Eng Commns, 66, 71-82 (1988).

17.9 Regenerative Heat Exchangers

1968-1979

Ramaswamy, E.R., and Gerhard, E.R., Performance estimate of a granular solids heater, Brit Chem Eng, 13(12), 1722-1725 (1968).

Jouhari, A.K., and Dey, D.N., Nomogram for cooling granular materials in beds, Brit Chem Eng, 17(2), 171 (1972).

Roy, D., and Gidaspow, D., Green's matrix representation of a crossflow regenerator, Chem Eng Sci, 27(4), 779-794 (1972).

Tanasawa, I., Thermal characteristics of storage-type heat exchangers, Int Chem Eng, 12(1), 143-162 (1972).
Balakrishnan, A.R., and Pei, D.C.T., Heat transfer in fixed beds, Ind Eng Chem Proc Des Dev, 13(4), 441-446 (1974).
El-Rifai, M.A., and Taymour, N.E., Temperature transients in fixed-bed heat regenerators, Chem Eng Sci, 29(8), 1687-1694 (1974).
Roy, D., and Gidaspow, D., Nonlinear coupled heat and mass exchange in crossflow regenerator, Chem Eng Sci, 29(10), 2101-2114 (1974).
Heggs, P.J., and Carpenter, K.J., The effect of fluid hold-up on the effectiveness of contraflow regenerators, Trans IChemE, 54, 232-238 (1976).
Li, C.H., and Finlayson, B.A., Heat transfer in packed beds, Chem Eng Sci, 32(9), 1055-1066 (1977).
Heggs, P.J., and Carpenter, K.J., Prediction of a dividing line between conduction and convection effects in regenerator design, Trans IChemE, 56, 86-90 (1978).
Balakrishnan, A.R., and Pei, D.C.T., Heat transfer in gas-solid packed bed systems, Ind Eng Chem Proc Des Dev, 18(1), 30-50, (1979).
Heggs, P.J., and Carpenter, K.J., A modification of the thermal regenerator infinite conduction model to predict the effects of intraconduction, Trans IChemE, 57, 228-236 (1979).

1980-1988

Heggs, P.J.; Bansal, L.S.; Bond, R.S., and Vazakas, V., Thermal regenerator design charts including intraconduction effects, Trans IChemE, 58, 265-270 (1980).
Willmott, A.J., and Duggan, R.C., Refined closed methods for contra-flow thermal regenerator problem, Int J Heat Mass Trans, 23(5), 655-662 (1980).
Gandhidasan, P.; Sriramulu, V., and Gupta, M.C., Heat and mass transfer in a solar regenerator, Chem Eng J, 21(1), 59-64 (1981).
Hinchcliffe, C., and Willmott, A.J., Lumped heat-transfer coefficients for thermal regenerators, Int J Heat Mass Trans, 24(7), 1229-1236 (1981).
Levenspiel, O., Design of long heat-regenerators by the dispersion model, Chem Eng Sci, 38(12), 2035-2046 (1983).
Ramachandran, P.A., and Dudukovic, M.P., Solution by triple collocation for periodic operation of heat regenerators, Comput Chem Eng, 8(6), 377-388 (1984).
Dudukovic, M.P., and Ramachandran, P.A., Evaluation of thermal efficiency of heat regenerators in periodic operation by approximate methods, Chem Eng Sci, 40(9), 1629-1640 (1985).
Dudukovic, M.P., and Ramachandran, P.A., Quick design and evaluation of heat regenerators, Chem Eng, 10 June, 63-72 (1985).
Delaunay, D.; Bransier, J., and Bardon, J.F., Modeling of cyclical latent heat storage, Int Chem Eng, 26(1), 82-90 (1986).
Gunn, D.J.; Ahmad, M.M., and Sabri, M.N., Radial heat transfer to fixed beds of particles, Chem Eng Sci, 42(9), 2163-2172 (1987).
Hill, A., and Willmott, A.J., A robust method for regenerative heat exchanger calculations, Int J Heat Mass Trans, 30(2), 241-250 (1987).

Krane, R.J., A Second Law analysis of the optimum design and operation of thermal energy storage systems, Int J Heat Mass Trans, 30(1), 43-58 (1987).
Voigt, B.; Mierke, T.; Ruhnau, R., and Moller, K., Heat storage: A materials problem, Chemtech, Dec, 746-749 (1987).
Atthey, D.R., Approximate thermal analysis for a regenerative heat exchanger, Int J Heat Mass Trans, 31(7), 1431-1442 (1988).
Baclic, B.S., Misinterpretations of the diabatic regenerator performances, Int J Heat Mass Trans, 31(8), 1605-1612 (1988).
Sklivaniotis, M.; Castro, J.A.A., and McGreavy, C., Characteristic features of parametric sensitivity in a fixed-bed heat exchanger, Chem Eng Sci, 43(7), 1517-1522 (1988).

17.10 Fired Heaters and Furnaces

1966-1977

Ellwood, P., and Danatos, S., Process furnaces, Chem Eng, 11 April, 151-174 (1966).
Roesler, F.C., Theory of radiative heat transfer in cocurrent tube furnaces, Chem Eng Sci, 22(10), 1325-1336 (1967).
Kuong, J., Nomogram for volumetric heat-transfer coefficient for kilns, Brit Chem Eng, 15(1), 109 (1970).
Von Wiesenthal, P., and Cooper, H.W., Guide to economics of fired heater design, Chem Eng, 6 April, 104-112 (1970).
Lucas, D.M., and Toth, H.E., Calculating the performance of a gas-fired coiled-tube air heater, Brit Chem Eng, 16(7), 598-600 (1971).
Sastri, S.R.S., and Narasimhan, K.S., Nomogram for heat loss through vertical furnace walls, Brit Chem Eng, 16(12), 1155 (1971).
Lucas, D.M., and Toth, H.E., Calculation of heat transfer in fire tubes of shell boilers, J Inst Fuel, 45, 521-527 (1972).
Sisson, W., Determining the air needed for combustion, Chem Eng, 10 July, 109 (1972).
Crisi, J.S., and Smith, G.R., How to expand fired heater tubes, Hyd Proc, 53(5), 123-126 (1974).
Lobo, W.E., Design of furnaces with flue gas temperature gradients, Chem Eng Prog, 70(1), 65-71 (1974).
Roffel, B., and Rijnsdorp, J.E., Dynamics and control of gas-liquid furnace, Chem Eng Sci, 29(10), 2083-2092 (1974).
Santoleri, J.J., Furnace spray nozzle selection, Chem Eng Prog, 70(9), 84-88 (1974).
Sisson, W., Combustion calculations for operators, Chem Eng, 10 June, 106-108 (1974).
Sissons, W., Operator calculations for reforming-furnace steam flows, Chem Eng, 30 Sept, 150-152 (1974).
Selcuk, N.; Siddall, R.G., and Beer, J.M., Comparison of mathematical models of radiative behaviour of an industrial heater, Chem Eng Sci, 30(8), 871-876 (1975).
Ganapathy, V., Holdup times in rotary kilns, Chem Eng, 20 Dec, 83 (1976).

Perara, W.G., and Rafique, K., Coking in a fired heater, Chem Engnr, Feb, 107-111 (1976).
Johnson, E.R., Firing oil on an ammonia reformer, Chem Eng Prog, 73(3), 97-99 (1977).
Lihou, D.A., Review of furnace design methods, Trans IChemE, 55, 225-242 (1977).

1978-1981

Bailey, T., and Wall, F.M., Ethylene furnace design, Chem Eng Prog, 74(7), 45-50 (1978).
Berman, H.L., Fired heaters, Chem Eng, 19 June, 98-104; 31 July, 87-96; 14 Aug, 129-140; 11 Sept, 165-169 (1978).
Griffin, J.J.; Kersey, B.R., and Eaton, S.J., Forced-draft firing in refinery heaters, Chem Eng Prog, 74(7), 57-61 (1978).
Saddler, J.W., Corrosion-resistant heater design, Chem Eng Prog, 74(1), 53-54 (1978).
Wimpress, N., Generalized method predicts fired-heater performance, Chem Eng, 22 May, 95-102 (1978).
Abdel-Halim, T.A., Program predicts radiant heat flux in direct-fired heaters, Chem Eng, 17 Dec, 87-91 (1979).
Tan, S.H., Combustion air requirements, Chem Eng, 27 Aug, 117-118 (1979).
Tscheng, S.H., and Watkinson, A.P., Convective heat transfer in a rotary kiln, Can JCE, 57, 433-533 (1979).
Gilbert, L.F., CO control heightens furnace efficiency, Chem Eng, 28 July, 69-73 (1980).
Mori, Y.; Yamada, Y., and Hijikata, K., Radiation effects on performances of radiative-gas heat exchangers at high temperatures, Int J Heat Mass Trans, 23(8), 1079-1090 (1980).
Neal, S.B.H.C.; Northover, E.W., and Preece, R.J., Measurement of radiant heat flux in large boiler furnaces, Int J Heat Mass Trans, 23(7), 1015-1032 (1980).
Vercammen, H.A.J., and Froment, G.F., Improved zone method using Monte Carlo techniques for simulation of radiation in industrial furnaces, Int J Heat Mass Trans, 23(3), 329-338 (1980).
Vijayaraghavan, D., Water formed during coal combustion, Chem Eng, 30 June, 140-142 (1980).
Antony, S.M., Check excess air for combustion quickly, Chem Eng, 14 Dec, 107-108 (1981).
Frankel, I., Shortcut calculation for fluegas volume, Chem Eng, 1 June, 88-89 (1981).
Guruz, H.K., and Bac, N., Mathematical modelling of rotary cement kilns by the zone method, Can JCE, 59, 540-548 (1981).
Jenkins, B.G., and Moles, F.D., Modelling of heat transfer from a large enclosed flame in a rotary kiln, Trans IChemE, 59, 17-25 (1981).

1982-1984

Arora, V.K., Thermal balance for direct-fired heaters, Chem Eng, 15 Nov, 152 (1982).

Ganapathy, V., Chart to estimate furnace parameters, Hyd Proc, Feb, 106-107 (1982).
Vancini, C.A., Program calculates flame temperature, Chem Eng, 22 March, 133-136 (1982).
Ganapathy, V., Quick estimate of heat loss through dampers, Proc Engng, Feb, 53 (1983).
Hottel, H.C., Flux distribution around tubes in radiant section of processing furnaces, Ind Eng Chem Fund, 22(2), 153-163 (1983).
Kumar, M., and Limpe, A.T., Design heaters for coal slurry flows, Hyd Proc, Nov, 207-209 (1983).
Lovejoy, G.R., and Clark, I.M., Furnace safety systems, Plant/Opns Prog, 2(1), 13-21 (1983).
Scholand, E., Calculation of radiant heat transfer in direct-fired tubed furnaces, Int Chem Eng, 23(4), 600-611 (1983).
Solomon, J., Process heating, Chem Engnr, April, 24-26 (1983).
Swanson, K., An advanced combustion control system, Plant/Opns Prog, 2(1), 30-34 (1983).
van Dongen, F.G., Heat transfer in gas-fired furnaces, J Inst Energy, 56, 184-189 (1983).
Yonezawa, M.; Amano, T.; Maruta, T., and Wall, F., New radiant coil furnace technology, Chem Eng Prog, 79(12), 50-55 (1983).
Lefebvre, A.H., Flame radiation in gas turbine combustion chambers, Int J Heat Mass Trans, 27(9), 1493-1510 (1984).
Nutcher, P.B., Forced-draft low-NOx burners applied to process fired heaters, Plant/Opns Prog, 3(3), 168-173 (1984).

1985-1988

Arora, V.K., Check fired heater performance, Hyd Proc, May, 85-87 (1985).
de Salis, C., Flow measurement in fired heaters, Chem Engnr, Oct, 49-51 (1985).
Gomes, V.G., Controlling fired heaters, Chem Eng, 7 Jan, 63-68 (1985).
Nogay, R., and Prasad, A., Better design method for fired heaters, Hyd Proc, Nov, 91-95 (1985).
Docherty, P., and Tucker, R.J., The influence of wall emissivity on furnace performance, J Inst Energy, March, 35-37 (1986).
Mahajani, V.V.; Kamat, A.B., and Mokashi, S.M., Recovering heat in fired heaters, Chem Eng, 18 Aug, 91-95 (1986).
Hayhurst, A.N., and Nedderman, R.M., The burning of a liquid oil droplet, Chem Eng Educ, 21(3), 126-129, 149-152 (1987).
Kurz, G., and Guthoff, H., Heat transfer with submerged combustors, Proc Engng, Jan, 31-33 (1987).
Marks, W.L., Coal-fired industrial steam boilers, J Inst Energy, March, 8-14 (1987).
Carvalho, M.G.; Oliveira, P., and Semiao, V., Three-dimensional modelling of an industrial glass furnace, J Inst Energy, 61, 143-156 (1988).
Dugwell, D.R., and Oakley, D.E., Model of heat transfer in tunnel kilns used for firing refractories, Int J Heat Mass Trans, 31(11), 2381-2390 (1988).
Garg, A., and Ghosh, H., Fired-heater design specifications, Chem Eng, 18 July, 77-80 (1988).

Yung, C.N.; Keith, T.G., and de Witt, K.J., Numerical prediction of cold turbulent flow in combustor configurations with different centerbody flame holders, Chem Eng Commns, 65, 61-78 (1988).

CHAPTER 18

REACTORS

18.1 Catalysis

1967-1972

Krasilnikova, M.K., and Topchieva, K.V., Catalytic properties of yttrium oxide, Int Chem Eng, 7(1), 40-45 (1967).

Rozhdestvenskaya, Z.B., et al., Phase composition of Raney nickel-molybdenum catalyst, Int Chem Eng, 7(1), 9-14 (1967).

Narsimhan, G.; Brahme, P.H., and Sengupta, D., Effects of added magnesium on activity of Raney nickel catalyst, Brit Chem Eng, 13(8), 1141-1142 (1968).

Shiba, T., Catalysis in organic synthesis: A review, Int Chem Eng, 8(3), 556-563 (1968).

Various, Engineering aspects of catalysis (feature report), Chem Eng, 29 July, 126-142 (1968).

Habermehl, R., Optimizing industrial catalysts, Chem Eng Prog, 65(10), 55-58 (1969).

Mukhlenov, I.P., Catalytic processes in a fluidized catalyst bed, Int Chem Eng, 9(1), 37-42 (1969).

Prettre, M., Review of contact catalysis, Int Chem Eng, 9(1), 107-118 (1969).

Henningsen, J., and Nielson, M.B., Catalytic reforming, Brit Chem Eng, 15(11), 1433-1436 (1970).

Zeis, L.A., and Heinz, E., Catalyst tubes in primary reformer furnaces, Chem Eng Prog, 66(7), 68-73 (1970).

Farkas, A., Homogeneous catalysis, Hyd Proc, 50(5), 137-140; 50(6), 137-142 (1971).

Heinemann, H., Homogeneous and heterogeneous catalysis, Chemtech, May, 286-291 (1971).

Shumskii, V.M., Static model of catalytic cracking, Int Chem Eng, 11(1), 64-66 (1971).

Thomas, W.J., The use of bifunctional catalysts in packed bed tubular reactors, Trans IChemE, 49, 204-214 (1971).

Various, Catalytic reforming (special report), Hyd Proc, 50(2), 85-102 (1971).

Venuto, P.B., Zeolite cracking catalysts, Chemtech, April, 215-224 (1971).

Gussow, S.; Higginson, G.W., and Schwint, I.A., Cracking with new zeolite catalyst, Hyd Proc, 51(6), 116-120 (1972).

Koches, C.F., and Smith, S.B., Reactivate powdered carbon, Chem Eng, 1 May, 46-47 (1972).

1973-1975

Hughes, R., and Shettigar, U.R., Regeneration of coked catalysts, Trans IChemE, 51, 192-198 (1973).

Jerocki, B., et al., Catalytic activity of molecular sieves, Int Chem Eng, 13(2), 201-206 (1973).

Pittman, C.U., and Evans, G.O., Polymer-bound catalysts and reagents, Chemtech, Sept, 560-566 (1973).

Richardson, J.T., SNG catalyst technology, Hyd Proc, 52(12), 91-95 (1973).

Sampath, B.S., and Hughes, R., Regeneration of beds of coked catalyst particles, Proc Tech Int, Jan, 39-43 (1973).

Sinfelt, J.H., Catalytic specificity (review paper), AIChEJ, 19(4), 673-683 (1973).
Trimm, D.L., Design and development of industrial catalysts, Chem & Ind, 3 Nov, 1012-1018 (1973).
Various, Catalysts in industry (topic issue), Chem Eng Prog, 69(5), 59-76 (1973).
Weisz, P.B., Zeolites: New horizons in catalysis, Chemtech, Aug, 498-505 (1973).
Burns, R.A.; Lippert, R.B., and Kerr, R.K., Choose catalyst objectively for Claus process, Hyd Proc, 53(11), 181-186 (1974).
Higginson, G.W., Making catalysts: An overview, Chem Eng, 30 Sept, 98-104 (1974).
Sampath, B.S., and Hughes, R., Regeneration of coked catalysts, Chemical Processing, April, 9-12; May, 13-17 (1974).
Andrews, S.P.S., Catalysts for the production of ammonia, Chem Engnr, Nov, 664-666 (1975).
Dehmlow, E.V., Phase-transfer catalysis, Chemtech, April, 210-218 (1975).
Michalska, Z.M., and Webster, D.E., Supported homogeneous catalysts, Chemtech, Feb, 117-122 (1975).
Reis, T., Coking, desulfurization and calcination, Hyd Proc, 54(4), 145-150; 54(6), 97-104 (1975).
Sampath, B.S.; Ramachandran, P.A., and Hughes, R., A comparison of models for the non-isothermal regeneration of coked catalyst pellets, Trans IChemE, 53, 234-241 (1975).
Various, Catalyst testing (topic issue), Chem Eng Prog, 71(1), 31-38, 44-47 (1975).

1976-1979

Andrew, S.P.S., The development of a non-precious metal based catalyst for NOx removal from automobile exhaust, Trans IChemE, 54, 196-198 (1976).
Bayer, E., and Schurig, V., New class of catalysts, Chemtech, March, 212-214 (1976).
Cromeans, J.S., and Fleming, H.W., Improved ammonia plant catalysts, Chem Eng Prog, 72(2), 56-58 (1976).
England, R., and Thomas, W.J., The significance and measurement of surface diffusion coefficients in catalysis, Trans IChemE, 54, 115-118 (1976).
Katzman, H.; Pandolfi, L.; Pedersen, L.A., and Libby, W.F., Lead-tolerant auto exhaust catalysts, Chemtech, June, 369-371 (1976).
Pierson, W.R., Sulfuric acid generation by automotive catalysts, Chemtech, May, 332-337 (1976).
Trimm, D.L., The deactivation and reactivation of nickel-based catalysts, Trans IChemE, 54, 119-123 (1976).
Baker, R.T.K., Catalysts in action, Chem Eng Prog, 73(4), 97-99 (1977).
Clarke, A.; Lloydlangston, J., and Thomas, W.J., An experimental investigation of a copper catalyst for the liquid phase hydrogenation of olefins, Trans IChemE, 55, 93-103 (1977).
Ford, W.D.; Reineman, R.C.; Vasalos, I.A., and Fahrig, R.J., Operating catalytic crackers for maximum profit, Chem Eng Prog, 73(4), 92-96 (1977).

Gold, R.F., et al., Testing olefin-polymerization catalysts, Chem Eng, 31 Jan, 119-121 (1977).
Boudart, M., Kinetics and catalysis, Chemtech, April, 230-234 (1978).
Farkas, A., What's new in catalysis, Hyd Proc, 57(5), 213-218 (1978).
Haensel, V., Catalytic concepts, Chemtech, July, 414-416 (1978).
Andrew, S.P.S., Designing and making catalysts, Chemtech, March, 180-184 (1979).
Lundberg, W.C., Extending catalyst life, Chem Eng Prog, 75(6), 81-86 (1979).
Sermon, P.A., Catalysis in the petrochemical economy, Chem & Ind, 20 Oct, 681-690 (1979).
Trimm, D.L., Designing catalysts, Chemtech, Sept, 571-577 (1979).

1980-1981

Delmon, B., Catalysts: New mechanistic model explaining synergy in hydrotreating catalysts, Int Chem Eng, 20(4), 639-642 (1980).
Gent, C.W., and Ward, S.A., Catalysts in coal-based ammonia plants, Chem Engnr, Feb, 85-90, 94 (1980).
Goodman, D.R., The economics of catalyst operation, Chem Engnr, Feb, 91-94 (1980).
Oudar, J., Sulfur adsorption and poisoning of metallic catalysts, Catalysis Reviews, 22(2), 171-196 (1980).
Starks, C., Selecting a phase-transfer catalyst, Chemtech, Feb, 110-117 (1980).
Various, Catalysts in use (topic issue), Chem Eng Prog, 76(6), 35-64 (1980).
Hegedus, L.L., and McCabe, R.W., Catalyst poisoning, Catalysis Reviews, 23(3), 377-476 (1981).
Henrici-Olive, G., and Olive, S., Mechanism for Ziegler-Natta catalysis, Chemtech, Dec, 746-752 (1981).
Lewis, D.A., and Kenney, C.N., Niobium disulphide as an isomerisation and hydrogenation catalyst in presence of hydrogen sulphide, Trans IChemE, 59, 186-195 (1981).
Reuben, B., and Sjoberg, K., Phase-transfer catalysis in industry, Chemtech, May, 315-320 (1981).
Satterfield, C.N., Industrial heterogeneous catalysis: An overview, Chemtech, Oct, 618-624 (1981).
Upson, L.L., Evaluating equilibrium catalyst tests, Hyd Proc, 60(11), 253-258 (1981).
Various, Papers from 1st Berkeley conference 'Catalysis and Surface Science', topics: Ammonia synthesis; Homogeneous Catalysis by Transition Metals; Ethylene Oxidation; Hydrogenation of Carbon Monoxide; Catalysis Reviews, 23(1/2), 1-328 (1981).
Zanker, A., Best catalyst regeneration cycle, Hyd Proc, 60(8), 95-97 (1981).

1982-1983

Corma, A., and Wojciechowski, B.W., Cracking catalyst design, Can JCE, 60, 11-16 (1982).
Dupin, T., and Voirin, R., Catalyst enhances Claus operations, Hyd Proc, Nov, 189-191 (1982).
Haggin, J.., Designing industrial catalysts, C&E News, 15 Nov, 11-17 (1982).
Herrington, D.R., Amphora catalysts, Chemtech, Jan, 42-47 (1982).

Jacobs, P.A., Acid zeolites: An attempt to develop unifying concepts, Catalysis Reviews, 24(3), 415-440 (1982).
Tucci, E.R., Use catalytic combustion for low-heating value gases, Hyd Proc, March, 159-166 (1982).
Various, Symposium on catalysis, Can JCE, 60, 1-60 (1982).
Whyte, T.E., and Dalla Betta, R.A., Zeolite advances in the chemical and fuel industries: A technical perspective, Catalysis Reviews, 24(4), 567-598 (1982).
Wohlfarht, K., Design of catalyst pellets, Chem Eng Sci, 37(2), 283-290 (1982).
Wolf, E.E., and Alfani, F., Catalysts deactivation by coking, Catalysis Reviews, 24(3), 329-372 (1982).
Palluzi, R.P.,and Matty, L., High-accuracy mass balancing in catalyst screening, Chem Eng Prog, 79(8), 41-45 (1983).
Randhava, R.; Goltermann, D.S., and Treworgy, E.D., Advanced configurations for catalyst research, Chem Eng Prog, 79(11), 52-58 (1983).
Rounthwaite, D.P., Improved ammonia plant catalysts, Plant/Opns Prog, 2(2), 127-131 (1983).
Sachtler, W.M.H., What makes a catalyst selective? Chemtech, July, 434-447 (1983).
Sokolsky, D.V., New hydrogenating catalysts for commercially important reactions, Periodica Polytechnica, 27, 37-48 (1983).
Somorjai, G.A., and Davis, S.M., Surface science of heterogeneous catalysis, Chemtech, Aug, 502-511 (1983).
Various, Papers from the 8th Canadian Symposium on Catalysis, Can JCE, 61(1), 21-49 (1983).
Various, Fluid catalytic cracking (special report), Hyd Proc, Feb, 35-42 (1983).

1984-1985

Bartholomew, C.H., Catalyst deactivation, Chem Eng, 12 Nov, 96-112 (1984).
Jahnig, C.E.; Martin, H.Z., and Campbell, D.L., Development of fluid catalytic cracking, Chemtech, Feb, 106-112 (1984).
Orchin, M., Homogeneous catalysis: Theory and experiment, Catalysis Reviews, 26(1), 59-80 (1984).
Oyekunle, L.O., and Hughes, R., Metal deposition in residuum hydrodesulphurisation catalysts, CER&D, 62, 339-343 (1984).
Plank, C.J., Invention of zeolite cracking catalysts, Chemtech, April, 243-249 (1984).
Prasad, R.; Kennedy, L.A., and Ruckenstein, E., Catalytic combustion, Catalysis Reviews, 26(1), 1-58 (1984).
Sinfelt, J.H., Reflections on catalysis, Chem & Ind, 4 June, 403-406 (1984).
Tesoriero, A., Predict particle size distribution from fluid catalytic cracking beds, Hyd Proc, Nov, 139-141 (1984).
Various, Papers from 2nd Berkeley 'Catalysis and Surface Science' Conference: Chemicals from Methanol; Hydrotreating of Hydrocarbons; Catalyst Preparation; Monomers and Polymers; Photocatalysis; Catalysis Reviews, 26(3/4), 300-728 (1984).
Wells, P., Progress in heterogeneous catalysis, Proc Engng, July, 25-28 (1984).
Boitiaux, J.P.; Cosyns,, J.; Derrien, M., and Leger, G., Newest hydrogenation catalysts, Hyd Proc, March, 51-59 (1985).

Carberry, J.J.; Tipnis, P., and Schmitz, R., Generate meaningful catalytic kinetics, Chemtech, May, 316-319 (1985).

Corma, A., and Wojciechowski, B.W., The chemistry of catalytic cracking, Catalysis Reviews, 27(1), 29-150 (1985).

Dewing, J.; Spencer, M.S., and Whittam, T.V., Synthesis, characterization and catalytic properties of Nu-1, Fu-1, and related zeolites, Catalysis Reviews, 27(3), 461-514 (1985).

Grove, D., Problems with homogeneous catalysis, Proc Engng, Nov, 89-90 (1985).

Kittrell, J.R.; Tam, P.S., and Eldridge, J.W., Predict catalyst poisoning rate, Hyd Proc, Aug, 63-67 (1985).

Komiyama, M., Design and preparation of impregnated catalysts, Catalysis Reviews, 27(2), 341-372 (1985).

Lachman, I.M., and McNally, R.N., Monolithic honeycomb supports for catalysis, Chem Eng Prog, 81(1), 29-31 (1985).

Martin, C., and Rives, V., Textural properties of vanadium/titanium oxides catalysts obtained from vanadium oxychloride, Ads Sci Tech, 2(4), 241-252 (1985).

Wachs, I.E.; Saleh, R.Y.; Chan, S.S., and Chersich, C., Supporting the catalyst, Chemtech, Dec, 756-761 (1985).

1986-1987

Brunovska, A., et al., Modelling of catalyst pellet deactivation: An irreversible chemisorption model, Ads Sci Tech, 3(3), 201-216 (1986).

Chen, N.Y., and Garwood, W.E., Industrial application of shape-selective catalysts, Catalysis Reviews, 28(2/3), 185-264 (1986).

Doughty, P.W.; Harrison, G., and Lawson, G.J., Hydrocracking of a coal extract with various catalysts, Fuel, 65(7), 937-944 (1986).

Hutchings, G.J., Crushing strength as a diagnostic test for catalysts, J Chem Tech Biotechnol, 36(6), 255-258 (1986).

Stinson, S.C., Process catalysts, C&E News, 17 Feb, 27-50 (1986).

Waller, F.J., Catalysis with metal cation-exchanged resins, Catalysis Reviews, 28(1), 1-12 (1986).

Biswas, J., and Do, D.D., A unified theory of coking deactivation in a catalyst pellet, Chem Eng J, 36(3), 175-192 (1987).

Burwell, R.L., Catalysis: A retrospective, Chemtech, Oct, 586-592 (1987).

Dougherty, R.C., and Verykios, X.E., Nonuniformly activated catalysts, Catalysis Reviews, 29(1), 101-150 (1987).

Fulton, J.W., Catalyst engineering (7 part feature report), Chem Eng, 17 Feb, 118-124; 12 May, 97-101; 7 July, 59-63; 13 Oct, 71-77 (1986); 19 Jan, 107-108; 11 May, 59-60; 14 Sept, 99-101 (1987).

Matsumoto, D.K., and Satterfield, C.N., Effects of poisoning a fused magnetite Fischer-Tropsch catalyst with dibenzothiophene, Energy & Fuels, 1(2), 203-211 (1987).

Matyi, R.J.; Schwartz, L.H., and Butt, J.B., Particle size, particle size distribution, and related measurements of supported metal catalysts, Catalysis Reviews, 29(1), 41-100 (1987).

Pawlicki, P.C., and Schmitz, R.A., Spatial effects on supported catalysts, Chem Eng Prog, 83(2), 40-45 (1987).

Pfefferle, L.D., and Pfefferle, W.C., Catalysis in combustion, Catalysis Reviews, 29(2/3), 219-267 (1987).
Takasaki, Y.; Komatsu, K., and Tachikawa, N., Kinetic method of evaluating a catalyst, Int Chem Eng, 27(4), 678-686 (1987).
Yee, H.A.; Palmer, H.J., and Chen, S.H., Solid-liquid phase transfer catalysis, Chem Eng Prog, 83(2), 33-39 (1987).
Zielinski, J.M., and Petersen, E.E., Monte Carlo simulation of diffusion and chemical reaction in catalyst pores, AIChEJ, 33(12), 1993-1997 (1987).

1988

Ayo, D.B., and Susu, A.A., Analysis of pulse microcatalytic technique for catalyst with variable activity, Chem Eng Commns, 70, 1-14 (1988).
Bhatia, S.K., Steady state multiplicity and partial internal wetting of catalyst particles, AIChEJ, 34(6), 969-979 (1988).
Biswas, J.; Bickle, G.M.; Gray, P.G.; Do, D.D., and Barbier, J., Role of deposited poisons and crystallite surface structure in activity and selectivity of reforming catalysts, Catalysis Reviews, 30(2), 161-248 (1988).
Campbell, W.G.; Lynch, D.T., and Wanke, S.E., Monte Carlo simulation of supported metal catalyst sintering and redispersion, AIChEJ, 34(9), 1528-1538 (1988).
Carberry, J.J., The contributions of heterogeneous catalysis to catalytic reaction engineering, Chem Eng Prog, 84(2), 51-60 (1988).
Cavaterra, E., Catalysts to make dichloroethane, Hyd Proc, 67(12), 63-67 (1988).
Felder, W.; Madronich, S., and Olson, D.B., Oxidation kinetics of carbon blacks over 1300-1700K, Energy & Fuels, 2(6), 743-750 (1988).
Funk, G.A.; Harold, M.P., and Ng, K.M., Effectiveness of partially wetted catalyst for bimolecular reaction kinetics, AIChEJ, 34(8), 1361-1366 (1988).
Grau, R.J.; Cassano, A.E., and Baltanas, M.A., Catalysts and network modelling in vegetable oil hydrogenation processes, Catalysis Reviews, 30(1), 1-48 (1988).
Haggin, J., Molecular sieves as cracking catalysts, C&E News, 21 March, 22, 24 (1988).
Harold, M.P., Partially wetted catalyst performance in consecutive-parallel network, AIChEJ, 34(6), 980-995 (1988).
Haynes, H.W., Experimental evaluation of catalyst effective diffusivity, Catalysis Reviews, 30(4), 563-628 (1988).
Hirschon, A.S., and Laine, R.M., Bulk ruthenium as an HDN catalyst, Energy & Fuels, 2(3), 292-295 (1988).
Ho., T.C., Hydrodenitrogenation catalysis, Catalysis Reviews, 30(1), 117-160 (1988).
Irandoust, S., and Andersson, B., Monolithic catalysts for nonautomobile applications, Catalysis Reviews, 30(3), 341-392 (1988).
Leary, K.J.; Michaels, J.N., and Stacey, A.M., Temperature programmed catalyst desorption: Multisite and subsurface diffusion models, AIChEJ, 34(2), 263-271 (1988).

Limbach, K.W., and Wei, J., Effect of nonuniform activity on hydrometallation catalyst, AIChEJ, 34(2), 305-313 (1988).
Lindner, D.; Werner, M., and Schumpe, A., Hydrogen transfer in slurries of carbon supported catalyst (HPO process), AIChEJ, 34(10), 1691-1697 (1988).
Mallat, T., and Petro, J., Electrochemical studies on noble metal alloy catalysts, Periodica Polytechnica, 32(1), 187-192 (1988).
Martin, G.A., Quantitative approach to ensemble model of catalysis by metals, Catalysis Reviews, 30(4), 519-562 (1988).
Meisel, S.L., Catalysis research for the methanol-to-gasoline process, Chemtech, Jan, 32-37 (1988).
Parshall, G.W., and Nugent, W.A., Homogeneous catalysis for agrochemicals, flavors, and fragrances, Chemtech, June, 376-383 (1988).
Parshall, G.W., and Nugent, W.A., Making pharmaceuticals via homogeneous catalysis, Chemtech, March, 184-190; May, 314-320 (1988).
Polyanszky, E.; Mallat, T., and Petro, J., Electrochemical investigation of metal catalysts in liquid phase, Periodica Polytechnica, 32(1), 193-197 (1988).
Reyes, S.C., and Ho, T.C., Heat effects in gas sulfiding of hydroprocessing catalysts, AIChEJ, 34(2), 314-320 (1988).
Thomas, W.J., and Ullah, U., Effect of intraparticle diffusion on the catalytic selectivity of ethanol dehydration, CER&D, 66(2), 138-146 (1988).
Tsotsis, T.T.; Sane, R.C., and Lindstrom, T.H., Bifurcation behavior of catalyst reaction due to slowly varying parameter, AIChEJ, 34(3), 383-388 (1988).
Tungler, A., et al., Complex studies on industrial nickel catalyst, Periodica Polytechnica, 32(1), 175-180 (1988).

18.2 Reaction Engineering and Kinetics

1966-1969

Kittrell, J.R., and Mezaki, R., Obtaining reaction rates from integral reactor data, Brit Chem Eng, 11(12), 1538-1539 (1966).
Chaplits, D.N., et al., Dehydration of tert-butyl alcohol in novel reactor, Int Chem Eng, 7(4), 672-676 (1967).
Frye, C.G., and Mosby, J.F., Kinetics of hydrodesulfurization, Chem Eng Prog, 63(9), 66-70 (1967).
Huang, I.D., Two-point plot proves key to kinetic relations, Chem Eng, 13 Feb, 135-139 (1967).
Metcalfe, T.B., Kinetics of coke combustion in catalyst regeneration, Brit Chem Eng, 12(3), 388-389 (1967).
Nikolaevskii, A.N., and Kucher, R.V., Liquid-phase oxidation of propylene-butene mixtures, Int Chem Eng, 7(2), 207-212 (1967).
Yantovskii, S.A., et al., Conditions of safe oxidation of toluene by atmospheric oxygen, Int Chem Eng, 7(1), 144-149 (1967).
Brisk, M.L.; Day, R.L.; Jones, M., and Warren, J.B., Development of a stirred gas-solid reactor for the measurement of catalyst kinetics, Trans IChemE, 46, T3-10 (1968).
Corrigan, T.E., and Dean, M.J., Application of chemical reaction kinetics in process scale-up, Brit Chem Eng, 13(9), 1273-1276 (1968).

Mezaki, R., and Hill, W.J., Study of rate data using transformations, Brit Chem Eng, 13(3), 358-360 (1968).
Mutzenberg, A.B., and Giger, A., Chemical reactions in thin-film equipment, Trans IChemE, 46, T187-189 (1968).
Reisser, A., Improving the efficiency of batch operations, Chem Eng, 12 Feb, 117-123 (1968).
Aris, R., Stability criteria of chemical reaction engineering, Chem Eng Sci, 24(1), 149-170 (1969).
Danly, D.E., Electrolysis makes petrochemicals, Hyd Proc, 48(6), 159-165 (1969).
Draper, N.R.; Kanemasu, H., and Mezaki, R., Estimating rate constants, Ind Eng Chem Fund, 8(3), 423-427 (1969).
Kladko, M., Device collects kinetic data under difficult conditions, Chem Eng, 14 July, 136 (1969).
Kohler, K., Thermodynamic analysis of methanol synthesis, Brit Chem Eng, 14(11), 1548-1550 (1969).
Sharma, R.L.; Dutt, J., and Nirdosh, I., Diffusion in electrolytes: Convective diffusion in laminar flow, Brit Chem Eng, 14(9), 1220-1221 (1969).
Skarchenko, V.K., Review of oxidative dehydrogenation of hydrocarbons, Int Chem Eng, 9(1), 1-23 (1969).
Spedding, P.L., Chemical reactions in non-disruptive electric discharges, Chem Engnr, Jan, CE17-50 (1969).
Wong, K.F., and Eckert, C.A., Solvent design for chemical reactions, Ind Eng Chem Proc Des Dev, 8(4), 568-573 (1969).

1970-1971

Cinadr, B.F., and Schooley, A.T., How to treat solvents for solution polymerization, Chem Eng, 26 Jan, 125-129 (1970).
Farkas, A., What you should know about catalytic hydrocarbon oxidation, Hyd Proc, 49(7), 121-130 (1970).
Fournier, C.D., and Groves, F.R., Isothermal temperatures for reversible reactions, Chem Eng, 9 Feb, 121-125 (1970).
Guerreri, G., Determine ammonia recycle gas equilibrium, Hyd Proc, 49(12), 74-76 (1970).
Thomas, P.J., Rate data in process development, Brit Chem Eng, 15(4), 531-532 (1970).
Thomas, W.J., and Naik, S.C., Catalysed synthesis of carbon disulphide from pentane and octane, Trans IChemE, 48, T129-139 (1970).
Various, Liquid-phase oxidation (special report), Hyd Proc, 49(3), 101-118 (1970).
Wesselingh, J.A., and Vant Hoog, A.C., Oxidation of aqueous sulphite solutions: A model reaction for measurements in gas-liquid dispersions, Trans IChemE, 48, T69-76 (1970).
Azpitarte, J.L.; Marzo, L., and Camacho, F., Nomograph gives NO oxidation, Hyd Proc, 50(2), 107-111 (1971).
Barrett, D., Rotating mixer reactor for kinetic studies, Trans IChemE, 49, 80-82 (1971).
Bondi, A., Handling kinetics from trickle-phase reactors, Chemtech, March, 185-188 (1971).

Cooper, J.L., and Pitcher, E., Rate analysis: A general technique, Chemtech, Aug, 484-486 (1971).
Elhalwagi, M.M., An engineering concept of reaction rate, Chem Eng, 31 May, 75-78 (1971).
Foss, S.D., Estimates of chemical kinetic rate constants by numerical integration, Chem Eng Sci, 26(2), 485-486 (1971).
Harmer, D.E., and Ballantine, D.S., Applying radiation to chemical processing, Chem Eng, 3 May, 91-98 (1971).
Harmer, D.E., and Ballantine, D.S., Radiation processing, Chem Eng, 19 April, 98-116 (1971).
Kladko, M., Case study of a chemical kinetics engineering design problem, Chemtech, March, 141-147 (1971).

1972-1974

Boudart, M., Two-step catalytic reactions (review paper), AIChEJ, 18(3), 465-478 (1972).
Diaper, E.W.J., Ozone generation and use, Chemtech, June, 368-375; Aug, 498-504 (1972).
Levins, D.M., and Glastonbury, J.R., Particle-liquid hydrodynamics and mass transfer in a stirred vessel, Trans IChemE, 50, 32-41, 132-146 (1972).
Samuels, M.R., and Eliassen, J.D., Optimum fractional conversion for chemical reactions at equilibrium, Ind Eng Chem Proc Des Dev, 11(3), 383-386 (1972).
Schlegel, W.F., Polymer-plant engineering: Reaction and polymer recovery, Chem Eng, 20 March, 88-106 (1972).
Alper, E., The kinetics of oxidation of sodium sulphite solution, Trans IChemE, 51, 159-161 (1973).
Costa, P., and Trevissoi, C., Thermodynamic stability of chemical reactors, Chem Eng Sci, 28(12), 2195-2204 (1973).
McIver, R.G., and Ratcliffe, J.S., Rate constants in the liquid-phase chlorination of some chloroethanes, Trans IChemE, 51, 68-71 (1973).
Sampath, B.S., and Hughes, R., A review of mathematical models in single particle gas-solid non-catalytic reactions, Chem Engnr, Oct, 485-492, 497 (1973).
April, G.C., and Pike, R.W., Modeling complex chemical-reaction systems, Ind Eng Chem Proc Des Dev, 13(1), 1-6 (1974).
Bowen, J.H., Gas-solid, non-catalytic reaction models, Trans IChemE, 52, 282-284 (1974).
Brostrom, A., Optimisation of a reactor of votator type for sulphonation of dodecylbenzene, Trans IChemE, 52, 384-386 (1974).
Lazaridis, A., Stoichiometric relationships in the combustion of organics, Chem Eng, 25 Nov, 122 (1974).
McIver, R.G., and Ratcliffe, J.S., Reaction of chlorine and vinyl chloride in a bubble column, Trans IChemE, 52, 269-281 (1974).
Various, Reaction engineering needs, Chem Eng Prog, 70(7), 31-44 (1974).
Wong, A., Calculating percent displacement by flows through vessels with mixed contents, Chem Eng, 18 March, 116 (1974).

1975-1976

Blau, G.E.; Neely, W.B., and Branson, D.R., Ecokinetics: A study of the fate and distribution of chemicals in laboratory ecosystems, AIChEJ, 21(5), 854-861 (1975).

Brostrom, A., Gas-liquid contacting in a reactor of votator type, Trans IChemE, 53, 26-28 (1975).

Brostrom, A., Mathematical model for simulating the sulphonation of dodecylbenzene with gaseous sulphur trioxide in an industrial reactor of votator type, Trans IChemE, 53, 29-33 (1975).

El-Rifai, M.A.; Saleh, M.A., and El-Shishing, S.S., Stagewise partial pressure sulphonation of aromatics, Trans IChemE, 53, 93-96 (1975).

Fitzjohn, J.L., Electro-organic synthesis, Chem Eng Prog, 71(2), 85-91 (1975).

Graveland, A., and Heertjes, P.M., Removal of manganese from ground water by heterogeneous autocatalytic oxidation, Trans IChemE, 53, 154-164 (1975).

Greco, G.; Iorio, G.; Tola, G., and Waldram, S.P., Unsteady-state diffusion in porous solids, Trans IChemE, 53, 55-58 (1975).

Pexidr, V., Evaluation of integral kinetic data, Int Chem Eng, 15(2), 358-368 (1975).

Pritchard, D.J., and Bacon, D.W., Statistical assessment of chemical kinetic models, Chem Eng Sci, 30(5), 567-574 (1975).

Ruckenstein, E., and Vavanellos, T., Kinetics of solid-phase reactions, AIChEJ, 21(4), 756-763 (1975).

Stanbridge, D.W., Quick calculation for gas yields from steam reforming processes, Chem Eng, 22 Dec, 69-70 (1975).

Bissett, L.A., Calculating the effect of a temperature change on reaction rate, Chem Eng, 25 Oct, 151-152 (1976).

Holton, R.D., and Trimm, D.L., Mathematical modelling of heterogeneous-homogeneous reactions, Chem Eng Sci, 31(7), 549-562 (1976).

Iorio, G.; Greco, G., and Waldram, S.P., Unsteady state diffusion in long hollow cylinders of porous material: A transient shape factor, Trans IChemE, 54, 199-201 (1976).

Sunderland, P., An assessment of laboratory reactors for heterogeneously catalysed vapour phase reactors, Trans IChemE, 54, 135-138 (1976).

Weller, S.W., Experiment on saponification of acetamide in a batch reactor, Chem Eng Educ, 10(2), 74-75 (1976).

1977-1978

Altwicker, E.R., The role of inhibitors in the kinetics of sulphite oxidation, Trans IChemE, 55, 281-282 (1977).

Astarita, G., and Mashelkar, R.A., Heat and mass transfer in non-Newtonian fluids (review paper), Chem Engnr, Feb, 100-105 (1977).

Bissett, L., Equilibrium constants for shift reactions, Chem Eng, 24 Oct, 155-156 (1977).

Cartlidge, J.; McGrath, L., and Wilson, S.H., Mathematical modelling of olefin oxidation over oxide catalysts, Trans IChemE, 52, 222-227 (1974); 53, 117-124 (1975); 55, 164-170 (1977).

Cartlidge, J.; McGrath, L., and Wilson, S.H., Mathematical modelling of olefin oxidation over oxide catalysts, Trans IChemE, 52, 222-227 (1974); 53, 117-124 (1975); 55, 164-170 (1977).
Duhne, C.R., Calculating the approach to equilibrium, Chem Eng, 29 Aug, 96-98 (1977).
Hofmann, H., et al., Reaction engineering, Int Chem Eng, 16(1), 132-148; 16(4), 620-630 (1976); 17(1), 8-19; 17(2), 193-208; 17(3), 414-425 (1977).
Nakada, K., and Tanaka, M., Interpretation of the coalescence behaviour of dispersed droplets in a stirred flow tank in terms of quasi-chemical kinetics, Trans IChemE, 55, 143-148 (1977).
Ng, T.H., and Vermeulen, T., Kinetic analysis of multiple-reaction systems by reaction-path slope methods, Ind Eng Chem Fund, 16(1), 125-130 (1977).
Sundberg, D.C.; Carleson, T.E., and McCollister, R.D., Experiments in reaction engineering, Chem Eng Educ, 11(3), 118-121,139 (1977).
Attar, A., Chemistry, thermodynamics and kinetics of reactions of sulphur in coal-gas reactions: A review, Fuel, 57(4), 201-212 (1978).
Cooke, M.J., and Robson, B., Gas from coal, Chem Engnr, Oct, 729-732 (1978).
Pearson, R.G., Modern kinetics, Chemtech, Sept, 552-558 (1978).
Takahashi, P.K., Balancing chemical equations without chemistry, Chem Eng, 13 Feb, 120-122 (1978).
Verhoff, F.H., Quadratic programming technique applied to kinetic rate constant determination, Chem Eng Sci, 33(3), 263-270 (1978).

1979-1980

Cappelli, A., and Trambouze, P., Significance of chemical reaction engineering for industry, Int Chem Eng, 19(1), 12-21 (1979).
Gans, M., Choosing between air and oxygen for chemical processes, Chem Eng Prog, 75(1), 67-72 (1979).
Hampson, G.M., A simple solution to steam reforming equations, Chem Engnr, July, 523; Aug, 621 (1979).
Hilder, M.H., An algebraic expression for first-order reaction in laminar flow, Trans IChemE, 57, 143-144 (1979).
Horwitz, B.A., pH in the kinetic rate equation, Chem Eng, 7 May, 115 (1979).
Kuleli, O., and Basar, K., The effect of hydrogen partial pressure on the methane selectivity of hydrogenolysis reactions, Trans IChemE, 57, 70-72 (1979).
Shindo, A., and Maekima, T., Reactor models for syn gas processes, Chem Eng Prog, 75(9), 92-97 (1979).
Smith, W.R., and Missen, R.W., What is chemical stoichiometry? Chem Eng Educ, 13(1), 26-32 (1979).
Tamhankar, S.S., and Doraiswamy, L.K., Analysis of solid-solid reactions (review paper), AIChEJ, 25(4), 561-582 (1979).
Youngquist, G.R., A mass transfer experiment, Chem Eng Educ, 13(1), 20-25 (1979).
Bejerano, T.; Germain, S.; Goodridge, F., and Wright, A.R., Electrochemical production of propylene oxide in a small pilot plant, Trans IChemE, 58, 28-32 (1980).
Benson, S.W., Predicting chemical reactivity, Chemtech, Feb, 121-126 (1980).
Grange, P., Catalytic hydrodesulfurization, Catalysis Reviews, 21(2), 135-182 (1980).

Jenkins, J.H., and Stephens, T.W., Kinetics of catalytic reforming, Hyd Proc, 59(11), 163-167 (1980).

Karlsson, H.T., and Bjerle, I., Enhancement factor for instananeous irreversible gas-liquid reactions, Trans IChemE, 58, 138-140 (1980).

Kung, H.H., Methanol synthesis, Catalysis Reviews, 22(2), 235-260 (1980).

Neelakantan, K., and Gehlawat, J.K., Kinetics of absorption of oxygen in aqueous solutions of ammonium sulfite in stirred cells, Ind Eng Chem Fund, 19(1), 36-39 (1980).

Paspek, S.C.; Varma, A., and Carberry, J.J., Experiment to determine kinetics of gas-solid catalytic reactions, Chem Eng Educ, 14(2), 78-82,100 (1980).

Smith, W.R., The computation of chemical equilibria in complex systems, Ind Eng Chem Fund, 19(1), 1-10 (1980).

Urbanek, A., and Trela, M., Catalytic oxidation of sulfur dioxide, Catalysis Reviews, 21(1), 73-134 (1980).

Van Hook, J.P., Methane-steam reforming, Catalysis Reviews, 21(1), 1-52 (1980).

Various, Chemical reaction engineering (topic issue), Chem Eng Sci, 35(9), 1821-2076 (1980).

Zamaraev, K.I., and Parmon, V.N., Potential methods and perspectives of solar energy conversion via photocatalytic processes, Catalysis Reviews, 22(2), 261-324 (1980).

1981-1982

Bjornbom, P., Components in chemical stoichiometry, Ind Eng Chem Fund, 20(2), 161-164 (1981).

Carberry, J.J., Some aspects of catalytic reactor engineering (review paper), Trans IChemE, 59, 75-82 (1981).

Fowler, W.B., Mixing model tracks changing compositions, Chem Eng, 14 Dec, 110 (1981).

Glowinski, J., and Stocki, J., Estimation of kinetic parameters: Initial guess generation method, AIChEJ, 27(6), 1041-1043 (1981).

Holleran, E.M., Handling chemical kinetics, Chemtech, Aug, 500-503 (1981).

Kam, E.K.T., and Hughes, R., Model for direct reduction of iron ore by mixtures of hydrogen and carbon monoxide in moving bed, Trans IChemE, 59, 196-206 (1981).

Lavery, J.E., Analyzing rate-process data economically, Chem Eng, 9 March, 87-90 (1981).

Morton, W., and Goodman, M.G., Parametric oscillations in simple catalytic reaction models, Trans IChemE, 59, 253-259 (1981).

Pavko, A.; Misic, D.M., and Levec, J., Kinetics in three-phase reactors, Chem Eng J, 21(2), 149-154 (1981).

Sunderland, P., and Ahmed, I., Mass transfer in a vibration mixed reactor, Trans IChemE, 59, 95-99 (1981).

White, C.W., and Seider, W.D., Analysis of chemical reaction systems, Chem Eng Commns, 9(1), 159-174 (1981).

Cerveny, L., and Ruzicka, V., Competitive catalytic hydrogenation in liquid phase on solid catalysts, Catalysis Reviews, 24(4), 503-566 (1982).

Dry, M.E., Sasol's Fischer-Tropsch experience, Hyd Proc, Aug, 121-124 (1982).

Finkelstein, E., Feeding liquified gas to a gas-liquid reaction, Chem Eng, 18 Oct, 132-134 (1982).
Legan, R.W., Applications of ultraviolet light for catalysis of chemical reactions, Chem Eng, 25 Jan, 95-100 (1982).
Madsack, H.J., Make synthesis gas by residuum partial oxidation, Hyd Proc, July, 169-172 (1982).
Marsch, H.D., and Herbort, H.J., Produce synthesis gas by steam reforming natural gas, Hyd Proc, June, 101-105 (1982).
Meisen, A., and Kennard, M.L., Diethanolamine degradation mechanism, Hyd Proc, Oct, 105-108 (1982).
Mu, J., and Kloos, D., A bench-scale continuous-flow rotary reactor for studies of high temperature reactions, Trans IChemE, 60, 377-379 (1982).
Smith, D.W., Runaway reactions and thermal explosion, Chem Eng, 13 Dec, 79-84 (1982).
Smith, J.M., Thirty-five years of applied catalytic kinetics (review), Ind Eng Chem Fund, 21(4), 327-332 (1982).

1983

Berger, F.P., and Ziai, A., Optimisation of experimental conditions for electrochemical mass transfer measurements, CER&D, 61, 377-382 (1983).
Boreskov, G.K., and Matros, Y.S., Unsteady-state performance of heterogeneous catalytic reactions, Catalysis Reviews, 25(4), 551-590 (1983).
Bourne, J.R., and Rohani, S., Micro-mixing and the selective iodination of l-tyrosine, CER&D, 61, 297-302 (1983).
Chang, C.D., Hydrocarbons from methanol, Catalysis Reviews, 25(1), 1-118 (1983).
Furimsky, E., Chemistry of catalytic hydrodeoxygenation, Catalysis Reviews, 25(3), 421-458 (1983).
Horwitz, B.A., Investigating unsteady-state kinetics, Chem Eng, 5 Sept, 115-117 (1983).
Mu, J., and Perlmutter, D.D., Comparison of a thermogravimetric analyser with bench-scale rotary reactor studies, CER&D, 61, 198-201 (1983).
Sharma, M.M., Perspectives in gas-liquid reactions, Chem Eng Sci, 38(1), 21-28 (1983).
Sokolsky, D.V., Liquid-phase electrocatalytic hydrogenation, Periodica Polytechnica, 27, 49-56 (1983).
Too, J.R.; Fan, L.T., and Nassar, R., Markov chain models of complex chemical reactions in continuous flow reactors, Comput Chem Eng, 7(1), 1-12 (1983).
Waldram, S.P., Chemical reaction engineering, Chem Engnr, What is chemical reaction engineering? July, 28-31; An equilibrium calculation, Oct, 16-17; Design of an ideal plug flow reactor, Nov, 35,41,42 (1983).

1984

Eigenberger, G., Practical problems in modelling chemical reactions in fixed bed reactors, Chem Eng & Proc, 18(1), 55-66 (1984).

Falconer, J.L., and Britten, J.A., Kinetics and catalysis demonstrations, Chem Eng Educ, 18(3), 140-144 (1984).

Guo-Tai, Z., and Shau-Drang, H., Experiments in chemical reaction engineering, Chem Eng Educ, 18(1), 10-13 (1984).

Guo-Tai, Z., and Shau-Drang, H., Developing kinetic rate equations from laboratory data, Chem Eng Educ, 18(2), 64-65,69 (1984).

Hainsworth, J., Optimal design for polymer processes, Chem Engnr, Jan, 14-17 (1984).

Hosten, L.H., Kinetics of gas-liquid reactions, Periodica Polytechnica, 28(1), 19-39 (1984).

Mann, R., Developments in reaction engineering, Proc Engng, Nov, 33-38 (1984).

Patil, V.K.; Joshi, J.B., and Sharma, M.M., Solid-liquid mass transfer coefficient in mechanically agitated contactors, CER&D, 62, 247-254 (1984).

Srinivasan, P.; Devotta, S., and Watson, F.A., Decomposition of trichloro-fluoromethane (R11) under static and flow conditions, CER&D, 62, 57-59 (1984).

Srinivasan, P.; Patwardhan, V.R.; Devotta, S., and Watson, F.A., Thermal decomposition of trichloro-trifluoroethane (R113) under static and flow conditions, CER&D, 62, 266-268 (1984).

Waldram, S.P., Chemical reaction engineering, Chem Engnr, The non-ideal reactor, Feb, 14-16; Introducing economic considerations, March, 41-43 (1984).

Warmoeskerken, M.M.C.G.; van Houwelingen, M.C.; Frijlink, J.J., and Smith, J.M., Role of cavity formation in stirred gas-liquid-solid reactors, CER&D, 62, 197-200 (1984).

Wright, A.R., Application of modelling and computer simulation to pharmaceutical processes: The Williamson synthesis, CER&D, 62, 391-397 (1984).

Youngquist, G.R., Intensity function and reaction conversion, Ind Eng Chem Proc Des Dev, 23(4), 769-772 (1984).

1985

Birk, R.H.; Thomas, W.J., and Yue, P.L., Effect of mass transport and catalyst deactivation on the kinetics of ethanol dehydration on zeolite 13X, CER&D, 63, 338-344 (1985).

Bolzern, O., and Bourne, J.R., Rapid chemical reactions in a centrifugal pump, CER&D, 63, 275-282 (1985).

Ferrer, M.; DAvid, R., and Villermaux, J., Homogeneous oxidation of n-butane in self-stirred reactor, Chem Eng & Proc, 19(3), 119-128 (1985).

Ratkowsky, D.A., A statistically suitable general formulation for modelling catalytic chemical reactions, Chem Eng Sci, 40(9), 1623-1628 (1985).

Rod, V.; Sir, Z., and Gruberova, A., Phase transfer enhancement in metal extraction, CER&D, 63, 89-100 (1985).

Sedahmed, G.H., and Shemilt, L.W., Free convection mass transfer characteristics of vertical screens, CER&D, 63, 378-382 (1985).

Stephan, R.; Emig, G., and Hofmann, H., Kinetics of hydrodesulfurization of gas oil, Chem Eng & Proc, 19(6), 303-316 (1985).

Tine, C.B.D., Simple model for non-catalytic gas-solid reaction, CER&D, 63, 112-116 (1985).

1986

Barlow, R.C., Reduce fluid catalytic cracking unit fouling, Hyd Proc, 65(7), 37-39 (1986).

Boudart, M., Classical catalytic kinetics, Ind Eng Chem Fund, 25(4), 656-659 (1986).

Burghardt, A., Transport phenomena and chemical reactions in porous catalysts for multicomponent and multireaction systems, Chem Eng & Proc, 20(5), 229-244 (1986).

Choudhary, V.R.; Sansare, S.D., and Thite, G.A., Adsorption of reaction species for hydrogenation of o-nitrotoluene on copper chromite under catalytic conditions, J Chem Tech Biotechnol, 36(2), 53-60 (1986).

Lede, J.; Verzaro, F.; Antoine, B., and Villermaux, J., Flash pyrolysis of wood in cyclone reactor, Chem Eng & Proc, 20(6), 309-318 (1986).

Luggenhorst, H.J., and Westerterp, K.R., Low-pressure carbonylation of benzyl chloride, Chem Eng & Proc, 20(4), 221-228 (1986).

Luss, D., Steady-state multiplicity features of chemically reacting systems, Chem Eng Educ, 20(1), 12-17, 52-56 (1986).

Pitchai, R., and Klier, K., Partial oxidation of methane, Catalysis Reviews, 28(1), 13-88 (1986).

Prasannan, P.C.; Ramachandran, P.A., and Doraiswamy, L.K., Gas-solid reactions: A method of direct solution for solid conversion profiles, Chem Eng J, 33(1), 19-26 (1986).

Scott Folger, H., Elements of Chemical Reaction Engineering, Prentice-Hall, New Jersey (1986).

Sivakumar, S.; Chidambaram, M., and Shankar, H.S., Analysis of non-catalytic gas-solid reactions in a vertical transport reactor, Chem Eng J, 33(2), 103-108 (1986).

Skaates, J.M., A computer graphics approach to the use of the integral method in kinetics, Chem Eng Educ, 20(3), 136-137 (1986).

Subramanian, B., and McHugh, M.A., Reactions in supercritical fluids: A review, Ind Eng Chem Proc Des Dev, 25(1), 1-12 (1986).

Tabak, S.A.; Krambeck, F.J., and Garwood, W.E., Conversion of propylene and butylene over ZMS-5 catalyst, AIChEJ, 32(9), 1526-1531 (1986).

Turek, F.; Geike, R., and Lange, R., Liquid-phase hydrogenation of nitrobenzene in slurry reactor, Chem Eng & Proc, 20(4), 213-220 (1986).

Various, Chemical reaction engineering (papers from an international symposium), Chem Eng Sci, 41(4), 607-1370 (1986).

Weir, E.D.; Gravenstine, G.W., and Hoppe, T.F., Thermal runaways: Problems with agitation, Plant/Opns Prog, 5(3), 142-147 (1986).

1987

Adelman, D.J., and Burnet, G., Carbochlorination of metal oxides with phosgene, AIChEJ, 33(1), 64-69 (1987).

Agrawal, R.K., and Sivasubramanian, M.S., Integral approximations for nonisothermal kinetics, AIChEJ, 33(7), 1212-1214 (1987).

Bhaskarwar, A.N., Analysis of gas absorption accompanied by a zero-order chemical reaction in a liquid-foam film surrounded by limited gas pockets, J Chem Tech Biotechnol, 37(3), 183-188 (1987).

Bilbao, J.; Arandes, J.M.; Romero, A., and Olazar, M., Kinetic study of the regeneration of solid catalysts under internal diffusion restrictions, Chem Eng J, 35(2), 115-122 (1987).

Braun, R.L., and Burnham, A.K., Analysis of chemical reaction kinetics using a distribution of activation energies and simpler models, Energy & Fuels, 1(2), 153-161 (1987).

Choi, K.Y., and Lei, G.D., Modeling of free-radical polymerization of styrene by bifunctional initiators, AIChEJ, 33(12), 2067-2076 (1987).

Cronin, J.L., and Nolan, P.F., Laboratory techniques for quantitative study of thermal decompositions, Plant/Opns Prog, 6(2), 89-97 (1987).

Cropley, J.B., Systematic errors in recycle reactor kinetic studies, Chem Eng Prog, 83(2), 46-51 (1987).

Franke, M.D.; Ernst, W.R., and Myerson, A.S., Kinetics of dissolution of alumina in acidic solution, AIChEJ, 33(2), 267-273 (1987).

Gardiner, W.C.; Hwang, S.M., and Rabinowitz, M.J., Shock tube and modeling study of methyl radical in methane oxidation, Energy & Fuels, 1(6), 545-549 (1987).

Gosselink, J.W., and Stork, W.H.J., Simple multicomponent description of catalysed hydrodesulfurization of heavy gas oil in trickle-flow reactors, Chem Eng & Proc, 22(3), 157-162 (1987).

Hanlon, R.T., Effects of gas partial pressures on hydrodenitrogenation of pyridine, Energy & Fuels, 1(5), 424-431 (1987).

Helling, R.K., and Tester, J.W., Oxidation kinetics of carbon monoxide in supercritical water, Energy & Fuels, 1(5), 417-424 (1987).

Henderson, L.S., Stability analysis of polymerization in continuous stirred-tank reactors, Chem Eng Prog, 83(3), 42-50 (1987).

Karlsen, L.G., and Villadsen, J., Isothermal reaction calorimeters: Literature review and data treatment, Chem Eng Sci, 42(5), 1153-1174 (1987).

Kevrekidis, I.G., Numerical study of global bifurcations in chemical dynamics, AIChEJ, 33(11), 1850-1864 (1987).

Kimura, S., and Smith, J.M., Kinetics of sodium carbonate-sulfur dioxide reaction, AIChEJ, 33(9), 1522-1532 (1987).

Kocaefe, D.; Karman, D., and Steward, F.R., Interpretation of sulfonation rate of CaO, MgO, ZnO with SO2 and SO3, AIChEJ, 33(11), 1835-1843 (1987).

Kucznski, M., et al., Reaction kinetics for methanol synthesis from CO and hydrogen on copper catalyst, Chem Eng & Proc, 21(4), 179-192 (1987).

Kwon, K.C.; Vahdat, N., and Ayers, W.R., Chemical reaction experiment for the undergraduate laboratory, Chem Eng Educ, 21(1), 30-32 (1987).

Ma, Y.H., and Savage, L.A., Xylene isomerization using zeolites in gradientless reactor system, AIChEJ, 33(8), 1233-1240 (1987).

McConvey, I.F., et al., Electrochemical reaction with parallel reversible surface adsorption: Interpretations of kinetics of anodic oxidation of aniline and phenol to carbon dioxide, Chem Eng & Proc, 22(4), 231-236 (1987).

Mills, D.B.; Bar, R., and Kirwan, D.J., Effect of solids on oxygen transfer in agitated three-phase systems, AIChEJ, 33(9), 1542-1549 (1987).

Moller, P., and Papp, H., Heats of adsorption and reaction of carbon monoxide on iron-manganese oxide catalysts, Ads Sci Tech, 4(3), 176-184 (1987).
Morrison, P.W., and Reimer, J.A., Simple method to study gas phase reactions, AIChEJ, 33(12), 2037-2046 (1987).
Oliveres, M.; Alonso, M.; Recasens, F., and Puigjaner, L., Modelling and simulation of styrene-acrylonitrile emulsion polymerization kinetics, Chem Eng J, 34(1) 1-10 (1987).
Rawlings, J.B., and Ray, W.H., Emulsion polymerization reactor stability: Simplified model analysis, AIChEJ, 33(10), 1663-1677 (1987).
Roy, R., and Bhatia, S., Kinetics of esterification of benzyl alcohol with acetic acid catalysed by cation-exchange resin (Amberlyst-15), J Chem Tech Biotechnol, 37(1), 1-10 (1987).
Salmi, T., A program package for simulation of heterogeneous catalytic reactions in ideal reactors, Comput Chem Eng, 11(2), 83-94 (1987).
Shanks, B.H., and Bailey, J.E., Modeling of slow dynamics in oxidation of CO over supported silver, AIChEJ, 33(12), 1971-1976 (1987).
Snel, R., Olefins from syngas, Catalysis Reviews, 29(4), 361-445 (1987).
Spencer, N.D., and Pereira, C.J., Catalytic partial oxidation of methane to formaldehyde, AIChEJ, 33(11), 1808-1812 (1987).
Tolen, D.F., and Desai, P.H., Increase profits from fluid catalytic cracking units, Hyd Proc, 66(7), 49-51 (1987).
Tsai, C.Y., and Scaroni, A.W., Pyrolysis and combustion of bituminous coal fractions in entrained flow reactor, Energy & Fuels, 1(3), 263-270 (1987).
Various, Plenary papers from 9th International Symposium on Chemical Reaction Engineering, Chem Eng Sci, 42(5), 923-1088 (1987).
Vatcha, S.R., Directly determine reaction order, Chem Eng, 16 Feb, 195-196 (1987).
Wimmers, O.J., and Fortuin, J.M.H., Gas absorption enhancement by heterogeneously catalysed chemical reaction in stagnant liquid: Temperature rise and catalyst effectiveness, Chem Eng Commns, 62(1), 107-122 (1987).
Yagi, H., and Hikita, H., Gas absorption into a slurry accompanied by chemical reaction with solute from sparingly soluble particles, Chem Eng J, 36(3), 169-174 (1987).

1988

Alper, E., and Abu-Sharkh, B., Kinetics of absorption of oxygen into aqueous sodium sulfite: Order in oxygen, AIChEJ, 34(8), 1384-1386 (1988).
Apte, N.G., et al., Kinetic modeling of thermal decomposition of aluminum sulfate, Chem Eng Commns, 74, 47-62 (1988).
Arai, N., et al., Formation of volatile nitrogen monoxide and nitrogen via ammonia with low oxygen concentration, Int Chem Eng, 28(2), 344-353 (1988).
Asami, K., et al., Vapor-phase oxidative coupling of methane under pressure, Energy & Fuels, 2(4), 574-578 (1988).
Astarita, G., and Ocone, R., Lumping nonlinear kinetics, AIChEJ, 34(8), 1299-1309 (1988).
Berty, J.M., The state of kinetic model development, Chem Eng Prog, 84(2), 61-67 (1988).

Bhaskarwar, A.N., Application of method of kinetic invariants to description of dissolution accompanied by chemical reaction, Chem Eng Commns, 72, 25-34 (1988).

Bose, A.C.; Dannecker, K.M., and Wendt, J.O.L., Coal composition effects on mechanisms governing destruction of nitric oxide and other nitrogeneous species during fuel-rich combustion, Energy & Fuels, 2(3), 301-309 (1988).

Burghardt, A., and Aerts, J., Pressure changes due to diffusion with chemical reaction in porous pellet, Int Chem Eng, 28(4), 713-723 (1988).

Burghardt, A., and Aerts, J., Pressure changes during diffusion with chemical reaction in porous pellet, Chem Eng & Proc, 23(2), 77-88 (1988).

Cale, T.S., Effectiveness factor for single pellet during ethane hydrogenalysis, Chem Eng Commns, 70, 57-66 (1988).

Chang, Y.H., Mathematical model for gasification of single coal particle, Int Chem Eng, 28(3), 520-527 (1988).

Charpentier, J.C., New trends in gas-liquid reaction engineering, Int Chem Eng, 28(2), 285-299 (1988).

Chou, M.Y., and Ho, T.C., Continuum theory for lumping nonlinear reactions, AIChEJ, 34(9), 1519-1527 (1988).

Collins, N.A.; Debenedetti, P.G., and Sundaresan, S., Disproportionation of toluene over ZSM-5 under near-critical conditions, AIChEJ, 34(7), 1211-1214 (1988).

Cooper, D.A., and Ljungstrom, E.B., Decomposition of ammonia over quartz sand at 840-960 degC, Energy & Fuels, 2(5), 716-719 (1988).

Daniell, P.; Soltani-Ahmadi, A., and Kono, H.O., Reaction kinetics of sulphur dioxide-calcium oxide system: Pore-closure model, Powder Tech, 55(2), 75-86 (1988).

Dobbins, M.S., and Burnet, G., Carbochlorination of dispersed oxides in molten salt reactor, AIChEJ, 34(7), 1086-1093 (1988).

Donnelly, T.J.; Yates, I.C., and Satterfield, C.N., Analysis and prediction of product distributions of Fischer-Tropsch synthesis, Energy & Fuels, 2(6), 734-740 (1988).

Ellis, M.F.; Taylor, T.W.; Gonzalez, V., and Jensen, K.F., Estimation of molecular weight distribution in batch polymerization, AIChEJ, 34(8), 1341-1353 (1988).

Ernst, W.R., and Chen, M.S.K., Hydrolysis of carbonyl sulfide in gas-liquid reactor, AIChEJ, 34(1), 158-162 (1988).

Ernst, W.R.; Shoaei, M., and Forney, L.J., Selectivity of chloride-chlorate reaction system in various reactor types, AIChEJ, 34(11), 1927-1930 (1988).

Fischer-Calderon, P.E., et al., Explicit expressions for Michaelis-Menten kinetic parameters from direct linear plot, Chem Eng Commns, 69, 13-28 (1988).

Floess, J.K.; Longwell, J.P., and Sarofim, A.F., Intrinsic reaction kinetics of microporous carbons, Energy & Fuels, 2(6), 756-765 (1988).

France, J.E.; Shamsi, A., and Ahsan, M.Q., Oxidative coupling of methane over Perovskite-type oxides, Energy & Fuels, 2(2), 235-236 (1988).

Fuerstenau, D.W.; Rosenbaum, J.M., and You, Y.S., Electrokinetic behavior of coal, Energy & Fuels, 2(3), 246-252 (1988).

Funatsu, K., et al., Oxygen transfer in the water-jet vessel, Chem Eng Commns, 73, 121-140 (1988).

Gabitto, J., et al., Supercritical catalytic dehydrogenation of toluene, AIChEJ, 34(7), 1225-1228 (1988).

Givi, P., and McMurtry, P.A., Nonpremixed reaction in homogeneous turbulence: Direct numerical simulations, AIChEJ, 34(6), 1039-1042 (1988).

Hanlon, R.T., and Satterfield, C.N., Reactions of selected 1-olefins and ethanol added during Fischer-Tropsch synthesis, Energy & Fuels, 2(2), 196-205 (1988).

Hepburn, J.S., and Stenger, H.G., Nitric oxide reduction by alumina-supported rhodium palladium and platinum, Energy & Fuels, 2(3), 289-292 (1988).

Herrick, D.E., et al., Acceleration of chlorination of alumina using supercritical carbon tetrachloride, AIChEJ, 34(4), 669-671 (1988).

Hillewaert, L.P.; Dierickx, J.L., and Froment, G.F., Computer generation of reaction schemes and rate equations for thermal cracking, AIChEJ, 34(1), 17-24 (1988).

Hsu, C.L., and Hsieh, J.S., Reaction kinetics in oxygen bleaching, AIChEJ, 34(1), 116-122 (1988).

Ivie, J.J., and Forney, L.J., Numerical model of carbon black synthesis by benzene pyrolysis, AIChEJ, 34(11), 1813-1820 (1988).

Kobayashi, M.; Kobayashi, H., and Kanno, T., Two reaction path kinetics in CO oxidation on zinc oxide, Chem Eng Commns, 66, 23-28 (1988).

Kuo, M.C., and Chou, T.C., Benzaldehyde oxidation catalyzed by wall of tubular bubble column reactor, AIChEJ, 34(6), 1034-1038 (1988).

Lai, C.K.; Peters, W.A., and Longwell, J.P., Reduction of NO by coke over CaO, Energy & Fuels, 2(4), 586-589 (1988).

Lalvani, S.B., and Shami, M., Reaction rate studies of indirect electro-oxidation of pyrite slurries, Chem Eng Commns, 70, 215-226 (1988).

Lee, J.S., and Oyama, S.T., Oxidative coupling of methane to higher hydrocarbons, Catalysis Reviews, 30(2), 249-280 (1988).

Lee, P.I., et al., Analysis of temperature-programmed desorption from porous catalysts in flow system, Chem Eng Commns, 63, 205-224 (1988).

Lox, E.; Coenen, F.; Vermeulen, R., and Froment, G.F., A versatile bench-scale unit for kinetic studies of catalytic reactions, Ind Eng Chem Res, 27(4), 576-580 (1988).

Maheshwari, M., and Akella, L., Calculation of pre-exponential term in kinetic rate expression, Chem Eng Educ, 22(3), 150-152 (1988).

Mathe, T.; Tungler, A., and Petro, J., Heterogeneous catalysis and catalytic hydrogenation in gas phase by electrochemical methods, Periodica Polytechnica, 32(1), 181-186 (1988).

Matsuda, M., et al., Heat-release characteristics of packed bed of CaO during exothermic hydration, Int Chem Eng, 28(4), 642-652 (1988).

Mazumdar, B.K.; Banerjee, D.D., and Ghosh, G., Coal-zinc chloride reaction: An interpretation, Energy & Fuels, 2(2), 224-230 (1988).

Nikolov, V., et al., New approach to oxidation of o-xylene into phthalic anhydride in fixed bed of vanadium-titanium catalyst, Chem Eng & Proc, 24(3), 157-162 (1988).

Rastogi, A.; Svrcek, W.Y., and Behie, L.A., Novel microreactor with quench system for kinetic study of propane pyrolysis, AIChEJ, 34(9), 1417-1422 (1988).
Simons, G.A., Parameters limiting sulfonation by CaO, AIChEJ, 34(1), 167-170 (1988).
Sivakumar, S.; Chidambaram, M., and Shankar, H.S., Differentiation of deactivation disguised kinetics in transport reactors by wavefront analysis, Chem Eng Commns, 68, 213-220 (1988).
Spence, J.P., and Noronha, J.A., Reliable detection on runaway reaction precursors in liquid-phase reactions, Plant/Opns Prog, 7(4), 231-235 (1988).
Sundaram, K.M., and Fernandez, J.M., Effect of methane and hydrogen during thermal cracking of light hydrocarbons, AIChEJ, 34(2), 321-325 (1988).
Sundaram, K.M.; Katzer, J.R., and Bischoff, K.B., Modeling of hydroprocessing reactions, Chem Eng Commns, 71, 53-72 (1988).
Suresh, A.K.; Sridhar, T., and Potter, O.E., Mass transfer and solubility in autocatalytic oxidation of cyclohexane, AIChEJ, 34(1), 55-93 (1988).
Tai, Z.G., and Min, Z.J., Effect of pore diffusion on deactivating rates of second-order reaction, Chem Eng Commns, 65, 21-28 (1988).
Various, Papers from 10th International Symposium on Chemical Reaction Engineering (ISCRE-10), 29 Aug 1988 (topic issue) Chem Eng Sci, 43(8), 1725-2325 (1988).
Vleeschower, P.H.M., et al., Transient behavior of chemically reacting system in CSTR, AIChEJ, 34(10), 1736-1739 (1988).
Wachi, S.; Morikawa, H., and Inoue, H., Conversion distribution in diffusion-governed chlorination of poly vinyl chloride, AIChEJ, 34(10), 1683-1690 (1988).
Waletzko, N., and Schmidt, L.D., Modeling catalytic gauze reactors: HCN synthesis, AIChEJ, 34(7), 1146-1156 (1988).
Wojciechowski, B.W., Kinetics of Fischer-Tropsch synthesis, Catalysis Reviews, 30(4), 629-702 (1988).
Wojcik, M., New method for determining the intrinsic deactivation rate constant, AIChEJ, 34(4), 692-693 (1988).
Yagi, H.; Nagashima, S., and Hikita, H., Semibatch precipitation accompanying gas-liquid reaction, Chem Eng Commns, 65, 109-120 (1988).
Zhdanov, V.P.; Pavlicek, J., and Knor, Z., Preexponential factors for elementary surface processes, Catalysis Reviews, 30(4), 501-518 (1988).
Zygourakis, K., and Sandmann, C.W., Discrete structural models and their application to gas-solid reacting systems, AIChEJ, 34(12), 2030-2040 (1988).

18.3 Reactor Design

1966-1967

Mukherjee, S.P., and Doraiswamy, L.K., Reaction kinetics and reactor design, Brit Chem Eng, 11(11), 1380-1384 (1966).
Almasy, G.A.; Jedlovszky, P., and Pallai, I.M., Optimum design of ammonia synthesis reactor, Brit Chem Eng, 12(8), 1219-1222 (1967).

Bauer, J.L., and Corrigan, T.E., Advantages of packed-tower reactors, Chem Eng, 27 March, 111-114 (1967).
Bhat, G.N., and Srimathi, C.R., Evaluation of residence time for a gas-solids interacting moving bed system, Brit Chem Eng, 12(10), 1597-1599 (1967).
Bridgwater, J., and Carberry, J.J., Gas-liquid reactors, Brit Chem Eng, 12(1). 58-64; 12(2), 217-222 (1967).
Chandra, R., and Srinivasan, R., Nomogram for rate of production of metals in electrolytic cells, Brit Chem Eng, 12(9), 1415 (1967).
Dor, M.N., On-line computer control of fixed-bed chemical reactors, Brit Chem Eng, 12(6), 888-891 (1967).
Engh, T.A., On smoothing quality variations in combined plug-flow and back-mix tank systems, Trans IChemE, 45, T408-416 (1967).
Fainzilber, A.M., and Fridlender, N.A., Optimal conditions for chemical reactions in gas streams, Brit Chem Eng, 12(7), 1100-1101 (1967).
Fair, J.R., Designing gas-sparged reactors, Chem Eng, 3 July, 67-74; 17 July, 207-214 (1967).
Kats, M.B., and Genin, L.S., Longitudinal mixing of liquid in cocurrent sparged reactors sectionalized with sieve trays, Int Chem Eng, 7(2), 246-252 (1967).
Kim, K.J., et al., Heat and mass transfer in fixed and fluidized bed reactors, Int Chem Eng, 7(1), 79-86 (1967).
Kiraly, G., and Harmathy, L., New type high-surface rotary film reactor, Int Chem Eng, 7(3), 388-391 (1967).
Kyung-Jong, L., and Chun-Khil, L., Heat and mass transfer in fixed and fluidized bed reactors, Int Chem Eng, 7(3), 408-418 (1967).
Mukherjee, S.P., and Doraiswamy, L.K., Reaction kinetics and reactor design, Brit Chem Eng, 12(1), 70-74 (1967).
Rudkin, J., The loop reactor, Brit Chem Eng, 12(9), 1374-1377 (1967).
Schroeder, J., and Seweryniak, M., Design, operation, and planning of high-temperature flame reactors, Int Chem Eng, 7(2), 296-304 (1967).
Soboleski, J., Temperature profiles in synthesis converters, Brit Chem Eng, 12(7), 1085-1087 (1967).
Valentin, F.H.H., Mass transfer in agitated tanks, Brit Chem Eng, 12(8), 1213-1218 (1967).
Wright, B.S., Computer simulation of a plug-flow reactor for a complex reaction, Brit Chem Eng, 12(11), 1750-1752 (1967).

1968

Bakos, M., Recent developments in contact catalytic reactors, Int Chem Eng, 8(1), 148-153 (1968).
Corrigan, T.E., and Eggers, E.L., Applying the fluid particle concept to nonideal flow reactors, Brit Chem Eng, 13(1), 98-100 (1968).
Corrigan, T.E., and Shaefer, R., Frequency response for N stirred tanks with recycle, Brit Chem Eng, 13(5), 680-683 (1968).
Crider, J.E., and Foss, A.S., Analytic solution for dynamics of packed adiabatic chemical reactor, AIChEJ, 14(1), 77-84 (1968).
El Halwagi, M.M., Simple technique determines residence time in reactors, Chem Eng, 23 Sept, 250 (1968).

Hodossy, L., A thermally stable high-pressure reactor for furfuryl alcohol production, Brit Chem Eng, 13(9), 1277-1280 (1968).
Hussain, S.Z., and Kamath, N.R., Design of a cascade of semi-continuous reactors, Brit Chem Eng, 13(7), 987-990 (1968).
Johnston, J.D., Improving pilot-plant reactors for organic syntheses, Chem Eng, 2 Dec, 121-126 (1968).
Kim, K.J.; Kim. D.J.; Chun, K.S., and Choo, S.S., Review of heat and mass transfer in fixed and fluidized bed reactors, Int Chem Eng, 8(3), 472-490 (1968).
Levine, R., New design method for backmixed reactors, Chem Eng, 1 July, 62-66; 29 July, 145-150; 12 Aug, 167-171 (1968).
Luyben, W.L., Nonlinear feedforward control of chemical reactors, AIChEJ, 14(1), 37-45 (1968).
Mecklenburgh, J.C., and Hartland, S., Design of reactors with backmixing, Chem Eng Sci, 23(1), 57-86 (1968).
Ogunye, A.F., and Ray, W.H., Non-simple control policies for reactors with catalyst decay, Trans IChemE, 46, T225-231 (1968).
Schroeder, J., and Seweryniak, M., High-temperature flame reactors, Brit Chem Eng, 13(2), 235-241 (1968).
Smith, J.M., Reactor scaledown, Chem Eng Prog, 64(8), 78-82 (1968).
Yablonskii, G.S.; Kamenko, B.L.; Gelbshtein, A.I., and Slinko, M.G., Modelling the hydrochlorination of acetylene in a fixed catalyst bed, Int Chem Eng, 8(1), 6-12 (1968).
Yeager, E., Fuel cells, Chem Eng Prog, 64(9), 92-96 (1968).

1969

Austin, L.G., and Almaula, B.C., Fuel cells: A progress report, Chem Eng, 16 June, 85-91 (1969).
Binns, D.T.; Kantyka, T.A., and Welland, R.C., Design of a tubular reactor with optimum temperature profile, Trans IChemE, 47, T53-58 (1969).
Bush, S.F., The design and operation of single-phase jet-stirred reactors for chemical kinetic studies, Trans IChemE, 47, T59-72 (1969).
Caldwell, A.D., and Calderbank, P.H., Catalyst dilution for temperature control in packed tubular reactors, Brit Chem Eng, 14(9), 1199-1201 (1969).
Corrigan, T.E., Evaluation of the analytical equations for adiabatic tubular reactors, Brit Chem Eng, 14(1), 59 (1969).
Edge, R.F., and Thompson, B.H., Reactor design in relation to modern high-pressure processes for fuel-gas manufacture, Chem Engnr, April, CE130-136 (1969).
Evangelista, J.J.; Katz, S., and Shinnar, R., Scale-up criteria for stirred-tank reactors, AIChEJ, 15(6), 843-853 (1969).
Floyd, J.R., Uses of aluminous materials, Hyd Proc, 48(4), 137-142 (1969).
Huang, I.D., and Dauerman, L., Predicting optimum reaction times in complex kinetic systems, Brit Chem Eng, 14(7), 969-972 (1969).
Keairns, D.L., and Manning, F.S., Model simulation of adiabatic continuous-flow stirred-tank reactors, AIChEJ, 15(5), 660-665 (1969).
Kitzen, M.R.; Wall, F.M., and Cijfer, H.J., Gas oil pyrolysis in tubular reactors, Chem Eng Prog, 65(7), 71-76 (1969).

Koo, L., and Ziegler, E.N., Design of multistage reactors for nonlinear kinetics, Chem Eng Sci, 24(2), 217-222 (1969).
Koo, L., and Ziegler, E.N., Reactor design made simple, Chem Eng, 24 Feb, 91-95 (1969).
Minhas, S., and Carberry, J.J., Adiabatic catalytic reactor simulation: Oxidation of sulphur dioxide, Brit Chem Eng, 14(6), 799-802 (1969).
Narsimhan, G., Optimisation of adiabatic reactor sequence with heat exchanger cooling, Brit Chem Eng, 14(10), 1402-1404 (1969).
Peterson, R.P.; Smith, D.E., and May, D.R., Reactor design from reaction kinetics, Chem Eng, 11 Aug, 101-106 (1969).
Porubszky, I.; Simonyi, E., and Ladanyi, G., Mathematical models describing ammonia synthesis reactors, Brit Chem Eng, 14(4), 495-496 (1969).
Ramirez, W.F., and Turner, B.A., Dynamic modeling, stability, and control of continuous stirred-tank chemical reactor, AIChEJ, 15(6), 853-860 (1969).
Rao, A.M., and Corrigan, T.E., Bode diagrams for some chemical reactors, Brit Chem Eng, 14(10), 1406-1407 (1969).
Small, W.M., Scaleup problems in reactor design, Chem Eng Prog, 65(7), 81-82 (1969).
Ting, A.P., and Wan, S.W., Sizing CO shift converters, Chem Eng, 19 May, 185-191 (1969).
Valstar, J.M.; Bik, J.D., and van den Berg, P.J., Calculations and measurements on models for fixed-bed tubular reactor with homogeneous reaction, Chem Engnr, April, CE136-143 (1969).
Weber, A.P., Residence-time spectrum in continuous-flow reactors, Chem Eng, 3 Nov, 79-80 (1969).
Zaloudik, P., Mixed model for continuous stirred-tank reactor with viscous fluid from experimental age-distribution study, Brit Chem Eng, 14(5), 657-659 (1969).

1970

Avruj, A., and Maymo, J.A., Charts for semibatch reactor design, Brit Chem Eng, 15(6), 776-777 (1970).
Bongiorno, S.J.; Connor, J.M., and Walton, B.C., Module concept of reformer design, Chem Eng Prog, 66(2), 64-67 (1970).
Craven, P., Effect of gas maldistribution on catalyst performance, Brit Chem Eng, 15(7), 918-919 (1970).
Flower, J.R.; Brooks, J.R., and Waite, O.E., Reactor optimization using a parallel-logic analogue computer, Brit Chem Eng, 15(5), 647-652 (1970).
Fournier, C.D., and Groves, F.R., Rapid method for calculating reactor temperature profiles, Chem Eng, 15 June, 157-159 (1970).
Gupalo, Y.P., and Ryazantsev, Y.S., Steady-state operating regimes of isothermal chemical flow reactors, Int Chem Eng, 10(1), 64-66 (1970).
Nebrensky, J.; Kristofovic, F., and Kosik, K., Reactors with pneumatic mixing, Int Chem Eng, 10(2), 262-268 (1970).
Ogorodnik, I.M., and Kafarov, V.V., Determination of maximum value of coefficients for scaling-up heat exchange apparatus used as a chemical reactor, Int Chem Eng, 10(3), 413-415 (1970).

Panchenkov, G.M.; Lazyan, Y.I.; Kozlov, M.V., and Zhorov, Y.M., Mathematical model of regenerator in catalytic cracking unit, Int Chem Eng, 10(3), 409-413 (1970).
Pexidr, V.; Hlubucek, V., and Pasek, J., Use of a model in designing a reactor for benzene hydrogenation, Int Chem Eng, 10(2), 188-198 (1970).
Porter, K., Design and performance of a semi-continuous plug-flow reactor fluidised bed, Brit Chem Eng, 15(6), 788-791 (1970).
Reith, T., and Beek, W.J., Bubble coalescence rates in a stirred tank contactor, Trans IChemE, 48, T63-68 (1970).
Rietema, K., and Ottengraf, S.P.P., Laminar liquid circulation and bubble street formation in a gas-liquid system, Trans IChemE, 48, T54-62 (1970).
Various, Reactors (special report), Hyd Proc, 49(6), 79-96 (1970).
Yen, Y.C., Bigger reactors or more recycle, Hyd Proc, 49(1), 157-161 (1970).

1971
Bor, T., The static mixer as a chemical reactor, Brit Chem Eng, 16(7), 610-612 (1971).
Gillespie, G.R., and Kenson, R.E., Catalyst system for ammonia oxidation to nitric acid, Chemtech, Oct, 627-632 (1971).
Hattiangadi, U.S., Calculate reactor residence time, Chem Eng, 6 Sept, 104-106 (1971).
Horacek, J., and Puschaver, S., Economics of dimensionally stable anodes, Chem Eng Prog, 67(3), 71-74 (1971).
Lund, M.M., and Seagrave, R.C., Variable-volume operation of a stirred-tank reactor, Ind Eng Chem Fund, 10(3), 494-504 (1971).
Lund, M.M., and Seagrave, R.C., Optimal operation of a variable-volume stirred-tank reactor, AIChEJ, 17(1), 30-37 (1971).
Slinko, M.G., Mathematical modelling of chemical reactors, Brit Chem Eng, 16(4), 363-364 (1971).

1972
Eschenbrenner, G.P., and Wagner, G.A., New high-capacity ammonia converter, Chem Eng Prog, 68(1), 62-66 (1972).
Harano, Y., and Matsura, T., Problems in designing photochemical reactors, Int Chem Eng, 12(1), 131-143 (1972).
Hill, S., and Hill, J.M., Distribution of residence times in a series of stirred reactors of unequal volumes, J Appl Chem Biotechnol, 22, 1277-1282 (1972).
Joffe, B.L., and Sargent, R.W.H., The design of an on-line control scheme for a tubular catalytic reactor, Trans IChemE, 50, 270-282 (1972).
Leitman, R.H., and Ziegler, E.N., Stirred-tank reactor studies, Chem Eng J, 2(4), 252-260 (1971); 3(3), 245-255 (1972).
Luus, R., Important considerations in reactor design, AIChEJ, 18(2), 438-439 (1972).
Mack, W.A., Bulk polymerization in screw-conveyor reactors, Chem Eng, 15 May, 99-102 (1972).
Narayanan, T.K., Simple method of estimating number of stages in an adiabatic reactor, Brit Chem Eng, 17(5), 419 (1972).

Potapchuk, V.S., and Kartsynel, M.B., Mathematical modelling of a waste-heat boiler as a chemical reactor, Int Chem Eng, 12(2), 254-257 (1972).

Vinter, A.A.; Dorozhkina, L.N., and Gorodetskii, I.Y., Determination of interfacial area in cocurrent bubbling reactors sectioned by sieve trays, Int Chem Eng, 12(3), 392-396 (1972).

Zimmerer, R.I., Predicting the time to arrive at a new steady state, Chem Eng, 30 Oct, 136 (1972).

1973

Allen, D.W., and Yen, W.H., Methanator design and operation, Chem Eng Prog, 69(1), 75-79 (1973).

Barona, N., and Prengle, H.W., Reactor design for liquid-phase processes, Hyd Proc, 52(3), 63-79; 52(12), 73-89 (1973).

Beckmann, G., Design of large polymerization reactors, Chemtech, May, 304-310 (1973).

Berryman, J.E., and Himmelblau, D.M., Stochastic analysis of a well-mixed tank reactor with heat transfer, Ind Eng Chem Fund, 12(3), 310-317 (1973).

Caha, J.; Hlavacek, V.; Kubicek, M., and Marek, M., Optimization of adiabatic reactor, Int Chem Eng, 13(3), 466-473 (1973).

Crescitelli, S., and Nicoletti, B., Near optimal control of batch reactors. Chem Eng Sci, 28(2), 463-472 (1973).

El-Rifai, M.A., and Saleh, M.A., Steady-state analysis of a class of heterogeneous reactors, Trans IChemE, 51, 22-28 (1973).

Koetsier, W.T.; Thoenes, D., and Frankena, J.F., Mass transfer in a closed stirred gas-liquid contactor, Chem Eng J, 5(1), 61-76 (1973).

Markovitz, R.E., Improve heat transfer with electropolished clad reactors, Hyd Proc, 52(8), 117-118 (1973).

Marroquin, G., and Luyben, W.L., Practical control studies of batch reactors using realistic mathematical models, Chem Eng Sci, 28(4), 993-1004 (1973).

Naidel, R.W., Hydrogen chloride production by combustion in graphite vessels, Chem Eng Prog, 69(2), 53-56 (1973).

Pierru, A., and Alexandre, C., Optimize PVC reactor, Hyd Proc, 52(6), 97-100 (1973).

Smith, J.M., Heat transfer in fixed-bed reactors (a review), Chem Eng J, 5(2), 109-116 (1973).

Stepanek, J.B., and Shilimkan, R.V., Residence time distribution in an axially-dispersed plug flow with capacity, Trans IChemE, 51, 112-115 (1973).

Williams, A., Flame and plasma reactors: A review, Trans IChemE, 51, 199-231 (1973).

1974

Bailey, J.E., Periodic operation of chemical reactors (review paper), Chem Eng Commns, 1(3), 111-124 (1974).

Berty, J.M., Reactor for vapor-phase catalytic studies, Chem Eng Prog, 70(5), 78-84 (1974).

Greaves, M., and Allan, B.W., Steady-state and dynamic characteristics of flotation in a single cell, Trans IChemE, 52, 136-148 (1974).

Hazbun, E.A., and White, J.W., Simulation, design and analysis of reactors, Chem Eng Prog, 70(7), 83-85 (1974).

Kantyka, T.A., Reactor development and design: Application of mathematical models, Chem Engnr, March, 141-144 (1974).

Kao, Y.K., and Bankoff, S.G., Optimal design of a chemical reactor with fast and slow reactions, Chem Eng Commns, 1(3), 141-154 (1974).

Mecklenburgh, J.C., Backmixing and design: A review, Trans IChemE, 52, 180-192 (1974).

Murthy, A.K.S., Material balance around a chemical reactor, Ind Eng Chem Proc Des Dev, 12(3), 246-248 (1973); 13(4), 347-349 (1974).

Oh, S.H., and Schmitz, R.A., Control of tubular reactor with recycle, Chem Eng Commns, 1(4), 199-216 (1974).

1975

Albright, L.F., and Bild, C.G., Designing reaction vessels for polymerization, Chem Eng, 15 Sept, 121-128 (1975).

Apostolopoulos, G.P., Parametric pump as chemical reactor, Ind Eng Chem Fund, 14(1), 11-16 (1975).

Bonem, J.M., Troubleshooting polymerization reactors, Chem Eng, 9 June, 91-94 (1975).

Dannatt, P.C., The steam generating heavy water reactor, Chem Engnr, April, 232-234 (1975).

Fleischmann, M., and Jansson, R.E.W., Characterisation of electrochemical reactors, Chem Engnr, Oct, 603-606 (1975).

Gerrard, A.M., and Liddle, C.J., Optimise batch reactors, Processing, Oct, 25 (1975).

Goto, S., and Smith, J.M., Trickle-bed reactor performance, AIChEJ, 21(4), 706-720 (1975).

Hong, H.M., and MacGregor, J.F., Identification and direct digital stochastic control of CSTR, Can JCE, 53, 211-220 (1975).

Hosegood, S.B., The high temperature reactor (HTR) as an industrial heat source, Chem Engnr, April, 238-241 (1975).

Insch, G.M., The high temperature reactor (HTR) as a power source, Chem Engnr, April, 235-237 (1975).

King, C.J.H.; Lister, K., and Plimley, R.E., A novel bi-polar electrolytic flow cell for synthesis, Trans IChemE, 53, 20-25 (1975).

McKelvey, K.N.; Yieh, H.N.; Zakanycz, S., and Brodkey, R.S., Turbulent motion, mixing, and kinetics in a chemical reactor configuration, AIChEJ, 21(6), 1165-1176 (1975).

Radcliffe, S.W., and Hickman, R.G., Diffusive catalytic combustors, J Inst Fuel, 48, 208-214 (1975).

Satterfield, C.N., Trickle-bed reactors (review paper), AIChEJ, 21(2), 209-228 (1975).

Shyam, R.; Davidson, B., and Vieth, W.R., Mass transfer and biochemical reaction in enzyme membrane reactor systems, Chem Eng Sci, 30(7), 669-678 (1975).

Sutherland, K.S., The place of the fast reactor, Chem Engnr, April, 231, 234 (1975).

Valstar, J.M.; Van den Berg, P.J., and Oyserman, J., Comparison between two-dimensional fixed bed reactor calculations and measurements, Chem Eng Sci, 30(7), 723-728 (1975).
Various, Nuclear reactors (topic issue), Chem Engnr, April, 231-244 (1975).

1976

Andres, R.P., and Hile, L.R., Homogeneous reactor experiments, Chem Eng Educ, 10(1), 18-22 (1976).
Cooke, B., Computer control of batch reactors, Proc Engng, Dec, 65-66 (1976).
Hansen, K.W., and Jorgensen, S.B., Dynamic modelling of gas-phase catalytic fixed-bed reactor, Chem Eng Sci, 31(7), 579-598 (1976).
Ibberson, V.J., and Sen, M., Plasma jet reactor design for hydrocarbon processing, Trans IChemE, 54, 265-275 (1976).
Joshi, J.B., and Sharma, M.M., Mass transfer characteristics of horizontal sparged contactors, Trans IChemE, 54, 42-53 (1976).
Kim, S.S., and Cooney, D.O., Improved theoretical model for hollow-fibre enzyme reactors, Chem Eng Sci, 31(4), 289-294 (1976).
Leuteritz, G.M.; Reimann, P., and Vergeres, P., Loop reactors: Better gas/liquid contact, Hyd Proc, 55(6), 99-100 (1976).
Robertson, C.R.; Michaels, A.S., and Waterland, L.R., Molecular separation barriers and their application to catalytic reactor design, Sepn & Purif Methods, 5(2), 301-332 (1976).
Sinkule, J.; Votruba, J.; Hlavacek, V., and Hofmann, H., Modeling of chemical reactors, Chem Eng Sci, 31(1), 23-36 (1976).
Uppal, A.; Ray, W.H., and Poore, A.B., Classification of dynamic behavior of continuous stirred tank reactors: Influence of reactor residence time, Chem Eng Sci, 31(3), 205-214 (1976).
Van Riet, K.; Boom, J.M., and Smith, J.M., Power consumption, impeller coalescence and reciriculation in aerated vessels, Trans IChemE, 54, 124-131 (1976).
Wirges, H.P., and Shah, S.R., For a given kinetic duty, select optimum reactor quickly, Hyd Proc, 55(4), 135-138 (1976).
Zanker, A., Nomograph for hydrocarbon cracking times, Proc Engng, Oct, 90-91 (1976).

1977

Bruns, D.D., and Bailey, J.E., Nonlinear feedback control for operating a nonisothermal CSTR near an unstable steady state, Chem Eng Sci, 32(3), 257-264 (1977).
Hanika, J.; Sporka, K.; Ruzicka, V., and Pistek, R., Dynamic behaviour of an adiabatic trickle-bed reactor, Chem Eng Sci, 32(5), 525-528 (1977).
Hinger, K.J., and Blenke, H., Economic optimization of a chemical reactor, Int Chem Eng, 17(1), 28-35 (1977).
Hofmann, H., Hydrodynamics, transport phenomena, and mathematical models in trickle-bed reactors, Int Chem Eng, 17(1), 19-28 (1977).
Hsu, E.H., and Bacher, S., Interacting feedback control system for gas-liquid reactor, Ind Eng Chem Fund, 16(2), 259-263 (1977).
Juvekar, V.A., and Sharma, M.M., Some aspects of process design of gas/liquid reactors, Trans IChemE, 55, 77-92 (1977).

Nauman, E.B., Nonisothermal reactors: Theory and application of thermal time distributions, Chem Eng Sci, 32(4), 359-368 (1977).
Ritchie, B.W., and Tobgy, A.H., General population balance modelling of unpremixed feed-stream chemical reactors (review paper), Chem Eng Commns, 2(4), 249-264 (1977).
Sriram, K., and Mann, R., Dynamic gas disengagement: New technique for assessing behaviour of bubble columns, Chem Eng Sci, 32(6), 571-580 (1977).
Various, Computer design and control of batch reaction systems (topic issue), Chem & Ind, 2 April, 246-266; 16 April, 293-308 (1977).

1978

Braun, R.L., Low-temperature pilot-plant batch reactors, Chem Eng, 16 Jan, 129-134 (1978).
Brisk, M.L., and Barton, G.W., On-line optimal control of catalytic reactors: A simplified approach, Trans IChemE, 56, 113-119 (1978).
Brusset, H.; Depeyre, D.; Richard, C., and Richard, P., Dynamic programming for optimization of multistage reactors, Ind Eng Chem Proc Des Dev, 17(4), 355-358 (1978).
Devia, N., and Luyben, W.L., Reactors: Size versus stability, Hyd Proc, 57(6), 119-122 (1978).
Gianetto, A.; Baldi, G.; Specchia, V., and Sicardi, S., Hydrodynamics and solid-liquid contacting effectiveness in trickle-bed reactors (review paper), AIChEJ, 24(6), 1087-1104 (1978).
Hanley, T.R., and Mischke, R.A., Mixing model for a continuous-flow stirred-tank reactor, Ind Eng Chem Fund, 17(1), 51-58 (1978).
Hinkle, R.E., and Friedman, J., Controlling heat-transfer systems for glass-lined reactors, Chem Eng, 30 Jan, 101-104 (1978).
Ishikawa, T., and Keister, R.G., New advanced cracking reactor, Hyd Proc, 57(12), 109-113 (1978).
Litchfield, R.J., Conversion of a CSTR to multistage operation, Trans IChemE, 56, 67-72 (1978).
Luyben, W.L., and Melcic, M., Consider reactor control lags, Hyd Proc, 57(3), 115-117 (1978).
Morita, S., and Smith, J.M., Mass transfer and contacting efficiency in trickle-bed reactor, Ind Eng Chem Fund, 17(2), 113-120 (1978).
Quartulli, O.J., and Wagner, G.A., Horizontal ammonia converters, Hyd Proc, 57(12), 115-122 (1978).
Radosz, M., and Kramarz, J., Predicting catalytic reformer yield, Hyd Proc, 57(7), 201-203 (1978).
Schweich, D., and Villermaux, J., The chromatographic reactor: A new theoretical approach, Ind Eng Chem Fund, 17(1), 1-7 (1978).
Shah, Y.T.; Stiegel, G.J., and Sharma, M.M., Backmixing in gas-liquid reactors (review paper), AIChEJ, 24(3), 369-400 (1978).
Shettigar, U.R., and Venkateswaran, V., Simulation of a non-catalytic adiabatic fixed-bed reactor, Chem Eng J, 16(3), 165-170 (1978).

1979

Barona, N., Design CSTR reactors this way, Hyd Proc, 58(7), 179-194 (1979).

Basaran, M., and Mann, R., Variable flow start-up of a back-mixed reactor exhibiting multiple steady states, Trans IChemE, 57, 182-187 (1979).
Brooks, B.W., Start-up dynamic behaviour of a chemical reactor, Chem Eng Sci, 34(12), 1417-1419 (1979).
Corpstein, R.R.; Dove, R.A., and Dickey, D.S., Stirred-tank reactor design, Chem Eng Prog, 75(2), 66-74 (1979).
Deckwer, W.D., Modelling and dimensioning of bubble-column reactors, Int Chem Eng, 19(1), 21-33 (1979).
Degnan, T.F., and Wei, J., Cocurrent reactor-heat exchanger, AIChEJ, 25(2), 338-344 (1979).
Hearfield, F., Thermal properties of reactors and some instabilities, Chem Engnr, March, 156-160 (1979).
Herskowitz, M.; Carbonell, R.G., and Smith, J.M., Effectiveness factors and mass transfer in trickle-bed reactors, AIChEJ, 25(2), 272-283 (1979).
Horak, J., Direct and indirect modelling in design of a chemical reactor, Int Chem Eng, 19(3), 535-540 (1979).
Kovenklioglu, S., and DeLancey, G.B., Optimal insulation for tubular reactors, Chem Eng Sci, 34(1), 150-155 (1979).
Lacksonen, J.W., Material balance calculations with reaction, Chem Eng Educ, 13(2), 92-94 (1979).
Litchfield, R.J.; Campbell, K.S., and Locke, A., The application of several Kalman filters to the control of a real chemical reactor, Trans IChemE, 57, 113-120 (1979).
Luckenbach, E.C., How to update a catalytic cracking unit, Chem Eng Prog, 75(2), 56-61 (1979).
Petrov, P.S., et al., Gas hold-up in aerated reactors, Chem Eng J, 17(3), 169-172 (1979).
Plumb, K.C., Sampling system for stirred tank reactors, Chem Engnr, May, 361 (1979).
Ponzi, P.R., and Kaye, L.A., Effect of flow maldistribution on conversion and selectivity in radial flow fixed-bed reactors, AIChEJ, 25(1), 100-108 (1979).
Raghuram, S.; Shah, Y.T., and Tierney, J.W., Multiple steady states in a gas-liquid reactor, Chem Eng J, 17(1), 63-76 (1979).
Ramachandran, P.A., and Smith, J.M., Dynamic behaviour of trickle-bed reactors, Chem Eng Sci, 34(1), 75-92 (1979).
Ramachandran, P.A., and Smith, J.M., Mixing-cell method for design of trickle-bed reactors, Chem Eng J, 17(2), 91-100 (1979).
Roth, D.D.; Basaran, V., and Seagrove, R.C., Mixing effects in variable-volume chemical reactors, Ind Eng Chem Fund, 18(4), 376-383 (1979).
Schmitz, R.A.; Bautz, R.R.; Ray, W.H., and Uppal, A., Dynamic behavior of CSTR, AIChEJ, 25(2), 289-297 (1979).
Silva, J.M.; Wallman, P.H., and Foss, A.S., Multi-bed catalytic reactor control systems: Configuration development and experimental testing, Ind Eng Chem Fund, 18(4), 383-391 (1979).
Trambouze, P., Reactor scaleup philosophy, Chem Eng, 10 Sept, 122-130 (1979).
Wallman, P.H.; Silva, J.M., and Foss, A.S., Multivariable integral controls for fixed-bed reactors, Ind Eng Chem Fund, 18(4), 392-400 (1979).

1980

Ahmed, M., and Fahien, R.W., Tubular reactor design, Chem Eng Sci, 35(4), 889-904 (1980).
Booth, A.D.; Karmarkar, M.; Knight, K., and Potter, R.C.L., Design of emergency venting system for phenolic resin reactors, Trans IChemE, 58, 75-90, 276 (1980).
Botton, R.; Cosserat, D., and Charpentier, J.C., Operating zone and scale-up of mechanically stirred gas-liquid reactors, Chem Eng Sci, 35(1), 82-89 (1980).
Carra, S., and Santacesaria, E., Engineering aspects of gas-liquid catalytic reactions, Catalysis Reviews, 22(1), 75-140 (1980).
Chaudhari, R.V., and Ramachandran, P.A., Three-phase slurry reactors (a review), AIChEJ, 26(2), 177-201 (1980).
Chen, J.M., and Yang, R.T., Modelling and kinetic studies of a rotary kiln reactor, Trans IChemE, 58, 98-106 (1980).
Cukierman, A.L., and Lemcoff, N.O., Heat and mass transfer in a tank reactor stirred by gaseous jets, Chem Eng J, 19(2), 125-130 (1980).
Degnan, T.F., and Wei, J., The co-current reactor-heat exchanger, AIChEJ, 26(1), 60-67 (1980).
Eissa, S.H., Gas holdup in stirred reactors, Chem Eng, 10 March, 170 (1980).
Hearfield, F., Adipic acid reactor development: Benefits in energy and safety, Chem Engnr, Oct, 625-627, 633 (1980).
Hinduja, M.J.; Sundaresan, S., and Jackson, R., Crossflow model of dispersion in packed bed reactors, AIChEJ, 26(2), 274-281 (1980).
Hsia, M.A., and Tavlarides, L.L., A simulation model for homogeneous dispersions in stirred tanks, Chem Eng J, 20(3), 225-236 (1980).
Kulkarni, B.D., and Doraiswamy, L.K., Estimation of effective transport properties in packed bed reactors, Catalysis Reviews, 22(3), 431-484 (1980).
Langner, F.; Moritz, H.U., and Reichert, K.H., Reactor scale-up for polymerization in suspension, Chem Eng Sci, 35(1), 519-526 (1980).
Le Cardinal, G.; Germain, E.; Gelus, M., and Guillon, B., Design of a stirred batch polymerization reactor, Chem Eng Sci, 35(1), 499-505 (1980).
Mochizuki, S., Design and performance of liquid-phase oxidation process: Laboratory and pilot-plant studies, Chem Eng Commns, 4(4), 547-556 (1980).
Ramachandran, P.A., and Chaudhari, R.V., Estimation of batch time of a semi-batch three-phase slurry reactor, Chem Eng J, 20(1), 75-78 (1980).
Ramachandran, P.A., and Chaudhari, R.V., Predicting performance of three-phase catalytic reactors, Chem Eng, 1 Dec, 74-85 (1980).
Sadana, A., Optimum temperature operations in deactivating fixed-bed reactors, Chem Eng Commns, 4(1), 51-56 (1980).
Various, Design studies of reactor models (topic issue), Chem Eng Sci, 35(1), 341-388 (1980).
Various, Dynamics, stability and control of chemical reactors (topic issue), Chem Eng Sci, 35(1), 241-294 (1980).
Werther, J., Mathematical modelling of fluidized bed reactors, Int Chem Eng, 20(4), 529-542 (1980).

1981

Balakotaiah, V., and Luss, D., Analysis of multiplicity patterns of a CSTR, Chem Eng Commns, 13(1), 111-132 (1981).

Barreto, G.F.; Ferretti, O.A.; Farina, I.H., and Lemcoff, N.O., Optimization of the operating conditions of CO converters, Ind Eng Chem Proc Des Dev, 20(4), 594-603 (1981).

Brodkey, R.S., Fundamentals of turbulent motion, mixing and kinetics in reactor design (review paper), Chem Eng Commns, 8(1), 1-24 (1981).

Colton, J.W., Carbon deposition in gas reactions, Hyd Proc, 60(1), 177-184 (1981).

Hammer, H., Bubble column reactors with suspended solids: Fundamentals, design and uses, Int Chem Eng, 21(2), 173-180 (1981).

Joosten, G.E.H.; Hoogstraten, H.W., and Ouwerkerk, C., Flow stability of multitubular continuous polymerization reactors, Ind Eng Chem Proc Des Dev, 20(2), 177-182 (1981).

Joshi, J.B., and Shah, Y.T., Hydrodynamic and mixing models for bubble column reactors (review paper), Chem Eng Commns, 11(1), 165-200 (1981).

Kiparissides, C., and Ponnuswamy, S.R., Hierarchical control of a train of continuous polymerization reactors, Can JCE, 59, 752-759 (1981).

Lee, P.L.; Newell, R.B., and Agnew, J.B., A comparative study of control techniques applied to an unstable steady-state of a chemical reactor, Trans IChemE, 59, 105-111 (1981).

Nagel, O.; Hegner, B., and Kurten, H., Criteria for selection and design of gas/liquid reactors, Int Chem Eng, 21(2), 161-173 (1981).

Nauman, E.B., Residence time distributions and micromixing (review paper), Chem Eng Commns, 8(1), 53-132 (1981).

Patterson, G.K., Application of turbulence fundamentals to reactor modelling and scaleup (review paper), Chem Eng Commns, 8(1), 25-52 (1981).

Shinnar, R., Chemical reactor modelling for controller design, Chem Eng Commns, 9(1), 73-100 (1981).

Tavlarides, L.L., Modelling and scaleup of dispersed phase liquid-liquid reactors (review paper), Chem Eng Commns, 8(1), 133-164 (1981).

Trotta, A., and DelGiudice, S., Behaviour of tubular reactors of arbitrary cross-section, Trans IChemE, 59, 207-210 (1981).

Umoh, N.F.; Hughes, R., and Harriott, P., Temperature dynamics and start-up of a continuous-flow stirred-tank polymerisation reactor, Chem Eng J, 21(2), 85-100 (1981).

Waghmare, R.S., and Lim, H.C., Optimal operation of isothermal reactors, Ind Eng Chem Fund, 20(4), 361-368 (1981).

Waller, K.V., and Makila, P.M., Chemical reaction invariants and variants and their use in reactor modeling, simulation and control (a review), Ind Eng Chem Proc Des Dev, 20(1), 1-11 (1981).

Werther, J., and Hegner, B., Optimum operating conditions of industrial fluidized-bed reactors, Int Chem Eng, 21(4), 585-595 (1981).

1982

Bartholomew, C.H., Carbon deposition in steam reforming and methanation, Catalysis Reviews, 24(1), 67-112 (1982).

Bourne, J.R., and Gros, H., The influence of heat effects on the modelling and performance of staged gas-liquid reactors, Trans IChemE, 60, 97-107 (1982).

Christoffel, E.G., Laboratory reactors and heterogeneous catalytic processes, Catalysis Reviews, 24(2), 159-232 (1982).

Gerrens, H., How to select polymerization reactors, Chemtech, June, 380-383; July, 434-443 (1982).

Guy, J.L., Dynamic modeling of tubular reactor systems, Chem Eng, 23 Aug, 91-94 (1982).

Guy, J.L., Dynamic modeling of tank-type reactor systems, Chem Eng, 28 June, 97-100 (1982).

Hlavacek, V., and Kubicek, M., Calculation of hollow box reactors, Chem Eng Commns, 17(1), 1-12 (1982).

Joshi, J.B.; Pandit, A.B., and Sharma, M.M., Mechanically agitated gas-liquid reactors (review paper), Chem Eng Sci, 37(6), 813-844 (1982).

King, R.W., Reactor modelling, Proc Engng, Dec, 49-53 (1982).

Kovarik, F.S., and Butt, J.B., Reactor optimization in the presence of catalyst decay, Catalysis Reviews, 24(4), 441-502 (1982).

Shah, Y.T.; Kelkar, B.G.; Godbole, S.P., and Deckwer, W.D., Design parameter estimations for bubble column reactors: A review, AIChEJ, 28(3), 353-379 (1982).

Svoronos, S.; Aris, R., and Stephanopoulos, G., Behavior of two stirred tanks in series, Chem Eng Sci, 37(3), 357-366 (1982).

Tilton, J.N., and Russell, T.W.F., Designing gas-sparged vessels for mass transfer, Chem Eng, 29 Nov, 61-68 (1982).

Yang, R.Y.K., and Colton, C.K., Computer-aided analysis and design of an ATP regeneration reactor, Chem Eng Commns, 16(1), 227-244 (1982).

Zahradnik, J.; Kratochvil, J.; Kastanek, F., and Rylek, M., Energy effectiveness of bubble-column reactors with sieve tray and ejector-type gas distributors, Chem Eng Commns, 15(1), 27-40 (1982).

Zanker, A., Comparing reactor efficiencies. Proc Engng, July, 61 (1982).

Zardi, U., Review of developments in ammonia and methanol reactors, Hyd Proc, Aug, 129-133 (1982).

Zenz, F.A., Scaleup fluid bed reactors, Hyd Proc, Jan, 155-156 (1982).

1983

Abugaber, H., and Pennington, J., Tube crimping keeps reformers on stream, Hyd Proc, Feb, 51-53 (1983).

Armstrong, W.S., and Coe, B.F., Multi-product batch reactor control, Chem Eng Prog, 79(1), 56-61 (1983).

Berty, J.M., Scaledown of a methanol reactor, Chem Eng Prog, 79(7), 64-67 (1983).

Handman, S.E., and LeBlanc, J.R., The horizontal ammonia converter, Chem Eng Prog, 79(5), 56-61 (1983).

Herskowitz, M., and Smith, J.M., Trickle-bed reactors: A review, AIChEJ, 29(1), 1-18 (1983).

Karlsson, H.T., Integral evaluation of tubular reactor data, Chem Eng Commns, 21(1), 37-46 (1983).

Laurent, A., and Charpentier, J.C., Use of experimental laboratory-scale models in predicting performance of gas-liquid reactors, Int Chem Eng, 23(2), 265-275 (1983).

Lee, H.H., and Ruckenstein, E., Catalyst sintering and reactor design, Catalysis Reviews, 25(4), 475-550 (1983).

Noble, R.D.; Jacquot, R.G., and Baldwin, L.B., Laboratory experiment for transient response of a stirred vessel, Chem Eng Educ, 17(2), 70-72 (1983).

Paterson, W.R., and Carberry, J.J., Fixed-bed catalytic reactor modelling: Heat transfer problems, Chem Eng Sci, 38(1), 175-180 (1983).

Turner, J.C.R., Perspectives on residence-time distributions, Chem Eng Sci, 38(1), 1-4 (1983).

van Deemter, J.J., Reactor modelling: Effect of scale, Chem Engnr, March, 22-23 (1983).

Villermaux, J., and David, R., Recent advances in understanding micromixing phenomena in stirred reactors, Chem Eng Commns, 21(1), 105-122 (1983).

Wei, C.N., and Ochiai, S., Tracer-response techniques in evaluation of mixing in reactors, Plant/Opns Prog, 2(2), 123-127 (1983).

1984

Anon., Temperature control for batch processes, Chem Engnr, April, 29-31 (1984).

Berty, J.M., 20 years of recycle reactors in reaction engineering, Plant/Opns Prog, 3(3), 163-168 (1984).

Capuder, E., and Koloini, T., Gas hold-up and interfacial area in aerated suspensions of small particles, CER&D, 62, 255-260 (1984).

Duarte, S.I.P.; Ferretti, O.A., and Lemcoff, N.O., Comparison of two-dimensional models for fixed bed catalytic reactors, Chem Eng Sci, 39(6), 1017-1024 (1984).

Hoose, J., Use liquid nitrogen to cool reactors, Hyd Proc, Oct, 71-72 (1984).

Koh, P.T.L.; Andrews, J.R.G., and Uhlherr, P.H.T., Flocculation in stirred tanks, Chem Eng Sci, 39(6), 975-986 (1984).

Martin, G.Q., Multiphase mixing problems in a reactor, Chem Eng Prog, 80(6), 68-73 (1984).

Shaikh, A.A., and Carberry, J.J., Model of isothermal transport line (riser) and moving-bed catalytic reactor, CER&D, 62, 387-390 (1984).

Smith, R.E.; Humphreys, G.C., and Griffiths, G.W., Optimize large methanol plant reactors, Hyd Proc, May, 95-100 (1984).

Tan, C.S., and Smith, J.M., Effectiveness factors for endothermic reactions in fixed-bed reactors, Chem Eng Sci, 39(9), 1329-1338 (1984).

Tarmy, B.; Chang, M.; Coulaloglou, C., and Ponzi, P., Hydrodynamic characteristics of three phase reactors, Chem Engnr, Oct, 18-23 (1984).

Trimm, D.L., Control of coking, Chem Eng & Proc, 18(3), 137-148 (1984).

Wagner, M.C.; Humes, W.H., and Magnabosco, L.M., Fully automated catalytic cracking test unit, Plant/Opns Prog, 3(4), 222-227 (1984).

Watanabe, N.; Kurimoto, H., and Matsubara, M., Periodic control of continuous stirred tank reactors, Chem Eng Sci, 39(1), 31-36 (1984).

Westerterp, K.R., and Ptasinski, K.J., Safe design of cooled tubular reactors for exothermic, multiple reactions: Parallel reactions, Chem Eng Sci, 39(2), 235-252 (1984).

Whatley, M.J., and Pott, D.C., Adaptive gain improves reactor control, Hyd Proc, May, 75-78 (1984).

Yeow, Y.L., Dynamics of a stirred tank experiment, Chem Eng Educ, 18(2), 78-81 (1984).

1985

Asfour, A.F.A., Improved design of a simple tubular reactor experiment, Chem Eng Educ, 19(2), 84-87 (1985).

Billet, R.; Mackowiak, J., and Pajak, M., Hydraulics and mass transfer in filled tube columns, Chem Eng & Proc, 19(1), 39-48 (1985).

Chitra, S.P., and Govind, R., Synthesis of optimal series reactor structures for homogeneous reactions, AIChEJ, 31(2), 177-194 (1985).

Cluett, W.R.; Shah, S.L., and Fisher, D.G., Adaptive control of a batch reactor, Chem Eng Commns, 38(1), 67-78 (1985).

Collins, G.M.; Hess, R.K., and Akgerman, A., Effect of volatile liquid phase on trickle bed reactor performance, Chem Eng Commns, 35(1), 281-292 (1985).

Costa, C., and Rodrigues, A., Regeneration of polymeric adsorbents in a CSTR, Chem Eng Sci, 40(5), 707-714 (1985).

Fajner, D.; Magelli, F.; Nocentini, M., and Pasquali, G., Solids concentration profiles in mechanically stirred and staged column slurry reactor, CER&D, 63, 235-240 (1985).

Goto, S., and Kojima, Y., Oxidation of sulfur dioxide in different types of three-phase reactors, Chem Eng Commns, 34(1), 213-224 (1985).

Huang, Y.J.; Lee, P.I.; Schwarz, J.A., and Heydweiller, J.C., Analysis of CSTR approximation under transient operation, Chem Eng Commns, 39, 355-370 (1985).

Isberg, P., Safer nuclear reactors for steam heating, Processing, June, 36,39 (1985).

Jiricny, V., and Stanek, V., Versatile correlation of liquid hold-up in two-phase countercurrent trickle-bed column, Chem Eng Commns, 35(1), 253-266 (1985).

Kim, D.H., and Chang, K.S., Analytical solution for tubular reactor model with dispersion in both radial and axial directions, Chem Eng Commns, 35(1), 293-304 (1985).

Krishna, A.S., and Parkin, E.S., Modeling the regenerator in commercial fluid catalytic cracking units, Chem Eng Prog, 81(4), 57-62 (1985).

Litz, L.M., A novel gas-liquid stirred tank reactor, Chem Eng Prog, 81(11), 36-39 (1985).

Muroyama, K.; Mitani,. Y., and Yasunishi, A., Hydrodynamic characteristics and gas-liquid mass transfer in a draft-tube slurry reactor, Chem Eng Commns, 34(1), 87-98 (1985).

Nowobilski, P., and Takoudis, C.G., Chemical oscillations in differential reactors: Case study of ammonia oxidation on platinum, Chem Eng Commns, 33(1), 211-218 (1985).
Russell, T.W.F., Semiconductor chemical reactor engineering and photovoltaic unit operations, Chem Eng Educ, 19(2), 72-77, 106-108 (1985).
Shuler, M.L., Use of chemically structured models for bioreactors (review paper), Chem Eng Commns, 36(1), 161-190 (1985).
Suzuki, S.; Shimizu, K., and Matsubara, M., Parameter-space classification of dynamic behavior of continuous microbial flow reactor, Chem Eng Commns, 33(5), 325-336 (1985).
Vayenas, C.G.; Debenedetti, P.G.; Yentekakis, I., and Hegedus, L.L., Cross-flow, solid-state electrochemical reactors: A steady-state analysis, Ind Eng Chem Fund, 24(3), 316-324 (1985).
Walas, S.M., Chemical reactor data, Chem Eng, 14 Oct, 79-83 (1985).
Wansbrough, R.W., Modeling chemical reactors, Chem Eng, 5 Aug, 95-102 (1985).
Westerterp, K.R., and Overtoom, R.R.M., Safe design of cooled tubular reactors for exothermic multiple reactions (consecutive reactions), Chem Eng Sci, 40(1), 155-166 (1985).
Yoon, B.J., and Rhee, H.K., Study of high-pressure polyethylene tubular reactor, Chem Eng Commns, 34(1), 253-266 (1985).

1986
Akyurtlu, A.; Akyurtlu, J.F.; Hamrin, C.E., and Fairweather, G., Reformulation and numerical solution of the equations for a catalytic porous-wall gas-liquid reactor, Comput Chem Eng, 10(4), 361-365 (1986).
Arandes, J.M., and Bilbao, J., Optimum feed policies for isothermal fixed bed reactors with deactivating catalysts under cyclic operation, CER&D, 64(6), 461-466 (1986).
Chaudhari, R.V.; Shah, Y.T., and Foster, N.R., Novel gas-liquid-solid reactors, Catalysis Reviews, 28(4), 431-518 (1986).
Chidambaram, M., Transient behaviour of particles in flow bubble column slurry reactors, Comput Chem Eng, 10(2), 115-118 (1986).
Clark, K.N., and Foster, N.R., Application of neutron techniques to studies of reactor fluid dynamics, Chem Eng J, 34(1) 35-46 (1986).
Davis, M.E., and Watson, L.T., Mathematical modeling of annular reactors, Chem Eng J, 33(3), 133-142 (1986).
Friedel, L., and Purps, S., Models and design methods for sudden depressurization of gas/vapor-liquid reaction systems, Int Chem Eng, 26(3), 396-408 (1986).
Gianetto, A., and Silveston, P.L. (Eds), Multiphase Chemical Reactors: Theory, Design, Scale-up, Springer-Verlag, Berlin (1986).
Grossel, S.S., Design and sizing of knock-out drums/catchtanks for reactor emergency relief systems, Plant/Opns Prog, 5(3), 129-135 (1986).
Hallaile, M., and Merchuk, J.C., Operation policies for gas-liquid stirred tank reactor, Chem Eng Commns, 46, 179-196 (1986).
Hoo, K.A., and Kantor, J.C., Linear feedback equivalence and control of an unstable biological reactor, Chem Eng Commns, 46, 385-399 (1986).

Ingleby, S.; Rossiter, A.P., and Douglas, J.M., Economic evaluation of integrated reactor systems, CER&D, 64(4), 241-247 (1986).

Iordache, O., and Corbu, S., Random residence-time distribution, Chem Eng Sci, 41(8), 2099-2102 (1986).

Kim, S.H., and Hlavacek, V., Detailed dynamics of coupled continuous stirred tank reactors, Chem Eng Sci, 41(11), 2767-2778 (1986).

Leigh, A.N., and Preece, P.E., Development of inclined-plate jet reactor system, Plant/Opns Prog, 5(1), 40-44 (1986).

Liptak, B.G., Controlling and optimizing chemical reactors, Chem Eng, 26 May, 69-81 (1986).

Mandler, J.A.; Morari, M., and Seinfeld, J.H., Robust multivariable control system design for a fixed-bed reactor, Ind Eng Chem Fund, 25(4), 645-656 (1986).

Mankin, J.C., and Hudson, J.L., Dynamics of coupled nonisothermal continuous stirred tank reactors, Chem Eng Sci, 41(10), 2651-2662 (1986).

Mann, R.; Stavridis, E., and Djamarani, K., Experimental fixed-bed reactor dynamics for sulphur dioxide oxidation, CER&D, 64(4), 248-257 (1986).

Merchuk, J.C., Gas hold-up and liquid velocity in a two-dimensional air-lift reactor, Chem Eng Sci, 41(1), 11-16 (1986).

Middleton, J.C.; Pierce, F., and Lynch, P.M., Computations of flow fields and complex reaction yield in turbulent stirred reactors and comparison with experimental data, CER&D, 64(1), 18-22 (1986).

Narendranathan, T.J., Designing fermentation equipment, Chem Engnr, May, 23-31 (1986).

Nowobilski, P.J., and Takoudis, C.G., Periodic operation of chemical reactor systems: Are global improvements attainable? Chem Eng Commns, 40, 249-264 (1986).

Pagani, G., and Monforte, A.A., Gas-liquid reactor design: Formulation and solution of a simple mathematical model, Comput Chem Eng, 10(2), 97-106 (1986).

Pandit, A.B., and Joshi, J.B., Mass and heat transfer characteristics of three-phase sparged reactors (review paper), CER&D, 64(2), 125-157 (1986).

Pratsinis, S.E., et al., Aerosol reactor design: Effect of reactor geometry on powder production and vapor deposition, Powder Tech, 47(1), 17-24 (1986).

Pratsinis, S.E.; Friedlander, S.K., and Pearlstein, A.J., Stability and dynamics of a continuous stirred-tank aerosol reactor, AIChEJ, 32(2), 177-185 (1986).

Recasens, F.; Alonso, M., and Puigjaner, L., Copolymer reactor operation for uniform product composition: Analysis of CSTR, Chem Eng & Proc, 20(2), 85-94 (1986).

Rossiter, A.P., and Ingleby, S., Integrated reactor systems: Design for variable operating conditions, CER&D, 64(5), 365-371 (1986).

Schugerl, K., Reaction engineering fundamentals relating to design and operation of bioreactors, Int Chem Eng, 26(2), 204-231 (1986).

Siegel, M.H.; Merchuk, J.C., and Schugerl, K., Air-lift reactor analysis, AIChEJ, 32(10), 1585-1596 (1986).

Stankiewicz, A., et al., Hydraulic design of multitubular reactors with heat carrier flowing in parallel to tubes, Chem Eng & Proc, 20(2), 79-84 (1986).

Stefoglo, E.F., Experimental study of hydrogenation process in gas-liquid reactors on suspended catalyst, Chem Eng Commns, 41, 327-354 (1986).
Wiedmann, J.A., Flooding behavior of two and three-phase stirred reactors, Int Chem Eng, 26(2), 189-204 (1986).
Yeung, P.O.Y., and Chapman, T.W., Design calculations for countercurrent gas-liquid reactors, Comput Chem Eng, 10(3), 259-267 (1986).
Zehner, P., Momentum, mass and heat transfer in bubble columns, Int Chem Eng, 26(1), 22-36 (1986).

1987

Beaudry, E.G.; Dudukovic, M.P., and Mills, P.L., Trickle-bed reactors: Liquid diffusional effects in gas-limited reaction, AIChEJ, 33(9), 1435-1447 (1987).
Bellgardt, D.; Schoessler, M., and Werther, J., Lateral non-uniformities of solids and gas concentrations in fluidized bed reactors, Powder Tech, 53(3), 205-216 (1987).
Botton, R.; Cosserat, D., and Charpentier, J.C., A new laboratory apparatus for study of gas-liquid reactions and simulation of reactors: The gas-lift bubble-column, Int Chem Eng, 27(2), 243-258 (1987).
Bourne, J.R., and Dell'Ava, P., Micro- and macro-mixing in stirred tank reactors of different sizes, CER&D, 65(2), 180-186 (1987).
Brown, L.F., and Falconer, J.L., Simplifying chemical reactor design by using molar quantities instead of fractional conversion, Chem Eng Educ, 21(1), 24-29 (1987).
Bukur, D.B., and Zimmerman, W.H., Modeling of bubble column slurry reactors for multiple reactions, AIChEJ, 33(7), 1197-1206 (1987).
Chalbi, M.; Castro, J.A.' Rodrigues, A.E., and Zoulalian, A., Heat transfer parameters in fixed bed catalytic reactors, Chem Eng J, 34(2) 89-98 (1987).
Chisti, M.Y., and Moo-Young, M., Airlift reactors: Characteristics, applications and design considerations, Chem Eng Commns, 60, 195-242 (1987).
Cinar, A., et al., Vibrational control of exothermic reaction in CSTR: Theory and experiments, AIChEJ, 33(3), 353-365 (1987).
Deckwer, W.D., and Schumpe, A., State of the art and current trends in bubble columns, Int Chem Eng, 27(3), 405-423 (1987).
Dutta, N.N., and Raghavan, P., Mass transfer and hydrodynamic characteristics of loop reactors with downflow liquid jet ejector, Chem Eng J, 36(2), 111-122 (1987).
Flaxman, R.J., and Hallett, W.L.H., Flow and particle heating in an entrained flow reactor, Fuel, 66(5), 607-611 (1987).
Fletcher, P., Heat transfer coefficients for stirred batch reactor design, Chem Engnr, April, 33-37 (1987).
Gabitto, J.F., and Lemcoff, N.O., Local solid-liquid mass transfer coefficients in a trickle bed reactor, Chem Eng J, 35(2), 69-74 (1987).
Kafarov, V.V.; Vinarov, A.Y., and Gordeev, L.S., Modeling of bioreactors, Int Chem Eng, 27(4), 615-642 (1987).
Larocca, M., Reactor particle-size calculations, Chem Eng, 2 March, 73-74 (1987).

Lede, J., et al., Measurement of solid particle residence time in cyclone reactor: Comparison of four methods, Chem Eng & Proc, 22(4), 215-222 (1987).

Lee, J.P.; Balakotaiah, V., and Luss, D., Thermoflow multiplicity in packed-bed reactor, AIChEJ, 33(7), 1136-1154 (1987).

Leung, P.C.; Recasans, F., and Smith, J.M., Hydration of isobutene in trickle-bed reactor: Wetting efficiency and mass transfer, AIChEJ, 33(6), 996-1007 (1987).

Mallikarjun, R., and Nauman, E.B., Optimization in tubular reactor systems, Chem Eng Commns, 50, 93-102 (1987).

Mazzarino, I.; Sicardi, S., and Baldi, G., Hydrodynamics and solid-liquid contacting effectiveness in an upflow multiphase reactor, Chem Eng J, 36(3), 151-160 (1987).

Metchis, S.G., and Foss, A.S., Averting extinction in autothermal catalytic reactor operations, AIChEJ, 33(8), 1288-1299 (1987).

Mohammadi, N.A., and Rempel, G.L., Control, data acquisition and analysis of catalytic gas-liquid mini slurry reactors using a personal computer, Comput Chem Eng, 11(1), 27-35 (1987).

Nauman, E.B., Chemical Reactor Design, Wiley, New York (1987).

Ng, K.M., and Chu, C.F., Trickle-bed reactors, Chem Eng Prog, 83(11), 55-63 (1987).

O'Dowd, W., et al., Gas and solids behavior in baffled and unbaffled slurry bubble column, AIChEJ, 33(12), 1959-1970 (1987).

Ozturk, S.S., et al., Organic liquids in bubble column: Holdups and mass transfer coefficients, AIChEJ, 33(9), 1473-1480 (1987).

Pirkle, J.C., and Wachs, I.E., Activity profiling in catalytic reactors, Chem Eng Prog, 83(8), 29-34 (1987).

Ramos, A.L., and Pironti, F.F., Effective thermal conductivity in packed-bed radial-flow reactor, AIChEJ, 33(10), 1747-1750 (1987).

Ravella, A., and de Lasa, H., The pseudoadiabatic regime for catalytic fixed bed reactors (limiting operating conditions), Chem Eng J, 34(1) 47-54 (1987).

Rovero, G.; Sicardi, S.; Baldi, G.; Osella, M., and Conti, R., Hydrodynamics of the gas phase in stirred gas-liquid contactors, Chem Eng J, 36(3), 161-168 (1987).

Schugerl, K.; Lubbert, A.; Korte, T., and Diekmann, J., Measuring techniques for characterizing gas/liquid reactors, Int Chem Eng, 27(4), 583-597 (1987).

Skrobek, H.; Brugel, E., and Legel, D., Hydrodynamics of highly concentrated suspensions in jet-agitated loop reactor, Chem Eng & Proc, 21(2), 95-100 (1987).

Steiner, R., Operating characteristics of special bubble column reactors, Chem Eng & Proc, 21(1), 1-8 (1987).

Sundaresan, S., Mathematical modeling of pulsating flow in large trickle beds, AIChEJ, 33(3), 455-469 (1987).

Tanaka, M., and Izumi, T., Gas entrainment in stirred-tank reactors, CER&D, 65(2), 195-198 (1987).

Thomas, W.J., The catalytic monolith reactor, Chem & Ind, 4 May, 315-319 (1987).

Touzani, A.; Klvana, D., and Belanger, G., A mathematical model for the dehydrogenation of methylcyclohexane in a packed bed reactor, Can JCE, 65(1), 56-63 (1987).

Turek, F.; Geike, R., and Lange, R., Problems encountered in the scale-up of a gas-liquid reaction in a stirred reactor with suspended catalyst, Chem Eng J, 36(1), 51-58 (1987).

Vaporciyan, G.G., and Kadlec, R.H., Equilibrium-limited periodic separating reactors, AIChEJ, 33(8), 1334-1343 (1987).

Waldram, S.P., Some general observations on the recycle flow model, Chem Eng Sci, 42(11), 2741-2744 (1987).

Wallis, G.B., and Goettgens, J., One-dimensional transients in a bubble column, Chem Eng Commns, 62(1), 79-92 (1987).

Wong, C.W.; Wang, J.P., and Huang, S.T., Investigations of fluid dynamics in mechanically stirred-aerated slurry reactors, Can JCE, 65(3), 412-419 (1987).

1988

Achenie, L.K.E., and Biegler, L.T., Developing targets for performance index of a chemical-reactor network: Isothermal systems, Ind Eng Chem Res, 27(10), 1811-1822 (1988).

Akyurtlu, J.F.; Akyurtlu, A., and Hamrin, C.E., Performance of catalytic porous-wall three-phase reactor, Chem Eng Commns, 66, 169-188 (1988).

Anastasov, A.; Elenkov, D., and Nikolov, V., Model study of conventional fixed bed tubular reactor with catalyst layer on inside tube wall, Chem Eng & Proc, 23(4), 203-212 (1988).

Asolekar, S.R.; Deshpande, P.K., and Kumar, R., Model for foam-bed slurry reactor, AIChEJ, 34(1), 150-154 (1988).

Baldyga, J., and Bourne, J.R., Calculation of micromixing in inhomogeneous stirred-tank reactors, CER&D, 66(1), 33-38 (1988).

Biswas, J.; Bhaskar, G.V., and Greenfield, P.F., Stratified flow model for two-phase pressure drop prediction in trickle beds, AIChEJ, 34(3), 510-513 (1988).

Bourne, J.R., and Tovstiga, G., Micromixing and fast chemical reactions in a turbulent reactor, CER&D, 66(1), 26-32 (1988).

Caneba, G.T., and Densch, B., Intermittency in nonisothermal CSTR, AIChEJ, 34(2), 333-334 (1988).

Cassano, A.E., et al., The 'a priori' design of photochemical reactors: Theory and experiments, Int Chem Eng, 28(2), 241-248 (1988).

Cavatorta, O.N., and Bohm, U., Heat and mass transfer in gas sparging systems: Empirical correlations and theoretical models, CER&D, 66(3), 265-274 (1988).

Chen, F., and Pearlstein, A.J., Temperature distributions in laminar-flow tubular photoreactor, AIChEJ, 34(8), 1381-1383 (1988).

Chiao, L., and Rinker, R.G., Simulated behavior of autothermal reactor under relaxed steady-state operation, Chem Eng Commns, 73, 163-182 (1988).

Chidambaram, M., Periodic operation of isothermal plug-flow reactors for autocatalytic reactions, Chem Eng Commns, 69, 215-224 (1988).

Chisti, M.Y.; Halard, B., and Moo-Young, M., Liquid circulation in airlift reactors, Chem Eng Sci, 43(3), 451-457 (1988).

Claria, M.A.; Irazoqui, H.A., and Cassano, A.E., Design of photoreactor for ethane chlorination, AIChEJ, 34(3), 366-382 (1988).
Cozewith, C., Transient response of CFST polymerization reactors, AIChEJ, 34(2), 272-282 (1988).
Cutler, A.H.; Antal, M.J., and Jones, M., A critical evaluation of the plug-flow idealization of tubular-flow reactor data, Ind Eng Chem Res, 27(4), 691-697 (1988).
Damkohler, G., Influence of diffusion, fluid flow, and heat transport on yield in chemical reactors, Int Chem Eng, 28(1), 132-198 (1988).
Elnashie, S.S.; Mahfouz, A.T., and Elshishini, S.S., Digital simulation of industrial ammonia reactor, Chem Eng & Proc, 23(3), 165-178 (1988).
Feyo de Azevedo, S., and Howell, J.A., Second-order model for low-density polyethylene pipeline reactors, CER&D, 66(2), 128-137 (1988).
Feyo de Azevedo, S.; Rodrigues, A., and Wardle, A.P., Optimization of tubular fixed-bed catalytic reactors, Chem Eng J, 38(1), 9-16 (1988).
Genon, G., and Ruggeri, B., Stability conditions of the fed-batch reactor, J Chem Tech Biotechnol, 42(4), 261-276 (1988).
Hamer, J.W., and Richenberg, C.B., On-line optimizing control of packed-bed immobilized-cell reactor, AIChEJ, 34(4), 626-632 (1988).
Haque, M.W., et al., Bubble rise velocity in bubble columns employing non-Newtonian solutions, Chem Eng Commns, 73, 31-42 (1988).
Heras, J.M., and Viscido, L., Behavior of water on metal surfaces, Catalysis Reviews, 30(2), 281-338 (1988).
Herskowitz, M., and Hagan, P.S., Accurate one-dimensional fixed-bed reactor model based on asymptotic analysis, AIChEJ, 34(8), 1367-1372 (1988).
Jones, C.K.S.; Yang, R.Y.K., and White, E.T., A novel hollow-fiber reactor with reversible immobilization of lactase, AIChEJ, 34(2), 293-304 (1988).
Kawase, Y., and Moo-Young, M., Volumetric mass-transfer coefficients in aerated stirred-tank reactors with Newtonian and non-Newtonian media, CER&D, 66(3), 284-288 (1988).
Kheshgi, H.S., et al., Transients in tubular reactors: Comparison of one- and two-dimensional models, AIChEJ, 34(8), 1373-1375 (1988).
Kim, T.O., and Kang, W.K., Liquid mixing in CFSTR, Int Chem Eng, 28(4), 690-698 (1988).
Kodas, T.T., and Friedlander, S.K., Design of tubular flow reactors for monodisperse aerosol production, AIChEJ, 34(4), 551-557 (1988).
Kwalik, K.M., and Schork, F.J., Adaptive pole-placement control of continuous polymerization reactor, Chem Eng Commns, 63, 157-180 (1988).
Lee, J.P.; Balakotaiah, V., and Luss, D., Thermoflow multiplicity in packed-bed reactor, AIChEJ, 34(1), 37-44 (1988).
Levec, J.; Grosser, K., and Carbonell, R.G., Hysteretic behavior of pressure drop and liquid holdup in trickle beds, AIChEJ, 34(6), 1027-1030 (1988).
Lisa, R.E., Determine the endpoint of viscous reactions, Chem Eng, 14 March, 160-162 (1988).
Losada, M.E., and Pironti, F.F., Experimental determination and modelling of liquid-solid mass transfer coefficients in three-phase reactor, Chem Eng Commns, 71, 95-112 (1988).
Ogut, A., and Hatch, R.T., Oxygen transfer into Newtonian and non-Newtonian fluids in mechanically agitated vessels, Can JCE, 66(1), 80-86 (1988).

Pan, D.F.; Schnitzlein, K., and Hofmann, H., Design of the control scheme of a concentration-controlled recycle reactor, Ind Eng Chem Res, 27(1), 86-93 (1988).

Peters, P.E.; Schiffino, R.S., and Harriott, P., Heat transfer in packed-tube reactors, Ind Eng Chem Res, 27(2), 226-232 (1988).

Pinkle, C.; Ruziska, P.A., and Shulik, L.J., Circulating magnetically stabilized bed reactors, Chem Eng Commns, 67, 89-110 (1988).

Ravella, A., and de Lasa, H., Cooling exothermic catalytic fixed-bed reactors, Can JCE, 65(6), 1021-1026 (1988).

Richardson, J.T.; Paripatyadar, S.A., and Shen, J.C., Dynamics of sodium heat pipe reforming reactor, AIChEJ, 34(5), 743-752 (1988).

Rigopoulos, K.; Shu, X., and Cinar, A., Forced periodic control of exothermic CSTR with multiple input oscillations, AIChEJ, 34(12), 2041-2051 (1988).

Roizard, C., and Tondeur, D., Transient thermal behavior of fixed bed percolated by a gas: Experiments and modeling, Int Chem Eng, 28(2), 271-285 (1988).

Rutkowska, D.A.; Stankiewicz, A., and Leszczynski, Z., Simulation of the operational characteristics of the large-scale multitubular reactor for maleic anhydride production, Comput Chem Eng, 12(2/3), 171-176 (1988).

Salmon, P.M.; Libicki, S.B., and Robertson, C.R., Theoretical investigation of convective transport in hollow-fiber reactor, Chem Eng Commns, 66, 221-246 (1988).

Saxena, S.C., and Patel, D., Assessment of experimental techniques for measurement of bubble size in bubble-slurry reactor as applied to indirect coal liquefaction, Chem Eng Commns, 63, 87-128 (1988).

Sotirchos, S.V.; Crowley, J.A., and Yu, H.C., Adsorption/reaction device for gas-solid reaction studies, Chem Eng Commns, 71, 83-94 (1988).

Stanek, V., and Vychodil, P., Assessment of the role of thermally induced gas flow inhomogeneities in fixed bed reactors, Chem Eng & Proc, 24(4), 203-210 (1988).

Thullie, J.; Chiao, L., and Rinker, R.G., Forced oscillations in concentration in tubular fixed-bed catalytic reactor, Int Chem Eng, 28(4), 723-731 (1988).

Wensley, J.H., and Harclerode, C.S., Chemical reactor control by fault-tolerant computer, Manufacturing Chemist, June, 65-68 (1988).

Wu, J.J., et al., Evaluation and control of particle properties in aerosol reactors, AIChEJ, 34(8), 1249-1256 (1988).

Yaici, W.; Laurent, A.; Midoux, N., and Charpentier, J.C., Determination of gas-side mass transfer coefficients in trickle bed reactors in presence of aqueous or organic liquid phase, Int Chem Eng, 28(2), 299-306 (1988).

CHAPTER 19

DISTILLATION

19.1 Theory and Principles

1967-1969

Chen, N.H., Basic equation solves many mass-transfer problems, Chem Eng, 2 Jan, 93-99 (1967).
Eyre, D.V., Some aspects of distillation near the critical region, Chem Eng Prog, 63(7), 93-96 (1967).
Furzer, I.A., and Ho, G.E., Distillation in packed columns: Relationship between HTU and packed height, AIChEJ, 13(3), 614-615 (1967).
Howard, G.M., Degrees of freedom for unsteady-state distillation processes, Ind Eng Chem Fund, 6(1), 86-89 (1967).
King, R.W., Distillation of heat-sensitive materials, Brit Chem Eng, 12(4), 568-572; 12(5), 722-727 (1967).
Ridgway, K., and Butler, P.A., Equilibrium times in equilibrium stills, Brit Chem Eng, 12(7), 1095-1099 (1967).
Various, Distillation (topic issue), Chem Eng Prog, 63(9), 41-65 (1967).
Corrigan, T.E., and Miller, J.H., Chemical reaction in a distillation column, Ind Eng Chem Proc Des Dev, 7(3), 383-384 (1968).
Golding, J.A.; Graydon, W.F., and Johnson, A.I., Mass transfer from single bubbles under distillation conditions, Trans IChemE, 46, T172-176 (1968).
Leslie, D.C., Development of flashing flow from existing nucleation sites, Brit Chem Eng, 13(4), 512-519 (1968).
Atkinson, B.; Richardson, J.F., and Roberts, N.W., Recovery of solutes and solvents from solutions using immiscible fluids, Brit Chem Eng, 14(9), 1193-1196 (1969).
Deam, J.R., and Maddox, R.N., Calculating three-phase flashing, Hyd Proc, 48(7), 163-164 (1969).
Tassios, D., Choosing solvents for extractive distillation, Chem Eng, 10 Feb, 118-122 (1969).
Todd, D.B., and Maclean, D.C., Centrifugal vapour-liquid contacting, Brit Chem Eng, 14(11), 1565-1568 (1969).
Various, Azeotropic and extractive distillation (topic issue), Chem Eng Prog, 65(9), 43-68 (1969).

1970-1972

Bassyoni, A.A.; McDaniel, R., and Holland, C.D., Examination of use of mass transfer rate expressions in description of packed distillation columns, Chem Eng Sci, 25(3), 437-444 (1970).
Brown, B.T.; Clay, H.A., and Miles, J.M., Drying liquid hydrocarbons via fractional distillation, Chem Eng Prog, 66(8), 54-60 (1970).
Forsyth, J.S., Degrees of freedom in a simple fractionating column, Ind Eng Chem Fund, 9(3), 507-509 (1970).
Furzer, I.A., and Ho, G.E., Differential backmixing in distillation columns, Chem Eng Sci, 25(8), 1297-1300 (1970).
Guerreri, G., Distillation: A refresher, Brit Chem Eng, 15(9), 1160-1164 (1970).
Uyehara, H., and Hagihara, Y., Application of film theory to analysis of mass transfer in liquid-phase molecular distillation, Int Chem Eng, 10(3), 466-471 (1970).

Atroshchenko, L.S., and Voronina, S.M., Distillation and rectification processes in homogeneous magnetic fields, Int Chem Eng, 11(3), 522-525 (1971).
Lees, F.P., Continuous vaporisation for drying organic liquids, Brit Chem Eng, 16(8), 696-697 (1971).
Liddle, C.J., Use of the Antoine equation in distillation calculations, Brit Chem Eng, 16(2), 193-197 (1971).
Nelson, P.A., Countercurrent equilibrium stage separation with reaction, AIChEJ, 17(5), 1043-1049 (1971).
Peiffer, C.C.; McCormick, R.H., and Fenske, M.R., Multistage cocurrent contacting vapor-liquid equilibrium unit, Ind Eng Chem Proc Des Dev, 10(3), 380-384 (1971).
Tosi, M., Vapor-liquid equilibrium still for use at 1 mm Hg pressure, AIChEJ, 17(4), 854-856 (1971).
Azizov, A.G., Analogy between heat and mass transfer processes during bubbling on sieve trays, Int Chem Eng, 12(2), 249-252 (1972).
Cloete, C.E., and de Clerk, K., Distillation vs. chromatography: Comparison based on purity index, Sepn Sci, 7(4), 449-456 (1972).
Hakki, T.A., and Lamb, J.A., Simultaneous heat and mass transfer in a wetted-wall column, Trans IChemE, 50, 115-118 (1972).
Moens, F.P., and Bos, R.G., Surface renewal effects in distillation, Chem Eng Sci, 27(2), 403-408 (1972).
Pearce, R.L.; Protz, J.E., and Lyon, G.W., Azeotropic distillation process, Hyd Proc, 51(12), 79-81 (1972).
Zavorka, J., Approximate calculation of heat transfer coefficient between liquid and vapour phases in a rectification column, Chem Eng J, 4(1), 1-7 (1972).

1973-1974

Davis, E.J.; Hung, S.C., and Dunn, C.S., Simultaneous heat and mass transfer with liquid-film vacuum distillation, Chem Eng Sci, 28(8), 1519-1534 (1973).
Ellerbe, R.W., Batch distillation basics, Chem Eng, 28 May, 110-116 (1973).
Ervin, E.D., and Danner, R.P., Continuous foam fractionation of phenol, Sepn Sci, 8(2), 179-184 (1973).
Featherstone, W., Estimating still requirements for non-ideal systems, Proc Tech Int, April, 185-189 (1973).
Ibragimov, M.G.; Konstantinov, E.A., and Serafimov, L.A., Kinetics of distillation with inert component, Int Chem Eng, 13(4), 705-708 (1973).
Ponter, A.B.; Boyes, A.P., and Houlihan, R.N., Distillation at near terminal compositions: Wetting effect at atmospheric and reduced pressures, Chem Eng Sci, 28(2), 593-596 (1973).
Rang, E.R., Computation of differential distillations from true boiling data for hydrocarbon mixtures, Chem Eng Sci, 28(6), 1349-1354 (1973).
Tsubaki, M., and Hiraiwa, H., Distribution of non-key components in multicomponent distillation, Int Chem Eng, 13(1), 183-192 (1973).
Zuiderweg, F.J., Distillation: Science and business, Chem & Ind, 21 July, 662-669 (1973).
Zuiderweg, F.J., Distillation: Science and business, Chem Engnr, Sept, 404-410 (1973).
Ellerbe, R.W., Steam-distillation basics, Chem Eng, 4 March, 105-112 (1974).

Ellis, S.R.M., and Boyes, A.P., Research and developments in distillation, Trans IChemE, 52, 202-210 (1974).
Fonyo, Z., Thermodynamic analysis of rectification, Int Chem Eng, 14(1), 18-27; 14(2), 203-211 (1974).
Shufelt, R.C.; Peiffer, C.C., and McCormick, R.H., Vapor-liquid contacting in descending cocurrent flow, Ind Eng Chem Proc Des Dev, 13(2), 165-168 (1974).

1975-1977
Gorodetskii, E.T., and Klyus, I.P., Heat effect during distillation of liquids below their boiling points, Int Chem Eng, 15(4), 631-633 (1975).
Jaques, D., Equation for salt effect in liquid-vapor equilibrium at constant liquid composition, Can JCE, 53, 713-720 (1975).
Kayihan, F.; Sandall, O.C., and Mellichamp, D.A., Simultaneous heat and mass transfer in binary distillation, Chem Eng Sci, 30(11), 1333-1340 (1975).
Pinczewski, W.V.; Benke, N.D., and Fell, C.J.D., Phase inversion on sieve trays, AIChEJ, 21(6), 1210-1213 (1975).
Pohorecki, R., Application pf Danckwerts method to measurements of interfacial area and liquid-side mass-transfer coefficient on sieve trays, Int Chem Eng, 15(4), 647-658 (1975).
Tyreus, B.D.; Luyben, W.L., and Schiesser, W.E., Stiffness in distillation models and use of implicit integration method to reduce computation times, Ind Eng Chem Proc Des Dev, 14(4), 427-433 (1975).
Zanker, A., Quick calculation of foam densities, Chem Eng, 17 Feb, 114 (1975).
Sawatzky, H.; George, A.E.; Smiley, G.T., and Montgomery, D.S., Hydrocarbon-type separation of heavy petroleum fractions, Fuel, 55(1), 9-15 (1976).
Scully, D.B., Interpretation of distillation of a three-component mixture on a ternary diagram, Ind Eng Chem Fund, 15(4), 344-346 (1976).
Woicik, J.F., Equilibrium flash calculations with a pocket calculator, Chem Eng, 16 Aug, 89-93; 22 Nov, 184-186 (1976).
Farcasiu, M., Fractionation and structural characterization of coal liquids, Fuel, 56(1), 9-14 (1977).
Furter, W.F., Salt effect in distillation: A literature review, Can JCE, 55, 229-240 (1977).
Holve, W.A., Theoretical analysis of changes in relative volatility via reversible metalation reactions and application to fractionation, Ind Eng Chem Fund, 16(1), 56-60 (1977).
Kayihan, F.; Sandall, O.C., and Mellichamp, D.A., Simultaneous heat and mass transfer in binary distillation, Chem Eng Sci, 32(7), 747-754 (1977).
Krishna, R., Film-model analysis of non-equimolar distillation of multicomponent mixtures, Chem Eng Sci, 32(10), 1197-1204 (1977).
Parwardhan, S.R., Fractionation of petroleum bitumens, Fuel, 56(1), 40-44 (1977).

1978-1979
Boston, J.F., and Britt, H.I., A radically different formulation and solution of single-stage flash problem, Comput Chem Eng, 2(2), 109-122 (1978).

Cavalier, J.C., and Chornet, E., Fractionation of peat-derived bitumen into oil and asphaltenes, Fuel, 57(5), 304-308 (1978).
Hirose, Y.; Nagai, Y., and Tsuda, M., Calculation of temperature and component distribution profiles of a distillation column by differentiation with respect to an actual parameter, Int Chem Eng, 18(2), 258-266 (1978).
Honorat, A., and Sandall, O.C., Simultaneous heat and mass transfer in a packed binary distillation column, Chem Eng Sci, 33(6), 635-640 (1978).
Kuzniar, J.; Kubisa, R., and Pasko, Z., Mass transfer in two-component distillation on bubble trays with allowance for complete mixing of the liquid, Int Chem Eng, 18(4), 586-591 (1978).
Mamedov, G.E., Polynomial mathematical model for multicomponent distillation, Int Chem Eng, 18(1), 166-169 (1978).
Medina, A.G.; McDermott, C., and Ashton, N., Surface tension effects in binary and multicomponent distillation, Chem Eng Sci, 33(11), 1489-1494 (1978).
Wagner, H., Thermodynamic design of multicomponent rectifications, Int Chem Eng, 18(3), 386-395 (1978).
Burke, F.P.; Winschel, R.A., and Wooton, D.L., Liquid column fractionation: Method of solvent fractionation of coal liquifaction and petroleum products, Fuel, 58(7), 539-541 (1979).
Krishna, R., and Standart, G.L., Mass and energy transfer in multicomponent systems, Chem Eng Commns, 3(4), 201-276 (1979).
Mah, R.S.H., and Wodnik, R.B., Binary distillations and their idealizations, Chem Eng Commns, 3(2), 59-64 (1979).
Naka, Y.; Araki, M., and Takamatsu, T., New matrix method for composition profiles in distillation, Int Chem Eng, 19(1), 137-145 (1979).
Schlunder, E.U., Effect of diffusion on selectivity of entraining distillation, Int Chem Eng, 19(3), 373-380 (1979).
Various, Distillation (special report), Hyd Proc, 58(2), 89-103 (1979).

1980-1981

Arwikar, K.J., and Sandall, O.C., Liquid-phase mass-transfer resistance in a small-scale packed distillation column, Chem Eng Sci, 35(11), 2337-2344 (1980).
Gallun, S.E., and Holland, C.D., Modification of Broyden's method for solution of sparse systems: Application to distillation problems described by nonideal thermodynamic functions, Comput Chem Eng, 4(2), 93-100 (1980).
Hiranuma, M., Composition effect and thermal effect to overall mass-transfer coefficient in distillation processes, Ind Eng Chem Proc Des Dev, 19(2), 220-222 (1980).
Niedzwiecki, J.L.; Springer, R.D., and Wolfe, R.G., Multicomponent distillation in the presence of free water, Chem Eng Prog, 76(4), 57-58 (1980).
Rees, G.J., Centrifugal molecular distillation, Chem Eng Sci, 35(4), 837-846 (1980).
Weisenfelder, A.J., and Olson, R.E., Solving recycle streams in multicomponent distillation, Chem Eng Prog, 76(1), 40-43 (1980).
Conder, J.R., and Fruitwala, N.A., Comparison of plate numbers and column lengths in chromatography and distillation, Chem Eng Sci, 36(3), 509-516 (1981).

Ferraris, G.B., and Morbidelli, M., Distillation models for two partially immiscible liquids, AIChEJ, 27(6), 881-888 (1981).
Fournier, R.L., and Boston, J.F., Quasi-Newton algorithm for solving multiphase equilibrium flash problems, Chem Eng Commns, 8(4), 305-326 (1981).
Haas, J.R.; Gomez, A., and Holland, C.D., Generalization of theta methods of convergence for solving distillation and absorber-type problems, Sepn Sci Technol, 16(1), 1-24 (1981).
Krishna, R.; Salomo, R.M., and Rahman, M.A., Ternary mass transfer in a wetted-wall column, Trans IChemE, 59, 35-53 (1981).
Ross, B.A., and Seider, W.D., Simulation of three-phase distillation towers, Comput Chem Eng, 5(1), 7-20 (1981).
Various, Distillation (topic issue), Chem Eng Prog, 77(9), 33-62 (1981).

1982-1983

Franklin, N.L., and Wilkinson, M.B., Reversibility in the separation of multicomponent mixtures, Trans IChemE, 60, 276-282 (1982).
Gallun, S.E., and Holland, C.D., Gear's procedure for simultaneous solution of differential and algebraic equations with application to unsteady-state distillation problems, Comput Chem Eng, 6(3), 231-244 (1982).
Grevillot, G.; Bailly, M., and Tondeur, D., Thermofractionation: New class of separation processes combining concepts of distillation, chromatography, and the heat pump, Int Chem Eng, 22(3), 440-454 (1982).
Ito, A., and Asano, K., Simultaneous heat and mass transfer in binary distillation, Int Chem Eng, 22(2), 309-319 (1982).
McCandless, F.P., Extent of separation in continuous multistage binary distillation, Sepn Sci Technol, 17(12), 1361-1386 (1982).
Nelson, R.E., Vacuum pump aids ejectors, Hyd Proc, Dec, 95-96 (1982).
Porter, K.E., and Jenkins, J.D., Distillation update, Processing, May, 23, 25, 45 (1982).
Tierney, J.W., and Riquelme, G.D., Calculation methods for distillation systems with reaction, Chem Eng Commns, 16(1), 91-108 (1982).
Tuohey, P.G.; Pratt, H.R.C., and Yost, R.S., Binary and ternary distillation in a stirred mass transfer cell, Chem Eng Sci, 37(12), 1741-1750 (1982).
Wasek, K., Calculation of crude flashing in presence of steam by Newton-Raphson method, Chem Eng Commns, 16(1), 39-44 (1982).
Burghardt, A., and Warmuzinski, K., Diffusional methods of calculation for multicomponent systems in rectification columns: Models of equimolar and nonequimolar mass transfer, Int Chem Eng, 23(2), 342-351 (1983).
Burghardt, A.; Warmuzinski, K.; Buzek, J., and Pytlik, A., Diffusional models of multicomponent distillation and experimental verification, Chem Eng J, 26(2), 71-84 (1983).
Erdweg, K.J., Molecular and short-path distillation, Chem & Ind, 2 May, 342-345 (1983).
Fahmi, M.F., and Mostafa, H.A., Distillation with optimal vapour compression, CER&D, 61, 391-392 (1983).
McCandless, F.P., Single-stage contribution to overall separation in binary multistage distillation, Sepn Sci Technol, 18(9), 803-820 (1983).

Nelson, A.R.; Olson, J.H., and Sandler, S.I., Sensitivity of distillation process design and operation to VLE data, Ind Eng Chem Proc Des Dev, 22(3), 547-552 (1983).
Ruckenstein, E.; Hassink, W.J., and Gourisankar, S.M.V., Combined effect of diffusion and evaporation on molecular distillation of ideal binary liquid mixtures, Sepn Sci Technol, 18(6), 523-546 (1983).
Zuiderweg, F.J., Marangoni effect in distillation of alcohol-water mixtures, CER&D, 61, 388-390 (1983).

1984

Burghardt, A., and Warmuzinski, K., Diffusional methods of calculation for multicomponent systems in rectification columns: Model of simultaneous mass transfer, Int Chem Eng, 24(4), 742-752 (1984).
Chien, H.H.Y., KB method in two-phase flash calculations, Comput Chem Eng, 8(1), 61-64 (1984).
Cho, Y.S., and Joseph, B., Reduced-order models for separation columns, Comput Chem Eng, 8(2), 81-90 (1984).
Dhulesia, H., Equation fits ASTM distillations, Hyd Proc, Sept, 179-180 (1984).
Fair, J.R., and Humphrey, J.L., Distillation: Research needs, Sepn Sci Technol, 19(13), 943-962 (1984).
Peterson, R.J.; Grewal, S.S., and El-Wakil, M.M., Investigations of liquid flashing and evaporation due to sudden depressurization, Int J Heat Mass Trans, 27(2), 301-310 (1984).
Porter, K.E., and Jenkins, J.E., Distillation update, Processing, Oct, 34,37 (1984).
Smith, W., Effects of thermal distillation on the performance of differential-contact distillation columns, Chem Eng Sci, 39(6), 997-1004 (1984).
Takamatsu, T.; Hashimoto, I., and Tomita, S., Structural analysis of algebraic equations using directed graphical representation, with application to ternary distillation, Int Chem Eng, 24(1), 88-97 (1984).
Westman, K.R.; Lucia, A., and Miller, D.C., Flash and distillation calculations by Newton-like method, Comput Chem Eng, 8(3), 219-228 (1984).

1985

Anon., Distillation update, Processing, Aug, 43,47 (1985).
Barba, D.; Brandani, V., and diGiacomo, G., Hyperazeotropic ethanol salted-out by extractive distillation, Chem Eng Sci, 40(12), 2287-2292 (1985).
Bastos, J.C.; Soares, M.E., and Medina, A.G., Selection of solvents for extractive distillation, Ind Eng Chem Proc Des Dev, 24(2), 420-426 (1985).
Berg, L., and Yeh, A.I., Unusual behavior of extractive distillation (reversing volatility), AIChEJ, 31(3), 504-506 (1985).
Byrne, G.D., and Baird, L.A., Distillation calculations using a locally paramaterized continuation method, Comput Chem Eng, 9(6), 593-600 (1985).
Chimowitz, E.H.; Macchietto, S., and Anderson, T.F., Dynamic multicomponent distillation using local thermodynamic models, Chem Eng Sci, 40(10), 1974-1979 (1985).

Cotterman, R.L., and Prausnitz, J.M., Flash calculations for continuous or semicontinuous mixtures using an equation of state, Ind Eng Chem Proc Des Dev, 24(2), 434-443 (1985).

Grimma, G.A., Concentration maxima in ternary distillation at total reflux, Sepn Sci Technol, 20(2), 85-100 (1985).

Hsu, S.N., and Maa, J.R., Foam fractionation with synergism, Ind Eng Chem Proc Des Dev, 24(1), 38-41 (1985).

Ikari, A.; Hatate, Y., and Maruyama, A., Behavior of a minute amount of diethyl sulfide in hydrocarbon distillation, Chem Eng Commns, 34(1), 241-252 (1985).

Kosuge, H.; Johkoh, T., and Asano, K., Experimental studies of diffusion fluxes of ternary distillation of acetone-methanol-ethanol systems using wetted-wall column, Chem Eng Commns, 34(1), 111-122 (1985).

McCandless, F.P., Effect of changes in operating variables on stagewise and cumulative separation in binary multistage distillation, Sepn Sci Technol, 20(9), 665-686 (1985).

Porter, K., and Jenkins, J., Distillation now, Chem Engnr, Nov, 26-30 (1985).

Schuil, J.A., and Bool, K.K., Three-phase flash and distillation, Comput Chem Eng, 9(3), 295-300 (1985).

Stewart, W.E.; Levien., K.L., and Morari, M., Simulation of fractionation by orthogonal collocation, Chem Eng Sci, 40(3), 409-422 (1985).

Vahdat, N., and Sather, G.A., Prediction of multicomponent azeotrope composition and temperature, Chem Eng J, 31(2), 83-96 (1985).

Yoshida, H., and Yorizane, M., Characteristics of distillation columns for azeotropic mixtures, Int Chem Eng, 25(4), 680-688 (1985).

1986

Angel, S.; Marmur, A., and Kehat, E., Comparison of methods of prediction of vapor-liquid equilibria and enthalpy in a distillation simulation program, Comput Chem Eng, 10(2), 169-180 (1986).

Brown, R.S., and Holland, C.D., Prediction of the steady-state relative gains by use of the total reflux model, Comput Chem Eng, 10(3), 287-295 (1986).

Cuille, P.E., and Reklaitis, G.V., Dynamic simulation of multicomponent batch rectification with chemical reactions, Comput Chem Eng, 10(4), 389-398 (1986).

Hirose, Y.; Takahashi, M., and Tachibana, H., Extension of Kubicek algorithm and its application to distillation problems, Comput Chem Eng, 10(1), 1-5 (1986).

Konoshita, M., and Takamatsu, T., A powerful solution algorithm for single-stage flash problems, Comput Chem Eng, 10(4), 353-360 (1986).

Koshy, T.D., and Rukovena, F., Reflux and surface tension effects on distillation, Hyd Proc, 65(5), 64-66 (1986).

Lucia, A., Uniqueness of solutions to single-stage isobaric flash processes involving homogeneous mixtures, AIChEJ, 32(11), 1761-1770 (1986).

Pucci, A.; Mikitenko, P., and Asselineau, L., Three-phase distillation: Simulation and application to the separation of fermentation products, Chem Eng Sci, 41(3), 485-494 (1986).

Rodriguez-Patino, J.M., and Rosello, S.A., Rectification in a pool column in absence of surface effects, Int Chem Eng, 26(2), 271-278 (1986).

Wu, J.S., and Bishnoi, P.R., A method for steady-state simulation of multistage separation columns involving nonideal systems, Comput Chem Eng, 10(4), 343-351 (1986).

1987

Davies, B.; Ali, Z., and Porter, K.E., Distillation of systems containing two liquid phases, AIChEJ, 33(1), 161-163 (1987).

Kehlen, H., and Ratzsch, M.T., Complex multicomponent distillation calculations by continuous thermodynamics, Chem Eng Sci, 42(2), 221-232 (1987).

Kovach, J.W., and Seider, W.D., Heterogeneous azeotropic distillation: Experimental and simulation results, AIChEJ, 33(8), 1300-1314 (1987).

Lagar, G.; Paloschi, J., and Romagnoli, J.A., Numerical studies in solving dynamic distillation problems, Comput Chem Eng, 11(4), 383-394 (1987).

Ognisty, T.P., and Sakata, M., Multicomponent diffusion: Theory vs. industrial data, Chem Eng Prog, 83(3), 60-65 (1987).

Porter, K.E., and Jenkins, J.D., Review of distillation 1979-87, Gas Sepn & Purif, 1(1), 11-15 (1987).

Taffe, P., Catalytic distillation, Processing, Sept, 47,49 (1987).

Tan, T.C., New screening technique and classification of salts for the salt distillation of close-boiling and azeotropic solvent mixtures, CER&D, 65(5), 421-425 (1987).

Various, Distillation (special supplement), Chem Engnr, Sept, 1-28 (1987).

Veeranna, D.; Husain, A.; Subrahmanyam, S., and Sarkar, M.K., An algorithm for flash calculations using an equation of state, Comput Chem Eng, 11(5), 489-496 (1987).

1988

Barbosa, D., and Doherty, M.F., The simple distillation of homogeneous reactive mixtures, Chem Eng Sci, 43(3), 541-550 (1988).

Bhandarkar, M., and Ferron, J.R., Transport processes in thin liquid films during high-vacuum distillation, Ind Chem Eng Res, 27(6), 1016-1024 (1988).

Drinkenburg, B., Distillation: Sophisticate or dinosaur? Chem Engnr, July, 28-29 (1988).

Fair, J.R., Distillation: Whither, not whether, CER&D, 66(4), 363-370 (1988).

Franklin, N.L., Theory of multicomponent countercurrent cascades, CER&D, 66(1), 65-74 (1988).

Luyben, W.L., Multicomponent batch distillation, Ind Eng Chem Res, 27(4), 642-647 (1988).

McDowell, J.K., and Davis, J.F., Characterization of diffusion distillation for azeotropic separation, Ind Eng Chem Res, 27(11), 2139-2149 (1988).

19.2 Calculations and Design Methods

1967-1968

Apelblat, A., Method for calculating stagewise unit operations, Brit Chem Eng, 12(9), 1378-1380 (1967).

Hamza, H.H., and Said, A.S., Simplified graphical method for distillation calculations, Chem Eng, 5 June, 164-166 (1967).
King, R.W., Separation by fractional distillation under vacuum, Brit Chem Eng, 12(10), 1599-1603 (1967).
Lee, R., and Cova, D.R., Distillation 'F' factor calculated from nomograph, Chem Eng, 28 Aug, 166-168 (1967).
Aristovich, V.Y., and Levin, A.I., Method for calculating rectification of multicomponent mixtures on digital computers, Int Chem Eng, 8(1), 1-5 (1968).
Distefano, G.P., Mathematical modeling and numerical integration of multicomponent batch distillation equations, AIChEJ, 14(1), 190-199 (1968).
Goldman, M.R., and Robinson, E.R., Computer simulation of batch distillation processes, Brit Chem Eng, 13(12), 1713-1716 (1968).
Guerreri, G., Experimental determination of theoretical plates in distillation columns, Brit Chem Eng, 13(1), 83-85 (1968).
Hanson, C., Non-adiabatic binary distillation calculations, Brit Chem Eng, 13(8), 1134-1135 (1968).
Hengstebeck, R.J., Finding feedplates from plots, Chem Eng, 29 July, 143-144 (1968).
Hiranuma, M., and Kugo, M., Fractionation calculations of multicomponent complex mixtures, Int Chem Eng, 8(3), 539-545 (1968).
Hoffman, E.J., Flash calculations for petroleum fractions, Chem Eng Sci, 23(9), 957-964 (1968).
Lee, R.; Cova, D.R., and Mueller, N.F., Reflux ratio determined easily with electric timer, Chem Eng, 8 April, 178 (1968).
Liddle, C.J., Improved shortcut method for distillation calculations, Chem Eng, 21 Oct, 137-142 (1968).
Ripps, D.L., Minimum reflux quickly, Hyd Proc, 47(12), 84 (1968).
Sterbacek, Z., Calculation of number of trays allowing for backmixing and entrainment, Brit Chem Eng, 13(6), 820-823 (1968).
Waterman, W.W.; Frazier, J.P., and Brown, G.M., Compute best distillation feed-point, Hyd Proc, 47(6), 155-160 (1968).

1969

Bakowski, S., Determination of number of trays for binary distillation, Brit Chem Eng, 14(9), 1213-1214 (1969).
Brosens, J.R., Quicker estimate of distillation trays, Hyd Proc, 48(10), 102-104 (1969).
Guerreri, G., Problems with shortcut distillation design, Hyd Proc, 48(8), 137-142 (1969).
Hariu, O.H., and Sage, R.C., Crude split calculated by computer, Hyd Proc, 48(4), 143-148 (1969).
Hassett, N.J., Improved graphical method for number of distillation trays, Brit Chem Eng, 14(7), 952-953 (1969).
Hengstebeck, R.J., Improved shortcut for calculating difficult multicomponent distillations, Chem Eng, 13 Jan, 115-118 (1969).
Lebedev, Y.N.; Aleksandrov, I.A., and Zykov, D.D., Modelling mass transfer in binary rectification, Int Chem Eng, 9(4), 698-702 (1969).

Lockhart, F.J., Stagewise still calculations by temperature profiles, Chem Eng, 8 Sept, 131-132 (1969).
Lowry, R.P., and Van Winkle, M., Foaming and frothing related to system physical properties in a small perforated-plate distillation column, AIChEJ, 15(5), 665-670 (1969).
Mecklenburgh, J.C., and Hartland, S., Design methods for countercurrent flow with backmixing: Distillation with variable enthalpy, Chem Eng Sci, 24(8), 1259-1268 (1969).
Mecklenburgh, J.C., and Hartland, S., Design methods for countercurrent flow with backmixing: Distillation with constant molar enthalpy, Chem Eng Sci, 24(5), 899-909 (1969).
Neretnieks, I.; Ericson, I., and Eriksson, S., Modified McCabe-Thiele construction, Brit Chem Eng, 14(12), 1711-1712 (1969).
Robinson, E.R., and Goldman, M.R., Simulation of multicomponent batch distillation processes on a small digital computer, Brit Chem Eng, 14(6), 809-812 (1969).
Scheiman, A.D., Find minimum reflux by heat balance, Hyd Proc, 48(9), 187-194 (1969).
Serov, V.V., and Zykov, D.D., Determination of feed plate concentration in multicomponent distillation, Int Chem Eng, 9(2), 274-276 (1969).
Various, Distillation design (topic issue), Chem Eng Prog, 65(3), 33-63 (1969).

1970

Billingsley, D.S., Numerical solution of problems in multicomponent distillation at steady state, AIChEJ, 16(3), 441-445 (1970).
Boynton, G.W., Newton-Raphson iteration for distillation problems, Hyd Proc, 49(1), 153-156 (1970).
Goldman, M.R., Simulating multicomponent batch distillation, Brit Chem Eng, 15(11), 1450-1453 (1970).
Goldstein, R.P., and Stanfield, R.B., Flexible method for solution of distillation design problems using Newton-Raphson technique, Ind Eng Chem Proc Des Dev, 9(1), 78-84 (1970).
Hattiangadi, U.S., How to interpret a negative value of minimum reflux ratio, Chem Eng, 18 May, 178 (1970).
Hosoda, K.; Takashima, I., and Hatano, S., Determination of vapor concentration in distillation columns, Int Chem Eng, 10(4), 672-678 (1970).
Mapstone, G.E., Batch distillation with fractionation at constant reflux, Brit Chem Eng, 15(2), 220 (1970).
McDaniel, R.; Bassyoni, A.A., and Holland, C.D., Use of field-test results in modeling of packed distillation columns and packed absorbers, Chem Eng Sci, 25(4), 633-652 (1970).
Mijolaj, P.G., Computer aid for crude oil distillation unit designs, Brit Chem Eng, 15(5), 638-643 (1970).
Sarkany, G.; Rozsa, P., and Tettamanti, K., Analytical calculation of number of theoretical plates, Periodica Polytechnica, 14, 321-332 (1970).
Serov, V.V., and Zykov, D.D., Determination of concentration of mixture of components adjacent to a temperature interface on an optimal feed tray, Int Chem Eng, 10(3), 415-418 (1970).

Sugie, H., and Lu, B.C.Y., Determination of minimum reflux ratio for multicomponent distillation column with any number of side-cut streams, Chem Eng Sci, 25(12), 1837-1846 (1970).
Treybal, R.E., A simple method for batch distillation, Chem Eng, 5 Oct, 95-98 (1970).

1971

Bakowski, S., Determination of the number of trays in a distillation column for separation of multicomponent mixtures, Brit Chem Eng, 16(11), 1013-1016 (1971).
Billingsley, D.S., and Boynton, G.W., Iterative methods for solving problems in multicomponent distillation at the steady state, AIChEJ, 17(1), 65-68 (1971).
Featherstone, W., Azeotropic systems: Rapid method of still design, Brit Chem Eng, 16(12), 1121-1124 (1971).
Hughmark, G.A., Models for vapor-phase and liquid-phase mass transfer on distillation trays, AIChEJ, 17(6), 1295-1299 (1971).
Jakob, R.R., Estimate number of crude trays, Hyd Proc, 50(5), 149-152 (1971).
Kondratev, A.A., and Golechek, A.A., Design of distillation columns for rectification of semi-continuous (complex) mixtures, Int Chem Eng, 11(4), 635-638 (1971).
Madsen, N., Finding the right reflux ratio, Chem Eng, 1 Nov, 73-75 (1971).
Osborne, A., Calculation of unsteady-state multicomponent distillation using partial differential equations, AIChEJ, 17(3), 696-703 (1971).
Shih, Y.P.; Chou, T.C., and Chen, C.T., Best distillation feed points, Hyd Proc, 50(7), 93-96 (1971).
Van Winkle, M., and Todd, W.G., Optimum fractionation design by simple graphical methods, Chem Eng, 20 Sept, 136-148 (1971).
Various, Vacuum distillation (topic issue), Chem Eng Prog, 67(3), 49-70 (1971).

1972

Barnes, F.J.; Hanson, D.N., and King, C.J., Calculation of minimum reflux for distillation columns with multiple feeds, Ind Eng Chem Proc Des Dev, 11(1), 136-140 (1972).
Bell, R.L., Experimental determination of residence time distributions on commercial scale distillation trays using a fibre optic technique, AIChEJ, 18(3), 491-497 (1972).
Bell, R.L., Residence time and fluid mixing on commercial-scale sieve trays, AIChEJ, 18(3), 498-505 (1972).
Clark, C.R., Distillation column computer program, Brit Chem Eng, 17(1), 44-48 (1972).
Guerreri, G.; Peri, B., and Seneci, F., Comparing distillation designs, Hyd Proc, 51(12), 77-78 (1972).
Hobler, T., and Pawelczyk, R., Interfacial area in bubbling through sieve-plate slots, Brit Chem Eng, 17(7), 624-628 (1972).
Jadbabaie, M.J., and Taheri, M., Generalised computer program for short-cut distillation calculations, Proc Tech Int, 17(11), 869-871 (1972).
Koppel, P.M., Fast way to solve problems for batch distillations, Chem Eng, 16 Oct, 109-112 (1972).

Molokanov, Y.K., et al., Approximate method of calculating the basic parameters of multicomponent fractionation, Int Chem Eng, 12(2), 209-213 (1972).

Various, Distillation design (topic issue), Chem Eng Prog, 68(10), 56-72 (1972).

1973

Bardone, E.; Mori, P., and Ferroni, E., Vacuum design vs distillation tests, Hyd Proc, 52(12), 71-72 (1973).

Chien, H.H.Y., Rigorous calculation method for minimum stages in multicomponent distillation, Chem Eng Sci, 28(11), 1967-1974 (1973).

Jashnani, I.L., and Lemlich, R., Transfer units in foam fractionation, Ind Eng Chem Proc Des Dev, 12(3), 312-321 (1973).

Jelinek, J.; Hlavacek, V., and Kubicek, M., Multicomponent multistage rectification by differentiation with respect to an actual parameter, Chem Eng Sci, 28(8), 1555-1564 (1973).

Jelinek, J.; Hlavacek, V., and Kubicek, M., Multicomponent multistage separation calculations by relaxation method, Chem Eng Sci, 28(10), 1825-1832 (1973).

Luyben, W.L., Azeotropic tower design by graph, Hyd Proc, 52(1), 109-112 (1973).

Maas, J.H., Optimum-feed-stage location in multicomponent distillation, Chem Eng, 16 April, 96-98 (1973).

Rozsa, P., and Sarkany, G., Analytical determination of number of theoretical stages in binary rectification and countercurrent extraction for nonlinear operating lines, Periodica Polytechnica, 17, 335-358 (1973).

Serafimov, L.A., and Timofeev, V.S., Mathematical model of distillation process, Int Chem Eng, 13(4), 666-669 (1973).

Serov, V.V.; Abramenko, V.P., and Zykov, D.D., Approximate group method for number of theoretical trays in a multicomponent distillation column, Int Chem Eng, 13(3), 514-517 (1973).

Sofer, S.S., and Anthony, R.G., Practical guidelines for distillation design, Hyd Proc, 52(3), 93-94 (1973).

Stewart, R.R.; Weisman, E.; Goodwin, B.M., and Speight, C.E., Effect of design parameters in multicomponent batch distillation, Ind Eng Chem Proc Des Dev, 12(2), 130-136 (1973).

Titov, A.A., et al., Determination of separation coefficients for dilute solutions using rectification columns, Int Chem Eng, 13(1), 62-65 (1973).

Yamada, I., Computer-aided distillation calculations, Int Chem Eng, 13(1), 106-122 (1973).

1974

Bruin, S., and Freije, A.D., A simple liquid mixing model for distillation plates with stagnant zones, Trans IChemE, 52, 75-79 (1974).

Horvath, A.L., Redlich-Kwong equation of state: Review for chemical engineering calculations, Chem Eng Sci, 29(5), 1334-1340 (1974).

Hutchison, H.P., and Shewchuk, C.F., A computational method for multiple distillation towers, Trans IChemE, 52, 325-336 (1974).

Jedlovszky, P., Linear approximation: New method for complex distillation column computations, Chem Eng Sci, 29(1), 287-288 (1974).

Kubicek, M.; Hlaracek, V., and Jelinek, J., Solution of countercurrent separation processes: Distillation problems, Chem Eng Sci, 29(2), 435-442 (1974).
Lee, E.S., Estimation of minimum reflux in distillation and multipoint boundary value problems, Chem Eng Sci, 29(3), 871-876 (1974).
Medina, A.G.; McDermott, C., and Ashton, N., Effect of experimental error on calculation of number of stages for a given distillation separation, Chem Eng Sci, 29(12), 2279-2282 (1974).
Mostafa, H.A., Direct calculation of number of actual plates, Chem Eng Sci, 29(9), 1997-2000 (1974).
Ricker, N.L., and Grens, E.A., Calculation procedure for design problems in multicomponent distillation, AIChEJ, 20(2), 238-244 (1974).
Strangio, V.A., and Treybal, R.E., Reflux-stages relations for distillation, Ind Eng Chem Proc Des Dev, 13(3), 279-285 (1974).
van Zyl, O.J., and Judd, M.R., Smoothing the material balance around a debutaniser-gasoline splitter using a new boiling point characterisation procedure, Trans IChemE, 52, 234-236 (1974).
Zanker, A., Quick calculation for minimum theoretical plates, Chem Eng, 2 Sept, 96 (1974).

1975

Biddulph, M.W., Multicomponent distillation simulation: Distillation of air, AIChEJ, 21(2), 327-335 (1975).
Jelinek, J., and Hlavacek, V., Mathematical modelling of the rectification process, Int Chem Eng, 15(3), 488-494 (1975).
Lo, C.T., Method of temperature estimation in bubble point method of iterative distillation calculations, AIChEJ, 21(6), 1223-1225 (1975).
Martin, G.Q., Guide to predicting azeotropes, Hyd Proc, 54(11), 241-246 (1975).
Mostafa, H.A.A., Operating curves for binary distillation calculation, Chem Eng, 17 Feb, 114-116 (1975).
Titov, A.A., and Zelvenskii, Y.D., Methods for estimating the active interfacial surface area in a packed distillation column, Int Chem Eng, 15(2), 371-376 (1975).
Vadnai, S,; Almasy, G.; Ser, V., and Veress, G., Modelling of distillation in the petroleum industry, Int Chem Eng, 15(3), 465-475 (1975).
Zanker, A., Batch distillation nomograms, Processing, July, 15-16 (1975).

1976

Block, U., and Hegner, B., Development and application of a simulation model for three-phase distillation, AIChEJ, 22(3), 582-589 (1976).
Chou, A.; Fayon, A.M., and Bauman, B.L., Simulations provide blueprint for distillation operations, Chem Eng, 7 June, 131-138 (1976).
Featherstone, W., Rapid method of design for batch distillation, Processing, May, 25-26 (1976).
Holland, C.D., et al., Solve more distillation problems, Hyd Proc, 53(7), 148-156; 53(11), 176-180 (1974); 54(1), 101-108; 54(7), 121-128 (1975); 55(1), 137-144; 55(6), 125-131 (1976).

Mommessin, P.E., and Holland, C.D., Solve more distillation problems, Hyd Proc, July, 148-156; Nov, 176-180 (1974); Jan, 101-108; July, 121-128 (1975); Jan, 137-144 (1976).

Winter, P., and Matthew, I., Improved design of multicolumn distillation, Processing, July, 11-13 (1976).

Zanker, A., Minimum reflux from a nomograph, Chem Eng, 12 April, 156 (1976).

1977

Brannock, N.F.; Verneuil, V.S., and Wang, Y.L., Rigorous distillation simulation, Chem Eng Prog, 73(10), 83-87 (1977).

Douglas, J.M., Rule-of-thumb for minimum trays, Hyd Proc, 56(11), 291 (1977).

Edwards, J.B., and Jassim, H.J., An analytical study of the dynamics of binary distillation columns, Trans IChemE, 55, 17-28 (1977).

Frank, O., Shortcut methods for distillation design, Chem Eng, 14 March, 110-128 (1977).

Fredenslund, A., et al., Computerized design of multicomponent distillation columns using the UNIFAC group contribution method for calculation of activity coefficients, Ind Eng Chem Proc Des Dev, 16(4), 450-462 (1977).

Hanson, D.N., and Newman, J., Calculation of distillation columns at the optimum feed-plate location, Ind Eng Chem Proc Des Dev, 16(2), 223-227 (1977).

Kister, H.Z., and Doig, I.D., Distillation pressure increases throughput, Hyd Proc, 56(7), 132-136 (1977).

Mussche, M., and Verhoeye, L., Analysis of relative errors in measuring number of theoretical plates in continuous distillation column, J Appl Chem Biotechnol, 27, 465-478 (1977).

Shewchuk, C.F., Extension of the quasi-linear method for non-ideal distillation calculations, Trans IChemE, 55, 130-136, 284 (1977).

Tan, H., Determine ideal distillation stages using a pocket calculator, Chem Eng, 14 March, 154 (1977).

Vanek, T.; Hlavacek, V., and Kubicek, M., Calculation of separation columns by non-linear block successive relaxation methods, Chem Eng Sci, 32(8), 839-844 (1977).

Wade, H.L., and Ryskamp, C.J., Tray flooding sets crude throughput, Hyd Proc, 56(11), 281-285 (1977).

Wagner, H., and Blass, E., Approximate design for multicomponent rectification, Int Chem Eng, 17(2), 237-245 (1977).

Zanker, A., Nomograph replaces Gilliland plot, Hyd Proc, 56(5), 263-264 (1977).

1978

Bush, M.J., et al., Modular approach to distillation column simulation, Comput Chem Eng, 2(4), 161-168 (1978).

Chien, H.H.Y., Rigorous method for calculating minimum reflux rates in distillation, AIChEJ, 24(4), 606-613 (1978).

Eckert, E., and Hlavacek, V., Calculation of multicomponent distillation of non-ideal mixtures by short-cut method, Chem Eng Sci, 33(1), 77-82 (1978).

Economopoulos, A.P., A fast computer method for distillation calculations, Chem Eng, 24 April, 91-100 (1978).
Furzer, I.A., Modelling of mass transfer in plate columns with dual flow, Chem Eng Sci, 33(3), 349-356 (1978).
Kochergin, N.A., et al., Model of longitudinal mixing of liquid on various mass transfer trays, Int Chem Eng, 18(1), 163-166 (1978).
Loud, G.D., and Waggoner, R.C., Effects of interstage backmixing on design of multicomponent distillation columns, Ind Eng Chem Proc Des Dev, 17(2), 149-156 (1978).
Steiner, L.; Barendreght, H.P., and Hartland, S., Concentration profiles in packed distillation column, Chem Eng Sci, 33(3), 255-262 (1978).
Yamada, I., et al., Distillation calculations for binary system using plate efficiency, Int Chem Eng, 18(4), 689-692 (1978).

1979

Akashah, S.; Erbar, J.H., and Maddox, R.N., Optimum feed plate location for multicomponent distillation separations, Chem Eng Commns, 3(6), 461-468 (1979).
Drogaris, G., and Lockett, M.J., Deleterious effect of vapour entrainment under the downcomer determined by Standart's entrainment model, Chem Eng Commns, 3(4), 291-302 (1979).
Harker, J.H., Economic balance in distillation, Processing, April, 39-40 (1979).
Hunek, J.; Foldes, P., and Sawinsky, J., Calculation methods for reactive distillation, Int Chem Eng, 19(2), 248-259 (1979).
Jafarey, A.; Douglas, J.M., and McAvoy, T.J., Short-cut techniques for distillation column design and control, Ind Eng Chem Proc Des Dev, 18(2), 197-210 (1979).
Kister, H.Z., and Doig, I.D., Computational analysis of the effect of pressure on distillation column feed capacity, Trans IChemE, 57, 43-48 (1979).
Mansouri, S., Streamline flash computations with a calculator program, Chem Eng, 27 Aug, 99-101 (1979).
Nguyen, H.X., Boiling points set minimum trays, Hyd Proc, 58(6), 95-96 (1979).
Tavana, M., and Hanson, D.N., Exact calculation of minimum flows in distillation columns, Ind Eng Chem Proc Des Dev, 18(1), 154-156, (1979).
Yaws, C.L.; Fang, C.S., and Patel, P.M., Estimating recoveries in multicomponent distillation, Chem Eng, 29 Jan, 101-104 (1979).

1980

Chang, H.Y., Computer aids shortcut distillation design, Hyd Proc, 59(8), 79-82 (1980).
Fitzmorris, R.E., and Mah, R.S.H., Improving distillation column design using thermodynamic availability analysis, AIChEJ, 26(2), 265-273 (1980).
Hirose, Y.; Kawase, Y., and Funada, I., Distillation calculation for cases of specified heat duty by modification of existing computer programs, Ind Eng Chem Proc Des Dev, 19(3), 505-507 (1980).
Hirose, Y.; Kawase, Y.; Sampei, K., and Kawai, T., New approach to distillation column at total reflux, Comput Chem Eng, 4(2), 133-138 (1980).

Mommessin, P.E., and Holland, C.D., Solve more distillation problems, Hyd Proc, June, 125-131 (1976); May, 241-248; June, 181-188 (1977); April, 195-203 (1980).
Said, A.S., Simple analytic formula for number of theoretical plates in distillation, Sepn Sci Technol, 15(10), 1699-1708 (1980).
Tierney, J.W., and Abdelmalek, N.A., Algorithms for total reflux calculations, Ind Eng Chem Fund, 19(4), 404-410 (1980).
Von Rosenberg, D.U., and Hadi, M.S., Numerical solution of multicomponent packed-tower distillation problems, Chem Eng Commns, 4(2), 313-324 (1980).

1981
Cerda, J., and Westerberg, A.W., Shortcut methods for complex distillation columns, Ind Eng Chem Proc Des Dev, 20(3), 546-557 (1981).
Challis, H., Cost-saving distillation, Proc Engng, Sept, 45-49 (1981).
Chang, H.Y., Gilliland plot in one equation, Hyd Proc, 60(10), 146 (1981).
Fleischer, M.T., and Prett, D.M., Multicomponent distillation calculations using simplified techniques, Chem Eng Commns, 10(4), 243-260 (1981).
Harg, K., Modified Smith-Brinkley algorithm for high-purity distillation, Chem Eng Sci, 36(3), 626-628 (1981).
Holland, C.D.; Gallun, S.E., and Lockett, M.J., Modeling azeotropic and extractive distillations, Chem Eng, 23 March, 185-200 (1981).
Kesler, M., Shortcut program for multicomponent distillation, Chem Eng, 4 May, 85-88 (1981).
Ledanois, J.M., Programs solve Underwood's equation, Hyd Proc, 60(4), 231-233 (1981).
Marinos-Kouris, D.S., A short-cut method for multicomponent distillation, Chem Eng, 9 March, 83-86 (1981).
Nandakumar, K., and Andres, R.P., Minimum reflux conditions, AIChEJ, 27(3), 450-465 (1981).
Olujic, Z., Optimum reflux ratio, Chem Eng, 19 Oct, 184 (1981).
Pierucci, S.J., et al., T-method computes distillation, Hyd Proc, 60(9), 179-183 (1981).
Thomas, P.J., Dynamic simulation of multicomponent distillation processes, Ind Eng Chem Proc Des Dev, 20(1), 166-168 (1981).
Torres-Marchal, C., Graphical design for ternary distillation system, Chem Eng, 19 Oct, 134-155 (1981).
Waggoner, R.C., Distillation in packed beds, Hyd Proc, 60(12), 119-122 (1981).
Yaws, C.L.; Li, K.Y., and Fang, C.S., How to find the minimum reflux for multicomponent systems in multiple-feed columns, Chem Eng, 1 June, 63-65 (1981).
Yaws, C.L.; Li, K.Y., and Fang, C.S., How to find the minimum reflux for binary systems in multiple-feed columns, Chem Eng, 18 May, 153-156 (1981).

1982
Bravo, J.L., and Fair, J.R., Generalized correlation for mass transfer in packed distillation columns, Ind Eng Chem Proc Des Dev, 21(1), 162-170 (1982).

Chianese, A.; Campana, H., and Picciotti, M., Better distillation feed point, Hyd Proc, Jan, 133-137 (1982).

Govind, R., Analytical form of the Ponchon-Savarit method for systems with straight enthalpy-concentration phase lines, Ind Eng Chem Proc Des Dev, 21(3), 532-535 (1982).

Tolliver, T.L., and Waggoner, R.C.,, Approximate solutions for distillation rating and operating problems using the Smoker equations, Ind Eng Chem Fund, 21(4), 422-427 (1982).

Yorizane, M.; Yoshida, H.; Kawasaki, S., and Soewarno, N., Extractive distillation calculations for the Petlyuk and two-column models, Int Chem Eng, 22(3), 533-543 (1982).

Yoshifuku, I., Method for multicomponent distillation calculation problems, Int Chem Eng, 22(3), 528-533 (1982).

Zanker, A., Quick estimate of bubble-cap pressure drop, Proc Engng, Feb, 63 (1982).

1983

Alper, W.S., and Osborne, R.L., Graphic input for complex distillation problems, Chem Eng Prog, 79(10), 27-43 (1983).

Guy, J.L., Modeling batch distillation in multitray columns, Chem Eng, 10 Jan, 99-103 (1983).

Holland, C.D., et al., Solve more distillation problems, Hyd Proc, 56(5), 241-248; 56(6), 181-188 (1977); 59(4), 195-203; 59(7), 144-147 (1980); 60(1), 189-196; 60(7), 133-139 (1981); 62(11), 195-200 (1983).

Ingels, R.M., Program to evaluate separation processes, Chem Eng, 30 May, 73-75 (1983).

Mommessin, P.E., and Holland, C.D., Solve more distillation problems, Hyd Proc, July, 144-148 (1980); Jan, 189-196; July, 133-139 (1981); Nov, 195-200 (1983).

Pierucci, S.J.; Ranzi, E.M., and Biardi, G.E., Corrected flowrates estimation by using theta-convergence promoter for distillation columns, AIChEJ, 29(1), 113-123 (1983).

Prokopakis, G.J., and Seider, W.D., Dynamic simulation of azeotropic distillation towers, AIChEJ, 29(6), 1017-1029 (1983).

Russell, R.A., Simple method for distillation-column problems, Chem Eng, 17 Oct, 53-59 (1983).

Sadatomo, H., and Miyahara, K., Calculation procedure for multicomponent batch distillation, Int Chem Eng, 23(1), 56-65 (1983).

Walker, C.A., and Halpern, B.L., Distillation calculations with a programmable calculator, Chem Eng Educ, 17(2), 86-88 (1983).

Zanker, A., Nomograph for feed plate location, Proc Engng, Nov, 67 (1983).

Zomosa, A., Quick design of distillation columns for binary mixtures, Chem Eng, 24 Jan, 95-98 (1983).

1984

Anderson, N.J., and Doherty, M.F., Approximate model for binary azeotropic distillation design, Chem Eng Sci, 39(1), 11-20 (1984).

Billingsley, D.S., Accelerated convergence for Newton-type iterations applied to multicomponent distillation problems, AIChEJ, 30(4), 686-688 (1984).

Glinos, K., and Malone, M.F., Minimum reflux product distribution and lumping rules for multicomponent distillation, Ind Eng Chem Proc Des Dev, 23(4), 764-768 (1984).

Kuno, M., Enthalpy method for converging distillation calculations, Ind Eng Chem Proc Des Dev, 23(3), 443-449 (1984).

Kvaalen, E.T., and Wankat, P.C., Analysis of multicomponent and adiabatic countercurrent columns, Ind Eng Chem Fund, 23(1), 14-19 (1984).

Ledanois, J.M., and Olivera-Fuentes, C., Modified Ponchon-Savarit and McCabe-Thiele methods for distillation of two-phase feeds, Ind Eng Chem Proc Des Dev, 23(1), 1-6 (1984).

Ngai, I., Two programs for multicomponent distillation, Chem Eng, 20 Aug, 145-149 (1984).

Rice, V.L., Way to predict tray temperatures, Hyd Proc, Aug, 83-84 (1984).

Rooney, J.M., Simulating batch distillation, Chem Eng, 14 May, 61-64 (1984).

1985

Al-Ameeri, R.S., and Said, A.S., Simple formula for Gilliland correlation in multicomponent distillation, Sepn Sci Technol, 20(7), 565-576 (1985).

Arnold, V.E., Calculating minimum reflux, Chem Eng, 4 Feb, 59-62 (1985).

Barna, B.A., and Ginn, R.F., Tray estimates for low reflux, Hyd Proc, May, 115-116 (1985).

Fasesan, S.O., Weeping from distillation/absorption trays, Ind Eng Chem Proc Des Dev, 24(4), 1073-1080 (1985).

Fisher, W.R.; Doherty, M.F., and Douglas, J.M., Shortcut calculation of optimal recovery fractions for distillation columns, Ind Eng Chem Proc Des Dev, 24(4), 955-961 (1985).

Glinos, K., and Malone, M.F., Minimum vapor flows in distillation column with sidestream stripper, Ind Eng Chem Proc Des Dev, 24(4), 1087-1090 (1985).

Hirose, Y., and Tsuda, M., New algorithm for solving a block-tridiagonal equation containing a singular submatrix and application to distillation calculations, Int Chem Eng, 25(2), 295-301 (1985).

Kister, H.Z., Complex binary distillation, Chem Eng, 21 Jan, 97-104 (1985).

Kister, H.Z., Complex multicomponent distillation, Chem Eng, 13 May, 71-80 (1985).

Krishnamurthy, R., and Taylor, R., Simulation of packed distillation and absorption columns, Ind Eng Chem Proc Des Dev, 24(3), 513-524 (1985).

Van Dongen, D.B., and Doherty, M.F., Design and synthesis of homogeneous azeotropic distillations, Ind Eng Chem Fund, 24(4), 454-485 (1985).

Yamada, I., et al., Ev-matrix method for multicomponent distillation problems involving partial condensation and entrainment, Int Chem Eng, 25(3), 507-517 (1985).

1986

Chou, S.M., and Yaws, C.L., Reflux for multifeed distillation, Hyd Proc, 65(12), 41-42 (1986).

Chou, S.M.; Yaws, C.L., and Cheng, J.S., Application of factor method for minimum reflux: Multiple-feed distillation columns, Can JCE, 64(2), 254-259 (1986).

Doherty, M.F., and co-workers, Design and synthesis of homogeneous azeotropic distillations, Ind Eng Chem Fund, Part 1: 24(4), 454-463 (1985); Part 2: 23(4), 463-474 (1985); Part 3: 24(4), 474-485 (1985),; Part 4: 25(2), 269-279 (1986); Part 5: 25(2), 279-289 (1986).

Ellis, M.F.; Koshy, R.; Mijares, G.; Munoz, A.G., and Holland, C.D., Use of multipoint algorithms and continuation methods in the solution of distillation problems, Comput Chem Eng, 10(5), 433-443 (1986).

Gani, R.; Ruiz, C.A., and Cameron, I.T., A generalized model for distillation columns, Comput Chem Eng, 10(3), 181-211 (1986).

Myers, A.L., and Fichthorn, K., Distillation column design by method of McCabe-Thiele: Distman computer graphics program, Chem Eng Commns, 40, 195-206 (1986).

Neil, J., and Stone, S.J., Using CAD systems to prepare McCabe-Thiele diagrams, Chem Eng, 10 Nov, 135-136 (1986).

Saeger, R.B., and Bishnoi, P.R., A modified 'inside-out' algorithm for simulation of multistage multicomponent separation processes using the UNIFAC group-contribution method, Can JCE, 64(5), 759-767 (1986).

Swartz, C.L.E., and Stewart, W.E., Collocation approach to distillation column design, AIChEJ, 32(11), 1832-1838 (1986).

Tsuo, F.M.; Yaws, C.L., and Cheng, J.S., Minimum relfux for sidestream columns, Chem Eng, 21 July, 49-53 (1986).

Yamada, I.; Sawada, M.; Zhang, B.Q., and Haraoka, S., Rapid solution of operation-type multicomponent distillation problems by Ev-matrix method, Int Chem Eng, 26(3), 534-540 (1986).

1987

Ammar, M.N., and Renon, H., Isothermal flash problem: New methods for phase split calculations, AIChEJ, 33(6), 926-939 (1987).

Bauerle, G.L., and Sandall, O.C., Batch distillation of binary mixtures at minimum reflux, AIChEJ, 33(6), 1034-1036 (1987).

Choe, Y.S., and Luyben, W.L., Rigorous dynamic models of distillation columns, Ind Eng Chem Res, 26(10), 2158-2161 (1987).

Grosser, J.H.; Doherty, M.F., and Malone, M.F., Modeling of reactive distillation systems, Ind Eng Chem Res, 26(5), 983-990 (1987).

Lipowicz, M., Distillation-column design software, Chem Eng, 16 Feb, 167-170 (1987).

Nikolaides, I.P., and Malone, M.F., Approximate design of multiple-feed/side-stream distillation systems, Ind Eng Chem Res, 26(9), 1839-1845 (1987).

Rao, K.V., and Raviprasad, A., Quickly determine multicomponent minimum reflux ratio, Chem Eng, 12 Oct, 137-138 (1987).

Shoaei, M., and Tedder, D.W., Design calculations for multicomponent distillation by an improved shortcut method, CER&D, 65(3), 251-260 (1987).

Swartz, C.L.E., and Stewart, W.E., Finite-element steady-state simulation of multiphase distillation, AIChEJ, 33(12), 1977-1985 (1987).

1988

Barbosa, D., and Doherty, M.F., Design and minimum-reflux calculations for double-feed multicomponent reactive-distillation columns, Chem Eng Sci, 43(9), 2377-2390 (1988).

Barbosa, D., and Doherty, M.F., Design and minimum-reflux calculations for single-feed multicomponent reactive distillation columns, Chem Eng Sci, 43(7), 1523-1538 (1988).

Bauerle, G.L., and Sandall, O.C., Design of batch distillation columns for binary mixtures, Chem Eng Commns, 65, 155-160 (1988).

Cairns, B.P., and Furzer, I.A., Rapid-robust calculational procedures for multicomponent distillation problems, Chem Engng in Australia 13(2), 13-15 (1988).

Chou, S.M., and Yaws, C.L., Minimum reflux for complex distillations, Chem Eng, 25 April, 79-82 (1988).

Henry, B.D., Use of mathematical models in the prediction of optimum distillation sequences, Chem Eng J, 37(2), 115-122 (1988).

Hitch, D.M., and Rousseau, R.W., Simulation of multicomponent batch distillation, Ind Eng Chem Res, 27(8), 1466-1473 (1988).

Lee, S.Y., and Wong, D.S.H., Flexibility and optimality of distillation column design, AIChEJ, 34(1), 144-146 (1988).

Morris, C.G.; Sim, W.D.; Vysniauskas, T., and Svrcek, W.Y., Crude-tower simulation on a personal computer, Chem Eng Prog, 84(11), 63-68 (1988).

Muhrer, C.A.; Collura, M.A., and Luyben, W.L., Concentration profile inversion in distillation rating programs with tray efficiencies, Ind Eng Chem Res, 27(4), 716-718 (1988).

Nikolaides, I.P., and Malone, M.F., Approximate design and optimization of a thermally coupled distillation with prefractionation, Ind Eng Chem Res, 27(5), 811-818 (1988).

Ratzsch, M.T.; Kehlen, H., and Schumann, J., Flash calculations for a crude oil by continuous thermodynamics, Chem Eng Commns, 71, 113-126 (1988).

Venkataraman, S., and Lucia, A., Solving distillation problems by Newton-like methods, Comput Chem Eng, 12(1), 55-70 (1988).

Wachter, J.A.; Ko, T.K.T., and Andres, R.P., Minimum reflux behavior of complex distillation columns, AIChEJ, 34(7), 1164-1184 (1988).

19.3 Efficiency

1966-1969

Chidambaram, S., Nomogram for plate efficiencies, Brit Chem Eng, 11(4), 288 (1966).

Diener, D.A., Calculation of effect of vapor mixing on tray efficiency, Ind Eng Chem Proc Des Dev 6(4), 499-503 (1967).

Ellis, S.R.M., and Biddulph, M.W., Effect of surface tension characteristics on plate efficiencies, Trans IChemE, 45, T223-228 (1967).

Everitt, C.T., and Hutchison, H.P., The effect of composition on the performance of a wetted-wall rectification column, Trans IChemE, 45, T9-15 (1967).

Thomas, W.J., and Campbell, M., Mixing, efficiency and mass transfer studies in a sieve plate downcomer system, Trans IChemE, 45, T64-73 (1967).
Diener, D.A., and Gerster, J.A., Point efficiencies in distillation of acetone-methanol-water, Ind Eng Chem Proc Des Dev, 7(3), 339-345 (1968).
Foldes, P., and Evangelidi, I., Efficiency of Turbogrid-tray distillation columns, Brit Chem Eng, 13(9), 1291-1293 (1968).
Foldes, P., and Evangelidi, I., Efficiency of Turbogrid-tray distillation columns, Brit Chem Eng, 13(6), 832-834 (1968).
Ho, G.E., and Prince, R.G.H., Plate efficiency: Effect of bubble size distribution on liquid-phase efficiency, Chem Eng Sci, 23(8), 948-951 (1968).
Hutchison, H.P., and Lusis, M.A., Distillation efficiencies of binary and ternary mixtures in a wetted-wall column, Trans IChemE, 46, T158-166 (1968).
May, R.A., and Horn, F.J.M., Stage efficiency of periodically operated distillation column, Ind Eng Chem Proc Des Dev, 7(1), 61-64 (1968).
Rampacek, C.M., and Van Winkle, M., Efficiency study of jet trays in a 6-inch diameter laboratory column, Ind Eng Chem Proc Des Dev, 7(2), 313-318 (1968).
Aleksandrov, I.A., Mass-transfer efficiency in plate cross-flow, Int Chem Eng, 9(4), 691-694 (1969).
Bakowski, S., Efficiency of sieve-tray columns, Brit Chem Eng, 14(7), 945-949 (1969).
Burghardt, A., and Kedzierski, S., Effect of an inert gaseous component on tray efficiency of a rectifying column, Int Chem Eng, 9(3), 486-502 (1969).
Furzer, I.A., Effect of vapor distribution on distillation plate efficiencies, AIChEJ, 15(2), 235-239 (1969).
Lebedev, Y.N.; Aleksandrov, I.A., and Zykov, D.D., Mass-transfer efficiency for binary distillation, Int Chem Eng, 9(1), 97-99 (1969).
Rao, K.B., Nomogram for efficiency of turbogrid-tray distillation columns, Brit Chem Eng, 14(10), 1457 (1969).

1970-1973

Ashley, M.J., and Haselden, G.G., Calculation of plate efficiency under conditions of finite mixing in both phases in multiplate columns, and potential advantage of parallel flow, Chem Eng Sci, 25(11), 1665-1672 (1970).
Gulyaev, F.A., et al., Relative evaluation of efficiency of cellular packing for vacuum mass-transfer equipment, Int Chem Eng, 10(3), 403-405 (1970).
Ho, G.E., and Prince, R.G.H., Hausen plate efficiency for binary systems, Trans IChemE, 48, T101-106 (1970).
Holland, C.D., and McMahon, K.S., Comparison of vaporization efficiencies with Murphree-type efficiencies in distillation, Chem Eng Sci, 25(3), 431-436 (1970).
Kirsten, R.A., and Van Winkle, M., Efficiency of jet trays, Ind Eng Chem Proc Des Dev, 9(1), 100-105 (1970).
Sealey, C.J., Optimal design of laboratory distillation column for efficiency studies at finite reflux, Chem Eng Sci, 25(4), 561-568 (1970).
Sealey, C.J., Column for investigating the effect of reflux ratio on efficiency, Brit Chem Eng, 15(8), 1053-1054 (1970).

Boyes, A.P., and Ponter, A.B., Prediction of distillation column performance for surface tension positive and negative systems, Ind Eng Chem Proc Des Dev, 10(1), 140-143 (1971).

Butcher, K.L., and Medani, M.S., Sieve plate distillation efficiencies of the methanol-benzene system at high vapour pressure, Trans IChemE, 49, 225-239 (1971).

Onda, K.; Sada, E.; Takahashi, K., and Mukhtar, S.A., Plate and column efficiencies of continuous rectifying columns for binary mixtures, AIChEJ, 17(5), 1141-1152 (1971).

Standart, G.L., Comparison of Murphree-type efficiencies with vaporisation efficiencies, Chem Eng Sci, 26(6), 985-989 (1971).

Ashley, M.J., and Haselden, G.G., Effectiveness of vapour-liquid contacting on a sieve plate, Trans IChemE, 50, 119-124 (1972).

MacFarland, S.A.; Sigmund, P.M., and Van Winkle, M., Predict distillation efficiency, Hyd Proc, 51(7), 111-114 (1972).

Miskin, L.G.; Ozalp, U., and Ellis, S.R.M., Ternary component efficiencies, Brit Chem Eng, 17(2), 153-155 (1972).

Porter, K.E.; Lockett, M.J., and Lim, C.T., The effect of liquid channelling on distillation plate efficiency, Trans IChemE, 50, 91-101 (1972).

Todd, W.G., and Van Winkle, M., Correlation of valve-tray efficiency data, Ind Eng Chem Proc Des Dev, 11(4), 589-604 (1972).

Todd, W.G., and Van Winkle, M., Fractionation efficiency, Ind Eng Chem Proc Des Dev, 11(4), 578-588 (1972).

Young, G.C., and Weber, J.H., Murphree point efficiencies in multicomponent systems, Ind Eng Chem Proc Des Dev, 11(3), 440-446 (1972).

Asbjornsen, O.A., Stage efficiency in dynamic models of phase separation processes, Chem Eng Sci, 28(12), 2223-2231 (1973).

Katayama, H., and Imoto, T., Effect of vapor mixing on tray efficiency of distillation columns, Int Chem Eng, 13(4), 728-734 (1973).

Lockett, M.J.; Lim, C.T., and Porter, K.E., The effect of liquid channelling on distillation column eficiency in the absence of vapour mixing, Trans IChemE, 51, 61-67 (1973).

1974-1977

Bell, R.L., and Solari, R.B., Effect of nonuniform velocity fields and retrograde flow on distillation tray efficiency, AIChEJ, 20(4), 688-695 (1974).

Lashmet, P.K., and Szczepanski, S.Z., Efficiency uncertainty and distillation column overdesign factors, Ind Eng Chem Proc Des Dev, 13(2), 103-106 (1974).

Lim, C.T.; Porter, K.E., and Lockett, M.J., The effect of liquid channelling on two-pass distillation plate efficiency, Trans IChemE, 52, 193-201 (1974).

Standart, G.L., Some problems of predicting plate efficiencies, Chem Engnr, Nov, 716-718 (1974).

van der Veen, A.J.; Drinkenburg, A.A.H., and Moens, F.P., Influence of liquid flowrate on the efficiency of a sieve plate at constant vapour throughput, Trans IChemE, 52, 228-233 (1974).

Lockett, M.J.; Porter, K.E., and Bassoon, K.S., Effect of vapour mixing on distillation plate efficiency when liquid channelling occurs, Trans IChemE, 53, 125-130 (1975).

Anderson, R.H.; Garrett, G., and Van Winkle, M., Efficiency comparison of valve and sieve trays in distillation columns, Ind Eng Chem Proc Des Dev, 15(1), 96-100 (1976).

Bolles, W.L., Multipass flow distribution and mass transfer efficiency for distillation plates, AIChEJ, 22(1), 153-158 (1976).

Brambilla, A., Effect of vapour mixing on efficiency of large diameter distillation plates, Chem Eng Sci, 31(7), 517-524 (1976).

Lockett, M.J., and Safekourdi, A., Effect of liquid flow pattern on distillation plate efficiency, Chem Eng J, 11(2), 111-122 (1976).

Sirkar, K.K., Point and stage efficiencies in distillation, Sepn Sci, 11(4), 303-316 (1976).

Biddulph, M.W., Predicted comparisons of efficiency of large valve trays and large sieve trays, AIChEJ, 23(5), 770-771 (1977).

Biddulph, M.W., Tray efficiency is not constant, Hyd Proc, 56(10), 145-148 (1977).

Biddulph, M.W., and Ashton, N., Deducing multicomponent distillation efficiencies from industrial data, Chem Eng J, 14(1), 7-16 (1977).

Calderbank, P.H., and Pereira, J., Prediction of distillation plate efficiencies from froth properties, Chem Eng Sci, 32(12), 1427-1434 (1977).

Clark, C.R., Calculating accurate plate efficiency, Proc Engng, June, 61-66 (1977).

Garrett, G.R.; Anderson, R.H., and Van Winkle, M., Calculation of sieve and valve tray efficiencies in column scale-up, Ind Eng Chem Proc Des Dev, 16(1), 79-82 (1977).

Golding, J.A., and Mah, C.C., A comparison of contacting efficiencies and mass transfer coefficients for various contacting conditions, Trans IChemE, 55, 216-219 (1977).

Krishna, R.; Martinez, H.F.; Sreedhar, R., and Standart, G.L., Murphree point efficiencies in multicomponent systems, Trans IChemE, 55, 178-183 (1977).

Porter, K.E.; Safekourdi, A., and Lockett, M.J., Plate efficiency in the spray regime, Trans IChemE, 55, 190-195 (1977).

Standart, G.L., and Martinez, J., Effect of spray mixing on tray efficiency, Chem Eng Commns, 2(4), 223-232 (1977).

1978-1982

Cervanka, J.; Endrst, M., and Kolar, V., Plate efficiency in rectification of binary mixtures at reduced pressures with expanded-metal inserts, Int Chem Eng, 18(2), 318-325 (1978).

Lenoir, J.M., and Sakata, M., Correlation of vapor-liquid equilibria used to determine tray efficiencies, Ind Eng Chem Fund, 17(2), 71-84 (1978).

Medina, A.G.; Ashton, N., and McDermott, C., Murphree and vaporization efficiencies in multicomponent distillation, Chem Eng Sci, 33(3), 331-340 (1978).

Lockett, M.J.; Kirkpatrick, R.D., and Uddin, M.S., Froth regime point efficiency for gas-film controlled mass transfer on a two-dimensional sieve tray, Trans IChemE, 57, 25-34 (1979).

Medina, A.G.; Ashton, N., and McDermott, C., Hausen and Murphree efficiencies in binary and multicomponent distillation, Chem Eng Sci, 34(9), 1105-1112 (1979).

Medina, A.G.; McDermott, C., and Ashton, N., Prediction of multicomponent distillation efficiencies, Chem Eng Sci, 34(6), 861-866 (1979).

Mostafa, H.A., Effect of concentration on distillation plate efficiency, Trans IChemE, 57, 55-59, 216 (1979).

Wankat, P.C., and Hubert, J., Use of the vaporization efficiency in closed form solutions for separation columns, Ind Eng Chem Proc Des Dev, 18(3), 394-398 (1979).

Holland, C.D., Computing large negative or positive values for Murphree efficiencies, Chem Eng Sci, 35(10), 2235-2236; 35(11), 2371 (1980).

Lockett, M.J., and Dhulesia, H.A., Murphree plate efficiency with nonuniform vapour distribution, Chem Eng J, 19(3), 183-188 (1980).

Weiler, D.W.; Kirkpatrick, R.D., and Lockett, M.J., Effect of downcomer mixing on distillation tray efficiency, Chem Eng Prog, 77(1), 63-69 (1981).

Hoek, P.J., and Zuiderweg, F.J., Influence of vapor entrainment on distillation tray efficiency at high pressures, AIChEJ, 28(4), 535-541 (1982).

Kister, H.Z., and Doig, I.D., Effect of pressure on separation efficiency in packed distillation columns, Can JCE, 60, 155-159 (1982).

Neuberg, H.J., and Chuang, K.T., Mass-transfer modeling for GS heavy water plants: Point and tray efficiencies on GS sieve trays, Can JCE, 60, 504-515 (1982).

1983-1985

Fair, J.R.; Null, H.R., and Bolles, W.L., Scale-up of plate efficiency from laboratory Oldershaw data, Ind Eng Chem Proc Des Dev, 22(1), 53-58 (1983).

Lockett, M.J., and Ahmed, I.S., Tray and point efficiencies from a 0.6 metre diameter distillation column, CER&D, 61, 110-118 (1983).

Lockett, M.J., and Plaka, T., Effect of non-uniform bubbles in the froth on the correlation and prediction of point efficiencies, CER&D, 61, 119-124 (1983).

Lockett, M.J.; Rahman, M.A., and Dhulesia, H.A., Effect of entrainment on distillation tray efficiency, Chem Eng Sci, 38(5), 661-672 (1983).

Mohan, T.; Rao, K.K., and Rao, D.P., Effect of vapor maldistribution and entrainment on tray efficiency, Ind Eng Chem Proc Des Dev, 22(3), 376-385 (1983).

Patberg, W.B.; Koers, A.; Steenge, W.D.E., and Drinkenburg, A.A.H., Effectiveness of mass transfer in a packed distillation column in relation to surface tension gradients, Chem Eng Sci, 38(6), 917-924 (1983).

Zanker, A., Murphree efficiency by nomograph, Proc Engng, May, 67 (1983).

Chan, H., and Fair, J.R., Prediction of point efficiencies on sieve trays, Ind Eng Chem Proc Des Dev, 23(4), 814-827 (1984).

Lockett, M.J.; Rahman, M.A., and Dhulesia, H.A., Prediction of the effect of weeping on distillation tray efficiency, AIChEJ, 30(3), 423-431 (1984).

Plaka, T.; Ahmed, I.S., and Lockett, M.J., Scaling-up distillation tray efficiency under conditions of weeping or entrainment when liquid channelling also occurs, CER&D, 62, 191-193, 200 (1984).

Savkovic-Stevanovic, J., Murphree, Hausen, vaporization, and overall efficiencies in binary distillation of associated systems, Sepn Sci Technol, 19(4), 283-296 (1984).

Vital, T.J.; Grossel, S.S., and Olsen, P.I., Estimating separation efficiency, Hyd Proc, Oct, 55-56; Nov, 147-153; Dec, 75-78 (1984).

Kouri, R.J., and Sohlo, J.J., Effect of developing liquid flow patterns on distillation plate efficiency, CER&D, 63, 117-124 (1985).

Krishna, R., Model for prediction of point efficiencies for multicomponent distillation, CER&D, 63, 312-322 (1985).

1986-1988

Biddulph, M.W., and Dribika, M.M., Distillation efficiencies on a large sieve plate with small-diameter holes, AIChEJ, 32(8), 1383-1388 (1986).

Dribika, M.M., and Biddulph, M.W., Scaling-up distillation efficiencies, AIChEJ, 32(11), 1864-1875 (1986).

Rocha, J.A.; Humphrey, J.L., and Fair, J.R., Mass-transfer efficiency of sieve tray extractors, Ind Eng Chem Proc Des Dev, 25(4), 862-872 (1986).

Rose, L.M., Distillation design in practice: Plate efficiencies, Proc Engng, Feb, 71-73 (1986).

Russell, L., and Gerrard, M., Maximising distillation column efficiency, Proc Engng, June, 45-51 (1986).

Biddulph, M.W., Efficiencies in ternary sieve-tray distillation column, Gas Sepn & Purif, 1(2), 90-93 (1987).

Kalbassi, M.A., and Biddulph, M.W., A modified Oldershaw column for distillation efficiency measurements, Ind Eng Chem Res, 26(8), 1127-1132 (1987).

Biddulph, M.W., and Kalbassi, M.A., Distillation efficiencies for methanol/1-propanol/water, Ind Eng Chem Res, 27(11), 2127-2135 (1988).

Biddulph, M.W.; Kalbassi, M.A., and Dribika, M.M., Multicomponent efficiencies in two types of distillation column, AIChEJ, 34(4), 618-625 (1988).

Cervenka, J.; Endrst, M., and Kolar, V., Efficiency of internals (horizontal insertions) in rectification of binary mixtures at atmospheric pressure, Int Chem Eng, 15(1), 45-52 (1988).

Martin, A.; Bravo, V., and Perez, J., Influence of nature of system on distillation efficiency: n-Hexane/benzene and benzene/n-heptane systems, Int Chem Eng, 28(1), 125-132 (1988).

19.4 Column Design Data

1966-1967

Eduljee, H.E., Hydraulics of grid and sieve-type shower plates, Brit Chem Eng, 11(12), 1519-1522 (1966).

Eichel, F.G., Capacity of packed columns in vacuum distillations, Chem Eng, 12 Sept, 197-204 (1966).

Thangappan, R., Nomogram for vapour velocities in bubble-cap distillation plates, Brit Chem Eng, 11(6), 517 (1966).

Thibodeaux, L.J., and Murrill, P.W., Comparing packed and plate columns, Chem Eng, 18 July, 155-158 (1966).
Aerov, M.E., et al., Hydraulics of sieve trays at high liquid loadings, Int Chem Eng, 7(2), 235-238 (1967).
Bozhov, I., and Elenkov, D., Influence of surfactants on mass transfer as determined by liquid boundary layer on plates without downcomers, Int Chem Eng, 7(2), 316-321 (1967).
Chase, J.D., Sieve-tray design, Chem Eng, 31 July, 105-116; 28 Aug, 139-146 (1967).
Galluzzo, J.F., Installing a reflux splitter under positive pump pressure, Chem Eng, 13 Feb, 180 (1967).
Hoppe, K.; Kruger, G., and Ikier, H., The Kittel overflow tray, Brit Chem Eng, 12(9), 1381-1384 (1967).
Hoppe, K.; Kruger, G., and Ikier, H., Development of the Kittel tray, Brit Chem Eng, 12(5), 715-718 (1967).
Malyusov, V.A., et al., High-velocity cocurrent rectification in a tubular column, Int Chem Eng, 7(2), 264-268 (1967).
McGurl, G.V., and Maddox, R.N., Vapor pulsing in a sieve-plate laboratory distillation column, Ind Eng Chem Proc Des Dev, 6(1), 6-9 (1967).
Ponter, A.B.; Davies, G.A.; Beaton, W., and Ross, T.K., Wetting of packings in distillation: Influence of contact angle, Trans IChemE, 45, T345-352 (1967).
Popov, V.V., and Popova, L.M., Design of tray-type mass-transfer equipment, Int Chem Eng, 7(2), 197-199 (1967).
Redwine, D.A.; Flint, E.M., and Van Winkle, M., Froth and foam height studies on a small perforated-plate distillation column, Ind Eng Chem Proc Des Dev 6(4), 525-532 (1967).
Rodionov, A.I., and Vinter, A.A., Interfacial surface area on sieve trays by chemical method, Int Chem Eng, 7(3), 468-471 (1967).
Ross, T.K.; Campbell, D.U., and Wragg, A.A., An electrochemical investigation of the hydrodynamics of liquid films on grid packings, Trans IChemE, 45, T401-407 (1967).
Sharma, M.M., and Gupta, R.K., Mass transfer characteristics of plate columns without downcomer, Trans IChemE, 45, T169-175 (1967).
Sterbacek, Z., Hydrodynamics of perforated trays with downcomers, Brit Chem Eng, 12(10), 1577-1581 (1967).
Thomas, W.J., and Campbell, M., Hydraulic studies in a sieve plate downcomer system, Trans IChemE, 45, T53-63 (1967).
Todd, W.G., and Van Winkle, M., Entrainment and pressure drop with jet trays in an air-water system, Ind Eng Chem Proc Des Dev, 6(1), 95-101 (1967).
White, D.A., Drainage rates from sieve plate and bubble-cap columns at shutdown, Chem Eng Sci, 22(11), 1515-1517 (1967).

1968-1969

Calcaterra, R.J.; Nicholls, C.W., and Weber, J.H., Free and captured entrainment and plate spacing in a perforated tray column, Brit Chem Eng, 13(9), 1294-1297 (1968).

Castellucci, N.T., and Krouskop, N.C., New low-density ceramic material for packing fractional distillation columns, J Appl Chem, 18, 143-145, 373-376 (1968).
Fair, J.R., and Bolles, W.L., Design of distillation columns, Chem Eng, 22 April, 156-178 (1968).
Fane, A., and Sawistowski, H., Surface tension effects in sieve-plate distillation, Chem Eng Sci, 23(8), 943-945 (1968).
Porter, K.E., and Ashton, N., Prediction and correlation of pressure drop per theoretical plate in packed distillation columns, Brit Chem Eng, 13(11), 1561-1566 (1968).
Rao, M.V.R., Nomogram for optimum gas velocity in turbogrid trays, Brit Chem Eng, 13(2), 257 (1968).
Sterbacek, Z., Liquid dispersion coefficients on perforated plates with downcomers, Trans IChemE, 46, T167-171 (1968).
Wood, C.E., Tray selection for column temperature control, Chem Eng Prog, 64(1), 85-88 (1968).
Billet, R., Development and progress in design and performance of valve trays, Brit Chem Eng, 14(4), 489-493 (1969).
Kohara, S.; Tokuda, M., and Warisawa, K., Performance of distillation pipe-trays, Int Chem Eng, 9(1), 172-176 (1969).
Molokanov, Y.K.; Korablina, T.P.; Abushevich, I.Z., and Rogozina, L.P., Liquid entrainment between trays by gas flow, Int Chem Eng, 9(4), 603-606 (1969).
Rodionov, A.I., and Vinter, A.A., Effect of specific weight of gas phase on hydrodynamic operating characteristics of grid trays, Int Chem Eng, 9(1), 46-48 (1969).
Saletan, D.I., Specifying distillation overdesign factors, Chem Eng Prog, 65(5), 80-82 (1969).
Sargent, R.W.H., and Murtagh, B.A., Design of plate distillation columns for multicomponent mixtures, Trans IChemE, 47, T85-95 (1969).

1970-1972

Edgeworth-Johnstone, R., Optimum plate spacing in a distillation column, Chem Engnr, May, CE119 (1970).
Kharbanda, P.O., and Chu, J.C., Experimental studies with a sieve plate column, Brit Chem Eng, 15(6), 792-794 (1970).
Koltunova, L.N.; Aerov, M.E., and Bystrova, T.A., Factors affecting mass transfer on grid trays, Int Chem Eng, 10(4), 521-524 (1970).
Kupferberg, A., and Jameson, G.J., Pressure fluctuations in a bubbling system with special reference to sieve plates, Trans IChemE, 48, T140-150 (1970).
Neeld, R.K., and O'Bara, J.T., Jet trays vs bubble cap trays, Chem Eng Prog, 66(7), 53-59 (1970).
Rodinov, A.I.; Vinter, A.A.; Ulyanov, B.A., and Zenkov, V.V., Effect of liquid-phase viscosity on hydrodynamic operating behavior of grid trays, Int Chem Eng, 10(2), 166-169 (1970).
Various, Trays and packings (topic issue), Chem Eng Prog, 66(3), 39-65 (1970).
Interess, E., Practical limitations on tray design, Chem Eng, 15 Nov, 167-170 (1971).

Kharbanda, P.O., Sieve plates: Pressure drop and minimum vapour velocity, Chem Engnr, April, 169 (1971).
Nemunaitis, R.R., Sieve trays? Consider viscosity, Hyd Proc, 50(11), 235-239 (1971).
Panicker, P.K.N., Stripper-downcomer sizing to avoid liquid entrainment, Chem Eng, 17 May, 160-162 (1971).
Poberezkin, A.E.; Gerasimov, P.V., and Alekseev, V.P., Rectification in regular corrugated packings, Int Chem Eng, 11(3), 396-398 (1971).
Weiler, D.W.; Bonnet, F.W., and Leavitt, F.W., Slotted sieve trays, Chem Eng Prog, 67(9), 86-88 (1971).
Bartholomai, G.B.; Gardner, R.G., and Hamilton, W., Characteristics of a sieve plate with downcomer, Brit Chem Eng, 17(1), 48-50 (1972).
Billet, R., Gauze-packed columns for vacuum distillation, Chem Eng, 21 Feb, 68-71 (1972).
Chekhov, O.S.; Yamshchikov, I.N., and Khodak, V.S., Hydraulics of bubbling trays with stagnant zones, Int Chem Eng, 12(1), 40-43 (1972).
Eduljee, H.E., Sieve plates: Minimum vapour velocity, Chem Engnr, March, 123-124 (1972).
Ellis, S.R.M., Some vacuum distillation devices, Chem Engnr, March, 115-120 (1972).
Lessi, A., How weeping affects distillation, Hyd Proc, 51(3), 109-111 (1972).
Pinczewski, W.V., and Fell, C.J.D., The transition from froth-to-spray regime on commercially loaded sieve trays, Trans IChemE, 50, 102-108 (1972).
Pozin, L.S., et al., Hydraulic calculation of trays with disc valves, Int Chem Eng, 12(4), 692-695 (1972).
Pozin, L.S., et al., Hydraulic design of trays with disc valves, Int Chem Eng, 12(3), 501-503 (1972).
Thorngren, J.T., Valve-tray pressure drop, Ind Eng Chem Proc Des Dev, 11(3), 428-429 (1972).
Zorina, G.I.; Zharkova, L.E.; Kruglov, S.A., and Skoblo, A.I., Effect of sectioning the liquid stream on holdup on bubbling sieve trays, Int Chem Eng, 12(2), 299-302 (1972).

1973-1974

Ashley, M.J., and Haselden, G.G., The improvement of sieve tray performance by controlled vapour-liquid contacting, Trans IChemE, 51, 188-191 (1973).
Cervenka, J., and Kolar, V., Structure and height of gaseous-liquid mixtures on sieve plates without downcomers, Chem Eng J, 6(1), 45-50 (1973).
Ellis, S.R.M.; Parry, D.W., and Hands, C.H.G., Fractional distillation in different diameter columns operated at low column top pressure and low pressure gradient, Trans IChemE, 51, 56-60 (1973).
Jeronimo, M.A.S., and Sawistowski, H., Phase inversion correlation for sieve trays, Trans IChemE, 51, 265-266 (1973).
Loon, R.E.; Pinczewski, W.V., and Fell, C.J.D., Dependence of the froth-to-spray transition on sieve tray design parameters, Trans IChemE, 51, 374-376 (1973).
Various, Distillation tower design (topic issue), Chem Eng Prog, 69(10), 67-76 (1973).

Vystoropskaya, A.A., et al., Hydrodynamics of ring trays at high liquid loadings, Int Chem Eng, 13(4), 618-622 (1973).
Zanelli, S., and Del Bianco, R., Perforated plate weeping, Chem Eng J, 6(3), 181-194 (1973).
Aguayo, G.A., and Lemlich, R., Countercurrent foam fractionation at high rates of throughput by means of perforated plate columns, Ind Eng Chem Proc Des Dev, 13(2), 153-159 (1974).
Eckert, J.S., Hy-Pak: An improved packing for vacuum distillation, Chem Engnr, Nov, 712-713 (1974).
Ellis, S.R.M., and Bayley, D.P., KnitMesh redistributors improve HETP in packed columns, Chem Engnr, Nov, 714-716 (1974).
Glaksin, V.G.; Zaostrovskii, F.P., and Krasikov, A.N., Liquid entrainment from small spaces between bubble-cap trays, Int Chem Eng, 14(4), 710-713 (1974).
Jeronimo, M.A.S., and Sawistowski, H., Projection velocities of droplets in the spray regime of sieve plate operation, Trans IChemE, 52, 291-293 (1974).
Korotkov, P.I.; Isaev, B.N., Ektov, V .V., and Teteruk, V.G., Operating experiments on distillation columns with trays of various constructions, Int Chem Eng, 14(1), 102-105 (1974).
Lockett, M.J., and Spiller, G.T., Transition from spray to froth on sieve plates, Chem Eng Sci, 29(5), 1309-1311 (1974).
Mikhailenko, G.G.; Bolshakov, A.G.; Ennan, A.A., and Golivets, G.I., Resistance of perforated dual-flow trays at high loadings, Int Chem Eng, 14(4), 641-644 (1974).
Pinczewski, W.V., and Fell, C.J.D., Nature of the two-phase dispersion on sieve plates operating in the spray regime, Trans IChemE, 52, 294-299 (1974).
Pozin, L.S.; Berezhnaya, K.P.; Bystrova, T.A., and Aerov, M.E., Equations for hydraulic resistance and froth height on tubular-valve trays without overflow devices, Int Chem Eng, 14(4), 703-706 (1974).
Raskop, F., Preform-Kontakt plates: A new development, Chem Engnr, Nov, 709-712 (1974).
Shoukry, E.; Cermak, J., and Kolar, V., Hydrodynamics of sieve plates without downcomers, Chem Eng J, 8(1), 27-52 (1974).
Tarat, E.Y., et al., Efficiency of new designs of valve trays in rectification of binary mixtures, Int Chem Eng, 14(4), 638-641 (1974).
Tyutyunnikov, A.B., et al., Flapper-sieve tray for distillation and absorption columns, Int Chem Eng, 14(1), 45-48 (1974).

1975

Belina-Freundlich, D.; Respondek, J., and Szust, J., Testing of packing materials in a vacuum distillation column, Int Chem Eng, 15(3), 574-579 (1975).
Berkovskii, M.A.; Sheinman, V.A., and Egorov, M.M., Investigation of operation of S-shaped combination trays, Int Chem Eng, 15(2), 247-251 (1975).
Bunin, V.B., and Skoblo, A.I., Effect of fractional composition of circulating liquid on operation of condensation section of a vacuum column, Int Chem Eng, 15(1), 43-45 (1975).
Chimes, A.R., Demonstrated data for tower design, Chem Eng, 27 Oct, 137-138 (1975).

Davy, C.A.E., and Haselden, G.G., Prediction of pressure drop across sieve trays, AIChEJ, 21(6), 1218-1220 (1975).
Featherstone, W., Estimating the cost of distillation columns, Processing, May, 21-25 (1975).
Haselden, G.G., Scope for improving fractionation equipment, Chem Engnr, July, 439-441, 445 (1975).
Huber, M., and Meier, W., Packings for vacuum rectification, Proc Engng, Sept, 83 (1975).
Kosyan, V.K., et al., Evaluation of lateral sectionalization of a packed distillation column with the aid of a coaxial tube, Int Chem Eng, 15(4), 627-631 (1975).
Rao, D.P., Inclined fractionators and absorbers, Chem Eng, 28 April, 132-134 (1975).
Rodionov, A.I., and Kochetov, N.M., Longitudinal mixing of liquid on bubble-cap sieve trays, Int Chem Eng, 15(4), 605-607 (1975).
Sewell, A., Practical aspects of distillation column design, Chem Engnr, July, 442-445 (1975).
Singales, B., Design of reflux drums, Chem Eng, 3 March, 157-160; 29 Sept, 87-90 (1975).
Smith, V.C., and Delnicki, W.V., Optimum sieve tray design, Chem Eng Prog, 71(8), 68-73 (1975).
Steiner, L.; Ballmer, J.F., and Hartland, S., Structure of gas-liquid dispersions on perforated plates, Chem Eng J, 10(1), 35-40 (1975).
Volejnik, M., Determination of parameters of industrial rectification columns, Int Chem Eng, 15(3), 564-574 (1975).
Zednik, A., and Karas, I., Tray pressure drop with directed flow of phases, Int Chem Eng, 15(4), 669-675 (1975).

1976

Haug, H.F., Stability of sieve trays with high overflow weirs, Chem Eng Sci, 31(4), 295-308 (1976).
Lockett, M.J.; Spiller, G.T., and Porter, K.E., The effect of the operating regime on entrainment from sieve plates, Trans IChemE, 54, 202-204 (1976).
Metcalfe, R.S.; Barton, P., and McCormick, R.H., Low-pressure vapor-liquid trays, Hyd Proc, 55(4), 97-100 (1976).
Mikhailenko, G.G., et al., Spray carry-over in columns with grid plates, Int Chem Eng, 16(2), 374-376 (1976).
Piqueur, H., and Verhoeye, L., Research on valve trays: Hydraulic performance in air-water system, Can JCE, 54, 177-185 (1976).
Ponter, A.B.; Trauffler, P., and Vijayan, S., Effect of surfactant addition on packed distillation column performance, Ind Eng Chem Proc Des Dev, 15(1), 196-199 (1976).
Saradhy, Y.P., and Kumar, R., Drop formation at sieve plate distributor, Ind Eng Chem Proc Des Dev, 15(1), 75-82 (1976).
Thomas, W.J., and Haq, M.A., Performance of a sieve-tray with 0.375in. diameter perforations, Ind Eng Chem Proc Des Dev, 15(4), 509-518 (1976).
Yakhno, O.M., and Vorontsov, E.G., Pressure drop in liquid distributors of plate-type equipment, Int Chem Eng, 16(2), 370-374 (1976).

1977

Drery, W., State of development of rectification plates, Int Chem Eng, 17(3), 389-395 (1977).

Feintuch, H.M., Distillation distributor design, Hyd Proc, 56(10), 150-151 (1977).

Fell, C.J.D., and Pinczewski, W.V., New considerations in design and operation of high-capacity sieve trays, Chem Engnr, Jan, 45-49 (1977).

Furzer, I.A., and Duffy, G.J., Mass transfer on sieve plates without downcomers, Chem Eng J, 14(3), 217-224 (1977).

Gilath, C.; Cohen, H., and Wolf, D., Flow dynamics in distillation columns packed with Dixon rings as used in isotope separation, Can JCE, 55, 168-180 (1977).

Hajdu, H.; Mizsey, P., and Foldes, P., Resonance effects in the hydraulic transient behaviour of shower-tray distillation columns, Periodica Polytechnica, 21, 265-276 (1977).

Kister, H.Z., and Doig, I.D., Entrainment flooding prediction for tray columns, Hyd Proc, 56(9), 149-151 (1977).

Moscicka, I.; Badowska, I., and Morawska, B., Design of perforated plates, Int Chem Eng, 17(4), 677-691 (1977).

Payne, G.J., and Prince, R.G.H., The relationship between the froth and spray regimes, and the orifice processes occurring on perforated distillation plates, Trans IChemE, 55, 266-273 (1977).

Pinczewski, W.V., and Fell, C.J.D., Droplet sizes on sieve plates operating in the spray regime, Trans IChemE, 55, 46-52 (1977).

Sohlo, J., and Kinnunen, S., Dispersion and flow phenomena on a sieve plate, Trans IChemE, 55, 71-73 (1977).

1978

Economopoulos, A.P., Computer design of sieve trays and tray columns, Chem Eng, 4 Dec, 109-120 (1978).

Hajdu, H.; Borus, A., and Foldes, P., Vapor flow lag in distillation columns, Chem Eng Sci, 33(1), 1-8 (1978).

Lockett, M.J., and Uddin, M.S., Slotted distillation trays: Momentum transfer from a single slot, Trans IChemE, 56, 194-199 (1978).

Raymond, B., Experimental study of gas-liquid transfer rates using an inclined perforated tray without downcomer, Chem Eng Sci, 33(8), 1157-1160 (1978).

Ruff, K.; Pilhofer, T., and Mersmann, A., Ensuring flow through all the openings of perforated plates for fluid dispersion, Int Chem Eng, 18(3), 395-402 (1978).

Stichlmair, J., and Mersmann, A., Dimensioning plate columns for absorption and rectification, Int Chem Eng, 18(2), 223-237 (1978).

Thomas, W.J., and Ogboja, O., Hydraulic studies in sieve-tray columns, Ind Eng Chem Proc Des Dev, 17(4), 429-443 (1978).

Thorngren, J.T., Valve tray flooding generalized, Hyd Proc, 57(8), 111-113 (1978).

1979

Brenner, M., Minimizing cost of column overdesign, Chem Eng, 2 July, 108 (1979).

Counce, R.M., and Perona, J.J., Gas-liquid interfacial area of sieve plate with downcomers and 0.6% perforation, Ind Eng Chem Proc Des Dev, 18(3), 562-564 (1979).

Fabre, S.; Ponter, A.B., and Huber, M., Assessment of distillation packing performance from wetting data obtained under equilibrium conditions, Ind Eng Chem Proc Des Dev, 18(2), 266-269 (1979).

Larsen, J., and Kummel, M., Hydrodynamic model for controlled cycling in tray columns, Chem Eng Sci, 34(4), 455-462 (1979).

Strigle, R.F., and Rukovena, F., Packed distillation column design, Chem Eng Prog, 75(3), 86-91 (1979).

Takahashi, T.; Akagi, Y., and Kishimoto, T., Gas and liquid velocities at incipient liquid stagnation on a sieve tray, Int Chem Eng, 19(1), 113-118 (1979).

Volejnik, M., Determination of parameters for rectification columns: Industrial applications of bubble-cap trays, Int Chem Eng, 19(2), 220-227 (1979).

Wong, P.F.Y., and Kwan, W.K., A generalised method for predicting the spray-bubbling transition on sieve plates, Trans IChemE, 57, 205-209 (1979).

1980-1981

Dolan, M.J., and Strigle, R.F., Advances in distillation column design, Chem Eng Prog, 76(11), 78-83 (1980).

Furzer, I.A., Flow distribution and mass transfer in a plate column fitted with a manifold, Chem Eng Sci, 35(6), 1291-1306 (1980).

Gaston-Bonhomme, Y.; Chevalier, J.L., and Cunin, G., Influence of non-wetted fluored polymer packings in distillation columns, Chem Eng Sci, 35(5), 1163-1178 (1980).

Klein, G.F., Tray spacing equation, Chem Eng, 5 May, 139 (1980).

Lockett, M.J., and Uddin, M.S., Liquid-phase controlled mass transfer in froths on sieve trays, Trans IChemE, 58, 166-174 (1980).

Yabe, I., and Kunii, D., Flow pattern around a spherical-cap gas bubble, Int Chem Eng, 20(2), 203-211 (1980).

Zhou, Y.F.; Shi, J.F.; Wang, X.M., and Ye, Y.H., Hydrodynamic behaviour of Linde-type flow-guided sieve plates, Int Chem Eng, 20(4), 642-654 (1980).

Colwell, C.J., Clear liquid height and froth density on sieve trays, Ind Eng Chem Proc Des Dev, 20(2), 298-307 (1981).

Kawagoe, M.; Otake, T.; Kimura, S., and Noda, Y., Flow characteristics of a new movable-perforated-plate column, Int Chem Eng, 21(4), 612-621 (1981).

Kister, H.Z., and Doig, I.D., Studies of effect of pressure on distillation heat requirements, Chem Eng Commns, 11(1), 1-12 (1981).

Kister, H.Z.; Pinczewski, V.W., and Fell, C.J.D., Entrainment from sieve trays operating in the spray regime, Ind Eng Chem Proc Des Dev, 20(3), 528-532 (1981).

Lockett, M.J., The froth to spray transition on sieve trays, Trans IChemE, 59, 26-34 (1981).

Priestman, G.H., and Brown, D.J., The mechanism of pressure pulsations in sieve-tray columns, Trans IChemE, 59, 279-282 (1981).

1982

Ito, A., and Asano, K., Thermal effects in non-adiabatic binary distillation, Chem Eng Sci, 37(7), 1007-1014 (1982).

Klein, G.F., Simplified model calculates valve-tray pressure drop, Chem Eng, 3 May, 81-85 (1982).

Kozubenko, G.Y.; Anoshin, I.M., and Krinitskii, B.P., Sectionalized vortex rectification apparatus, Int Chem Eng, 22(2), 379-382 (1982).

Kumar, A., and Hartland, S., Prediction of drop size produced by a multiorifice distributor, Trans IChemE, 60, 35-39 (1982).

Pierucci, S.J.; Ranzi, E.M., and Biardi, G.E., Possibility of enlarging 'standard specifications' in distillation columns, Ind Eng Chem Proc Des Dev, 21(4), 604-606 (1982).

Pinczewski, W.V., and Fell, C.J.D., Froth to spray transition on sieve trays, Ind Eng Chem Proc Des Dev, 21(4), 774-776 (1982).

Raper, J.A.; Kearney, M.S.; Burgess, J.M., and Fell, C.J.D., Structure of industrial sieve tray froths, Chem Eng Sci, 37(4), 501-506 (1982).

Scali, C., and Zanelli, S., Characterization of oscillating phenomena on sieve plates of transfer columns, Chem Eng J, 25(2), 191-200 (1982).

Sohlo, J., and Kouri, R.J., Analysis of enhanced transverse dispersion on distillation plates, Chem Eng Sci, 37(2), 193-198 (1982).

Solari, R.B.; Saez, E., D'Apollo, I., and Bellet, A., Velocity distribution and liquid flow patterns on industrial sieve trays, Chem Eng Commns, 13(4), 369-384 (1982).

Yanagi, T., and Sakata, M., Performance of commercial scale 14% hole-area sieve tray, Ind Eng Chem Proc Des Dev, 21(4), 712-717 (1982).

Zuiderweg, F.J., Sieve trays: A state-of-the-art review, Chem Eng Sci, 37(10), 1441-1464 (1982).

1983-1984

Bennett, D.L.; Agrawal, R., and Cook, P.J., New pressure drop correlation for sieve tray distillation columns, AIChEJ, 29(3), 435-442 (1983).

Dhulesia, H., Operating flow regimes on the valve tray, CER&D, 61, 329-332 (1983).

Haq, M.A., Fluid dynamics on sieve trays, Hyd Proc, April, 165-168; Aug, 117-119 (1982); Feb, 55-58 (1983).

Miyahara, T.; Matsuba, Y., and Takahashi, T., Size of bubbles generated from perforated plates, Int Chem Eng, 23(3), 517-524 (1983).

Ponter, A.B., and Tsay, T., Sieve plate simulation study: Contact angle and frequency of emission of bubbles from a submerged orifice with liquid crossflow, CER&D, 61, 259-263 (1983).

Barduhn, A.J., Setting the pressure in distillation columns, Chem Eng Educ, 18(1), 38-41 (1984).

Bradford, M., and Durrett, D.G., Sizing distillation safety valves, Chem Eng, 9 July, 78-84 (1984).

Klein, G.F., Evaluating turndown of valve trays, Chem Eng, 17 Sept, 128-129 (1984).
Lieberman, N.P., Packing expands low-pressure fractionators, Hyd Proc, April, 143-145 (1984).
Raper, J.A.; Pinczewski, W.V., and Fell, C.J.D., Liquid passage on sieve trays operating in the spray regime, CER&D, 62, 111-116 (1984).
Spagnolo, D.A., and Chuang, K.T., Improving sieve tray performance with knitted mesh packing, Ind Eng Chem Proc Des Dev, 23(3), 561-565 (1984).
Zuiderweg, F.J.; Hofhuis, P.A.M., and Kuzniar, J., Flow regimes on sieve trays: Significance of emulsion flow regime, CER&D, 62, 39-47 (1984).

1985-1986

Bravo, J.L.; Rocha, J.A., and Fair, J.R., Mass transfer in gauze packings, Hyd Proc, Jan, 91-95 (1985).
Dhulesia, H., Clear liquid height on sieve and valve trays, CER&D, 62, 321-326 (1984); 63, 206-208 (1985).
Gorak, A., and Vogelpohl, A., Experimental study of ternary distillation in a packed column, Sepn Sci Technol, 20(1), 33-62 (1985).
Kolev, N., and Semkov, K., Influence of axial mixing in distillation columns with random packing, Chem Eng & Proc, 19(4), 175-182 (1985).
Ye, Y.; Shi, J., and Zhou, Y., Transition in regime of flow and point of inflection of rate of entrainment on a perforated plate, Int Chem Eng, 25(1), 176-182 (1985).
Zanetti, R.; Short, H., and Hope, A., Better column distributors, Chem Eng, 27 May, 22-27 (1985).
Coombs, S., New distillation packings, Processing, March, 23,25 (1986).
Lockett, M.J., Distillation Tray Fundamentals, Cambridge University Press, U.K. (1986).
Lockett, M.J., and Banik, S., Weeping from sieve trays, Ind Eng Chem Proc Des Dev, 25(3), 561-569 (1986).
Lygeros, A.I., and Magoulas, K.G., Column flooding and entrainment, Hyd Proc, 65(12), 43-44 (1986).
Melli, T.R.; Spekuljak, Z., and Cerro, R.L., Structural model for optimal geometrical design of packed distillation towers, Ind Eng Chem Proc Des Dev, 25(3), 612-618 (1986).
Muir, L.A., and Briens, C.L., Low-pressure-drop gas distributors for packed distillation columns, Can JCE, 64(6), 1027-1032 (1986).
Ozgen, C., and Somer, T.G., Performance of a sieve plate with small holes, Ind Eng Chem Proc Des Dev, 25(2), 375-380 (1986).
Solari, R.B., and Bell, R.L., Fluid-flow patterns and velocity distribution on commercial-scale sieve trays, AIChEJ, 32(4), 640-649 (1986).
Tassev, Z.; Stefanov, Z., and Kamenski, D., Longitudinal mixing of the liquid phase on a cocurrent valve tray with entrainment separators, Ind Eng Chem Proc Des Dev, 25(1), 181-184 (1986).
Zuiderweg, F., Influence of two-phase flow regimes on separation performance of sieve plates, Int Chem Eng, 26(1), 1-11 (1986).

1987-1988

Chynoweth, E., Structured packings for distillation, Proc Engng, June, 51 (1987).

Dribika, M.M., and Biddulph, M.W., Surface tension effects on a large rectangular tray with small diameter holes, Ind Eng Chem Res, 26(8), 1489-1495 (1987).

Fasesan, S.O., Hydraulic characteristics of sieve and valve trays, Ind Eng Chem Res, 26(10), 2114-2122 (1987).

McDermott, W.T.; Anselmo, K.J., and Chetty, A.S., Hydraulic behavior on a circular flow distillation tray, Comput Chem Eng, 11(5), 497-502 (1987).

Prado, M.; Johnson, K.L., and Fair, J.R., Bubble-to-spray transition on sieve trays, Chem Eng Prog, 83(3), 32-40 (1987).

Scheffe, R.D., and Weiland, R.H., Mass-transfer characteristics of valve trays, Ind Eng Chem Res, 26(1), 228-237 (1987).

Yoshida, H., Liquid flow over distillation column plates, Chem Eng Commns, 51, 261-276 (1987).

Chuang, K.T., and Miller, A.I., Performance of packings for water distillation, Can JCE, 66(3), 377-381 (1988).

Herron, C.C.; Kruelskie, B.K., and Fair, J.R., Hydrodynamics and mass transfer on three-phase distillation trays, AIChEJ, 34(8), 1267-1274 (1988).

Hufton, J.R.; Bravo, J.L., and Fair, J.R., Scale-up of laboratory data for distillation columns containing corrugated metal-type structured packing, Ind Eng Chem Res, 27(11), 2096-2101 (1988).

Kerlin, M., Spreadsheet tray hydraulic design, Hyd Proc, 67(3), 49-51 (1988).

Kister, H.Z., and Haas, J.R., Entrainment from sieve trays in froth regime, Ind Eng Chem Res, 27(12), 2331-2340 (1988).

Kler, S.C., and Lavin, J.T., Simulation of flow on distillation trays, Gas Sepn & Purif, 2(1), 34-40 (1988).

Muller, E.A.; Cavero, A., and Estevez, L.A., Improving flow patterns in distillation tray by modifying downcomer apron shape, Chem Eng Commns, 74, 195-208 (1988).

Prahl, W.H., Solving vacuum column liquid wetting problems, Chem Eng, 26 Sept, 92-93 (1988).

Salem, A.B., and Alsaygh, A.A., Add packing for better sieve trays, Hyd Proc, 67(5), 76G-76H (1988).

Various, Distillation design (feature report), Chem Eng, 26 Sept, 71-76 (1988).

Walraven, F.F.Y.; Van Rompay, P.V., and Luyten, R., Fast algorithm for three-phase distillation towers, Chem Eng & Proc, 24(3), 133-144 (1988).

19.5 Optimization

1967-1975

Billet, R., and Raichle, L., Optimizing method for vacuum rectification, Chem Eng, 13 Feb, 145-150; 27 Feb, 149-154 (1967).

Coward, I., Time-optimal problem in binary batch distillation, Chem Eng Sci, 22(4), 503-516; 22(12), 1881-1885 (1967).

Robinson, E.R., Optimisation of batch distillation operations, Chem Eng Sci, 24(11), 1661-1668 (1969).

Billet, R., Cost optimization of distillation towers, Chem Eng Prog, 66(1), 41-50 (1970).
Neretnieks, I., Optimisation of sieve tray columns, Brit Chem Eng, 15(2), 193-198 (1970).
Kropholler, H.W., and Lees, F.P., Coupling between liquid flow and liquid concentration responses in unsteady-state plate distillation and gas absorption column models, Chem Eng Sci, 26(1), 39-44 (1971).
Luyben, W.L., Practical aspects of optimal batch distillation design, Ind Eng Chem Proc Des Dev, 10(1), 54-59 (1971).
Mayur, D.N., and Jackson, R., Time-optimal problems in batch distillation for multicomponent mixtures and for columns with holdup, Chem Eng J, 2(3), 150-163 (1971).
Nishimura, H., and Hiraizumi, Y., Optimal system pattern for multicomponent distillation systems, Int Chem Eng, 11(1), 188-194 (1971).
Robinson, E.R., Time-optimal problem in binary batch distillation with recycled waste cut, Chem Eng J, 2(2), 135-136 (1971).
De Lorenzo, F., et al., Asymmetric behavior of distillation systems, Chem Eng Sci, 27(6), 1211-1222 (1972).
Van Winkle, M., and Todd, W.G., Minimizing distillation costs via graphical techniques, Chem Eng, 6 March, 105-112 (1972).
Pyatnichko, A.I.; Kryukov, V.A., and Cheglikov, A.G., Optimum fractionation processes in nonadiabatic columns, Int Chem Eng, 13(1), 17-20 (1973).
Bekassy-Molnar, E.; Foldes, P., and Kollar-Hunek, K., Joint optimization of the construction and operation at various pressures of plate distillation columns, Periodica Polytechnica, 18, 327-336 (1974).
Dickson, A.N., The optimisation of multi-stage flash distillation plant, Chem Engnr, Feb, 79-81 (1975).
Freshwater, D.C., and Henry, B.D., Optimal configuration of multicomponent distillation trains, Chem Engnr, Sept, 533-536 (1975).
Various, Distillation practices (topic issue), Chem Eng Prog, 71(6), 49-84 (1975).

1977-1988

Batyzhev, E.A., and Zykov, D.D., Conditions for optimum statics in fractionation of continuous mixtures, Int Chem Eng, 17(2), 322-326 (1977).
Kuznik, J., and Krzyzanowski, R., Dynamic properties of distillation columns, Int Chem Eng, 17(2), 360-367 (1977).
Waller, K.V., and Gustafsson, T.K., Optimal steady-state operation in distillation, Ind Eng Chem Proc Des Dev, 17(3), 313-317 (1978).
Al-Haq-Ali, N.S., and Holland, C.D., Determine distillation optimum, Hyd Proc, 58(7), 165-175; 58(8), 111-119 (1979).
Rees, G.J.; Scully, D.B., and Weldon, L.H.P., Optimisation of separating efficiency in the centrifugal molecular still, Chem Eng Sci, 34(2), 159-170 (1979).
Fleischer, M.T., and Prett, D.M., Simplified techniques for simulating complex columns, Chem Eng Prog, 77(2), 72-75 (1981).
Jones, E.A., and Mellbom, M.E., Fractionating column economics, Chem Eng Prog, 78(5), 52-55 (1982).

Melli, T.R., and Spekuljak, Z., Optimum design of packed distillation towers, Ind Eng Chem Proc Des Dev, 22(2), 230-236 (1983).
Ferre, J.A.; Castells, F., and Flores, J., Optimization of a distillation column with a direct vapor recompression heat pump, Ind Eng Chem Proc Des Dev, 24(1), 128-132 (1985).
Kaiser, V., and Gourlia, J.P., The ideal column concept: Applying exergy to distillation, Chem Eng, 19 Aug, 45-53 (1985).
Severance, W.A.N., Differential radiation scanning improves the visibility of liquid distribution in a column, Chem Eng Prog, 81(4), 48-51 (1985).
Hansen, T.T., and Jorgensen, S.B., Optimal control of binary batch distillation in tray or packed columns, Chem Eng J, 33(3), 151-156 (1986).
Kapoor, N.; McAvoy, T.J., and Marlin, T.E., Effect of recycle structure on distillation tower time constraints, AIChEJ, 32(3), 411-418 (1986).
Kler, S.C., and Lavin, J.T., Use of fibre-optic endoscope for visual investigation of weeping in cryogenic distillation columns, Gas Sepn & Purif, 1(2), 110-116 (1987).
Duprat, F.; Gassend, R., and Gau, G., Reactive distillation process optimization by empirical formulae construction, Comput Chem Eng, 12(11), 1141-1150 (1988).
Glinos, K., and Malone, M.F., Optimality regions for complex column alternatives in distillation systems, CER&D, 66(3), 229-240 (1988).
Kingsley, J.P., and Lucia, A., Simulation and optimization of three-phase distillation processes, Ind Eng Chem Res, 27(10), 1900-1910 (1988).
Lucia, A., and Kumar, A., Distillation optimization, Comput Chem Eng, 12(12), 1263-1266 (1988).
Rosendorf, P.; Kibicek, M., and Schongut, J., On-line optimization of a rectification column, Comput Chem Eng, 12(2/3), 199-204 (1988).

19.6 Operation

1967-1970

Badami, V.N., Constant reflux ratio maintained by simple splitter, Chem Eng, 8 May, 180-182 (1967).
Francis, R.C., and Berg, J.C., Effect of surfactants on a packed distillation column, Chem Eng Sci, 22(4), 685-692 (1967).
Peacock, D.G., Selection of test mixtures for distillation columns, Chem Eng Sci, 22(7), 957-962 (1967).
Furter, W.F., Salt effect in distillation, Chem Engnr, June, CE173-177 (1968).
Garvin, R.G., and Norton, E.R., Sieve-tray performance in a heavy-water plant, Chem Eng Prog, 64(3), 99-106 (1968).
Katzen, R., et al., A self-descaling distillation tower, Chem Eng Prog, 64(1), 79-84 (1968).
Nisenfeld, A.E., and Stravinski, C.A., Feedforward control for azeotropic distillations, Chem Eng, 23 Sept, 227-236 (1968).
Kolar, V., Structure of gas-liquid mixtures on sieve trays of separation columns, Chem Eng Sci, 24(8), 1285-1290 (1969).
Nisenfeld, A.E., Reflux or distillate: Which to control? Chem Eng, 6 Oct, 169-171 (1969).

Bollinger, R.E., Predicting fractionator dynamics by using a frequency domain solution technique, AIChEJ, 16(4), 673-678 (1970).
Gerster, J.A., and Scull, H.M., Performance of tray columns operated in the cycling mode, AIChEJ, 16(1), 108-111 (1970).
Howard, G.M., Unsteady-state behavior of multicomponent distillation columns, AIChEJ, 16(6), 1022-1033 (1970).
McLaren, D.B., and Upchurch, J.C., Guide to trouble-free distillation, Chem Eng, 1 June, 139-152 (1970).
Panicker, P.K.N., Siphon systems aid distillation, Chem Eng, 18 May, 180 (1970).
Tommasi, G., and Rice, P., Dynamics of packed tower distillation, Ind Eng Chem Proc Des Dev, 9(2), 234-243 (1970).

1971-1975

Shunta, J.P., and Luyben, W.L., Dynamic effects of temperature-control tray location in distillation columns, AIChEJ, 17(1), 92-96 (1971).
Sittel, C.N., and Fisher, G.T., New way to detect incipient plate-column flooding, Chem Eng, 27 Dec, 92 (1971).
Moens, F.P., Effect of composition and driving force on performance of packed distillation columns, Chem Eng Sci, 27(2), 275-294 (1972).
Singh, R., Hyperbolic distillations in multiple-section fractionator, Chem Eng Sci, 27(4), 677-684 (1972).
Weigand, W.A.; Jhawar, A.K., and Williams, T.J., Calculation method for response time to step inputs for approximate dynamic models of distillation columns, AIChEJ, 18(6), 1243-1252 (1972).
Brumbaugh, K.H., and Berg, J.C., Effect of surface active agents on sieve-plate distillation column, AIChEJ, 19(5), 1078-1080 (1973).
Stainthorp, F.P., and Searson, H.M., The dynamics of fractionating columns, Trans IChemE, 51, 42-55 (1973).
Biddulph, M.W., and Stephens, D.J., Oscillating behavior on distillation trays, AIChEJ, 20(1), 60-67 (1974).
Gallier, P.W., and McCune, L.C., Simple internal reflux control, Chem Eng Prog, 70(9), 71-76 (1974).
Waddington, W.; Kohler, H.K., and Brown, D.J., Vibration excitation of sieve plate columns by bubbling, Trans IChemE, 52, 381-383 (1974).
Biddulph, M.W., Oscillating behavior on distillation trays, AIChEJ, 21(1), 41-49 (1975).
Joseph, L.M., and Kadlec, R.H., Cyclic operation of a vapor-liquid separator, Ind Eng Chem Proc Des Dev, 14(2), 187-191 (1975).
Kemp, D.W., and Ellis, D.G., Computer control of fractionation plants, Chem Eng, 8 Dec, 115-118 (1975).
Pinczewski, W.V., and Fell, C.J.D., Oscillations on sieve trays, AIChEJ, 21(6), 1019-1021 (1975).
Pratt, C.F., and Hobbs, S.Y., Quick kill of foam on fractionator trays, Chem Eng, 12 May, 112 (1975).

1976-1979

Aleksandrov, N.A., et al., Sharp-rectification of close-boiling hydrocarbon mixtures, Int Chem Eng, 16(3), 464-469 (1976).

Furzer, I.A., and Duffy, G.J., Periodic cycling of plate columns, AIChEJ, 22(6), 1118-1125 (1976).
Komatsu, H., Reactive distillation by total reflux, Int Chem Eng, 16(3), 563-566 (1976).
Naka, Y., et al., Changes in distillate composition during ternary azeotropic batch distillations, Int Chem Eng, 16(2), 272-280 (1976).
Various, Distillation operating practices (topic issue), Chem Eng Prog, 72(9), 43-71 (1976).
Vrba, J., General balance calculation for multicomponent distillation, Int Chem Eng, 16(1), 110-118 (1976).
Steingaszner, P.; Balint, A., and Kojnok, M., Improving a furfural distillation plant, Periodica Polytechnica, 21, 59-72 (1977).
Castellano, E.N.; McCain, C.A., and Nobles, F.W., Digital control of a distillation system, Chem Eng Prog, 74(4), 56-60 (1978).
Duffy, G.J., and Furzer, I.A., Mass transfer on a single sieve plate column operated with periodic cycling, AIChEJ, 24(4), 588-598 (1978).
Furzer, I.A., et al., Periodic cycling of plate columns, Chem Eng Sci, 33(7), 897-912 (1978).
Kerkhof, L.H.J., and Vissers, A., Profit from optimum control in batch distillation, Chem Eng Sci, 33(7), 961-970 (1978).
Medani, M.S., and Soliman, M.A., Approximate analytical solutions for systems with axial dispersion: Steady-state behaviour of a binary distillation plate, Comput Chem Eng, 2(1), 53-60 (1978).
Nisenfeld, A.E., and Harbison, J., Benefits of better distillation control, Chem Eng Prog, 74(7), 88-90 (1978).
Shah, G.C., Troubleshooting distillation columns, Chem Eng, 31 July, 70-78 (1978).
Tedder, D.W., and Rudd, D.F., Parametric studies in industrial distillation, AIChEJ, 24(2), 303-334 (1978).
Chin, T.G., Guide to distillation pressure control methods, Hyd Proc, 58(10), 145-153 (1979).
Furzer, I.A., Periodic cycling of plate columns, AIChEJ, 25(4), 600-609 (1979).

1980-1984

Goss, D.W., and Furzer, I.A., Mass transfer in periodically cycled plate columns containing multiple sieve plates, AIChEJ, 26(4), 663-669 (1980).
Hoyos, G.H., and Sommerfeld, J.T.S., Save energy by reducing reflux ratio, Chem Eng, 10 March, 168 (1980).
Baron, G.; Wajc, S., and Lavie, R., Stepwise periodic distillation, Chem Eng Sci, 35(4), 859-866 (1980); 36(11), 1819-1828 (1981).
Kennedy, J.P., Sequential control of continuous distillation, Chem Eng Prog, 77(11), 33-37 (1981).
Various, Aids to better distillation (special report), Hyd Proc, 60(2), 91-107 (1981).
Longwell, E.J., Control system design for distillation columns, Chem Eng Prog, 78(9), 63-66 (1982).

Matsuda, A.; Honda, K.; Okada, K., and Munakata, T., Performance of wetted-wall distillation columns under reduced pressure, Int Chem Eng, 22(3), 495-503 (1982).

Rathore, R.N.S., Reusing energy lowers fuel needs of distillation towers, Chem Eng, 14 June, 155-159 (1982).

Drew, J.W., Distillation column startup, Chem Eng, 14 Nov, 221-226 (1983).

Guy, J.L., Dynamic modeling of continuous distillation and emergency systems, Chem Eng, 21 March, 65-69 (1983).

Dubeau, Y., Use temperature column indicators to locate distillation-column liquid level, Chem Eng, 10 Dec, 98 (1984).

Rose, L.M., and Hyka, J., Analysis of operating distillation-column data, Ind Eng Chem Proc Des Dev, 23(3), 429-437 (1984).

1985-1988

Abu-Eishah, S.I., and Luyben, W.L., Design and control of a two-column azeotropic distillation system, Ind Eng Chem Proc Des Dev, 24(1), 132-140 (1985).

Ayral, T.E., Control improves fractionator performance, Hyd Proc, April, 87-88 (1985).

Fisher, W.R.; Doherty, M.F., and Douglas, J.M., Effect of overdesign on operability of distillation columns, Ind Eng Chem Proc Des Dev, 24(3), 593-598 (1985).

Matsubara, M.; Watanabe, N., and Kurimoto, H., Binary periodic distillation scheme with enhanced energy recovery, Chem Eng Sci, 40(5), 715-722, 755-758 (1985).

Szonyi, L., and Furzer, I.A., Periodic cycling of distillation columns using a new tray design, AIChEJ, 31(10), 1707-1713 (1985).

Yamada, I., et al., Minimum temperature of heat source necessary for multi-effect distillation, Int Chem Eng, 25(3), 527-534 (1985).

Basta, N., Facelift for distillation, Chem Eng, 2 March, 14-16 (1987).

Kapoor, N., and McAvoy, T.J., An analytical approach to approximate dynamic modeling of distillation towers, Ind Eng Chem Res, 26(12), 2473-2482 (1987).

Kister, H.Z., and Hower, T.C., Unusual operating histories of gas processing and olefins plant columns, Plant/Opns Prog, 6(3), 151-161 (1987).

Toftegard, B., and Jorgensen, S.B., Design algorithm for periodic cycled binary distillation columns, Ind Eng Chem Res, 26(5), 1041-1043 (1987).

Velasco, J.R.G.; Ortiz, M.A.G.; Pelayo, J.M.C., and Marcos, J.A.G., Improvements in batch distillation startup, Ind Eng Chem Res, 26(4), 745-750 (1987).

Yoshida, H., and Yorizane, M., Comparative characteristics of distillation columns with open steam and reboilers, Int Chem Eng, 27(1), 117-121 (1987).

Anon., Alcohol column revamp increases capacity, Chem Eng, 10 Oct, 108-110 (1988).

Priestman, G.H., and Brown, D.J., Flow induced vibration and damage in sieve-tray columns, Chem Eng Commns, 63, 181-192 (1988).

19.7 Control and Instrumentation

1967-1969

Block, B., Control of batch distillations, Chem Eng, 16 Jan, 147-150 (1967).

Cadman, T.W.; Rothfus, R.R., and Kermode, R.I., Design and effectiveness of feedforward control systems for multicomponent distillation columns, Ind Eng Chem Fund, 6(3), 421-431 (1967).

Davison, E.J., Control of a distillation column with pressure variation, Trans IChemE, 45, T229-250 (1967).

Hutchinson, A.W., and Shelton, R.J., Measurement of dynamic characteristics of full-scale plant using random perturbating signals: An application to a refinery distillation column, Trans IChemE, 45, T334-342 (1967).

Schrodt, V.N., et al., Plant-scale study of controlled cyclic distillation, Chem Eng Sci, 22(5), 759-768 (1967).

Tetlow, N.J.; Groves, D.M., and Holland, C.D., Generalized model for dynamic behavior of a distillation column, AIChEJ, 13(3), 476-485 (1967).

Thorogood, R.M., Dynamic response of air rectification column to interruptions of feed flow, Chem Eng Sci, 22(11), 1457-1474 (1967).

Brosilow, C.B., and Handley, K.R., Optimal control of a distillation column, AIChEJ, 14(3), 467-472 (1968).

Luyben, W.L., Feed plate manipulation in distillation column feedforward control, Ind Eng Chem Fund, 7(3), 502-508 (1968).

Marino, P.A.; Perna, A.J., and Stutzman, L.F., Sinusoidal and pulse response of a plate distillation column by reflux upset, AIChEJ, 14(6), 866-869 (1968).

Burman, L.K., and Maddox, R.N., Dynamic control of distillation columns, Ind Eng Chem Proc Des Dev, 8(4), 433-438 (1969).

Levy, R.E.; Foss, A.S., and Grens, E.A., Response modes of a binary distillation column, Ind Eng Chem Fund, 8(4), 765-776 (1969).

Luyben, W.L., Distillation, feedforward control with intermediate feedback control trays, Chem Eng Sci, 24(6), 997-1008 (1969).

Luyben, W.L., Feedback control of distillation columns by double differential temperature control, Ind Eng Chem Fund, 8(4), 739-744 (1969).

Pohjola, V.J., and Norden, H.V., Process dynamics of binary distillation, Chem Eng Sci, 24(11), 1687-1698 (1969).

1970-1973

Luyben, W.L., Distillation decoupling, AIChEJ, 16(2), 198-203 (1970).

Robinson, E.R., Optimal control of industrial batch distillation column, Chem Eng Sci, 25(6), 921-928 (1970).

Shoneman, K.F., and Gerster, J.A., Feedback control of an enriching column, AIChEJ, 16(6), 1080-1086 (1970).

Wahl, E.F., and Harriott, P., Understanding and prediction of dynamic behavior of distillation columns, Ind Eng Chem Proc Des Dev, 9(3), 398-407 (1970).

Luyben, W.L., Control of distillation columns with sharp temperature profiles, AIChEJ, 17(3), 713-718 (1971).

Niederlinski, A., Two-variable distillation control: Decouple or not decouple, AIChEJ, 17(5), 1261-1263 (1971).

Shunta, J.P., and Luyben, W.L., Studies of sampled-data control of distillation columns, Ind Eng Chem Fund, 10(3), 486-493 (1971).
Speicher, E.J., and Luyben, W.L., Experimental studies of feed plate manipulation for distillation column feedforward control, Ind Eng Chem Fund, 10(1), 147-149 (1971).
Svrcek, W.Y., and Wilson, H.W., Case history of a distillation column control scheme, Chem Eng Prog, 67(2), 45-51 (1971).
Changlai, Y., and Ward, T.J., Decoupling control of distillation column, AIChEJ, 18(1), 225-227 (1972).
Hu, Y.C., and Ramirez, W.F., Application of modern control theory to distillation columns, AIChEJ, 18(3), 479-486 (1972).
Merluzzi, P., and Brosilow, C.B., Nearly optimal control of a pilot-plant distillation column, AIChEJ, 18(4), 739-744 (1972).
Mizuno, H.; Watanabe, Y.; Nishimura, Y., and Matsubara, M., Asymmetric properties of continuous distillation column dynamics, Chem Eng Sci, 27(1), 129-136 (1972).
Shunta, J.P., and Luyben, W.L., Sampled-data noninteracting control for distillation columns, Chem Eng Sci, 27(6), 1325-1336 (1972).
Beaverstock, M.C., and Harriott, P., Experimental closed-loop control of a distillation column, Ind Eng Chem Proc Des Dev, 12(4), 401-407 (1973).
Sittel, C.N., and Fisher, G.T., Transient response of a distillation column plate, Sepn Sci, 8(4), 419-472 (1973).

1974-1979

Alekseev, Y.A., and Gorban, V.A., Control of multicomponent distillation, Int Chem Eng, 14(1), 118-122 (1974).
Lee, E.S., Quasilinearization, parameter estimation, and distillation column design, Chem Eng Commns, 1(5), 249-260 (1974).
Paul, R.N., Reboiler liquid closes level-control circuit, Chem Eng, 13 May, 128 (1974).
Pike, D.H., and Thomas, M.E., Optimal control of a continuous distillation column, Ind Eng Chem Proc Des Dev, 13(2), 97-102 (1974).
Svrcek, W.Y., and Ritter, R.A., Dynamic response of binary distillation column, Chem Eng Sci, 29(11), 2253-2257 (1974).
Chan, A.L., and Talbot, F.D., Optimal two-point control of binary distillation column, Can JCE, 53, 91-100 (1975).
Luyben, W.L., Steady-state energy conservation aspects of distillation column control-system design, Ind Eng Chem Fund, 14(4), 321-325 (1975).
Matsubara, M., et al., Periodic control of continuous distillation processes, Chem Eng Sci, 30(9), 1075-1084 (1975).
Tyreus, B., and Luyben, W.L., Control of binary distillation column with sidestream drawoff, Ind Eng Chem Proc Des Dev, 14(4), 391-398 (1975).
Joseph, B.; Brosilow, C.B.; Howell, J.C., and Kerr, W.R.D., Distillation control by multiple temperature measurements, Hyd Proc, 55(3), 127-131 (1976).
Shinskey, F.G., Energy-conserving control systems for distillation units, Chem Eng Prog, 72(5), 73-78 (1976).
Skrokov, M.R., The benefits of microprocessor control for distillation, Chem Eng, 11 Oct, 133-139 (1976).

Furzer, I.A., Discrete residence-time distribution of distillation column operated with microprocessor-controlled periodic cycle, Can JCE, 56, 747-755 (1978).

Jafarey, A., and McAvoy, T.J., Degeneracy of decoupling in distillation columns, Ind Eng Chem Proc Des Dev, 17(4), 485-490 (1978).

Meyer, C.; Seborg, D.E., and Wood, R.K., Experimental application of time delay compensation techniques to distillation column control Ind Eng Chem Proc Des Dev, 17(1), 62-67 (1978).

Doherty, M.F., and Perkins, J.D., Dynamics of distillation processes, Chem Eng Sci, 33(3), 281-302 (1978); 33(5), 569-579 (1978); 34(12), 1401-1414 (1979).

Jafarey, A.; McAvoy, T.J., and Douglas, J.M., Analytical relationships for the relative gain for distillation control, Ind Eng Chem Fund, 18(2), 181-187 (1979).

McAvoy, T.J., Steady-state decoupling of distillation columns, Ind Eng Chem Fund, 18(3), 269-273 (1979).

Meyer, C.B.G.; Wood, R.K., and Seborg, D.E., Experimental evaluation of analytical and Smith predictors for distillation column control, AIChEJ, 25(1), 24-32 (1979).

Multala, R., et al., Computer-controlled pilot-plant distillation unit and studies on multicomponent distillation, Comput Chem Eng, 3, 47-52 (1979).

Roffel, B., and Fontein, H.J., Constraint control of distillation processes, Chem Eng Sci, 34(8), 1007-1018 (1979).

Tyreus, B.D., Multivariable control system design for an industrial distillation column, Ind Eng Chem Proc Des Dev, 18(1), 177-182 (1979).

1980-1982

DiBiano, R.J., Improve reliability of advanced distillation control, Hyd Proc, 59(10), 117-120 (1980).

Jafarey, A.; McAvoy, T.J., and Douglas, J.M., Steady-state noninteracting controls for distillation columns: An analytical study, Ind Eng Chem Proc Des Dev, 19(1), 114-117 (1980).

Taiwo, O., Application of the method of inequalities to the multivariable control of binary distillation columns, Chem Eng Sci, 35(4), 847-858 (1980).

Weischedel, K., and McAvoy, T.J., Feasibility of decoupling in conventionally controlled distillation columns, Ind Eng Chem Fund, 19(4), 379-384 (1980).

Dahlquist, S.A., Control of a distillation column using self-tuning regulators, Can JCE, 59, 118-127 (1981).

Doukas, N.P., and Luyben, W.L., Control of an energy-conserving prefractionator/sidestream column distillation system, Ind Eng Chem Proc Des Dev, 20(1), 147-153 (1981).

Kim, Y.S., and McAvoy, T.J., Steady-state analysis of interaction between pressure and temperature or composition loops in a single distillation tower, Ind Eng Chem Fund, 20(4), 381-388 (1981).

Morris, C.G., and Svrcek, W.Y., Dynamic simulation of multicomponent distillation, Can JCE, 59, 382-387 (1981).

Shinskey, F.G., Predict distillation column response using relative gains, Hyd Proc, 60(5), 196-200 (1981).

Takamatsu, T.; Kawachi, K., and Watanabe, F., Design of decoupling control system for binary distillation column with uncertain parameters, Int Chem Eng, 21(1), 40-50 (1981).

Doherty, M.F., and Perkins, J.D., Dynamics of distillation processes, Chem Eng Sci, 37(3), 381-392 (1982).

Gordon, L.M., Practical evaluation of relative gains: The key to designing dual composition controls, Hyd Proc, Dec, 87-92 (1982).

Hammarstrom, L.G.; Waller, K.V., and Fagervik, K.C., Modeling accuracy for multivariable distillation control, Chem Eng Commns, 19(1), 77-90 (1982).

Patke, N.G., and Deshpande, P.B., Experimental evaluation of inferential systems for distillation control, Chem Eng Commns, 13(4), 343-360 (1982).

Patke, N.G.; Deshpande, P.B., and Chou, A.C., Evaluation of inferential and parallel cascade schemes for distillation control, Ind Eng Chem Proc Des Dev, 21(2), 266-272 (1982).

Pierucci, S., and Ranzi, E., T-method for distillation columns, Chem Eng Commns, 14(1), 1-22 (1982).

Rinne, R.; Sunell, H.; Latour, P.R., and Paynter, K.K., Experience with distillation unit computer control, Hyd Proc, March, 141-148 (1982).

1983-1985

Defaye, G.; Caralp, L., and Jouve, P., Application of a simple deterministic predictive control algorithm for a distillation column, Chem Eng J, 27(3), 161-166 (1983).

Fagervik, K.C.; Waller, K.V., and Hammarstrom, L.G., Two-way or one-way decoupling in distillation, Chem Eng Commns, 21(4), 235-250 (1983).

Fuentes, C., and Luyben, W.L., Control of high-purity distillation columns, Ind Eng Chem Proc Des Dev, 22(3), 361-366 (1983).

Kisakurek, B., Predictive model for dynamic distillation, Chem Eng Commns, 20(1), 63-80 (1983).

Ogunnaike, B.A.; Lemaire, J.P.; Morari, M., and Ray, W.H., Advanced multivariable control of a pilot-plant distillation column, AIChEJ, 29(4), 632-640 (1983).

Takamatsu, T., et al., Decoupling control of multicomponent distillation column using pseudobinary model, Int Chem Eng, 23(4), 699-707 (1983).

Frey, R.M.; Doherty, M.F.; Douglas, J.M., and Malone, M.F., Controlling thermally linked distillation columns, Ind Eng Chem Proc Des Dev, 23(3), 483-490 (1984).

Rys, R.A., Advanced control techniques for distillation columns, Chem Eng, 10 Dec, 75-81 (1984).

Yu, C.C., and Luyben, W.L., Use of multiple temperatures for control of multicomponent distillation columns, Ind Eng Chem Proc Des Dev, 23(3), 590-597 (1984).

Chiang, T.P., and Luyben, W.L., Incentives for dual composition control in single and heat-integrated binary distillation columns, Ind Eng Chem Fund, 24(3), 352-359 (1985).

Elaahi, A., and Luyben, W.L., Control of an energy-conservative complex configuration of distillation columns for four-component separations, Ind Eng Chem Proc Des Dev, 24(2), 368-376 (1985).

Glinos, K., and Malone, M.F., Steady-state control of sidestream distillation columns, Ind Eng Chem Proc Des Dev, 24(3), 608-613 (1985).
Haskins, D.E.,; Tolfo, F., and Chauvin, L., Group methods for advanced column control compared, Hyd Proc, May, 93-96 (1985).
Marchetti, J.L.; Benallou, A.; Seborg, D.E., and Mellichamp, D.A., Pilot-scale distillation facility for digital-computer control research, Comput Chem Eng, 9(3), 301-310 (1985).
Shimizu, K., and Matsubara, M., Directions of disturbances and modelling errors on the control quality in distillation systems, Chem Eng Commns, 37(1), 67-92 (1985).
Shimizu, K.; Holt, B.R.; Morari, M., and Mah, R.H.S., Assessment of control structures for binary distillation columns with secondary reflux and vaporization, Ind Eng Chem Proc Des Dev, 24(3), 852-858 (1985).
Stanley, G.T., and McAvoy, T.J., Dynamic energy conservation aspects of distillation control, Ind Eng Chem Fund, 24(4), 439-443 (1985).
Stathaki, A.; Mellichamp, D.A., and Seborg, D.E., Dynamic simulation of a multicomponent distillation column with asymmetric dynamics, Can JCE, 63(3), 510-518 (1985).
Tsogas, A., and McAvoy, T., Gain scheduling for composition control of distillation columns, Chem Eng Commns, 37(1), 275-292 (1985).
Waller, K.V.; Wikman, K.E., and Gustafsson, S.E., Decoupler design and control system tuning by INA for distillation composition control, Chem Eng Commns, 35(1), 149-174 (1985).

1986-1987

Alatiqi, I.M., and Luyben, W.L., Control of a complex sidestream column/stripper distillation configuration, Ind Eng Chem Proc Des Dev, 25(3), 762-767 (1986).
Chien, I.L.; Mellichamp, D.A., and Seborg, D.E., Multivariable self-tuning control strategy for distillation columns, Ind Eng Chem Proc Des Dev, 25(3), 595-600 (1986).
Davis, J.; James, D.; Catt, M., and Adair, J., Computer control of a desulfurizer fractionator, Chem Eng Prog, 82(3), 62-66 (1986).
Gani, R.; Romagnoli, J.A., and Stephanopoulos, G., Control studies in extractive distillation process: Simulation and measurement structure, Chem Eng Commns, 40, 281-302 (1986).
Stanton, B.D., Distillation towers: Designing front-end control schemes, Chem Eng, 24 Nov, 87-92 (1986).
Tolliver, T.L., Improving column control to reduce distillation operating cost, Chem Eng, 24 Nov, 99-101 (1986).
Baker-Counsell, J., Modelling and control of distillation processes, Proc Engng, Jan, 34-35 (1987).
Biddulph, M.W., How stable are split-flow distillation columns? Proc Engng, Feb, 61-64 (1987).
Christensen, F.M., and Jorgensen, S.B., Optimal control of binary batch distillation with recycled waste cut, Chem Eng J, 34(2) 57-64 (1987).
Iftikhar, N.Z.; Ramulu, C.P.; Husain, A., and Deshpande, P.B., RGA analysis for distillation control, Chem Eng Commns, 51, 19-46 (1987).

Levien, K.L., and Morari, M., Internal model control of coupled distillation columns, AIChEJ, 33(1), 83-98 (1987).
McDonald, K.A., and McAvoy, T.J., Application of dynamic matrix control to moderate and high purity distillation towers, Ind Eng Chem Res, 26(5), 1011-1018 (1987).
Mizsey, P.; Hajdu, H., and Foldes, P., How construction affects column control, Hyd Proc, 66(2), 53-56 (1987).
Skogestad, S., and Morari, M., LV-control of a high-purity distillation column, Chem Eng Sci, 43(1), 33-48 (1987).
Skogestad, S., and Morari, M., Control configuration selection for distillation columns, AIChEJ, 33(10), 1620-1635 (1987).
Takamatsu, T.; Hashimoto, I., and Hashimoto, Y., Selection of manipulated variables to minimize interaction in multivariate control of a distillation column, Int Chem Eng, 27(4), 669-678 (1987).
Vu, L.D.; Gadkari, P.B., and Govind, R., Analysis of ternary distillation column sequences, Sepn Sci Technol, 22(7), 1659-1691 (1987).
Waller, K.V., and Finnerman, D.H., Using sums and differences to control distillation, Chem Eng Commns, 56, 253-268 (1987).
Yasuoka, H.; Nakanishi, E., and Kunigita, E., Design of on-line startup system for distillation column based on simple algorithm, Int Chem Eng, 27(3), 466-473 (1987).

1988

Arkun, Y., and Morgan, C.O., Use of the structured singular value for robustness analysis of distillation column control, Comput Chem Eng, 12(4), 303-306 (1988).
Bozenhardt, H.F., Modern control tricks solve distillation problems, Hyd Proc, 67(6), 47-50 (1988).
Chang, Y.A., and Seader, J.D., Simulation of continuous reactive distillation by homotopy-continuation method, Comput Chem Eng, 12(12), 1243-1257 (1988).
Chiang, T.P., and Luyben, W.L., Comparison of the dynamic performance of three heat-integrated distillation configurations, Ind Eng Chem Res, 27(1), 99-105 (1988).
Coser, R.J., Distillation cut-point calculations with modern distributed control systems, Hyd Proc, 67(4), 35-37 (1988).
Floudas, C.A., and Anastasiadis, S.H., Synthesis of distillation sequences with several multicomponent feed and product streams, Chem Eng Sci, 43(9), 2407-2420 (1988).
Galindez, H., and Fredenslund, A., Simulation of multicomponent batch distillation processes, Comput Chem Eng, 12(4), 281-288 (1988).
Georgiou, A.; Georgakis, C., and Luyben, W.L., Nonlinear dynamic matrix control for high-purity distillation columns, AIChEJ, 34(8), 1287-1298 (1988).
Haggblom, K.E., and Waller, K.V., Transformations and consistency relations of distillation control structures, AIChEJ, 34(10), 1634-1648 (1988).
Karim, M.N., and Lee, G.K.F., Design of robust control systems for distillation columns, Chem Eng Commns, 68, 81-98 (1988).

Lu, S.M., and Huang, R.H., Microcomputer control of distillate streams from pot distillation of wines, Chem Eng Commns, 74, 73-84 (1988).

McDonald, K.A.; Palazoglu, A., and Bequette, B.W., Impact of model uncertainty descriptions for high-purity distillation control, AIChEJ, 34(12), 1996-2004 (1988).

Mountziaris, T.J., and Georgiou, A., Design of robust noninteracting controllers for high-purity binary distillation columns, Ind Eng Chem Res, 27(8), 1450-1460 (1988).

Paules, G.E., and Floudas, C.A., Synthesis of flexible distillation sequences for multiperiod operation, Comput Chem Eng, 12(4), 267-280 (1988).

Skogestad, S., and Morari, M., Dynamic behavior of distillation columns, Ind Eng Chem Res, 27(10), 1848-1863 (1988).

Takamatsu, T., et al., Relationship between interaction in multivariable control system of distillation column and column characteristics, Int Chem Eng, 28(3), 435-447 (1988).

Waller, K.V., et al., Experimental comparison of four control structures for two-point control of distillation, Ind Eng Chem Res, 27(4), 624-630 (1988).

Waller, K.V.; Haggblom, K.E.; Sandelin, P.M., and Finnerman, D.H., Disturbance sensitivity of distillation control structures, AIChEJ, 34(5), 853-858 (1988).

19.8 Systems, Sequences and Applications

1967-1972

Aerov, M.E.; Vostrikova, V.N., and Gurovich, R.E., Rectification drying of hydrocarbons, Int Chem Eng, 7(2), 279-281 (1967).

Gaumer, L.S., Flash column performance in a helium recovery plant, Chem Eng Prog, 63(5), 72-77 (1967).

Husain, A., Distillation columns with side streams or multiple feeds, Brit Chem Eng, 12(2), 226-229 (1967).

Latimer, R.E., Distillation of air (state-of-the-art review), Chem Eng Prog, 63(2), 35-59; 63(3), 89-90 (1967).

Huggett, R., and King, P.J., Helical coil distillation columns, Trans IChemE, 46, T101-105 (1968).

Taylor, J.H., Systems design for centrifugal molecular distillation, Chem Eng, 26 Aug, 109-112 (1968).

Armstrong, M., and Schofield, A.E., The design of air separation distillation columns using a computer, Chem Engnr, May, CE184-189 (1969).

Frank, J.C.; Geyer, G.R., and Kehde, H., Styrene-ethylbenzene vacuum distillation with sieve trays, Chem Eng Prog, 65(2), 79-86 (1969).

Hajiev, S.N., Rectification of methylchloro-silanes, Ind Eng Chem Proc Des Dev, 9(2), 229-233 (1970).

Wolf, D., and Cohen, H., Connecting distillation columns in cascade for long-period operation, Chem Eng, 14 Dec, 180-182 (1970).

Broughton, D.B., and Uitti, K.D., Estimate tower for naphtha cuts, Hyd Proc, 50(10), 109-112 (1971).

Burgess, M.P., Mathematical model of heavy water extraction and distillation, AIChEJ, 17(3), 529-535 (1971).

Clancy, G.M., and Townsend, R.W., Ethylene plant fractionation, Chem Eng Prog, 67(2), 41-44 (1971).
Mehta, D.D., and Pan, W.W., Methanol distillation, Hyd Proc, 50(2), 115-120 (1971).
Taylor, D.L., and Edmister, W.C., Solutions for distillation processes treating petroleum fractions, AIChEJ, 17(6), 1324-1329 (1971).
Peiffer, C.C.; Metcalfe, R.S.; Kopko, R.J., and McCormick, R.H., A 20-stage cocurrent contacting vapor-liquid equilibrium unit, Ind Eng Chem Proc Des Dev, 11(4), 525-529 (1972).

1973-1979

Davies, B.; Jenkins, J.D., and Jeffreys, G.V., The continuous trans-esterification of ethyl alcohol and butyl acetate in a sieve plate column, Trans IChemE, 51, 267-280 (1973).
Jelinek, J.; Hlavacek, V., and Krivsky, Z., Computation of two interlinked countercurrent separation columns, Chem Eng Sci, 28(10), 1833-1838 (1973).
Packer, L.G.; Ellis, S.R.M., and Soares, L.D.J., Miniature equilibrium still, Chem Eng Sci, 28(2), 597-600 (1973).
Bairamov, R.B.; Toiliev, K., and Mukhammetdurdyeva, O., Greenhouse solar distillation unit combined with heat pipes (unsteady-state conditions), Int Chem Eng, 15(3), 454-456 (1975).
Olsson, B., and Svensson, S.G., Formalin distillation: Vapour-liquid equilibria and tray efficiencies, Trans IChemE, 53, 97-105 (1975).
Yost, R.W., Application of modern distillation equipment in analytical chemistry, Sepn & Purif Methods, 4(1), 1-22 (1975).
Kubicek, M.; Hlavacek, V., and Prochaska, F., Global modular Newton-Raphson technique for simulation of an interconnected plant applied to complex rectification columns, Chem Eng Sci, 31(4), 277-284 (1976).
Prenosil, J.E., Multicomponent steam distillation: Comparison between digital simulation and experiment, Chem Eng J, 12(1), 59-68 (1976).
Ritchey, K.J.; Canfield, F.B., and Challand, T.B., Heavy-oil distillation via computer simulation, Chem Eng, 2 Aug, 79-88 (1976).
Fair, J.R., Advances in distillation system design, Chem Eng Prog, 73(11), 78-83 (1977).
Kern, R., Layout arrangements for distillation columns, Chem Eng, 15 Aug, 153-160 (1977).
Mah, R.S.H.; Nicholas, J.J., and Wodnik, R.B., Distillation with secondary reflux and vaporization: A comparative evaluation, AIChEJ, 23(5), 651-658 (1977).
Sutton, T.L., and MacGregor, J.F., Analysis and design of binary vapour-liquid equilibrium experiments, Can JCE, 55, 602-620 (1977).
Darton, R.C.; van Grinsven, P.F.A., and Simon, M.M., Development in steam stripping of sour water, Chem Engnr, Dec, 923-927 (1978).
Doukas, N., and Luyben, W.L., Economics of alternative distillation configurations for separation of ternary mixtures, Ind Eng Chem Proc Des Dev, 17(3), 272-281 (1978).

Rasquin, E.A.; Lynn, S., and Hanson, D.N., Vacuum steam stripping of volatile, sparingly soluble organic compounds from water streams, Ind Eng Chem Fund, 17(3), 170-174 (1978).

Ruheman, M., Review of cryogenic developments with particular reference to chemical engineering, Chem Engnr, May, 362-365 (1978).

Sargent, R.W.H., and Sullivan, G.R., Development of feed changeover policies for refinery distillation units, Ind Eng Chem Proc Des Dev, 18(1), 113-124, (1979).

Zacchi, G., and Aly, G., Simulation and design of distillation units for treatment of sulfite pulping condensates to recover methanol and furfural, Can JCE, 57, 311-325 (1979).

1980-1984

Black, C., Distillation modeling of ethanol recovery and dehydration processes for ethanol and gasohol, Chem Eng Prog, 76(9), 78-85 (1980).

Noworyta, A.W., Distillation of heat-sensitive substances, Chem Eng J, 19(1), 75-82 (1980).

Economopoulos, A.P., Data inputs and interchange in integrated solution/design tower packages, Ind Eng Chem Proc Des Dev, 20(2), 219-223 (1981).

Englund, S.M., Monomer removal from latex by steam stripping, Chem Eng Prog, 77(8), 55-59 (1981).

Hsu, H., and Tierney, J.W., Exact calculation methods for systems of interlinked columns, AIChEJ, 27(5), 733-739 (1981).

Carta, R.; Kovacic, A., and Tola, G., Separation of ethylbenzene-xylenes mixtures by distillation: Possible application of vapour recompression concept, Chem Eng Commns, 19(1), 157-166 (1982).

Holmes, A.S.; Ryan, J.M.; Price, B.C., and Styring, R.E., Cryogenic process improves acid gas distillation, Hyd Proc, May, 131-136 (1982).

Naka, Y.; Terashita, M., and Takamatsu, T., Thermodynamic approach to multicomponent distillation system synthesis, AIChEJ, 28(5), 812-820 (1982).

Berg, L., Separation of benzene and toluene from close boiling nonaromatics by extractive distillation, AIChEJ, 29(6), 961-966 (1983).

Pibouleau, L.; Said, A., and Domenech, S., Synthesis of optimal and near-optimal distillation sequences by a bounding strategy, Chem Eng J, 27(1), 9-20 (1983).

Prokopakis, G.J., and Seider, W.D., Feasible specifications in azeotropic distillation, AIChEJ, 29(1), 49-60 (1983).

Ramshaw, C., 'Higee' distillation: An example of process intensification, Chem Engnr, Feb, 13-14 (1983).

Schmitt, D., and Vogelpohl, A., Distillation of ethanol-water solutions in presence of potassium acetate, Sepn Sci Technol, 18(6), 547-554 (1983).

Zhi, W.X., and Wankat, P.C., Continuous multicomponent fractionation by parametric pumping, Ind Eng Chem Fund, 22(2), 172-176 (1983).

Berg, L., and Yeh, A.I., Separation of isopropyl ether from methyl ethyl ketone by extractive distillation, Chem Eng Commns, 29(1), 283-290 (1984).

Berg, L., and Yeh, A.I., Separation of methyl acetate from methanol by extractive distillation, Chem Eng Commns, 30(1), 113-118 (1984).

Berg, L.; Ratanapupech, P., and Yeh, A.I., Packed rectification columns for extractive distillation, AIChEJ, 30(5), 845-849 (1984).
Canfield, F.B., Computer simulation of the parastillation process, Chem Eng Prog, 80(2), 58-62 (1984).
Kumar, S.; Wright, J.D., and Taylor, P.A., Modelling and dynamics of an extractive distillation column, Can JCE, 62(6), 780-789 (1984).
Lynd, L.R., and Grethlein, H.E., IHOSR/extractive distillation for ethanol separation, Chem Eng Prog, 80(11), 59-62 (1984).

1985-1986

Berg, L.; Yeh, A.I., and Ratanapupech, P., Recovery of ethyl acetate by extractive distillation, Chem Eng Commns, 39, 193-200 (1985).
Duckett, M., and Ruhemann, M., Cryogenic gas separation, Chem Engnr, Dec, 14-17 (1985).
Gilnos, K., and Malone, M.F., Design of sidestream distillation columns, Ind Eng Chem Proc Des Dev, 24(3), 822-828 (1985).
Gomez-Munoz, A., and Seader, J.D., Synthesis of distillation trains by thermodynamic analysis, Comput Chem Eng, 9(4), 311-342 (1985).
Gooding, C.H., and Bahouth, F.J., Membrane-aided distillation of azeotropic solutions, Chem Eng Commns, 35(1), 267-280 (1985).
Malone, M.F.; Glinos, K.; Marquez, F.E., and Douglas, J.M., Simple analytical criteria for the sequencing of distillation columns, AIChEJ, 31(4), 683-689 (1985).
O'Neill, P.S.; Wisz, M.W.; Ragi, E.G.; Page, E.H., and Antonelli, R., Vapor recompression systems with high efficiency components, Chem Eng Prog, 81(7), 57-62 (1985).
Strigle, R.F., Distillation of light hydrocarbons in packed columns, Chem Eng Prog, 81(4), 67-71 (1985).
Terrill, D.L.; Sylvestre, L.F., and Doherty, M.F., Separation of closely boiling mixtures by reactive distillation, Ind Eng Chem Proc Des Dev, 24(4), 1062-1073 (1985).
Berg, L., and Yeh, A.I., Breaking of ternary acetate-alcohol-water azeotropes by extractive distillation, Chem Eng Commns, 48, 93-102 (1986).
Fullarton, D., and Schlunder, E.U., Diffusion distillation: New separation process for azeotropic mixtures, Chem Eng & Proc, 20(5), 255-270 (1986).
Glinos, K.N.; Nikolaides, I.P., and Malone, M.F., New complex column arrangements for ideal distillation, Ind Eng Chem Proc Des Dev, 25(3), 694-699 (1986).
Lynn, S., and Hanson, D.N., Multieffect extractive distillation for separating aqueous azeotropes, Ind Eng Chem Proc Des Dev, 25(4), 936-942 (1986).
Pettit, D.R., Performance of absorption and distillation columns in a lunar environment, Chem Eng Commns, 46, 111-123 (1986).
Pibouleau, L., and Domenech, S., Discrete and continuous approaches to the optimal synthesis of distillation sequences, Comput Chem Eng, 10(5), 479-491 (1986).
Ulowetz, M.A., New distillation pilot plant for packed column designs, Chem Eng Prog, 82(11), 41-45 (1986).

1987-1988

Barker, P.E.; Bhambra, K.S.; Alsop, R.M., and Gibbs, R., Fractionation of dextran using ethanol, CER&D, 65(5), 390-395 (1987).

Barnette, D.T., and Sommerfeld, J.T., Discrete-event simulation of a sequence of multicomponent batch distillation columns, Comput Chem Eng, 11(4), 395-398 (1987).

Berg, L., Recovery of methyl t-butyl ether from close-boiling impurities by extractive distillation, Chem Eng Commns, 52, 105-108 (1987).

Berg, L., and Yeh, A.I., Separation of m-xylene from o-xylene by extractive distillation, Chem Eng Commns, 54, 149-160 (1987).

Gostoli, C.; Sarti, G.C., and Matulli, S., Low-temperature distillation through hydrophobic membranes, Sepn Sci Technol, 22(2), 855-872 (1987).

Jubin, R.T.; Counce, R.M.; Holland, W.D.; Groenier, W.S., and North, E.D., A simplified process for the extractive distillation of nitric acid, Ind Eng Chem Res, 26(5), 990-997 (1987).

Lee, F.M., and Coombs, D.M., Two-liquid-phase extractive distillation for aromatics recovery, Ind Eng Chem Res, 26(3), 564-574 (1987).

Shoemaker, J.D., and Jones, E.M., Cumene by catalytic distillation, Hyd Proc, 66(6), 57-58 (1987).

Weiss, S., and Arlt, R., Modelling of mass transfer in extractive distillation, Chem Eng & Proc, 21(2), 107-114 (1987).

Zaidi, S.A.H., Extractive distillation of propanol-water mixtures employing less-soluble and sparingly-soluble salts as separating agents, J Chem Tech Biotechnol, 38(2), 85-88 (1987).

Bamopoulos, G.; Nath, R., and Motard, R.L., Heuristic synthesis of nonsharp separation sequences, AIChEJ, 34(5), 763-780 (1988).

Berg, L., Separation of lower boiling alcohols by extractive distillation, Chem Eng Commns, 66, 1-22 (1988).

Franklin, N.L., Counterflow cascades, CER&D, 66(1), 47-64 (1988).

Gadkari, P.B., and Govind, R., Analytical screening criterion for sequencing of distillation columns, Comput Chem Eng, 12(12), 1199-1214 (1988).

Lee, F.M., and Coombs, D.M., Two-liquid-phase extractive distillation for upgrading the octane number of the catalytically cracked gasoline, Ind Eng Chem Res, 27(1), 118-123 (1988).

Long, J.J.C., and Burney, A.H., Effect of various schemes in multicomponent distillation system, Chem Eng & Proc, 23(1), 33-40 (1988).

Mahapatra, A.; Gaikar, V.G., and Sharma, M.M., New strategies in extractive distillation: Use of aqueous solutions of hydrotropes and organic bases as solvent for organic acids, Sepn Sci Technol, 23(4), 429-436 (1988).

Wagler, R.M., and Douglas, P.L., Method for design of flexible distillation sequences, Can JCE, 66(4), 579-590 (1988).

Yeh, A.I.; Berg, L., and Warren, K.J., Separation of acetone-methanol mixture by extractive distillation, Chem Eng Commns, 68, 69-80 (1988).

19.9 Energy Integration

1976-1985

Geyer, G.R., and Kline, P.E., Energy conservation schemes for distillation processes, Chem Eng Prog, 72(5), 49-51 (1976).

Petterson, W.C., and Wells, T.A., Energy-saving schemes in distillation, Chem Eng, 26 Sept, 78-86 (1977).

Bannon, R.P., and Marple, S., Heat recovery in hydrocarbon distillation, Chem Eng Prog, 74(7), 41-45 (1978).

Mix, T.J.; Dweck, J.S.; Weinburg, M., and Armstrong, R.C., Energy conservation in distillation, Chem Eng Prog, 74(4), 49-55 (1978).

Stephenson, R.M., and Anderson, T.F., Energy conservation in distillation, Chem Eng Prog, 76(8), 68-71 (1980).

Luyben, W.L., Energy consumption in purge distillation columns, Chem Eng Prog, 77(10), 78-82 (1981).

Quadri, G.P., Use heat pump for propylene-propane rectification, Hyd Proc, 60(2), 119-126; 60(3), 147-150 (1981).

Fuentes, C., and Luyben, W.L., Comparison of energy models for distillation columns, Ind Eng Chem Fund, 21(3), 323-325 (1982).

Thiagarajan, N.; Ilgner, H.; Heck, G., and Lienerth, I., Minimum energy pure methanol (distillation process), Hyd Proc, March, 89-91 (1984).

Becker, F.E., and Zakak, A.I., Recovering energy by mechanical vapor recompression, Chem Eng Prog, 81(7), 45-49 (1985).

Cheng, H.C., and Luyben, W.L., Heat-integrated distillation columns for ternary separations, Ind Eng Chem Proc Des Dev, 24(3), 707-713 (1985).

1986-1988

Bingzhen, C., and Westerberg, A.W., Structural flexibility for heat-integrated distillation columns, Chem Eng Sci, 41(2), 355-378 (1986).

Kattan, M.K., and Douglas, P.L., A new approach to thermal integration of distillation sequences, Can JCE, 64(1), 162-170 (1986).

Lynd, L.R., and Grethlein, H.E., Distillation with intermediate heat pumps and optimal sidestream return, AIChEJ, 32(8), 1347-1359 (1986).

Meszaros, I., and Fonyo, Z., A new bounding strategy for synthesizing distillation schemes with energy integration, Comput Chem Eng, 10(6), 545-550 (1986).

Pritchard, C.L., Distillation with vapour compression, Chem Eng Educ, 20(3), 132-135 (1986).

Strigle, R.F., and Fukuyo, K., Cut C4 recovery costs: Less reboiler duty for packed distillation columns, Hyd Proc, 65(6), 47-48 (1986).

Fidkowski, Z., and Krolikowski, L., Minimum energy requirements of thermally coupled distillation systems, AIChEJ, 33(4), 643-653 (1987).

Glenchur, T., and Govind, R., Study on a continuous heat-integrated distillation column, Sepn Sci Technol, 22(12), 2323-2338 (1987).

Isla, M.A., and Cerda, J., Simultaneous synthesis of distillation trains and heat exchanger networks, Chem Eng Sci, 42(10), 2455-2464 (1987).

Meili, A., and Stuecheli, A., Distillation columns with direct vapor recompression, Chem Eng, 16 Feb, 133-143 (1987).

Isla, M.A., and Cerda, J., Heuristic method for synthesis of heat-integrated distillation systems, Chem Eng J, 38(3), 161-178 (1988).

CHAPTER 20

ABSORPTION AND COOLING TOWERS

20.1 Absorber Design

1967

Buchanan, J.E., Holdup in irrigated ring-packed towers below the loading point, Ind Eng Chem Fund, 6(3), 400-407 (1967).

Gildenblat, I.A.; Gurova, N.M., and Ramm, V.M., Effect of wetting distribution and packed column height on absorption efficiency in columns with various ring packings, Int Chem Eng, 7(1), 149-154 (1967).

Jameson, G.J., Predictions from a model for liquid distribution in packed columns, Trans IChemE, 45, T74-81 (1967).

Kozicki, W.; Hsu, C.J., and Tiu, C., Non-Newtonian flow through packed beds and porous media, Chem Eng Sci, 22(4), 487-502 (1967).

Krell, E., Wettability of packings and effect on mass transfer, Brit Chem Eng, 12(4), 562-567 (1967).

Metcalfe, T.B., and Newton, W.M., Characterising parameter for flow through packed beds, Brit Chem Eng, 12(6), 892-893 (1967).

Rajagopalan, R., and Laddha, G.S., Liquid flow through fixed and fluidised beds, Brit Chem Eng, 12(6), 894-896 (1967).

Rama Iyer, S., and Murti, P.S., Longitudinal mixing in slat tray columns, Brit Chem Eng, 12(4), 573-575 (1967).

Reiss, L.P., Cocurrent gas-liquid contacting in packed columns, Ind Eng Chem Proc Des Dev 6(4), 486-499 (1967).

Ridgway, K., and Tarbuck, K.J., Random packing of spheres, Brit Chem Eng, 12(3), 384-388 (1967).

Ross, T.K.; Campbell, D.U., and Wragg, A.A., An electrochemical investigation of the hydrodynamics of liquid films on grid packings, Trans IChemE, 45, T401-407 (1967).

Sharma, M.M., and Gupta, R.K., Mass transfer characteristics of plate columns without downcomer, Trans IChemE, 45, T169-175 (1967).

Shulman, H.L., and Mellish, W.G., Performance of packed columns, AIChEJ, 13(6), 1137-1140 (1967).

Susskind, H., and Becker, W., Pressure drop in geometrically ordered packed beds of spheres, AIChEJ, 13(6), 1155-1159 (1967).

Sweeney, D.E., Correlation for pressure drop in two-phase cocurrent flow in packed beds, AIChEJ, 13(4), 663-669 (1967).

Thomas, W.J., and Campbell, M., Hydraulic studies in a sieve plate downcomer system, Trans IChemE, 45, T53-63 (1967).

Thomas, W.J., and Campbell, M., Mixing, efficiency and mass transfer studies in a sieve plate downcomer system, Trans IChemE, 45, T64-73 (1967).

Turpin, J.L., and Huntington, R.L., Prediction of pressure drop for two-phase two-component concurrent flow in packed beds, AIChEJ, 13(6), 1196-1202 (1967).

Vidwans, A.D., and Sharma, M.M., Gas-side mass-transfer coefficient in packed columns, Chem Eng Sci, 22(4), 673-684 (1967).

Yen, I.K., Predicting packed-bed pressure drop, Chem Eng, 13 March, 173-176 (1967).

1968

Chung, S.F., and Wen, C.Y., Longitudinal dispersion of liquid flowing through fixed and fluidized beds, AIChEJ, 14(6), 857-866 (1968).

Dutkai, E., and Ruckenstein, E., Liquid distribution in packed columns, Chem Eng Sci, 23(11), 1365-1374 (1968).

Gunn, D.J., Mixing in packed and fluidised beds, Chem Engnr, June, CE153-172 (1968).

Handley, D., and Heggs, P.J., Momentum and heat transfer mechanisms in regular shaped packings, Trans IChemE, 46, T251-267 (1968).

Iczkowski, R.P., Displacement of liquids from random sphere packings, Ind Eng Chem Fund, 7(4), 572-576 (1968).

Jackson, G.S., and Marchello, J.M., Correlation of gravitational force for absorption in packed columns, Ind Eng Chem Proc Des Dev, 7(3), 359-361 (1968).

Jhaveri, A.S., and Sharma, M.M., Effective interfacial area in a packed column, Chem Eng Sci, 23(7), 669-676 (1968).

Lees, F.P., Frequency response of a packed gas-absorption column, Chem Eng Sci, 23(2), 97-108 (1968).

Lieberman, N.P., Bottleneck removal in a reboiled absorber system, Chem Eng Prog, 64(7), 58-60 (1968).

Mostafa, H.A., Curved operating lines for absorption of multicomponent systems, Brit Chem Eng, 13(5), 671-673 (1968).

Porter, K.E.; Barnett, V.D., and Templeman, J.J., Liquid flow in packed columns, Trans IChemE, 46, T69-94 (1968).

Rotstein, E., et al., Polygonal helicoids as packing, Brit Chem Eng, 13(12), 1730-1733 (1968).

Thomas, B.L., Nomograph for gas-absorption efficiency, Chem Eng, 26 Aug, 138 (1968).

Yoshida, F., and Arakawa, S.I., Pressure dependence of liquid-phase mass-transfer coefficients, AIChEJ, 14(6), 962-963 (1968).

1969

Beskin, L.Z.; Streltsov, V.V., and Demshin, V.Y., Mass transfer in an absorber with hydrodynamic mixing, Int Chem Eng, 9(1), 88-91 (1969).

Buchanan, J.E., Pressure gradient and liquid holdup in irrigated packed towers, Ind Eng Chem Fund, 8(3), 502-511 (1969).

Carpenter, R.E., Is graphite packing worth the cost? Hyd Proc, 48(4), 149-150 (1969).

Cetinbudaklar, A.G., and Jameson, G.J., Mechanism of flooding in vertical countercurrent two-phase flow, Chem Eng Sci, 24(11), 1669-1680 (1969).

Chand, P., Pressure drop across fixed beds of spherical particles, Brit Chem Eng, 14(3), 329-330 (1969).

Coggan, G.C., and Bourne, J.R., Design of gas absorbers with heat effects, Trans IChemE, 47, T96-106, T160-165 (1969).

Coughlin, R.W., Effect of liquid-packing surface interaction on gas absorption and flooding in a packed column, AIChEJ, 15(5), 654-659 (1969).

Eastwood, J.; Matzen, E.J.P.; Young, M.J., and Epstein, N., Random loose porosity of packed beds, Brit Chem Eng, 14(11), 1542-1545 (1969).

Formisano, F.A., Method quickly troubleshoots packed-column problems, Chem Eng, 3 Nov, 108-110 (1969).
Freedman, W., and Davidson, J.F., Hold-up and liquid circulation in bubble columns, Trans IChemE, 47, T251-262 (1969).
Gilath, C.; Naphthali, L.M., and Resnick, W., Transient response of a packed column to changes in liquid and gas flowrates, Ind Eng Chem Proc Des Dev, 8(3), 324-333 (1969).
Gunn, D.J., Theory of axial and radial dispersion in packed beds, Trans IChemE, 47, T351-359 (1969).
Gunn, D.J., and Pryce, C., Dispersion in packed beds, Trans IChemE, 47, T341-350 (1969).
Huckabay, H.K., and Garrison, R.L., Packed tower transfer rate by graphs, Hyd Proc, 48(6), 153-158 (1969).
Kalika, P.W., Effect of water recirculation and steam plumes on scrubber design, Chem Eng, 28 July, 133-138 (1969).
Markin, A., and Sommerfeld, J.T., Packing pressure drop by computer, Hyd Proc, 48(9), 206-208 (1969).
Mehta, D., and Hawley, M.C., Wall effect in packed columns, Ind Eng Chem Proc Des Dev, 8(2), 280-282 (1969).
Pasiuk,-Bronikowska, W., Liquid-film coefficient for physical absorption and effective interfacial area in sieve-plate column by the chemical method, Chem Eng Sci, 24(7), 1139-1148 (1969).
Prahl, W.H., Pressure drop in packed columns, Chem Eng, 11 Aug, 89-96 (1969).
Rubac, R.E.; McDaniel, R., and Holland, C.D., Packed distillation columns and absorbers at steady-state operation, AIChEJ, 15(4), 568-575 (1969).
Sharma, M.M.; Mashelkar, R.A., and Mehta, V.D., Mass transfer in plate columns, Brit Chem Eng, 14(1), 70-76 (1969).
Strek, E.; Werner, J., and Paniuticz, A., Effect of free cross-section on flooding point of packing on a grid, Int Chem Eng, 9(3), 464-470 (1969).
Surowiec, A.J., Estimate reboiled absorbers, Hyd Proc, 48(9), 211-214 (1969).
Svonava, M.; Palka, J., and Gregor, M., Advantages of foam-type scrubbing operation, Int Chem Eng, 9(3), 480-485 (1969).
Umeda, T., Design of an absorber-stripper system, Ind Eng Chem Proc Des Dev, 8(3), 308-317 (1969).
Walker, G.J., Design sour-water strippers quickly, Hyd Proc, 48(6), 121-124 (1969).

1970

Bennett, A., and Goodridge, F., Hydrodynamic and mass transfer studies in packed absorption columns, Trans IChemE, 48, T232-244 (1970).
Camerinelli, I., Sizing dehydrator-absorbers, Hyd Proc, 49(2), 103-104 (1970).
England, R., and Gunn, D.J., Dispersion, pressure drop, and chemical reaction in packed beds of cylindrical particles, Trans IChemE, 48, T265-275 (1970).
Furzer, I.A., and Michell, R.W., Liquid-phase dispersion in packed beds with two-phase flow, AIChEJ, 16(3), 380-385 (1970).
Kafarov, V.V., and Falin, V.A., Procedure for designing absorption processes with chemical reactions in section bubbling reactors, Int Chem Eng, 10(3), 418-422 (1970).

Mashelkar, R.A., Bubble columns, Brit Chem Eng, 15(10), 1297-1304 (1970).
McDaniel, R., and Holland, C.D., Modeling of packed absorbers at unsteady-state operation, Chem Eng Sci, 25(8), 1283-1296 (1970).
Mehta, D.S., and Calvert, S., Performance of a porous-plate column, Brit Chem Eng, 15(6), 781-785 (1970).
Mehta, K.C., and Sharma, M.M., Mass transfer in spray columns, Brit Chem Eng, 15(11), 1440-1444; 15(12), 1556-1558 (1970).
Popov, V.V.; Gerasimov, A.D., and Filippov, E.P., Efficiency of mass-transfer equipment, Brit Chem Eng, 15(4), 535 (1970).
Prahl, W.H., Liquid density distorts packed column correlation, Chem Eng, 2 Nov, 109-112 (1970).
Reith, T., Interfacial area and scaling-up of gas-liquid contactors, Brit Chem Eng, 15(12), 1559-1563 (1970).
Sycheva, A.M., et al., Resistance of a packed bed during radially directed flow, Int Chem Eng, 10(1), 66-70 (1970).
Zanker, A., Spray-tower volume fractions calculated with nomograph, Chem Eng, 30 Nov, 96 (1970).

1971

Buchanan, J., Liquid feed distribution in packed towers, Chem Eng Sci, 26(5), 746-749 (1971).
Buchanan, J.E., Axial dispersion and backmixing in packed columns in gas-liquid operations, AIChEJ, 17(3), 746-747 (1971).
Chen, B.H., Cylindrical screen packings, Brit Chem Eng, 16(2), 197-199 (1971).
Danckwerts, P.V., and Rizvi, S.F., The design of gas absorbers, Trans IChemE, 49, 124-127 (1971).
Goto, S.; Kitai, A., and Ozaki, A., Operating conditions for preventing back flow in a perforated-plate column with cocurrent gas-liquid flow, Int Chem Eng, 11(3), 542-547 (1971).
Gunn, D.J., Axial dispersion in packed beds, Trans IChemE, 49, 109-113 (1971).
Hoppe, K., Developments in column equipment, Brit Chem Eng, 16(9), 807-813 (1971).
Klimecek, R.; Krivsky, Z., and Veverka, V., Absorber packed with orientated helices, Brit Chem Eng, 16(11), 1018-1020 (1971).
Kovacs, K., Certain problems in absorption of multicomponent gas mixtures, Int Chem Eng, 11(4), 645-654 (1971).
Roche, E.C., General design algorithm for multistage countercurrent equilibrium processes, Brit Chem Eng, 16(9), 821-824 (1971).
Rowland, C.H., and Grens, E.A., Design absorbers using real stages, Hyd Proc, 50(9), 201-204 (1971).
Shulman, H.L.; Mellish, W.G., and Lyman, W.H., Performance of packed columns, AIChEJ, 17(3), 631-640 (1971).
Van der Merwe, D.F., and Gauvin, W.H., Velocity and turbulence measurements of air flow through a packed bed, AIChEJ, 17(3), 519-528 (1971).
Watson, J.S., and Cochran, H.D., Simple method for estimating effect of axial backmixing on countercurrent column performance, Ind Eng Chem Proc Des Dev, 10(1), 83-85 (1971).

1972

Gardner, R.G., Heat-transfer characteristics of an integrally-cooled sieve plate, Brit Chem Eng, 17(7), 619-624 (1972).

Kafarov, V.V.; Shestopalov, V.V., and Belkov, V.P., Longitudinal mixing of liquid in a plate column with sieve trays for absorption of nitric oxide, Int Chem Eng, 12(2), 257-260 (1972).

Karanth, N.G., and Kuloor, N.R., Optimisation of inlet temperature of absorbent in sulphur trioxide absorber, Proc Tech Int, 17(11), 863-864 (1972).

Krauze, R., and Serwinski, M., Evaluation of hydrodynamic similarity of fluid flow through a packed bed with fractionation in a packed column, Int Chem Eng, 12(4), 679-685 (1972).

Litz, W.J., Design of gas distributors, Chem Eng, 13 Nov, 162-166 (1972).

MacDonald, J.O.S., Developments in column internals, Brit Chem Eng, 17(7), 631, 633 (1972).

Michell, R.W., and Furzer, I.A., Trickle flow in packed beds, Trans IChemE, 50, 334-342 (1972).

Miyauchi, T., et al., Mass-transfer coefficients in packed beds at low Reynolds numbers, Int Chem Eng, 12(2), 360-366; 373-378 (1972).

Sakata, N., and Prados, J.W., Dynamics of a packed-bed gas absorber by pulse response technique, AIChEJ, 18(3), 572-581 (1972).

Szekely, J., and Mendrykowski, J., Flooding criteria for liquids with high density and high interfacial tension, Chem Eng Sci, 27(5), 959-964 (1972).

Thirkell, H., Reduce carbon dioxide absorption costs by combining processes, Hyd Proc, 51(1), 115-118 (1972).

Wankat, P.C., Graphical methods for nonisothermal absorption, Ind Eng Chem Proc Des Dev, 11(2), 302-307 (1972).

Zanker, A., Finding linear velocities through packed columns, Chem Eng, 25 Dec, 84-85 (1972).

Zarycki, R., Effect of gas flow velocity on liquid distribution in packed column, Int Chem Eng, 12(1), 88-92 (1972).

Zenz, F.A., Designing gas-absorption towers, Chem Eng, 13 Nov, 120-138 (1972).

1973

Bekirov, T.M., et al., Design of industrial absorption system with recycle, Int Chem Eng, 13(3), 441-443 (1973).

Brignole, E.A.; Zacharonek, G., and Mangosio, J., Liquid distribution in packed columns, Chem Eng Sci, 28(5), 1225-1230 (1973).

Egberongbe, S.A., Design of compact mass-transfer packing for maximum efficiency, Proc Engng, Feb, 82-85 (1973).

Gianetto, A.; Specchia, V., and Baldi, G., Absorption in packed towers with concurrent downward high-velocity flows, AIChEJ, 19(5), 916-922 (1973).

Hughmark, G.A., Gas-phase mass transfer in packed columns, AIChEJ, 19(6), 1258-1259 (1973).

Juvekar, V.A., and Sharma, M.M., Chemical methods to determine liquid-side mass transfer coefficient and effective interfacial area in gas-liquid contactors, Chem Eng Sci, 28(3), 976-978 (1973).

Mandelbaum, J.A., and Bohm, U., Mass transfer in packed beds at low Reynolds numbers, Chem Eng Sci, 28(2), 569-576 (1973).
Onda, K.; Takeuchi, H.; Maeda, Y., and Takeuchi, N., Liquid distribution in a packed column, Chem Eng Sci, 28(9), 1677-1694 (1973).
Pritchard, D.W., and Tiley, P.F., Construction and operation of laboratory differential vapour-absorption column, Chem Eng Sci, 28(10), 1839-1846 (1973).
Raal, J.D., and Khurana, M.K., Design of packed absorbers with large heat effects, Proc Tech Int, June, 267-269 (1973).
Ruziska, P.A., Packings for hot carbonate absorption systems, Chem Eng Prog, 69(2), 67-70 (1973).
Sahay, B.N., and Sharma, M.M., Absorption in packed bubble columns, Chem Eng Sci, 28(12), 2245-2256 (1973).
Sahay, B.N., and Sharma, M.M., Effective interfacial area and liquid and gas-side mass-transfer coefficients in a packed column, Chem Eng Sci, 28(1), 41-48 (1973).
Sisson, W., How to determine MEA circulation rates, Chem Eng, 26 Nov, 98-100 (1973).
Sobotka, V., Modelling of absorption column for nitric acid manufacture, Int Chem Eng, 13(4), 718-728 (1973).
Tamir, A., and Taitel, Y., Absorption mass transfer in presence of axial diffusion in convective-diffusive flows, Chem Eng Sci, 28(11), 1921-1930 (1973).
Tichy, J., Liquid hold-up in gas-liquid countercurrent flow through a packed bed, Chem Eng Sci, 28(2), 655-658 (1973).
Uchida, S., and Wen, C.Y., Gas absorption by alkaline solutions in venturi scrubber, Ind Eng Chem Proc Des Dev, 12(4), 437-443 (1973).
Ufford, R.C., and Perona, J.J., Liquid-phase mass transfer with concurrent flow through packed towers, AIChEJ, 19(6), 1223-1226 (1973).
Zanker, A., Specific surface and voids fraction for tower packings, Chem Eng, 3 Sept, 126 (1973).

1974

Andrieu, J., Pressure drop below the load zone in a raschig ring packed column with countercurrent air-water flows, Chem Eng J, 7(3), 257-260 (1974).
Bridgwater, J., and Scott, A.M., Statistical models of packing: Application to gas absorption and solids mixing, Trans IChemE, 52, 317-324 (1974).
Brooks, P.C., Flooding packed towers for more capacity, Chem Eng, 18 Feb, 152 (1974).
Colquhoun-Lee, I., and Stepanek, J., Mass transfer in single phase flow in packed beds, Chem Engnr, Feb, 108-111 (1974).
Guerreri, G., and King, C.J., Designing falling film absorbers, Hyd Proc, 53(1), 131-136 (1974).
Hills, J.H., Radial non-uniformity of velocity and voidage in a bubble column, Trans IChemE, 52, 1-9 (1974).
Hopke, S.W., and Lin, C.J., Improve absorber predictions by better equations of state, Hyd Proc, 53(6), 136-142 (1974).
Hutton, B.E.T., and Leung, L.S., Cocurrent gas-liquid flow in packed columns, Chem Eng Sci, 29(8), 1681-1686 (1974).

Hutton, B.E.T.; Leung, L.S.; Brooks, P.C., and Nicklin, D.J., Flooding in packed columns, Chem Eng Sci, 29(2), 493-500 (1974).

Kabakov, M.I., and Matusevich, A.A., Effect of initial distribution of wetting liquid on efficiency of packed absorbers, Int Chem Eng, 14(3), 435-440 (1974).

Knight, M.W.; McDermott, P.G., and Cooper, D., Composite structures for hot gas scrubbers, Chem Eng Prog, 70(9), 89 (1974).

Linek, V.; Stoy, V.; Machon, V., and Krivsky, Z., Increasing the effective interfacial area in plastic packed absorption columns, Chem Eng Sci, 29(9), 1955-1960 (1974).

Oorts, A.J., and Hellinckx, L.J., Modified time-delay model for flow in packed columns, Chem Eng J, 7(2), 147-154 (1974).

Puranik, S.S., and Vogelpohl, A., Effective interfacial area in irrigated packed columns, Chem Eng Sci, 29(2), 501-508 (1974).

Sakol, S.L., and Schwartz, R.A., Construction materials for wet scrubbers, Chem Eng Prog, 70(8), 63-68 (1974).

Shende, B.W., and Sharma, M.M., Mass transfer in cocurrent packed columns, Chem Eng Sci, 29(8), 1763-1772 (1974).

Specchia, V.; Sicardi, S., and Gianetto, A., Absorption in packed towers with concurrent upward flow, AIChEJ, 20(4), 646-653 (1974).

Veverka, V., and Krivsky, Z., Distribution of liquid in an absorber packed with orientated helices, Chem Eng Commns, 1(5), 217-220 (1974).

Woodburn, E.T., Gas-phase axial mixing at extremely high irrigation rates in a large packed absorption tower, AIChEJ, 20(5), 1003-1009 (1974).

1975

Danckwerts, P.V., and Alper, E., Design of gas absorbers, Trans IChemE, 53, 34-40 (1975).

Eckert, J.S., How tower packings behave, Chem Eng, 14 April, 70-76 (1975).

Goto, S.; Levec, J., and Smith, J.M., Mass transfer in packed beds with two-phase flow, Ind Eng Chem Proc Des Dev, 14(4), 473-478 (1975).

Lockett, M.J., and Kirkpatrick, R.D., Ideal bubbly flow and actual flow in bubble columns, Trans IChemE, 53, 267-273 (1975).

Mashelkar, R.A., and Ramachandran, P.A., Longitudinal dispersion in circulation dominated bubble columns, Trans IChemE, 53, 274-277 (1975).

Merchuk, J.C., The Danckwerts-Gillham method of gas absorber design, AIChEJ, 21(4), 815-817 (1975).

Miyauchi, T., and Kikuchi, T., Axial dispersion in packed beds, Chem Eng Sci, 30(3), 343-348 (1975).

Nemunaitis, R.R., Heat transfer in packed towers, Chem Eng Prog, 71(8), 60-67 (1975).

Prchlik, J., et al., Liquid distribution in columns and reactors filled with irrigated dumped packing, Int Chem Eng, 15(1), 125-136 (1975).

Saada, M.Y., Fluid mechanics of cocurrent two-phase flow in packed beds: Pressure drop and liquid holdup studies, Periodica Polytechnica, 19, 317-338 (1975).

Sylvester, N.D., and Pitayagulsarn, P., Mass transfer for two-phase cocurrent downflow in a packed bed, Ind Eng Chem Proc Des Dev, 14(4), 421-427 (1975).

Szekely, J., and Poveromo, J.J., Flow maldistribution in packed beds, AIChEJ, 21(4), 769-775 (1975).
Verma, S.L., and Delancey, G.B., Thermal effects in gas absorption, AIChEJ, 21(1), 96-102 (1975).
Zanker, A., Simplified calculations for packed towers, Chem Eng, 29 Sept, 100-101 (1975).

1976
Achwal, S.K., and Stepanek, J.B., Holdup profiles in packed beds, Chem Eng J, 12(1), 69-76 (1976).
Alper, E., and Danckwerts, P.V., Laboratory scale-model of a complete packed-column absorber, Chem Eng Sci, 31(7), 599-608 (1976).
Charpentier, J.C., Recent progress in two-phase gas-liquid mass transfer in packed beds, Chem Eng J, 11(3), 161-182 (1976).
Endrst, M.; Cervenka, J., and Kolar, V., Absorption on horizontal expanded-metal packings, Int Chem Eng, 16(1), 118-121 (1976).
Hills, J.H., Operation of a bubble column at high throughputs, Chem Eng J, 12(2), 89-100 (1976).
Leung, L.S., New method for flooding packed towers, Chem Eng, 20 Dec, 88 (1976).
Mathur, V.K., and Wellek, R.M., Effect of axial dispersion on interphase mass transfer in packed absorption columns, Can JCE, 54, 90-100 (1976).
Miller, J.D., and Rehm, T.R., Packed column mass-transfer coefficients for concurrrent and countercurrent flow, Chem Eng Educ, 10(2), 84-89,102-104 (1976).
Miyauchi, T.; Kataoka, H., and Kikuchi, T., Gas-film coefficient of mass transfer in low Peclet number region for sphere packed beds, Chem Eng Sci, 31(1), 9-14 (1976).
Sicardi, S., and Baldi, G., Model for mass transfer in packed towers: Gas-phase controlling resistance, Chem Eng Sci, 31(8), 651-656 (1976).
Tanoka, Y., and Inoue, I., Radial mixing in a packed column, Int Chem Eng, 16(3), 479-485 (1976).
Zanker, A., Determine optimum diameter of packed columns, Processing, June, 18 (1975); April, 10 (1976).

1977
Alonso, J.R.F., Calculate droplet size for scrubber design, Hyd Proc, 56(1), 141-142 (1977).
Badr El-Din, A.A.; El-Halwagi, M.M., and Saleh, M.A., Liquid flow distribution in two-phase countercurrent packed column, Chem Eng Sci, 32(3), 343-346 (1977).
Douglas, J.M., Quick estimates for design of plate-type gas absorbers, Ind Eng Chem Fund, 16(1), 131-138 (1977).
Dunn, W.E.; Vermeulen, T.; Wilke, C.R., and Word, T.T., Longitudinal dispersion in packed gas-absorption columns, Ind Eng Chem Fund, 16(1), 116-124 (1977).
Gardiner, M.C.S., New multiphase contacting device, Processing, Aug, 37-39 (1977).

Groenhof, H.C., Scaling-up of packed columns, Chem Eng J, 14(3), 181-204 (1977).

Imura, H.; Kusuda, H., and Funatsu, S., Flooding velocity in countercurrent annular two-phase flow, Chem Eng Sci, 32(1), 79-88 (1977).

Klykov, M.V.; Rogozin, V.I.; Svinukhov, A.G., and Panchenkov, G.M., Interfacial contact area in countercurrent mass-transfer apparatus with netted packing, Int Chem Eng, 17(1), 112-115 (1977).

Linek, V.; Krivsky, Z., and Hudec, P., Effective interfacial area in plastic-packed absorption columns, Chem Eng Sci, 32(3), 323-326 (1977).

Manieh, A.A., Comparing equilibrium stages with transfer units, Chem Eng, 9 May, 163-164 (1977).

Meier, W.; Stoecker, W.D., and Weinstein, B., Performance of a new high-efficiency packing, Chem Eng Prog, 73(11), 71-77 (1977).

Mottola, A.C., Diffusivities streamline wet scrubber design, Chem Eng, 19 Dec, 77-80 (1977).

Nguyen, V.T., and Spedding P.L., Holdup in two-phase, gas-liquid flow, Chem Eng Sci, 32(9), 1003-1022 (1977).

Picciotti, M., Design quench-water towers, Hyd Proc, 56(6), 163-170 (1977).

Pillai, K.K., Voidage variation at the wall of a packed bed of spheres, Chem Eng Sci, 32(1), 59-62 (1977).

Simpson, S.G., and Lynn, S., Vacuum-spray stripping of sparingly soluble gases from aqueous solutions, AIChEJ, 23(5), 666-679 (1977).

Specchia, V., and Baldi, G., Pressure drop and liquid holdup for two-phase concurrent flow in packed beds, Chem Eng Sci, 32(5), 515-524 (1977).

Stockar, U., and Wilke, C.R., Rigorous and short-cut design calculations for gas absorption involving large heat effects, Ind Eng Chem Fund, 16(1), 88-103 (1977).

Thibodeaux, L.J., et al., Mass transfer units in single and multiple stage packed bed, cross-flow devices, Ind Eng Chem Proc Des Dev, 16(3), 325-330 (1977).

1978

Bemer, G.G., and Kalis, G.A.J., New method to predict hold-up and pressure drop in packed columns, Trans IChemE, 56, 200-204 (1978).

Bemer, G.G., and Zuiderweg, F.J., Radial liquid spread and maldistribution in packed columns under different wetting conditions, Chem Eng Sci, 33(12), 1637-1644 (1978).

Bhavaraju, S.M.; Russell, T.W.F., and Blanch, H.W., Design of gas sparged devices for viscous liquid systems, AIChEJ, 24(3), 454-466 (1978).

Botton, R.; Cosserat, D, and Charpentier, J.C., Influence of column diameter and high gas throughputs on the operation of a bubble column, Chem Eng J, 16(2), 107-116 (1978).

Buffham, B.A., and Rathor, M.N., The influence of viscosity on axial mixing in trickle flow in packed beds, Trans IChemE, 56, 266-273 (1978).

Calvert, S., Field evaluation of fine particle scrubbers, Chem Engnr, June, 485-490 (1978).

Colquhoun-Lee, I., and Stepanek, J.B., Solid-liquid mass transfer in two phase co-current upward flow in packed beds, Trans IChemE, 56, 136-144 (1978).

Eastham, I.E.; van den Broek, D., and Miertschin, G.N., Advances in tower packing design, Chem Eng Prog, 74(4), 61-65 (1978).
Feintuch, H.M., and Treybal, R.E., Design of adiabatic packed towers for gas absorption and stripping, Ind Eng Chem Proc Des Dev, 17(4), 505-513 (1978).
Gunn, D.J., Liquid distribution and redistribution in packed columns, Chem Eng Sci, 33(9), 1211-1232 (1978).
Hatton, T.A., and Woodburn, E.T., Mixing and mass transfer at high liquid rates in steady-state counterflow operation of packed columns, AIChEJ, 24(2), 187-192 (1978).
Ho, S.P.; Fisher, J.A., and Im, U.K., Modeling and simulation of oil/water quench system in an olefin unit, Ind Eng Chem Proc Des Dev, 17(1), 82-87 (1978).
Hodge, F.G., High performance alloys make wet scrubbers work, Chem Eng Prog, 74(10), 84-88 (1978).
Hoppe, K.; Keller, J., and Krell, L., A new high-performance packing for gas/liquid contact, Chem Engnr, Feb, 110-114 (1978).
Legrys, G.A., Power demand and mass transfer capability of mechanically agitated gas-liquid contactors and their relationship to air-lift fermenters, Chem Eng Sci, 33(1), 83-86 (1978).
Linek, V.; Benes, P.; Sinkule, J., and Krivsky, Z., Simultaneous determination of mass transfer coefficient and of gas and liquid axial dispersions and holdups in a packed absorption column by dynamic response method, Ind Eng Chem Fund, 17(4), 298-305 (1978).
Mottola, A.C., and Fellinger, L.L., Packed scrubber design, Chem Eng Prog, 74(10), 94-95 (1978).
Nguyen, H.X., Computer program expedites packed-tower design, Chem Eng, 20 Nov, 181-184 (1978).
Shah, Y.T.; Ratway, C.A., and McIlvried, H.G., Back-mixing characteristics of a bubble column with vertically suspended tubes, Trans IChemE, 56, 107-112 (1978).
Stichlmair, J., and Mersmann, A., Dimensioning plate columns for absorption and rectification, Int Chem Eng, 18(2), 223-237 (1978).

1979

Alexander, B.F.; Shah, Y.T., and Wilson, J.H., Radial dispersion in vertically suspended packed baskets inside a bubble column, Trans IChemE, 57, 252-255 (1979).
Alper, E., Measurement of effective interfacial area in a packed column absorber by chemical methods, Trans IChemE, 57, 64-66 (1979).
Brown, D.E., and Halsted, D.J., Liquid-phase mixing model for stirred gas-liquid contactor, Chem Eng Sci, 34(6), 853-860 (1979).
Burghardt, A., and Bartelmus, G., Experimental determination of longitudinal dispersion in two-phase flow through packing, Chem Eng Sci, 34(3), 405-412 (1979).
Counce, R.M., and Perona, J.J., Gaseous nitrogen oxide absorption in a sieve-plate column, Ind Eng Chem Fund, 18(4), 400-406 (1979).
Dixon, A.G., and Cresswell, D.L., Theoretical prediction of effective heat-transfer parameters in packed beds, AIChEJ, 25(4), 663-676 (1979).

Douglas, J.M., Equations spur design of plate-type gas absorbers, Chem Eng, 13 Aug, 135-139 (1979).
Johnston, I.W., and Pollitt, R.R., Horizontal two-phase flow of liquid water/water vapour through packed beds, Trans IChemE, 57, 256-261 (1979).
Kuk, M.S., Key design variables in packed towers, Chem Eng Prog, 75(5), 68-71 (1979).
Lock, J., Column packings, Processing, Oct, 51, 53, 55 (1979).
McCarthy, J.E., Select the right scrubber, Hyd Proc, 58(8), 129-133 (1979).
Meister, D.; Post, T.; Dunn, I.J., and Bourne, J.R., Design and characterization of a multistage mechanically stirred column absorber, Chem Eng Sci, 34(12), 1367-1374 (1979).
Mink, W.H., Calculator program aids quench-tower design, Chem Eng, 3 Dec, 95-98 (1979).
Mitchell, M.G., and Perona, J.J., Gas-liquid interfacial areas for high-porosity tower packings in concurrent downward flow, Ind Eng Chem Proc Des Dev, 18(2), 316-318 (1979).
Nguyen, H.X., Quicker answers for packed beds, Hyd Proc, 58(3), 115-117 (1979).
Nguyen, H.X., Calculating actual plates in absorbers and strippers, Chem Eng, 9 April, 113-117 (1979).
Refre, A.E., and Hellinckx, L.J., Flow conditions in a packed bed through local measurements, Chem Eng J, 18(1), 1-12 (1979).
Roes, A.W.M., and van Swaaij, W.P.M., Hydrodynamic behaviour of a gas-solid countercurrent packed column at trickle flow, Chem Eng J, 17(2), 81-90 (1979).
Roes, A.W.M., and van Swaaij, W.P.M., Mass transfer in a gas-solid packed column at trickle flow, Chem Eng J, 18(1), 13-38 (1979).
Sawant, S.B.; Pangarkar, V.G., and Joshi, J.B., Gas hold-up and mass transfer characteristics of packed bubble columns, Chem Eng J, 18(2), 143-150 (1979).
Trasi, P., and Khang, S.J., Residence-time-distribution studies on gas-liquid countercurrent packed column with intermittent voids along the axis, Ind Eng Chem Fund, 18(3), 256-260 (1979).
Wild, N.H., Calculator program for sour-water-stripper design, Chem Eng, 12 Feb, 103-113 (1979).
Wolff, H.J.; Radeke, K.H., and Gelbin, D., Heat and mass transfer in packed beds, Chem Eng Sci, 34(1), 101-108 (1979).
Zanker, A., Nomograph for packed tower and column calculations, Proc Engng, Nov, 139-141 (1979).

1980

Albrecht, J.J.; Kershenbaum, L.S., and Pyle, D.L., Identification and linear multivariable control in an absorption-desorption pilot plant, AIChEJ, 26(3), 496-504 (1980).
Burghardt, A., and Bartelmus, G., Experimental studies of longitudinal dispersion in two-phase flow in packed columns, Int Chem Eng, 20(1), 117-136 (1980).

Clements, L.D., and Schmidt, P.C., Dynamic liquid holdup and pressure drop in two-phase cocurrent downflow in packed beds (air-silicone oil systems), AIChEJ, 26(2), 314-319 (1980).

Cornelissen, A.E., Simulation of absorption of hydrogen sulphide and carbon dioxide into aqueous alkanolamines in tray and packed columns, Trans IChemE, 58, 242-251 (1980).

Davies, J.T., Interfacial effects of gas transfer to liquids, Chem & Ind, 1 March, 189-193 (1980).

Dolejs, V., and Lecjaks, Z., Pressure drop in flow of Newtonian liquid through a fixed randomly packed bed of spherical particles, Int Chem Eng, 20(3), 466-473 (1980).

Field, R.W., and Davidson, J.F., Axial dispersion in bubble columns, Trans IChemE, 58, 228-236 (1980).

Gunn, D.J., Theory of liquid-phase dispersion in packed columns, Chem Eng Sci, 35(12), 2405-2414 (1980).

Hruby, J., Variation of pressure fluctuations and Murphree efficiency on a slot tray without weirs in an absorption column, Int Chem Eng, 20(4), 673-681 (1980).

Hughmark, G.A., Mass transfer and flooding in wetted-wall and packed columns, Ind Eng Chem Fund, 19(4), 385-389 (1980).

Khoury, F.M., Simulate absorbers by successive iteration, Chem Eng, 29 Dec, 51-53 (1980).

Leva, M., Performance of a new tower packing, Chem Eng Prog, 76(9), 73-77 (1980).

Mahajani, V.V., and Sharma, M.M., Mass transfer in packed columns with different packings, Chem Eng Sci, 35(4), 941-948 (1980).

Mangers, R.J., and Ponter, A.B., Liquid-phase resistance to mass transfer in a laboratory absorption column packed with glass and PTFE rings, Chem Eng J, 19(2), 139-152 (1980).

Mangers, R.J., and Ponter, A.B., Effect of viscosity on liquid-film resistance to mass transfer in packed column, Ind Eng Chem Proc Des Dev, 19(4), 530-537 (1980).

Merchuk, J.C., Mass transfer characteristics of a column with small plastic packings, Chem Eng Sci, 35(3), 743-745 (1980).

Pancuska, V.I., Calculator program for designing packed towers, Chem Eng, 5 May, 113-114 (1980).

Savage, D.W.; Astarita, G., and Joshi, S., Chemical absorption and desorption of carbon dioxide from hot carbonate solutions, Chem Eng Sci, 35(7), 1513-1522 (1980).

Ter Ver, K.J.R.; Van der Klooster, H.W., and Drinkenburg, A.A.H., Influence of liquid distribution on the efficiency of a packed column, Chem Eng Sci, 35(3), 759-761 (1980).

Thibodeaux, L.J., Fluid dynamic observations on packed crossflow cascade tower at high loadings, Ind Eng Chem Proc Des Dev, 19(1), 33-40 (1980).

Uchida, S.; Suzuki, T., and Maejima, H., Flooding condition of turbulent contact absorber, Can JCE, 58, 406-409 (1980).

Waggoner, R.C.; Calvin, S.J., and Mills, T.K., Solve absorption and leaching problems on a hand calculator, Chem Eng, 14 July, 119-123 (1980).

Wild, G., and Schlunder, E.U., Multicomponent gas absorption in bubble columns, Chem Eng Sci, 35(1), 506-511 (1980).

1981

Kister, H.Z., Column internals, Chem Eng, 19 May, 138-142; 28 July, 79-83; 8 Sept, 119-123; 17 Nov, 283-285; 29 Dec, 55-60 (1980); 9 Feb, 107-109; 6 April, 97-100 (1981).

Knaebel, K.S., Simplified sparger design, Chem Eng, 9 March, 116-118 (1981).

Liapis, A.I., and McAvoy, T.J., Transient solutions for a class of hyperbolic countercurrent distributed heat and mass transfer systems, Trans IChemE, 59, 89-94 (1981).

Mink, W.H., Hole-area distribution for liquid spargers, Chem Eng, 17 Nov, 277-281 (1980); 6 April, 93-95 (1981).

Pancuska, V.I., Packed-bed flow computed, Hyd Proc, 60(3), 103-104 (1981).

Patil, V.K., and Sharma, M.M., Packed tube columns: Hydrodynamics and effective interfacial area (Pall rings and multifilament wire-gauze packings), Can JCE, 59, 606-613 (1981).

Patwardhan, V.S., and Shrotri, V.R., Mass-transfer coefficient between static and dynamic hold-ups in packed column, Chem Eng Commns, 10(6), 349-356 (1981).

Rizzuti, L.; Augugliaro, V., and Cascio, G.L., Influence of liquid viscosity on effective interfacial area in packed columns, Chem Eng Sci, 36(6), 973-978 (1981).

Sarma, H., How to size gas scrubbers, Hyd Proc, 60(9), 251-255 (1981).

Shirato, M., et al., Gravitational drainage of granular packed bed, Int Chem Eng, 21(2), 294-303 (1981).

Skold, J.O., Energy savings in cooling tower packings, Chem Eng Prog, 77(10), 48-53 (1981).

1982

Bolles, W.L., and Fair, J.R., Improved mass-transfer model enhances packed-column design, Chem Eng, 12 July, 109-116 (1982).

Diab, S., and Maddox, R.N., Absorption, Chem Eng, 27 Dec, 38-56 (1982).

Fedkiw, P.S., and Newman, J., Mass-transfer coefficients in packed beds at very low Reynolds numbers, Int J Heat Mass Trans, 25(7), 935-944 (1982).

Horner, B.; Abbenseth, R.J., and Dialer, K., Prediction of mass-transfer coefficients in absorption from the measured turbulence properties of the liquid, Int Chem Eng, 22(2), 226-234 (1982).

Joshi, J.B., and Sharma, M.M., A circulation cell model for bubble columns, Trans IChemE, 57, 244-251 (1979); 60, 255-256 (1982).

Lewis, D.A.; Field, R.W.; Xavier, A.M., and Edwards, D., Heat transfer in bubble columns, Trans IChemE, 60, 40-47 (1982).

Liu, C.P.; McCarthy, G.E., and Tien, C.L., Flooding in vertical gas-liquid countercurrent flow through multiple short paths, Int J Heat Mass Trans, 25(9), 1301-1312 (1982).

Mahajani, V.V., Figuring packed-tower diameter, Chem Eng, 20 Sept, 132 (1982).

Miconnet, M.; Guigon, P., and Large, J.F., Scrubbing of acid gases in columns with fixed or mobile packings, Int Chem Eng, 22(1), 133-142 (1982).

Morsi, B.I., et al., Hydrodynamics and interfacial areas in downward cocurrent gas-liquid flow through fixed beds, Int Chem Eng, 22(1), 142-152 (1982).

Niranjan, K.; Sawant, S.B.; Joshi, J.B., and Pangarkar, V.G., Countercurrent absorption using wire gauze packings, Chem Eng Sci, 37(3), 367-374 (1982).

Ponter, A.B., and Au-Yeung, P.H., Estimation of liquid-film mass-transfer coefficients for columns randomly packed with partially wetted rings, Can JCE, 60, 94-99 (1982).

Ponter, A.B., and Yekta-Fard, M., Prediction of flooding in columns packed with polymer packings, Chem Eng Sci, 37(10), 1587-1589 (1982).

Thomas, W.J., and Ogboja, O., Mass transfer studies on sieve trays with one-inch diameter perforations, Ind Eng Chem Proc Des Dev, 21(2), 217-222 (1982).

Ward, H.C., and Sommerfeld, J.T., New equation for flooding, Hyd Proc, Oct, 99-100 (1982).

1983

Andrew, S.P.S., Liquid-film limited physical solution of gases in packed absorbers, Chem Eng Sci, 38(1), 9-20 (1983).

Au-Yeung, P.H., and Ponter, A.B., Estimation of liquid-film mass-transfer coefficients for randomly packed absorption columns (review paper), Can JCE, 61(4), 481-493 (1983).

Chen, G.K.; Kitterman, L., and Shieh, J., High-efficiency packing for product separation, Chem Eng Prog, 79(9), 46-49 (1983).

Chen, G.K.; Kitterman, L., and Shieh, J.H., Performance of high-efficiency packing, Chem Eng Prog, 79(11), 49-51 (1983).

Chen, J.J.J.; Leung, Y.C., and Spedding, P.L., Simple method of measuring voidage in a bubble column, CER&D, 61, 325-328 (1983).

Niranjan, K.; Pangarkar, V.G., and Joshi, J.B., Estimate tower pressure drop, Chem Eng, 27 June, 67 (1983).

Patil, V.K., and Sharma, M.M., Hydrodynamics and mass transfer characteristics of co-current downflow packed-tube columns, Can JCE, 61(4), 509-516 (1983).

Rao, V.G.; Ananth, M.S., and Varma, Y.B.G., Hydrodynamics of two-phase cocurrent downflow through packed beds, AIChEJ, 29(3), 467-483 (1983).

Verduijn, W.D., Corrosion of a carbon dioxide absorber tower wall, Plant/Opns Prog, 2(3), 153-160 (1983).

Won, K.W., Sour-water stripper efficiency, Plant/Opns Prog, 2(2), 108-114 (1983).

1984

Albright, M.A., Packed tower distributors tested, Hyd Proc, Sept, 173-177 (1984).

Chen, G.K., Packed column internals, Chem Eng, 5 March, 40-51 (1984).

Chou, T.S., Liquid distribution in a trickle bed with redistribution screens placed in the column, Ind Eng Chem Proc Des Dev, 23(3), 501-505 (1984).

Echarte, R.; Campana, H., and Brignole, E.A., Effective areas and liquid-film mass-transfer coefficients in packed columns, Ind Eng Chem Proc Des Dev, 23(2), 349-354 (1984).

Escoe, A.K., Simple method reduces vortex-induced tower vibration, Hyd Proc, Oct, 81-82 (1984).

Fadel, T.M., Selecting packed column auxiliaries, Chem Eng, 23 Jan, 71-76 (1984).

Furzer, I.A., Liquid dispersion in packed columns, Chem Eng Sci, 39(7), 1283-1314 (1984).

Hixson, T.J., Packed-column design on pocket calculator, Chem Eng, 6 Feb, 95-98 (1984).

Horner, G., Selecting suitable materials for tower packings and internals, Chem Engnr, Nov, 21-24 (1984).

Kelly, R.M.; Rousseau, R.W., and Ferrell, J.K., Design of packed adiabatic absorbers: Physical absorption of acid gases in methanol, Ind Eng Chem Proc Des Dev, 23(1), 102-109 (1984).

Kshirsagar, S.V., and Pangarkar, V.G., Mass transfer characteristics of slotted plastic tower packings, Chem Eng J, 28(3), 179-182 (1984).

Kumar, P.; Devotta, S., and Holland, F.A., Effect of flow ratio on performance of experimental absorption cooling system, CER&D, 62, 194-196 (1984).

Lewis, D.A.; Nicol, R.S., and Thompson, J.W., Measurement of bubble sizes and velocities in gas-liquid dispersions, CER&D, 62, 334-336 (1984).

Linek, V.; Petricek, P.; Benes, P., and Braun, R., Effective interfacial area and liquid side mass transfer coefficients in absorption columns packed with hydrophilised and untreated plastic packings, CER&D, 62, 13-21 (1984).

O'Brien, N.G., and Porter, H.F., Design of gas-liquid contactors, Chem Eng Prog, 80(5), 44-46 (1984).

Ogboja, O., Liquid dispersion and efficiency on a sieve tray with inch diameter perforations, CER&D, 62, 53-56 (1984).

Pinilla, E.A.; Diaz, J.M., and Coca, J., Mass transfer and axial dispersion in a spray tower for gas-liquid contacting, Can JCE, 62(5), 617-622 (1984).

Prasad, C.C., and Prasad, B.V.R.K., Experimental studies on combined state and parameter estimation for a packed column, Chem Eng Sci, 39(1), 185-187 (1984).

Stockar, U.V., and Cevey, P.F., Influence of physical properties of the liquid on axial dispersion in packed columns, Ind Eng Chem Proc Des Dev, 23(4), 717-724 (1984).

Tosun, G., Cocurrent downflow of nonfoaming gas-liquid systems in packed beds, Ind Eng Chem Proc Des Dev, 23(1), 29-39 (1984).

Uchida, S., and Tsuchuja, K., Simulation of spray drying absorber for removal of HCl in flue gas from incinerators. Ind Eng Chem Proc Des Dev, 23(2), 300-307 (1984).

Viswanathan, S.; Gnyp, A.W., and Pierre, C.C.S., Examination of gas-liquid flow in a venturi scrubber, Ind Eng Chem Fund, 23(3), 303-308 (1984).

Vital, T.J.; Grossel, S.S., and Olsen, P.I., Estimating separation efficiency, Hyd Proc, Oct, 55-56; Nov, 147-153; Dec, 75-78 (1984).

1985

Bonsignore, D., et al., Mass transfer in plunging jet absorbers, Chem Eng & Proc, 19(2), 85-94 (1985).

Choe, D.K., and Lee, W.K., Liquid-phase dispersion in packed column with countercurrent two-phase flow, Chem Eng Commns, 34(1), 295-304 (1985).

Dharwadkar, S.V., and Sawant, S.B., Mass transfer and hydrodynamic characteristics of tower packings larger than 25mm nominal size, Chem Eng J, 31(1), 15-22 (1985).

Han, N.W.; Bhakta, J., and Carbonell, R.G., Longitudinal and lateral dispersion in packed beds: Effect of column length and particle size distribution, AIChEJ, 31(2), 277-288 (1985).

Hile, R.L., Absorber/stripper design with a programmable calculator, Chem Eng, 29 April, 53-56 (1985).

Horner, G., Selecting internals for packed columns, Proc Engng, May, 79,81 (1985).

Horsthemke, A., and Schroder, J.J., Wettability of industrial surfaces: Contact angle measurements and thermodynamic analysis, Chem Eng & Proc, 19(5), 277-286 (1985).

Hsu, S.L., Packing pressure drop estimated, Hyd Proc, July, 89-90 (1985).

Ju, D.P., Effect of packing height on mass-transfer coefficient in a packed column, Int Chem Eng, 25(4), 703-710 (1985).

Krebs, C., Gas-side mass transfer in irrigated packed columns, Chem Eng & Proc, 19(2), 95-102; 19(3), 129-142 (1985).

Kulbe, B.; Hoppe, K., and Keller, J., Development and application of column packings for direct heat transfer, Int Chem Eng, 25(3), 474-480 (1985).

Levec, J., and Carbonell, R.G., Longitudinal and lateral thermal dispersion in packed beds, AIChEJ, 31(4), 581-602 (1985).

Lewis, D.A., and Davidson, J.F., Pressure drop for bubbly gas-liquid flow through orifice plates and nozzles, CER&D, 63, 149-156 (1985).

Marini, L.; Clement, K.; Georgakis, C., and Svenson, M.M., Experimental and theoretical investigation of an absorber-stripper pilot plant under nonequilibrium conditions, Ind Eng Chem Fund, 24(3), 296-301 (1985).

Saez, A.E., and Carbonell, R.G., Hydrodynamic parameters for gas-liquid cocurrent flow in packed beds, AIChEJ, 31(1), 52-62 (1985).

Svenson, M.M.; Georgakis, C., and Evans, L.B., Steady-state and dynamic modeling of a gas absorber-stripper system, Ind Eng Chem Fund, 24(3), 288-295 (1985).

Visvanathan, C., and Leung, L.S., Design of a fluidized bed scrubber, Ind Eng Chem Proc Des Dev, 24(3), 677-683 (1985).

Zanetti, R.; Short, H., and Hope, A., Better column distributors, Chem Eng, 27 May, 22-27 (1985).

1986

Ahn, B.J.; Zoulalian, A., and Smith, J.M., Axial dispersion in packed beds with large wall effects, AIChEJ, 32(1), 170-174 (1986).

Bravo, J.L.; Rocha, J.A., and Fair, J.R., Pressure drop in structured packings, Hyd Proc, 65(3), 45-49 (1986).

Daraktschiev, R.; Boev, A., and Kolev, N., Determination of distribution coefficient of horizontal expanded sheet packing, Chem Eng & Proc, 20(2), 73-78 (1986).

Gerrard, M.; Puc, G., and Simpson, E., Optimize the design of wire-mesh separators, Chem Eng, 10 Nov, 91-93 (1986).
Govindarao, V.M.H., and Froment, G.F., Voidage profiles in packed beds of spheres, Chem Eng Sci, 41(3), 533-540 (1986).
Haque, M.W., and Nigam, K.D.P., and Joshi, J.B., Optimum gas sparger design for bubble columns with a low height:diameter ratio, Chem Eng J, 33(2), 63-70 (1986).
Hitch, D.M.; Rousseau, R.W., and Ferrell, J.K., Simulation of continuous-contact separation processes: Multicomponent adiabatic absorption, Ind Eng Chem Proc Des Dev, 25(3), 699-705 (1986).
Hoek, P.J.; Wesselingh, J.A., and Zuiderweg, F.J., Small scale and large scale liquid maldistribution in packed columns, CER&D, 64(6), 431-449 (1986).
Hughmark, G.A., Packed column efficiency fundamentals, Ind Eng Chem Fund, 25(3), 405-410 (1986).
Kolff, S.W., Corrosion of carbon dioxide absorber tower, Plant/Opns Prog, 5(2), 65-72 (1986).
Krishnamurthy, R., and Taylor, R., Absorber simulation and design using a nonequilibrium stage model, Can JCE, 64(1), 96-105 (1986).
Leye, L.D., and Froment, G.F., Rigorous simulation and design of columns for gas absorption and chemical reaction, Comput Chem Eng, 10(5), 493-515 (1986).
Llorens, J.; Mans, C., and Costa, J., Design of absorption columns in the presence of surfactants, Ind Eng Chem Proc Des Dev, 25(1), 305-308 (1986).
Monat, J.P.; McNulty, K.J.; Michelson, I.S., and Hansen, O.V., Accurate evaluation of Chevron mist eliminators, Chem Eng Prog, 82(12), 32-39 (1986).
Nagy, E., et al., Determination of gas-liquid interfacial area of a perforated plate operating with crossflow, Int Chem Eng, 26(4), 637-647 (1986).
Pettit, D.R., Performance of absorption and distillation columns in a lunar environment, Chem Eng Commns, 46, 111-123 (1986).
Schubert, C.N.; Lindner, J.R., and Kelly, R.M., Experimental methods for measuring static liquid holdup in packed columns, AIChEJ, 32(11), 1920-1923 (1986).
Skomorokov, V.B.; Kirillov, V.A., and Baldi, G., Simulation of liquid hydrodynamics in cocurrent two-phase upward flow through a packed bed, Chem Eng J, 33(3), 169-174 (1986).
Som, S.K., and Biswas, G., Dispersion of spray from swirl nozzles, Chem Eng & Proc, 20(4), 191-200 (1986).
Spagnolo, D.A., and Chuang, K.T., Hydraulic, mass transfer and heat transfer performance comparison between ordered bed packing and sieve trays, Can JCE, 64(1), 62-67 (1986).
Spedding, P.L., and Jones, M.T., Mass transfer coefficients in a packed tower: Height, end and solute concentration effects, Chem Eng J, 33(1), 1-18 (1986).
Stephenson, J.L., and Stewart, W.E., Optical measurements of porosity and fluid motion in packed beds, Chem Eng Sci, 41(8), 2161-2170 (1986).

Valderrama, J.O., and Yanez, M.A., A simple and accurate model for the residence time distribution in liquid trickle flow through packed beds, Chem Eng J, 33(2), 109-112 (1986).

Volkov, S.A., et al., Nonuniformity of packed beds and its influence on longitudinal dispersion, Chem Eng Sci, 41(2), 389-398 (1986).

Zioudas, A.P., and Dadach, Z., Absorption rates of carbon dioxide and hydrogen sulphide in sterically hindered amines, Chem Eng Sci, 41(2), 405-409 (1986).

1987

Billet, R., Performance of low-pressure-drop packings, Chem Eng Commns, 54, 93-118 (1987).

Bright, R.L., and Leister, D.A., Gas treaters need clean amines, Hyd Proc, 66(12), 47-48 (1987).

Dylag, M., and Maszek, L., Flat wire-mesh demisters for gases, Int Chem Eng, 27(2), 358-370, 27(4), 737-742 (1987).

Keil, Z.O., and Russell, T.W.F., Design of commercial-scale gas-liquid contactors, AIChEJ, 33(3), 488-496 (1987).

Klingman, K.J., and Lee, H.H., Alternating flow model for mass and heat dispersion in packed beds, AIChEJ, 33(3), 366-381 (1987).

Kolev, N.; Istatkova, E., and Darakchiev, R., A redistribution layer for packed columns intended for high flow velocities, Chem Eng & Proc, 21(2), 77-82 (1987).

Kunesh, J.G., Practical tips on tower packing, Chem Eng, 7 Dec, 101-105 (1987).

Kunesh, J.G.; Lahm, L., and Yanagi, T., Commercial scale experiments that provide insight on packed tower distributors, Ind Eng Chem Res, 26(9), 1845-1851 (1987).

Marchot, P.; Crine, M., and L'Homme, G.A., Two-phase flow through a packed bed: A stochastic model based on percolation concepts, Chem Eng J, 36(3), 141-150 (1987).

McNulty, K.J.; Monat, J.P., and Hansen, O.V., Performance of commercial Chevron mist eliminators, Chem Eng Prog, 83(5), 48-55 (1987).

Niederkruger, M., and Yuksel, M.L., Direct measurement of surface temperature of falling films, Chem Eng & Proc, 21(1), 33-40 (1987).

Sai, P.S.T., and Varma, Y.B.G., Pressure drop in gas-liquid downflow through packed beds, AIChEJ, 33(12), 2027-2036 (1987).

Stanek, V., and Vychodil, P., Mathematical model and assessment of thermally induced gas flow inhomogeneities in fixed beds, Chem Eng & Proc, 22(2), 107-116 (1987).

Thompson, R.E., and King, C.J., Energy conservation in regenerated chemical absorption processes, Chem Eng & Proc, 21(3), 115-130 (1987).

Tien, C.L., and Hunt, M.L., Boundary-layer flow and heat transfer in porous beds, Chem Eng & Proc, 21(2), 53-64 (1987).

Tsotsas, E., and Martin, H., Review of thermal conductivity of packed beds, Chem Eng & Proc, 22(1), 19-38 (1987).

Vatcha, S.R., Relating transfer units and theoretical stages, Chem Eng, 9 Nov, 101-103 (1987).

Wronski, S., and Molga, E., Axial dispersion in packed beds: Effect of particle size non-uniformities, Chem Eng & Proc, 22(3), 123-126 (1987).

Yu, W.C., and Astarita, G., Design of packed towers for selective chemical absorption, Chem Eng Sci, 42(3), 425-434 (1987).

1988

Apte, V.B.; Wall, T.F., and Truelove, J.S., Gas flows in raceways formed by high velocity jets in a two-dimensional packed bed, CER&D, 66(4), 357-362 (1988).

Badssi, A., et al., Influence of pressure on gas-liquid interfacial area and gas-side mass transfer coefficient of laboratory column with crossflow sieve plates, Chem Eng & Proc, 23(2), 89-98 (1988).

Buchanan, J.E., Operating hold-up on film-type packings, AIChEJ, 34(5), 870-872 (1988).

Butcher, C., Structured packings, Chem Engnr, Aug, 25-30 (1988).

Dolejs, V., and Machac, I., Pressure drop for liquid flow in fixed bed of particles, Int Chem Eng, 28(4), 739-746 (1988).

Grosser, K.; Carbonell, R.G., and Sundaresan, S., Onset of pulsing in two-phase cocurrent downflow through a packed bed, AIChEJ, 34(11), 1850-1860 (1988).

Kunesh, J.G., Recent developments in packed columns, Can JCE, 65(6), 907-912 (1988).

Kuthan, K., and Broz, Z., Mass transfer in liquid films during absorption, Chem Eng & Proc, 24(4), 221-232 (1988).

Lindner, J.R.; Schubert, C.N., and Kelly, R.M., Influence of hydrodynamics on physical and chemical gas adsorption in packed columns, Ind Eng Chem Res, 27(4), 636-642 (1988).

Nicol, R.S., and Davidson, J.F., Effect of surfactants on the gas hold-up in circulating bubble columns, CER&D, 66(2), 159-164 (1988).

Nicol, R.S., and Davidson, J.F., Gas hold-up in circulating bubble columns, CER&D, 66(2), 152-158 (1988).

Pohorecki, R., and Moniuk, W., Plate efficiency for absorption with chemical reaction, Chem Eng J, 39(1), 27-46 (1988).

Rousseau, R.W., and Staton, J.S., Analyzing chemical absorbers and strippers, Chem Eng, 18 July, 91-95 (1988).

Rubio, F.C., et al., Pressure drop from friction and liquid holdup in two-phase ascending gas-liquid flow in packed columns, Int Chem Eng, 28(4), 627-634 (1988).

Spagnolo, D.A.; Plaice, E.L.; Neuburg, H.J., and Chuang, K.T., Heat-transfer modelling of sieve trays, Can JCE, 66(3), 367-376 (1988).

Tsotsas, E., and Schlunder, E.U., Axial dispersion in packed beds with fluid flow, Chem Eng & Proc, 24(1), 15-32 (1988).

Velaga, A., et al., Packed crisscross-flow cascade tower efficiencies for methanol-water separation, Ind Eng Chem Res, 27(8), 1481-1487 (1988).

Yih, S.M., and Kuo, C.C., Performance of wetted-wall sheet bundle column, Chem Eng Commns, 71, 239-250 (1988).

Yih, S.M., and Kuo, C.C., Design and testing of new type of falling film gas-liquid contacting device, AIChEJ, 34(3), 499-501 (1988).

Ziolkowska, I., and Ziolkowski, D., Fluid flow inside packed beds, Chem Eng & Proc, 23(3), 137-164 (1988).

20.2 Absorption Reactions

1967-1970

Brian, P.L.T.; Vivian, J.E., and Matiatos, D.C., Criterion for supersaturation in simultaneous gas absorption and desorption, Chem Eng Sci, 22(1), 7-10 (1967).

Calderbank, P.H., Gas absorption from bubbles: A review, Chem Engnr, Oct, CE209-233 (1967).

Danckwerts, P.V., and McNeil, K.M., The absorption of carbon dioxide into aqueous amine solutions and the effects of catalysis, Trans IChemE, 45, T32-49 (1967).

Marushkin, B.K., Calculation of circulating absorbent composition in absorption of hydrocarbon gases, Int Chem Eng, 7(2), 341-346 (1967).

Nienow, A.W., Transfer processes with a high mass flux, Brit Chem Eng, 12(11), 1737-1743 (1967).

Pritchard, C.L., and Biswas, S.K., Mass transfer from drops in forced convection, Brit Chem Eng, 12(6), 879-885 (1967).

Whitt, F.R., Absorption of gases by liquids in agitated vessels, Brit Chem Eng, 12(4), 554-557 (1967).

Barrere, C.A., and Deans, H.A., Investigation of absorption-reaction of carbon dioxide in liquid diethanolamine by direct chromatographic perturbation method, AIChEJ, 14(2), 280-285 (1968).

Broughton, D.B., The Molex absorption process, Chem Eng Prog, 64(8), 60-65 (1968).

Petrovic, L.J., and Thodos, G., Mass transfer in flow of gas through packed beds, Ind Eng Chem Fund, 7(2), 274-280 (1968).

Volgin, B.P.; Efimova, T.F., and Gofman, M.S., Absorption of sulfur dioxide by ammonium sulfite-bisulfite solution in a venturi scrubber, Int Chem Eng, 8(1), 113-118 (1968).

Markovs, J.; Lee, M.N.Y., and Nasser, B.E., Determining impurities in absorption process streams, Chem Eng Prog, 65(5), 68-74 (1969).

Thibodeaux, L.J., Continuous crosscurrent mass transfer in packed towers, Chem Eng, 2 June, 165-170 (1969).

Thomas, W.J., and Nicholl, E.McK., Interfacial turbulence accompanying absorption with reaction, Trans IChemE, 47, T325-331 (1969).

Gunn, D.J., and Saleem, A., Absorption and liquid-phase oxidation of sulphur dioxide, Trans IChemE, 48, T46-53 (1970).

Mashelkar, R.A., and Sharma, M.M., Mass transfer in bubble and packed bubble columns, Trans IChemE, 48, T162-172 (1970).

Sharma, M.M., and Danckwerts, P.V., Chemical methods of measuring interfacial area and mass transfer coefficients in two-fluid systems, Brit Chem Eng, 15(4), 522-528 (1970).

Sweny, J.W., and Valentine, J.P., Selexol: Physical solvent for gas treatment-purification, Chem Eng, 7 Sept, 54-56 (1970).

1971-1975

Davis, J.C., Sulfur dioxide absorbed from tail gas with sodium sulfite, Chem Eng, 29 Nov, 43-45 (1971).

Hamilton, W., and Bartholomai, G.B., Mass transfer with heat transfer, Brit Chem Eng, 16(12), 1133-1134 (1971).

Hawkes, E.N., and Mago, B.F., Prevent MEA-carbon dioxide unit corrosion, Hyd Proc, 50(8), 109-112 (1971).

Ramachandran, P.A., and Sharma, M.M., Simultaneous absorption of two gases, Trans IChemE, 49, 253-280 (1971).

Bakowski, S., Mass transfer in packed columns, Proc Tech Int, 17(10), 789-792 (1972).

Billet, R., Absorption of nitrogen oxides, Brit Chem Eng, 17(9), 705-708; 17(10), 807 (1972).

Rosner, D.E., and Epstein, M., Effects of interface kinetics, capillarity and solute diffusion on bubble growth rates in highly supersaturated liquids, Chem Eng Sci, 27(1), 69-88 (1972).

Hulswitt, C.E., Adiabatic and falling-film absorption of hydrogen chloride, Chem Eng Prog, 69(2), 50-52 (1973).

Kado, T., and Himmelblau, D.M., Stochastic analysis of a countercurrent two-phase absorption or extraction column, Ind Eng Chem Proc Des Dev, 12(3), 321-328 (1973).

Moo-Young, M., and Shoda, M., Gas absorption rates at the free surface of a flowing water stream, Ind Eng Chem Proc Des Dev, 12(4), 410-414 (1973).

Bourne, J.R.; Stockar, U.V., and Coggan, G.C., Gas absorption with heat effects, Ind Eng Chem Proc Des Dev, 13(2), 115-132 (1974).

La Nauze, R.D., and Harris, I.J., Gas bubble formation at elevated system pressures, Trans IChemE, 52, 337-348 (1974).

Stepanek, J.B., and Shilimkan, R.V., Enhancement factor in absorption with instantaneous irreversible chemical reaction at high mass transfer rates, Trans IChemE, 52, 313-316 (1974).

Jelenek, J., and Hlavacek, V., Mathematical modelling of absorption and stripping processes, Int Chem Eng, 15(4), 664-669 (1975).

Richardson, I.M.J., and O'Connell, J.P., Some generalizations about processes to absorb acid gases and mercaptans, Ind Eng Chem Proc Des Dev, 14(4), 467-470 (1975).

Strelzoff, S., Choosing the optimum carbon dioxide-removal system, Chem Eng, 15 Sept, 115-120 (1975).

1976-1979

Hozawa, M.; Shoji, K., and Tadaki, T., Effect of Rayleigh instability on gas-absorption rate, Int Chem Eng, 16(2), 341-346 (1976).

Kirillov, V.A., and Nasamanyan, M.A., Mass transfer processes between liquid and packing in a three-phase fixed bed, Int Chem Eng, 16(3), 538-543 (1976).

Lee, J.I.; Otto, F.D., and Mather, A.E., Equilibrium between carbon dioxide and aqueous monoethanolamine solutions, J Appl Chem Biotechnol, 26, 541-549 (1976).

Thomas, W.J., and Ray, M.S., Physical absorption and surface resistance in a flowing liquid, Chem Eng Commns, 2(2), 135-147 (1976).
Thomas, W.J.; Ray, M.S., and Palmer, E.W., Physical absorption of carbon dioxide in water flowing in an inclined cell, Chem Eng Commns, 2(2), 121-134 (1976).
Zanker, A., Nomograph for Hatta number for gas absorption, Proc Engng, Sept, 100-101 (1976).
Adler, P., Correct absorber lean oil cuts operating costs, Chem Eng, 5 Dec, 125-130 (1977).
Hikita, H.; Asai, S.; Ishikawa, H., and Honda, M., Kinetics of reactions of carbon dioxide with mono-, di-, and triethanolamine by rapid mixing method, Chem Eng J, 13(1), 7-12 (1977).
Mann, R., and Moyes, H., Exothermic gas absorption with chemical reaction, AIChEJ, 23(1), 17-23 (1977).
Rizzuti, L.; Augugliaro, V., and Marrucci, G., Ozone absorption in aqueous phenol solutions, Chem Eng J, 13(3), 219-224 (1977).
Sada, E., et al., Absorption of carbon dioxide in aqueous solutions of sodium phenoxide, Chem Eng J, 13(1), 41-44 (1977).
Sada, E.; Kumazawa, H,, and Butt, M.A., Absorption of carbon dioxide into aqueous solutions of ethylenediamine: Effect of interfacial turbulence, Chem Eng J, 13(3), 213-218 (1977).
Sada, E.; Kumazawa, H., and Butt, M.A., Simultaneous absorption of three reacting gases, Chem Eng J, 13(3), 225-232 (1977).
Vidaurri, F.C., and Kahre, L.C., Controlled absorption of hydrogen sulfide from sour gas, Hyd Proc, 56(11), 333-337 (1977).
Hatta, S., Rate of absorption of gases by liquids, Int Chem Eng, 18(3), 443-476 (1978).
Hikita, H.; Asai, S., and Nose, H., Absorption of sulfur dioxide into water, AIChEJ, 24(1), 147-149 (1978).
Ouwerkerk, C., Design for selective hydrogen sulfide absorption, Hyd Proc, 57(4), 89-94 (1978).
Picciotti, M., Optimize caustic scrubbing systems, Hyd Proc, 57(5), 201-209 (1978).
Shilimkan, R.V., and Stepanek, J.B., Mass transfer in cocurrent gas-liquid flow, Chem Eng Sci, 33(12), 1675-1680 (1978).
Watanabe, H., Voidage function in particulate fluid systems, Powder Tech, 19, 217-225 (1978).
Alper, E., Physical absorption of a gas in laboratory models of a packed column, AIChEJ, 25(3), 545-547 (1979).
Billet, R., and Mackowiak, J., Liquid-phase mass transfer in absorption packed columns, Chem Eng Commns, 3(1), 1-14 (1979).
Butwell, K.F.; Kubek, D.J., and Sigmund, P.W., Amine guard III absorption system, Chem Eng Prog, 75(2), 75-81 (1979).
Danckwerts, P.V., Reaction of carbon dioxide with ethanolamines, Chem Eng Sci, 34(4), 443-446 (1979).
DeLancey, G.B., and Wu, K.O., Absorption of carbon dioxide by ammonia solutions, Chem Eng Sci, 34(1), 148-150 (1979).
Hikita, H.; Asai, S.; Katsu, Y., and Ikuno, S., Absorption of carbon dioxide into aqueous MEA solutions, AIChEJ, 25(5), 793-780 (1979).

Ly, L.N.; Carbonell, R.G., and McCoy, B.J., Diffusion of gases through surfactant films: Interfacial resistance to mass transfer, AIChEJ, 25(6), 1015-1024 (1979).

Rivas, O.R., and Prausnitz, J.M., Sweetening of sour natural gases by mixed-solvent absorption, AIChEJ, 25(6), 975-984 (1979).

1980-1982

Bettelheim, J.; Foster, P.M., and Kyte, W.S., The effect of condensation upon transfer rates with application to flue-gas washing plants, Trans IChemE, 58, 1-8 (1980).

Counce, R.M., and Perona, J.J., Mathematical model for nitrogen oxide absorption in a sieve-plate column, Ind Eng Chem Proc Des Dev, 19(3), 426-431 (1980).

Pittaway, K.R., and Thibodeaux, L.J., Measurement of the oxygen desorption rate in a single-stage cross-flow packed tower, Ind Eng Chem Proc Des Dev, 19(1), 40-46 (1980).

Ruether, J.A.; Yang, C.S., and Hayduk, W., Particle mass transfer during cocurrent downward gas-liquid flow in packed beds, Ind Eng Chem Proc Des Dev, 19(1), 103-107 (1980).

Taylor, R., and Webb, D.R., Stability of the film model for multicomponent mass transfer, Chem Eng Commns, 6(1), 175-190 (1980).

Horwitz, B.A., Scrub organics with organics, Chem Eng, 1 June, 87 (1981).

Krishna, R., Binary and multicomponent mass transfer at high transfer rates, Chem Eng J, 22(3), 251-258 (1981).

Patwardhan, V.S., Gas-liquid reactions in packed beds: Effectiveness of static hold-up in presence of gas-side resistance, Can JCE, 59, 483-486 (1981).

Charpentier, J.C., What's new in absorption with chemical reaction (review paper)? Trans IChemE, 60, 131-156 (1982).

Glushchenko, V.I., and Kirichuk, E.D., Mathematical model of absorption columns for production of nitric acid, Int Chem Eng, 22(1), 181-187 (1982).

Hills, J.H.; Abbott, C.J., and Westall, L.J., A simple apparatus for the measurement of mass transfer from gas bubbles to liquids, Trans IChemE, 60, 369-372 (1982).

Krishna, R., A turbulent-film model for multicomponent mass transfer, Chem Eng J, 24(2), 163-172 (1982).

Lee, Y.H., and Luk, S., Oxygen absorption in a stirred tank, Ind Eng Chem Fund, 21(4), 428-434 (1982).

Lefers, J.B., and van den Berg, P.J., Absorption of nitrogen oxides into diluted and concentrated nitric acid, Chem Eng J, 23(2), 211-222 (1982).

Meisen, A., and Kennard, M.L., Diethanolamine degradation mechanism, Hyd Proc, Oct, 105-108 (1982).

Neelakanten, K., and Gehlawat, J.K., New chemical systems for the determination of liquid-side mass-transfer coefficient and effective interfacial area in gas-liquid contactors, Chem Eng J, 24(1), 1-6 (1982).

Sada, E.; Katoh, S.; Yoshii, H., and Yasuda, K., Rates of gas absorption into molten salts, Ind Eng Chem Fund, 21(1), 43-46 (1982).

Sandall, O.C.; Hanna, O.T., and Valeri, F.J., Heat effects for physical and chemical absorption in turbulent liquid films, Chem Eng Commns, 16(1), 135-148 (1982).
Wolfer, W., Helpful hints for physical solvent absorption, Hyd Proc, Nov, 193-197 (1982).
Won, Y.S., and Mills, A.F., Correlation of effects of viscosity and surface tension on gas absorption rates into freely falling turbulent liquid films, Int J Heat Mass Trans, 25(2), 223-230 (1982).

1983
Blauwhoff, P.M.M.; Versteeg, G.F., and van Swaaij, W.P.M., Study of reaction between carbon dioxide and alkanolamines in aqueous solutions, Chem Eng Sci, 38(9), 1411-1430 (1983).
Carta, G., and Pigford, R.L., Absorption of nitric oxide in nitric acid and water, Ind Eng Chem Fund, 22(3), 329-335 (1983).
Counce, R.M., and Perona, J.J., Scrubbing of gaseous nitrogen oxides in packed towers, AIChEJ, 29(1), 26-32 (1983).
Glaves, P.S.; McKee, R.L.; Kensell, W.W., and Kobayashi, R., Pick your data for carbon dioxide dehydration carefully, Hyd Proc, Nov, 213-215 (1983).
Grossman, G., Simultaneous heat and mass transfer in film absorption under laminar flow, Int J Heat Mass Trans, 26(3), 357-372 (1983).
Keaton, M.M., and Bourke, M.J., Activated carbon system cuts foaming and amine losses, Hyd Proc, Aug, 71-73 (1983).
Nicholas, D.M.; Wilkins, J.T., and Li, T.C., Optimize acid gas removal, Hyd Proc, Sept, 123-129 (1983).
Pandya, J.D., Adiabatic gas absorption and stripping with chemical reaction in packed towers, Chem Eng Commns, 19(4), 343-362 (1983).
Patil, V.K., and Sharma, M.M., Solid-liquid mass transfer coefficients in bubble columns up to one metre diameter, CER&D, 61, 21-28 (1983).

1984
Barth, D.; Tondre, C., and Delpuech, J.J., Reactions of carbon dioxide with MDEA and DEA, Chem Eng Sci, 39(12), 1753-1758 (1984).
Blauwhoff, P.M.M.; Versteeg, G.F., and van Swaaij, W.P.M., Study of reaction between carbon dioxide and alkanolamines in aqueous solutions, Chem Eng Sci, 39(2), 207-226 (1984).
Carnell, P., and Starkey, P., Gas desulphurisation offshore (absorption systems, including 'Higee'), Chem Engnr, Nov, 30-34 (1984).
Chakma, A., and Meisen, A., Diethanolamine properties by computer, Hyd Proc, Oct, 79-80 (1984).
Daviet, G.R.; Sundermann, R.; Donnelly, S.T., and Bullin, J.A., Switch to MDEA absorption raises capacity, Hyd Proc, May, 79-82 (1984).
Gazzi, L.; D'Ambra, R.; DiCintio, R.; Rescalli, C., and Vetere, A., Treat high acid gases, Hyd Proc, July, 99-103 (1984).
Grossman, G., and Heath, M.T., Simultaneous heat and mass transfer in absorption of gases in turbulent liquid films, Int J Heat Mass Trans, 27(12), 2365-2376 (1984).

Haimour, N., and Sandall, O.C., Absorption of carbon dioxide into aqueous methyl-diethanolamine, Chem Eng Sci, 39(12), 1791-1796 (1984).
Herwig, J.; Schleppinghoff, B., and Schulwitz, S., New low energy absorption process for MTBE and TAME, Hyd Proc, June, 86-88 (1984).
Holmes, J.W.; Spears, M.L., and Bullin, J.A., Sweetening LPG with amines, Chem Eng Prog, 80(5), 47-50 (1984).
Lee, S.Y., and Tankin, R.S., Study of liquid spray (water) in a condensable environment (steam), Int J Heat Mass Trans, 27(3), 363-374 (1984).
Lee, S.Y., and Tankin, R.S., Study of liquid spray (water) in a non-condensable environment (air), Int J Heat Mass Trans, 27(3), 351-362 (1984).
Lee, Y.H.; Luk, S., and Sirdeshpande, G., Mechanism of carbon dioxide absorption and desorption, Chem Eng Commns, 28(1), 111-116 (1984).
Mohamed, R.S., and Klinzing, G.E., Absorption of nitrogen dioxide and sulfur dioxide in methanol, Can JCE, 62(1), 99-102 (1984).
Mortko, R.A., Remove hydrogen sulfide selectively, Hyd Proc, June, 78-82 (1984).
Pauley, C.R., Carbon dioxide recovery from flue gas, Chem Eng Prog, 80(5), 59-62 (1984).
Rocha, F.A.N., and Guedes de Carvalho, J.R.F., Absorption during gas injection through a submerged nozzle, CER&D, 62, 303-314 (1984).
Sada, E.; Kumazawa, H., and Lee, C.H., Chemical absorption of carbon dioxide and sulfur dioxide into aqueous concentrated slurries of calcium hydroxide, Chem Eng Sci, 39(1), 117-120 (1984).
Various, Treating acid and sour gas (topic issue), Chem Eng Prog, 80(10), 27-77 (1984).
Zanker, A., Nomograph for mass transfer time and efficiency, Proc Engng, June, 73 (1984).

1985

Blauwhoff, P.M.M., and van Swaaij, W.P.M., Simultaneous mass transfer of hydrogen sulphide and carbon dioxide with complex chemical reactions in aqueous di-isopropanolamine solution, Chem Eng & Proc, 19(2), 67-84 (1985).
Blauwhoff, P.M.M., et al., Absorber design in sour natural gas treatment plants: Impact of process variables on operation and economics, Chem Eng & Proc, 19(1), 1-26 (1985).
Chakravarty, T.; Phukan, U.K., and Weiland, R.H., Reaction of acid gases with mixtures of amines, Chem Eng Prog, 81(4), 32-36 (1985).
Chang, C.S., and Rochelle, G.T., Sulfur dioxide absorption into sodium hydroxide and sodium sulfite aqueous solutions, Ind Eng Chem Fund, 24(1), 7-11 (1985).
Jou, F.Y.; Otto, F.D., and Mather, A.E., Equilibria of hydrogen sulfide and carbon dioxide in triethanolamine solutions, Can JCE, 63(1), 122-125 (1985).
Knaff, G., and Schlunder, E.U., Competitive physical absorption of gases into water, Chem Eng & Proc, 19(4), 191-198 (1985).
Lal, D.; Otto, F.D., and Mather, A.E., Solubility of hydrogen sulfide and carbon dioxide in a diethanolamine solution at low partial pressures, Can JCE, 63(4), 681-685 (1985).

Mahiout, S., and Vogelpohl, A., Absorption of oxygen by aqueous glycerol solutions and squalane, Chem Eng & Proc, 19(4), 221-226 (1985).
Schulze, G., and Schlunder, E.U., Effect of multicomponent diffusion on mass transfer during absorption of single gas bubbles, Chem Eng & Proc, 19(5), 257-266 (1985).
Schulze, G., and Schlunder, E.U., Physical absorption of single gas bubbles in degassed and preloaded water, Chem Eng & Proc, 19(1), 27-38 (1985).
Siddique, Q.M., The changing face of gas purification, Chem Engnr, June, 26-28 (1985).
Thurner, F., and Schlunder, E.U., Wet-bulb temperature of binary mixtures, Chem Eng & Proc, 19(6), 337-344 (1985).
Various, Amine inhibiting (special report), Hyd Proc, May, 70-75 (1985).

1986

Alper, E., and Ozturk, S., The effect of activated carbon loading on oxygen absorption into aqueous sodium sulphide solutions in a slurry reactor, Chem Eng J 32(2), 127-130 (1986).
Carta, G., Scrubbing of nitrogen oxides with nitric acid solutions, Chem Eng Commns, 42, 157-170 (1986).
Edgerton, M.E.; Byrne, G.D., and Ho, W.S., Numerical calculation of the simultaneous absorption of two gases with reversible chemical reactions, Comput Chem Eng, 10(6), 551-556 (1986).
Gazzi, L.; Rescalli, C., and Sguera, O., Selefining process: A new route for selective hydrogen sulfide removal, Chem Eng Prog, 82(5), 47-49 (1986).
Haynes, H.W., A note on diffusive mass transport, Chem Eng Educ, 20(1), 22-27 (1986).
Horner, B.; Viebahn, U., and Dialer, K., Mass transfer in a turbulent liquid during absorption, Chem Eng Sci, 41(7), 1723-1734 (1986).
Sada, E.; Kumazawa, H.; Osawa, Y.; Matsuura, M., and Han, Z.Q., Reaction kinetics of carbon dioxide with amines in non-aqueous solvents, Chem Eng J, 33(2), 87-96 (1986).
Sheppard, S.V., Ionizing wet scrubber for air pollution control, Chem Eng Prog, 82(2), 40-43 (1986).
Spedding, P.L.; Jones, M.T., and Lightsey, G.R., Ammonia absorption into water in a packed tower, Chem Eng J, 32(3), 151-164 (1986).
Spedding, P.L.; Munro, P.A., and Jones, M.T., Ammonia absorption into water in a packed tower, Chem Eng J, 32(2), 65-76 (1986).
Weisweiler, W.; Blumhofer, R., and Westermann, T., Absorption of nitrogen monoxide in aqueous solutions containing sulfite and transition-metal chelates, Chem Eng & Proc, 20(3), 155-166 (1986).
Witte, I., and Kind, R., Modelling of absorption of sulphur dioxide with aqueous sodium carbonate, Chem Eng & Proc, 20(4), 183-190 (1986).

1987

Bhattacharya, A.; Gholap, R.V., and Chaudhari, R.V., Gas absorption with bimolecular (1,1 order) reaction, AIChEJ, 33(9), 1507-1513 (1987).
Cooney, D.O., and Olsen, D.P., Absorption of sulfur dioxide and hydrogen sulfide in small-scale venturi scrubbers, Chem Eng Commns, 51, 291-306 (1987).

Heisel, M.P., and Marold, F.J., New gas scrubber removes hydrogen sulfide, Hyd Proc, 66(4), 35-37 (1987).

Hitch, D.M.; Rousseau, R.W., and Ferrell, J.K., Simulation of continuous-contact separation processes: Unsteady-state, multicomponent, adiabatic absorption, Ind Eng Chem Res, 26(6), 1092-1100 (1987).

Knapp, H.; Zeck, S., and Langhorst, R., Phase equilibria for design of gas wash systems: Experimental techniques, Chem Eng & Proc, 21(1), 25-32 (1987).

Krishna, R., Physical significance of the mass transfer coefficient, Chem Eng J, 35(1) 67-68 (1987).

Linek, V.; Vacek, V., and Benes, P., A critical review and experimental verification of the correct use of the dynamic method for the determination of oxygen transfer in aerated agitated vessels, Chem Eng J, 34(1) 11-34 (1987).

Miller, D.N., Mass transfer in nitric acid absorption, AIChEJ, 33(8), 1351-1358 (1987).

Rocha, F.A.N., and Guedes de Carvalho, J.R.F., Absorption during gas injection through a submerged nozzle, CER&D, 65(3), 279-284 (1987).

Rousseau, R.W.; Ferrell, J.K., and Staton, J.S., Conditioning coal gas with aqueous solutions of potassium carbonate: Model development and testing, Gas Sepn & Purif, 1(1), 44-54 (1987).

Yih, S.M., and Lai, H.C., Simultaneous absorption of carbon dioxide and hydrogen sulfide in hot carbonate solutions in a packed absorber-stripper unit, Chem Eng Commns, 51, 277-290 (1987).

Yih, S.M., and Sun, C.C., Simultaneous absorption of hydrogen sulphide and carbon dioxide into potassium carbonate solution with or without amine promoters, Chem Eng J, 34(2) 65-72 (1987).

Yu, W.C., and Astarita, G., Selective absorption of hydrogen sulphide in tertiary amine solutions, Chem Eng Sci, 42(3), 419-424 (1987).

Yuksel, M.L., and Schlunder, E.U., Heat and mass transfer in non-isothermal absorption of gases in falling liquid films, Chem Eng & Proc, 22(4), 193-214 (1987).

1988

Altwicker, E.R., and Lindhjem, C.E., Absorption of gases into drops, AIChEJ, 34(2), 329-332 (1988).

Crawford, D.B., and Counce, R.M., Depletion of aqueous nitrous acid in packed towers, Sepn Sci Technol, 23(12), 1573-1594 (1988).

Mahajani, V.V., and Joshi, J.B., Kinetics of reactions between carbon dioxide and alkanolamines (review paper), Gas Sepn & Purif, 2(2), 50-64 (1988).

Moravec, P., and Stanek, V., Counter-current absorption of oxygen by frequency response technique, Chem Eng & Proc, 24(2), 93-104 (1988).

Newman, B.L., and Carta, G., Mass transfer in absorption of nitrogen oxides in alkaline solutions, AIChEJ, 34(7), 1190-1199 (1988).

Roberts, B.E., and Mather, A.E., Solubility of carbon dioxide and hydrogen sulfide in hindered amine solution, Chem Eng Commns, 64, 105-112 (1988).

Sada, E.; Kumazawa, H., and Yoshikawa, Y., Simultaneous removal of nitric oxide and sulfur dioxide by absorption into aqueous mixed solutions, AIChEJ, 34(7), 1215-1220 (1988).

Sanyal, D.; Vasishtha, N., and Saraf, D.N., Modeling of carbon dioxide absorber using hot carbonate process, Ind Eng Chem Res, 27(11), 2149-2156 (1988).

Selby, G.W., and Counce, R.M., Aqueous scrubbing of dilute nitrogen oxide gas mixtures, Ind Eng Chem Res, 27(10), 1917-1922 (1988).

Sen, A.K., Perturbation analysis of gas absorption with chemical reaction: A nonvolatile liquid reactant, Chem Eng Commns, 65, 223-230 (1988).

Singh, M., and Bullin, J.A., Determination of rate constants for reaction between diglycolamine and carbonyl sulphide, Gas Sepn & Purif, 2(3), 131-137 (1988).

Sweeney, C.W.; Ritter, T.J., and McGinley, E.B., Strategy for screening physical solvents, Chem Eng, 20 June, 119-125 (1988).

Tseng, P.C.; Ho, W.S., and Savage, D.W., Carbon dioxide absorption into promoted carbonate solutions, AIChEJ, 34(6), 922-931 (1988).

Versteeg, G.F., and van Swaaij, W.P.M., Absorption of carbon dioxide and hydrogen sulphide in aqueous alkanolamine solutions using fixed-bed reactor with cocurrent downflow operation in pulsing flow regime, Chem Eng & Proc, 24(3), 163-176 (1988).

Weiss, H., Rectisol wash for purification of partial oxidation gases, Gas Sepn & Purif, 2(4), 171-176 (1988).

Wilson, D.J., Fine-bubble aeration: Mathematical modeling of time-dependent operation, Sepn Sci Technol, 23(14), 2211-2230 (1988).

Yih, S.M., and Lii, C.W., Absorption of NO and sulfur dioxide in Fe(II)-EDTA solutions, Chem Eng Commns, 73, 43-66 (1988).

20.3 Dehumidifiers

1967-1985

Stewart, R.R., and Bruley, D.F., Thermal dynamics of a distributed-parameter nonadiabatic humidification process, AIChEJ, 13(4), 793-796 (1967).

Shleinikov, V.M., and Sobolev, M.A., Water-evaporation process for cooling air in scrubbers of air separation plants, Int Chem Eng, 12(4), 662-669 (1972).

Barrett, E.C., and Dunn, S.G., Design of direct contact humidifiers and dehumidifiers using tray columns, Ind Eng Chem Proc Des Dev, 13(4), 353-358 (1974).

Nori, S., and Ishii, T., Humidification of air with warm water in countercurrent packed beds. Chem Eng Sci, 37(3), 487-490 (1982).

Fahim, M.A.; Al-Ameeri, R.S., and Wakao, N., Equations for calculating temperatures in adiabatic air-water contact towers, Chem Eng Commns, 36(1), 1-8 (1985).

Larson, R.S., Design of recirculated-liquid gas cooler/humidifier, Ind Eng Chem Proc Des Dev, 24(4), 1023-1026 (1985).

20.4 Desorption

1969-1988

Lees, F.P., Desorption into a dry gas for drying organic liquids, Brit Chem Eng, 14(2), 173-174 (1969).
Liddle, C.J., How to design desorption systems based on pressure reduction, Chem Eng, 13 July, 87-94 (1970).
McLachlan, C.N.S., and Danckwerts, P.V., Desorption of carbon dioxide from aqueous potash solutions, Trans IChemE, 50, 300-309 (1972).
Thomas, W.J., and Ray, M.S., A study of gas desorption from static liquid pools, Chem Eng J, 9(1), 71-82 (1975).
Shah, Y.T., and Sharma, M.M., Desorption with or without chemical reaction, Trans IChemE, 54, 1-41 (1976).
Thuy, L.T., and Weiland, R.H., Mechanisms of gas desorption from aqueous solution, Ind Eng Chem Fund, 15(4), 286-293 (1976).
Cable, M., and Cardew, G.E., Kinetics of desorption with interfacial resistance and concentration-dependent diffusivity, Chem Eng Sci, 32(5), 535-542 (1977).
Krabach, M.H.; Donnelly, R.G.; Reber, S.A., and Davis, G.J., Kinetics of seawater deaeration by gas sparging in a cross-flow flume, Ind Eng Chem Fund, 16(4), 430-439 (1977).
Lin, W.C.; Rice, P.A.; Cheng, Y.S., and Barduhn, A.J., Vacuum stripping of refrigerants in water sprays, AIChEJ, 23(4), 409-415 (1977).
Mitsutake, H., and Sakai, M., Desorption of air from liquid in gas absorption, AIChEJ, 23(4), 599-601 (1977).
Astarita, G., and Savage, D.W., Theory of chemical desorption, Chem Eng Sci, 35(3), 649-656 (1980).
Bronikowska, W.P., and Rudzinski, K.J., Mathematical model of bubble gas desorption from liquids, Chem Eng Sci, 35(1), 512-518 (1980).
Bronikowska, W.P., and Rudzinski, K.J., Gas desorption from liquids, Chem Eng Sci, 36(7), 1153-1160 (1981).
Weiland, R.H.; Rawal, M., and Rice, R.G., Stripping of carbon dioxide from MEA solutions in a packed column, AIChEJ, 28(6), 963-973 (1982).
Avedesian, M.M.; Spira, P., and Kanduth, H., Stripping of HCN in a packed tower, Can JCE, 61(6), 801-806 (1983).
Mahajani, V.V., and Danckwerts, P.V., Stripping of carbon dioxide from amine-promoted potash solutions at 100 degC, Chem Eng Sci, 38(2), 321-328 (1983).
Mangnall, K., Deaeration and its adaption to water flood systems (application to oil recovery), Chem Engnr, Nov, 44-48 (1984).
Merchuk, J.C., and Herskowitz, M., Desorption with chemical reaction in contacting devices, Chem Eng Commns, 29(1), 79-88 (1984).
Cho, J.S., and Wakao, N., Stripping of benzene and toluene in bubble aeration tank, Chem Eng Commns, 56, 139-148 (1987).
Hoogendoorn, G.C., et al., Desorption of volatile electrolytes in a tray column (sour-water stripping), CER&D, 66(6), 483-502 (1988).

20.5 Cooling Towers

1966-1975

Davis, D.S., Nomogram for capacities of cooling towers, Brit Chem Eng, 11(5), 360 (1966).

Paige, P.M., Costlier cooling towers require a new approach to water-systems design, Chem Eng, 3 July, 93-98 (1967).

DeMonbrun, J.R., Factors to consider in selecting a cooling tower, Chem Eng, 9 Sept, 106-116 (1968).

Furzer, I.A., The natural draught cooling tower, Brit Chem Eng, 13(9), 1287-1290 (1968).

Vouyoucalos, S., Cross-flow cooling towers analysed, Brit Chem Eng, 13(7), 1004-1006 (1968).

Furzer, I.A., Natural draft cooling tower, Ind Eng Chem Proc Des Dev, 7(4), 555-565 (1968); 8(4), 599 (1969).

Hawkins, P., The thermal and functional design of natural draught cooling towers, Chem Engnr, Sept, 328-333 (1971).

Kuehmsted, A.M., Operation and maintenance of cooling towers, Chem Eng, 3 May, 112-115 (1971).

Kunesch, A.M., Mechanical draught cooling towers, Chem Engnr, Sept, 337-342 (1971).

Various, Cooling towers (topic issue), Chem Eng Prog, 67(7), 39-76 (1971).

Butler, P., Natural-draught cooling towers, Proc Engng, Oct, 102-103 (1972).

Wnek, W.J., and Snow, R.H., Design of cross-flow cooling towers and ammonia stripping towers, Ind Eng Chem Proc Des Dev, 11(3), 343-349 (1972).

Zanker, A., Estimating cooling tower costs from operating data, Chem Eng, 12 June, 118-119 (1972).

Friar, F., Cooling-tower basin design, Chem Eng, 22 July, 122-124 (1974).

Burger, R., Cooling tower drift elimination, Chem Eng Prog, 71(7), 73-76 (1975).

Caplan, F., Quick calculation of cooling tower blowdown and makeup, Chem Eng, 7 July, 110 (1975).

Elgawhary, A.W., Spray cooling system design, Chem Eng Prog, 71(7), 83-87 (1975).

Furzer, I.A., Radial air distribution functions in water cooling towers, Chem Eng Sci, 30(3), 349-351 (1975).

Inazumi, H., and Kageyama, S., Successive graphical method of design of crossflow cooling tower, Chem Eng Sci, 30(7), 717-722 (1975).

Jordan, D.R.; Bearden, M.D., and McIlhenny, W.F., Cooling tower blowdown concentration by electrodialysis, Chem Eng Prog, 71(7), 77-82 (1975).

Maze, R.W., Air cooler or cooling tower? Chem Eng, 6 Jan, 106-114 (1975).

Sussman, S., Facts on water use in cooling towers, Hyd Proc, 54(7), 147-153 (1975).

1977-1983

Anon., Cooling tower maintenance, Processing, Jan, 21 (1977).

Reed, D.T.; Klen, E.F., and Johnson, D.A., Side-stream softening reduces cooling tower blowdown, Hyd Proc, 56(11), 339-342 (1977).

Wigley, S., Steel cooling towers, Proc Engng, Feb, 45-46 (1977).

Haupt, T., Cycle control cuts cooling-tower costs, Chem Eng, 11 Sept, 161-163 (1978).
Kunesch, T., Environmental aspects of cooling tower selection, Proc Engng, Nov, 86-91 (1978).
Kunesch, T., Cooling tower selection, Proc Engng, Sept, 164-167 (1978).
Meytsar, J., Estimate cooling tower requirements easily, Hyd Proc, 57(11), 238-239 (1978).
Puckorius, P.R., Proper startup protects cooling-tower systems, Chem Eng, 2 Jan, 101-104 (1978).
Brooke, J.M., Zero cooling tower blowdown? Hyd Proc, 58(7), 211-214 (1979).
Burger, R., Cooling tower retrofit, Chem Eng Prog, 75(3), 78-81 (1979).
Veazey, J.A., Protecting cooling towers from overpressure, Chem Eng Prog, 75(7), 73-77 (1979).
Thomas, W.B., Keys to successful cooling tower operation, Hyd Proc, 59(5), 203-204 (1980).
Burrows, S., Cooling-tower design, Proc Engng, Nov, 85-87 (1981).
Rogers, A.N.; Weekes, M.C.; May, S.C., and Houle, E.H., Treatment of cooling tower blowdown, Chem Eng Prog, 77(7), 31-38 (1981).
Burger, R., Cooling towers can make money, Chem Eng Prog, 78(2), 84-87 (1982).
Gardner, B.R., and Winter, R.J., Development of assisted draught cooling towers, J Inst Energy, 55, 102-107 (1982).
Johnson, J.D., Variable fan speeds cut cooling-tower operating costs, Chem Eng, 8 Aug, 95-96 (1983).
Reeves, G.G., Cooling water chemistry for plant design, Hyd Proc, Sept, 60B-60HH (1983).
Various, Cooling tower developments (topic issue), Chem Eng Prog, 79(12), 23-38 (1983).

1984-1988

Burger, R., Cooling towers: The overlooked money maker, Plant/Opns Prog, 3(3), 184-188 (1984).
Campagne, W.V.L., and McDonough, L.J., How cooling towers affect process energy savings, Hyd Proc, June, 103-107 (1984).
Lefevre, M.R., Reducing water consumption in cooling towers, Chem Eng Prog, 80(7), 55-62 (1984).
Millington, P., Evaporative cooling-tower design, Proc Engng, Oct, 53-57 (1984).
Burger, R., Retrofit your cooling towers to save energy and money, Hyd Proc, July, 67-70 (1985).
van Dijk, H., Cooling tower control, Proc Engng, Oct, 57-60 (1985).
Warner, J.D., Timing a cooling tower, Chem Eng, 30 Sept, 95-98; 28 Oct, 71-73 (1985).
Burger, R., Retrofitting cuts cooling-tower costs, Chem Eng, 18 Aug, 117-119 (1986).
Guzdial, C.J., Energy-efficient location of blowdown, Chem Eng, 12 May, 123-124 (1986).
McDonald, C.M., Cooling tower maintenance forestalls problems, Chem Eng, 4 Aug, 85-88; 1 Sept, 95-101 (1986).

Burger, R., Getting rid of asbestos: Removal from cooling towers, Chem Eng, 22 June, 167-170 (1987).

Noor, A., and Allen, G.W., Cooling tower calculations, Chem Eng, 22 June, 173-174 (1987).

Ogboja, O., A procedure for computer-aided design of water-cooling towers, Chem Eng J, 35(1) 43-50 (1987).

Smith, M., Understanding cooling towers: Route to energy savings, Plant/Opns Prog, 6(4), 181-184 (1987).

Kaguei, S.; Nishio, M., and Wakao, N., Parameter estimation for packed cooling tower operation using a heat input-response technique, Int J Heat Mass Trans, 31(12), 2579-2586 (1988).

Reidenbach, R., Spray treatment procedure improves cooling system efficiency and reliability, Plant/Opns Prog, 7(3), 209-214 (1988).

CHAPTER 21

LIQUID-LIQUID EXTRACTION

21.1 Theory and Data

1967-1969

Bakker, C.A.P., and Beek, W.J., Influence of driving force in liquid-liquid extraction, Chem Eng Sci, 22(11), 1349-1356 (1967).

Guy, K.W.A.; Malanowski, S.K., and Rowlinson, J.S., Liquid methane as extractive solvent for separation of hydrocarbons, Chem Eng Sci, 22(5), 801-803 (1967).

Hanson, C.; Patel, A.N., and Chang-Kakoti, D.K., Studies on the separation of ortho and para-chloronitrobenzene by solvent extraction, J Appl Chem, 16, 341-344 (1966); 17, 169-170 (1967).

Sprow, F.B., Distribution of drop-sizes produced in turbulent liquid-liquid dispersion, Chem Eng Sci, 22(3), 435-442 (1967).

Ward, J.P., and Knudsen, J.G., Turbulent flow of unstable liquid-liquid dispersions: Drop sizes and velocity distributions, AIChEJ, 13(2), 356-365 (1967).

Barton, P.; McCormick, R.H., and Fenske, M.R., Ammonia: A versatile liquid extraction solvent, Ind Eng Chem Proc Des Dev, 7(3), 366-371 (1968).

Brown, A.H., Mechanism of liquid coalescence, Brit Chem Eng, 13(12), 1719-1721 (1968).

Hanson, C.; Patel, A.N., and Chang-Kakoti, D.K., Separation of ortho and para-chlorotoluene by solvent extraction, J Appl Chem, 18, 89-91 (1968).

Hartland, S., The coalescence of a liquid drop at a liquid-liquid interface, Trans IChemE, 46, T275-282 (1968).

Humphrey, J.L., and Van Winkle, M., Polar compounds as extractive solvents for hydrocarbon systems, Ind Eng Chem Proc Des Dev, 7(4), 581-585 (1968).

Scheele, G.F., and Meister, B.J., Drop formation at low velocities in liquid-liquid systems, AIChEJ, 14(1), 9-19 (1968).

Schindler, H.D., and Treybal, R.E., Continuous-phase mass-transfer coefficients for liquid extraction in agitated vessels, AIChEJ, 14(5), 790-798 (1968).

Sharma, M.M., and Nanda, A.K., Extraction with second order reaction, Trans IChemE, 46, T44-52 (1968).

Blaschke, G., and Schugerl, K., New method for measurement of concentration of solute at interface of two liquids, Chem Eng Sci, 24(10), 1543-1552 (1969).

Gross, B., and Hixson, A.N., Interferometric study of interfacial turbulence accompanying liquid-liquid mass transfer, Ind Eng Chem Fund, 8(2), 296-302 (1969).

McCoy, B.J., and Madden, A.J., Drop size in stirred liquid-liquid systems via encapsulation, Chem Eng Sci, 24(2), 416-419 (1969).

Nakamura, A., Effect of salts on liquid-liquid equilibria, Int Chem Eng, 9(3), 521-525 (1969).

1970-1972

Hanson, C.; Patel, A.N., and Chang-Kakoti, D.K., Separation of thiophen from benzene by solvent extraction, J Appl Chem, 19, 320-323 (1969); 20, 42-44 (1970).

Kessler, D.P., and York, J.L., Characteristics of inclusions in the dispersed phase of liquid-liquid suspensions, AIChEJ, 16(3), 369-374 (1970).

Liljenzin, J.O., and Reinhardt, H., Theoretical model for behavior of drops in a centrifuge, Ind Eng Chem Fund, 9(2), 248-251 (1970).

Mizrahi, J., and Barnea, E., Effects of solid additives on the formation and separation of emulsions, Brit Chem Eng, 15(4), 497-503 (1970).

Mudge, L.K., and Heideger, W.J., Interferometric determination of effect of surface active agents on liquid-liquid mass transfer rates, AIChEJ, 16(4), 602-608 (1970).

Raj, P., et al., Laboratory studies on solvent extraction of low-temperature tar oils by aqueous sodium salicylate, J Appl Chem, 20, 252-255 (1970).

Anwar, M.M.; Hanson, C., and Pratt, M.W.T., Dissociation extraction, Trans IChemE, 49, 95-100 (1971).

Lang, S.B., and Wilke, C.R., Hydrodynamic mechanism for coalescence of liquid drops, Ind Eng Chem Fund, 10(3), 329-352 (1971).

Mok, Y.I., and Treybal, R.E., Continuous-phase mass transfer coefficients for liquid extraction in agitated vessels, AIChEJ, 17(4), 916-920 (1971).

Niinimaki, L., and Orjans, J.R., Solvent extraction scores in making nitrophosphates, Chem Eng, 11 Jan, 63-65 (1971).

Paul, R.N., and Johny, C.J., Nitrobenzene as an extraction solvent for pyridine, Brit Chem Eng, 16(12), 1135-1134 (1971).

Skelland, A.H.P., and Minhas, S.S., Dispersed-phase mass transfer during drop formation and coalescence in liquid-liquid extraction, AIChEJ, 17(6), 1316-1324 (1971).

Bailes, P.J., and Winward, A., Progress in liquid-liquid extraction, Trans IChemE, 50, 240-258 (1972).

Brown, D.E., and Pitt, K., Drop-size distribution of stirred non-coalescing liquid-liquid system, Chem Eng Sci, 27(3), 577-584 (1972).

Davies, G.A.; Jeffreys, G.V.; Ali, F., and Afzal, M., Coalescence in droplet dispersion bands: An unusual surface effect, Chem Engnr, Oct, 392-396 (1972).

Flett, D.S., Recent advances in solvent extraction of metals, Chem Engnr, Dec, 465-471 (1972).

Godfrey, N.B., Solvent selection via miscibility number, Chemtech, June, 359-363 (1972).

Grossman, G., Determination of droplet size distribution in liquid-liquid dispersions, Ind Eng Chem Proc Des Dev, 11(4), 537-542 (1972).

Izard, J.A., Prediction of drop volumes in liquid-liquid systems, AIChEJ, 18(3), 634-638 (1972).

Komasawa, I.; Saito, T., and Otake, T., Mass transfer and turbulence at a liquid-liquid interface, Int Chem Eng, 12(2), 345-352 (1972).

Miachon, J.P., Solvent extraction unit reduces residence times, Proc Engng, Dec, 79-80 (1972).

Mlynek, Y., and Resnick, W., Drop sizes in an agitated liquid-liquid system, AIChEJ, 18(1), 122-127 (1972).

Moore, F.L., Liquid-liquid extraction of mercury with high-molecular-weight amines from iodide and bromide solutions, Sepn Sci, 7(5), 505-512 (1972).

Mumford, C.J., and Thomas, R.J., Coalescer aids in liquid-liquid settlers, Proc Engng, Dec, 54-58 (1972).

Paul, R.N., and Pai, M.U., Solvent for pyridine extraction, Brit Chem Eng, 17(1), 69-70 (1972).

Shah, S.T.; Wasan, D.T., and Kintner, R.C., Passage of a liquid drop through a liquid-liquid interface, Chem Eng Sci, 27(5), 881-894 (1972).

Shinde, V.M., Solvent extraction separation of selected transition metals with mesityl oxide, Sepn Sci, 7(2), 97-104 (1972).

1973-1974

Anwar, M.M.; Hanson, C.; Patel, A.N., and Pratt, M.W.T., Dissociation extraction: Multi-stage extraction, Trans IChemE, 51, 151-158 (1973).

Grishunin, A.V., et al., Separation of gasoline from mother liquor by countercurrent liquid extraction method, Int Chem Eng, 13(1), 122-126 (1973).

Jones, W.T., and Payne, V., New solvent to extract aromatics, Hyd Proc, 52(3), 91-92 (1973).

McLaughlin, C.M., and Rushton, J.H., Interfacial areas of liquid-liquid dispersions from light transmission measurements, AIChEJ, 19(4), 817-822 (1973).

Otto, W.; Streicher, R., and Schugerl, K., Influence of surface active agents on mass transfer across liquid-liquid interfaces, Chem Eng Sci, 28(10), 1777-1788 (1973).

Perez de Ortiz, E.S., and Sawistowski, H., Interfacial stability of binary liquid-liquid systems, Chem Eng Sci, 28(11), 2051-2070 (1973).

Shah, B.H., and Ramkrishna, D., Population balance model for mass transfer in lean liquid-liquid dispersions, Chem Eng Sci, 28(2), 389-400 (1973).

Traher, A., and Kirwan, D.J., Interferometric measurement of concentrations near an interface during liquid-liquid mass transfer, Ind Eng Chem Fund, 12(2), 244-246 (1973).

Zabel, T.; Hanson, C., and Ingham, J., The influence of system purity, drop separation and heat transfer on the terminal velocity of falling drops in liquid-liquid systems, Trans IChemE, 51, 162-164 (1973).

Arrowsmith, A., and Foster, P.J., The reduction of drag on drops falling in a stream in a liquid-liquid system, Trans IChemE, 52, 211-212 (1974).

Kirkpatrick, R.D., and Lockett, M.J., Influence of approach velocity on bubble coalescence, Chem Eng Sci, 29(12), 2363-2374 (1974).

Mecklenburgh, J.C., Backmixing and design: A review, Trans IChemE, 52, 180-192 (1974).

Ramkrishna, D., Drop-breakage in agitated liquid-liquid dispersions, Chem Eng Sci, 29(4), 987-992 (1974).

Smyth, T.N., Measurement of drop size in liquid-liquid dispersions, Chem Eng Sci, 29(2), 583-588 (1974).

Zabel, T.; Hanson, C., and Ingham, J., Direct contact heat transfer to single freely falling liquid drops, Trans IChemE, 52, 307-312 (1974).

1975-1976

Barford, R.A., Modern countercurrent distribution, Sepn & Purif Methods, 4(2), 351-397 (1975).

Ejaz, M., Extraction separation studies of uranium (VI) by amine oxides, Sepn Sci, 10(4), 425-446 (1975).

Hanson, C., and Ismail, H.A.M., Solubility and distribution data for benzene and toluene between aqueous and organic phases, J Appl Chem Biotechnol, 25, 319-326 (1975).

Lawson, G.J., Solvent extraction of metals from chloride solutions, J Appl Chem Biotechnol, 25, 949-958 (1975).

McManamey, W.J.; Multani, S.K.S., and Davies, J.T., Molecular diffusion and liquid-liquid mass transfer in stirred transfer cells, Chem Eng Sci, 30(12), 1536-1539 (1975).

Medir, M., and Mackay, D., Extraction of phenol from water with mixed solvents, Can JCE, 53, 274-280 (1975).

Mhaskar, R.D., and Sharma, M.M., Extraction with reaction in both phases, Chem Eng Sci, 30(8), 811-818 (1975).

Park, J.Y., and Blair, L.M., Effect of coalescence on drop size distribution in agitated liquid-liquid dispersion, Chem Eng Sci, 30(9), 1057-1064 (1975).

Perez de Ortiz, E.S., and Sawistowski, H., Stability analysis of liquid-liquid systems under conditions of simultaneous heat and mass transfer, Chem Eng Sci, 30(12), 1527-1529 (1975).

Sethy, A., and Cullinan, H.T., Transport of mass in ternary liquid-liquid systems, AIChEJ, 21(3), 571-582 (1975).

Standart, G.L.; Cullinan, H.T.; Paybarah, A., and Louizos, N., Ternary mass transfer in liquid-liquid extraction, AIChEJ, 21(3), 554-559 (1975).

Thomas, W.J.; Ismail, S.I.A., and Palmer, E.W., Interferometric studies of interfacial mass transfer in a liquid-liquid system, Chem Eng Commns, 2(1), 87-102 (1975).

Tunescu, R.C., and Duinea, N.M., Easy test for aromatic solvent extraction, Hyd Proc, 54(9), 160-161 (1975).

Bailes, P.J.,; Hanson, C., and Hughes, M.A., Liquid-liquid extraction: A review, Chem Eng, 19 Jan, 86-100; 10 May, 115-120; 30 Aug, 86-94 (1976).

Bulicka, J., and Prochazka, J., Mass transfer beyween two turbulent liquid phases, Chem Eng Sci, 31(2), 137-146 (1976).

Cullinan, H.T., and Ram, S.K., Mass transfer in ternary liquid-liquid system, Can JCE, 54, 156-165 (1976).

Iqbal, M.; Ejaz, M.; Chaudhri, S.A., and Ahmad, R., Extraction separation studies of gold by 2-hexylpyridine, Sepn Sci, 11(3), 255-278 (1976).

Rawat, B.S., and Gulati, I.B., Liquid-liquid equilibrium studies for separation of aromatics, J Appl Chem Biotechnol, 26, 425-435 (1976).

Various, Diluents in the solvent extraction of metals (topic issue), Chem & Ind, 6 March, 170-189 (1976).

Vijayan, S.; Furrer, M., and Ponter, A.B., Effect of temperature on coalescence of liquid drops at liquid-liquid interfaces, Can JCE, 54, 269-279 (1976).

Walia, D.S., and Vir, D., Interphase mass transfer during drop or bubble formation, Chem Eng Sci, 31(7), 525-534 (1976).

1977-1978

Bohnet, M., Separation of immiscible fluid dispersions, Int Chem Eng, 17(3), 395-409 (1977).

Coulaloglou, C.A., and Tavlarides, L.L., Description of interaction processes in agitated liquid-liquid dispersions, Chem Eng Sci, 32(11), 1289-1298 (1977).

Grace, J.R.; Wairegi, T., and Nguyen, T.H., Shapes and velocities of single drops and bubbles moving freely through immiscible liquids, Trans IChemE, 54, 167-173 (1976), 55, 285 (1977).

Landau, J.; Gomaa, H.G., and Taweel, A.M.A., Measurement of large interfacial areas by light attenuation, Trans IChemE, 55, 212-215 (1977).

Mattila, T.K., and Lehto, T.K., Nitrate removal from waste solutions by solvent extraction, Ind Eng Chem Proc Des Dev, 16(4), 469-472 (1977).

McDonald, C.W., and Bajwa, R.S., Removal of toxic metal ions from metal-finishing wastewater by solvent extraction, Sepn Sci, 12(4), 435-446 (1977).

Miller, J.W., and Harper, D.O., Enhancement of liquid extraction by instantaneous irreversible reaction, Can JCE, 55, 534-540 (1977).

Omran, N.M., and Foster, P.J., The terminal velocity of a chain of drops or bubbles in a liquid, Trans IChemE, 55, 171-177 (1977).

Otake, T.; Tone, S.; Nakao, K., and Mitsuhashi, Y., Coalescence and breakup of bubbles in liquids, Chem Eng Sci, 32(4), 377-384 (1977).

Takeuchi, H., and Numata, Y., Effect of interfacial turbulence on liquid-liquid mass transfer rates, Int Chem Eng, 17(3), 468-474 (1977).

Whewell, R.J., Kinetics of solvent extraction of copper, Chem & Ind, 17 Sept, 755-760 (1977).

Karabelas, A.J., Droplet size spectra generated in turbulent pipe flow of dilute liquid/liquid dispersions, AIChEJ, 24(2), 170-180 (1978).

McDonald, C.W., and Butt, N., Solvent extraction studies of zinc with Alamine-336 in aqueous chloride and bromide media, Sepn Sci Technol, 13(1), 39-46 (1978).

1979-1980

Baird, M.H.I., and Ho, M.K., Liquid-liquid extraction in laminar slug flow, Can JCE, 57, 467-480 (1979).

Brooks, B.W., Drop size distributions in an agitated liquid-liquid dispersion, Trans IChemE, 57, 210-212 (1979).

Hanson, C., Solvent extraction, Chem Eng, 7 May, 83-87 (1979).

Kawano, Y., and Nakashio, F., Effect of an interfacial reaction on extraction rate, Int Chem Eng, 19(1), 131-137 (1979).

Kikic, I.; Alessi, P., and Lapasin, R., Butyronitrile-C8H16O as a double solvent for hydrocarbon separation, Chem Eng J, 18(1), 39-46 (1979).

McManamey, W.J., Sauter mean and maximum drop diameters of liquid-liquid dispersions in turbulent, agitated vessels at low dispersed-phase holdup, Chem Eng Sci, 34(3), 432-434 (1979).

Mukhopadhyay, M., Thermodynamic method based upon theory of regular solutions for selection of solvents and process conditions for aromatics extraction, J Chem Tech Biotechnol, 29, 634-641 (1979).

Ottino, J.M., Ranz, W.E., and Macosko, C.W., Lamellar model for analysis of liquid-liquid mixing, Chem Eng Sci, 34(6), 877-890 (1979).

Sagert, N.H., and Quinn, M.J., Coalescence of n-hexane droplets in aqueous solution of n-alcohols, Can JCE, 57, 29-39 (1979).

Scholtens, B.J.R.; Bruin, S., and Bijsterbosch, B.H., Liquid-liquid mass transfer and its retardation by macromolecular adsorption, Chem Eng Sci, 34(5), 661-670 (1979).

Smith, I.K., and Garmendia, L.A., The effects of an electrostatic field and air stream on water spray droplet size, Trans IChemE, 57, 60-63 (1979).

Snyder, L., Solvents and solubility, Chemtech, Dec, 750-754 (1979).

Arashmid, M., and Jeffreys, G.V., Analysis of phase inversion characteristics of liquid-liquid dispersions, AIChEJ, 26(1), 51-55 (1980).

Grosjean, P.R.L., and Sawistowski, H., Liquid-liquid mass transfer accompanied by instantaneous chemical reaction, Trans IChemE, 58, 59-65 (1980).

Kikic, I.; Alessi, P., and Cividini, A., Solvent extraction of furfural from aqueous solutions, Can JCE, 58, 119-122 (1980).

Molag, M.; Joosten, G.E.H., and Drinkenburg, A.A.H., Droplet breakup and distribution in stirred immiscible two-liquid systems, Ind Eng Chem Fund, 19(3), 275-281 (1980).

Voigtlander, R.; Blaschke, H.G.; Halwachs, W., and Schugerl, K., Investigation of mass transfer across the interface of two concurrent laminar flowing liquids in a horizontal cylindrical channel, Chem Eng Sci, 35(5), 1211-1222 (1980).

Wankat, P.C., Calculations for separations with three phases, Ind Eng Chem Fund, 19(4), 358-363 (1980).

1981-1982

Al-Saadi, A.N., and Jeffreys, G.V., Esterification of butanol in a two-phase liquid-liquid system, AIChEJ, 27(5), 754-773 (1981).

Austin, D.G., and Jeffreys, G.V., Coalescence phenomena in liquid-liquid systems, J Chem Tech Biotechnol, 31, 475-488 (1981).

Bailes, P.J., and Larkai, S.K.L., An experimental investigation into the use of high voltage D.C. fields for liquid phase separation, Trans IChemE, 59, 229-237 (1981).

Coppus, J.H.C., and Rietema, K., Mass transfer from spherical cap bubbles: Contribution of the bubble rear, Trans IChemE, 59, 54-63 (1981).

Luks, K.D., and Kohn, J.P., LNG liquid-liquid immiscibility, Hyd Proc, 60(9), 257-258 (1981).

Monaghan, J.; Hill, G.A., and Soucey, W.G., Formulae for mass-median and mass-mean drop diameters, Can JCE, 59, 776-780 (1981).

Rod, V.; Lukesova, and Sir, Z., Kinetics of extraction by solvation, Chem Eng Commns, 11(4), 281-290 (1981).

Thornton, J.D., and Anderson, T.J., Surface renewal phenomena in liquid-liquid droplet systems with and without mass transfer, Int J Heat Mass Trans, 24(11), 1847-1848 (1981).

Various, Solvent extraction of biological molecules (topic issue), Chem & Ind, 16 May, 355-362 (1981).

Aguirre, F.J.; Klinzing, G.E.; Chiang, S.H., and Jing, W.K., Temperature measurements during mass transfer in partially miscible liquid-liquid systems, Chem Eng Commns, 17(1), 117-122 (1982).

Bailes, P.J., and Larkai, S.K.L., Liquid phase separation in pulsed D.C. fields, Trans IChemE, 60, 115-121 (1982).

Barnes, J.E., and Edwards, J.D., Solvent extraction for precious metal extraction, Chem & Ind, 6 March, 151-155 (1982).

Eyal, A., and Baniel, A., Liquid-liquid extraction of strong mineral acids by organic acid-base couples, Ind Eng Chem Proc Des Dev, 21(2), 334-337 (1982).

Freeman, R.L., and Tavlarides, L.L., Study of interfacial kinetics for liquid-liquid systems, Chem Eng Sci, 35(3), 559-566 (1980); 37(10), 1547-1556 (1982).

Greminger, D.C.; Burns, G.P.; Lynn, S.; Hanson, D.N., and King, C.J., Solvent extraction of phenols from water, Ind Eng Chem Proc Des Dev, 21(1), 51-54 (1982).

Hirata, A.; Hattori, T., and Nishimura, K., Multiple-solvent, single-stage extraction calculations, Int Chem Eng, 22(1), 99-107 (1982).

Leeper, S.A., and Wankat, P.C., Gasohol production by extraction of ethanol from water using gasoline as solvent, Ind Eng Chem Proc Des Dev, 21(2), 331-334 (1982).

Lewis, D.A., and Davidson, J.F., Bubble splitting in shear flow, Trans IChemE, 60, 283-291 (1982).

Mersmann, A., and Grossmann, H., Dispersion of immiscible liquids in agitated vessels, Int Chem Eng, 22(4), 581-591 (1982).

Rod, V., and Misek, T., Stochastic modelling of dispersion formation in agitated liquid-liquid systems, Trans IChemE, 60, 48-53 (1982).

Various, Carbon dioxide in solvent extraction (topic issue), Chem & Ind, 19 June, 385-405 (1982).

1983-1984

Ajawin, L.A.; de Ortiz, E.S.P., and Sawistowski, H., Extraction of zinc by di(2-ethylhexyl) phosphoric acid, CER&D, 61, 62-66 (1983).

Bailes, P.J.; Godfrey, J.C., and Slater, M.J., Liquid-liquid extraction test systems, CER&D, 61, 321-324 (1983).

Bapat, P.M.; Tavlarides, L.L., and Smith, G.W., Monte Carlo simulation of mass transfer in liquid-liquid dispersions, Chem Eng Sci, 38(12), 2003-2014 (1983).

Maljkovic, D., and Maljkovic, D., Three liquid phases in extraction of hydrochloric acid with mixture of di-isopropyl ether and n-pentyl alcohol, Solv Extn & Ion Exch, 1(2), 281-298 (1983).

McDowell, W.J.; Michelson, D.C.; Moyer, B.A., and Coleman, C.F., A source of solvent extraction information, Solv Extn & Ion Exch, 1(1), 1-4 (1983).

Nandi, B.; Das, N.R., and Bhattacharyya, S.N., Solvent extraction of zirconium and hafnium, Solv Extn & Ion Exch, 1(1), 141-202 (1983).

Rod, V., Wei-Yang, F., and Hanson, C., Evaluation of mass transfer and backmixing parameters in extraction columns from measured solute concentrations, CER&D, 61, 290-296 (1983).

Takahashi, K.; Kamiya, H.; Uchida, S., and Takeuchi, H., Extraction of copper by LIX65N in a vessel and a multistage column under stirred conditions, Solv Extn & Ion Exch, 1(2), 311-336 (1983).

Wisniak, J., Liquid-liquid phase splitting, Chem Eng Sci, 38(6), 969-978 (1983).

Bailes, P.J., and Larkai, S.K.L., Influence of phase ratio on electrostatic coalescence of water-in-oil dispersions, CER&D, 62, 33-38 (1984).

Bapat, P.M., and Tavlarides, L.L., Phase separation techniques for liquid dispersions, Ind Eng Chem Fund, 23(1), 120-123 (1984).

Bhattacharyya, S.N., and Ganguly, B.N., Solvent-extraction separation of niobium and tantalum (review paper), Solv Extn & Ion Exch, 2(4), 699-740 (1984).

Burkholz, A., Droplet separators, Int Chem Eng, 24(4), 618-629 (1984).

Croker, J.R., and Bowrey, R.G., Liquid extraction of furfural from aqueous solution, Ind Eng Chem Fund, 23(4), 480-484 (1984).

Danesi, P.R., Relative importance of diffusion and chemical reactions in liquid-liquid extraction kinetics, Solv Extn & Ion Exch, 2(1), 29-44 (1984).

Grilc, V.; Golob, J., and Modic, R., Drop coalescence in liquid-liquid dispersions by flow through glass fibre beds, CER&D, 62, 48-52 (1984).

Humphrey, J.L.; Rocha, J.A., and Fair, J.R., The essentials of extraction, Chem Eng, 17 Sept, 76-95 (1984).

Miller, J.D., and Mooiman, M.B., Review of new developments in amine solvent extraction systems for hydrometallurgy, Sepn Sci Technol, 19(11), 895-910 (1984).

Munson, C.L., and King, C.J., Factors influencing solvent selection for extraction of ethanol from aqueous solutions, Ind Eng Chem Proc Des Dev, 23(1), 109-115 (1984).

Nakaike, Y.; Mizukoshi, T.; Aonuma, T., and Tadaki, T., Mass transfer accompanied by interfacial turbulence during droplet formation in a liquid-liquid system, Int Chem Eng, 24(3), 527-536 (1984).

Rahman, M.; Mikitenko, P., and Asselineau, L., Solvent extraction of aromatics from middle distillates: Equilibria prediction method by group contribution, Chem Eng Sci, 39(11), 1543-1558 (1984).

Schulz, W.W., and Navratil, J.D., Development and applications of bifunctional organophosphorus liquid-liquid extraction reagents, Sepn Sci Technol, 19(11), 927-942 (1984).

Steiner, L.; Berger, J., and Hartland, S., Mass transfer rates in liquid-liquid extraction with partially miscible solvents, Solv Extn & Ion Exch, 2(4), 553-579 (1984).

Wisniak, J., Liquid-liquid phase splitting, Chem Eng Sci, 39(1), 111-116; 39(6), 967-974 (1984).

Yagodin, G.A., and Tarasov, V.V., Interfacial phenomena in liquid-liquid extraction, Solv Extn & Ion Exch, 2(2), 139-178 (1984).

1985

Aguirre, F.J.; Klinzing, G.E., and Chiang, S.H., Use of image intensification for mass transfer studies in liquid-liquid systems, Int J Heat Mass Trans, 28(10), 1891-1898 (1985).

Beaver, W.H., and Turpin, J.L., Ternary extraction modeling using the Redlich-Kister expansion, Chem Eng Commns, 33(5), 277-286 (1985).

Cooney, D.O., and Jin, C.L., Solvent extraction of phenol from aqueous solution in hollow fiber device, Chem Eng Commns, 37(1), 173-192 (1985).

Gradon, L., and Orlicki, D., Separation of liquid mixtures in the freezing-out process: Mathematical description and experimental verification, Int J Heat Mass Trans, 28(11), 1983-1990 (1985).

Horwitz, E.P., et al., The TRUEX process: Extraction of transuranic elements from nitric acid wastes using modified Purex solvent, Solv Extn & Ion Exch, 3(1), 75-110 (1985).

Joos, F.M., and Snaddon, R.W.L., Frequency dependence of electrically enhanced emulsion separation, CER&D, 63, 305-311 (1985).

Kandil, A.T.; El-Atrash, A.M., and El-Medani, S.M., Solvent extraction of Cu(2+) by benzeneazo-2-naphthol, Solv Extn & Ion Exch, 3(4), 453-472 (1985).

Lee, J.M., and Soong, Y., Effects of surfactants on the liquid-liquid dispersions in agitated vessels, Ind Eng Chem Proc Des Dev, 24(1), 118-121 (1985).

Riquelme, G.D.; Nunez, P.M., and Triday, J.L., Mathematical model for metal solvent extraction systems (selectivity effect), Chem Eng Commns, 39, 1-22 (1985).

Ruiz, F.; Marcilla, A.; Ancheta, A.M., and Caro, J.A., Purification of wet process phosphoric acid by solvent extraction with isoamyl alcohol, Solv Extn & Ion Exch, 3(3), 331-356 (1985).

Schlichting, E.; Halwachs, W., and Schugerl, K., Reactive extraction of salicylic acid and D,L-phenylalanine in bench-scale pulsed sieve plate column, Chem Eng & Proc, 19(6), 317-328 (1985).

Sebba, F., Predispersed solvent extraction, Sepn Sci Technol, 20(5), 331-334 (1985).

Shuler, R.G., et al., Extraction of cesium and strontium from acidic high-activity nuclear waste using Purex-process compatible organic extractant, Solv Extn & Ion Exch, 3(5), 567-604 (1985).

Thornton, J.D.; Anderson, T.J.; Javed, K.H., and Achwal, S.K., Surface phenomena and mass transfer interactions in liquid-liquid systems, AIChEJ, 31(7), 1069-1076 (1985).

Yorulmaz, Y., and Karpuzcu, F., Sulpholane versus diethylene glycol in recovery of aromatics, CER&D, 63, 184-190 (1985).

1986

Bensalem, A.; Steiner, L., and Hartland, S., Effect of mass transfer on drop size in Karr column, Chem Eng & Proc, 20(3), 129-136 (1986).

Cockrem, M.C.M.; Meyer, R.A., and Lightfoot, E.N., Recovery on n-butanol from dilute solution by extraction, Sepn Sci Technol, 21(10), 1059-1074 (1986).

Dalingaros, W.; Kumar, A., and Hartland, S., Effect of physical properties and dispersed-phase velocity on drop size produced at multi-nozzle distributor, Chem Eng & Proc, 20(2), 95-102 (1986).

Dekker, M.; Van Riet, K.; Weijers, S.R.; Baltussen, W.A.; Laane, C., and Bijsterbosch, B.H., Enzyme recovery by liquid-liquid extraction using reversed micelles, Biochem Eng J, 33(2), B27-B34 (1986).

Dreisinger, D.B., and Cooper, W.C., Use of rotating diffusion cell technique for study of solvent extraction kinetics, Solv Extn & Ion Exch, 4(1), 135-148 (1986).

Gaikar, V.G., and Sharma, M.M., Extractive separations with hydrotropes, Solv Extn & Ion Exch, 4(4), 839-846 (1986).

Gasparini, G.M., and Grossi, G., Long-chain distributed aliphatic amides as extracting agents in industrial applications of solvent extraction, Solv Extn & Ion Exch, 4(6), 1233-1272 (1986).

Giavedoni, M.D., and Deiber, J.A., Model of mass transfer through a fluid-fluid interface, Chem Eng Sci, 41(7), 1921-1925 (1986).

Hanzevack, E.L., Concentration determination in liquid-liquid flow systems by laser image processing, Chem Eng Prog, 82(1), 47-50 (1986).
Jeelani, S.A.K., and Hartland, S., Variation of drop size and hold-up in dense-packed dispersion, Chem Eng & Proc, 20(5), 271-276 (1986).
Kreysa, G., and Woebcken, C., Mass transfer in liquid-liquid systems, Chem Eng Sci, 41(2), 307-316 (1986).
Martin, K.A.; Horwitz, E.P., and Ferraro, J.R., Infrared studies of bifunctional extractants, Solv Extn & Ion Exch, 4(6), 1149-1170 (1986).
Mehra, Y., Using extraction to treat hydrocarbon gases, Chem Eng, 27 Oct, 53-55 (1986).
Ruiz, F., and Gomis, V., Correlation of quaternary liquid-liquid equilibrium data using UNIQUAC, Ind Eng Chem Proc Des Dev, 25(1), 216-221 (1986).
Sebba, F., Polyaphrons in process engineering, Chem Engnr, March, 12-14 (1986).
Yamada, H.; Adachi, K.; Fujii, Y., and Mizuta, M., Comparison of benzoic acid with decanoic acid as extracting agents, Solv Extn & Ion Exch, 4(6), 1109-1120 (1986).

1987

Babcock, R.E.; Beaver, W.H., and Turpin, J.L., Recovery of fatty acid from a solvated hydrocarbon mixture applicable to the Beaver-Herter solvent extraction process, Chem Eng Commns, 57, 245-262 (1987).
Chaiko, D.J., and Osseo-Asare, K., The air-water interface as a model system for solvent extraction monolayer studies, Solv Extn & Ion Exch, 5(2), 277-286 (1987).
Chaiko, D.J., and Osseo-Asare, K., Aqueous solubility of solvent extraction reagents: A monolayer dissolution method, Solv Extn & Ion Exch, 5(2), 287-300 (1987).
Chiarizia, R., and Horwitz, E.P., Solvent extraction of selected organic complexants by Truex process solvent, Solv Extn & Ion Exch, 5(1), 175-194 (1987).
Goklen, K.E., and Hatton, T.A., Liquid-liquid extraction of low molecular weight proteins by selective solubilization in reversed micelles, Sepn Sci Technol, 22(2), 831-842 (1987).
Gupte, P.A., and Danner, R.P., Prediction of liquid-liquid equilibria with UNIFAC: A critical evaluation, Ind Eng Chem Res, 26(10), 2036-2042 (1987).
Harvala, T.; Alkio, M., and Komppa, V., Extraction of tall oil with supercritical carbon dioxide, CER&D, 65(5), 386-389 (1987).
Honda, H.; Mano, T.; Taya, M.; Shimizu, K.; Matsubara, M., and Kobayashi, T., A general framework for the assessment of extractive fermentations, Chem Eng Sci, 42(3), 493-498 (1987).
Jurkiewicz, K., Ion flotation and solvent extraction of ferric thiocyanate complexes, Sepn Sci Technol, 22(12), 2381-2402 (1987).
Martin, G., Extraction from viscous polymer solutions, Chem Eng Prog, 83(8), 54-59 (1987).
Ramirez, F.M.; Jimenez-Reyes, M., and Maddock, A.G., Third-phase formation in the solvent extraction system: Ferrous chloride-HCl-water-Dipe, Solv Extn & Ion Exch, 5(3), 533-560 (1987).

Sereno, A.M.; Anderson, T.F., and Medina, A.G., Dynamic simulation of liquid-liquid operations using simple non-linear models, Comput Chem Eng, 11(2), 177-185 (1987).

Shimizu, K., and Matsubara, M., A solvent screening criterion for multicomponent extractive fermentation, Chem Eng Sci, 42(3), 499-504 (1987).

1988

Abbott, N.L., and Hatton, T.A., Liquid-liquid extraction for protein separations, Chem Eng Prog, 84(8), 31-41 (1988).

Cambridge, V.J.; Constant, W.D., and Wolcott, J.M., Measurement of liquid-liquid interfacial shear viscosity, Chem Eng Commns, 70, 137-156 (1988).

Egan, B.Z.; Lee., D.D., and McWhirter, D.A., Solvent extraction and recovery of ethanol from aqueous solutions, Ind Eng Chem Res, 27(7), 1330-1332 (1988).

Eubank, P.T., and Barrufet, M.A., Simple algorithm for calculation of phase separation, Chem Eng Educ, 22(1), 36-41 (1988).

Farag, I.H., and Peshori, D.L., Computer-aided graphics of liquid-liquid extraction, Chem Eng Commns, 65, 29-38 (1988).

Golding, J.A., and Barclay, C.D., Equilibrium characteristics for extraction of cobalt and nickel into di(2-ethylhexyl)phosphoric acid, Can JCE, 66(6), 970-979 (1988).

Guilinger, T.R.; Grislingas, A.K., and Erga, O., Phase inversion behavior of water-kerosene dispersions, Ind Chem Eng Res, 27(6), 978-982 (1988).

Guimaraes, M.M.L., and Cruz-Pinto, J.J.C., Mass transfer and dispersed-phase mixing in liquid-liquid systems, Comput Chem Eng, 12(11), 1075-1082 (1988).

Likidis, Z., and Schugerl, K., Reextraction of Penicillin G and V from organic phase in bench-scale Karr column after reactive extraction by LA-2, Chem Eng & Proc, 23(1), 61-64 (1988).

Mukherjee, D.; Buswas, M.N., and Mitra, A.K., Hydrodynamics of liquid-liquid dispersion in ejectors and vertical two-phase flow, Can JCE, 66(6), 896-907 (1988).

Nolan, B.T., and McTernan, W.F., Application of solvent sublation to simultaneous removal of emulsified coal tar and dissolved organics, Chem Eng Commns, 63, 1-16 (1988).

Patil, T.A.; Sawant, S.B.; Joshi, J.B., and Sikdar, S.K., Mass-transfer coefficients in two-phase aqueous extraction, Biochem Eng J., 39(1), B1-B6 (1988).

Ruiz, F.; Gomis, V., and Botella, R.F., Extraction of ethanol from aqueous solution, Ind Eng Chem Res, 27(4), 648-650 (1988).

Salem, A.B., Modeling phase inversion in a mixer-settler, Sepn Sci Technol, 23(1), 261-272 (1988).

Schugerl, K., et al., Reactive extraction, Int Chem Eng, 28(3), 393-406 (1988).

Tanigawa, M.; Nishiyama, S.; Tsuruya, S., and Masai, M., Solvent extraction of alkali metals by crown ethers, Chem Eng J, 39(3), 157-168 (1988).

Various, Solvent extraction processes and advances (symposium papers), Sepn Sci Technol, 23(12), 1191-1472 (1988).

Zhang, Y., and Muhammed, M., Removal of nitric acid from concentrated phosphoric acid by extraction with methyl isobutyl ketone in aromatic diluents, Solv Extn & Ion Exch, 6(6), 993-1006 (1988).
Zhang, Y., and Muhammed, M., Extraction of phosphoric acid from nitrate solutions with isoamyl alcohol, Solv Extn & Ion Exch, 6(6), 973-992 (1988).

21.2 Equipment Design

1967

Angelino, H.; Alran, C.; Boyadzhiev, L., and Mukherjee, S.P., Efficiency of a pulsed extraction column with rotary agitators, Brit Chem Eng, 12(12), 1893-1895 (1967).
Baird, M.H.I., and Garstang, J.H., Power consumption and gas hold-up in a pulsed column, Chem Eng Sci, 22(12), 1663-1674 (1967).
Belter, P.A., and Speaker, S.M., Controlled-cycle operations applied to extraction processes, Ind Eng Chem Proc Des Dev, 6(1), 36-42 (1967).
Burns, P.E., and Hanson, C., Transient response of a multistage mixer-settler, Brit Chem Eng, 12(1), 75-77 (1967).
Chen, H.T., and Middleman, S., Drop size distribution in agitated liquid-liquid systems, AIChEJ, 13(5), 989-995 (1967).
Fernandes, J.B., and Sharma, M.M., Effective interfacial area in agitated liquid-liquid contactors, Chem Eng Sci, 22(10), 1267-1282 (1967).
Gelperin, N.I., et al., Determination of interfacial surface area in liquid extraction processes by sedimentation method, Int Chem Eng, 7(2), 231-235 (1967).
Greskovich, E.J.; Barton, P., and Hersh, R.E., Heat transfer in liquid-liquid spray towers, AIChEJ, 13(6), 1160-1166 (1967).
Hartland, S., The optimization of forward and back extraction, Trans IChemE, 45, T82-93 (1967).
Hartland, S., and Wise, G.D., Back-mixing in a Morris contactor, Trans IChemE, 45, T353-359 (1967).
Hughmark, G.A., Liquid-liquid spray column drop size, holdup and continuous-phase mass transfer, Ind Eng Chem Fund, 6(3), 408-413 (1967).
Krishnaiah, M.M.; Pai, M.U.; Rao, M.V.R., and Sastri, S.R.S., Performance of a rotating-disc contactor with perforated rotors, Brit Chem Eng, 12(5), 719-721 (1967).
Letan, R., and Kehat, E., Mechanics of a spray column, AIChEJ, 13(3), 443-449 (1967).
McAllister, R.A.; Groenier, W.S., and Ryon, A.D., Correlation of flooding in pulsed perforated-plate extraction columns, Chem Eng Sci, 22(7), 931-944 (1967).
Smith, S., Use of reflux in solvent extraction, Brit Chem Eng, 12(9), 1361-1363 (1967).
Sprow, F.B., Drop size distributions in strongly coalescing agitated liquid-liquid systems, AIChEJ, 13(5), 995-998 (1967).
Susanov, E.Y., et al., Use of sieve-tray columns for liquid extractions, Int Chem Eng, 7(3), 423-428 (1967).

1968

Doninger, J.E., and Stevens, W.F., Dynamic behavior of a packed liquid-extraction column, AIChEJ, 14(4), 591-599 (1968).

Doyle, C.M.; Doyle, W.G.P.; Rauch, E.H., and Lowry, C.D., Centrifugal liquid-liquid extractors, Chem Eng Prog, 64(12), 68-74 (1968).

Eubanks, I.D., and Lowe, J.T., Steady-state solvent extraction calculations for curium recovery, Ind Eng Chem Proc Des Dev, 7(2), 172-177 (1968).

Fernandes, J.B., and Sharma, M.M., Air-agitated liquid-liquid contactors, Chem Eng Sci, 23(1), 9-16 (1968).

Hanson, C., Solvent extraction: A review, Chem Eng, 26 Aug, 76-98; 9 Sept, 135-142 (1968).

Lovland, J., Graphical solution of cyclic extraction, Ind Eng Chem Proc Des Dev, 7(1), 65-67 (1968).

Lowe, J.T., Calculation of transient behavior of solvent extraction processes, Ind Eng Chem Proc Des Dev, 7(3), 362-366 (1968).

Mecklenburgh, J.C., and Hartland, S., Two-phase countercurrent extraction with high backmixing, Chem Eng Sci, 23(12), 1421-1430 (1968).

Misek, T., Operating conditions in mechanical liquid extractors, Int Chem Eng, 8(3), 439-442 (1968).

Mumford, C.J., Advances in equipment for liquid-liquid extraction, Brit Chem Eng, 13(7), 981-986 (1968).

Murty, R.K., and Rao, C.V., Perforated-plate liquid-liquid extraction towers, Ind Eng Chem Proc Des Dev, 7(2), 166-172 (1968).

1969

Bell, R.L., and Babb, A.L., Holdup and axial distribution of holdup in a pulsed sieve-plate solvent extraction column, Ind Eng Chem Proc Des Dev, 8(3), 392-400 (1969).

Gelperin, N.I.; Pebalk, V.L., and Shashkova, M.N., Mass transfer and longitudinal mixing in a horizontal-tubular multisectional extractor, Int Chem Eng, 9(3), 391-395 (1969).

Hartland, S., Optimum operation of an existing forward and back extractor, Chem Eng Sci, 24(7), 1075-1082 (1969).

Keey, R.B., and Glen, J.B., Area-free mass transfer coefficients for liquid extraction in a continuously worked mixer, AIChEJ, 15(6), 942-947 (1969).

Lelli, U.; Gatta, A., and Pasquali, G., Heat transfer in multistage mixer columns, Chem Eng Sci, 24(8), 1203-1212 (1969).

Mecklenburgh, J.C., and Hartland, S., Design methods for countercurrent flow with backmixing: Solvent extraction with partially miscible solvents, Chem Eng Sci, 24(7), 1063-1074 (1969).

Rahn, R.W., and Smutz, M., Development of a laboratory scale continuous multistage extractor, Ind Eng Chem Proc Des Dev, 8(3), 289-293 (1969).

Roche, E.C., Rigorous solution of multicomponent multistage liquid-liquid extraction problems, Brit Chem Eng, 14(10), 1393-1397 (1969).

Sebenik, R.F., and Smutz, M., Optimization of a solvent extraction plant to process monazite rare earth nitrates, Ind Eng Chem Proc Des Dev, 8(2), 225-231 (1969).

Thomas, W.J., Oscillating baffle contactor: A new liquid-liquid extractor, Trans IChemE, 47, T304-314 (1969).
Thomas, W.J., and Chui, Y.T., Efficiency and axial dispersion in an oscillating baffle contactor, Trans IChemE, 47, T315-321 (1969).
Wellek, R.M., et al., Liquid-extraction column with reciprocated wire mesh packing, Ind Eng Chem Proc Des Dev, 8(4), 515-527 (1969).
Zanker, A., Nomograph for determining extraction efficiency, Chem Eng, 24 Feb, 132 (1969).

1970

Burrill, K.A., and Woods, D.R., Separation of two immiscible liquids in a hydrocyclone, Ind Eng Chem Proc Des Dev, 9(4), 545-552 (1970).
Cadman, T.W., and Hsu, C.K., Dynamics and control of multistage liquid extraction, Trans IChemE, 48, T209-226 (1970).
Foster, H.R.; McKee, R.E., and Babb, A.L., Transient holdup behavior of pulsed sieve-plate solvent extraction column, Ind Eng Chem Proc Des Dev, 9(2), 272-278 (1970).
Grigar, K.; Prochazka, J., and Landau, J., Loss in pressure associated with drop formation in liquid-liquid systems, Chem Eng Sci, 25(11), 1773-1784 (1970).
Hartland, S., and Mecklenburgh, J.C., Effect of backmixing on the concepts of height of transfer unit and stage efficiency, Brit Chem Eng, 15(2), 216-219 (1970).
Henton, J.E., and Cavers, S.D., Continuous-phase axial dispersion in liquid-liquid spray towers, Ind Eng Chem Fund, 9(3), 384-392 (1970).
Jeffreys, G.V.; Davies, G.A., and Pitt, K., Rate of coalescence of dispersed-phase in laboratory mixer/settler, AIChEJ, 16(5), 823-831 (1970).
Kafarov, V.V.; Vygon, V.G., and Gordeev, L.S., Turbulent diffusion in continuous phase of a pulsating extractor, Int Chem Eng, 10(3), 341-345 (1970).
Massoumi, A.; Edrissi, M., and Hedrick, C.E., Solvent extraction using dialysis through artificial phase boundaries, J Appl Chem, 20, 357-360 (1970).
Miller, J.W., and Eastburn, F.J., Individual film resistances in liquid extraction in a one inch i.d. York-Scheibel column, Ind Eng Chem Proc Des Dev, 9(4), 601-604 (1970).
Misek, T., and Marek, J., Asymmetric rotating disc extractor, Brit Chem Eng, 15(2), 202-207 (1970).
Puranik, S.A., and Sharma, M.M., Effective interfacial area in packed liquid extraction columns, Chem Eng Sci, 25(2), 257-266 (1970).
Sallaly, M., and Reynier, J.P., Dispersed phase dynamics in packed columns for cocurrent liquid-liquid downflow, Chem Eng Sci, 25(11), 1709-1718 (1970).
Spalding, W.M., Determination of tie lines in liquid systems, Brit Chem Eng, 15(1), 62 (1970).
Thomas, W.J., Performance of a rotating disc contactor using carbon tetrachloride-acetic acid-water system, J Appl Chem, 20, 261-273 (1970).
Thomas, W.J., and Weng, P.K.J., Oscillating baffle column fitted with stator rings, J Appl Chem, 20, 21-29, 48-68 (1970).

1971

Chen, E.C., and Chon, W.Y., Effects of ultrasonic vibrations on the droplet size in spray columns, Brit Chem Eng, 16(10), 919-921 (1971).

Haug, H.F., A correlation for backmixing in multistage agitated contactors, AIChEJ, 17(3), 585-589 (1971).

Heertjes, P.M.; de Nie, L.H., and de Vries, H.J., Drop formation in liquid-liquid systems, Chem Eng Sci, 26(2), 441-460 (1971).

Johnson, T.R.; Pierce, R.D.,; Teats, F.G., and Johnston, E.F., Behavior of countercurrent liquid-liquid columns with a liquid metal, AIChEJ, 17(1), 14-18 (1971).

Kehat, E., and Letan, R., Role of wakes in the mechanism of extraction in spray columns, AIChEJ, 17(4), 984-990 (1971).

McFerrin, A.R., and Davison, R.R., Effect of surface phenomena on a solvent extraction process, AIChEJ, 17(5), 1021-1027 (1971).

Michot, G., and Wiegandt, H.F., Countercurrent liquid-liquid extraction in rotating multiple helix, Ind Eng Chem Proc Des Dev, 10(4), 586-592 (1971).

Nemunaitis, R.R.; Eckert, J.S.; Foote, E.H., and Rollison, L.R., Packed liquid-liquid extractors, Chem Eng Prog, 67(11), 60-67 (1971).

Pratt, H.R.C., Axial dispersion in differential extraction columns with 'perfect' mass transfer, Ind Eng Chem Fund, 10(1), 170-174 (1971).

Prochazka, J.; Landau, J.; Souhrada, F., and Heyberger, A., Reciprocating-plate liquid-liquid extraction column, Brit Chem Eng, 16(1), 42-44 (1971).

Rod, V., Model describing longitudinal hold-up profiles in mixed extraction columns, Brit Chem Eng, 16(7), 617-619 (1971).

Witt, P.A., and Forbes, M.C., By-product recovery via solvent extraction, Chem Eng Prog, 67(10), 90-94 (1971).

Ziolkowski, Z., and Pajak, M., Mass transfer and longitudinal displacement in extraction column with rotating disks (Peclet numbers in continuous and disperse phases), Int Chem Eng, 11(3), 530-536 (1971).

1972

Aly, G., and Wittenmark, B., Dynamic behaviour of mixer-settlers, J Appl Chem Biotechnol, 22, 1165-1184 (1972).

Burge, D.A., and Clements, W.C., Distributed-parameter model for liquid extraction, Chem Eng Sci, 27(8), 1537-1548 (1972).

Cinelli, E.; Noe, S., and Paret, G., Solvent extraction of aromatics, Hyd Proc, 51(4), 141-144 (1972).

Davies, G.A.; Jeffreys, G.V., and Azfal, M., New packings for coalescence and separation of liquid dispersions, Brit Chem Eng, 17(9), 709-714 (1972).

De, A.K., and Ray, U.S., Solvent extraction behavior of transitional metals with liquid ion-exchangers, Sepn Sci, 6(1), 25-35 (1971); 7(4), 409-424 (1972).

Dunn, I.J., and Ingham, J., Digital simulation of extraction columns with backmixing, Chem Eng Sci, 27(9), 1751-1753 (1972).

Hutton, A.E., and Holland, C.D., Use of field tests in modelling of liquid-liquid extraction columns, Chem Eng Sci, 27(5), 919-936 (1972).

Ingham, J., Back-mixing in a multi-mixer liquid-liquid extraction column, Trans IChemE, 50, 372-385 (1972).

Jeffreys, G.V.; Mumford, C.J., and Herridge, M.H., Optimisation of a liquid extraction process, J Appl Chem Biotechnol, 22, 319-334 (1972).

Kagan, S.Z.; Trukhanov, V.G., and Ogai, Y.A., High-efficiency rotor extractor for liquid-liquid systems (hydrodynamic characteristics and mass-transfer rates), Int Chem Eng, 12(3), 396-401 (1972).

Lelli, U.; Magelli, F., and Sama, C., Backmixing in multistage mixer columns, Chem Eng Sci, 27(5), 1109-1118 (1972).

Lo, T.C., and Karr, A.E., Development of a laboratory-scale reciprocating-plate extraction column, Ind Eng Chem Proc Des Dev, 11(4), 495-501 (1972).

Noel, D.E., and Meloan, C.E., Some empirical correlations in solvent extraction, Sepn Sci, 7(1), 75-84 (1972).

Perrut, M., and Loutaty, R., Drop size in a liquid-liquid dispersion, Chem Eng J, 3(3), 286-293 (1972).

Reynier, J.P., and Rojey, A., Statistical model for dispersed-phase axial mixing, Chem Eng J, 3(2), 187-195 (1972).

Rojey, A., Efficiency of a multistage process, Chem Eng Sci, 27(11), 1901-1908 (1972).

Sheikh, A.R.; Ingham, J., and Hanson, C., Axial mixing in a Graesser raining bucket liquid-liquid contactor, Trans IChemE, 50, 199-207 (1972).

Simonis, H., Solvent extraction unit for difficult separations, Proc Engng, Nov, 110-112 (1972).

Todd, D.B., Improving performance of centrifugal extractors, Chem Eng, 24 July, 152-158 (1972).

Vedaujan, S., et al., Performance characteristics of spray columns, AIChEJ, 18(1), 161-168 (1972).

Wagle, R.D., and Dwivedi, M.C., Calculate furfural solvent extraction yields, Hyd Proc, 51(6), 136-138 (1972).

Watson, J.S., and McNeese, L.E., Axial dispersion in packed columns during countercurrent flow, Ind Eng Chem Proc Des Dev, 11(1), 120-121 (1972).

Zeitlin, M.A., and Tavlarides, L.L., Prediction of mass-transfer coefficients, local concentrations and binary and ternary mass-transfer rates for extractors, Ind Eng Chem Proc Des Dev, 11(4), 532-537 (1972).

1973

Aly, G., and Ottertun, H., Dynamic behaviour of mixer-settlers, J Appl Chem Biotechnol, 23, 643-660 (1973).

Baird, M.H.I., and Lane, S.J., Drop size and holdup in reciprocating plate extraction column, Chem Eng Sci, 28(3), 947-958 (1973).

Bruin, S., Geometry effects on axial mixing of the continuous phase in rotating disc contactors, Trans IChemE, 51, 355-360 (1973).

Casto, M.G.; Hoh, Y.C.; Smutz, M., and Bautista, R.G., Dynamic considerations for a multistage mixer-settler extractor and stripper, Ind Eng Chem Proc Des Dev, 12(4), 432-433 (1973).

Chen, B.H., Performance of screen-packed liquid-liquid extraction tower, Ind Eng Chem Proc Des Dev, 12(1), 115-119 (1973).

Fandeev, M.A.; Gelperin, N.I., and Nazarov, P.S., Packed extraction columns with cyclic operating conditions, Int Chem Eng, 13(4), 692-695 (1973).

Furzer, I.A., Periodic cycling of plate columns, Chem Eng Sci, 28(1), 296-299 (1973).

Golovko, S.N., and Zadorskii, V.M., Capacities of packed-pulsed extraction columns, Int Chem Eng, 13(3), 419-423 (1973).
Henton, J.E.; Fish, L.W., and Cavers, S.D., Liquid-liquid spray towers: Continuous-phase Peclet numbers, Ind Eng Chem Fund, 12(3), 365-372 (1973).
Hudson, B., and Wankat, P.C., Two-dimensional cross-flow extraction, Sepn Sci, 8(5), 599-612 (1973).
Jones, D.A., and Wilkinson, W.L., Dynamic response of a multiple-mixer column to flowrate disturbances, Chem Eng Sci, 28(8), 1577-1590 (1973).
Jones, D.A., and Wilkinson, W.L., Dynamic response of multiple-mixer solvent extraction column to concentration disturbances, Chem Eng Sci, 28(2), 539-552 (1973).
Kagan, S.Z.; Veisbein, B.A.; Trukhanov, V.G., and Muzychenko, L.A., Longitudinal mixing and its effect on mass transfer in pulsed-screen extractors, Int Chem Eng, 13(2), 217-221 (1973).
Mizrahi, J., and Barnea, E., Compact settler for separation of liquid-liquid dispersions, Proc Engng, Jan, 60-65 (1973).
Plummer, M.A., and Pouska, G.A., Operating two solvent extractors together, Hyd Proc, 52(6), 91-94 (1973).
Skelland, A.H.P., and Conger, W.L., A rate approach to design of perforated-plate extraction columns, Ind Eng Chem Proc Des Dev, 12(4), 448-454 (1973).
Sreenivasan, K., and Viswanath, D.S., Mass transfer coefficients in mixer-settlers, J Appl Chem Biotechnol, 23, 160-174 (1973).
Vyazovkin, E.S., and Nikolaev, N.A., Liquid-droplet motion in vortex-type mass-transfer equipment, Int Chem Eng, 13(2), 320-324 (1973).
Wankat, P.C., Cycling zone extraction, Sepn Sci, 8(4), 473-500 (1973).
Wankat, P.C., Liquid-liquid extraction by parametric pumping, Ind Eng Chem Fund, 12(3), 372-381 (1973).
Warwick, G.C.I., Liquid-liquid contacting equipment, Chem & Ind, 5 May, 403-408 (1973).
Weinstein, B., and Treybal, R.E., Dispersed-phase holdup in baffled mixing vessels, AIChEJ, 19(4), 851-852 (1973).
Weinstein, B., and Treybal, R.E., Liquid-liquid contacting in unbaffled agitated vessels, AIChEJ, 19(2), 304-312 (1973).
Yeheskel, J., and Kehat, E., Wake phenomena in a liquid-liquid fluidized bed, AIChEJ, 19(4), 720-728 (1973).

1974

Barton, R.L., Sizing liquid-liquid phase separators empirically, Chem Eng, 8 July, 111 (1974).
Brown, D.E., and Pitt, K., Effect of impeller geometry on drop break-up in stirred liquid-liquid contactor, Chem Eng Sci, 29(2), 345-348 (1974).
Chan, K.W., and Baird, M.H.I., Wall friction in oscillating liquid columns, Chem Eng Sci, 29(10), 2093-2100 (1974).
Dale, E.B., and Furzer, I.A., Application of Zakian's method to solve dynamics of periodically cycled plate column, Chem Eng Sci, 29(12), 2378-2380 (1974).

Hafez, M.M., and Prochazka, J., Dynamic effects in vibrating-plate and pulsed extractors, Chem Eng Sci, 29(8), 1745-1762 (1974).

Ingham, J., and Dunn, I.J., Digital simulation of stagewise processes with backmixing, Chem Engnr, June, 354-367 (1974).

Kerkhof, P.J., and Thijssen, H.A.C., Simple model describing effect of axial mixing on countercurrent mass exchange, Chem Eng Sci, 29(6), 1427-1434 (1974).

Lamey, S.C., and Maloy, J.T., Ultrapurification of aromatic hydrocarbons through liquid-liquid extraction with sulfuric acid, Sepn Sci, 9(5), 391-400 (1974).

Misek, T., Design and normal flows through a column extractor, Int Chem Eng, 14(1), 107-112 (1974).

Misek, T., Simple method of extraction calculations and application in complex industrial systems, Int Chem Eng, 14(1), 60-65 (1974).

Moryakov, V.S., and Nikolaev, N.A., Liquid-phase mass transfer in vortex-type equipment, Int Chem Eng, 14(2), 297-300 (1974).

Novitskii, V.S.; Bulatov, S.N., and Gryaznov, V.V., Graphical method of optimizing equipment design in extraction processes using sieve-column extractors, Int Chem Eng, 14(3), 447-451 (1974).

Pan, S.C., New type of continuous countercurrent extractor with two continuous phases, Sepn Sci, 9(3), 227-248 (1974).

Peel, D., and Coggan, G.C., Further work on the transurface contactor: A device for liquid extraction by filmwise contact, Trans IChemE, 52, 67-74 (1974).

Rimmer, B.F., Refining of gold from precious metal concentrates by liquid-liquid extraction, Chem & Ind, 19 Jan, 63-66 (1974).

Ruskan, R.P., Effects of surfactant on a mechanically agitated extraction column, Ind Eng Chem Proc Des Dev, 13(3), 203-208 (1974).

Various, Liquid-liquid extraction and ion exchange in analytical chemistry (topic issue), Chem & Ind, 17 Aug, 639-647 (1974).

1975

Barnea, E., and Mizrahi, J., Separation mechanism of liquid-liquid dispersions in a deep-layer gravity settler, Trans IChemE, 53, 61-92 (1975).

Chartres, R.H., and Korchinsky, W.J., Modelling of liquid-liquid extraction columns: Predicting the influence of drop size distribution, Trans IChemE, 53, 247-254 (1975).

Drew, J.W., Design for solvent recovery, Chem Eng Prog, 71(2), 92-99 (1975).

El-Rifai, M.A., Composition dynamics in multi-mixer-settler extractive reaction batteries, Chem Eng Sci, 30(1), 79-88 (1975).

Landau, J.; Dim, A., and Shemilt, L.W., Dynamic behavior of reciprocating plate columns, Can JCE, 53, 9-15 (1975).

Miyanami, K.; Tojo, K.; Yano, T., and Miyaji, K., Drop size distributions and holdups in a multistage vibrating disk column, Chem Eng Sci, 30(11), 1415-1420 (1975).

Pratt, H.R.C., Simplified analytical design method for differential extractors with backmixing, Ind Eng Chem Proc Des Dev, 14(1), 74-80 (1975).

Reinhardt, H., Solvent extraction recovery of metal waste, Chem & Ind, 1 March, 210-214 (1975).

Sigales, B., Design of settling drums, Chem Eng, 23 June, 141-144 (1975).

Skelland, A.H.P., and Shah, A.V., Extraction with oscillating droplets in a perforated-plate column, Ind Eng Chem Proc Des Dev, 14(4), 379-384 (1975).

Verma, R.P., and Sharma, M.M., Mass transfer in packed liquid-liquid extraction columns, Chem Eng Sci, 30(3), 279-292 (1975).

Watson, J.S.; McNeese, L.E.; Day, J., and Carroad, P.A., Flooding rates and holdup in packed liquid-liquid extraction columns, AIChEJ, 21(6), 1080-1086 (1975).

Wilkinson, D.; Mumford, C.J., and Jeffreys, G.V., Phase separation of primary dispersions in beds packed with spherical packings, AIChEJ, 21(5), 910-917 (1975).

1976

Arthayukti, W.; Muratet, G., and Angelino, H., Longitudinal mixing in dispersed phase in pulsed perforated-plate columns, Chem Eng Sci, 31(12), 1193-1198 (1976).

Boyadzhiev, L., and Angelov, G., Multicomponent extraction in prescence of backmixing, Int Chem Eng, 16(3), 427-432 (1976).

Coulaloglou, C.A., and Tavlarides, L.L., Drop size distributions and coalescence frequencies in liquid-liquid dispersions in flow vessels, AIChEJ, 22(2), 289-297 (1976).

Eckert, J.S., Liquid-liquid extraction variables defined, Hyd Proc, 55(3), 117-124 (1976).

Erving, W.J., and Chen, B.H., Axial mixing in liquid-extraction spray column, Can JCE, 54, 636-645 (1976).

Ioannou, J.; Hafez, M., and Hartland, S., Mass transfer and power consumption in reciprocating plate extractors, Ind Eng Chem Proc Des Dev, 15(3), 469-471 (1976).

Jelinek, J., and Hlavacek, V., Calculation of multistage multicomponent liquid-liquid extraction by relaxation method, Ind Eng Chem Proc Des Dev, 15(4), 481-484 (1976).

Johnson, R.A.; Gibson, W.E., and Libby, D.R., Performance of liquid-liquid cyclones, Ind Eng Chem Fund, 15(2), 110-115 (1976).

Karr, A.E., and Lo, T.C., Scaleup of large diameter reciprocating-plate extraction columns, Chem Eng Prog, 72(11), 68-70 (1976).

Kim, S.D., and Baird, M.H.I., Effect of hole size on hydrodynamics of reciprocating perforated-plate extraction column, Can JCE, 54, 235-245 (1976).

Kim, S.D., and Baird, M.H.I., Axial dispersion in reciprocating-plate extraction column, Can JCE, 54, 81-90 (1976).

Korchinsky, W.J., and Azimzadeh, S., Improved stagewise model of countercurrent-flow liquid-liquid contactors, Chem Eng Sci, 31(10), 871-876 (1976).

Laddha, G.S., et al., Performance characteristics of liquid-phase spray columns, AIChEJ, 22(3), 456-462 (1976).

Lelli, U.; Magelli, F., and Pasquali, G., Multistage mixer column: Fluid-dynamic studies, Chem Eng Sci, 31(4), 253-256 (1976).

Magiera, J.; Tal, B., and Zadlo, J., Hydrodynamic and kinetic problems of mass transfer in rotating-disc extraction columns, Int Chem Eng, 16(4), 744-751 (1976).

Pratt, H.R.C., Simplified analytical design method for differential extractors with backmixing, Ind Eng Chem Proc Des Dev, 14(1), 74-80 (1975); 15(1), 34-41 (1976).

Pratt, H.R.C., Simplified analytical design method for stagewise extractors with backmixing, Ind Eng Chem Proc Des Dev, 15(4), 544-548 (1976).

Prochazka, J., and Jiricny, V., Contribution to theory of fractional liquid extraction, Chem Eng Sci, 31(3), 179-186 (1976).

Tojo, K.; Miyanami, K., and Yano, T., Axial mixing in a multistage vibrating disc column with countercurrent liquid-liquid flow, Chem Eng J, 11(2), 101-104 (1976).

Various, Solvent extraction in the pharmaceutical industry (topic issue), Chem & Ind, 21 Aug, 677-685 (1976).

1977

Breuer, M.E.; Yoon, C.Y.; Jones, D.P., and Nurry, M.J., Countercurrent controlled cycle liquid-liquid extraction, Chem Eng Prog, 73(6), 95-96 (1977).

Drown, D.C., and Thompson, W.J., Fluid mechanic considerations in liquid-liquid settlers, Ind Eng Chem Proc Des Dev, 16(2), 197-206 (1977).

El-Rifai, M.A.; El Nashaie, S.S.E., and Kafafi, A.A., Analysis of a countercurrent tallow-splitting column, Trans IChemE, 55, 59-63, 283 (1977).

Golob, J., and Modic, R., Coalescence of liquid-liquid dispersions in gravity settlers, Trans IChemE, 55, 207-211 (1977).

Goto, S., and Matsubara, M., Liquid-liquid extraction parametric pumping with reversible reaction, Ind Eng Chem Fund, 16(2), 193-200 (1977).

Landau, J., and Chin, M., Contactor for studying mass transfer with reaction in liquid-liquid systems, Can JCE, 55, 161-170 (1977).

Mills, T.K., and Waggoner, R.C., Effect of bi-directional recycle streams on extractor operations, Comput Chem Eng, 1(3), 191-196 (1977).

Various, Solvent extraction (topic issue), Chem & Ind, 3 Sept, 705-737 (1977).

Various, Solvent extraction in the petrochemicals industry (topic issue), Chem & Ind, 15 Jan, 65-73 (1977).

Wang, P.S.M.; Ingham, J., and Hanson, C., Performance of a Graesser raining bucket liquid-liquid contactor, Trans IChemE, 55, 196-201 (1977).

1978

Barnea, E., Flooding conditions in a deep-layer liquid-liquid settler, Trans IChemE, 56, 73-76 (1978).

Chartres, R.H., and Korchinsky, W.J., Drop size and extraction efficiency measurements in a pilot plant rotating disc contactor, Trans IChemE, 56, 91-95 (1978).

Chopra, S.J., and Mukhopadhyay, P.K., Liquid-liquid extraction process for removal of aromatics from kerosine, Hyd Proc, 57(2), 113-117 (1978).

Clarke, S.I., and Sawistowski, H., Phase inversion of stirred liquid-liquid dispersions under mass transfer conditions, Trans IChemE, 56, 50-55 (1978).

Hafez, M.M., and Baird, M.H.I., Power consumption in a reciprocating plate extraction column, Trans IChemE, 56, 229-238 (1978).

Horvath, M.; Steiner, L., and Hartland, S., Prediction of drop diameter, holdup, and backmixing coefficients in liquid-liquid spray columns, Can JCE, 56, 9-20 (1978).

Johnson, K.L., and Miller, J.W., A guide for regulating liquid-liquid extraction columns, Chem Eng, 13 March, 106-108 (1978).

Khemangkorn, V.; Molinier, J., and Angelino, H., Influence of mass transfer direction on efficiency of a pulsed perforated-plate column, Chem Eng Sci, 33(4), 501-508 (1978).

Komasawa, I., and Ingham, J., Effect of system properties on performance of liquid-liquid extraction columns, Chem Eng Sci, 33(3), 341-348; 33(4), 479-486; 33(5), 541-546 (1978).

Laddha, G.S.; Degaleesan, T.E., and Kannappan, R., Hydrodynamics and mass transport in rotary disk contactors, Can JCE, 56, 137-145 (1978).

Logsdail, D., Developments in liquid-liquid extraction, Proc Engng, Oct, 40-43 (1978).

Milyakh, S.V., and Ivanova, V.I., Increasing the operating efficiency of extraction towers, Int Chem Eng, 18(1), 112-115 (1978).

Murakami, A.; Misonou, A., and Inoue, K., Dispersed-phase holdup and backmixing in a rotating-disc extraction column, Int Chem Eng, 18(1), 16-26 (1978).

Ramamoorthy, P., and Treybal, R.E., Drop coalescence in liquid-liquid fluidized beds, AIChEJ, 24(6), 985-992 (1978).

Rao, K.V.K.; Jeelani, S.A.K., and Balasubramanian, G.R., Backmixing in pulsed-plate columns, Can JCE, 56, 120-129 (1978).

Reissinger, K.H., and Schroter, J., Selection criteria for liquid-liquid extractors, Chem Eng, 6 Nov, 109-118 (1978).

Sawinsky, J., Axial mixing in the continuous phase of a mixing-extractor column, Int Chem Eng, 18(1), 169-176 (1978).

Sawinsky, J.; Pekovits, L., and Hunek, J., Effect of axial mixing on dynamics of extractor column, Int Chem Eng, 18(2), 325-330 (1978).

Sharma, R.N., and Baird, M.H.I., Solvent extraction of copper in reciprocating-plate column, Can JCE, 56, 310-320 (1978).

Skelland, A.H.P., and Lee, J.M., Agitator speeds in baffled vessels for uniform liquid-liquid dispersions, Ind Eng Chem Proc Des Dev, 17(4), 473-478 (1978).

Skelland, A.H.P., and Seksaria, R., Minimum impeller speeds for liquid-liquid dispersion in baffled vessels, Ind Eng Chem Proc Des Dev, 17(1), 56-61 (1978).

Steiner, L.; Horvath, M., and Hartland, S., Mass transfer between two liquid phases in spray column at unsteady state, Ind Eng Chem Proc Des Dev, 17(2), 175-182 (1978).

Various, Solvent extraction equipment (topic issue), Chem & Ind, 7 Oct, 745-764 (1978).

Waggoner, R.C., and Burkhart, L.E., Nonequilibrium computations for multistage extractors, Comput Chem Eng, 2(4), 169-176 (1978).

Walia, D.S., Concentration profiles in a liquid-liquid spray column, Chem Eng J, 16(3), 185-192 (1978).

1979

Blumberg, R., Design of solvent extraction systems, Sepn & Purif Methods, 8(1), 45-72 (1979).

Bulicka, J., and Prochazka, J., Ternary liquid-liquid mass transfer in turbulent flow, Chem Eng Commns, 3(4), 325-338 (1979).

Graham, E.E.; Ooi, K.L., and Odell, M.H., Drop-size distributions for non-coalescing liquid-liquid systems in sieve plate columns, Chem Eng J, 18(3), 189-196 (1979).

Groenier, W.S.; Rainey, R.H., and Watson, S.B., Analysis of transient and steady-state operation of countercurrent liquid-liquid solvent extraction process, Ind Eng Chem Proc Des Dev, 18(3), 385-390 (1979).

Hafez, M.M.; Baird, M.H.I., and Nirdesh, I., Flooding and axial dispersion in reciprocating-plate extraction columns, Can JCE, 57, 150-161 (1979).

Hoffer, M.S., and Resnick, W., A study of agitated liquid-liquid dispersions, Trans IChemE, 57, 1-14 (1979).

Murray, D.J., Equipment development in solvent extraction, J Chem Tech Biotechnol, 29, 367-378 (1979).

Roszak, J., and Gawronski, R., Cocurrent liquid-liquid extraction in fluidized beds, Chem Eng J, 17(2), 101-110 (1979).

Tojo, K.; Miyanami, K.; Minami, I., and Yano, T., Power dissipation in a vibrating disc column, Chem Eng J, 17(3), 211-218 (1979).

Various, Papers from IMM/SCI conference on solvent extraction, J Chem Tech Biotechnol, 29, 193-272 (1979).

Wijffels, J.B., and Rietema, K., Flow patterns and axial mixing in liquid-liquid spray columns, Trans IChemE, 50, 224-239 (1972); 57, 84-93, 147-155 (1979).

1980

Angelov, G., and Boyadzhiev, L., Hydrodynamic study of stagewise controlled cycling extraction column, Chem Eng Commns, 4(1), 77-88 (1980).

Bott, T.R., Supercritical gas extraction, Chem & Ind, 15 March, 228-232 (1980).

Cruz-Pinto, J.J.C., amd Lorchinsky, W.J., Experimental confirmation of the influence of drop size distribution on liquid-liquid extraction column performance, Chem Eng Sci, 35(10), 2213-2220 (1980).

Jackson, P.J., and Agnew, J.B., Model-based scheme for on-line optimization of liquid extraction process, Comput Chem Eng, 4(4), 241-250 (1980).

Karr, A.E.; Gebert, W., and Wang, M., Extraction of whole fermentation broth with Karr reciprocating-plate extraction column, Can JCE, 58, 249-252 (1980).

Kumar, A.; Vohra, D.K., and Hartland, S., Sedimentation of droplet dispersions in countercurrent spray columns, Can JCE, 58, 154-159 (1980).

Peker, S.; Comden, M., and Atagunduz, G., Effect of interfacial instabilities and hydrodynamic interaction on liquid-liquid mass transfer, Chem Eng Sci, 35(8), 1679-1686 (1980).

Robbins, L.A., Liquid-liquid extraction as a pretreatment process for wastewater, Chem Eng Prog, 76(10), 58-61 (1980).

Sarkar, S.; Mumford, C.J., and Phillips, C.R., Liquid-liquid extraction with interphase chemical reaction in agitated columns, Ind Eng Chem Proc Des Dev, 19(4), 665-679 (1980).

Sarkar, S.; Phillips, C.R.; Mumford, C.J., and Jeffreys, G.V., Mechanisms of phase inversion in rotary agitated columns, Trans IChemE, 58, 43-50 (1980).

Steiner, L., and Hartland, S., Agitated liquid-liquid extraction, Chem Eng Prog, 76(12), 60-62 (1980).

Tojo, K.; Miyanami, K., and Yano, T., Design method and performance characteristics of a multistage vibrating-disc column (MVDC) extractor, Ind Eng Chem Proc Des Dev, 19(3), 459-465 (1980).

Various, Solvent extraction (symposium papers), Sepn Sci Technol, 15(4), 799-1034 (1980).

Venkatarama, J.; Degaleesan, T.E., and Laddha, G.S., Continuous-phase axial mixing in rotaty disk contactors, Can JCE, 58, 206-211 (1980).

1981

Bailes, P.J., Solvent extraction in an electrostatic field, Ind Eng Chem Proc Des Dev, 20(3), 564-570 (1981).

Bailes, P.J.; Godfrey, J.C., and Slater, M.J., Designing liquid-liquid extraction equipment, Chem Engnr, July, 331-333 (1981).

Bender, E.; Berger, P.; Leuckel, W., and Wolf, D., Operating characteristics of pulsed packed columns for liquid-liquid extraction, Int Chem Eng, 21(1), 29-40 (1981).

Berry, W.W., Recovery of uranium from phosphoric acid by solvent extraction, Chem Eng Prog, 77(2), 76-82 (1981).

Bhave, R.R., and Sharma, M.M., Liquid-liquid reactors: Operation under refluxing conditions and measurement of effective interfacial area, Trans IChemE, 59, 161-169 (1981).

Cruz-Pinto., J.J.C., and Korchinsky, W.J., Drop breakage in countercurrent liquid-liquid extraction columns, Chem Eng Sci, 36(4), 687-694 (1981).

Duyckaerts, G., et al., Liquid-liquid extraction: Standard procedures and international collaboration, Sepn Sci Technol, 16(8), 937-942 (1981).

Flett, D.S., Some recent developments in the application of liquid extraction in hydrometallurgy, Chem Engnr, July, 321-324 (1981).

Hiby, J.W., Definition and measurement of degree of mixing in liquid mixtures, Int Chem Eng, 21(2), 197-205 (1981).

Ikari, A., and Hatate, H., Plate efficiency for minor component in Oldershaw-type column, Int Chem Eng, 21(3), 453-459 (1981).

Joshi, J.B., Axial mixing in multiphase contactors: A unified correlation, Trans IChemE, 58, 155-165 (1980); 59, 138-143 (1981).

Kehat, E., and Ghitis, B., Simulation of an extraction column, Comput Chem Eng, 5(3), 171-180 (1981).

Pilhofer, T., Optimum design of unpulsed sieve-plate extraction columns, Chem Eng Commns, 11(4), 241-254 (1981).

Pratt, H.R.C., and Garg, M.O., Scale-up of backmixed liquid extraction columns, Ind Eng Chem Proc Des Dev, 20(3), 489-492 (1981).

Purarelli, C., How to size a gravity settler with an internal weir, Chem Eng, 29 June, 112-114 (1981).
Sawinsky, J., Effect of longitudinal mixing on separating efficiency of countercurrent extraction columns, Int Chem Eng, 21(1), 121-129 (1981).
Sawinsky, J., and Hunek, J., Methods for investigating backmixing in the continuous phase of multiple-mixer extraction columns, Trans IChemE, 59, 64-66 (1981).
Scuffham, J.B., Combined mixer settler: A Davy McKee concept, Chem Engnr, July, 328-330 (1981).
Spencer, J.L.; Steiner, L., and Hartland, S., Model-based analysis of data from countercurrent liquid-liquid extraction processes, AIChEJ, 27(6), 1008-1016 (1981).
Stowe, L.R., and Shaeiwitz, J.A., Hydrodynamics and mass transfer characteristics of a liquid/liquid stirred cell, Chem Eng Commns, 11(1), 17-26 (1981).
Various, Liquid-liquid extraction (symposium papers), Sepn Sci Technol, 16(9), 1113-1298 (1981).
Zanker, A., Separation in a continuous gravity decanter, Proc Engng, Sept, 91-93 (1981).
Zhang, S.H.; Ni, X.D., and Su, Y.F., Hydrodynamics, axial mixing and mass transfer in rotating disc contactors, Can JCE, 59, 573-583 (1981).

1982

Aly, G., Dynamic behaviour of mixer-settlers, Trans IChemE, 60, 240-244 (1982).
Baird, M.H.I., and Krovvidi, K.R., Liquid-liquid extraction by a suspended slug, Can JCE, 60, 569-574 (1982).
Balla, L., and Sisak, C., Hydrodynamics of liquid phase in bubble and sieve-plate column cascades with selective partitions, Int Chem Eng, 22(4), 736-744 (1982).
Danesi, P.R., et al., ARMOLLEX: Apparatus for solvent extraction kinetic measurements, Sepn Sci Technol, 17(7), 961-968 (1982).
Geankoplis, C.J.; Sapp, J.B.; Arnold, F.C., and Marroquin, G., Axial dispersion coefficients of the continuous phase in liquid-liquid spray towers, Ind Eng Chem Fund, 21(3), 306-311 (1982).
Hellyar, K.G., and de Filippi, R.P., Extraction processes using solvents near their thermodynamic critical point, Chem Engnr, April, 136-138 (1982).
Katoh, N., Dynamic programming approach to design of solvent extraction process in fast-breeder reactor fuel reprocessing, Chem Eng Commns, 17(1), 31-42 (1982).
Mason, G.S., Decanting without interface control, Chem Eng, 20 Sept, 129-130 (1982).
Rachez, D.; Delaveau, G.; Grevillot, G., and Tondeur, D., Stagewise liquid-liquid extraction parametric pumping: Equilibrium analysis and experiments, Sepn Sci Technol, 17(4), 589-620 (1982).
Salem, A.B., and Jeffreys, G.V., Find optimum temperature for countercurrent extraction, Hyd Proc, Oct, 93-97 (1982).
Schulz, L., and Pilhofer, T., Tray efficiency in unpulsed sieve-tray extraction columns, Int Chem Eng, 22(1), 61-68 (1982).

Taylor, P.A.; Baird, M.H.I., and Kusuma, I., Computer control of holdup in a reciprocating plate extraction column, Can JCE, 60, 556-565 (1982).

Waggoner, R.C., and Eye, J.M., Quin-diagonal algorithm describing mass transfer and backmixing in extraction, Comput Chem Eng, 6(4), 265-270 (1982).

1983

Bonnet, J.C., and Jeffreys, G.V., Measurement of concentration profiles in a liquid-liquid extraction column, J Chem Tech Biotechnol, 33A(4), 176-186 (1983).

Cruz-Pinto, J.J.C., and Korchinsky, W.J., Exact solutions of Newman and Handlos-Baron model equations for countercurrent flow extraction, Comput Chem Eng, 7(1), 19-26 (1983).

Hatton, T.A.; Lightfoot, E.N., Cahn, R.P., and Li, N.N., An internal recycle mixer for solvent extraction, Ind Eng Chem Fund, 22(1), 27-35 (1983).

Heyberger, A.; Kratky, M.,and Prochazka, J., Parameter evaluation of solvent extractor with backmixing, Chem Eng Sci, 38(8), 1303-1308 (1983).

Hossain, K.T.; Sarkar, S.; Mumford, C.J., and Phillips, C.R., Hydrodynamics of mixer-settlers, Ind Eng Chem Proc Des Dev, 22(4), 553-563 (1983).

Hsia, M.A., and Tavlarides, L.L., Simulation analysis of drop breakage, coalescence and micromixing in liquid-liquid stirred tanks, Chem Eng J, 26(3), 201-216 (1983).

Kumar, A., and Hartland, S., Correlations for dispersed phase hold-up in pulsed sieve-plate liquid-liquid extraction columns, CER&D, 61, 248-252 (1983).

Pratt, H.R.C., Generalized design equations for liquid-liquid extractors, Solv Extn & Ion Exch, 1(4), 669-688 (1983).

Rao, N.V.R.; Srinivas, N.S., and Varma, Y.B.G., Dispersed phase holdup and drop size distributions in reciprocating plate columns, Can JCE, 61(2), 168-177 (1983).

Sovova, H., Model of dispersion hydrodynamics in a vibrating plate extractor, Chem Eng Sci, 38(11), 1863-1872 (1983).

1984

Angelov, G.; Boydzhiev, L., and Kyutchoukov, G., Separator for liquid-liquid dispersions, Chem Eng Commns, 25(1), 311-320 (1984).

Baird, M.H.I., and Shen, Z.J., Holdup and flooding in reciprocating plate extraction columns, Can JCE, 62(2), 218-227 (1984).

Blass, E., and Zimmermann, H., Mathematical simulation and experimental determination of unsteady behaviour of liquid-pulsated sieve-plate column in liquid-liquid extraction, Int Chem Eng, 24(2), 214-227 (1984).

Caminos, A.A.; Gani, R., and Brignole, E.A., Liquid-liquid extractor model based on UNIFAC, Comput Chem Eng, 8(2), 127-136 (1984).

Fair, J.R., and Humphrey, J.L., Liquid-liquid extraction: Possible alternative to distillation, Solv Extn & Ion Exch, 2(3), 323-352 (1984).

Fair, J.R.; Rocha, A., and Humphrey, J.L., Efficiency of crossflow sieve-tray extractors, Solv Extn & Ion Exch, 2(7), 985-1008 (1984).

Gaonkar, A.G., and Neuman, R.D., Purity considerations and interfacial behavior of solvent extraction systems, Sepn & Purif Methods, 13(2), 141-152 (1984).

Kumar, A., and Hartland, S., Correlation for drop-size in liquid-liquid spray columns, Chem Eng Commns, 31(1), 193-208 (1984).

Malmary, G.; Molinier, J.; Mankowski, G., and Lenzi, J., Metals separation by liquid extraction, Chem Eng Educ, 18(2), 88-91 (1984).

Noth, H., and Mersmann, A., Heat transfer phenomena in liquid-liquid spray columns, Int J Heat Mass Trans, 27(11), 2015-2024 (1984).

Pajak, M., and Kaczmarski, K., Hydrodynamics of two-phase drops-liquid systems, Chem Eng Commns, 26(1), 173-192 (1984).

Pietzsch, W., and Pilhofer, T., Calculation of the drop size in pulsed sieve-plate extraction columns, Chem Eng Sci, 39(6), 961-966 (1984).

Pratt, H.R.C., Droplet coalescence and breakage rates in liquid extraction columns, Solv Extn & Ion Exch, 2(4), 521-552 (1984).

Rao, N.V.R., and Baird, M.H.I., Liquid extraction laboratory studies using a combined mixer-settler, Can JCE, 62(4), 497-506 (1984).

Rod, V., Unconventional separations of metals by liquid-liquid extraction, Chem Eng J, 29(2), 77-84 (1984).

Sato, T.; Maeda, T.; Mizuno, Y., and Nakamura, T., Some methods for analysis of solvent extraction data and their applications, Solv Extn & Ion Exch, 2(6), 755-764 (1984).

Schlip, R., and Blass, E., Flooding capacity of perforated plates in rotating liquid-liquid systems, Chem Eng Commns, 28(1), 85-98 (1984).

Vasudevan, T.V., and Sharma, M.M., Process design of liquid-liquid reactors, Ind Eng Chem Proc Des Dev, 23(2), 400-406 (1984).

1985

Asai, S.; Hatanaka, J., and Kuroi, M., Continuous-phase mass transfer in laminar liquid-liquid slug flow, Chem Eng J, 30(3), 133-140 (1985).

Bapat, P.M., and Tavlarides, L.L., Mass transfer in a liquid-liquid CFSTR, AIChEJ, 31(4), 659-666 (1985).

Bensalem, A.; Steiner, L., and Hartland, S., Effect of mass transfer on flooding and holdup in a Karr column, Solv Extn & Ion Exch, 3(5), 697-722 (1985).

Berger, R., and Walter, K., Flooding in pulsed sieve plate extractors, Chem Eng Sci, 40(12), 2175-2184 (1985).

Bonnet, J.C., and Jeffreys, G.V., Hydrodynamics and mass transfer characteristics of a Scheibel extractor, AIChEJ, 31(5), 788-801 (1985).

Delaine, J., Separating oil from water offshore, Chem Engnr, Nov, 31-34 (1985).

Gaikar, V.G., and Sharma, M.M., Dissociation extraction: Prediction of separation factor and selection of solvent, Solv Extn & Ion Exch, 3(5), 679-696 (1985).

Gillett, G.A., and Rowden, G.A., Combined mixer-settler design, Chem & Ind, 2 Sept, 583-589 (1985).

Horvath, M., and Hartland, S., Mixer-settler-extraction column: Mass transfer efficiency and entrainment, Ind Eng Chem Proc Des Dev, 24(4), 1220-1225 (1985).

Hughes, M.A., and Parker, N., Computer study of liquid-liquid stagewise calculations in typical and new countercurrent contacting, J Chem Tech Biotechnol, 35A(5), 255-262 (1985).

Kamath, M.S., and Rau, M.G.S., Prediction of operating range of rotor speeds for rotating disc contactors, Can JCE, 63(4), 578-584 (1985).

Kaul, A., and Van Wormer, K.A., Effects of internal stage recycle on efficiency and performance of a mixer-settler, Ind Eng Chem Proc Des Dev, 24(3), 636-646 (1985).

Korchinsky, W.J., and Young, C.H., Computational techniques for liquid-liquid extraction column model-parameter estimation using the forward-mixing model, J Chem Tech Biotechnol, 35A(7), 347-357 (1985).

Krishna, R.; Low, C.Y.; Newsham, D.; Fuentes, C.G., and Standart, G.L., Ternary mass transfer in liquid-liquid extraction, Chem Eng Sci, 40(6), 893-904 (1985).

Kumar, A., and Hartland, S., Gravity settling in liquid-liquid dispersions, Can JCE, 63(3), 368-376 (1985).

Palagyi, S., Theoretical efficiency of pulsed polyurethane foam column separations, Solv Extn & Ion Exch, 3(4), 517-530 (1985).

Phillips, C., Hydraulic behaviour of pulsed columns for solvent extraction, Chem & Ind, 2 Sept, 577-582 (1985).

Sankey, B.M., New-lubricants extraction process, Can JCE, 63(1), 3-7 (1985).

Shen, Z.J.; Rao, N.V.R., and Baird, M.H.I., Mass transfer in a reciprocating-plate extraction column: Effects of mass transfer direction and plate material, Can JCE, 63(1), 29-36 (1985).

Slater, M.J., Liquid-liquid extraction column design, Can JCE, 63(6), 1004-1005 (1985).

Wardius, D.S., and Hatton, T.A., Model for liquid membrane extraction with instantaneous reaction in cascaded mixers, Chem Eng Commns, 37(1), 159-172 (1985).

Weatherley, L., Future trends in downstream solvent extraction, Chem Engnr, May, 25-28 (1985).

Zhang, S.H.; Yu, S.C.; Zhou, Y.C., and Su, Y.F., Model for liquid-liquid extraction column performance: Influence of drop size distribution on extraction efficiency, Can JCE, 63(2), 212-226 (1985).

1986

Bailes, P.J., Gledhill, J.; Godfrey, J.C., and Slater, M.J., Hydrodynamic behaviour of packed rotating disc and Kuhni liquid/liquid extraction columns, CER&D, 64(1), 43-55 (1986).

Batey, W.; Lonie, S.J.; Thompson, P.J., and Thornton, J.D., Dynamics of pulsed plate extraction columns, CER&D, 64(5), 396-403 (1986).

Bocangel, J., Design of liquid-liquid gravity separators, Chem Eng, 17 Feb, 133-135 (1986).

Dalingaros, W., and Hartland, S., Effect of drop size and physical properties on dispersion height in the separating section of a liquid-liquid extraction column, Can JCE, 64(6), 925-930 (1986).

Hopkins, M.R., and Ng, K.M., Liquid-liquid relative permeability: Network models and experiments, Chem Eng Commns, 46, 253-280 (1986).

Hsu, H.W., Separations by liquid centrifugations, Ind Eng Chem Fund, 25(4), 588-593 (1986).

Jeelani, S.A.K., and Hartland, S., Prediction of dispersion height in liquid-liquid gravity settlers from batch settling data, CER&D, 64(6), 450-460 (1986).

Korchinsky, W.J., and Al-Husseini, R., Liquid-liquid extraction column (rotating disc contactor): Model parameters from drop size distribution and solute concentration measurements, J Chem Tech Biotechnol, 36(9), 395-409 (1986).

Kumar, A., and Hartland, S., Prediction of drop size in pulsed perforated-plate extraction columns, Chem Eng Commns, 44, 163-182 (1986).

Kyu, K.B., and Kwang, K., Behavior of dispersed phase in liquid-liquid extraction in fluidized beds, Chem Eng Commns, 41, 101-120 (1986).

Le Lann, M.V.; Najim, K., and Casamatta, G., Generalized predictive control of a pulsed liquid-liquid extraction column, Chem Eng Commns, 48, 237-254 (1986).

Louvar, J.F., and Capraro, M.A., Liquid-liquid centrifuges (performance characteristics), Chem Eng, 27 Oct, 137-140 (1986).

Penner, L.R.; Siemens, R.E., and Galvan, G.J., Anaerobic extraction of cobalt in pulse column, Solv Extn & Ion Exch, 4(2), 345-360 (1986).

Ruiz, F.; Marcilla, A., and Gomis, V., Method for equilibrium-stage calculations in liquid-liquid extraction, Ind Eng Chem Proc Des Dev, 25(3), 631-634 (1986).

Seibert, A.F.; Humphrey, J.L., and Fair, J.R., Efficiency of a controlled-cycle extractor, Solv Extn & Ion Exch, 4(5), 1049-1072 (1986).

Stevens, G.W., and Pratt, H.R.C., Droplet coalescence and breakage rates in a packed liquid-extraction column, AIChEJ, 32(12), 2079-2082 (1986).

Thew, M., Hydrocyclone redesign for liquid-liquid separation, Chem Engnr, July, 17-23 (1986).

Wichterlova, J.; Drohos, J.; Cermak, J., and Rod, V., Measurement of holdup and drop size distribution in extraction column using continuous sampling by capillary tube, Chem Eng Commns, 40, 1-16 (1986).

Yoshida, H., et al., Effect of baffle on tray under oscillating condition, Chem Eng Commns, 43, 263-274 (1986).

1987

Asai, S., and Tanaka, H., Continuous-phase mass transfer in a laminar suspended slug for a liquid-liquid system, Chem Eng J, 34(3) 165-168 (1987).

Batey, W.; Lonie, S.J.; Thompson, P.J., and Thornton, J.D., Dynamics of pulsed-plate extraction columns, Solv Extn & Ion Exch, 5(4), 573-596 (1987).

Cooney, D.O., and Poufos, M.G., Liquid-liquid extraction in a hollow-fiber device, Chem Eng Commns, 61, 159-168 (1987).

Das, P.K., et al., Effect of mass transfer on droplet breakup in stirred liquid-liquid dispersions, AIChEJ, 33(11), 1899-1902 (1987).

Eldridge, R.B.; Humphrey, J.L., and Fair, J.R., Continuous phase mixing on crossflow extraction sieve trays, Sepn Sci Technol, 22(2), 1121-1134 (1987).

Fan, Z.; Oloidi, J.O., and Slater, M.J., Liquid-liquid extraction column design data acquisition for short columns, CER&D, 65(3), 243-250 (1987).

Jenkins, J., Pulsed columns in the nuclear industry, Chem Engnr, Oct, 16-19 (1987).

Kumar, A., and Hartland, S., Prediction of dispersed-phase holdup in rotating disc extractors, Chem Eng Commns, 56, 87-106 (1987).

Lahiere, R.J.; Humphrey, J.L., and Fair, J.R., Mass transfer in countercurrent supercritical extraction, Sepn Sci Technol, 22(2), 379-394 (1987).

Leonard, R.A., Electronic worksheets for calculation of stagewise solvent extraction processes, Sepn Sci Technol, 22(2), 535-556 (1987).

Najim, K.; Lann, M.U.L., and Casamatta, G., Learning control of a pulsed liquid-liquid extraction column, Chem Eng Sci, 42(7), 1619-1628 (1987).

O'Quinn, L.N., and van Brunt, V., Structural aspects of hydrometallurgical solvent extraction, Sepn Sci Technol, 22(2), 467-486 (1987).

Prasad, R., and Sirkar, K.K., Microporous membrane solvent extraction, Sepn Sci Technol, 22(2), 619-640 (1987).

Scott, T.C., Visualization of flow fields and interfacial phenomena in liquid-liquid solvent extraction, Sepn Sci Technol, 22(2), 503-512 (1987).

Seibert, A.F.; Humphrey, J.L., and Fair, J.R., Evaluation of packings for use in liquid-liquid extraction columns, Sepn Sci Technol, 22(2), 281-314 (1987).

Skelland, A.H.P., and Ramsay, G.G., Minimum agitator speeds for complete liquid-liquid dispersion, Ind Eng Chem Res, 26(1), 77-82 (1987).

Stamatoudis, M., and Tavlarides, L.L., The effect of continuous-phase viscosity on the unsteady state behaviour of liquid-liquid agitated dispersions, Chem Eng J, 35(2), 137-144 (1987).

Tavlarides, L.L.; Bae, J.H., and Lee, C.K., Solvent extraction, membranes and ion exchange in hydrometallurgical dilute metals separation, Sepn Sci Technol, 22(2), 581-618 (1987).

Thompson, P.J., Solvent extraction equipment development at the Dounreay Nuclear Power Development Establishment, CER&D, 65(5), 371-374 (1987).

Thornton, J., Liquid-liquid extraction: A neglected technology? Proc Engng, Aug, 32-33 (1987).

Various, Liquid-liquid extraction principles (topic issue), Chem & Ind, 16 March, 174-196 (1987).

Warner, J., and Harris, I.J., Design of multicomponent solvent extraction systems, CER&D, 65(3), 261-266 (1987).

Yih, S.M.; Wu, Y.M.; Pan, R.K.; Wu, Y.F., and Chen, T.F., Holdup measurement in a Scheibel extraction column, Solv Extn & Ion Exch, 5(2), 353-366 (1987).

1988

Alper, E., Effective interfacial area in the RTL extractor from rates of extraction with chemical reaction, CER&D, 66(2), 147-151 (1988).

Baird, M.H.I.; Rohatgi, A., and He, W., Rising-film extractor, CER&D, 66(2), 121-127 (1988).

Bandyopadhyay, N.; Ray, P., and Dutta, B.K., Gas holdup in bubble column with immiscible liquid mixtures, Can JCE, 66(6), 995-999 (1988).

Godfrey, J.C., et al., Continuous-phase axial mixing in pulsed sieve-plate liquid-liquid extraction columns, CER&D, 66(5), 445-457 (1988).

Haman, J., and Misek, T., Analysis of extractor performance, Chem Eng J, 39(1), 1-16 (1988).

Hazel, G.; Karr, A., and Cusak, R., Mixing theory and reality for successful solvent extraction, Proc Engng, June, 31-33 (1988).
Horng, J.S.; Lu, D., and Hoh, Y.C., Interfacial effects in a multistage mixer-settler operation, J Chem Tech Biotechnol, 42(4), 277-288 (1988).
Hussain, A.A.; Liang, T.B., and Slater, M.J., Characteristic velocity of drops in liquid-liquid extraction pulsed sieve-plate column, CER&D, 66(6), 541-554 (1988).
Janosi, T., and Hunek, J., Axial mixing in extraction column, Int Chem Eng, 28(4), 731-739 (1988).
Jeelani, S.A.K., and Hartland, S., Dynamic response of gravity settlers to changes in dispersion throughput. AIChEJ, 34(2), 335-340 (1988).
Karr, A.E., and Ramanujan, S., Scaleup and performance of 5ft (1.52 m) diameter reciprocating-plate extraction column, Solv Extn & Ion Exch, 6(2), 221-232 (1988).
Kim, S.D.; Yu, Y.H., and Han, P.W., Phase holdups and liquid-liquid extraction in three-phase fluidized beds, Chem Eng Commns, 68, 57-68 (1988).
Kirou, V.I.; Tavlarides, L.L.; Bonnet, J.C., and Tsouris, C., Flooding, holdup, and drop size measurements in a multistage column extractor, AIChEJ, 34(2), 283-292 (1988).
Korchinsky, W.J., and Ismail, A.M., Mass-transfer parameters in rotating-disc contactors: Influence of column diameter, J Chem Tech Biotechnol, 43(2), 147-158 (1988).
Kumar, A., and Hartland, S., Mass transfer in a Kuhni extraction column, Ind Eng Chem Res, 27(7), 1198-1203 (1988).
Kumar, A., and Hartland, S., Prediction of dispersed phase hold-up in pulsed perforated-plate extraction columns, Chem Eng & Proc, 23(1), 41-60 (1988).
Misek, T.; Haman, J., and Rod, V., Analysis of extractor performance, Chem Eng J, 39(2), 69-88 (1988).
Najim, K., and Le Lann, M.V., Multivariable learning control of an extractor, Chem Eng Sci, 43(7), 1539-1546 (1988).
Najim, K., and Le Lann, M.V., Control of pulsed liquid-liquid extraction column based on multilevel system of automata, Chem Eng Commns, 70, 107-126 (1988).
Nataraj, S.; Wehrum, W.L., and Wankat, P.C., Continuous, regenerative, two-dimensional extraction, Ind Eng Chem Res, 27(4), 650-657 (1988).
Prasad, R., and Sirkar, K.K., Dispersion-free solvent extraction with microporous hollow-fibre modules, AIChEJ, 34(2), 177-188 (1988).
Seibert, A.F., and Fair, J.R., Hydrodynamics and mass transfer in spray and packed liquid-liquid extraction columns, Ind Eng Chem Res, 27(3), 470-481 (1988).
Seibert, A.F., and Moosberg, D.G., Performance of spray, sieve tray and packed contactors for high pressure extraction, Sepn Sci Technol, 23(12), 2049-2064 (1988).
Steiner, L., and Hartland, S., Sensitivity of countercurrent extraction column response to changes in operating parameters, Chem Eng & Proc, 23(4), 193-202 (1988).

Tawfik, W.Y.; Eckles, A.J., and Tedder, D.W., Performance correlations for reciprocating-plate extraction columns, Solv Extn & Ion Exch, 6(4), 563-584 (1988).
Various, Solvent extraction technology (topic issue), Chem & Ind, 17 Oct, 642-662 (1988).
Wilson, D.J., Countercurrent extraction: Mass transfer kinetics and time-dependent behavior, Sepn Sci Technol, 23(1), 133-152 (1988).

21.3 Supercritical Extraction

1981-1988

Williams, D.F., Extraction with supercritical gases (review paper), Chem Eng Sci, 36(11), 1769-1788 (1981).
Various, Supercritical gases in extraction and chromatography (topic issue), Sepn Sci Technol, 17(1), 1-288 (1982).
Logsdail, D.H., Applications and prospects for supercritical extraction, Proc Engng, Sept, 32-35 (1983).
Shimshick, E.J., Extraction with supercritical carbon dioxide, Chemtech, June, 374-375 (1983).
Hoyer, G.G., Extraction with supercritical fluids, Chemtech, July, 440-448 (1985).
Korner, J.P., Design and construction of full-scale supercritical gas extraction plants, Chem Eng Prog, 81(4), 63-66 (1985).
Nelson, S.R., and Roodman, R.G., Residuum oil supercritical extraction process, Chem Eng Prog, 81(5), 63-68 (1985).
Chimowitz, E.H., and Pennisi, K.J., Process synthesis concepts for supercritical gas extraction in the crossover region, AIChEJ, 32(10), 1665-1676 (1986).
Josten, H., and Hartmann, H., Liquified gases as medium for extractive separation of liquid mixtures, Chem Eng & Proc, 21(4), 217-228 (1987).
Knaff, G., and Schlunder, E.U., Diffusion coefficients of naphthalene and caffeine in supercritical carbon dioxide, Chem Eng & Proc, 21(2), 101-106 (1987).
Rathkamp, P.J.; Bravo, J.L., and Fair, J.R., Evaluation of packed columns in supercritical extraction processes, Solv Extn & Ion Exch, 5(3), 367-392 (1987).
Scholsky, K.M., Process polymers with supercritical fluids, Chemtech, Dec, 750-757 (1987).
Debenedetti, P.G., and Kumar, S.K., Molecular basis of temperature effects in supercritical extraction, AIChEJ, 34(4), 645-657 (1988).
Kim, S., and Johnston, P., Adjustment of selectivity of Diels-Alder reaction network using supercritical fluids, Chem Eng Commns, 63, 49-60 (1988).
Schaeffer, S.T.; Zalkow, L.H., and Teja, A.S., Supercritical extraction of crotalaria spectabilis in cross-over region, AIChEJ, 34(10), 1740-1742 (1988).
Schmitt, W.J., and Reid, R.C., Solubility of paraffinic hydrocarbons and their derivatives in supercritical carbon dioxide, Chem Eng Commns, 64, 155-176 (1988).

Tan, C.S.; Liang, S.K., and Liou, D.C., Fluid-solid mass transfer in a supercritical fluid extractor, Chem Eng J, 38(1), 17-22 (1988).

Tan, C.S., and Wu, Y.C., Supercritical fluid distribution in packed column, Chem Eng Commns, 68, 119-132 (1988).

Triday, J., and Smith, J.M., Dynamic behavior of supercritical extraction of kerogen from shale, AIChEJ, 34(4), 658-668 (1988).

CHAPTER 22

ADSORPTION

22.1 Theory

1967-1969

Anon., Adsorption in packed beds, Brit Chem Eng, 11(1), 12; 11(2), 96; 11(5), 364 (1966).

Carter, J.W., Adsorption, solids drying and membrane permeation, Brit Chem Eng, 11(7), 718-720 (1966).

Hoory, S.E., and Prausnitz, J.M., Monolayer adsorption of gas mixtures on homogeneous and heterogeneous solids, Chem Eng Sci, 22(7), 1025-1034 (1967).

Lee, J.H., Statistical mechanical theory of adsorption life, Int Chem Eng, 7(1), 63-72 (1967).

Meyer, O.A, and Weber, T.W., Nonisothermal adsorption in fixed beds, AIChEJ, 13(3), 457-465 (1967).

Niac, C., Kinetic method for determination of adsorption isotherms, Int Chem Eng, 7(4), 666-672 (1967).

Pan, C.Y., and Basmadjian, D., Constant-pattern adiabatic fixed-bed adsorption, Chem Eng Sci, 22(3), 285-298 (1967).

Satterfield, C.N., and Frabetti, A.J., Sorption and diffusion of gaseous hydrocarbons in synthetic mordenite, AIChEJ, 13(4), 731-738 (1967).

Stuart, F.X., and Camp, D.T., Comparison of kinetic and diffusional models for packed bed adsorption, Ind Eng Chem Fund, 6(1), 156-158 (1967).

Carter, J.W., Developments in adsorption processes, Brit Chem Eng, 13(2), 229-234 (1968).

Carter, J.W., Isothermal and adiabatic adsorption in fixed beds, Trans IChemE, 46, T213-222 (1968).

Payne, H.K.; Sturdevant, G.A., and Leland, T.W., Improved two-dimensional equation of state to predict adsorption of pure and mixed hydrocarbons, Ind Eng Chem Fund, 7(3), 363-374 (1968).

Punwani, D.; Chi, C.W., and Wasan, D.T., Dynamic sorption by hygroscopic salts, Ind Eng Chem Proc Des Dev, 7(3), 410-415 (1968).

Schneider, P., and Smith, J.M., Adsorption rate constants from chromatography, AIChEJ, 14(5), 762-771 (1968).

Cerro, R.L., and Smith, J.M., Effects of heat release and nonlinear equilibrium on transient adsorption, Ind Eng Chem Fund, 8(4), 796-802 (1969).

Eberly, P.E., Diffusion studies in zeolites and related solids by gas chromatography, Ind Eng Chem Fund, 8(1), 25-30 (1969).

Eteson, D.C., and Zwiebel, I., Hybrid computer solution of the simple fixed bed adsorption model, AIChEJ, 15(1), 124-126 (1969).

Hawtin, P.; Dawson, R.W., and Roberts, J., The diffusion of gases through graphite, Trans IChemE, 47, T109-113 (1969).

Kurbanaliev, T.G.; Shabataev, S.A., and Rasulov, A.M., Determination of activity of adsorbent in short-cycle adsorption systems, Int Chem Eng, 9(3), 451-453 (1969).

Pigford, R.L.; Baker, B., and Blum, D.E., An equilibrium theory of the parametric pump, Ind Eng Chem Fund, 8(1), 144-149 (1969).

Sweed, N.H., and Wilhelm, R.H., Parametric pumping (The stop-go method), Ind Eng Chem Fund, 8(2), 221-231 (1969).

Van Ness, H.C., Adsorption of gases on solids, Ind Eng Chem Fund, 8(3), 464-473 (1969).

1970-1972

Cooper, R.S., and Liberman, D.A., Fixed-bed adsorption kinetics with pore diffusion control, Ind Eng Chem Fund, 9(4), 620-623 (1970).

Gonzalez, A.J., and Holland, C.D., Adsorption of multicomponent mixtures by solid adsorbents, AIChEJ, 16(5), 718-724 (1970).

Nemeth, E.J., and Stuart, E.B., Pore-diffusion mechanisms during vapor-phase adsorption, AIChEJ, 16(6), 999-1004 (1970).

Pan. C.Y., and Basmadjian, D., Analysis of adiabatic sorption of single solutes in fixed beds: Pure thermal wave formation and its practical implications, Chem Eng Sci, 25(11), 1653-1664 (1970).

Baker, B., and Pigford, R.L., Cycling-zone adsorption: Quantitative theory and experimental results, Ind Eng Chem Fund, 10(2), 283-292 (1971).

Clough, P.S.; Dollimore, D., and Nicklin, T., Comparison of isothermal adsorption data by technique of absolute adsorption, J Appl Chem Biotechnol, 21, 137-138 (1971).

Forrester, S.D., and Giles, C.H., Gas-solid adsorption isotherms: A historical survey to 1918, Chem & Ind, 24 July, 831-839 (1971).

Gupta, R., and Sweed, N.H., Equilibrium theory of cycling-zone adsorption, Ind Eng Chem Fund, 10(2), 280-283 (1971).

Pan, C.Y., and Basmadjian, D., Equilibrium theory analysis of adiabatic sorption of single solutes in fixed beds, Chem Eng Sci, 26(1), 45-58 (1971).

Safanov, M.S., Criterion for formation of steady-state sorption front, Sepn Sci, 6(1), 35-42 (1971).

Yoon, S.M., and Kunii, D., Physical adsorption in a moving bed of fine adsorbents, Ind Eng Chem Proc Des Dev, 10(1), 64-70 (1971).

Brunet, J., Three-dimensional treatment of adsorption, Chem Eng Sci, 27(4), 685-694 (1972).

Chen, H.T.; Rak, J.L.; Stokes, J.D., and Hill, F.B., Separations via continuous parametric pumping, AIChEJ, 18(2), 356-361 (1972).

Chen, J.W.; Cunningham, F.L., and Buege, J.A., Computer simulation of plant-scale multicolumn adsorption processes under periodic countercurrent operation, Ind Eng Chem Proc Des Dev, 11(3), 430-434 (1972).

Cooney, D.O., and Strusi, F.P., Analytical description of fixed-bed sorption of two Langmuir solutes under nonequilibrium conditions, Ind Eng Chem Fund, 11(1), 123-126 (1972).

Friederich, R.O., and Mullins, J.C., Adsorption equilibria of binary hydrocarbon mixtures on homogeneous carbon black at 25 degC, Ind Eng Chem Fund, 11(4), 439-445 (1972).

Garg, D.R., and Ruthven, D.M., Effect of concentration dependence of diffusivity on zeolitic sorption curves, Chem Eng Sci, 27(2), 417-424 (1972).

Ma, Y.H., and Mancel, C., Diffusion studies of carbon dioxide, nitric oxide, nitrogen dioxide, and sulfur dioxide on molecular sieve zeolites by gas chromatography, AIChEJ, 18(6), 1148-1153 (1972).

Rhee, H.K.; Heerdt, E.D., and Amundson, N.R., Analysis of adiabatic adsorption columns, Chem Eng J, 1(2), 241; 1(3), 279 (1970); 3(1), 22-34; 3(2), 121-135 (1972).

1973-1974

Blaisdell, C.T., and Kammermeyer, K., Countercurrent and cocurrent gas separation, Chem Eng Sci, 28(6), 1249-1256 (1973).

Chen, H.T.; Reiss, E.H.; Stokes, J.D., and Hill, F.B., Separations via semicontinuous parametric pumping, AIChEJ, 19(3), 589-595 (1973).

Danner, R.P., and Wenzel, L.A., Gas-mixture adsorption on molecular sieves, AIChEJ, 19(4), 870 (1973).

Fleck, R.D.; Kirwan, D.J., and Hall, K.R., Mixed-resistance diffusion kinetics in fixed-bed adsorption under constant pattern conditions, Ind Eng Chem Fund, 12(1), 95-99 (1973).

Furusawa, T., and Smith, J.M., Dynamics of packed-bed adsorbers using the cell model, Ind Eng Chem Fund, 12(3), 388-390 (1973).

Furusawa, T., and Smith, J.M., Diffusivities from dynamic adsorption data, AIChEJ, 19(2), 401-403 (1973).

Garg, D.R., and Ruthven, D.M., Fixed-bed sorption behavior of gases with nonlinear equilibria, AIChEJ, 19(4), 852-853 (1973).

Hashimoto, N., and Smith, J.M., Macropore diffusion in molecular sieve pellets by chromatography, Ind Eng Chem Fund, 12(3), 353-359 (1973).

Hori, Y., and Kobayashi, R., Thermodynamic properties of adsorbate for high-pressure multilayer adsorption, Ind Eng Chem Fund, 12(1), 26-30 (1973).

Jury, S.H., and Horng, J.S., Molecular sieve 4A water vapor sorption therm, AIChEJ, 19(2), 371-372 (1973).

Kocirik, M.; Zikanova, A., and Dubsky, J., Numerical solution of the adsorption kinetics with a nonlinear isotherm, Ind Eng Chem Fund, 12(4), 440-443 (1973).

Kyte, W.S., Freundlich isotherm for nonlinear adsorption in fixed beds, Chem Eng Sci, 28(10), 1853-1856 (1973).

Myers, A.L., Adsorption of gas mixtures on molecular sieves, AIChEJ, 19(3), 666-667 (1973).

Ruthven, D.M.; Loughlin, K.F., and Holborow, K.A., Multicomponent sorption equilibrium in molecular sieve zeolites, Chem Eng Sci, 28(3), 701-710 (1973).

Sircar, S., and Myers, A.L., Surface potential theory of multilayer adsorption from gas mixtures, Chem Eng Sci, 28(2), 489-500 (1973).

Bowen, J.H., Gas-solid, non-catalytic reaction models, Trans IChemE, 52, 282-284 (1974).

Chen, H.T.; Lin, W.W.; Stokes, J.D., and Fabisiak, W.R., Separation of multicomponent mixtures via thermal parametric pumping, AIChEJ, 20(2), 306-310 (1974).

Cooney, D.O., Numerical investigation of adiabatic fixed-bed adsorption, Ind Eng Chem Proc Des Dev, 13(4), 368-373 (1974).

Gilliland, E.R.; Baddour, R.F.; Perkinson, G.P., and Sladek, K.J., Diffusion on surfaces, Ind Eng Chem Fund, 13(2), 95-105 (1974).

Gregory, R.A., Comparison of parametric pumping with conventional adsorption, AIChEJ, 20(2), 294-300 (1974).

Kocirik, M., and Zikanova, A., Analysis of adsorption kinetics in materials with polydisperse pore structure, Ind Eng Chem Fund, 13(4), 347-350 (1974).

Meir, D., and Lavie, R., Continuous cyclic zone adsorption, Chem Eng Sci, 29(5), 1133-1138 (1974).

Wankat, P.C., Cyclic separation processes, Sepn Sci, 9(2), 85-116 (1974).

Weber, T.W., and Chakravorti, R.K., Pore and solid diffusion models for fixed-bed adsorbers, AIChEJ, 20(2), 228-238 (1974).

Zwiebel, I.; Gariepy, R.L., and Schnitzer, J.J., Fixed-bed desorption behavior of gases with nonlinear systems, AIChEJ, 18(6), 1139-1147 (1972); 20(5), 915-923 (1974).

1975-1976

Basmadjian, D.; Ha, K.D., and Pan, C.Y., Nonisothermal desorption by gas purge of single solutes in fixed-bed adsorbers, Ind Eng Chem Proc Des Dev, 14(3), 328-347 (1975).

Choi, P.S.K.; Fab, L.T., and Hsu, H.H., Modeling and simulation of adiabatic adsorber, Sepn Sci, 10(6), 701-722 (1975).

Garg, D.R., and Ruthven, D.M., Linear driving force approximations for diffusion controlled adsorption in molecular sieve columns, AIChEJ, 21(1), 200-202 (1975).

Greco, G.; Iorio, G.; Tola, G., and Waldram, S.P., Unsteady-state diffusion in porous solids, Trans IChemE, 53, 55-58 (1975).

Karger, J., and Bulow, M., Theoretical prediction of uptake behaviour in adsorption kinetics of binary gas mixtures using irreversible thermodynamics, Chem Eng Sci, 30(8), 893-896 (1975).

Lunde, P.J., and Kester, F.L., Chemical and physical gas adsorption in finite multimolecular layers, Chem Eng Sci, 30(12), 1497-1506 (1975).

Camero, A.A., and Sweed, N.H., Separation of nonlinearly sorbing solutes by parametric pumping, AIChEJ, 22(2), 369-376 (1976).

Chihara, K.; Suzuki, M., and Kawazoe, K., Effect of heat generation on measurement of adsorption rate by gravimetric method, Chem Eng Sci, 31(6), 505-507 (1976).

Dore, J.C., and Wankat, P.C., Multicomponent cycling zone adsorption, Chem Eng Sci, 31(10), 921-928 (1976).

England, R., and Thomas, W.J., The significance and measurement of surface diffusion coefficients in catalysis, Trans IChemE, 54, 115-118 (1976).

Ferrell, J.K.; Rousseau, R.W., and Branscome, M.R., Development and testing of mathematical model for complex adsorption beds, Ind Eng Chem Proc Des Dev, 15(1), 114-122 (1976).

Gidaspow, D.; Dharia, D., and Leung, L., Gas purification by porous solids with structural changes, Chem Eng Sci, 31(5), 337-344 (1976).

Hsu, C.C.; Rudzinski, W., and Wojciechowski, B.W., Three-dimensional mobile adsorption: Evaluation of density distribution in adsorbed phases, Chem Eng Sci, 31(12), 1123-1130 (1976).

Nagy, L.G., Study of porous adsorbents by isotopic molecular exchange method, Periodica Polytechnica, 20, 25-36 (1976).

Rice, R.G., Progress in parametric pumping, Sepn & Purif Methods, 5(1), 139-188 (1976).

Ruthven, D.M., Sorption and diffusion in molecular sieve zeolites, Sepn & Purif Methods, 5(2), 189-246 (1976).

Svedberg, U.G., Numerical solution of multicolumn adsorption processes under periodic countercurrent operation, Chem Eng Sci, 31(5), 345-354 (1976).

Szirmay, L., Relative diffusivities from breakthrough curves through exchange adsorption, Sepn Sci, 11(2), 159-170 (1976).

Tien, C.; Hsieh, J.S.C, and Turian, R.M., Application of h-transformation for solution of multicomponent adsorption in fixed bed, AIChEJ, 22(3), 498-505 (1976).

Waksmundzki, A., et al., Application of gas-adsorption chromatography data to investigation of adsorptive properties of adsorbents, Sepn Sci, 11(1), 29-38; 11(4), 411-416 (1976).

1977-1978

Allen, T., and Burevski, D., Adsorption of gases on microporous carbons, Powder Tech, 18, 139-148 (1977).

Bye, G.C., and Chigbo, G.O., Analysis of some adsorption isotherms on calcium silicate hydrates, J Appl Chem Biotechnol, 27, 48-54 (1977).

Grevillot, G., and Tondeur, D., Equilibrium staged parametric pumping, AIChEJ, 22(6), 1055-1063 (1976); 23(6), 840-851 (1977).

Jaroniec, M.; Borowko, M., and Rudzinski, W., Gonzalez-Holland model for adsorption of gas mixtures, AIChEJ, 23(4), 605-607 (1977).

Klaus, R.; Aiken, R.C., and Rippin, D.W.T., Simulated binary isothermal adsorption on activated carbon in periodic countercurrent column operation, AIChEJ, 23(4), 579-586 (1977).

Ma, Y.H., and Lee, T.Y., Diffusion of binary gas mixtures in zeolite X pellets, Ind Eng Chem Fund, 16(1), 44-48 (1977).

Ozil, P., and Bonnetain, L., Dynamical adsorption in fixed bed, Chem Eng Sci, 32(3), 303-310 (1977).

Shah, D.B., and Ruthven, D.M., Measurement of zeolitic diffusivities and equilibrium isotherms by chromatography, AIChEJ, 23(6), 804-809 (1977).

Various, Adsorption and ion exchange (topic issue), Chem Eng Prog, 73(10), 44-64 (1977).

von Rosenberg, D.U.; Chambers, R.P., and Swan, G.A., Numerical solution of surface-controlled fixed-bed adsorption, Ind Eng Chem Fund, 16(1), 154-157 (1977).

Wankat, P.C., Fractionation by cycling zone adsorption, Chem Eng Sci, 32(11), 1283-1288 (1977).

Carleton, F.B.; Kershenbaum, L.S., and Wakeham, W.A., Adsorption in non-isobaric fixed beds, Chem Eng Sci, 33(9), 1239-1246 (1978).

Chihara, K.; Suzuki, M., and Kawazoe, K., Adsorption rate on molecular sieving carbon by chromatography, AIChEJ, 24(2), 237-246 (1978).

Chung, I.J., and Hsu, H.W., Analysis of packed bed adsorption, Chem Eng Sci, 33(3), 399-403 (1978).

DiGiano, F.A.; Baldauf, G.; Frick, B., and Sontheimer, H., Simplified competitive equilibrium adsorption model, Chem Eng Sci, 33(12), 1667-1674 (1978).

Dyer, A., Separation of closely related systems by molecular sieve zeolites, Sepn Sci Technol, 13(6), 501-516 (1978).

Glandt, E.D.; Myers, A.L., and Fitts, D.D., Physical adsorption of gases on graphetized carbon black, Chem Eng Sci, 33(12), 1659-1666 (1978).

Liapis, A.I., and Rippin, D.W.T., Simulation of binary adsorption in activated carbon columns, Chem Eng Sci, 33(5), 593-600 (1978).

Nelson, W.C.; Silarski, D.F., and Wankat, P.C., Continuous flow equilibrium-staged model for cycling zone adsorption, Ind Eng Chem Fund, 17(1), 32-38 (1978).

Ozil, P., and Bonnetain, L., Theoretical prediction of temperature profile in adsorbent fixed-bed, Chem Eng Sci, 33(9), 1233-1238 (1978).

Razavi, M.S.; McCoy, B.J., and Carbonell, R.G., Moment theory of breakthrough curves for fixed-bed adsorbers and reactors, Chem Eng J, 16(3), 211-222 (1978).

Szirmay, L., Dynamic behaviour and relative mass transfer coefficients of a porous adsorbent through exchange adsorption, Trans IChemE, 56, 101-106 (1978).

Wankat, P.C., Continuous recuperative-mode parametric pumping, Chem Eng Sci, 33(6), 723-734 (1978).

Weber, T.W., Batch adsorption for pore diffusion with film resistance and irreversible isotherm, Can JCE, 56, 187-197 (1978).

Wiedemann, K.; Roethe, A.; Radeke, K.H., and Gelbin, D., Modelling of adsorption-desorption breakthrough curves using statistical moments, Chem Eng J, 16(1), 19-26 (1978).

1979

Gelbin, D., and Bunke, G., Equilibrium adsorption dynamics in the cyclic steady state with concave isotherms, Chem Eng J, 17(3), 191-200 (1979).

Karger, K.J., and Zikanova, A., Influence of heat generated on adsorption in bidisperse adsorbents, J Chem Tech Biotechnol, 29, 339-345 (1979).

Kiselev, A.V., Molecular statistical study of thermodynamic characteristics of hydrocarbon adsorption on zeolites, J Chem Tech Biotechnol, 29, 673-685 (1979).

Lee, L.K.; Yucel, H., and Ruthven, D.M., Kinetics of adsorption in bi-porous molecular sieves, Can JCE, 57, 65-80 (1979).

Liapis, A.I., and Rippin, D.W.T., Simulation of binary adsorption in continuous countercurrent operation and comparison with other operating modes, AIChEJ, 25(3), 455-460 (1979).

Liaw, C.H., et al., New solution to kinetics of fixed-bed adsorption, AIChEJ, 25(2), 376-381 (1979).

Yang, R.T., and Liu, R.T., Gaseous diffusion in carbon with particular reference to graphite, Ind Eng Chem Proc Des Dev, 18(2), 245-249 (1979).

1980

Basmadjian, D., Rapid procedures for prediction of fixed-bed adsorber behavior, Ind Eng Chem Proc Des Dev, 19(1), 129-144 (1980).

Danner, R.P.; Nicoletti, M.P., and Al-Ameeri, R.S., Determination of gas mixture adsorption equilibria by the tracer-pulse technique, Chem Eng Sci, 35(10), 2129-2134 (1980).

Gelbin, D., and Fiedler, K., Concentration dependence of diffusion coefficients in zeolites, AIChEJ, 26(3), 510-513 (1980).

Harwell, J.H.; Liapis, A.I.; Litchfield, R., and Hanson, D.T., Non-equilibrium model for fixed-bed multicomponent adiabatic adsorption, Chem Eng Sci, 35(11), 2287-2296 (1980).

Hayhurst, D.T., Gas adsorption by some natural zeolites, Chem Eng Commns, 4(6), 729-736 (1980).

Kelly, J.F., and Fuller, O.M., An evaluation of a method for investigating sorption and diffusion in porous solids, Ind Eng Chem Fund, 19(1), 11-17 (1980).

Liapis. A.I., and Litchfield, R.J., Ternary adsorption in columns, Chem Eng Sci, 35(11), 2366-2370 (1980).

Moharir, A.S.; Kunzru, D., and Saraf, D.N., Theoretical prediction of sorption curves for molecular sieves, Chem Eng Sci, 35(6), 1435-1442 (1980).

Mor, L.; Mor, L.A.,; Sideman, S., and Brandes, J.M., Time dependent packed bed adsorption of a chemically-bound adsorbate, Chem Eng Sci, 35(3), 725-736 (1980).

Oberoi, A.S.; Fuller, O.M., and Kelly, J.F., Methods of interpreting transient response curves from dynamic sorption experiments, Ind Eng Chem Fund, 19(1), 17-21 (1980).

Ranade, M.G., and Evans, J.W., Reaction between gas and solid in nonisothermal packed bed: Simulation and experiments, Ind Eng Chem Proc Des Dev, 19(1), 118-123 (1980).

Ruthven, D.M., and Kumar, R., An experimental study of single-component and binary adsorption equilibria by a chromatographic method, Ind Eng Chem Fund, 19(1), 27-32 (1980).

Ruthven, D.M.; Lee, K.L., and Yucel, H., Kinetics of non-isothermal sorption in molecular sieve crystals, AIChEJ, 26(1), 16-23 (1980).

Schweich, D.; Villermaux, J., and Sardin, M., Introduction to nonlinear theory of adsorptive reactors, AIChEJ, 26(3), 477-486 (1980).

Suwanayuen, S., and Danner, R.P., Gas adsorption isotherm equation based on vacancy solution theory, AIChEJ, 26(1), 68-83 (1980).

Tan, H.K.S., Kinetics of fixed-bed sorption processes, Chem Eng, 24 March, 117-119 (1980).

Wang, M.L.; Liou, C.T., and Chang, R.Y., Numerical technique for solving partial differential equations with applications to adsorption process, Comput Chem Eng, 4(2), 85-92 (1980).

Wilson, D.J. and Clarke, A.N., Theory of adsorption by activated carbon, Sepn Sci Technol, 14(3), 227-242; 14(5), 415-430 (1979); 15(1), 1-22 (1980).

1981

Al-Sahhaf, T.A.; Sloan, E.D., and Hines, A.L., Application of the modified potential theory to the adsorption of hydrocarbon vapors on silica gel, Ind Eng Chem Proc Des Dev, 20(4), 658-662 (1981).

Belfort, G., Similarity of adsorbed solution and potential theories for adsorption from a bulk phase onto a solid surface, AIChEJ, 27(6), 1021-1022 (1981).

Chihara, K.; Smith, J.M., and Suzuki, M., Regeneration of powdered activated carbon, AIChEJ, 27(2), 213-225 (1981).

Goto, S.; Goto, M., and Teshima, H., Simplified evaluations of mass-transfer resistances from batch-wise adsorption and ion-exchange data, Ind Eng Chem Fund, 20(4), 368-375 (1981).

Jacob, P., and Tondeur, D., Non-isothermal gas adsorption in fixed beds, Chem Eng J, 22(3), 187-202 (1981).
Knox, D., and Dadyburjor, D.B., Bounds for acceptable values of adsorption entropy, Chem Eng Commns, 11(1), 99-112 (1981).
Liapis, A.I., and Litchfield, R.J., Off-diagonal terms of the effective pore diffusivity matrix, Trans IChemE, 59, 122-124 (1981).
Okazaki, M.; Tamon, H., and Toei, R., Adsorbed gas flow through porous media, AIChEJ, 27(2), 262-277 (1981).
Ortlieb, H.J.; Bunke, G., and Gelbin, D., Separation efficiency in the cyclic steady-state for periodic countercurrent adsorption, Chem Eng Sci, 36(6), 1009-1016 (1981).
Paderewski, M.; Majkut, A., and Jedrzejak, A., Simplified model of adiabatic fixed-bed adsorption, Int Chem Eng, 21(1), 129-135 (1981).
Peel, R.G., and Benedek, A., Simplified driving-force model for activated carbon adsorption, Can JCE, 59, 688-692 (1981).
Peel, R.G.; Benedek, A., and Crowe, C.M., Branched-pore kinetic model for activated carbon adsorption, AIChEJ, 27(1), 26-32 (1981).
Rasmuson, A., Exact solution of model for diffusion and transient adsorption in particles and longitudinal dispersion in packed beds, AIChEJ, 27(6), 1032-1035 (1981).
Rice, R.G., Adsorptive distillation, Chem Eng Commns, 10(1), 111-126 (1981).
Ruthven, D.M., and Lee, L.K., Kinetics of nonisothermal sorption, AIChEJ, 27(4), 654-663 (1981).
Sircar, S., and Gupta, R., Semi-empirical adsorption equation for single component gas-solid equilibria, AIChEJ, 27(5), 806-812 (1981).
Vanderschuren, J., Plate efficiency of multistage fluidized-bed adsorbers, Chem Eng J, 21(1), 1-10 (1981).
Yang, R.T., and Wong, C., Role of surface diffusion in Langmuir-Hinshelwood mechanism, Chem Eng Commns, 11(4), 317-326 (1981).

1982

Calligaris, M.B., and Tien, C., Species grouping in multicomponent adsorption calculations, Can JCE, 60, 772-780 (1982).
Frey, D.D., Model of adsorbent behavior applied to use of layered beds in cycling zone adsorption, Sepn Sci Technol, 17(13), 1485-1498 (1982).
Friday, D.K., and LeVan, M.D., Solute condensation in adsorption beds during thermal regeneration, AIChEJ, 28(1), 86-91 (1982).
Karger, J., et al., Importance of dimension variation in determining the limiting steps in adsorption kinetics, J Chem Tech Biotechnol, 32, 376-381 (1982).
Knopf, F.C., and Rice, R.G., Adsorptive distillation: Optimum solids profiles, Chem Eng Commns, 15(1), 109-124 (1982).
Kumar, R.; Duncan, R.C., and Ruthven, D.M., Chromatographic study of diffusion of single components and binary mixtures of gases in 4A and 5A zeolites, Can JCE, 60, 493-499 (1982).
Linek, F., and Dudukovic, M.P., Representation of breakthrough curves for fixed-bed adsorbers and reactors using moments of the impulse response, Chem Eng J, 23(1), 31-36 (1982).
Mathews, A.P., Analytical solution for fixed bed adsorption with variable input concentration, Chem Eng Commns, 15(5), 313-322 (1982).

Moharir, A.S.; Kunzru, D., and Saraf, D.N., Sorption of non-uniform zeolite crystals for various bulk concentration profiles, Chem Eng Commns, 18(1), 15-28 (1982).

Morbidelli, M.; Servida, A.; Storti, G., and Carra, S., Simulation of multicomponent adsorption beds: Model analysis and numerical solution, Ind Eng Chem Fund, 21(2), 123-131 (1982).

Rice, R.G., Approximate solutions for batch, packed tube and radial flow adsorbers: Comparison with experiment, Chem Eng Sci, 37(1), 83-92 (1982).

Rousar, I., and Ditl, P., Kinetic characteristics of batch adsorber or ion exchange device operated under nonisothermal conditions, Chem Eng Commns, 18(5), 341-354 (1982).

Sacco, A.; Chung, B., and Aksoy, Y., Nondestructive method to measure residual adsorption capacity of charcoal filters, Chem Eng Commns, 17(1), 43-56 (1982).

Urano, K.; Yamamoto, E., and Takeda, H., Regeneration rates of granular activated carbons containing adsorbed organic matter, Ind Eng Chem Proc Des Dev, 21(1), 180-185 (1982).

Wilson, D.J., Theory of adsorption by activated carbon, Sepn Sci Technol, 17(11), 1281-1292 (1982).

1983

Aris, R., Interpretation of sorption and diffusion data in porous solids, Ind Eng Chem Fund, 22(1), 150-151 (1983).

Bac, N.; Sacco, A., and Hammarstrom, J.L., Measurement of residual adsorption capacity of charcoal filters under conditions of variable humidity, Chem Eng Commns, 24(4), 205-214 (1983).

Cooney, D.O., and Hines, A.L., Extractive purification on activated carbon, Ind Eng Chem Proc Des Dev, 22(2), 208-211 (1983).

Coppola, A.P., and Levan, M.D., Adsorption with axial diffusion in shallow beds, Chem Eng Sci, 38(7), 991-998 (1983).

Do, D.D., Adsorption in porous solids having bimodal pore size distribution, Chem Eng Commns, 23(1), 27-56 (1983).

Garza, G., and Rosales, M.A., Adsorption and diffusion rate parameters from dynamic gravimetric techniques, Ind Eng Chem Proc Des Dev, 22(1), 168-169 (1983).

Gelbin, D.; Bunke, G.; Wolff, H.J., and Neinass, J., Adsorption separation efficiency in the cyclic steady state, Chem Eng Sci, 38(12), 1993-2002 (1983).

Knaebel, K.S., and Pigford, R.L., Equilibrium and dissipative effects in cycling zone adsorption, Ind Eng Chem Fund, 22(3), 336-346 (1983).

Raghavan, N.S., and Ruthven, D.M., Numerical simulation of a fixed-bed adsorption column by orthogonal collocation method, AIChEJ, 29(6), 922-925 (1983).

Ruthven, D.M., Axial-dispersed plug-flow model for continuous countercurrent adsorbers, Can JCE, 61(6), 881-883 (1983).

Sheindorf, C.; Rebhun, M., and Sheintuch, M., Prediction of breakthrough curves from fixed-bed adsorbers with Freundlich-type multisolute isotherm Chem Eng Sci, 38(2), 335-342 (1983).

Sircar, S., and Kumar, R., Adiabatic adsorption of bulk binary gas mixtures: Analysis by constant pattern model, Ind Eng Chem Proc Des Dev, 22(2), 271-280 (1983).

Sircar, S., and Kumar, R., Adsorption of dilute adsorbate: Effects of small changes in column temperature, Ind Eng Chem Proc Des Dev, 22(2), 280-287 (1983).

Sircar, S.; Kumar, R., and Anselmo, K.J., Effects of column nonisothermality or nonadiabaticity on the adsorption breakthrough curves, Ind Eng Chem Proc Des Dev, 22(1), 10-15 (1983).

Viswanathan, K.; Khakhar, D., and Rao, D.S., Fluidized-bed adsorber modelling and experimental study, Chem Eng Commns, 20(1), 235-252 (1983).

Yoshida, H., and Ruthven, D.M., Dynamic behaviour of an adiabatic adsorption column, Chem Eng Sci, 38(6), 877-884 (1983).

1984

Aharoni, C., Review of kinetics of adsorption: The S-shaped z-t plot, Ads Sci Tech, 1(1), 1-29 (1984).

Al-Ameeri, R.S., and Danner, R.P., Improved tracer-pulse method for measurement of gas adsorption equilibria, Chem Eng Commns, 26(1), 11-24 (1984).

Berzins, A.R., et al., Isothermal chemisorption upon oxide-supported platinum, Ads Sci Tech, 1(1), 51-76 (1984).

Birnholtz, H.; Nir, A.; Lotan, N., and Aharoni, C., Surface diffusion as rate determining step in activated chemisorption, Can JCE, 62(2), 233-240 (1984).

Carbonell, R.G., and Whitaker, S., Adsorption and reaction at a catalytic surface, Chem Eng Sci, 39(7), 1319-1321 (1984).

Friday, D.K., and LeVan, M.D., Thermal regeneration of adsorption beds: Equilibrium theory for solute condensation, AIChEJ, 30(4), 679-682 (1984).

Friedrich, S., and Gelbin, D., Model of constant-volume adsorption kinetics allowing for two internal resistances, Chem Eng Sci, 39(5), 912-915 (1984).

Hills, J.H., and Pirzada, I.M., Examination of the accuracy of an approximate solution to isothermal packed bed adsorption with a linear isotherm, Chem Eng Sci, 39(5), 919-923 (1984).

Kadlec, O., Mechanisms of volume filling of micropores during adsorption of vapours, Ads Sci Tech, 1(2), 133-150 (1984).

Kaul, B.K., Correlation and prediction of adsorption isotherm data for pure and mixed gases, Ind Eng Chem Proc Des Dev, 23(4), 711-716 (1984).

Kumar, R., and Sircar, S., Adiabatic sorption of bulk or dilute single adsorbate from an inert gas: Effect of gas-solid mass and heat transfer coefficients, Chem Eng Commns, 26(4), 339-354 (1984).

Lee, T.V.; Huang, J.C., and Madey, R., Separation-factor method for analysis of ideal binary mixtures in gas-solid adsorption, Sepn Sci Technol, 19(2), 157-172 (1984).

Linares-Solano, A., et al., The n-nonane preadsorption method applied to activated carbons, Ads Sci Tech, 1(2), 123-132 (1984).

Mansour, A.R., et al., Numerical solution of general nonequilibrium multicomponent adsorption model, Sepn Sci Technol, 19(8), 479-496 (1984).

Mansour, A.R.; Shahalam, A.B., and Darwish, N., Comprehensive study of parameters influencing performance of multicomponent adsorption in fixed beds, Sepn Sci Technol, 19(13), 1087-1112 (1984).

Mehrotra, A.K., and Tien, C., Further work in species grouping in multicomponent adsorption calculation, Can JCE, 62(5), 632-643 (1984).

Radcliffe, D.F., Lumped parameter models of adsorption kinetics in fixed beds, Chem Eng Commns, 25(1), 183-192 (1984).

Raghavan, N.S., and Ruthven, D.M., Dynamic behaviour of an adiabatic adsorption column, Chem Eng Sci, 39(7), 1201-1212 (1984).

Srivastava, R.K., and Joseph, B., Simulation of packed-bed separation processes using orthogonal collocation, Comput Chem Eng, 8(1), 43-50 (1984).

Urano, K., and Yamamoto, E., Adsorption of organic vapors on activated carbon, Ind Eng Chem Proc Des Dev, 23(4), 665-669 (1984).

Valenzuela, D., and Myers, A.L., Gas adsorption equilibria, Sepn & Purif Methods, 13(2), 153-183 (1984).

Wang, S.C.P., and Tien, C., Interaction between adsorption and bacterial activity in granular activated carbon columns, AIChEJ, 30(5), 786-801 (1984).

Wankat, P.C., New adsorption methods, Chem Eng Educ, 18(1), 20-25, 44-48 (1984).

1985

Basmadjian, D., and Karayannopoulos, C., Rapid procedures for prediction of fixed-bed adsorber behavior, Ind Eng Chem Proc Des Dev, 24(1), 140-149 (1985).

Blasinski, H., and Krauze, S.M., Method of moments for analyzing and predicting outlet curve for adsorption process, Int Chem Eng, 25(1), 182-191 (1985).

Bobyleva, M.S., et al., Relation between structure of molecules of nitrogen-containing heterocyclic compounds and their adsorption on graphitized carbon black, Ads Sci Tech, 2(3), 165-176 (1985).

Cagliostro, D.E.; Changtai, W., and Smith, J.M., Gas adsorption on carbon-containing fabrics, Ind Eng Chem Proc Des Dev, 24(2), 377-381 (1985).

Cochran, T.W.; Kabel, R.L., and Danner, R.P., Vacancy solution theory of adsorption using Flory-Huggins activity coefficient equations, AIChEJ, 31(2), 268-277 (1985).

Cochran, T.W.; Kabel, R.L., and Danner, R.P., Vacancy solution model of adsorption, AIChEJ, 31(12), 2075-2081 (1985).

Do, D.D., Discrete cell model of fixed-bed adsorbers with rectangular adsorption isotherms, AIChEJ, 31(8), 1329-1337 (1985).

Evteeva, V.A., et al., Adsorption from binary gas mixtures and its theoretical description, Ads Sci Tech, 2(3), 153-164 (1985).

Friedrich, S., and Gelbin, D., Simplified mathematical model describing activation of carbon, Chem Eng & Proc, 19(3), 143-150 (1985).

Ghosh, A.K., and Sridhar, T., Non-dissociative adsorption on activated carbon: A statistical rate theory of interfacial turbulence, Can JCE, 63(5), 784-788 (1985).

Hyun, S.H., and Danner, R.P., Gas adsorption isotherms by use of perturbation chromatography, Ind Eng Chem Fund, 24(1), 95-101 (1985).

Hyun, S.H., and Danner, R.P., Adsorption equilibrium constants and intraparticle diffusivities in molecular sieves by tracer-pulse chromatography, AIChEJ, 31(7), 1077-1085 (1985).

Jacob, P., and Tondeur, D., Nonisothermal gas adsorption in fixed beds, Chem Eng J, 22(3), 187-201 (1981); 26(1), 41-58 (1983); 31(1), 23-38 (1985).

Knaff, G., and Schlunder, E.U., Experimental confirmation of Graham's law of diffusion up to pore diameters of 2 micrometre, Chem Eng & Proc, 19(3), 167-174 (1985).

Liang, S., and Weber, W.J., Parameter evaluation for modeling multicomponent mass transfer in fixed-bed adsorbers, Chem Eng Commns, 35(1), 49-65 (1985).

Mansour, A.R.; Shahalam, A.B., and Sotari, M.A., Parametric sensitivity study of multicomponent adsorption in agitated tanks, Sepn Sci Technol, 20(1), 1-20 (1985).

McKay, G., and Bino, M.J., Application of two-resistance mass transfer model to adsorption systems, CER&D, 63, 168-174 (1985).

Mehta, S.D., and Danner, R.P., An improved potential theory method for predicting gas-mixture adsorption equilibria, Ind Eng Chem Fund, 24(3), 325-330 (1985).

Nemeth, J.; Vasanits, E.V., and Virag, T., Mathematical modelling of countercurrent adsorption, Int J Heat Mass Trans, 28(4), 859-866 (1985).

O'Brien, J.A., and Myers, A.L., Rapid calculations of multicomponent adsorption equilibria from pure isotherm data, Ind Eng Chem Proc Des Dev, 24(4), 1188-1191 (1985).

Palancz, B., Modelling and simulation of heat and mass transfer in a packed bed of solid particles having high diffusion resistance, Comput Chem Eng, 9(6), 567-582 (1985).

Pfeifer, H., et al., Concentration dependence of intracrystalline self-diffusion in zeolites, Ads Sci Tech, 2(4), 229-240 (1985).

Rudzinski, W., et al., Simple adsorption equation for adsorption of nonionic surfactants on hydrophilic surfaces of silica, Ads Sci Tech, 2(4), 207-218 (1985).

Ruthven, D.M., and Wong, F., Generalized statistical model for the prediction of binary adsorption equilibria in zeolites, Ind Eng Chem Fund, 24(1), 27-32 (1985).

Sircar, S., New isotherm for multilayer adsorption of vapours on non-porous adsorbents, Ads Sci Tech, 2(1), 23-30 (1985).

Sircar, S., and Kumar, R., Equilibrium theory for adiabatic desorption of bulk binary gas mixtures by purge, Ind Eng Chem Proc Des Dev, 24(2), 358-364 (1985).

Smith, D.M., and Keller, J.F., Nonlinear sorption effects on the determination of diffusion/sorption parameters, Ind Eng Chem Fund, 24(4), 497-499 (1985).

Wojciechowski, B.W.; Hsu, C.C., and Rudzinski, W., Adsorption from multicomponent gas mixtures on the heterogeneous surfaces of solid catalysts, Can JCE, 63(5), 789-794 (1985).

Wojsz, R., and Rozwadowski, M., Thermodynamical analysis of adsorption isotherms measured for microporous adsorbents, Chem Eng Sci, 40(1), 105-110 (1985).

1986

Adschiri, T., and Furusawa, T., Relation between carbon dioxide reactivity of coal char and BET surface area, Fuel, 65(7), 927-931 (1986).

Biswas, J.; Do, D.D.; Greenfield, P.F., and Smith, J.M., Importance of finite adsorption rate in the evaluation of adsorption and diffusion parameters in porous catalysts, AIChEJ, 32(3), 493-496 (1986).

Blasinski, H., and Krauze, S.M., Method of moments for analyzing and predicting outlet curve for adsorption process: Estimation of outlet time and bed height, Int Chem Eng, 26(2), 340-348 (1986).

Cen, P.L., and Yang, R.T., Analytic solution for adsorber breakthrough curves with bidisperse sorbents (zeolites), AIChEJ, 32(10), 1635-1641 (1986).

Ching, C.B., and Ruthven, D.M., Experimental study of a simulated countercurrent adsorption system, Chem Eng Sci, 41(12), 3063-3072 (1986).

Chitra, S.P., and Govind, R., Application of a group contribution method for predicting adsorbability on activated carbon, AIChEJ, 32(1), 167-169 (1986).

Do, D.D., Analysis of a batch adsorber with rectangular adsorption isotherms, Ind Eng Chem Fund, 25(3), 321-326 (1986).

Do, D.D., and Rice, R.G., Validity of the parabolic velocity profile assumption in adsorption studies, AIChEJ, 32(1), 149-154 (1986).

Fairbridge, C.; Ng, S.H., and Palmer, A.D., Fractal analysis of gas adsorption on Sycrude coke, Fuel, 65(12), 1759-1762 (1986).

Fiedler, K., and Grauert, B., Monte Carlo simulation of thermodynamic function in zeolites, Ads Sci Tech, 3(3), 181-188 (1986).

Gelbin, D., et al., Breakthrough curves for single solutes in beds of activated carbon with a broad pore-size distribution, Chem Eng Sci, 41(3), 541-554 (1986).

Hidajat, K.; Ching, C.B., and Ruthven, D.M., Simulated countercurrent adsorption processes: A theoretical analysis of the effect of subdividing the adsorbent bed, Chem Eng Sci, 41(11), 2953-2956 (1986).

High, M.S., and Danner, R.P., Treatment of gas-solid adsorption data by the error-in-variables method, AIChEJ, 32(7), 1138-1145 (1986).

Kumar, R., Column dynamics for multicomponent adsorption: Constant pattern formation, Sepn Sci Technol, 21(10), 1039-1046 (1986).

Lee, C.S., and O'Connell, J.P., Statistical mechanical model for adsorption and flow of pure and mixed gases in porous media with homogeneous surfaces, AIChEJ, 32(1), 96-122 (1986).

Lee, T.V.; Rothstein, D., and Madey, R., Moment analysis of time-dependent transmission of step-function input of radioactive gas through adsorber bed, Sepn Sci Technol, 21(6), 689-700 (1986).

Matteson, M.J., and Schirmer, W., Adsorption of gases at evaporating and condensing water surfaces, Chem Eng J, 33(1), 27-38 (1986).

Moon, H., and Lee, W.K., A lumped model for multicomponent adsorptions in fixed beds, Chem Eng Sci, 41(8), 1995-2004 (1986).

Nakahara, T., Calculation of adsorption equilibria for the binary gaseous mixtures on a heterogeneous surface, Chem Eng Sci, 41(8), 2093-2098 (1986).

Palekar, M.G., and Rajadhyaksha, R.A., Sorption in zeolites, Chem Eng Sci, 40(7), 1085-1092 (1985); 41(3), 463-468 (1986).

Palekar, M.G., and Rajadhyaksha, R.A., Sorption accompanied by chemical reaction on zeolites, Catalysis Reviews, 28(4), 371-429 (1986).

Patwardhan, V.S., and Tien, C., Effect of particle stratification on the performance of fluidized adsorption beds, AIChEJ, 32(2), 321-324 (1986).

Rousar, I., and Ditl, P., Numerical simulation of multicomponent isobaric adsorption in fixed-bed columns, Ads Sci Tech, 3(2), 49-60 (1986).

Sircar, S., and Kumar, R., Column dynamics for adsorption of bulk binary gas mixtures on activated carbon, Sepn Sci Technol, 21(9), 919-940 (1986).

Sircar, S., and Myers, A.L., Characteristic adsorption isotherm for adsorption of vapors on heterogeneous adsorbents, AIChEJ, 32(4), 650-656 (1986).

Stanley-Wood, N.G.; Sadeghnejad, G.R., and York, P., Adsorption potential characterisation of modified celluloses, Powder Tech, 46(2), 195-200 (1986).

Talu, O., and Zwiebel, I., Multicomponent adsorption equilibria of nonideal mixtures, AIChEJ, 32(8), 1263-1276 (1986).

Tan, C.S., Pseudo-steady-state approximation for a packed-bed adsorber, Chem Eng Sci, 41(11), 2956-2958 (1986).

Tien, C., Incorporation of IAS theory in multicomponent adsorption calculations, Chem Eng Commns, 40, 265-280 (1986).

Utrilla, J.R., and Garcia, M.F., Effect of carbon-oxygen and carbon-nitrogen surface complexes on adsorption of cations by activated carbons, Ads Sci Tech, 3(4), 293-302 (1986).

Wittkopf, H., and Brauer, P., Thermodynamic formulations of excess and absolute values of adsorption on solid surfaces: One and two phase approaches, Ads Sci Tech, 3(4), 271-292 (1986).

1987

Aharoni, C., Adsorption by nonhomogeneous porous solids: Effect of adsorption energy gradient on surface flow, AIChEJ, 33(2), 303-306 (1987).

Allen, S.J., Equilibrium adsorption isotherms for peat, Fuel, 66(9), 1171-1175 (1987).

Altshuller, D., et al., Analysis of countercurrent adsorber, Chem Eng Commns, 52, 311-330 (1987).

Aracil, J., Use of factorial design of experiments in the determination of adsorption equilibrium constants (methyl iodide on charcoals), J Chem Tech Biotechnol, 38(3), 143-152 (1987).

Burganos, V.N., and Sotirchos, S.V., Diffusion in pore networks: Effective medium theory and smooth field approximation, AIChEJ, 33(10), 1678-1689 (1987).

Chen, T.L., and Hsu, J.T., Prediction of breakthrough curves by fast Fourier transform method, AIChEJ, 33(8), 1387-1390 (1987).

Davis, M.M., and LeVan, M.D., Equilibrium theory for complete adiabatic adsorption cycles, AIChEJ, 33(3), 470-479 (1987).

Jaroniec, M., and Madey, R., Gas adsorption on structurally heterogeneous microporous solids, Sepn Sci Technol, 22(12), 2367-2380 (1987).

Kapoor, A., and Yang, R.T., Roll-up in fixed-bed multicomponent adsorption under pore-diffusion limitation, AIChEJ, 33(7), 1215-1217 (1987).

Mansour, A.R., Comparison of equilibrium and nonequilibrium models in simulation of multicomponent sorption processes, Sepn Sci Technol, 22(4), 1219-1234 (1987).

Marutovsky, R.M., and Bulow, M., Sorption kinetics of multicomponent gaseous and liquid mixtures on porous sorbents (review paper), Gas Sepn & Purif, 1(2), 66-76 (1987).

McKay, G.; Otterburn, M.S., and Aga, J.A., Pore diffusion and external mass transport during dye adsorption on to Fuller's earth and silica, J Chem Tech Biotechnol, 37(4), 247-256 (1987).

Pesaran, A.A., and Mills, A.F., Moisture transport in silica gel packed beds, Int J Heat Mass Trans, 30(6), 1037-1060 (1987).

Talu, O., and Kabel, R.L., Isosteric heat of adsorption and vacancy solution model, AIChEJ, 33(3), 510-514 (1987).

Tine, C.B.D., Single pellet model application to a two component fixed bed adsorber, CER&D, 65(2), 199-206 (1987).

Van Deventer, J.S.J., Parametric sensitivity study of adsorption in periodic countercurrent cascade of stirred tanks, Sepn Sci Technol, 22(7), 1737-1760 (1987).

Wankat, P.C., Intensification of sorption processes, Ind Eng Chem Res, 26(8), 1579-1586 (1987).

Wittkopf, H., and Brauer, P.., Statistical thermodynamic calculations for adsorption data of gases: Two-phase approach, Ads Sci Tech, 4(4), 251-274 (1987).

1988

Al Duri, B., and McKay, G., Basic dye adsorption on carbon using a solid-phase diffusion model, Chem Eng J, 38(1), 23-32 (1988).

Annesini, M.C.; Gironi, F., and Marrelli, L., Multicomponent adsorption of continuous mixtures, Ind Eng Chem Res, 27(7), 1212-1217 (1988).

Bhatia, S.K., Combined surface and pore volume diffusion in porous media, AIChEJ, 34(7), 1094-1105 (1988).

Biyani, P., and Goochee, C.F., Nonlinear fixed-bed sorption when mass transfer and sorption are controlling, AIChEJ, 34(10), 1747-1751 (1988).

Buso, A.; Paratella, A., and Trotta, A., Solution of dynamic adsorption beds using the finite element method, Comput Chem Eng, 12(2/3), 247-252 (1988).

Crittenden, B., Selective adsorption. Chem Engnr, Sept, 21-24 (1988).

Do, D.D., Asymmetry of adsorption and desorption in microporous solids, Chem Eng Commns, 74, 123-136 (1988).

Do, D.D., and Nguyen, T.S., A power law adsorption model and its significance, Chem Eng Commns, 72, 171-186 (1988).

Doong, S.J., and Yang, R.T., A simple potential-theory model for predicting mixed-gas adsorption, Ind Eng Chem Res, 27(4), 630-635 (1988).

Filippov, L.K., Theoretical basis of adsorption processes for separation of multicomponent mixtures, Gas Sepn & Purif, 2(3), 138-143 (1988).

Haas, O.W.; Kapoor, A., and Yang, R.T., Confirmation of heavy-component rollup in diffusion-limited fixed-bed adsorption, AIChEJ, 34(11), 1913-1916 (1988).

Harriott, P., and Cheng, A.T.Y., Kinetics of spent activated carbon regeneration, AIChEJ, 34(10), 1656-1662 (1988).

Huang, C.C., and Fair, J.R., Adsorption and desorption of multiple adsorbates in a fixed bed, AIChEJ, 34(11), 1861-1877 (1988).

Jasra, R.V., and Bhat, S.G.T., Adsorptive bulk separations by zeolite molecular sieves, Sepn Sci Technol, 23(10), 945-990 (1988).

Karger, J., et al., NMR study of mass transfer in granulated molecular sieves, AIChEJ, 34(7), 1185-1189 (1988).

Klotz, W.L., and Rousseau, R.W., Anomalous mass transfer for vapor adsorption on activated carbon, AIChEJ, 34(8), 1403-1406 (1988).

Kluge, G., and Nagel, G., Modelling of non-isothermal multicomponent adsorption in adiabatic fixed beds, Chem Eng Sci, 43(10), 2885-2889 (1988).

Le Van, M.D., et al., Fixed-bed adsorption of gases: Effect of velocity variations on transition types, AIChEJ, 34(6), 996-1005 (1988).

Moon, H., and Tien, C., Adsorption of gas mixtures on adsorbents with hetergeneous surfaces, Chem Eng Sci, 43(11), 2967-2980 (1988).

O'Brien, J.A., and Myers, A.L., Comprehensive technique for equilibrium calculations in adsorbed mixtures: Generalized Fast-IAS method, Ind Eng Chem Res, 27(11), 2085-2091 (1988).

Rabo, J.A., New advances in molecular sieve science and technology, Periodica Polytechnica, 32(4), 211-234 (1988).

Rota, R., et al., Generalized statistical model for multicomponent adsorption equilibria on zeolites, Ind Eng Chem Res, 27(5), 845-851 (1988).

Rudisill, E.N., and Le Van, M.D., Multicomponent adsorption equilibrium: Henry's law limit for pore-filling models, AIChEJ, 34(12), 2080-2082 (1988).

Schlunder, E.U., et al., Desorption of a binary mixture from activated carbon and the resulting separation effect, Chem Eng Sci, 43(9), 2391-2398 (1988).

Smith, D.M.; Ross, S.B., and Ciftcioglu, M., Calculation of continuous pore size distributions from adsorption isotherms, Powder Tech, 55(3), 225-228 (1988).

Sotirchos, S.V., and Burganos, V.N., Analysis of multicomponent diffusion in pore networks, AIChEJ, 34(7), 1106-1118 (1988).

Storti, G., et al., Adsorption separation processes: Countercurrent and simulated countercurrent operations, Comput Chem Eng, 12(5), 475-482 (1988).

Talu, O., and Myers, A.L., Rigorous thermodynamic treatment of gas adsorption, AIChEJ, 34(11), 1887-1893 (1988).

Valenzuela, D.P.; Myers, A.L.; Talu, O., and Zwiebel, I., Adsorption of gas mixtures: Effect of energetic heterogeneity, AIChEJ, 34(3), 397-402 (1988).

Whitaker, S., Diffusion in packed beds of porous particles, AIChEJ, 34(4), 679-683 (1988).
Wotzak, G.P.; Kim, H.I., and Koronich, E., Transient adsorption by use of dynamic balance method, Chem Eng Commns, 69, 53-64 (1988).
Yoshida, H.; Kataoka, T., and Ruthven, D.M., Dynamic behaviour of an adiabatic adsorption column, Chem Eng Sci, 43(7), 1647-1656 (1988).
Young, B.D., and van Vliet, B.M., Effect of surface roughness on fluid-to-particle mass transfer in a packed adsorber bed, Int J Heat Mass Trans, 31(1), 27-34 (1988).

22.2 Design Data

1967-1972

Lerch, R.G., and Ratkowsky, D.A., Optimum allocation of adsorbent in stagewise adsorption operations, Ind Eng Chem Fund, 6(2), 308-310; 6(3), 480 (1967).
Todes, O.M., and Lezin, Y.S., Dynamics of continuous adsorption and desorption in fluidized bed, Int Chem Eng, 7(3), 447-450; 7(4), 577-581 (1967).
Chen, J.W.; Buege, J.A.; Cunningham, F.L., and Northam, J.I., Scale-up of a column adsorption process by computer simulation, Ind Eng Chem Proc Des Dev, 7(1), 26-31 (1968).
Dayan, J., and Levenspiel, O., Longitudinal dispersion in packed beds of porous adsorbing solids, Chem Eng Sci, 23(11), 1327-1334 (1968).
Shulman, H.L.; Youngquist, G.R.; Marsh, J.L., and Mehta, V.K., Development of a continuous countercurrent fluid-solids adsorber, Ind Eng Chem Proc Des Dev, 7(4), 493-496 (1968).
Tikonova, N.M., Gas flow distribution in packed beds, Int Chem Eng, 8(1), 135-138 (1968).
Carter, J.W., Scale-up in the design of fixed-bed adsorption plant, Brit Chem Eng, 14(3), 303-307 (1969).
Lucas, J.P., and Ratkowsky, D.A., Optimization in various multistage adsorption operations, Ind Eng Chem Fund, 8(3), 576-581 (1969).
Mehta, D.S., and Calvert, S., Calculating actual plates in continuous sorptions, Brit Chem Eng, 14(11), 1563 (1969).
Symoniak, M.F., Correlation for sizing adsorption systems, Chem Eng, 6 Oct, 172-174 (1969).
Zwiebel, I., Fixed-bed adsorption with variable gas velocity due to pressure drop, Ind Eng Chem Fund, 8(4), 803-807 (1969).
Alonso, J.R.F., Simple breakthrough curves yield more information about fixed-bed adsorbers, Brit Chem Eng, 16(7), 602-603 (1971).
Alonsono, J.R.F., Design of fixed bed adsorbers, Brit Chem Eng, 17(4), 326; 17(6), 517-522 (1972).
Jeffreson, C.P., Prediction of breakthrough curves in packed beds, AIChEJ, 18(2), 409-420 (1972).
Johnston, W.A., Designing fixed-bed adsorption columns, Chem Eng, 27 Nov, 87-92 (1972).

King, P.J., and Denizman, E., Channelling in porous beds, Chem Engnr, May, 177-181 (1972).
Kowler, D.E., and Kadlec, R.H., Optimal control of a periodic adsorber, AIChEJ, 18(6), 1207-1219 (1972).
Lavie, R., and Reilly, M.J., Limit cycles in fixed adsorption beds operated in alternating modes, Chem Eng Sci, 27(10), 1835-1844 (1972).
Rimmer, P.G., and Bowen, J.H., Design of fixed bed sorbers using a quadratic driving force equation, Trans IChemE, 50, 168-175 (1972).
Webber, D.A., Adsorptive removal of carbon dioxide from air at intermediate low temperatures, Chem Engnr, Jan, 18-23 (1972).

1973-1979

Garg, D.R., and Ruthven, D.M., Theoretical prediction of breakthrough curves for molecular sieve adsorption columns, Chem Eng Sci, 28(3), 791-806 (1973).
Hutchins, R.A., Design of activated-carbon systems, Chem Eng, 20 Aug, 133-138 (1973).
Lukchis, G.M., Design of adsorption systems, Chem Eng, 11 June, 111-116; 9 July, 83-87; 6 Aug, 83-90 (1973).
Zanker, A., Space rates for fixed-bed adsorption columns, Chem Eng, 26 Nov, 102 (1973).
Garg, D.R., and Ruthven, D.M., Performance of molecular sieve adsorption columns: Macropore diffusion control, Chem Eng Sci, 29(9), 1961-1968 (1974).
Garg, D.R., and Ruthven, D.M., Performance of molecular sieve adsorption columns: Systems with micropore diffusion control, Chem Eng Sci, 29(2), 571-582 (1974).
Bunke, G., and Gelbin, D., Effects of cyclic operation on adsorber performance, Chem Eng Sci, 30(10), 1301-1304 (1975).
Carter, J.W., Regeneration of fixed adsorber beds, AIChEJ, 21(2), 380-382 (1975).
Garg, D.R., and Ruthven, D.M., Performance of molecular sieve adsorption columns: Combined effects of mass transfer and longitudinal diffusion, Chem Eng Sci, 30(9), 1192-1195 (1975).
Papadatos, K.; Svrcek, W.Y., and Bergougnou, M.A., Holdup dynamics of single-stage gas-solid fluidized bed adsorber, Can JCE, 53, 686-695 (1975).
Ruthven, D.M.; Garg, D.R., and Crawford, R.M., Performance of molecular sieve adsorption columns: Non-isothermal systems, Chem Eng Sci, 30(8), 803-810 (1975).
Mostafa, H.A., and Said, A.S., Theoretical-plate concept for fixed-bed adsorption and ion-exchange, Trans IChemE, 54, 132-134 (1976).
Bunke, G., and Gelbin, D., Breakthrough curves in cyclic steady state for adsorption systems with concave isotherms, Chem Eng Sci, 33(1), 101-108 (1978).
Ruthven, D.M., Prediction of desorption times for fixed-bed adsorbers, AIChEJ, 24(3), 540-542 (1978).
Ikeda, K., Performance of nonisothermal fixed-bed adsorption column with nonlinear isotherms, Chem Eng Sci, 34(7), 941-950 (1979).

Sung, E.; Han, C.D., and Rhee, H.K., Optimal design of multistage adsorption-bed systems, AIChEJ, 25(1), 87-100 (1979).

1980-1984

Moharir, A.S.; Kunzru, D., and Saraf, D.N., Effect of adsorbent particle size distribution on breakthrough curves for molecular sieve columns, Chem Eng Sci, 35(8), 1795-1802 (1980).

Robinson, K.S., and Thomas, W.J., The adsorption of methane/ethane mixtures on a molecular sieve, Trans IChemE, 58, 219-227 (1980).

Weber, W.J., and Liu, K.T., Determination of mass-transport parameters for fixed-bed adsorbers, Chem Eng Commns, 6(1), 49-60 (1980).

Weiss, A.; Freund, T., and Biron, E., A method to determine the residual capacity of an adsorber, Chem Engnr, April, 213-215 (1980).

Werling, K., and Wimmerstedt, R., A kinetic study of the adsorption and desorption process of two commercial molecular sieves, Chem Eng Sci, 35(8), 1783-1786 (1980).

Coppola, A.P., and Levan, M.D., Adsorption with axial diffusion in deep beds. Chem Eng Sci, 36(6), 967-972 (1981).

Radeke, K.H.; Ortlieb, H.J., and Gelbin, D., Evaluating breakthrough curves with the method of moments for systems obeying the Langmuir isotherm, Chem Eng Sci, 36(1), 11-18 (1981).

Resnick, W., and Golt, M., Particle-to-gas mass-transfer measurements and coefficients in fixed beds at low Reynolds numbers, Int J Heat Mass Trans, 24(3), 387-394 (1981).

Hunt, L., Adsorptive gas-generators, Proc Engng, Aug, 30-31 (1982).

Liberti, L., and Passino, R., Simplified method for calculating cyclic exhaustion-regeneration operations in fixed-bed adsorbers, Ind Eng Chem Proc Des Dev, 21(2), 197-203 (1982).

Marcussen, L., and Vinding, C., Comparison of experimental and predicted breakthrough curves for adiabatic adsorption in fixed beds, Chem Eng Sci, 37(2), 299-318 (1982).

Balzli, M.W.; Liapis, A.I., and Rippin, D.W.T., Applications of mathematical modelling to simulation of multicomponent adsorption in activated carbon columns, Trans IChemE, 56, 145-156 (1978); CER&D, 61, 393 (1983).

Ray, M.S., Separation and purification of gases using solid adsorbents (review paper), Sepn Sci Technol, 18(2), 95-120 (1983).

Sacco, A.; Bac, N.; Hammarstrom, J.L., and Chung, B., Prediction of residual capacity in thin adsorbers, Can JCE, 61(5), 665-671 (1983).

Umehara, T.; Harriott, P., and Smith, J.M., Regeneration of activated carbon, AIChEJ, 29(5), 732-741 (1983).

Hymore, K., and Laguerie, C., Analysis and modelling of operation of counterflow multistage fluidized bed adsorber for drying moist air, Chem Eng & Proc, 18(5), 255-268 (1984).

Tan, H.K.S., Programs for fixed-bed sorption, Chem Eng, 24 Dec, 57-61 (1984).

1985-1988

Casey, J.T., and Liapis, A.I., Fixed bed sorption with recycle, CER&D, 62, 315-320, 344-350 (1984); 63, 398-402 (1985).

Ching, C.B., and Ruthven, D.M., Experimental study of a simulated countercurrent adsorption system, Chem Eng Sci, 40(6), 877-892; 40(8), 1411-1418 (1985).

Friday, D.K., and LeVan, M.D., Hot purge-gas regeneration of adsorption beds with solute condensation: Experimental studies, AIChEJ, 31(8), 1322-1328 (1985).

Kaguei, S.; Yu, Q., and Wakao, N., Thermal waves in an adsorption column: Parameter estimation, Chem Eng Sci, 40(7), 1069-1076 (1985).

Morbidelli, M., et al., Role of the desorbent in bulk adsorption separations, Chem Eng Sci, 40(7), 1155-1168 (1985).

Sircar, S., and Myers, A.L., Gas adsorption operations: Equilibrium, kinetics, column dynamics and design, Ads Sci Tech, 2(2), 69-88 (1985).

Rajniak, P., and Ilavsky, J., Mathematical and experimental modelling of adsorption in fixed bed, Ads Sci Tech, 3(4), 233-252 (1986).

Chow, D.K., Optimal process parameters for fixed bed adsorption, Can JCE, 65(5), 871-876 (1987).

Dolesjs, V., and Machac, I., Pressure drop in flow of fluid through fixed bed of particles, Int Chem Eng, 27(4), 730-737 (1987).

Kler, S.C., and Lavin, J.T., Computer simulation of gas distribution in large-shallow-packed adsorbers, Gas Sepn & Purif, 1(1), 55-61 (1987).

Mohilla, R.; Argyelan, J., and Szolcsanyi, P., Rapid method for calculating breakthrough curve of gas-cleaning adsorbers, Int Chem Eng, 27(4), 723-730 (1987).

Awum, F.; Narayan, S., and Ruthven, D.M., Measurement of intracrystalline diffusivities in NaX zeolite by liquid chromatography, Ind Eng Chem Res, 27(8), 1510-1515 (1988).

Filippov, L.K., Modelling of adsorption separation of gas mixtures with allowance for thermal effects accompanying adsorption, Gas Sepn & Purif, 2(3), 144-150; 2(4), 190-195 (1988).

Jedrzejak, A., and Paderewski, M., Adsorption-desorption cycles in fixed bed of adsorbent, Int Chem Eng, 28(4), 707-713 (1988).

McKay, G., and Al-Duri, B., Prediction of bisolute adsorption isotherms using single-component data for dye adsorption onto carbon, Chem Eng Sci, 43(5), 1133-1143 (1988).

Suckow, M., Modelling of large-scale adsorptive gas separation processes, Gas Sepn & Purif, 2(3), 151-158; 2(4), 196-204 (1988).

22.3 Adsorbents

1969-1979

Gribkova, L.V.; Sarakhov, A.I.; Tverdokhleb, N.A., and Ryabikova, A.I., Adsorption and mechanical properties of molded zeolites containing clay, Int Chem Eng, 9(3), 503-505 (1969).

Koshelev, V.S., and Byk, S.S., Regeneration of NaA and KA synthetic zeolites by propylene, Int Chem Eng, 10(1), 21-23 (1970).

Green, D.W.; Hardy, R.G.; Beri, P., and Vickburg, C.D., Make activated carbon from coke, Hyd Proc, 50(1), 105-108 (1971).

Ruthven, D.M., and Loughlin, K.F., Effect of crystallite shape and size distribution on diffusion measurements in molecular sieves, Chem Eng Sci, 26(5), 577-584 (1971).

Borthakur, P.C., et al., Quality assessment of molecular sieve zeolites by thermal analysis, J Appl Chem Biotechnol, 23, 415-418 (1973).

Barrer, R.M., and Harding, D.A., Optimization and modification of Offretite sorbents, Sepn Sci, 9(3), 195-210 (1974).

Burwell, R.L., Modified silica gel as adsorbents and catalysts, Chemtech, June, 370-377 (1974).

Parkash, S., Chemistry of activated carbon, Chem & Ind, 1 June, 445-449 (1974).

DeJohn, P.B., Granular activated carbons from lignite and coal, Chem Eng, 28 April, 113-116 (1975).

Matsumura, Y., Production of acidified active carbon by wet oxidation, and its carbon structure, J Appl Chem Biotechnol, 25, 39-56 (1975).

Various, Activated carbon regeneration (topic issue), Chem Eng Prog, 71(5), 80-91 (1975).

Hancil, V., Regeneration of granulated active carbon, Int Chem Eng, 16(4), 663-668 (1976).

Matsumoto, Z., and Numasaki, K., Regenerate granular carbon, Hyd Proc, 55(5), 157-160 (1976).

Trimm, D.L., The deactivation and reactivation of nickel-based catalysts, Trans IChemE, 54, 119-123 (1976).

Burfield, D.R.; Gan, G.H., and Smithers, R.H., Molecular sieves: Desiccants of choice, J Appl Chem Biotechnol, 28, 23-30 (1978).

Gorbaty, M.L., Effect of drying on the adsorptive properties of subbituminous coal, Fuel, 57(12), 796-797 (1978).

Robson, H., Synthesizing zeolites, Chemtech, March, 176-180 (1978).

Suzuki, M.; Misic, D.; Koyama, O., and Kawazoe, K., Thermal regeneration of spent activated carbons, Chem Eng Sci, 33(3), 271-280 (1978).

Zanitsch, R.H., and Lynch, R.T., Selecting a thermal regeneration system for activated carbon, Chem Eng, 2 Jan, 95-100 (1978).

Durie, R.A., and Schafer, H.N.S., Production of active carbon from brown coal in high yields, Fuel, 58(6), 472-476 (1979).

Roques, M., and Bastick, M., Some structural characteristics of microporous carbons deduced from adsorption and diffusion data, Fuel, 58(8), 561-564 (1979).

1980-1984

Dixon, D.R., Magnetic adsorbents: Properties and applications, J Chem Tech Biotechnol, 30, 572-578 (1980).

Linares-Solano, A., et al., Active carbons from almond shells as adsorbents in gas and liquid phases, J Chem Tech Biotechnol, 30, 65-72 (1980).

Seko, M.; Miyake, T., and Inada, K., Sieves for mixed xylenes separation, Hyd Proc, 59(1), 133-138 (1980).

Barrer, R.M., Sorption in porous crystals: Equilibria and their interpretation, J Chem Tech Biotechnol, 31, 71-85 (1981).

Juntgen, H.; Knoblauch, K., and Harder, K., Carbon molecular sieves: Production from coal and application in gas separation, Fuel, 60(9), 832-838 (1981).

Knoblauch, K.; Richter, E., and Jungten, H., Application of active coke in flue gas purification, Fuel, 60(9), 832-838 (1981).

Ordoyno, N.F., and Rowan, S.M., Adsorptive properties of carbons prepared by the pyrolysis of binary mixtures of textile fibres, J Chem Tech Biotechnol, 31, 415-423 (1981).

Wilson, J., Active carbons from coals, Fuel, 60(9), 823-831 (1981).

Idiculla, R., and Seshadri, S.K., Desiccant characteristics of coconut pith, J Chem Tech Biotechnol, 32, 1065-1072 (1982).

Wilkins, C.S.H., New uses for activated carbon, Chem Engnr, Oct, 15, 23 (1983).

Gobet, J., and Kovats, E., Specific surface area of hydrated silicon dioxide measured by chemisorption, Ads Sci Tech, 1(2), 111-122 (1984).

Gobet, J., and Kovats, E., Preparation of hydrated silicon dioxide for reproducible chemisorption experiments, Ads Sci Tech, 1(1), 77-92 (1984).

Goodboy, K.P., and Fleming, H.L., Trends in adsorption with aluminas, Chem Eng Prog, 80(11), 63-68 (1984).

Miura, K., and Hashimoto, K., Model representing change of pore structure during activation of carbonaceous materials, Ind Eng Chem Proc Des Dev, 23(1), 138-145 (1984).

Various, Zeolites for industry (topic issue), Chem & Ind, 2 April, 237-272 (1984).

1985-1988

Bohra, J.N., and Sing, K.S.W., Micropore structure of carbonized rayon yarn by nitrogen adsorption and nonane preadsorption, Ads Sci Tech, 2(2), 89-96 (1985).

Goworek, J., Adsorption properties of hydrated and thermally modified silica gel, Ads Sci Tech, 2(3), 195-206 (1985).

Lopez-Peinado, A., et al., Porous texture characterization of coals and chars, Ads Sci Tech, 2(1), 31-38 (1985).

Shah, D.B., and Hayhurst, D.T., Research on molecular sieve technology, Chem Eng Educ, 19(4), 198-202 (1985).

Hudec, P., et al., Possibility of using t-plots obtained from nitrogen adsorption for valuation of zeolites, Ads Sci Tech, 3(3), 159-166 (1986).

Lohse, U., et al., Dependence of adsorption capacity and thermal stability of Y-zeolites upon Si/Al ratio, Ads Sci Tech, 3(3), 149-158 (1986).

Lohse, U., et al., Adsorption properties of zeolites prepared by different processes, Ads Sci Tech, 3(3), 173-180 (1986).

Lohse, U.; Noack, M., and Jahn, E., Adsorption properties of the AlPO4-5 molecular sieve, Ads Sci Tech, 3(1), 19-24 (1986).

Lopez-Gonzalez, J.D., et al., Preparation of active carbons from olive wood, Ads Sci Tech, 2(4), 263-270 (1985); 3(1), 41-48 (1986).

Stach, H., et al., Influence of pore diameter on adsorption behaviour of nonpolar molecules on silica adsorbents, Ads Sci Tech, 3(4), 261-270 (1986).

Thamm, H., et al., Calorimetric investigation of adsorption properties of microporous alumino-phosphate, Ads Sci Tech, 3(4), 217-220 (1986).

Tipnis, P.R., and Harriott, P., Thermal regeneration of activated carbons, Chem Eng Commns, 46, 11-28 (1986).
Buczek, B., and Czepirski, L., Improvement of capacity for active carbons, Ads Sci Tech, 4(4), 217-223 (1987).
Dubinin, M.M., and Kadlec, O., Microporous adsorbents with limiting development of micropores, Ads Sci Tech, 4(1), 45-52 (1987).
Weisz, P.B., Zeolites: The remarkable active site, Chemtech, June, 368-373 (1987).
Chen, N.Y., and Degnan, T.F., Industrial catalytic applications of zeolites, Chem Eng Prog, 84(2), 32-41 (1988).
Chiang, P.C., and You, J.H., Use of sewage-sludge for manufacturing adsorbents, Can JCE, 65(6), 921-926 (1988).
Costa, E.; de Lucas, A.; Uguina, M.A., and Ruiz, J.C., Synthesis of 4A zeolite from calcined kaolins for use in detergents, Ind Eng Chem Res, 27(7), 1291-1296 (1988).
Iovtchev, K.; Nikolov, L., and Elenkov, D., Experimental study of new mechanically resistant adsorbent, Can JCE, 66(6), 1013-1016 (1988).
Ramdas, S., Computer graphics and zeolite chemistry, Chem Eng Prog, 84(2), 68 (1988).
Ruthven, D.M., Zeolites as selective adsorbents, Chem Eng Prog, 84(2), 42-50 (1988).
Schork, J.M., and Fair, J.R., Steaming of activated carbon beds, Ind Eng Chem Res, 27(8), 1545-1548 (1988).
Van Vliet, B.M., and Weber, W.J., Particle surface roughness effects on interfacial mass transfer dynamics of microporous adsorbents, Chem Eng Commns, 68, 165-176 (1988).
Vaughan, D.E.W., Synthesis and manufacture of zeolites, Chem Eng Prog, 84(2), 25-31 (1988).

22.4 PSA and Cyclic Systems

1967-1979

Alexis, R.W., Upgrading hydrogen via heatless adsorption, Chem Eng Prog, 63(5), 69-71 (1967).
Dellosso, L., and Winnick, J., Mixed-gas adsorption and vacuum desorption of carbon dioxide on molecular sieve, Ind Eng Chem Proc Des Dev, 8(4), 469-482 (1969).
Fair, J.R., Sorption processes for gas separation, Chem Eng, 14 July, 90-110 (1969).
Stewart, H.A., and Heck, J.L., Pressure swing adsorption, Chem Eng Prog, 65(9), 78-83 (1969).
Turnock, P.H., and Kadlec, R.H., Separation of nitrogen and methane via periodic adsorption, AIChEJ, 17(2), 335-342 (1971).
van der Vlist, E., Oxygen and nitrogen enrichment in air by cycling zone adsorption, Sepn Sci, 6(5), 727-732 (1971).
Raghuraman, K.S., and Johansen, T., Hydrogen by PSA process, Processing, Oct, 10-11 (1975).

Wankat, P.C., Multicomponent cycling zone separations, Ind Eng Chem Fund, 14(2), 96-102 (1975).
Wankat, P.C.; Dore, J.C., and Nelson, W.C., Cycling zone separations, Sepn & Purif Methods, 4(2), 215-266 (1975).
Heck, J.L., and Johansen, T., PSA process improves large-scale hydrogen production, Hyd Proc, 57(1), 175-177 (1978).
Narraway, R., PSA gas generators, Processing, Jan, 29 (1978).
Wood, R., Nitrogen from PSA process, Proc Engng, June, 44-47 (1978).
Corr, F.; Dropp, F., and Rudelstorfer, E., PSA produces low-cost high-purity hydrogen, Hyd Proc, 58(3), 119-122 (1979).
Gay, P., PSA vs piped gas, Proc Engng, Feb, 41,43 (1979).

1980-1984

Foo, S.C.; Bergsman, K.H., and Wankat, P.C., Thermal-mode cycling zone adsorption for multicomponent separations, Ind Eng Chem Fund, 19(1), 86-93 (1980).
Hill, F.B., Recovery of weakly adsorbed impurity by pressure swing adsorption, Chem Eng Commns, 7(1), 37-44 (1980).
Jacob, P., and Tondeur, D., Nonisothermal adsorption: Separation of gas mixtures by modulation of feed temperature, Sepn Sci Technol, 15(8), 1563-1578 (1980).
Thomas, W.J., Gas separation by adsorption, Chem & Ind, 3 May, 366-372 (1980).
Chan, Y.N.I.; Hill, F.B., and Wong, Y.W., Equilibrium theory of a pressure swing adsorption process, Chem Eng Sci, 36(2), 243-252 (1981).
Hill, F.B.; Wong, Y.W., and Chan, Y.N.I., Temperature swing adsorption for hydrogen isotope separation, AIChEJ, 28(1), 1-6 (1982).
Wong, Y.W., and Hill, F.B., Separation of hydrogen isotopes via single-column pressure swing adsorption, Chem Eng Commns, 15(5), 343-356 (1982).
Carter, J.W., and Wyszynski, M.L., Pressure swing adsorption drying of compressed air, Chem Eng Sci, 38(7), 1093-1100 (1983).
Fernandez, G.F., and Kenney, C.N., Modelling of the pressure swing air separation process, Chem Eng Sci, 38(6), 827-834 (1983).
Knaebel, K.S., and Hill, F.B., Analysis of gas purification by pressure swing adsorption: Priming the parametric pump, Sepn Sci Technol, 18(12), 1193-1220 (1983).
Tsai, M.C.; Wang, S.S., and Yang, R.T., Temperature-swing adsorption for hydrogen-methane separation, AIChEJ, 29(6), 966-975 (1983).
Watson, A.M., Use pressure swing adsorption for lowest cost hydrogen, Hyd Proc, March, 91-95 (1983).
Rieke, R.D., Cycling zone adsorption: Variable-feed mode of operation, Sepn Sci Technol, 19(4), 261-282 (1984).

1985-1986

Anon., Applications of pressure-swing adsorption, Proc Engng, Sept, 67-71 (1985).
Cen, P., and Yang, R.T., Separation of five-component gas mixture by pressure swing adsorption, Sepn Sci Technol, 20(9), 725-748 (1985).

Cen, P.L.; Chen, W.N., and Yang, R.T., Ternary gas mixture separation by pressure swing adsorption, Ind Eng Chem Proc Des Dev, 24(4), 1201-1208 (1985).
Cheng, H.C., and Hill, F.B., Separation of helium-methane mixtures by pressure swing adsorption, AIChEJ, 31(1), 95-102 (1985).
Costa, C., and Rodrigues, A., Design of cyclic fixed-bed adsorption procesess, AIChEJ, 31(10), 1645-1665 (1985).
Knaebel, K.S., and Hill, F.B., Pressure swing adsorption: Development of an equilibrium theory for gas separations, Chem Eng Sci, 40(12), 2351-2360 (1985).
Platt, D., and Lavie, R., Pressure cyclic zone adsorption, Chem Eng Sci, 40(5), 733-740 (1985).
Raghavan, N.S.; Hassan, M.M., and Ruthven, D.M., Numerical simulation of a PSA system, AIChEJ, 31(3), 385-392; 31(12), 2008-2025 (1985).
Tondeur, D., and Wankat, P.C., Gas purification by pressure swing adsorption, Sepn & Purif Methods, 14(2), 157-212 (1985).
Tsai, M.C.; Wang, S.S.; Yang, R.T., and Desai, N.J., Temperature-swing separation of hydrogen-methane mixture, Ind Eng Chem Proc Des Dev, 24(1), 57-62 (1985).
Yang, R.T., and Doong, S.J., Gas separation by pressure swing adsorption: A pore-diffusion model for bulk separation, AIChEJ, 31(11), 1829-1842 (1985).
Carta, G., and Pigford, R.L., Analytical solution for cycling-zone adsorption, Chem Eng Sci, 41(3), 511-518 (1986).
Cen, P., and Yang, R.T., Bulk gas separation by pressure swing adsorption, Ind Eng Chem Fund, 25(4), 758-768 (1986).
Cen, P.L., and Yang, R.T., Separation of binary gas mixture into two high-purity products by new pressure-swing adsorption cycle, Sepn Sci Technol, 21(9), 845-864 (1986).
Crabb, K.S.; Perona, J.J.; Byers, C.H., and Watson, J.S., Vacuum sorption pumping studies with pure gases on molecular sieves, AIChEJ, 32(2), 255-262 (1986).
Doong, S.J., and Yang, R.T., Parametric study of pressure swing adsorption process for gas separation: Criterion for pore diffusion limitation, Chem Eng Commns, 41, 163-180 (1986).
Doong, S.J., and Yang, R.T., Bulk separation of multicomponent gas mixtures by pressure swing adsorption, AIChEJ, 32(3), 397-410 (1986).
Kayser, J.C., and Knaebel, K.S., Pressure swing adsorption: Experimental study of an equilibrium theory, Chem Eng Sci, 41(11), 2931-2938 (1986).
Pritchard, C.L., and Simpson, G.K., Design of an oxygen concentrator using the rapid pressure-swing adsorption principle, CER&D, 64(6), 467-472 (1986).
Raghaven, N.S.; Hassan, M.M., and Ruthven, D.M., Numerical simulation of a PSA system using a pore diffusion model, Chem Eng Sci, 41(11), 2787-2794 (1986).
Ray, M.S., Pressure swing adsorption: A review of UK patent literature, Sepn Sci Technol, 21(1), 1-38 (1986).
Underwood, R.P., Model of a pressure-swing adsorption process for nonlinear adsorption equilibrium, Chem Eng Sci, 41(2), 409-412 (1986).

1987-1988

Doong, S.J., and Yang, R.T., Comparison of gas separation performance by different pressure swing adsorption cycles, Chem Eng Commns, 54, 61-72 (1987).

Doong, S.J., and Yang, R.T., Bidisperse pore diffusion model for zeolite pressure swing adsorption, AIChEJ, 33(6), 1045-1049 (1987).

Hachiya, K.; Takeda, K., and Yasunaga, T., Pressure-jump method to adsorption-desorption kinetics, Ads Sci Tech, 4(1), 25-44 (1987).

Lu, X.; Rothstein, D.; Madey, R., and Huang, J.C., Pressure swing adsorption for system with Freundlich isotherm, Sepn Sci Technol, 22(6), 1547-1556 (1987).

Shin, H.S., and Knaebel, K.S., Pressure swing adsorption: A theoretical study of diffusion-induced separations, AIChEJ, 33(4), 654-662 (1987).

Basta, N., New developments in pressure swing adsorption, Chem Eng, 26 Sept, 26-31 (1988).

Chiang, A.S.T., Arithmetic of PSA process scheduling, AIChEJ, 34(11), 1910-1913 (1988).

Chiang, A.S.T.; Hwong, M.Y.; Lee, T.Y., and Cheng, T.W., Oxygen enrichment by pressure swing adsorption, Ind Eng Chem Res, 27(1), 81-86 (1988).

Chynoweth, E., PSA on-site nitrogen generation, Processing, Feb, 30 (1988).

Davis, M.M.; McAvoy, R.L., and Le Van, M.D., Periodic states for thermal swing adsorption of gas mixtures, Ind Eng Chem Res, 27(7), 1229-1235 (1988).

Farooq, S.; Hassan, M.M., and Ruthven, D.M., Heat effects in pressure swing adsorption systems, Chem Eng Sci, 43(5), 1017-1033 (1988).

Kapoor, A., and Yang, R.T., Separation of hydrogen-lean mixtures for high-purity hydrogen by vacuum swing adsorption, Sepn Sci Technol, 23(1), 153-178 (1988).

Kapoor, A., and Yang, R.T., Optimization of a pressure swing adsorption cycle, Ind Eng Chem Res, 27(1), 204-206 (1988).

Kayser, J.C., and Knaebel, K.S., Integrated steps in pressure swing adsorption cycles, Chem Eng Sci, 43(11), 3015-3022 (1988).

Lu, X.; Rothstein, D.; Madey, R., and Huang, J.C., Pressure swing adsorption for a system with a Langmuir isotherm, Sepn Sci Technol, 23(4), 281-292 (1988).

Matz., M.J., and Knaebel, K.S., Pressure swing adsorption: Effects of incomplete purge, AIChEJ, 34(9), 1486-1492 (1988).

Rousar, I., and Ditl, P., Optimization of pressure swing adsorption equipment, Chem Eng Commns, 70, 67-106 (1988).

Schork, J.M., and Fair, J.R., Parametric analysis of thermal regeneration of adsorption beds, Ind Eng Chem Res, 27(3), 457-469 (1988).

Shin, H.S., and Knaebel, K.S., Pressure swing adsorption: Experimental study of diffusion-induced separation, AIChEJ, 34(9), 1409-1416 (1988).

Sircar, S., Separation of methane and carbon dioxide gas mixtures by pressure swing adsorption, Sepn Sci Technol, 23(6), 519-530 (1988).

Sircar, S., and Kratz, W.C., Pressure-swing adsorption process for production of 23-50% oxygen-enriched air, Sepn Sci Technol, 23(4), 437-450 (1988).

Sircar, S., and Kratz, W.C., Simultaneous production of hydrogen and carbon dioxide from steam reformer off-gas by pressure swing adsorption, Sepn Sci Technol, 23(14), 2397-2416 (1988).
Toftegard, B., and Jorgensen, S.B., Stationary profiles for periodic cycled separation columns, Ind Eng Chem Res, 27(3), 481-485 (1988).
Wiessner, F.G., Basics and industrial applications of pressure swing adsorption for gas separations, Gas Sepn & Purif, 2(3), 115-119 (1988).

22.5 Systems and Applications

1967-1969

Haydel, J.J., and Kobayashi, R., Adsorption equilibria in the methane-propane-silica gel system at high pressures, Ind Eng Chem Fund, 6(4), 546-554 (1967).
Levinson, S.Z., and Orochko, D.I., Staged countercurrent multisectional contactors for continuous adsorptive treatment of petroleum products, Int Chem Eng, 7(4), 649-654 (1967).
Mathur, B.C., and Banerjee, P.K., Solid desicants for vapour-phase dehydration, Brit Chem Eng, 11(11), 1388-1389 (1966); 12(3), 381-384 (1967).
Schumacher, W.J., and York, R., Separation of hydrocarbons in fixed beds of molecular sieves, Ind Eng Chem Proc Des Dev, 6(3), 321-327 (1967).
Silbernagel, D.R., New uses for molecular sieves in olefin plants, Chem Eng Prog, 63(4), 99-102 (1967).
Carter, J.W., Some aspects of the prediction of performance of a large air-dryer, Trans IChemE, 46, T222-224 (1968).
Collins, J.J., Where to use molecular sieves, Chem Eng Prog, 64(8), 66-71 (1968).
Fukunaga, P.; Hwang, K.C.; Davis, S.H., and Winnick, J., Mixed-gas adsorption and vacuum desorption of carbon dioxide on molecular sieve, Ind Eng Chem Proc Des Dev, 7(2), 269-275 (1968).
Masukawa, S., and Kobayashi, R., Correlations for adsorption of a binary gas mixture on a heterogeneous adsorbent: Methane-ethane-silica gel system, AIChEJ, 14(5), 740-746 (1968).
Rimpel, A.E.; Camp, D.T.; Kostecki, J.A., and Canjar, L.N., Kinetics of physical adsorption of propane from helium on fixed beds of activated alumina, AIChEJ, 14(1), 19-24 (1968).
Shen, J., and Smith, J.M., Adsorption isotherms for benzene-hexane mixtures, Ind Eng Chem Fund, 7(1), 100-105 (1968).
Shen, J., and Smith, J.M., Rates of adsorption in the benzene-hexane system, Ind Eng Chem Fund, 7(1), 106-114 (1968).
Wilhelm, R.H.; Rice, A.W.; Rolke, R.W., and Sweed, N.H., Parametric pumping, Ind Eng Chem Fund, 7(3), 337-349 (1968).
Asher, W.J.; Campbell, M.L.; Epperly, W.R., and Robertson, J.L., Desorb n-paraffins from molecular sieves with ammonia, Hyd Proc, 48(1), 134-138 (1969).
Avery, D.A., and Boiston, D.A., The recovery of solvents from gaseous effluents, Chem Engnr, Jan, CE8-11 (1969).

Danner, R.P., and Wenzel, L.A., Adsorption of CO-nitrogen, CO-oxygen, and oxygen-nitrogen mixtures on synthetic zeolites, AIChEJ, 15(4), 515-520 (1969).

Emery, D.L., Drying natural gas with alumina, Hyd Proc, 48(3), 130-132 (1969).

Kazakova, E.A.; Khiterer, R.Z., and Bomshtein, V.E., Purification of exhaust gases from nitric acid plants, Brit Chem Eng, 14(5), 667-668 (1969).

Layton, L., and Youngquist, G.R., Sorption of sulfur dioxide by ion exchange resins, Ind Eng Chem Proc Des Dev, 8(3), 317-324 (1969).

Ozawa, Y., Regeneration of coked catalyst in adiabatic fixed beds at lower temperatures, Ind Eng Chem Proc Des Dev, 8(3), 378-383 (1969).

Ponder, T.C., Adsorption systems for alkane recovery, Hyd Proc, 48(10), 141 (1969).

Romberg, E., and Shorrock, J.C., Determination of adsorption equilibria for the helium-methane-krypton system on active charcoal using a dynamic method, Trans IChemE, 47, T3-10 (1969).

1970-1971

Allen, T., and Patel, R.M., Adsorption of alcohols on finely divided powders, J Appl Chem, 20, 165-171 (1970).

Broughton, D.B.; Neuzil, R.W.; Pharis, J.M., and Brearley, C.S., Parex process for recovering paraxylene, Chem Eng Prog, 66(9), 70-75 (1970).

Chi, C.W., and Wasan, D.T., Fixed-bed adsorption drying, AIChEJ, 16(1), 23-31 (1970).

DiNapoli, R.N., Adsorption systems for LNG gas pretreatment, Hyd Proc, 49(12), 93-96 (1970).

Jenczewski, T.J., and Myers, A.L., Separation of gas mixtures by pulsed adsorption, Ind Eng Chem Fund, 9(2), 216-221 (1970).

Kehat, E., and Heineman, M., Desorption of normal paraffins from 5A molecular sieve, Ind Eng Chem Proc Des Dev, 9(1), 72-78 (1970).

Kidnay, A.J., and Hiza, M.J., Purification of helium gas by physical adsorption at 76K, AIChEJ, 16(6), 949-954 (1970).

Kotb, A.K., Adsorption of sulphur dioxide on coal, J Appl Chem, 20, 147-152 (1970).

LaPlante, L.J., and Symoniak, M.F., Molecular sieves for protein-from-paraffins process, Hyd Proc, 49(12), 77-82 (1970).

Marcussen, L., Kinetics of water adsorption on porous alumina, Chem Eng Sci, 25(9), 1487-1500 (1970).

Michelson, K.J., and Price, C.D., Molecular sieve pre-drying, Chem Eng Prog, 66(5), 73-74 (1970).

Petukhov, S.S.; Tumanov, A.I., and Trokhina, G.A., Combined process for removal of impurities from air using synthetic zeolites, Int Chem Eng, 10(3), 405-409 (1970).

Phillips, J.R., Desorption of normal paraffins from 5A molecular sieves, Ind Eng Chem Proc Des Dev, 9(3), 484-485 (1970).

Snyder, C.F., and Chao, K.C., Heat of adsorption of light hydrocarbons and their mixtures on activated carbon, Ind Eng Chem Fund, 9(3), 437-443 (1970).

Tan, V.A., et al., Continuous fluid bed adsorber with centrifugal separation of the solid phase, Brit Chem Eng, 15(10), 1295-1296 (1970).

Tomassi, W., Effect of temperature variations on the state of chlorine in adsorption layer on activated carbon, Int Chem Eng, 10(3), 372-374 (1970).
Alexis, R.W., and Dailey, L.W., Molecular sieve driers, Hyd Proc, 50(5), 145-148 (1971).
Barnebey, H.L., Activated charcoal in the petrochemical industry, Chem Eng Prog, 67(11), 45-48 (1971).
Barrere, C.A., Feed-gas drying with molecular sieves, Hyd Proc, 50(8), 126-128 (1971).
Gonzalez, A.J., and Holland, C.D., Adsorption equilibria of light hydrocarbon gases on activated carbon and silica, AIChEJ, 17(2), 470-475 (1971).
Hales, G.E., Gas dehydration using molecular sieves, Hyd Proc, 50(6), 151-154 (1971).
Hales, G.E., Drying reactive fluids with molecular sieves, Chem Eng Prog, 67(11), 49-53 (1971).
Kondis, E.F., and Dranoff, J.S., Kinetics of isothermal sorption of ethane on 4A molecular sieve pellets, Ind Eng Chem Proc Des Dev, 10(1), 108-114 (1971).
McCarthy, W.C., Adsorption of light hydrocarbons from nitrogen with activated carbon, Ind Eng Chem Proc Des Dev, 10(1), 13-18 (1971).
Reikert, L., Rates of sorption and diffusion of hydrocarbons in zeolites, AIChEJ, 17(2), 446-454 (1971).
Ruthven, D.M., and Loughlin, K.F., Sorption and diffusion of n-butane in Linde 5A molecular sieve, Chem Eng Sci, 26(8), 1145-1154 (1971).
Thomas, W.J., and Lombardi, J.L., Binary adsorption of benzene-toluene mixtures, Trans IChemE, 49, 240-250 (1971).
Thomas, W.J., and Qureshi, A.R., Adsorption of toluene in fixed beds of active carbons, Trans IChemE, 49, 60-69 (1971).

1972-1973

Brooking, H.L., and Walton, D.C., The specification of molecular sieve adsorption systems, Chem Engnr, Jan, 13-18 (1972).
Butts, T.J.; Gupta, R., and Sweed, N.H., Parametric pumping separations of multicomponent mixtures: An equilibrium theory, Chem Eng Sci, 27(5), 855-866 (1972).
Carter, J.W., and Husain, H., Adsorption of carbon dioxide in fixed beds of molecular sieves, Trans IChemE, 50, 69-75 (1972).
Colley, C.R., Adsorbent selection for gas drying, Brit Chem Eng, 17(3), 229-233 (1972).
Fernbacher, J.M., and Wenzel, L.A., Adsorption equilibria at high pressures in the helium-nitrogen-activated carbon system, Ind Eng Chem Fund, 11(4), 457-465 (1972).
Harper, C., A molecular sieve plant as the main helium purifier for a high temperature, gas cooled nuclear reactor, Chem Engnr, July, 271-276 (1972).
Joithe, W.; Bell, A.T., and Lynn, S., Removal and recovery of NOx from nitric acid plant tail gas by adsorption on molecular sieves, Ind Eng Chem Proc Des Dev, 11(3), 434-439 (1972).

Loughlin, K.F., and Ruthven, D.M., Sorption and diffusion of ethane in type A zeolites, Chem Eng Sci, 27(7), 1401-1408 (1972).
Malkin, L.S., et al., Mass transfer during drying of oils by NaA synthetic zeolites, Int Chem Eng, 12(2), 331-335 (1972).
Maslan, F., and Aberth, E.R., Low-temperature adsorption of hydrogen on activated carbon, Sepn Sci, 7(5), 601-605 (1972).
Patrick, R.R.; Schrodt, J.T., and Kermode, R.I., Thermal parametric pumping of air-sulfur dioxide, Sepn Sci, 7(4), 331-344 (1972).
Carter, J.W., and Barrett, D.J., Comparative study for fixed bed adsorption of water vapour by activated alumina, silica gel, and molecular sieve adsorbents, Trans IChemE, 51, 75-81 (1973).
El-Rifai, M.A.; Saleh, M.A., and Youssef, H.A., Steam regeneration of a solvents adsorber, Chem Engnr, Jan, 36-38 (1973).
Gregg, S.J.; Nashed, S., and Malik, M.T., Adsorption of water vapour on a microporous carbon black, Powder Tech, 7, 15-19 (1973).
Ma, Y.H., and Roux, A.J., Multicomponent rates of sorption of sulfur dioxide and carbon dioxide on sodium mordenite, AIChEJ, 19(5), 1055-1059 (1973).
Miller, W.C., Adsorption cuts SO2, NOx, Hg, Chem Eng, 6 Aug, 62-63 (1973).
Otani, S., Adsorption separates xylenes, Chem Eng, 17 Sept, 106-107 (1973).
Various, Removal of organics by adsorption (topic issue), Chem & Ind, 1 Sept, 823-831 (1973).

1974-1975

Bond, A., Compressed-air drying systems, Proc Engng, March, 52-53 (1974).
Bourgeois, S.V.; Groves, F.R., and Wehe, A.H., Analysis of fixed-bed sorption: Flue gas desulfurization, AIChEJ, 20(1), 94-103 (1974).
Carter, J.W., and Husain, H., Simultaneous adsorption of carbon dioxide and water vapour by fixed beds of molecular sieves, Chem Eng Sci, 29(1), 267-274 (1974).
Chen, H.T.; Park, J.A., and Rak, J.L., Equilibrium parametric pumps, Sepn Sci, 9(1), 35-46 (1974).
Chiu, H.M.; Hashimoto, N., and Smith, J.M., Chromatographic studies of adsorption of nitric oxide on activated carbon, Ind Eng Chem Fund, 13(3), 282-285 (1974).
Donnet, J.B.; Kobel, L., and Sevenster, A., Helium adsorption on solid surfaces, Ind Eng Chem Fund, 13(1), 83-86 (1974).
Weaver, K., and Hamrin, C.E., Separation of hydrogen isotopes by heatless adsorption, Chem Eng Sci, 29(9), 1873-1882 (1974).
Aharoni, C.; Neuman, M., and Notea, A., Adsorption of lead chloride vapors, Ind Eng Chem Proc Des Dev, 14(4), 417-421 (1975).
Cummings, W.P., Save energy in adsorption, Hyd Proc, 54(2), 97-98 (1975).
Garg, D.R., and Ruthven, D.M., Sorption of carbon dioxide in Davison 5A molecular sieve, Chem Eng Sci, 30(4), 436-437 (1975).
Nandi, S.P., and Walker, P.L., Carbon molecular sieves for concentration of oxygen from air, Fuel, 54(3), 169-178 (1975).
Schunder, E.U., et al., Scale-up of activated carbon columns for water purification based on batch-test results, Chem Eng Sci, 30(5), 529-548 (1975).

Subbotin, A.I., et al., Dynamics of adsorption of methanol-containing air-vapor mixture, Int Chem Eng, 15(3), 392-394 (1975).
Wang, L.K., et al., Treatment of industrial effluents by activated carbon, J Appl Chem Biotechnol, 25, 475-502 (1975).

1976-1979
Chen, H.T.; Rastog, A.K.; Kim, C.Y., and Rak, J.L., Nonequilibrium parametric pumps, Sepn Sci, 11(4), 335-346 (1976).
Hsu, H.H.; Wang, K.B., and Fan, L.T., Gaseous pollutant removal by single bed cyclic adsorber with synchronous thermal contact, Sepn Sci, 11(2), 109-132 (1976).
Kiovsky, J.R.; Koradia, P.B., and Hook, D.S., Molecular sieves for sulfur dioxide removal, Chem Eng Prog, 72(8), 98-103 (1976).
Nandi, S.P., and Walker, P.L., Separation of oxygen and nitrogen using 5A zeolite and carbon molecular sieves, Sepn Sci, 11(5), 441-454 (1976).
Neretnieks, I., Analysis of some adsorption experiments with activated carbon, Chem Eng Sci, 31(11), 1029-1036 (1976).
Ruthven, D.M., Sorption of oxygen, nitrogen, carbon monoxide, methane, and binary mixtures of these gases in 5A molecular sieve, AIChEJ, 22(4), 753-759 (1976).
Ruthven, D.M., and Doetsch, I.H., Diffusion of hydrocarbons in 13X zeolite, AIChEJ, 22(5), 882-886 (1976).
Sun, Y.C., and Killat, G.R., Adsorption for vapor control, Hyd Proc, 55(9), 241-242 (1976).
Ammons, R.D.; Dougherty, N.A., and Smith, J.M., Adsorption of methylmercuric chloride on activated carbon, Ind Eng Chem Fund, 16(2), 263-269 (1977).
Koradia, P.B., and Kiovsky, J.R., Drying chlorinated hydrocarbons using zeolites, Chem Eng Prog, 73(4), 105-106 (1977).
Allen, T., and Burevski, D., Models of adsorption of sulphur dioxide on powdered adsorbents, Powder Tech, 21, 91-96 (1978).
Larsen, J.W.; Kennard, L., and Kuemmerle, E.W., Thermodynamics of adsorption of organic compounds on the surface of Bruceton coal measured by gas chromatography, Fuel, 57(5), 309-313 (1978).
Adler, M.S., and Johnson, D.R., A flexible butylene separation process, Chem Eng Prog, 75(1), 77-79 (1979).
Danner, R.P., and Choi, E.C.F., Mixture adsorption equilibria of ethane and ethylene on 13X molecular sieves, Ind Eng Chem Fund, 17(4), 248-253 (1978); 18(3), 300 (1979).
Martin, M.; Cuellar, J., and Galan, M.A., Adsorption characteristics of n-propanol on silica gel by gas-chromatography, Chem Eng Sci, 34(5), 691-696 (1979).
Parmele, C.S.; O'Connell, W.L., and Basdekis, H.S., Vapor-phase adsorption cuts pollution, recovers solvents, Chem Eng, 31 Dec, 58-70 (1979).
Rice, R.G.; Foo, S.C., and Gough, G.G., Limiting separations in parametric pumps, Ind Eng Chem Fund, 18(2), 117-123 (1979).
Ruthven, D.M., and Kumar, R., Chromatographic study of diffusion of nitrogen, methane and their binary mixtures in 4A molecular sieve, Can JCE, 57, 342-352 (1979).

Scamehorn, J.F., Removal of vinyl chloride from gaseous streams by adsorption on activated carbon, Ind Eng Chem Proc Des Dev, 18(2), 210-217 (1979).

Yang, R.T., and Liu, R.T., Preferential adsorption of methane over hydrogen on certain coals, Ind Eng Chem Fund, 18(3), 299-300 (1979).

Yang, R.T., and Shen, M.S., Calcium silicates as regenerative sorbents for hot-gas desulfurization, AIChEJ, 25(5), 811-819 (1979).

1980-1981

Andrieu, J., and Smith, J.M., Rate parameters for adsorption of carbon dioxide in beds of carbon particles, AIChEJ, 26(6), 944-948 (1980).

Johansson, R., and Neretnieks, I., An experimental study of adsorption on activated carbon in countercurrent flow, Chem Eng Sci, 35(4), 979-986 (1980).

McKay, G., and Allen, S.J., Surface mass transfer processes using peat as adsorbent for dyestuffs, Can JCE, 58, 521-526 (1980).

Rolniak, P.D., and Kobayashi, R., Adsorption of methane and methane-carbon dioxide mixtures at elevated pressures and ambient temperatures on 5A and 13X molecular sieves by tracer perturbation chromatography, AIChEJ, 26(4), 616-625 (1980).

Skalsky, M., and Farrell, P.C., Uptake characteristics of selected biochemicals on coated and uncoated activated charcoal, Trans IChemE, 58, 91-97 (1980).

Snape, E., and Lynch, F.E., Metal hydrides adsorb hydrogen, Chemtech, Sept, 578-583; Dec, 768-773 (1980).

Wankat, P.C., and Partin, L.R., Process for recovery of solvent vapors with activated carbon, Ind Eng Chem Proc Des Dev, 19(3), 446-451 (1980).

Andrieu, J., and Smith, J.M., Adsorption rates for sulfur dioxide and hydrogen sulfide in beds of activated carbon, AIChEJ, 27(5), 840-842 (1981).

Basmadjian, D., and Wright, D.W., Non-isothermal sorption of ethane-carbon dioxide mixtures in beds of 5A molecular sieves, Chem Eng Sci, 36(5), 937-940 (1981).

Costa, E.; Sotelo, J.L.; Calleja, G., and Marron, C., Adsorption of binary and ternary hydrocarbon gas mixtures on activated carbon, AIChEJ, 27(1), 5-14 (1981).

Fraenkel, D., Hydrogen adsorption by zeolites, Chemtech, Jan, 60-65 (1981).

Fritz, W.; Merk, W., and Schlunder, E.U., Simultaneous adsorption of organic solutes in water by activated carbon, Int Chem Eng, 21(3), 384-399 (1981).

Gala, H.B.; Kara, M.; Sung, S.; Chiang, S.H., and Klinzing, G.E., Hydrogen permeation through nickel in gas and liquid phase, AIChEJ, 27(1), 159-162 (1981).

Kincal, S., and Culfaz, A., Sorption of sulfur dioxide on hydrogen-form and aluminium-deficient mordenites, Sepn Sci Technol, 16(3), 229-236 (1981).

Reich, R.; Ziegler, W.T., and Rogers, K.A., Adsorption of methane, ethane, and ethylene gases and their binary and ternary mixtures and carbon dioxide on activated carbon (212-301K and pressures to 35 atmospheres), Ind Eng Chem Proc Des Dev, 19(3), 336-344 (1980); 20(1), 175 (1981).

Sandrock, G.D., and Huston, E.L., How metals adsorb hydrogen, Chemtech, Dec, 754-762 (1981).

Various, Gas handling (special report), Processing, July, 15-21, 48 (1981).

1982-1983

Cavalletto, G., and Smith, J.M., Adsorption of ethylene in beds of activated carbon, AIChEJ, 28(6), 1039-1040 (1982).

Chihara, K.; Suzuki, M., and Smith, J.M., Cyclic regeneration of activated carbon in fluidized beds, AIChEJ, 28(1), 129-134 (1982).

Ezell, E.L., and Gelo, J.F., Cut cracked gas drying costs, Hyd Proc, May, 191-193 (1982).

Ghezelayagh, H., and Gidaspow, D., Micro-macropore model for sorption of water on silica gel in a dehumidifier, Chem Eng Sci, 37(8), 1181-1198 (1982).

Kumar, R., Effect of Freon-12 exposure on sieving property of 4A zeolite, Can JCE, 60, 577-580 (1982).

Liapis, A.I., and Crosser, O.K., Comparison of model predictions with non-isothermal sorption data for ethane-carbon dioxide mixtures in beds of 5A molecular sieves, Chem Eng Sci, 37(6), 958-962 (1982).

Mahajan, P.O.; Youssef, A., and Walker, P.L., Surface-modified carbons for drying of gas streams, Sepn Sci Technol, 17(8), 1019-1026 (1982).

Matsuda, S.; Kamo, T.; Imahashi, J., and Nakajima, F., Adsorption and oxidative desorption of hydrogen sulfide by molybdenum oxide-titanium dioxide, Ind Eng Chem Fund, 21(1), 18-22 (1982).

Milewski, M., and Berak, J.M., Effect of adsorbent preparation parameters on selectivity for xylene isomers separation, Sepn Sci Technol, 17(2), 369-374 (1982).

Moseman, M.H., and Bird, G., Desiccant dehydration of natural gasoline, Chem Eng Prog, 78(2), 78-83 (1982).

Santacesaria, E., et al., Separation of xylenes on Y zeolites, Ind Eng Chem Proc Des Dev, 21(3), 440-457 (1982).

Tien, C., and Wang, S.C., Dynamics of adsorption columns with bacterial growth outside adsorbents, Can JCE, 60, 363-376 (1982).

Barrow, J.A., Proper design saves energy (adsorption dehydrators), Hyd Proc, Jan, 117-120 (1983).

Capes, P., Solvent-recovery adsorption system, Proc Engng, Jan, 33 (1983).

McKay, G., Adsorption of acidic and basic dyes onto activated carbon in fluidised beds, CER&D, 61, 29-36 (1983).

Milewski, M., et al., Efficiency of dynamic tests used to evaluate adsorbents for separation of C8 isomers, Sepn Sci Technol, 18(2), 187-194 (1983).

Miura, K.; Nakanishi, A., and Hashimoto, K., Methyl bromide vapor adsorption by activated carbon, Ind Eng Chem Proc Des Dev, 22(3), 469-477 (1983).

Santangelo, J.G., and Chen, G.T., Metal hydrides for hydrogen gas purification, Chemtech, Oct, 621-623 (1983).

Sorial, G.A.; Granville, W.H., and Daly, W.O., Adsorption equilibria for oxygen and nitrogen gas mixtures on 5A molecular sieves, Chem Eng Sci, 38(9), 1517-1524 (1983).

Wang, S.S., and Yang, R.T., Multicomponent separation by cyclic processes: A process for combined hydrogen/methane separation and acid gas removal in coal conversions, Chem Eng Commns, 20(1), 183-190 (1983).

Zuech, J.L.; Hines, A.L., and Sloan, E.D., Methane adsorption on 5A molecular sieve (4-690kPa), Ind Eng Chem Proc Des Dev, 22(1), 172-174 (1983).

1984

Abd El-Salaam, K.M., et al., Physical adsorption studies on mixed vanadium oxide catalysts, Ads Sci Tech, 1(2), 169-176 (1984).

Andrews, G.F., and Elkcechen, S., Solid adsorbents in batch fermentations, Chem Eng Commns, 29(1), 139-152 (1984).

Carrott, P.J.M., and Sing, K.S.W., Adsorption of nitrogen on precipitated and pyrogenic silicas, Ads Sci Tech, 1(1), 30-40 (1984).

Galan, M.A., Adsorption characteristics of nitric oxide on various adsorbents, Chem Eng J, 28(2), 105-114 (1984).

Glanz, P., and Findenegg, G.H., Adsorption of gas mixtures of propene and propane on graphitized carbon black, Ads Sci Tech, 1(1), 41-50 (1984).

Ishikawa, T., et al., Adsorption of organic halides on silver-impregnated silica gels, Ads Sci Tech, 1(1), 93-102 (1984).

Ladisch, M.R.; Voloch, M.; Hong, J.; Blenkowski, P., and Tsao, G.T., Cornmeal adsorber for dehydrating ethanol vapours, Ind Eng Chem Proc Des Dev, 23(3), 437-443 (1984).

Lee, M.H.; Petty, L.E.; Wilson, R.H., and Galvin, C., The ultra low temperature reaction adsorption process, Chem Eng Prog, 80(5), 33-38 (1984).

Lee, T.V., et al., Adsorption of binary mixtures of ethane and acetylene on activated carbon, Sepn Sci Technol, 19(1), 1-20 (1984).

Lopez-Garzon, F.J., et al., High-temperature adsorption of hydrocarbons prepared from olive stones, Ads Sci Tech, 1(1), 103-110 (1984).

Lynch, D.T., The use of adsorption/desorption models to describe the forced periodic operation of catalytic reactors, Chem Eng Sci, 39(9), 1325-1328 (1984).

McKay, G., and Allen, S.J., Pore diffusion model for dye adsorption onto peat in batch adsorbers, Can JCE, 62(3), 340-345 (1984).

Mills, B., and Rothery, E., Gas drying, Chem Engnr, April, 19-23 (1984).

Morbidelli, M.; Storti, G.; Carra, S.; Niederjaufner, G., and Pontoglio, A., Study of a separation process using a molecular sieve for chlorotoluene isomers, Chem Eng Sci, 39(3), 383-394 (1984).

Ruthven, D.M.; Tezel, F.H.; Devgun, J.S., and Sridhar, T.S., Adsorptive separation of krypton from nitrogen, Can JCE, 62(4), 526-534 (1984).

Yue, P.L., and Olaofe, O., Kinetic analysis of the catalytic dehydration of alcohols over zeolites, CER&D, 62, 81-91 (1984).

Yue, P.L., and Olaofe, O., Molecular sieving effects of zeolites in the dehydration of alcohols, CER&D, 62, 167-172 (1984).

1985

Baba, Y.; Nakao, T.; Inoue, K., and Nakamori, I., Adsorption mechanism of iodine on activated carbon, Sepn Sci Technol, 20(1), 21-32 (1985).

Bottani, E.J., and Cascarini, L.E., Physical adsorption of carbon dioxide on graphite: Monolayer and multilayer regions, Ads Sci Tech, 2(4), 253-262 (1985).

Carlson, N.W., and Dranoff, J.S., Adsorption of ethane by 4A zeolite pellets, Ind Eng Chem Proc Des Dev, 24(4), 1300-1302 (1985).

Costa, E.; Calleja, G., and Domingo, F., Adsorption of gaseous hydrocarbons on activated carbon: Characteristic kinetic curve, AIChEJ, 31(6), 982-991 (1985).

Emesh, I.T.A., and Gay, I.D., Adsorption and NMR studies of ethene, ethane and carbon monoxide on Zn and Cd exchanged A-zeolites, J Chem Tech Biotechnol, 35A(3), 115-120 (1985).

Jean, G., et al., Selective removal of nitrogeneous-type compounds from fuels using zeolites, Sepn Sci Technol, 20(7), 541-564 (1985).

Joshi, J.B.; Mahajani, V.V., and Juvekar, V.A., Adsorption of NOx gases (review paper), Chem Eng Commns, 33(1), 1-92 (1985).

Lao, M.Z., and Ye, Z.H., Phase equilibrium of adsorption separation of xylene isomers with type-Y zeolites, Chem Eng Commns, 35(1), 89-100 (1985).

Lee, T.V.; Madey, R., and Huang, J.C., Adsorption equilibria for ethane and propane gas mixtures on activated carbon, Sepn Sci Technol, 20(5), 461-480 (1985).

Santacesaria, E.; Gelosa, D.; Danise, P., and Carra, S., Separation of xylenes on Y zeolites in the vapor phase, Ind Eng Chem Proc Des Dev, 24(1), 78-92 (1985).

Tamon, H., and Toei, R., Solar-powered adsorber dehumidifier, Ind Eng Chem Proc Des Dev, 24(2), 450-457 (1985).

Wang, J.H., and Smith, J.M., Thermal regeneration of the phenol-carbon system, AIChEJ, 31(3), 496-498 (1985).

Yang, R.T., and Saunders, J.T., Adsorption of gases on coals and heat-treated coals at elevated temperature and pressure, Fuel, 64(5), 616-626 (1985).

Yumura, M., and Furimsky, E., Comparison of adsorbents for hydrogen sulfide removal at high temperatures, Ind Eng Chem Proc Des Dev, 24(4), 1165-1168 (1985).

Zagorevskaya, E.V., et al., Adsorption of polychlorocarbons on carbon blacks by gas chromatography, Ads Sci Tech, 2(4), 219-228 (1985).

1986

Bulow, M.; Schlodder, H., and Struve, P., Sorption uptake of molecular mobility of n-paraffins in ZSM-5 type zeolite, Ads Sci Tech, 3(4), 229-232 (1986).

Cornel, P.; Sontheimer, H.; Summers, R.S., and Roberts, P.V., Sorption of dissolved organics from aqueous solution by polystyrene resins, Chem Eng Sci, 41(7), 1791-1810 (1986).

Furlong, D.N.; Sing, K.S.W., and Parfitt, G.D., Adsorption of water vapour on rutile and silica-coated rutile, Ads Sci Tech, 3(1), 25-32 (1986).

Garg, D.R., and Yon, C.M., Adsorptive heat recovery drying system, Chem Eng Prog, 82(2), 54-60 (1986).

Ghosh, A.K., and Agnew, J.B., Adsorption of acetylene, hydrogen chloride and vinyl chloride on activated carbons: Kinetics and thermodynamics, using transient response technique, Chem Eng Commns, 40, 169-182 (1986).

Heiti, R.V., and Thodos, G., Water sorption from air by celite-supported calcium chloride, AIChEJ, 32(7), 1169-1175 (1986).

Janchen, J., and Stach, H., n-Decane adsorption on silicon dioxide adsorbents, Ads Sci Tech, 3(1), 3-10 (1986).

Roethe, K.P., et al., Adsorption of aromatics on silica gel, Ads Sci Tech, 3(2), 65-74 (1986).

Schoellner, R., and Mueller, U., Influence of mono- and bivalent cations in 4A zeolites on adsorptive separation of ethene and propene from crack-gases, Ads Sci Tech, 3(3), 167-172 (1986).

Van Deventer, J.S.J., Competitive equilibrium adsorption of metal cyanides on activated carbon, Sepn Sci Technol, 21(10), 1025-1038 (1986).

1987

Al-Damkhi, A.M.; Al-Ameeri, R.S.; Jeffreys, G.V., and Mumford, C.J., Optimal separation of n-paraffins from Kuwait kerosene using a molecular sieve adsorbent, J Chem Tech Biotechnol, 37(4), 215-228 (1987).

Arnold, M., Hydrogen adsorption by metals, Proc Engng, Aug, 24-25 (1987).

Beevers, A., Adsorption for biotech separations, Processing, Sept, 41,43 (1987).

Beschmann, K.; Kokotailo, G.T., and Riekert, L., Kinetics of sorption of aromatics in zeolite ZSM-5, Chem Eng & Proc, 22(4), 223-230 (1987).

Bottani, E.J., et al., Argon and nitrogen physical adsorption on boron nitride, Ads Sci Tech, 4(1), 121-130 (1987).

Braslaw, J.; Golovoy, A., and Beckwith, E.C., Adsorption of automotive paint solvents on activated carbon, Chem Eng Commns, 51, 321-333 (1987).

Carton, A.; Gonzalez, G.; Torre, A.I., and Cabezas, J.L., Separation of ethanol-water mixtures using 3A molecular sieve, J Chem Tech Biotechnol, 39(2), 125-132 (1987).

Chou, C.L., Dynamic modelling of water vapor adsorption by activated alumina, Chem Eng Commns, 56, 211-228 (1987).

Costa, E.; Calleja, G., and Marijuan, L., Adsorption of phenol and p-nitrophenol on activated carbon, Ads Sci Tech, 4(1), 59-78 (1987).

Guo, D.; Venkat, C., and Weiss, A.H., Cyclic adsorption of styrene and of plasticizer on BPL carbon, Ads Sci Tech, 4(1), 15-24 (1987).

McAllister, J., and Parsons, P., Meeting hydrocarbon dewpoint requirements: Moisture removal using adsorbents, Chem Engnr, April, 24-26 (1987).

McKay, G., and Bino, M.J., Adsorption of pollutants on to activated carbon in fixed beds, J Chem Tech Biotechnol, 37(2), 81-94 (1987).

Michele, H., Purification of flue gases by dry sorbents, Int Chem Eng, 27(2), 183-197 (1987).

Miller, G.W.; Knaebel, K.S., and Ikels, K.G., Equilibria of nitrogen, oxygen, argon, and air in molecular sieve 5A, AIChEJ, 33(2), 194-201 (1987).

O'Shea, S., et al., Gas adsorption phenomena in evacuated tubular solar collectors, Ads Sci Tech, 4(4), 275-285 (1987).

Richter, E.; Knoblauch, K., and Juntgen, H., Mechanisms and kinetics of sulphur dioxide adsorption and NOx reduction on active coke, Gas Sepn & Purif, 1(1), 35-43 (1987).

Simons, G.A.; Garman, A.R., and Boni, A.A., Kinetic rate of sulfur dioxide sorption by CaO, AIChEJ, 33(2), 211-217 (1987).

Singh, V.S., and Pandey, B.P., Adsorption of petroleum sulfonate TRS 10-80 on a mineral sand, J Chem Tech Biotechnol, 37(1), 45-58 (1987).

Yang, R.T., Gas Separation by Adsorption Processes, Butterworth Publishing Co., Massachusetts (1987).

1988

Abdallah, K.; Grenier, P.; Sun, L.M., and Meunier, F., Nonisothermal adsorption of water by synthetic NaX zeolite pellets, Chem Eng Sci, 43(10), 2633-2644 (1988).

Banerjee, B.D., Spacing of fissuring network and rate of desorption of methane from coals, Fuel, 67(11), 1584-1586 (1988).

Chiang, P.C.; Lou, J.C., and Tseng, S.K., Development of a procedure for selecting the optimum adsorbent for ABS removal, Can JCE, 65(6), 913-920 (1988).

De-Xin, Z.; Zhi-Jing, X., and Zhen, F., Prediction of breakthrough curves of oxygen-nitrogen coadsorption system on molecular sieves, Gas Sepn & Purif, 2(4),184-189 (1988).

Dexin, Z., and Youfan, G., Prediction of breakthrough curves in system of methane, ethane, and carbon dioxide coadsorption on 4A molecular sieve, Gas Sepn & Purif, 2(1), 28-33 (1988).

Duprat, F.; Gassend, R., and Gau, G., Inductive adsorption: A new method of isomer separation, Ind Eng Chem Res, 27(5), 831-836 (1988).

Ferraz, M.C.A., Preparation of activated carbon for air pollution control, Fuel, 67(9), 1237-1241 (1988).

George, N., and Davies, J.T., Parameters affecting adsorption of microorganisms on activated charcoal cloth, J Chem Tech Biotechnol, 43(3), 173-186 (1988).

Gollakota, S.V., and Chriswell, C.D., Study of an adsorption process using silicate for sulfur dioxide removal from combustion gases, Ind Eng Chem Res, 27(1), 139-143 (1988).

Henning, K.D., et al., Impregnated activated carbon for mercury removal, Gas Sepn & Purif, 2(1), 20-22 (1988).

Joshi, S., and Fair, J.R., Adsorptive drying of toluene, Ind Eng Chem Res, 27(11), 2078-2084 (1988).

Kim, H.C., and Woo, S.I., Preparation of oxygen-enriched air by selective adsorption with platinum crystallites supported on alumina, Ind Eng Chem Res, 27(11), 2135-2139 (1988).

Kobuke, Y., et al., Composite fiber adsorbent for rapid uptake of Uranyl from seawater, Ind Eng Chem Res, 27(8), 1461-1466 (1988).

Koo, Y.M., and Wankat, P.C., Modeling of size-exclusion parametric pumping, Sepn Sci Technol, 23(4), 413-428 (1988).

Koske, P.H.; Ohlrogge, K., and Peinemann, K.V., Uranium recovery from seawater by adsorption, Sepn Sci Technol, 23(12), 1929-1940 (1988).

Kuo, S.L., and Hines, A.L., Adsorption of chlorinated hydrocarbon pollutants on silica gel, Sepn Sci Technol, 23(4), 293-304 (1988).

Le Van Mao, R., and McLaughlin, G.P., Ethylene recovery from low-grade gas stream by adsorption on zeolites and controlled desorption, Can JCE, 66(4), 686-690 (1988).

McCormick, R.L., et al., Surface acidity studied temperature-programmed desorption of tert-butylamine, Energy & Fuels, 2(6), 740-743 (1988).

McKay, G., Fluidized bed adsorption of pollutants onto activated carbon, Chem Eng J, 39(2), 87-96 (1988).

McKay, G., and McAleavey, G., Ozonation and carbon adsorption in three-phase fluidized bed for colour removal from peat water, CER&D, 66(6), 531-536 (1988).
Murakami, M., and Nomura, M., Catalyst-adsorption type purifiers for ultra-pure gas supply, Gas Sepn & Purif, 2(2), 95-102 (1988).
Niklasson, C., and Anderson, B., Adsorption and reaction of hydrogen and deuterium on nickel/silicon dioxide catalyst, Ind Eng Chem Res, 27(8), 1370-1376 (1988).
Nirdosh, I.; Tremblay, W.B.; Muthuswami, S.V., and Johnson, C.R., Adsorption-desorption studies on the radium-silica system, Can JCE, 65(6), 927-933 (1988).
Rybolt, T.R., and English, K.J., Virial analysis of methane adsorption in 5A zeolite, AIChEJ, 34(7), 1207-1210 (1988).
Shadman, F., and Dombek, P.E., Enhancement of sulfur dioxide sorption on lime by structure modifiers, Can JCE, 66(6), 930-935 (1988).
Shah, D.B., et al., Sorption and diffusion of benzene in HZSM-5 and silicalite crystals, AIChEJ, 34(10), 1713-1717 (1988).
Sircar, S., Air fractionation by adsorption, Sepn Sci Technol, 23(14), 2379-2396 (1988).
Sowerby, B., and Crittenden, B.D., Experimental comparison of type A molecular sieves for drying the ethanol-water azeotrope, Gas Sepn & Purif, 2(2), 77-83 (1988).
Sowerby, B., and Crittenden, B.D., Vapour-phase separation of alcohol-water mixtures by adsorption onto silicalite, Gas Sepn & Purif, 2(4),177-183 (1988).
Tan, C.S., and Liou, D.C., Desorption of ethyl acetate from activated carbon by supercritical carbon dioxide, Ind Chem Eng Res, 27(6), 988-991 (1988).
van der Merwe, P.F., and van Deventer, J.S.J., Influence of oxygen on adsorption of metal cyanides on activated carbon, Chem Eng Commns, 65, 121-138 (1988).
Yan, T.Y., Effects of moisture in separation of C8 aromatics using medium-pore zeolites, Ind Eng Chem Res, 27(9), 1665-1668 (1988).

22.6 Liquid-Phase Adsorption

1966-1973

Enneking, J.C., How activated carbon recovers gas liquids, Hyd Proc, 45(10), 189-192 (1966).
Gehrhardt, H.M., and Kyle, B.G., Fixed-bed liquid-phase drying with molecular sieve adsorbent, Ind Eng Chem Proc Des Dev, 6(3), 265-267 (1967).
Morton, E.L., and Murrill, P.W., Analysis of liquid-phase adsorption fractionation in fixed beds, AIChEJ, 13(5), 965-972 (1967).
Roberts, P.V., and York, R., Adsorption of normal paraffins from binary liquid solutions by 5A molecular sieve adsorbent, Ind Eng Chem Proc Des Dev 6(4), 516-525 (1967).
Miller, C.O.M., and Clump, C.W., Liquid-phase adsorption study of rate of diffusion of phenol from aqueous solution into activated carbon, AIChEJ, 16(2), 169-172 (1970).

Erskine, D.B., and Schuliger, W.G., Activated carbon processes for liquids, Chem Eng Prog, 67(11), 41-44 (1971).
Misic, D.M., and Smith, J.M., Adsorption of benzene in carbon slurries, Ind Eng Chem Fund, 10(3), 380-389 (1971).
Sircar, S., and Myers, A.L., Thermodynamic consistency test for adsorption from binary liquid mixtures on solids, AIChEJ, 17(1), 186-190 (1971).
Kim, Y.S., and Zeitlin, H., Separation of zinc and copper from seawater by adsorption colloid flotation, Sepn Sci, 7(1), 1-12 (1972).
Radke, C.J., and Prausnitz, J.M., Adsorption of organic solutes from dilute aqueous solution on activated carbon, Ind Eng Chem Fund, 11(4), 445-451 (1972).
Radke, C.J., and Prausnitz, J.M., Thermodynamics of multi-solute adsorption from dilute liquid solutions, AIChEJ, 18(4), 761-768 (1972).
Sircar, S.; Novosad, J., and Myers, A.L., Adsorption from liquid mixtures on solids, Ind Eng Chem Fund, 11(2), 249-254 (1972).
Butler, P., Activated carbon for organics removal from aqueous solutions, Proc Engng, June, 134-135 (1973).
Furusawa, T., and Smith, J.M., Mass-transfer adsorption rates in slurries by chromatography, Ind Eng Chem Fund, 12(3), 360-364 (1973).
Furusawa, T., and Smith, J.M., Fluid-particle and intraparticle mass-transport adsorption rates in slurries, Ind Eng Chem Fund, 12(2), 197-203 (1973).
Minka, C., and Myers, A.L., Adsorption from ternary liquid mixtures on solids, AIChEJ, 19(3), 453-459 (1973).
Sircar, S., and Myers, A.L., Prediction of adsorption at liquid-solid interface from adsorption isotherms of pure unsaturated vapors, AIChEJ, 19(1), 159-166 (1973).

1974-1977

Dernini, S.; DeSantis, R., and Pasquinucci, A., Liquid-phase adsorption of xylenes by molecular sieves, Chemical Processing, June, 20, 22 (1974).
Furusawa, T., and Smith, J.M., Intraparticle mass transport in slurries by dynamic adsorption studies, AIChEJ, 20(1), 88-93 (1974).
Komiyama, H.; Furusawa, T., and Smith, J.M., Effectiveness factors for adsorption in slurries, Ind Eng Chem Fund, 13(3), 293-296 (1974).
Malati, M.A., and Mazza, R.J., Mechanism of adsorption of alkali metal ions on silica, Powder Tech, 9, 107-110 (1974).
Deineko, P.S., et al., Zeolite activation for liquid-phase adsorptive separation of hydrocarbons, Int Chem Eng, 15(2), 339-341 (1975).
Prober, R.; Pyeha, J.J., and Kidon, W.E., Interaction of activated carbon with dissolved oxygen, AIChEJ, 21(6), 1200-1204 (1975).
Skrivanek, J., and Hostomsky, J., Model of isothermal adsorption in a cascade of continuous flow stirred-tank reactors, Chem Eng Commns, 2(1), 109-114 (1975).
de Rosset, A.J.; Neuzil, R.W., and Korous, D.J., Liquid column chromatography as a predictive tool for continuous countercurrent adsorptive separations, Ind Eng Chem Proc Des Dev, 15(2), 261-266 (1976).
Foti, G., and Nagy, L.G., Critical evaluation of the linear approximation of the individual adsorption isotherms of binary liquid mixtures on solid surfaces, Periodica Polytechnica, 20, 107-114 (1976).

Niiyama, H., and Smith, J.M., Adsorption of nitric oxide in aqueous slurries of activated carbon, AIChEJ, 22(6), 961-970 (1976).

Alexander, F., and McKay, G., Kinetics of the removal of basic dye from effluent using silica, Chem Engnr, April, 243-246 (1977).

Hsieh, J.S.C.; Turian, R.M., and Tien, C., Multicomponent liquid-phase adsorption in fixed beds, AIChEJ, 23(3), 263-275 (1977).

Myers, A.L., and Moser, F., Slurry sorption separations, Chem Eng Sci, 32(5), 529-534 (1977).

Niiyama, H., and Smith, J.M., Adsorption rates of oxygen in aqueous slurries of activated carbon, AIChEJ, 23(4), 592-596 (1977).

Rizzo, J.L., and Shepherd, A.R., Treating industrial wastewater with activated carbon, Chem Eng, 3 Jan, 95-100 (1977).

1978-1980

Anon., Carbon regeneration in wastewater treatment, Processing, July, 36 (1978).

Chester, A., Wastewater treatment by carbon adsorption, Processing, Feb, 69-70 (1978).

Jossens, L., et al., Thermodynamics of multi-solute adsorption from dilute aqueous solutions, Chem Eng Sci, 33(8), 1097-1106 (1978).

Mahajan, O.P.; Youssef, A., and Walker, P.L., Surface-treated activated carbon for removal of ammonia from water, Sepn Sci Technol, 13(6), 487-500 (1978).

Ramachandran, P.A., and Smith, J.M., Adsorption of hydrogen sulfide in a slurry reactor, Ind Eng Chem Fund, 17(1), 17-23 (1978).

Brunovska, A., and Brunovsky, P., Optimal temperature control of a stirred adsorber, Chem Eng Sci, 34(3), 379-386 (1979).

Kars, R.L.; Best, R.J., and Drinkenburg, A.A.H., Sorption of propane in slurries of active carbon in water, Chem Eng J, 17(3), 201-210 (1979).

Hutchins, R.A., Liquid-phase adsorption: Maximising performance, Chem Eng, 25 Feb, 101-110 (1980).

Mahajan, O.P.; Castilla, C.M., and Walker, P.L., Surface-treated activated carbon for removal of phenol from water, Sepn Sci Technol, 15(10), 1733-1752 (1980).

McKay, G., Surface mass transfer during acid dye removal from water using carbon, Chem Engnr, April, 219-221, 224 (1980).

McKay, G., and Poots, V.J.P., Kinetics and diffusion processes in colour removal from effluent using wood as an adsorbent, J Chem Tech Biotechnol, 30, 279-292 (1980).

Sylvester, N.D., and Dianat, S., Oxygen absorption and adsorption in aqueous slurries of powdered activated carbon, Ind Eng Chem Proc Des Dev, 19(1), 199-201 (1980).

1981-1983

Bhattacharyya, D., et al., Separation of toxic organotin compounds from aqueous solution by adsorption, Sepn Sci Technol, 16(5), 495-504 (1981).

Fritz, W.; Merk, W., and Schlunder, E.U., Competitive adsorption of two dissolved organics onto activated carbon, Chem Eng Sci, 36(4), 721-758 (1981).

Gupta, R.K.; Kunzru, D., and Saraf, D.N., Liquid-phase adsorption of binary and ternary systems of n-paraffins on LMS-5A, Ind Eng Chem Fund, 20(1), 28-34 (1981).

Hasanain, M.A., and Hines, A.L., Application of the adsorption potential theory to adsorption of carboxylic acids from aqueous solutions onto a macroreticular resin, Ind Eng Chem Proc Des Dev, 20(4), 621-625 (1981).

Jaroniec, M., and Derylo, A., Simple relationships for predicting multi-solute adsorption from dilute aqueous solutions, Chem Eng Sci, 36(6), 1017-1020 (1981).

McKay, G., Design models for adsorption systems in wastewater treatment, J Chem Tech Biotechnol, 31, 717-731 (1981).

McKay, G., and McConvey, I.F., External mass transfer of basic and acidic dyes on wood, J Chem Tech Biotechnol, 31, 401-408 (1981).

Perineau, F., and Gaset, A., Adsorption of ionic and non-ionic surfactants by wool charring waste, J Chem Tech Biotechnol, 31, 395-400 (1981).

Choudhary, V.R., and Vaidya, S.H., Adsorption of copper nitrate from solution on silica gel, J Chem Tech Biotechnol, 32, 888-892 (1982).

Ghim, Y.S., and Chang, H.N., Adsorption characteristics of glucose and fructose in ion-exchange resin columns, Ind Eng Chem Fund, 21(4), 369-374 (1982).

Hashimoto, K.; Miura, K., and Watanabe, T., Kinetics of thermal regeneration reaction of activated carbons used in wastewater treatment, AIChEJ, 28(5), 737-746 (1982).

Heitkamp, D., and Wagener, K., Uranium recovery by adsorption from seawater, Ind Eng Chem Proc Des Dev, 21(4), 781-784 (1982).

Kaneko, S., Adsorption of several dyes from aqueous solutions on silica-containing complex-oxide gels, Sepn Sci Technol, 17(13), 1499-1510 (1982).

Mansour, A.; von Rosenberg, D.U., and Sylvester, N.D., Numerical solution of liquid-phase multicomponent adsorption in fixed beds, AIChEJ, 28(5), 765-772 (1982).

McKay, G., Adsorption of dyestuffs from aqueous solutions with activated carbon, J Chem Tech Biotechnol, 32, 759-780 (1982).

Pedram, E.; Hines, A.L.; Poulson, R.E., and Cooney, D.O., Adsorption of organics from a true in-situ oil-shale retort water on activated carbon in packed beds, Chem Eng Commns, 15(5), 291-304 (1982).

Pedram, E.O.; Hines, A.L., and Cooney, D.O., Kinetics of adsorption of organics from an above-ground oil-shale retort water, Chem Eng Commns, 19(1), 167-176 (1982).

Perineau, F.; Molinier, J., and Gaset, A., Adsorption of ionic dyes on charred plant material, J Chem Tech Biotechnol, 32, 749-758 (1982).

Wang, S.C., and Tien, C., Multicomponent liquid-phase adsorption in fixed beds, AIChEJ, 28(4), 565-573 (1982).

Asteimer, L., et al., Development of sorbers for recovery of uranium from seawater, Sepn Sci Technol, 18(4), 307-340 (1983).

Borwanker, R.P., and Wasan, D.T., Kinetics of adsorption of surface active agents at gas-liquid surfaces, Chem Eng Sci, 38(10), 1637-1650 (1983).

Keaton, M.M., and Bourke, M.J., Activated carbon system cuts foaming and amine losses, Hyd Proc, Aug, 71-73 (1983).

McKay, G., Adsorption of dyestuffs from aqueous solutions using activated carbon, J Chem Tech Biotechnol, 33A(4), 196-218 (1983).

Sutikno, T., and Himmelstein, K.J., Desorption of phenol from activated carbon by solvent regeneration, Ind Eng Chem Fund, 22(4), 420-425 (1983).

1984

Baba, Y., et al., Adsorption equilibrium of copper and cadmium on silica gel from ammoniacal solutions, Sepn Sci Technol, 19(6), 417-428 (1984).

Belfort, G., et al., Selective adsorption of organic homologues onto activated carbon from dilute aqueous solutions, AIChEJ, 30(2), 197-207 (1984).

Blasinski, H., and Kazmierczak, J., Surface diffusion on active carbon in adsorption from solutions, Chem Eng Commns, 25(1), 351-362 (1984).

Broughton, D.B., Production-scale adsorptive separations of liquid mixtures by simulated moving-bed technology, Sepn Sci Technol, 19(11), 723-732 (1984).

Jaroniec, M.; Dabrowski, A., and Toth, J., Multilayer single-solute adsorption from dilute solutions on energetically heterogeneous solids., Chem Eng Sci, 39(1), 65-70 (1984).

Larson, A.C., and Tien, C., Multicomponent liquid-phase adsorption in batch, Chem Eng Commns, 27(5), 339-379 (1984).

Mathews, A.P., and Weber, W.J., Modeling and parameter evaluation for adsorption in slurry reactors, Chem Eng Commns, 25(1), 157-172 (1984).

McKay, G., Adsorption of basic dye onto silica from aqueous solution-solid diffusion model, Chem Eng Sci, 39(1), 129-138 (1984).

McKay, G., Batch adsorption of dyestuffs from aqueous solution using activated carbon, Chem Eng J, 27(3), 187-196 (1983); 28(2), 95-104 (1984).

McKay, G., Two-resistance mass transfer models for adsorption of dyestuffs from aqueous solutions using activated carbon, J Chem Tech Biotechnol, 34A(6), 294-310 (1984).

McKay, G., Analytical solution for adsorption of basic dye on silica, AIChEJ, 30(4), 692-697 (1984).

McKay, G., Mass transfer processes during the adsorption of solutes in aqueous solutions in batch and fixed bed adsorbers, CER&D, 62, 235-246 (1984).

McKay, G.; Allen, S.J.; McConvey, I.F., and Walters, J.H.R., External mass transfer and homogeneous solid-phase diffusion effects during adsorption of dyestuffs, Ind Eng Chem Proc Des Dev, 23(2), 221-226 (1984).

Milestone, N.B., and Bibby, D.M., Adsorption of alcohols from aqueous solution by ZSM-5, J Chem Tech Biotechnol, 34A(2), 73-79 (1984).

Rivera-Utrilla, J., Adsorption of radionuclides on activated carbons from aqueous solutions, J Chem Tech Biotechnol, 34A(5), 243-250 (1984).

Tsezos, M., and Noh, S.H., Extraction of uranium from seawater using biological-origin adsorbents, Can JCE, 62(4), 559-561 (1984).

1985

Asfour, H.M.; Nassar, M.M.; Fadali, O.A., and El-Geundi, M.S., Adsorption of textile dyes, J Chem Tech Biotechnol, 35A(1), 21-35 (1985).

Baba, Y.; Doi, S.; Inoue, K., and Nakamori, I., Adsorption rate of copper (II) from aqueous ammonium nitrate solution on silica gel, Solv Extn & Ion Exch, 3(5), 741-752 (1985).

Bui, S.; Verykios, X., and Mutharasan, R., Removal of ethanol from fermentation broths by adsorption, Ind Eng Chem Proc Des Dev, 24(4), 1209-1213 (1985).

Cloutier, J.N.; Leduy, A., and Ramalho, R.S., Peat adsorption of herbicide 2,4-D from wastewaters, Can JCE, 63(2), 250-257 (1985).

Desai, N.J., and Do, D.D., Adsorption of organic solutes into activated carbons (batch studies), Chem Eng Commns, 39, 101-126 (1985).

Friedrich, M.; Seidel, A., and Gelbin, D., Measuring adsorption rates from an aqueous solution, AIChEJ, 31(2), 324-327 (1985).

Hashimoto, K.; Miura, K., and Kyotani, S., Regeneration of activated carbons used in wastewater treatment by a moving-bed regenerator, AIChEJ, 31(12), 1986-1996 (1985).

Hinds, B.C., and McCoy, B.J., Perturbation theory and moments for nonlinear Freundlich desorption in continuous-flow well-stirred vessel, Chem Eng Commns, 37(1), 265-274 (1985).

Jayaraj, K., and Tien, C., Characterization of adsorption affinity of unknown substances in aqueous solutions, Ind Eng Chem Proc Des Dev, 24(4), 1230-1239 (1985).

Kosmulski, M.; Jaroniec, M., and Szczypa, J., Liquid-solid interfaces: Kinetics of isotope exchange, Ads Sci Tech, 2(2), 97-120 (1985).

Laszlo, K.; Nagy, L.G.; Foti, G., and Schay, G., Investigation of HPLC packings by liquid mixture adsorption, Periodica Polytechnica, 29(2), 73-86 (1985).

Malmary, G.; Perineau, F.; Molinier, J., and Geset, A., Continuous process for dye removal from liquid effluents using carbonised wool waste, J Chem Tech Biotechnol, 35A(8), 431-437 (1985).

McConvey, I.F., and McKay, G., Mass transfer model for adsorption of basic dyes on woodmeal in agitated batch adsorbers, Chem Eng & Proc, 19(5), 267-276 (1985).

McKay, G., Adsorption of dyestuffs from aqueous solutions using activated carbon, AIChEJ, 31(2), 335-339 (1985).

McKay, G., and McConvey, I.F., Adsorption of acid dye onto woodmeal by solid diffusional mass transfer, Chem Eng & Proc, 19(6), 287-296 (1985).

Sircar, S., Adsorption of dilute hydrocarbon solutes from aqueous solutions on hetergeneous adsorbents, Ads Sci Tech, 2(1), 1-8 (1985).

1986

Akser, M.; Wan, R.Y., and Miller, J.D., Gold adsorption from alkaline aurocyanide solution by neutral polymeric adsorbents, Solv Extn & Ion Exch, 4(3), 531-546 (1986).

Bindal, R.C., and Misra, B.M., Separation of binary liquid systems by sorption: Comparison with pervaporation, Sepn Sci Technol, 21(10), 1047-1058 (1986).

Ching, C.B.; Ho, C., and Ruthven, D.M., An improved adsorption process for the production of high-fructose syrup, AIChEJ, 32(11), 1876-1880 (1986).

Dabrowski, A.; Jaroniec, M., and Oscik, J., Surface activity coefficients in adsorption at liquid-solid interface, Ads Sci Tech, 3(4), 221-228 (1986).

Garbacz, J.K.; Biniak, S., and Swiatkowski, A., Description of adsorption of iodine from non-aqueous solutions on active carbon, Ads Sci Tech, 3(2), 61-64 (1986).

Goworek, J., Adsorption of alcohols from multicomponent solutions onto silica gel, Ads Sci Tech, 3(3), 141-148 (1986).
Hidajat, K.; Ching, C.B., and Ruthven, D.M., Numerical simulation of a semi-continuous countercurrent adsorption unit for fructose-glucose separation, Biochem Eng J, 33(3), B55-B62 (1986).
Lee, H.S., and Ihm, S.K., Statistical thermodynamic analysis on the sorption selectivity for binary liquid mixture on solid surface, Chem Eng Commns, 48, 21-34 (1986).
Omichi, H., et al., New type of amidoxime-group-containing adsorbent for recovery of uranium from seawater, Sepn Sci Technol, 20(2), 163-178 (1985); 21(6), 563-574 (1986).
Seidel,, A., et al., Equilibrium adsorption of two-component organic solutes from aqueous solutions on activated carbon, Ads Sci Tech, 3(3), 189-200 (1986).
Sircar, S., and Myers, A.L., Liquid adsorption operations: Equilibrium, kinetics, column dynamics, and applications, Sepn Sci Technol, 21(6), 535-562 (1986).
Tsezos, D.E.; Baird, M.H.I., and Shemilt, L.W., Adsorptive treatment with microbial biomass of 226 Ra-containing wastewaters, Biochem Eng J, 32(2), B29-B38 (1986).
Utrilla, J.R.; Garcia, M.A.F.; Mingorance, M.D., and Toledo, I.B., Adsorption of lead on activated carbons, J Chem Tech Biotechnol, 36(2), 47-52 (1986).
Van Deventer, J.S.J., Kinetic model for reversible adsorption of gold cyanide on activated carbon, Chem Eng Commns, 44, 257-274 (1986).

1987

Al-Zaid, K., et al., Adsorption of aromatic compounds and mixtures from solutions on molecular sieve 13X, Ads Sci Tech, 4(3), 185-210 (1987).
Cooney, D.O., An experiment in liquid-phase adsorption fundamentals, Chem Eng Educ, 21(4), 200-203 (1987).
Garbacz, J.K., et al., Theoretical equations for adsorption from binary non-electrolyte solutions with limited miscibility, Ads Sci Tech, 3(4), 253-260 (1986); 4(1), 105-112 (1987).
Gomez-Jimenez, L.; Rodriguez, A.G.; Gonzalez, J.L., and Guijosa, A.N., Study of kinetics of adsorption by activated carbons of 2,4,5-trichlorophenoxyacetic acid from aqueous solution, J Chem Tech Biotechnol, 37(4) 271-280 (1987); 38(1), 1-14 (1987).
Gonzalez-Pradas, E., et al., Adsorption of para-substituted anilines from cyclohexane solution on sepiolite, Ads Sci Tech, 4(1), 79-86 (1987).
Gonzalez-Pradas, E.; Sanchez, M.V.; Viciana, M.S.; Rey-Bueno, F.D., and Rodriguez, A.G., Adsorption of thiram from aqueous solution on activated carbon and sepiolite, J Chem Tech Biotechnol, 39(1), 19-28 (1987).
Goworek, J., Adsorption of alcohols from ternary solutions in benzene and n-heptane on silica gel, Ads Sci Tech, 4(1), 113-120 (1987).
Goworek, J., Effects of surface heterogeneity in adsorption from binary solutions on modified silica gel, Ads Sci Tech, 4(4), 224-229 (1987).
Helmy, A.K.; Ferreiro, E.A., and DeBussetti, S.G., Adsorption of acid on charcoal, Ads Sci Tech, 4(3), 211-216 (1987).

Khare, S.K.; Panday, K.K.; Srivastava, R.M., and Singh, V.N., Removal of Victoria Blue from aqueous solution by fly ash, J Chem Tech Biotechnol, 38(2), 99-104 (1987).
King, C.J., et al., Use of adsorbents for recovery of acetic acid from aqueous solutions, Sepn & Purif Methods, 16(1), 31-102 (1987).
Kumar, S.; Upadhyay, S.N., and Upadhya, Y.D., Removal of phenols by adsorption on fly ash, J Chem Tech Biotechnol, 37(4), 281-290 (1987).
Kuo, J.F., et al., Kinetics of adsorption of organics from water produced during in-situ tar sands experiments, Chem Eng Commns, 50, 201-212 (1987).
McConvey, I.F., Sorption of vitamin B12 from aqueous solution, CER&D, 65(3), 231-233 (1987).
McKay, G., Mass transport processes for adsorption of dyestuffs onto chitin, Chem Eng & Proc, 21(1), 41-52 (1987).
McKay, G., and Al Duri, B., Simplified model for equilibrium adsorption of dyes from mixtures using activated carbon, Chem Eng & Proc, 22(3), 145-156 (1987).
Price, P.E., and Danner, R.P., Prediction of multicomponent liquid adsorption equilibria, AIChEJ, 33(4), 551-557 (1987).

1988

Friedrich, M.; Seidel, A., and Gelbin, D., Kinetics of adsorption of phenol and indole from aqueous solutions on activated carbons, Chem Eng & Proc, 24(1), 33-38 (1988).
Lee, C.S., and Belfort, G., Thermodynamics of multiorganic solute adsorption from dilute aqueous solution, Ind Chem Eng Res, 27(6), 951-955 (1988).
McKay, G., and Al Duri, B., Branched-pore model applied to adsorption of basic dyes on carbon, Chem Eng & Proc, 24(1), 1-14 (1988).
Omichi, H.; Katakai, A., and Okamoto, J., Simulation of adsorption of uranium from seawater using liquid-film mass transfer controlling model, Sepn Sci Technol, 23(10), 1133-1144 (1988).
Omichi, H.; Katakai, A., and Okamoto, J., Effect of ultrasonic irradiation on recovery of uranium from seawater with adsorbents, Sepn Sci Technol, 23(14), 2445-2450 (1988).
Price, P.E., and Danner, R.P., Extension and evaluation of the Minka and Myers theory of liquid adsorption, Ind Eng Chem Res, 27(3), 506-512 (1988).
Sircar, S., and Myers, A.L., Determination of surface area and pore volume of adsorbents from adsorption isotherms of binary liquid mixtures, Chem Eng Sci, 43(12), 3259-3263 (1988).

CHAPTER 23

MEMBRANE-TYPE SEPARATION PROCESSES

23.1 Membrane Materials

1974-1988

McKinney, R., Ionically modified aromatic polyamide membranes, Sepn & Purif Methods, 3(1), 87-110 (1974).

Ross, W.D., Open pore polyurethane: A new separation medium, Sepn & Purif Methods, 3(1), 111-132 (1974).

Allan, A., PTFE membrane laminates, Chem Engnr, Oct, 374-376 (1982).

Kesting, R.E., Synthetic Polymeric Membranes, 2nd Edn., Wiley, New York (1985).

Lloyd, D.R., Materials Science of Synthetic Membranes, ACS, Washington, D.C. (1985).

Johnson, J.S., Materials for membranes, Chem Eng, 18 Aug, 121-123 (1986).

Lee, K.H., and Khang, S.J., New silicon-based material formed by pyrolysis of silicon rubber and its properties as a membrane, Chem Eng Commns, 44, 121-132 (1986).

Leenaars, A.F.M.; Keizer, K., and Burggraaf, A.J., Porous alumina membranes, Chemtech, Sept, 560-564 (1986).

Zanetti, R., Ceramics make strong bid for tough membrane uses, Chem Eng, 9 June, 19-22 (1986).

Freitas, R.F.S., and Cussler, E.L., Temperature-sensitive gels as size-selective absorbents, Sepn Sci Technol, 22(2), 911-920 (1987).

Johnson, J.S., Making membranes, Chemtech, Dec, 742-745 (1987).

Koresh, J.E., and Sofer, A., Molecular-sieve carbon permselective membrane, Sepn Sci Technol, 18(8), 723-734 (1983); 22(2), 973-982 (1987).

Marr, R., et al., Emulsion liquid membranes, Chem Eng & Proc, 20(6), 319-330 (1986); 21(2), 83-94 (1987).

McCaffrey, R.R., et al., Inorganic membrane technology, Sepn Sci Technol, 22(2), 873-888 (1987).

Meares, P., The manufacture of microporous membranes and their structure and properties, J Chem Tech Biotechnol, 37(3), 189-194 (1987).

Haggin, J., New generation of membranes developed for industrial separations, C&E News, 6 June, 7-16 (1988).

Neogi, P., Nonlinear elastodiffusion at small deformations in polymer membranes, Chem Eng Commns, 68, 185-196 (1988).

Okubo, T., and Inoue, H., Improvement of surface transport property of porous glass membranes by surface modification, AIChEJ, 34(6), 1031-1033 (1988).

23.2 Membrane Separation Theory

1968-1975

Kaufmann, T.G., and Leonard, E.F., Mechanism of interfacial mass transfer in membrane transport, AIChEJ, 14(3), 421-425 (1968).

Kaufmann, T.G., and Leonard, E.F., Studies of intramembrane transport, AIChEJ, 14(1), 110-117 (1968).

Osborn, J.C., and Bennion, D.N., Mass transfer of binary electrolytes in membranes at high concentrations, Ind Eng Chem Fund, 10(2), 273-280 (1971).

Steyn, W.M., and de Wet, W.J., Separation of cations using glass membranes, Sepn Sci, 7(5), 457-464 (1972).

Walawender, W.P., and Stern, S.A., Analysis of membrane separation parameters, Sepn Sci, 7(5), 553-584 (1972).

Reusch, C.F., and Cussler, E.L., Selective membrane transport, AIChEJ, 19(4), 736-741 (1973).

Batra, V.K., and Dullien, F.A.L., Measuring permeability of porous media, Chem Eng, 2 Sept, 92 (1974).

Bollenbeck, P.H., and Ramirez, W.F., Use of Rayleigh interferometer for membrane transport studies, Ind Eng Chem Fund, 13(4), 385-393 (1974).

Cussler, E.L., and Evans, D.F., Design of liquid membrane separations, Sepn & Purif Methods, 3(2), 399-422 (1974).

Hwang, S.T.; Choi, C.K., and Kammermeyer, K., Gaseous transfer coefficients in membranes, Sepn Sci, 9(6), 461-478 (1974).

Lee, C.H., Permeation properties in laminated membranes, Sepn Sci, 9(6), 479-486 (1974).

Pan, C.Y., and Habgood, H.W., Analysis of single-stage gaseous permeation membrane process, Ind Eng Chem Fund, 13(4), 323-331 (1974).

Schultz, J.S.; Goddard, J.D., and Suchdeo, S.R., Facilitated transport via carrier-mediated diffusion in membranes (review paper), AIChEJ, 20(3), 417-445; 20(4), 625-645 (1974).

Caracciolo, F.; Cussler, E.L., and Evans, D.F., Membranes with common ion pumping, AIChEJ, 21(1), 160-167 (1975).

Donaldson, T.L., and Quinn, J.A., Carbon dioxide transport through enzymatically active synthetic membranes, Chem Eng Sci, 30(1), 103-116 (1975).

Fang, S.M.; Stern, S.A., and Frisch, H.L., 'Free volume' model of permeation of gas and liquid mixtures through polymeric membranes, Chem Eng Sci, 30(8), 773-780 (1975).

Matulevicius, E.S., and Li, N.N., Facilitated transport through liquid membranes, Sepn & Purif Methods, 4(1), 73-96 (1975).

Steele, R.D., and Halligan, J.E., Hydrodynamic effects in liquid membrane transfer, Sepn Sci, 10(4), 461-470 (1975).

Stevenson, J.F.; Von Deak, M.A.; Weinberg, M., and Schuette, R.W., Unsteady-state method for measuring permeability of small tubular membranes, AIChEJ, 21(6), 1192-1199 (1975).

1976-1980

Min, S.; Duda, J.L.; Notter, R.H., and Vrentas, J.S., Interferometric technique for study of steady-state membrane transport, AIChEJ, 22(1), 175-182 (1976).

Paul, D.R., Solution-diffusion model for swollen membranes, Sepn & Purif Methods, 5(1), 33-50 (1976).

Doshi, M.R.; Gill, W.N., and Kabadi, V.N., Optimal design of hollow-fiber modules, AIChEJ, 23(5), 765-768 (1977).

Gilroy, K.; Brighton, E., and Gaylor, J.D.S., Fluid vortices and mass transfer in a curved-channel artificial membrane lung, AIChEJ, 23(1), 106-115 (1977).
Mohanty, K.K., and Sikar, K.K., Stage and point efficiencies in barrier separation processes, Chem Eng Sci, 32(2), 97-108 (1977).
Stern, S.A.; Onorato, F.J., and Libove, C., Permeation of gases through hollow silicone rubber fibres: Effect of fiber elasticity on gas permeability, AIChEJ, 23(4), 567-578 (1977).
Volkel, W.; Wandrey, C., and Schugerl, K., Determination of selective mass transfer rates across liquid membranes by measurement of permeation coefficients, Sepn Sci, 12(4), 425-434 (1977).
Alessi, P.; Kikic, I.; Canepa, B., and Costa, P., Inversion of selectivity in liquid membrane permeation process, Sepn Sci Technol, 13(7), 613-624 (1978).
Pan, C.Y., and Habgood, H.W., Gas separation by permeation, Can JCE, 56, 197-220 (1978).
Huang, T.C., and Lian, P.H., Interdiffusion of counterions in a cation-exchange membrane, Ind Eng Chem Fund, 18(3), 221-226 (1979).
Leung, P.S., Mathematical simulation for membrane filtration systems, Sepn Sci Technol, 14(2), 167-174 (1979).
Notter, R.H.; Tam, Y.M., and Min, S., Steady nonionic countergradient transport through membranes by coupled diffusion, AIChEJ, 25(3), 469-478 (1979).
Stern, S.A., and Leone, S.M., Separation of krypton and xenon by selective permeation, AIChEJ, 26(6), 881-901 (1980).

1981-1984

Deen, W.M.; Bohrer, M.P., and Epstein, N.B., Effects of molecular size and configuration on diffusion in microporous membranes, AIChEJ, 27(6), 952-959 (1981).
Desai, N., and Stroeve, P., Application of an electrodic technique to study transport across monolayers, Chem Eng Commns, 11(1), 113-122 (1981).
Lee, C.H., Effects of flow dispersion on hollow-fiber module performance, Sepn Sci Technol, 16(1), 81-86 (1981).
Light, W.G., and Tran, T.V., Improvement of thin-channel design for pressure-driven membrane systems, Ind Eng Chem Proc Des Dev, 20(1), 33-40 (1981).
Noda, I., and Gryte, C.C., Multistage membrane separation processes, AIChEJ, 27(6), 904-912 (1981).
Do, D.D., Transient response of a reactive membrane bounded by two finite reservoirs, Chem Eng J, 25(2), 201-210 (1982).
Soles, E.; Smith, J.M., and Parrish, W.R., Gas transport through polyethylene membranes, AIChEJ, 28(3), 474-479 (1982).
Stroeve, P., and Varanasi, P.P., Transport processes in liquid membranes: Double emulsion separation systems, Sepn & Purif Methods, 11(1), 29-70 (1982).
Bohrer, M.P., Diffusional boundary layer resistance for membrane transport, Ind Eng Chem Fund, 22(1), 72-78 (1983).
Kemp, N.J., and Noble, R.D., Heat transfer effects in facilitated transport liquid membranes, Sepn Sci Technol, 18(12), 1147-1166 (1983).

Various, Permeation-membrane separation (special report), Hyd Proc, Aug, 43-62 (1983).
Boucif, N.; Majumdar, S., and Sirkar, K.K., Series solutions for a gas permeator with countercurrent and cocurrent flow, Ind Eng Chem Fund, 23(4), 470-480 (1984).
Izatt, R.M., et al., Cation selectivity in a toluene emulsion membrane system, Solv Extn & Ion Exch, 2(3), 459-478 (1984).
Pintauro, P.N., and Bennion, D.N., Mass transport of electrolytes in membranes, Ind Eng Chem Fund, 23(2), 230-243 (1984).
Rangarajan, R.; Mazid, M.A.,; Matsuura, T., and Sourirajan, S., Permeation of pure gases under pressure through asymmetric porous membranes (membrane characterization and prediction of performance), Ind Eng Chem Proc Des Dev, 23(1), 79-87 (1984).

1985-1986

Haraya, K.; Hakuta, T., and Yoshitome, H., Calculation of a serial fed gas permeator system, Sepn Sci Technol, 20(5), 403-422 (1985).
Meldon, J.H.; Kang, Y.S., and Sung, N.H., Analysis of transient permeation through a membrane with immobilizing chemical reaction, Ind Eng Chem Fund, 24(1), 61-64 (1985).
Perrin, J.E., and Stern, S.A., Modeling of permeators with two different types of polymer membranes, AIChEJ, 31(7), 1167-1177 (1985).
Rautenbach, R., and Dahm, W., Separation of multicomponent mixtures by gas permeation, Chem Eng & Proc, 19(4), 211-220 (1985).
Shindo, Y.; Hakuta, T.; Yoshitome, H., and Inoue, H., Calculation methods for multicomponent gas separation by permeation, Sepn Sci Technol, 20(5), 445-460 (1985).
Hernandez, A., et al., Experimentally fitted and simple model for pores in nuclepore membranes, Sepn Sci Technol, 21(6), 665-678 (1986).
Kontturi, K., and Pajari, H., Effect of membrane thickness on countercurrent electrolysis in a thin porous membrane, Sepn Sci Technol, 21(10), 1089-1100 (1986).
Mohan, K., and Govind, R., Analysis of a cocurrent membrane reactor, AIChEJ, 32(12), 2083-2086 (1986).
Neogi, P.; Kim, M., and Yang, Y., Diffusion in solids under strain, with emphasis on polymer membranes, AIChEJ, 32(7), 1146-1157 (1986).
Pan, C.Y., Gas separation by high-flux asymmetric hollow-fiber membrane, AIChEJ, 32(12), 2020-2027 (1986).
Park, J.K., and Chang, H.N., Flow distribution in the fibre lumen side of a hollow-fibre module, AIChEJ, 32(12), 1937-1947 (1986).
Rice, R.G., and Howell, S.W., Elastic and flow mechanics for membrane spargers, AIChEJ, 32(8), 1377-1382 (1986).
Yang, M.C., and Cussler, E.L., Designing hollow-fibre contactors, AIChEJ, 32(11), 1910-1916 (1986).

1987-1988

Aris, R., and Cussler, E.L., A general theory of anisotropic membranes, Chem Eng Commns, 58, 3-16 (1987).

Deen, W.M., Hindered transport of large molecules in liquid-filled pores, AIChEJ, 33(9), 1409-1425 (1987).
Gorissen, H., Temperature changes involved in membrane gas separations, Chem Eng & Proc, 22(2), 63-68 (1987).
Haraya, K., et al., Concentration polarization phenomenon on surface of gas separation membrane, Sepn Sci Technol, 22(5), 1425-1438 (1987).
Jem, K.M.; Aris, R., and Cussler, E.L., Anisotropic membrane transport, Chem Eng Commns, 55, 5-18 (1987).
Kao, Y.K., and Yan, Z., Dynamic modeling and simulation of simple membrane permeators, Chem Eng Commns, 59, 343-370 (1987).
Noble, R.D., Overview of membrane separations, Sepn Sci Technol, 22(2), 731-744 (1987).
Pons, M.N., Monte Carlo model of microporous plane membranes, Chem Eng J, 35(3), 201-210 (1987).
Barkey, D.P., Determination of breakthrough curves for selective-membrane sorption processes, Chem Eng Commns, 72, 213-220 (1988).
Basaran, O.A., and Auvil, S.R., Asymptotic analysis of gas separation by membrane module, AIChEJ, 34(10), 1726-1731 (1988).
Borwanker, R.P., et al., Analysis of effect of internal phase leakage on liquid membrane separations, AIChEJ, 34(5), 753-762 (1988).
Evangelista, F., An improved analytical method for the design of spiral-wound modules, Chem Eng J, 38(1), 33-40 (1988).
Gottschlich, D.E.; Roberts, D.L., and Way, J.D., Theoretical comparison of facilitated transport and solution-diffusion membrane modules for gas separation, Gas Sepn & Purif, 2(2), 65-71 (1988).
Grau, R.J.; Cassano, A.E., and Irazoqui, H.A., Mass transfer through permeable walls: New integral equation approach for cylindrical tubes with laminar flow, Chem Eng Commns, 64, 47-66 (1988).
Leighton, D.T., and McCready, M.J., Shear enhanced transport in oscillatory liquid membranes, AIChEJ, 34(10), 1709-1712 (1988).
Noble, R.D., Relationship of system properties to performance in facilitated transport systems, Gas Sepn & Purif, 2(1), 16-19 (1988).
Prakash, S., and Gaddis, J.L., An exponential transformation to assist in solution of polarization boundary layers in membrane separation, Comput Chem Eng, 12(4), 345-350 (1988).
Sengupta, A.; Basu, R., and Sirkar, K.K., Separation of solutes from aqueous solutions by contained liquid membranes, AIChEJ, 34(10), 1698-1708 (1988).
Shere, A.J., and Cheung, H.M., Modeling of leakage in liquid surfactant membrane systems, Chem Eng Commns, 68, 143-164 (1988).

23.3 Membrane Separation Applications

1966-1975

Friedlander, H.Z., and Rickles, R.N., Membrane separation processes, Chem Eng, 28 Feb, 111-116; 28 March, 121-124; 25 April, 163-168; 23 May, 153-156; 6 June, 145-148; 20 June, 217-224 (1966).

Kubica, J., and Kucharski, M., Separation of liquid mixtures by permeation, Int Chem Eng, 7(2), 321-326 (1967).
Kucharski, M., and Stelmaszek, J., Separation of liquid mixtures by permeation, Int Chem Eng, 7(4), 618-622 (1967).
Buckles, R.G.; Merrill, E.W., and Gilliland, E.R., Analysis of oxygen absorption in a tubular membrane oxygenator, AIChEJ, 14(5), 703-708 (1968).
Carter, J.W., Developments in membrane processes, Brit Chem Eng, 13(4), 533-536 (1968).
Carter, J.W., Developments in membrane processes, Brit Chem Eng, 13(5), 674-676 (1968).
Gantzel, P.K., and Merten, U., Gas separations with high-flux cellulose acetate membranes, Ind Eng Chem Proc Des Dev, 9(2), 331-332 (1970).
Li, N.N., Separation of hydrocarbons by liquid membrane permeation, Ind Eng Chem Proc Des Dev, 10(2), 215-221 (1971).
Paul, D.R., Membrane separation of gases using steady cyclic operation, Ind Eng Chem Proc Des Dev, 10(3), 375-379 (1971).
Stavenger, P.L., Putting semipermeable membranes to work, Chem Eng Prog, 67(3), 30-36 (1971).
Lacey, R.E., Membrane separation processes, Chem Eng, 4 Sept, 56-74 (1972).
Jagur-Grodzinski, J.; Marian, S., and Vofsi, D., Mechanism of selective permeation of ions through 'solvent polymeric membranes', Sepn Sci, 8(1), 33-44 (1973).
Cook, R.L., and Tock, R.W., Aqueous membranes for separation of gaseous mixtures, Sepn Sci, 9(3), 185-194 (1974).
Steele, R.D., and Halligan, J.E., Factors influencing liquid membrane mass transfer, Sepn Sci, 9(4), 299-312 (1974).
Tock, R.W.; Cheung, J.Y., and Cook, R.L., Dioxane-water transport through Nylon-6 membranes, Sepn Sci, 9(5), 361-380 (1974).
Haas, F.W., and Tock, R.W., Permeation of permanent gases through liquid membranes, Sepn Sci, 10(6), 723-730 (1975).
Thorman, J.M.; Rhim, H., and Hwang, S.T., Gas separation by diffusion through silicone rubber capillaries, Chem Eng Sci, 30(7), 751-754 (1975).
Waterland, L.R.; Robertson, C.R., and Michaels, A.S., Enzymatic catalysis using asymmetric hollow fibre membranes, Chem Eng Commns, 2(1), 37-48 (1975).

1976-1980

Tang, T.E., and Hwang, S.T., Mass transfer of dissolved gases through tubular membranes, AIChEJ, 22(6), 1000-1006 (1976).
Antonson, C.R.; Gardner, R.J.; King, C.F., and Ko, D.Y., Analysis of gas separation by permeation in hollow fibres, Ind Eng Chem Proc Des Dev, 16(4), 463-469 (1977).
Gardner, R.J.; Crane, R.A., and Hannan, J.F., Hollow fiber permeator for separating gases, Chem Eng Prog, 73(10), 76-78 (1977).
Gutman, R.G., The design of membrane separation plant, Chem Engnr, July, 510-513, 521-523 (1977).
Sirkar, K.K., Separation of gaseous mixtures with asymmetric dense polymeric membranes, Chem Eng Sci, 32(10), 1137-1146 (1977).

Walch, A., Industrial applications of membrane technology, Int Chem Eng, 17(3), 425-430 (1977).
Casamatta, G.; Bouchez, D., and Angelino, H., Liquid membrane separation: Modelling and development of a continuous countercurrent pilot scale contactor, Chem Eng Sci, 33(2), 145-152 (1978).
Lee, K.H.; Evans, D.F., and Cussler, E.L., Selective copper recovery with two types of liquid membranes, AIChEJ, 24(5), 860-868 (1978).
Sirkar, K.K., Skin thickness of dense asymmetric or composite membranes for maximum extent of separation of permanent gas mixtures, Sepn Sci Technol, 13(2), 165-173 (1978).
Peeters, H.; Vanderstraeten, P., and Verhoeye, L., Permeation of hydrocarbons through a polyethylene membrane, J Chem Tech Biotechnol, 29, 581-590 (1979).
Alessi, P.; Kikic, I., and Visalberghi, M.O., Liquid membrane permeation for the separation of C8 hydrocarbons, Chem Eng J, 19(3), 221-228 (1980).
Fritzsche, A.K., Permeation of phenol through ethylene copolymer hollow fibers, Sepn Sci Technol, 15(6), 1323-1338 (1980).
Halwachs, W., and Schugerl, K., Liquid membranes: A promising extraction process, Int Chem Eng, 20(4), 519-529 (1980).
Hwang, S.T., and Thorman, J.M., The continuous membrane column, AIChEJ, 26(4), 558-566 (1980).
Kuehne, D.L., and Friedlander, S.K., Selective transport of sulfur dioxide through polymer membranes, Ind Eng Chem Proc Des Dev, 19(4), 609-623 (1980).
Spear, M., Hydrogen recovery by membrane separators, Proc Engng, Sept, 66-67 (1980).
Ukihashi, H., A membrane for electrolysis, Chemtech, Feb, 118-120 (1980).
Various, Membrane separations (symposium series), Sepn Sci Technol, 15(4), 1035-1204 (1980).

1981-1982

Berry, R.I., Membranes separate gas, Chem Eng, 13 July, 63-67 (1981).
Danesi, P.R., et al., Mass-transfer rate through liquid membranes: Interfacial chemical reactions and diffusion as simultaneous permeability controlling factors, Sepn Sci Technol, 16(2), 201-212 (1981).
Frankenfeld, J.W.; Cahn, R.P., and Li, N.N., Extraction of copper by liquid membranes, Sepn Sci Technol, 16(4), 385-402 (1981).
Lee, C.H., Separation of liquid through polymer membrane: Benzene and cyclohexane system, Sepn Sci Technol, 16(1), 25-30 (1981).
Li, N.N., Encapsulation and separation by liquid surfactant membranes, Chem Engnr, July, 325-327 (1981).
Achwal, S.K., Transfer of dissolved gases through tubular membrane, Can JCE, 60, 443-446 (1982).
Bock, J., and Valint, P.L., Uranium extraction from wet phosphoric acid using a liquid membrane, Ind Eng Chem Fund, 21(4), 417-422 (1982).
Bollinger, W.A.; MacLean, D.L., and Narayan, R.S., Membrane separation systems for oil refining and production, Chem Eng Prog, 78(10), 27-32 (1982).

Chaiko, D.J., and Asare, K.O., Characterization of liquid membrane supports, Sepn Sci Technol, 17(15), 1659-1680 (1982).
Coady, A.B., and Davis, J.A., Carbon dioxide recovery by gas permeation, Chem Eng Prog, 78(10), 44-49 (1982).
Cooper, A.R., and Van Derveer, D.S., Membrane testing using polymeric dyes, Sepn Sci Technol, 17(4), 621-624 (1982).
Fox, J.L., Developments in membrane technology, C&E News, 8 Nov, 7-12 (1982).
Imai, M.; Furusaki, S., and Miyauchi, T., Separation of volatile materials by gas membranes, Ind Eng Chem Proc Des Dev, 21(3), 421-426 (1982).
Marr, R., and Kopp, A., Liquid membrane technology: Survey of phenomena, mechanisms, and models, Int Chem Eng, 22(1), 44-61 (1982).
Mazur, W.H., and Chan, M.C., Membranes for natural gas sweetening, Chem Eng Prog, 78(10), 38-43 (1982).
Meldon, J.H.; Stroeve, P., and Gregoire, C.E., Facilitated transport of carbon dioxide (review paper), Chem Eng Commns, 16(1), 263-300 (1982).
Schell, W.J., and Houston, C.D., Spiral-wound permeators for purification and recovery, Chem Eng Prog, 78(10), 33-37 (1982).
Schell, W.J., and Houston, C.D., Process gas with selective membranes, Hyd Proc, Sept, 249-252 (1982).
Trouve, G.; Malher, E.; Colinart, P., and Renon, H., Liquid membranes for mass transfer in emulsions, Chem Eng Sci, 37(8), 1225-1234 (1982).
Various, Membrane separation techniques: Plant and equipment survey, Processing, April, 33-37 (1982).

1983

Anon., Membranes for gas separations, Processing, Nov, 32 (1983).
Cianetti, C., and Danesi, P.R., Facilitated transport of nitric acid through supported liquid membrane containing a tertiary amine as carrier, Solv Extn & Ion Exch, 1(3), 565-584 (1983).
Danesi, P.R.; Cianetti, C., and Horwitz, E.P., Acid extraction by supported liquid membranes containing basic carriers, Solv Extn & Ion Exch, 1(2), 299-310 (1983).
Dickson, J.M.; Babai-Pirouz, M., and Lloyd, D.R., Aromatic hydrocarbon-water separations by pressure-driven membrane separation process, Ind Eng Chem Proc Des Dev, 22(4), 625-632 (1983).
Forssell, P., and Kontturi, K., Experimental verification of separation of ions using countercurrent electrolysis in a thin porous membrane, Sepn Sci Technol, 18(3), 205-214 (1983).
Gokalp, M.; Hodgson, K.T., and Cussler, E.L., Selective electrorefining with liquid membranes, AIChEJ, 29(1), 144-149 (1983).
Izatt, R.M., et al., Metal separations using emulsion liquid membranes, Sepn Sci Technol, 18(12), 1113-1130 (1983).
Kuo, Y., and Gregor, H.P., Acetic acid extraction by solvent membrane, Sepn Sci Technol, 18(5), 421-440 (1983).
Matson, S.L.; Lopez, J., and Quinn, J.A., Separation of gases with synthetic membranes, Chem Eng Sci, 38(4), 503-524 (1983).
Matson, S.L.; Lopez, J., and Quinn, J.A., Gas separation with synthetic membranes (review paper), Chem Eng Sci, 38(4), 503-524 (1983).

Noble, R.D., Shape factors in facilitated transport through membranes, Ind Eng Chem Fund, 22(1), 139-144 (1983).
Pan, C.Y., Gas separation by permeators with high-flux asymmetric membranes, AIChEJ, 29(4), 545-552 (1983).
Teramoto, M., et al., Mechanism and modeling of copper permeation in membranes, Sepn Sci Technol, 18(10), 871-892; 18(11), 985-998 (1983).
Various, Flowthrough porous electrodes (feature report), Chem Eng, 21 Feb, 57-67 (1983).

1984

Applegate, L.E., Membrane separation processes, Chem Eng, 11 June, 64-89 (1984).
Bateman, B.R.; Way, J.D., and Larson, K.M., Gas-flux measurement through immobilized liquid membranes, Sepn Sci Technol, 19(1), 21-32 (1984).
Belfort, G., Synthetic Membrane Processes, Academic Press, Florida (1984).
Bollinger, W.A.; Long, S.P., and Metzger, T.R., Optimizing hydrocracker hydrogen membrane separators, Chem Eng Prog, 80(5), 51-57 (1984).
Danesi, P.R., Separation of metal species by supported liquid membranes, Sepn Sci Technol, 19(11), 857-894 (1984).
Gutman, R., and Leaver, G., Membrane separation techniques for biotechnology, Proc Engng, June, 37-40 (1984).
Macasek, F.; Rajec, P.; Kopunec, R., and Mikulaj, V., Membrane extraction in preconcentration of some uranium fission products, Solv Extn & Ion Exch, 2(2), 227-252 (1984).
Mitrovic, M.; Radovanovic, F., and Knezic, L., Dual-membrane separation, Chem Eng J, 28(1), 53-64 (1984).
Mohnot, S.M., and Cussler, E.L., Microporous membrane reactors, Chem Eng Sci, 39(3), 569-578 (1984).
Schendel, R.L., Using membranes for separation of acid gases and hydrocarbons, Chem Eng Prog, 80(5), 39-43 (1984).

1985

Chern, R.T.; Koros, W.J., and Fedkiw, P.S., Simulation of a hollow-fibre gas separator: Effects of process and design variables, Ind Eng Chem Proc Des Dev, 24(4), 1015-1022 (1985).
Cooley, T.E., and Dethloff, W.L., Field tests show membrane gas processing attractive, Chem Eng Prog, 81(10), 45-50 (1985).
Gooding, C.H., Applications of membrane technology, Chemtech, June, 348-354 (1985).
Hsu, E.C., and Li, N.N., Membrane recovery in liquid membrane separation processes, Sepn Sci Technol, 20(2), 115-130 (1985).
Le, M.S., and Billigheimer, P.J., Membranes in downstream processing, Chem Engnr, July, 48-53 (1985).
Mazid, M.A.; Rangarajan, R.; Matsuura, T., and Sourirajan, S., Separation of hydrogen-methane gas mixtures by permeation through porous cellulose acetate membranes, Ind Eng Chem Proc Des Dev, 24(4), 907-913 (1985).
McGregor, W.C. (Ed.), Membrane Separations in Biotechnology, Marcel Dekker, New York (1985).

McReynolds, K.B., A new membrane air-separation system, Chem Eng Prog, 81(6), 27-29 (1985).
Qi, Z., and Cussler, E.L., Hollow fiber gas membranes, AIChEJ, 31(9), 1548-1553 (1985).
Schulz, G., Gas separation in one- or two-step membrane processes, Chem Eng & Proc, 19(5), 235-242 (1985).
Short, H.C., and Skole, R., Membrane-cell developments ease electrochemistry scaleup, Chem Eng, 4 March, 41-43 (1985).
Sweeney, M.J., Membrane-based liquid separation systems, Chem Eng Prog, 81(1), 32-35 (1985).
Yamashiro, H.; Hirajo, M.; Schell, W.J., and Maitland, C.F., Plant uses membrane separation, Hyd Proc, Feb, 87-89 (1985).

1986

Baker, R.W., and Blume, I., Permselective membranes separate gases, Chemtech, April, 232-238 (1986).
Basta, N., Use electrodialytic membranes for waste recovery, Chem Eng, 3 March, 42-43 (1986).
Brennan, M.S.; Fane, A.G., and Fell, C.J.D., Natural gas separation using supported liquid membranes, AIChEJ, 32(9), 1558-1560 (1986).
Danesi, P.R., and Rickert, P.G., Performance of hollow-fiber supported liquid membranes for Co-Ni separations, Solv Extn & Ion Exch, 4(1), 149-164 (1986).
Gabler, R., and Messinger, S., Scaling up membrane filter systems, Chemtech, Oct, 616-621 (1986).
Meares, P., Industrial use of membranes, Chem Engnr, Feb, 38-40 (1986).
Perrin, J.E., and Stern, S.A., Separation of helium-methane mixture in permeators with two types of polymer membranes, AIChEJ, 32(11), 1889-1901 (1986).
Roberts, D.L., and Ching, G.D., Recovery of freon gases with silicon rubber membranes, Ind Eng Chem Proc Des Dev, 25(4), 971-974 (1986).
Stookey, D.J.; Patton, C.J., and Malcolm, G.L., Membranes separate gases selectively, Chem Eng Prog, 82(11), 36-40 (1986).
Weber, W.F., and Bowman, W., Membranes replacing other separation technologies, Chem Eng Prog, 82(11), 23-28 (1986).

1987

Anon., Inorganic membrane for filtration, Processing, Sept, 37 (1987).
Anon., Unique membrane system spurs gas separations, Chem Eng, 30 Nov, 62-66 (1987).
Anon., World lead for USSR gas membrane technology, Chem Engnr, July, 25 (1987).
Baird, R.S.; Bunge, A.L., and Noble, R.D., Batch extraction of amines using emulsion liquid membranes: Importance of reaction reversibility, AIChEJ, 33(1), 43-53 (1987).
Bellobono, I.R., et al., Transport of oxygen through facilitated-liquid membrane immobilized by photografting onto cellulose, Gas Sepn & Purif, 1(2), 103-106 (1987).
Blume, R., Preparing ultrapure water, Chem Eng Prog, 83(12), 55-57 (1987).

Boey, S.C.; Garcia del Cerro, M.C., and Pyle, D.L., Extraction of citric acid by liquid membrane extraction, CER&D, 65(3), 218-223 (1987).

Buchalter, E.M.; Hofman, D.L.; Craig, W.M.; Birkill, R.S., and Smit, J.J., Supported liquid membrane technology applied to the recovery of useful isotopes from reactor pool water, CER&D, 65(5), 381-385 (1987).

Chowdhury, J., Membranes set to tackle larger separation tasks, Chem Eng, 28 Sept, 14-17 (1987).

Dworzak, W.R., and Naser, A.J., Pilot-scale evaluation of supported liquid-membrane extraction, Sepn Sci Technol, 22(2), 677-690 (1987).

Gostoli, C.; Sarti, G.C., and Matulli, S., Low-temperature distillation through hydrophobic membranes, Sepn Sci Technol, 22(2), 855-872 (1987).

Haggin, J., High-temperature membrane separates gases, C&E News, 4 May, 27 (1987).

Haraya, K., et al., Review of gas-separation membrane development in Japanese C1 chemistry project, Gas Sepn & Purif, 1(1), 3-10 (1987).

Hoffmann, H.; Scheper, T.; Schugerl, K., and Schmidt, W., Use of membranes to improve bioreactor performance, Biochem Eng J, 34(1) B13-B19 (1987).

Meldrum, A., Hollow-fibre membrane bioreactors, Chem Engnr, Oct, 28-31 (1987).

Prasad, R., and Sirkar, K.K., Microporous membrane solvent extraction, Sepn Sci Technol, 22(2), 619-640 (1987).

Prasad, R., and Sirkar, K.K., Solvent extraction with microporous hydrophilic and composite membranes. AIChEJ, 33(7), 1057-1066 (1987).

Rautenbach, R., and Dahm, W., Gas permeation: Module design and arrangement, Chem Eng & Proc, 21(3), 141-150 (1987).

Sengupta, A., and Sirkar, K.K., Ternary gas mixture separation in two-membrane permeators, AIChEJ, 33(4), 529-539 (1987).

Teramoto, M., et al., Development of spiral-type supported liquid membrane module for separation and concentration of metal ions, Sepn Sci Technol, 22(11), 2175-2202 (1987).

1988

Agrawal, R., et al., Membrane/cryogenic hybrid processes for hydrogen purification, Gas Sepn & Purif, 2(1), 9-15 (1988).

Chiou, J.S., and Paul, D.R., Gas permeation in a dry Nafion membrane, Ind Eng Chem Res, 27(11), 2161-2165 (1988).

Dahuron, L., and Cussler, E.L., Protein extractions with hollow fibres, AIChEJ, 34(1), 130-136 (1988).

Del Cerro, C., and Boey,, D., Liquid membrane extraction, Chem & Ind, 7 Nov, 681-687 (1988).

DiMartino, S.P.; Glazer, J.L.; Houston, C.D., and Schott, M.E., Hydrogen/carbon monoxide separation with cellulose acetate membranes, Gas Sepn & Purif, 2(3), 120-125 (1988).

Evangelista, F., and Jonsson, G., Optimal design and performance of spiral wound modules, Chem Eng Commns, 72, 69-94 (1988).

Fouda, A.E.; Matsuura, T., and Lui, A., Permeation of gas mixtures (carbon dioxide/methane) in cellulose acetate membranes, Sepn Sci Technol, 23(12), 2175-2190 (1988).

Gardner, J.B., Research on membranes for gas separation, Gas Sepn & Purif, 2(3), 114-115 (1988).

Godbole, G.C.; Klinzing, G.E., and Brainard, A.J., Facilitated transport of nitric oxide through immobilized liquid membranes, Sepn Sci Technol, 23(1), 215-226 (1988).

Haggin, J., Membrane separations for small-scale industrial processing, C&E News, 11 July, 25-32 (1988).

Haggin, J., Membrane-based compressed air dryer, C&E News, 18 July, 35 (1988).

Haraya, K.; Obata, K.; Hakuta, T., and Yoshitome, H., Performance of gas separator with high-flux polyimide hollow-fiber membrane, Sepn Sci Technol, 23(4), 305-320 (1988).

Igawa, M., Ion separation by charge-mosaic membrane system, Sepn & Purif Methods, 17(2), 141-155 (1988).

Jin, M.; Sikdar, S.K., and Bischke, S.D., Glycine permeation through Na+, Ag+ and Cs+, forms of perfluorosulfonated ion-exchange membranes, Sepn Sci Technol, 23(14), 2293-2308 (1988).

Kirkkopru, A.; Noble, R.D., and Bunge, A.L., Amine phase partitioning using emulsion liquid membranes, Chem Eng Commns, 64, 207-216 (1988).

Kontturi, K., and Westerberg, L.M., Countercurrent electrolysis in a cell where porous membranes have been connected in series with ion-exchange membranes, Sepn Sci Technol, 23(1), 227-242 (1988).

Langsam, M.; Anand, M., and Karwacki, E.J., Substitute propyne polymers for gas separation membranes, Gas Sepn & Purif, 2(4), 162-170 (1988).

Lenski, U.; Passlack, J., and Staude, E., Tranport measurements through membranes by drying, Chem Eng Commns, 69, 65-80 (1988).

Majumdar, S.; Guha, A.K., and Sirkar, K.K., New liquid membrane technique for gas separation, AIChEJ, 34(7), 1135-1145 (1988).

Nakagawa, T.; Saito, T.; Asakawa, S., and Saito, Y., Polyacetylene derivatives as membranes for gas separation, Gas Sepn & Purif, 2(1), 3-8 (1988).

Pellegrino, J.J.; Nassimbene, R., and Noble, R.D., Facilitated transport of carbon dioxide through highly swollen ion-exchange membranes: Effect of hot glycerine pretreatment, Gas Sepn & Purif, 2(3), 126-130 (1988).

Prasad, R., and Sirkar, K.K., Dispersion-free solvent extraction with microporous hollow-fibre modules, AIChEJ, 34(2), 177-188 (1988).

Saito, K., et al., Recovery of uranium from seawater using amidoxime hollow fibres, AIChEJ, 34(3), 411-416 (1988).

Scott, K., and Winnick, J., Electrochemical membrane cell design for sulphur dioxide separation from flue gas, Gas Sepn & Purif, 2(1), 23-27 (1988).

Shackleton, R., Ceramic membrane filters, Chem Engnr, Feb, 15-16 (1988).

Shiyao, B., et al., Surface force-pore flow model in predicting separation and concentration of polyhydric alcohols in aqueous solutions using cellulose acetate membranes, Sepn Sci Technol, 23(1), 77-90 (1988).

Sidhoum, M.; Sengupta, A., and Sirkar, K.K., Asymmetric cellulose acetate hollow fibres: Studies in gas permeation, AIChEJ, 34(3), 417-425 (1988).

Tanigaki, M.; Shiode, T.; Ueda, M., and Eguchi, W., Facilitated transport of zinc chloride through hollow-fiber supported liquid-membrane, Sepn Sci Technol, 23(10), 1145-1182 (1988).

Tremblay, A.Y., et al., Simplex method to characterise dry cellulose acetate membranes for gas separations, Can JCE, 66(6), 1027-1030 (1988).
Uragami, T.; Yoshida, F., and Sugihara, M., Studies on synthesis and permeabilities of special polymer membranes, Sepn Sci Technol, 23(10), 1067-1082 (1988).
Various, Advances in membrane separations (symposium papers), Sepn Sci Technol, 23(12), 1595-1852 (1988).
Walke, L., et al., Recovery of carbon dioxide from flue gas using an electrochemical membrane, Gas Sepn & Purif, 2(2), 72-76 (1988).

23.4 Ion Exchange

1967-1969

Copeland, J.P.; Henderson, C.L., and Marchello, J.M., Influence of resin selectivity on film diffusion-controlled ion exchange, AIChEJ, 13(3), 449-452 (1967).
Gilwood, M.E., Saving capital and chemicals with countercurrent ion exchange, Chem Eng, 18 Dec, 83-88 (1967).
Klein, G.; Tondeur, D., and Vermeulen, T., Multicomponent ion exchange in fixed beds, Ind Eng Chem Fund, 6(3), 339-364 (1967).
Solt, G.S., Continuous countercurrent ion exchange (the CI process), Brit Chem Eng, 12(10), 1582-1586 (1967).
Tallmadge, J.A., Ion exchange treatment of mixed electroplating wastes, Ind Eng Chem Proc Des Dev, 6(4), 419-423 (1967).
Various, Desalination (topic issue), Chem Eng Prog, 63(1), 53-103 (1967).
Wallace, R.M., Concentration and separation of ions by Donnan membrane equilibrium, Ind Eng Chem Proc Des Dev, 6(4), 423-431 (1967).
George, D.R.; Riley, J.M., and Ross, J.R., Potassium recovery by chemical precipitation and ion exchange, Chem Eng Prog, 64(5), 96-99 (1968).
Hall, G.R.; Streat, M., and Creed, G.R.B., Ion exchange in nuclear chemical processes, Trans IChemE, 46, T53-59 (1968).
Lifshutz, N., and Dranoff, J.S., Inversion of concentrated sucrose solutions in fixed beds of ion exchange resin, Ind Eng Chem Proc Des Dev, 7(2), 266-269 (1968).
Michalson, A.W., High quality water via ion exchange, Chem Eng Prog, 64(10), 67-73 (1968).
Tuichiev, I.S.; Rizaev, N.U.; Merenkov, K.V., and Yusipov, M.M., Hydrodynamic properties of ion exchange resins during fluidization, Int Chem Eng, 8(2), 221-223 (1968).
Turner, J.C.R., and Snowdon, C.B., Liquid-side mass-transfer coefficients in ion exchange: Nernst-Planck model, Chem Eng Sci, 23(3), 221-230; 23(9), 1099-1104 (1968).
Turner, J.C.R.; Snowdon, C.B.; Jones, D.C., and Ward, J.W.C., Estimation of ion-exchange equilibrium diagrams involving weakly dissociated electrolytes, Trans IChemE, 46, T232-235 (1968).
Copeland, J.P., and Marchello, J.M., Film-diffusion controlled ion-exchange with a selective resin, Chem Eng Sci, 24(9), 1471-1474 (1969).

Kadlec, V., and Matejka, Z., Mixed-bed deionisation by weak electrolyte ion-exchange resins regenerated in situ by carbon dioxide, J Appl Chem, 19, 352-355 (1969).

Kuong, J.F., Maximising ion-exchanger throughput, Chem Eng, 15 Dec, 160 (1969).

Pollio, F.X.; Kunin, R., and Petralia, J.W., Treat sour water by ion exchange, Hyd Proc, 48(5), 124-126 (1969).

1970-1973

Campbell, D.O., and Buxton, S.R., Rapid ion exchange separations, Ind Eng Chem Proc Des Dev, 9(1), 89-99 (1970).

McGovern, T.J., and Dranoff, J.S., Sucrose inversion by partially deactivated ion-exchange resin beds, AIChEJ, 16(4), 536-538 (1970).

Streat, M., and Brignal, W.J., Representation of ternary ion exchange equilibria, Trans IChemE, 48, T151-155 (1970).

Turner, J.C.R., and Snowdon, C.B., Liquid-side mass transfer coefficients in ion exchange, Chem Eng Sci, 25(11), 1673-1678 (1970).

Weber, O.W.; Miller, I.F., and Gregor, H.P., Absorption of carbon dioxide by weak-base ion exchange resins, AIChEJ, 16(4), 609-614 (1970).

Colwell, C.J., and Dranoff, J.S., Nonlinear equilibrium and axial mixing effects in intraparticle diffusion-controlled sorption by ion-exchange resin beds, Ind Eng Chem Fund, 10(1), 65-70 (1971).

Danes, F., Batch process application to ion-exchange unit operation, Chem Eng Sci, 26(8), 1277-1288 (1971).

Gardiner, W.C., and Munoz, F., Mercury removal from waste effluent via ion exchange, Chem Eng, 23 Aug, 57-59 (1971).

Gondo, S.; Itai, M., and Kusunoki, K., Computational and experimental studies on a moving ion-exchange bed, Ind Eng Chem Fund, 10(1), 140-146 (1971).

Kunin, R., and Downing, D.G., Ion-exchange system boasts more pulling power, Chem Eng, 28 June, 67-69 (1971).

Dodds, J.A., and Tondeur, D., Design of cyclic fixed-bed ion-exchange operations, Chem Eng Sci, 27(6), 1267-1282; 27(12), 2291-2298 (1972).

Golden, L.S., and Irving, J., Osmotic and mechanical strength of ion-exchange resins, Chem & Ind, 4 Nov, 837-844 (1972).

Holliday, D.C., Continuous ion exchange: Design and development, Chem & Ind, 16 Sept, 717-723 (1972).

Parker, K.J., Ion exchange in the sugar industry, Chem & Ind, 21 Oct, 782-790 (1972).

Presedo, H.; Baker, B.L., and Gibbons, J.H., Hydraulic design of membrane ion exchanger, Ind Eng Chem Proc Des Dev, 11(4), 529-532 (1972).

Qureshi, M.; Qureshi, S.Z.; Gupta, J.P., and Rathore, H.S., Progress in ion-exchange studies on insoluble salts of polybasic metals, Sepn Sci, 7(6), 615-630 (1972).

Conrard, P.; Caude, M., and Rosset, R., Separation of close species on ion exchangers, Sepn Sci, 7(5), 465-490 (1972); 8(1), 1-10; 8(2), 269-278 (1973).

Lal, B.B., and Douglas, W.J.M., Techniques for measuring sorption of water by ion-exchange resin spheres, Ind Eng Chem Fund, 12(3), 381-384 (1973).

Letan, R., Continuous ion-exchanger, Chem Eng Sci, 28(3), 981-985 (1973).

Meares, P., Characteristics and uses of ion exchange membranes, Chem & Ind, 1 Dec, 103-107 (1973).

Millar, J.R., Fundamentals of ion exchange, Chem & Ind, 5 May, 409-413 (1973).

Moore, J.H., and Schechter, R.S., Liquid ion-exchange membranes, AIChEJ, 19(4), 741-747 (1973).

Williams, R.C., Ion exchange resins in power stations, Chem & Ind, 19 May, 465-470 (1973).

1974-1976

Bull, P.S.; Evans, J.V., and Nicholson, F.D., Condensate polishing performance of powdered ion-exchange resins, J Appl Chem Biotechnol, 24, 475-486 (1974).

Caude, M.; Conrard, P., and Rosset, R., Displacement development on ion exchangers, Sepn Sci, 9(4), 269-286 (1974).

Dodds, J.A., and Tondeur, D., Design of cyclic fixed-bed ion-exchange operations, Chem Eng Sci, 29(2), 611-620 (1974).

Lal, B.B., and Douglas, W.J.M., Equilibrium water sorption and volumetric behavior of ion-exchange resin spheres, Ind Eng Chem Fund, 13(3), 223-227 (1974).

Various, Liquid-liquid extraction and ion exchange in analytical chemistry (topic issue), Chem & Ind, 17 Aug, 639-647 (1974).

Bolto, B.A., Sirotherm ion-exchange desalination, Chemtech, May, 303-307 (1975).

Braud, C., and Selegny, E., Interrelation of swelling and selectivity of ion-exchange resins, Sepn Sci, 9(1), 13-26 (1974); 10(1), 47-110; 10(2), 175-244; 10(3), 331-358 (1975).

Farkas, E.J., and Himsley, A., Fundamental aspects of behavior of ion exchange equipment, Can JCE, 53, 575-585 (1975).

Kataoka, T.; Nishiki, T., and Ueyama, K., Mass transfer with liquid anion exchange, Chem Eng J, 10(3), 189-196 (1975).

Prengle, H.W., et al., Recycle waste water by ion exchange, Hyd Proc, 54(4), 173-184 (1975).

Vermeer, D.J.; Lynn, S., and Vermeulen, T., Cation-exchange column behavior in desalination process with regenerant recovery, Ind Eng Chem Proc Des Dev, 14(3), 290-297 (1975).

Kadlec, V., and Hubner, P., Ion exchange deionisation with recirculation of regenerant by heat, Chem & Ind, 4 Sept, 744-746 (1976).

Kataoka, T.; Nishiki, T., and Ueyama, K., Simultaneous mass transfer of acid and ions in a liquid anion exchanger, Chem Eng J, 12(3), 233-238 (1976).

Mostafa, H.A., and Said, A.S., Theoretical-plate concept for fixed-bed adsorption and ion-exchange, Trans IChemE, 54, 132-134 (1976).

Roland, L.D., Ion exchange: Operational advantages of continuous plants, Processing, Jan, 11-12 (1976).

Slater, M.J., and Lucas, B.H., Flow patterns and mass transfer rates in fluidized-bed ion-exchange equipment, Can JCE, 54, 264-270 (1976).

Smirnov, N.N., Mathematical models of ion-exchange process, Int Chem Eng, 16(2), 234-240 (1976).

Weatherley, L.R., and Turner, J.C.R., Ion-exchange kinetics: Comparison between a macroporous and a gel resin, Trans IChemE, 54, 89-94 (1976).

1977-1979

Holl, W., and Sontheimer, H., Ion exchange kinetics of the protonation of weak-acid ion-exchange resins, Chem Eng Sci, 32(7), 755-762 (1977).

Pusch, W., Ion-exchange membranes, Int Chem Eng, 17(1), 62-75 (1977).

Various, Novel ion exchangers (topic issue), Chem & Ind, 6 Aug, 634-652 (1977).

Danesi, P.R., and Chiarizia, R., Mass transfer rate with liquid ion exchangers, J Appl Chem Biotechnol, 28, 581-598 (1978).

De, A.K., and Sen, A.K., Synthetic inorganic ion-exchangers, Sepn Sci Technol, 13(6), 517-540 (1978).

Hubner, P., and Kadlec, V., Kinetic behavior of weak-base anion exchangers, AIChEJ, 24(1), 149-154 (1978).

Marra, R.A., and Cooney, D.O., Multicomponent sorption operations: Bed shrinking and swelling in an ion-exclusion case, Chem Eng Sci, 33(12), 1597-1602 (1978).

Smith, R.P., and Woodburn, E.T., Prediction of multicomponent ion exchange equilibria for ternary systems from binary systems data, AIChEJ, 24(4), 577-587 (1978).

Wiley, J.R., Decontamination of alkaline radioactive waste by ion exchange, Ind Eng Chem Proc Des Dev, 17(1), 67-71 (1978).

Abe, M., and Kasai, K., Synthetic inorganic ion-exchange materials, Sepn Sci Technol, 14(10), 895-908 (1979).

Agarwal, J.C., and Klumpar, I.V., Role of liquid ion exchange in processing of complex solutions, J Chem Tech Biotechnol, 29, 730-740 (1979).

Erickson, K.L., and Rase, H.F., Fixed-bed ion exchange with differing ionic mobilities and nonlinear equilibria, Ind Eng Chem Fund, 18(4), 312-317 (1979).

Goto, S.; Sato, N., and Teshima, H., Periodic operation for desalting water with thermally regenerable ion-exchange resin, Sepn Sci Technol, 14(3), 209-218 (1979).

Gupta, A.R., Isotope effects in ion-exchange equilibria in aqueous and mixed solvent systems, Sepn Sci Technol, 14(9), 843-858 (1979).

Hadzismajlovic, D.E., et al., Mass transfer in liquid spout-fluid beds of ion exchange resin, Chem Eng J, 17(3), 227-236 (1979).

Knaebel, K.S.; Cobb, D.D.; Shih, T.T., and Pigford, R.L., Ion-exchange rates in bifunctional resins, Ind Eng Chem Fund, 18(2), 175-180 (1979).

Various, Ion exchange in the water industry (topic issue), Chem & Ind, 3 March, 142-165 (1979).

1980-1981

Brown, J.M., and Wilson, D.J., Macroreticular resin columns, Sepn Sci Technol, 15(8), 1533-1555 (1980).

Burfield, D.R., and Smithers, R.H., Desiccant efficiency in solvent drying: Applications of cationic exchange resins, J Chem Tech Biotechnol, 30, 491-496 (1980).

Calmon, C., Explosion hazards of using nitric acid in ion-exchange equipment, Chem Eng, 17 Nov, 271-274 (1980).
Kennedy, D.C., Predict sorption of metals on ion-exchange resins, Chem Eng, 16 June, 106-118 (1980).
MacLean, G.T., Effect of synthetic flocculant on ion-exchange resin, Sepn Sci Technol, 15(8), 1555-1563 (1980).
Omatete, O.O.; Clazie, R.N., and Vermeulen, T., Column dynamics of ternary ion exchange, Chem Eng J, 19(3), 229-250 (1980).
Soldatov, V.S., and Bichkova, V.A., Ternary ion-exchange equilibria, Sepn Sci Technol, 15(2), 89-110 (1980).
Various, Advances in ion-exchange water treatment (topic issue), Chem & Ind, 20 Sept, 712-743 (1980).
Dyer, A.; Enamy, H., and Townsend, R.P., Plotting and interpretation of ion-exchange isotherms in zeolite systems, Sepn Sci Technol, 16(2), 173-184 (1981).
Gomez-Vaillard, R.; Kershenbaum, L.S., and Streat, M., Performance of continuous, cyclic ion-exchange reactors, Chem Eng Sci, 36(2), 307-326 (1981).
Matsuda, H.; Yamamoto, T.; Goto, S., and Teshima, H., Periodic operation for desalination with thermally regenerable ion-exchange resins (dynamic studies), Sepn Sci Technol, 16(1), 31-42 (1981).
Nigam, P.C.; Singh, D., and Sharma, R.N., Studies on ion-exclusion phenomena, Ind Eng Chem Proc Des Dev, 20(2), 182-188 (1981).
Rahman, K., and Streat, M., Mass transfer in liquid fluidized beds of ion exchange particles, Chem Eng Sci, 36(2), 293-306 (1981).
Raman, M.S., Polymer resins for water treatment, Chemtech, April, 252-255 (1981).
Rice, R.G., and Foo, S.C., Continuous desalination using cyclic mass-transfer ion exchange with bifunctional resins, Ind Eng Chem Fund, 20(2), 150-155 (1981).
Various, Pharmaceutical applications of ion exchange and solvent extraction (topic issue), Chem & Ind, 3 Oct, 677-690 (1981).

1982

Clifford, D., Multicomponent ion-exchange calculations for selected ion separations, Ind Eng Chem Fund, 21(2), 141-153 (1982).
Graham, E.E., and Dranoff, J.S., Application of Stefan-Maxwell equations to diffusion in ion exchangers, Ind Eng Chem Fund, 21(4), 360-369 (1982).
Graham, E.E., and Fook, C.F., Rate of protein absorption and desorption on cellulosic ion exchangers, AIChEJ, 28(2), 245-250 (1982).
Husain, S.W., et al., Synthesis and ion-exchange properties of lanthanum tungstate, Sepn Sci Technol, 17(7), 935-944 (1982).
Koff, F.W.; Sifniades, S., and Tunick, A.A., Ion-exchange process for recovery of hydroxylamine from Raschig synthesis mixtures, Ind Eng Chem Proc Des Dev, 21(2), 204-216 (1982).
Kojima, T., et al., Fundamental study on recovery of copper with a cation-exchange membrane, Can JCE, 60, 642-658 (1982).
Novosad, J., and Myers, A.L., Thermodynamics of ion exchange as an adsorption process, Can JCE, 60, 500-503 (1982).

Pelosi, P., and McCarthy, J., Preventing fouling of ion-exchange resins, Chem Eng, 9 Aug, 75-78; 6 Sept, 125-128 (1982).
Rao, M.G., and Gupta, A.K., Ion exchange processes accompanied by ionic reactions, Chem Eng J, 24(2), 181-190 (1982).
Reschke, M.; Halwachs, W., and Schugerl, K., Ion pair extraction of water soluble dyes, Chem Eng Sci, 37(10), 1529-1538 (1982).
Rousar, I., and Ditl, P., Kinetic characteristics of batch adsorber or ion exchange device operated under nonisothermal conditions, Chem Eng Commns, 18(5), 341-354 (1982).
Schenk, H.J., et al., Development of sorbers for recovery of uranium from seawater, Sepn Sci Technol, 17(11), 1293-1308 (1982).
Slater, M.J., The relative sizes of fixed bed and continuous countercurrent flow ion exchange equipment, Trans IChemE, 60, 54-58 (1982).
Various, Ion exchange in the petrochemical industry (topic issue), Chem & Ind, 21 Aug, 561-573 (1982).

1983

Abe, M., and Hayashi, K., Synthetic inorganic ion-exchange materials, Solv Extn & Ion Exch, 1(1), 97-112 (1983).
Barba, D.; del Re, G., and Foscolo, P.U., Numerical simulation of multicomponent ion-exchange operations, Chem Eng J, 26(1), 33-40 (1983).
Bobman, M.H.; Golden, T.C., and Jenkins, R.G., Ion exchange in selected low-rank coals: Equilibrium and kinetics, Solv Extn & Ion Exch, 1(4), 791-826 (1983).
Boundy, T.M., and Hubbard, D.W., Conduction and mass transfer in reinforced cation exchange membranes, Sepn Sci Technol, 18(12), 1131-1146 (1983).
Choppin, G.R., and Ohene-Aniapam, F., Equilibrium sorption of Am(III), Ce(III), and Eu(III), on Biorex 70 ion-exchange resin, Solv Extn & Ion Exch, 1(3), 585-596 (1983).
Fujine, S.; Saito, K.; Shiba, K., and Itoi, T., Liquid mixing in a large-sized column of ion exchange, Solv Extn & Ion Exch, 1(1), 113-126 (1983).
Goto, M.; Hayashi, N., and Goto, S., Separation of electrolyte and nonelectrolyte by ion retardation resin, Sepn Sci Technol, 18(5), 475-484 (1983).
Manning, M.J., and Melsheimer, S.S., Binary and ternary ion-exchange equilibria with a perfluorosulfonic acid membrane, Ind Eng Chem Fund, 22(3), 311-317 (1983).
Turner, J.C.R., and Murphy, T.K., A CSTR method for determining ion-exchange equilibria, Chem Eng Sci, 38(1), 147-154 (1983).
Various, Uses of ion exchange in the food industry (topic issue), Chem & Ind, 7 Nov, 804-824 (1983).

1984

Costa, E.; Lucas, A., and Gonzalez, M.E., Ion exchange: Determination of interdiffusion coefficients, Ind Eng Chem Fund, 23(4), 400-405 (1984).
Jenkins, I.L., Ion exchange in the atomic energy industry with particular reference to actinide and fission product separation (review paper), Solv Extn & Ion Exch, 2(1), 1-28 (1984).

Jepson, B.E., and Shockey, G.C., Calcium hydroxide isotope effect in calcium isotope enrichment by ion exchange, Sepn Sci Technol, 19(2), 173-182 (1984).

Klein, G., Calculation of ideal or empirically modified mass-action equilibria in heterovalent multicomponent ion exchange, Comput Chem Eng, 8(3), 171-178 (1984).

Tsuji, M., and Abe, M., Synthetic inorganic ion-exchange materials, Solv Extn & Ion Exch, 2(2), 253-274 (1984).

van der Meer, A.P.; Woerde, H.M., and Wesselingh, J.A., Mass transfer in countercurrent ion-exchange plate column, Ind Eng Chem Proc Des Dev, 23(4), 660-664 (1984).

1985

Egawa, H.; Nonaka, T., and Maeda, H., Studies of selective adsorption resins, Sepn Sci Technol, 20(9), 653-664 (1985).

Law, H.H.; Wilson, W.L., and Gabriel, N.E., Separation of gold cyanide ion from anion-exchange resins, Ind Eng Chem Proc Des Dev, 24(2), 236-238 (1985).

Mathur, J.N., and Khopkar, P.K., Ion exchange behaviour of chelating resin Dowex A-1 with actinides and lanthanides, Solv Extn & Ion Exch, 3(5), 753-762 (1985).

Riveros, P.A., and Cooper, W.C., Extraction of silver from cyanide solutions with ion-exchange resins, Solv Extn & Ion Exch, 3(3), 357-376 (1985).

van der Walt, T.N.; Strelow, F.W.E., and Verheij, R., Influence of cross-linkage on distribution coefficients and anion exchange behaviour of some elements in hydrochloric acid, Solv Extn & Ion Exch, 3(5), 723-740 (1985).

Wildhagen, G.R.S.; Qassim, R.Y.; Rajagopal, K., and Rahman, K., Effective liquid-phase diffusivity in ion exchange, Ind Eng Chem Fund, 24(4), 423-432 (1985).

Yoshida, H.; Kataoka, T., and Ikeda, S., Intraparticle mass transfer in bidispersed porous ion exchanger, Can JCE, 63(3), 422-435 (1985).

1986

Frey, D.D., Prediction of liquid-phase mass-transfer coefficients in multicomponent ion exchange: Comparison of matrix, film-model, and effective-diffusivity methods, Chem Eng Commns, 47, 273-294 (1986).

Geldart, R.W.; Yu, Q.; Wankat, P.C., and Wang, N., Improving elution and displacement ion-exchange chromatography by adjusting eluent and displacer affinities, Sepn Sci Technol, 21(9), 873-886 (1986).

Golden, L., Industrial use of ion exchange resins, Chem Engnr, Oct, 31-34 (1986).

Jackson, M.B., and Pilkington, N.H., Effect of the degree of crosslinking on the selectivity of ion-exchange resins, J Chem Tech Biotechnol, 36(2), 88-94 (1986).

Lefevre, L.J., Ion exchange: Problems and troubleshooting, Chem Eng, 7 July, 73-75 (1986).

Strelow, F.W.E., Influence of resin loading on cation exchange distribution coefficients of some elements in hydrochloric acid, Solv Extn & Ion Exch, 4(6), 1193-1208 (1986).

Ushio, S., End of the line for Japan's mercury cells: Change to ion-exchange membrane cells for chlor-alkali production, Chem Eng, 18 Aug, 12N (1986).

Wilson, D.J., Modeling of ion-exchange column operation, Sepn Sci Technol, 21(8), 767-788; 21(10), 991-1008 (1986).

Yoshida, H., and Kataoka, T., Recovery of amine and ammonia by ion exchange method: Comparison of ligand sorption and ion exchange accompanied by neutralization reaction, Solv Extn & Ion Exch, 4(6), 1171-1192 (1986).

Yoshida, H.; Kataoka, T., and Fujikawa, S., Kinetics in a chelate ion exchanger, Chem Eng Sci, 41(10), 2517-2530 (1986).

Yu, Q., and Wang, N.H.L., Multicomponent interference phenomena in ion exchange columns, Sepn & Purif Methods, 15(2), 127-158 (1986).

1987

Higgins, I.R., and Denton, M.S., CSA continuous countercurrent ion exchange technology, Sepn Sci Technol, 22(2), 997-1016 (1987).

Huang, T.C.; Huang, Y.C., and Tsai, F.N., Intraparticle diffusion-controlled kinetics of phenol adsorption on ion exchange resins, Chem Eng Commns, 56, 77-86 (1987).

Kataoka, T., and Yoshida, H., Dynamics in a thermally regenerable ion exchange column, Chem Eng J, 36(1), 41-50 (1987).

Kataoka, T., et al., Liquid-side ion exchange mass transfer in ternary system, AIChEJ, 33(2), 202-210 (1987).

Mikhail, E.M., and Misak, N.Z., Ion exchange characteristics of ceric tungstate: Kinetics of exchange, J Chem Tech Biotechnol, 39(4), 219-230 (1987).

Misak, N.Z., and Mikhail, E.M., Ion-exchange characteristics of a new manganese oxide, Solv Extn & Ion Exch, 5(5), 939-976 (1987).

Tavlarides, L.L.; Bae, J.H., and Lee, C.K., Solvent extraction, membranes and ion exchange in hydrometallurgical dilute metals separation, Sepn Sci Technol, 22(2), 581-618 (1987).

Various, Ultrapure water by ion exchange (topic issue), Chem & Ind, 16 Feb, 104-118 (1987).

Way, J.D., et al., Facilitated transport of carbon dioxide in ion exchange membranes, AIChEJ, 33(3), 480-487 (1987).

Yan, T.Y., and Shu, P., Regeneration of ion-exchange resin in nonaqueous media, Ind Eng Chem Res, 26(4), 753-755 (1987).

1988

Biscans, B.; Riba, J.P., and Couderc, J.P., Continuous equipment for ion exchange in fluidized bed: Prospects and problems, Int Chem Eng, 28(2), 248-257 (1988).

Bolden, W.B., and Groves, F.R., Batch sorption by lignand exchange: Determination of intraparticle diffusivity, Chem Eng Commns, 64, 125-136 (1988).

Geckler, K.E.; Shkinev, V.M., and Spivakov, B.Y., Liquid-phase polymer-based retention: A new method for selective ion separation, Sepn & Purif Methods, 17(2), 105-140 (1988).
Haas, C.N., Existence of ternary interactions in ion exchange, AIChEJ, 34(4), 702-703 (1988).
Huang, T.C., and Cho, L.T., Adsorption of phenol on anion exchange resins in presence of p-nitrophenol, Chem Eng Commns, 74, 169-178 (1988).
Hwang, Y.L.; Helfferich, F.G., and Leu, R.J., Multicomponent equilibrium theory for ion-exchange columns involving reactions, AIChEJ, 34(10), 1615-1626 (1988).
Kataoka, T., and Yoshida, H., Kinetics of ion exchange accompanied by neutralization reaction, AIChEJ, 34(6), 1020-1026 (1988).
Miyai, Y.; Ooi, K., and Katoh, S., Recovery of lithium from seawater using a new type of ion-sieve adsorbent based on magnesium-manganese oxide, Sepn Sci Technol, 23(1), 179-192 (1988).
Mustafa, S.; Hussain, S.Y., and Ali, H., Ion exchange sorption of phosphate, Solv Extn & Ion Exch, 6(4), 725-738 (1988).
Riveros, P.A., and Cooper, W.C., Kinetic aspects of ion exchange extraction of gold, silver, and base-metal cyano complexes, Solv Extn & Ion Exch, 6(3), 479-504 (1988).
Sengupta, A.K., and Lim, L., Modeling chromate ion-exchange processes, AIChEJ, 34(12), 2019-2029 (1988).
Solt, G.S., The basis of deionization plant design, Chem Engnr, Jan, 14-15 (1988).
Solt, G.S.; Nowosielski, A.W., and Feron, P., Predicting performance of ion-exchange columns, CER&D, 66(6), 524-530 (1988).
Staniewski, J.W.; Latto, B., and Hamielec, A.E., Sorption of water by poly (sodium acrylate) resin from organic solutions and mixtures, CER&D, 66(4), 371-377 (1988).
Taffe, P., Compact water-deionizer unit, Processing, July, 23-26 (1988).
Thonchk, N.K., et al., Extraction of thiocyanate ions from coal gasification effluents by ion exchange, CER&D, 66(6), 503-517 (1988).
Various, Ion exchange and chromatographic separations (symposium papers), Sepn Sci Technol, 23(12), 1853-1928 (1988).

23.5 Reverse Osmosis

1967-1969

Kimura, S., and Souirirajan, S., Analysis of data for reverse osmosis with porous cellulose acetate membranes, AIChEJ, 13(3), 497-503 (1967).
Rosenfeld, J., and Loeb, S., Turbulent region performance of reverse osmosis desalination tubes, Ind Eng Chem Proc Des Dev, 6(1), 122-127 (1967).
Sherwood, T.K.; Brian, P.L.T., and Fisher, R.E., Desalination by reverse osmosis, Ind Eng Chem Fund, 6(1), 2-12 (1967).
Srinivasan, S.; Tien, C., and Gill, W.N., Simultaneous development of velocity and concentration profiles in reverse osmosis systems, Chem Eng Sci, 22(3), 417-434 (1967).

Carter, J.W.; Psaras, G., and Hoyland, G., Demineralisation of aqueous solutions by reverse osmosis, Trans IChemE, 46, T265-269 (1968).
Kimura, S., and Sourirajan, S., Concentration polarization effects in reverse osmosis using porous cellulose acetate membranes, Ind Eng Chem Proc Des Dev, 7(1), 41-48 (1968).
Kimura, S., and Sourirajan, S., Mass-transfer coefficients for reverse-osmosis design, Ind Eng Chem Proc Des Dev, 7(4), 539-547 (1968).
Miller, E.F., Reverse-osmosis desalting, Chem Eng, 18 Nov, 153-158 (1968).
Shor, S.M., and Thodos, G., Reverse osmosis for water recovery from aqueous salt solutions, J Appl Chem, 18, 322-326 (1968).
Agrawal, J.P., and Sourirajan, S., Specification, selectivity and performance of porous cellulose acetate membranes in reverse osmosis, Ind Eng Chem Proc Des Dev, 8(4), 439-449 (1969).
Kimura, S.; Sourirajan, S., and Ohya, H., Stagewise reverse osmosis process design, Ind Eng Chem Proc Des Dev, 8(1), 79-89 (1969).
Lukavyi, L.S., and Dytnerskii, Y.I., Separation of homogeneous liquid mixtures by reverse osmosis, Int Chem Eng, 9(3), 458-464 (1969).
Mattson, R.J., and Tomsic, V.J., Improved water quality by reverse osmosis, Chem Eng Prog, 65(1), 62-68 (1969).
Ohya, H., and Sourirajan, S., General equations for reverse-osmosis process design, AIChEJ, 15(6), 829-836 (1969).

1970-1974
Banfield, D.L., Reverse osmosis costs, Chem & Ind, 14 March, 348-351 (1970).
Sharples, A., Introduction to reverse osmosis, Chem & Ind, 7 March, 322-328 (1970).
Worley, N., Reverse osmosis plant, Chem & Ind, 14 March, 352-357 (1970).
Anon., Hollow-fibre reverse osmosis for water purification, Chem Eng, 29 Nov, 54-59 (1971).
Matsuura, T., and Sourirajan, S., Reverse osmosis separation of some organic solutes in aqueous solution using porous cellulose acetate membranes, Ind Eng Chem Proc Des Dev, 10(1), 102-108 (1971).
Dejmek, P., Prediction of reverse osmosis apparatus performance, Chem Eng Sci, 27(8), 1577-1582 (1972).
Grover, J.R., and Delve, M.H., Operating experience with a 23 m3/day reverse osmosis pilot plant, Chem Engnr, Jan, 24-29 (1972).
Sharples, A., Structure and behaviour of reverse osmosis membranes, Chem Engnr, Jan, 34-37 (1972).
Gill, W.N., and Bansal, B., Analysis and design of hollow-fiber reverse osmosis systems, AIChEJ, 19(4), 823-831 (1973).
Grover, J.R., Reverse osmosis, Chem & Ind, 21 April, 369-371 (1973).
Kaup, E.C., Design factors in reverse osmosis, Chem Eng, 2 April, 46-55 (1973).
Alfani, F., and Drioli, E., Rejection coefficient variation in reverse osmosis process: Theory and experiments, Chem Eng Sci, 29(11), 2197-2204 (1974).
Butler, P., Reverse osmosis and deionisation for very pure water, Proc Engng, April, 83 (1974).

Carter, J.W.; Hoyland, G., and Hasting, A.P.M., Concentration polarization in reverse osmosis flow systems under laminar conditions: Effect of surface roughness and fouling, Chem Eng Sci, 29(7), 1651-1658 (1974).

Derzansky, L.J., and Gill, W.N., Mechanisms of brine-side mass transfer in a horizontal reverse-osmosis tubular membrane, AIChEJ, 20(4), 751-761 (1974).

Johnson, A.R., Interferometric measurements of polarization effects in reverse osmosis, AIChEJ, 20(5), 966-974 (1974).

Kennedy, T.J.; Merson, R.L., and McCoy, B.J., Improving permeation flux by pulsed reverse osmosis, Chem Eng Sci, 29(9), 1927-1932 (1974).

Leightell, B., Application of reverse osmosis in water treatment, Chem & Ind, 1 June, 437-440 (1974).

1975-1979

Carter, J.C., and De, S.C., Calculation of product flux and purity in reverse osmosis, Trans IChemE, 53, 16-19 (1975).

Dandavati, M.S.; Doshi, M.R., and Gill, W.N., Hollow-fiber reverse osmosis: Experiments and analysis of radial flow systems, Chem Eng Sci, 30(8), 877-886 (1975).

Doshi, M.R.; Gill, W.N., and Subramanian, R.S., Unsteady reverse osmosis or ultrafiltration in a tube, Chem Eng Sci, 30(12), 1467-1476 (1975).

Goldstein, W.E., and Verhoff, F.H., Investigation of anomalous osmosis and thermoosmosis, AIChEJ, 21(2), 229-238 (1975).

Thayer, W.L.; Pageau, L., and Sourirajan, S., Improved mass-transfer coefficient in reverse osmosis by oscillatory flow, Can JCE, 53, 422-430 (1975).

Kutowy, O.; Matsuura, T., and Sourirajan, S., Permeation characteristics of cellulose acetate propionate reverse-osmosis membranes, Can JCE, 54, 364-370 (1976).

Sastri, V.S., and Ashbrook, A.W., Reverse osmosis performance of cellulose acetate membranes in uranium separation from dilute solutions, Sepn Sci, 11(4), 361-377 (1976).

Sastri, V.S., and Ashbrook, A.W., Performance of some reverse osmosis membranes and application to separation of metals in acid mine-water, Sepn Sci, 11(2), 133-146 (1976).

Morris, R.M., Desalination of sea water, Chem & Ind, 6 Aug, 653-658 (1977).

Sastri, V.S., Reverse osmosis performance of cellulose acetate membranes for zinc separation from dilute solutions, Sepn Sci, 12(3), 257-270, (1977).

Spatz, D.D., Reclamation via reverse osmosis, Chemtech, Nov, 696-699 (1977).

Matsuura, T., and Sourirajan, S., Characterization of membrane material, specification of membranes, and predictability of membrane performance in reverse osmosis, Ind Eng Chem Proc Des Dev, 17(4), 419-428 (1978).

Pepper, D., Reverse osmosis for preconcentration, Chem Engnr, Dec, 916-918 (1978).

Sastri, V.S., Mechanism of separation of metal ions by reverse osmosis, Sepn Sci Technol, 13(6), 541-546 (1978).

Sastri, V.S., Reverse osmosis separation of nickel from dilute solutions, Sepn Sci Technol, 13(6), 475-486 (1978).

Kabadi, V.N.; Doshi, M.R., and Gill, W.N., Radial-flow hollow-fiber reverse osmosis experiments and theory, Chem Eng Commns, 3(4), 339-366 (1979).

Muratova, N.G., et al., Wastewater treatment by reverse osmosis, Int Chem Eng, 19(2), 350-352 (1979).

Sastri, V.S., Reverse osmosis separation of metal ions in acid mine-water, Sepn Sci Technol, 14(8), 711-720 (1979).

1980-1982

Baxter, A.G., et al., Reverse osmosis concentration of flavor components in apple-juice and grape-juice waters, Chem Eng Commns, 4(4), 471-484 (1980).

Subramanian, K.S.; Malaiyandi, M., and Sastri, V.S., Reverse osmosis separation of thiosalts from mining effluents, Sepn Sci Technol, 15(5), 1205-1212 (1980).

Malaiyandi, M., and Sastri, V.S., Reverse osmosis separation of sulfate, nitrate and ammonia from mining effluents, Sepn Sci Technol, 16(4), 371-376 (1981).

Matsuura, T., and Sourirajan, S., Reverse osmosis transport through capillary pores under the influence of surface forces, Ind Eng Chem Proc Des Dev, 20(2), 273-282 (1981).

Sirkar, K.K., and Rao, G.H., Approximate design equations and alternate design methodologies for tubular reverse-osmosis desalination, Ind Eng Chem Proc Des Dev, 20(1), 116-127 (1981).

Soltanieh, M., and Gill, W.N., Review of reverse osmosis membranes and transport models, Chem Eng Commns, 12(4), 279-364 (1981).

Yeager, H.L.; Matsuura, T., and Sourirajan, S., Some characteristics of aromatic polyamide-hydrazide (1:1) copolymer membranes for reverse osmosis transport, Ind Eng Chem Proc Des Dev, 20(3), 451-456 (1981).

Lin, C.C., and Tsai, G.J., Reverse osmosis in unstirred batch system, Sepn Sci Technol, 17(5), 727-738; 17(6), 839-848 (1982).

Malaiyandi, M.; Matsuura, T., and Sourirajan, S., Predictability of membrane performance for mixed-solute reverse-osmosis systems, Ind Eng Chem Proc Des Dev, 21(2), 277-282 (1982).

Malaiyandi, M.; Shah, S.M., and Sastri, V.S., Reverse osmosis separation of some organic acids from their aqueous solutions, Sepn Sci Technol, 17(8), 1065-1074 (1982).

Min, B.R., and Gill, W.N., Reverse osmosis batch cells for membrane transport studies, Chem Eng Commns, 17(1), 251-260 (1982).

Sirkar, K.K.; Dang, P.T., and Rao, G.H., Approximate design equations for reverse-osmosis desalination by spiral-wound modules, Ind Eng Chem Proc Des Dev, 21(3), 517-527 (1982).

Soltanieh, M., and Gill, W.N., Analysis and design of hollow fiber reverse osmosis systems, Chem Eng Commns, 18(5), 311-330 (1982).

Sourirajan, S., and Matsuura, T., Science of reverse osmosis, Chem Engnr, Oct, 359-368, 376 (1982).

1983-1984

Anon., Membrane separations for RO and UF, Processing, Jan, 13,15 (1983).

Chan, K.; Matsuura, T., and Sourirajan, S., Reverse osmosis separations of free radicals in aqueous solutions, Sepn Sci Technol, 18(3), 223-228 (1983).

Farnand, B.A.; Talbot, F.D.F.; Matsuura, T., and Sourirajan, S., Reverse osmosis separations, Ind Eng Chem Proc Des Dev, 22(2), 179-187 (1983).

Lin, C.C., and Chiang, M.K., Mathematical model for composite membranes for reverse osmosis, Sepn Sci Technol, 18(3), 229-238 (1983).

Salt, G., Membrane processes for RO and UF: A survey, Processing, Nov, 11-18 (1983).

Steven, J.H., Separations based on electrodialysis, reverse osmosis and ultrafiltration, Chem & Ind, 2 May, 346-349 (1983).

Anon., Reverse osmosis and ultrafiltration, Processing, April, 21-24 (1984).

Farnand, B.A., et al., Reverse osmosis separations of some inorganic solutes in ethanol solutions using cellulose acetate membranes, Sepn Sci Technol, 19(1), 33-50 (1984).

Matsuura, T.; Tweddle, T.A., and Sourirajan, S., Predicting performance of reverse osmosis membranes, Ind Eng Chem Proc Des Dev, 23(4), 674-684 (1984).

Mazid, M.A., Mechanisms of transport through reverse osmosis membranes (review paper), Sepn Sci Technol, 19(6), 357-374 (1984).

Prasad, R., and Sirkar, K.K., Analytical design equations for multisolute reverse osmosis processes by tubular modules, Ind Eng Chem Proc Des Dev, 23(2), 320-329 (1984).

Rangarajan, R.; Baxter, A.G.; Matsuura, T., and Sourirajan, S., Predictability of membrane performance for mixed solute reverse osmosis systems, Ind Eng Chem Proc Des Dev, 23(2), 367-374 (1984).

1985-1986

Evangelista, F., Short-cut method for design of reverse osmosis desalination plants, Ind Eng Chem Proc Des Dev, 24(1), 211-223 (1985).

Gooding, C.H., Reverse osmosis and ultrafiltration solve separation problems, Chem Eng, 7 Jan, 56-62 (1985).

Gupta, S.K., Analytical design equations for reverse osmosis systems, Ind Eng Chem Proc Des Dev, 24(4), 1240-1244 (1985).

Kim, S.S.; Chang, H.N., and Ghim, Y.S., Separation of fructose and glucose by reverse osmosis, Ind Eng Chem Fund, 24(4), 409-412 (1985).

Kurokawa, Y.; Mukaigawara, H., and Saito, S., Reverse osmosis separation of several alcohols from aqueous solution, Chem Eng Commns, 36(1), 333-342 (1985).

Matsuura, T., and Sourirajan, S., Reverse osmosis for separation of mixed uni-univalent electrolytes in aqueous solutions, Ind Eng Chem Proc Des Dev, 24(2), 297-303 (1985).

Prasad, R., and Sirkar, K.K., Design of multicomponent reverse osmosis processes with spiral-wound modules, Ind Eng Chem Proc Des Dev, 24(2), 350-358 (1985).

Rangarajan, R.; Mazid, M.A.; Matsuura, T., and Sourirajan, S., Predictability of membrane performance for mixed-solute reverse osmosis systems, Ind Eng Chem Proc Des Dev, 24(4), 977-985 (1985).

Sekino, M., et al., Reverse osmosis modules for water desalination, Chem Eng Prog, 81(12), 52-56 (1985).

Stone, M., Membrane developments for reverse osmosis and ultrafiltration, Processing, Oct, 9-20 (1985).
Cross, J., Combined reverse osmosis and deionisation techniques for water purification, Processing, June, 20-21 (1986).
Evangelista, F., Improved graphical-analytical method for the design of reverse-osmosis plants, Ind Eng Chem Proc Des Dev, 25(2), 366-374 (1986).
Taffe, P., Membranes for reverse osmosis, Processing, Nov, 35-43 (1986).

1987-1988

Evangelista, F., Approximate design method for reverse osmosis plants equipped with imperfectly rejecting membranes, Ind Eng Chem Res, 26(6), 1109-1117 (1987).
Farnand, B.A.; Talbot, F.D.F.; Matsuura, T., and Sourirajan, S., Reverse osmosis separations with cellulose acetate membranes, Ind Eng Chem Res, 26(6), 1080-1087 (1987).
Gupta, S.K., Design and analysis of radial-flow hollow-fibre reverse-osmosis system, Ind Eng Chem Res, 26(11), 2319-2323 (1987).
McCray, S.B, and Ray, R.J., Concentration of synfuel process condensates by reverse osmosis, Sepn Sci Technol, 22(2), 745-762 (1987).
Palanki, S., and Gupta, S.K., Analytical design equations for multisolute reverse osmosis systems, Ind Eng Chem Res, 26(12), 2449-2454 (1987).
Slater, C.S., and Paccione, J.D., A reverse osmosis system for an advanced separation process laboratory, Chem Eng Educ, 21(3), 138-143 (1987).
Wakeman, R., and Gallagher, P., Membranes for RO and UF, Processing, Nov, 51-59 (1987).
Aly, S.E., Reverse-osmosis desalination by waste-heat adsorption power cycle, J Inst Energy, March, 33-37 (1988).
De Pinho, M.N., et al., Reverse osmosis separation of glucose-ethanol-water system by cellulose acetate membranes, Chem Eng Commns, 64, 113-124 (1988).
Farnand, B.A., et al., Fabrication and performance of cellulose reverse-osmosis membranes, Chem Eng Commns, 66, 57-70 (1988).
Muldowney, G.P., and Punzi, V.L., Comparison of solute rejection models in reverse osmosis membranes for water-sodium chloride-cellulose acetate, Ind Eng Chem Res, 27(12), 2341-2352 (1988).
Rautenbach, R., and Janisch, I., Reverse osmosis for separation of organics from aqueous solutions, Chem Eng & Proc, 23(2), 67-76 (1988).

23.6 Ultrafiltration

1967-1975

Nakano, Y.; Tien, C., and Gill, W.N., Hyperfiltration application of nonlinear convective diffusion, AIChEJ, 13(6), 1092-1098 (1967).
Michaels, A.S., Membrane ultrafiltration separations, Chem Eng Prog, 64(12), 31-43 (1968).
Goldsmith, R.L., Macromolecular ultrafiltration with microporous membranes, Ind Eng Chem Fund, 10(1), 113-120 (1971).

Blatt, W.F., et al., Rapid salt exchange by coupled ultrafiltration and dialysis in anisotropic hollow fibers, Sepn Sci, 7(3), 271-284 (1972).

Forbes, F., Considerations in optimisation of ultrafiltration, Chem Engnr, Jan, 29-34 (1972).

Kozinski, A.A., and Lightfoot, E.N., Protein ultrafiltration, AIChEJ, 18(5), 1030-1040 (1972).

Porter, M.C., and Michaels,, A.S., Membrane ultrafiltration, Chemtech, Jan, 56-63; April, 248-254; July, 440-445; Oct, 633-637 (1971); Jan, 56-61 (1972).

Grieves, R.B.; Bhattacharyya, D.; Schomp, W.G., and Bewley, J.L., Membrane ultrafiltration of a nonionic surfactant, AIChEJ, 19(4), 766-774 (1973).

Harriott, P., Mechanism of partial rejection by ultrafiltration membranes, Sepn Sci, 8(3), 291-302 (1973).

McDonald, D.P., Ultrafiltration units, Proc Engng, Jan, 76-79 (1973).

Bhattacharyya. D.; Garrison, K.A.; Jumawan, A.B., and Grieves, R.B., Membrane ultrafiltration of a nonionic surfactant and inorganic salts from complex aqueous suspensions: Design for water reuse, AIChEJ, 21(6), 1057-1065 (1975).

Huffman, W.J.; Ward, R.M., and Harshman, R.C., Effects of forced and natural convection during ultrafiltration of protein-saline solutions and whole blood in thin channels, Ind Eng Chem Proc Des Dev, 14(2), 166-170 (1975).

Various, Filtration/separation (topic issue), Chem Eng Prog, 71(12), 37-80 (1975).

1976-1980

Pace, G.W., et al., Effect of temperature on flux from stirred ultrafiltration cell, Sepn Sci, 11(1), 65-78 (1976).

Klinkowski, P.R., Ultrafiltration: An emerging unit-operation, Chem Eng, 8 May, 164-173 (1978).

Tanny, G.B., Dynamic membranes in ultrafiltration and reverse osmosis, Sepn & Purif Methods, 7(2), 183-220 (1978).

Bhattacharyya, D., et al., Charged membrane ultrafiltration of multisalt systems, Sepn Sci Technol, 14(3), 193-208 (1979).

Bhattacharyya, D., et al., Ultrafiltration characteristics of oil-detergent-water systems: Membrane fouling mechanisms, Sepn Sci Technol, 14(6), 529-550 (1979).

Cooper, A.R., and Van Derveer, D.S., Characterization of ultrafiltration membranes by polymer transport measurements, Sepn Sci Technol, 14(6), 551-556 (1979).

Leung, W.F., and Probstein, R.F., Low polarization in laminar ultrafiltration of macromolecular solutions, Ind Eng Chem Fund, 18(3), 274-278 (1979).

Papenfuss, H.D.; Gross, J.F., and Thorson, S.T., Analytical study of ultrafiltration in a hollow-fiber artificial kidney, AIChEJ, 25(1), 170-179 (1979).

Shen, J.J.S., and Probstein, R.F., Turbulence promotion and hydrodynamic optimization in ultrafiltration process, Ind Eng Chem Proc Des Dev, 18(3), 547-554 (1979).

Michaels, A.S., Analysis and prediction of sieving curves for ultrafiltration membranes: A universal correlation? Sepn Sci Technol, 15(6), 1305-1322 (1980).

Trettin, D.R., and Doshi, M.R., Limiting flux in ultrafiltration of macromolecular solutions, Chem Eng Commns, 4(4), 507-522 (1980).

Trettin, D.R., and Doshi, M.R., Ultrafiltration in an unstirred batch cell, Ind Eng Chem Fund, 19(2), 189-194 (1980).

1981-1985

Doshi, M.R., and Trettin, D.R., Ultrafiltration of colloidal suspensions and macromolecular solutions in an unstirred batch cell, Ind Eng Chem Fund, 20(3), 221-229 (1981).

Michaels, A.S., Ultrafiltration, Chemtech, Jan, 36-43 (1981).

Klein, W., and Hoelz, W., Crossflow microfiltration in chemical processes, Chem Engnr, Oct, 369-373 (1982).

Sarbolouki, M.N., General diagram for estimating pore size of ultrafiltration and reverse osmosis membranes, Sepn Sci Technol, 17(2), 381-386 (1982).

Zuk, J.S.; Rucka, M., and Rak, J., Dye recovery by low pressure ultrafiltration, Chem Eng Commns, 19(1), 67-76 (1982).

Le, M.S., and Howell, J.A., Model for the effects of adsorbents and cleaners on ultrafiltration membrane structure, CER&D, 61, 191-197 (1983).

Radovich, J.M., and Behnam, B., Concentration ultrafiltration and diafiltration of albumin with an electric field, Sepn Sci Technol, 18(3), 215-222 (1983).

Short, J.L., Industrial applications of hollow fibre ultrafiltration, Chem Engnr, Aug, 47-51 (1983).

Bertera, R.; Steven, H., and Metcalfe, M., Development studies of crossflow microfiltration, Chem Engnr, June, 10-14 (1984).

Le, M.S., and Howell, J.A., Alternative model for ultrafiltration, CER&D, 62, 373-380 (1984).

Modolo, R.; Vittori, O., and Rumeau, M., Hydrodynamic behavior of water-tri-n-butylphosphate emulsions during ultrafiltration, Sepn Sci Technol, 19(4), 297-306 (1984).

Chong, R.; Jelen, P., and Wong, W., Effect of cleaning agents on noncellulosic ultrafiltration membrane, Sepn Sci Technol, 20(5), 393-402 (1985).

Mullon, C.; Radovich, J.M., and Behnam, B., Semiempirical model for electroultrafiltration-diafiltration, Sepn Sci Technol, 20(1), 63-72 (1985).

Schlumpf, J.P., and Quemeneur, F., Ultrafiltration of an alkylbenzene sulfonate, Int Chem Eng, 25(2), 240-246 (1985).

Vigo, F.; Uliana, C., and Lupino, P., Performance of rotating module in oily emulsions ultrafiltration, Sepn Sci Technol, 20(2), 213-230 (1985).

Warashina, T.; Hashino, Y., and Kobayashi, T., Hollow-fiber ultrafiltration, Chemtech, Sept, 558-561 (1985).

1986-1988

Bemberis, I., and Neely, K., Ultrafiltration as a competitive unit process, Chem Eng Prog, 82(11), 29-35 (1986).

Kovasin, K.K.; Hughes, R.R., and Hill, C.G., Optimization of an ultrafiltration-diafiltration process using dynamic programming, Comput Chem Eng, 10(2), 107-114 (1986).

Mahenc, J.; Lafaille, J.P., and Sanchez, V., Estimation of performance of hollow-fibre ultrafiltration modules, Int Chem Eng, 26(4), 660-671 (1986).
Nunes, S.P., et al., A new centrifigal ultrafiltration device, Sepn Sci Technol, 21(8), 823-830 (1986).
Vigo, F., and Uliana, C., Influence of vorticity at membrane surface on performances of ultrafiltration rotating module, Sepn Sci Technol, 21(4), 367-382 (1986).
Davis, R.H., and Birdsell, S.A., Hydrodynamic model and experiments for crossflow microfiltration, Chem Eng Commns, 49, 217-234 (1987).
Lafaille, J.P.; Sanchez, V., and Mahenc, J., Mass transfer in hollow-fibre ultrafiltration modules, Int Chem Eng, 27(2), 258-268 (1987).
Ohtani, T.; Watanabe, A.; Hoshino, C., and Kimura, S., Rejection properties of dynamic membranes of ovalbumin in ultrafiltration, Int Chem Eng, 27(2), 295-304 (1987).
Poyen, S., et al., Improvement of flux of permeate in ultrafiltration by turbulence promoters, Int Chem Eng, 27(3), 441-448 (1987).
Zuk, J.S., and Rucka, M., Resistance of a gel layer during ultrafiltration of casein solution, Chem Eng Commns, 54, 85-92 (1987).
Bedwell, W.B.; Yates, S.F.; Brubaker, I.M., and Uban, S., Crossflow microfiltration-fouling mechanisms studies, Sepn Sci Technol, 23(6), 531-548 (1988).
Christian, S.D., et al., Micellar-enhanced ultrafiltration of chromate anion from aqueous streams, AIChEJ, 34(2), 189-194 (1988).
Frederick, W.J., et al., Solute rejection in ultrafiltration of polydisperse organics from natural products, Chem Eng Commns, 68, 197-212 (1988).
Gill, W.N.; Wiley, D.E.; Fell, C.J.D., and Fane, A.G., Effect of viscosity on concentration polarization in ultrafiltration, AIChEJ, 34(9), 1563-1567 (1988).
Matsumoto, Y.; Nakao, S., and Kimura, S., Crossflow ceramic microfiltration of polymer solutions, Int Chem Eng, 28(4), 677-684 (1988).
Mazid, M.A., Separation and fractionation of macromolecular solutions by ultrafiltration, Sepn Sci Technol, 23(14), 2191-2210 (1988).
Nonaka, M., Macrokinetic modeling of membrane microfiltration processes, Sepn Sci Technol, 23(4), 387-412 (1988).

23.7 Dialysis

1967-1974

Bloch, R.; Finkelstein, A.; Kedem, O., and Vofsi, D., Metal-ion separation by dialysis through solvent membranes, Ind Eng Chem Proc Des Dev, 6(2), 231-237 (1967).
Guccione, E., How the artificial kidney cleans blood, Chem Eng, 30 Jan, 94-96 (1967).
Oldenburg, C.C., Dialysis for separating solutes of different molecular weights, Ind Eng Chem Proc Des Dev, 6(1), 111-114 (1967).
Kelman, S., and Grieves, R.B., Electrodialytic transport of inorganic and organic anions, J Appl Chem, 18, 20-24 (1968).

Yamane, R.; Ichikawa, M.; Mizutani, Y., and Onoue, Y., Concentrated brine production from sea water by electrodialysis using ion exchange membranes, Ind Eng Chem Proc Des Dev, 8(2), 159-165 (1969).

Wendt, R.P., et al., Measurements of membrane permeabilities using a rotating batch dialyzer, Ind Eng Chem Fund, 10(3), 406-411 (1971).

Klein, E.; Smith, J.K.; Wendt, R.P., and Shyamkant, V., Solute separations from water by dialysis, Sepn Sci, 7(3), 285-292 (1972); 8(5), 585-592 (1973).

Baker, J.A.,; Osburn, J.O., and Lawton, R.L., Concentration profiles in hollow-fiber dialyzers, Sepn Sci, 9(5), 411-422 (1974).

Chang, T.M.S., Comparison of semipermeable microcapsules and standard dialysers for separations, Sepn & Purif Methods, 3(2), 245-262 (1974).

Cooney, D.O.; Kim, S.S., and Davis, E.J., Analyses of mass transfer in hemodialyzers for laminar blood flow and homogeneous dialysate, Chem Eng Sci, 29(8), 1731-1738 (1974).

1975-1982

Lee, C.H., and Perry, E., Mathematical analysis of membrane separation parameters for liquid-liquid dialysis in single or multiple stages, Sepn Sci, 10(1), 21-32 (1975).

Rhee, C.T., Electrodialisis using ion-exchange membrane, Int Chem Eng, 15(2), 280-286 (1975).

Isaacson, M.S., and Sonin, A.A., Sherwood number and friction factor correlations for electrodialysis systems, with application to process optimization, Ind Eng Chem Proc Des Dev, 15(2), 313-321 (1976).

Ng, P.; Lundblad, J., and Mitra, G., Optimization of solute separation by diafiltration, Sepn Sci, 11(5), 499-502 (1976).

Giddings, J.C.; Yang, F.J., and Myers, M.N., Flow field-flow fractionation channel as a versatile pressure dialysis and ultrafiltration cell, Sepn Sci, 12(5), 499-510 (1977).

Patel, R.D.; Lang, K.C., and Miller, I.F., Polarization in ion-exchange membrane electrodialysis, Ind Eng Chem Fund, 16(3), 340-348 (1977).

Lake, M.A., and Melsheimer, S.S., Mass transfer characterization of Donnan dialysis, AIChEJ, 24(1), 130-137 (1978).

Noda, I., and Gryte, C.C., Mass transfer in regular arrays of hollow fibres in countercurrent dialysis, AIChEJ, 25(1), 113-122 (1979).

Urano, K.; Kawabata, M.; Yamada, N., and Masaki, Y., Selectivity of ion transport in desalination by electrodialysis, Ind Eng Chem Proc Des Dev, 19(1), 59-64 (1980).

Wang, N.H.L.; Kessler, D.P., and Ash, S.R., Mass transfer characteristics of sorbent-based reciprocating dialyzer, Chem Eng Commns, 5(5), 347-366 (1980).

Makai, A.J., and Turner, J.C.R., Electrodialysis at high current density using a laboratory stack, Trans IChemE, 60, 88-96 (1982).

1983-1988

Goncalves, M.C., et al., Pervaporation and dialysis of water-ethanol solutions using silicone rubber membranes, Sepn Sci Technol, 18(10), 893-904 (1983).

Huang, T.C., and Yu, T.M., Ion-exchange membrane electrodialysis of sugar solution, Chem Eng J, 26(2), 119-126 (1983).
Huang, T.C.; Yu, I.Y., and Lin, S.B., Ionic mass transfer rate of copper-sulphate in electrodialysis, Chem Eng Sci, 38(11), 1873-1880 (1983).
Bollinger, J.M., and Adams, R.A., Electrofiltration of ultrafine aqueous dispersions, Chem Eng Prog, 80(11), 54-58 (1984).
Strathmann, H., Electrodialysis and its application in the chemical process industry, Sepn & Purif Methods, 14(1), 41-66 (1985).
Taffe, P., Membranes for electrodialysis, Processing, March, 27-29 (1987).
Abbas, M., and Tyagi, V.P., Mass transfer in a circular conduit dialyzer when ultrafiltration is coupled with dialysis, Int J Heat Mass Trans, 31(3), 591-602 (1988).
Davis, J.C.; Valus, R.J., and Lawrence, E.G., Affinity dialysis: Method of continuous, rapid ion separation using dialysis membranes and selective water-soluble polymers as extractants, Sepn Sci Technol, 23(10), 1039-1066 (1988).
Gering, K.L., and Scamehorn, J.F., Use of electrodialysis to remove heavy metals from water, Sepn Sci Technol, 23(14), 2231-2268 (1988).
Niwa, M., et al., Concentration of aqueous solutions of ethanol by piezodialysis, Int Chem Eng, 28(3), 469-477 (1988).
Sakai, K.; Ohashi, H., and Naitoh, A., Effects of blood contact on the properties of tubular dialysis membranes, Biochem Eng J, 38(1), B1-B6 (1988).

23.8 Chromatography

1967-1973

Lai, C.L., and Roth, J.A., Dynamic simulation of gas chromatographic column, Chem Eng Sci, 22(10), 1299-1304 (1967).
Ryan, J.M.; Timmins, R.S., and O'Donnell, J.F., Production-scale gas chromatography, Chem Eng Prog, 64(8), 53-59 (1968).
Cooke, J.P., Understanding a gas chromatograph, Chem Eng, 10 March, 134-144 (1969).
Timmins, R.S.; Mir, L., and Ryan, J.M., Large-scale chromatography: New separation tool, Chem Eng, 19 May, 170-178 (1969).
Heines, V., A history of chromatography, Chemtech, May, 280-285 (1971).
Hsu, H.W., Optimum adsorbent volume in liquid adsorption chromatography, Sepn Sci, 6(5), 645-652 (1971).
Maldacker, T.A., and Rogers, L.B., Effect of loading on separation efficiency using steric exclusion chromatography, Sepn Sci, 6(6), 747-758 (1971).
Mir, L., Comparison of static bed and moving bed chromatography, Sepn Sci, 6(4), 515-536 (1971).
Moreland, A.K., and Rogers, L.B., Effects of slow mass transfer using microporous adsorbents in gas-solid chromatography, Sepn Sci, 6(1), 1-24 (1971).
Smuts, T.W.; Jordaan, J.T., and Pretorius, V., Phenomenological plate height equation for packed chromatographic columns, Sepn Sci, 6(5), 653-684 (1971).

Various, Gel permeation chromatography (topic issue), Sepn Sci, 6(1), 47-164; 6(2), 207-330 (1971).
Cloete, C.E., and de Clerk, K., Distillation vs. chromatography: Comparison based on purity index, Sepn Sci, 7(4), 449-456 (1972).
Buys, T.S., and de Clerk, K., Effect of temperature on production rate in chromatography, Sepn Sci, 8(5), 551-566 (1973).
Johnson, J.F.; Macphail, M.G.; Cooper, A.R., and Bruzzone, A.R., Effect of column length on chromatographic fractionation of polymers, Sepn Sci, 8(5), 577-584 (1973).
Martin, J.R., and Johnson, J.F., Cost-efficiency comparisons of some polymer chromatographic fractionation techniques, Sepn Sci, 8(5), 619-622 (1973).
Metzger, V.G.; Barford, R.A., and Rothbart, H.L., Chromatography and countercurrent distribution, Sepn Sci, 8(2), 143-160 (1973).
Ouano, A.C., and Barker, J.A., Computer simulation of linear gel permeation chromatography, Sepn Sci, 8(6), 673-700 (1973).
Stevens, B., Chromatographic refining unit, Proc Engng, March, 82-84 (1973).
Weiss, G.H., and Dishon, M., Resolution in nonuniform chromatographic systems, Sepn Sci, 8(3), 337-344 (1973).

1974-1978

Kirchner, J.G., Thin-layer chromatography, Chemtech, Feb, 79-82 (1974).
Nikelly, J.G., Porous-layer open-tubular gas chromatography columns, Sepn & Purif Methods, 3(2), 423-441 (1974).
Scott, C.D., High-pressure ion exchange chromatography applied to separation of complex biochemical mixtures, Sepn & Purif Methods, 3(2), 263-298 (1974).
Singhal, R.P., Separation and analysis of nucleic acids and their constituents by ion-exclusion and ion-exchange column chromatography, Sepn & Purif Methods, 3(2), 339-398 (1974).
Various, Chromatographic separations (topic issue), Sepn & Purif Methods, 3(1), 1-86, 133-244 (1974).
Whitlock, L.R., and Siggia, S., Fusion reaction gas chromatography, Sepn & Purif Methods, 3(2), 299-338 (1974).
Pawlowski, L., and Zytomirski, S., Influence of ion exchange capacity and total concentration of solution of ions of different valency on their chromatographic separation, Sepn Sci, 10(1), 33-38 (1975).
Rendell, M., Future of large-scale chromatography, Proc Engng, April, 66-70 (1975).
Cooper, A.R., and Lynn, T.R., Coiled high-efficiency liquid chromatography columns, Sepn Sci, 11(1), 39-44 (1976).
Ito, Y., and Bowman, R.L., Foam countercurrent chromatography, Sepn Sci, 11(3), 201-206 (1976).
Lee, H.L., and Lightfoot, E.N., Preliminary report on ultrafiltration-induced polarization chromatography: Analog of field-flow fractionation, Sepn Sci, 11(5), 417-440 (1976).
Sussman, M.V., Continuous chromatography (review paper), Chemtech, April, 260-264 (1976).
Talmon, Y., and Rubin, E., Chromatographic separation by foam, Sepn Sci, 11(6), 509-533 (1976).

Pauls, R.E., and Rogers, L.B., Comparisons of methods for calculating retention and separation of chromatographic peaks, Sepn Sci, 12(4), 395-415 (1977).

Pauls, R.E., et al., Experimental variables in recycle gas chromatography, Sepn Sci, 12(3), 289-306, (1977).

Umbreit, G.R., Chromatographic anomalies, Chemtech, Feb, 101-106 (1977).

Barker, P.E.; Ellison, F.J., and Hatt, B.W., Countercurrent chromatographic unit for continuous fractionation of dextran, Ind Eng Chem Proc Des Dev, 17(3), 302-309 (1978).

1980-1983

Curtis, M.A., et al., Liquid chromatographic fractionations of mixtures of polystyrene oligomers, Sepn Sci Technol, 15(7), 1413-1428 (1980).

Takahashi, T., and Gill, W.N., Hydrodynamic chromatography, Chem Eng Commns, 5(5), 367-380 (1980).

Various, Chromatographic processes (symposium papers), Sepn Sci Technol, 15(3), 587-696; 15(4), 697-798 (1980).

Annino, R.., Chromatographs can run on air, Chemtech, Aug, 482-487 (1981).

Barker, P.E., and Chuah, C.H., A sequential chromatographic process for the separation of glucose/fructose mixtures, Chem Engnr, Aug, 389-393 (1981).

Huang, J.C.; Forsythe, R., and Madey, R., Gas-solid chromatography of methane-helium mixtures: Transmission of step increase in concentration of methane through activated carbon adsorber bed at 25 degC, Sepn Sci Technol, 16(5), 475-486 (1981).

Moharir, A.S.; Saraf, D.N., and Kunzru, D., Effect of crystal size distribution on chromatographic peaks in molecular sieve columns, Chem Eng Commns, 11(6), 377-386 (1981).

Phillips, J.B.; Wright, N.A., and Burke, M.F., Probabilistic approach to digital simulation of chromatographic processes, Sepn Sci Technol, 16(8), 861-884 (1981).

Said, A.S., Theory of nonlinear chromatography, Sepn Sci Technol, 16(2), 113-134 (1981).

Various, Advances in chromatography (topic issue), Chem & Ind, 17 Oct, 710-732 (1981).

Huang, J.C., et al., Gas-solid chromatography of methane-helium mixtures: Moment analysis of breakthrough curves, Sepn Sci Technol, 17(12), 1417-1424 (1982).

Altshuller, D., Design equations and transient behaviour of the countercurrent moving-bed chromatographic reactor, Chem Eng Commns, 19(4), 363-376 (1983).

Barker, P.E.; England, K., and Vlachogiannis, G., Mathematical model for the fractionation of dextran on a semi-continuous countercurrent simulated moving bed chromatograph, CER&D, 61, 241-247 (1983).

Begovich, J.M.; Byers, C.H., and Sisson, W.G., A high-capacity pressurized continued chromatograph, Sepn Sci Technol, 18(12), 1167-1192 (1983).

Shih, C.K., et al., Large-scale liquid chromatography separation system, Chem Eng Prog, 79(10), 53-57 (1983).

1984-1987

Bailly, M., and Tondeur, D., Reversibility and performances in productive chromatography, Chem Eng & Proc, 18(6), 293-302 (1984).

Bonmati R., et al., Industrial gas chromatography process applied to essential oils, Sepn Sci Technol, 19(2), 113-156 (1984).

Miller, G.H., and Wankat, P.C., Moving port chromatography: A method of improving preparative chromatography, Chem Eng Commns, 31(1), 21-44 (1984).

Scott, F., Larger high-pressure liquid-chromatography systems, Proc Engng, Feb, 26-31 (1984).

Walton, H.F., Counter-ion effects in partition chromatography, Sepn Sci Technol, 19(11), 849-856 (1984).

Kamiyanagi, K., and Furusaki, S., Analysis of chromatography by transfer functions, Int Chem Eng, 25(2), 301-308 (1985).

Sommer, C.C., et al., Recycle gas chromatography using coarse packings, Sepn Sci Technol, 20(7), 523-540 (1985).

Ecknig, W., and Polster, H.J., Supercritical chromatography of paraffins on a molecular sieve: Analytical and preparative scale, Sepn Sci Technol, 21(2), 139-156 (1986).

Jun, S.H., and Ruckenstein, E., Separation of multicomponent mixture of proteins by potential barrier chromatography, Sepn Sci Technol, 21(2), 111-138 (1986).

Arve, B.H., and Liapis, A.I., Modeling and analysis of affinity chromatography in finite bath, AIChEJ, 33(2), 179-193 (1987).

Row, K.H., and Lee, W.K., Separation of close-boiling components using new chromatographic method, Sepn Sci Technol, 22(7), 1761-1778 (1987).

Smith, R.D., et al., Solubilities in supercritical fluids: Application of chromatographic measurement methods, Sepn Sci Technol, 22(2), 1065-1086 (1987).

Various, Preparative-scale chromatography (topic issue), Sepn Sci Technol, 22(8), 1791-2110 (1987).

1988

Barker, P.E., and Ganetsos, G., Chemical and biochemical separations using preparative and large-scale batch and continuous chromatography, Sepn & Purif Methods, 17(1), 1-66 (1988).

Forsythe, R., et al., Gas-solid chromatography: Longitudinal and intraparticle diffusion of acetylene in activated carbon, Sepn Sci Technol, 23(14), 2319-2328 (1988).

Howard, A.J.; Carta, G., and Byers, C.H., Separation of sugars by continuous annular chromatography, Ind Eng Chem Res, 27(10), 1873-1882 (1988).

Kawasaki, T., Specification of general theory of quasi-static linear gradient chromatography, Sepn Sci Technol, 23(14), 2365-2378 (1988).

Keum, D.K., and Lee, W.K., Simulation of moving feed port chromatography by rate model with mass transfer effect, Sepn Sci Technol, 23(14), 2349-2364 (1988).

Sanders, S.J., et al., Modeling the separation of amino acids by ion-exchange chromatography, Chem Eng Prog, 84(8), 47-54 (1988).

Sisson, W.G.; Begovich, J.M.; Byers, C.H., and Scott, C.D., Continuous chromatography, Chemtech, Aug, 498-502 (1988).
Takeda, K., et al., Equilibrium principle of displacement chromatography, Sepn Sci Technol, 23(14), 2329-2348 (1988).
Wankat, P.C., and Koo, Y.M., Scaling rules for isocratic elution chromatography, AIChEJ, 34(6), 1006-1019 (1988).
Ward, K.J.; Kaliaguine, S.C.; Tanguy, P.A., and Jean, G., Numerical simulation of a chromatograph column: Linear case, Ind Eng Chem Res, 27(8), 1474-1480 (1988).

23.9 Miscellaneous Separations

1969-1975

Slater, M.J., Review of continuous counter-current contactors for liquids and particulate solids, Brit Chem Eng, 14(1), 41-46 (1969).
Featherstone, W., and Cox, T., Separation of aqueous-organic mixtures by pervaporation, Brit Chem Eng, 16(9), 817-819 (1971).
Various, Electrophoresis techniques (topic issue), Sepn Sci, 7(6), 659-817 (1972).
Giddings, J.C., Parameters for optimum separations in field-flow fractionation, Sepn Sci, 8(5), 567-576 (1973).
Pugh, O., Desalination (a review), Proc Tech Int, Jan, 57-59 (1973).
Myers, M.N.; Caldwell, K.D., and Giddings, J.C., Study of retention in thermal field-flow fractionation, Sepn Sci, 9(1), 47-70 (1974).
Various, New methods of separation (symposium papers), Sepn Sci, 9(6), 491-562 (1974).
Giddings, J.C., et al., Nonequilibrium plate height for field-flow fractionation in ideal parallel plate columns, Sepn Sci, 10(4), 447-460 (1975).
Giddings, J.C.; Yang, F.J.F., and Myers, M.N., Application of sedimentation field-flow fractionation to biological particles, Sepn Sci, 10(2), 133-150 (1975).
Hur, Y.; Barrall, E.M., and Johnson, J.F., Preparatory column fractionation of polymers: Selection of solvent gradient, Sepn Sci, 10(6), 731-740 (1975).

1977-1988

Giddings, J.C.; Yang, F.J., and Myers, M.N., Criteria for concentration field-flow fractionation, Sepn Sci, 12(4), 381-394 (1977).
Krishnamurthy, S., and Subramanian, R.S., Exact analysis of field-flow fractionation, Sepn Sci, 12(4), 347-380 (1977).
Wakeham, W.A., and Mason, E.A., Diffusion through multiperforate laminae (a review), Ind Eng Chem Fund, 18(4), 301-305 (1979).
Various, Field-flow fractionation (topic issue), Sepn Sci Technol, 16(6), 549-744 (1981).
Lo, Y.S.; Gidaspow, D., and Wasan, D.T., Separation of colloidal particles from nonaqueous media by crossflow electrofiltration, Sepn Sci Technol, 18(12), 1323-1350 (1983).
Giddings, J.C., Review of field-flow fractionation, Sepn Sci Technol, 19(11), 831-848 (1984).

Naumann, R.J., and Rhodes, P.H., Thermal considerations in continuous flow electrophoresis, Sepn Sci Technol, 19(1), 51-76 (1984).
Thormann, W., Principles of isotachophoresis and dynamics of isotachophoretic separation of two components, Sepn Sci Technol, 19(8), 455-468 (1984).
Various, Field-flow fractionation (topic issue), Sepn Sci Technol, 19(10), 629-722 (1984).
Lock, J., Pervaporation using membranes, Processing, March, 17,19 (1986).
Rautenbach, R., and Albrecht, R., Pervaporation and gas permeation: Fundamentals of process design, Int Chem Eng, 27(1), 10-25 (1987).
Schwarzbach, J.; Nilles, M., and Schlunder, E.U., Microconvection in porous media during pervaporation of liquid mixture, Chem Eng & Proc, 22(3), 163-176 (1987).
Giddings, J.C., Field-flow fractionation, C&E News, 10 Oct, 34-45 (1988).
Gobie, W.A., and Ivory, C.F., Recycle continuous-flow electrophoresis: Zero-diffusion theory, AIChEJ, 34(3), 474-482 (1988).
Rolchigo, P.M., and Graves, D.J., Analytical and preparative electrophoresis in nonuniform electric field, AIChEJ, 34(3), 483-492 (1988).
Tomida, T., and McCoy, B.J., Separations in conventional, hyperlayer, and steric field-flow fractionation, AIChEJ, 34(2), 341-346 (1988).

CHAPTER 24

FLUIDIZATION

24.1 Theory and Equipment Design

1967

Akopyan, L.A., et al., Process driving force in fluidized-bed equipment, Int Chem Eng, 7(3), 418-421 (1967).

Anderson, T.B., and Jackson, R., Fluid-mechanical description of fluidized beds, Ind Eng Chem Fund, 6(4), 527-539 (1967).

Chidambaram, S., Nomogram for superficial gas rate at minimum fluidisation, Brit Chem Eng, 12(8), 1251 (1967).

Davies. L., and Richardson, J.F., Gas interchange between bubbles and continuous phase in a fluidised bed, Brit Chem Eng, 12(8), 1223-1226 (1967).

Fridland, M.I., Mass-transfer coefficients in fluidized systems, Int Chem Eng, 7(4), 598-604 (1967).

Halwagi, M.M., and Gomezplata, A., Investigation of solids distribution, mixing, and contacting characteristics of gas-solid fluidized beds, AIChEJ, 13(3), 503-512 (1967).

Kiselnikov, V.N., et al., Problem of electrostatic phenomena in fluidized bed, Int Chem Eng, 7(3), 428-432 (1967).

Knudsen, I.E., and Olsen, W.F., Direct indication of particle size in fluidized beds, Chem Eng, 10 April, 244-246 (1967).

Landrock, A.H., Coating of plastic resin particles in a fluidized bed, Chem Eng Prog, 63(2), 67-74 (1967).

Mireur, J.P., and Bischoff, K.B., Mixing and contacting models for fluidized beds, AIChEJ, 13(5), 839-845 (1967).

Singh, B., and Bhat, G.N., Minimum fluidising velocity for hematite-air systems, Brit Chem Eng, 12(2), 242 (1967).

Zinoveva, A.P., and Bashkirova, S.G., Efficiency of contacting between heterogeneous phases and degree of chemical conversion in fluidized bed, Int Chem Eng, 7(3), 381-385 (1967).

1968

Anderson, T.B., and Jackson, R., Fluid-mechanical description of fluidized beds, Ind Eng Chem Fund, 7(1), 12-21 (1968).

Botterill, J.S.M., Progress in fluidisation, Brit Chem Eng, 13(8), 1121-1126 (1968).

Capes, C.E., and McIlhinney, A.E., Pseudoparticulate expansion of screen-packed gas-fluidized beds, AIChEJ, 14(6), 917-922 (1968).

Carlos, C.R., and Richardson, J.F., Solids movement in liquid fluidised beds, Chem Eng Sci, 23(8), 813-832 (1968).

De Nevers, N., Bubble-driven fluid circulations, AIChEJ, 14(2), 222-226 (1968).

Gunn, D.J., Mixing in packed and fluidised beds, Chem Engnr, June, CE153-172 (1968).

Hovmand, S., and Davidson, J.F., Chemical conversion in a slugging fluidised bed, Trans IChemE, 46, T190-203 (1968).

Kozin, V.E., and Baskakov, A.P., Investigation of the grid zone of a fluidized bed above cap-type distributors, Int Chem Eng, 8(2), 257-260 (1968).

Kunii, D., and Levenspiel, O., Bubbling bed model for kinetic processes in fluidized beds, Ind Eng Chem Proc Des Dev, 7(4), 481-492 (1968).
Kunii, D., and Levenspiel, O., Bubbling fluidized-bed model, Ind Eng Chem Fund, 7(3), 446-452 (1968).
Latham, R.; Hamilton, C., and Potter, O.E., Backmixing and chemical reaction in fluidised beds, Brit Chem Eng, 13(5), 666-671 (1968).
Luss, D., and Amundson, N.R., Stability of batch catalytic fluidized beds, AIChEJ, 14(2), 211-221 (1968).
Nauman, E.B., and Collinge, C.N., Theory and measurement of contact time distributions in gas-fluidized beds, Chem Eng Sci, 23(11), 1309-1326 (1968).
Paul, R.J.A.; Al-Naimi, T.T., and Gupta, D.K.D., Stability of coupled fluid beds, Brit Chem Eng, 13(5), 683-684 (1968).
Sandblom, H., Pulse technique for investigating solids mixing in a fluidised bed, Brit Chem Eng, 13(5), 677-679 (1968).
Stewart, P.S.B., Isolated bubbles in fluidised beds: Theory and experiment, Trans IChemE, 46, T60-66 (1968).
Winter, O., Density and pressure fluctuations in gas fluidized beds, AIChEJ, 14(3), 426-434 (1968).
Woollard, I.N.M., and Potter, O.E., Solids mixing in fluidized beds, AIChEJ, 14(3), 388-391 (1968).

1969

Anderson, T.B., and Jackson, R., Fluid-mechanical description of fluidized beds, Ind Eng Chem Fund, 8(1), 137-144 (1969).
Godard, K., and Richardson, J.F., Correlation of data for minimum fluidising velocity and bed expansion in particulately fluidised systems, Chem Eng Sci, 24(2), 363-368 (1969).
Godard, K., and Richardson, J.F., Bubble velocities and bed expansions in freely bubbling fluidised beds, Chem Eng Sci, 24(4), 663-670 (1969).
Grace, J.R., and Harrison, D., Behaviour of freely bubbling fluidised beds, Chem Eng Sci, 24(3), 497-508 (1969).
Horsler, A.G.; Lacey, J.A., and Thompson, B.H., High-pressure fluidized beds, Chem Eng Prog, 65(10), 59-64 (1969).
Howard, A.J.; Morris, P.J.; Ward, K.A., and Pyle, D.L., The flow of particles from pressurised fluidised beds, Chem Engnr, Oct, CE364-366 (1969).
Lyall, E., Photography of bubbling fluidised beds, Brit Chem Eng, 14(4), 501-506 (1969).
Matsen, J.M.; Hovmand, S., and Davidson, J.F., Expansion of fluidized beds in slug flow, Chem Eng Sci, 24(12), 1743-1754 (1969).
Soroko, V.E.; Mikhalev, M.F., and Mukhlenov, I.P., Calculation of minimum height of space above the bed in fluidized-bed contact equipment, Int Chem Eng, 9(2), 280-281 (1969).

1970

Efremov, G.I., and Vakhrushev, I.A., Hydrodynamics of three-phase fluidized beds, Int Chem Eng, 10(1), 37-42 (1970).
Geldart, D., Size and frequency of bubbles in two- and three-dimensional gas-fluidized beds, Powder Tech, 4, 41-55 (1970).

Geldart, D., Design of fluidised bed chemical reactors, Chem Engnr, June, CE147-155 (1970).
Gelperin, N.I.; Ainshtein, V.G., and Lapshenkov, G.I., Efflux of solid particles of granular materials from fluidized beds, Int Chem Eng, 10(3), 401-403 (1970).
Gelperin, N.I.; Kvasha, V.B., and Komarov, A.S., Design of sectioned equipment with fluidized beds and rotating redistribution, Int Chem Eng, 10(2), 257-262 (1970).
Gliddon, B.J., and Cranfield, R.R., Gas particle heat-transfer coefficients in packed beds at Reynolds numbers between 2 and 100, Brit Chem Eng, 15(4), 481-482 (1970).
Kunz, R.G., Minimum fluidization velocity: Fluid cracking catalysts and spherical glass beads, Powder Tech, 4, 156-162 (1970).
Morina, I.M., et al., Solid-phase mixing in a sectioned fluidized-bed apparatus, Int Chem Eng, 10(1), 27-30 (1970).
Sycheva, T.N., and Donat, E.V., Effect of height and fractional composition of a fluidized bed on carry-over of granular material, Int Chem Eng, 10(2), 172-176 (1970).
Tone, S.; Suda, J., and Otake, T., Characteristics of flow of solids through a feed pipe in stirred-moving and fluidized beds, Int Chem Eng, 10(4), 659-666 (1970).
Vail, Y.K.; Manakov, N.K., and Marshilin, V.V., Gas content of three-phase fluidized beds, Int Chem Eng, 10(2), 244-248 (1970).
Vance, S.W., and Lang, R.W., Versatility of fluid-bed dryers, Chem Eng Prog, 66(7), 92-93 (1970).
Yoon, S.M., and Kunii, D., Gas flow and pressure drop through moving beds, Ind Eng Chem Proc Des Dev, 9(4), 559-565 (1970).

1971

Argyriou, D.T.; List, H.L., and Shinnar, R., Bubble growth by coalescence in gas fluidized beds, AIChEJ, 17(1), 122-130 (1971).
Bratu, E., and Jinescu, G.I., Effect of vertical vibrations on fluidisation, Brit Chem Eng, 16(8), 691-695 (1971).
Creasy, D.E., Gas fluidisation at pressures and temperatures above ambient, Brit Chem Eng, 16(7), 605-610 (1971).
Goodridge, F.; Holden, D.I.; Murray, H.D., and Plimley, R.F., Fluidized-bed electrodes, Trans IChemE, 49, 128-141 (1971).
Hovmand, S.; Freedman, W., and Davidson, J.F., Chemical reaction in a pilot-scale fluidized bed, Trans IChemE, 49, 149-162 (1971).
McGrath, L., and Streatfield, R.E., Bubbling in shallow gas-fluidized beds of large particles, Trans IChemE, 49, 70-79 (1971).
Merry, J.M.D., Penetration of a horizontal gas jet into a fluidized bed, Trans IChemE, 49, 189-195 (1971).
Rao, K.B., and Murti, P.S., Nomogram for total height of fluidised-bed reactors, Brit Chem Eng, 16(2), 162-163 (1971).
Reh, L., Fluidized bed processing, Chem Eng Prog, 67(2), 58-63 (1971).
Shine, N.B., Fluidized-bed reactor for hydrogen cyanide, Chem Eng Prog, 67(2), 52-57 (1971).

Zenz, F.A., Determine attrition in fluidized beds, Hyd Proc, 50(2), 103-105 (1971).

1972

Capes, C.E., and McIlhinney, A.E., Expansion of particulate mixtures in screen-packed gas-fluidized beds, Trans IChemE, 50, 1-5 (1972).

Cherepanov, G.P., Theory of fluidization, Ind Eng Chem Fund, 11(1), 9-19 (1972).

Clift, R., and Grace, J.R., The coalescence of bubble chains in fluidised beds, Trans IChemE, 50, 364-371 (1972).

Do, H.T.; Grace, J.R., and Clift, R., Particle ejection and entrainment from fluidised beds, Powder Tech, 6, 195-200 (1972).

Fryer, C., and Potter, O.E., Bubble size variation in two-phase models of fluidized bed reactors, Powder Tech, 6, 317-322 (1972).

Geldart, D., Effect of particle size and size distribution on behaviour of gas-fluidised beds, Powder Tech, 6, 201-215 (1972).

Geldart, D., and Cranfield, R.R., Gas fluidization of large particles, Chem Eng J, 3(2), 211-231 (1972).

Geldart, D., and Kelsey, J.R., Use of capacitance probes in gas fluidised beds, Powder Tech, 6, 45-50 (1972).

Ghosal, S.K., and Mukherjea, R.N., Momentum transfer in solid-liquid batch fluidised beds, Brit Chem Eng, 17(3), 248-250; 17(4), 341-342 (1972).

Letan, R., and Elgin, J.C., Fluid mixing in particulate fluidized beds, Chem Eng J, 3(2), 136-144 (1972).

Leung, L.S., Design of gas distributors and prediction of bubble size in large gas-solids fluidized beds, Powder Tech, 6, 189-193 (1972).

Page, R.E., and Harrison, D., Size distribution of gas bubbles leaving a three-phase fluidized bed, Powder Tech, 6, 245-249 (1972).

Rajeevalochanan, V., and Ibrahim, S.H., Nomogram for minimum fluidization velocity, Brit Chem Eng, 17(9), 723 (1972).

Ritzmann, H.; Hoffmann, G., and Schugerl, K., Mixing processes in liquid fluidised beds, Powder Tech, 6, 225-230 (1972).

Rowe, P.N., and Everett, D.J., Fluidised bed bubbles viewed by X-rays, Trans IChemE, 50, 42-60 (1972).

Rowe, P.N.; Nienow, A.W., and Agbim, A.J., The mechanisms by which particles segregate in gas fluidised beds, Trans IChemE, 50, 310-333 (1972).

Singh, B.; Fryer, C., and Potter, O.E., Solids motion caused by a bubble in a fluidized bed, Powder Tech, 6, 239-244 (1972).

Strijbos, S., Motion and distribution of large particles suspended in a fluidized bed, Powder Tech, 6, 337-342 (1972).

Zenz, F.A., and Smith, R., Fines in equilibrium in fluidised beds, Hyd Proc, 51(2), 104-106 (1972).

Ziegler, E.N., Interstitial fluid-bed reactor, Chemtech, Nov, 690-694 (1972).

1973

Alfredson, P.G., and Doig, I.D., Behaviour of pulsed fluidised beds, Trans IChemE, 51, 232-246 (1973).

Chatlynne, C.J., and Resnick, W., Determination of flow patterns for unsteady-state flow of granular materials, Powder Tech, 8, 177-182 (1973).
Cheremisinoff, P.N., and Rao, B., Fluid-bed free-board design, Proc Tech Int, March, 121-124 (1973).
Doheim, M.A., Mathematical model for iron ore reduction in a continuous multi-stage fluidised system, J Appl Chem Biotechnol, 23, 375-388 (1973).
Geldart, D., Types of gas fluidization, Powder Tech, 7, 285-292 (1973).
Ghar, R.N.; Kumar, A., and Gupta, P.S., Quick method for degree and time of fluid-bed mixing, Proc Engng, March, 78-79 (1973).
Heertjes, P.M., and Jessurun, R.M.M., Nickel and cobalt from iron-laterites using a fluidised bed, Trans IChemE, 51, 293-301 (1973).
Manieh, A.A., Sensing the height of a gas-fluidized bed, Chem Eng, 6 Aug, 112 (1973).
Merry, J.M.D., and Davidson, J.F., 'Gulf stream' circulation in shallow fluidised beds, Trans IChemE, 51, 361-368 (1973).
Nienow, A.W.; Rowe, P.N., and Agbim, A.J., Liquid-like properties of gas fluidised beds, Trans IChemE, 51, 260-264 (1973).
Singh, B.; Rigby, G.R., and Callcott, T.G., Measurement of minimum fluidisation velocities at elevated temperatures, Trans IChemE, 51, 93-96 (1973).
Takahashi, H., and Yanai, H., Flow profile and void fraction of granular solids in a moving bed, Powder Tech, 7, 205-214 (1973).
Verloop, J., and Heertjes, P.M., Onset of fluidization, Powder Tech, 7, 161-168 (1973).

1974

Abrahami, S.N., and Resnick, W., Fluidised bed behaviour near incipient fluidisation in a three-dimensional bed, Trans IChemE, 52, 80-87 (1974).
Baeyens, J., and Geldart, D., Investigation of slugging fluidized beds, Chem Eng Sci, 29(1), 255-266 (1974).
Broughton, J., The influence of bed temperature and particle size distribution on incipient fluidisation behaviour, Trans IChemE, 52, 105-107 (1974).
Cranfield, R.R., and Geldart, D., Large particle fluidization, Chem Eng Sci, 29(4), 935-948 (1974).
Darton, R.C., and Harrison, D., The rise of single gas bubbles in liquid fluidised beds, Trans IChemE, 52, 301-306 (1974).
De Jong, J.A.H., and Nomden, J.F., Homogeneous gas-solid fluidization, Powder Tech, 9, 91-97 (1974).
Fuchs, W., and Yavorsky, P.M., New method for measuring bed levels, Chem Eng, 24 June, 145-148 (1974).
Gibilaro, L.G., and Rowe, P.N., Model for a segregating gas fluidized bed, Chem Eng Sci, 29(6), 1403-1412 (1974).
Grace, J.R., and Clift, R., Two-phase theory of fluidization, Chem Eng Sci, 29(2), 327-334 (1974).
Harrison, D.; Aspinall, P.N., and Elder, J., Suppression of particle elutriation from a fluidised bed, Trans IChemE, 52, 213-216 (1974).
Heertjes, P.M., and Jessurun, R.M.M., Influence of size of granules of a lateritic ore in a fluidised bed on the rate of conversion of nickel and cobalt, Trans IChemE, 52, 53-57 (1974).

Huthwaite, J.A., Examples of the use and specification of fluid bed dryers, Chem Engnr, May, 295-297, 303 (1974).
Martinola, F., The Lewatit fluidised bed process, Chem Engnr, Jan, 25-27, 34 (1974).
Medlin, J.; Wong, H.W., and Jackson, R., Fluid-mechanical description of fluidized beds, Ind Eng Chem Fund, 13(3), 247-259 (1974).
Naveh, E., and Resnick, W., Particle size segregation in baffled fluidised beds, Trans IChemE, 52, 58-66 (1974).
Nishinaka, M.; Morooka, S., and Kato, Y., Longitudinal dispersion of solid particles in fluid bed with horizontal baffles, Powder Tech, 9, 1-6 (1974).
Otero, A.R., and Munoz, R.C., Fluidized-bed gas distributors of bubble-cap type, Powder Tech, 9, 279-286 (1974).
Rooney, N.M., and Harrison, D., Spouted beds of fine particles, Powder Tech, 9, 227-230 (1974).
Werther, J., Bubbles in gas fluidised beds, Trans IChemE, 52, 149-169 (1974).

1975
Barnea, E., and Mednick, R.L., Correlation for minimum fluidisation velocity, Trans IChemE, 53, 278-281 (1975).
Baskakov, A.P.; Malykh, G.A., and Shishko, I.I., Separation of materials in equipment with fluidized bed and continuous charging and discharging, Int Chem Eng, 15(2), 286-289 (1975).
Beran, Z., and Lutcha, J., Optimising particle residence time in a fluidised bed drier, Chem Engnr, Nov, 678-681 (1975).
Broadhurst, T.E., and Becker, H.A., Onset of fluidization and slugging in beds of uniform particles, AIChEJ, 21(2), 238-247 (1975).
Damronglerd, S.; Couderc, J.P., and Angelino, H., Mass transfer in particulate fluidisation, Trans IChemE, 53, 175-180 (1975).
Darton, R.C., and Harrison, D., Gas and liquid hold-up in three-phase fluidisation, Chem Eng Sci, 30(5), 581-586 (1975).
Kmiec, A., Modelling of solid-liquid fluidized beds, Chem Eng J, 9(3), 251-254 (1975).
Mori, S., and Wen, C.Y., Estimation of bubble diameter in gaseous fluidized beds, AIChEJ, 21(1), 109-115 (1975).
Nguyen, X.T., and Bergougnou, M.A., Approximate correlation for fluidized bed density within the slug flow regime, Can JCE, 53, 102-110 (1975).
Rowe, P.N., and Yacono, C., The distribution of bubble size in gas fluidised beds, Trans IChemE, 53, 59-60 (1975).
Roy, G.K., and Gupta, P.S., Semi-fluidization design study, Processing, Jan, 10-11 (1975).
Slater, M.J., Assessment of fluidised-bed ion-exchange equipment, J Appl Chem Biotechnol, 25, 367-378 (1975).
Varma, Y.B.G., Pressure drop of the fluid and flow patterns of the phases in multistage fluidisation, Powder Tech, 12, 167-174 (1975).
Walker, B.V., The effective rate of gas exchange in a bubbling fluidised bed, Trans IChemE, 53, 255-266 (1975).
Yates, J.G., Fluidised bed reactors, Chem Engnr, Nov, 671-677 (1975).
Yuan, E.K.C.; Pulsifer, A.H., and Wheelock, T.D., Arcing at electrode surfaces in fluidised beds, Trans IChemE, 53, 41-43 (1975).

1976

Bhattacharya, S.C., and Harrison, D., Heat transfer in a pulsed fluidised bed, Trans IChemE, 54, 281-286 (1976).

Brea, F.M.; Edwards, M.F., and Wilkinson, W.L., Flow of non-Newtonian slurries through fixed and fluidised beds, Chem Eng Sci, 31(5), 329-336 (1976).

Fryer, C., and Potter, O.E., Experimental investigation of models for fluidized-bed catalytic reactors, AIChEJ, 22(1), 38-47 (1976).

Germain, B., and Claudel, B., Fluidization at mean pressures less than 30 torr, Powder Tech, 13, 115-121 (1976).

Haribabu, P.; Sarkar, M.K., and Subba-Rao, D., Formation of bubbles in gas fluidized beds, Can JCE, 54, 451-460 (1976).

Hills, J.H., and Darton, R.C., The rising velocity of a large bubble in a bubble swarm, Trans IChemE, 54, 258-264 (1976).

La Nauze, R.D., Circulating fluidised bed, Powder Tech, 15, 117-127 (1976).

Merry, J.M.D., Fluid and particle entrainment into vertical jets in fluidized beds, AIChEJ, 22(2), 315-323 (1976).

Rowe, P.N., Prediction of bubble size in a gas fluidised bed, Chem Eng Sci, 31(4), 285-288 (1976).

Rowe, P.N., and Nienow, A.W., Particle mixing and segregation in gas fluidised beds: A review, Powder Tech, 15, 141-147 (1976).

Werther, J., Convective solids transport in large diameter gas-fluidized beds, Powder Tech, 15, 155-167 (1976).

Whitehead, A.B., et al., Fluidization studies in large gas-solid systems, Powder Tech, 14, 61-70 (1976); 15, 77-87 (1976).

Yerushalmi, J.; Turner, D.H., and Squires, A.M., The fast fluidized bed, Ind Eng Chem Proc Des Dev, 15(1), 47-53 (1976).

1977

Beeckmans, J.M., and Minh, T., Separation of mixed granular solids using fluidized countercurrent cascade principle, Can JCE, 55, 493-505 (1977).

Coppus, J.H.C.; Rietema, K., and Ottengraf, S.P.P., Wake phenomena behind spherical-cap bubbles and solid spherical-cap bodies, Trans IChemE, 55, 122-129 (1977).

Darton, R.C.; La Nauze, R.D.; Davidson, J.F., and Harrison, D., Bubble growth due to coalescence in fluidised beds, Trans IChemE, 55, 274-280 (1977).

Denloye, A.O.O., and Botterill, J.S.M., Heat transfer in flowing packed beds, Chem Eng Sci, 32(5), 461-466 (1977).

Dwivedi, P.N., and Upadhyay, S.N., Particle-fluid mass transfer in fixed and fluidized beds, Ind Eng Chem Proc Des Dev, 16(2), 157-165 (1977).

Fouda, A.E., and Capes, C.E., Hydrodynamic particle volume and fluidized bed expansion, Can JCE, 55, 386-395 (1977).

Garside, J., and Al-Dibouni, M.R., Velocity-voidage relationships for fluidization and sedimentation in solid-liquid systems, Ind Eng Chem Proc Des Dev, 16(2), 206-214 (1977).

Hengl, G.; Hiquily, N., and Couderc, J.P., New distributor for gas fluidization, Powder Tech, 18, 277-278 (1977).

Hsiung, T.H., and Thodos, G., Mass transfer in gas-fluidized beds: Measurement of actual driving forces, Chem Eng Sci, 32(6), 581-592 (1977).
Hsiung, T.H., and Thodos, G., Expansion characteristics of gas-fluidized beds, Can JCE, 55, 221-230 (1977).
Nguyen, H.V.; Whitehead, A.B., and Potter, O.E., Gas backmixing, solids movement, and bubble activities in large scale fluidized beds, AIChEJ, 23(6), 913-922 (1977).
Patel, R.D., and Simpson, J.M., Heat transfer in aggregative and particulate liquid-fluidized beds, Chem Eng Sci, 32(1), 67-74 (1977).
Rigby, G.R.; Callcott, T.G.; Singh, B., and Evans, B.R., New distributor for gas fluidised beds, Trans IChemE, 55, 68-70 (1977).
Saxena, S.C., and Vogel, G.J., The measurement of incipient fluidisation velocities in a bed of coarse dolomite at temperature and pressure, Trans IChemE, 55, 184-189 (1977).
Strumillo, C., and Kudra, T., Interfacial area in three-phase fluidized beds, Chem Eng Sci, 32(2), 229-232 (1977).
Thiel, W.J., and Potter, O.E., Slugging in fluidized beds, Ind Eng Chem Fund, 16(2), 242-247 (1977).
Wallis, G.B., Simple correlation for fluidisation and sedimentation, Trans IChemE, 55, 74-75 (1977).
Whitehead, A.B.; Dent, D.C., and McAdam, J.C.H., Fluidization studies in large gas-solid systems, Powder Tech, 18, 231-237 (1977).
Yates, J.G., and Rowe, P.N., A model for chemical reaction in the freeboard region above a fluidised bed, Trans IChemE, 55, 137-142 (1977).
Zanker, A., Nomograph for fluidization velocity, Proc Engng, March, 92-93 (1977).
Zanker, A., Nomograph for fluidization velocities of solid mixtures, Proc Engng, Sept, 132-133 (1977).
Zenz, F.A., Effect of flow phenomena on fluidized bed design, Chem Eng, 19 Dec, 81-91 (1977).

1978

Atimtay, A., and Cakaloz, T., Investigation of gas mixing in fluidized beds, Powder Tech, 20, 1-7 (1978).
Baker, C.G.J., and Geldart, D., Investigation of slugging characteristics of large particles, Powder Tech, 19, 177-187 (1978).
Baker, C.G.J.; Armstrong, E.R., and Bergougnou, M.A., Heat transfer in three-phase fluidized beds, Powder Tech, 21, 195-204 (1978).
Behie, L.A.; Voegelin, B.E., and Bergougnou, M.A., Design of fluid bed grids using orifice equation, Can JCE, 56, 404-410 (1978).
Botterill, J.S.M., and Denloye, A.O.O., Theoretical model of heat transfer to a packed or quiescent fluidized bed, Chem Eng Sci, 33(4), 509-516 (1978).
Crooks, M.J., and Schade, H.W., Fluidized bed granulation of a microdose pharmaceutical powder, Powder Tech, 19, 103-108 (1978).
Doganoglu, Y.; Jog, V.; Thambimuthu, K.V., and Clift, R., Removal of fine particulates from gases in fluidised beds, Trans IChemE, 56, 239-248 (1978).
Eleftheriades, C.M., and Judd, M.R., Design of downcomers joining gas-fluidized beds in multistage systems, Powder Tech, 21, 217-225 (1978).

Goedecke, R.; Schugerl, K., and Todt, J., Influence of the sorption process on gas residence times distribution in bench scale fluidized beds, Powder Tech, 21, 227-244 (1978).

Kmiec, A., Particle distributions and dynamics of particle movement in solid-liquid fluidized beds, Chem Eng J, 15(1), 1-12 (1978).

Lehmann, J., and Schugerl, K., Investigation of gas mixing and gas distributor performance in fluidized beds, Chem Eng J, 15(2), 91-110 (1978).

Masson, H., Solid-circulation studies in a gas-solid fluid bed, Chem Eng Sci, 33(5), 621-623 (1978).

Muzyka, D.; Beeckmans, J.M., and Jeffs, A., Solids separation in countercurrent fluidized cascade: Jetsam-rich mixtures at total reflux, Can JCE, 56, 286-295 (1978).

Rowe, P.N., Design considerations for a fluidised-bed chemical reactor, Chem & Ind, 17 June, 424-426 (1978).

Sathiyamoorthy, D., and Rao, C.S., Gas distributors in fluidised beds, Powder Tech, 20, 47-52 (1978).

Sen Gupta, P., and Vaid, R.P., Minimum fluidization velocities in beds of mixed solids, Can JCE, 56, 292-300 (1978).

Singh, B., Theory of slugging lifters, Powder Tech, 21, 81-89 (1978).

Tanaka, I., and Shinohara, H., Estimation of column height of fluidized bed, Int Chem Eng, 18(2), 276-279 (1978).

Thiel, W.J., and Potter, O.E., Mixing of solids in slugging gas fluidized beds, AIChEJ, 24(4), 561-569 (1978).

Vaux, W.G., and Newby, R.A., Wear on tubes by jet impingement in a fluidized bed, Powder Tech, 19, 79-88 (1978).

Xavier, A.M.; Lewis, D.A., and Davidson, J.F., The expansion of bubbling fluidised beds, Trans IChemE, 56, 274-280 (1978).

Yerushalmi, J., and Cankurt, N.T., High-velocity fluid beds, Chemtech, Sept, 564-572 (1978).

1979

Al-Dibouni, M.R., and Garside, J., Particle mixing and classification in liquid fluidised beds, Trans IChemE, 57, 94-103 (1979).

Allahwala, S.A., and Potter, O.E., Rise velocity equation for isolated bubbles and for isolated slugs in fluidized beds, Ind Eng Chem Fund, 18(2), 112-116 (1979).

Burli, A.B.; Senthilnathan, P.R., and Subramanian, N., Axial dispersion of liquids in fluidized beds: Effect of internals, Can JCE, 57, 648-658 (1979).

Chen, T.Y.; Walawender, W.P., and Fan, L.T., Moving-bed solids flow between two fluidized beds, Powder Tech, 22, 89-96 (1979).

Christiansen, O.B., Polymer drying and cooling systems, Chem Eng Prog, 75(11), 58-64 (1979).

Darton, R.C., A bubble growth theory of fluidised bed reactors, Trans IChemE, 57, 134-138 (1979).

El-Temtamy, S.A., and Epstein, N., Contraction or expansion of three-phase fluidized beds containing fine/light solids, Can JCE, 57, 520-530 (1979).

El-Temtamy, S.A.; El-Sharnoubi, Y.O., and El-Halwagi, M.M., Liquid dispersion in gas-liquid fluidized beds, Chem Eng J, 18(2), 151-168 (1979).

Fan, L.T., and Chang, Y., Mixing of large particles in two-dimensional gas fluidized beds, Can JCE, 57, 88-100 (1979).
Fan, L.T., and Fan, L.S., Simulation of catalytic fluidized bed reactors, Chem Eng Sci, 34(2), 171-180 (1979).
Fan, L.T.; Too, J.R.; Lai, F.S., and Akao, Y., Studies on multicomponent solids mixing and mixtures, Powder Tech, 22, 205-213; 23, 99-113; 24, 73-89 (1979).
Fieldes, R.B.; Burdett, N.A., and Davidson, J.F., Reaction of sulphur dioxide with limestone particles in a fluidised bed, Trans IChemE, 57, 276-280 (1979).
Ganapathy, V., Nomograph for fluidized bed variables, Proc Engng, Oct, 110-111 (1979).
Ganapathy, V., Estimating superficial-velocity boundaries in fluidized beds, Chem Eng, 7 May, 116-117 (1979).
Geldart, D.; Cullinan, J.; Georghiades, S.; Gilvray, D., and Pope, D.J., Effect of fines on entrainment from gas fluidised beds, Trans IChemE, 57, 269-275 (1979).
Grace, J.R., and Tuot, J., A theory for cluster formation in vertically conveyed suspensions of intermediate density, Trans IChemE, 57, 49-54 (1979).
Jodra, L.G.; Aragon, J.G., and Corella, J., Fluidized beds with internal screens, Int Chem Eng, 19(4), 654-671 (1979).
Qureshi, A.E., and Creasy, D.E., Fluidised-bed gas distributors, Powder Tech, 22, 113-119 (1979).
Rowe, P.N.; MacGillivray, H.J., and Cheesman, D.J., Gas discharge from an orifice into a gas fluidised bed, Trans IChemE, 57, 194-199 (1979).
Sathiyamoorthy, D., amd Rao, C.S., Multi-orifice plate distributors in gas fluidized beds: Model for distributor design, Powder Tech, 24, 215-223 (1979).
Saxena, S.C.; Chatterjee, A., and Patel, R.C., Effect of distributors on gas-solid fluidization, Powder Tech, 22, 191-198 (1979).
Vrbata, O., and Schugerl, K., Gas residence time distributions in single and multistage fluidized beds with porous particles , Powder Tech, 22, 179-185; 24, 207-214 (1979).
Wood, R.M., The significance of particles entrained in bubbles on the products of fluidised bed gasification, Trans IChemE, 57, 213-214 (1979).
Yacono, C.; Rowe, P.N., and Angelino, H., Analysis of flow distribution between phases in a gas-fluidized bed, Chem Eng Sci, 34(6), 789-800 (1979).
Yerushalmi, J., and Cankurt, N.T., Further studies of the regimes of fluidization, Powder Tech, 24, 187-205 (1979).

1980

Abrahamsen, A.R., and Geldart, D., Behaviour of gas-fluidized beds of fine powders, Powder Tech, 26, 35-65 (1980).
Allen, M.L., The effect of particle proximity on the drag force generated at high Reynolds numbers, Trans IChemE, 58, 187-194 (1980).
Doheim, M.A., and Collinge, C.N., Contact time distribution in fluidized-bed reactors, Chem Eng J, 19(1), 39-56 (1980).

Fakhimi, S., and Harrison, D., The voidage fraction near a horizontal tube immersed in a fluidised bed, Trans IChemE, 58, 125-131 (1980).

Ganguly, U.P., Direct method for prediction of expanded bed height in liquid-solid fluidization, Can JCE, 58, 559-563 (1980).

Grewal, N.S., and Saxena, S.C., Comparison of commonly used correlations for minimum fluidization velocity of small solid particles, Powder Tech, 26, 229-234 (1980).

Irani, R.K.; Kulkarni, B.D., and Doraiswamy, L.K., Analysis of complex reaction schemes in a fluidized bed: Application of the Kunii-Levenspiel model, Ind Eng Chem Proc Des Dev, 19(1), 24-30 (1980).

Ishida, M.; Shirai, T., and Nishiwaki, A., Measurement of velocity and direction of flow of solid particles in a fluidized bed, Powder Tech, 27, 1-6 (1980).

Kitic, D., and Brea, F.M., Influence of moisture on fluidising velocities in batch fluid bed drying, Trans IChemE, 58, 208-210 (1980).

Knight, M.J.; Rowe, P.N.; MacGillivray, H.J., and Cheesman, D.J., On measuring the density of finely divided porous particles such as fluid bed cracking catalyst, Trans IChemE, 58, 203-207 (1980).

Kono, H., A new concept for three-phase fluidized beds, Hyd Proc, 59(1), 123-129 (1980).

Kuhne, J., and Wippern, D., Application of linear three-phase model to a fluidized-bed reactor, Can JCE, 58, 527-530 (1980).

McKay, G., and McLain, H., The fluidisation of cuboid particles, Trans IChemE, 58, 107-115 (1980).

Molerus, O., Coherent representation of pressure drop in fixed beds and of bed expansion for particulate fluidized beds, Chem Eng Sci, 35(6), 1331-1340 (1980).

Morooka, S.; Kawazuishi, K., and Kato, Y., Holdup and flow pattern of solid particles in freeboard of gas-solid fluidized bed with fine particles, Powder Tech, 26, 75-82 (1980).

Nienow, A.W., and Naimer, N.S., Continuous mixing of two particulate species of different density in a gas fluidised bed, Trans IChemE, 58, 181-186 (1980).

Parulekar, S.J., and Shah, Y.T., Steady-state behavior of gas-liquid-solid fluidized-bed reactors, Chem Eng J, 20(1), 21-34 (1980).

Singh, A.N., and Sen Gupta, P., Expansion behaviour of liquid-solid fluidized bed systems, Can JCE, 58, 116-119 (1980).

Singh, A.N.; Kesavan, S., and Sen Gupta, P., Rate of mass transfer in a semi-fluidized bed, Int J Heat Mass Trans, 23(3), 279-282 (1980).

1981

Allahwala, S.A.; Singh, B., and Potter, O.E., Bubble frequency and distribution in fluidized beds, Chem Eng Commns, 11(4), 255-280 (1981).

Alvarez-Cuenca, M., and Nerenberg, M.A., Plug-flow model for mass transfer in three-phase fluidized beds and bubble columns, Can JCE, 59, 739-745 (1981).

Cooper, P.F., The use of biological fluidised beds for treatment of domestic and industrial wastewaters, Chem Engnr, Aug, 373-376 (1981).

Epstein, N., Three-phase fluidization: Some knowledge gaps (review paper), Can JCE, 59, 649-657 (1981).

Geldart, D.; Baeyens, J.; Pope, D.J., and Van de Wijer, P., Segregation in beds of large particles at high velocities, Powder Tech, 30, 195-205 (1981).

Joshi, J.B., Axial mixing in multiphase contactors: A unified correlation, Trans IChemE, 58, 155-165 (1980); 59, 138-143 (1981).

King, D.F.; Mitchell, F.R.G., and Harrison, D., Dense phase viscosities of fluidised beds at elevated pressures, Powder Tech, 28, 55-58 (1981).

Li, Y., et al., Rapid fluidization, Int Chem Eng, 21(4), 670-679 (1981).

Martin, B.L.A.; Kolar, Z., and Wesselingh, J.A., The falling velocity of a sphere in a swarm of different spheres, Trans IChemE, 59, 100-104 (1981).

Pandey, D.K.; Upadhyay, S.N., and Kumar, S., Mass transfer between particles and fluid in fluidized beds of large particles, Int J Heat Mass Trans, 24(7), 1221-1228 (1981).

Rowe, P.N., and Masson, H., Interaction of bubbles with probes in gas fluidised beds, Trans IChemE, 59, 177-187 (1981).

Sathiyamoorthy, D., and Rao, C.S., Choice of distributor to bed pressure drop ratio in gas fluidised beds, Powder Tech, 30, 139-143 (1981).

Shieh, W.K.; Mulcahy, L.T., and LaMotta, E.J., Fluidised bed biofilm reactor effectiveness factor expressions, Trans IChemE, 59, 129-133 (1981).

Sitnal, O., Solids mixing in a fluidized bed with horizontal tubes, Ind Eng Chem Proc Des Dev, 20(3), 533-538 (1981).

Vaid, R.P., and Sen Gupta, P., Velocities of entrainment in liquid fluidized beds of mixed solids, Can JCE, 59, 35-41 (1981).

1982

Avidan, A.A., and Yerushalmi, J., Bed expansion in high velocity fluidization, Powder Tech, 32, 223-232 (1982).

Balsam, G., and Staffin, H.K., Fluidized-bed cleaning, Chem Eng Prog, 78(7), 81-83 (1982).

Bauer, W., and Werther, J., Role of gas distribution in fluidized-bed chemical-reactor design, Chem Eng Commns, 18(1), 137-148 (1982).

Botterill, J.S.M.; Teoman, Y., and Yuregir, K.R., Effect of operating temperature on velocity of minimum fluidization, bed voidage and general behaviour, Powder Tech, 31, 101-110 (1982).

Botterill, J.S.M.; Teoman, Y., and Yuregir, K.R., Effect of temperature on fluidized bed behaviour, Chem Eng Commns, 15(1), 227-238 (1982).

Bull, M.A.; Sterritt, R.M., and Lester, J.N., The effect of organic loading on the performance of anaerobic fluidised beds treating high strength wastewaters, Trans IChemE, 60, 373-376 (1982).

Burgess, J.M.; Fane, A.G., and Fell, C.J.D., Application of an electroresistivity probe technique to a two-dimensional fluidised bed, Trans IChemE, 60, 249-252, 383-384 (1982).

Corella, J., and Bilbao, R., Fluid dynamic study of a new type of solid-gas contactor: The fluidized/fixed or fluidized bed, Ind Eng Chem Proc Des Dev, 21(4), 545-550 (1982).

Dencs, B., and Ormos, Z., Particle formation from solution in a gas fluidized bed, Powder Tech, 31, 85-99 (1982).

Fasso, L.; Chao, B.T., and Soo, S.L., Measurement of electrostatic charges and concentration of particles in the freeboard of a fluidized bed, Powder Tech, 33, 211-221 (1982).

Ganguly, U.P., Elutriation characteristics of solids from liquid fluidized bed systems, Can JCE, 60, 466-474 (1982).
Geldart, D., Survey of current world-wide research in gas fluidization (May 1981), Powder Tech, 31, 1-25 (1982).
King, D.F., and Harrison, D., The dense phase of a fluidised bed at elevated temperatures, Trans IChemE, 60, 26-30 (1982).
Krishnaiah, K., and Varma, Y.B.G., Pressure drop, solids concentration and mean holding time in multistage fluidisation, Can JCE, 60, 346-352 (1982).
Nielson, R.H.; Harnby, N, and Wheelock, T.D., Mixing and circulation in fluidized beds of flour, Powder Tech, 32, 71-86 (1982).
Pechey, R., Applications of fluidised beds, Proc Engng, Jan, 38-41 (1982).
Rao, K.V.K., and Prakash, S.G., Some aspects of liquid-solid fluidized beds, Can JCE, 60, 859-862 (1982).
Sobreiro, L.E.L., and Monteiro, J.L.F., Effect of pressure on fluidized bed behaviour, Powder Tech, 33, 95-100 (1982).
Tatterson, G.B., Effect of draft tubes on circulation and mixing times, Chem Eng Commns, 19(1), 141-148 (1982).
Yue, P.L., and Kolaczkowski, J.A., Multiorifice distributor design for fluidised beds, Trans IChemE, 60, 164-170 (1982).
Zenz, F.A., Scaleup fluid bed reactors, Hyd Proc, Jan, 155-156 (1982).

1983

Casal, J., and Puigjaner, L., Segregation and apparent minimum fluidization velocity in particulate fluidization, Chem Eng Commns, 23(1), 125-136 (1983).
Geldart, D., Survey of current world-wide research in gas fluidisation (Jan 1982-May 1983), Powder Tech, 36, 149-180 (1983).
Geldart, D., and Haesebrouck, M., Studies on the intermittent discharge of coarse solids from fluidised beds, CER&D, 61, 224-232 (1983).
Gidaspow, D.; Seo, Y.C., and Ettehadieh, B., Hydrodynamics of fluidization: Experimental and theoretical bubble sizes in a two-dimensional bed with a jet, Chem Eng Commns, 22(5), 253-272 (1983).
Glass, D.H., and Mojtahedi, W., Measurement of fluidised-bed bubbling properties using a fibre-optic light probe, CER&D, 61, 37-44 (1983).
Ho, T.C.; Yutani, N.; Fan, L.T., and Walawender, W.P., Onset of slugging in gas-solid fluidized beds with large particles, Powder Tech, 35, 249-257 (1983).
Juma, A.K.A., and Richardson, J.F., Segregation and mixing in liquid fluidized beds, Chem Eng Sci, 38(6), 955-968 (1983).
Martin, P.D., On the 'particulate' and 'delayed bubbling' regimes in fluidisation, CER&D, 61, 318-320 (1983).
Martin, P.D., and Davidson, J.F., Flow of powder through an orifice from a fluidised bed, CER&D, 61, 162-166 (1983).
McKay, G., Adsorption of acidic and basic dyes onto activated carbon in fluidised beds, CER&D, 61, 29-36 (1983).
Pillay, P.S., and Varma, Y.B.G., Pressure drop and solids holding time in multistage fluidisation, Powder Tech, 35, 223-231 (1983).

Viswanathan, K.; Khakhar, D., and Rao, D.S., Fluidized-bed adsorber modelling and experimental study, Chem Eng Commns, 20(1), 235-252 (1983).
Webb, C.; Black, G.M., and Atkinson, B., Liquid fluidisation of highly porous particles, CER&D, 61, 125-134 (1983).
Weimer, A.W., and Clough, D.E., Rise velocity of slugs in fluidized beds, Chem Eng Commns, 21(1), 175-182 (1983).

1984

Chen, P., and Pei, D.C.T., Fluidization characteristics of fine particles, Can JCE, 62(4), 464-468 (1984).
Chitester, D.C.; Kornosky, R.M.; Fan, L.S., and Danko, J.P., Characteristics of fluidization at high pressure, Chem Eng Sci, 39(2), 253-262 (1984).
Colakyan, M., and Levenspiel, O., Elutriation from fluidized beds, Powder Tech, 38, 223-232 (1984).
Dry, R.J.; Judd, M.R., and Shingles, T., Bubble velocities in fluidized beds of fine, dense powders, Powder Tech, 39, 69-75 (1984).
Geldart, D.; Harnby, N., and Wong, A.C., Fluidization of cohesive powders, Powder Tech, 37, 25-38 (1984).
Glicksman, L.R., Scaling relationships for fluidized beds, Chem Eng Sci, 39(9), 1373-1380 (1984).
Joshi, J.B., Solid-liquid fluidised beds: Some design aspects (review paper), CER&D, 61, 143-161 (1983); 62, 190, 271-272, 398 (1984).
Patwardhan, V.S., and Tien, C., Distribution of solid particles in liquid fluidized beds, Can JCE, 62(1), 46-54 (1984).
Raufast, C.R., Process details of BP Chimie's LLDPE fluidised beds, Hyd Proc, May, 105-108 (1984).
Rowe, P.N., Effect of pressure on minimum fluidisation velocity, Chem Eng Sci, 39(1), 173-174 (1984).
Saxena, S.C.; Mathur, A., and Sharma, G.K., Bubble dynamics and elutriation studies in gas-fluidized beds, Chem Eng Commns, 29(1), 35-62 (1984).
Shieh, W.K., and Chen, C.Y., Biomass hold-up correlations for a fluidised bed biofilm reactor, CER&D, 62, 133-136 (1984).
Siegell, J.H., High temperature defluidization, Powder Tech, 38, 13-22 (1984).
Song, J.C.; Fan, L.T., and Yutani, N., Fault detection of fluidized bed distributor by pressure fluctuation signal, Chem Eng Commns, 25(1), 105-116 (1984).
Stanley, C.K.; Bridgwater, J., and Botterill, J.S.M., Solids flow between interconnected shallow fluidized beds, Chem Eng Sci, 39(12), 1797-1806 (1984).
Stubington, J.F.; Barrett, D., and Lowry, G., Bubble size measurements and correlation in a fluidised bed at high temperatures, CER&D, 62, 173-178 (1984).
Van der Meer, A.P.; Blanchard, C., and Wesselingh, J.A., Mixing of particles in liquid fluidised beds, CER&D, 62, 214-222 (1984).
Yates, J.G., Applications of gas fluidization, Proc Engng, July, 33-36 (1984).
Zenz, F.A.; Zenz, F.E., and Zenz, J.A., Riser design considerations under slug flow conditions, Powder Tech, 38, 205-210 (1984).

1985

Abed, R., Characterization of hydrodynamic nonuniformity in large fluidized beds, Ind Eng Chem Fund, 24(1), 78-82 (1985).

Agarwal, P.K., Bubble characteristics in gas fluidised beds, CER&D, 63, 323-337 (1985).

Baskakov, A.P.; Tuponogov, V.G., and Philippovsky, N.F., Uniformity of fluidization on a multi-orifice gas distributor, Can JCE, 63(6), 886-890 (1985).

Beeckmans, J.M., Correlation of data on segregation in gas-fluidized beds, Chem Eng Sci, 40(4), 675-677 (1985).

Bin, A.; Warych, J., and Komorowski, R., Batch granulation in a fluidised bed, Powder Tech, 41, 1-11 (1985).

Bonsu, A.K., and Meisen, A., Fluidized bed Claus reactor studies, Chem Eng Sci, 40(1), 27-38 (1985).

Dry, R.J., and Judd, M.R., Fluidised beds of fine, dense powders: Scale-up and reactor modelling, Powder Tech, 43, 41-53 (1985).

Fan, L.S.; Kawamura, T.; Chitester, D.C., and Kornosky, R.M., Experimental observation of nonhomogeneity in a liquid/solid fluidized bed of small particles, Chem Eng Commns, 37(1), 141-158 (1985).

Feng, D.; Chen, H., and Whiting, W.B., Effects of distributor design on fluidized-bed hydrodynamic behaviour, Chem Eng Commns, 36(1), 317-332 (1985).

Geldart, D., and Baeyens, J., Design of distributors for gas-fluidized beds, Powder Tech, 42, 67-78 (1985).

Glicksman, L.R., and McAndrews, G., Effect of bed width on hydrodynamics of large particle fluidized beds, Powder Tech, 42, 159-167 (1985).

Grace, J.R., and Hosny, N., Forces on horizontal tubes in gas fluidised beds, CER&D, 63, 191-198 (1985).

Hiraoka, S.; Kim, K.C.; Shin, S.H., and Fan, L.T., Properties of pressure fluctuations in gas-solids fluidized bed under a free bubbling condition, Powder Tech, 45, 245-265 (1985).

Jianhong, Y., and Schugerl, K., Development of relationship for solid recirculation in highly expanded (fast) circulating fluidized bed, Chem Eng & Proc, 19(6), 297-302 (1985).

Kawase, Y., and Ulbrecht, J.J., Mass and momentum transfer with non-Newtonian fluids in fluidized beds, Chem Eng Commns, 32(1), 263-288 (1985).

Kuramoto, M.; Furusawa, T., and Kunii, D., Development of new system for circulating fluidized particles within a single vessel, Powder Tech, 44, 77-84 (1985).

Muramoto, T., et al., Measurement of flow of solid particles in gas fluidized bed, Chem Eng Commns, 35(1), 193-202 (1985).

Muroyama, K., and Fan, L.S., Fundamentals of gas-liquid-solid fluidization (a review), AIChEJ, 31(1), 1-34 (1985).

Nguyen-Tien, K.; Patwari, A.N.; Schumpe, A., and Deckwer, W.D., Gas-liquid mass transfer in fluidized particle beds, AIChEJ, 31(2), 194-201 (1985).

Olsen, K.W., Recent advances in fluid-bed agglomerating and coating technology, Plant/Opns Prog, 4(3), 135-139 (1985).

Puncocher, M.; Drahos, J.; Cermak, J., and Selucky, K., Evaluation of minimum fluidizing velocity in gas fluidized bed from pressure fluctuations, Chem Eng Commns, 35(1), 81-88 (1985).

Rahman, K., and Streat, M., Mass transfer in fixed and fluidized beds, Chem Eng Sci, 40(9), 1783-1785 (1985).

Schruben, J.S., and Vaux, W.G., Attrition in bubbling zone of steady-state fluidized bed, Chem Eng Commns, 33(5), 337-348 (1985).

Shi, Y.F., and Fan, L.T., Lateral mixing of solids in gas-solid fluidized beds with continuous flow of solids, Powder Tech, 41, 23-28 (1985).

Thomas, C.R., and Yates, J.G., Expansion index for biological fluidised beds, CER&D, 63, 67-70 (1985).

Yang, Z.; Tung, Y., and Kwauk, M., Characterizing fluidization by bed collapsing method, Chem Eng Commns, 39, 217-232 (1985).

Yue, P.L., and Birk, R.H., Fluidised bed studies of dehydration of ethanol over zeolite catalyst, CER&D, 63, 250-258 (1985).

Yutani, N., and Fan, L.T., Mixing of randomly moving particles in liquid-solid fluidized beds, Powder Tech, 42, 145-152 (1985).

1986

Bellgardt, D., and Werther, J., Novel method of investigation of particle mixing in gas-solid systems, Powder Tech, 48(2), 173-180 (1986).

Chen, D.; Zhao, L., and Yang, G., Expansion of large-scale fluidized beds with internals, Int Chem Eng, 26(1), 155-160 (1986).

Cheremisinoff, N.P., Review of experimental methods for studying the hydrodynamics of gas-solid fluidized beds, Ind Eng Chem Proc Des Dev, 25(2), 329-351 (1986).

Dry, R.J., Radial concentration profiles in a fast fluidised bed, Powder Tech, 49(1), 37-44 (1986).

Dry, R.J., and Potter, O.E., Improved throughput and gas-solids contact in a fluidised bed: Concept of a bubble collector, Powder Tech, 46(1), 13-22 (1986).

Erdesz, K., and Mujumdar, A.S., Comparison of hydrodynamic aspects of conventional and vibrofluidized beds, Powder Tech, 46(2), 167-172 (1986).

Forster, C.F.; Boyes, A.P.; Hay, B.A., and Butt, J.A., An aerobic fluidised bed reactor for wastewater treatment, CER&D, 64(6), 425-430 (1986).

Geldart, D., (Ed.), Gas Fluidization Technology, Wiley, New York (1986).

Geldart, D., and Rhodes, M., Developments in fluidisation, Chem Engnr, Feb, 30-32 (1986).

Gyure, D.C., and Clough, D.E., State estimation of bubble frequency and velocity in a bubbling fluidized bed, Chem Eng Commns, 46, 365-384 (1986).

Herskowitz, M., and Merchuk, J.C., A loop three-phase fluidized bed reactor, Can JCE, 64(1), 57-61 (1986).

Hoffmann, A.C., and Yates, J.G., Experimental observations of fluidized beds at elevated pressures, Chem Eng Commns, 41, 133-150 (1986).

Horio, M.; Nonaka, A.; Sawa, Y., and Muchi, I., New similarity rule for fluidized bed scale-up, AIChEJ, 32(9), 1466-1482 (1986).

Kono, H.O.; Chiba, S.; Ells, T., and Suzuki, M., Characterization of emulsion phase in fine particle fluidized beds, Powder Tech, 48(1), 51-58 (1986).
Lee, P.L.; Chong, Y.O., and Leung, L.S., Experimental investigation of quality control of fluidized beds, Powder Tech, 46(1), 77-80 (1986).
Lin, S.C.; Arastoopour, H., and Kono, H., Experimental and theoretical study of multistage fluidized-bed reactor, Powder Tech, 48(2), 125-140 (1986).
Lucas, A.; Arnaldos, J.; Casal, J., and Puigjaner, L., High-temperature incipient fluidization in mono- and poly-disperse systems, Chem Eng Commns, 41, 121-132 (1986).
Mathur, A.; Saxena, S.C., and Zhang, Z.F., Hydrodynamic characteristics of gas-fluidized beds over a broad temperature range, Powder Tech, 47(3), 247-256 (1986).
Mihail, R., and Straja, S., A theoretical model concerning bubble size distributions, Chem Eng J, 33(2), 71-78 (1986).
Morl, L.; Krell, L.; Kunne, H.J., and Kliefoth, J., Calculation of angle of repose on fluidized-bed plates, Int Chem Eng, 26(2), 231-236 (1986).
Noda, K.; Uchida, S.; Makino, T., and Kamo, H., Minimum fluidization velocity of binary mixture of particles with large size ratio, Powder Tech, 46(2), 149-154 (1986).
Patel, K.; Nienow, A.W., and Milne, I.P., Attrition of urea in a gas-fluidized bed, Powder Tech, 47(3), 257-262 (1986).
Patwari, A.N., et al., Three-phase fluidized beds with viscous liquid: Hydrodynamics and mass transfer, Chem Eng Commns, 40, 49-66 (1986).
Reh, L., The circulating fluid bed reactor: Main features and applications, Chem Eng & Proc, 20(3), 117-128 (1986).
Rios, G.M.; Tran, K.D., and Masson, H., Free object motion in a gas fluidized bed, Chem Eng Commns, 47, 247-272 (1986).
Saunders, J.H., Particle entrainment from rotating fluidized beds, Powder Tech, 47(3), 211-218 (1986).
Schuart, L., et al., Computer-aided development of quotations for fluidized-bed equipment, Int Chem Eng, 26(3), 419-423 (1986).
Siegell, J.H.; Dupre, G.D., and Pirkle, C., Chromatographic separations in a crossflow magnetically stabilized fluidized bed, Chem Eng Prog, 82(11), 57-61 (1986).
Yutani, N.; Ototake, N., and Fan, L.T., Stochastic analysis of fluctuations in local void fraction of gas-solids fluidized bed, Powder Tech, 48(1), 31-38 (1986).

1987

Agarwal, P.K., Effect of bed diameter on bubble growth and incipient slugging in gas fluidised beds, CER&D, 65(4), 345-354 (1987).
Baron, T., et al., Electrostatic effects on entrainment from a fluidized bed, Powder Tech, 53(1), 55-68 (1987).
Beeckmans, J.M., and Stahl, B., Mixing segregation kinetics in a strongly segregated gas-fluidized bed, Powder Tech, 53(1), 31-38 (1987).
Bejcek, V., et al., Bubble size above an isolated gas jet penetrating a fluidized bed, Chem Eng Commns, 62(1), 303-314 (1987).
Bilbao, R.; Lezaun, J., and Abanades, J.C., Fluidization velocities of sand/straw binary mixtures, Powder Tech, 52(1), 1-6 (1987).

Biswas, J., and Leung, L.S., Applicability of choking correlations for fast-fluid bed operation, Powder Tech, 51(2), 179-180 (1987).

Bolton, L.W., and Davidson, J.F., Dense-phase circulating fluidised beds, Chem Eng Commns, 62(1), 31-52 (1987).

Carsky, M., et al., Binary system fluidized bed equilibrium, Powder Tech, 51(3), 237-242 (1987).

Carter, B.; Ghadiri, M.; Clift, R., and Jury, A.W., Behaviour of large jetsam particles in fluidised beds, Powder Tech, 52(3), 263-266 (1987).

Chan, I.H.; Sishtla, C., and Knowlton, T.M., Effect of pressure on bubble parameters in gas-fluidized beds, Powder Tech, 53(3), 217-236 (1987).

Chen, Y.M., Fundamentals of centrifugal fluidized bed, AIChEJ, 33(5), 722-728 (1987).

Chong, Y.O., et al., Fluidization quality control in tall beds using variance of pressure fluctuations, Powder Tech, 53(3), 237-246 (1987).

Currier, R., and Herman, M.F., Non-Markovian model for particle motion in fluidized beds, Chem Eng Commns, 56, 203-210 (1987).

Dry, R.J., Radial particle size segregation in fast fluidised bed, Powder Tech, 52(1), 7-16 (1987).

Dry, R.J.; Christensen, I.N., and White, C.C., Gas-solids contact efficiency in high-velocity fluidised bed, Powder Tech, 52(3), 243-250 (1987).

Fan, L.S., et al., Hydrodynamics of gas-liquid-solid fluidization under high gas hold-up conditions, Powder Tech, 53(3), 285-294 (1987).

Fan, L.S.; Kitano, K., and Kreischer, B.E., Hydrodynamics of gas-liquid-solid annular fluidization, AIChEJ, 33(2), 225-231 (1987).

Fan, L.S.; Ramesh, T.S.; Tang, W.T., and Long, T.R., Gas-liquid mass transfer in a two-stage draft tube gas-liquid-solid fluidized bed, Chem Eng Sci, 42(3), 543-554 (1987).

Flemmer, R.L.C., and Clark, N.N., Wave velocity based on new equation of state for fluidized beds, Powder Tech, 50(1), 77-78 (1987).

Foscolo, P.U., et al., Expansion characteristics of gas fluidized beds of fine catalysts, Chem Eng & Proc, 22(2), 69-78 (1987).

Gibilaro, L.G., et al., Review of applications of fluid-particle interaction model to predictions of fluidised bed behaviour, Chem Eng Commns, 62(1), 17-30 (1987).

Glicksman, L.R., and Piper, G.A., Particle density distribution in freeboard of fluidized bed, Powder Tech, 53(3), 179-186 (1987).

Glicksman, L.R.; Lord, W.K., and Sakagami, M., Bubble properties in large-particle fluidized beds, Chem Eng Sci, 42(3), 479-492 (1987).

Guo, F., Gas flow and mixing behavior in fine-powder fluidized bed, AIChEJ, 33(11), 1895-1898 (1987).

Ho, T.C.; Yau, S.J., and Hopper, J.R., Hydrodynamics of semi-fluidization in gas-solid systems, Powder Tech, 50(1), 25-34 (1987).

Horio, M., and Nonaka, A., Generalized bubble diameter correlation for gas-solid fluidized beds, AIChEJ, 33(11), 1865-1872 (1987).

Hu, T.T., and Wu, J.Y., Characteristics of biological fluidized bed in magnetic field, CER&D, 65(3), 238-242 (1987).

Jacob, K.V., and Weimer, A.W., High-pressure particulate expansion and minimum bubbling of fine carbon powders, AIChEJ, 33(10), 1698-1706 (1987).

Johnsson, J.E.; Grace, J.R., and Graham, J.J., Fluidized-bed reactor model verification on industrial scale, AIChEJ, 33(4), 619-627 (1987).

Kai, T., et al., Change in bubble behavior for different fluidizing gases in a fluidized bed, Powder Tech, 51(3), 267-272 (1987).

Kao, J.; Pfeffer, R., and Tardos, G.I., Partial fluidization in rotating fluidized beds, AIChEJ, 33(5), 858-861 (1987).

Kathuria, D.G., and Saxena, S.C., Variable-thickness two-dimensional bed for investigating gas-solid fluidized bed hydrodynamics, Powder Tech, 53(2), 91-96 (1987).

Kono, H.O., et al., Quantitative criteria for emulsion phase characterization and transition between particulate and bubbling fluidization, Powder Tech, 52(1), 69-76 (1987).

Kono, H.O.; Soltani, A., and Suzuki, M., Kinetic forces of solid particles in coarse-particle fluidized beds, Powder Tech, 52(1), 49-58 (1987).

Krambeck, F.J., et al., Predicting fluid-bed reactor efficiency using adsorbing gas tracers, AIChEJ, 33(10), 1727-1732 (1987).

Lee, S.L.P., and de Lasa, H.I., Phase holdups in three-phase fluidized beds, AIChEJ, 33(8), 1359-1370 (1987).

Mydlarz, J., Prediction of the packed bed height in liquid-solid semi-fluidization of homogeneous mixtures, Chem Eng J, 34(3) 155-158 (1987).

Nieh, S., and Yang, G., Particle flow pattern in freeboard of vortexing fluidized bed, Powder Tech, 50(2), 121-132 (1987).

Nienow, A.W., and Killick, R.C., Binding agents to enhance fluidised bed gas cleaning, Powder Tech, 50(3), 267-274 (1987).

Nienow, A.W.; Naimer, N.S., and Chiba, T., Segregation/mixing in fluidised beds of different size particles, Chem Eng Commns, 62(1), 53-66 (1987).

Noordergraaf, I.W., Fluidization and slugging in large-particle systems, Powder Tech, 52(1), 59-68 (1987).

Puncochar, M.; Drahos, J., and Cermak, J., Porosity in gas-solid fluidized bed at onset of slugging, Powder Tech, 51(2), 183-184 (1987).

Ray, Y.C.; Jiang, T.S., and Jiang, T.L., Particle population model for fluidized bed with attrition, Powder Tech, 52(1), 35-48 (1987).

Ray, Y.C.; Jiang, T.S., and Wen, C.Y., Particle attrition phenomena in fluidized bed, Powder Tech, 49(3), 193-206 (1987).

Rhodes, M.J., and Geldart, D., A model for the circulating fluidized bed, Powder Tech, 53(3), 155-162 (1987).

Saberian, M.B., et al., Hydrodynamic study of gas-liquid-solid fluidized-bed reactors, Int Chem Eng, 27(3), 423-441 (1987).

Salam, T.F., and Gibbs, B.M., Solid circulation between fluidized beds using jet pumps, Powder Tech, 52(2), 107-116 (1987).

Salam, T.F., and Gibbs, B.M., Gas and solid discharge from fluidized bed using jet pump, Powder Tech, 50(2), 111-120 (1987).

Saxena, S.C.; Mathur, A., and Zhang, Z.F., Incipient fluidization at different temperatures and powder characterization, AIChEJ, 33(3), 500-502 (1987).

Siegell, J.H., Liquid-fluidized magnetically stabilized beds, Powder Tech, 52(2), 139-148 (1987).

Stievenart, P., et al., Development of sandwiched fluidized beds, Chem Eng Commns, 62(1), 269-284 (1987).

Sung, J.S., and Burgess, J.M., Laser-based method for bubble parameter measurement in two-dimensional fluidised beds, Powder Tech, 49(2), 165-176 (1987).

Yang, J.S.; Liu, Y.A., and Squires, A.M., Simple light-probe method for quantitative measurements of particle volume-fractions in fluidized beds, Powder Tech, 49(2), 177-188 (1987).

Yang, J.S.; Liu, Y.A., and Squires, A.M., Pressure drop across shallow fluidized beds: Theory and experiment, Powder Tech, 53(2), 79-90 (1987).

Yang, W.C. (Ed.), Fluidization and fluid/particle systems (various papers), Powder Tech, 53(3), 153-300 (1987).

Yang, W.C., and Keairns, D.L., Fine particle residence time in a jetting fluidized bed, Powder Tech, 53(3), 169-178 (1987).

Yates, J.G., and Ruiz-Martinez, R.S., Interaction between horizontal tubes and gas bubbles in fluidised bed, Chem Eng Commns, 62(1), 67-78 (1987).

Yoshioka, S., et al., Circulatory flow of particles in a 0.96m diameter fluidized bed, Int Chem Eng, 27(2), 281-288 (1987).

Zhang, M.C., and Yang, R.Y.K., Scaling laws for bubbling gas-fluidized bed dynamics, Powder Tech, 51(2), 159-166 (1987).

1988

Arastoopour, H.; Huang, C.S., Weil, S.A., Fluidization behavior of particles under agglomerating conditions, Chem Eng Sci, 43(11), 3063-3076 (1988).

Arters, D.C.; Fan, L.S., and Kim, B.C., Mass transfer and bed expansion of solid-slurry fluidized beds, AIChEJ, 34(7), 1221-1224 (1988).

Atkinson, C.M., and Clark, N.N., Novel probe system for gas sampling from fluidized beds, Powder Tech, 54(1), 59-70 (1988).

Baron, T.; Briens, C.L., and Bergougnou, M.A., Measurement of flux of clusters ejected from a fluidized bed, Powder Tech, 55(2), 115-126 (1988).

Barreto, G.F.; Mazza, G.D., and Yates, J.G., Significance of bed collapse experiments in characterization of fluidized beds of fine powders, Chem Eng Sci, 43(11), 3037-3048 (1988).

Basta, N., Fluidized bed developments, Chem Eng, 9 May, 30-33 (1988).

Bilbao, R.; Lezaun, J.; Menendez, M., and Abanades, J.C., Model of mixing-segregation for straw/sand mixtures in fluidized beds, Powder Tech, 56(3), 149-156 (1988).

Briens, C.L.; Bergougnou, M.A., and Baron, T., Prediction of entrainment from gas-solid fluidized beds, Powder Tech, 54(3), 183-196 (1988).

Briens, C.L.; Tyagi, A.K., and Bergougnou, M.A., Pressure drop through multiorifice gas distributors in fluidized bed columns, Can JCE, 66(5), 740-748 (1988).

Burns, M.A., and Graves, D.J., Structural studies of liquid-fluidized magnetically fluidized bed, Chem Eng Commns, 67, 315-330 (1988).

Daw, C.S., and Frazier, C., Quantitative analysis of binary solids segregation in large-particle gas-fluidized beds, Powder Tech, 56(3), 165-178 (1988).

Donsi, G.; Ferrari, G., and Formisani, B., Segregation mechanism of percolating fines in coarse-particle fluidized beds, Powder Tech, 55(2), 153-158 (1988).

Drahos, J., et al., Characterization of flow regime transitions in a circulating fluidized bed, Powder Tech, 56(1), 41-48 (1988).

Drahos, J., et al., Diagnostics of slugging in fluidized beds, Powder Tech, 55(4), 285-288 (1988).
Drahos, J.; Cermak, J., and Schugerl, K., Characterization of axial nonuniformity in fluidized bed by amplitude of local pressure drop fluctuations, Chem Eng Commns, 65, 49-60 (1988).
Ermakova, A.; Kuzmin, V.A., and Umbetov, A.S., Mass transfer from individual particles to a liquid in three-phase systems of fixed and fluidized beds: Derivation of generalized mass-transfer equation, Int Chem Eng, 28(3), 559-570 (1988).
Ettehadieh, B.; Yang, W.C., and Haldipur, G.B., Motion of solids, jetting and bubbling dynamics in large jetting fluidized bed, Powder Tech, 54(4), 243-254 (1988).
Gbordzoe, E.A.M., et al., Gas transfer between central jet and large two-dimensional gas-fluidized bed, Powder Tech, 55(3), 207-222 (1988).
Geuzens, P., and Thoenes, D., Magnetically stabilized fluidization, Chem Eng Commns, 67, 217-242 (1988).
Gibilaro, L.G.; Di Felice, R., and Foscolo, P.U., Minimum bubbling voidage and the Geldart classification for gas-fluidised beds, Powder Tech, 56(1), 21-30 (1988).
Glicksman, L.R., Scaling relationships for fluidized beds, Chem Eng Sci, 43(6), 1419-1421 (1988).
Jacob, K.V., and Weimer, A.W., Normal bubbling of fine carbon powders in high-pressure fluidized beds, AIChEJ, 34(8), 1395-1397 (1988).
Krishna, R., Simulation of industrial fluidized bed reactor using a bubble growth model, CER&D, 66(5), 463-469 (1988).
Lancia, A., et al., Detection of transition from slugging to turbulent flow regimes in fluidized beds using capacitance probes, Powder Tech, 56(1), 49-56 (1988).
Levy, E.K; Chen, H.K.; Radcliff, R., and Caram, H.S., Analysis of gas flow through erupting bubbles in gas-fluidized bed, Powder Tech, 54(1), 45-58 (1988).
Muller, C., and Flament, G., Monitoring of gas-solid reaction in fluidized bed by measuring pressure drop, Int Chem Eng, 28(1), 62-75 (1988).
Ruder, Z., et al., Voidage and expansion in oil-shale ash fluidized beds with non-uniform particle size distribution, Powder Tech, 56(2), 69-82 (1988).
Sciazko, M.; Raczek, J., and Bandrowski, J., Model of gas flow above bubbling fluidized bed: Prediction of splash zone height, Chem Eng & Proc, 24(1), 49-56 (1988).
Siegell, J.H., Applications of crossflow magnetically stabilized fluidized beds, Chem Eng Commns, 67, 43-54 (1988).
Stevens, J.G., et al., Magnetically stabilized fluidized beds with time-varying magnetic fields, Powder Tech, 56(2), 119-128 (1988).
Vadivel, R., and Saxena, S.C., Particle dynamics near surface of horizontal tube in gas-fluidized bed, AIChEJ, 34(11), 1919-1921 (1988).
Webster, G.H., and Perona, J.J., Liquid mixing in tapered fluidized bed, AIChEJ, 34(8), 1398-1402 (1988).
Wu, C.S., and Whiting, W.B., Interacting jets in fluidized beds, Chem Eng Commns, 73, 1-18 (1988).

Yasuda, M.; Yasukawa, S., and Sekine, T., Experimental study of current-potential curves for concentric-cylindrical fluidized-bed electrode, Int Chem Eng, 28(2), 306-314 (1988).
Yu, Y.H., and Kim, S.D., Bubble characteristics in radial direction of three-phase fluidized beds, AIChEJ, 34(12), 2069-2072 (1988).
Zimmels, Y., Generalized approach to flow through fixed beds, fluidization and hindered sedimentation, Chem Eng Commns, 67, 19-42 (1988).

24.2 Spouted Beds

1967-1977

Rao, M.V.R., Nomograph determines fluid velocity needed for spouting, Chem Eng, 3 July, 122 (1967).
Balakrishan, S., and Krishnan, V., Nomogram for spout diameter in a spouted bed, Brit Chem Eng, 13(1), 111 (1968).
Lefroy, G.A., and Davidson, J.F., The mechanics of spouted beds, Trans IChemE, 47, T120-128 (1969).
Zanker, A., Nomograph for calculating the superficial fluid velocity in spouted beds, Chem Engnr, Oct, CE351-353 (1970).
Bridgwater, J., and Mathur, K.B., Prediction of spout diameter in a spouted bed: A theoretical model, Powder Tech, 6, 183-187 (1972).
McNab, G.S., Prediction of fluid-bed spout diameter, Brit Chem Eng, 17(6), 532 (1972).
Mann, U., and Crosby, E.J., Cycle-time distribution measurements in spouted beds, Can JCE, 53, 579-590 (1975).
Grbavcic, Z.B., et al., Fluid flow pattern, minimum spouting velocity and pressure drop in spouted beds, Can JCE, 54, 33-40 (1976).
Lim, C.J., and Mathur, K.B., Flow model for gas movement in spouted beds, AIChEJ, 22(4), 674-680 (1976).
Vajda, T., Calculation of residence-time distribution in spouted beds, Int Chem Eng, 16(3), 543-552 (1976).
Asenjo, J.A.; Munoz, R., and Pyle, D.L., Transition from fixed to spouted bed, Chem Eng Sci, 32(2), 109-118 (1977).
Littman, H., et al., Theory for predicting maximum spoutable height in a spouted bed, Can JCE, 55, 497-505 (1977).
Piccinini, N., et al., Segregation phenomenon in spouted beds, Can JCE, 55, 122-130 (1977).
Suciu, G.C., and Patrascu, M.H., Phase distribution and residence time in a spouted bed, AIChEJ, 23(3), 312-318 (1977).
Zanker, A., Designing spouted beds, Chem Eng, 21 Nov, 207-209 (1977).

1978-1984

Cook, H.H., and Bridgwater, J., Segregation in spouted beds, Can JCE, 56, 636-645 (1978).
Epstein, N.; Lim, C.J., and Mathur, K.B., Data and models for flow distribution and pressure drop in spouted beds, Can JCE, 56, 436-445 (1978).
Grace, J.R., and Mathur, K.B., Height and structure of fountain region above spouted beds, Can JCE, 56, 533-540 (1978).

Robinson, T., and Waldie, B., Particle cycle times in spouted bed of polydisperse particles, Can JCE, 56, 632-640 (1978).
Suciu, G.C., and Patrascu, M., Particle circulation in a spouted bed, Powder Tech, 19, 109-114 (1978).
Littman, H., et al., Prediction of maximum spoutable height and average spout/inlet tube diameter ratio in spouted beds of spherical particles, Can JCE, 57, 684-694 (1979).
King, D.F., and Harrison, D., Minimum spouting velocity of a spouted bed at elevated pressure, Powder Tech, 26, 103-107 (1980).
Kmiec, A., Hydrodynamics of flows and heat transfer in spouted beds, Chem Eng J, 19(3), 189-200 (1980).
Mele, D., and Martinez, J., Calculator program aids design of spouted beds, Chem Eng, 20 Oct, 137-139 (1980).
Ishikura, T.; Shinohara, H., and Funatsu, K., Minimum spouting velocity for binary mixtures of particles, Can JCE, 60, 697-699 (1982).
Ishikura, T.; Tanaka, I., and Shinohara, H., Elutriation of fines from a continuous spouted bed containing a gas-solid system: Effect of column height on the fines elutriation rate, Int Chem Eng, 22(2), 346-355 (1982).
Various, Papers from the 2nd International Symposium on Spouted Beds, Can JCE, 61(3), 265-477 (1983).
Fan, L.S.; Hwang, S.J., and Matsuura, A., Hydrodynamic behaviour of a draft tube gas-liquid-solid spouted bed, Chem Eng Sci, 39(12), 1677-1688 (1984).

1985-1988

Sutanto, W.; Epstein, N., and Grace, J.R., Hydrodynamics of spout-fluid beds, Powder Tech, 44, 205-212 (1985).
Hwang, S.J., and Fan, L.S., Some design considerations of a draft tube gas-liquid-solid spouted bed, Chem Eng J, 33(1), 49-56 (1986).
Law, L., et al., Aerodynamic and solids circulation measurements in a slotted spouted bed of grains, Powder Tech, 46(2), 141-148 (1986).
Passos, M.L., et al., Pressure drop in slotted spouted beds of grains: Comparison of data with models, Powder Tech, 52(2), 131-138 (1987).
Pavarini, P.J., and Coury, J.R., Granulation of an insoluble powder in a spouted bed, Powder Tech, 53(2), 97-104 (1987).
Sit, S.P., and Grace, J.R., Interphase mass transfer from spouts connecting the distributor plate to forming bubbles in fluidized beds, Chem Eng Commns, 62(1), 315-332 (1987).
Sullivan, C.; Benkrid, A., and Caram, H., Prediction of solids circulation patterns in spouted bed, Powder Tech, 53(3), 257-272 (1987).
Wu, S.W.M.; Lim, C.J., and Epstein, N., Hydrodynamics of spouted beds at elevated temperatures, Chem Eng Commns, 62(1), 251-268 (1987).
Berruti, F.; Muir, J.R., and Behie, L.A., Solids circulation in spout-fluid bed with draft tube, Can JCE, 66(6), 919-923 (1988).
Matthew, M.C.; Morgan, M.H., and Littman, H., Hydrodynamics within draft-tube spouted bed system, Can JCE, 66(6), 908-918 (1988).

24.3 Heat Transfer in Fluidized Beds

1967-1971

Legler, B.M., Feed injection for heated fluidized beds, Chem Eng Prog, 63(2), 75-82 (1967).

Wilkinson, G.T., and Norman, J.R., Heat transfer to a suspension of solids in a gas, Trans IChemE, 45, T314-318 (1967).

Botterill, J.S.M., and Butt, M.H.D., Achieving high heat-transfer rates in fluidised beds, Brit Chem Eng, 13(7), 1000-1004 (1968).

Petrie, J.C.; Freeby, W.A., and Buckham, J.A., Heat exchange in fluidized beds, Chem Eng Prog, 64(7), 45-51 (1968).

Gelperin, N.I.; Ainshtein, V.G., and Korotyanskaya, L.A., Heat transfer between a fluidized bed and staggered bundles of horizontal tubes, Int Chem Eng, 9(1), 137-142 (1969).

Botterill, J.S.M., Heat transfer to gas-fluidised beds, Powder Tech, 4, 19-26 (1970).

Botterill, J.S.M.; Chandrasekhar, R., and Van der Kolk, M., Heat transfer and pressure loss for the flow of a fluidised solid across banks of tubes, Brit Chem Eng, 15(6), 769-772 (1970).

Roots, W.K., and Tulunay, E., Reduce fluid-bed operating costs by heat exchange, Hyd Proc, 49(5), 161-164 (1970).

Wright, S.J.; Hickman, R., and Ketley, H.C., Heat transfer in fluidised beds of wide size spectrum at elevated temperatures, Brit Chem Eng, 15(12), 1551-1554 (1970).

Brea, F.M., and Hamilton, W., Heat transfer in liquid fluidised beds with a concentric heater, Trans IChemE, 49, 196-203 (1971).

Gelperin, N.I.; Einstein, V.G., and Toskubayev, I.N., Heat-transfer coefficient between a surface and a fluid bed, Brit Chem Eng, 16(10), 922 (1971).

1972-1981

Baskakov, A.P., and Suprun, V.M., Determination of convective component of heat-transfer coefficient for a gas in a fluidized bed, Int Chem Eng, 12(1), 53-55; 12(2), 324-327 (1972).

Botterill, J.S.M., and Desai, M., Limiting factors in gas-fluidised bed heat transfer, Powder Tech, 6, 231-238 (1972).

Baskakov, A.P., et al., Heat transfer to objects immersed in fluidized beds, Powder Tech, 8, 273-282 (1973).

Zabrodsky, S.S.; Antonishin, N.V., and Parnas, A.L., Fluidized bed-to-surface heat transfer, Can JCE, 54, 52-60 (1976).

Howe, W.C., and Aulisio, C., Control variables in fluidized-bed steam generation, Chem Eng Prog, 73(7), 69-73 (1977).

Denloye, A.O.O., and Botterill, J.S.M., Bed to surface heat transfer in a fluidized bed of large particles, Powder Tech, 19, 197-203 (1978).

Cole, W.S., and Suo, M., Waste heat recovery with fluidized beds, Chem Eng Prog, 75(12), 38-42 (1979).

Stubington, J.F., Heat transfer between a freely-rising slug and a fluidised bed, Trans IChemE, 58, 195-202 (1980).

Ganapathy, V., Nomograph for heat-transfer coefficient in fluidised beds, Proc Engng, Jan, 53 (1981).

Grewal, N.S., Generalized correlation for heat transfer between a gas-solid fluidized bed of small particles and immersed staggered array of horizontal tubes, Powder Tech, 30, 145-154 (1981).

Hoelen, Q.E., and Stemerding, S., Heat transfer in a fluidized bed, Powder Tech, 30, 161-184 (1981).

Hussein, F.D.; Maitra, P.P., and Jackson, R., Heat transfer to a fluidized bed using a particulate heat transfer medium, Ind Eng Chem Proc Des Dev, 20(3), 511-519 (1981).

Martin, H., Fluid-bed heat exchangers: A new model for particle convective energy transfer, Chem Eng Commns, 13(1), 1-22 (1981).

Zabrodsky, S.S., et al., Heat transfer in a large-particle fluidized bed with immersed in-line and staggered bundles of horizontal smooth tubes, Int J Heat Mass Trans, 24(4), 571-580 (1981).

1982-1986

Ganzha, V.L.; Upadhyay, S.N., and Saxena, S.C., Mechanistic theory for heat transfer between fluidized beds of large particles and immersed surfaces, Int J Heat Mass Trans, 25(10), 1531-1540 (1982).

Borodulya, V.A., and Kovensky, V.I., Radiative heat transfer between a fluidized bed and a surface, Int J Heat Mass Trans, 26(2), 277-288 (1983).

Botterill, J.S.M.; Teoman, Y., and Yuregir, K.R., Factors affecting heat transfer between gas-fluidized beds and immersed surfaces, Powder Tech, 39, 177-189 (1984).

Martin, H., Heat transfer between gas fluidized beds of solid particles and surfaces of immersed heat exchanger elements, Chem Eng & Proc, 18(3), 157-170; 18(4), 199-224 (1984).

Rossi, R.A., Indirect heat transfer in fluidised beds, Chem Eng, 15 Oct, 95-102 (1984).

Chen, P., and Pei, D.C.T., Model of heat transfer between fluidized beds and immersed surfaces, Int J Heat Mass Trans, 28(3), 675-682 (1985).

Kang, Y.; Suh, I.S., and Kim, S.D., Heat transfer characteristics of three-phase fluidized beds, Chem Eng Commns, 34(1), 1-14 (1985).

Kubie, J., Heat transfer between gas fluidized beds and immersed surfaces, Int J Heat Mass Trans, 28(7), 1345-1354 (1985).

Virr, M.J., and Williams, H.W., Heat recovery by shallow fluidized beds, Chem Eng Prog, 81(7), 50-56 (1985).

Anon., Fluid bed heat exchanger to minimise fouling, Chem Engnr, Feb, 28 (1986).

Barreto, G.F.; Lancia, A., and Volpicelli, G., Heat transfer and fluid dynamic characteristics of gas-fluidized beds under pressure, Powder Tech, 46(2), 155-166 (1986).

Farag, I.H.; Karri, S.B.R.; Breault, R., and Tsai, K.Y., Graphical solution of multistage fluidized-bed heat exchanger, Chem Eng Commns, 48, 331-348 (1986).

Koutchoukali, M.S.; Laguerie, C., and Najim, K., Modeling of transient and steady-state operation of fluidized bed reactor with adaptive control of temperature, Chem Eng Commns, 44, 197-218 (1986).

St John, B., Economics of atmospheric fluidized-bed boilers, Chem Eng, 8 Dec, 157-159 (1986).

Weimer, A.W., and Jacob, K.V., Bed-to-surface heat transfer for vertical U-tubes in high-pressure slugging hydrogen-fluidized bed of carbon-supported catalyst powder, Powder Tech, 48(3), 247-252 (1986).

White, T.R.; Mathur, A., and Saxena, S.C., Effect of vertical tube diameter on heat transfer coefficient in gas-fluidized beds, Chem Eng J, 32(1), 1-14 (1986).

1987-1988

Biyikli, S.; Tuzla, K., and Chen, J.C., Freeboard heat transfer in high-temperature fluidized beds, Powder Tech, 53(3), 187-194 (1987).

Deffenbaugh, D.M., and Green, S.T., Transient heat transfer and bubble dynamics in a pressurized fluidized bed, Int J Heat Mass Trans, 30(10), 2151-2160 (1987).

Ho, T.C.; Wang, R.C., and Hopper, J.R., Characteristics of grid zone heat transfer in gas-solid fluidized bed, AIChEJ, 33(5), 843-847 (1987).

Kolar, A.K., and Sastri, V.M.K., Extended surface heat transfer in fluidized beds, Chem Eng & Proc, 22(1), 1-18 (1987).

Mathur, A., and Saxena, S.C., Total and radiative heat transfer to immersed surface in gas-fluidized bed, AIChEJ, 33(7), 1124-1135 (1987).

Renzhang, Q.; Wendi, H.; Yunsheng, X., and Dechang, L., Experimental research of radiative heat transfer in fluidized beds, Int J Heat Mass Trans, 30(5), 827-832 (1987).

Turton, R.; Colakyan, M., and Levenspiel, O., Heat transfer from fluidized beds to immersed fine wires, Powder Tech, 53(3), 195-204 (1987).

Basu, P., and Nag, P.K., An investigation into heat transfer in circulating fluidized beds, Int J Heat Mass Trans, 30(11), 2399-2410 (1988).

El-Halwagi, A.M., et al., Mathematical modeling of fluidized bed heat regenerators, Chem Eng Commns, 72, 121-140 (1988).

Filtris, Y.; Flamant, G., and Hatzik, P., Wall-to-fluidized bed radiative heat transfer analysis using the particle model, Chem Eng Commns, 72, 187-200 (1988).

Grewal, N.S., and Zimmerman, A.T., Heat transfer from horizontal tube immersed in liquid-solid fluidized bed, Powder Tech, 54(2), 137-146 (1988).

Kurosaki, Y.; Ishiguro, H., and Takahashi, K., Fluidization and heat-transfer characteristics around a horizontal heated circular cylinder immersed in a gas fluidized bed, Int J Heat Mass Trans, 31(2), 349-358 (1988).

Magiliotou, M.; Chen, Y.M., and Fan, L.S., Bed-immersed object heat transfer in three-phase fluidized bed, AIChEJ, 34(6), 1043-1047 (1988).

24.4 Fluidized-Bed Combustion

1968-1979

Bowling, K.M., and Waters, P.L., Fuel processing in fluidised beds, Brit Chem Eng, 13(8), 1127-1133 (1968).

Buckham, J.A., et al., Fluidized bed combustion, Chem Eng Prog, 64(7), 52-57 (1968).

Avedesian, M.M., and Davidson, J.F., Combustion of carbon particles in a fluidised bed, Trans IChemE, 51, 121-131 (1973).
Dickey, B.R.; Grimmett, E.S., and Kilian, D.C., Waste heat disposal via fluidized beds, Chem Eng Prog, 70(1), 60-64 (1974).
Locke, H.B., and Lunn, H.G., Clean heat and power cycles using fluidised combustion, Chem Engnr, Nov, 667-670 (1975).
Basu, P., Burning rate of carbon in fluidized beds, Fuel, 56(4), 390-392 (1977).
Steedman, W.G.; Corder, W.C., and Meeks, H.C., Solids segregation and flow in a coal gasification burner, Chem Eng Prog, 73(1), 69-70 (1977).
Altenkirch, R.A.; Peck, R.E., and Chen, S.L., Fluidized bed feeding of pulverized coal, Powder Tech, 20, 189-196 (1978).
Bishop, R.J., Deposition/corrosion in coal-fired fluidized beds, Energy World, Oct, 5-8 (1978).
McKenzie, E.C., Burning coal in fluidized beds, Chem Eng, 14 Aug, 116-127 (1978).
Pritchard, A.B., and Caplin, P.B., Fluid bed combustion systems, Energy World, Aug, 16-18 (1978).
Gay, P., Fluidised bed boilers, Proc Engng, March, 61,63 (1979).
Golan, L.P., et al., Particle size effects in fluidized bed combustion, Chem Eng Prog, 75(7), 63-72 (1979).
Leung, L.S., and Smith, I.W., The role of fuel reactivity in fluidized-bed combustion, Fuel, 58(5), 354-360 (1979).
Scarborough, C.E., et al., Fluidized-bed combustion at high heat-flux levels, Chem Eng Prog, 75(7), 41-46 (1979).
Van Dyk, G.C., Mechanical design of fluid bed combustors, Chem Eng Prog, 75(12), 46-51 (1979).
Various, Fluidized-bed combustion (topic issue), J Inst Energy, 52, 51-99 (1979).

1980-1984

Anon., Fluidized bed combustion, Processing, April, 49-51 (1980).
Highley, J., Design and control of fluidized-bed fired boilers, J Inst Energy, 53, 208-216 (1980).
Rajan, R.R., and Wen, C.Y., A comprehensive model for fluidized bed coal combustors, AIChEJ, 26(4), 642-655 (1980).
Roscoe, J.C.; Witkowski, A.R., and Harrison, D., The temperature of coke particles in a fluidised combustor, Trans IChemE, 58, 69-72 (1980).
Stanmore, B.R., and Jung, K., The burnout rates of brown coal char particles in fluidised bed combustors, Trans IChemE, 58, 66-68 (1980).
Ganapathy, V., Predict performance of fluid bed combustors, Hyd Proc, 60(11), 269-270 (1981).
Gill, D.W., The potential of fluidised-bed combustion for emission control, Chem Engnr, June, 278-280 (1981).
Kawabata, J.I., et al., Performance of a pressurized two-stage fluidized gasification process for production of low-Btu gas from coal char, Chem Eng Commns, 11(6), 335-346 (1981).
Pillai, K.K., Burning rates of coal in fluidized-bed combustion, Fuel, 60(2), 163-164 (1981).

Ross, I.B.; Patel, M.S., and Davidson, J.F., The temperature of burning carbon particles in fluidised beds, Trans IChemE, 59, 83-88 (1981).

Various, Fluidized-bed combustion (topic issue), J Inst Energy, 54, 38-58, 94-102 (1981).

Allen, P.M.; French, M.T.; Watters, J.C., and Cunningham, R.D., Pneumatic feeder/splitter system for fluidized bed combustors, Chem Eng Commns, 18(5), 331-340 (1982).

Ross, I.B., and Davidson, J.F., The combustion of carbon particles in a fluidised bed, Trans IChemE, 60, 108-114 (1982).

Cooke, M.J., Fluidized bed combustion for burning biomass, Energy World, March, 2-7 (1984).

Fitzgerald, T.; Bushnell, D.; Crane, S., and Shieh, Y.C., Testing of cold scaled bed modeling for fluidized-bed combustors, Powder Tech, 38, 107-120 (1984).

Green, R.C.; Paterson, N.P., and Summerfield, I.R., Fluidized bed gasification for industrial applications, Energy World, July, 7-10 (1984).

Turnbull, E., et al., Effect of pressure on combustion of char in fluidised beds, CER&D, 62, 223-234 (1984).

Various, Fluidized bed combustion (topic issue), Chem Eng Prog, 80(1), 35-67 (1984).

1985-1986

Datta, A.B.; Nandi, S.S., and Bhaduri, D., Fluidized bed combustion of coal, Fuel, 64(4), 564-567 (1985).

Grewal, N.S.; Sorenson, E.S., and Goblirsch, G., Heat transfer to horizontal tubes in a pilot-scale fluidized-bed combustor burning low-rank coals, Chem Eng Commns, 39, 43-68 (1985).

Jones, L., and Glicksman, L.R., Experimental investigation of gas flow in a scale model of a fluidized-bed combustor, Powder Tech, 45, 201-213 (1985).

La Nauze, R.D., Mass transfer considerations in fluidised-bed combustion with particular reference to the influence of system pressure, CER&D, 63, 219-229 (1985).

La Nauze, R.D., Fundamentals of coal combustion in fluidised beds (review paper), CER&D, 63, 3-33 (1985).

Miccio, M., and Salatino, P., Computations of performance of fluidized coal combustors, Powder Tech, 43, 163-167 (1985).

Stubington, J.F., Comparison of techniques for measuring the temperature of char particles burning in a fluidised bed, CER&D, 63, 241-249 (1985).

Taeed, O., Systems-based methodology applied to fluidized-bed combustion, Energy World, Feb, 7-10 (1985).

Basu, P., Design considerations for calculating fluidized bed combustors, J Inst Energy, Dec, 179-183 (1986).

La Nauze, R.D., and Jung, K., Mass transfer relationships in fluidized-bed combustors, Chem Eng Commns, 43, 275-286 (1986).

1987-1988

Agarwal, P.K., and Gaissmaier, A.E., Combustion of coal volatiles in gas fluidized beds, CER&D, 65(5), 431-441 (1987).

Botterill, J.S.M., Technological aspects of fluidised-bed combustor design, Energy World, May, 2-6 (1987).
Gendreau, R.J., and Raymond, D.L., Assessment of burning waste fuel in circulating fluidized-bed boilers, Plant/Opns Prog, 6(1), 46-51 (1987).
Grewal, N.S., et al., Heat transfer to horizontal tubes immersed in a fluidized-bed combustor, Powder Tech, 52(2), 149-160 (1987).
Ho, T.C., et al., Dynamic simulation of shallow-jetting fluidized-bed coal combustor, Powder Tech, 53(3), 247-256 (1987).
La Nauze, R.D., A review of the fluidised bed combustion of biomass, J Inst Energy, June, 66-76 (1987).
Massimilla, L., and Salatino, P., Theoretical approach to characterization of carbon attrition in fluidized-bed combustor, Chem Eng Commns, 62(1), 285-302 (1987).
Riley, R.K., and Judd, M.R., Measurement of char-steam gasification kinetics for design of fluidised-bed coal gasifier containing draft tube, Chem Eng Commns, 62(1), 151-160 (1987).
Weiss, V., et al., Mathematical modelling of circulating fluidized bed reactors by reference to solids decomposition reaction and coal combustion, Chem Eng & Proc, 22(2), 79-90 (1987).
Weluda, K.D., et al., Bubble-phase behavior in high-temperature fluidized-bed reactor with coal combustion, Chem Eng Commns, 62(1), 161-187 (1987).
Wilkie, F., Steam generation using a fluidised bed combustor, Proc Engng, March, 61-63 (1987).
Zhao, J.; Lim, C.J., and Grace, J.R., Coal burnout times in spouted and spout-fluid beds, CER&D, 65(5), 426-430 (1987).
Achara, N.; Horsley, M.E.; Purvis, M.R.I., and Teague, R.H., Zonal heat transfer rates in a fluidized bed combustor, Int J Heat Mass Trans, 31(3), 577-582 (1988).
Cooper, D.A., and Ljungstrom, E.B., In-bed gas measurements in a commercial 16 MW fluidized bed combustor, Chem Eng J, 39(3), 139-146 (1988).
Dawes, S.G.; Gibbs, G.B., and Highley, J., The British Coal/CEGB project on pressurised fluidised-bed combustion, J Inst Energy, March, 17-26 (1988).
Lim, C.J., et al., Spouted, fluidized and spout-fluid bed combustion of bituminous coals, Fuel, 67(9), 1211-1217 (1988).
Saffer, M.; Ocampo, A., and Laguerie, C., Gasification of coal in fluidized bed in presence of water and oxygen: Experimental study and reactor modeling, Int Chem Eng, 28(1), 46-62 (1988).
Schouten, J.C.; Valkenburg, P.J.M., and van den Bleek, C.M., Segregation in slugging FBC large-particle system, Powder Tech, 54(2), 85-98 (1988).
Thyn, J.; Kolar, Z.; Martens, W., and Korving, A., Gas flow in a pressurized fluidized-bed combustor, Powder Tech, 56(3), 157-164 (1988).

CHAPTER 25

CRYSTALLIZATION

25.1 Theory and Data

1967-1968

Barduhn, A.J., Desalination by crystallization, Chem Eng Prog, 63(1), 98-103 (1967).
Botsaris, G.D.; Mason, E.A., and Reid, R.C., Incorporation of ionic impurities in crystals growing from solution: Lead ions in potassium chloride crystals, AIChEJ, 13(4), 764-768 (1967).
Cayey, N.W., and Estrin, J., Secondary nucleation in agitated magnesium sulfate solutions, Ind Eng Chem Fund, 6(1), 13-20 (1967).
Han, C.D., Determination of crystal growth rate by analog computer simulation, Chem Eng Sci, 22(4), 611-618 (1967).
Harriott, P., Growth of ice crystals in a stirred tank, AIChEJ, 13(4), 755-759 (1967).
Mullin, J.W., and Amatavivadhana, A., Growth kinetics of ammonium- and potassium-dihydrogen phosphate crystals, J Appl Chem, 17, 151-156 (1967).
Razumovskii, L.A., and Streltsov, V.V., Salt crystallization process from solutions in fluidized beds, Int Chem Eng, 7(4), 604-608 (1967).
Abegg, C.F.; Stevens, J.D., and Larson, M.A., Crystal-size distribution in continuous crystallizers when growth rate is size dependent, AIChEJ, 14(1), 118-122 (1968).
Amin, A.B., and Larson, M.A., Crystallization of calcium sulfate from phosphoric acid, Ind Eng Chem Proc Des Dev, 7(1), 133-137 (1968).
Han, C.D., Evaluation of some of the kinetic parameters in crystallization, Chem Eng Sci, 23(4), 321-330 (1968).
Hidalgo, A.F., and Orr, C., Homogeneous nucleation of sodium chloride solutions, Ind Eng Chem Fund, 7(1), 79-83 (1968).
Larson, M.A.; Timm, D.C., and Wolff, P.R., Effect of suspension density on crystal size distribution, AIChEJ, 14(3), 448-452 (1968).
Mitsuda, H.; Miyake, K., and Nakai, T., Experimental study of seed crystal formation, Int Chem Eng, 8(4), 733-739 (1968).
Mullin, J.W., and Garside, J., Crystallization of aluminium potassium sulphate, Trans IChemE, 45, T285-295 (1967); 46, T11-18 (1968).
Penney, W.R., Crystallization on a constant temperature surface, AIChEJ, 14(4), 661-662 (1968).
Randolph, A.D.; Deepak, C., and Iskander, M., Narrowing of particle size distributions in staged vessels with classified product removal, AIChEJ, 14(5), 827-830 (1968).
Rohatgi, P.K.; Jain, S.M., and Adams, C.M., Dendritic crystallization of ice from aqueous solutions, Ind Eng Chem Fund, 7(1), 72-79 (1968).
Savage, H.R.; Butt, J.B., and Tallmadge, J.A., Kinetics of reaction and crystallization in condensed phases: Aqueous potassium dipicrylamine system, AIChEJ, 14(2), 266-274 (1968).
Timm, D.C., and Larson, M.A., Effect of nucleation kinetics on dynamic behavior of continuous crystallizer, AIChEJ, 14(3), 452-457 (1968).
Watanabe, A.; Kawakami, T., and Moroto, S., Crystallization of spent pickling liquor with acetone, Int Chem Eng, 8(2), 369-374 (1968).

1969-1970

Albertins, R., and Powers, J.E., Column crystallizer purification of material with eutectic forming impurities, AIChEJ, 15(4), 554-560 (1969).

Aldcroft, D.; Bye, G.C., and Hughes, C.A., Crystallisation processes in aluminium hydroxide gels, J Appl Chem, 19, 167-172 (1969).

Chiu, S.Y.; Fan, L.T., and Akins, R.G., Experimental study of ice-making operation in inversion desalination freezing process, Ind Eng Chem Proc Des Dev, 8(3), 347-356 (1969).

Estrin, J.; Sauter, W.A., and Karshina, G.W., Size-dependent crystal growth-rate expressions, AIChEJ, 15(2), 289-290 (1969).

Kirwan, D.J., and Pigford, R.L., Crystallization kinetics of pure and binary melts, AIChEJ, 15(3), 442-449 (1969).

Liu, S.L., Continuous process of zeolite A crystallization, Chem Eng Sci, 24(1), 57-64 (1969).

Randolph, A.D., Effect of crystal breakage on crystal size distribution in a mixed-suspension crystallizer, Ind Eng Chem Fund, 8(1), 58-63 (1969).

Rubin, B., Growth of single crystals by controlled diffusion in silica gel, AIChEJ, 15(2), 206-208 (1969).

Sutherland, D.N., Crystallisation-dissolution cycling of sucrose crystals, Chem Eng Sci, 24(1), 192-193 (1969).

Botsaris, G.D., and Denk, E.G., Growth rates of aluminium potassium sulfate crystals in aqueous solutions, Ind Eng Chem Fund, 9(2), 276-283 (1970).

Canning, T.F., Interpreting population-density data from crystallizers, Chem Eng Prog, 66(7), 80-85 (1970).

Dikshit, R.C., and Chivate, M.R., Separation of ortho and para nitrochlorobenzenes by extractive crystallisation, Chem Eng Sci, 25(2), 311-318 (1970).

Gates, W.C., and Powers, J.E., Determination of mechanisms causing and limiting separations by column crystallization, AIChEJ, 16(4), 648-658 (1970).

Ishibashi, T., and Tani, Y., Crystallization of Nylon 6, Int Chem Eng, 10(2), 294-303 (1970).

Mullin, J.W.; Amatavivadhana, A., and Chakraborty, M., Crystal habit modification studies with ammonium and potassium dihydrogen phosphate, J Appl Chem, 20, 153-158 (1970).

Mullin, J.W.; Chakraborty, M., and Mehta, K., Nucleation and growth of ammonium sulphate crystals from aqueous solution, J Appl Chem, 20, 367-371 (1970).

Murthy, A.S.A., and Mahadevappa, D.S., Distribution of isomorphous salts between aqueous and solid phases in fractional crystallization, Ind Eng Chem Proc Des Dev, 9(2), 260-263 (1970).

Randolph, A.D., Crystallization problems, Chem Eng, 4 May, 80-96 (1970).

Randolph, A.D., and Rajagopal, K., Direct measurement of crystal nucleation and growth rate kinetics in backmixed crystal slurry, Ind Eng Chem Fund, 9(1), 165-171 (1970).

White, E.T., and Lawrence, J., Variation of volume-surface mean size for growing particles, Powder Tech, 4, 104-107 (1970).

Zanker, A., Nomogram for theoretical yield calculation of a crystallisation process, Brit Chem Eng, 15(11), 1465 (1970).

1971-1972

Cheng, C.T., and Pigford, R.L., Purity of crystals grown from binary organic melts, Ind Eng Chem Fund, 10(2), 220-228 (1971).

Dikshit, R.C., and Chivate, M.R., Selectivity of solvent for extractive crystallisation, Chem Eng Sci, 26(5), 719-728 (1971).

Garside, J., Concept of effectiveness factors in crystal growth, Chem Eng Sci, 26(9), 1425-1432 (1971).

Hsu, A.C.T., Nucleation and crystal proliferation kinetics, AIChEJ, 17(6), 1311-1315 (1971).

Lei, S.J.; Shinnar, R., and Katz, S., Stability and dynamic behavior of continuous crystallizer with fines trap, AIChEJ, 17(6), 1459-1470 (1971).

Mullin, J.W., Crystallization: A study in molecular engineering, Chem & Ind, 27 Feb, 237-242 (1971).

Sun, Y.C., Water: Key to new crystallization process for purifying organics, Chem Eng, 12 July, 87-90 (1971).

Vasile, C., Study of crystallization and melting of amorphocrystalline polymers by differential thermal analysis, Int Chem Eng, 11(4), 698-709 (1971).

Clontz, N.A., and McCabe, W.L., Effects of diffusion, surface growth resistance and particle solubility on early growth of magnesium sulfate crystals from solution, Chem Eng Sci, 27(2), 307-314 (1972).

Clontz, N.A.; Johnson, R.T.; McCabe, W.L., and Rousseau, R.W., Growth of magnesium sulfate heptahydrate crystals from solution, Ind Eng Chem Fund, 11(3), 368-373 (1972).

Fitzgerald, T.J., and Yang, T.C., Size distribution for crystallization with continuous growth and breakage, Ind Eng Chem Fund, 11(4), 588-590 (1972).

Mullin, J.W., The measurement of supersaturation, Chem Engnr, May, 186-187, 193 (1972).

Murthy, A.S.A., and Mahadevappa, D.S., Distribution of isomorphous salts between aqueous and solid phases in fractional crystallization, Ind Eng Chem Proc Des Dev, 11(2), 201-203 (1972).

Randolph, A.D., and Cise, M.D., Nucleation kinetics of the potassium sulfate-water system, AIChEJ, 18(4), 798-807 (1972).

Rees, N.W.; Frew, J.A.; Batterham, R.J., and Thornton, G.J., Optimal control theory applied to batch crystallization of sugar, Chem Eng J, 3(3), 301-303 (1972).

Rogers, J.F., and Creasy, D.E., Growth calculation of polydisperse crystal suspensions, Powder Tech, 6, 263-269 (1972).

Youngquist, G.R., and Randolph, A.D., Secondary nucleation in a class II system: Ammonium sulfate-water, AIChEJ, 18(2), 421-429 (1972).

Zanker, A., Nomograph for determination of viscosity of liquid-solid suspensions, Chem Engnr, Feb, 76-77 (1972).

1973

Bennett, R.C.; Fiedelman, H., and Randolph, A.D., Crystallizer influenced nucleation, Chem Eng Prog, 69(7), 86-93 (1973).

Butler, P., New techniques in crystallization, Proc Engng, April, 92-93 (1973).

Ching, W., and McCartney, E.R., Use of a membrane electrode to study crystallisation of calcium sulphate from aqueous solution, J Appl Chem Biotechnol, 23, 441-456 (1973).
Edie, D.D., and Kirwan, D.J., Impurity trapping during crystallization from melts, Ind Eng Chem Fund, 12(1), 100-106 (1973).
Frew, J.A., Optimal control of batch raw-sugar crystallization, Ind Eng Chem Proc Des Dev, 12(4), 460-467 (1973).
Garside, J.; Mullin, J.W., and Das, S.N., Importance of crystal shape in crystal growth rate determinations, Ind Eng Chem Proc Des Dev, 12(3), 369-371 (1973).
Ishii, T., Multi-particle crystal growth rates in vertical cones, Chem Eng Sci, 28(5), 1121-1128 (1973).
Jones, A.G., and Mullin, J.W., Crystallisation kinetics of potassium sulphate in a draft-tube agitated vessel, Trans IChemE, 51, 302-308 (1973).
Kehhia, B., and Foster, P.J., Calculation of growth of polydisperse crystal suspensions, Powder Tech, 8, 191-192 (1973).
Liu, Y.A., Crystal-size intensity function and interpretation of population-density data from crystallizers, AIChEJ, 19(6), 1254-1257 (1973).
Liu, Y.A., and Botsaris, G.D., Impurity effects in continuous-flow mixed-suspension crystallizers, AIChEJ, 19(3), 510-516 (1973).
Miller, R.D., Porous phase barrier in crystallization, Sepn Sci, 8(5), 521-536 (1973).
Mullin, J.W., Solution concentration and supersaturation units and their conversion factors, Chem Engnr, June, 316-317 (1973).
Ottens, E.P.K., and de Jong, E.J., Model for secondary nucleation in a stirred-vessel cooling crystallizer, Ind Eng Chem Fund, 12(2), 179-184 (1973).
Randolph, A.D.; Beer, G.L., and Keener, J.P., Stability of the class II classified-product crystallizer with fines removal, AIChEJ, 19(6), 1140-1149 (1973).
Strickland-Constable, R.F., Secondary nucleation, Chem Engnr, Dec, 603-604 (1973).
Sung, C.Y.; Estrin, J., and Youngquist, G.R., Secondary nucleation of magnesium sulfate by fluid shear, AIChEJ, 19(5), 957-962 (1973).
Wey, J.S., and Estrin, J., Modeling the batch crystallization process: Ice-brine system, Ind Eng Chem Proc Des Dev, 12(3), 236-246 (1973).

1974

Bamforth, A.W., Crystals and crystalliser systems, Chem Engnr, July, 455-457 (1974).
Bauer, L.G.; Rousseau, R.W., and McCabe, W.L., Influence of crystal size on the rate of contact nucleation in stirred-tank crystallizers, AIChEJ, 20(4), 653-659 (1974).
Chien, H.H.Y., and Larsen, A.H., Calculation of multistage multiphase reacting systems in crystallization, Ind Eng Chem Proc Des Dev, 13(3), 299-303 (1974).
Desai, R.M.; Rachow, J.W., and Timm, D.C., Collision breeding in cooling crystallization, AIChEJ, 20(1), 43-50 (1974).
Evans, T.W.; Margolis, G., and Sarofim, A.F., Mechanisms of secondary nucleation in agitated crystallizers, AIChEJ, 20(5), 950-966 (1974).

Garside, J.; Mullin, J.W., and Das, S.N., Growth and dissolution kinetics of potassium sulfate crystals in an agitated vessel, Ind Eng Chem Fund, 13(4), 299-305 (1974).

Jones, A.G., and Mullin, J.W., Programmed cooling crystallization of potassium sulphate solutions, Chem Eng Sci, 29(1), 105-118 (1974).

Kane, S.G.; Evans, T.W.; Brian, P.L.T., and Sarofim, A.F., Determination of kinetics of secondary nucleation in batch crystallizers, AIChEJ, 20(5), 855-862 (1974).

Konak, A.R., New model for surface reaction-controlled growth of crystals from solution, Chem Eng Sci, 29(7), 1537-1544 (1974).

Konak, A.R., Surface reaction-controlled dissolution of crystals in a solvent or solution, Chem Eng Sci, 29(8), 1785-1788 (1974).

Mullin, J.W., Recommended symbols for industrial crystallisation, Chem Engnr, July, 458-459, 464 (1974).

Mullin, J.W., and Ang, H.M., Crystal size measurement: Comparison of sieving and Coulter counter techniques, Powder Tech, 10, 153-156 (1974).

Omran, A.M., and King, C.J., Kinetics of ice crystallization in sugar solutions and fruit juices, AIChEJ, 20(4), 795-803 (1974).

Phillips, V.R., and Epstein, N., Growth of nickel sulfate in a laboratory-scale fluidized-bed crystallizer, AIChEJ, 20(4), 678-687 (1974).

Razumovskii, L.A.; Slivchenko, E.S., and Shvedov, Y.P., Crystal growth rates without aggregation, Int Chem Eng, 14(2), 199-203 (1974).

Rogers, J.F., and Creasy, D.E., Crystallisation of pentaerythritol, J Appl Chem Biotechnol, 24, 171-180 (1974).

Rosen, H.N., Importance of slip velocity in determining growth and nucleation kinetics in continuous crystallization, AIChEJ, 20(2), 388-390 (1974).

Sato, T., Crystallisation of gelatinous aluminium hydroxide, J Appl Chem Biotechnol, 24, 187-198 (1974).

Shadman-Yazdi, F., and Petersen, E.E., Co-crystallization of isomorphic substances, Chem Eng Sci, 29(1), 191-196 (1974).

Toussaint, A.G., and Donders, A.J.M., Mixing criterion in crystallization by cooling, Chem Eng Sci, 29(1), 237-246 (1974).

Vlahakis, J.G., and Barduhn, A.J., Growth rate of ice crystals in flowing water and salt solutions, AIChEJ, 20(3), 581-591 (1974).

1975-1976

Gnyra, B., and Brown, N., Coarsening of Bayer alumina trihydrate by crystallization modifiers, Powder Tech, 11, 101-105 (1975).

Jancic, S., and Garside, J., Determination of crystallization kinetics from crystal size distribution data, Chem Eng Sci, 30(10), 1299-1301 (1975).

Rosner, D.E., and Epstein, M., Simultaneous kinetic and heat transfer limitations in crystallization of highly undercooled melts, Chem Eng Sci, 30(5), 511-521 (1975).

Rousseau, R.W.; McCabe, W.L., and Tai, C.Y., Stability of nuclei generated by contact nucleation, AIChEJ, 21(5), 1017-1019 (1975).

Shen, C.Y., Production of crystalline pyrophosphoric acid and its salts, Ind Eng Chem Proc Des Dev, 14(1), 80-85 (1975).

Shiloh, K.; Sideman, S., and Resnick, W., Crystallization in a dispersed phase, Can JCE, 53, 157-168 (1975).

Tai, C.Y.; McCabe, W.L., and Rousseau, R.W., Contact nucleation of various crystal types, AIChEJ, 21(2), 351-358 (1975).
Wells, G.L.; Makin, A.C.I., and Scarlett, P., Process stages for mother liquor separation, Trans IChemE, 53, 136-142 (1975).
Yu, K.M., and Douglas, J.M., Self-generated oscillations in continuous crystallizers, AIChEJ, 21(5), 917-930 (1975).
Arkenbout, G.J., Continuous fractional crystallization, Chemtech, Sept, 596-599 (1976).
Duncan, A.G., and Phillips, R.H., Crystallisation by direct contact cooling and its application to p-xylene production, Trans IChemE, 54, 153-159 (1976).
Garside, J., and Jancic, S.J., Growth and dissolution of potash alum crystals in subsieve size range, AIChEJ, 22(5), 887-894 (1976).
Garside, J.; Phillips, V.R., and Shah, M.B., Size-dependent crystal growth, Ind Eng Chem Fund, 15(3), 230-233 (1976).
Grossman, G., Melting, freezing, and channeling phenomena in ice counterwashers, AIChEJ, 22(6), 1033-1042 (1976).
Halfon, A., and Kaliaguine, S., Alumina trihydrate crystallization, Can JCE, 54, 160-175 (1976).
Randolph, A.D., and Sikdar, S.K., Creation and survival of secondary crystal nuclei, Ind Eng Chem Fund, 15(1), 64-71 (1976).
Sarig, S.; Leshem, R., and Ben-Yosef, N., Effect of fluoride ions on crystallization of calcium hydrogen phosphate (DCP), Chem Eng Sci, 31(11), 1061-1064 (1976).
Sikdar, S.K., and Randolph, A.D., Secondary nucleation of two fast growth systems in a mixed suspension crystallizer: Magnesium sulfate and citric acid-water systems, AIChEJ, 22(1), 110-117 (1976).
Stocking, J.H., and King, C.J., Secondary nucleation of ice in sugar solutions and fruit juices, AIChEJ, 22(1), 131-140 (1976).
Tare, J.P., and Chivate, M.R., Selection of solvent for adductive crystallization, Chem Eng Sci, 31(10), 893-900 (1976).

1977-1978

Helt, J.E., and Larson, M.A., Effects of temperature on crystallization of potassium nitrate by direct measurement of supersaturation, AIChEJ, 23(6), 822-830 (1977).
Ishii, T.; Fujita, S., and Johnson, A.I., Continuous multi-particle crystal growth in vertical cones, Can JCE, 55, 177-185 (1977).
Juzaszek, P., and Larson, M.A., Influence of fines dissolving on crystal size distribution in MSMPR crystallizer, AIChEJ, 23(4), 460-468 (1977).
Kallungal, J.P., and Barduhn, A.J., Growth rate of an ice crystal in subcooled pure water, AIChEJ, 23(3), 294-303 (1977).
Ladenburger, M.A.; Gunn, R.J., and Bauer, L.G., Contact nucleation of para-dichlorobenzene, AIChEJ, 23(1), 124-125 (1977).
Myerson, A.S., and Kirwan, D.J., Impurity trapping during dendritic crystal growth, Ind Eng Chem Fund, 16(4), 414-425 (1977).
Randolph, A.D., and White, E.T., Modeling size dispersion in the prediction of crystal-size distribution, Chem Eng Sci, 32(9), 1067-1076 (1977).
Randolph, A.D.; Beckman, J.R., and Kraljevich, Z.I., Crystal size distribution dynamics in a classified crystallizer, AIChEJ, 23(4), 500-520 (1977).

Whitehead, B.D., Crystallization and drying of polyethylene terephthalate, Ind Eng Chem Proc Des Dev, 16(3), 341-346 (1977).
Arkenbout, G.J., Progress in continuous fractional crystallization, Sepn & Purif Methods, 7(1), 99-134 (1978).
Bourne, J.R.; Davey, R.J., and McCulloch, J., Growth kinetics of hexamethylene tetramine crystals from a water/acetone solution, Chem Eng Sci, 33(2), 199-204 (1978).
Garside, J., and Jancic, S.J., Prediction and measurement of crystal size distributions for size-dependent growth, Chem Eng Sci, 33(12), 1623-1630 (1978).
Janse, A.H., and de Jong, E.J., Width of the meta-stable zone, Trans IChemE, 56, 187-193 (1978).
Jongenelen, F.C.H., and Heertjes, P.M., Crystallization on a vertical wall, Chem Eng Sci, 33(1), 47-64 (1978).
Khambaty, S., and Larson, M.A., Crystal regeneration and crystal growth of small crystals in contact nucleation, Ind Eng Chem Fund, 17(3), 160-165 (1978).
Kraljevich, Z.I., and Randolph, A.D., Design-oriented model of fines dissolving, AIChEJ, 24(4), 598-606 (1978).
Lee, H.H., Determination of birth and growth rate of secondary nuclei: SSBCR crystallizer, AIChEJ, 24(3), 535-537 (1978).
Lodaya, K.D.; Lahti, L.E., and Jones, M.L., Nucleation kinetics of urea crystallization in water, Ind Eng Chem Proc Des Dev, 16(3), 294-297 (1977); 17(4), 576-577 (1978).
Miller, A.G., Determination of particle size distribution of salt crystals in aqueous slurries, Powder Tech, 21, 275-284 (1978).
O'Dell, F.P., and Rousseau, R.W., Magma density and dominant size for size-dependent crystal growth, AIChEJ, 24(4), 738-741 (1978).
Rivera, T., and Randolph, A.D., Model for precipitation of pentaerythritol tetranitrate, Ind Eng Chem Proc Des Dev, 17(2), 182-188 (1978).
Sarig, S.; Edelman, N.; Glasner, A., and Epstein, J.A., Effect of supersaturation on crystal characteristics of potassium chloride, J Appl Chem Biotechnol, 28, 663-667 (1978).
Shadman, F., and Randolph, A.D., Nucleation and growth rates of ammonium chloride in organic media, AIChEJ, 24(5), 782-788 (1978).
Zanker, A., Nomograph for particle size distribution after crystallization, Proc Engng, Oct, 130-131 (1978).

1979-1980

Ang, H.M., and Mullin, J.W., Crystal growth rate determinations from desupersaturation measurements: Nickel ammonium sulphate hexahydrate, Trans IChemE, 57, 237-243 (1979).
Chivate, M.R.; Palwe, B.G., and Tavare, N.S., Effect of seed concentration in a batch dilution crystallizer, Chem Eng Commns, 3(3), 127-134 (1979).
Fernandez-Lozano, J.A., and Wint, A., Production of potassium sulphate by an ammoniation process, Chem Engnr, Oct, 688-690 (1979).
Garside, J., and Jancic, S.J., Measurement and scale-up of secondary nucleation kinetics for potash alum-water system, AIChEJ, 25(6), 948-958 (1979).

Garside, J.; Rusli, I.T., and Larson, M.A., Origin and size distribution of secondary nuclei, AIChEJ, 25(1), 57-64 (1979).
Heist, J.A., Freeze crystallization, Chem Eng, 7 May, 72-82 (1979).
Mullin, J.W., and Jancic, S.J., Interpretation of metastable zone widths, Trans IChemE, 57, 188-193 (1979).
Ramshaw, C., Industrial crystallisation research, Chem Engnr, Oct, 691-694 (1979).
Viola, M.S., and Botsaris, G.D., Simulation of nucleation kinetics: Application to secondary nucleation, Chem Eng Sci, 34(7), 993-1000 (1979).
Bourne, J.R., amd Zabelka, M., Influence of gradual classification on continuous crystallization, Chem Eng Sci, 35(3), 533-542 (1980).
DeSilva, R.L.; Falanges, E., and Creasy, D.E., Crystal growth rate determination from desupersaturation measurements, Trans IChemE, 58, 135-137 (1980).
Garside, J., and Davey, R.J., Secondary contact nucleation: Kinetics, growth and scale-up (review paper), Chem Eng Commns, 4(4), 393-424 (1980).
Ghez, R., Expansions in time for solution of one-dimensional Stefan problems of crystal growth, Int J Heat Mass Trans, 23(4), 425-432 (1980).
Ishii, T., and Randolph, A.D., Stability of high-yield MSMPR crystallizer with size-dependent growth rate, AIChEJ, 26(3), 507-510 (1980).
Karpinski, P.H., Crystallization as a mass transfer phenomenon, Chem Eng Sci, 35(11), 2321-2324 (1980).
Mukhopadhyay, S.C., and Epstein, M.A.F., Computer model for crystal size distribution control in a semi-batch evaporative crystallizer, Ind Eng Chem Proc Des Dev, 19(3), 352-358 (1980).
Mullin, J.W., Nucleation and crystal growth, Chem & Ind, 3 May, 372-377 (1980).
Mullin, J.W., and Whiting, M.J.L., Succinic acid crystal growth rates in aqueous solution, Ind Eng Chem Fund, 19(1), 117-121 (1980).
Rousseau, R.W., Crystallization: Review of recent developments, Chemtech, Sept, 566-571 (1980).
Rousseau, R.W., and O'Dell, F.P., Separation of multiple solutes by selective nucleation, Ind Eng Chem Proc Des Dev, 19(4), 603-608 (1980).
Rousseau, R.W., and O'Dell, F.P., Moments of crystal size distributions in systems with selective crystal removal or size dependent crystal growth, Chem Eng Commns, 6(4), 293-304 (1980).
Sarig, S., and Mullin, J.W., Size reduction of crystals in slurries by the use of crystal habit modifiers, Ind Eng Chem Proc Des Dev, 19(3), 490-494 (1980).
Viola, M.S., and Botsaris, G.D., Simulation of nucleation kinetics, Chem Eng Commns, 7(1), 1-16 (1980).
Wey, J.S., and Terwilliger, J.P., Effect of temperature on suspension crystallization processes, Chem Eng Commns, 4(2), 297-306 (1980).

1981
Brecevic, L., and Garside, J., Measurement of crystal size distributions in the micrometer size range, Chem Eng Sci, 36(5), 867-870 (1981).
Garside, J., and Tavare, N.S., Non-isothermal effectiveness factors for crystal growth, Chem Eng Sci, 36(5), 863-866 (1981).

Ishii, T., and Fujita, S., Crystallization from supersaturated cupric sulfate solutions in a batchwise stirred tank, Chem Eng J, 21(3), 255-260 (1981).

Jensen, R.A., Improving crystallizer performance by electropolishing, Chem Eng Prog, 77(7), 74-75 (1981).

Larson, M.A., Analysis of secondary nucleation, Chem Eng Commns, 12(1), 161-170 (1981).

Lieberman, A., On-line particle counting for crystallizer operation control, Chem Eng Prog, 77(11), 50-53 (1981).

Melikhov, I.V., and Berliner, L.B., Simulation of batch crystallization, Chem Eng Sci, 36(6), 1021-1034 (1981).

Randolph, A.D., and Puri, A.D., Effect of chemical modifiers on borax crystal growth, nucleation and habit, AIChEJ, 27(1), 92-99 (1981).

Randolph, A.D.; White, E.T., and Low, C.C.D., On-line measurement of fine-crystal response to crystallizer disturbances, Ind Eng Chem Proc Des Dev, 20(3), 496-503 (1981).

Rousseau, R.W., and Parks, R.M., Size-dependent growth of magnesium sulfate heptahydrate, Ind Eng Chem Fund, 20(1), 71-76 (1981).

Scrutton, A., and Grootscholten, P.A.M., A study of the dissolution and growth of sodium chloride crystals, Trans IChemE, 59, 238-246 (1981).

Slaminko, P., and Myerson, A.S., Effect of crystal size on occlusion formation during crystallization from solution, AIChEJ, 27(6), 1029-1031 (1981).

Toyokura, K.; Uchiyama, M.; Hirasawa, I., and Kawai, M., Effect of collisions between impeller blades and crystals on secondary nucleation rate of potash-alum system, Int Chem Eng, 21(2), 269-276 (1981).

Wang, M.L.; Huang, H.T., and Estrin, J., Secondary nucleation of citric acid, AIChEJ, 27(2), 312-315 (1981).

1982

Dabir, B.; Peters, R.W., and Stevens, J.D., Precipitation kinetics of magnesium hydroxide in a scaling system, Ind Eng Chem Fund, 21(3), 298-305 (1982).

Garside, J.; Gibilaro, L.G., and Tavare, N.S., Evaluation of crystal growth kinetics from a desupersaturation curve using initial derivatives, Chem Eng Sci, 37(11), 1625-1628 (1982).

Nyvlt, J., and Broul, M., Kinetic exponents in crystallization and the accuracy of their determination, Int Chem Eng, 22(3), 543-549 (1982).

Rousseau, R.W., and Howell, T.R., Comparison of simulated crystal size distribution control systems based on nuclei density and supersaturation, Ind Eng Chem Proc Des Dev, 21(4), 606-610 (1982).

Sarig, S., and Mullin, J.W., Effect of trace impurities on calcium sulphate precipitation, J Chem Tech Biotechnol, 32, 525-531 (1982).

Schliephake, D., Processes for crystallization from solution, Int Chem Eng, 22(3), 415-426 (1982).

Scrutton, A.; Grootscholten, P.A.M., and de Jong, E.J., Effect of impeller draft tube clearance on the crystallisation kinetics of sodium chloride, Trans IChemE, 60, 345-351 (1982).

Swinney, L.D.; Stevens, J.D., and Peters, R.W., Calcium carbonate crystallization kinetics, Ind Eng Chem Fund, 21(1), 31-36 (1982).

Tavare, N.S., and Garside, J., Determination of the Peclet number for crystal growth, Chem Eng J, 25(2), 229-232 (1982).

Wey, J.S., and Jagannathan, R., Determination of growth kinetics of polyhedral crystals, AIChEJ, 28(4), 697-698 (1982).
White, G.; Estrin, J.; Youngquist, G.R., and Glasser, M.L., Nonhomogeneous nucleation from solution, Chem Eng Commns, 15(1), 239-256 (1982).

1983
Akulichev, V.A., and Bulanov, V.A., Crystallization nuclei in liquid in a sound field, Int J Heat Mass Trans, 26(2), 289-300 (1983).
Berglund, K.A.; Kaufman, E.L., and Larson, M.A., Growth of contact nuclei of potassium nitrate, AIChEJ, 29(5), 867-869 (1983).
Bubnik, Z., and Kadlec, P., Growth rate of sucrose crystals, Chem Eng Commns, 19(4), 263-272 (1983).
Garside, J., Tailoring of crystal growth, Chem Engnr, May, 66-68 (1983).
Habshi, F.; Naito, K., and Awadalla, F.T., Crystallisation of impurities from black phosphoric acid at high temperatures, J Chem Tech Biotechnol, 33A(5), 261-265 (1983).
Kawashima, Y.; Naito, M.; Lin, S.Y., and Takenaka, H., Kinetics of spherical crystallization of sodium theophylline monohydrate, Powder Tech, 34, 255-260 (1983).
Sohnel, O., Metastable zone of solutions, CER&D, 61, 186-190 (1983).
Wang, M.L., and Chang, R.Y., Legendre function approximations of ordinary differential equation and application to continuous crystallization processes, Chem Eng Commns, 22(1), 115-125 (1983).

1984
Berglund, K.A., and Larson, M.A., Modeling of growth rate dispersion of citric acid monohydrate in continuous crystallizers, AIChEJ, 30(2), 280-287 (1984).
Chang, R.Y., and Wang, M.L., Shifted Legendre function approximation to differential equations: Application to crystallization processes, Comput Chem Eng, 8(2), 117-126 (1984).
Kawashima, Y; Okumura, M., and Takenaka, H., Effects of temperature on spherical crystallization of salicylic acid, Powder Tech, 39, 41-47 (1984).
Melikhov, I.V.; Vukovic, Z.; Bacic, S., and Lazic, S., Some characteristics of chemical crystallization in water purification processes, Chem Eng Sci, 39(12), 1707-1714 (1984).
Mullin, J.W., and Williams, J.R., Comparison between indirect and direct contact cooling methods for the crystallisation of potassium sulphate, CER&D, 62, 296-302 (1984).
Yagi, H., et al., Crystallization of calcium carbonate accompanying chemical absorption, Ind Eng Chem Fund, 23(2), 153-158 (1984).

1985
Beer, W.F., and Hock, K.L., Design criteria for crystallization of an organic salt from aqueous solution, Chem Eng & Proc, 19(1), 49-56 (1985).
Brown, D.J., and Felton, P.G., Direct measurement of concentration and size for particles of different shapes using laser light diffraction, CER&D, 63, 125-132 (1985).

Budz, J.; Karpinski, P.H., and Nuruc, Z., Effect of temperature on crystallization and dissolution processes in a fluidized bed, AIChEJ, 31(2), 259-268 (1985).

Girolami, M.W., and Rousseau, R.W., Size-dependent crystal growth, AIChEJ, 31(11), 1821-1828 (1985).

Grootscholten, P.A.M.; Scrutton, A., and de Jong, E.J., Influence of crystal breakage on kinetics of sodium chloride crystallisation, CER&D, 63, 34-42 (1985).

Huang, J.S., and Barduhn, A.J., Effect of natural convection on ice crystal growth rates in salt solutions, AIChEJ, 31(5), 747-752 (1985).

Karpinski, P.H., Importance of the two-step crystal growth model, Chem Eng Sci, 40(4), 641-646 (1985).

Larson, M.A.; White, E.T.; Ramanarayanan, K.A., and Berglund, K.A., Growth rate dispersion in MSMPR crystallizers, AIChEJ, 31(1), 90-94 (1985).

Mathis-Lilley, J.J., and Berglund, K.A., Contact nucleation from aqueous potash alum solutions, AIChEJ, 31(5), 865-867 (1985).

Shirai, Y.; Nakanishi, K.; Matsuno, R., and Kamikubo, T., Kinetics of ice crystallization in batch crystallizers, AIChEJ, 31(4), 676-682 (1985).

Shmidt, L., and Shmidt, J., Mechanism of crystallization in agitated solutions, Chem Eng Commns, 36(1), 233-250 (1985).

Tavare, N.S., Crystal growth rate dispersion, Can JCE, 63(3), 436-442 (1985).

Wey, J.S., Analysis of batch crystallization processes, Chem Eng Commns, 35(1), 231-253 (1985).

1986

Berglund, K.A., Summary of recent research on growth rate dispersion of contact nuclei, Chem Eng Commns, 41, 355-360 (1986).

Chang, J.C.S., and Brna, T.G., Gypsum crystallization for limestone FGD, Chem Eng Prog, 82(11), 51-56 (1986).

Garside, J., and Tavare, N.S., Recent advances in industrial crystallisation research, CER&D, 64(2), 77-79 (1986).

Jensen, B.E., Graph paper for crystallizer calculations, Chem Eng, 8 Dec, 161-164 (1986).

Jones, A.G.; Budz, J., and Mullin, J.W., Crystallization kinetics of potassium sulfate in an MSMPR agitated vessel, AIChEJ, 32(12), 2002-2009 (1986).

Kotzev, A.; Resnick, W., and Lavie, R., Evaluation of discrete approximations to continuous size distributions for control of crystallization processes, Chem Eng Sci, 41(12), 3045-3052 (1986).

Myerson, A.S.; Decker, S.E., and Weiping, F., Solvent selection and batch crystallization, Ind Eng Chem Proc Des Dev, 25(4), 925-929 (1986).

Rohani, S., Dynamic study and control of crystal size distribution in a KCl crystallizer, Can JCE, 64(1), 112-116 (1986).

Shirai, Y.; Sakai, K.; Nakanishi, K., and Matsuno, R., Analysis of ice crystallization in continuous crystallizers based on a particle size-dependent growth rate model, Chem Eng Sci, 41(9), 2241-2246 (1986).

Tavare, N.S., and Garside, J., Simultaneous estimation of crystal nucleation and growth kinetics from batch experiments, CER&D, 64(2), 109-118 (1986).

Webster, G., and Tavare, N.S., Dissolution and dispersion characteristics of potassium chloride crystals from size distributions in a batch dissolver, Chem Eng Sci, 41(9), 2446-2448 (1986).

Wynn, N., Use melt crystallization for higher purities, Chem Eng, 28 April, 26-27 (1986).

1987

Azoury, R.; Robertson, W.G., and Garside, J., Generation of supersaturation using reverse osmosis, CER&D, 65(4), 342-344 (1987).

Blem, K.E., Generation and growth of secondary ammonium dihydrogen phosphate nuclei, AIChEJ, 33(4), 677-680 (1987).

Davey, R.J., Looking into crystal chemistry, Chem Engnr, Dec, 24-27 (1987).

David, R.; Klein, J.P., and Villermaux, J., Crystallization and precipitation engineering, Chem Eng Sci, 43(1), 59-78 (1987).

Elankovan, P., and Berglund, K.A., Contact nucleation from aqueous dextrose solutions, AIChEJ, 33(11), 1844-1849 (1987).

Gaikar, V.G., and Sharma, M.M., Dissociation extractive crystallization, Ind Eng Chem Res, 26(5), 1045-1048 (1987).

Hsu, J.P., and Fan, L.T., Transient analysis of crystallization: Effect of size-dependent growth rate, Chem Eng Commns, 56, 19-40 (1987).

Huang, T.C.; Juang, R.S., and Huang, T.H., Mass transfer of sodium chloride in simulated crystallizing pond during the rainfall period, J Chem Tech Biotechnol, 39(2), 93-106 (1987).

Jones, A.G., and Chianese, A., Fines destruction during batch crystallization, Chem Eng Commns, 62(1), 5-16 (1987).

Karagodin, M.A., Mathematical model of sugar crystallization, Periodica Polytechnica, 31(3), 169-186 (1987).

Laudise, R.A., Hydrothermal crystallization, C&E News, 28 Sept, 30-43 (1987).

Liang, B.M.; Hartel, R.W., and Berglund, K.A., Contact nucleation in sucrose crystallization, Chem Eng Sci, 42(11), 2723-2728 (1987).

Liang, B.M.; Hartel, R.W., and Berglund, K.A., Growth rate dispersion in seeded batch sucrose crystallization, AIChEJ, 33(12), 2077-2079 (1987).

Qian, R.Y., et al., Crystallization kinetics of potassium chloride from brine and scale-up criterion, AIChEJ, 33(10), 1690-1697 (1987).

Saska, M., and Myerson, A.S., Crystal aging and crystal habit of terephthalic acid, AIChEJ, 33(5), 848-852 (1987).

Shiau, L.D., and Berglund, K.A., Growth kinetics of fructose crystals formed by contact nucleation, AIChEJ, 33(6), 1028-1033 (1987).

Tavare, N.S., Simulation of Ostwald ripening in reactive batch crystallizer, AIChEJ, 33(1), 152-156 (1987).

Wakao, H.; Hiraguchi, H., and Ishii, T., A simulation of crystallization from aqueous supersaturated solutions in a batch isothermal stirred tank, Chem Eng J, 35(3), 169-178 (1987).

Zumstein, R.C., and Rousseau, R.W., Growth rate dispersion by initial growth rate distributions and growth rate fluctuations, AIChEJ, 33(1), 121-129 (1987).

Zumstein, R.C., and Rousseau, R.W., Growth rate dispersion in batch crystallization with transient conditions, AIChEJ, 33(11), 1921-1925 (1987).

1988

Brown, R.A., Theory of transport processes in single crystal growth from the melt (review paper), AIChEJ, 34(6), 881-911 (1988).

Hounslow, M.J.; Ryall, R.L., and Marshall, V.R., Discretized population balance for nucleation, growth, and aggregation, AIChEJ, 34(11), 1821-1832 (1988).

Hsu, J.P.; Fan, L.T., and Chou, S.T., Transient analysis of crystallization: Effect of initial size distribution, Chem Eng Commns, 69, 95-114 (1988).

Kishimoto, S., and Naruse, M., Process development for bundling crystallization of aspartame, J Chem Tech Biotechnol, 43(1), 71-82 (1988).

Ring, T.A., Crystallization and powder precipitation, Chemtech, Jan, 60-64 (1988).

Sohnel, O., and Mullin, J.W., Role of time in metastable zone width determinations, CER&D, 66(6), 537-540 (1988).

Sohnel, O.; Mullin, J.W., and Jones, A.G., Crystallization and agglomeration kinetics in batch precipitation of strontium molybdate, Ind Eng Chem Res, 27(9), 1721-1727 (1988).

Steemson, M.L., and White, E.T., Numerical modelling of steady-state continuous crystallization processes using piecewise cubic spline functions, Comput Chem Eng, 12(1), 81-90 (1988).

25.2 Equipment Design

1967-1969

McKay, D.L., and Goard, H.W., Crystal purification column with cyclic solids movement, Ind Eng Chem Proc Des Dev, 6(1), 16-21 (1967).

Sherwin, M.B.; Shinnar, R., and Katz, S., Dynamic behavior of well-mixed isothermal crystallizer, AIChEJ, 13(6), 1141-1154 (1967).

Betts, W.D.; Freeman, J.W., and McNeil, D., Continuous multistage fractional crystallization, J Appl Chem, 18, 180-187 (1968).

Han, C.D., and Shinnar, R., Steady-state behavior of crystallizers with classified product removal, AIChEJ, 14(4), 612-619 (1968).

Koziol, K., and Bandrowski, J., Multiple crystallization with interstage evaporation, Int Chem Eng, 8(3), 469-472 (1968).

Li, N.N., Surfactant system for eliminating drop sticking in a dewaxing crystallization tower, Ind Eng Chem Proc Des Dev, 7(2), 239-243 (1968).

Streltsov, V.V., and Razumovskii, L.A., Particle aggregation in fluidized bed crystallization, Int Chem Eng, 8(4), 729-732 (1968).

Armstrong, R.M., The scraped-shell crystalliser, Brit Chem Eng, 14(5), 647-649 (1969).

Bolsaitis, P., Continuous column crystallization, Chem Eng Sci, 24(12), 1813-1826 (1969).

Han, C.D., Control study on isothermal mixed crystallizers, Ind Eng Chem Proc Des Dev, 8(2), 150-158 (1969).

Messing, T., Development in industrial crystallisation, Brit Chem Eng, 14(5), 641-644 (1969).

Nyvlt, J., Evaluation and characteristics of particle-size composition of products from agitated crystallizers, Int Chem Eng, 9(3), 422-426 (1969).
Player, M.R., Mathematical analysis of column crystallization, Ind Eng Chem Proc Des Dev, 8(2), 210-217 (1969).
Razumovskii, L.A., and Streltsov, V.V., Dispersivity characteristics of crystals during crystallization of salts from solutions in fluidized beds, Int Chem Eng, 9(1), 100-102 (1969).
Svalov, G.N., Calculation of drum crystallizers, Int Chem Eng, 9(4), 606-608 (1969).

1970-1972

Henry, J.D., and Powers, J.E., Experimental and theoretical investigation of continuous flow column crystallization, AIChEJ, 16(6), 1055-1063 (1970).
Hill, S., Residence time distribution in continuous crystallisers, J Appl Chem, 20, 300-304 (1970).
Mullin, J.W., and Garside, J., Velocity-voidage relationships in the design of suspended bed crystallisers, Brit Chem Eng, 15(6), 773-775 (1970).
Mullin, J.W., and Nyvlt, J., Design of classifying crystallizers, Trans IChemE, 48, T7-14 (1970).
Nyvlt, J., and Mullin, J.W., Periodic behavior of continuous crystallizers, Chem Eng Sci, 25(1), 131-148 (1970).
Orcutt, J.C., and Carey, T.P,, Simulation of liquid mixing in a freezer-crystallizer vessel, Ind Eng Chem Proc Des Dev, 9(1), 58-63 (1970).
Svalov, G.V., Estimating drum crystallisers, Brit Chem Eng, 15(9), 1153 (1970).
Margolis, G.; Sherwood, T.K.; Brian, P.L.T., and Sarofim, A.F., Performance of continuous well-stirred ice crystallizer, Ind Eng Chem Fund, 10(3), 439-452 (1971).
Mullin, J.W., and Nyvlt, J., Programmed cooling of batch crystallizers, Chem Eng Sci, 26(2), 369-378 (1971).
Timm, D.C., and Cooper, T.R., Crystallization: Kinetics and design considerations, AIChEJ, 17(2), 285-288 (1971).
Witte, J.F., and Voncken, R.M., Two case histories in the design of crystallisers, Brit Chem Eng, 16(8), 681-685; 16(11), 1061,1076 (1971).
Duncan, A.G., and West, C.D., Prevention of incrustation on crystallizer heat exchangers by ultrasonic vibration, Trans IChemE, 50, 109-114 (1972).
Wiegandt, H.F.; Von Berg, R.L., and Leinroth, J.P., Piston beds for crystal washing, Ind Eng Chem Proc Des Dev, 11(3), 404-414 (1972).

1973-1974

Anchus, B.E., and Ruckenstein, E., Stability of a well-stirred isothermal crystallizer, Chem Eng Sci, 28(2), 501-514 (1973).
Arkenbout, G.J.; van Kuijk, A., and Smit, W.M., New simple crystallisation column, Chem & Ind, 3 Feb, 139-142 (1973).
Larson, M.A., and Garside, J., Crystallizer design techniques using the population balance, Chem Engnr, June, 318-328 (1973).
Letan, R., Direct-contact cooled crystallizer, Ind Eng Chem Proc Des Dev, 12(3), 300-305 (1973).
Lieb, E.B., and Osmers, H.R., Effect of diffusion on crystal size distribution in continuous crystallizer, Chem Eng Commns, 1(1), 13-20 (1973).

Moyers, C.G., and Randolph, A.D., Crystal-size distribution and interaction with crystallizer design (review paper), AIChEJ, 19(6), 1089-1104 (1973).
Ramshaw, C., and Parker, I.B., Crystalliser design: Models of steady-state operation, Trans IChemE, 51, 82-92 (1973).
Ajinkya, M.B., and Ray, W.H., Optimal operation of crystallization processes, Chem Eng Commns, 1(4), 181-186 (1974).
Bamforth, A.W., Types of crystallisers, Chem Engnr, July, 443-445 (1974).
Bauer, L.G.; Larson, M.A., and Dallons, V.J., Contact nucleation of magnesium sulphate crystals in continuous MSMPR crystallizer, Chem Eng Sci, 29(5), 1253-1262 (1974).
Houghton, J., Pilot plant crystallisation, Chem Engnr, July, 450-454 (1974).
Jones, A.G., Optimal operation of batch cooling crystallizer, Chem Eng Sci, 29(5), 1075-1088 (1974).
Molinari, J.G.D., and Dodgson, B.V., Theory and practice related to the Brodie purifier, Chem Engnr, July, 460-464 (1974).
Moyers, C.G., and Olson, J.H., Dense-bed column crystallizer, AIChEJ, 20(6), 1118-1124 (1974).
Ramshaw, C., Secondary nucleation in mechanically agitated crystallisers, Chem Engnr, July, 446-449, 457 (1974).
Wiker, S.L., and Anshus, B.E., Characteristics of non-isothermal loop crystallizer, Chem Eng Sci, 29(7), 1575-1584 (1974).

1975-1979

Bochenek, W., and Drury, M.D., A plastic film crystalliser in the chemical industry, Chem Engnr, July, 435-436, 438 (1975).
Chivate, M.R., and Tavare, N.S., Growth rate measurements in DTB crystallizer, Chem Eng Sci, 30(3), 354-355 (1975).
Streltsov, V.V., et al., Mathematical modelling of crystallization equipment, Int Chem Eng, 15(2), 216-219 (1975).
Nienow, A.W., The effect of agitation and scale-up on crystal growth rates and on secondary nucleation, Trans IChemE, 54, 205-207 (1976).
Wey, J.S., and Terwilliger, J.P., Design considerations for a multistage cascade crystallizer, Ind Eng Chem Proc Des Dev, 15(3), 467-469 (1976).
Mullin, J.W., and Sohnel, O., Expressions of supersaturation in crystallization studies, Chem Eng Sci, 32(7), 683-686 (1977).
Singh, D., Modified crystallizer design for specialized applications, AIChEJ, 23(3), 400-401 (1977).
Tavare, N.S., and Chivate, M.R., Analysis of batch evaporative crystallizers, Chem Eng J, 14(3), 175-180 (1977).
Kay, J., Continuous fractional crystallizer, Processing, Dec, 25-28 (1978).
Larson, M.A., Guidelines for selecting a crystallizer, Chem Eng, 13 Feb, 90-102 (1978).
McNeil, T.J.; Weed, D.R., and Estrin, J., Modeling laboratory batch crystallizers, AIChEJ, 24(4), 728-731 (1978).
Mullin, J.W., and Broul, M., Laboratory-scale evaporative crystalliser, Chem & Ind, 1 April, 226-228 (1978).
Randolph, A.D., and Tan, C.S., Numerical design techniques for staged classified recycle crystallizers: Examples of continuous alumina and sucrose crystallizers, Ind Eng Chem Proc Des Dev, 17(2), 189-200 (1978).

Tavare, N.S., and Chivate, M.R., Growth rate correlation for potassium sulphate crystals in a fluidized-bed crystallizer, Chem Eng Sci, 33(9), 1290-1293 (1978).
Tavare, N.S., and Chivate, M.R., CSD analysis from single-stage and two-stage cascade MSCPR crystallizers, Can JCE, 56, 758-770 (1978).
Armstrong, A.J., Cooling crystallisation and flow patterns in scraped-surface crystallisers, Chem Engnr, Oct, 685-687 (1979).
Nagashima, Y., and Maeda, S., Mixing characteristics of fluid in a horizontal crystallizer, Int Chem Eng, 19(1), 126-131 (1979).
Neelakantan, P.S., and Mukesh, D., Computer modelling of a continuous evaporative crystallizer, Ind Eng Chem Proc Des Dev, 18(1), 56-59, (1979).
Nyvlt, J., and Broul, M., Crystallization using recycle of mother liquor, Int Chem Eng, 19(3), 547-553 (1979).
Tavare, N.S., and Chivate, M.R., Growth and dissolution kinetics of potassium sulphate crystals in a fluidised bed crystalliser, Trans IChemE, 57, 35-42 (1979).

1980-1983

Barone, J.P.; Furth, W., and Loynaz, S., Simplified derivation of the general population balance equation for a seeded continuous flow crystallizer, Can JCE, 58, 137-139 (1980).
Bourne, J.R., and Hungerbuehler, K., An experimental study of the scale-up of a well-stirred crystalliser, Trans IChemE, 58, 51-58 (1980).
Garside, J., and Shah, M.B., Crystallization kinetics from MSMPR crystallizers (a review), Ind Eng Chem Proc Des Dev, 19(4), 509-514 (1980).
Morari, M., Optimal operation of batch crystallizers, Chem Eng Commns, 4(2), 167-172 (1980).
Moritoki, M., Method of fractional crystallization under high pressure, Int Chem Eng, 20(3), 394-402 (1980).
Saxer, K., and Papp, A., The MWB crystallization process, Chem Eng Prog, 76(4), 64-66 (1980).
Tavare, N.S.; Garside, J., and Chivate, M.R., Analysis of batch crystallizers, Ind Eng Chem Proc Des Dev, 19(4), 653-665 (1980).
Sowul, L., and Epstein, M.A.F., Crystallization kinetics of sucrose in CMSMPR evaporative crystallizer, Ind Eng Chem Proc Des Dev, 20(2), 197-203 (1981).
Grootscholten, P.A.M.; De Leer, B.G.M.; De Jong, E.J., and Asselbergs, C.J., Factors affecting secondary nucleation rate of sodium chloride in an evaporative crystallizer, AIChEJ, 28(5), 728-737 (1982).
Jensen, R.A., Process changes to improve crystallization, Chem Eng Prog, 78(10), 66-69 (1982).
Tavare, N.S., and Garside, J., Estimation of crystal growth and dispersion parameters using pulse response techniques in batch crystallisers, Trans IChemE, 60, 334-344 (1982).
Zacek, S.; Nyvlt, J.; Garside, J., and Nienow, A.W., A stirred tank for continuous crystallization studies, Chem Eng J, 23(1), 111-114 (1982).

Bazhal, I.G.; Mihailik, V.A., and Trebin, L.I., Account of heat of sucrose phase transformations in designing and exploitation of crystallization equipment, Chem Eng Commns, 21(1), 23-28 (1983).

Garside, J., Industrial crystallization, Chem & Ind, 4 July, 509-515 (1983).

Toyokura, K., et al., Growth rate of needle crystals of gypsum in fluidized-bed crystallizer, Int Chem Eng, 23(1), 65-72 (1983).

1984-1985

Akoglu, K.; Tavare, N.S., and Garside, J., Dynamic simulation of a non-isothermal MSMPR crystallizer, Chem Eng Commns, 29(1), 353-368 (1984).

Bennett, R.C., Advances in industrial crystallization techniques, Chem Eng Prog, 80(3), 89-95 (1984).

Chang, R.Y., and Wang, M.L., Modeling the batch crystallization process via shifted Legendre polynomials, Ind Eng Chem Proc Des Dev, 23(3), 463-468 (1984).

Chien, H.H.Y., Simulation of countercurrent equilibrium crystallizer, Ind Eng Chem Proc Des Dev, 23(2), 279-283 (1984).

de Jong, E.J., Development of crystallizers, Int Chem Eng, 24(3), 419-432 (1984).

Grootscholten, P.A.M.; Brekel, L.D.M., and Jong, E.J., Effect of scale-up on secondary nucleation kinetics for the sodium chloride-water system, CER&D, 62, 179-189 (1984).

King, R.J., Mechanical vapor recompression crystallizers, Chem Eng Prog, 80(7), 63-69 (1984).

Mersmann, A., Design and scale-up of crystallizers, Int Chem Eng, 24(3), 401-419 (1984).

Palwe, B.G.; Chivate, M.R., and Tavare, N.S., Growth kinetics of potassium sulphate crystals in a DTB agitated crystallizer, Chem Eng Sci, 39(5), 903-905 (1984).

Ring, T.A., Continuous precipitation of monosized particles with a packed bed crystallizer, Chem Eng Sci, 39(12), 1731-1734 (1984).

Shock, R., Advances in crystalliser design, Proc Engng, Sept, 43-49 (1984).

Chen, M.R., and Larson, M.A., Crystallization kinetics of calcium nitrate tetrahydrate from MSMPR crystallizer, Chem Eng Sci, 40(7), 1287-1294 (1985).

Garside, J., Industrial crystallization from solution (review article), Chem Eng Sci, 40(1), 3-26 (1985).

Garside, J., Industrial crystallization from solution (review paper), Chem Eng Sci, 40(1), 3-26 (1985).

Garside, J., and Tavare, N.S., Mixing, reaction and precipitation: Limits of micromixing in an MSMPR crystallizer, Chem Eng Sci, 40(8), 1485-1494 (1985).

Palwe, B.G.; Chivate, M.R., and Tavare, N.S., Growth kinetics of ammonium nitrate crystals in a draft-tube, baffled, agitated batch crystallizer, Ind Eng Chem Proc Des Dev, 24(4), 914-919 (1985).

Tavare, N.S., and Garside, J., Multiplicity in continuous MSMPR crystallizers, AIChEJ, 31(7), 1121-1135 (1985).

Tavare, N.S.; Shah, M.B., and Garside, J., Crystallization and agglomeration kinetics of nickel ammonium sulphate in MSMPR crystallizer, Powder Tech, 44, 13-18 (1985).

Winter, B., and Georgi, H., Extended crystallizer model for sizing and optimization of crystallizer cascades, Int Chem Eng, 25(4), 611-617 (1985).

1986-1988

Berglund, K.A., and Murphy, V.G., Modeling growth rate dispersion in a batch sucrose crystallizer, Ind Eng Chem Fund, 25(1), 174-177 (1986).

Farmer, R.W., and Beckman, J.R., Particle size improvement by a countercurrent tower crystallizer, AIChEJ, 32(7), 1099-1107 (1986).

Moyers, C.G., Industrial crystallization for ultrapure products, Chem Eng Prog, 82(5), 42-46 (1986).

Shin, Y.J.; Yun, C.H., and Lee, C.S., Start-up dynamics of a CMSMPR crystallizer, Int Chem Eng, 26(2), 348-355 (1986).

Tavare, N.S., Crystallization kinetics from transients of an MSMPR crystallizer, Can JCE, 64(5), 752-758 (1986).

Tavare, N.S., Mixing in continuous crystallizers, AIChEJ, 32(5), 705-732 (1986).

Tavare, N.S.; Garside, J., and Larson, M.A., Crystal size distribution from a cascade of MSMPR crystallizers with magma recycle, Chem Eng Commns, 47, 185-200 (1986).

Karagodin, M.A., The optimization of vacuum-pan crystallization, Periodica Polytechnica, 31(3), 161-168 (1987).

Karel, M., and Nyvlt, J., Fluidized-bed apparatus for crystal growth rate measurements and testing with copper sulphate, Chem Eng Commns, 61, 319-326 (1987).

Mashingaidze, T.A.; Garside, J., and Tavare, N.S., Continuous mixed-suspension mixed-product removal dissolver, Chem Eng Commns, 62(1), 221-232 (1987).

Randolph, A.D.; Chen, L., and Tavana, A., Feedback control of CSD in a KCl crystallizer with a fines dissolver, AIChEJ, 33(4), 583-591 (1987).

Shiau, L.D., and Berglund, K.A., Model for a cascade crystallizer in the presence of growth rate dispersion, Ind Eng Chem Res, 26(12), 2515-2522 (1987).

Tai, C.Y.; Chen, C.Y., and Wu, J.F., Crystal dissolution and growth in a lean fluidized-bed crystallizer, Chem Eng Commns, 56, 329-340 (1987).

Tavare, N.S., Batch crystallizers: A review, Chem Eng Commns, 61, 259-318 (1987).

Wang, M.L.; Horng, H.N., and Chang, R.Y., Modeling of crystallization systems via generalized orthogonal polynomials method, Chem Eng Commns, 57, 197-214 (1987).

White, E.T., and Randolph, A.D., Graphical solution of the material balance constraint for MSMPR crystallizers, AIChEJ, 33(4), 686-689 (1987).

Wohlk, W., and Hofmann, G., Types of crystallizers, Int Chem Eng, 27(2), 197-205 (1987).

Bennett, R.C., Matching crystallizer to material, Chem Eng, 23 May, 118-127 (1988).

Hedrick, R.H., Checking crystallizer performance, Chem Eng, 7 Nov, 116-118 (1988).
Mayrhofer, B., and Nyvlt, J., Programmed cooling of batch crystallizers, Chem Eng & Proc, 24(4), 217-220 (1988).
Mersmann, A., Design of crystallizers, Chem Eng & Proc, 23(4), 213-228 (1988).
Mohanty, R., et al., Characterizing the product crystals from a mixing tee process, AIChEJ, 34(12), 2063-2068 (1988).
Tayler, C., Improving batch crystalliser performance, Proc Engng, Nov, 36-38 (1988).

CHAPTER 26

DRYING

26.1 Theory

1967-1970

Aerov, M.E.; Vostrikova, V.N., and Gurovich, R.E., Rectification drying of hydrocarbons, Int Chem Eng, 7(2), 279-281 (1967).

Keey, R.B., Interpreting mixing with isotropic turbulence theory, Brit Chem Eng, 12(7), 1081-1085 (1967).

Sandall, O.C.; King, C.J., and Wilke, C.R., Relationship between transport properties and rates of freeze-drying of poultry meat, AIChEJ, 13(3), 428-438 (1967).

Various, Drying (feature report), Chem Eng, 19 June, 167-214 (1967).

Dyer, D.F., and Sunderland, J.E., Role of convection in drying, Chem Eng Sci, 23(9), 965-970 (1968).

Davidson, J.F.; Robson, M.W.L., and Roesler, F.C., Drying of granular solids subjected to alternating boundary conditions, Chem Eng Sci, 24(5), 815-828 (1969).

Harmathy, T.Z., Simultaneous moisture and heat transfer in porous systems with particular reference to drying, Ind Eng Chem Fund, 8(1), 92-103 (1969).

Oliver, E.D., Crossflow drying of hot solids with cool air, Chem Eng, 14 July, 132-134 (1969).

Peck, R.E., and Kauh, J.Y., Evaluation of drying schedules, AIChEJ, 15(1), 85-88 (1969).

Rulkens, W.H., and Thijssen, H.A.C., Numerical solution of diffusion equations with strongly variable diffusion coefficients: Calculation of flavour loss in drying food liquids, Trans IChemE, 47, T292-298 (1969).

Bailey, G.H.; Slater, I.W., and Eisenklam, P., Dynamic equations and solutions for particles undergoing mass transfer, Brit Chem Eng, 15(7), 912-916 (1970).

Capes, C.E., Efflorescence and the drying of agglomerates, Powder Tech, 4, 77-82 (1970).

Garside, J.; Lord, L.W., and Reagan, R., Drying of granular fertilizers, Chem Eng Sci, 25(7), 1133-1148 (1970).

Keey, R.B., Estimation of drying-flux profiles in continuously worked dryers, Chem Eng Sci, 25(5), 897-899 (1970).

Labutin, V.A.; Golubev, L.G., and Kotov, A.I., Heat transfer and kinetics of drying process during filtration regime, Int Chem Eng, 10(1), 56-58 (1970).

Lykov, A.V., Drying capillary-porous colloidal materials, Int Chem Eng, 10(4), 599-604 (1970).

1971-1973

Glasser, L.S.D., and Lee, C.K., Drying of sodium silicate solutions, J Appl Chem Biotechnol, 21, 127-133 (1971).

Peck, R.E.; Max, D.A., and Ahluwalia, M.S., Predicting drying times for thin materials, Chem Eng Sci, 26(2), 389-404 (1971).

Ashworth, J.C., and Keey, R.B., Evaporation of moisture from wet surfaces, Chem Eng Sci, 27(10), 1797-1806 (1972).

Bluestein, P.M., and Labuza, T.P., Kinetics of water vapor sorption in a model freeze-dried food, AIChEJ, 18(4), 706-712 (1972).

Butler, P., Low residence time annular dryer avoids product degradation, Proc Engng, Nov, 119 (1972).

Chandrasekaran, S.K., and King, C.J., Volatiles retention during drying of food liquids, AIChEJ, 18(3), 520-526 (1972).

Das, B.P., and Narasimhan, K.S., Moisture retention capacity of granular materials, Proc Tech Int, 17(11), 894-895 (1972).

Emmett, R.C., and Dahlstrom, D.A., Steam drying of filter cake, Chem Eng Prog, 68(1), 51-55 (1972).

Fox, E.C., and Thomson, W.J., Coupled heat and mass transport in unsteady sublimation drying, AIChEJ, 18(4), 792-797 (1972).

Fritze, H., Dry gelatinized starch produced on different types of drum dryers, Ind Eng Chem Proc Des Dev, 12(2), 142-148 (1973).

Kiselnikov, V.N., and Talanov, N.M., Drying kinetics of fiberboard with convective supply of heat, Int Chem Eng, 13(3), 415-419 (1973).

Lykov, A.V.; Kuts, P.S., and Olshanskii, A.I., Kinetics of heat transfer in process of drying moist materials, Int Chem Eng, 13(4), 639-643 (1973).

Pievskii, I.M., Convective heat and mass transfer in drying and evaporation equipment, Int Chem Eng, 13(3), 511-514 (1973).

Sieniutycz, S., Thermodynamic approach to fluidized drying and moistening optimization, AIChEJ, 19(2), 277-285 (1973).

1974-1977

Dutta, S., and Shirai, T., Kinetics of drying and decomposition of calcium hydroxide, Chem Eng Sci, 29(9), 2000-2003 (1974).

Harbert, F.C., Automatic control of drying processes: Moisture measurement and control by temperature difference method, Chem Eng Sci, 29(3), 888-891 (1974).

Kawala, Z., High-vacuum sublimation and sublimation drying, Int Chem Eng, 14(1), 70-74 (1974).

Various, Drying gases and liquids (feature report), Chem Eng, 16 Sept, 92-110 (1974).

Audu, T.O.K., and Jeffreys, G.V., The drying of drops of particulate slurries, Trans IChemE, 53, 165-172 (1975).

Evans, A.A., and Keey, R.B., Moisture diffusion coefficient of a shrinking clay on drying, Chem Eng J, 10(2), 127-134 (1975).

Evans, A.A., and Keey, R.B., Definition and variation of diffusion coefficients when drying capillary-porous materials, Chem Eng J, 10(2), 135-144 (1975).

Gibson, R., Radiation drying, Proc Engng, Sept, 94-96 (1975).

Kisakurek, B.; Peck, R.E., and Cakaloz, T., Generalized drying curves for porous solids, Can JCE, 53, 53-60 (1975).

Novak, L.T., and Coulman, G.A., Mathematical models for drying of rigid porous materials, Can JCE, 53, 60-70 (1975).

King, R., Determining solids moisture content, Proc Engng, Aug, 49-50 (1976).

Murthy, S.S.; Murthy, M.V.K., and Ramachandran, A., Low-intensity convective drying of non-hygroscopic porous materials of high initial moisture content, Chem Eng Sci, 31(11), 975-984 (1976).

Van der Lijn, J., 'Constant rate period' during drying of shrinking spheres, Chem Eng Sci, 31(10), 929-936 (1976).

Zanker, A., Drying time of non-porous solids, Processing, May, 38 (1976).
Belcher, D.W., Spray drying of war gas residue, Chem Eng Prog, 73(7), 101-104 (1977).
Pham, Q.T., and Keey, R.B., Evaporation of a spray in the jet zone from a nozzle atomiser, Trans IChemE, 55, 114-121 (1977).
Whitaker, S., Toward a diffusion theory of drying, Ind Eng Chem Fund, 16(4), 408-414 (1977).

1978-1981

Long, G.E., Spraying theory and practice, Chem Eng, 13 March, 73-77 (1978).
Rotstein, E., and Cornish, A.R.H., Prediction of sorptional equilibrium relationship for drying of foodstuffs, AIChEJ, 24(6), 956-966 (1978).
Taylor, R.; Boardman, C.F.B., and Wallis, R.G., Sterile freeze-drying in an unclean environment, J Appl Chem Biotechnol, 28, 213-214 (1978).
Wakabayashi, K.; Yamaguchi, S.; Matsumoto, T., and Mita, T., Liquid moisture movement during the drying process in a bed of fine particles, Int Chem Eng, 18(3), 508-514 (1978).
Shishido, I.; Suzuki, M.; Endo, A., and Ohtani, S., Critical moisture content and characteristic drying curve of granular material, Int Chem Eng, 19(3), 491-498 (1979).
Various, Drying techniques (topic issue), Chem Eng Prog, 75(4), 37-60 (1979).
Abdul Majeed, P.M., Analysis of heat transfer during hydrair cooling of spherical food products, Int J Heat Mass Trans, 24(2), 323-334 (1981).
Chen, C.C., and Wang, C.C., Closed-cycle drying for coated film with organic solvent on continuous sheet materials, Chem Eng Commns, 10(6), 331-348 (1981).
Fortes, M., and Okos, M.R., Drying of hygroscopic products, AIChEJ, 27(2), 255-262 (1981).
Murty, K.N., Quick estimate of spray-nozzle mean drop size, Chem Eng, 27 July, 96-98 (1981).
Rajan, T.S., and Ibrahim, S.H., Through-circulation drying of vanadium pentoxide, Can JCE, 59, 642-647 (1981).

1982-1984

Balazs, T.; Parti, M.; Topar, J., and Banoczy, J., Drying and shrinkage, Int Chem Eng, 22(4), 744-750 (1982).
Beukema, K.J.; Bruin, S., and Schenk, J., Heat and mass transfer during cooling and storage of agricultural products, Chem Eng Sci, 37(2), 291-298 (1982).
Crabble, P.G., and Fryer, C., Drying and storage of Solanum plant material, J Chem Tech Biotechnol, 32, 559-565 (1982).
Cross, A.D.; Jones, P.L., and Lawton, J., Simultaneous energy and mass transfer in radio-frequency fields, Trans IChemE, 60, 67-78 (1982).
Hubble, P.E., Consider microwave drying, Chem Eng, 4 Oct, 125-127 (1982).
Mahajan, P.O.; Youssef, A., and Walker, P.L., Surface-modified carbons for drying of gas streams, Sepn Sci Technol, 17(8), 1019-1026 (1982).
Milojevic, D.Z., and Stefanovic, M.S., Convective drying of thin and deep beds of grain, Chem Eng Commns, 13(4), 261-270 (1982).
Zakarlan, J.A., and King, C.J., Volatiles loss in the nozzle zone during spray drying of emulsions, Ind Eng Chem Proc Des Dev, 21(1), 107-113 (1982).

Hallstrom, A., and Wimmerstedt, R., Drying of porous granular materials, Chem Eng Sci, 38(9), 1507-1516 (1983).
Root, W.L., Indirect drying of solids, Chem Eng, 2 May, 52-64 (1983).
Agarwal, P.K.; Genetti, W.E., and Lee, Y.Y., Pseudo steady-state receding core model for drying with shrinkage of low ranked coals, Chem Eng Commns, 27(1), 9-22 (1984).
Burfield, D.R.; Hefter, G.T., and Koh, D.S.P., Desicant efficiency in solvent and reagent drying: Molecular sieve column drying of 95% ethanol, J Chem Tech Biotechnol, 34A(4), 187-194 (1984).
Dunhill, S., Product drying and solvent recovery, Proc Engng, Oct, 61-63 (1984).
Etzel, M.R., and King, C.J., Loss of volatile trace organics during spray drying, Ind Eng Chem Proc Des Dev, 23(4), 705-710 (1984).
Kuramae, M., Prediction of drying characteristics of coarse granular bed using relationship between capillary suction pressure and moisture content, Int Chem Eng, 24(1), 113-119 (1984).
Schlunder, E.U., and Mollekopf, N., Vacuum contact drying of free flowing mechanically agitated particulate material, Chem Eng & Proc, 18(2), 93-112 (1984).
Sieniutycz, S., Lumped parameter modelling and an introduction to optimization of one-dimensional nonadiabatic drying systems, Int J Heat Mass Trans, 27(11), 1971-1984 (1984).
Walsh, J.J., Drying of temperature-sensitive solids, Plant/Opns Prog, 3(2), 74-80 (1984).

1985-1986

Kobari, M.; Shimizu, Y.; Endo, M., and Inazumi, H., Contact drying of fibrous sheet material, Ind Eng Chem Proc Des Dev, 24(1), 188-194 (1985).
Plumb, O.A.; Spolek, G.A., and Olmstead, B.A., Heat and mass transfer in wood during drying, Int J Heat Mass Trans, 28(9), 1669-1678 (1985).
Androutsopoulos, G.P., and Linardos, T.J., Effects of drying upon lignite macropore structure, Powder Tech, 47(1), 9-16 (1986).
Carr-Brion, K., Ways to measure water content on-line, Proc Engng, March, 55-59 (1986).
Hallstrom, A., Alternating boundary conditions in drying, Chem Eng Sci, 41(9), 2225-2234 (1986).
Heimann, F.; Thurner, F., and Schlunder, E.U., Intermittent drying of porous materials containing binary mixtures, Chem Eng & Proc, 20(3), 167-174 (1986).
Huang, C.L.D., and Miller, P.L., Drying light-weight concrete slabs, Chem Eng Commns, 49, 1-12 (1986).
Keey, R.B., and Kecheng, M., The humidity potential coefficient in drying, CER&D, 64(2), 119-124 (1986).
Li, Y.H., and Skinner, J.L., Deactivation of dried subbituminous coal, Chem Eng Commns, 49, 81-98 (1986).
Li, Y.H., and Skinner, J.L., Development and validation of a process simulator drying subbituminous coal, Chem Eng Commns, 49, 99-118 (1986).
Luu, D.V., and Benner, S.M., Hygroscopic moisture transfer in fibrous material, Chem Eng Commns, 48, 317-330 (1986).

Mujumdar, A.S. (Ed.), Drying of Solids: Recent International Developments, Wiley, New York (1986).

Rajan, T.S., and Narayanaswamy, S., A novel curve fitting technique applied to drying curves, Can JCE, 64(6), 1033-1034 (1986).

Romero, A.; Bilbao, J., and Aguayo, A.T., Vacuum drying of silica gel, Int Chem Eng, 26(1), 90-97 (1986).

Stanish, M.A.; Schajer, G.S., and Kayihan, F., Mathematical model of drying for hygroscopic porous media, AIChEJ, 32(8), 1301-1311 (1986).

Thurner, F., and Schlunder, E.U., Drying of porous materials wetted with binary mixtures, Chem Eng & Proc, 20(1), 9-26 (1986).

Thurner, F., and Wischniewski, M., Drying of polyvinyl alcohol in closed-circuit dryer, Chem Eng & Proc, 20(1), 43-52 (1986).

Tsotsas, E., and Schlunder, E.U., Vacuum contact drying of free-flowing mechanically agitated multigranular packings, Chem Eng & Proc, 20(6), 339-350 (1986).

Tsotsas, E., and Schlunder, E.U., Contact drying of mechanically agitated particulate material in presence of inert gas, Chem Eng & Proc, 20(5), 277-288 (1986).

1987-1988

Kuster, U., Recycling enhances slurry drying, Chem Eng, 9 Nov, 87-89 (1987).

Tsotsas, E., and Schlunder, E.U., Vacuum contact drying of mechanically agitated beds: Influence of hygroscopic behaviour on drying rate curve, Chem Eng & Proc, 21(4), 199-208 (1987).

Arnaud, G., and Fohr, J.P., Slow drying simulation in thick layers of granular products, Int J Heat Mass Trans, 31(12), 2517-2527 (1988).

Crapiste, G.H.; Whitaker, S., and Rotstein, E., Drying of cellular material, Chem Eng Sci, 43(11), 2919-2936 (1988).

Dutta, S.K.; Nema, V.K., and Bhardwaj, R.K., Drying behaviour of spherical grains, Int J Heat Mass Trans, 31(4), 855-862 (1988).

Heimann, F., and Schlunder, E.U., Vacuum contact drying of mechanically granular beds wetted with binary mixture, Chem Eng & Proc, 24(2), 75-92 (1988).

Nasrallah, S.B., and Perre, P., Detailed study of a model of heat and mass transfer during convective drying of porous media, Int J Heat Mass Trans, 31(5), 957-968 (1988).

Schlunder, E.U., Mechanism of constant-drying-rate period and relevance to diffusion-controlled catalytic gas-phase reactions, Chem Eng Sci, 43(10), 2685-2688 (1988).

Shibata, H.; Mada, J., and Shinohara, H., Steam drying of sintered-glass-bead spheres, Ind Eng Chem Res, 27(12), 2353-2362, 2385-2387 (1988).

Wang, B.X., and Yu, W.P., Method for evaluation of heat and mass transport properties of moist porous media, Int J Heat Mass Trans, 31(5), 1005-1010 (1988).

Young, G.C., Simple method for drying conditions, Chem Eng, 20 June, 107-108 (1988).

26.2 Equipment Design

1967-1969

Clark, W.E., Fluid-bed drying, Chem Eng, 13 March, 177-184 (1967).

Krivsky, Z., and Vanecek, V., Calculation for a fluid bed dryer by computer, Brit Chem Eng, 12(12), 1886-1889 (1967).

Masters, K., and Mohtadi, M.F., Study of centrifugal atomisation and spray drying, Brit Chem Eng, 12(12), 1890-1892 (1967).

Romankov, P.G., and Rashkovskaya, N.B., Modern drying techniques in the chemical industry, Int Chem Eng, 7(2), 212-220 (1967).

Wisniak, J.; Fertilio, A., and Freed, C., Tray drying of sand: Properties and critical moisture, Brit Chem Eng, 12(10), 1590-1592 (1967).

Keey, R.B., Batch drying with air recirculation, Chem Eng Sci, 23(11), 1299-1308 (1968).

Kovats, F., Increase capacity of rotary dryers, Chem Eng, 3 June, 126 (1968).

Masters, K., and Mohtadi, M.F., Study of centrifugal atomisation and spray drying, Brit Chem Eng, 13(1), 88-89; 13(2), 242-244 (1968).

Rosenberg, R.E., Sampling fluid-bed batch dryers in operation, Chem Eng, 26 Aug, 136 (1968).

Anon., Pneumatic dryers survey, Brit Chem Eng, 14(9), 1225-1228 (1969).

Bakhshi, N.N., and Chai, C.Y., Fluidized bed drying of sodium sulfate solutions, Ind Eng Chem Proc Des Dev, 8(2), 275-279 (1969).

Baltas, L., and Gauvin, W.H., Performance predictions for a cocurrent spray dryer, AIChEJ, 15(5), 764-779 (1969).

Moncman, E., Average retention time of material in a fluo-solids dryer, Int Chem Eng, 9(3), 471-476 (1969).

Ziolkowski, Z., and Kmiec, A., Drying aqueous sodium salicylate solutions in a spray dryer, Int Chem Eng, 9(4), 668-673 (1969).

1970-1972

Bukareva, M.F.; Chlenov, V.A., and Mikhailov, N.V., Drying of finely dispersed powders in vibro-fluidized beds, Int Chem Eng, 10(3), 384-386 (1970).

Shakova, N.A.; Dmitrenko, E.V., and Zhelonkin, V.G., Drying polymer materials in a fluidized bed, Int Chem Eng, 10(4), 597-599 (1970).

Thygeson, J.R., and Grossmann, E.D., Optimization of continuous through-circulation dryer, AIChEJ, 16(5), 749-754 (1970).

Warman, K.G., and Reichel, A.J., Development of a freeze-drying process for food, Chem Engnr, May, CE134-139 (1970).

Coggan, G.C., Innovation following computer studies in design of drying plant, Chem Eng J, 2(1), 55-62 (1971).

Kuong, J.F., Drying-kiln design calculations, Brit Chem Eng, 16(2), 180-182 (1971).

Lyne, C.W., Review of spray drying, Brit Chem Eng, 16(4), 370-373 (1971).

Paris, J.R.; Ross, P.N.; Dastur, S.P., and Morris, R.L., Modeling of air flow pattern in a countercurrent spray-drying tower, Ind Eng Chem Proc Des Dev, 10(2), 157-164 (1971).

Various, Drying (symposium papers), Chem Engnr, Feb, 51-80 (1971).

Wormald, D., and Burnell, E.M.W., Design of fluidised bed driers and coolers, Brit Chem Eng, 16(4), 376-380 (1971).
Chirife, J., and Gardner, R.G., Mass transfer in through-circulation drying of packed beds, Ind Eng Chem Proc Des Dev, 11(2), 312-313 (1972).
Parker, J., and Smith, H.M., Design and construction of a freeze dryer incorporating improved standards of biological safety, J Appl Chem Biotechnol, 22, 925-932 (1972).
Sutherland, K.S., Introduction to pneumatic drying, Brit Chem Eng, 17(1), 55-61 (1972).
Various, Drying and dryer design (topic issue), Proc Engng, Sept, 84-106 (1972).

1973-1974
Janda, F., Dimensions of disc spray dryers with intensive circulation of drying medium, Int Chem Eng, 13(4), 649-658 (1973).
Longmore, A.P., Freeze drying, Chem & Ind, 20 Oct, 970-975 (1973).
MacDonald, J.O.S., Progress in drying, Proc Tech Int, March, 127-133; April, 203-206 (1973).
Martin, C.G., Some comments on dryer design, Chem Engnr, July, 367-369, 376 (1973).
Moles, F.D.; Watson, D., and Lain, P.B., Aerodynamics of rotary cement kilns, J Inst Fuel, 46, 353-362 (1973).
Pearce, K.W., Heat transfer model for rotary kilns, J Inst Fuel, 46, 363-372 (1973).
Thorpe, G.R.; Wint, A., and Coggan, G.C., Mathematical modelling of industrial pneumatic driers, Trans IChemE, 51, 339-348 (1973).
Butler, P., Fluid-bed dryers, Proc Engng, March, 56-59 (1974).
Greenfield, P.F., Cyclic-pressure freeze drying, Chem Eng Sci, 29(10), 2115-2124 (1974).
Huthwaite, J.A., Examples of the use and specification of fluid bed dryers, Chem Engnr, May, 295-297, 303 (1974).
Inazumi, H., and Suzuki, T., Dehumidification of moist air in a fluidized bed, Int Chem Eng, 14(4), 768-776 (1974).
Jones, P.L.; Lawton, J., and Parker, I.M., High frequency paper drying, Trans IChemE, 52, 121-135, 386 (1974).
Lees, F.P., Design of plant for drying organic liquids, Chem & Ind, 5 Jan, 25-35 (1974).
Parti, M., and Palancz, B., Mathematical model for spray drying, Chem Eng Sci, 29(2), 355-362 (1974).
Pictor, J., Spray drying and granulation, Proc Engng, June, 66-67 (1974).
Pye, E.W., Vibrating-bed dryers, Proc Engng, March, 61-63 (1974).
Sulg, E.O.; Mitev, D.T.; Raskovskaya, N.B., and Romankov, P.G., Modelling and design of vortex-bed equipment for drying processes, Int Chem Eng, 14(4), 700-703 (1974).
White, A., Batch-type fluid-bed driers, Chemical Processing, Aug, 17-19 (1974).

1975-1976

Backhurst, J.R., and Harker, J.H., Drying and incineration of pig manure, Chem Engnr, July, 449-452 (1975).

Barr, D.J., Closed-circuit drying, Proc Engng, March, 49 (1975).

Beran, Z., and Lutcha, J., Optimising particle residence time in a fluidised bed drier, Chem Engnr, Nov, 678-681 (1975).

Jouhari, A.K., and Dey, D.N., Determination of variables of rotary kilns, coolers and driers, Processing, Feb, 11 (1975).

Katta, S., and Gauvin, W.H., Fundamental aspects of spray drying, AIChEJ, 21(1), 143-152 (1975).

Ma, Y.H., and Peltre, P.R., Freeze dehydration by microwave energy, AIChEJ, 21(2), 335-350 (1975).

Various, Drying developments, Proc Engng, March, 50-55 (1975).

Barr, D.J., Two-stage drying cuts pollution costs, Proc Engng, Aug, 47 (1976).

Beveridge, G.S.G., and Lyne, C.W., Continuous vacuum band dryer, Chem Engnr, July, 513-515, 521 (1976).

Gauvin, W.H., and Katta, S., Basic concepts of spray dryer design, AIChEJ, 22(4), 713-724 (1976).

Keey, R.B., and Pham, Q.T., Behaviour of spray dryers with nozzle atomisers, Chem Engnr, July, 516-521 (1976).

Lee, D.A., Drying and calcining ammonium diuranate, Chem Engnr, July, 522-524 (1976).

McIntosh, M.J., Mathematical model of drying in a brown-coal mill system, Fuel, 55(1), 47-58 (1976).

Mortensen, S., Spray-fluidised dryer, Proc Engng, Aug, 44-45 (1976).

Peter, U., Fluidized-bed dryer and cooler, Int Chem Eng, 16(4), 587-590 (1976).

1977-1978

Ang, T.K.; Pei, D.C.T., and Ford, J.D., Microwave freeze drying: An experimental investigation, Chem Eng Sci, 32(12), 1477-1490 (1977).

Brink, H.J.; Coch, J.; Fischer, J., and Behrens, F., Drying powders and fine-grained products in centrifugal flash dryer, Int Chem Eng, 17(1), 80-84 (1977).

Comings, E.W., et al., Spray drying of Serratia marcescens in a jet spray dryer, Ind Eng Chem Fund, 16(1), 12-18 (1977).

Kawamura, Y.; Shinagawa, H., and Kiyama, H., Vacuum-puff drying of highly viscous heat-sensitive material, Int Chem Eng, 17(2), 271-276 (1977).

Mrowiec, M., and Ciesielczyk, W., Fluidized bed dryers for paste materials, Int Chem Eng, 17(2), 373-379 (1977).

Sunavala, P.D., Mathematical analysis of temperature distribution of gas and charge in dry-process rotary cement kilns, J Inst Fuel, 50, 23-32 (1977).

Updegrove, L.B., Batch drying in a Nauta mixer, Chem Eng Prog, 73(4), 107-112 (1977).

Various, Methods for analyzing drying equipment (feature report), Chem Eng, 17 Jan, 104-123 (1977).

White, A., Economics of high-speed drying, Processing, Sept, 72 (1977).

Ade-John, A.O., and Jeffreys, G.V., Flow visualisation and residence time in a spray drier, Trans IChemE, 56, 36-42 (1978).

Caloine, R., and Clayton, C.G.A., Drying and handling of polyester granules, Chem Engnr, March, 185-188 (1978).
Christiansen, O.B., Closed-loop spray drying systems, Chem Eng Prog, 74(1), 83-86 (1978).
Cook, E.M., Influence of solvent properties on dryer design, Chem Eng Prog, 74(4), 75-78 (1978).
Janda, F., Calculation of height of drying tower with atomizing nozzle for a single substance, Int Chem Eng, 18(2), 310-318 (1978).
Kmiec, A.; Mielczarski, S., and Pajakowska, J., Experimental study on hydrodynamics of a pneumatic flash dryer, Powder Tech, 20, 67-74 (1978).
Various, Dryers and drying (feature report), Processing, Aug, 41-59 (1978).

1979-1980

Liapis, A.I., and Litchfield, R.J., Optimal control of a freeze dryer, Chem Eng Sci, 34(7), 975-982 (1979).
Litchfield, R.J., and Liapis, A.I., Adsorption-sublimation model for a freeze dryer, Chem Eng Sci, 34(9), 1085-1090 (1979).
Noble, R.D., Laboratory experiment on freeze drying of fruits and vegetables, Chem Eng Educ, 13(3), 142-144 (1979).
Purcell, J.G., Practical rotary cascading dryer design, Chem Engnr, July, 496-497, 500 (1979).
Reay, D., Theory in the design of dryers, Chem Engnr, July, 501-503, 506 (1979).
Snowman, J.W., Freeze drying applications, Proc Engng, Feb, 37,39 (1979).
van Brakel, J., The choice and design of dryers, Chem Engnr, July, 493-495 (1979).
Various, Freeze drying of foodstuffs (topic issue), Chem & Ind, 21 July, 461-468 (1979).
Wijlhuizen, A.E.; Kerkhof, P.J., and Bruin, S., Theoretical study of the inactivation of phosphatase during spray drying of skim milk, Chem Eng Sci, 34(5), 651-660 (1979).
Basmadjian, D., Throughflow drying and conditioning of beds of moist porous solids, AIChEJ, 26(4), 625-634 (1980).
Foster, B.S., Automatic conical mixer-drier, Chem Eng Prog, 76(12), 57-59 (1980).
Herron, D., and Hummel, D., How to select polymer drying equipment, Chem Eng Prog, 76(1), 44-52 (1980).
Hoebink, J.H., and Rietema, K., Drying granular solids in a fluidized bed, Chem Eng Sci, 35(10), 2135-2140; 35(11), 2257-2266 (1980).
Kieckbusch, T.G., and King, C.J., Volatiles loss during atomization in spray drying, AIChEJ, 26(5), 718-725 (1980).
Kitic, D., and Brea, F.M., Influence of moisture on fluidising velocities in batch fluid bed drying, Trans IChemE, 58, 208-210 (1980).
Masters, K., Foodstuffs: Where spray drying plays a role, Chem Engnr, March, 145-148 (1980).
Mu, J., and Perlmutter, D.D., Mixing of granular solids in a rotary dryer, AIChEJ, 26(6), 928-934 (1980).

Rowbotham, D.W., and Beveridge, G.S.G., Design and operation of glass-lined conical dryer blenders with reference to the safety aspects, Chem Engnr, Part 1: Feb, 95-97; Part 2: March, 155-160 (1980).
Various, Dryers and drying (feature report), Processing, Feb, 25-43 (1980).

1981-1983

Anon., Advances in freeze drying, Processing, March, 17-19 (1981).
Branston, J.J., and Marsh, G.S., Application of computer control to a staple fibre drier, Chem Engnr, March, 106-108 (1981).
Gauvin, W.H., Novel approach to spray drying using plasmas of water vapour, Can JCE, 59, 697-704 (1981).
Lee, D.A., Classifying and selecting spray driers, Chem Eng Prog, 77(3), 34-38 (1981).
Litchfield, R.J.; Farhadpour, F.A., and Liapis, A.I., Cyclical pressure freeze drying, Chem Eng Sci, 36(7), 1233-1238 (1981).
Various, Dryers and drying (special report), Processing, June, 21-27 (1981).
Litchfield, R.J., and Liapis, A.I., Optimal control of a freeze dryer, Chem Eng Sci, 37(1), 45-56 (1982).
Murty, K.N., Predict spray-nozzle performance, Chem Eng, 8 March, 118 (1982).
Tuma, V., Feed and discharge problems for continuous contact driers, Proc Engng, Feb, 47-49 (1982).
Various, Selection and design of dryers (topic issue), Processing, Sept, 25-34 (1982).
Audu, T.O.K., Optimum parameters for rotary drying, Chem Eng J, 26(2), 157-164 (1983).
Baker, C., Selection of drying equipment, Proc Engng, July, 43-44 (1983).
Goffredi, R.A., and Crosby, E.J., Limiting analytical relationships for prediction of spray drier performance, Ind Eng Chem Proc Des Dev, 22(4), 665-672 (1983).
Lee, D.A., Finding thermal dryer capabilities, Chem Eng, 12 Dec, 87 (1983).
Palancz, B., Mathematical model for continuous fluidized bed drying, Chem Eng Sci, 38(7), 1045-1060 (1983).
Tsamopoulos, J.A., and Georgakis, C., Effect of relative size on dynamics of fluidized bed dryers, Chem Eng Commns, 23(4), 343-362 (1983).
Various, Industrial drying practices (topic issue), Chem Eng Prog, 79(4), 37-76 (1983).

1984-1985

Afacan, A., and Masliyah, J., Tray drying of solids, Chem Eng Educ, 18(3), 132-135 (1984).
Agarwal, P.K.; Genetti, W.E.,; Lee, Y.Y., and Prasad, S.N., Model for drying during fluidized-bed combustion of wet low-rank coals, Fuel, 63(7), 1020-1027 (1984).
Anon., More effective dryers, Processing, Aug, 31 (1984).
Bahu, R., and Baker, C., Dryers and drying: A survey, Processing, March, 25-31 (1984).
Becker, H.A.; Douglas, P.L., and Ilias, S., Optimization of industrial grain dryer systems, Can JCE, 62(6), 738-745 (1984).

Forbes, J.F.; Jacobson, B.A.; Rhodes, E., and Sullivan, G.R., Model-based control strategies for commercial grain drying systems, Can JCE, 62(6), 773-779 (1984).
Martin, H., and Saleh, A.H., Drying of fine granular material in a pneumatic drier, Int Chem Eng, 24(1), 14-23 (1984).
Matsumoto, S., and Pei, D.C.T., Mathematical analysis of pneumatic drying of grains, Int J Heat Mass Trans, 27(6), 843-856 (1984).
Mills, B., and Rothery, E., Gas drying, Chem Engnr, April, 19-23 (1984).
Palancz, B., Analysis of solar-dehumidification drying, Int J Heat Mass Trans, 27(5), 647-656 (1984).
Tamir, A.; Elperin, I., and Luzzatto, K., Drying in a new two-impinging-streams reactor, Chem Eng Sci, 39(1), 139-146 (1984).
van't Land, C.M., Selection of industrial dryers, Chem Eng, 5 March, 53-61 (1984).
Ashworth, J.C., Optimizing drying costs, Processing, May, 12-23 (1985).
Escott, J., Design of fluid-bed dryers, Proc Engng, Oct, 77,79 (1985).
Lang, W., Contact dryers revisited, Chem Engnr, Feb, 37-39 (1985).
MacKay, I., 'Whither spray drying?' Chem Engnr, July, 19-20 (1985).
Peacock, E.M., Heaterless dryers are not heatless, Chem Eng, 11 Nov, 199-203 (1985).
Toering, W., Installation and start-up experiences of synthesis gas dryers in existing ammonia plant, Plant/Opns Prog, 4(3), 127-132 (1985).

1986-1988

Anon., Dryers survey, Processing, July, 21-27 (1986).
Ashworth, J., Drying high cost materials, Chem Engnr, Sept, 30-34 (1986).
Frey, D.D., and King, C.J., Effect of surfactants on mass transfer during spray drying, AIChEJ, 32(3), 437-443 (1986).
Garg, D.R., and Yon, C.M., Adsorptive heat recovery drying system, Chem Eng Prog, 82(2), 54-60 (1986).
Groves, F.R., and Wu, T.H., Dynamic mathematical model of rotary dryer: Method of characteristics, Chem Eng Commns, 49, 35-50 (1986).
Kamke, F.A., and Wilson, J.B., Computer simulation of a rotary drier, AIChEJ, 32(2), 263-275 (1986).
Keey, R.B., Some current developments in drying, CER&D, 64(2), 83-88 (1986).
Lai, F.S.; Chen, Y., and Fan, L.T., Modelling and simulation of a continuous fluidized bed drier, Chem Eng Sci, 41(9), 2419-2430 (1986).
Morl, L., Growth of granules in fluidized-bed drying, accounting for nuclei formation, Int Chem Eng, 26(2), 236-243 (1986).
Morl, L., and Kunne, H.J., Granulate growth during unsteady-state operation in liquid-sprayed fluidized beds, Int Chem Eng, 26(3), 423-428 (1986).
Pearson, W.K., Advanced batch pilot plant for unloading and drying system, Chem Eng Prog, 82(4), 25-29 (1986).
Strumke, M., and Morl, L., Discontinuous drying of granular materials with high initial moisture content in fluidized beds, Int Chem Eng, 26(4), 629-637 (1986).
Viswanathan, K., Model for continuous drying of solids in fluidized/spouted beds, Can JCE, 64(1), 87-95 (1986).

Viswanathan, K.; Lyall, M.S., and Raychaudhuri, B.C., Spouted bed drying of agricultural grains, Can JCE, 64(2), 223-232 (1986).
Lambrana, J.I., and Diaz, J.M., Heat programming to improve efficiency in a batch freeze-drier, Biochem Eng J, 35(3), B23-B30 (1987).
Alden, M.; Torkington, P., and Strutt, A.C.R., Control and instrumentation of fluidized-bed drier using temperature-difference technique, Powder Tech, 54(1), 15-26 (1988).
Butcher, C., Fluid-bed dryers, Chem Engnr, July, 16-18 (1988).
Cook, E.M., and DuMont, H.D., Improving dryer performance, Chem Eng, 9 May, 70-78 (1988).
Daud, W.R., and Armstrong, W.D., Residence time distribution of the drum dryer, Chem Eng Sci, 43(9), 2399-2406 (1988).
Holmes, J.G.; Hedman, B.A., and Salama, S.Y., Overview of industrial drying needs and competing technologies, Plant/Opns Prog, 7(3), 199-203 (1988).
McCormick, P.Y., The key to drying solids, Chem Eng, 15 Aug, 113-122 (1988).
Papadakis, S.E., and King, C.J., Air temperature and humidity profiles in spray drying, Ind Eng Chem Res, 27(11), 2111-2123 (1988).
Thomas, B.; Liu, Y.A.; Mason, M.O., and Squires, A.M., Vibrated beds for drying, Chem Eng Prog, 84(6), 65-75 (1988).
Zbicinski, I., Mathematical modelling of spray drying, Comput Chem Eng, 12(2/3), 209-214 (1988).

26.3 Energy Considerations

1974-1988

Debrand, S., Heat transfer during a flash drying process, Ind Eng Chem Proc Des Dev, 13(4), 396-404 (1974).
Butler, P., Dryer system design for energy savings, Proc Engng, March, 44-47 (1975).
Hodgett, D.L., Efficient drying using heat pumps, Chem Engnr, July, 510-512 (1976).
Reay, D., Energy conservation in industrial drying, Chem Engnr, July, 507-509 (1976).
Barr, J.D., Energy saving in drying, Processing, April, 33 (1977).
Holland, C.R., and McKay, G., Spray drier energy savings in a food processing plant, Energy World, Dec, 8-12 (1981).
Kragh, O.T., and Kraglund, A., Heat recovery in dryers, Chem Engnr, April, 149-153, 158 (1981).
Root, W.L., and Cook, E.M., Energy use in paddle driers, Chem Eng Prog, 77(3), 31-33 (1981).
Anon., Energy economy trends in dryer design, Processing, Jan, 17,19 (1983).
Toei, R., et al., Heat transfer in an indirect-heat agitated dryer, Chem Eng & Proc, 18(3), 149-156 (1984).
Coumans, W.J., and Willeboer, W., Simple calculation model for radiant heat transfer in a paper dryer, Chem Eng & Proc, 21(1), 15-24 (1987).
Maclean, R., Energy saving techniques in the evaporation and drying of distillery effluent, Chem Engnr, July, 29-34 (1987).

Deakin, A., and Kemp, I., Pinch technology for dryer efficiency, Processing, Oct, 49-56 (1988).
Gauden, C.G., Air knives for energy efficiency in drying, Energy World, Dec, 7-10 (1988).
McCormick, P.Y., Reducing heat losses in dryers, Chem Eng, 12 Sept, 107-108 (1988).

CHAPTER 27

MISCELLANEOUS

27.1 Gasification

1973-1982

Hashimoto, K., and Silveston, P.L., Gasification, AIChEJ, 19(2), 259-277 (1973).

Various, Coal gasification (topic issue), Chem Eng Prog, 69(3), 31-66 (1973).

Perry, H., Coal conversion technology, Chem Eng, 22 July, 88-102 (1974).

Various, Coal processing (topic issue), Chem Eng Prog, 71(4), 61-92 (1975).

Various, Coal processing (topic issue), Chem Eng Prog, 72(8), 51-79 (1976).

Various, Coal processing (topic issue), Chem Eng Prog, 73(6), 49-88 (1977).

Yen, L.C.; Frith, J.F.S.; Chao, K.C., and Lin, H.M., Data deficiency hampers coal-gasification plant design, Chem Eng, 9 May, 127-130 (1977).

Cooke, M.J., and Robson, B., Gas from coal, Chem Engnr, Oct, 729-732 (1978).

Peel, R.B., Coal utilisation and research in Brazil, Chem Engnr, Oct, 736-738 (1978).

Various, Coal processing technology (topic issue), Chem Eng Prog, 74(8), 43-102 (1978).

Various, Coal processing technology (topic issue), Chem Eng Prog, 75(6), 33-80 (1979).

Stephens, D.R.; Brandenburg, C.F., and Burwell, E.L., Synfuel trends: Underground coal gasification, Chem Eng Prog, 76(4), 89-94 (1980).

Various, Synfuel trends (topic issue), Chem Eng Prog, 76(3), 43-92 (1980).

Chiaramonte, G.R., and Sharma, R., Upgrade coke by gasification (Tosca process), Hyd Proc, Sept, 255-257 (1982).

Pechey, R., Coal gasification developments, Proc Engng, March, 27-31 (1982).

Theis, K.A., and Nitschke, E., Make syngas from lignite, Hyd Proc, Sept, 233-237 (1982).

Various, Gas processing handbook (special report), Hyd Proc, April, 85-164 (1982).

Various, Syngas production (special report), Hyd Proc, March, 89-116 (1982).

Various, Synfuels (special report), Hyd Proc, June, 77-96 (1982).

1983-1988

Various, Coal processing practices (topic issue), Chem Eng Prog, 79(5), 31-51 (1983).

Jeske, H.O., Charge-gas compressors in coal gasification and olefin plants, Chem Eng & Proc, 18(2), 113-122 (1984).

Bastick, M.; Perrot, J.M., and Weber, J., General characteristics of coal gasification, Int Chem Eng, 26(2), 243-257 (1986).

Graff, R.A., and Brandes, S.D., Modification of coal by subcritical steam: Pyrolysis and extraction yields, Energy & Fuels, 1(1), 84-89 (1987).

Ohtsuka, Y.; Tamai, Y., and Tomita, A., Iron-catalyzed gasification of brown coal at low temperatures, Energy & Fuels, 1(1), 32-36 (1987).

Ross, D.S., et al., Coal conversion in carbon monoxide-water, Energy & Fuels, 1(3), 287-294 (1987).

Serio, M.A., et al., Kinetics of volatile product evolution in coal pyrolysis: Experiment and theory, Energy & Fuels, 1(2), 138-152 (1987).

Takarada, T., et al., New utilization of NaCl as catalyst precursor for catalytic gasification of low-rank coal, Energy & Fuels, 1(3), 308-310 (1987).

Various, Pyrolysis in petroleum exploration geochemistry (symposium papers), Energy & Fuels, 1(6), 451-476 (1987).
Ballal, G.; Amundson, N.R., and Zygourakis, K., Pore structural effects in catalytic gasification of coal chars, AIChEJ, 34(3), 426-434 (1988).
Brown, B.W., et al., Measurement and prediction of entrained-flow gasification processes, AIChEJ, 34(3), 435-446 (1988).
Furimsky, E.; Sears, P., and Suzuki, T., Iron-catalyzed gasification of char in carbon dioxide, Energy & Fuels, 2(5), 634-639 (1988).
Hasatani, M.; Matsuda, H., and Kataoka, A., Desulfurization of low-calorie gas produced by solid-fuel gasification in packed bed of iron oxide pellets, Int Chem Eng, 28(3), 497-504 (1988).
Lacey, J.A., Gasification: A key to the clean use of coal, Energy World, Feb, 3-7, 9 (1988).
Niksa, S., Rapid coal devolatilization as an equilibrium flash distillation, AIChEJ, 34(5), 790-802 (1988).
Oh, M.S., et al., Ammonia evolution during oil shale pyrolysis, Energy & Fuels, 2(1), 100-105 (1988).
Suzuki, T.; Nakajima, S., and Watanabe, Y., Catalytic activity of rare-earth compounds for steam and carbon dioxide gasification of coal, Energy & Fuels, 2(6), 848-854 (1988).
Various, Coal pyrolysis: Mechanisms and modeling (symposium papers), Energy & Fuels, 2(4), 361-437 (1988).
Wilson, J.S.; Halow, J., and Ghate, M.R., Gasification: Key to chemicals from coal, Chemtech, Feb, 123-128 (1988).

27.2 Liquefaction

1976-1988

O'Hara, J.B., Coal liquefaction, Hyd Proc, 55(11), 221-226 (1976).
Various, Filtering coal liquids (topic issue), Chem Eng Prog, 77(12), 52-72 (1981).
Camier, R.J., and Stevens, S., Prospects for brown coal liquefaction in Australia, Proc Econ Int, 4(1), 11-17; 4(2), 6-12 (1983).
Various, Coal liquefaction (topic issue), Chem Eng Prog, 80(11), 29-53 (1984).
Detrez, C., and Houzelot, J.L., Review of design and scale-up of coal hydroliquefaction processes, Int Chem Eng, 26(3), 458-468 (1986).
Chiba, K.; Tagaya, H., and Saito, N., Liquefaction of Yallourn coal by binary system solvent, Energy & Fuels, 1(4), 338-343 (1987).
Ruether, J.A., et al., Effect of water and hydrogen partial pressures during direct liquefaction in catalyzed systems with a low solvent:coal ratio, Energy & Fuels, 1(2), 198-203 (1987).
Shin, S.C.; Baldwin, R.M., and Miller, R.L., Coal reactivity in direct hydrogenation liquefaction processes: Measurement and correlation with coal properties, Energy & Fuels, 1(4), 377-380 (1987).
Suzuki, T.; Ando, T., and Watanabe, Y., Kinetic studies on hydroliquefaction of coals using organometallic complexes, Energy & Fuels, 1(3), 294-300 (1987).

Tagaya, H., et al., Low-temperature coal liquefaction using n-butylamine as solvent, Energy & Fuels, 1(5), 397-401 (1987).
Liden, A.G.; Berruti, F., and Scott, D.S., Kinetic model for production of liquids from flash pyrolysis of biomass, Chem Eng Commns, 65, 207-222 (1988).
Miyake, M., and Stock, L.M., Factors governing successful coal solubilization through C-alkylation, Energy & Fuels, 2(6), 815-819 (1988).
Moritomi, H., et al., Mechanism of semicoke formation during coal liquefaction, Energy & Fuels, 2(4), 529-534 (1988).
Nagaishi, H., et al., Evaluation of coal reactivity for liquefaction based on kinetic characteristics, Energy & Fuels, 2(4), 522-529 (1988).
Sato, K., et al., Effect of coal particle and spray droplet sizes on combustion characteristics of coal-water mixtures, Powder Tech, 54(2), 127-136 (1988).
Shadle, L.J.; Neill, P.H., and Given, P.H., Dependence of coal liquifaction behaviour on coal characteristics, Fuel, 67(11), 1459-1476 (1988).
Toda, M., et al., Influence of particle size distribution of coal on fluidity of coal-water mixtures, Powder Tech, 55(4), 241-246 (1988).
Vassallo, A.M.; Wilson, M.A., and Attalla, M.I., Promotion of coal liquefaction by iodomethane, Energy & Fuels, 2(4), 539-548 (1988).

27.3 Gas Processing

1967-1979

Various, Liquified natural gas (topic issue), Chem Eng Prog, 63(6), 51-72 (1967).
Nelson, W.L., Natural gas liquifaction, Brit Chem Eng, 13(4), 509-511 (1968).
Various, Natural gas processing (special report), Hyd Proc, 48(4), 93-114 (1969).
Ring, T.A.; Mann, W.L., and Tse, Y.S., Innovations in hydrogen plant design, Chem Eng Prog, 66(12), 59-64 (1970).
Various, Natural gas processing (special report), Hyd Proc, 49(4), 85-108 (1970).
Various, NG/SNG handbook, Hyd Proc, 50(4), 93-168 (1971).
Bresler, S.A., and Ireland, J.D., Substitute natural gas: Processes, equipment and costs, Chem Eng, 16 Oct, 94-108 (1972).
Various, NG/SNG developments (special report), Hyd Proc, 51(4), 89-113 (1972).
Various, Liquified natural gas (topic issue), Chem Eng Prog, 68(9), 53-84 (1972).
Various, Substitute natural gas (topic issue), Chem Eng Prog, 68(12), 39-54 (1972).
Various, NG/LNG/SNG handbook, Hyd Proc, 52(4), 87-132 (1973).
Various, Gas processing developments (special report), Hyd Proc, 53(4), 69-91 (1974).
Various, Gas processing handbook, Hyd Proc, 54(4), 79-138 (1975).
Lehman, L.M., and Van Baush, E.H., Cryogenic purification of hydrogen, Chem Eng Prog, 72(1), 44-49 (1976).
Various, Gas processing (special report), Hyd Proc, 55(4), 69-92 (1976).

Various, Gas processing (special report), Hyd Proc, 56(4), 93-144 (1977).
Various, Gas processing developments (special report), Hyd Proc, 57(4), 89-124 (1978).
Various, Gas processing handbook 1979, Hyd Proc, 58(4), 99-170 (1979).

1980-1988

Bassett, L.C., and Natarajan, R.S., Hydrogen: Buy it or make it? Chem Eng Prog, 76(3), 93-97 (1980).
Various, Gas processing developments (special report), Hyd Proc, 59(4), 103-140 (1980).
Various, Gas processing developments (special report), Hyd Proc, 60(4), 109-152 (1981).
Various, Synfuels processing (topic issue), Chem Eng Prog, 77(5), 39-98 (1981).
Various, Synfuels (feature report), Hyd Proc, 60(6), 119-138 (1981).
Anon., 1983 LNG world overview, Hyd Proc, June, 140P-140T (1983).
Fluor, J.R., Natural gas: Global challenges and choices, Hyd Proc, April, 174D-174Q (1983).
Various, Gas processing developments (special report), Hyd Proc, April, 67-87 (1983).
Various, Gas process handbook 1984 (special report), Hyd Proc, April, 51-114 (1984).
Johnson, J.E., and Walter, F.B., Gas processing needs for enhanced oil recovery, Hyd Proc, Oct, 62-66 (1985).
Various, Gas processing developments (special report), Hyd Proc, April, 67-81 (1985).
Wall, J., Gas process handbook, Hyd Proc, 65(4), 75-106 (1986).
Haggin, J., Methane-to-gasoline in New Zealand, C&E News, 22 June, 22-25 (1987).
Lucadamo, G.A.; Bernhard, D.P., and Rowles, H.C., Improved ethylene and LPG recovery through dephlegmator technology, Gas Sepn & Purif, 1(2), 94-102 (1987).
Various, Oil and natural gas production including enhanced oil recovery and reservoir simulation (special topic issue), CER&D, 65, 1-112 (1987).
Thomas, E.R., and Denton, R.D., Conceptual studies of carbon dioxide-natural gas separation using controlled freeze zone process, Gas Sepn & Purif, 2(2), 84-89 (1988).
Various, Oil and natural gas production (topic issue), CER&D, 66(4), 289-344 (1988).
Various, Gas Process Handbook (special report), Hyd Proc, 67(4), 51-80 (1988).

27.4 Petrochemicals

1967-1976

Phillips, R.F., Petrochemicals (feature report), Chem Eng, 22 May, 153-180 (1967).
Various, Hydrocracking (topic issue), Chem Eng Prog, 63(5), 51-68 (1967).

Various, Petrochemical developments (special report), Hyd Proc, 47(11), 127-202 (1968).
Stobaugh, R.B., et al., Petrochemical guide (common chemicals), Hyd Proc, 45(10), 143-157 (1966); 49(6), 97-101 (1970); 49(7), 131-135; 49(8), 117-119; 47(9), 275-280; 49(10), 105-115 (1970).
Various, Petrochemical handbook, Hyd Proc, 50(11), 113-228 (1971).
Ockerbloom, N.E., Xylenes and higher aromatics, Hyd Proc, 50(7), 112-114; 50(8), 113-116; 50(9), 162-166; 50(10), 101-103; 50(12), 101-105 (1971); 51(1), 93-99 (1972); 51(2), 101-103; 51(4), 114-118 (1972).
Various, Petrochemical developments (special report), Hyd Proc, 51(11), 69-146 (1972).
Stobaugh, R.B., et al., Petrochemical guide (common chemicals), Hyd Proc, 50(1), 109-120 (1971); 51(5), 153-161 (1972); 52(1), 99-108 (1973); 52(2), 99-110 (1973).
Various, Petrochemical processes handbook, Hyd Proc, 52(11), 89-200 (1973).
Various, Petrochemical developments (special report), Hyd Proc, 53(11), 97-166 (1974).
Various, Petrochemical handbook, Hyd Proc, 54(11), 97-224 (1975).
Baba, T.B., and Kennedy, J.R., Ethylene and its coproducts, Chem Eng, 5 Jan, 116-128 (1976).
Various, Petrochemical developments (special report), Hyd Proc, 55(11), 93-185 (1976).

1977-1988

Chambers, L.E., and Singer, H., Economics of naptha versus gas cracking, Chem Engnr, July, 500-505 (1977).
Hatch, L.F., and Matar, S., From hydrocarbons to petrochemicals (particular chemicals and processes), Hyd Proc, 56(5), 191-196; 56(6), 189-194; 56(7), 191-201; 56(8), 151-155; 56(9), 165-173; 56(10), 153-163; 56(11), 349-357 (1977).
Various, Petrochemical handbook 1977, Hyd Proc, 56(11), 115-242 (1977).
Hatch, L.F., and Matar, S., From hydrocarbons to petrochemicals (particular chemicals and processes), Hyd Proc, 57(1), 135-139; 57(3), 129-139; 57(4), 155-166; 57(6), 149-162; 57(8), 153-165; 57(11), 291-301 (1978).
Various, Petrochemical developments (special report), Hyd Proc, 57(11), 99-184 (1978).
Various, Petrochemical handbook 1979, Hyd Proc, 58(11), 115-257 (1979).
Various, Petrochemical feedstocks (topic issue), Chem Eng Prog, 76(12), 31-56 (1980).
Various, Petrochemical developments 1980, Hyd Proc, 59(11), 93-153 (1980).
Various, Petrochemical handbook 1981, Hyd Proc, 60(11), 117-244 (1981).
Various, Petrochemical developments (special report), Hyd Proc, Nov, 87-141 (1982).
Various, Petrochemical handbook (special report), Hyd Proc, Nov, 67-162 (1983).
Various, Petrochemical developments (special report), Hyd Proc, Nov, 83-120 (1984).
Various, Petrochemical handbook (special report), Hyd Proc, Nov, 115-178 (1985).

Anon., Petrochemical handbook, Hyd Proc, 66(11), 59-90 (1987).
Greek, B.F., Petrochemical production in 1987, C&E News, 9 Feb, 9-13 (1987).
Various, Petrochemical developments: A special report, Hyd Proc, 67(11), 57-84 (1988).

27.5 Chemical Processes

1966-1970

Albright, L.F., Modern chemical technology (polymerization processes for ethylene, PVC, etc), Chem Eng, 28 March, 119-120; 25 April, 169-172; 9 May, 161-164; 6 June, 149-156; 4 July, 119-126; 15 Aug, 143-150; 12 Sept, 205-210; 10 Oct, 209-215; 21 Nov, 127-131; 19 Dec, 113-120 (1966); 16 Jan, 169-174; 13 Feb, 159-164; 27 March, 123-130; 10 April, 219-226; 8 May, 151-158; 5 June, 145-152; 3 July, 85-92; 17 July, 197-202; 14 Aug, 165-171; 11 Sept, 197-202; 9 Oct, 249-256; 6 Nov, 251-259; 4 Dec, 179-187 (1967).
Mantell, C.L., Electro-organic processing, Chem Eng, 5 June, 128-135 (1967).
Various, Fertilizer trends (topic issue), Chem Eng Prog, 63(10), 37-73 (1967).
Various, Sulfur and sulfuric acid plants (topic issue), Chem Eng Prog, 64(11), 47-92 (1968).
Various, The pulp and paper industry (topic issue), Chem Eng Prog, 64(8), 15-34 (1968).
Various, Phosphoric and phosphate processes (topic issue), Chem Eng Prog, 64(5), 49-88 (1968).
Dundas, P.H., and Thorpe, M.L., Economics and technology of chemical processing with electric-field plasmas, Chem Eng, 30 June, 123-128 (1969).
Edwards, R.H., Modern production methods for sulphuric and phosphoric acids, Brit Chem Eng, 14(6), 795-798; 14(7), 955-957 (1969).
Tucker, J.F., Ethylene liquefaction: A case history, Hyd Proc, 48(2), 99-102 (1969).
Dundas, P.H., and Thorpe, M.L., Titanium dioxide production by plasma processing, Chem Eng Prog, 66(10), 66-71 (1970).
Williams, V.C., Cryogenics, Chem Eng, 16 Nov, 92-108 (1970).

1971-1980

Kuhn, A., Modern trends in electrochemical industry, Brit Chem Eng, 16(2), 149-153 (1971).
Tomilov, A.P., and Fioshin, M.Y., Industrial electrolysers for organic syntheses, Brit Chem Eng, 16(2), 154-159 (1971).
Various, Chemicals: Products, manufacturing and distribution (deskbook issue), Chem Eng, 8 Oct, 9-107 (1973).
Austin, G.T., Industrially significant organic chemicals, Chem Eng, 21 Jan, 127-132; 18 Feb, 125-128; 18 March, 87-92; 15 April, 86-90; 29 April, 143-150; 27 May, 101-106; 24 June, 149-156; 22 July, 107-116; 5 Aug, 96-100 (1974).

Manitius, A.; Kurcyusz, E., and Kawecki, W., Mathematical model of the aluminum oxide rotary kiln, Ind Eng Chem Proc Des Dev, 13(2), 134-142 (1974).
Cronan, C.S., The Australian CPI, Chem Eng, 13 Oct, 119-122 (1975).
Various, Synthetic detergents (special report), Hyd Proc, 54(3), 71-92 (1975).
Various, Air vs oxygen for commercial oxidation processes (special report), Hyd Proc, 55(3), 69-100 (1976).
Various, Pulp and paper technology (topic issue), Chem Eng Prog, 72(6), 45-71 (1976).
Douglas, J.R., and Davis, C.H., Fertilizer supply and demand, Chem Eng, 18 July, 88-94 (1977).
Netzer, D., and Moe, J., Ammonia from coal, Chem Eng, 24 Oct, 129-132 (1977).
Stiles, A.B., Methanol: Past, present, and speculation on the future (review paper), AIChEJ, 23(3), 362-375 (1977).
Anon., Chemicals from coal: Planning for the 1990s and beyond, Proc Econ Int, 1(2), 12-15 (1979).
Various, Vinyl chloride monomer (special report), Hyd Proc, 58(3), 75-93 (1979).
Various, Polyvinyl chloride (special report), Hyd Proc, 59(3), 39-50 (1980).
Various, New techniques in pulp and paper technology (topic issue), Chem Eng Prog, 76(2), 43-67 (1980).

1981-1988

Various, Downstream methanol processing (special report), Hyd Proc, 60(3), 71-100 (1981).
Anon., The Australian chemical industry in the 1980s, Proc Econ Int, 3(1), 33-39 (1982).
Anon., Natural products as feedstocks for the chemical industry, Proc Econ Int, 3(1), 55-62 (1982).
Various, Developments in emulsion polymerisation (topic issue), Chem & Ind, 21 March, 220-236 (1983).
Anon., Distillery industry survey, Processing, Nov, 9-11 (1984).
Bayliss, M., Advances in pulp and paper processing, Proc Engng, June, 21-24 (1984).
Layman, P.L., Detergent chemistry, C&E News, 23 Jan, 17-49 (1984).
Ozero, B.J., and Procelli, J.V., Can developments keep ethylene oxide viable, Hyd Proc, March, 55-61 (1984).
Canterford, J.H., Hydrometallurgy, Chem Eng, 28 Oct, 41-48 (1985).
Springmann, H., Cryogenics: Principles and applications, Chem Eng, 13 May, 58-67 (1985).
Anon., Hydrocarbon Processing Industries construction boxscore (supplement), Hyd Proc, 67(10), October (1988).
Atkinson, N., Electrochemical synthesis of organics, Proc Engng, Nov, 49-50 (1988).
Greek, B.F., Detergents (product report), C&E News, 25 Jan, 21-53 (1988).
Mahi, P., The end of the Bayer process? Chem & Ind, 18 July, 445-451 (1988).
Nichols, J., and Thomas, P., Fine chemicals in the 21st century, Chem Engnr, March, 21-24 (1988).

Reisch, M.S., US dye producers, C&E News, 25 July, 7-14 (1988).
Various, Italian Chemical Engineering and Processing Handbook (1989), Chem Eng, 19 Dec, I1-I156 (1988).

27.6 Refining

1968-1979

Various, Refining processes handbook, Hyd Proc, 47(9), 135-238 (1968).
Hatch, L.F., A chemical view of refining, Hyd Proc, 48(2), 77-88 (1969).
Various, Refining processes developments (special report), Hyd Proc, 48(9), 131-155 (1969).
Dosher, J.R., Trends in petroleum refining, Chem Eng, 10 Aug, 96-112 (1970).
Various, Refining-petrochemicals interface (special report), Hyd Proc, 49(7), 83-98 (1970).
Various, Refining processes handbook, Hyd Proc, 49(9), 163-274 (1970).
Various, Refining processes developments (special report), Hyd Proc, 50(9), 123-161 (1971).
Various, Refining processes handbook, Hyd Proc, 51(9), 111-222 (1972).
Various, Refining processes developments (special report), Hyd Proc, 52(9), 107-146 (1973).
Various, Refining processes handbook, Hyd Proc, 53(9), 103-214 (1974).
Various, Refining developments (special report), Hyd Proc, 54(9), 93-159 (1975).
Various, Refining processes handbook (special report), Hyd Proc, 55(9), 103-230 (1976).
Various, Refining developments (special report), Hyd Proc, 56(9), 85-145 (1977).
Various, Refining process handbook 1978, Hyd Proc, 57(9), 97-224 (1978).
Various, Refining developments (special report), Hyd Proc, 58(9), 103-165 (1979).

1980-1988

Various, Refining process handbook 1980, Hyd Proc, 59(9), 93-220 (1980).
Various, Processing heavy crudes (topic issue), Chem Eng Prog, 77(12), 29-51 (1981).
Various, Processing heavy crude oil (topic issue), Chem Eng Prog, 77(2), 31-64 (1981).
Various, Refining developments (special report), Hyd Proc, 60(9), 103-157 (1981).
Anon., Refining process handbook (special report), Hyd Proc, Sept, 101-212 (1982).
Hyne, J.B., Don't produce Carsul, Hyd Proc, Sept, 241-244 (1982).
Various, Refining developments (special report), Hyd Proc, Sept, 77-111 (1983).
Various, Refining process handbook (special report), Hyd Proc, Sept, 67-146 (1984).
Various, Refining developments (special report), Hyd Proc, Sept, 67-95 (1985).
Anon., Refining process handbook, Hyd Proc, 65(9), 83-114 (1986).

Various, Refining developments (special report), Hyd Proc, 66(9), 59-70 (1987).
Hansen, J.H., et al., Thermodynamic model for predicting wax formation in crude oils, AIChEJ, 34(12), 1937-1942 (1988).
Mukherjee, N.L., Shale-oil hydroprocessing to produce synthetic jet fuels, Chem Eng Commns, 63, 193-204 (1988).
Various, 1988 Refining Handbook, Hyd Proc, 67(9), 61-91 (1988).

27.7 Offshore/Subsea Operations

1968-1988

Hamilton, R.W., and Schreiner, H.R., Engineering and medical problems of ocean exploitation, Chem Eng, 17 June, 263-270 (1968).
Mero, J.L., Exploiting seafloor minerals, Chem Eng, 1 July, 73-80 (1968).
Various, Offshore engineering (topic issue), Chem Engnr, June, 356-371 (1975).
Various, Economics of North Sea oil and gas (topic issue), Chem Engnr, June, 387-402 (1977).
Archer, J.S., and Wilson, D.C., Reservoir simulation in the development of North Sea oil fields, Chem Engnr, July, 565-567 (1978).
Buckley, P.S.; Dawe, R.A., and Grist, D.M., Chemical requirements for offshore oil recovery, Chem Engnr, July, 571-573 (1978).
Holding, J., What's ahead for process design offshore, Chem Engnr, Aug, 539-542 (1980).
Lang, J.R.A., Offshore oil production systems: The tethered buoyant platform, Chem Engnr, Aug, 543-545 (1980).
Various, Developments in offshore process plant, Chem Engnr, Sept, 28-31 (1983).
Delaine, J., Separating oil from water offshore, Chem Engnr, Nov, 31-34 (1985).
Waldie, B., Chemical engineering offshore, Chem Engnr, June, 17-19 (1986).
Head, J., and Rumley, J., Design of production facilities for floating offshore platforms, Chem Engnr, Jan, 17-21 (1987).
Head, J., and Rumley, J., Production design for floating platforms, Chem Engnr, July, 17-22 (1987).
Steward, P., Subsea separation and transport, Chem Engnr, Jan, 22-25 (1988).

27.8 Oil Recovery and Processes

1978-1988

Brown, C.E., Physiochemical aspects of enhanced oil recovery, Chem & Ind, 18 Nov, 875-881 (1978).
McConnell, A.J.; McKay, G.; Murphy, W.R., and Williams, J.D., Optimization study of operation of light distillate reforming gas plant, Energy World, April, 8-11 (1978).
Thurlow, G.G., Oil from coal, Chem Engnr, Oct, 733-735 (1978).
Various, Shale oil (topic issue), Chem Eng Prog, 75(9), 64-91 (1979).
Anon., Technical and economic status of enhanced oil recovery, Proc Econ Int, 1(4), 17-24 (1980).

Chiwetelu, C.; Hornof, V., and Neale, G.H., Enhanced oil recovery using lignosulphonate-petroleum sulphonate mixtures, Trans IChemE, 60, 177-182 (1982).
Davis, G.B., and Hill, J.M., Some theoretical aspects of oil recovery from fractured reservoirs, Trans IChemE, 60, 352-358 (1982).
Goddin, C.S., Enhanced oil recovery using carbon dioxide, Hyd Proc, May, 125-130 (1982).
Various, Lubes for the future (special report), Hyd Proc, Feb, 75-89 (1982).
Aquilo, A.; Alder, J.S.; Freeman, D.N., and Voorhoeve, R.J.H., Focus on C1 chemistry, Hyd Proc, March, 57-65 (1983).
Scanlan, T., Soviet oil and gas in the 1980s, Hyd Proc, June, 140C-140P (1983).
Stegemeier, G.L., Application of research to steam injection processes for oil recovery, Chem Eng Prog, 81(4), 25-31 (1985).
Philp, R.P., Geochemistry in the search for oil, C&E News, 10 Feb, 28-43 (1986).
Various, Advances in oil shale chemistry (symposium papers), Energy & Fuels, 1(6), 477-528 (1987).
Puig, J.E., et al., Roles of co-surfactant and co-solvent in surfactant waterflooding, Chem Eng Commns, 65, 169-186 (1988).

27.9 Minerals Extraction and Electrochemistry

1975-1988

Various, Cell design in electrowinning and electrorefining (topic issue), Chem & Ind, 19 April, 326-336; 3 May, 370-390 (1975).
Dotson, R.L., Modern electrochemical technology, Chem Eng, 17 July, 106-118 (1978).
Various, Hydrometallurgy (topic issue), Chem & Ind, 20 June, 406-431 (1981).
Various, Recovery of valuable materials from dilute sources (symposium papers), Sepn Sci Technol, 16(9), 971-1112 (1981).
Habashi, F., Hydrometallurgy (feature report), C&E News, 8 Feb, 46-58 (1982).
Burkin, A.R., Hydrometallurgy 1952-1982, Chem & Ind, 19 Sept, 690-695 (1983).
Eisebe, J.A.; Colombo, A.E., and McClelland, G.E., Recovery of gold and silver from ores by hydrometallurgical processing, Sepn Sci Technol, 18(12), 1081-1094 (1983).
Nguyen, V.V.; Pinder, G.F.; Gray, W.G., and Botha, J.F., Numerical simulation of uranium in-situ mining, Chem Eng Sci, 38(11), 1855-1862 (1983).
Various, A decade of electrochemical technology (topic issue), Chem & Ind, 7 March, 187-191 (1983).
Various, Electrohydrometallurgy (topic issue), Chem & Ind, 1 Sept, 569-577 (1986).
Goodridge, F., Some aspects of electrochemical engineering, J Chem Tech Biotechnol, 38(2), 127-142 (1987).
Hughes, D., Reviving electrochemistry, Chem Engnr, Sept, 17-18 (1987).
Hollitzer, E., and Sartori, P., Electrochemical fluorination (review), Int Chem Eng, 28(2), 221-231 (1988).

Rosensweig, R.E., Introduction to ferrohydrodynamics, Chem Eng Commns, 67, 1-18 (1988).

27.10 Miscellaneous Process Technology

1967-1988

Flinn, J.E., and Nack, H., Microencapsulation techniques, Chem Eng, 4 Dec, 171-178 (1967).

Various, Paper coatings (topic issue), Chem Eng Prog, 63(3), 45-69 (1967).

Lih, M.M., Color technology, Chem Eng, 12 Aug, 146-156 (1968).

Vance, R.W., Future cryogenic applications, Brit Chem Eng, 13(10), 1439-1440 (1968).

Ligi, J.J., Processing with cryogenics, Hyd Proc, 48(5), 137-140; 48(6), 118-120 (1969).

Price, F.C., Iron and steel today, Chem Eng, 11 Aug, 76-88 (1969).

Bonet, C., Thermal plasma processing, Chem Eng Prog, 72(12), 63-69 (1976).

Leesley, M.E., Predicting the combustion rate of pulverised fuel from its initial size distribution, AIChEJ, 23(4), 520-528 (1977).

Norman, P., and Fells, I., Contamination of plasma arcs by electrode vapour, Trans IChemE, 55, 149-152 (1977).

Waldie, B., and Hancock, R., Effect of a cool central gas feed on temperatures and velocities in an induction plasma torch, Trans IChemE, 56, 178-186 (1978).

Coleby, M.G., and Davies, G.J., Effect of some process variables on the thermal decomposition of molybdenite in an induction-coupled argon plasma, Trans IChemE, 57, 128-133 (1979).

Various, Isotope separations (symposium papers), Sepn Sci Technol, 15(3), 370-586 (1980).

Various, Separation methods in fossel fuel utilization (symposium papers), Sepn Sci Technol, 16(10), 1545-1666 (1981).

Davies, T.W., Density reduction of kaolinite by flash heating, CER&D, 63, 82-88 (1985).

Kudela, L., and Sampson, M.J., Understanding sublimation technology, Chem Eng, 23 June, 93-98 (1986).

Liu, G.E., and Law, C.K., Combustion of coal-water slurry droplets, Fuel, 65(2), 171-176 (1986).

Goodger, E.M., and Eissa, A.F.M., Spontaneous ignition research: A review of experimental data, J Inst Energy, June, 84-94 (1987).

Papachristodoulou, G., and Trass, O., Coal slurry fuel technology (a review), Can JCE, 65(2), 177-201 (1987).

Park, J.K., and Phillips, J.A., Ammonia catalyzed organosolv delignification of poplar, Chem Eng Commns, 65, 187-206 (1988).

Various, Advances in soot chemistry (symposium papers), Energy & Fuels, 2(4), 438-504 (1988).

27.11 Space Technology

1969-1987

Dallaire, E.E., Advances in engineering from space technology, Chem Eng, 25 Aug, 119-124 (1969).
Motz, L., Cosmic chemical engineering, Chem Eng, 1 April, 86-94 (1974).
Waldron, R.D.; Erstfeld, T.E., and Criswell, D.R., The role of chemical engineering in space manufacturing, Chem Eng, 12 Feb, 80-94 (1979).
Santandrea, R.P., Industrial processing in space, Chem Eng Prog, 81(11), 33-35 (1985).
Roble, R.G., Chemistry in the thermosphere and ionosphere, C&E News, 16 June, 23-38 (1986).
Zenz, F.A., The particulate nature of the universe, Chem Eng Commns, 62(1), 123-134 (1987).

27.12 Mass Transfer and Transport Processes

1970-1980

Suciu, D.G.; Smigelschi, O., and Ruckenstein, E., Effect of surface tension sinks on the behaviour of stagnant liquid films, Trans IChemE, 48, T176-177 (1970).
Hsu, C.K., Transfer function of multistage separation processes, Trans IChemE, 49, 251-252 (1971).
Hills, J.H., The rise of a large bubble through a swarm of smaller ones, Trans IChemE, 53, 224-233 (1975).
Smith, I.K., and Garmendia, L.A., The effects of an electrostatic field and air stream on water jet spray angle, Trans IChemE, 53, 173-174 (1975).
Giddings, J.C., Basic approaches to separation: Analysis and classification of methods according to underlying transport characteristics, Sepn Sci Technol, 13(1), 3-24 (1978).
Wong, P.F.Y.; Ko, N.W.M., and Yip, P.C., Mass transfer from large diameter vibrating cylinders, Trans IChemE, 56, 214-216 (1978).
Ferreira, A.B., and Jansson, R.E.W., Mass transfer to co-rotating discs: Further studies on the rotating electrolyser, Trans IChemE, 57, 262-268 (1979).
Giddings, J.C., Basic approaches to separation: Steady-state zones and layers, Sepn Sci Technol, 14(10), 871-882 (1979).
Sandoval-Robles, J.G.; Riba, J.P., and Couderc, J.P., Mass transfer around a sphere, Trans IChemE, 58, 132-134 (1980).
Slattery, J.C., Interfacial transport phenomena (review paper), Chem Eng Commns, 4(2), 149-166 (1980).

1981-1987

Various, Use of surfactants in separation processes (symposium papers), Sepn Sci Technol, 16(10), 1429-1544 (1981).
Corry, W.D.; Seaman, G.V.F., and Szafron, D.A., New method for evaluating effectiveness of separation processes, Sepn Sci Technol, 17(13), 1469-1484 (1982).

Seil, H., Discrete mass balance for acid-dipping, Chem Eng, 15 Nov, 150-152 (1982).
Berger, F.P., and Ziai, A., Optimisation of experimental conditions for electrochemical mass transfer measurements, CER&D, 61, 377-382 (1983).
Mahiout, S., and Vogelpohl, A., Mass transfer in high viscosity media, Chem Eng & Proc, 18(4), 225-232 (1984).
Bright, A., Minimum drop volume in liquid jet breakup, CER&D, 63, 59-66, 403-404 (1985).
Greaves, M., and Patel, K., Flow of polymer solution in porous media, CER&D, 63, 199-202 (1985).
Tung, H.H., and Mah, R.S.H., Modeling liquid mass transfer in Higee separation process, Chem Eng Commns, 39, 147-154 (1985).
Ramshaw, C., Opportunities for exploiting centrifugal fields, Chem Engnr, June, 17-21 (1987).
Tondeur, D., Unifying concepts in nonlinear unsteady processes, Chem Eng & Proc, 21(4), 167-178; 22(2), 91-106 (1987).

1988

Adebekun, A.K., and Schork, F.J., Stability of some chemical processes via the theory of positive systems, Chem Eng Commns, 69, 43-52 (1988).
Brauer, H., Mass transfer machines: Better equipment design in processing industries, Int Chem Eng, 28(2), 207-221 (1988).
Chebi, R.; Rice, P.A., and Schwartz, J.A., Heat dissipation in microelectronic systems using phase change materials with natural convection, Chem Eng Commns, 69, 1-12 (1988).
Chen, H.T., and Chen, C.K., Natural convection of non-Newtonian fluids in porous medium on horizontal surface, Chem Eng Commns, 69, 29-42 (1988).
Dagan, Z., and Maldarelli, C., Surface tension retardation of Marangoni waves, Chem Eng Commns, 64, 83-104 (1988).
Farag, I.H.; Audet, D.M., and Allam, T.A., Remote spectral sensing of temperature profiles in flames, Chem Eng Commns, 72, 221-232 (1988).
Ho, K.L., and Chang, H.C., Nonlinear doubly-diffusive Marangoni instability, AIChEJ, 34(5), 705-722 (1988).
Lin, I.J., and Shalom, A.L., Static equilibrium build-up model for dielectrophoretic capture on a single rod, Chem Eng Commns, 67, 55-68 (1988).
Mersmann, A.; Voit, H., and Zeppenfeld, R., Do we need mass transfer machines? Int Chem Eng, 28(1), 1-14 (1988).
Moissis, A.A., and Zahn, M., Boundary value problems in electrofluidized and magnetically stabilized beds, Chem Eng Commns, 67, 181-204 (1988).
Morin, T.J.; Chapman, R., and Hawley, M., Flow calorimetry of electrical discharges, Chem Eng Commns, 73, 183-204 (1988).
Osorio, C., and Onken, U., Influence of liquid properties on determination of volumetric mass-transfer coefficient with the hydrazine method, Chem Eng Commns, 68, 43-56 (1988).
Rai, B.N., et al., Forced convective mass transfer in annuli, Chem Eng Commns, 68, 15-30 (1988).

Rosner, D.E., and Liang, B., Experimental studies of deposition rates in presence of alkali sulfate vapor scavenging by submicron particles in combustion gas boundary layers, Chem Eng Commns, 64, 27-46 (1988).
Tsebers, A., and Blums, E., Long-range magnetic forces in two-dimensional hydrodynamics of magnetic fluid pattern formation, Chem Eng Commns, 67, 69-88 (1988).
Zanwar, S.S., and Pangarkar, V.G., Solid-liquid mass transfer in packed beds: Enhancement due to ultrasound, Chem Eng Commns, 68, 133-142 (1988).

27.13 Foams

1968-1987

Lemlich, R., Foam fractionation, Chem Eng, 16 Dec, 95-102 (1968).
Kouloheris, A.P., Foam destruction and inhibition, Chem Eng, 27 July, 143-146 (1970).
Hartland, S., and Barber, A.D., A model for a cellular foam, Trans IChemE, 52, 43-52 (1974).
Barber, A.D., and Hartland, S., The collapse of cellular foams, Trans IChemE, 53, 106-111 (1975).
Grieves, R.B., Foam separations: A review, Chem Eng J, 9(2), 93-106 (1975).
Steiner, L.; Hunkeler, R., and Hartland, S., Behaviour of dynamic cellular foams, Trans IChemE, 55, 153-163 (1977).
Burley, R.W.; Nutt, C.W., and Bayat, M.G., Studies of fluid displacement by foam in porous media, CER&D, 62, 92-100 (1984).
Ali, J.; Burley, R.W., and Nutt, C.W., Foam enhanced oil recovery from sand packs, CER&D, 63, 101-111 (1985).
Castanier, L.M., and Brigham, W.E., Selecting foaming agents for enhanced oil recovery steam injection improvement, Chem Eng Prog, 81(6), 37-40 (1985).
Koczo, K.; Ludanyi, B., and Racz, G.Y., Foaminess and foam stability of surfactant solutions, Periodica Polytechnica, 31(1/2), 83-92 (1987).
Kouloheris, A.P., Foam: Friend or foe? Chem Eng, 26 Oct, 88-97 (1987).

27.14 Miscellaneous Topics

1968-1988

Various, Patents (feature report), Chem Eng, 26 Feb, 138-154 (1968).
Various, Fast, functional writing (feature report), Chem Eng, 30 June, 104-122 (1969).
Fair, J.R., History of The Chemical Engineers' Handbook, Chem Eng, 18 Feb, 129-132 (1974).
Hougen, O.A., Seven decades of chemical engineering, Chem Eng Prog, 73(1), 89-104 (1977).
Moon, G.B., A law primer for the chemical engineer, Chem Eng, 10 April, 114-120 (1978).
Wall, C.G., The chemical engineer in petroleum engineering, Chem Engnr, July, 562-564 (1978).

Kemp, M.K., and Joyce, J., Whizbang chemistry, Chemtech, April, 210-215 (1979).
Conkling, J.A., Chemistry of fireworks, C&E News, 29 June, 24-32 (1981).
Daus, D.G., What can be patented? Chem Eng, 23 Aug, 74-76 (1982).
Michaels, A.S., Frontiers of chemical engineering, Chem Eng Commns, 17(1), 99-106 (1982).
Witherell, C.E., The products-liability threat, Chem Eng, 24 Jan, 72-87 (1983).
Torres-Marchal, C., Do-it-yourself technical Russian, Chem Eng, 4 Feb, 49-51 (1985).
Szabadvary, F., The history of chemical laboratory equipment, Periodica Polytechnica, 30(1/2), 77-96 (1986).
Chittenden, D.H., 'How to solve it' - revisited!: Engineering problem-solving approach, Chem Eng, 16 March, 89-92 (1987).
Morris, M., Creative thinking: The new technology, Chem Eng, 19 Jan, 127-129 (1987).
Levenspiel, O., Danckwerts memorial lecture: Chemical Engineering's Grand Adventure, CER&D, 66(5), 387-395 (1988).
Various, History of chemical engineering (special issue), Chem Engnr, Dec, 17-77 (1988).

Index